Taylor Expansions

$$e^{ax} = \sum_{n=0}^{\infty} \frac{(ax)^n}{n!} = 1 + ax + \frac{1}{2!}(ax)^2 + \frac{1}{3!}(ax)^3 + \cdots$$

$$\sin ax = \sum_{n=0}^{\infty} (-1)^n \frac{(ax)^{2n+1}}{(2n+1)!} = ax - \frac{1}{3!}(ax)^3 + \frac{1}{5!}(ax)^5 - \cdots$$

$$\cos ax = \sum_{n=0}^{\infty} (-1)^n \frac{(ax)^{2n}}{(2n)!} = 1 - \frac{1}{2!}(ax)^2 + \frac{1}{4!}(ax)^4 - \cdots$$

$$\sinh ax = \sum_{n=0}^{\infty} \frac{(ax)^{2n+1}}{(2n+1)!} = ax + \frac{1}{3!}(ax)^3 + \frac{1}{5!}(ax)^5 + \cdots$$

$$\cosh ax = \sum_{n=0}^{\infty} \frac{(ax)^{2n}}{(2n)!} = 1 + \frac{1}{2!}(ax)^2 + \frac{1}{4!}(ax)^4 + \cdots$$

$$\ln(1+x) = \sum_{n=1}^{\infty} (-1)^{n+1} \frac{x^n}{n} = x - \frac{x^2}{2} + \frac{x^3}{3} - \cdots, \ |x| \leq 1, \ x \neq -1$$

$$(1+x)^{\alpha} = 1 + \alpha x + \frac{\alpha(\alpha-1)}{2!}x^2 + \frac{\alpha(\alpha-1)(\alpha-2)}{3!}x^3 + \cdots, \qquad |x| < 1$$

Vector Identities

For vector fields $\vec{A}$, $\vec{B}$ and scalar fields f, g:

$$\vec{A} \times \vec{B} = -\vec{B} \times \vec{A}$$

$$\vec{A} \cdot (\vec{B} \times \vec{C}) = \vec{B} \cdot (\vec{C} \times \vec{A}) = \vec{C} \cdot (\vec{A} \times \vec{B})$$

$$\vec{A} \times (\vec{B} \times \vec{C}) = \vec{B}(\vec{A} \cdot \vec{C}) - \vec{C}(\vec{A} \cdot \vec{B}) \qquad [\text{``BAC-CAB''}]$$

$$(\vec{A} \times \vec{B}) \cdot (\vec{C} \times \vec{D}) = (\vec{A} \cdot \vec{C})(\vec{B} \cdot \vec{D}) - (\vec{A} \cdot \vec{D})(\vec{B} \cdot \vec{C})$$

$$(\vec{A} \times \vec{B}) \times (\vec{C} \times \vec{D}) = [(\vec{A} \times \vec{B}) \cdot \vec{D}]\vec{C} - [(\vec{A} \times \vec{B}) \cdot \vec{C}]\vec{D}$$

$$\vec{\nabla}(fg) = f\vec{\nabla}g + g\vec{\nabla}f$$

$$\vec{\nabla} \cdot (f\vec{A}) = f(\vec{\nabla} \cdot \vec{A}) + \vec{\nabla}f \cdot \vec{A}$$

$$\vec{\nabla} \times (f\vec{A}) = f(\vec{\nabla} \times \vec{A}) + \vec{\nabla}f \times \vec{A}$$

$$\vec{\nabla} \times \vec{\nabla}f = 0$$

$$\vec{\nabla} \cdot (\vec{\nabla} \times \vec{A}) = 0$$

$$\vec{\nabla} \cdot (\vec{A} \times \vec{B}) = (\vec{\nabla} \times \vec{A}) \cdot \vec{B} - \vec{A} \cdot (\vec{\nabla} \times \vec{B})$$

$$\vec{\nabla} \times (\vec{A} \times \vec{B}) = (\vec{B} \cdot \vec{\nabla})\vec{A} - \vec{B}(\vec{\nabla} \cdot \vec{A}) + \vec{A}(\vec{\nabla} \cdot \vec{B}) - (\vec{A} \cdot \vec{\nabla})\vec{B}$$

$$\vec{\nabla} \times (\vec{\nabla} \times \vec{A}) = \vec{\nabla}(\vec{\nabla} \cdot \vec{A}) - \nabla^2 \vec{A}$$

$$\vec{\nabla}(\vec{A} \cdot \vec{B}) = \vec{A} \times (\vec{\nabla} \times \vec{B}) + \vec{B} \times (\vec{\nabla} \times \vec{A}) + (\vec{A} \cdot \vec{\nabla})\vec{B} + (\vec{B} \cdot \vec{\nabla})\vec{A}$$

Mathematical Methods and Physical Insights

An Integrated Approach

Mathematics instruction is often more effective when presented in a physical context. Schramm uses this insight to help develop students' physical intuition as he guides them through the mathematical methods required to study upper-level physics. Based on the undergraduate Math Methods course he has taught for many years at Occidental College, the text encourages a symbiosis through which the physics illuminates the math, which in turn informs the physics. Appropriate for both classroom and self-study use, the text begins with a review of useful techniques to ensure students are comfortable with prerequisite material. It then moves on to cover vector fields, analytic functions, linear algebra, function spaces, and differential equations. Written in an informal and engaging style, it also includes short supplementary digressions ("By the Ways") as optional boxes showcasing directions in which the math or physics may be explored further. Extensive problems are included throughout, many taking advantage of Mathematica, to test and deepen comprehension.

Alec J. Schramm is a professor of physics at Occidental College, Los Angeles. In addition to conducting research in nuclear physics, mathematical physics, and particle phenomenology, he teaches at all levels of the undergraduate curriculum, from courses for non-majors through general relativity and relativistic quantum mechanics. After completing his Ph.D., he lectured at Duke University and was a KITP Scholar at the Kavli Institute for Theoretical Physics at UC Santa Barbara. He is regularly nominated for awards for his physics teaching and clear exposition of complex concepts.

Mathematical Methods and Physical Insights

An Integrated Approach

ALEC J. SCHRAMM

Occidental College, Los Angeles

CAMBRIDGE
UNIVERSITY PRESS

University Printing House, Cambridge CB2 8BS, United Kingdom

One Liberty Plaza, 20th Floor, New York, NY 10006, USA

477 Williamstown Road, Port Melbourne, VIC 3207, Australia

314–321, 3rd Floor, Plot 3, Splendor Forum, Jasola District Centre,
New Delhi – 110025, India

103 Penang Road, #05–06/07, Visioncrest Commercial, Singapore 238467

Cambridge University Press is part of the University of Cambridge.

It furthers the University's mission by disseminating knowledge in the pursuit of
education, learning, and research at the highest international levels of excellence.

www.cambridge.org
Information on this title: www.cambridge.org/highereducation/isbn/9781107156418
DOI: 10.1017/9781316661314

First published 2022

Printed in the United Kingdom by TJ Books Limited, Padstow Cornwall 2022

A catalogue record for this publication is available from the British Library.

ISBN 978-1-107-15641-8 Hardback

Additional resources for this publication at www.cambridge.org/schramm

To Eitan, whose wit and humor eases my load
To Tirtza, who leads by example to show me the way
To Aviya, who teaches that obstacles need not block my road
To Keshet, whose refracted sunshine colors each day

And to Laurel, whose smile lights my life

Contents

List of BTWs

Preface

You're holding in your hands a book on mathematical methods in the physical sciences. Wait — don't put it down just yet! I know, I know: there are already many such books on the shelves of scientists and engineers. So why another one?

Motivations

As with many textbooks, this one has its origins in the classroom — specifically, a math methods course I teach annually for third-year physics and engineering students. The project began, simply enough, with the realization that many students have trouble integrating functions beyond polynomials and elementary trig. At first I thought this merely reflected gaps in their background. After all, it's not unusual for standard calculus courses to largely skip once-invaluable topics such as trig substitution. To be sure, there's a cogent argument to be made that many of the integration techniques taught back in my college days are passé in the age of such resources as Wolfram Alpha. One unfortunate consequence, however, is that students too often see an integral as a black-box calculator, rather than as a mathematical statement in its own right. And indeed, I soon realized that the difficulties I was encountering had less to do with students' integration abilities and more with their basic comprehension. So one year I wrote a handout for the first week of the semester, in the hope that learning a few standard techniques would help demystify integration. I was astounded by the overwhelmingly positive response. Moreover, despite the relatively quick pace, students became notably less intimidated by integrals common in upper-level physics and engineering. Though they did not always master the techniques, most students began to develop a solid intuition for both the physics and mathematics involved — which was the goal, after all. This early handout evolved into Chapter 8, "Ten Integration Techniques and Tricks."

Encouraged by this modest success — and having identified other areas of similar weakness — I began to develop a set of notes on an array of topics. Although upper-level students have already seen most of this material in other courses, I've found that even if they can push the symbols around, their understanding is often wanting. And once again, the results were gratifying. These notes form the backbone of Part I of this book, which is intended as a review with an eye toward strengthening intuition.

As I continued writing handouts for various topics covered in the course, a book began to emerge. Along the way I found myself influenced by and reacting to the many texts which often stress formality over pragmatism, or are too recipe-driven to instill insight. Neither approach, in my view, gets it quite right. All too often, students lost in the abstract find refuge in rote symbol manipulation; emphasizing manipulation alone, however, is woefully inadequate. The ability to play the notes doesn't make you a musician.[1] So it's a delicate balance. An effective program must be more than a collection of theorems, proofs, or How-To's, and should be illustrated with physical examples whose context is familiar to the intended audience.

The approach taken here is not meant to be rigorous; indeed, many of us were attracted to the physical sciences more by the playfulness and fun of mathematics than its formality. For the most part, derivations and proofs are reserved for situations in which they aid comprehension. My goal has been to leverage that delight we've all experienced when the depth and inherent beauty of a mathematical concept or technique reveal themselves — not just the endorphin rush of understanding, but the empowering command that comes with intuition.

Content and Organization

Of course, one has to start somewhere. It's impractical to go all the way back to an introduction of slopes and derivatives, or the development of the integral from a Riemann sum. For such topics, many excellent texts are readily available. As this book is aimed at upper-level physics and engineering students, the presumed background includes standard introductory physics through the sophomore level. A background in multivariable calculus and linear algebra is also assumed — though some of this material is reviewed in Parts II and IV.

Much of the coverage is similar to that of other math methods texts (as indeed it must be), so there's more material than is generally covered in two semesters.[2] But the design is intended to give instructors flexibility without sacrificing coherence. The book is separated into Parts. Each begins with a prelude introduction, has an interlude to try to flesh out intuition, and ends with a coda to tie together many of the ideas in a physical context. Some of the interludes and codas can be assigned as reading rather than presented during class time without loss of continuity. Aside from Part I (see below), the Parts were constructed to build upon one another pedagogically, with properties and representations of vectors as a common theme. Even so, the Parts are relatively self-contained.

- Part I: With rare exception, instructors of math methods courses cite the same central challenge: the uneven preparation of students. This creates an enormous pedagogical hurdle on both sides of the white board. Addressing this obstacle is the motivation underlying Part I, which, as its title suggests, students Just Gotta' Know.[3] I've found that starting the semester with a potpourri of topics helps smooth out students' variation in background. In fact, despite the unit's staccato character, students across the preparation spectrum have clearly derived benefit — as well as, I daresay, a sense of empowerment. I usually cover this material over the first 2–3 weeks of the semester, though one could envision covering some sections only as needed as the term progresses.
- Part II: This unit is largely an introduction to grad, div, and curl. For completeness, a review of general line and surface integrals is given — but even here, the goal is to prepare for the theorems of Gauss and Stokes. The primary paradigm is the behavior of electric and magnetic fields. Though the assumption is that students will have had a typical first-year E&M course, I've tried to make these examples largely self-contained. The Coda discusses simply connected regions, and includes an example of a circuit with non-trivial topology.

---

[1] As I often ask my students, would you want to fly on a plane designed by someone who knows how to push around symbols, but with a weak grasp of their meaning? I'm reminded of my 7th grade math teacher, Mr. Ford, who used to berate students with the line "It's not for you to understand — it's for you to do!" I cannot in good conscience recommend his airplane designs.

[2] Good general references include Arfken and Weber (1995); Boas (2005); Kusse and Westwig (1998); Mathews and Walker (1970), and Wong (1991).

[3] Because of its potpourri nature, I wanted to call it "Unit 0," but my editors wisely advised that this would be seen as too quirky.

- Part III: As a natural extension of the discussion in the previous unit, this one begins by reformulating $\mathbb{R}^2$ vector fields as functions in the complex plane. Note that Part I only discusses the rudiments of complex numbers — more algebra than calculus. The topics addressed in this unit include Laurent series, Cauchy's integral formula, and the calculus of residues. The Interlude considers conformal mapping, and the Coda the connection between analyticity and causality. Instructors who prefer to cover these subjects at another time can choose to skip this unit on a first pass, since little of the material is needed later.

- Part IV: This unit turns to linear algebra, and as such the focus shifts from vectors as functions in $\mathbb{R}^2$ and $\mathbb{R}^3$, to basic mathematical entities which do little more than add and subtract. Over the course of the unit more structure is introduced — most notably, the inner product. Complete orthonormal bases are emphasized, and Dirac's bra-ket notation is introduced. Orthogonal, unitary, and Hermitian matrices are discussed. An Interlude on rotations is complemented by an appendix presenting different approaches to rotations in $\mathbb{R}^3$. We then turn to the eigenvalue problem and diagonalization. The Coda focuses on normal modes, serving as a natural bridge to Part V.

- Entr'acte: Following the developments in the previous unit, the Entr'acte classifies vectors in terms of their behavior under rotation. This naturally leads to discussions of tensors, metrics, and general covariance.

- Part V: Here the results of Part IV are extended from finite- to infinite-dimensional vector space. Orthogonal polynomials and Fourier series are derived and developed as examples of complete orthogonal bases. Taking the continuum limit leads to Fourier transforms and convolution, and then to Laplace transforms. The Interlude introduces spherical harmonics and Bessel functions as examples of orthogonal functions beyond sine and cosine — the goal being to develop familiarity and intuition for these bases in advance of tackling their defining differential equations in Part VI. The Coda is an introduction to signal processing, with an emphasis on Fourier analysis and the convolution theorem.

- Part VI: After a relatively brief discussion of first-order differential equations, this unit turns to second-order ODEs and PDEs. The orthogonal polynomials of Part V are rederived by series solutions. Initial and boundary value problems, as well as Green's function solutions are introduced. The unitarity and hermiticity of differential operators are examined in an Interlude on the Sturm–Liouville problem, and the Coda provides an introduction to quantum mechanical scattering as an application of Green's functions.

It's a challenge to bring students up to speed while simultaneously lifting and enticing them to the next level. Towards this end, sprinkled throughout the text are asides called "By The Ways" (BTWs). Averaging about a page in length, these digressions are what I hope are delightful ("geeky-cool") forays into the tangential — sometimes math, sometimes physics. Though the array of BTW topics is necessarily rather idiosyncratic, they were chosen for beauty, charm, or insight — and in the best instances, all three. Since the BTWs are not required for understanding of the mainstream material, students can choose to skip over them (though they'll miss much of the fun).

One of my professors liked to say "The best professor is the worst professor." By this he meant that ultimately one must be self-taught — and that an ineffective lecturer is often required to compel students to learn on their own. Though personally I think he was just making excuses for poor lecture prep, it is true that students cannot master the material without working problems themselves. So each chapter includes extensive homework exercises, many written to take advantage of the capabilities of Mathematica.[4] (Online solutions are available for instructors.) The objective, of course, is to recruit and retain every student's best teacher.

---

[4] Trott (2004) is a good reference source.

Inspiration and Aspiration

Ultimately, my own experiences as a student of physics and math motivate the guiding principle: the physics illuminates and encourages the math, which in turn further elucidates the physics. But it doesn't come easy; effort is required. As I often tell students, walking through an art museum or listening to music will not elicit the same depth of wonder and delight without having first studied art or music. So too, appreciation of the inherent harmonies between math and physics favors the prepared mind. Moreover, such leveraging of the physics/math symbiosis is truly a life-long exercise. But it is my hope that this approach will enhance students' appreciation for the inherent beauty of this harmony, inspiring them to continue their pursuit into the wonders and mysteries of physical science.

Acknowledgements

I want to begin by thanking my parents, who sadly are no longer here to show this book to their friends. To my brothers, from whom I've learned so much on a remarkably wide array of subjects. And most especially to Laurel and our children, whose love and support I could not do without.

To the incredible professionals Vince Higgs, Heather Brolly, and especially Melissa Shivers at Cambridge University Press, my deep appreciation for their guidance, advice, and seemingly inexhaustible patience. And to David Hemsley, for his sharp eye and good humor. It truly has been a pleasure working and bantering with all of them.

Mr. Ford and the above-referenced professor notwithstanding, I've had the good fortune of learning from some incredible teachers and colleagues, in both formal and informal settings. But I've also learned from students. In particular, I want to thank the many who have taken my courses over the years using early versions of the manuscript. Special shout-outs of gratitude to Olivia Addington, Claire Bernert, Ian Convy, Emily Duong, Alejandro Fernandez, Leah Feuerman, Michael Kwan, Sebastian Salazar, Hunter Weinreb, and Hedda Zhao. Their (mostly constructive) criticisms, suggestions, and perspective have greatly improved the text. Without Jason Detweiler this book may never have seen the light of day. And deepest gratitude and appreciation to my friend and colleague Jochen Rau, who read much of the text, prevented many embarrassing errors — and taught me the word "betriebsblind."

Any remaining errors, of course, are mine alone; they tried their best.

Part I

Things You Just Gotta' Know

The noblest pleasure is the joy of understanding.

Leonardo da Vinci

$$\int f(x)\,dx$$

1 Prelude: Symbiosis

We begin with some basic tools. Many of these will undoubtedly be review; perhaps a few are new. Still other techniques may be familiar but never mastered. Indeed, it is not uncommon to possess an ability to push symbols around without a firm appreciation of their meaning. And sometimes symbol-pushing camouflages that lack of understanding — even to oneself. This seldom ends well.

Moreover, it's important to realize that mathematical concepts and techniques are more than mere tools: they can be a source of physical insight and intuition. Conversely, physical systems often inspire an appreciation for the power and depth of the math — which in turn can further elucidate the physics. This sort of leveraging infuses much of the book. In fact, let's start with three brief illustrations of the symbiotic relationship between math and physics — examples we'll return to often.

Example 1.1 Oscillations

One of the most important equations in physics is

$$\frac{d^2x}{dt^2} = -\omega^2 x, \tag{1.1}$$

which describes simple harmonic motion with frequency $\omega = 2\pi f$. The solution can be written

$$x(t) = A\cos\omega t + B\sin\omega t, \tag{1.2}$$

where the amplitudes of oscillation A and B can be determined from initial values of position and speed. The significance of this equation lies in the ubiquity of oscillatory motion; moreover, many of the most common differential equations in physics and engineering are in some sense generalizations of this simple expression. These "variations on a theme" consider the effects of additional terms in (1.1), resulting in systems rich both mathematically and physically. But even the simple variation

$$\frac{d^2x}{dt^2} = +\omega^2 x \tag{1.3}$$

brings new possibilities and insight. Although differing from (1.1) by a mere sign, this seemingly trivial mathematical distinction yields a solution which is completely different physically:

$$x(t) = Ae^{-\omega t} + Be^{+\omega t}. \tag{1.4}$$

The mathematics is clearly hinting at an underlying relationship between oscillation and exponential growth and decay. Appreciating this deeper connection requires we take a step back to consider *complex numbers*. Given that position $x(t)$ is real, this may seem counterintuitive. But as we'll see in Part I and beyond, the broader perspective provided by allowing for complex numbers generates physical insight.

Example 1.2 **Angular Momentum and Moment of Inertia**

Recall that angular momentum is given by

$$\vec{L} = \vec{r} \times \vec{p} = I\vec{\omega}, \qquad (1.5)$$

where the angular velocity $|\vec{\omega}| = v/r$, and I is the moment of inertia. In introductory mechanics, one is told that moment of inertia is a scalar quantity which measures the distribution of mass around the axis of rotation. Thus (1.5) clearly shows that angular momentum $\vec{L}$ is always parallel to the angular velocity $\vec{\omega}$. Right?

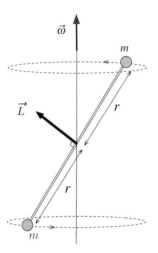

Figure 1.1 Angular momentum of a dumbbell.

Wrong. Perhaps the simplest counter-example is a massless rod with two identical weights at either end, spinning as shown in Figure 1.1 with angular velocity $\vec{\omega}$ pointing vertically (Figure 1.1). With the origin at the center of the rod, each mass contributes equally to the total angular momentum. Since the momentum of each is perpendicular to its position vector $\vec{r}$, the angular momentum is easily found to have magnitude

$$L = |\vec{r} \times \vec{p}| = 2mvr = 2mr^2\omega. \qquad (1.6)$$

This much is certainly correct. The problem, however, is that $\vec{L}$ is not parallel to $\vec{\omega}$! Thus either $\vec{L} = I\vec{\omega}$ holds only in special circumstances, or I cannot be a simple scalar. The resolution requires expanding our mathematical horizons into the realm of *tensors* — in which scalars and vectors are special cases. In this unit we'll introduce notation to help manage this additional level of sophistication; a detailed discussion of tensors is presented in the Entr'acte.

Example 1.3 **Space, Time, and Spacetime**

The Special Theory of Relativity is based upon two postulates put forward by Einstein in 1905:

 I. The laws of physics have the same form in all inertial reference frames.
 II. The speed of light in vacuum is the same in all inertial reference frames.

The first postulate is merely a statement of Galileo's principle of relativity; the second is empirical fact. From these postulates, it can be shown that time and coordinate measurements in two frames with relative speed V along their common x-axis are related by the famous *Lorentz transformations*,

$$t' = \gamma(t - Vx/c^2) \qquad y' = y$$
$$x' = \gamma(x - Vt) \qquad z' = z, \tag{1.7}$$

where $\gamma \equiv (1 - V^2/c^2)^{-1/2}$. At its simplest, special relativity concerns the relationship between measurements in different inertial frames.

Intriguingly, we can render the Lorentz transformations more transparent by rescaling the time as $x_0 \equiv ct$, so that it has the same dimensions as x; the second postulate makes this possible, since it promotes the speed of light to a universal constant of nature. To make the notation consistent, we'll also define $(x_1, x_2, x_3) \equiv (x, y, z)$. Then introducing the dimensionless parameter $\beta \equiv V/c$, (1.7) become

$$x'_0 = \gamma(x_0 - \beta x_1) \qquad x'_2 = x_2$$
$$x'_1 = \gamma(x_1 - \beta x_0) \qquad x'_3 = x_3. \tag{1.8}$$

In this form, the equal footing of space and time, a hallmark of relativity, is manifest. It also clearly displays that transforming from one frame to another "mixes" space and time. Curiously, this is very similar to the way a rotation mixes up position coordinates. Upon rotation of coordinate axes, a vector's new components (x', y') are a simple linear combination of the old components (x, y). Could it be that relativity similarly "rotates" space and time?

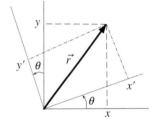

In fact, Lorentz transformations can be understood as rotations in a single, unified mathematical structure called *spacetime* — also known as *Minkowski space* after the mathematician who first worked out this geometric interpretation. Recognition of the fundamental nature of spacetime represents one of the greatest paradigm shifts in the history of science: the nearly 300 year-old notion of a three-dimensional universe plus an external time parameter is replaced by a single *four*-dimensional structure. Deeper insight into the mathematics of rotations will heighten the appreciation for this shift.

2 | Coordinating Coordinates

2.1 Position-Dependent Basis Vectors

Working with a vector-valued function $\vec{V}(\vec{r})$ often involves choosing two coordinate systems: one for the functional dependence, and one for the basis used to decompose the vector. So a vector in the plane, say, can be rendered on the cartesian basis $\{\hat{\imath}, \hat{\jmath}\}$ using either cartesian or polar coordinates,[1]

$$\vec{V}(\vec{r}) = \hat{\imath}\, V_x(x, y) + \hat{\jmath}\, V_y(x, y) \tag{2.1a}$$

$$= \hat{\imath}\, V_x(\rho, \phi) + \hat{\jmath}\, V_y(\rho, \phi), \tag{2.1b}$$

where $V_x \equiv \hat{\imath} \cdot \vec{V}$ and $V_y \equiv \hat{\jmath} \cdot \vec{V}$. But we could just as easily choose the polar basis $\{\hat{\rho}, \hat{\phi}\}$,

$$\vec{V}(\vec{r}) = \hat{\rho}\, V_\rho(\rho, \phi) + \hat{\phi}\, V_\phi(\rho, \phi) \tag{2.1c}$$

$$= \hat{\rho}\, V_\rho(x, y) + \hat{\phi}\, V_\phi(x, y), \tag{2.1d}$$

with $V_\rho \equiv \hat{\rho} \cdot \vec{V}$ and $V_\phi \equiv \hat{\phi} \cdot \vec{V}$. For instance, the position vector $\vec{r}$ can be written

$$\vec{r} = \hat{\imath}\, x + \hat{\jmath}\, y \tag{2.2a}$$

$$= \hat{\imath}\, \rho \cos\phi + \hat{\jmath}\, \rho \sin\phi \tag{2.2b}$$

$$= \hat{\rho}\, \rho \tag{2.2c}$$

$$= \hat{\rho}\, \sqrt{x^2 + y^2}. \tag{2.2d}$$

To date, you have probably had far more experience using cartesian components. At the risk of stating the obvious, this is because the cartesian unit vectors $\hat{\imath}, \hat{\jmath}$ never alter direction — each is parallel to itself everywhere in space. This is not true in general; in other systems *the directions of the basis vectors depend on their spatial location*. For example, at the point $(x, y) = (1, 0)$, the unit vectors in polar coordinates are $\hat{\rho} \equiv \hat{\imath}$, $\hat{\phi} \equiv \hat{\jmath}$, whereas at $(0, 1)$ they are $\hat{\rho} \equiv \hat{\jmath}$, $\hat{\phi} \equiv -\hat{\imath}$ (Figure 2.1). A position-dependent basis vector is not a trivial thing.

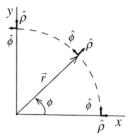

Figure 2.1 The polar basis.

Because the cartesian unit vectors are constant, they can be found by taking derivatives of the position $\vec{r} = x\hat{\imath} + y\hat{\jmath}$ — for instance, $\hat{\jmath} = \partial\vec{r}/\partial y$. We can leverage this to find $\hat{\rho}$ and $\hat{\phi}$ as functions of position, beginning with the chain rule in polar coordinates,

$$d\vec{r} = \frac{\partial\vec{r}}{\partial\rho}\, d\rho + \frac{\partial\vec{r}}{\partial\phi}\, d\phi. \tag{2.3}$$

---

[1] $V(x, y)$ is generally a different function than $V(\rho, \phi)$; we use the same symbol for simplicity.

Now if we hold ϕ fixed but vary ρ by $d\rho$, the resulting displacement $d\vec{r}$ must be in the $\hat{\rho}$-direction. In other words, $\partial \vec{r}/\partial \rho$ is a vector tangent to curves of constant ϕ. Similarly, $\partial \vec{r}/\partial \phi$ is a vector tangent to curves of constant ρ, and points in the $\hat{\phi}$-direction. (See Problem 6.36.) To calculate unit tangent vectors, we decompose the position vector $\vec{r}$ along cartesian axes but with components expressed in polar coordinates,

$$\vec{r} = x\hat{i} + y\hat{j} = \rho \cos \phi \, \hat{i} + \rho \sin \phi \, \hat{j}. \tag{2.4}$$

Then we find

$$\hat{\rho} \equiv \frac{\partial \vec{r}/\partial \rho}{|\partial \vec{r}/\partial \rho|} = \cos \phi \, \hat{i} + \sin \phi \, \hat{j} \tag{2.5a}$$

and

$$\hat{\phi} \equiv \frac{\partial \vec{r}/\partial \phi}{|\partial \vec{r}/\partial \phi|} = -\sin \phi \, \hat{i} + \cos \phi \, \hat{j} \,. \tag{2.5b}$$

Note that the fixed nature of the cartesian basis is crucial in this derivation. As a quick consistency check, not only are these vectors normalized, $\hat{\rho} \cdot \hat{\rho} = \hat{\phi} \cdot \hat{\phi} = 1$, they are also orthogonal, $\hat{\rho} \cdot \hat{\phi} = 0$. And as expected, they are position-dependent; direct differentiation of (2.5) gives

$$\frac{\partial \hat{\rho}}{\partial \phi} = \hat{\phi} \qquad \frac{\partial \hat{\phi}}{\partial \phi} = -\hat{\rho}, \tag{2.6a}$$

and also, as expected,

$$\frac{\partial \hat{\rho}}{\partial \rho} = 0 \qquad \frac{\partial \hat{\phi}}{\partial \rho} = 0. \tag{2.6b}$$

We can trivially extend these results to cylindrical coordinates (ρ, ϕ, z), since $\hat{k} \equiv \hat{z}$ is a constant basis vector.

Example 2.1   Velocity in the Plane

Consider the velocity $\vec{v} = d\vec{r}/dt$. Along cartesian axes, this is just

$$\vec{v} \equiv \frac{d\vec{r}}{dt} = \frac{d}{dt}(x\hat{i}) + \frac{d}{dt}(y\hat{j}) = \frac{dx}{dt}\hat{i} + \frac{dy}{dt}\hat{j}, \tag{2.7}$$

while on the polar basis we have instead

$$\frac{d\vec{r}}{dt} = \frac{d}{dt}(\rho\hat{\rho}) = \frac{d\rho}{dt}\hat{\rho} + \rho\frac{d\hat{\rho}}{dt}$$

$$= \frac{d\rho}{dt}\hat{\rho} + \rho\left(\frac{\partial\hat{\rho}}{\partial\rho}\frac{d\rho}{dt} + \frac{\partial\hat{\rho}}{\partial\phi}\frac{d\phi}{dt}\right), \tag{2.8}$$

where the chain rule was used in the last step. Of course, (2.7) and (2.8) are equivalent expressions for the same velocity. As it stands, however, (2.8) is not in a particularly useful form: if we're going to use the polar basis, $\vec{v}$ should be clearly expressed in terms of $\hat{\rho}$ and $\hat{\phi}$, not their derivatives. This is where (2.6) come in, allowing us to completely decompose the velocity on the polar basis:

$$\vec{v} = \frac{d\rho}{dt}\hat{\rho} + \rho\frac{d\phi}{dt}\frac{\partial\hat{\rho}}{\partial\phi}$$

$$\equiv \dot{\rho}\,\hat{\rho} + \rho\dot{\phi}\,\hat{\phi}, \tag{2.9}$$

where we have used the common "dot" notation to denote time derivative, $\dot{u} \equiv du/dt$. The forms of the radial and tangential components,

$$v_\rho = \dot{\rho}, \quad v_\phi = \rho \dot{\phi} \equiv \rho\, \omega, \tag{2.10}$$

should be familiar; $\omega \equiv \dot{\phi}$, of course, is the angular velocity.

Example 2.2   **Acceleration in the Plane**

In cartesian coordinates, a particle in the plane has acceleration

$$\vec{a} \equiv \frac{d\vec{v}}{dt} = \frac{d^2\vec{r}}{dt^2} = \ddot{x}\hat{\imath} + \ddot{y}\hat{\jmath}, \tag{2.11}$$

where $\ddot{u} \equiv d^2u/dt^2$. The same calculation in polar coordinates is more complicated, but can be accomplished without difficulty: differentiating (2.9) and using (2.6a) and (2.6b) gives

$$\begin{aligned}
\vec{a} &= \left(\ddot{\rho} - \rho\dot{\phi}^2\right)\hat{\rho} + \left(\rho\ddot{\phi} + 2\dot{\rho}\dot{\phi}\right)\hat{\phi} \\
&\equiv \left(\ddot{\rho} - \rho\omega^2\right)\hat{\rho} + \left(\rho\alpha + 2\dot{\rho}\omega\right)\hat{\phi},
\end{aligned} \tag{2.12}$$

where $\alpha \equiv \ddot{\phi}$ is the angular acceleration. The $\ddot{\rho}$ and $\rho\alpha$ terms have straightforward interpretations: changing speed in either the radial or tangential directions results in acceleration. The $-\rho\omega^2\hat{\rho}$ term is none other than centripetal acceleration (recall that $\rho\omega^2 = v_\phi^2/\rho$); herein is the elementary result that acceleration can occur even at constant speed. The fourth term, $2\dot{\rho}\omega\hat{\phi}$, is the Coriolis acceleration observed within a non-inertial frame rotating with angular velocity $\omega\,\hat{\phi}$.

Though a little more effort, the procedure that led to (2.6a) and (2.6b) also works for spherical coordinates r, θ, and ϕ. Starting with

$$d\vec{r} = \frac{\partial\vec{r}}{\partial r}\,dr + \frac{\partial\vec{r}}{\partial\theta}\,d\theta + \frac{\partial\vec{r}}{\partial\phi}\,d\phi \tag{2.13}$$

and

$$\vec{r} = r\sin\theta\,\cos\phi\,\hat{\imath} + r\sin\theta\,\sin\phi\,\hat{\jmath} + r\cos\theta\,\hat{k}, \tag{2.14}$$

we use the tangent vectors to define

$$\hat{r} \equiv \frac{\partial\vec{r}/\partial r}{|\partial\vec{r}/\partial r|} \qquad \hat{\theta} \equiv \frac{\partial\vec{r}/\partial\theta}{|\partial\vec{r}/\partial\theta|} \qquad \hat{\phi} \equiv \frac{\partial\vec{r}/\partial\phi}{|\partial\vec{r}/\partial\phi|}. \tag{2.15}$$

As you'll verify in Problem 2.1, this leads to

$$\hat{r} = \hat{\imath}\sin\theta\cos\phi + \hat{\jmath}\sin\theta\sin\phi + \hat{k}\cos\theta \tag{2.16a}$$

$$\hat{\theta} = \hat{\imath}\cos\theta\cos\phi + \hat{\jmath}\cos\theta\sin\phi - \hat{k}\sin\theta \tag{2.16b}$$

$$\hat{\phi} = -\hat{\imath}\sin\phi + \hat{\jmath}\cos\phi, \tag{2.16c}$$

so that

$$\partial\hat{r}/\partial\theta = \hat{\theta} \qquad\qquad \partial\hat{r}/\partial\phi = \sin\theta\,\hat{\phi} \tag{2.17a}$$

$$\partial\hat{\theta}/\partial\theta = -\hat{r} \qquad\qquad \partial\hat{\theta}/\partial\phi = \cos\theta\,\hat{\phi} \tag{2.17b}$$

$$\partial\hat{\phi}/\partial\theta = 0 \qquad\qquad \partial\hat{\phi}/\partial\phi = -\sin\theta\,\hat{r} - \cos\theta\,\hat{\theta}. \tag{2.17c}$$

And of course, all the unit vectors have vanishing radial derivatives.

Once again, a quick consistency check verifies that $\hat{r}, \hat{\theta}$, and $\hat{\phi}$ form an *orthonormal* system — that is, that they are unit vectors ("normalized"),

$$\hat{r} \cdot \hat{r} = \hat{\theta} \cdot \hat{\theta} = \hat{\phi} \cdot \hat{\phi} = 1,$$

and they're mutually orthogonal,

$$\hat{r} \cdot \hat{\theta} = \hat{\theta} \cdot \hat{\phi} = \hat{\phi} \cdot \hat{r} = 0.$$

Moreover, the so-called "right-handedness" of cartesian coordinates, $\hat{\imath} \times \hat{\jmath} = \hat{k}$, bequeathes an orientation to spherical coordinates,[2]

$$\hat{\imath} \times \hat{\jmath} = \hat{k} \iff \hat{r} \times \hat{\theta} = \hat{\phi}.$$

Similarly, in cylindrical coordinates $\hat{\rho} \times \hat{\phi} = \hat{k}$. (Caution: ρ denotes the cylindrical radial coordinate, r the spherical. But it's very common to use r for both — as we will often do. So pay attention to context!)

> **Example 2.3** **A First Look at Rotations**
>
> There's a simple and handy way to convert back and forth between these coordinate bases. A mere glance at (2.5) and (2.16) reveals that they can be expressed in matrix form,
>
> $$\begin{pmatrix} \hat{\rho} \\ \hat{\phi} \end{pmatrix} = \begin{pmatrix} \cos\phi & \sin\phi \\ -\sin\phi & \cos\phi \end{pmatrix} \begin{pmatrix} \hat{\imath} \\ \hat{\jmath} \end{pmatrix}, \tag{2.18}$$
>
> and
>
> $$\begin{pmatrix} \hat{r} \\ \hat{\theta} \\ \hat{\phi} \end{pmatrix} = \begin{pmatrix} \sin\theta\cos\phi & \sin\theta\sin\phi & \cos\theta \\ \cos\theta\cos\phi & \cos\theta\sin\phi & -\sin\theta \\ -\sin\phi & \cos\phi & 0 \end{pmatrix} \begin{pmatrix} \hat{\imath} \\ \hat{\jmath} \\ \hat{k} \end{pmatrix}. \tag{2.19}$$
>
> The inverse transformations $\{\hat{\rho}, \hat{\phi}\} \to \{\hat{\imath}, \hat{\jmath}\}$ and $\{\hat{r}, \hat{\theta}, \hat{\phi}\} \to \{\hat{\imath}, \hat{\jmath}, \hat{k}\}$ can be found either by inverting (2.5) and (2.16) algebraically, or equivalently by finding the inverse of the matrices in (2.18) and (2.19). But this turns out to be very easy, since (as you can quickly verify) the inverses are just their transposes. Thus
>
> $$\begin{pmatrix} \hat{\imath} \\ \hat{\jmath} \end{pmatrix} = \begin{pmatrix} \cos\phi & -\sin\phi \\ \sin\phi & \cos\phi \end{pmatrix} \begin{pmatrix} \hat{\rho} \\ \hat{\phi} \end{pmatrix}, \tag{2.20}$$
>
> and
>
> $$\begin{pmatrix} \hat{\imath} \\ \hat{\jmath} \\ \hat{k} \end{pmatrix} = \begin{pmatrix} \sin\theta\cos\phi & \cos\theta\cos\phi & -\sin\phi \\ \sin\theta\sin\phi & \cos\theta\sin\phi & \cos\phi \\ \cos\theta & -\sin\theta & 0 \end{pmatrix} \begin{pmatrix} \hat{r} \\ \hat{\theta} \\ \hat{\phi} \end{pmatrix}. \tag{2.21}$$
>
> In Part III, we'll see that $R^{-1} = R^T$ is a defining feature of rotations in $\mathbb{R}^n$.

2.2 Scale Factors and Jacobians

As a vector, $d\vec{r}$ can be decomposed on any basis. As you'll show in Problem 2.3, the chain rule — together with (2.4) and (2.5) — leads to the familiar line element in cylindrical coordinates,

$$d\vec{r} = \hat{\rho}\,d\rho + \hat{\phi}\,\rho\,d\phi + \hat{k}\,dz. \tag{2.22}$$

---

[2] A left-handed system has $\hat{\jmath} \times \hat{\imath} = \hat{k}$ and $\hat{\theta} \times \hat{r} = \hat{\phi}$.

The extra factor of ρ in the second term is an example of a *scale factor*, and has a simple geometric meaning: at a distance ρ from the origin, a change of angle $d\phi$ leads to a change of distance $\rho\,d\phi$. Such scale factors, usually denoted by h, neatly summarize the geometry of the coordinate system. In cylindrical coordinates,

$$h_\rho \equiv 1, \qquad h_\phi \equiv \rho, \qquad h_z \equiv 1. \tag{2.23}$$

In spherical, the scale factors are

$$h_r \equiv 1, \qquad h_\theta \equiv r, \qquad h_\phi \equiv r\sin\theta, \tag{2.24}$$

corresponding to the line element

$$d\vec{r} = \hat{r}\,dr + \hat{\theta}\,r\,d\theta + \hat{\phi}\,r\sin\theta\,d\phi. \tag{2.25}$$

So in spherical coordinates, a change of $d\theta$ does not by itself give the displacement; for that we need an additional factor of r. Indeed, neither $d\theta$ nor $d\phi$ has the same units as $d\vec{r}$; scale factors, which are generally functions of position, remedy this.

BTW 2.1 Curvilinear Coordinates I

Cylindrical and spherical coordinates are the two most common $\mathbb{R}^3$ examples of *curvilinear coordinates*. Cartesian coordinates, of course, are notable by their straight axes and constant unit vectors; by contrast, curvilinear systems have curved "axes" and position-dependent unit vectors. The procedure we used to derive the cylindrical and spherical unit vectors and scale factors can be generalized to any set of curvilinear coordinates.

Let's simplify the notation by denoting cartesian coordinates as x_i, where $i = 1$, 2, or 3; for curvilinear coordinates, we'll use u_i with unit vectors $\hat{e}_i$. (So in cylindrical coordinates, we have the assignments $\hat{e}_1 \to \hat{\rho}$, $\hat{e}_2 \to \hat{\phi}$, $\hat{e}_3 \to \hat{k}$, and in spherical $\hat{e}_1 \to \hat{r}$, $\hat{e}_2 \to \hat{\theta}$, $\hat{e}_3 \to \hat{\phi}$.)

Given the functions $x_i = x_i(u_1, u_2, u_3)$ relating the two coordinate systems, we want a general expression for the unit vectors $\hat{e}_i$ and their derivatives. Once again we begin with the chain rule,

$$d\vec{r} = \frac{\partial \vec{r}}{\partial u_1}\,du_1 + \frac{\partial \vec{r}}{\partial u_2}\,du_2 + \frac{\partial \vec{r}}{\partial u_3}\,du_3, \tag{2.26}$$

where, as before, the position vector $\vec{r} \equiv (x_1, x_2, x_3)$. Now the vector $\partial\vec{r}/\partial u_i$ is tangent to the u_i coordinate axis (or perhaps more accurately, the u_i coordinate curve), and so defines a coordinate direction $\hat{e}_i$. These unit vectors can be decomposed on the cartesian basis as

$$\hat{e}_i \equiv \frac{1}{h_i}\frac{\partial\vec{r}}{\partial u_i} = \frac{1}{h_i}\left(\hat{i}\,\frac{\partial x_1}{\partial u_i} + \hat{j}\,\frac{\partial x_2}{\partial u_i} + \hat{k}\,\frac{\partial x_3}{\partial u_i}\right), \tag{2.27a}$$

where the scale factors h_i are determined by the condition $\hat{e}_i \cdot \hat{e}_i = 1$,

$$h_i \equiv \left|\frac{\partial\vec{r}}{\partial u_i}\right| = \sqrt{\left(\frac{\partial x_1}{\partial u_i}\right)^2 + \left(\frac{\partial x_2}{\partial u_i}\right)^2 + \left(\frac{\partial x_3}{\partial u_i}\right)^2}. \tag{2.27b}$$

In general, the $\hat{e}_i$ and the h_i are both functions of position $\vec{r}$ — that's how we got into this mess in the first place. Instead of staying fixed, the triplet of unit vectors rotates as we move around in space.

Note that a scale factor is given by the ratio of position to curvilinear coordinate. So even though the coordinates u_i may not have units of distance, the scale factors guarantee that the displacement does,

$$d\vec{r} = \hat{e}_1 h_1 du_1 + \hat{e}_2 h_2 du_2 + \hat{e}_3 h_3 du_3. \tag{2.28}$$

We saw this for cylindrical and spherical coordinates in (2.22) and (2.25).

The scale factors are relevant not just for calculations of distance. Suppose we want to find the area of some region S in $\mathbb{R}^2$. In cartesian coordinates this is conceptually straightforward: divide the region into little rectangles with sides Δx, Δy and add up the area of these rectangles. The more rectangles we cram into the region, the better our sum approximates the region's true area; in the limit in which the number of these rectangles goes to infinity we get the exact area of S. Of course, in this limit the sides of the rectangles become infinitesimally small and the sum becomes an integral, $\sum_S \Delta x \Delta y \longrightarrow \int_S dx\, dy$.

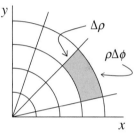

This much is simple. But what if we choose to calculate the area in, say, polar coordinates? One might naively try the same approach and assert $\int_S dx\, dy = \int_S d\rho\, d\phi$ — in other words, that the infinitesimal measure $dx\, dy$ is equivalent to $d\rho\, d\phi$. This, however, cannot be correct, if only because the units of these two expressions don't match. But a glance at (2.22) shows what's going on. If we cut up S into little rectangles along curves of constant ρ and ϕ (Figure 2.2), the area of each little rectangle is not $\Delta\rho\Delta\phi$ but rather $\Delta\rho \cdot \rho\Delta\phi$; the extra factor is the product $h_\rho h_\phi$. Thus the area element $dx\, dy$ is actually transformed into the familiar $\rho\, d\rho\, d\phi$. Though area on the surface of a sphere deforms differently than inside a circle, the same basic thing

Figure 2.2 Polar scale factors.

occurs in spherical coordinates; in this case, we'd find $h_\theta h_\phi = r^2 \sin\theta$. Similarly, the volume element requires the product of all three h's — so in spherical coordinates, $d\tau = r^2 \sin\theta\, dr\, d\theta\, d\phi$. As before, the extra factors ensure the correct units for area and volume.

Example 2.4 **Solid Angle**

Consider a circle. A given angle θ subtends an arc length s which grows linearly with the radius r of the circle. Thus the angle can be fully specified by the ratio $\theta = s/r$. This in fact is the definition of a radian; as a ratio of distances it is dimensionless. Since the entire circle has arc length (circumference) $2\pi r$, there are 2π radians in a complete circle.

Now consider a sphere. A given *solid angle* Ω subtends an area A ("two-dimensional arc") which grows *quadratically* with the radius of the sphere. Thus the solid angle can be fully specified by the ratio $\Omega = A/r^2$ (Figure 2.3). This in fact is the definition of a square radian, or *steradian*; as a ratio of areas it is dimensionless. Since the entire sphere has surface area $4\pi r^2$, there are 4π steradians in a complete sphere.[3] Note that in both cases, the radial line is perpendicular to the circular arc or the spherical surface.

An infinitesimal solid angle $d\Omega = dA/r^2$ can be expressed in terms of the scale factors h_θ and h_ϕ,

$$d\Omega = dA/r^2 = h_\theta h_\phi\, d\theta\, d\phi / r^2 = \sin\theta\, d\theta\, d\phi. \tag{2.29}$$

---

[3] 4π steradians $\approx 41,253$ degrees^2.

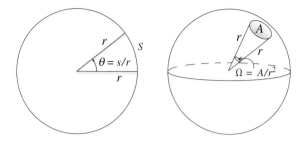

Figure 2.3 Definitions of radian and steradian.

So in terms of solid angle, the surface area of a sphere is just

$$A = \int r^2 \, d\Omega = r^2 \int_0^\pi \sin\theta \, d\theta \int_0^{2\pi} d\phi = 4\pi r^2, \tag{2.30}$$

where, since r is constant on the spherical surface, it can be taken out of the integral. This confirms that there are 4π steradians in a complete sphere — or equivalently, that the surface area of the unit sphere is 4π.

Implicit in these geometric constructions is that the new coordinates are orthogonal; this allows us to treat infinitesimal regions as rectangles (or cubes) with areas (or volumes) given by the product of the lengths of the sides. Scale factors of non-orthogonal systems will not so easily lead to the correct infinitesimal elements.

Often, however, all that is needed are the *products* of the scale factors, in which case deriving the individual h's is not the best way to determine the integration measures da or $d\tau$. Instead, when transforming from one system of coordinates (x, y) to another system (u, v) — even non-orthogonal ones — area and volume elements can be found directly from the *Jacobian* $|J|$, defined in two dimensions by

$$da = dx \, dy = |J| \, du \, dv, \tag{2.31}$$

where $|J|$ is the determinant of the Jacobian matrix[4]

$$J = \frac{\partial(x, y)}{\partial(u, v)} \equiv \begin{pmatrix} \partial x/\partial u & \partial x/\partial v \\ \partial y/\partial u & \partial y/\partial v \end{pmatrix}. \tag{2.32}$$

For orthogonal systems, the Jacobian is the product of the scale factors. (Verification of this is a worthwhile calculation, so we relegate it to Problem 2.10.)

Example 2.5 **From Cartesian to Polar**

For polar coordinates, $x = \rho\cos\phi$, $y = \rho\sin\phi$, and (2.32) gives

$$|J| = \left| \frac{\partial(x, y)}{\partial(\rho, \phi)} \right| = \begin{vmatrix} \cos\phi & -\rho\sin\phi \\ \sin\phi & \rho\cos\phi \end{vmatrix} = \rho, \tag{2.33}$$

as expected. (Does it matter that we assigned $u \equiv \rho$ and $v \equiv \phi$ rather than vice-versa?)

---

[4] Sometimes J is defined to be the determinant itself, so $dx \, dy = J \, du \, dv$.

Example 2.6 **From Cartesian to Skew**

What about the change from orthogonal $\{\hat{\imath}, \hat{\jmath}\}$ to a non-orthogonal (sometimes called "skew") system $\{\hat{u}, \hat{v}\}$ given by, say, $u = x$ and $v = x + y$? Then

$$|J| = \left| \frac{\partial(x, y)}{\partial(u, v)} \right| = \begin{vmatrix} 1 & 0 \\ -1 & 1 \end{vmatrix} = 1, \tag{2.34}$$

so that $da = dx\,dy = du\,dv$. Note that were we to find the h_i individually from (2.27), we'd find $h_u = \sqrt{2}$, $h_v = 1$ — in other words, $h_u h_v \neq |J|$. But then, $\{\hat{u}, \hat{v}\}$ is not an orthogonal system, $\hat{u} \cdot \hat{v} \neq 0$, so the product of the scale factors shouldn't be expected to lead to the correct integration measure.

The procedure summarized by (2.31) and (2.32) is fundamental, and applies whenever an invertible coordinate change is made — in other words, whenever $|J| \neq 0$. Moreover, it readily generalizes to $\mathbb{R}^n$,

$$\int dx_1 dx_2 \cdots dx_n = \int \left| \frac{\partial(x_1, \ldots, x_n)}{\partial(u_1, \ldots, u_n)} \right| du_1 du_2 \cdots du_n. \tag{2.35}$$

The notation is intended to be suggestive of the chain rule.

As a quick aside, you should be(come) familiar with the volume element notation

$$d\tau \equiv d^n x \equiv d^n r \tag{2.36}$$

in n-dimensional space. In particular, the expression $d^n x$ does *not* presume a choice of the coordinate system. So, for instance, in $\mathbb{R}^3$

$$d^3 x = dx_1\, dx_2\, dx_3 = \rho\, d\rho\, d\phi\, dz = r^2 \sin\theta\, dr\, d\theta\, d\phi. \tag{2.37}$$

All three are equivalent, each with its appropriate Jacobian factor.

BTW 2.2 The n-Dimensional Sphere S^n

The circle and sphere are one- and two-dimensional examples of the locus of points satisfying

$$S^n : x_1^2 + x_2^2 + \cdots + x_{n+1}^2 = R^2, \tag{2.38}$$

for some constant radius R. The so-called n-sphere S^n is an n-dimensional spherical surface in $\mathbb{R}^{n+1}$. The circle is S^1, conventionally parametrized by the single angle ϕ,

$$x = R\cos\phi \qquad y = R\sin\phi, \tag{2.39}$$

where $0 \leq \phi < 2\pi$. The surface S^2 is parametrized by the familiar

$$x = R\sin\theta\cos\phi \qquad y = R\sin\theta\sin\phi \qquad z = R\cos\theta, \tag{2.40}$$

where $0 \leq \phi < 2\pi$ and $0 \leq \theta \leq \pi$. One can visualize S^2 as the two-dimensional surface in $\mathbb{R}^3$ which emerges by rotating S^1 around a diameter; since only a half-turn is required to generate the sphere, θ need only go up to π.

Though impossible to picture, by similarly rotating S^2 a half-turn out of $\mathbb{R}^3$ through an angle ψ one can generate S^3, the three-dimensional spherical surface of a ball in four-dimensional space,

$$x = R \sin \psi \sin \theta \cos \phi$$
$$y = R \sin \psi \sin \theta \sin \phi$$
$$z = R \sin \psi \cos \theta$$
$$w = R \cos \psi. \tag{2.41}$$

This is easily shown to satisfy (2.38) for $n = 3$. We still have $0 \leq \phi < 2\pi$, but both θ and ψ run from 0 to π.

For non-constant $R \to r$, (2.41) defines spherical (sometimes called "hyperspherical") coordinates in $\mathbb{R}^4$. Though perhaps a bit tedious, finding the Jacobian determinant is a straightforward calculation,

$$|J| = r^3 \sin^2 \psi \sin \theta. \tag{2.42}$$

We can then identify $d\Omega_3 \equiv \sin^2 \psi \sin \theta$ as the three-dimensional element of solid angle, with a total of

$$\int d\Omega_3 = \int \sin^2 \psi \sin \theta \, d\psi d\theta d\phi = 2\pi^2 \tag{2.43}$$

cubic radians. The hypersurface area of the three-sphere, then, is $2\pi^2 R^3$. Similarly, the volume of the four-dimensional ball is

$$\int r^3 \, dr \, d\Omega_3 = \frac{1}{2}\pi^2 R^4, \tag{2.44}$$

As we continue to step up in dimension, it's convenient to define

$$\theta_0 \equiv \pi/2 - \varphi \qquad \theta_1 \equiv \theta \qquad \theta_2 \equiv \psi, \dots, \tag{2.45}$$

where all the $\theta_{i>0}$ run from 0 to π. Then the parametrization of S^3 takes the satisfying form

$$x_1 = R \sin \theta_2 \sin \theta_1 \sin \theta_0$$
$$x_2 = R \sin \theta_2 \sin \theta_1 \cos \theta_0$$
$$x_3 = R \sin \theta_2 \cos \theta_1$$
$$x_4 = R \cos \theta_2. \tag{2.46}$$

The emerging pattern is sufficient to generalize for all n, with Jacobian matrix

$$|J| = R^n \sin^{n-1} \theta_{n-1} \sin^{n-2} \theta_{n-2} \cdots \sin \theta_1$$

$$= R^n \prod_{m=1}^{n-1} \sin^m \theta_m = R^n \prod_{m=1}^{n} \sin^{m-1} \theta_{m-1}, \tag{2.47}$$

where $\prod$ denotes the product, and the last expression includes $\sin^0 \theta_0 = 1$ for aesthetic balance.

Problems

2.1 Verify (2.5) and (2.16).

2.2 Draw a sketch verifying the relationships in (2.6).

2.3 Verify the line elements in (2.22) and (2.25).

2.4 Find the matrix which expresses the spherical coordinates r, θ, ϕ in terms of the cylindrical coordinates ρ, ϕ, z, and vice-versa.

2.5 Demonstrate that the basis vectors of spherical coordinates, $\{\hat{r}, \hat{\theta}, \hat{\phi}\}$ are in fact unit vectors and that the system is right handed. Do the same for cylindrical coordinates $\{\hat{\rho}, \hat{\phi}, \hat{k}\}$.

2.6 Write in cartesian, cylindrical, and spherical coordinates equations for the following surfaces:

(a) the unit sphere centered at the origin
(b) a cylinder of unit radius centered on the z-axis.

2.7 Consider a line through the origin at angles with respect to the three coordinate axes given by α, β, and γ. More useful are the line's so-called *direction cosines*, i.e., $\cos \alpha$, $\cos \beta$, and $\cos \gamma$.

(a) Show that $\cos^2 \alpha + \cos^2 \beta + \cos^2 \gamma = 1$.
(b) Given a second line with direction cosines $\cos \alpha'$, $\cos \beta'$, and $\cos \gamma'$, show that the angle θ between the lines is given by

$$\cos \theta = \cos \alpha \cos \alpha' + \cos \beta \cos \beta' + \cos \gamma \cos \gamma'.$$

2.8 Consider two vectors $\vec{r}$ and $\vec{r}'$ with spherical coordinates (r, θ, ϕ) and (r', θ', ϕ'). Show that the angle γ between them is given by

$$\cos \gamma = \cos \theta \cos \theta' + \sin \theta \sin \theta' \cos (\phi - \phi').$$

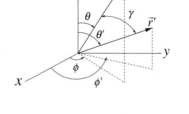

2.9 Follow the steps outlined in Example 2.2 to verify that in polar coordinates acceleration in the plane is given by (2.12)

$$\vec{a} \equiv \ddot{\vec{r}} = \left(\ddot{\rho} - \rho \omega^2 \right) \hat{\rho} + \left(\rho \alpha + 2 \dot{\rho} \omega \right) \hat{\phi},$$

where $\omega \equiv \dot{\phi} = d\phi/dt$ and $\alpha \equiv \ddot{\phi} = \dot{\omega} = d^2\phi/dt^2$.

2.10 For a change of coordinates $u = u(x, y)$ and $v = v(x, y)$, show that if $\hat{u} \cdot \hat{v} = 0$, then the Jacobian is the product of the scale factors, $|J| = h_u h_v$. This is easier to do starting with $|J|^2$ rather the $|J|$. Two basic properties of determinants makes things even easier: first, the determinant of a matrix is the same as that of its transpose; second, the determinant of a product is the product of the determinants. Use these to show that $|J|^2 = |JJ^T| = (h_u h_v)^2$.

2.11 Use the cross product, together with properties of $\hat{\imath}$, $\hat{\jmath}$, and $\hat{k}$, to find the normal to the surface of a sphere.

2.12 With the direction given by the unit normal to the surface, find the vector-valued area elements $d\vec{a}$:

(a) in cylindrical coordinates:

(i) on a surface of constant ρ

  (ii) on a surface of constant ϕ
  (iii) on a surface of constant z.

(b) In spherical coordinates:

  (i)   on a surface of constant r
  (ii) on a surface of constant θ
  (iii) on a surface of constant ϕ.

2.13 Motion in a central potential is confined to a plane, courtesy of the antisymmetry of the cross product (Example 4.6). Thus polar coordinates (r, ϕ) are the natural choice for analyzing orbits in an inverse-square force $\vec{F} = -k\hat{r}/r^2$.

(a) Decompose Newton's second law $m\ddot{\vec{r}} = \vec{F}$ to find equations for radial motion $\ddot{r}$ and angular motion $\ddot{\phi}$.

(b) Show that the $\ddot{\phi}$ equation implies conservation of angular momentum ℓ. This is the implication of Kepler's second law ("equal areas in equal times").

(c) Find an expression for the kinetic energy $\frac{1}{2}mv^2$ in terms of ℓ. What form does it take for circular orbits?

2.14 **Gauss' Law**
The net electric flux Φ of an electric field $\vec{E}$ through a surface S is given by adding up all the contributions $d\Phi = \vec{E} \cdot \hat{n}\, dA$ from each area element dA on the surface with unit normal $\hat{n}$. Consider a point charge q enclosed by a surface S of arbitrary shape.

(a) Show that $d\Phi = \frac{q}{4\pi\epsilon_0} \frac{\cos\theta\, dA}{r^2}$, where θ is the angle between the electric field and the normal to the surface. Draw a sketch, identifying $\hat{n}$ and $\hat{r}$ — and remember that since S is of arbitrary shape, dA is not necessarily perpendicular to a radial line from the origin.

(b) Show on your sketch (or draw a new one) that the component of the *vector-valued* area $d\vec{A} \equiv dA\,\hat{n}$ perpendicular to the radial line is $dA \cos\theta$.

(c) Use the definition of solid angle to show that the net flux $\Phi = \oint_S \vec{E} \cdot \hat{n}\, dA$ is given by Gauss' law,

$$\Phi = q/\epsilon_0.$$

2.15 Use the Jacobian to derive the form of the volume element in spherical coordinates.

2.16 Elliptical coordinates in the plane are defined by

$$x = a\cosh u \cos\phi \qquad y = a\sinh u \sin\phi,$$

where $u \geq 0, 0 \leq \phi < 2\pi$, and constant a. Just as polar coordinates are useful when a system has circular symmetry, elliptical coordinates are appropriate when the symmetry is elliptical.

(a) Show that curves of constant u are ellipses with foci at $\pm a$, while curves of constant ϕ are hyperbolas. Use Mathematica's `ParametricPlot` command to create a plot of the coordinates curves with $a = 1$. Identify the curve $u = 0$.

(b) Find the scale factors h_u and h_ϕ.

2.17 Oblate spheroidal coordinates are defined by

$$x = a\cosh u \sin\theta \cos\phi \qquad y = a\cosh u \sin\theta \sin\phi \qquad z = a\sinh u \cos\theta,$$

where $u > 0, 0 \leq \theta \leq \pi, 0 \leq \phi < 2\pi$, and constant a. Just as spherical coordinates are useful when a system has circular or spherical symmetry, oblate spheroidal coordinates are appropriate when the symmetry is elliptical or hyperbolic.

(a) Use the identity $\cosh^2 u - \sinh^2 u = 1$ to show that surfaces of constant u are ellipsoids. (In fact, since $\sinh^2 u < \cosh^2 u$, they form *oblate spheroids*.) Similarly, show that surfaces of constant θ are hyperboloids.

(b) Show that the scale factors for oblate spheroidal coordinates are $h_u = h_\theta = a\sqrt{\sinh^2 u + \cos^2 \theta}$, and $h_\phi = a \cosh u \sin \theta$. [Note: $\frac{d}{du} \sinh u = + \cosh u$, $\frac{d}{du} \cosh u = + \sinh u$.]

(c) Find the Jacobian $|J|$ of the transformation from cartesian to oblate spheroidal coordinates.

(d) What are the area elements da for surfaces of constant u? constant θ?

2.18 Following the procedure used to derive (2.5) and (2.16) (see also (2.27) in BTW 2.1), express the vectors $\hat{e}_u$, $\hat{e}_\theta$ and $\hat{e}_\phi$ for oblate spheroidal coordinates on the cartesian basis $\{\hat{i}, \hat{j}, k\}$, and show that they form an orthonormal set — that is, each is a unit vector, perpendicular to the other two. (You may use the scale factors given in Problem 2.17.)

2.19 Show that the four-dimensional line element in hyperspherical coordinates, (2.41) with $R \rightarrow r$, is

$$d\vec{r} = \hat{r}\, dr + \hat{\psi}\, r\, d\psi + \hat{\theta}\, r\sin\psi\, d\theta + \hat{\phi}\, r\sin\psi \sin\theta\, d\phi. \tag{2.48}$$

Demonstrate that the unit vectors are orthogonal, and that the product of the scale factors gives the Jacobian, (2.42).

2.20 Verify the Jacobian determinant (2.42) for the hyperspherical coordinates of (2.41). [With patience and fortitude, you can calculate this directly. But you'll find the determinant a bit more manageable if you take the coordinates x_i in reverse order, so that the top row of J contains the derivatives of x_4 — two of which are zero. Also, if you're clever you'll be able to factor the 4×4 determinant into a sum over 3×3 determinants — each of which is the Jacobian of conventional spherical coordinates in $\mathbb{R}^3$, whose determinant you already know.]

3 Complex Numbers

3.1 Representations

Complex numbers can be expressed in cartesian coordinates along the real and imaginary axes (Figure 3.1),

$$z = a + ib. \tag{3.1}$$

The same complex number can also be represented in polar coordinates,

$$z = r\cos\varphi + ir\sin\varphi \equiv re^{i\varphi}, \tag{3.2}$$

where we have invoked Euler's ("oy-ler") indispensable identity

$$e^{ix} = \cos x + i\sin x. \tag{3.3}$$

Associated with any complex number z is its *complex conjugate* z^*,

$$z^* \equiv a - ib = re^{-i\varphi}. \tag{3.4}$$

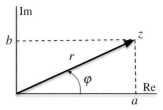

Figure 3.1 The complex plane.

Geometrically, z^* is the reflection of z across the real axis. The need for both z and z^* arises from the fact that complex numbers live in two-dimensional space, and thus encode two independent pieces of information. Using polar coordinates in the complex plane, these are

$$r^2 = a^2 + b^2 = zz^* \equiv |z|^2 \qquad \varphi = \tan^{-1}(b/a) \equiv \arg(z). \tag{3.5}$$

In standard parlance, $|z| = r$ is called the "modulus" or "magnitude" of z. Why the absolute value symbol? Because just as for real numbers, $|z|$ measures distance from the origin. Note that although z^2 can be negative, $|z|^2 \geq 0$.

The "argument" or "phase" of z is $\arg(z) \equiv \varphi$. Real numbers have phase $\varphi = 0$ or π; an imaginary number has $\varphi = \pi/2$ or $3\pi/2$. For any φ, a complex number of the form $e^{i\varphi}$ has unit modulus,

$$|e^{i\varphi}|^2 = (\cos\varphi + i\sin\varphi)(\cos\varphi - i\sin\varphi) = \cos^2\varphi + \sin^2\varphi = 1, \tag{3.6}$$

or, more simply,

$$|e^{i\varphi}|^2 = e^{i\varphi}e^{-i\varphi} = 1. \tag{3.7}$$

A complex number of unit magnitude is often called a "pure phase"; the equation $z = e^{i\varphi}$ describes the unit circle in the complex plane.

Example 3.1 *i* is Geometry!

Ask someone what i is, and you're likely to hear "it's just $\sqrt{-1}$." But ask someone what i *does*, and you'll often get an unsatisfactory response. The problem is that most of us have been introduced to complex numbers *algebraically* rather than *geometrically*. We already know part of the story: a complex number $z = a + ib$ represents a point (a, b) in $\mathbb{R}^2$. Thus adding and subtracting complex numbers is equivalent to vector addition in the plane. But what about multiplying vectors? Although one can dot and cross vectors, neither operation returns another vector in $\mathbb{R}^2$. Indeed, neither of those operations can be inverted since division is not part of their algebraic structure. Complex numbers introduce something new: the ability to multiply and divide points in the plane — an operation with a straightforward geometric interpretation:

$$Z \equiv z_2 z_1 = r_2 e^{i\varphi_2} r_1 e^{i\varphi_1} = r_2 r_1 e^{i(\varphi_2 + \varphi_1)}. \tag{3.8}$$

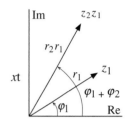

Thus z_2 acts on z_1, *rotating* z_1 by φ_2 and *magnifying* it by r_2. Since multiplication of complex numbers is commutative, we can equivalently say that z_1 operates on z_2. Either way, the result is the new complex number number Z. In particular, $i = e^{i\pi/2}$ is a 90° counter-clockwise rotation. Thus the algebraic relation $i^2 = -1$ has a simple geometric explanation: two 90° rotations are equivalent to a single 180° rotation. Similarly, $(-1)^2 = +1$ because rotation by π is its own inverse.

In this way, algebraic equations can be read geometrically. Consider, for example,

$$z^3 - z^2 + 2 = 0, \tag{3.9}$$

which has solutions

$$z_1 = \sqrt{2}\, e^{i\pi/4} = 1 + i \qquad z_2 = z_1^* \qquad z_3 = e^{i\pi} = -1. \tag{3.10}$$

We can interpret an expression like (3.9) as an instruction for taking linear combinations of a given solution vector. The z^3 term, for instance, can be understood as z^2 operating on z. For solution z_1, this gives

$$z_1^3 = z_1^2 z_1 = 2e^{i\pi/2} z_1 = 2\sqrt{2}\, e^{3i\pi/4} = -2 + 2i. \tag{3.11a}$$

So z_1^2 rotates z_1 by 90° and doubles its length. The next term rotates z_1 counter-clockwise by $5\pi/4 = 225°$ (since -1 is a π rotation) and magnifies it by $\sqrt{2}$,

$$-z_1^2 = e^{i\pi} z_1 z_1 = -2e^{i5\pi/4} z_1 = -2e^{i3\pi/2} = 0 - 2i. \tag{3.11b}$$

Adding these results satisfies (3.9). The procedure for $z_2 \equiv z_1^*$ is the same, though the rotations are clockwise. Finally, real z_3 requires only π rotations.

Although the polar representation is arguably more useful, the more-familiar cartesian real and imaginary parts of a complex number z are linear combinations of z and z^*,

$$\Re e(z) \equiv \frac{1}{2}(z + z^*) = a \qquad \Im m(z) \equiv \frac{1}{2i}(z - z^*) = b. \tag{3.12}$$

Unless otherwise specified by either statement or context, one often presumes numbers to be complex; real and imaginary numbers are special cases. A simple test is to compare z with z^*: if $z = z^*$ then z is real; if $z = -z^*$, it is imaginary.

Example 3.2 **Real, Imaginary, or Complex?**

(a) $z_1 = i^i$: It certainly looks imaginary, but since $1/i = -i$

$$z_1^* = (-i)^{-i} = (1/i)^{-i} = +i^i = z_1,$$

it's real. In fact, from (3.3) we discover $i = e^{i\pi/2}$, so $i^i = e^{-\pi/2} \approx 0.21$.

(b) $z_2 = \sin(i\varphi)$: This too looks imaginary — and indeed,

$$z_2^* = \sin(-i\varphi) = -\sin(i\varphi),$$

since sine is an odd function, $\sin(-x) = -\sin(x)$. So z_2 is imaginary.

(c) $z_3 = z \pm z^*$: It's easy to verify that $z + z^*$ is always real, $z - z^*$ always imaginary.

(d) $z_4 = (z-z^*)/(z+z^*)$: The numerator is imaginary, the denominator real, so the ratio is imaginary. You should show that $z_4 = i \tan \varphi$, where $\varphi = \arg(z)$.

(e) $z_5 = z/(z + z^*)$: There is no simple relationship between z_5 and z_5^*, so the most one can say is that z_5 is complex.

3.2 Euler's Formula and Trigonometry

In the first volume of his justly famous *Lectures on Physics*, Richard Feynman (2011) calls Euler's formula (3.3) "one of the most remarkable, almost astounding, formulas in all of mathematics" because it provides the connection between algebra, trigonometry, and analysis.[1] You're asked to verify it in Problems 3.3 and 6.25.

To get a sense for Feynman's admiration, we note three immediate and useful consequences of Euler's formula:

1. Since $e^{inx} = (e^{ix})^n$,

$$\cos nx + i \sin nx = (\cos x + i \sin x)^n. \tag{3.13}$$

This is *de Moivre's theorem*, and is very useful for finding trig identities. For example:

$$\cos(2\varphi) + i\sin(2\varphi) = (\cos \varphi + i \sin \varphi)^2$$
$$= \left(\cos^2 \varphi - \sin^2 \varphi\right) + 2i \sin \varphi \cos \varphi. \tag{3.14}$$

Matching real parts and imaginary parts, we find

$$\cos(2\varphi) = \cos^2 \varphi - \sin^2 \varphi \qquad \sin(2\varphi) = 2 \sin \varphi \cos \varphi. \tag{3.15}$$

For higher powers the binomial theorem (Section 5) can be used to expand the left-hand side of (3.13), leading to multiple-angle trig identities.

2. We can express cosine and sine as combinations of the real and imaginary parts of e^{ix}, yielding a form that is often more amenable to mathematical manipulation:

$$\cos x = \frac{1}{2}\left(e^{ix} + e^{-ix}\right), \quad \sin x = \frac{1}{2i}\left(e^{ix} - e^{-ix}\right). \tag{3.16}$$

---

[1] The relation $e^{i\pi} + 1 = 0$ is often cited for connecting five fundamental constants in mathematics.

Note that in this form, $\cos x$ and $\sin x$ are manifestly even and odd, respectively — that is, $\cos(-x) = +\cos(x)$, $\sin(-x) = -\sin(x)$.

3. For integer n, $e^{in\pi} = (-1)^n$. Thus $e^{i\pi} = -1$ and $e^{2i\pi} = +1$. Similarly, $e^{in\pi/2} = i^n$.

By connecting $e^{i\varphi}$ to trigonometry, Euler's formula encapsulates the inherent periodicity of the complex plane. For instance, for integer m,

$$z^{1/m} = \left(re^{i\varphi}\right)^{1/m} = r^{1/m}\,e^{i(\varphi + 2\pi n)/m}, \quad n = 0, 1, \ldots, m-1. \tag{3.17}$$

Thus there are m *distinct* roots; for $r = 1$, they are called the "m roots of unity." Euler's formula also underlies a physical interpretation for complex numbers. Consider a time-dependent phase $\varphi = \omega t$. For real ω, we can depict $Ae^{i\omega t}$ as a vector of length A rotating counterclockwise with frequency ω. The real part — the shadow it casts on the real axis — is undergoing simple harmonic motion with the same frequency, $\Re e[Ae^{i\omega t}] = A\cos\omega t$. The complex conjugate rotates clockwise, but has the same real part — and so describes the same oscillation. The imaginary part also describes oscillation, though with a $\pi/2$ phase shift. The real and imaginary parts thus provide two independent pieces of information. This allows $e^{i\omega t}$ to be interpreted as the complex representation of harmonic oscillation's two fundamental degrees of freedom, the amplitude of the motion and the phase as a function of time (Problem 3.20).

Example 3.3   Complex Waveforms

If $e^{i\omega t}$ describes oscillation, then $e^{i(kx-\omega t)} = e^{ik(x-vt)}$ is a complex representation of a wave traveling in x with speed ω/k. This is consistent with the general idea that a wave is simultaneously simple harmonic in both space and time, with the relation $\omega/k = v$ the connection between the two oscillations. Moreover, it's easy to see that the sum of two identical waves traveling in opposite directions form a standing wave,

$$e^{i(kx-\omega t)} + e^{-i(kx+\omega t)} = 2e^{-i\omega t}\cos kx, \tag{3.18}$$

in which each point x oscillates harmonically in time.

The waveform

$$A\,e^{i(\vec{k}\cdot\vec{r} - \omega t)} \tag{3.19}$$

is called a *plane wave* because the wavefronts are (infinite) planes perpendicular to the direction of propagation $\hat{k}$. For the one-dimensional case $e^{i(kx-\omega t)}$, $\hat{k} = \hat{\imath}$ and the wavefronts span y and z. The energy of a wave is always proportional to the square of the wave amplitude — in this case, $|A|^2$. So parallel planes in space have the same energy passing through them; no energy is lost in wave propagation.

Plane waves, however, do not describe more physically realistic situations such as circular wavefronts of water on the surface of a lake. Although overall energy must still be conserved (at least in the absence of damping), the energy *density* must decrease as the wavefronts spread out from the source. Since a circle's circumference grows with r, the energy density must fall like $1/r$ — and hence circular waves have the form

$$\frac{A}{\sqrt{r}}\,e^{i(\vec{k}\cdot\vec{r} - \omega t)} = \frac{A}{\sqrt{r}}\,e^{i(kr - \omega t)}, \tag{3.20}$$

describing propagation in the radial direction, $\hat{k} = \hat{r}$. (Note that in this example, $\vec{r} = x\hat{\imath} + y\hat{\jmath} = r\hat{r}$ is the position in the plane.) Spherical waves are described in a similar fashion — though since the energy must fall with distance at the same rate as the surface area of a sphere grows, spherical waves are given by

$$\frac{A}{r} e^{i(\vec{k}\cdot\vec{r}-\omega t)} = \frac{A}{r} e^{i(kr-\omega t)}, \tag{3.21}$$

with r the radial coordinate in $\mathbb{R}^3$.

Of course all of these waveforms can be expressed as cosines and sines, but the complex exponential form is often easier to work with. And in quantum mechanics, which requires complex wavefunctions, it's largely unavoidable.

Euler's formula also allows us to ascribe meaning to trigonometric functions of imaginary arguments — a far cry indeed from ratios of the sides of a triangle. Writing $x = iy$, (3.16) become

$$\cos iy = \frac{1}{2}\left(e^{-y} + e^{+y}\right) \qquad \sin iy = \frac{1}{2i}\left(e^{-y} - e^{+y}\right). \tag{3.22}$$

Curiously, $\cos iy$ is purely real and $\sin iy$ purely imaginary; throwing in an extra factor of i to the latter yields the real functions cosh and sinh (pronounced "cosh" and "sinch"),[2]

$$\cosh y \equiv \cos iy = \frac{1}{2}\left(e^y + e^{-y}\right), \quad \sinh y \equiv -i\sin iy = \frac{1}{2}\left(e^y - e^{-y}\right). \tag{3.23}$$

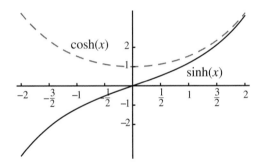

Figure 3.2 $\sinh x$ and $\cosh x$.

Like cos and sin, cosh and sinh are manifestly even and odd, respectively — as Figure 3.2 reveals. Indeed, the similarity of (3.23) and (3.16) clearly motivates the names of these functions. But why the "h"? Well, recall that $\cos\theta$ and $\sin\theta$ give the horizontal and vertical coordinates of a point on the unit circle, $\cos^2\theta + \sin^2\theta = 1$; for this reason they are often called "circular functions." For imaginary arguments, though, this becomes

$$\cos^2(iy) + \sin^2(iy) = 1 \implies \cosh^2 y - \sinh^2 y = 1. \tag{3.24}$$

This is the equation of a hyperbola — hence cosh and sinh are called hyperbolic trig functions.

In terms of sinh and cosh, Euler's formula becomes

$$e^{\pm y} = \cosh y \pm \sinh y, \tag{3.25}$$

---

[2] Remembering where to put the extra i relating sin and sinh is a common pitfall. One remedy is to think of the relationship between cos ↔ cosh and sin ↔ sinh as an extension of their even/odd properties:

$$\cos(\pm iy) = \cosh y, \quad \sin(\pm iy) = \pm i \sinh y.$$

and de Moivre's theorem, (3.13), reads

$$(\cosh y + \sinh y)^n = \cosh ny + \sinh ny, \tag{3.26}$$

from which hyperbolic trig identities can be derived. Not surprisingly, trigonometric identities and their hyperbolic analogs differ by no more than a few sign flips.

Example 3.4 Damped Oscillations

For imaginary ω, the complex function $e^{i\omega t}$ is transformed into a real exponential. So instead of bounded oscillations we get exponential growth and decay. The astounding algebraic similarity between these radically different physical behaviors can be ascribed to a single sign in the defining differential equation: for real ω, $\ddot{x} = -\omega^2 x$ gives bounded, oscillatory motion, while $\ddot{x} = +\omega^2 x$ yields exponential growth and decay (see Example 1.1).

Consider an oscillator with natural frequency ω_0 moving through a viscous medium. For relatively small viscosity, energy is slowly drained from the oscillator, so we expect oscillation with steadily decreasing amplitude; for large enough viscosity, there should be no oscillation at all. But how can we account for such disparate behaviour from a single expression of Newton's second law?

An oscillator's restoring force is commonly written $F_0 = -m\omega_0^2 x$ (recall that for a mass on a spring, $\omega_0^2 = k/m$). We'll assume that the applied damping is proportional to the speed of the oscillator, $F_d = -bv$, with the constant b characterizing the viscosity of the medium. (Note the signs: the restorative force responsible for oscillation is always in the direction opposite the displacement x, whereas the resistive force always acts against the motion v.) Then Newton's second law can be rendered

$$m\ddot{x} = -m\omega_0^2 x - bv \implies \ddot{x} + 2\beta\dot{x} + \omega_0^2 x = 0, \tag{3.27}$$

where $\beta \equiv b/2m$ has units of frequency. The solution to this second-order differential equation is (Problem 3.20)

$$x(t) = e^{-\beta t}\left(Ae^{i\Omega t} + Be^{-i\Omega t}\right)$$
$$= Ce^{-\beta t}\cos\left(\Omega t + \delta\right), \tag{3.28}$$

where $\Omega^2 \equiv \omega_0^2 - \beta^2$, and the constants A, B, C, and δ are determined from initial conditions x_0 and v_0. Real Ω results in oscillation with steadily decreasing amplitude — so-called *underdamped oscillation* (Figure 3.3a). The condition $\beta < \omega_0$, then, is what is meant by "small" damping.

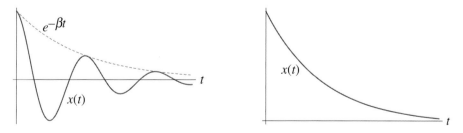

Figure 3.3 Underdamped and overdamped oscillations.

On the other hand, if the resistive force is large, $\beta > \omega_0$, then $\Omega = i\omega \equiv i\sqrt{\omega_0^2 - \beta^2}$ is imaginary and (3.28) becomes

$$x(t) = e^{-\beta t}\left(Ae^{-\omega t} + Be^{+\omega t}\right)$$

$$= Ce^{-\beta t}\cosh(\omega t + \delta). \tag{3.29}$$

This is known as *overdamped oscillation* (Figure 3.3b). As expected, for large damping no oscillation whatsoever occurs – although, depending on initial conditions, an overdamped system could over-shoot equilibrium and then reverse before coming to rest. A third, subtly different solution

$$x(t) = (At + B)e^{-\beta t} \tag{3.30}$$

describes *critical damping* when $\beta = \omega_0$, as we'll see in Chapter 39.

Note that all three of (3.28) — (3.30) are solutions to the differential equation (3.27). Indeed, the *calculus* is the same either way; the behavior of the solution depends upon the relative size of the physical parameters.

Example 3.5 **Forced Oscillations**

All three types of oscillation in Example 3.4 are "transient" in that they quickly decay as damping removes energy from the system. Energy can be pumped back in to achieve a "steady state" by subjecting the oscillator to an external force $F(t) = mf(t)$; this adds an x-independent *source term* to the differential equation,

$$\frac{d^2x}{dt^2} + 2\beta\frac{dx}{dt} + \omega_0^2 x = f(t). \tag{3.31}$$

If we take the driving force to be sinusoidal,

$$f(t) = f_0 \cos\omega t = f_0\,\Re e\left[e^{i\omega t}\right], \tag{3.32}$$

then surely its periodicity is directly reflected in the behavior of the oscillator. So we'll assume the steady-state solution has the complex form

$$z(t) = C(\omega)e^{i\omega t}, \tag{3.33}$$

where $x = \Re e\,z$, and see what emerges. Inserting (3.33) into the differential equation gives

$$z(t) = \frac{f_0\,e^{i\omega t}}{\left(\omega_0^2 - \omega^2\right) + 2i\beta\omega}. \tag{3.34}$$

Each term in the denominator corresponds to a term in (3.31), each representing different physics: $+\omega_0^2$ the oscillator, $-\omega^2$ the system's inertia, and $2i\beta\omega$ the damping — all with different phases from one another. Only the ω_0 term is in phase with the driving force; when it dominates the other two, the oscillator response z and the driving force are in phase. Similarly, when either the inertia or damping is dominant, the oscillator and driving force are out of phase by π and $\pi/2$, respectively. The frequency ω for which the oscillator achieves its largest amplitude response is the one which maximizes

$$|C(\omega)| = \frac{f_0}{\sqrt{\left(\omega_0^2 - \omega^2\right)^2 + 4\beta^2\omega^2}}. \tag{3.35}$$

Straightforward calculation reveals that this happens at the *resonant frequency*

$$\omega_{\text{res}}^2 = \omega_0^2 - 2\beta^2, \tag{3.36}$$

and so can only occur for underdamped systems with $\beta < \omega_0/\sqrt{2}$. Indeed, we intuitively expect smaller damping to result in greater oscillator response. Moreover, the smaller the damping the more precisely $\omega^2 - \omega_0^2$ must cancel when ω is in the vicinity of ω_{res} — and hence the narrower the range of frequencies which result in large amplitude. For $\omega \approx \omega_{res}$, even small f_0 can give rise to a large-amplitude response.

These insights were much simpler to uncover by working with complex z. Solving for $x(t)$ is easier too: writing $z(t) = Ce^{i\omega t} \equiv |C|e^{i(\omega t - \varphi)}$ and doing a bit of algebra are all that's needed to find the physical displacement and relative phase (Problem 3.21),

$$x(t) = \frac{f_0 \cos(\omega t - \varphi)}{\sqrt{\left(\omega_0^2 - \omega^2\right)^2 + 4\beta^2\omega^2}}, \qquad \tan\varphi = \frac{2\beta\omega}{\omega_0^2 - \omega^2}. \qquad (3.37)$$

BTW 3.1 Special Relativity and Hyperbolic Geometry

As we discussed in Example 1.3, the Lorentz transformations of special relativity can be written

$$x_0' = \gamma(x_0 - \beta x_1)$$
$$x_1' = \gamma(x_1 - \beta x_0), \qquad (3.38)$$

where $\beta = V/c$ and $\gamma = 1/\sqrt{1 - \beta^2}$. These equations are very similar to the way a rotation mixes up the coordinates as in (2.18), except that a normal rotation transforms along a circle, leaving constant the magnitude of the vector $x_0^2 + x_1^2$. But in relativity, it's the invariant interval $s^2 \equiv x_0^2 - x_1^2$ which is the same in all frames — as you can verify directly from the Lorentz transformations. The form of s^2 clearly implies that (3.38) must describe a hyperbolic "rotation." Let's introduce the parameter Φ such that

$$\cosh\Phi \equiv \gamma, \qquad \sinh\Phi \equiv \gamma\beta, \qquad (3.39)$$

which, of course, satisfies $\cosh^2\Phi - \sinh^2\Phi = 1$. Then the Lorentz transformations take the form

$$\begin{pmatrix} x_0' \\ x_1' \end{pmatrix} = \begin{pmatrix} \cosh\Phi & -\sinh\Phi \\ -\sinh\Phi & \cosh\Phi \end{pmatrix} \begin{pmatrix} x_0 \\ x_1 \end{pmatrix}. \qquad (3.40)$$

Comparison with (2.18) allows an interpretation as a rotation through an imaginary angle.

Problems

3.1 Which of the following are pure phases?

(a) $e^{-i\pi}$
(b) π^{ie}
(c) $(-\pi)^{ie}$
(d) $i^{e\pi}$
(e) $(-i)^{e\pi^i}$

3.2 Use both cartesian and polar representations to compare $|z|^2$ and $|z^2|$. Generalize to $|z^n|$.

3.3 For $z = \cos\theta + i\sin\theta$, express $dz/d\theta$ in terms of z to establish that $z = e^{i\theta}$.

3.4 By considering the expression $|(a + ib)(c + id)|^2$, prove that if two integers can be expressed as the sum of two squares, then so can their product, i.e., if $M = a^2 + b^2$ and $N = c^2 + d^2$, then $MN = p^2 + q^2$.

3.5 (a) Write both $i + \sqrt{3}$ and $i - \sqrt{3}$ in the form $R\,e^{i\theta}$, where $R \equiv |z|$. Write θ as a rational multiple of π.

(b) Find the square roots of $2i$ and $2 + 2i\sqrt{3}$.

3.6 Demonstrate whether the following are real, imaginary, or complex. If real or imaginary, render into a form in which this is (more?) obvious.

(a) $(-1)^{1/i}$

(b) $(z/z^*)^i$

(c) $z_1 z_2^* - z_1^* z_2$

(d) $\displaystyle\sum_{n=0}^{N} e^{in\theta}$

(e) $\displaystyle\sum_{n=-N}^{N} e^{in\theta}$

3.7 Simplify:

(a) $(1+i)^4$

(b) $\frac{1+2i}{2-i}$

(c) $\left|\frac{1+4i}{4+i}\right|$

(d) $\left|\frac{(3+4i)^4}{(3-4i)^3}\right|$.

3.8 Use the product $(2+i)(3+i)$ to establish the relation $\pi/4 = \tan^{-1}(1/2) + \tan^{-1}(1/3)$.

3.9 Use complex exponentials to show that:

(a) $\cos(a+b) + \cos(a-b) = 2\cos a \cos b$

(b) $\sin(a+b) + \sin(a-b) = 2\sin a \cos b$.

3.10 Prove that:

(a) $e^{i\alpha} + e^{i\beta} = 2\cos\left[\frac{\alpha-\beta}{2}\right]e^{\frac{i(\alpha+\beta)}{2}}$

(b) $e^{i\alpha} - e^{i\beta} = 2i\sin\left[\frac{\alpha-\beta}{2}\right]e^{\frac{i(\alpha+\beta)}{2}}$.

3.11 For two complex numbers $z_1 = r_1 e^{i\phi_1}$ and $z_2 = r_2 e^{i\phi_2}$:

(a) Find the magnitude and argument of the product $z = z_1 z_2$. Give a geometrical interpretation for z.

(b) Find the magnitude and argument of the sum $z = z_1 + z_2$. Show how the magnitude of z is related to the law of cosines.

3.12 For complex number $z = a + ib$, show that $\tan^{-1}(b/a) = \frac{1}{2i}\ln\left(\frac{a+ib}{a-ib}\right)$.

3.13 Show that $\frac{d^n}{dt^n}\left[e^{at}\sin bt\right] = (a^2 + b^2)^{n/2} e^{at}\sin\left[bt + n\tan^{-1}(b/a)\right]$.

3.14 Use de Moivre's theorem to recast each of the following as a sum of powers of $\sin\theta$ and/or $\cos\theta$:

(a) $\cos 3\theta$ (b) $\sin 3\theta$.

3.15 Find all the solutions to the equation $z^6 = 1$. Write each in the form $A + iB$ and plot them in the complex plane. [Hint: Write $z = Re^{i\theta}$.]

3.16 Find all the solutions of $\sqrt[i]{i}$.

3.17 Find all the zeros of (a) $\sin z$ and (b) $\cosh z$.

3.18 Show that $(z = x + iy)$:

(a) $\sinh z = \sinh x \cos y + i \cosh x \sin y$
(b) $\cosh z = \cosh x \cos y + i \sinh x \sin y$
(c) $|\sinh z|^2 = \sinh^2 x + \sin^2 y$
(d) $|\cosh z|^2 = \sinh^2 x + \cos^2 y$

(e) $\tanh z = \frac{\sinh 2x + i \sin 2y}{\cosh 2x + \cos 2y}$.

3.19 For any two complex numbers z_1 and z_2, prove algebraically that

$$|z_1| - |z_2| \leq |z_1 + z_2| \leq |z_1| + |z_2|$$

and give a geometric interpretation.

3.20 As a second-order differential equation, the general solution of $\frac{d^2}{dx^2} f(x) = -k^2 f(x)$ has two constants (which are ultimately determined from boundary conditions). We know that sine and cosine each solve; as a linear equation, the general solution can thus be written

$$f(x) = C_1 \cos kx + C_2 \sin kx,$$

with constants $C_{1,2}$. Of course, sines and cosines are merely phase-shifted versions of one another, so we should also be able to express the solution as

$$A \cos (kx + \alpha) \qquad \text{or} \qquad B \sin (kx + \beta),$$

with constants A, α or B, β. Given the close relationship between oscillations and exponentials, we might also expect to be able to write this as

$$D_1 e^{ikx} + D_2 e^{-ikx},$$

or even

$$\text{Re}(F e^{ikx}),$$

where a complex $F = a + ib$ still encodes two constants. Show that all these expressions are equivalent, and determine all the constants in terms of C_1 and C_2. What's the relationship between D_1 and D_2 if $f(x)$ is real?

3.21 Verify the displacement x and phase φ of the forced harmonic oscillator, (3.37), from the complex solution (3.34).

3.22 Consider a polynomial $P_n(z) = \sum_{k=0}^{n} a_k z^k$.

(a) Prove that if all the coefficients a_k are real, the complex roots always occur in complex conjugate pairs — i.e., that if z_0 is a root, z_0^* must be as well. [Hint: Rewrite P_n in polar form.]
(b) Show that if a polynomial is of odd order, it must have at least one real root.
(c) Verify that a paired linear factor of the form $(z - z_0)(z - z_0^*)$ is a quadratic with real coefficients.

4 Index Algebra

In general, any vector $\vec{A}$ can be represented as a one-dimensional array of components, e.g., $(A_1, A_2, \ldots, A_n)$. More abstractly, we can dispense with the enumeration of components and simply denote the vector as A_i, where i can take any value from 1 to n. The beauty of this notation is that since the value of i is not specified, A_i can represent not just a single component but also the *totality* of components. In other words, if the value of the index i is arbitrary, A_i denotes any of the components — or, conceptually, all the components simultaneously.

4.1 Contraction, Dummy Indices, and All That

Consider matrix multiplication. You may think the rule for the product of two matrices is written

$$
C = AB = \begin{pmatrix} A_{11} & \cdots & A_{1m} \\ A_{21} & \cdots & A_{2m} \\ \vdots & \ddots & \vdots \\ A_{n1} & \cdots & A_{nm} \end{pmatrix} \begin{pmatrix} B_{11} & \cdots & B_{1p} \\ B_{21} & \cdots & B_{2p} \\ \vdots & \ddots & \vdots \\ B_{m1} & \cdots & B_{mp} \end{pmatrix},
$$

where A is $n \times m$ and B is $m \times p$ (so C is thus $n \times p$). But note that the actual *recipe* for multiplying a row of A with a column of B is nowhere to be found — unless, of course, you explicitly write out each element of C in terms of A and B. But that's rather cumbersome, especially for large matrices. The rule for matrix multiplication is best given in index notation, in which a matrix is a two-indexed object C_{ij},

$$
C = AB \iff C_{ij} = (AB)_{ij} = \sum_k A_{ik} B_{kj},
$$

where k runs from 1 to m. Similarly, $D = ABC$ is given by

$$
D_{ij} = \sum_{k, \ell} A_{ik} B_{k\ell} C_{\ell j}.
$$

Not only is this more compact than writing out the full matrix product, matrix notation is simply not an option for arrays of numbers with more than two indices.[1]

---

[1] Since summed indices always appear twice in an expression, the very common "Einstein summation convention" dispenses with the explicit $\sum$ altogether. Thus $C_{ij} \equiv A_{ik} B_{kj} = (AB)_{ij}$.

Example 4.1 **2 × 2 Matrices**

For 2 × 2 matrices $A = \begin{pmatrix} A_{11} & A_{12} \\ A_{21} & A_{22} \end{pmatrix}$ and $B = \begin{pmatrix} B_{11} & B_{12} \\ B_{21} & B_{22} \end{pmatrix}$,

$$C_{ij} = \sum_{k=1}^{2} A_{ik}B_{kj} = A_{i1}B_{1j} + A_{i2}B_{2j} \tag{4.1}$$

$$\implies C = \begin{pmatrix} A_{11}B_{11} + A_{12}B_{21} & A_{11}B_{12} + A_{12}B_{22} \\ A_{21}B_{11} + A_{22}B_{21} & A_{21}B_{12} + A_{22}B_{22} \end{pmatrix}. \tag{4.2}$$

Manipulating expressions using index notation can often be simpler than using vector or matrix notation. In particular, although matrices do not generally commute,

$$AB \neq BA, \tag{4.3}$$

the products of their *elements* certainly do,

$$(AB)_{ij} = \sum_{k} A_{ik}B_{kj} = \sum_{k} B_{kj}A_{ik}. \tag{4.4}$$

So how does the lack of commutivity, (4.3), manifest in index notation? Carefully compare expression (4.4) for AB with the analogous one for BA,

$$(BA)_{ij} = \sum_{k} B_{ik}A_{kj} = \sum_{k} A_{kj}B_{ik}, \tag{4.5}$$

and pay close attention to how the index notation distinguishes the two.

When working with index notation, the "net" number of indices denotes the type of object you have. Thus

$$\sum_{i} B_i C_i, \qquad \sum_{ijk} B_{ijk} C_{jik}, \qquad \sum_{ijkl} B_{ij} C_{ik} D_{lk} E_{jl} \tag{4.6a}$$

with no unsummed indices are all single numbers — that is, scalars — whereas

$$\sum_{j} B_{ij} C_j, \qquad \sum_{jk} B_{ijk} C_{jk}, \qquad \sum_{jkl} B_{kj} C_{jkil} D_l \tag{4.6b}$$

with one net index are all vectors.

Note that summed indices are completely summed over. In the scalar examples in (4.6a), for instance, i doesn't take just one value — rather, it ranges over *all* its values. Since it disappears once the sum is done, i is said to be a "contracted index." Indeed, the actual index label used in a sum is completely arbitrary and irrelevant; it is called a *dummy index*. Thus[2]

$$\sum_{i} B_i C_i = \sum_{j} B_j C_j \tag{4.7}$$

---

[2] This is the same thing that occurs in a definite integral. For instance,

$$\int_0^\infty e^{-ax^2} dx = \int_0^\infty e^{-ay^2} dy = \sqrt{\pi/a}$$

independent of x and y; they are dummy variables.

are completely equivalent. Similarly,

$$A_i = \sum_j B_{ij} C_j = \sum_\ell B_{i\ell} C_\ell. \tag{4.8}$$

Since A_i has a single index, i must be the sole surviving index emerging from the sum on the right; it is not a dummy index. So in order for an equation to make sense, the net number of indices and their labels must match on both sides.

Take care with the order of the indices. It matters. For instance

$$\sum_j B_{ij} C_j \neq \sum_j B_{ji} C_j. \tag{4.9}$$

The expression on the left is the matrix B times a column vector; on the right, it's the transpose matrix $B_{ij}^T = B_{ji}$ which acts on the column vector C_j. (Equivalently, $\sum_j C_j B_{ji}$ can be construed as a row vector C_j acting on B_{ji} from the left.)

Symmetric and antisymmetric matrices in index notation satisfy the conditions

$$S = S^T \longleftrightarrow S_{ij} = S_{ji} \quad \text{(symmetric)}$$

$$A = -A^T \longleftrightarrow A_{ij} = -A_{ji} \quad \text{(antisymmetric)}. \tag{4.10}$$

Note that the diagonal elements A_{ii} of an antisymmetric matrix are all zero.

Example 4.2 Derivatives

How do we take derivatives of an expression like $\sum_j A_j B_j$? Well, a derivative with respect to a specific vector component picks out a single term in the sum:

$$\frac{d}{dA_i}\left(\sum_j A_j B_j\right) = B_i, \tag{4.11}$$

as is readily verified by expanding the sum and assigning i to a particular value between 1 and n. On the other hand, a time derivative requires the product rule:

$$\frac{d}{dt}\left(\sum_j A_j B_j\right) = \sum_j \left(\dot{A}_j B_j + A_j \dot{B}_j\right). \tag{4.12}$$

The time derivative of a scalar function like $\sqrt{\sum_j A_j B_j}$ uses the chain rule in the usual way — though the deployment of dummy indices is a bit more subtle:

$$\frac{d}{dt}\sqrt{\sum_j A_j B_j} = \frac{1}{2\sqrt{\sum_j A_j B_j}}\sum_k \left(\dot{A}_k B_k + A_k \dot{B}_k\right). \tag{4.13}$$

Note that the result is a scalar. As it must be.

4.2 Two Special Tensors

Physics is replete with indexed objects called *tensors* — of which vectors are the most familiar examples. We'll postpone until Chapter 29 a discussion of tensors and their properties; for now we merely introduce two tensors which are particularly useful in algebraic manipulations.

The *Kronecker delta* is the index-notation version of the identity matrix $\mathbb{1}$,

$$\delta_{ij} = \begin{cases} 1, & i = j \\ 0, & i \neq j \end{cases}. \tag{4.14}$$

Clearly, it's symmetric: $\delta_{ij} = \delta_{ji}$. Here i and j range from 1 to the dimension n of the space — so in two dimensions

$$\delta_{ij} \longrightarrow \begin{pmatrix} 1 & 0 \\ 0 & 1 \end{pmatrix}, \tag{4.15}$$

whereas for $n = 3$,

$$\delta_{ij} \longrightarrow \begin{pmatrix} 1 & 0 & 0 \\ 0 & 1 & 0 \\ 0 & 0 & 1 \end{pmatrix}. \tag{4.16}$$

An important feature of the Kronecker delta is that it "collapses" a sum,

$$\sum_i A_i \delta_{ij} = A_j. \tag{4.17}$$

In other words, as i sweeps through its values, only for $i = j$ is a non-zero contribution made to the sum.

Example 4.3 **Vector Label or Vector Index?**

The Kronecker delta often emerges from the properties of a set of n vectors. Take, for example, the set of unit vectors $\hat{e}_\ell$ with $\ell = 1$ to n. In this case, the index ℓ is *not* a vector index, but simply a label to denote which unit vector we're talking about. Instead, the "hat" marks $\hat{e}_\ell$ as a vector; its components are denoted $(\hat{e}_\ell)_i$. For instance, the cartesian set $\{\hat{e}_1, \hat{e}_2, \hat{e}_3\} \equiv \{\hat{\imath}, \hat{\jmath}, \hat{k}\}$ has components $(\hat{e}_1)_i = \delta_{1i}$, $(\hat{e}_2)_i = \delta_{2i}$, $(\hat{e}_3)_i = \delta_{3i}$.

The Kronecker also emerges in relationship *amongst* the $\hat{e}_\ell$. As unit vectors, we have $\hat{e}_\ell \cdot \hat{e}_\ell = 1$ for all ℓ; if they are also orthogonal, then $\hat{e}_\ell \cdot \hat{e}_m = 0$, for all $\ell \neq m$. In other words,

$$\hat{e}_\ell \cdot \hat{e}_m = \delta_{\ell m} \text{ for all } \ell, m. \tag{4.18}$$

This is the succinct mathematical statement of the *orthonormality* of the set $\{\hat{e}_\ell\}$.

The *Levi-Civita epsilon* ("levee-cheeVEEta") is the index-notation version of a completely antisymmetric matrix. But unlike δ_{ij}, the number of indices is equal to the dimensionality of the space. This means that it can be written in standard matrix notation only for $n = 2$,

$$\epsilon_{ij} = \begin{cases} +1, & i = 1, j = 2 \\ -1, & i = 2, j = 1 \\ 0, & i = j \end{cases} \longrightarrow \begin{pmatrix} 0 & 1 \\ -1 & 0 \end{pmatrix}. \tag{4.19}$$

In three dimensions, the Levi-Civita epsilon has 27 components — but most of them are zero,

$$\epsilon_{ijk} = \begin{cases} +1, & \{ijk\} = \{123\}, \{231\}, \text{ or } \{312\} \\ -1, & \{ijk\} = \{132\}, \{321\}, \text{ or } \{213\} \\ 0, & \text{any two indices the same} \end{cases}. \tag{4.20}$$

The signs of the six non-zero elements depend on whether the indices are an even or odd permutation of {123} — that is, how many adjacent indices must be flipped in {123} to achieve a particular arrangement {ijk}. (This is why the Levi-Civita epsilon is often referred to as the permutation symbol.) In three dimensions, the even permutations are equivalent to so-called "cyclic" permutations: like keys arranged clockwise on a ring, 1 is followed by 2 is followed by 3 is followed by 1 ... (Figure 4.1). The odd permutations are "anti-cyclic." Note that odd permutations are all even permutations of *one another*.

Figure 4.1 Cyclic order.

For a right-handed set of orthonormal vectors, it is straightforward to show

$$\hat{e}_i \times \hat{e}_j = \sum_\ell \epsilon_{ij\ell} \hat{e}_\ell. \tag{4.21a}$$

Dotting both sides with $\hat{e}_k$ and using (4.18) to collapse the sum gives

$$\left(\hat{e}_i \times \hat{e}_j\right) \cdot \hat{e}_k = \epsilon_{ijk}. \tag{4.21b}$$

If this looks completely foreign to you, try applying it to $\hat{\imath}$, $\hat{\jmath}$, and $\hat{k}$.

The antisymmetry of ϵ_{ijk} means that reversing any two adjacent indices reverses its sign; so, for example, from (4.21a) we rediscover the familiar

$$\hat{e}_i \times \hat{e}_j = \sum_\ell \epsilon_{ij\ell} \hat{e}_\ell$$

$$= -\sum_\ell \epsilon_{ji\ell} \hat{e}_\ell = -\hat{e}_j \times \hat{e}_i. \tag{4.22}$$

In a similar way, the contraction of ϵ_{ijk} with a symmetric matrix $T_{ij} = T_{ji}$ vanishes:

$$A_k \equiv \sum_{ij} \epsilon_{ijk} T_{ij} = \sum_{ij} \epsilon_{ijk} T_{ji} \quad \text{[symmetry of } T_{ij}\text{]}$$

$$= \sum_{ij} \epsilon_{jik} T_{ij} \quad \text{[relabeling dummy indices, } i \leftrightarrow j\text{]}$$

$$= -\sum_{ij} \epsilon_{ijk} T_{ij} \quad \text{[antisymmetry of } \epsilon_{ijk}\text{]}$$

$$= -A_k. \tag{4.23}$$

Thus $A_k \equiv 0$.

It may come as little surprise to learn that δ_{ij} and ϵ_{ijk} can be related. Two useful identities are (Problems 4.9–4.10)

$$\sum_{jk} \epsilon_{mjk} \epsilon_{njk} = 2\delta_{mn} \tag{4.24}$$

$$\sum_\ell \epsilon_{mn\ell} \epsilon_{ij\ell} = \delta_{mi}\delta_{nj} - \delta_{mj}\delta_{ni}. \tag{4.25}$$

Note that the first is symmetric under $m \leftrightarrow n$, and the second antisymmetric under $m \leftrightarrow n$ and $i \leftrightarrow j$ (but symmetric under both simultaneously).

4.3 Common Operations and Manipulations

Consider the multiplication of two vectors, $\vec{A} = \sum_i A_i \hat{e}_i$ and $\vec{B} = \sum_i B_i \hat{e}_i$. In three dimensions, there are nine possible products $A_i B_j$ amongst their components. The dot product uses orthonormality, (4.18), to pick out the three that are symmetric,

$$\vec{A} \cdot \vec{B} = \sum_{i,j} A_i B_j (\hat{e}_i \cdot \hat{e}_j) = \sum_{i,j} A_i B_j \delta_{ij} = \sum_i A_i B_i, \tag{4.26}$$

which as a scalar has no remaining index. The other six products $A_i B_j$ can be similarly combined, this time using ϵ_{ijk} and (4.21a),

$$\vec{A} \times \vec{B} = \sum_{i,j} A_i B_j (\hat{e}_i \times \hat{e}_j) = \sum_{i,j,k} \epsilon_{ijk} A_i B_j \hat{e}_k. \tag{4.27}$$

Expressing the vector $\vec{A} \times \vec{B}$ itself in index notation,

$$(\vec{A} \times \vec{B})_\ell \equiv (\vec{A} \times \vec{B}) \cdot \hat{e}_\ell = \left(\sum_{i,j,k} \epsilon_{ijk} A_i B_j \hat{e}_k \right) \cdot \hat{e}_\ell$$

$$= \sum_{i,j,k} \epsilon_{ijk} A_i B_j \delta_{k\ell} = \sum_{i,j} \epsilon_{ij\ell} A_i B_j. \tag{4.28}$$

One need never work through these formal manipulations again — just use:

$$\vec{A} \cdot \vec{B} = \sum_i A_i B_i \tag{4.29a}$$

and

$$\vec{A} \times \vec{B} \rightarrow (\vec{A} \times \vec{B})_\ell = \sum_{ij} \epsilon_{ij\ell} A_i B_j. \tag{4.29b}$$

Example 4.4 Dot Products and Unit Vectors

Using index notation, a unit vector like $\hat{r} \equiv \vec{r}/|\vec{r}|$ can be rendered

$$(\hat{r})_i = \frac{r_i}{|\vec{r}|} = \frac{r_i}{\sqrt{\sum_k r_k^2}}, \tag{4.30}$$

where r_i is the ith component of $\vec{r}$ and $(\hat{r})_i$ the ith component of $\hat{r}$. Note that k is completely summed away. Now let's take the time derivative of $\hat{r}$:

$$\frac{d(\hat{r})_i}{dt} = \frac{d}{dt} \left(\frac{r_i}{\sqrt{\sum_k r_k^2}} \right)$$

$$= \frac{v_i}{\sqrt{\sum_k r_k^2}} - r_i \frac{\sum_j 2 r_j v_j}{2 \left(\sqrt{\sum_k r_k^2} \right)^3}$$

$$= \frac{v_i}{r} - r_i \frac{\left(\sum_j r_j v_j \right)}{r^3}, \tag{4.31}$$

where $r \equiv |\vec{r}|$ and we've used (4.13) with $v_i \equiv \dot{r}_i$. Thus[3]

$$\frac{d\hat{r}}{dt} = \frac{\vec{v}}{r} - \frac{\vec{r} \cdot \vec{v}}{r^3} \vec{r} = \frac{1}{r^2} \left[r\vec{v} - (\hat{r} \cdot \vec{v}) \vec{r} \right]. \tag{4.32}$$

The beauty of (4.32) is that it uses the physically relevant vectors $\vec{r}$ and $\vec{v}$ rather than coordinate frame unit vectors. In fact, (4.32) holds for $\hat{r}$ in any number of dimensions.

Example 4.5 **The Antisymmetry of the Cross Product**

As with (4.22), the antisymmetry of ϵ_{ijk} leads directly to a familiar result

$$(\vec{A} \times \vec{B})_\ell = \sum_{ij} \epsilon_{ij\ell} A_i B_j$$

$$= -\sum_{ij} \epsilon_{ji\ell} A_i B_j = -(\vec{B} \times \vec{A})_\ell. \tag{4.33}$$

If this holds for arbitrary component ℓ, then it must hold for all components. Thus $\vec{A} \times \vec{B} = -\vec{B} \times \vec{A}$.

Example 4.6 **Cross Products and Central Forces**

Any radially directed force whose magnitude depends only on the distance r from the source is called a *central force*, $\vec{F} = f(r)\hat{r}$. Associated with such a force is a potential energy U that depends only on distance, $U(\vec{r}) = U(r)$, with $f = -\partial U / \partial r$.[4] The most important central force is the familiar inverse-square law,

$$f(r) = -\frac{k}{r^2} \qquad U(r) = -\frac{k}{r}. \tag{4.34}$$

For positive k, the force is attractive — which is always the case for gravity. By contrast, electrostatic forces have $k < 0$ for two like-sign electric charges.

Fundamental to all central forces is that they conserve angular momentum,

$$\frac{d\vec{L}}{dt} = \frac{d}{dt}(\vec{r} \times \vec{p}) = (\dot{\vec{r}} \times \vec{p}) + (\vec{r} \times \dot{\vec{p}}) = 0. \tag{4.35}$$

Both terms vanish due to the antisymmetry of the cross product, (4.33) — the first because $\vec{p} = m\dot{\vec{r}}$, the second because the force is central, $\dot{\vec{p}} \equiv \vec{F} \sim \hat{r}$. And a constant angular momentum vector implies that the motion occurs in a plane perpendicular to $\vec{L}$. Thus the antisymmetry of the cross product leads immediately to a crucial physical insight: *orbits in a central potential are confined to a plane*.

BTW 4.1 Beyond Three Dimensions

The Kronecker delta always has two indices, δ_{ij}. Thus the dot product takes essentially the same form regardless of the dimensionality of the space N,

---

[3] This can be found more directly without index notation by simply noting $dr/dt \equiv v_r = \hat{r} \cdot \vec{v}$.

[4] As we'll see in Part II, a conservative force is the negative gradient of the potential energy, $\vec{F} = -\vec{\nabla} U$.

$$\vec{A} \cdot \vec{B} = \sum_{i,j=1}^{N} A_i B_j \delta_{ij} = \sum_{i=1}^{N} A_i B_i. \tag{4.36}$$

The result of this symmetric combination is always an index-less object — that is, a scalar. By contrast, the antisymmetric combination of vectors in three dimensions,

$$\sum_{jk} \epsilon_{ijk} A_j B_k, \tag{4.37}$$

is a vector; this, of course, is the cross product. But as we saw in (4.19) and (4.20), the number of indices on the Levi-Civita tensor is equal to the dimensionality of the space. So in two dimensions the antisymmetric combination of two vectors is a scalar,

$$\sum_{i,j} \epsilon_{ij} A_i B_j, \tag{4.38}$$

whereas in four dimensions we get a two-indexed tensor

$$\sum_{i,j} \epsilon_{ijk\ell} A_i B_j. \tag{4.39}$$

In N dimensions, the antisymmetric product yields a tensor with $N - 2$ indices. So the familiar cross product combining two vectors to produce another vector only exists in three dimensions.

You likely learned to calculate cross products as a 3×3 determinant. In fact, if we consider the rows of a matrix M as three-component vectors

$$M = \begin{pmatrix} \longleftarrow & \vec{M}_1 & \longrightarrow \\ \longleftarrow & \vec{M}_2 & \longrightarrow \\ \longleftarrow & \vec{M}_3 & \longrightarrow \end{pmatrix} \tag{4.40}$$

so that $(\vec{M}_i)_j \equiv M_{ij}$, its determinant can be written as a "triple scalar product,"

$$\det[M] = \sum_{\ell mn} \epsilon_{\ell mn} M_{1\ell} M_{2m} M_{3n}$$

$$= \sum_{\ell} M_{1\ell} \left(\sum_{mn} \epsilon_{\ell mn} M_{2m} M_{3n} \right)$$

$$= \vec{M}_1 \cdot \left(\vec{M}_2 \times \vec{M}_3 \right). \tag{4.41}$$

Since the determinant is unchanged under cyclic permutations of the rows, we can also write this as

$$\epsilon_{ijk} \det[M] = \sum_{\ell mn} \epsilon_{\ell mn} M_{i\ell} M_{jm} M_{kn}, \tag{4.42}$$

which doesn't require that we specify row vectors.

Far simpler than the determinant is the *trace* of a matrix, the sum of its diagonal elements. As the determinant is to ϵ_{ijk} and cross product, the trace is to δ_{ij} and the dot product,

$$\text{tr}[M] \equiv \sum_{ij} M_{ij} \delta_{ij} = \sum_{i} \left[\sum_{j} M_{ij} \delta_{ij} \right] = \sum_{i} M_{ii}. \tag{4.43}$$

Note in particular that the trace of the identity matrix $\mathbb{1}$ is the dimensionality of the space,

$$\text{tr}[\mathbb{1}] = \sum_{ij} \delta_{ij}\delta_{ij} = \sum_i \delta_{ii} = n. \tag{4.44}$$

Example 4.7   Cyclic Permutations of the Trace

$$\begin{aligned}
\text{Tr}(ABC) &= \sum_{i,j} \left(\sum_{k,\ell} A_{ik}B_{k\ell}C_{\ell j} \right) \delta_{ij} = \sum_{i,k,\ell} A_{ik}B_{k\ell}C_{\ell i} \\
&= \sum_{i,k,\ell} B_{k\ell}C_{\ell i}A_{ik} = \text{Tr}(BCA) \\
&= \sum_{i,k,\ell} C_{\ell i}A_{ik}B_{k\ell} = \text{Tr}(CAB). \tag{4.45}
\end{aligned}$$

For ease of reference, a compilation of these and other common vector identities are collected inside the front cover; all of them are straightforward to verify using index notation (Problem 4.11).

BTW 4.2   Integration Measures

Vector indices can help streamline the notation and derivation of integration measures in terms of unit vectors $\hat{e}_i$ and scale factors h_i as in (2.27). A fairly intuitive approach begins with the tangent vectors to the coordinate curves,

$$\frac{\partial \vec{r}}{\partial u_i} = h_i \hat{e}_i. \tag{4.46}$$

This leads to the expression for vector displacement given in (2.28); from this we can find the net *scalar* displacement ds,

$$(ds)^2 = d\vec{r} \cdot d\vec{r} = \sum_{ij} h_i h_j (\hat{e}_i \cdot \hat{e}_j) \, du_i du_j, \tag{4.47}$$

which holds even for non-orthogonal systems.

For the two-dimensional area element, we turn to the cross product. Recall that $\vec{A} \times \vec{B}$ gives the area of the parallelogram delineated by $\vec{A}$ and $\vec{B}$; the vector-valued area element in the infinitesimal ij-plane, then, can be obtained by crossing the i and j tangent vectors,

$$da_{ij} = \left| \frac{\partial \vec{r}}{\partial u_i} \times \frac{\partial \vec{r}}{\partial u_j} \right| du_i du_j = \left| \hat{e}_i \times \hat{e}_j \right| h_i h_j \, du_i du_j. \tag{4.48}$$

Similarly, the volume of the parallelepiped described by vectors $\vec{A}$, $\vec{B}$, and $\vec{C}$ is given by the triple product $\vec{A} \cdot (\vec{B} \times \vec{C})$ — so the three-dimensional volume element can be rendered as

$$d\tau = \hat{e}_1 \cdot (\hat{e}_2 \times \hat{e}_3) \, h_1 h_2 h_3 \, du_1 \, du_2 \, du_3. \tag{4.49}$$

Though these forms for ds, da_{ij}, and $d\tau$ may seem unfamiliar, take solace in noting that they simplify for orthogonal systems, for which $\hat{e}_i \cdot \hat{e}_j = \delta_{ij}$:

$$(ds)^2 = \sum_i h_i^2 \, (du_i)^2 \tag{4.50a}$$

$$d\vec{a} = \left(\sum_k \epsilon_{ijk} \hat{e}_k \right) h_i h_j \, du_i du_j \tag{4.50b}$$

$$d\tau = h_1 h_2 h_3 \, du_1 \, du_2 \, du_3, \tag{4.50c}$$

where we have used (4.21a) to write the area element explicitly as a vector in the $\hat{e}_k$-direction. The expressions in (4.50) easily reproduce the familiar results for cartesian, cylindrical, and spherical coordinates — as well as the less-familiar coordinates systems listed in Appendix A.

4.4 The Moment of Inertia Tensor

You may recall two definitions for angular momentum: on the one hand, a particle at $\vec{r}$ with momentum $\vec{p}$ has $\vec{L} = \vec{r} \times \vec{p}$ relative to the origin. But there's also the familiar $\vec{L} = I\vec{\omega}$, where $\vec{\omega}$ is the angular velocity and $I \sim mr^2$ is the so-called *moment of inertia*. Are these two definitions equivalent? Is one more fundamental, in some sense, than the other? It would seem so, since $\vec{L} = I\vec{\omega}$ appears to apply only when $\vec{L}$ is parallel to $\vec{\omega}$. But to properly answer the question requires a complete expression for I.

Consider a rigid object made up of many individual masses m_α. (We'll use α to distinguish the different masses, reserving $i, j,$ and k for vector indices.) Each one of these particles has angular momentum $\vec{L}_\alpha = \vec{r}_\alpha \times \vec{p}_\alpha$ (Figure 4.2). The total angular momentum of the extended body, then, is the (vector) sum over all the masses in the object,

$$\vec{L} = \sum_\alpha \vec{r}_\alpha \times \vec{p}_\alpha. \tag{4.51}$$

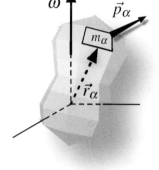

Now just as there's no momentum without velocity, there's no angular momentum without angular velocity; since $\vec{p}$ is directly proportional to $\vec{v}$, we fully expect $\vec{L}$ to be proportional to $\vec{\omega}$. This is easily verified: recalling $\vec{v} = \vec{\omega} \times \vec{r}$, each mass has momentum

$$\vec{p}_\alpha = m_\alpha \vec{v}_\alpha = m_\alpha \vec{\omega} \times \vec{r}_\alpha,$$

Figure 4.2 $\vec{L}_\alpha = \vec{r} \times \vec{p}_\alpha$

so that (4.51) becomes

$$\vec{L} = \sum_\alpha m_\alpha \vec{r}_\alpha \times (\vec{\omega} \times \vec{r}_\alpha). \tag{4.52}$$

For a continuous body the mass elements m_α are infinitesimal, and the sum becomes an integral over $dm \equiv \rho \, d\tau$, where $\rho(\vec{r})$ is the mass density. Thus

$$\vec{L} = \int \vec{r} \times (\vec{\omega} \times \vec{r}) \rho \, d\tau, \tag{4.53}$$

where $\vec{r} = (x, y, z)$ is the position of the mass element $\rho(\vec{r}) d\tau$.

Now to identify the moment of inertia from within the angular momentum, we need to factor out $\vec{\omega}$ from (4.53). This may appear impossible, given how the components of $\vec{r}$ and $\vec{\omega}$ are entangled. Clever wielding of index notation, however, can do the trick. First apply the "BAC-CAB" identity (see inside front cover) to get

$$\vec{L} = \int \left[r^2 \vec{\omega} - \vec{r}(\vec{r} \cdot \vec{\omega}) \right] \rho \, d\tau, \tag{4.54}$$

which in index notation is

$$L_i = \int \left(r^2 \omega_i - r_i \sum_j r_j \omega_j \right) \rho \, d\tau. \tag{4.55}$$

Using $\omega_i = \sum_j \omega_j \delta_{ij}$, we can factor out ω_j; then reversing the order of sum and integral yields

$$L_i = \int \sum_j \left[\delta_{ij} r^2 - r_i r_j \right] \omega_j \, \rho \, d\tau$$

$$= \sum_j \left[\int \left(r^2 \delta_{ij} - r_i r_j \right) \rho \, d\tau \right] \omega_j \equiv \sum_j I_{ij} \omega_j. \tag{4.56}$$

Thus we find the *moment of inertia tensor*

$$I_{ij} \equiv \int \left(r^2 \delta_{ij} - r_i r_j \right) \rho \, d\tau, \tag{4.57}$$

which is completely independent of $\vec{\omega}$. So whereas the relation between linear momentum and velocity is determined by the mass of the object, the relation between angular momentum and angular velocity is regulated by the *distribution* of mass within the object. The moment of inertia quantifies this distribution. The utility of $L_i = \sum_j I_{ij} \omega_j$ is that it distinguishes kinematics from dynamics: the moment of inertia tensor is a property of a rigid body, independent of its motion.

So we see that the two definitions of angular momentum are in fact equivalent — but only if I is understood as a *matrix*, rather than a simple number proportional to mr^2. Indeed, I_{ij}'s off-diagonal terms are what account for situations in which $\vec{L}$ is not parallel to $\vec{\omega}$.

Example 4.8   Just A Moment!

Consider a dumbbell with two equal point masses connected by a massless rod with orientation θ, ϕ, rotating around the z-axis as shown (Example 1.2). The angular momentum $\vec{L}$ is given by

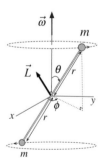

$$\vec{L} = \sum_\alpha \vec{r}_\alpha \times \vec{p}_\alpha = 2m\,\vec{r} \times (\vec{\omega} \times \vec{r})$$

$$= 2mr^2 \omega \sin\theta \hat{L}, \tag{4.58}$$

where the $\sin\theta$ emerges from $\vec{\omega} \times \vec{r}$, and

$$\hat{L} \equiv (-\cos\theta \cos\phi, \, -\cos\theta \sin\phi, \, \sin\theta), \tag{4.59}$$

as you can verify from the cross products. How does this compare to the equation $\vec{L} = I\vec{\omega}$? In this example, the integral of (4.57) reduces to a discrete sum over the two masses,

$$I_{ij} = \sum_\alpha m_\alpha \left(r_\alpha^2 \delta_{ij} - r_{\alpha,i} r_{\alpha,j} \right), \tag{4.60}$$

from which we find diagonal elements (Problem 4.18)

$$I_{11} = 2mr^2(1 - \sin^2\theta\cos^2\phi)$$

$$I_{22} = 2mr^2(1 - \sin^2\theta\sin^2\phi) \qquad (4.61)$$

$$I_{33} = 2mr^2(1 - \cos^2\theta) = 2mr^2\sin^2\theta,$$

and off-diagonal elements $I_{ij} = I_{ji}$

$$I_{12} = I_{21} = -2mr^2\sin^2\theta\sin\phi\cos\phi$$

$$I_{13} = I_{31} = -2mr^2\sin\theta\cos\theta\cos\phi \qquad (4.62)$$

$$I_{23} = I_{32} = -2mr^2\sin\theta\cos\theta\sin\phi .$$

This looks tedious, but since $\vec{\omega}$ is aligned along z, the matrix multiplication is easy:

$$\vec{L} = \begin{pmatrix} & & \\ & I_{ij} & \\ & & \end{pmatrix}\begin{pmatrix} 0 \\ 0 \\ \omega \end{pmatrix} = 2mr^2\omega\sin\theta\begin{pmatrix} -\cos\theta\cos\phi \\ -\cos\theta\sin\phi \\ \sin\theta \end{pmatrix}, \qquad (4.63)$$

which is identical to (4.58)–(4.59).

Problems

4.1 Show that any matrix T_{ij} can be written as a sum of a symmetric and an antisymmetric matrix.

4.2 Show that the transpose of the product is the product of the transposes: $(AB)^T = B^T A^T$.

4.3 Let Λ be a diagonal matrix with numbers $\lambda^{(j)}$ along the diagonal, so that $\Lambda = \begin{pmatrix} \lambda^{(1)} & 0 & \cdots & 0 \\ 0 & \lambda^{(2)} & \cdots & 0 \\ \vdots & & \ddots & \vdots \\ 0 & 0 & \cdots & \lambda^{(n)} \end{pmatrix}$, or $\Lambda_{ij} = \lambda^{(j)}\delta_{ij}$. Note that the superscript on the λ's is not an index, but rather labels the diagonal elements of Λ. Use index notation to show that an arbitrary matrix M does not commute with Λ, $M\Lambda \neq \Lambda M$, unless all the $\lambda^{(j)}$ are the same.

4.4 Explicitly demonstrate that $(\vec{A} \times \vec{B})_\ell = \sum_{jk} \epsilon_{\ell jk}A_j B_k$.

4.5 For any 3×3 matrix M, verify that $\det M = \sum_{ijk} \epsilon_{ijk}M_{1i}M_{2j}M_{3k}$.

4.6 Show that for right-handed orthonormal vectors $\hat{e}_i$:

(a) $\hat{e}_i \times \hat{e}_j = \sum_k \epsilon_{ijk}\hat{e}_k$
(b) $(\hat{e}_i \times \hat{e}_j) \cdot \hat{e}_k = \epsilon_{ijk}$.

4.7 Use (4.50) to find all the line, area, and volume elements in both cylindrical and spherical coordinates.

4.8 Use (4.46) and (4.50) to find the line and volume elements for hyperspherical coordinates of (2.41).

4.9 Verify the identity $\sum_\ell \epsilon_{mn\ell}\epsilon_{ij\ell} = \delta_{mi}\delta_{nj} - \delta_{mj}\delta_{ni}$ by stepping through the sum over ℓ and using the antisymmetry of ϵ_{ijk}.

4.10 Use the identity in the previous problem to work out the following sums:

(a) $\sum_{jk} \epsilon_{mjk}\epsilon_{njk}$
(b) $\sum_{ijk} \epsilon_{ijk}\epsilon_{ijk}$.

4.11 Use index algebra to verify the following identities:

(a) $\vec{A} \times \vec{B} = -\vec{B} \times \vec{A}$
(b) $\vec{A} \cdot (\vec{A} \times \vec{B}) = 0$
(c) $\vec{A} \cdot (\vec{B} \times \vec{C}) = \vec{B} \cdot (\vec{C} \times \vec{A}) = \vec{C} \cdot (\vec{A} \times \vec{B})$
(d) $\vec{A} \times (\vec{B} \times \vec{C}) = \vec{B}(\vec{A} \cdot \vec{C}) - \vec{C}(\vec{A} \cdot \vec{B})$ ["BAC-CAB"]
(e) $(\vec{A} \times \vec{B}) \cdot (\vec{C} \times \vec{D}) = (\vec{A} \cdot \vec{C})(\vec{B} \cdot \vec{D}) - (\vec{A} \cdot \vec{D})(\vec{B} \cdot \vec{C})$.

4.12 For $r^2 = x^2 + y^2 + z^2 = \sum_i x_i^2$:

(a) Show that $\frac{\partial}{\partial x_i}(1/r) = -x_i/r^3$.

(b) Argue that symmetry dictates the two-derivative expression $\frac{\partial}{\partial x_i}\frac{\partial}{\partial x_j}(1/r)$ to have the form $a\,\delta_{ij} + b\,x_i x_j$, for scalars a and b.

(c) Take the trace of both sides of this proposed form to find $3a + br^2 = 0$. (The operator $\sum_i \partial_i^2 \equiv \nabla^2$ is called the *Laplacian*.)

(d) Work out the derivatives for $i \neq j$ to find b, and thence that

$$\frac{\partial}{\partial x_i}\frac{\partial}{\partial x_j}\left(\frac{1}{r}\right) = \frac{3x_i x_j - \delta_{ij} r^2}{r^5}. \tag{4.64}$$

[Though applicable in most cases, this expression is not quite correct. The missing subtlety is addressed in Problem 15.28.]

4.13 Indices are not used exclusively to denote vector components; they can also keep track of the number of items in a sum. We saw an example of this in Section 4.4, where the index α was used to keep track of individual mass elements m_α. In a similar vein, consider the electrostatic potential energy U of an arbitrary collection of point charges q_i. For a two-charge configuration, the energy is

$$\frac{1}{4\pi\epsilon_0} \frac{q_i q_j}{|\vec{r}_i - \vec{r}_j|} = q_i V(\vec{r}_i),$$

where $V(\vec{r}_i)$ is the voltage at $\vec{r}_i$ due to the charge at $\vec{r}_j$. Note that the indices on $\vec{r}_i$ and $\vec{r}_j$ are not vector indices (as the arrow-on-top confirms), but rather denote the location of charges. Show that summing over all such pairs in a collection of N charges yields the total energy

$$U = \frac{1}{2}\sum_{i=1}^{N} q_i V(\vec{r}_i),$$

where now $V(\vec{r}_i)$ is the voltage at $\vec{r}_i$ due to *all* the other charges in the configuration. [Hint: You'll need to start with a double sum — one for all the charges i, the other over all the other charges $j \neq i$ which comprise $V(\vec{r}_i)$.]

4.14 For real symmetric matrices $A = A^T$, the Rayleigh quotient is the scalar function

$$\rho(\vec{v}) = \frac{v^T A v}{v^T v}.$$

Use $\partial\rho/\partial v_i = 0$ to show that ρ's extremal values occur when $\vec{v}$ satisfies the *eigenvalue equation* $A\vec{v} = \rho\vec{v}$.

4.15 The moment of inertia tensor can be written

$$I_{ij} \equiv \int \mathcal{I}_{ij}\,\rho\,d\tau,$$

where $\mathcal{I}_{ij} \equiv r^2\delta_{ij} - r_i r_j$. Write out the matrix $\mathcal{I}$ in cartesian coordinates. (When integrated to get I, the diagonal elements are called the *moments of inertia*, and the off-diagonal elements are the *products of inertia*.)

4.16   A uniform solid cube of mass M and sides of length a has it edges parallel to the coordinate axes. Use (4.57) to find the moment of inertial tensor when the coordinate origin is:

(a) at the center of the cube

(b) at one corner of the cube.

In both cases, find axes through the origin for which $\vec{L}$ and $\vec{\omega}$ are parallel. (Only for such axes is torque-free, constant-ω motion possible.)

4.17   Using index notation:

(a) Show that the moment of inertia I_{ij} is symmetric. How many independent elements does it have? In general, how many independent elements does a real $n \times n$ symmetric matrix have?

(b) Show that the diagonal elements of I_{ij} must be non-negative.

(c) Find the trace of I_{ij}.

4.18   Verify (4.61) and (4.62).

5 Brandishing Binomials

5.1 The Binomial Theorem

The binomial theorem allows one to express the power of a sum as a sum of powers,

$$(x+y)^n = \sum_{m=0}^{n} \binom{n}{m} x^{n-m} y^m, \tag{5.1}$$

where the integer $n \geq 0$, and the so-called *binomial coefficients* are given by

$$\binom{n}{m} \equiv \frac{n!}{m!\,(n-m)!}. \tag{5.2}$$

For example (recall that $0! = 1$),

$$(x+y)^5 = \binom{5}{0} x^5 y^0 + \binom{5}{1} x^4 y^1 + \binom{5}{2} x^3 y^2 + \binom{5}{3} x^2 y^3 + \binom{5}{4} x^1 y^4 + \binom{5}{5} x^0 y^5$$

$$= x^5 + 5x^4 y + 10x^3 y^2 + 10x^2 y^3 + 5xy^4 + y^5. \tag{5.3}$$

Expanding an expression such as this using the binomial theorem is often easier than direct multiplication — but it's even easier using *Pascal's triangle*,

$$
\begin{array}{ccccccccccc}
 & & & & & 1 & & & & & \\
 & & & & 1 & & 1 & & & & \\
 & & & 1 & & 2 & & 1 & & & \\
 & & 1 & & 3 & & 3 & & 1 & & \\
 & 1 & & 4 & & 6 & & 4 & & 1 & \\
1 & & 5 & & 10 & & 10 & & 5 & & 1 \\
\end{array}
$$

which can be continued indefinitely in an obvious fashion: each entry in the triangle is given by the sum of the two numbers diagonally above it. A little mental calculation leads to the discovery that its rows are the binomial coefficients. The sixth row, for instance, is made up of the values of $\binom{5}{m}$, $m = 0$ to 5 found in (5.3). (This is usually labeled the fifth row, with the triangle's apex corresponding to $n = 0$.)

Inspection of Pascal's triangle reveals several symmetries amongst the binomial coefficients — perhaps the two most apparent being

$$\binom{n}{m} = \binom{n}{n-m} \quad \text{and} \quad \binom{n}{m} + \binom{n}{m+1} = \binom{n+1}{m+1}. \tag{5.4}$$

The first is clear from the definition in (5.2); the second requires a few lines of algebra to derive directly. Also note that the sum of the nth row of Pascal's triangle is

$$\sum_{m=0}^{n} \binom{n}{m} = 2^n, \tag{5.5}$$

which, for example, is the total number of possible outcomes of n coin flips.

In any event, the origin of the binomial coefficients in (5.3) is not hard to appreciate. If you try to calculate $(x + y)^5$ by lining up 5 factors of $(x + y)$, it becomes clear that there are multiple paths from which a term such as, say, $x^3 y^2$ emerges. With 5 terms to multiply, how many different ways can you choose 3 of them to be x? The answer is $\binom{5}{3} = 10$. More generally, the expression $\binom{n}{m}$ — often read "n choose m" — gives the number of ways m items can be picked out from a group of n. For instance, what's the probability of flipping a fair coin 6 times and getting 2 tails? Generating the next row of Pascal's triangle shows that 4 heads, 2 tails occurs 15 times out of a total of $1 + 6 + 15 + 20 + 15 + 6 + 1 = 2^6 = 64$ possible outcomes, so the answer is 15/64, or 23.4%. Another way is to note that the probability of getting 2 tails is 0.5^2, the probability of getting 4 heads is 0.5^4 — but since there are 15 ways for this specific combination to occur, the overall probability is $15 \times (0.5)^2 \times (0.5)^4 = 0.234$. In general, given probability p for getting one of two mutually exclusive outcomes, the probability of getting m of them in n tries is

$$P(m) = \binom{n}{m} p^m (1 - p)^{n-m}. \tag{5.6}$$

This is the famous *binomial distribution*.

Example 5.1 **Fair is Foul and Foul is Fair**

When flipping a set of fair coins, the most likely outcome (or "state") is one with half heads and half tails. So if instead we find a state composed of, say, at least 80% heads or at least 80% tails, should we become suspicious that something was rigged? What is the probability of such a result using fair coins?

To be concrete, consider a set of 10 fair coins. The probability of flipping them and getting at least 8 heads is found from (5.6) with $p = 0.5$:

$$P(8) + P(9) + P(10) = \left[\binom{10}{8} 2^{-8} 2^{-2} + \binom{10}{9} 2^{-9} 2^{-1} + \binom{10}{10} 2^{-10} 2^0 \right]$$

$$= (45 + 10 + 1) 2^{-10} = \frac{56}{1024} = 0.055. \tag{5.7}$$

The probability of getting 8 or more of *either* heads or tails, then, is 0.11 — about 1/9. Small, but certainly not insignificant. The reason is the small sample size. If instead of at least 8 heads or tails from 10 coin flips we asked for at least 80 from a sample of 100, the probability plummets to about 10^{-9}.

Now imagine that we have 10 students, each with a set of 10 fair coins. If all the students flip their coins, what is the probability of at least one of them ending up with at least 8 heads or 8 tails? Note that there are only two outcomes or states — either a student's set has 8 or more heads/tails (State A, $p = 0.11$), or it doesn't (State B, $1 - p = 0.89$). Each set of coins represents one trial. Thus the binomial distribution, (5.6), applies with $n = 10$ and $p = 0.11$. Summing over all possible outcomes with at least one occurrence of State A,

$$\text{Prob}(\geq 8 \text{ H/T}) = \sum_{1}^{10} \binom{10}{m} p^m (1 - p)^{10-m} = 0.688 \approx \frac{2}{3}. \tag{5.8}$$

If we ask about heads alone, $p = 0.055$ and the probability of getting 8 or more is still a respectable 43%. A most unexpected large result — but it doesn't mean something in the state is rotten. Once again, our intuition presumes a larger sample size. And indeed, if we repeat the exercise with 100 set of 100 coins, the probability of getting more than 80% heads or tails is only about 1 in 10^7.

Example 5.2 **Happy Birthday(s)!**

Consider the following party trick: what's the probability that in a room of N people at least two share the same birthday? Given the length of a year, it seems unlikely that the odds would be substantial for groups smaller than about 200 people. But let's work it out.

It's actually easier to calculate the probability $P_u(N)$ that each person has a unique birthday, and then just subtract that answer from one. (For simplicity, we'll ignore leap-year complications.) Imagine filling the room one by one. The first person has 365 choices for a unique birthday. The second person only has 364, the third 363, etc. So the probability of N people all having unique birthdays is

$$P_u(N) = \frac{365}{365} \cdot \frac{364}{365} \cdot \frac{363}{365} \cdot \frac{362}{365} \cdots \frac{365 - (N - 1)}{365}$$

$$= 1 \cdot \left(1 - \frac{1}{365}\right) \left(1 - \frac{2}{365}\right) \left(1 - \frac{3}{365}\right) \cdots \left(1 - \frac{(N - 1)}{365}\right)$$

$$= \prod_{m=1}^{N-1} \left(1 - \frac{m}{365}\right). \tag{5.9}$$

A little effort reveals that $P_u(N)$ drops below 50% at the surprisingly low value $N = 23$. In other words, in a room of only 23 people, the odds are slightly better than 50% that two of them share a birthday!

Binomial coefficients hold the key to understanding this counter-intuitive result. We've allowed *any* of the available 365 days to be doubly represented — and in a group of 23 people, there are $\binom{23}{2} = 253$ possible pairs. Had we specified instead that at least one of the other 22 people must share the birthday of the first person picked, (5.6) tells us that the odds become

$$\sum_{m=1}^{22} \binom{22}{m} \left(\frac{1}{365}\right)^m \left(\frac{364}{365}\right)^{22-m} = 0.059,$$

a reassuringly small 6%.

BTW 5.1 Fun with Pascal's Triangle

Pascal's triangle can be used to find expressions for common sums. The outermost, "zeroth" diagonal of the triangle is made up of all 1's. Boring. The first diagonal, though, is the set of integers, in ascending order — which, for reasons that will soon be clear, we write

$$n\text{th integer:} \quad \binom{n+0}{1} = \frac{n}{1!}. \tag{5.10}$$

The second diagonal is composed of the so-called "triangular" numbers, which is the number of dots that can be arranged in a two-dimensional triangular grid (Figure 5.1),

Figure 5.1 Triangular numbers.

nth triangular number: $\qquad \dbinom{n+1}{2} = \dfrac{1}{2}n(n+1)\,.$ (5.11)

If we stack these triangles to form tetrahedra, we get the "tetrahedral" numbers, which are found in the third diagonal,

nth tetrahedral number: $\qquad \dbinom{n+2}{3} = \dfrac{1}{3!}n(n+1)(n+2).$ (5.12)

The next diagonal gives the number of dots in a stack of tetrahedra that form a four-dimensional triangular array, sometimes called a pentatope,

nth pentatope number: $\qquad \dbinom{n+3}{4} = \dfrac{1}{4!}n(n+1)(n+2)(n+3).$ (5.13)

Equations (5.11)–(5.13) can be verified with the defining sum used to generate Pascal's triangle. In addition, each element of a diagonal contains the running sum of the numbers in the diagonal above it. So the triangular numbers, which lie beneath the diagonal of integers, are simply given by the running sum of consecutive integers; from (5.11), then, we have the useful expression

$$\sum_{k=1}^{n} k = \frac{n(n+1)}{2}.$$ (5.14)

In the same way, the running sum of triangular numbers yields the tetrahedral numbers.

The formula (5.14) recalls a famous, if apocryphal, story told about Gauss. It seems that his elementary school teacher told the students to add up all the integers from 1 to 100; Gauss, it is said, returned with the correct answer in seconds! How did he do it? He may have grouped the integers into complementary pairs, each summing to 101: 1+100, 2+99, 3+98, ... Since there are 50 such pairs, the total is just $50{\cdot}101 = 5050$. But Gauss may have solved the problem visually by representing each integer from 1 to n by a row of dots, and then stacking the rows into a right triangle of side n. Two such triangles can be aligned to form an $n \times (n+1)$ rectangle. And the area of the rectangle is just twice the number of dots in the triangle.

Continuing this geometric mode of thought, just as two right triangles can be aligned to form a square, so too the sum of consecutive triangular numbers yields a perfect square, $\binom{n}{2} + \binom{n+1}{2} = n^2$. Similarly, the sum of two consecutive tetrahedral numbers gives the so-called pyramidal number: 1, 5, 14, 30, ... (perhaps most-commonly seen in a stacked display of oranges at the market). Since each layer of the pyramid is a square, the nth pyramidal number is just $\sum_1^n k^2$. Then using (5.12) to add the nth and $(n-1)$st tetrahedral numbers yields the useful formula

$$\sum_{k=1}^{n} k^2 = \frac{n(n+1)(2n+1)}{3!}.$$ (5.15)

5.2 Beyond Binomials

A slightly different view emerges by expressing the binomial coefficient (5.2) as

$$\binom{n}{i\,j} \equiv \frac{n!}{i!\,j!}, \tag{5.16}$$

where $n = i + j$. Recall that a set of n items has $n!$ permutations — but since there are only two *types* of items (i's and j's), some of these permutations are indistinguishable. To find the number of *distinct* permutations — the combinations — we need to divide out the $i!$ and $j!$ permutations of identical elements. This is what (5.16) does. For example, in the string *aabbabaa*, $i = 5$ is the number of a's, $j = 3$ is the number of b's, and $n = i + j = 8$. Then the number of distinct permuations of a's and b's is $\binom{8}{5\,3} \equiv \binom{8}{5} = \binom{8}{3} = 8!\,/5!\,3! = 8 \cdot 7 = 56$.

It's straightforward to expand (5.16) from two to three possible outcomes and obtain what are sometimes called *trinomial* coefficients,

$$\binom{n}{i\,j\,k} \equiv \frac{n!}{i!\,j!\,k!}, \tag{5.17}$$

where $i + j + k = n$. For instance, the number of distinct combinations of the letters *abcaba* is

$$\binom{6}{3\,2\,1} \equiv \frac{6!}{3!\,2!\,1!} = 60.$$

Just as the binomial coefficients emerge in (5.1), so-called trinomial coefficients arise in the expansion

$$(x + y + z)^n = \sum_{ijk} \binom{n}{i\,j\,k} x^i y^j z^k, \tag{5.18}$$

for positive $n \equiv i + j + k$. We can also introduce a trinomial distribution. Consider a collection of a's, b's, and c's with probabilities p_a, p_b, and p_c of selecting each at random. Assuming $\sum_i p_i = 1$, the probability that n random choices yields i a's, j b's, and k c's is

$$P(i,j,k) = \binom{n}{i\,j\,k} p_a^i\, p_b^j\, p_c^k, \tag{5.19}$$

which reduces to the binomial distribution (5.6) for $k = 0$. (5.2) and (5.19) easily generalize using "multinomial" coefficients,

$$\binom{n}{k_1\ k_2\ \cdots\ k_m} \equiv \frac{n!}{k_1!\,k_2!\cdots k_m!}, \qquad n = \sum_i k_i. \tag{5.20}$$

So, for example, the number of distinct combinations of the 12 letters in *blaise pascal* are

$$\binom{12}{1\,2\,3\,1\,2\,1\,1\,1} \equiv \frac{12!}{1!\,2!\,3!\,1!\,2!\,1!\,1!\,1!} = 19,958,400.$$

Multinomial coefficients can be generated using higher-dimensional forms of Pascal's triangle. For trinomial coefficients, Pascal's *tetrahedron* can be viewed as an exploded stack of his triangles,

$$
\begin{array}{ccccccccccccc}
 & & & & & & & & & & 1 & & \\
 & & & & & & 1 & & & 4 & & 4 & \\
1 & & 1 & & 2 & 2 & & 3 & 3 & & 6 & 12 & 6 & \cdots. \\
 & 1 & 1 & 1 & 2 & 1 & 3 & 6 & 3 & 4 & 12 & 12 & 4 \\
 & & & & & & 1 & 3 & 3 & 1 & 4 & 6 & 4 & 1 \\
\end{array}
$$

Notice that each face of the tetrahedron is just Pascal's triangle.

Rather than increasing the number of choices, what happens to the binomial distribution as the number of trials n increases? As displayed graphically in the series of plots in Figure 5.2, the binomial distribution "smooths out" into the well-known bell curve. These graphs represent, say, the probabilities of obtaining m 3's in n rolls of a fair six-sided die. So in one roll of the die, there's a 5/6 probability of getting no 3's, a 1/6 probability of getting one 3; in two rolls, there's only 1/36 probability of getting two 3's; etc.

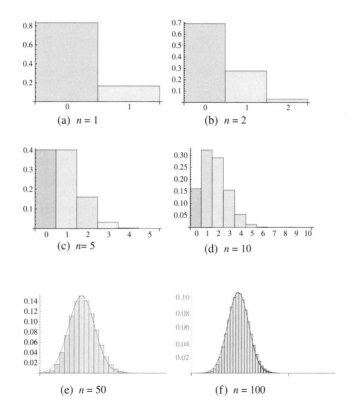

Figure 5.2 The $p = 1/6$ binomial distribution in the large n limit (note that the scales differ).

On the last two graphs is superimposed a bell curve, given by the function

$$
f(x) = \frac{1}{\sqrt{2\pi\sigma^2}} e^{-(x-\mu)^2/2\sigma^2}, \tag{5.21}
$$

were μ is the mean and σ the standard deviation. This functional form is known as a *Gaussian*. Of course the binomial distribution is defined only for integer values of m and n, whereas a Gaussian distribution is defined for continuous x. But for large n, moving slightly along the horizontal axis has little effect on the probability — at least in the vicinity of the mean. In other words, for a large number of trials, n essentially becomes a continuous variable. As you'll show in Problem 5.9, the binomial distribution has mean np and standard deviation $\sqrt{np(1-p)}$; in the large n and np limit, a Gaussian emerges with just these values. The emergence of a Gaussian distribution for a large number of rolls of a fair die is an example of the so-called central limit theorem.

Problems

5.1 Use the definition of the derivative together with the binomial theorem to rediscover the well-known result for the derivative of a polynomial.

5.2 Verify each of the following algebraically. Then write down the first few rows of Pascal's triangle and circle numbers that demonstrate the relation.

(a) $\dbinom{n}{m} + \dbinom{n}{m+1} = \dbinom{n+1}{m+1}$

(b)
$$\binom{n}{m} = \sum_{k=0}^{m} \binom{n-k-1}{m-k}$$
$$= \binom{n-1}{m} + \binom{n-2}{m-1} + \binom{n-3}{m-2} + \cdots + 1$$

[Hint: Use the identity in (a) recursively.]

(c) Find $\sum_{m=0}^{n} \binom{n}{m}$. What about $\sum_{m=0}^{n} (-1)^m \binom{n}{m}$? [Hint: Consider $(1 \pm 1)^n$.]

(d) Show $\sum_{m=0}^{n} \binom{n}{2m+1} = \sum_{m=0}^{n} \binom{n}{2m} = 2^{n-1}, n > 0$. [Note: $\binom{n}{k} \equiv 0, k > n$.]

5.3 Prove the Leibniz rule

$$\left(\frac{d}{dx}\right)^n (uv) = \sum_{m=0}^{n} \binom{n}{m} \left[\left(\frac{d}{dx}\right)^m u \right] \left[\left(\frac{d}{dx}\right)^{n-m} v \right]. \tag{5.22}$$

Then, using induction:

(a) show that the rule holds for $n = 1$
(b) show that if the rule holds for some n, it also holds for $n + 1$. [Hint: Use the recursion relation in Problem 5.2a.]

5.4 Use de Moivre's theorem and the binomial theorem to show that for some integer m,

$$2^m \sin\left(\frac{m\pi}{2}\right) = \binom{2m}{1} - \binom{2m}{3} + \binom{2m}{5} - \cdots + (-1)^{m+1} \binom{2m}{2m-1}.$$

5.5 Use de Moivre's theorem for $\sin(5\varphi)$ to show that $\sin^2(\pi/5) = \frac{1}{8}(5 - \sqrt{5})$. What's $\sin^2(2\pi/5)$?

5.6 Combining de Moivre's theorem with the binomial theorem allows for a fairly straightforward derivation of useful identities. Show that for integer N,

$$\sin(N\varphi) = \sum_{k=0}^{N} \binom{N}{k} \sin\left(\frac{(N-k)\pi}{2}\right) \cos^k \varphi \sin^{N-k} \varphi \tag{5.23a}$$

and

$$\cos(N\varphi) = \sum_{k=0}^{N} \binom{N}{k} \cos\left(\frac{(N-k)\pi}{2}\right) \cos^k \varphi \sin^{N-k} \varphi. \tag{5.23b}$$

[Hint: Isolate the i in your binomial theorem expansion, and apply de Moivre's theorem to its polar form $i = e^{i\pi/2}$.] Work out these identities for $N = 2$ and 3, and comment on the practical implications of the $\sin\left(\frac{(N-k)\pi}{2}\right)$ and $\cos\left(\frac{(N-k)\pi}{2}\right)$ terms in (5.23).

5.7 Use $\cos x = \frac{1}{2}\left(e^{ix} + e^{-ix}\right)$ and the binomial theorem to show that

$$\cos^n x = \frac{1}{2^n}\sum_{k=0}^{n}\binom{n}{k}\cos[(n-2k)x]. \tag{5.24}$$

[The symmetry $\binom{n}{k} = \binom{n}{n-k}$ together with the even parity of the cosine implies that this sum can be reduced to half the range from 0 to n — but at the expense of distinguishing odd and even n. See Problem 5.8.]

5.8 Tedious use of (5.23) would yield identities for powers of $\sin x$ and $\cos x$ as sums over sines and cosines of multiple angles. But it's easier to find the identities directly.

(a) If you've worked Problem 5.7, you can start from (5.24) — otherwise, use $\cos x = \frac{1}{2}\left(e^{ix} + e^{-ix}\right)$ and the binomial theorem to show that for even and odd powers

$$\cos^{2N} x = \frac{1}{2^{2N}}\binom{2N}{N} + \frac{1}{2^{2N-1}}\sum_{k=0}^{N-1}\binom{2N}{k}\cos[2(N-k)x], \tag{5.25a}$$

and

$$\cos^{2N+1} x = \frac{1}{2^{2N}}\sum_{k=0}^{N}\binom{2N+1}{k}\cos\left[2\,(N-k+1/2)\,x\right]. \tag{5.25b}$$

Write out these identities for $N = 1$.

(b) Similarly, show

$$\sin^{2N} x = \frac{1}{2^{2N}}\binom{2N}{N} + \frac{1}{2^{2N-1}}\sum_{k=0}^{N-1}(-1)^{N+k}\binom{2N}{k}\cos[2(N-k)x], \tag{5.26a}$$

and

$$\sin^{2N+1} x = \frac{1}{2^{2N}}\sum_{k=0}^{N}(-1)^{N+k}\binom{2N+1}{k}\sin[2(N-k+1/2)x]. \tag{5.26b}$$

Write out these identities for $N = 1$.

5.9 Any proper probability distribution $P(m)$ must be *normalized*,

$$\sum_{m=0}^{N} P(m) = 1.$$

Then the *mean* μ of a probability distribution is given by the weighted sum

$$\mu \equiv \sum_{m=0}^{N} mP(m) \equiv \langle m \rangle.$$

This is often called the "expected value" or the "expectation value." The mean of the distribution, however, does not say anything about the width. For that there's the *variance* σ^2, the average of the squared deviation from the mean,

$$\sigma^2 \equiv \sum_{m=0}^{N} (m - \mu)^2 P(m) \equiv \langle (m - \mu)^2 \rangle.$$

The *standard deviation* σ is the square-root of the variance.

(a) Show that for the binomial distribution $\mu = Np$. [Hint: Note that in the definition of μ the $m = 0$ term makes no contribution, so that the sum can formally begin at $m = 1$. Then change variables to $r = m - 1$ and use the normalization condition.]

(b) Show that the standard deviation can be expressed as the average of the square minus the square of the average,

$$\sigma^2 = \sum_{m=0}^{N} m^2 P(m) - \left(\sum_{m=0}^{N} mP(m) \right)^2 \equiv \langle m^2 \rangle - \langle m \rangle^2 .$$

Why don't we just define σ directly by $\sum_{m=0}^{N}(m - \mu)P(m) \equiv \langle m - \mu \rangle$?

(c) Show that for the binomial distribution, $\sigma^2 = Np(1 - p)$.

5.10   Consider the binomial distribution $P(m) = \binom{N}{m} p^m q^{N-m}$, where $q = 1 - p$.

(a) Use the large-n Stirling approximation, $n! \approx \sqrt{2\pi n}\, n^n e^{-n}$ (Problem 8.32) in the binomial coefficient to show that, for large N,

$$P(m) \approx \sqrt{\frac{N}{2\pi m(N - M)}} \left(\frac{Np}{m} \right)^m \left(\frac{Nq}{N - m} \right)^{N-m}.$$

(b) Use $x = m - Np$, which measures the deviation from the mean, to replace m and $N - m$ everywhere by $x + Np$ and $Nq - x$.

(c) Take the logarithm and expand the logs on the right through second order for large N and large Np. (In this limit, x and m become continuous variables.)

(d) Simplify your expression to find the Gaussian

$$P(m) = \frac{1}{\sqrt{2\pi\sigma^2}} e^{-(m-\mu)^2/2\sigma^2},$$

where $\mu = Np$ and $\sigma^2 = Npq$.

6 Infinite Series

The binomial theorem (5.1) is a series with a finite number of terms. And the sum of a finite number of finite numbers always yields a finite number. By contrast, an *infinite* series $S = \sum_{k=0}^{\infty} a_k$ may or may not converge to a finite result — even if all the a_k are finite.

An infinite series can be considered as the limit of a *sequence* of terms $S_0, S_1, S_2, \ldots$, where

$$S_N = \sum_{k=0}^{N} a_k \tag{6.1}$$

is the Nth *partial sum*. A series converges if the sequence of partial sums has a finite limit.

Example 6.1 Geometric Series

A series of the form $S_N = \sum_{k=0}^{N} r^k$ which has a common ratio r between consecutive terms is called a *geometric series*. It is one of the more common examples of a series, and also happens to be one of the simplest to analyze. Simply subtract rS_N from S_N and solve:

$$S_N = 1 + r + r^2 + r^3 + \cdots + r^N$$

$$rS_N = r + r^2 + r^3 + r^4 + \cdots + r^{N+1}$$

$$\leadsto (1-r)S_N = 1 - r^{N+1} \quad \implies \quad S_N = \frac{1 - r^{N+1}}{1 - r}. \tag{6.2}$$

This holds for all finite values of r — even the superficially troublesome value $r = 1$, as straightforward application of l'Hôpital's rule confirms.

A "closed" from for the infinite geometric series can be found from the partial sums in (6.2):

$$S \equiv \lim_{N \to \infty} S_N = \frac{1}{1 - r} \quad \textit{only if } r < 1. \tag{6.3}$$

Thus, for example, we easily find that $1 + \frac{1}{2} + \frac{1}{4} + \cdots = \frac{1}{1-1/2} = 2$.[1] As we'll see, the series actually converges for complex r with magnitude less than 1; for $|r| \geq 1$, however, the infinite series *diverges*.

---

[1] An infinite number of mathematicians walk into a bar. The first one orders a pint of beer, the second a half-pint, the third a quarter-pint. As they continue in this fashion, the bartender calmly pours exactly two pints of beer and says "You mathematicians need to know your limits!"

Example 6.2 **Of Prime Importance**

The *Riemann zeta function* $\zeta(s)$ is commonly expressed as the infinite series

$$\zeta(s) \equiv \sum_{k=1}^{\infty} \frac{1}{k^s}, \quad \Re e(s) > 1 \tag{6.4}$$

and appears in such disparate areas as statistics and number theory. A very intriguing alternative form is as a *product* over the prime numbers,

$$\zeta(s) = \prod_{\{p\}} \frac{1}{1 - p^{-s}}, \quad \{p\} = \text{the set of primes.} \tag{6.5}$$

The proof is both instructive and straightforward — and relies on the convergence of geometric series. Denoting individual primes by p_i, we start by rewriting (6.5) as an infinite product of geometric series,

$$\prod_{\{p\}} \frac{1}{1 - p^{-s}} = \frac{1}{1 - p_1^{-s}} \cdot \frac{1}{1 - p_2^{-s}} \cdot \frac{1}{1 - p_3^{-s}} \cdots$$

$$= \left[\sum_{k=0}^{\infty} \left(\frac{1}{p_1^s} \right)^k \right] \cdot \left[\sum_{k=0}^{\infty} \left(\frac{1}{p_2^s} \right)^k \right] \cdot \left[\sum_{k=0}^{\infty} \left(\frac{1}{p_3^s} \right)^k \right] \cdots, \tag{6.6}$$

where (6.3) applies since $p_i^{-s} < 1$. Multiplying the sums out, we get

$$\prod_{\{p\}} \frac{1}{1 - p^{-s}} = \left(1 + \frac{1}{p_1^s} + \frac{1}{p_1^{2s}} + \frac{1}{p_1^{3s}} + \cdots \right) \cdot \left(1 + \frac{1}{p_2^s} + \frac{1}{p_2^{2s}} + \frac{1}{p_2^{3s}} + \cdots \right) \cdots$$

$$= 1 + \sum_i \frac{1}{p_i^s} + \sum_{i \geq j} \frac{1}{(p_i \, p_j)^s} + \sum_{i \geq j \geq k} \frac{1}{(p_i \, p_j \, p_k)^s} + \cdots, \tag{6.7}$$

where the restrictions on the sums prevent multiple counting of cross-terms arising from the product. The first sum in (6.7) contributes the reciprocals of all the primes (raised to the s power), the second sum is over all integers which can be factored into two primes, the third over all integers which can be written as the product of three primes, etc. Now according to the *Fundamental Theorem of Arithmetic*, every positive integer greater than 1 can be factored into a unique product of prime numbers. But that's exactly what appears in each of the denominators of (6.7). Thus every integer occurs in a denominator exactly once, so that

$$\prod_{\{p\}} \frac{1}{1 - p^{-s}} = 1 + \frac{1}{2^s} + \frac{1}{3^s} + \frac{1}{4^s} + \cdots = \sum_{k=1}^{\infty} \frac{1}{k^s} = \zeta(s). \tag{6.8}$$

6.1 Tests of Convergence

Conceptually, the notion of convergence is fairly straightforward: as the partial sums S_N include more and more terms (that is, as N grows), their values become increasingly indistinguishable from S. In other words, if you set a precision goal of ϵ, there comes a point N_0 in a convergent sequence beyond which all the subsequent terms $S_{N>N_0}$ are within ϵ of S. Formally one says that for any given number $\epsilon > 0$, no matter how small, there is always some number N_0 in a convergent sequence such that

$$|S - S_N| < \epsilon \tag{6.9}$$

for all $N > N_0$. Intuitive though this definition may be, it requires knowing S — even before you've established that the series even has a finite limit. And anyway, often the more important question is merely whether or not a series converges, not the value it converges to. So what we need are tests for convergence that focus only on the terms a_k in the sum.

Consider series with non-negative real elements, $a_k \geq 0$. Then if the a_k grow for large k, the series will diverge. What's required is that they decrease *monotonically* — that is, $a_{k+1} < a_k$ for all k. (Actually they only need to decrease for all k above some N, since a finite number of non-monotonic terms will not affect the question of convergence.) We must also have $\lim_{k\to\infty} a_k = 0$ — otherwise the infinite sum will surely diverge. The geometric series with $r < 1$ satisfies both conditions. But so too does $\zeta(1) = \sum_k 1/k$, even though this series diverges (as we'll soon see). Thus the requirement that the a_k monotonically decrease to zero does not guarantee convergence (necessary but not sufficient).

The issue is not merely that the a_k decrease monotonically, but whether they fall off *fast* enough. Factorials, for example, grow very quickly, so we might reasonably expect the series $\sum \frac{1}{k!}$ to converge. As for the zeta function, we just claimed that $\zeta(1)$ diverges — but a moment's consideration shows that $\zeta(s) \to 1$ as s grows to infinity. So there must be some finite value of $s > 1$ for which the terms begin to fall off fast enough the render the series convergent. There are many tests of convergence which establish just what is "fast enough" for any given series; we'll look at a few of the most common.

1. **Ratio Test:** A series $S = \sum_k a_k$ converges if the ratio of consecutive terms is less than 1,

$$r = \lim_{k\to\infty} \frac{a_{k+1}}{a_k} < 1. \tag{6.10}$$

At first glance, this condition seems insufficient since there is no consideration of how fast the a_k fall. But we already know that the geometric series converges for any ratio less then 1. So no matter the precise value of $r < 1$, there is always a larger value (still less than 1, of course) for which the geometric series converges. And if a series with a larger ratio converges, then certainly the smaller-r series S must converge (and do so more rapidly). By the same token, if $r = 1$ the test is inconclusive.

Note that the ratio of consecutive terms can be greater than 1 for a *finite* number of terms — hence r in (6.10) is defined in the $k \to \infty$ limit.

Example 6.3 $\sum_{k=1}^{\infty} 1/k!$

Using the ratio test, we have

$$r = \lim_{k\to\infty} \frac{1/(k+1)!}{1/k!} = \lim_{k\to\infty} \frac{k!}{(k+1)!}$$
$$= \lim_{k\to\infty} \left(\frac{1}{k+1}\right)$$
$$= 0.$$

So, as expected, the series converges.

Example 6.4 $\zeta(s) = \sum_{k=1}^{\infty} \frac{1}{k^s}$

The ratio test gives

$$r = \lim_{k\to\infty} \left[\frac{1/(k+1)^s}{1/k^s}\right] = \lim_{k\to\infty} \left[\frac{k}{k+1}\right]^s = 1.$$

Thus the ratio test is inconclusive. We'll have to try other tests to determine whether, and for what values of s, $\zeta(s)$ converges. (Note that it is the limit as $k \to \infty$ that counts here: the fact that $\frac{k}{k+1} < 1$ for finite k is not relevant.)

2. **Comparison Test:** In establishing the ratio test, we appealed to perhaps the most basic approach of all: comparison with a known convergent series. Specifically, if another series $\sum_k b_k$ is known to converge, then S converges if $a_k \leq b_k$. This condition, though, need not hold for all k; as before, if $a_k > b_k$ only for a *finite* number of terms, S will still be finite.

Example 6.5 $\sum_{k=1}^{\infty} 1/k!$

If we compare this series

$$\sum_{k=1}^{\infty} \frac{1}{k!} = 1 + \frac{1}{2} + \frac{1}{6} + \frac{1}{24} + \cdots \tag{6.11a}$$

with the geometric series

$$\sum_{k=1}^{\infty} \frac{1}{2^k} = \frac{1}{2} + \frac{1}{4} + \frac{1}{8} + \frac{1}{16} + \cdots \tag{6.11b}$$

we see that aside from the first three terms, each term in the geometric series is larger than the corresponding term in the $1/k!$ series. Although the first three will certainly affect the *value* of the sum, since they are finite in number they cannot affect its *convergence*. Thus since the geometric series converges, the series $1/n!$ must converge as well.

Example 6.6 **The Harmonic Series, $\zeta(1)$**

The Riemann zeta function for $s = 1$ is called the *harmonic series*,

$$\zeta(1) = 1 + \frac{1}{2} + \frac{1}{3} + \frac{1}{4} + \frac{1}{5} + \frac{1}{6} + \cdots. \tag{6.12a}$$

We can use the comparison test by grouping terms as follows:

$$\zeta(1) = \sum_{k=1}^{\infty} \frac{1}{k} = 1 + \left(\frac{1}{2}\right) + \left(\frac{1}{3} + \frac{1}{4}\right) + \left(\frac{1}{5} + \frac{1}{6} + \frac{1}{7} + \frac{1}{8}\right) + \left(\frac{1}{9} + \cdots + \frac{1}{16}\right) + \cdots$$

$$> 1 + \left(\frac{1}{2}\right) + \left(\frac{1}{4} + \frac{1}{4}\right) + \left(\frac{1}{8} + \frac{1}{8} + \frac{1}{8} + \frac{1}{8}\right) + \left(\frac{1}{16} + \cdots + \frac{1}{16}\right) + \cdots$$

$$= 1 + \left(\frac{1}{2}\right) + \left(\frac{1}{2}\right) + \left(\frac{1}{2}\right) + \left(\frac{1}{2}\right) + \cdots. \tag{6.12b}$$

Clearly, then, the harmonic series *diverges*.

Example 6.7 $\zeta(2)$

So $\zeta(1)$ diverges; what about $\zeta(2)$? Noting that $k^2 \geq k(k + 1)/2$ for $k \geq 1$, we can use the comparison test as follows:

$$\zeta(2) = \sum_{k=1}^{\infty} \frac{1}{k^2} = 1 + \frac{1}{4} + \frac{1}{9} + \frac{1}{16} + \cdots$$

$$< 1 + \frac{1}{3} + \frac{1}{6} + \frac{1}{20} + \cdots$$

$$= \sum_{k=1}^{\infty} \frac{2}{k(k + 1)}. \tag{6.13a}$$

But this new series not only converges, it is an example of a simple-to-sum "telescoping" series:

$$\sum_{k=1}^{\infty} \frac{1}{k(k+1)} = \sum_{k=1}^{\infty} \left(\frac{1}{k} - \frac{1}{k+1} \right)$$

$$= \left[\left(1 - \frac{1}{2} \right) + \left(\frac{1}{2} - \frac{1}{3} \right) + \left(\frac{1}{3} - \frac{1}{4} \right) + \left(\frac{1}{4} - \frac{1}{5} \right) + \cdots \right]$$

$$= \left[1 + \left(-\frac{1}{2} + \frac{1}{2} \right) + \left(-\frac{1}{3} + \frac{1}{3} \right) + \left(-\frac{1}{4} + \frac{1}{4} \right) + \cdots \right]$$

$$= 1. \tag{6.13b}$$

Thus from (6.13a) we see not merely that $\zeta(2)$ converges, but that $\zeta(2) < 2$. In fact, $\zeta(2) = \pi^2/6$.

3. **Integral Test:** Rather than compare a given infinite sum with a second known series, if we think of the a_k as functions of integers k, we can also compare $\sum_k a_k$ to the integral $\int a(x)\, dx$. Since we're presuming that past some point N the $a_{k>N}$ are monotonically decreasing, we get the upper bound

$$a_k = \int_k^{k+1} a_k\, dx \leq \int_k^{k+1} a(x)\, dx. \tag{6.14}$$

Then

$$\sum_{k=N}^{\infty} a_k \leq \sum_{k=N}^{\infty} \int_k^{k+1} a(x)\, dx = \int_N^{\infty} a(x)\, dx. \tag{6.15}$$

So if the integral converges, so too does the sum. A similar argument can be made replacing $\int_k^{k+1} \to \int_{k-1}^{k}$ in (6.14), resulting in a lower bound for the sum. Thus if the integral diverges, so too the sum.

Example 6.8 $\zeta(s)$ **Revisited**

$$\zeta(s) = \sum_{k=1}^{\infty} \frac{1}{k^s} \longrightarrow \int \frac{dx}{x^s} = \begin{cases} \frac{x^{1-s}}{1-s}, & s \neq 1 \\ \ln x, & s = 1 \end{cases} \tag{6.16}$$

To properly compare with the infinite sum, we take the limit $x \to \infty$. Thus we see once again that the harmonic series ($s = 1$) diverges, whereas $\zeta(s)$ converges for all $s > 1$.

BTW 6.1 Why Should the Harmonic Series be Divergent?

If you try to explicitly evaluate the sum in (6.12a), you'll find it grows very slowly: even with 100,000 terms, the sum is only a little larger than 12 — and the next term only contributes an additional 10^{-6}. So it's not unreasonable to expect the infinite sum to converge to some finite value. But even though the harmonic series is the sum of ever-decreasing terms, these terms do not drop off quickly enough to compensate for their infinite number. And this, in the end, is what determining the convergence of a series is really all about.

   In some sense, the harmonic series is just "barely" divergent — indeed, as the integral test in (6.16) shows, $\zeta(1+\epsilon)$ *is* convergent for any small $\epsilon > 0$. The slow, unceasing growth of the partial sum $\zeta_N(1) \equiv \sum_{k=1}^{N} \frac{1}{k}$ is actually logarithmic in N. (So, for instance, the partial sum with a $100{,}000$ terms is about 12.1; but the partial sum with $N = 10^6$, despite having 10 times as many terms, is merely 14.4 — a difference of only $\ln 10 \approx 2.3$.) We can extract a finite result from $\zeta(1)$ if we *subtract off* this logarithmic divergence,

$$\gamma \equiv \lim_{N \to \infty} \left(\sum_{k=1}^{N} \frac{1}{k} - \ln N \right) \approx 0.5772, \tag{6.17}$$

where γ is called the Euler–Mascheroni constant (and has nothing to do with the γ of relativity). This can be used in reverse to quickly estimate $\sum_{k=1}^{N} \frac{1}{k}$ for *finite* N. For example, rather than tediously summing the first 100 terms in the harmonic series, we can quickly estimate $\sum_{k=1}^{100} \frac{1}{k} \approx \gamma + \ln 100 = 5.182$ — which is accurate to better than 0.1%.

While we're in the neighborhood, it's worth noting that for $s = 1$, the product-over-primes form in (6.5) can be rendered $\zeta(1) = \prod \frac{p}{p-1}$, which is finite only if there are a finite number of primes. But since we know that $\zeta(1)$ diverges, there must be infinitely many prime numbers!

BTW 6.2 Cauchy Sequences

As we've seen, the formal statement of convergence given in (6.9) has limited utility. Cauchy ("ko-shee") devised a more useful criterion which concerns only the elements of a sequence without reference to an unknown limiting value. He reasoned that beyond a certain point in any sequence with a finite asymptotic limit, there will be less and less "room" for the remaining (infinite) terms to crowd into, so that all elements will necessarily have to cluster together. In other words, the elements of a convergent sequence should eventually become arbitrarily close to one another — not just consecutive elements S_n and S_{n+1}, but *any* two elements S_n and S_m. So given a precision goal of ϵ, there must come a point N in the sequence beyond which all subsequent terms $S_{n>N}$ are within ϵ of one another. Formally one says that for any given number $\epsilon > 0$, no matter how small, there is always some number N in a convergent sequence such that

$$|S_m - S_n| < \epsilon \tag{6.18}$$

for all $n, m > N$. A sequence that satisfies this condition is called a *Cauchy sequence*. In common parlance, one says the sequence is "cauchy." Being cauchy is what "fast enough" means.

Consider once again the harmonic series. Even though *consecutive* partial sums become arbitrarily close, the same is not true of all partial sums. For $m > n$, we have

$$|S_m - S_n| = \left| \sum_{k=1}^{m} \frac{1}{k} - \sum_{k=1}^{n} \frac{1}{k} \right| = \sum_{k=n+1}^{m} \frac{1}{k}$$

$$\geq \int_{n+1}^{m} \frac{dx}{x} = \ln \left(\frac{m}{n+1} \right). \tag{6.19}$$

Now as $m \to \infty$, this blows up; it is certainly not within *any* finite precision. The harmonic series is logarithmically divergent.

For sequences of real numbers, all convergent sequences are Cauchy sequences. However, there are intriguing examples in which a Cauchy sequence could converge in a *different space* than that in which the partial sums belong. Consider the sequence of partial sums $S_n = (1 + 1/n)^n$. Although each S_n is rational, the sequence converges to e, which is decidedly not rational. Similarly, the rational sequence $\{1., 1.4, 1.414, \ldots\}$ converges to the irrational $\sqrt{2}$. In these examples, the sequences converge, but not in the space of the individual terms. They are "cauchy" in the reals, but not in the space of rationals.

4. **Absolute Convergence:** Convergence is a little trickier if the a_k vary in sign or, more generally, are complex. One approach is to apply the ratio test to the *magnitudes* $|a_k|$. If $r = \lim_{k \to \infty} |a_{k+1}/a_k| < 1$, the series is said to be *absolutely convergent*. And any series which is absolutely convergent is also convergent — after all, the sum of all positive $|a_k|$ should be larger than the sum of phase-varying a_k. (Just as when adding arrows $\vec{a}$ in the plane, $\sum_k |\vec{a}_k| \geq |\sum_k \vec{a}_k|$.)

> **Example 6.9** **The Geometric Series with $r < 0$**
>
> The ratio test clearly shows that the geometric series converges for $0 \leq r < 1$; but in fact, since the series is absolutely convergent the expansion is valid for all r between -1 and 1.

5. **Alternating Series:** An alternating series $\sum_k (-1)^k a_k$ may be convergent even if it isn't absolutely convergent. This is not unreasonable, since an alternating series will not grow as quickly as its absolute cousin. For $a_k > 0$, converges as long as (*i*) the a_k decrease for large k and (*ii*) $\lim_{k \to \infty} a_k = 0$. (That is, if beyond a certain point the a_k decrease to zero monotonically, the alternating series is cauchy.)

> **Example 6.10** **The Alternating Harmonic Series $S = \sum_{k=1}^{\infty} (-1)^{k+1}/k$**
>
> Both (*i*) and (*ii*) above are satisfied, so this series converges despite the fact that the harmonic series does not. In other words, the alternating harmonic series is convergent, but not absolutely convergent. Computing the partial sums $S_N \equiv \sum_{k=1}^{N} (-1)^{k+1}/k$ for finite N, one can show $S = \lim_{N \to \infty} S_N = \ln 2$.

BTW 6.3 Conditional Convergence

A convergent series which is not absolutely convergent is said to be "conditionally convergent." Conditional convergence can be tricky. As an example, consider again the alternating harmonic series $S = \sum_{k=1}^{\infty} (-1)^{k+1}/k = \ln 2$. If we re-arrange the terms, grouping the first positive term with the first two negatives and so on, we get

$$\ln 2 = \left(1 - \frac{1}{2} - \frac{1}{4}\right) + \left(\frac{1}{3} - \frac{1}{6} - \frac{1}{8}\right) + \cdots$$

$$= \sum_{k=1}^{\infty} \left(\frac{1}{2k-1} - \frac{1}{4k-2} - \frac{1}{4k}\right) = \sum_{k=1}^{\infty} \frac{1}{4k(2k-1)}$$

$$= \frac{1}{2} \sum_{k=1}^{\infty} \left(\frac{1}{2k-1} - \frac{1}{2k}\right) = \frac{1}{2} \sum_{k=1}^{\infty} \frac{(-1)^{k+1}}{k} = \frac{1}{2} \ln 2 \, ,$$

a clear absurdity. In fact, by rearranging the order of the terms we can actually make a conditionally convergent series converge to any value whatsoever — or even render it divergent! This bit of legerdemain is the essence of the Riemann rearrangement theorem. Even without delving too deeply into its details, it's not hard to see why one must take care when dealing with conditionally convergent series. Consider the positive and negative terms in a series separately,

$$S \equiv \sum_{k}^{\infty} a_k - \sum_{k}^{\infty} b_k,$$

where both the a_k and b_k are positive. Clearly, the full series converges absolutely if both $\sum a_k$ and $\sum b_k$ converge. If the full series diverges, then one of $\sum a_k$ and $\sum b_k$ must diverge. But in a conditionally convergent series *both* $\sum a_k$ and $\sum b_k$ diverge. Thus a conditionally convergent series is essentially the *difference* of divergent quantities.

6.2 Power Series

What if we have a series like $S(x) = \sum_{k=0}^{\infty} a_k x^k$, which is a function of x? Now (absolute) convergence depends not just on the coefficients a_k but on the value of x for which the sum is being evaluated. Applying the ratio test gives

$$\lim_{k\to\infty}\left|\frac{a_{k+1}x^{k+1}}{a_k x^k}\right| = |x|\lim_{k\to\infty}\left|\frac{a_{k+1}}{a_k}\right| = |x|r. \tag{6.20}$$

So the sum converges only for $|x|r < 1$ or, defining $R \equiv 1/r$,

$$|x| < R = \lim_{k\to\infty}\left|\frac{a_k}{a_{k+1}}\right|, \tag{6.21}$$

where R is called the *radius of convergence*.

Example 6.11   **Finding the Radius of Convergence**

- $\frac{1}{1-x} = 1 + x + x^2 + \cdots \implies R = 1$, so the series converges for $|x| < 1$.
- $e^x = \sum_{k=0}^{\infty} \frac{x^k}{k!} \implies R = \lim_{k\to\infty}(k+1)!\,/k! = \infty$, so this series converges for all finite x. Given the close relationship between e^x and trigonometry, this implies that the series for $\sin x$, $\cos x$, $\sinh x$, and $\cosh x$ are also convergent for finite x.
- $\ln(1+x) = x - \frac{1}{2}x^2 + \frac{1}{3}x^3 - \cdots \implies R = 1$, so the series converges for $|x| < 1$. In fact, it also converges at $x = 1$, where it becomes the alternating harmonic series of Example 6.10 — which, as we claimed, converges to $\ln(2)$. On the other hand, $x = -1$ gives the (regular) harmonic series, so the series is divergent there.
- $\tan x = x + \frac{x^3}{3} + \frac{2x^5}{15} + \frac{17x^7}{315} + \cdots$. This series doesn't seem to follow a simple pattern, nor is it terribly useful beyond the approximation $\tan x \approx x + \frac{x^3}{3}$ for small x. But it can be written in terms of the so-called Bernoulli numbers B_k,[2]

$$\tan x = \sum_{k=1}^{\infty}(-1)^{k-1}\frac{2^{2k}(2^{2k}-1)B_{2k}}{(2k)!}\,x^{2k-1}.$$

This looks a mess, but for present purposes all we need to know about B_{2k} is

$$|B_{2k}| = \frac{2(2k)!}{(2\pi)^{2k}}\,\zeta(2k),$$

where ζ is the Riemann zeta function. Then, since $\zeta(n) \to 1$ for large n, the radius of convergence of $\tan x$ is easy to find from (6.21):

$$|x|^2 < \frac{\pi^2}{4} \implies R = \frac{\pi}{2}.$$

As expected.

---

[2] Bernoulli numbers are the coefficients in the expansion $\frac{x}{e^x-1} = \sum_{k=0}^{\infty} B_k \frac{x^k}{k!}$.

BTW 6.4 Why Is It Called the *Radius* of Convergence?

Wouldn't "interval" seem more appropriate? To answer this, let's examine two closely related series:

$$f(x) = 1 + x^2 + x^4 + \cdots = \sum_{k=0}^{\infty} x^{2k}, \qquad (6.22)$$

$$g(x) = 1 - x^2 + x^4 - \cdots = \sum_{k=0}^{\infty} (-1)^k x^{2k}. \qquad (6.23)$$

The ratio test shows that (6.22) converges only for $|x| < 1$. In fact, as a geometric series we have

$$f(x) = \frac{1}{1 - x^2}, \qquad (6.24)$$

clearly showing that the function blows up at $x = \pm 1$. The same, however, cannot be said of (6.23); this geometric series sums to

$$g(x) = \frac{1}{1 + x^2}, \qquad (6.25)$$

but is not singular at $x = \pm 1$. Yet straightforward evaluation shows that the series in (6.23) breaks down for $|x| > 1$, converging to (6.25) *only* for $|x| < 1$ (Figure 6.1). But there appears to be no discernible explanation for $\sum_{k=0}^{\infty} (-1)^k x^{2k}$ having any finite radius of convergence at all!

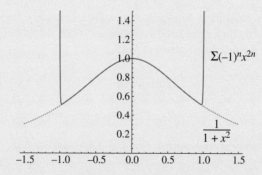

Figure 6.1 Comparison of $1/(1 + x^2)$ and the series $\sum_n (-1)^n x^{2n}$.

When confronted with such an apparent conflict, it often pays to take a step back to broaden your scope. In this case, though we're considering real-valued power series, note that f and g are special cases of the complex function $h(z) = 1/(1 - z^2)$: h reduces to f on the real axis (i.e., when z is real), and to g on the imaginary axis. The singularities of h are at ± 1 *and* $\pm i$; all four lie on a circle in the complex plane. So even though our series expansions are real, with regard to convergence what matters is the distance to the nearest singularity *in the complex plane*.

This example illustrates the basic result: a power series expansion of a function will converge everywhere from the center of the series to the nearest singularity in the complex plane. Hence the term "*radius* of convergence." (We'll demonstrate why this is so in Section 19.2.)

A power series is in some sense a generalization of an Nth-order polynomial — which is just a truncated power series whose coefficients $a_{k>N}$ are all zero. Yet despite not have a highest power, power series are notable and useful in that they can often be treated as polynomials. Two series can be added, multiplied, and even divided, with the result valid in their common region of convergence (see Problem 6.18). They can also be differentiated term-by-term, resulting in power series expansions for df/dx and $\int f\, dx$ with the same radius of convergence. The same, alas, cannot be said of more general function series, $f(x) = \sum_k a_k u_k(x)$, even when the functions $u_k(x)$ are reasonably well-behaved. We'll address this question in Chapter 34.

6.3 Taylor Series

A Taylor expansion is a power series which gives the value of a function at x in terms of the function and its derivatives at x_0:

$$f(x) = \sum_{n=0}^{\infty} \frac{(x - x_0)^n}{n!} f^{(n)}(x_0), \tag{6.26}$$

where

$$f^{(n)}(x_0) \equiv \frac{d^n}{dx^n} f(x)\bigg|_{x=x_0}.$$

Thus if the series converges, the value of $f(x)$ and all its derivatives at *one point* allow the reconstruction of $f(x)$ *everywhere*.[3] The infinite power series is sometimes referred to as an "open" representation of the function $f(x)$. Note that the Taylor series of a polynomial has a finite number of terms — and is just the polynomial itself. Similarly, truncating a function's Taylor series approximates the function as a polynomial. Such a truncated polynomial $P_n(x)$ is the best nth-order approximation to $f(x)$ which has the same the first n derivatives of f at $x = x_0$,

$$P_n(x) = f(x_0) + f'(x_0)\,(x - x_0) + \frac{1}{2!} f''(x_0)\,(x - x_0)^2$$
$$+ \frac{1}{3!} f'''(x_0)(x - x_0)^3 + \cdots + \frac{1}{n!} f^{(n)}(x_0)(x - x_0)^n. \tag{6.27}$$

Within the radius of convergence R, these polynomials are the partial sums which converge to an infinitely differentiable function $f(x)$. In fact, for $|x| < R$, every power series is a Taylor series of some $f(x)$.

Example 6.12 **exp(x)**

The Taylor series for $f(x) = e^x$ is particularly important. And because all its derivatives at $x_0 = 0$ evaluate to one, it's also particularly easy to work out:

$$e^x = 1 + x + \frac{1}{2!}x^2 + \frac{1}{3!}x^3 + \cdots = \sum_{n=0}^{\infty} \frac{x^n}{n!}. \tag{6.28}$$

In addition to e^x, other frequently used Taylor series include $\sin x$, $\cos x$, and $\ln(1 + x)$; for convenience, these and others are listed inside the front cover. If any look unfamiliar, take a moment with (6.26) to work them out for yourself. They are well worth having on your fingertips.

---

[3] A Taylor series around $x_0 = 0$ is called a Maclaurin series.

Example 6.13 **The Ubiquity of Harmonic Motion**

Springs, pendulums, electronic circuitry, molecular bonds …. Oscillatory motion seems to pop up everywhere you look. And Taylor series can help us understand why. Consider a potential energy $U(x)$ with a stable equilibrium at x_0. Assuming its derivatives exist to all orders, we can Taylor expand U around x_0,

$$U(x) = U(x_0) + (x - x_0)U'(x_0) + \frac{1}{2!}(x - x_0)^2 U''(x_0) + \frac{1}{3!}(x - x_0)^3 U'''(x_0) + \cdots, \qquad (6.29)$$

where $U^{(n)}(x_0) \equiv \frac{d^n U}{dx^n}\big|_{x=x_0}$. Now $U(x_0)$ is a constant, so although it contributes to the energy, it has no dynamic consequences. Moreover, since x_0 is an equilibrium point, the derivative $U'(x)$ vanishes there. So the first term that affects the motion is $\frac{1}{2!}(x - x_0)^2 U''(x_0)$ — which, since x_0 is a stable equilibrium, must be positive. Now for $x - x_0$ small enough, the higher-order terms in the Taylor series pale in size compared with the second-derivative term. Thus the potential energy can be fruitfully approximated as

$$U(x) \approx \frac{1}{2}m\omega^2(x - x_0)^2, \qquad (6.30)$$

where $m\omega^2 \equiv U''(x_0)$. In other words, for positions close enough to a stable equilibrium, a system behaves as a harmonic oscillator.

Example 6.14 **Keeping the Piece(s)!**

Any smooth function can be expanded in a Taylor series, but sometimes it's easier to expand it piecemeal rather than calculate all the derivatives needed for a Taylor expansion. When expanding this way, *it is crucial to expand the various parts consistently to the same order.*

This is best demonstrated with an example. Although $\exp[x/(1 - x)]$ can certainly be put directly into a Taylor expansion, it's easier to successively expand its more-familiar pieces:

$$\exp\left[\frac{x}{1 - x}\right] = 1 + \frac{x}{1 - x} + \frac{1}{2!}\left[\frac{x}{1 - x}\right]^2 + \frac{1}{3!}\left[\frac{x}{1 - x}\right]^3 + \cdots$$

$$= 1 + x(1 + x + x^2 + \cdots) + \frac{x^2}{2!}(1 + x + x^2 + \cdots)^2$$

$$+ \frac{x^3}{3!}(1 + x + x^2 + \cdots)^3 + \cdots$$

$$= 1 + x(1 + x + x^2) + \frac{x^2}{2!}(1 + 2x) + \frac{x^3}{3!}(1) + \mathcal{O}(x^4)$$

$$= 1 + x + \frac{3}{2}x^2 + \frac{13}{6}x^3 + \mathcal{O}(x^4), \qquad (6.31)$$

which is valid for $|x| < 1$. In the first line we've expanded e^y where $y = \frac{x}{1-x}$ (legitimate for all $x \neq 1$), and in the second we've expanded $(1 - x)^{-n}$ (which is valid for $|x| < 1$).

Example 6.15 **Functions of Operators**

A matrix placed to the left of a column vector can be construed as an instruction to transform the vector into another vector. This is an example of an *operator*. The derivative d/dx is another example, in this case taking a function and returning another function. Generally speaking, operators can be manipulated much as any other algebraic object, though there are some restrictions (it makes little sense, say, to add matrices of different dimension).

Functions of operators are defined via their Taylor expansion. For example, for some matrix M

$$e^M \equiv \mathbb{1} + M + \frac{1}{2!}M^2 + \cdots,$$ (6.32)

where $\mathbb{1}$ is the identity matrix. Thus e^M is a matrix.

Consider the matrix $\sigma_3 = \begin{pmatrix} 1 & 0 \\ 0 & -1 \end{pmatrix}$. Noting that $\sigma_3^2 = \mathbb{1}$ and $\sigma_3^3 = \sigma_3$, we can easily evaluate $e^{i\sigma_3\theta}$:

$$e^{i\sigma_3\theta} = \mathbb{1} + i\theta\sigma_3 - \frac{\theta^2}{2!}\mathbb{1} - i\frac{\theta^3}{3!}\sigma_3 + \frac{\theta^4}{4!}\mathbb{1} + i\frac{\theta^5}{5!}\sigma_3 - \cdots$$

$$= \mathbb{1}\left(1 - \frac{\theta^2}{2!} + \frac{\theta^4}{4!} - \cdots\right) + i\sigma_3\left(\theta - \frac{\theta^3}{3!} + \frac{\theta^5}{5!} - \cdots\right)$$

$$= \mathbb{1}\cos\theta + i\sigma_3\sin\theta$$

$$= \begin{pmatrix} e^{i\theta} & 0 \\ 0 & e^{-i\theta} \end{pmatrix}.$$ (6.33)

Example 6.16 **Found in Translation**

What happens when we take a function and add a constant to its argument? The effect of this *translation*, if small, can be determined from the function's slope,

$$f(x + \Delta x) \approx f(x) + \Delta x \frac{df}{dx}$$

$$= \left(1 + \Delta x \frac{d}{dx}\right)f(x).$$ (6.34)

Clearly this is only exact for a linear function. Still, the smaller the Δx, the more accurate the formula. Moreover, we should be able to construct any *finite* translation by repeated application of an *infinitesimal* one. Mathematically, this is accomplished by breaking up some finite distance a into N small pieces, $\Delta x = a/N$, which become infinitesimal in the large-N limit. Successive application of (6.34) then gives (Problem 6.15)

$$f(x + a) = \lim_{N \to \infty}\left(1 + \frac{a}{N}\frac{d}{dx}\right)^N f(x)$$

$$= e^{a\frac{d}{dx}}f(x).$$ (6.35)

The operator $\hat{T}(a) \equiv e^{a\,d/dx}$ is called the *translation operator*; d/dx is called the *generator* of translation. (In quantum mechanics, the translation operator is written $\hat{T} = e^{ia\hat{p}/\hbar}$, allowing the physical identification of the generator of translations, $\hat{p} = \frac{\hbar}{i}\frac{d}{dx}$, as the momentum operator.)

How are we to understand this exponential expression for $\hat{T}$? Well, according to our rule for a function of an operator,

$$e^{a\frac{d}{dx}}f(x) = \left(1 + a\frac{d}{dx} + \frac{a^2}{2!}\frac{d^2}{dx^2} + \frac{a^3}{3!}\frac{d^3}{dx^3} + \cdots\right)f(x).$$ (6.36)

Note that the first two terms give the linear approximation of (6.34). The exact value of the function at the point $x + a$ is given by an infinite sum of powers of a and derivatives of f evaluated at x — this sounds like a Taylor expansion! Indeed, we can derive this directly from (6.26). First, rewrite (6.26) in terms of the displacement, that is, eliminate x_0 in favor of $a \equiv x - x_0$. Then make the substitution $x \to x + a$ — that is, translate by a — to find

$$f(x+a) = \sum_{n=0}^{\infty} \frac{a^n}{n!} f^{(n)}(x)$$

$$= \sum_{n=0}^{\infty} \frac{1}{n!} \left(a \frac{d}{dx} \right)^n f(x), \tag{6.37}$$

which is just (6.36).

Example 6.17   Higher Dimensions

In the form of (6.37), it's not hard to generalize a Taylor expansion to functions of N independent variables with vector displacement $\vec{a} = (a_1, \ldots, a_N)$:

$$f(\vec{r} + \vec{a}) = \sum_{n=0}^{\infty} \frac{1}{n!} \left[\sum_{m=1}^{N} a_m \frac{\partial}{\partial x_m} \right]^n f(\vec{r})$$

$$\equiv \sum_{n=0}^{\infty} \frac{1}{n!} \left(\vec{a} \cdot \vec{\nabla} \right)^n f(\vec{r}), \tag{6.38}$$

where $\vec{\nabla} = (\partial/\partial x_1, \ldots, \partial/\partial x_N)$ is the *gradient operator* in cartesian coordinates.

For instance, in two dimensions, $f(\vec{r}) = f(x, y)$, $\vec{a} = (a_x, a_y)$, and $\vec{\nabla} = (\partial_x, \partial_y)$. Then (6.38) gives

$$f(x + a_x, y + a_y) = f(x, y) + (a_x \partial_x + a_y \partial_y)f(x, y) + \frac{1}{2}(a_x \partial_x + a_y \partial_y)^2 f(x, y) + \cdots$$

$$= \left[1 + (a_x \partial_x + a_y \partial_y) + \frac{1}{2} \left(a_x^2 \partial_x^2 + 2a_x a_y \partial_x \partial_y + a_y^2 \partial_y^2 \right) + \cdots \right] f(x, y). \tag{6.39}$$

Note that the derivative operators are easily found from Pascal's triangle; thus the third-order term is just

$$\frac{1}{3!}(a_x \partial_x + a_y \partial_y)^3 = \frac{1}{3!} \left[a_x^3 \partial_x^3 + 3(a_x \partial_x)^2(a_y \partial_y) + 3(a_x \partial_x)(a_y \partial_y)^2 + a_y^3 \partial_y^3 \right]$$

$$= \frac{1}{3!} \left[a_x^3 \partial_x^3 + 3a_x^2 a_y \partial_x^2 \partial_y + 3a_x a_y \partial_x \partial_y^2 + a_y^3 \partial_y^3 \right]. \tag{6.40}$$

The Taylor expansion around $x_0 = 0$ of a function of the form $f(x) = (1 + x)^\alpha$,

$$(1 + x)^\alpha = 1 + \alpha x + \frac{\alpha(\alpha - 1)}{2!} x^2 + \frac{\alpha(\alpha - 1)(\alpha - 2)}{3!} x^3 + \cdots, \tag{6.41}$$

is called the *binomial expansion*, convergent for $|x| < 1$. The power of x gives the *order*: in (6.41), αx is called the first-order term, $\frac{1}{2!}\alpha(\alpha - 1) x^2$ the second-order term, etc. Sometimes they are referred to as first- and second-order *corrections* to the so-called "zeroth-order" term, $x^0 = 1$. Note that the expansion in (6.41) *terminates* if α is a positive integer n:

$$(1+x)^n = \sum_{m=0}^{n} \frac{n!}{m!\,(n-m)!}\, x^m \equiv \sum_{m=0}^{n} \binom{n}{m} x^m,$$

which is just the binomial theorem, (5.1), for $a = 1, b = x$.

The binomial expansion is a crucial analytic tool for both algebraic and numerical calculations, as the next several examples seek to demonstrate.

Example 6.18   Pull Out a Calculator!

We can use the binomial expansion to approximate $\frac{1}{1+10^{-\beta}} \approx 1 - 10^{-\beta}$ for large β. This first-order formula is accurate at the 0.01% level only for $\beta \geq 1.5$. Including the second-order term, $\frac{(-1)(-2)}{2!}(10^{-\beta})^2 = +10^{-2\beta}$, we can achieve 0.01% accuracy down to $\beta = 1$. Even smaller β? Include higher orders!

Example 6.19   Special Relativity (Analytic)

In special relativity, we discover the formula $E = \gamma mc^2$, where $\gamma = (1 - v^2/c^2)^{-1/2}$. How can we give this formula a physical interpretation? Since $v < c$, we can use a binomial expansion to see if we recognize any of the terms:

$$E = \gamma mc^2$$

$$= mc^2 \left[1 - \frac{1}{2} \left(\frac{-v^2}{c^2} \right) + \frac{(-1/2)(-3/2)}{2!} \left(\frac{-v^2}{c^2} \right)^2 + \cdots \right]$$

$$= mc^2 + \frac{1}{2}mv^2 \left[1 + \frac{3}{4} \frac{v^2}{c^2} + \frac{5}{8} \left(\frac{v^2}{c^2} \right)^2 + \cdots \right]. \tag{6.42}$$

In the non-relativistic limit (i.e., $v \ll c$), the term in square brackets is effectively equal to one, leaving us with the usual expression for kinetic energy plus a new term which only depends on the mass m — essentially a previously unknown contribution to potential energy. We conclude that $E = \gamma mc^2$ must be the total energy.

Note that we did not take the $v \to 0$ limit (or equivalently, $c \to \infty$). We're interested in the relative sizes of the contributions to the energy for $v < c$. Actually taking the limit would have destroyed all sense of scale, obliterating the information we wanted to extract.

Example 6.20   Special Relativity (Numerical)

Kinetic energy, being the energy of motion, is by definition that portion of E which depends on v. From the previous example then, we find that the *exact* expression for kinetic energy is $(\gamma - 1)mc^2$. As (6.42) shows, the largest relativistic contribution to the non-relativistic $\frac{1}{2}mv^2$ is $\frac{3}{8} \frac{mv^4}{c^2}$ — or equivalently, the first-order *fractional* correction is $\frac{3}{4} \frac{v^2}{c^2}$. How important is this correction? It depends. Consider a fast-ball thrown at 100 mph $\approx$ 45 m/s. The mass of a baseball is about 145 g, so $\frac{1}{2}mv^2 \approx 147$ J. The first-order correction is about $\frac{v^2}{c^2} \sim 10^{-14}$. So you'd have to have a precision of at least one part in 10^{14} for relativistic effects to even begin to become noticeable. By contrast, take an electron, $mc^2 = 0.511$ MeV, moving at $\frac{1}{2}c$. The "zeroth order" kinetic energy is $\frac{1}{2}mv^2 \approx$ 63.9 keV. But the first-order correction is almost 19%, $\frac{3v^2}{4c^2} = 0.188$, so the kinetic energy to first order is $63.9(1 + 0.188) = 75.9$ keV. The second-order correction will add about another 2.5 keV. To get a sense for how quickly the series converges, compare this with the limit $(\gamma - 1)mc^2 = 79.1$ keV.

Example 6.21 Series Expansion of the Logarithm

Consider the integral

$$\ln(1 + x) = \int_0^x \frac{du}{1 + u}. \tag{6.43}$$

If $|x| < 1$, we can expand the integrand and integrate term-by-term to obtain a power series for the logarithm:

$$\ln(1 + x) = \int_0^x \frac{du}{1 + u} = \int_0^x \left(1 - u + u^2 - u^3 + \cdots\right) du$$

$$= x - \frac{1}{2}x^2 + \frac{1}{3}x^3 - \frac{1}{4}x^4 + \cdots, \tag{6.44}$$

or, if you prefer to use Σ's,

$$\ln(1 + x) = \int_0^x \left(\sum_{n=0}^{\infty}(-1)^n u^n\right) du$$

$$= \sum_{n=0}^{\infty}(-1)^n \int_0^x u^n \, du = \sum_{n=0}^{\infty}(-1)^n \frac{x^{n+1}}{n + 1}. \tag{6.45}$$

This is identical to the Taylor series of $\ln(1 + x)$ around the origin.

Example 6.22 Happy Birthday(s) — Revisited

Although (5.9) in Example 5.2 is not a terribly difficult expression to evaluate (at least with a computer), a little clever algebra results in a much simpler expression when N is small. Taking the log to turn the product into a sum,

$$\ln P_u = \ln \left[\prod_{m=1}^{N-1} \left(1 - \frac{m}{365}\right)\right]$$

$$= \sum_{m=1}^{N-1} \ln \left(1 - \frac{m}{365}\right) \approx \sum_{m=1}^{N-1} \left(-\frac{m}{365}\right) = -\frac{1}{365} \sum_{m=1}^{N-1} m, \tag{6.46}$$

where we've used $\ln(1 + x) \approx x$ for small x. Then (5.14) gives $\ln P_u \approx -\frac{1}{730} N(N - 1)$. So our approximate expression for the probability of unique birthdays is

$$P_u(N) \approx e^{-N(N-1)/730}, \tag{6.47}$$

which is a much simpler result than (5.9). Now virtually no effort is required to verify that only 23 people are needed to achieve even odds for a shared birthday. More importantly, the exponential form of the approximation provides an intuitive understanding of the behavior of $P_u(N)$ — something not so easily grasped from (5.9).

Still, (6.47) is only an approximation; how does it compare to the exact result? The second-order term in the $\ln(1+x)$ expansion, $-x^2/2$, gives an error for (or if you prefer, a correction to) (6.46) of

$$\delta(\ln P_u) = -\sum_{m=1}^{N-1} \frac{1}{2}\left(\frac{-m}{365}\right)^2 = -\frac{1}{2}\frac{1}{365^2}\sum_{m=1}^{N-1} m^2$$

$$= -\frac{1}{12}\frac{1}{365^2} N(N-1)(2N-1), \tag{6.48}$$

where we have used (5.15). But the error in $\ln(P)$ is the fractional error in P itself, that is,

$$\delta P_u/P_u = \delta(\ln P_u) = -\frac{1}{12}\frac{1}{365^2} N(N-1)(2N-1). \tag{6.49}$$

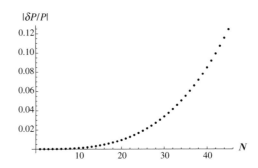

Figure 6.2 The growth of the fractional error in P_u with N.

For $N=23$, the second-order fractional correction is -0.014 — a mere 1.4%. As shown in Figure 6.2, the error doesn't reach 10% until about $N = 42$; including the third-order correction to (6.47) drops the overall error back down to about 1.2%.

Problems

6.1   Use geometric series to prove that $0.9999\ldots = 1$.

6.2   Use infinite series to find the numerical value of $\sqrt{2\sqrt{2\sqrt{2.\ldots}}}$.

6.3   Use both the ratio and comparison tests to determine whether the following series are convergent:

(a) $\sum_1^\infty \frac{1}{\sqrt{n}}$

(b) $\sum_1^\infty \frac{1}{n2^n}$

(c) $\sum_1^\infty \frac{2n^2+1}{n^4 i}$

(d) $\sum_2^\infty \frac{1}{\ln n}$.

6.4   The beauty of the comparison test is that the basic idea is really rather intuitive — but to use it you must come up with a suitable series for comparison. By contrast, the ratio test is easier to wield (though we derived its validity from the comparison test).

(a) Use the comparison test with the known convergence properties of the geometric series for $1/(1-x)$ to derive the ratio test for a series $\sum_n a_n$.

(b) Instead of the series for $1/(1-x)$, use the known convergence properties of the Riemann zeta function to derive a "new and improved" ratio test for $\sum_n a_n$. [Hint: Expand the ratio test for $\zeta(s)$ and retain the first-order term in $1/n$.]

(c) Apply the "new and improved" ratio test to determine the convergence of:

(i) $\sum_1^\infty \frac{1}{\sqrt{n}}$

(ii) $\sum_1^\infty \frac{2n^2+1}{n^\pi}$.

6.5 Find the radius of convergence for the following series:

(a) $\sum_1^\infty \frac{x^{2n}}{2^n n^2}$

(b) $\sum_1^\infty \frac{x^{3n}}{n}$

(c) $\sum_1^\infty \frac{n^{1/n}}{x^n}$

(d) $\sum_1^\infty \frac{(x-1)^n}{nx^n}$

(e) $\sum_1^\infty n! \, x^n$.

6.6 Curiously, the Riemann ζ function can be used to approximate the gamma function Γ (see Example 8.20). For what range of x does the series

$$\ln \Gamma(1+x) = -\gamma x + \sum_{n=2}^\infty (-1)^n \zeta(n) \frac{x^n}{n}$$

converge? (γ is the Euler–Mascheroni constant, see (6.17).)

6.7 A power series converges everywhere within the radius of convergence, and diverges everywhere beyond the radius of convergence. What about *on* the radius of convergence? Find the radius of convergence R of each of the following power series, and then determine their convergence *at R*.

(a) $S_1 = \sum_{n=0}^\infty x^n$

(b) $S_2 = \sum_{n=1}^\infty x^n/n$

(c) $S_3 = \sum_{n=1}^\infty x^n/n^2$

6.8 Find closed expressions for the finite series $S_1 = \sum_{k=1}^{n+1} \cos k\theta$ and $S_2 = \sum_{k=1}^{n+1} \sin k\theta$. Evaluate at $\theta = \frac{2\pi}{n+1}$ and interpret geometrically.

6.9 Show that $\sum_{n=-N}^N e^{inx} = \frac{\sin[(N+1/2)x]}{\sin(x/2)}$.

6.10 Show that:

(a) $\sum_{n \text{ odd}}^\infty \sin\left(\frac{n\pi x}{a}\right) = \frac{1}{2\sin(\pi x/a)}, \ 0 < x < a$

(b) $\sum_{k=1}^N \cos(2k-1)\theta = \frac{\sin 2N\theta}{2\sin\theta}$.

6.11 Prove that

$$\sum_{k=0}^\infty \frac{\cos(kx)}{r^k} = \frac{r(r-\cos x)}{r^2 - 2r\cos x + 1}.$$

6.12 Your old car takes a long time to warm up. Whenever you go someplace, you can only drive the first mile at 1 mph, the second mile at 2 mph, the third mile at 3 mph, etc. Approximately how long will it take you drive 100 miles?

6.13 Taking the derivative of a power series does not change its radius of convergence, and sometimes it's easier to find the radius of convergence for the derivative series. As an example, consider the series for the arctangent,

$$\arctan(z) = z - \frac{z^3}{3} + \frac{z^5}{5} - \frac{z^7}{7} + \cdots . \tag{6.50}$$

(a) Find the closed form for the derivative of this series.
(b) Use your result in (a) to find the radius of convergence for the derivative series.
(c) Verify your result by finding the radius of convergence directly from (6.50).

6.14 Expand the function $\ln(\sin x)$ as follows. Start with $\sin x = \sqrt{\frac{1}{2}(1 - \cos 2x)}$ and substitute $1 - \cos 2x = \frac{1}{2}(1 - e^{2ix})(1 - e^{-2ix})$. Then use the series expansion for $\ln(1 - z)$ to obtain the series $-\ln(\sin x) = \ln(2) + \sum_{n=1}^{\infty} \frac{\cos 2nx}{n}$. Use this result to sum the alternating harmonic series of Example 6.10.

6.15 Use the binomial expansion to show that

$$e^x = \lim_{n \to \infty} (1 + x/n)^n. \tag{6.51}$$

6.16 Use (6.51) to verify $\det[e^M] = e^{\text{Tr}[M]}$ for a matrix M.

6.17 Evaluate $\int \frac{dx}{1+x^2}$ by expanding the integrand in a binomial series and integrating term by term. The resulting series is a representation of what well-known function? Over what range of x is this series valid?

6.18 Two power series expanded around the same center can be manipulated in the same ways as polynomials.

(a) *Add* the power series for $1/(1 \pm x)$ centered at $x = 0$ to obtain the power series for $1/(1 - x^2)$.
(b) *Multiply* the power series for $1/(1 \pm x)$ centered at $x = 0$ to obtain the power series for $1/(1 - x^2)$.
(c) What about *division*? For power series $A(x) = \sum_n a_n x^n$ and $B(x) = \sum_n b_n x^n$, write $C(x) = A(x)/B(x) \equiv \sum_n c_n x^n$. To find the c_n, multiply both sides by B and work out the product of the series B and C — that is, multiply out the known coefficients b_n with the unknown c_n. By uniqueness, the coefficients of this product must equal the known coefficients of A. In other words, the coefficients of x^n on both sides must match for all n. This can get messy in the general case — so just verify the procedure for the simple case $A(x) = 1$, $B(x) = e^x$. Check that your answer is, of course, the series for e^{-x}.

6.19 Most of our examples of expansions have been around the origin — for instance, $1/(1 - x) = \sum_{n=0}^{\infty} x^n$ for $|x| < 1$. In order to find a series representation outside this range, we need to shift the center of the expansion.

(a) Expand $f(x) = \frac{1}{a-x}$ around some arbitrary point k by adding and subtracting k in the denominator and expanding in the shifted variable $x' = x - k$. Describe the circle of convergence of this series.
(b) Use this result to express $g(x) = \frac{1}{1-x^2}$ as the sum of two series centered at k. [Hint: Decompose $g(x)$ into a sum of fractions of the form $f(x)$.] Describe the circle of convergence of this series.
(c) We can do this for complex functions as well. Express $h(z) = \frac{1}{1+z^2}$ as the sum of two series centered at k. Describe the circle of convergence of this series.
(d) Restrict $h(z)$ to the real axis and assume k is real. Use the result in (c) to show that

$$\frac{1}{1+x^2} = \sum_{n=0}^{\infty} \frac{\sin\left[(n+1)\phi\right]}{\left(\sqrt{1+k^2}\right)^{n+1}} (x-k)^n,$$

where $\phi = \arg(i - k) \equiv \tan^{-1}(-1/k)$.

6.20 For power series $A(x) = \sum_{n=0}^{\infty} a_n x^n$ and $B(x) = \sum_{m=0}^{\infty} b_m x^m$, show that the product $C(x) = A(x)B(x) = \sum_{\ell=0}^{\infty} c_\ell x^\ell$ has coefficients

$$c_\ell = (a * b)_\ell \equiv \sum_{n=0}^{\ell} a_n b_{\ell-n}. \tag{6.52}$$

This is called the *Cauchy product* or the discrete *convolution* of the coefficients a_n and b_n. Note that since $AB = BA$, it must be that $(a * b)_\ell = (b * a)_\ell$.

6.21 It's not at all obvious from their Taylor expansions that sine and cosine are periodic — or that π figures in them at all. And yet we know it must. Derive the following properties of $\sin\varphi$ and $\cos\varphi$ using only their Taylor expansions:

(a) Show $\sin 0 = 0$ and $\cos 0 = 1$, and $(\sin\varphi)' = \cos\varphi$, $(\cos\varphi)' = -\sin\varphi$.

(b) Use the convolution (Cauchy product) from Problem 6.20 to show that

$$\sin^2\varphi = \frac{1}{2}\left[1 - \sum_{\ell=0}^{\infty} (-1)^\ell \frac{(2\varphi)^{2\ell}}{(2\ell)!}\right] \qquad \cos^2\varphi = \frac{1}{2}\left[1 + \sum_{\ell=0}^{\infty} (-1)^\ell \frac{(2\varphi)^{2\ell}}{(2\ell)!}\right].$$

[Hint: Set $x = 1$ and use (6.52) to solve for $c_\ell = c_\ell(\varphi)$. You'll also find results from Problem 5.2 to be helpful.]

(c) Use the results of (a) and (b) to argue that there must be a number $\xi > 0$ such that $\sin\xi = 1$ when $\cos\xi = 0$.

(d) Use the Cauchy product to show that $\sin(\varphi + \alpha) = \sin\varphi\cos\alpha + \cos\varphi\sin\alpha$ and $\cos(\varphi + \alpha) = \cos\varphi\cos\alpha - \sin\varphi\sin\alpha$. Then demonstrate that sin and cos have periodicity 4ξ.

(e) In light of the expansions in (b), deduce that $\xi = \pi/2$.

6.22 (a) Find all the singularities of $\frac{z}{e^z-1}$ in the complex plane. ($z = 0$ is not a singularity; why?)

(b) What is the radius of convergence of the power series

$$\frac{x}{e^x - 1} = \sum_{n=0}^{\infty} B_n x^n / n! .$$

(The coefficients B_n are called the Bernoulli numbers.)

6.23 Determine whether the following series are absolutely convergent or conditionally convergent. Search through the various examples of power series that we've seen to compute the values of the convergent ones.

(a) $\sum_{n=0}^{\infty} \frac{1}{n!}$

(b) $\sum_{n=0}^{\infty} \frac{(-1)^n}{n!}$

(c) $\sum_{n=0}^{\infty} \frac{(-1)^n}{n+1}$

(d) $\sum_{n=0}^{\infty} \frac{(-1)^n}{2n+1}$

6.24 Work out the Taylor expansions for both $\sinh x$ and $\cosh x$ around the origin.

6.25 Verify that $e^{i\theta} = \cos\theta + i\sin\theta$ by comparing the Taylor expansions of e, sin, and cos.

6.26 Expand the function $f(x) = \frac{\sin x}{(\cosh x + 2)}$ in a Taylor series around the origin going up to x^3. Use a calculator to find $f(0.1)$ from this series, and compare it to the answer obtained using the exact expression.

6.27 Since $\ln z$ cannot be expanded in a Maclauren series (why?), construct a Taylor expansion around $z = 1$, and then let $z \to z + 1$ to find the expansion of $\ln(1 + z)$. Use $z = i$ to find infinite series for both $\ln(2)$ and $\pi/4$.

6.28 We can use our familiarity with e^x to derive a power series expansion for the logarithm. Let $f(x) = \ln(1+x)$ for real x, so that $(1 + x) = e^{f(x)}$.

(a) Using the result of Problem 6.15, we can approximate

$$1 + \frac{f}{n} \approx (1+x)^{1/n}.$$

Use the binomial expansion to obtain an approximate series for $f(x)$. Then take the $n \to \infty$ limit to obtain an exact power series for $\ln(1+x)$.

(b) The power series around $x = 0$ is necessarily of the form $f(x) = \sum_{n=1} c_n x^n$ (note that $c_0 = 0$). Use this in the Taylor expansion of e^f around $f = 0$ and solve for c_n, $n = 1$ to 4, by equating powers of x.

6.29 Remember L'Hôpital's rule? For functions $f(x)$ and $g(x)$ such that $f(x_0) = g(x_0) = 0$, use Taylor expansions to prove

$$\lim_{x \to x_0} \frac{f(x)}{g(x)} = \lim_{x \to x_0} \frac{f'(x)}{g'(x)}.$$

6.30 The sagging shape of a free-hanging flexible chain suspended at its ends $\pm L$ is $y(x) = a\cosh(x/a)$, called a *catenary* (from the Latin for "chain").

(a) Show that a is the y coordinate of the lowest point on the chain. Large a, then, corresponds to a chain with small sag.

(b) Most people intuitively expect the shape of a free-hanging chain to be parabolic. Show that this is not a bad guess for small sag.

6.31 Expand in a Taylor series a two-dimensional function $f(x,y)$ through third order. Identify the emerging pattern.

6.32 Data sets $\{f_n\}$ measured at discrete positions x_n are often samplings approximating an underlying continuous function $f(x)$. Similarly, modeling space as a discrete lattice is a common technique in numerical calculations. In both cases, one needs to be able to use discrete values to approximate derivatives of the continuous function $f(x)$. There are three standard approaches:

forward difference: $\left(\dfrac{df}{dx}\right)_n \approx \dfrac{f_{n+1} - f_n}{a}$

backward difference: $\left(\dfrac{df}{dx}\right)_n \approx \dfrac{f_n - f_{n-1}}{a}$

central difference: $\left(\dfrac{df}{dx}\right)_n \approx \dfrac{f_{n+1} - f_{n-1}}{2a}$

where a is the lattice spacing.

(a) Show that the forward and backward differences are accurate to order $\mathcal{O}(a)$, whereas the central difference is accurate to $\mathcal{O}(a^2)$.

(b) Use these Taylor expansions to show that to $\mathcal{O}(a^2)$,

$$\left(\frac{d^2 f}{dx^2}\right)_n = \frac{f_{n+1} - 2f_n + f_{n-1}}{a^2}.$$

Verify that this can also be obtained by calculating $\frac{\partial}{\partial x}\left(\frac{\partial f}{\partial x}\right)$ using forward difference for one derivative, backward difference for the other.

(c) Use central differences to show that at lattice site (m,n) on a two-dimensional square lattice of side a,

$$\left(\frac{\partial^2 f}{\partial x \partial y}\right)_{m,n} = \frac{1}{4a^2}\left(f_{m+1,n-1} - f_{m-1,n+1} - f_{m+1,n+1} + f_{m-1,n-1}\right).$$

6.33 A *spherical triangle* is a triangle on the surface of a sphere whose sides are all segments of great circles. A right spherical triangle with sides a, b, and c and angle $\pi/2$ between b and c obeys the relation

$$\cos(a/r) = \cos(b/r)\cos(c/r),$$

where r is the radius of the sphere. Find the limit of this expression as the radius goes to infinity, and comment on the result.

6.34 Consider a sphere of radius a.

(a) Show that the distance between two infinitesimally separated points on the surface of the sphere is given by the line element

$$ds^2 = a^2 d\Omega^2 = a^2 \left(d\theta^2 + \sin^2\theta \, d\phi\right).$$

(b) Show that the distance s between two points on the surface of a sphere can be expressed as (Problem 2.8)

$$s = a\gamma = a\cos^{-1}\left[\cos\theta\cos\theta' + \sin\theta\sin\theta'\cos(\phi - \phi')\right].$$

(c) For $\theta' = \theta + d\theta$ and $\phi' = \phi + d\phi$, show that $d\gamma^2 = d\Omega^2$ to lowest non-trivial order in $d\theta$ and $d\phi$.

6.35 Show that the distance between two points $\vec{r}_1$ and $\vec{r}_2$ can be approximated to $\mathcal{O}(1/|\vec{r}_1|)$ as

$$|\vec{r}_1 - \vec{r}_2| \approx |\vec{r}_1| - \hat{r}_1 \cdot \vec{r}_2, \qquad |\vec{r}_1| \gg |\vec{r}_2|,$$

where $\hat{r} = \vec{r}/|\vec{r}|$.

6.36 The position vector can be expanded on a basis as $\vec{r} = \sum_i u_i \vec{e}_i$. In three-dimensional cartesian coordinates, $u_1 \equiv x, u_2 \equiv y, u_3 \equiv z$, and $\vec{e}_1 \equiv \hat{i}$, $\vec{e}_2 \equiv \hat{j}$, and $\vec{e}_3 \equiv \hat{k}$. These directions are parallel to the coordinate axes — and as constant vectors are easily found by taking derivatives of position with respect to the relevant coordinate, $\vec{e}_i \equiv \partial\vec{r}/\partial x_i$. In curvilinear coordinates, however, the coordinate "axes" are curves, so a basis vector is not parallel but rather *tangent* to the coordinate curve at a point P. Thus curvilinear basis vectors are position dependent. By calculating the difference between position vectors $\vec{r}(P)$ and $\vec{r}(P + du_i)$, show that these n tangent vectors are still given by the derivatives $\vec{e}_i = \partial\vec{r}/\partial u_i$.

6.37 The force of gravity on an object m a distance h above the Earth's surface is

$$F = GmM_E/(R_E + h)^2,$$

where M_E and R_E are the Earth's mass and radius, respectively. Find an approximate, lowest-order, numerical value for the object's acceleration when it is near the surface, i.e., for $h \ll R_E$. How high does h need to be (in units of R_E) to make a 10% correction to this result? Is the correction positive or negative?

6.38 The electric field along the x-axis of a uniformly charged disk of radius a and total charge Q is given by

$$\vec{E}(x) = \frac{Q}{2\pi\epsilon_0 a^2}\left[1 - \frac{x}{\sqrt{x^2 + a^2}}\right]\hat{i}.$$

Find the limiting algebraic form of $\vec{E}(x)$ for $x \gg a$. Interpret the result physically.

6.39 A constant force mg acts along the x-axis on a particle of mass m. Including the effects of relativity, the displacement of the particle is

$$x = \frac{c^2}{g}\left\{\left[1 + \left(\frac{gt}{c}\right)^2\right]^{1/2} - 1\right\},$$

where c is the speed of light. Find an approximate expression for the particle's displacement for small time t. Discuss the physical implications of your result.

6.40 Using the relativistic expression $E^2 = p^2c^2 + m^2c^4$, expand the kinetic energy $T = E - mc^2$ in powers of p to find the first-order relativistic correction. How large must the momentum be before the non-relativistic expression yields a 25% error? (Express the answer in units of the particle's rest energy mc^2.)

6.41 In special relativity, a particle of mass m has energy and momentum

$$E = \frac{mc^2}{\sqrt{1 - \beta^2}}, \qquad p = \frac{m\beta c}{\sqrt{1 - \beta^2}}$$

with $\beta = v/c$ and c the speed of light. Verify that $E^2 = p^2c^2 + m^2c^4$. Expand E and p keeping terms up to β^4 and show that the identity holds to this order.

6.42 You may recall that a pendulum of length L is only simple harmonic for small amplitudes (i.e, only for small amplitudes is the period independent of the amplitude). Application of Newton's second law shows that the exact expression for the period of a pendulum is

$$T = 4\sqrt{\frac{L}{g}} \int_0^{\pi/2} \frac{d\phi}{\sqrt{1 - k^2 \sin^2 \phi}},$$

where $k = \sin(\alpha/2)$ and α is the initial angular displacement. (Integrals of this sort are known as *elliptic integrals*.) Show that in the limit of small α the pendulum is indeed simple harmonic to lowest order. Expand the integrand as a series in $k^2 \sin^2 \phi$ and determine the lowest-order correction. For $\alpha = \pi/3$, what is the fractional change in T due to this term?

6.43 (a) Use the Taylor series for e^x and/or $\ln(1 + x)$ to prove that for numbers a and b, $e^{a+b} = e^a e^b$.
(b) Verify that for matrices A and B, $e^{A+B} = e^A e^B$ only holds if A and B commute, i.e., only if $AB = BA$. Show that to second order, $e^{A+B} = e^A e^B e^{-\frac{1}{2}[A,B]}$, where the commutator $[A, B] \equiv AB - BA$. [The full identity is known as the Baker–Campbell–Hausdorff formula.]
(c) Show that the Trotter formula

$$\lim_{n \to \infty} \left(e^{iAt/n} e^{iBt/n} \right)^n = e^{i(A+B)t}$$

holds independent of whether or not A and B commute.

6.44 Exponentiate the sub-diagonal matrix $M = \begin{pmatrix} 0 & 0 & 0 & 0 \\ 1 & 0 & 0 & 0 \\ 0 & 2 & 0 & 0 \\ 0 & 0 & 3 & 0 \end{pmatrix}$ and interpret your result. [This is a lot easier than it may look.]

6.45 Consider two matrices

$$\sigma_1 = \begin{pmatrix} 0 & 1 \\ 1 & 0 \end{pmatrix}, \quad \sigma_2 = \begin{pmatrix} 0 & -i \\ i & 0 \end{pmatrix}.$$

Calculate both $e^{i\sigma_1 \theta}$ and $e^{i\sigma_2 \theta}$, simplifying as much as possible.

6.46 Use the translation operator to derive a closed-form expression for $f(t + \tau)$, where $f(t) = \sin \omega t$.

7 Interlude: Orbits in a Central Potential

As we saw in Example 4.6, motion in a central potential $U(r)$ is confined to a plane, courtesy of the antisymmetry of the cross product. Thus the central force on an object depends only on the two-dimensional distance $r = \sqrt{x^2 + y^2}$ — making polar coordinates (r, ϕ) the natural choice with which to analyze orbits.

The motion of a object of mass m under the gravitational attraction of a central body of mass M is governed by Kepler's laws:

I. Orbits are conic sections of eccentricity ϵ with the origin at one focus,

$$r(\phi) = \frac{r_0}{1 + \epsilon \cos \phi}. \tag{7.1}$$

The simplest case is $\epsilon = 0$, which is the circular orbit $r(\phi) = r_0$. As you'll verify in Problem 7.1, $r(\phi)$ describes an ellipse for $0 < \epsilon < 1$, a parabola for $\epsilon = 1$, and a hyperbola for $\epsilon > 1$. This is a generalized form of Kepler's first law; his original statement was concerned with planets, and so only explicitly referred to ellipses with semi-major axis $a = r_0/(1 - \epsilon^2)$.

II. Angular momentum is perpendicular to the orbital plane, and has magnitude

$$\ell \equiv mr^2 \dot{\phi} = \text{constant}. \tag{7.2}$$

This is the physics behind Kepler's famous "equal areas in equal time" statement.

III. For $\epsilon < 1$, square of the period is proportional to the cube of the semi-major axis,

$$T^2 = \left(\frac{4\pi^2}{GM}\right) a^3. \tag{7.3}$$

The first two of these laws date back to 1609, the third to 1619; all three were determined from observational data. In *Principia*, published in 1687, Newton derived them from his law of gravity. In this Interlude, we'll explore different derivations of the first two laws, making use of some of the tools discussed thus far in Part I.

7.1 The Runge–Lenz Vector

The orbit equation (7.1) can be derived using vector methods. Consider the time derivative of $\vec{p} \times \vec{L}$,

$$\frac{d}{dt}\left(\vec{p} \times \vec{L}\right) = \left(\frac{d\vec{p}}{dt} \times \vec{L}\right) + \left(\vec{p} \times \frac{d\vec{L}}{dt}\right). \tag{7.4}$$

The second term, of course, vanishes by conservation of angular momentum. Recognizing $d\vec{p}/dt$ as the force $f(r)\hat{r}$, we apply the BAC-CAB identity (see inside the front cover) to the first term to find

$$\frac{d}{dt}(\vec{p} \times \vec{L}) = f(r)\hat{r} \times (\vec{r} \times m\vec{v})$$
$$= mf(r)\left[(\hat{r} \cdot \vec{v})\vec{r} - r\vec{v}\right]. \tag{7.5}$$

This may look familiar — in fact, (4.32)

$$\frac{d\hat{r}}{dt} = \frac{1}{r^2}\left[r\vec{v} - (\hat{r} \cdot \vec{v})\,\vec{r}\right] \tag{7.6}$$

reveals

$$\frac{d}{dt}(\vec{p} \times \vec{L}) = -mr^2 f(r)\frac{d\hat{r}}{dt}. \tag{7.7}$$

For a central force of the form $f(r) = -k/r^2$, we can integrate (7.7) to find

$$\vec{p} \times \vec{L} = mk\hat{r} + \vec{A}, \tag{7.8}$$

where $\vec{A}$ is the vector constant of integration — a constant that gives a new conserved quantity known as the *Runge–Lenz vector*,[1]

$$\vec{A} \equiv (\vec{p} \times \vec{L}) - mk\hat{r}. \tag{7.9}$$

To learn what $\vec{A}$ is physically, let's dot it with $\vec{L}$. The cyclic identity for $\vec{A} \cdot (\vec{B} \times \vec{C})$ reveals

$$\vec{L} \cdot \vec{A} = \vec{L} \cdot (\vec{p} \times \vec{L}) - mk\,\vec{L} \cdot \hat{r} = 0. \tag{7.10}$$

Thus we discover that $\vec{A}$ is perpendicular to the angular momentum — in other words, it points somewhere in the plane of the orbit. We can find the angle it makes with the planet by dotting it with $\vec{r}$, again using the cyclic identity:

$$\vec{r} \cdot \vec{A} = \vec{r} \cdot (\vec{p} \times \vec{L}) - mk\,\vec{r} \cdot \hat{r}$$
$$= \ell^2 - mk\,r, \tag{7.11}$$

where $|\vec{L}| = \ell$, as before. Writing the left-hand side $\vec{r} \cdot \vec{A} = rA\cos\phi$, we rediscover the form of the orbit equation (7.1),

$$\frac{1}{r} = \frac{mk}{\ell^2}\left(1 + \frac{A}{mk}\cos\phi\right). \tag{7.12}$$

Thus the angle $\vec{A}$ makes with $\vec{r}$ is the angle the planet makes with the perihelion. Direct comparison with (7.1) reveals that $|\vec{A}| = mk\epsilon$. For this reason, $\vec{A}$ is sometimes called the eccentricity vector, pointing from the origin to the perihelion. This conserved vector, then, encodes both the shape and orientation of the orbit (Biedenharn and Louck, 1981; Taylor, 2005).

---

[1] Note that $\vec{A}$ is conventional notation, and is distinct from the area enclosed by an orbit.

7.2 Orbits in the Complex Plane

Let's derive the first two laws by formulating the problem in the complex plane, so that orbital motion is described by a single complex function

$$z(t) = x + iy = re^{i\phi}, \tag{7.13}$$

where $r(t)$ is the distance from the source (which we place at the origin), and the argument $\phi(t)$ is the angular position. Taking time derivatives will start to reveal advantages of this formulation:

$$\dot{z} = (\dot{r} + ir\dot{\phi})\, e^{i\phi} \tag{7.14}$$

and

$$\ddot{z} = \left[(\ddot{r} - r\dot{\phi}^2) + i(2\dot{r}\dot{\phi} + r\ddot{\phi})\right]e^{i\phi}. \tag{7.15}$$

These are just (2.9) and (2.12) — derived here a lot quicker — with

$$\hat{r} \longrightarrow e^{i\phi} = \frac{z}{|z|} \qquad \hat{\phi} \longrightarrow ie^{i\phi} = i\frac{z}{|z|}. \tag{7.16}$$

Since $i = e^{i\pi/2}$, these representations of $\hat{r}$ and $\hat{\phi}$ are indeed perpendicular. Note too that they are consistent with (2.6a).

With only an $\hat{r} \leftrightarrow z/|z|$ component, a central force can be represented in the complex plane as $\vec{F} \longrightarrow f(r)e^{i\phi}$ for real f, so that the equation of motion becomes

$$f(r)e^{i\phi} = m\ddot{z}. \tag{7.17}$$

Of course, real and imaginary parts must balance separately. Using (7.15) and (7.17), the imaginary part satisfies

$$0 = 2\dot{r}\dot{\phi} + r\ddot{\phi} = \frac{1}{r}\frac{d}{dt}(r^2\dot{\phi}). \tag{7.18}$$

Thus the quantity $r^2\dot{\phi}$ is a constant of the motion. This, of course, is just the conservation of angular momentum $\ell = mr^2\dot{\phi}$ — which, as noted earlier, holds for any central force. As for the real part, with an attractive inverse-square force $f(r) = -k/r^2$, (7.17) can be written

$$m\ddot{z} = -\frac{k}{r^2}e^{i\phi} = -\frac{mk}{\ell}\dot{\phi}\, e^{i\phi} = i\frac{mk}{\ell}\frac{d}{dt}e^{i\phi} = i\frac{mk}{\ell}\frac{d}{dt}\frac{z}{|z|}. \tag{7.19}$$

Invoking angular momentum conservation shows

$$\frac{d}{dt}\left(i\ell\dot{z} + \frac{kz}{|z|}\right) = 0, \tag{7.20}$$

revealing that

$$Ae^{i\phi_0} \equiv -\left(i\ell\dot{z} + \frac{kz}{|z|}\right) \tag{7.21}$$

is a conserved quantity. We've written this complex-valued constant in polar form, so that A is real; the minus sign is for later convenience. The real part of (7.21) gives (Problem 7.2)

$$A\cos(\phi - \phi_0) = \ell^2/mr - k \tag{7.22}$$

or

$$\frac{1}{r} = \frac{m}{\ell^2} \left[k + A \cos(\phi - \phi_0) \right]. \tag{7.23}$$

This is the orbit equation (7.12) with perihelion at $\phi = \phi_0$. Which means that $\vec{A}$ is the Runge–Lenz vector of (7.9).

7.3 The Anomalies: True, Mean, and Eccentric

Solving for $r(\phi)$ yields a catalog of possible two-body orbits. But observational astronomy prefers the function $\phi(t)$, since one can then determine in what direction to look for an object at any time t. Solving for $\phi(t)$, however, turns out to be a difficult problem, and over the centuries a great deal of mathematical effort has been applied to it. Physically speaking, the challenge arises from the fact that a planet does not orbit at constant speed. Angular momentum conservation (7.2) implies that $v \equiv r\dot{\phi}$ varies as $1/r$, so a planet moves faster when closer to the sun.

Solving for $\phi(t)$ — traditionally known as the *true anomaly*[2] — begins with the conservation of angular momentum and the orbit (7.1), giving

$$\dot{\phi} = \frac{\ell}{mr^2} = C \left(1 + \epsilon \cos \phi \right)^2. \tag{7.24}$$

Note that C has units of inverse-time and, as a constant, must reflect the characteristic time scale of the system. For bound orbits ($\epsilon < 1$), that scale is of course the period T — and from Kepler's third law (7.3), one can show $C = (1 - \epsilon^2)^{-3/2} 2\pi/T$. So for a circular orbit ($\epsilon = 0$), (7.24) reduces to the familiar constant angular speed $2\pi/T$. But, in general, $n \equiv 2\pi/T$ is the *average* angular speed — usually referred to as the *mean motion*,

$$n = \frac{2\pi}{T} = \sqrt{\frac{GM}{a^3}}. \tag{7.25}$$

The quantity $\mu \equiv nt$, called the *mean anomaly*, is the angle of a fictitious point moving in a circle of radius a with uniform angular speed n (Figure 7.1). Only for circular orbits is a satellite's true angular position $\phi(t)$ equivalent to $\mu(t)$.

To find a general expression for $\phi(t)$, we begin by separating the variables in (7.24) — that is, we put all the t dependence on one side, all the ϕ on the other. This leads to the innocent-looking expression

$$\mu \equiv \frac{2\pi t}{T} = \left(1 - \epsilon^2\right)^{3/2} \int \frac{d\phi}{(1 + \epsilon \cos \phi)^2}. \tag{7.26}$$

Innocent perhaps, but this is a challenging integral — which we'll leave for Problem 8.31. (You're welcome.) Once computed, you'll find

$$\mu = 2 \tan^{-1} \left[\sqrt{\frac{1 - \epsilon}{1 + \epsilon}} \tan (\phi/2) \right] - \epsilon \sqrt{1 - \epsilon^2} \frac{\sin \phi}{1 + \epsilon \cos \phi}. \tag{7.27}$$

---

[2] The observed positions of planets often deviated slightly from expectations — hence the use of "anomaly" rather than "angle."

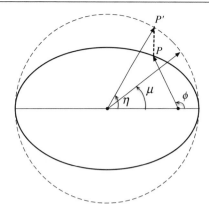

Figure 7.1 The true anomaly ϕ, eccentric anomaly η, and mean anomaly μ. The true anomaly locates the satellite at point P in its elliptical orbit, whereas the eccentric anomaly tracks the corresponding point P' on the so-called auxiliary circle. Thus ϕ and η sweep through the orbit together at the same non-constant rate, faster than μ at perihelion, slower at aphelion. The three angles coincide only at perihelion and aphelion.

Traditionally, the connection between the mean anomaly μ and the true anomaly ϕ is given by a third angle originally introduced by Kepler — the so-called *eccentric anomaly* η,

$$\tan(\eta/2) \equiv \sqrt{\frac{1-\epsilon}{1+\epsilon}} \tan(\phi/2). \tag{7.28}$$

Looking at the first term in (7.27), this seems a sensible choice. And for circular orbits ($\epsilon = 1$), η reduces to ϕ. But the real benefit is discovering that

$$\sin\eta = \sqrt{1-\epsilon^2}\,\frac{\sin\phi}{1+\epsilon\cos\phi}. \tag{7.29}$$

Thus (7.27) becomes *Kepler's equation* (see Figure 7.2)

$$\mu = \eta - \epsilon\sin\eta. \tag{7.30}$$

So suppose we want to know how long it takes a satellite to get from ϕ_1 to ϕ_2. We can use (7.28) to find the corresponding eccentric anomalies $\eta_{1,2}$, and Kepler's equation to find the mean anomalies $\mu_{1,2}$. Then since $\mu_2 = \mu_1 + n\Delta t$, we obtain the time Δt between the true anomalies $\phi_{1,2}$.

What we really want, however, is the function $\phi(t)$. And for that, we need to invert Kepler's equation to find $\eta(\mu)$. For circular orbits, this is not a problem: $\epsilon = 0$ gives $\eta = \mu$. But for non-zero eccentricity, Kepler's is a transcendental equation with no closed-form solution; a

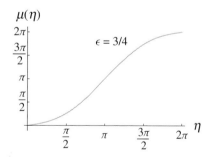

Figure 7.2 Kepler's equation.

series solution is our only recourse. Since bound orbits necessarily have eccentricities less than 1, let's formally write $\eta = \eta(\mu,\epsilon)$ and try doing a Taylor expansion in powers of ϵ around zero with coefficients which are functions of μ,

$$\eta(\mu,\epsilon) = \sum_{k=0} c_k(\mu)\epsilon^k = \mu + \epsilon\,\frac{d\eta}{d\epsilon}\bigg|_{\epsilon=0} + \frac{\epsilon^2}{2!}\frac{d^2\eta}{d\epsilon^2}\bigg|_{\epsilon=0} + \cdots. \tag{7.31}$$

The first derivative can be found directly from Kepler's equation, (7.30),

$$0 = d\eta - d\epsilon \sin \eta - \epsilon \cos \eta \, d\eta \quad \implies \quad \frac{d\eta}{d\epsilon} = \frac{\sin \eta}{1 - \epsilon \cos \eta}, \tag{7.32}$$

so that $c_1(\mu) = \frac{d\eta}{d\epsilon}\big|_{\epsilon=0} = \sin \mu$. Higher-order derivatives are straightforward (Problem 7.5). Through third order, we find

$$\eta(\mu, \epsilon) = \mu + \epsilon \sin \eta$$

$$= \mu + \epsilon \sin \mu + \frac{1}{2}\epsilon^2 (2 \sin \mu \cos \mu) + \frac{1}{3!}\epsilon^3 (6 \sin \mu - 9 \sin^3 \mu) + \cdots$$

$$= \mu + \epsilon \sin \mu + \frac{\epsilon^2}{2} \sin 2\mu + \frac{\epsilon^3}{8}(3 \sin 3\mu - \sin \mu) + \cdots. \tag{7.33}$$

If it converges, then this series representation of $\eta(\mu)$ can be used to track a satellite of eccentricity ϵ to arbitrary precision at any time $t = (T/2\pi) \mu$.

Unfortunately, the radius of convergence of (7.33) is only 0.663, known as the Laplace limit; thus only for eccentricities less than about 2/3 does the series converge. Still, most planets and dwarf planets in our solar system have $\epsilon \lesssim 0.1$ (amongst the planets, Mercury tops out at $\epsilon \simeq 0.21$). On the other hand, Sedna has eccentricity 0.85 and Halley's Comet 0.97. So we'd really rather have a fully convergent series for all $\epsilon < 1$. We'll revisit this problem in Example 35.3.

Problems

7.1 Show by converting to cartesian coordinates that (7.1) describes conic sections. [Note that r_0 is a constant, not a coordinate. You may find it helpful to complete the square.]

7.2 (a) Derive the orbit equation (7.23) from the real part of (7.21).
   (b) Show that after multiplying both sides by $dt/d\phi$, the imaginary part leads to the same result.

7.3 We found the complex Runge–Lenz vector, (7.21), by using the polar representation $z = re^{i\phi}$. Ideally, one would like to find A without choosing a representation — that is, without the need to refer to r and ϕ (except, perhaps, to identify something physically).

   (a) Show that the complex equation of motion is $m\ddot{z} = -kz/|z|^3$.
   (b) Show that the difference between $z^*\ddot{z}$ and its complex conjugate vanishes. Construct a conserved quantity J from this result. Argue that the functional form of J lends itself to a physical interpretation — and then verify this using the polar representation.
   (c) Construct the simplest complex function W whose imaginary part Q is proportional to J. Show that the real part $P = |z|\frac{d}{dt}|z|$.
   (d) Multiply $W = P + iQ$ by the equation of motion, and manipulate to discover a conserved quantity: the complex Runge–Lenz vector. [Hint: Multiply $m\ddot{z}$ with Q, $-kz/|z|^3$ with W and P.]

7.4 Recasting mechanics problems in the complex plane is fine for two-dimensional problems. But how do we translate vector operations such as the dot and cross products?

   (a) For complex numbers A and B representing two-dimensional vectors $\vec{a}$ and $\vec{b}$, respectively, show that $A^*B = \vec{a} \cdot \vec{b} + i(\vec{a} \times \vec{b})$. Thus the dot and cross products can be distilled from the real and imaginary parts of A^*B. How is the right-hand rule reflected in this?
   (b) Use the correspondence $\vec{r} \to z$ and $\vec{p} \to m\dot{z}$ and the polar representation to find the angular momentum $\vec{r} \times \vec{p}$.
   (c) The complex plane, of course, cannot accommodate vectors pointing out of the plane; the complex representation regards vectors like angular momentum as scalars. This means that we cannot calculate

something like $\vec{p} \times \vec{L}$ directly. The $BAC - CAB$ identity, however, only involves vectors in the plane. Use it to evaluation $\vec{p} \times \vec{L}$ in complex notation, and show that it's consistent with (7.21).

7.5 (a) Verify the terms in the Taylor expansion of (7.33).
(b) This expansion in powers of ϵ can be generated more easily using an iterative scheme:

$$\eta = \mu + \mathcal{O}(\epsilon)$$
$$= \mu + \epsilon \sin\left[\mu + \mathcal{O}(\epsilon)\right] = \mu + \epsilon \sin\mu + \mathcal{O}(\epsilon^2)$$
$$= \mu + \epsilon \sin\left[\mu + \epsilon \sin\mu + \mathcal{O}(\epsilon^2)\right] = \cdots$$
$$= \cdots,$$

where the expansion for η at each step is used as the argument of sine in the following step, and this sine is then expanded in powers of ϵ to the same order. Use this procedure to reproduce (7.33).

7.6 (a) The terms in (7.33) are equivalent to derivatives of powers of $\sin\eta$ — and in general, for a function $f(\eta)$ for which $\eta = \mu + \epsilon f(\eta)$ is invertible,

$$\left.\frac{d^m \eta}{d\epsilon^m}\right|_{\epsilon=0} = \frac{d^{m-1}}{d\mu^{m-1}} f(\mu)^m.$$

Verify this for $m = 1, 2$, and 3.
(b) With $\lfloor \alpha \rfloor \equiv$ the integer part of $\alpha > 0$, use (5.26) to show that

$$\left(\frac{d}{dx}\right)^{m-1} \sin^m x = \frac{1}{2^{m-1}} \sum_{k=0}^{\lfloor m/2 \rfloor} (-1)^k \binom{m}{k} (m - 2k)^{m-1} \sin\left[(m - 2k)x\right]$$

holds for all integer $m \geq 1$. (The "floor" function $\lfloor \alpha \rfloor$ gives the greatest integer less than or equal to α, regardless of sign. Its complement the ceiling function, $\lceil \alpha \rceil$, is the smallest integer greater than or equal to α.)

8 Ten Integration Techniques and Tricks

Integration is summation. But an integral is more than just an instruction to calculate, it's a statement about a system. Indeed, the process of evaluating an integral frequently leads to greater insight into any underlying physics. Even with easy access to both symbolic and numeric computing routines, the ability to manipulate and evaluate integrals remains a fundamental skill. In this section we discuss 10 common techniques and tricks which should be in everyone's toolbox.[1]

8.1 Integration by Parts

Using the product rule, we have

$$\int_a^b (uv)' = \int_a^b u'v + \int_a^b uv'$$

$$\implies \int_a^b uv' = \int_a^b (uv)' - \int_a^b u'v = (uv)|_a^b - \int_a^b u'v. \tag{8.1}$$

Example 8.1 $\int x\, e^{ax} dx$

Let $u = x$, $dv = e^{ax} dx$; then

$$\int x\, e^{ax} dx = \frac{1}{a} x\, e^{ax} - \int \frac{1}{a} e^{ax} dx$$

$$= \frac{1}{a} x\, e^{ax} - \frac{1}{a^2} e^{ax} = \frac{1}{a^2} e^{ax} (ax - 1), \tag{8.2}$$

as is easily checked by differentiation.

Example 8.2 $\int \ln(x)\, dx$

Integration by parts can be used to integrate functions whose derivatives are easier to handle, using the following trick:

$$\int f(x)\, dx = \int 1 \cdot f(x)\, dx$$

$$= \int \frac{d}{dx} [x] \cdot f(x)\, dx = xf(x) - \int xf'(x)dx. \tag{8.3}$$

Since $d(\ln x)/dx = 1/x$, we find that $\int \ln(x)\, dx = x\ln x - x$, as is easily verified by direct differentiation.

---

[1] We happily concede the distinction between a "trick" and a "technique" to be a matter of taste.

8.2 Change of Variables

For $x = x(u)$, $dx = (dx/du)du$, so that

$$\int_{x_1}^{x_2} f(x)dx = \int_{u(x_1)}^{u(x_2)} f(x(u)) \frac{dx}{du} \, du. \tag{8.4}$$

Note that in more than one dimension, the scale factor dx/du generalizes to the Jacobian.

Example 8.3 $I_1 = \int_0^\infty x \, e^{-ax^2} dx$

Let $u = x^2$, so $du = 2x \, dx$; then

$$I_1 = \int_0^\infty x \, e^{-ax^2} dx = \frac{1}{2} \int_0^\infty e^{-au} du = \frac{1}{2a}. \tag{8.5}$$

Example 8.4 $d(\cos \theta)$

The θ integral in (2.30) is easily evaluated with the variable change $u = \cos \theta$,

$$\int_0^\pi \sin \theta \, d\theta = \int_{-1}^1 d(\cos \theta) = \int_{-1}^1 du = 2. \tag{8.6}$$

This substitution is useful anytime the integrand's θ dependence occurs only in terms of $\cos \theta$ — for example, $f(\theta) = \frac{1}{2}(3 \cos^2 \theta - 1) \longleftrightarrow g(u) = \frac{1}{2}(3u^2 - 1)$. Then

$$\int_0^\pi f(\theta) \sin \theta \, d\theta = \int_{-1}^1 g(\cos \theta) \, d(\cos \theta) = \int_{-1}^1 g(u) \, du. \tag{8.7}$$

Such $\cos \theta$ dependence is actually very common.

Example 8.5 **Trig Substitution**

Consider the integral

$$\int_0^a \frac{x}{x^2 + a^2} \, dx. \tag{8.8}$$

Note that the only scale in the problem is a, which has dimensions of x. But the integral itself is dimensionless — and there's no way to construct a dimensionless number from a single dimensional parameter. (As a definite integral, of course, the result cannot have an x dependence.) Thus the answer must be a pure number, independent of a. To find it, let $x = a \tan \theta$; then

$$dx = \frac{dx}{d\theta} d\theta = a \sec^2 \theta \, d\theta \tag{8.9}$$

and

$$\theta(x = 0) = 0, \quad \theta(x = a) = \pi/4. \tag{8.10}$$

Thus

$$\int_0^a \frac{x \, dx}{x^2 + a^2} = \int_0^{\pi/4} \frac{a^2 \tan \theta \sec^2 \theta \, d\theta}{a^2(\tan^2 \theta + 1)}$$

$$= \int_0^{\pi/4} \tan \theta \, d\theta = -\ln(\cos \theta)\Big|_0^{\pi/4} = \ln(\sqrt{2}) = \frac{1}{2}\ln(2). \tag{8.11}$$

Example 8.6 **Rational Functions of sin and cos**

Integrands that are rational functions of $\sin\theta$ and $\cos\theta$ can be handled with the half-angle trig substitution

$$u = \tan(\theta/2), \tag{8.12}$$

so that $\sin(\theta/2) = u/\sqrt{1+u^2}$ and $\cos(\theta/2) = 1/\sqrt{1+u^2}$. You should use this to verify

$$d\theta = \frac{2\,du}{1+u^2}, \tag{8.13}$$

and, using half-angle identities, that

$$\sin\theta = \frac{2u}{1+u^2} \qquad \cos\theta = \frac{1-u^2}{1+u^2}. \tag{8.14}$$

So, for example,

$$\int \frac{d\theta}{1+\cos\theta} = \int \frac{2\,du/(1+u^2)}{1+(1-u^2)/(1+u^2)}$$
$$= \int \frac{2\,du}{(1+u^2)+(1-u^2)} = \int du = \tan(\theta/2) = \frac{\sin\theta}{1+\cos\theta}, \tag{8.15}$$

where the identity in the last step follows from (8.14) and (8.12). If that seemed too easy, then consider this:

$$\int \frac{1}{1-\sin\theta}\,d\theta = \int \frac{1}{1-2u/(1+u^2)}\frac{2\,du}{1+u^2}$$
$$= 2\int \frac{du}{(1-u)^2} = \frac{2}{(1-u)} = \frac{2}{1-\tan(\theta/2)}, \tag{8.16}$$

up to a constant of integration.

Example 8.7 **Dimensionless Integrals**

In physics and engineering, one is often confronted with integrals over dimensional quantities, such as distance, time, energy, or electric charge. But computer integration routines and integral tables (showing my age here) usually require recasting an integral into a sum over dimensionless variables. Moreover, sometimes an abundance of physical constants and variables can obscure both the mathematics and the physics. Thus it's very often a good idea to scale the integration variable to make it dimensionless. As an example, consider a box of photons in equilibrium at temperature T. The energy per volume for each wavelength λ is given by *Planck's law*,

$$u(\lambda) = \frac{8\pi hc}{\lambda^5(e^{hc/\lambda kT} - 1)}, \tag{8.17}$$

where h is Planck's constant, k is Boltzmann's constant, and c the speed of light. The total energy density is found by integrating this over all wavelengths,

$$U = \int_0^\infty u(\lambda)\,d\lambda = 8\pi hc \int_0^\infty \frac{d\lambda}{\lambda^5(e^{hc/\lambda kT} - 1)}. \tag{8.18}$$

This integral is challenging — even without the integration variable appearing in the *denominator* of the exponent. We could sidestep that problem by rewriting the integral not as a sum over wavelength but over frequency $\nu = c/\lambda$. However, the expression is most transparent when expressed in terms of the dimensionless variable $x \equiv h\nu/kT = hc/\lambda kT$; then, as is readily verified, (8.18) becomes

$$U = 8\pi hc \left(\frac{kT}{hc}\right)^4 \int_0^\infty \frac{x^3\, dx}{e^x - 1}. \tag{8.19}$$

Now all the physics is encapsulated by the constants in front of the integral — most notably, that the energy density scales as T^4. (This is the Stefan–Boltzmann law.) The remaining integral is just a dimensionless number. And it's in a form ripe for using a computer or looking up in a table. Or you could try to estimate it; plotting $x^3/(e^x - 1)$, you might decide 5 is a pretty good guess for the area under the curve. The exact value is $\pi^4/15 \approx 6.5$. (A method for evaluating this integral analytically is given in Example 8.20.)

8.3 Even/Odd

One day in college, my friend David and I were approached by someone seeking help with a particular integral. (This was in medieval times before such aides as Wolfram Alpha.) It looked complicated, something like $\int_{-\infty}^{\infty} \frac{\sin^3 x \cos^3 x}{x^4 - 3x^2 + 5} e^{-ax^2}\, dx$. David, far more clever than I, immediately said "Oh, it's just zero!" How did he do it — and so fast?

Functions for which $f(-x) = \pm f(x)$ are said to have *definite parity*. Specifically (Figure 8.1),

• any function with $f(-x) = +f(x)$ is called *even*
• any function with $f(-x) = -f(x)$ is called *odd*.

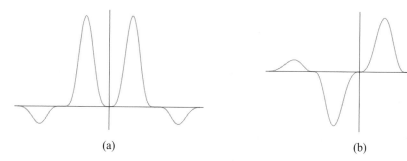

(a)   (b)

Figure 8.1 Functions with definite parity: (a) even (b) odd.

Over a symmetric interval, an integrand with definite parity makes identical-magnitude contributions to the summation for both $x > 0$ and $x < 0$. Thus:

$$\int_{-a}^{a} f(x)dx = \begin{cases} 2\int_0^a f(x)dx, & \text{if } f(x) \text{ is even} \\ 0, & \text{if } f(x) \text{ is odd.} \end{cases} \tag{8.20}$$

Example 8.8 $\int_{-\infty}^{\infty} \sin^3 x \cos^3 x\, e^{-ax^2}/(x^4 - 3x^2 + 5)\, dx$

Though this looks formidable, note that $\sin^3 x$ is an odd function, while $\cos^3 x$, e^{-ax^2}, and $x^4 - 3x^2 + 5$ are all even functions. Thus the entire integrand is *odd*. This is what David immediately recognized — and since the integral is over a symmetric interval, he knew that the contribution for $x > 0$ precisely offsets that for $x < 0$. Thus he trivially concluded

$$\int_{-\infty}^{\infty} \frac{\sin^3 x \cos^3 x}{x^4 - 3x^2 + 5} e^{-ax^2}\, dx = 0. \tag{8.21}$$

Example 8.9   **Integrating Absolute Values**

The function $f(x) = |x|$ is an even function; thus

$$\int_{-a}^{a} |x|\, dx = 2 \int_{0}^{a} x\, dx, \tag{8.22}$$

where in the second integral we drop the absolute value sign since x is always positive over the region of integration.

For any function $f(x)$, $f(|x|)$ is even. When the interval is not symmetric, however, to integrate $f(|x|)$ one must break up the integration into intervals in which x does not change sign. Thus

$$\int_{-a}^{b} f(|x|)\, dx = \int_{-a}^{0} f(-x)\, dx + \int_{0}^{b} f(x)\, dx. \tag{8.23}$$

Compare, for example, the following integrals for $f(x) = x^5 - 1$:

(a) $\int_{-1}^{2} f(x)\, dx = 15/2$

(b) $\int_{-1}^{2} f(|x|)\, dx = \int_{-1}^{0}[(-x)^5 - 1]\, dx + \int_{0}^{2}[x^5 - 1]\, dx = 47/6$

(c) $\int_{-1}^{2} |f(x)|\, dx = \int_{-1}^{1}(1 - x^5)\, dx + \int_{1}^{2}(x^5 - 1)\, dx = 23/2.$

Note the fundamental relation

$$\int |f(x)|\, dx \geq \left| \int f(x)\, dx \right|, \tag{8.24}$$

which is supported by comparison of integrals 8.9 and 8.9. We saw the discrete version of this when discussing absolute convergence. As in that case, the relation is easy to understand: the integral on the left is a sum over a strictly positive integrand, while on the right the integrand can vary in sign and phase.

BTW 8.1 The Cauchy Principal Value

Integrals over an infinite interval, deemed "improper," are defined as

$$\int_{-\infty}^{b} f(x)\, dx = \lim_{a \to \infty} \int_{-a}^{b} f(x)\, dx \quad \text{and} \quad \int_{a}^{\infty} f(x)\, dx = \lim_{b \to \infty} \int_{a}^{b} f(x)\, dx, \tag{8.25}$$

assuming the limits exist. Often this seems a mere formality — for example,

$$\int_{-\infty}^{\infty} \frac{dx}{1 + x^2} = \lim_{a,b \to \infty} \int_{-a}^{b} \frac{dx}{1 + x^2}$$

$$= \lim_{b \to \infty} \tan^{-1}(b) - \lim_{a \to \infty} \tan^{-1}(-a) = \pi, \tag{8.26}$$

since each limit exists. But consider

$$\lim_{a,b \to \infty} \int_{-a}^{b} x\, dx = \lim_{b \to \infty} \frac{1}{2} b^2 - \lim_{a \to \infty} \frac{1}{2} a^2 = \infty - \infty, \tag{8.27}$$

whatever that means. To be sure, as an odd integrand over a symmetric interval we expect the integral to vanish — but the difference of infinities is ill-defined. For instance, if we let $b = 2a$ we'll get

$$\lim_{a\to\infty} \int_{-a}^{2a} x\,dx = \lim_{a\to\infty} \left(2a^2 - \frac{1}{2}a^2\right) = \infty, \tag{8.28}$$

whereas $a = 2b$ would give $-\infty$. The problem is not difficult to resolve: since the area under the curve of an odd integrand only vanishes for finite $a = b$, so too the improper integral should be evaluated with a single symmetric limit,

$$\lim_{a\to\infty} \int_{-a}^{a} x\,dx = \lim_{a\to\infty} \left(\frac{1}{2}a^2 - \frac{1}{2}a^2\right) = 0. \tag{8.29}$$

Taking symmetric limits yields what is known as the *Cauchy principal value* (or just "principal value") of the integral. It's not so much that the other results are wrong; rather, the integral is ambiguous without appending a limiting procedure — and the symmetric limit best aligns with expectations. Moreover, when the integral is unambiguous, it agrees with the principal value.

A similar ambiguity occurs for improper integrals which pass through a point of discontinuity. For $f(x)$ discontinuous at $x = c$, we can define the integral as

$$\int_{a}^{b} f(x)\,dx = \lim_{\epsilon_1\to 0} \int_{a}^{c-\epsilon_1} f(x)\,dx + \lim_{\epsilon_2\to 0} \int_{c+\epsilon_2}^{b} f(x)\,dx, \tag{8.30}$$

with $\epsilon_{1,2} > 0$. A straightforward non-trivial example is

$$\int_{0}^{1} \frac{dx}{\sqrt{x}} = \lim_{\epsilon_2\to 0} \int_{0+\epsilon_2}^{1} \frac{dx}{\sqrt{x}} = \lim_{\epsilon_2\to 0} 2\sqrt{x}\,\Big|_{\epsilon_2}^{1} = 2. \tag{8.31}$$

So the singularity at the origin notwithstanding, the area under the curve is finite. But consider

$$\int_{-1}^{1} \frac{dx}{x}. \tag{8.32}$$

Though the integrand has a singularity at $x = 0$, the integrand is odd and the interval symmetric — so we expect the area under the curve to vanish. However,

$$\int_{-1}^{1} \frac{dx}{x} = \lim_{\epsilon_1\to 0} \int_{-1}^{-\epsilon_1} \frac{dx}{x} + \lim_{\epsilon_2\to 0} \int_{\epsilon_2}^{1} \frac{dx}{x}$$

$$= \lim_{\epsilon_1\to 0} \ln|x|\,\Big|_{-1}^{-\epsilon_1} + \lim_{\epsilon_2\to 0} \ln|x|\,\Big|_{\epsilon_2}^{1}, \tag{8.33}$$

which again yields the dicey expression $-\infty + \infty$. Taking a symmetric limit, however, yields the integral's principal value

$$\lim_{\epsilon\to 0} \left(\int_{-1}^{\epsilon} \frac{dx}{x} + \int_{\epsilon}^{1} \frac{dx}{x}\right) = 0. \tag{8.34}$$

By taking symmetric limits, we can extend intuition to cases where the improper integral is not well defined. (Mathematica has the integration option PrincipalValue $\to$ True.)

Another way to gain insight into the principal value is to consider the integral

$$\int_{-1}^{1} \frac{g(x)}{x}\,dx. \tag{8.35}$$

If $g(x)$ is a smooth function, we can Taylor expand it around zero to get

$$\int_{-1}^{1} \frac{g(x)}{x}\,dx = g(0)\int_{-1}^{1}\frac{dx}{x} + g'(0)\int_{-1}^{1}dx + \frac{1}{2!}g''(0)\int_{-1}^{1}x\,dx + \cdots . \tag{8.36}$$

All but the first term can be integrated in familiar fashion. That problematic term is an odd function over a symmetric interval, so we choose a limiting procedure which sends it to zero; what's left is the integral's principal value. In physics parlance, discarding the divergent piece in this way is an example of "renormalization" of the integral.

8.4 Products and Powers of Sine & Cosine

(a) Integrals of the form $\int \sin(\alpha x)\cos(\beta x)\,dx$ can be dispensed with using the familiar multi-angle formulas,

$$\sin(\alpha \pm \beta) = \sin\alpha\cos\beta \pm \cos\alpha\sin\beta \tag{8.37a}$$

$$\cos(\alpha \pm \beta) = \cos\alpha\cos\beta \mp \sin\alpha\sin\beta . \tag{8.37b}$$

Thus the integrand can be recast as

$$\sin(\alpha)\cos(\beta) = \frac{1}{2}\left[\sin(\alpha - \beta) + \sin(\alpha + \beta)\right] . \tag{8.38a}$$

Similarly, for products of sines and products of cosines,

$$\sin(\alpha)\sin(\beta) = \frac{1}{2}\left[\cos(\alpha - \beta) - \cos(\alpha + \beta)\right] \tag{8.38b}$$

$$\cos(\alpha)\cos(\beta) = \frac{1}{2}\left[\cos(\alpha - \beta) + \cos(\alpha + \beta)\right] . \tag{8.38c}$$

A single example should suffice.

Example 8.10

Using (8.38a),

$$\int \sin(3x)\cos(2x)\,dx = \frac{1}{2}\int\left[\sin(5x) + \sin(x)\right]dx \tag{8.39}$$

$$= -\left[\frac{1}{2}\cos(x) + \frac{1}{10}\cos(5x)\right], \tag{8.40}$$

plus an arbitrary constant (of course).

(b) Integrals of the form $\int \sin^m(x)\cos^n(x)\,dx$ for integer m and n fall into two subcategories. (The general case of non-integer powers is considered in Problem 8.21.)

(i) **m and/or n odd:** Since an odd integer has the form $2k+1$, straightforward substitution reduces the integrand to a polynomial:

$$\int \sin^m(x)\cos^{2k+1}(x)\,dx = \int \sin^m(x)\cos^{2k}(x)\,\cos(x)\,dx \tag{8.41}$$

$$= \int u^m(1 - u^2)^k\,du. \tag{8.42}$$

Similarly,

$$\int \sin^{2k+1}(x) \cos^n(x)\, dx = \int \sin^{2k}(x) \cos^n(x)\, \sin(x)\, dx \tag{8.43}$$

$$= -\int (1 - u^2)^k u^n\, du. \tag{8.44}$$

The idea is simple: group the "extra" odd power with dx to form du. For n odd, let $u = \sin x$; with m odd, $u = \cos x$. If both are odd, take your pick.

Example 8.11

$$\int \sin^3 x \cos^2 x\, dx = \int \sin^2 x \cos^2 x\, \sin x\, dx$$

$$= -\int (1 - u^2) u^2\, du = -\frac{1}{3} u^3 + \frac{1}{5} u^5$$

$$= \frac{1}{5} \cos^5 x - \frac{1}{3} \cos^3 x + C. \tag{8.45}$$

Similarly,

$$\int \sin^2 x \cos^3 x\, dx = \int \sin^2 x \cos^2 x\, \cos x\, dx$$

$$= \int u^2 (1 - u^2)\, du = \frac{1}{3} u^3 - \frac{1}{5} u^5$$

$$= \frac{1}{3} \sin^3 x - \frac{1}{5} \sin^5 x + C. \tag{8.46}$$

(ii) m, n **both even:** The easiest approach is often to use the trig identities,[2]

$$\left. \begin{array}{c} \sin^2(x) \\ \cos^2(x) \end{array} \right\} = \frac{1}{2} \left[1 \mp \cos(2x) \right]. \tag{8.47}$$

Then, with $m = 2k$ and $n = 2\ell$,

$$\int \sin^{2k}(x) \cos^{2\ell}(x)\, dx = \frac{1}{2^{k+\ell}} \int [1 - \cos(2x)]^k [1 + \cos(2x)]^\ell\, dx. \tag{8.48}$$

Example 8.12

The simplest example is the fairly common case $m = 0, n = 2$ or vice-versa,

$$\int \left. \begin{array}{c} \sin^2(x) \\ \cos^2(x) \end{array} \right\}\, dx = \frac{1}{2} \int [1 \mp \cos(2x)] = \frac{x}{2} \mp \frac{1}{4} \sin(2x) + C. \tag{8.49}$$

The definite integral of either $\sin^2(x)$ or $\cos^2(x)$ over a half-period is thus $\pi/2$.

---

[2] Mnemonic: for small angles, $\sin x < \cos x$ — so $\sin^2 x$ has the minus sign.

Example 8.13

Consider $m = n = 2$,

$$\int \sin^2(x)\cos^2(x)\,dx = \frac{1}{4}\int [1 - \cos(2x)][1 + \cos(2x)]\,dx$$

$$= \frac{1}{4}\int \left[1 - \cos^2(2x)\right] = \frac{1}{4}\int \sin^2(2x)\,dx$$

$$= \frac{1}{8}\left[x - \frac{1}{4}\sin(4x)\right] + C, \tag{8.50}$$

where we used (8.49). When $m = n$, we can also use $\sin(2x) = 2\sin(x)\cos(x)$ (which is just (8.38a) with $\alpha = \beta$):

$$\int \sin^2(x)\cos^2(x)\,dx = \int \left[\frac{1}{2}\sin(2x)\right]^2 dx = \frac{1}{8}\int [1 - \cos(4x)]\,dx$$

$$= \frac{1}{8}\left[x - \frac{1}{4}\sin(4x)\right] + C. \tag{8.51}$$

8.5 Axial and Spherical Symmetry

Integrals that are difficult in one coordinate system may be simpler in another. This is particularly true if the integral exhibits some sort of symmetry. For example, an axially symmetric function is constant everywhere in a fixed circle centered on an axis. If we label this as the z-axis, then the function is independent of ϕ both in cylindrical and spherical coordinates (where the symmetry is often called cylindrical and azimuthal, respectively). For instance, if a volume integral encircles the z-axis, we can write $d^3x = 2\pi\rho\,d\rho\,dz = 2\pi r^2\sin\theta\,drd\theta$. So the proper choice of coordinate system allows a three-dimensional integration to be trivially reduced down to just two. Similarly, a function f with spherical symmetry takes the same value over the surface of a sphere, which is to say $f(\vec{r}) = f(r)$. This independence of both θ and ϕ allows a three-dimensional integral over a sphere to be immediately reduced to one dimension:

$$f(\vec{r}) = f(r) \implies \int f(\vec{r})\,d\tau = \int f(r)\,r^2 dr d\Omega = 4\pi\int f(r)\,r^2 dr. \tag{8.52}$$

As with the integrand, oftentimes the region of integration has symmetry which can be exploited by a shrewd choice of coordinate system — as demonstrated in the following examples.

Example 8.14 **Hemispherical Distribution of Charge**

The total charge on a surface S with a charge density $\sigma(\vec{r})$ requires evaluation of the integral

$$Q = \int_S \sigma(\vec{r})\,dA. \tag{8.53}$$

Consider a hemispherical surface with $z > 0$ and radius R. Then for charge density $\sigma = k\,(x^2 + y^2)/R^2$, the integrand is axially symmetric, and either cylindrical or spherical coordinates would make sense. But since the surface S is spherically symmetric, the integral is easier to evaluate in spherical coordinates:

$$Q = \int_S \sigma(\vec{r})\,dA = \frac{k}{R^2}\int_S (x^2 + y^2)\,dA \tag{8.54}$$

$$= \frac{k}{R^2} \int_S (R^2 \sin^2 \theta \cos^2 \phi + R^2 \sin^2 \theta \sin^2 \phi)\, R^2\, d\Omega \quad [r \to R \text{ since evaluating } on\ S]$$

$$= kR^2 \int_S \sin^2 \theta\ \sin \theta\ d\theta d\phi \quad [\text{axial symmetry} \Rightarrow \text{ absence of } \phi \text{ dependence}]$$

$$= 2\pi kR^2 \int_0^{\pi/2} \sin^3 \theta\ d\theta = \frac{4\pi kR^2}{3}. \quad [\theta \in [0, \pi/2] \text{ since } S \text{ is a hemisphere}]$$

Example 8.15 $\Phi(\vec{r}) = \int \frac{d^3 k}{(2\pi)^3} \frac{e^{-i\vec{k}\cdot\vec{r}}}{k^2}$

In cartesian coordinates $d^3 k = dk_x dk_y dk_z$, so

$$\Phi(\vec{r}) = \int_{-\infty}^{\infty} \frac{dk_x}{2\pi} \int_{-\infty}^{\infty} \frac{dk_y}{2\pi} \int_{-\infty}^{\infty} \frac{dk_z}{2\pi} \frac{\exp[-i(k_x x + k_y y + k_z z)]}{k_x^2 + k_y^2 + k_z^2}. \tag{8.55}$$

This expression of the integral is difficult to evaluate because the denominator prevents it being factored into a product of three one-dimensional integrals. But in spherical coordinates, $d^3 k = k^2 dk d\Omega$, and this same integral becomes

$$\Phi(\vec{r}) = \frac{1}{(2\pi)^3} \int_0^{\infty} k^2 dk \int_{-1}^{1} d(\cos\theta) \int_0^{2\pi} d\phi\, \frac{e^{-ikr\cos\theta}}{k^2}, \tag{8.56}$$

where we've chosen the k_z-axis to be in the direction of $\vec{r}$, so that the k-space spherical coordinate θ is the angle between $\vec{k}$ and $\vec{r}$. The ϕ integral is trivial — an example of azimuthal symmetry. Moreover, the pesky denominator cancels with the Jacobian's k^2, leaving the straightforward expression

$$\Phi(\vec{r}) = \frac{1}{(2\pi)^2} \int_0^{\infty} dk \int_{-1}^{1} d(\cos\theta)\, e^{-ikr\cos\theta}$$

$$= \frac{1}{(2\pi)^2} \int_0^{\infty} dk \frac{1}{(-ikr)} \left[e^{-ikr} - e^{+ikr} \right]$$

$$= \frac{1}{(2\pi)^2} \int_0^{\infty} dk \frac{1}{(-ikr)} [-2i \sin(kr)]$$

$$= \frac{1}{2\pi^2} \int_0^{\infty} dk \frac{\sin(kr)}{kr}. \tag{8.57}$$

We'll evaluate the remaining k integral in Example 8.16.

8.6 Differentiation with Respect to a Parameter

In Example 8.1 on integration-by-parts, a is a parameter with units $[x^{-1}]$, but with no actual dependence on x. Thus as a purely formal expression we can write

$$\int x\, e^{ax} dx = \frac{\partial}{\partial a} \left(\int e^{ax} dx \right)$$

$$= \frac{\partial}{\partial a} \left[\frac{1}{a} e^{ax} \right]$$

$$= -\frac{1}{a^2} e^{ax} + \frac{1}{a} x\, e^{ax} = \frac{1}{a^2} e^{ax} (ax - 1). \tag{8.58}$$

We can also *introduce* a parameter, as the next example illustrates.

Example 8.16 $\int_0^\infty \frac{\sin x}{x}\, dx$

Since $|\sin x| \le 1$, the integral has no convergence difficulties for large x. The lower limit, however, looks as if it may be problematic — but recalling $\lim_{x\to 0}(\sin x/x) = 1$, we can be assured that the integral converges. The $1/x$, though, makes evaluating it a challenge. So let's introduce a parameter α and define a function $J(\alpha)$ by

$$J(\alpha) \equiv \int_0^\infty e^{-\alpha x}\, \frac{\sin x}{x}\, dx, \quad \alpha \ge 0. \tag{8.59}$$

The value we're after is $J(0)$. How does this help? Well, taking the derivative of J with respect to α rids ourselves of that pesky x in the denominator, resulting in a more tractable integral:

$$\frac{dJ}{d\alpha} = -\int_0^\infty e^{-\alpha x} \sin x\, dx = -\frac{1}{1+\alpha^2}, \tag{8.60}$$

as you can verify by, say, substituting $\sin x \to e^{ix}$ and taking the imaginary part at the end. Now that we've done the x-integration, we integrate in α to find J,

$$J(\alpha) = -\int \frac{d\alpha}{1+\alpha^2} = -\tan^{-1}\alpha + \pi/2, \tag{8.61}$$

where we've chosen the integration constant $\pi/2$ to ensure that $J(\infty) = 0$. Then $J(0)$ is simply

$$\int_0^\infty \frac{\sin x}{x}\, dx = \frac{\pi}{2}. \tag{8.62}$$

Introducing α made the problem tractable — though rather than a single integral to evaluate, we needed two. This is the cost imposed by the venerable "No Free Lunch Theorem." (An easier technique is discussed in Example 21.13.)

8.7 Gaussian Integrals

Try to evaluate $I_0 = \int_0^\infty e^{-ax^2}\, dx$ for $\Re e(a) > 0$ by any of the methods above and you'll have trouble. The simple change of variables $u = x^2$, for instance, would work if the integrand had an additional factor of x as in Example 8.3. But it doesn't. So what does one do?

First let's try a little dimensional analysis. The exponent of course must be dimensionless, which requires a to have dimension $[x^{-2}]$. And as a definite integral, I_0 has no actual x dependence — so $I_0 = I_0(a)$. Then since I_0 has dimensions of x, we conclude that $I_0 \sim a^{-1/2}$.

To find the precise result, we use a little sleight of hand:

$$\begin{aligned}
I_0^2 &= \left(\frac{1}{2}\int_{-\infty}^{+\infty} e^{-ax^2}\, dx\right)\left(\frac{1}{2}\int_{-\infty}^{+\infty} e^{-ay^2}\, dy\right) \quad \text{[note the use of dummy variables]}\\
&= \frac{1}{4}\int_{-\infty}^{+\infty}\int_{-\infty}^{+\infty} dx\,dy\, e^{-a(x^2+y^2)}\\
&= \frac{1}{4}\int_0^{2\pi} d\theta \int_0^\infty r\, dr\, e^{-ar^2} \quad \text{[changing to polar coordinates]}\\
&= \frac{\pi}{2}\int_0^\infty dr\, r\, e^{-ar^2}\\
&= \frac{\pi}{2}\cdot\frac{1}{2}\int_0^\infty du\, e^{-au} \quad \text{[using the substitution } u = r^2]\\
&= \frac{\pi}{4a}. \tag{8.63}
\end{aligned}$$

Therefore

$$I_0 \equiv \int_0^\infty e^{-ax^2}\, dx = \frac{1}{2}\sqrt{\frac{\pi}{a}}. \tag{8.64}$$

So dimensional analysis correctly gave us the crucial $a^{-1/2}$ dependence. Moreover, as the only parameter in the integral, a provides a scale — that is, it gives meaning to the notion of large and small x and their relative contributions to the integral.

It's fair to say that the answer is far more important than the technique, if only because Gaussian integrals $I_n \equiv \int_0^\infty x^n e^{-ax^2}\, dx$ are so common. Indeed, once we know I_0 and I_1 (Example 8.3), we can easily derive the entire family of Gaussian integrals for $n \geq 2$ by differentiating $I_{n=0,1}$ with respect to a.

Example 8.17 $I_2 = \int_0^\infty x^2 e^{-ax^2}\, dx$

Dimensional analysis tells us that $I_2 \sim a^{-3/2}$; differentiation with respect to a gives the precise relationship:

$$\int_0^\infty x^2 e^{-ax^2}\, dx = -\frac{d}{da} \int_0^\infty e^{-ax^2}\, dx = -\frac{dI_0}{da}$$

$$= -\frac{d}{da}\left(\frac{1}{2}\sqrt{\frac{\pi}{a}}\right) = +\frac{1}{4}\sqrt{\frac{\pi}{a^3}}. \tag{8.65}$$

Convention Alert: Be aware that Gaussian integrals appear in many different forms. First, we've defined the integration running from zero to infinity, but $\pm\infty$ is arguably just as common. (Care to speculate why we've chosen 0 as the lower limit?) More insidious is the location and form of the parameter a. A plot of e^{-ax^2} reveals the famous bell-shaped curve; $1/a$ is a measure of its width. So it's common to find the integrand expressed to give a units of x, such as e^{-x^2/a^2} or $e^{-x^2/2a^2}$. It even sometimes appears as $e^{-x^2/a}$. So while it's helpful have (8.64) at your fingertips, be careful to make adjustments appropriate for the integral you're given.

BTW 8.2 The Error Function

The trick used to evaluate the integral in (8.64) works because we are integrating over all space: only with infinite limits is the integral of $dx\, dy$ over a square equivalent to the integral of $r\, dr\, d\theta$ over a circle. For finite limits, the integral must be evaluated numerically; the results define the *error function* erf(x),

$$\text{erf}(x) \equiv \frac{2}{\sqrt{\pi}} \int_0^x e^{-t^2}\, dt, \tag{8.66}$$

as shown in Figure 8.2. The $2/\sqrt{\pi}$ out front serves to normalize the function so that erf$(\infty) = 1$. As defined here, erf(x) gives the fraction of the total area under the Gaussian between $-x$ and x. (Be aware that there are other closely related functions and conventions.)

So what does the area under a Gaussian have to do with error? Every laboratory measurement is fraught with uncertainties; the challenge is to minimize errors as much as possible, and use statistics to manage whatever remains. Now a measurement which is subject only to small, random errors will yield values scattered with the distribution of a bell curve — in other words, in accordance with the normal distribution in (5.21),

$$f(z) = \frac{1}{\sqrt{2\pi\sigma^2}} e^{-(z-\mu)^2/2\sigma^2}. \tag{8.67}$$

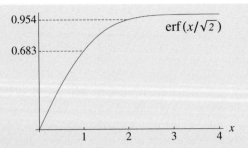

Figure 8.2 The error function.

This distribution is maximum at $z = \mu$, which is thus the best estimate of the expected value of z; the quality of that estimate is quantified by the standard deviation σ. Oftentimes the mean μ represents a theoretical prediction, which one then tries to confirm experimentally by teasing it out of the data.

The normal distribution gives the relative probability of measuring values of z scattered only by random error; the probability for a measurement to fall within a particular range (a,b) is given by the integral $\int_a^b f(z)dz$. The probability of falling within n standard deviations σ of the mean μ is

$$\frac{1}{\sqrt{2\pi\sigma^2}} \int_{\mu-n\sigma}^{\mu+n\sigma} e^{-(z-\mu)^2/2\sigma^2}\, dz = \text{erf}\left(\frac{n}{\sqrt{2}}\right), \tag{8.68}$$

where we evaluated the integral with the variable change $t \to (z - \mu)/\sqrt{2}\sigma$.

So the probability that a measurement will be within one standard deviation of the mean is about erf$(1/\sqrt{2})$, approximately 68%. On the other hand, the probability that a measurement could vary from the expected value μ by as much or more than 2σ is only $1 - 95.4\% = 4.6\%$. Should your measurement deviate this much, there may be other sources of error — or you may have evidence discrediting a prediction that the true value of the mean is μ.

8.8 Completing the Square

The integral

$$\int_{-\infty}^{\infty} e^{-ax^2-bx}\, dx$$

is certainly Gaussian-esque, but the linear term in the exponent makes things more complicated. We can massage it into an explicit Gaussian by completing the square:

$$ax^2 + bx = a\left(x^2 + \frac{b}{a}x + \frac{b^2}{4a^2}\right) - \frac{b^2}{4a}$$

$$= a\left(x + \frac{b}{2a}\right)^2 - \frac{b^2}{4a}. \tag{8.69}$$

Then

$$\int_{-\infty}^{\infty} e^{-ax^2 - bx} \, dx = \int_{-\infty}^{\infty} \exp\left[-a\left(x + \frac{b}{2a}\right)^2 + \frac{b^2}{4a}\right] dx$$

$$= e^{b^2/4a} \int_{-\infty}^{\infty} e^{-a\left(x + \frac{b}{2a}\right)^2} dx$$

$$= e^{b^2/4a} \int_{-\infty}^{\infty} e^{-ay^2} \, dy \qquad [\text{using } y = x + \frac{b}{2a}]$$

$$= \sqrt{\frac{\pi}{a}} \, e^{b^2/4a}. \tag{8.70}$$

Example 8.18 $\int_{-\infty}^{\infty} x \, e^{-ax^2 - bx} \, dx$

Differentiating with respect to b, we can replace this integral with

$$-\frac{d}{db} \int_{-\infty}^{\infty} e^{-ax^2 - bx} \, dx = -\frac{d}{db}\left(\sqrt{\frac{\pi}{a}} \, e^{b^2/4a}\right) = -\sqrt{\frac{\pi}{a}} \frac{b}{2a} \, e^{b^2/4a}. \tag{8.71}$$

Example 8.19 $\int dx/\sqrt{ax^2 + bx + c}$

Though this may look intimidating, the denominator is just the square root of a quadratic. So given the well-known result $\frac{d}{dx} \sin^{-1}(x) = \frac{1}{\sqrt{1-x^2}}$, we might expect this integral to be an inverse sine for $a < 0$. To massage it into this canonical form, complete the square:

$$ax^2 + bx + c = a\left[x^2 + \frac{bx}{a} + \frac{c}{a}\right]$$

$$= a\left[x^2 + \frac{bx}{a} + \left(\frac{b^2}{4a^2} - \frac{b^2}{4a^2}\right) + \frac{c}{a}\right] = a\left[\left(x + \frac{b}{2a}\right)^2 - \frac{q^2}{4a^2}\right], \tag{8.72}$$

where we've defined $q^2 \equiv b^2 - 4ac$. Then

$$\int \frac{dx}{\sqrt{ax^2 + bx + c}} = \frac{1}{\sqrt{a}} \int \frac{dx}{\sqrt{\left(x + \frac{b}{2a}\right)^2 - \frac{q^2}{4a^2}}}$$

$$= \frac{1}{\sqrt{a}} \int \frac{dy}{\sqrt{y^2 - 1}} = \frac{1}{\sqrt{-a}} \int \frac{dy}{\sqrt{1 - y^2}}, \tag{8.73}$$

where we've made the change of variable $y = (2a/q)(x + \frac{b}{2a}) = (2ax + b)/q$. So, as expected,

$$\int \frac{dx}{\sqrt{ax^2 + bx + c}} = \frac{1}{\sqrt{-a}} \sin^{-1}\left(\frac{2ax + b}{q}\right), \tag{8.74}$$

up to an arbitrary constant. The precise form, of course, will depend on the signs and relative magnitudes of the parameters a, b, and c. The arcsine, for instance, is real for $q^2 > 0$; it's not too difficult to verify that $a > 0$, $q^2 < 0$ results in an inverse sinh instead. Finally, since $\frac{d}{dy}\cos^{-1}(-y) = \frac{1}{\sqrt{1-y^2}}$ for real y, we can also write

$$\int \frac{dx}{\sqrt{ax^2 + bx + c}} = \frac{1}{\sqrt{-a}}\cos^{-1}\left(-\frac{2ax + b}{q}\right), \tag{8.75}$$

when $(2ax + b)/q$ is real.

8.9 Expansion of the Integrand

An integral that appears formidable in closed form may be tractable when the integrand is expanded; sometimes, the integrated pieces can be identified or their sum recast into closed form.

Example 8.20 **The Gamma Function**

Consider the integral $\int_0^\infty \frac{x^{s-1}}{e^x-1}\,dx$. What makes it difficult to evaluate is the denominator — but note that near the upper end of the interval, the integrand goes like $x^{s-1}e^{-x}$. So let's make this asymptotic form manifest,

$$\frac{x^{s-1}}{e^x - 1} = \frac{x^{s-1}\,e^{-x}}{1 - e^{-x}}. \tag{8.76}$$

Since the integration is over $x \geq 0$, we can now expand the integrand to effectively raise the denominator into the numerator:

$$\int_0^\infty \frac{x^{s-1}}{e^x - 1}dx = \int_0^\infty \frac{e^{-x}\,x^{s-1}}{1 - e^{-x}}dx \qquad \text{[denominator} \sim 1 - y, \ y < 1]$$

$$= \int_0^\infty e^{-x}\left[\sum_{n=0}^\infty (e^{-x})^n\right] x^{s-1}\,dx \qquad \text{[geometric series]}$$

$$= \sum_{n=0}^\infty \int_0^\infty e^{-(n+1)x} x^{s-1}dx. \qquad \text{[switching order of } \int \text{ and } \sum]$$

This is all pretty standard fare. Now if we can do the integral for arbitrary n, only the sum will remain. In this particular case, however, a judicious change of variables will allow us to extricate the sum from the integral altogether — in effect, factoring the expression. You should verify that with $m \equiv n + 1$ and $u = mx$, the integral becomes

$$\int_0^\infty \frac{x^{s-1}}{e^x - 1}dx = \left(\sum_{m=1}^\infty \frac{1}{m^s}\right)\left(\int_0^\infty e^{-u}\,u^{s-1}\,du\right). \tag{8.77}$$

So we've replaced one unknown integral with another integral and an infinite sum — this is progress? Actually it is, because these are both well-known functions (Table 8.1). The sum is the Riemann zeta function $\zeta(s)$ (6.4), and the integral is the *gamma function* Γ,

$$\Gamma(s) \equiv \int_0^\infty e^{-u}\,u^{s-1}\,du. \tag{8.78}$$

Table 8.1 Gamma and zeta functions

s	$\Gamma(s)$	$\zeta(s)$
3/2	$\sqrt{\pi}/2$	2.612
2	1	$\pi^2/6$
5/2	$3\sqrt{\pi}/4$	1.341
3	2	1.202
7/2	$15\sqrt{\pi}/8$	1.127
4	6	$\pi^4/90$

Thus

$$\int_0^\infty \frac{x^{s-1}}{e^x - 1} dx = \zeta(s)\Gamma(s). \tag{8.79}$$

Incidentally, for a real *integer* $s > 0$, $\Gamma(s)$ is not really new: it's easy to show that $\Gamma(1) = 1$ and

$$\Gamma(s+1) = s\Gamma(s), \quad s \in \text{integers.} \tag{8.80}$$

Thus $\Gamma(s) = (s-1)!$. In other words, the Gamma function $\Gamma(s)$ is essentially the continuation of the factorial beyond positive integers. So, for instance, should you come across something bizarre like $\left(-\frac{1}{2}\right)!$, interpret it as $\left(-\frac{1}{2}\right)! = \left(\frac{1}{2} - 1\right)! = \Gamma\left(\frac{1}{2}\right) = \sqrt{\pi}$. For positive integer n, the following half-integer formulas can be useful:

$$\Gamma\left(n + \frac{1}{2}\right) = \frac{\sqrt{\pi}}{2^n}(2n-1)!! = \frac{\sqrt{\pi}}{2^{2n}}\frac{(2n)!}{n!} \tag{8.81a}$$

$$\Gamma\left(-n + \frac{1}{2}\right) = (-1)^n \frac{2^n\sqrt{\pi}}{(2n-1)!!} = (-1)^n\sqrt{\pi}\, 2^{2n}\frac{n!}{(2n)!}, \tag{8.81b}$$

where the "double factorial" $(2n-1)!! \equiv (2n-1) \cdot (2n-3)\cdots 5 \cdot 3 \cdot 1 = \frac{(2n)!}{2^n n!}$. (By definition, $(-1)!! \equiv 1$, so the formulas agree for $n = 0$.)

BTW 8.3 Asymptotic Expansions

Since we have no analytic technique for evaluating $\text{erf}(x) = \frac{2}{\sqrt{\pi}}\int_0^x \exp(-t^2)dt$, approximations are very useful. Following the lead of the previous example, we can expand the exponential in (8.66) and integrate term-by-term,

$$\text{erf}(x) = \frac{2}{\sqrt{\pi}}\int_0^x \left[1 - t^2 + \frac{1}{2}t^4 - \frac{1}{3!}t^6 + \cdots\right]dt$$

$$= \frac{2}{\sqrt{\pi}}\left(x - \frac{x^3}{3} + \frac{x^5}{2 \cdot 5} - \frac{x^7}{3! \cdot 7} + \cdots\right). \tag{8.82}$$

Although this series converges for all x, as a practical matter it is only useful for small values. For example, only for $x \lesssim 0.5$ do the first two terms yield better than 1% accuracy.

So what do we do if, say, we want to integrate the tails of the Gaussian — that is, the area under the curve between $|x|$ and $\pm\infty$? This area is, of course, complementary to that of $\text{erf}(x)$ — in fact, it is called the complementary error function $\text{erfc}(x)$,

$$\text{erfc}(x) \equiv 1 - \text{erf}(x) = \frac{2}{\sqrt{\pi}} \int_x^\infty e^{-t^2} \, dt, \tag{8.83}$$

where we've used $\text{erf}(\infty) = 1$.

Rather than expanding around 0 as we did in (8.82), it seems reasonable to expand around x — but now with infinity as the upper integration limit, we need to be more clever. To be useful for large x the expansion should be in powers of $1/x$. We can obtain such an expansion if we make the change of variables $u = t^2$,

$$\text{erfc}(x) = \frac{2}{\sqrt{\pi}} \int_{x^2}^\infty e^{-u} \frac{du}{2\sqrt{u}}, \tag{8.84}$$

and expand $1/\sqrt{u}$ in a Taylor series around x^2:

$$\frac{1}{\sqrt{u}} = \frac{1}{x} - \frac{1}{2x^3}(u - x^2) + \frac{3}{4x^5} \frac{(u - x^2)^2}{2!} - \frac{15}{8x^7} \frac{(u - x^2)^3}{3!} + \cdots . \tag{8.85}$$

The u-integration is now straightforward, yielding a series expansion for erfc,

$$\text{erfc}(x) = \frac{2}{\sqrt{\pi}} e^{-x^2} \left(\frac{1}{2x} - \frac{1}{2^2 x^3} + \frac{3!!}{2^3 x^5} - \frac{5!!}{2^4 x^7} + \cdots + (-1)^{n+1} \frac{(2n-3)!!}{2^n x^{2n-1}} + \cdots \right). \tag{8.86}$$

An expansion such as (8.86) in inverse powers of x is called an *asymptotic expansion*.

Success? Well, yes. Sort of. The series seems to work well; for example, the first term alone yields better than 1% accuracy down to $x \approx 7$ — and the larger the value of x, the better the approximation. But casual inspection of the series reveals something very disturbing: the radius of convergence is zero! In other words, the series does not converge for *any* value of x! For instance, the asymptotic expansion of $\text{erfc}(x)$ for $x = 2$ improves only through $n = 4$, at which point the terms start to grow — so including further terms in the sum yields increasingly poorer approximations. The value of n where this turn-around occurs depends upon x; the greater x, the more terms can be included until the best approximation for $\text{erfc}(x)$ is achieved. This is a general feature of asymptotic series — and not at all what we're accustomed to. Indeed, the distinction between a convergent and an asymptotic series for $f(x)$ is their limits:

- a convergent series $\sum_n^N c_n x^n$ approaches $f(x)$ as $N \to \infty$ for fixed x
- an asymptotic series $\sum_n^N c_n/x^n$ approaches $f(x)$ as $x \to \infty$ for fixed N.

8.10 Partial Fractions

Consider the integral $\int \frac{dx}{1-x^2}$. Since $1 - x^2 = (1-x)(1+x)$ we can decompose the integrand into a sum of simpler integrands,

$$\frac{1}{1 - x^2} = \frac{\alpha}{1 - x} + \frac{\beta}{1 + x} . \tag{8.87}$$

A little algebra gives $\alpha = \beta = \frac{1}{2}$, so that

$$\int \frac{dx}{1 - x^2} = \frac{1}{2} \int \left(\frac{dx}{1 - x} + \frac{dx}{1 + x} \right)$$

$$= -\frac{1}{2} \ln |1 - x| + \frac{1}{2} \ln |1 + x| + C = \frac{1}{2} \ln \left| \frac{1 + x}{1 - x} \right| + C. \tag{8.88}$$

This is a very simple example of the method of *partial fractions*. Suppose we wish to integrate the rational function $r(x) = f(x)/g(x)$. If the degree of $f(x)$ is less than the degree of $g(x)$, then we can write $r(x)$ as a sum of fractions having simpler denominators. To do this, we need the factors of $g(x)$. In our example, $1 - x^2$ has only two linear factors. More generally, any polynomial with real coefficients can be expressed as a product of real linear and quadratic factors, $(ax - b)$ and $(cx^2 + dx + e)$. In the partial fraction decomposition, to each linear factor $(ax - b)^k$ of $g(x)$ there corresponds the sum

$$\frac{C_1}{ax - b} + \frac{C_2}{(ax - b)^2} + \cdots + \frac{C_k}{(ax - b)^k} \tag{8.89}$$

and to each quadratic factor $(cx^2 + dx + e)^\ell$ the sum

$$\frac{A_1 x + B_1}{cx^2 + dx + e} + \frac{A_2 x + B_2}{(cx^2 + dx + e)^2} + \cdots + \frac{A_\ell x + B_\ell}{(cx^2 + dx + e)^\ell}. \tag{8.90}$$

Equating the sum of equations (8.89) and (8.90) to the original function $r(x)$ allows one to find the constants A, B, and C.

In equation (8.87), both factors are linear with $k = 1$: $C_1 \equiv \alpha$ is the constant associated with the factor $(1 - x)$, $C_1 \equiv \beta$ the constant associated with $(1 + x)$. But let's demonstrate with a more complicated example.

Example 8.21 $\int \frac{4 - 2x}{(x^2 + 1)(x - 1)^2} \, dx$

We have one quadratic factor and a double linear factor, so

$$\frac{4 - 2x}{(x^2 + 1)(x - 1)^2} = \frac{Ax + B}{x^2 + 1} + \frac{C}{x - 1} + \frac{D}{(x - 1)^2}, \tag{8.91}$$

where, for notational simplicity, we use C and D rather than C_1 and C_2. Multiplying both sides by the common denominator gives

$$4 - 2x = (Ax + B)(x - 1)^2 + C(x - 1)(x^2 + 1) + D(x^2 + 1). \tag{8.92}$$

Setting $x = 1$ instantly reveals that $D = 1$. The other three constants can be found by equating the coefficients on each side power-by-power in x: rewriting the equation as a polynomial in x (and using $D = 1$) we get

$$4 - 2x = (A + C)x^3 + (-2A + B - C + 1)x^2 + (A - 2B + C)x + (B - C + 1). \tag{8.93}$$

Then we must have

$$\text{order } x^0 : 4 = B - C + 1$$
$$\text{order } x^1 : -2 = A - 2B + C$$
$$\text{order } x^2 : 0 = -2A + B - C + 1$$
$$\text{order } x^3 : 0 = A + C.$$

Solving gives $A = 2, B = 1, C = -2$, so that

$$\frac{4 - 2x}{(x^2 + 1)(x - 1)^2} = \frac{2x + 1}{x^2 + 1} - \frac{2}{x - 1} + \frac{1}{(x - 1)^2}. \tag{8.94}$$

Now the integral can be done more easily.

BTW 8.4 The Fundamental Theorem of Algebra

The method of partial fractions relies on the fact that any polynomial with real coefficients can be expressed as a product of real linear and quadratic factors. You might find this curious: can't a quadratic be expressed as a product of two linear factors? So why not fully factor the polynomial into linear pieces?

That any nth-order polynomial $P_n(z)$ can be factored into a product of n linear pieces is essentially the content of the *Fundamental Theorem of Algebra*,

$$P_n(z) = \sum_{k=0}^{n} a_k z^k = (z - z_1)(z - z_2) \cdots (z - z_n), \tag{8.95}$$

where the n values z_i are the *roots* of the polynomial, i.e., $P_n(z_i) = 0$. (The roots are not necessarily distinct.) So indeed, any polynomial can be fully factored into a product of linear pieces; what the theorem does not guarantee is that all these roots are *real*. However, it's not hard to show that if the coefficients a_k are real, the complex roots always occur in complex conjugate pairs — in other words, if z_0 is a root, so is z_0^* (see Problem 3.22). All these paired linear factors in equation (8.95) multiply to give quadratics with real coefficients. The method of partial fractions, then, seeks to keep everything real by decomposing a polynomial with real coefficients into linear and quadratic factors.

BTW 8.5 The Mean Value Theorem

We've saved this for last since it's less a technique than a basic property of integrals. But the theorem is of sufficient importance to merit inclusion. So consider a continuous function $f(x)$ defined on a closed interval $[a, b]$. The Mean Value Theorem for Integrals simply states that somewhere within the closed interval there's at least one point c where $f(c)$ is the average of $f(x)$ over the interval,

$$\frac{1}{(b - a)} \int_a^b f(x) dx = f(c), \quad a < c < b. \tag{8.96}$$

In other words, the area of the rectangle, $(b - a)f(c)$, equals the area under the curve $f(x)$ between a and b (Figure 8.3).

More generally, for continuous functions $f(x)$ and $g(x)$ on a closed interval $[a, b]$ with $g(x) > 0$ on the open interval (a, b), there is a point c between a and b such that

$$\int_a^b f(x)g(x) dx = f(c) \int_a^b g(x) dx, \tag{8.97}$$

which is (8.96) with $g(x) = 1$. Note that the theorem doesn't tell you what c is; nevertheless, this result can be very useful for analyzing or estimating integrals.

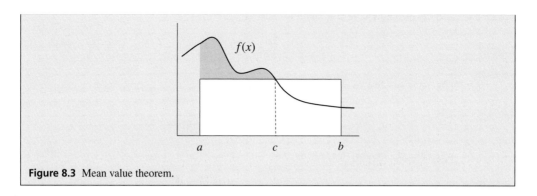

Figure 8.3 Mean value theorem.

Problems

8.1 Evaluate the following integrals; be sure to explain your method.

(a) $\int_0^1 2^x dx$

(b) $\int_{-\infty}^{\infty} e^{-x^2/2} e^x dx$

(c) $\int_{-\infty}^{\infty} e^{i\alpha x^2} dx$ [Express your answer in polar form, $re^{i\varphi}$.]

(d) $\int_{-a}^{a} \sin x\, e^{-ax^2} dx$

(e) $\int_0^1 e^{\sqrt{x}} dx$

(f) $\int dx/\sqrt{1+x^2}$

(g) $\int \tanh x\, dx$

(h) $\int_0^{2\pi} \sin(4\phi)\, e^{i\cos\phi}\, d\phi$ [Think: Change of variables and parity.]

(i) $\int \tan^{-1} x\, dx$ [Hint: Write $\tan^{-1}(x) = 1 \cdot \tan^{-1}(x)$.]

(j) $\int_S z^2\, dA$, with z is the usual cartesian coordinate, and $S = 1/8$ of the surface of the unit sphere centered at the origin.

8.2 Use the symmetry amongst the integration variables x, y, and z to show that

$$J_3 \equiv \int_0^1 \int_0^1 \int_0^1 dx\, dy\, dz \left(\frac{x+y-z}{x+y+z}\right) = 1/3.$$

Generalize to find an expression for

$$J_n = \int_0^1 \cdots \int_0^1 dx_1 \cdots dx_n \left(\frac{\sum_{i=1}^{n-1} x_i - x_n}{\sum_{i=1}^{n} x_i}\right).$$

8.3 Evaluate the integral

$$\int_0^a \frac{x}{x^2+a^2}\, dx$$

using the substitution $x = a \sinh u$.

8.4 Evaluate $\int \sin^2 x \cos^4 x\, dx$.

8.5 Evaluate each of these integrals twice, once with the substitution $u = \sin x$, and again with $u = \cos x$. Does it bother you that the methods give different results? Explain.

(a) $\int \sin x \cos x\, dx$

(b) $\int \sin^5 x \cos^3 x\, dx$

8.6   Use the substitution $t = \frac{1}{\sqrt{2}} \sinh x$ to show

$$\int \sqrt{2t^2 + 1} \, dt = \frac{1}{2\sqrt{2}} \sinh^{-1}(\sqrt{2}t) + \frac{1}{2} t \sqrt{2t^2 + 1}.$$

8.7   Use the results of Example 8.16 to show that

$$\int_{-\infty}^{\infty} \frac{e^{ikx}}{x} \, dx = i\pi \, \text{sgn}(k),$$

where the *signum* function is defined as

$$\text{sgn}(k) \equiv \begin{cases} +1, & k > 0 \\ 0, & k = 0 \\ -1, & k < 0 \end{cases}.$$

8.8   Evaluate the following integrals, explaining your method clearly.

(a) $\int_0^{\infty} dx \, \frac{x}{e^x - 1}$

(b) $\int_0^{\infty} dx \, \frac{x}{e^x + 1}$

(c) $\int_0^{\infty} dx \, \frac{x^n}{e^x - 1}$

(d) $\int_0^{\infty} dx \, \frac{x^n}{e^x + 1}$

8.9   (a) Show that for $\alpha > 0$,

$$\int_0^{\infty} \frac{\alpha x^{s-1}}{e^x - \alpha} \, dx = \zeta_\alpha(s)\Gamma(s),$$

where

$$\zeta_\alpha(s) = \sum_{n=1}^{\infty} \frac{\alpha^n}{n^s}$$

is called the *polylogarithm*. The Riemann zeta function $\zeta(s) = \zeta_1(s)$. For what range of real α and s does $\zeta_\alpha(s)$ converge?

(b) Though we chose the notation $\zeta_\alpha(s)$ to emphasize its relation to the Riemann zeta function, the polylogarithm is more commonly denoted as $\text{Li}_s(\alpha)$, with independent variable α and parameter s. The name of the function comes from recursive relationships such as

$$\frac{\partial}{\partial \log x} \text{Li}_s(x) = x \frac{\partial}{\partial x} \text{Li}_s(x) = \text{Li}_{s-1}(x).$$

Verify that this is so. Use this, the related integral recursion, and/or the series representation to find closed expressions for $\text{Li}_0(x)$ and $\text{Li}_{\pm 1}(x)$.

8.10   Evaluate the integral

$$\int_{-\infty}^{\infty} dx \, \frac{x^2 e^x}{(e^x + 1)^2}.$$

[Hint: First demonstrate that the integrand is even. Then use integration by parts and results from Problem 8.8.]

8.11   Show that $\int_0^{\infty} e^{-x^4} dx = \Gamma(\frac{5}{4})$.

8.12   (a) Derive the recursion relation $\Gamma(z + 1) = z\Gamma(z)$, $\Re e(z) > 0$ using integration by parts.

(b) Show that for integer $z > 0$, $\Gamma(z + 1) = z!$

(c) Show that for integer $z \leq 0$, $1/\Gamma(z) = 0$.

8.13   Calculate $\Gamma(1/2)$. Use the result to find $\Gamma(3/2)$ and $\Gamma(-1/2)$.

8.14   Consider the family of functions $I_n(a) \equiv \int_0^\infty x^n \, e^{-ax^2} dx$. Use I_0 and I_1 to find I_3 and I_4.

8.15   Let $J_n(a) \equiv \int_{-\infty}^\infty x^n \, e^{-x^2/2a^2} dx$. Without integrating:

(a) find J_0
(b) show that $J_{2n}/J_0 = a^{2n}(2n-1)(2n-3)\cdots 5\cdot 3\cdot 1 \equiv a^{2n}(2n-1)!!$.

8.16   Consider the error function, $\operatorname{erf}(x) = \frac{2}{\sqrt{\pi}} \int_0^x e^{-t^2} dt$.

(a) Use integration by parts to evaluate $\int \operatorname{erf}(x) \, dx$.
(b) Evaluate $\int_0^\infty \operatorname{erf}(\sqrt{qx}) \, e^{-px} dx$, where $p > 0$, $q > 0$.

8.17   Derive the first term in the expansion (8.86) of the complementary error function $\operatorname{erfc}(x)$ in (8.83) using integration by parts with the assignments $u = 1/t$ and $dv = e^{-t^2} t \, dt$. In a similar fashion, find the second term in (8.86). Assuming you've found the nth term in the expansion, what integration-by-parts assignments should you make to find the $n+1$st?

8.18   Evaluate $\int_0^\infty e^{-(x^2+1/x^2)} dx$. [Hint: "Gaussian-ize" the integrand by recasting it in the form $e^{-(x-1/x)^2}$. Then investigate what you can do with the substitution $y = 1/x$.]

8.19   Gammas and Gaussians: $\Gamma(n)$ vs. I_n

(a) Manipulate the Γ function integral to verify $\Gamma(n+1) = 2I_{2n+1}(1) = n!$
(b) Similarly verify that $\Gamma(n+\frac{1}{2}) = 2I_{2n}(1) = \sqrt{\pi} \, (2n-1)!!/2^n$,
   where the "double factorial" $(2n-1)!! \equiv (2n-1)(2n-3)\cdots 5\cdot 3\cdot 1 = \frac{(2n)!}{2^n n!}$.

8.20   Gammas and Betas

With appropriate changes of integration variables, show that the product of gamma functions,

$$\frac{\Gamma(r)\Gamma(s)}{\Gamma(r+s)} = B(r,s),$$

where

$$B(r,s) \equiv \int_0^1 t^{r-1}(1-t)^{s-1} dt = \int_0^\infty \frac{u^{r-1}}{(1+u)^{r+s}} du$$

is the *beta function*.

8.21   (a) Adapt our method for evaluating I_0 to the product $\Gamma(r)\Gamma(s)$ to show that

$$B(r,s) = \frac{\Gamma(r)\Gamma(s)}{\Gamma(r+s)} = 2\int_0^{\pi/2} \sin^{2r-1}\theta \, \cos^{2s-1}\theta \, d\theta,$$

for $\Re e(r) > 0$, $\Re e(s) > 0$.
(b) Verify the formulas

$$\int_0^{\pi/2} \sin^{2n} x \, dx = \int_0^{\pi/2} \cos^{2n} x \, dx = \frac{\pi}{2}\frac{(2n-1)!!}{(2n)!!}$$

$$\int_0^{\pi/2} \sin^{2n+1} x \, dx = \int_0^{\pi/2} \cos^{2n+1} x \, dx = \frac{(2n)!!}{(2n+1)!!},$$

for integer n. [Hint: You'll need to show that $(2n)!! = 2^n n!$.]

8.22   In (2.30), we directly calculated the surface area of the unit sphere; let's do it again in a rather indirect fashion.

(a) Use cartesian coordinates to evaluate over all space $\int d^3x\, e^{-r^2}$, where $r^2 = x_1^2 + x_2^2 + x_3^2$.

(b) Recast the integral in spherical coordinates, and evaluate the r integral.

(c) Use the results in (a) and (b) to solve *algebraically* for the remaining angular integral, $\int d\Omega$.

8.23 Let's generalize Problem 8.22 to find the "area" of the unit n-dimensional sphere S^n in $\mathbb{R}^{n+1}$.

(a) Using cartesian coordinates, evaluate over all space $J_n \equiv \int d^{n+1}x\, e^{-r^2}$, where $r^2 = \sum_{i=1}^{n+1} x_i^2$.

(b) In $\mathbb{R}^3$, going from cartesian to spherical coordinates introduces the Jacobian $r^2 dr d\Omega$. This generalizes to $r^n dr d\Omega_n$, where $d\Omega_n$ is the n-dimensional element of solid angle in $n+1$ dimensions. Use this to recast J_n in spherical coordinates, and then evaluate the r integral.

(c) Compare your results in (a) and (b) to discover the general expression for the surface area of the unit sphere in n-dimensions, S^n. Verify that your formula reproduces the known results for S^1 (circle) and S^2 (sphere). What's the surface area of S^3?

(d) Find the volume of the ball in n-dimensions, B^n. Verify that your formula reproduces the known results for B^2 (disk) and B^3 (ball). [Hint: Now that you know the surface area of S^n, introduce a radial coordinate s to your result in part (c) to write an expression for the volume of a shell of radius s and thickness ds. Then integrate.]

8.24 To evaluate the integral

$$ J = \lim_{n\to\infty} \int_0^1 \int_0^1 \cdots \int_0^1 \frac{n}{x_1 + x_2 + \cdots + x_n}\, dx_1\, dx_2 \cdots dx_n, $$

introduce $\exp[-a(x_1 + x_2 + \cdots x_n)/n]$ into the integrand, and differentiate with respect to a parameter (see Example 8.16).

8.25 Integrate by parts repeatedly to show that for integer n,

$$ \frac{1}{\pi} \int_0^\pi \cos^{2n}(x)\, dx = \frac{(2n-1)!!}{(2n)!!}. $$

What about $\frac{1}{\pi} \int_0^\pi \sin^{2n}(x)\, dx$?

8.26 Evaluate the integral

$$ \int_0^\infty e^{-ax} \sin kx\, dx $$

in three different ways:

(a) Integration by parts. (You'll have to do it twice and use some algebra.)

(b) Express $\sin kx$ in terms of complex exponentials and integrate.

(c) Replace $\sin kx$ with e^{ikx}.

8.27 Use the result of the previous problem to evaluate $\int_0^\infty xe^{-ax} \sin kx\, dx$ and $\int_0^\infty xe^{-ax} \cos kx\, dx$.

8.28 Evaluate the integral

$$ I = \int_0^{\pi/2} \ln(\sin x)\, dx $$

(without integrating!) as follows:

(a) Show that $I = \int_{\pi/2}^\pi \ln(\sin x)\, dx = \int_0^{\pi/2} \ln(\cos x)\, dx$.

(b) Express $2I$ as the sum of log sine and log cosine functions, and rewrite it in terms of $\sin 2x$.

(c) Solve algebraically for I. (Check that the overall sign is what you'd expect.)

8.29   Use partial fractions to integrate $\int \frac{1}{x^4 - 1}$.

8.30   Consider the integral

$$I(\alpha) = \frac{1}{\pi} \int_{-\infty}^{\infty} \frac{e^{i\alpha u}}{1 + u^2} \, du, \quad \alpha > 0.$$

(a) Show that $I = \frac{2}{\pi} \int_0^{\infty} \frac{\cos \alpha u}{1 + u^2} \, du$. What's $I(0)$?

(b) Show that although $I'(\alpha)$ is convergent, $I'(0)$ can't be obtained by evaluating the integrand at $\alpha = 0$. Instead, use partial fractions to find

$$I'(\alpha) = -\frac{2}{\pi} \int_0^{\infty} \frac{\sin(\alpha u)}{u} + \frac{2}{\pi} \int_0^{\infty} \frac{\sin(\alpha u)}{u(1 + u^2)}.$$

So what's $I'(0)$? (See Example 8.16.)

(c) Use the result of (b) to show that $I''(\alpha) = I(\alpha)$. (What's wrong with just taking two derivatives of I?) Then use $I(0)$ and $I'(0)$ to find $I(\alpha)$.

(d) Find an expression for $I(\alpha)$ valid for both positive and negative α.

8.31   Perform the integral in (7.26) to find Kepler's equation as expressed in (7.27). Follow these steps:

(i) Use the substitution $u = \tan \phi / 2$ (Example 8.6).

(ii) To help reduce clutter, introduce the parameter $\beta^2 \equiv (1 - \epsilon)/(1 + \epsilon)$, which is the ratio of perihelion to aphelion.

(iii) Use a partial fraction decomposition to express the integrand as a sum of integrands.

(iv) Integrate each piece, using trig substitutions as warranted.

(v) Express the result in terms of the original angle ϕ to find Kepler's equation.

8.32   **Method of Steepest Descent**

Consider an integral of the form

$$\int_{-\infty}^{\infty} e^{-af(x)} \, dx, \quad \Re e(a) > 0.$$

If a is large, the integral is dominated by contributions around the minimum $f(x_0)$. Since the first derivative term vanishes at an extremum, the Taylor expansion around x_0,

$$f(x) = f(x_0) + \frac{1}{2} f''(x_0) \, (x - x_0)^2 + \cdots,$$

allows us to approximate the integral as a Gaussian,

$$\int_{-\infty}^{\infty} e^{-af(x)} \, dx \approx e^{-af(x_0)} \int_{-\infty}^{\infty} e^{-\frac{a}{2} f''(x_0)(x - x_0)^2} \, dx = e^{-af(x_0)} \sqrt{\frac{2\pi}{af''(x_0)}}.$$

As an example, derive the Stirling approximation: $n! \approx \sqrt{2\pi n}\, n^n e^{-n}$ for large n. To do this, we need to approximate the integral representation of $n!$ (the gamma function),

$$n! = \Gamma(n + 1) = \int_0^{\infty} x^n e^{-x} dx,$$

for large n. Note that in this limit, expanding the integrand directly won't help since a truncated expansion is only valid for *small* x. But sketching $x^n e^{-x}$ shows that it becomes more and more tightly peaked as n grows, suggesting that a possible approximation scheme is to expand the integrand around this maximum.

(a) Show that the integrand peaks at $x = n$.

(b) Rewrite the integrand in purely exponential form, $e^{f(x)}$. Since we want to be able to truncate our expansion, expand $f(x)$ around its maximum, keeping only terms through x^2. The higher-order terms will contribute relatively little to the integral.

(c) Clean up the integrand as much as possible, and argue that extending the lower integration limit to $-\infty$ is not unreasonable. Evaluate the integral, and enjoy the glow of a derivation well done.

The Stirling approximation appears frequently in statistical mechanics where logarithms of large numbers are common: $\ln(n!) \approx n \ln n - n + \frac{1}{2} \ln(2\pi n)$.

9 The Dirac Delta Function

9.1 The Infinite Spike

Consider a function $\delta(x)$ implicitly defined by the integral

$$\int_{-\infty}^{\infty} f(x)\delta(x-a)\,dx = f(a),\tag{9.1}$$

for any continuous, well-behaved function f. At first glance, this expression seems absurd: how can integrating an arbitrary function $f(x)$ over all x return the value of f at a single point?

Let's parse the integral to see if we can make sense of it. First, of the infinity of x values sampled, only one, $x = a$, is singled out by the integration. Although one might imagine an odd situation in which the contribution for $x < a$ exactly cancels that for $x > a$, this is certainly not possible for *arbitrary* $f(x)$. More fundamentally, since dx is infinitesimal, the contribution at a single point a should be vanishingly small — which would make the entire integral vanish!

The only way to avoid this outcome is for $\delta(0)$ to be infinite, but vanish for all $x \neq a$. In other words, $\delta(x-a)$ must be an infinite spike at $x = a$ (Figure 9.1),

$$\delta(x-a) = \begin{cases} 0, & x \neq a \\ \infty, & x = a \end{cases}.\tag{9.2}$$

But this alone is not quite enough. In order to be consistent, we must also require

$$\int_{-\infty}^{\infty} \delta(x)\,dx = 1.\tag{9.3}$$

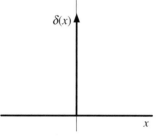

Figure 9.1 Dirac delta function.

We can now understand (9.1) as follows: as the integral sweeps through an infinity of x-values, the integrand is non-zero only at $x = a$. But here, the infinity of $\delta(0)$ is moderated by the infinitesimal dx in such a way that their product is unity. So the integral is $f(a)$. (If you prefer, you could assert that since δ is an infinitely thin, infinitely high spike, $f(x)\delta(x-a) \equiv f(a)\delta(x-a)$; the $f(a)$ can then be taken out of the integral in (9.1), leaving only the condition in (9.3).)

$\delta(x)$ is known as the *Dirac delta function*. Despite the name, it is not a function in the usual sense (Lighthill, 1958). A true function cannot have a single-point, infinite discontinuity. Indeed, from a strict mathematical perspective, delta functions only make sense under an integral, as in (9.1). This is important to remember, because algebraic relations and identities are often presented with the integration implied.

The delta function comes in many different guises called "representations"; many of the most common representations are listed inside the back cover. Some relate $\delta(x)$ to a differential equation, others

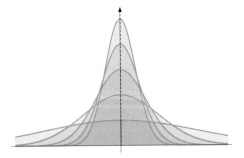

Figure 9.2 The Dirac delta as a limit of Gaussians.

render it as an infinite sum over a family of functions. The most common way to *think* of a δ-function, however, is as the limit of a sequence of functions — specifically, the limit of a sequence of peaked curves which become increasingly narrow and high in such a way that the total area under each curve is always unity. For example, one common representation utilizes the sequence of Gaussians $\frac{1}{\sqrt{\pi}}\, n\, e^{-n^2 x^2}$ which integrate to one for all n. As n grows, the familiar bell-shaped curve becomes taller and narrower (Figure 9.2). Only in the $n \to \infty$ limit, however, do we finally obtain the delta function $\delta(x)$.[1]

Example 9.1   Impulse

How do you analyze the force imparted on a kicked ball? The force is clearly time dependent, growing rapidly from zero when contact is made, peaking, and then rapidly dropping back to zero. A mathematical expression for $F(t)$ is difficult to come by; the manifestation of the force, however, is still given by Newton's second law,

$$I \equiv \int F(t)\, dt = \Delta p, \tag{9.4}$$

where Δp is the change in the ball's momentum. I, called the *impulse*, is all that is needed to calculate the overall motion of the ball; the details of the foot-on-ball interaction are not necessary.

A common idealization is to imagine that the impulsive force acts over shorter and shorter time intervals, though always with the same I — that is, always imparting the same change in the ball's momentum. Thus we successively replace the force with a sequence of ever-narrower, ever-taller functions, all of which have the same area under the curve. In the limit, the change in the ball's momentum — the foot-and-ball interaction itself — occurs at a single instant of time t_0. Not terribly realistic, perhaps, but certainly a reasonable approximation if the time over which the force acts is too short to be of any relevance. In this limit, the impulsive force becomes a delta function

$$F(t) = \Delta p\, \delta(t - t_0), \tag{9.5}$$

which clearly satsifies (9.4). When $\Delta p = 1$, this is often called a unit impulse.

A completely different but useful integral representation of the delta function is

$$\delta(x) \equiv \frac{1}{2\pi} \int_{-\infty}^{\infty} e^{ikx}\, dk. \tag{9.6}$$

Physically, this is a superposition of an infinite number of plane waves e^{ikx} of equal magnitude. So we can try to understand (9.6) by arguing that these waves are all out of phase with one another, and so

---

[1] This limit of functions is not a function, so the sequence is not cauchy. See BTW 6.2.

interfere destructively everywhere — except of course at $k = 0$, where the integral obviously diverges. Fine. But what about the 2π? At first, you might think a multiplicative constant is irrelevant; after all, the only values $\delta(x)$ takes are zero and infinity. But remember, strictly speaking a delta function only makes sense under an integral, and here a multiplicative constant can be crucial to ensure that the representation of δ is properly normalized — i.e., that $\int \delta(x)\,dx = 1$. In Problem 9.11, you'll use (9.1) and integration by parts to verify the factor of 2π.

9.2 Properties of the Delta Function

1. Normalization: The integral of $\delta(x)$ over all of space is unity,

$$\int_{-\infty}^{\infty} \delta(x)\,dx = 1. \tag{9.7}$$

For that matter, any range of integration will do as long as the zero of the delta function's argument is included,

$$\int_{x_1}^{x_2} \delta(x - a)\,dx = \begin{cases} 1, & x_1 < a < x_2 \\ 0, & \text{else} \end{cases}. \tag{9.8}$$

2. Sieve: The delta function acts as a sieve,

$$\int_{-\infty}^{\infty} f(x)\,\delta(x)\,dx = f(0), \tag{9.9}$$

sifting out of an integral a particular value of f. More generally,[2]

$$\int_{x_1}^{x_2} f(x)\,\delta(x - a)\,dx = \begin{cases} f(a), & x_1 < a < x_2 \\ 0, & \text{else} \end{cases}. \tag{9.10}$$

Note that (9.7) and (9.8) are merely special cases of (9.9) and (9.10).

These equations are sometimes cited as the *operational* definition of $\delta(x)$. More specifically, validating a proposed representation of the delta function requires verification of (9.9) — or equivalently, (9.1). For instance, the limit of a sequence of functions $\phi_k(x)$ is a valid representation for $\delta(x)$ only if

$$\lim_{k \to \infty} \int_{-\infty}^{\infty} f(x)\,\phi_k(x)\,dx = f(0). \tag{9.11}$$

Example 9.2 **Delta as a Limit of Rectangles**

Consider the function

$$\phi_k(x) = \begin{cases} k/2, & |x| < 1/k \\ 0, & |x| > 1/k \end{cases}.$$

This family of rectangular functions has unit area, independent of k (Figure 9.3),

$$\int_{-\infty}^{\infty} \phi_k(x)\,dx = \frac{k}{2} \int_{-1/k}^{1/k} dx = 1.$$

---

[2] Compare this to the action of a Kronecker delta under a discrete sum, $\sum_i f(x_i)\delta_{ij} = f(x_j)$.

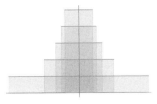

Figure 9.3 The family of functions $\phi_k(x)$.

To be a valid representation, (9.11) must be satisfied. Invoking the mean value theorem from BTW 8.5,

$$\lim_{k\to\infty} \int_{-\infty}^{\infty} f(x)\phi_k(x)\,dx = \lim_{k\to\infty} \frac{k}{2} \int_{-1/k}^{1/k} f(x)\,dx$$

$$= \lim_{k\to\infty} f(c_k) \left(\frac{k}{2} \int_{-1/k}^{1/k} dx \right) = \lim_{k\to\infty} f(c_k), \qquad (9.12)$$

where (8.97) was used in the second step. Now the theorem dictates that c_k be somewhere within the integration interval, $|c_k| < 1/k$ — which means $c_k \to 0$ as $k \to \infty$. Thus

$$\lim_{k\to\infty} \int_{-\infty}^{\infty} f(x)\phi_k(x)\,dx = \lim_{k\to\infty} f(c_k) = f(0). \qquad (9.13)$$

So the representation is valid.

The defining relations (9.9) and (9.10) allow the δ-function to be thought of as a *distribution* or weight function which emphasizes the contribution of $f(x)$ at some points more than at others — just as, say, a charge density $\lambda(x)$ is larger at some points than others, giving those points greater weight in an integral like $Q = \int \lambda\,dx$. The Dirac delta, though, is an extreme case: it places an infinite weight on the point(s) at which its argument goes to zero.

Note that just as a spatial density in one dimension has units of stuff-per-length, the distribution $\delta(x)$ has units $1/x$. (Indeed, how else could the integral in (9.1) have the units of f?) More generally, the units of a delta function are the reciprocal of the units of its argument.

3. More Complicated Arguments: What if we have $\delta(ax)$, for some constant a? At first, you wouldn't expect any effect: the argument of $\delta(ax)$ still goes to 0 at $x = 0$. Once again, though, we must be more careful. The easiest way to see that $\delta(x)$ and $\delta(ax)$ differ is by examining one of the representations. For instance, we established in Example 9.2 that the limit of the sequence of rectangular functions $\phi_k(x)$ is a valid representation of $\delta(x)$. But clearly $\phi_k(ax)$ does not obey (9.7). A similar observation can be made for any representation.

The key to finding the correct relationship is to change the integration variable in (9.10) to be the same as the argument of the delta; for $\delta(ax)$, that means integrating over $y \equiv ax$:

$$\int_{-\infty}^{\infty} f(x)\,\delta(ax)\,dx = \int_{-\infty}^{\infty} f(y/a)\,\delta(y)\,\frac{1}{a}\,dy = \frac{1}{a}f(0)$$

$$\implies \delta(ax) = \frac{1}{a}\delta(x). \qquad (9.14)$$

Accounting for the case of negative a gives (see Problem 9.4)

$$\delta(ax) = \frac{1}{|a|}\delta(x). \qquad (9.15)$$

Note that, as expected, $\delta(ax)$ has units $(ax)^{-1}$.

For a more complicated argument $\delta(g(x))$, the variable change $y = g(x)$ gives

$$\int_{y_1}^{y_2} f(y)\,\delta(y)\,dy = \int_{x_1}^{x_2} f(g(x))\,\delta(g(x))\left|\frac{dg}{dx}\right|dx. \tag{9.16}$$

In order for the right-hand side to correctly integrate to $f(0)$, it must be that

$$\delta(g(x)) = \frac{1}{|dg/dx|_{x_0}}\,\delta(x - x_0), \tag{9.17}$$

where x_0 is a root of g, $g(x_0) \equiv 0$. Note that (9.15) is a special case of this result. If there are several zeros of g within the integration interval, this generalizes to the useful rule

$$\delta(g(x)) = \sum_i \frac{1}{|g'(x_i)|}\,\delta(x - x_i), \tag{9.18}$$

where the x_i are roots of $g(x)$. Of course, this presumes that $g'(x_i) \neq 0$.

Example 9.3 $\delta(x^2 - a^2)$

Direct application of (9.18) gives

$$\delta(x^2 - a^2) = \frac{1}{2a}\big[\delta(x - a) + \delta(x + a)\big], \tag{9.19}$$

for $a > 0$. So, for example,

$$\int_{-\infty}^{\infty} dx\, x^3\, \delta(x^2 - a^2) = \frac{1}{2a}\int_{-\infty}^{\infty} dx\, x^3\big[\delta(x - a) + \delta(x + a)\big]$$
$$= \frac{1}{2a}\big[a^3 + (-a)^3\big] = 0, \tag{9.20}$$

as you'd expect of an odd integrand and a symmetric interval. On the other hand,

$$\int_0^{\infty} dx\, x^3\, \delta(x^2 - a^2) = \frac{1}{2a}\int_0^{\infty} dx\, x^3\big[\delta(x - a) + \delta(x + a)\big]$$
$$= \frac{1}{2a}(a^3) = \frac{1}{2}a^2. \tag{9.21}$$

Example 9.4 $\delta(\cos\theta)$

The zeros of $\cos\theta$ are odd multiples of $\pi/2$, $\theta_n = (2n+1)\frac{\pi}{2} = (n+\frac{1}{2})\pi$. But of course, $\sin\theta_n = \pm1$. Thus

$$\delta(\cos\theta) = \sum_{n=-\infty}^{\infty} \frac{1}{|\sin\theta_n|}\delta(\theta - \theta_n) = \sum_{n=-\infty}^{\infty} \delta(\theta - \theta_n). \tag{9.22}$$

Often, the limits of integration collapse this infinite sum to only a term or two, for example

$$\int_0^{\pi} d\theta\, e^{i\theta}\, \delta(\cos\theta) = \int_0^{\pi} d\theta\, e^{i\theta} \sum_{n=-\infty}^{\infty} \delta(\theta - \theta_n)$$
$$= \int_0^{\pi} d\theta\, e^{i\theta}\, \delta(\theta - \pi/2) = e^{i\pi/2} = i. \tag{9.23}$$

4. Derivatives of $\delta(x)$: Thinking of the delta function as an infinite spike, it makes little sense to talk of its derivative. Considering its behavior under an integral, however, integration by parts yields

$$\int_{-\infty}^{\infty} f(x)\delta'(x)\,dx = f(x)\delta(x)\Big|_{-\infty}^{\infty} - \int_{-\infty}^{\infty} f'(x)\delta(x)\,dx$$

$$= -\int_{-\infty}^{\infty} f'(x)\delta(x)\,dx = -f'(0). \tag{9.24}$$

Thus

$$f(x)\delta'(x) = -f'(x)\delta(x). \tag{9.25}$$

This easily generalizes to

$$f(x)\delta^{(n)}(x) = (-1)^n f^{(n)}(x)\delta(x), \tag{9.26}$$

where in this context, $\delta^{(n)}(x) \equiv \left(\frac{d}{dx}\right)^n \delta(x)$.

5. Relation to the Heaviside Step Function, $\Theta(x)$: A mere glance at a sketch of the step function

$$\Theta(x) = \begin{cases} 0, & x < 0 \\ 1, & x > 0 \end{cases}, \tag{9.27}$$

shown in Figure 9.4, should suggest to you (the proof is Problem 9.2) that

$$\frac{d\Theta}{dx} = \delta(x). \tag{9.28}$$

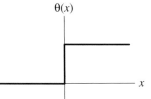

Figure 9.4 Step function.

Since $\delta(x)$ has units of x^{-1}, the step function must be dimensionless.

6. Dirac δ in Higher Dimensions: In higher dimensions,

$$\int_{\text{all space}} \delta(\vec{r})\,d\tau = 1, \tag{9.29}$$

or, consistent with (9.10),

$$\int_V f(\vec{r})\,\delta(\vec{r} - \vec{a})\,d\tau = \begin{cases} f(\vec{a}), & \vec{a} \in V \\ 0, & \text{else} \end{cases}. \tag{9.30}$$

An n-dimensional delta function is often written as $\delta^{(n)}(\vec{r})$ — which should not be confused with the nth derivative. When the dimensionality is unambiguous, though, it is common to omit the superscript and write $\delta(\vec{r})$, leaving only the vector-valued argument to indicate $n > 1$.

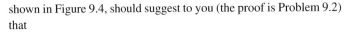

Point Charge Distribution

The voltage of a point charge q at a position $\vec{a}$ is given by Coulomb's law[3]

$$\Phi(r) = \frac{q}{4\pi\epsilon_0} \frac{1}{|\vec{r} - \vec{a}|}, \tag{9.31}$$

[3] We use Φ rather than V to avoid confusion with volume.

with $\Phi = 0$ at infinity. For a continuous charge distribution $\rho(\vec{r}\,')$, we can consider separately each element of volume $d\tau\,'$ centered at $\vec{r}\,'$ with infinitesimal charge $dq = \rho(\vec{r}\,')d\tau\,'$; each such element has voltage $d\Phi$ given by (9.31) with $q \to dq$ and $\vec{a} \to \vec{r}\,'$,

$$d\Phi(r) = \frac{1}{4\pi\epsilon_0} \frac{\rho(\vec{r}\,')\,d\tau\,'}{|\vec{r} - \vec{r}\,'|}. \tag{9.32}$$

The total voltage at some point $\vec{r}$ due to all these infinitesimal charges is then simply given by superposition — that is, the integral over the volume of charge

$$\Phi(\vec{r}) = \frac{1}{4\pi\epsilon_0} \int_V \frac{\rho(\vec{r}\,')}{|\vec{r} - \vec{r}\,'|}\, d\tau\,'. \tag{9.33}$$

This expression should hold for any physically acceptable charge configuration. But what about for a single charge q at $\vec{r}\,' = \vec{a}$ — what's the correct expression for the charge density of a point charge? In other words, what function ρ in (9.33) will reproduce (9.31)?

Physical intuition is very helpful here: a point charge crams a finite charge q into a single point. So the density of a point charge must be infinite at that point, and zero everywhere else — in other words, a delta function. Thus we posit

$$\rho(\vec{r}\,') \equiv q\,\delta(\vec{r}\,' - \vec{a}), \tag{9.34}$$

and indeed,

$$\frac{q}{4\pi\epsilon_0} \int_V \frac{\delta(\vec{r}\,' - \vec{a})}{|\vec{r} - \vec{r}\,'|}\, d\tau\,' = \frac{q}{4\pi\epsilon_0} \frac{1}{|\vec{r} - \vec{a}|}, \tag{9.35}$$

which is (9.31).

Integrals such as (9.33) are conceived and constructed using physical considerations — but to actually *do* the integral one often has to rewrite it as the product of one-dimensional integrals. How is this done when the integrand contains a delta function? Using the defining condition (9.29), the answer is straightforward in cartesian coordinates:

$$\delta(\vec{r} - \vec{r}_0) \equiv \delta(x - x_0)\,\delta(y - y_0)\,\delta(z - z_0). \tag{9.36}$$

What does $\delta(\vec{r})$ look like in non-cartesian coordinates? Probably the easiest way to answer this is to appreciate that (9.29) must be independent of coordinate system. Recall that under coordinate change $\{x, y, z\} \to \{u, v, w\}$, the volume element transforms as

$$d\tau = dx\,dy\,dz = |J|\,du\,dv\,dw, \tag{9.37}$$

where J is the Jacobian of the transformation. But since $\delta(\vec{r} - \vec{r}_0)$ integrates to 1,

$$\int_{\text{all space}} \delta(\vec{r} - \vec{r}_0)\,d\tau = \int_{\text{all space}} \delta(\vec{r} - \vec{r}_0)\,|J|\,du\,dv\,dw = 1, \tag{9.38}$$

we must conclude that[4]

$$\delta(\vec{r} - \vec{r}_0) = \frac{1}{|J|}\,\delta(u - u_0)\delta(v - v_0)\delta(w - w_0). \tag{9.39}$$

---

[4] This is the generalization to higher dimensions of our derivation of (9.18).

So in cylindrical coordinates we have

$$\delta(\vec{r} - \vec{r}_0) = \frac{1}{r}\,\delta(r - r_0)\,\delta(\phi - \phi_0)\,\delta(z - z_0), \tag{9.40}$$

and in spherical

$$\delta(\vec{r} - \vec{r}_0) = \frac{1}{r^2 \sin\theta}\,\delta(r - r_0)\,\delta(\theta - \theta_0)\,\delta(\phi - \phi_0)$$

$$= \frac{1}{r^2}\,\delta(r - r_0)\,\delta(\cos\theta - \cos\theta_0)\,\delta(\phi - \phi_0). \tag{9.41}$$

An n-dimensional δ-function can always be written as the product of n one-dimensional δ's with the appropriate Jacobian.

Example 9.6 **Mass m at $x = y = -z = 1$**

Clearly,

$$\rho(\vec{r}) = m\,\delta(x - 1)\delta(y - 1)\delta(z + 1). \tag{9.42a}$$

In cylindrical coordinates, the mass is located at $r = \sqrt{2}$, $\phi = \pi/4$, giving

$$\rho(\vec{r}) = \frac{m}{r}\delta(r - \sqrt{2})\,\delta(\phi - \pi/4)\,\delta(z + 1). \tag{9.42b}$$

In spherical coordinates, $r = \sqrt{3}$, $\cos\theta = -1/\sqrt{3}$, and $\phi = \pi/4$, so

$$\rho(\vec{r}) = \frac{m}{r^2}\,\delta(r - \sqrt{3})\,\delta\left(\cos\theta + 1/\sqrt{3}\right)\delta(\phi - \pi/4). \tag{9.42c}$$

Example 9.7 **Line Charge Distribution**

The delta and theta functions can also be used for continuous charge or mass distributions. Consider a uniform line charge centered along the z-axis with total charge Q and length 2ℓ so that $\lambda_0 = Q/2\ell$. To express this as a volume charge density $\rho(\vec{r})$ in spherical coordinates, we need to confine the non-zero contributions to $\theta = 0$ and π; we can do this with a factor

$$\delta(\cos\theta - 1) + \delta(\cos\theta + 1). \tag{9.43}$$

This keeps us on the z-axis, but doesn't restrict the charge to lie between $-\ell$ and ℓ. To do that, we can use the step function (9.27),

$$\Theta(\ell - |z|) = \Theta(\ell - r|\cos\theta|). \tag{9.44}$$

(Be careful not to confuse the coordinate θ with the step function Θ.) Then

$$\rho(\vec{r}) = \frac{\lambda_0}{r^2}\,\Theta(\ell - r|\cos\theta|)\,\delta(\phi)\,[\delta(\cos\theta - 1) + \delta(\cos\theta + 1)]$$

$$= \frac{\lambda_0}{2\pi r^2}\,\Theta(\ell - r|\cos\theta|)\,[\delta(\cos\theta - 1) + \delta(\cos\theta + 1)]. \tag{9.45}$$

Let's check. First, it has the correct units of charge/volume (since Θ is dimensionless). Integrating over all space,

$$\int \rho(\vec{r})\, d\tau = \frac{\lambda_0}{2\pi} \int_0^\infty r^2 dr \int_{-1}^1 d(\cos\theta)$$

$$\times \int_0^{2\pi} d\phi\, \frac{1}{r^2}\, \Theta(\ell - r|\cos\theta|)\, [\delta(\cos\theta - 1) + \delta(\cos\theta + 1)]$$

$$= \lambda_0 \int_0^\infty dr \int_{-1}^1 d(\cos\theta)\, \Theta(\ell - r|\cos\theta|)\, [\delta(\cos\theta - 1) + \delta(\cos\theta + 1)]$$

$$= \lambda_0 \int_0^\infty dr\, 2\Theta(\ell - r) = 2\lambda_0 \int_0^\ell dr = 2\lambda_0\ell = Q. \tag{9.46}$$

Example 9.8 Surface Charge Distribution

Consider a uniform surface charge distributed on a disk of radius a so that $\sigma_0 = Q/\pi a^2$. To express it as a volume density, the natural choice is cylindrical coordinates. Confining the charge to the xy, plane requires a factor of $\delta(z)$; indeed,

$$\sigma_0\, \delta(z) \tag{9.47}$$

has the correct units for a volume charge density. But how do we restrict the distribution to $r < a$? A factor like $\delta(r-a)$ only restricts the charge to the circle $r = a$, not the entire disk. We could use a step function, as in the previous example. This time, however, let's do something a little more clever — essentially taking advantage of the differential relationship (9.28). Introduce a dummy variable r' and the expression

$$\int_0^a \delta(r' - r)dr' = \begin{cases} 1, & 0 < r < a \\ 0, & \text{else} \end{cases}. \tag{9.48}$$

This dimensionless expression correctly confines us to the disk. Thus

$$\rho(\vec{r}) = \sigma_0\, \delta(z) \int_0^a \delta(r' - r)dr'. \tag{9.49}$$

As a final check, let's integrate:

$$\int_{\text{all space}} \rho(\vec{r})d\tau = \frac{Q}{\pi a^2} \int_{-\infty}^\infty \delta(z)\, dz \int_0^{2\pi} d\phi \int_0^\infty r\, dr \int_0^a dr'\delta(r' - r)$$

$$= \frac{2Q}{a^2} \int_0^\infty r\, dr \int_0^a dr'\delta(r' - r)$$

$$= \frac{2Q}{a^2} \int_0^a dr' \int_0^\infty dr\, r\, \delta(r' - r)$$

$$= \frac{2Q}{a^2} \int_0^a dr'r' = Q. \tag{9.50}$$

So it works.

Problems

9.1   Evaluate the following integrals:

(a) $\int_{-2}^{3} [x^3 - (2x+5)^2] \, \delta(x-1) \, dx$

(b) $\int_{0}^{3} (5x^2 - 3x + 4) \, \delta(x+2) \, dx$

(c) $\int_{0}^{1} \cos x \, \delta(x - \pi/6) \, dx$

(d) $\int_{-\pi}^{\pi} \ln(\sin x + 2) \, \delta(x + \pi/2) \, dx$

(e) $\int_{-1}^{1} (x^3 - 3x^2 + 2) \, \delta(x/7) \, dx$

(f) $\int_{-1}^{1} (x-1) \, e^{x^2} \, \delta(-3x) \, dx$

(g) $\int_{-\pi}^{\pi} 4x^2 \cos^{-1} x \, \delta(2x-1) \, dx$

(h) $\int_{p}^{\infty} \delta(x+q) \, dx$

(i) $\int_{0}^{2b} x \, \delta[(x^2 - b^2)(x - b/2)] \, dx$

(j) $\int_{-\pi/2}^{\pi/2} e^x \, \delta(\tan x) \, dx.$

9.2   Let $\Theta(x)$ be the step function of (9.27),

$$\Theta(x) = \begin{cases} 1, & x > 0 \\ 0, & x < 0. \end{cases}$$

Show that $d\Theta/dx = \delta(x)$.

9.3   Show that the sequence of Gaussians

$$\phi_n(x) = \frac{n}{\sqrt{\pi}} \, e^{-n^2 x^2}$$

gives a valid representation of the delta function in the $n \to \infty$ limit.

9.4   Verify the identity

$$\delta(ax) = \frac{1}{|a|} \delta(x),$$

being careful to justify the use of the absolute value.

9.5   Determine the parity of $\delta(x)$ — i.e., is it odd/even/neither?

9.6   Verify the identity

$$\delta(x^2 - a^2) = \frac{1}{2a} \big[\delta(x+a) + \delta(x-a) \big]$$

by examining the behavior of $x^2 - a^2$ in the vicinity of $x = \pm a$. Note that I am not asking you to *apply* the identity in (9.18); rather, I want you to *verify* it in this particular case. Start by arguing that the integral $\int_{-\infty}^{\infty} f(x) \, \delta(x^2 - a^2) \, dx$ only has contributions in infinitesimal regions around $\pm a$. Then factor $x^2 - a^2$ into linear pieces, simplify as appropriate in the vicinity of $\pm a$, and use the results of Problem 9.4.

9.7   The energy–momentum relationship in special relativity,

$$E^2 = p^2 c^2 + m^2 c^4,$$

identifies mass as a *Lorentz invariant*: though different inertial observers of the same system may measure different values for energy and momentum, they will all agree on the quadratic combination $E^2 - p^2 c^2$.

(For instance, no matter the speed of a proton whizzing by, measurements of its energy and momentum will always return a mass of $mc^2 = 938$ MeV.) The four-dimensional integral

$$\int d^3p \int_0^\infty dE \; \delta(E^2 - p^2c^2 - m^2c^4)$$

is a sum over all E and momentum $\vec{p}$ subject to the relativistic energy–momentum constraint. Use this expression to show that

$$\int \frac{d^3p}{2\sqrt{p^2c^2 + m^2c^4}}$$

is a Lorentz invariant.

9.8   **Limit Representations** In the expression $\delta(g(x))$, it makes sense that the delta function picks out values $x = x_0$ such that $g(x_0) = 0$. But our derivation of (9.18) notes makes clear that this formula only applies to functions for which $g'(x_0) \neq 0$. In other words, x_0 must be a simple root. So then what do you do if you have a function like $g(x) = (x-a)^2$ which has a double root at $x = a$? Deriving a general rule is a bit cumbersome. Instead we can use a sequence of functions $\phi_n(x)$ which provides a representation of the δ-function in the $n \to \infty$ limit. As an example, take $g(x) = x^2$ and use any *two* of the three limit representations given inside the back cover to calculate

$$\int_{-\infty}^{\infty} |x| \, \delta(x^2) \, dx.$$

Of course, the different representations must give the same result.

9.9   Express $x^n \frac{d^n}{dx^n} \delta(x)$ in terms of $\delta(x)$.

9.10   Write an expression for the mass density ρ (mass/volume) of a uniform circular ring of radius R and total mass M sitting in the xy-plane:

(a) in cylindrical coordinates
(b) in spherical coordinates.

9.11   In Problem 8.7 we found

$$\int_{-\infty}^{\infty} \frac{e^{ikx}}{k} \, dk = i\pi \; \mathrm{sgn}(x),$$

where the so-called *signum* function is

$$\mathrm{sgn}(x) \equiv \begin{cases} +1, & x > 0 \\ 0, & x = 0 \\ -1, & x < 0. \end{cases}$$

Note that for $x \neq 0$, $\mathrm{sgn}\, x = |x|/x$. Use this, together with integration by parts, to verify the representation

$$\delta(x) \equiv \frac{1}{2\pi} \int_{-\infty}^{\infty} e^{ikx} \, dk \qquad (9.51)$$

by integrating it with an arbitrary function $f(x)$ from a to b.

9.12   **Lagrange Inversion**
Given $\mu(\eta) = \eta - \epsilon f(\eta)$, we want to find the inverse function $\eta(\mu)$. (Kepler's equation (7.30) is the special case $f(\eta) = \sin \eta$.) As a transcendental equation, however, the best we can do is a series solution. We'll derive one such series in several steps, using several Part I tools.

First, a couple of conditions on the function $f(\eta)$: as with the mean and eccentric anomalies, we take $\mu = 0$ when $\eta = 0$. This gives $f(0) = 0$. Then to be assured an inverse exists, we assume $\epsilon f'(\eta) < 1$ for positive η. This renders $\mu(\eta) = \eta - \epsilon f(\eta)$ a strictly monotonically increasing function, mapping $0 \leq \eta < \infty$ one-to-one to $0 \leq \mu < \infty$.

(a) Noting that μ is given by the defining expression $\mu = \eta - \epsilon f(\eta)$, show that

$$\eta(\mu) = \int \delta(t - \eta(\mu))\, t\, dt = \int \delta[(t - \epsilon f(t)) - \mu]\left(1 - \epsilon f'(t)\right) t\, dt,$$

where $f'(t) = \partial f/\partial t$. Note that μ is a fixed parameter in the integral.

(b) Use the δ function representation of (9.51) to write this expression as a double integral over both t and k.

(c) Use $e^{a+b} = e^a e^b$ to separate the $f(t)$ portion of the exponential, and replace that separated piece with its Taylor expansion in powers of $f(t)$.

(d) Use differentiation with respect to the parameter μ to remove the k-dependence of the Taylor series from the integral.

(e) Use (9.51) again to recover a delta function, and use it to evaluate the t integral. Your fully integrated expression should be the infinite sum

$$\eta(\mu) = \sum_{m=0}^{\infty} \frac{\epsilon^m}{m!} \left(\frac{\partial}{\partial\mu}\right)^m \left[\mu f(\mu)^m \left(1 - \epsilon f'(\mu)\right)\right].$$

(f) Explicity remove $m = 0$ from the first sum, and shift the sum $m \to m - 1$ in the second, so that both begin at 0. Clever use of the product rule in the second sum should ultimately reveal the series

$$\eta(\mu) = \mu + \sum_{m=1}^{\infty} \frac{\epsilon^m}{m!} \left(\frac{\partial}{\partial\mu}\right)^{m-1} f(\mu)^m.$$

(g) So we have a series for η; what about for a general function $g(\eta)$? Emend the derivation above to find the *Lagrange inversion theorem*,

$$\eta(\mu) = \mu + \sum_{m=1}^{\infty} \frac{\epsilon^m}{m!} \left(\frac{\partial}{\partial\mu}\right)^{m-1} \left[\frac{\partial g(\mu)}{\partial\mu} f(\mu)^m\right]. \tag{9.52}$$

The earlier result is the special case $g(\eta) = \eta$.

10 Coda: Statistical Mechanics

The behavior of a non-relativistic classical system is governed by Newtonian mechanics; that of a quantum system, by Schrödinger mechanics. Though in principle this is true independent of the size of the system, in practice neither mechanics is up to the task of investigating a system with many particles. In particular, a macroscopic system with some 10^{23} particles is completely intractable with only the tools of Newton or Schrödinger analysis. Instead, statistical methods must be introduced to demonstrate how the microscopic rules of classical and quantum mechanics determine the properties of a macroscopic system with well-defined characteristics such as temperature, energy, and pressure.

This theoretical framework is called *statistical mechanics*, and several topics discussed in Part I are its basic tools. A proper introduction to "stat mech" requires an entire course all its own; here we have the more modest goal of presenting some of its tools in a physical context — for example, keeping track of indices, infinite sums, hyperbolic trig, and integration techniques. We'll also see how consternation over a divergent series can lead to a remarkable physical discovery.[1]

10.1 The Partition Function

One cannot do much better than to start by citing what Feynman calls "the key principle of statistical mechanics": if a system in equilibrium can be in one of N states, then the probability of the system being in state s with energy E_s is

$$\mathcal{P}_s = \frac{1}{Z} e^{-\beta E_s}, \tag{10.1}$$

where $e^{-\beta E_s}$ is the famous[2] Boltzmann factor, $\beta \equiv 1/kT$ the so-called "inverse temperature," and k is Boltzmann's constant. The *partition function* Z is defined as the sum over all states,

$$Z = \sum_{s=1}^{N} e^{-\beta E_s}, \tag{10.2}$$

ensuring that the probability in (10.1) properly sums to unity,

$$\sum_{s} \mathcal{P}_s = \frac{1}{Z} \sum_{s} e^{-\beta E_s} = 1. \tag{10.3}$$

Note that for small T/large β, only low energy states will have much chance of being occupied. But as the temperature increases and β falls, the probabilities become more uniformly distributed amongst the various states.

---

[1] Much of this Coda is influenced by Feynman (1972); see also Baierlein (1999) and Rau (2017).
[2] Or at least famous amongst those who know of it.

Although Z is introduced by the necessity of having a normalized probability disttibution, the partition function can be used to calculate thermodynamic properties arising from the distribution of the underlying states s. For instance, differentiation with respect to the parameter β will give the system's average energy $U \equiv \langle E \rangle$ as a function of temperature,

$$U \equiv \frac{1}{Z} \sum_s E_s e^{-\beta E_s} = -\frac{1}{Z} \frac{\partial}{\partial \beta} \sum_s e^{-\beta E_s}$$

$$= -\frac{1}{Z} \frac{\partial Z}{\partial \beta} = -\frac{\partial \log Z}{\partial \beta}. \tag{10.4}$$

It is largely due to manipulations of this sort that the inverse temperature β is preferred over $1/kT$.

Example 10.1 The Maxwell Speed Distribution

Consider a monatomic ideal gas. We know that the temperature T is a measure of the average speed $\langle v \rangle$ of the molecules in the gas — but what precisely is the relationship between the two? Just how does a vast cauldron of particle positions and momenta lead to a physical interpretation of temperature?

The mathematical expression for the distribution, $D(v)$, is defined so that the probability $\mathcal{P}(v_1, v_2)$ of a molecule of gas having speed between v_1 and v_2 is

$$\mathcal{P}(v_1, v_2) = \int_{v_1}^{v_2} D(v) \, dv. \tag{10.5}$$

The greater D is for some value of v, the greater the probability of finding a molecule in the gas with that speed. Our goal is to find the function $D(v)$, and from there the relationship between temperature and average speed.

Classical states are specified by a particle's position and velocity, so the partition sum becomes an integral. The relative probability $d\mathcal{P}$ of finding a molecule with energy E_s somewhere within a volume d^3x of space and $d^3v = dv_x dv_y dv_z$ of "velocity space" is

$$d\mathcal{P} \sim e^{-E_s/kT} d^3x \, d^3v. \tag{10.6}$$

The energy of an ideal gas is entirely kinetic, $E_s = \frac{1}{2}mv^2$, so we can immediately integrate over space to get an overall factor of the volume — which we'll discard, as it has no bearing on the distribution of speeds within the gas. Since the energy E_s is dependent only on a particle's speed rather than its velocity, the remaining integral is spherically symmetric, so $d^3v = v^2 dv \, d\Omega = 4\pi v^2 dv$. Then the probability that a molecule has a speed v within an interval dv is

$$d\mathcal{P} = D(v) \, dv = \frac{1}{Z} e^{-mv^2/2kT} 4\pi v^2 dv. \tag{10.7}$$

The condition $\int d\mathcal{P} = 1$ gives partition function

$$Z = 4\pi \int_0^\infty e^{-mv^2/2kT} v^2 dv. \tag{10.8}$$

This integral is a Gaussian; from Example 8.17 we get

$$Z = 4\pi \int_0^\infty e^{-mv^2/2kT} v^2 dv = \left(\frac{2\pi kT}{m} \right)^{3/2}. \tag{10.9}$$

As a consistency check, you can use (10.4) to find the well-known result for the average energy of an ideal gas of N molecules, $U = \frac{3}{2}NkT$.

Putting everything together, we find that the probability (10.5) is

$$P(v_1, v_2) = 4\pi \left(\frac{m}{2\pi kT}\right)^{3/2} \int_{v_1}^{v_2} v^2 e^{-mv^2/2kT} dv. \tag{10.10}$$

Thus we identify

$$D(v) = 4\pi \left(\frac{m}{2\pi kT}\right)^{3/2} v^2 e^{-mv^2/2kT}, \tag{10.11}$$

which is known as the *Maxwell speed distribution* (Figure 10.1).

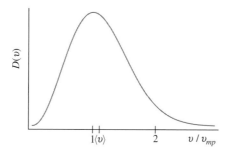

Figure 10.1 The Maxwell speed distribution.

A distribution function is meant to be integrated, giving appropriate weights to the various elements in the sum. For example, the average molecular speed $\langle v \rangle$ within the gas is given by

$$\langle v \rangle = \int_0^\infty v D(v) dv = 4\pi \left(\frac{a}{\pi}\right)^{3/2} I_3(a), \tag{10.12}$$

where $a = m/2kT$. Since

$$I_3(a) = -\frac{d}{da} I_1(a) = \frac{1}{2a^2}, \tag{10.13}$$

we get $\langle v \rangle = \sqrt{8kT/\pi m}$. This is the sought-after relationship between temperature and average speed. Note, incidentally, that $D(v)$ is asymmetric — quadratic for small v, a decaying exponential for large. Thus the *most probable* speed v_{mp}, where $D(v)$ reaches a maximum, is not the average; a straightforward calculation reveals $v_{mp} = v_{max} = \sqrt{2kT/m} \approx 0.89\langle v \rangle$.

Example 10.2 The Einstein Solid

Atoms in a crystal lattice are often modeled as connected by springs, each vibrating harmonically around its equilibrium position. Einstein, in a 1907 paper, was the first to apply quantum mechanics to this picture, making the simplifying assumptions that all atoms oscillate with the same frequency, independently of their neighbors. In other words, each oscillator is considered to be decoupled from the rest of the lattice, which it sees as a large reservoir with some fixed temperature T.

In quantum mechanics, the harmonic oscillator is characterized by equally spaced energies

$$\varepsilon_a = (a + 1/2)\hbar\omega, \qquad a = 0, 1, 2, \dots, \tag{10.14}$$

where ω is the natural frequency of the oscillator. (We use ε rather than E to emphasize that this is the energy of a single particle.) With $\varepsilon_0 = \frac{1}{2}\hbar\omega$ denoting the ground state ("zero point") energy, the partition function Z_1 of a single oscillator in thermal equilibrium with temperature $T = 1/k\beta$ is given by a geometric series,

$$Z_1 = \sum_{a=0}^{\infty} e^{-\beta\varepsilon_a} = e^{-\beta\varepsilon_0} \sum_{a=0}^{\infty} e^{-2\beta a\varepsilon_0} = \frac{e^{-\beta\varepsilon_0}}{1 - e^{-2\beta\varepsilon_0}} = \frac{1}{2\sinh(\beta\varepsilon_0)}. \tag{10.15}$$

Using (10.4), we get the particle's average energy,

$$u = \frac{\varepsilon_0}{\tanh(\beta\varepsilon_0)}. \tag{10.16}$$

The lower-case u indicates that this is for a single oscillator; for a solid of N identical oscillators, this needs to be multiplied by N. An additional factor to account for oscillations in three dimensions then gives the average energy

$$U = \frac{3N\varepsilon_0}{\tanh(\beta\varepsilon_0)}. \tag{10.17}$$

Einstein compared this result to data by calculating the heat capacity

$$C = \frac{\partial U}{\partial T} = -k\beta^2 \frac{\partial U}{\partial \beta}$$
$$= +3Nk \left(\frac{\beta\varepsilon_0}{\sinh\beta\varepsilon_0}\right)^2 = 3Nk \left[\frac{T_E/2T}{\sinh(T_E/2T)}\right]^2, \tag{10.18}$$

where in the last step we defined the Einstein temperature,

$$T_E \equiv \hbar\omega/k = 2\varepsilon_0/k. \tag{10.19}$$

This defines a natural scale characteristic of the crystal, giving meaning to the notion of large and small T. At high temperatures the heat capacity becomes the T-independent constant $3Nk$, in agreement with the classical result. But that classical result, called the Dulong–Petit law, asserts that the heat capacity is constant for *all* T, whereas data show that $C \to 0$ as T goes to zero — which *is* consistent with (10.18). Einstein's model was the first to reproduce this sort of low-T behavior, and as such was an important early demonstration of the fundamental nature of the quantum hypothesis. (In 1912, Debye improved upon Einstein's model both by allowing different frequencies, and by incorporating interactions between neighboring lattice sites — for which he received the 1936 Nobel prize for chemistry.)

10.2 The Chemical Potential

One of the challenges of statistical mechanics is to determine the multi-particle partition function from a single-particle partition function Z_1. Because the oscillators in the Einstein solid are decoupled from one another, the full N-particle partition function Z_N is the product of N single-particle partition functions, $Z_N = Z_1^N$. Since Z and U are related by a *logarithmic* derivative, (10.4), we did this implicitly when we multiplied the single-particle energy u by N to get the average energy U of the Einstein solid.

This relatively simple construction of Z_N is possible not only because the oscillators are independent. The fact that they are localized around their individual lattice sites provides an additional attribute with which they can be distinguished from one another — even if they are all the same type of atom. In a gas, however, we wouldn't be able to tell them apart. This is not merely an issue of keeping track of some 10^{23} particles; rather, in quantum mechanics identical particles are *fundamentally indistinguishable*. So, for instance, a configuration in which particle 1 has energy ε_a and particle 2 energy ε_b is the same quantum state as that with particle 1 having energy ε_b and particle 2 energy ε_a. As a result, both configurations only make one contribution to the partition function's sum over states. So a quantum system of N identical, non-interacting particles is determined not by identifying each particle's energy individually, but rather by specifying the *number* of particles n_a with energy ε_a, so that

$$E_s = \sum_a n_a \varepsilon_a. \tag{10.20}$$

Then the partition function's sum over states becomes sums over all possible values of the n_a's,

$$Z = \sum_s e^{-\beta E_s} = \sideset{}{'}\sum_{n_1, n_2, \ldots} \exp\left(-\beta \sum_a n_a \varepsilon_a\right), \tag{10.21}$$

where the notation $\sum'$ indicates that the sum is constrained by having a fixed number of particles, $N = \sum_a n_a$.

The restriction on the sum in (10.21) is difficult to implement directly. A clever way to impose it is to imagine a bottle of gas which can exchange particles with a large reservoir of particles. We don't assume this is free diffusion, but rather that it costs an energy μ, called the *chemical potential*, for a particle to pass from the bottle to the reservoir. The larger μ, the larger the number of particles N in the bottle. Their relationship is implicit in the condition that the number of particles in the bottle is the sum of the average $\langle n_a \rangle$ in each state a (called the *occupation numbers*),

$$N = \sum_a n_a = \sum_a \langle n_a \rangle. \tag{10.22}$$

Rather than the awkward constrained sum of (10.21), we'll fix the particle number by finding the value of μ which satisfies (10.22).

The explicit relationship between particle number and chemical potential can be derived from the partition function. Uniformly shifting the energies $\varepsilon_a \to \varepsilon_a - \mu$, (10.21) becomes

$$\begin{aligned}
Z &= \sum_{n_1, n_2, \ldots} \exp\left(-\beta \sum_a n_a(\varepsilon_a - \mu)\right) \\
&= \sum_{n_1, n_2, \ldots} e^{-\beta n_1(\varepsilon_1 - \mu)} e^{-\beta n_2(\varepsilon_2 - \mu)} \cdots \\
&= \left(\sum_{n_1} e^{-\beta n_1(\varepsilon_1 - \mu)}\right)\left(\sum_{n_2} e^{-\beta n_2(\varepsilon_2 - \mu)}\right) \cdots.
\end{aligned} \tag{10.23}$$

Notice that the restriction on the sum in (10.21) has been removed. (Mathematically, the chemical potential is an example of a *Lagrange multiplier*.)

Now there are two types of quantum particles with very different collective properties. We'll explore here the case of *bosons*; these are particles whose spins are integer multiples of $\hbar$, such as pions or alpha particles. It turns out that for bosons, n can be any non-negative integer. Thus each of the sums in (10.23) is an infinite geometric series,[3]

$$\sum_n e^{-\beta n(\varepsilon - \mu)} = \frac{1}{1 - e^{-\beta(\varepsilon - \mu)}}. \tag{10.24}$$

---

[3] By contrast, *fermions* can only have n equal 0 or 1 — see Problem 10.5.

Then the full N-particle boson partition function is the much-leaner infinite product

$$Z = \prod_a \frac{1}{1 - e^{-\beta(\varepsilon_a - \mu)}}. \tag{10.25}$$

With Z in hand, we can find the occupation numbers $\langle n_a \rangle$. In a calculation similar to that in (10.4), the logarithmic derivative of Z with respect to ε_a for fixed a gives (Problem 10.2)

$$\langle n_a \rangle_{\mathrm{B}} = \frac{1}{e^{+\beta(\varepsilon_a - \mu)} - 1}, \tag{10.26}$$

where the subscript indicates that this is for bosons. So from (10.22) we find

$$N = \sum_a \frac{1}{e^{+\beta(\varepsilon_a - \mu)} - 1}. \tag{10.27}$$

Having found this expression, inspection of (10.25) reveals that N itself can be found more directly with a logarithmic derivative of Z (Problem 10.2),

$$N = \frac{1}{\beta} \frac{\partial \log Z}{\partial \mu}. \tag{10.28}$$

Remember, we introduced the chemical potential in order to accommodate the restriction $\sum_a n_a = N$. So if are able to calculate $N(\mu)$ from (10.28), then we can choose μ to yield the required fixed number of particles in the bottle of gas. As we'll see (and as you may expect), that value of μ will depend on temperature.

10.3 The Ideal Boson Gas

Let's apply this machinery to an ideal gas of zero-spin identical bosons. Particles in an ideal gas are non-interacting, so $\varepsilon = p^2/2m$ is entirely kinetic. Then (10.25) gives

$$Z = \prod_a \frac{1}{1 - e^{-\beta(p_a^2/2m - \mu)}}. \tag{10.29}$$

Since we only need logarithmic derivatives of Z, we can go ahead and take the log to convert the unwieldy product to a more-tractable sum,

$$\log Z = -\sum_a \log\left(1 - e^{-\beta(p_a^2/2m - \mu)}\right). \tag{10.30}$$

To compute the sum, we need to know the momenta p_a. Let's take our gas to be in a cubic box of side L. Then the quantum mechanical states in each orthogonal direction are standing waves with $\lambda = 2L/n$, where n is a positive integer. Using de Broglie's $p = h/\lambda = \hbar k$ for each direction, we get momentum states

$$\vec{p}_a = \frac{\pi \hbar}{L}\left(n_x, n_y, n_z\right) \longrightarrow \varepsilon_a = \frac{p_a^2}{2m} = \frac{\pi^2 \hbar^2}{2mL^2}\left(n_x^2 + n_y^2 + n_z^2\right). \tag{10.31}$$

So the sum over states is a sum over the triplet of positive integers, $a = \{n_x, n_y, n_z\}$. Note that the spacing of these quantum states becomes smaller and smaller as the size of the box grows; since we're talking about atoms in a macroscopic fixed volume, it's therefore reasonable to replace the discrete sum with a continuous integral. This is accomplished by noting each one-dimensional momentum interval has width $\Delta p = \pi \hbar / L$, so that

$$\log Z = -\sum \log \left(1 - e^{-\beta(p_a^2/2m - \mu)}\right) \frac{V}{(\pi\hbar)^3} \Delta p_x \Delta p_y \Delta p_z, \tag{10.32}$$

where $V = L^3$. Then in the large-volume limit,

$$\log Z = -V \int \log \left(1 - e^{-\beta(p^2/2m - \mu)}\right) \frac{d^3p}{(2\pi\hbar)^3}. \tag{10.33}$$

The extra factor of 2^3 has been introduced because the sum is only over positive n_x, n_y, n_z — that is, over 1/8 of the space — whereas the integral ranges over all of momentum space.

Invoking (10.28), we need to calculate the integral

$$N = \frac{1}{\beta} \frac{\partial \log Z}{\partial \mu} = V \int \frac{e^{-\beta(p^2/2m - \mu)}}{1 - e^{-\beta(p^2/2m - \mu)}} \frac{d^3p}{(2\pi\hbar)^3}. \tag{10.34}$$

Given that the integrand is a function only of p^2, this is best evaluated in spherical coordinates. After dispatching the angular integrals, we're left with

$$N = \frac{V}{2\pi^2\hbar^3} \int_0^\infty \frac{e^{-\beta(p^2/2m - \mu)} p^2 \, dp}{1 - e^{-\beta(p^2/2m - \mu)}} = \frac{Vm^{3/2}}{\sqrt{2}\,\pi^2\hbar^3} \int_0^\infty \frac{\sqrt{\varepsilon}\, d\varepsilon}{e^{+\beta(\varepsilon - \mu)} - 1}, \tag{10.35}$$

where $\varepsilon = p^2/2m$. This integral has the form considered in Example 8.20; after working through Problem 10.3, you'll find

$$N = V \left(\frac{m}{2\pi\hbar^2\beta}\right)^{3/2} \zeta_\alpha(3/2) = V \left(\frac{mkT}{2\pi\hbar^2}\right)^{3/2} \zeta_\alpha(3/2), \tag{10.36}$$

where we've used $\Gamma(3/2) = \sqrt{\pi}/2$, and $\zeta_\alpha(s) \equiv \mathrm{Li}_s(\alpha)$ is the polylogarithm of Problem 8.9,

$$\zeta_\alpha(s) = \sum_{n=1}^\infty \frac{\alpha^n}{n^s}, \qquad \alpha \equiv e^{+\beta\mu}. \tag{10.37}$$

So we've apparently achieved our goal: for a given N and density $\rho = N/V$, we can sweep through a range of T values and use

$$\zeta_\alpha(3/2) = \left(\frac{2\pi\hbar^2}{mk}\right)^{3/2} \rho \, T^{-3/2} \tag{10.38}$$

to solve for $\mu = \frac{1}{\beta} \log \alpha$ at each step. The result, shown in Figure 10.2, reveals that $\mu < 0$ for T greater than some critical value T_c (defined below).

But something's clearly not right: though (10.37) and (10.38) dictate that $\alpha = e^{\beta\mu}$ grows as T falls, $\zeta_\alpha(3/2)$ only converges for $\alpha \leq 1$. So our calculation in fact *fails* for temperatures below the critical value

$$T_c = \frac{2\pi\hbar^2}{mk} \left(\frac{\rho}{2.612}\right)^{2/3}, \tag{10.39}$$

where we've used (10.38) with $\zeta_1(3/2) = 2.612$ (see Table 8.1). This makes no sense; somewhere there's an error. As we'll now see, the problem is that we've focused on μ, whereas the relevant quantity is $\varepsilon_a - \mu$.

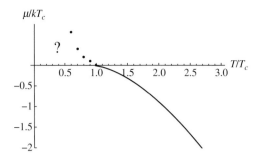

Figure 10.2 The chemical potential $\mu(T)$ for an ideal boson gas. The question mark indicates its unknown behavior below T_c.

Let's apply some physical intuition. Since a state ε_a can accommodate any number of bosons, then as $T \to 0$ we expect all N particles to end up in the ground state ε_1; from (10.26), this is the condition

$$\langle n_1 \rangle_{T \to 0} = \lim_{\beta \to \infty} \frac{1}{e^{+\beta(\varepsilon_1 - \mu)} - 1} = N. \tag{10.40}$$

Now regardless of temperature, $\langle n \rangle$ must be positive or zero. So it must always be true that $\varepsilon - \mu > 0$. But since N is a large number, (10.40) requires $e^{+\beta(\varepsilon_1 - \mu)} - 1 \ll 1$. Taken together, we see that for small temperature, μ must be less than, but very close to, ε_1. Thus for low T we can replace the exponential with its first-order expansion $e^x \approx 1 + x$, so that

$$\langle n_1 \rangle_{T \to 0} = \frac{1}{\beta(\varepsilon_1 - \mu)} = N, \tag{10.41}$$

and therefore for small T (Figure 10.3)

$$\mu(T) = \varepsilon_1 - \frac{1}{N\beta} = \varepsilon_1 - \frac{kT}{N}. \tag{10.42}$$

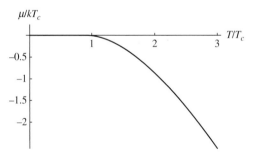

Figure 10.3 The chemical potential for an ideal boson gas. On this scale, $\mu(T < T_c) \approx 0$.

Given the rapid growth of e^x, for sufficiently low temperature (large β) no other energy $\varepsilon_{a>1}$ makes a contribution comparable to that of the ground state. Let's run some numbers to get a sense of scale. In an experiment with a gas of ^{87}Rb atoms in a box $\sim 10^{-5}$ m on a side, (10.31) gives a ground state energy of about 7.3×10^{-14} eV, and the first excited state twice that. But with $N \sim 2 \times 10^4$ in the box, T_c is about 40 nK; at that temperature, $\varepsilon_1 - \mu \approx 6 \times 10^{-16}$, whereas $\varepsilon_2 - \mu \sim \varepsilon_2$. That's a two order of magnitude disparity — and the lower the temperature, the more stark the contrast.

This reveals where our mistake lies: replacing a discrete sum with an integral as we did in (10.33) is only valid when the contributions to the sum are closely spaced. While that continues to be the case above the ground state, at low temperature the near-equality of ε_1 and μ requires that $\langle n_1 \rangle$ be discretely accounted for — so (10.35) must be replaced by

$$N = \langle n_1 \rangle + \sum_{a \geq 2} \langle n_a \rangle = \langle n_1 \rangle + \frac{V m^{3/2}}{\sqrt{2}\,\pi^2 \hbar^3} \int_{\varepsilon_2}^{\infty} \frac{\sqrt{\varepsilon}\,d\varepsilon}{e^{+\beta(\varepsilon-\mu)} - 1}. \tag{10.43}$$

Because μ is much less than all energies $\varepsilon_{a \geq 2}$, we introduce no substantive error by setting $\mu = 0$ in the integral; similarly, the factor of $\sqrt{\varepsilon}$ allows us to replace the lower limit with zero. Then, in contrast to (10.35)–(10.36), we find that as the temperature falls below T_c,

$$N = \langle n_1 \rangle + 2.612 \left(\frac{mkT}{2\pi \hbar^2} \right)^{3/2} V. \tag{10.44}$$

In terms of the critical temperature (10.39), the occupation number for the ground state is

$$\langle n_1 \rangle = N \left[1 - \left(\frac{T}{T_c} \right)^{3/2} \right], \quad T < T_c. \tag{10.45}$$

So as T falls below T_c, a large fraction of the N particles abruptly drop to the ground state: for $T/T_c = 0.99$, 1.5% of the total have energy ε_1, but at the lower temperature $T/T_c = 0.9$, it's 15%, This phenomenon, where below some critical temperature a considerable fraction of the particles coalesce in the ground state, is called *Bose–Einstein condensation*. Einstein predicted this behavior in 1924, but experimental verification did not come until 1995 — a technical tour-de-force for which Cornell, Wieman, and Ketterle shared the 2001 Nobel prize in physics.

Problems

10.1 Verify (10.16) for the single-particle average energy in the Einstein solid.

10.2 (a) Find a derivative expression for $\langle n \rangle$ from (10.21). Then use it to verify (10.26).
  (b) Verify(10.28).

10.3 Evaluate the integral (10.35) to find the expression for N in (10.36).

10.4 Though (10.33) for the ideal boson gas fails for small T, it does correctly describe a classical ideal gas in the large T limit. We'll demonstrate this by extracting the expression for the average energy U in this limit.

  (a) The average energy can be derived from the partition function Z using

$$U = -\frac{\partial \log Z}{\partial \beta} + \mu N,$$

  which is just (10.4) plus the contribution from the chemical potential. Show that this gives U proportional to $\zeta_\alpha(5/2)$, where $\alpha = e^{\beta\mu}$, and $\zeta_\alpha(s)$ is the polylogarithm function of (10.37).
  (b) Use (10.38) to argue that in the classical limit of high temperature and/or low density, the classical expression $U = \frac{3}{2}NkT$ emerges.

10.5 **The Difference a Sign Makes**
  Any number of bosons can occupy a single-particle state. By contrast, *fermions* obey the Pauli exclusion principle, limiting n to either 0 or 1. (Fermions are particles whose spins are half-integer multiples of $\hbar$.) Thus unlike bosons, fermions cannot form a condensate. How does this manifest mathematically?

(a) Show that the fermion occupation number is given by

$$\langle n_a \rangle_\mathrm{F} = \frac{1}{e^{+\beta(\varepsilon_a - \mu)} + 1}.$$

(b) Though the difference between the expressions $\langle n_a \rangle_\mathrm{B}$ and $\langle n_a \rangle_\mathrm{F}$ may appear mathematically minor, the sign flip precludes the formation of a condensate. Instead, an ideal Fermi gas at $T = 0$ has particles stacked in ever-higher single particle energy states; the energy of the highest-energy particle is the system's *Fermi energy* ε_F. Sketch $\langle n_a \rangle_\mathrm{F}$ as a function of ε_a at $T = 0$ and identify ε_F. How is the Fermi energy related to the chemical potential $\mu(T = 0)$?

(c) The sign flip allows the sum over states to be replaced by an integral without difficulty. Adapt (10.35) to find

$$\epsilon_\mathrm{F} = \frac{\hbar^2}{2m} \left(\frac{3\pi^2 N}{V} \right)^{2/3},$$

including an overall factor of 2 to account for both up and down spin states.

Part II

The Calculus of Vector Fields

Mathematics is the music of reason.

James Joseph Sylvester

11 Prelude: Visualizing Vector Fields

One of the most fundamental concepts of mathematical physics is the vector. Whether in the perhaps mundane but powerful second law of Newton; as the abstract fields of gravity, electricity, and magnetism; or the complex and often perplexing wavefunction of quantum mechanics, vectors appear innumerable times in mathematics, physics, and engineering. Over the course of this book, we will discuss different mathematical manifestations and representations of vectors, including vector-valued functions of space and time, column vectors obeying the rules of matrix manipulation, and even functions such as Gaussians, which can be regarded as vectors in an infinite-dimensional space.

In this Part, we explore the calculus of vector fields in the familiar confines of $\mathbb{R}^3$ (good physics references include Griffiths, 2017 and Jackson, 1998). Unlike a scalar field $f(\vec{r})$, which assigns to every point a number (say, the mass or temperature at $\vec{r}$), a vector-valued field $\vec{F}(\vec{r})$ assigns both a number *and* a direction. These numbers and directions can be easily depicted by drawing arrows from representative points.

Example 11.1 $\vec{F}(x,y,z) = \hat{\imath}x + \hat{\jmath}y + \hat{k}z$

Allowing the relative lengths of the arrows to represent the field strength at each point, we easily appreciate that this vector field explodes out from the origin. Though $\vec{F}$ may appear to be algebraically three-dimensional, because of its clear spherical symmetry it is essentially a one-dimensional field — as can be demonstrated by changing to spherical coordinates,

$$\vec{F}(x,y,z) = \hat{\imath}x + \hat{\jmath}y + \hat{k}z = \vec{r} = r\hat{r}, \qquad (11.1)$$

where $\hat{r} \equiv \vec{r}/|\vec{r}|$. So this field has only a radial component. Moreover, in this form it's obvious that the magnitude of $\vec{F}$ depends not on separate values of x, y, and z, but only on the distance r from the origin. Ultimately, this dimensional reduction is possible because $\hat{r}$ is not constant.

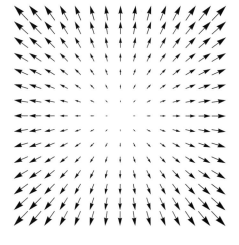

Although depicting vector fields using scaled arrows emphasizes the mapping of a vector to each point in space, you're probably more accustomed to the so-called "streamline" representation. Thinking of the vector function as representing the flow of water, streamlines indicate the overall velocity field: the direction of flow is given by the tangent to the streamline at a point, while the speed of the flow is represented by the density of the streamlines. Consider, for instance, water flowing in a pipe of varying cross section, as

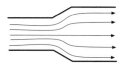

Figure 11.1
Incompressible flow.

depicted in Figure 11.1. Streamlines quickly and intuitively show that water speed is faster where the pipe is narrower. Incidentally, note that streamlines can never cross, since otherwise the field would have two values at that point.

Water is "incompressible" — that is, for any given region in the velocity field, the amount of water flowing in per unit time must equal the amount flowing out. Thus (in a way we'll later make more precise) these streamlines exemplify some form of conservation of water. In fact, the only way to introduce or remove water is with spigots or drains — often called sources or sinks. So streamlines can only begin on sources and end on sinks (or extend to infinity); the number of lines emerging from a source indicates the relative strength of the source.[1]

Consider N streams of water emerging symmetrically from a single spherical sprinkler head (Figure 11.2).[2] Every one of these N streams will also flow through imaginary nested spheres centered on the sprinkler. So the product of a sphere's surface area A with the density σ of streamlines passing through the surface is just the number of streamlines, $N = \sigma A$. Of course N must be the same for all of these nested spheres, so since the surface area of a sphere scales like $A \sim r^2$, the density of the lines must fall as $\sigma \sim 1/r^2$. This gives the velocity field of the water.

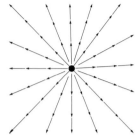

Figure 11.2 The field $\hat{r}/r^2$.

This should sound a lot like the flux of electric field from a point charge. Gauss' law, recall, stipulates that all of the electric field lines emerging from a source must also pass through any surrounding sphere — in other words, that the electric flux through a surface depends only on the source within. Then once again, since the surface area of a sphere scales like r^2, the flux is constant only if the electric field falls as $1/r^2$. So the field of a positive point charge at the origin,

$$\vec{E}(\text{pt charge}) = \frac{\hat{i}x + \hat{j}y + \hat{k}z}{(x^2 + y^2 + z^2)^{3/2}} = \frac{\vec{r}}{r^3} = \frac{\hat{r}}{r^2}, \tag{11.2}$$

has the same behavior as the velocity field of our spherically symmetric sprinkler head. Indeed, the direction of $\vec{E}$ is tangent to the streamlines, and the strength of the field is indicated by the density of the curves. In common parlance, these continuous curves are what are usually meant by the term "field lines" or "lines of force." But we can go further. Since all static $\vec{E}$ field configurations can be assembled by the superposition of point charges, all physical electrostatic fields can be represented this way — even for $\vec{E}$ fields which don't scale as $1/r^2$. For example, the field lines of two equal and opposite charges — the familiar "spider diagram" of a *dipole* (Figure 11.3) — is easily interpreted. The individual field vector at each point is the tangent to the streamline at

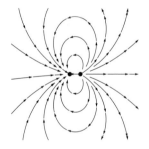

Figure 11.3 Dipole field.

that point, and the strength of the field diminishes with distance (in this case as $1/r^3$). It's all very intuitive.

Since the density of field lines diverges at sources, a streamline representation is usually only straightforward in regions between sources. For a discrete number of point sources, such as the dipole, this is not a problem. But for a continuous source distribution, streamlines can often be difficult to use. As we'll see, fields like the one in Example 11.1 depend on continuous sources — and thus new streamlines emerge at every point within the source region. For these cases, it's easier to depict the fields as individual vectors assigned to each point in space, with the length of a vector reflecting the strength of the field at that point.

---

[1] Sinks, of course, are just negative sources — so without confusion we'll use "source" for both.

[2] The flat page constrains the depiction to two dimensions, which can lead to confusion; remember that Figure 11.2 represents a field in three dimensions.

Example 11.2 $\vec{G}(x,y) = (-\hat{i}y + \hat{j}x)/\sqrt{x^2 + y^2}$

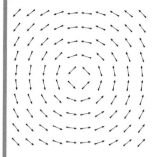

This vector field swirls around the origin, but is undefined there. It has the property that the field strength is constant everywhere. In fact it's a normalized field,

$$\vec{G} \cdot \vec{G} = \frac{1}{x^2 + y^2}\left(y^2 + x^2\right) = 1, \tag{11.3}$$

independent of x and y. $\vec{G}$'s constant magnitude, together with its clear lack of sources, makes the "arrow at each point" depiction essentially identical to a streamline representation of closed concentric circles.

Even were the field to vary in strength, a tidier streamline representation can still be useful. For example, for $\vec{G}/r$, with $r^2 = x^2 + y^2$, we merely space out the closed loops so that their density drops like $1/r$. These are the field lines of the magnetic field of a steady current moving through the origin in the positive z-direction. Algebraically $\vec{G}$ may appear to be two-dimensional, but looking at the field shows that it is essentially a one-dimensional field. This can be demonstrated by changing to polar coordinates,

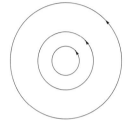

$$\vec{G}(x,y) = \frac{-\hat{i}y + \hat{j}x}{\sqrt{x^2 + y^2}} = \hat{\phi}. \tag{11.4}$$

So the field has only a tangential component. Moreover, we now see that the field depends not on separate values of x and y, but only on the angle ϕ around the origin, $\tan\phi = y/x$. Ultimately, this dimensional reduction is possible because $\hat{\phi}$ is not constant.

Example 11.3 **Finding the Streamlines**

A streamline $\vec{s}(t) = (x(t), y(t), z(t))$ is the directed curve a small particle follows in the velocity field $\vec{v}$ of a fluid. The tangent to the streamline at every point P is the vector field $\vec{v}(P)$ at that point. Thus a streamline $\vec{s}(t)$ — also called an *integral curve* — can be found by solving the differential equation

$$\frac{d\vec{s}}{dt} = \vec{v}\left(\vec{s}(t)\right) \equiv \vec{v}(x(t), y(t), z(t)). \tag{11.5}$$

Consider the two-dimensional $\vec{G}(x,y)$ from the previous example as the velocity field. The streamline must have tangent $\dot{\vec{s}} = (\dot{x}, \dot{y})$ such that

$$\frac{dx}{dt} = G_x = -\frac{y}{\sqrt{x^2 + y^2}} \qquad \frac{dy}{dt} = G_y = +\frac{x}{\sqrt{x^2 + y^2}}. \tag{11.6}$$

These can be solved by dividing the differential equations,

$$\frac{dy/dt}{dx/dt} = \frac{dy}{dx} = -\frac{x}{y} \quad \Longrightarrow \quad \frac{1}{2}y^2 = -\frac{1}{2}x^2 + k,$$

where k is the constant of integration. So, as expected, the streamlines are concentric circles — specifically, curves of constant $f(x,y) = x^2 + y^2$.

Note that, in a sense, a vector field depends on the underlying coordinates *twice*: once to provide the unit vectors (or basis) upon which to decompose the field, and once to determine the functional dependence of the components. So there is no reason, say, that $\hat{\imath}$ need be multiplied by a function only of x, nor $\hat{r}$ by a function only of r. (Indeed, the swirl or circulation of $\vec{G}$ comes from the y-dependence of the $\hat{\imath}$-component and the x-dependence of the $\hat{\jmath}$-component; by contrast, $\vec{F}$ has no swirl at all, only a radial surge or flux.)

We don't even have to use the same coordinate system for both the basis and the functional dependence. For instance, it is perfectly acceptable to express the field $\vec{F}$ in mixed coordinates,

$$\vec{F} = \sqrt{x^2 + y^2 + z^2}\,\hat{r}$$
$$= r\sin\theta\cos\phi\,\hat{\imath} + r\sin\theta\sin\phi\,\hat{\jmath} + r\cos\theta\,\hat{k}. \tag{11.7}$$

We used the mixed coordinates of (2.4) in our derivation of $\{\hat{\rho},\hat{\phi}\}$ as functions of $\{\hat{\imath},\hat{\jmath}\}$ in Part I, taking advantage of the vanishing derivatives of the cartesian basis vectors. Similarly, integrating a vector decomposed into cartesian components allows the constants $\{\hat{\imath},\hat{\jmath},\hat{k}\}$ to be taken out of the integral. This does not, however, preclude the components from being expressed in another coordinate system for which the integration may be easier.

Problems

11.1 Sketch the field lines for the following vector fields in two dimensions:

(a) $\hat{\rho}/\rho$, where $\rho^2 = x^2 + y^2$
(b) $y\hat{\imath} + x\hat{\jmath}$.

11.2 Find the parameterized streamlines $\vec{s}(t)$ for the field $\vec{v} = (-y, x)$.

11.3 Construct the streamlines $y(x)$ for $\vec{E} = xy^2\hat{\imath} + x^2 y\hat{\jmath}$ and $\vec{B} = x^2 y\hat{\imath} - xy^2\hat{\jmath}$. Show that they are mutually perpendicular by comparing their slopes.

11.4 Consider the electric field of a point charge, $\vec{E}(\vec{r}) = \hat{r}/r^2$, and the integral

$$\int_V \vec{E}(\vec{r})\,d^3x,$$

where the volume V is the hemisphere $x^2 + y^2 + z^2 = R^2$ with $z > 0$.

(a) Express (but do not evaluate) the definite integral first in spherical coordinates using spherical components, and then in cartesian coordinates using cartesian components.
(b) Now express the definite integral in spherical coordinates but using cartesian components. Explain why this choice is more practical for evaluating the integral than either of the "pure" forms in part (a).
(c) Use symmetry arguments to deduce the integrals of the $\hat{\imath}$- and $\hat{\jmath}$-components. Then explicitly evaluate these integrals to verify your expectations.
(d) Integrate the z-component. Why does the symmetry argument used in (c) not apply to the z-component?

11.5 Use the vector-valued integral

$$\vec{R}_{\text{cm}} = \frac{1}{M}\int_S \vec{r}\,dm$$

to find the center of mass of a hollow half-pipe with length ℓ, radius R, mass M, and uniform surface density σ. Take the origin at the center of the pipe.

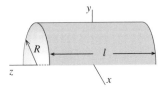

12 Grad, Div, and Curl

12.1 The Del Operator

Consider a scalar function $f(\vec{r})$. According to the chain rule, the value of the function an infinitesimal distance $d\vec{r} \equiv (dx, dy, dz)$ away changes by the infinitesimal amount

$$df = \left(\frac{\partial f}{\partial x}\right) dx + \left(\frac{\partial f}{\partial y}\right) dy + \left(\frac{\partial f}{\partial z}\right) dz. \tag{12.1}$$

We can introduce notation to make this look suspiciously like a dot product:

$$df = \left(\frac{\partial f}{\partial x}, \frac{\partial f}{\partial y}, \frac{\partial f}{\partial z}\right) \cdot (dx, dy, dz) \tag{12.2}$$

or

$$df = \vec{\nabla} f \cdot d\vec{r}, \tag{12.3}$$

where we have defined

$$\vec{\nabla} f \equiv \left(\frac{\partial f}{\partial x}, \frac{\partial f}{\partial y}, \frac{\partial f}{\partial z}\right) = \hat{\imath} \frac{\partial f}{\partial x} + \hat{\jmath} \frac{\partial f}{\partial y} + \hat{k} \frac{\partial f}{\partial z}. \tag{12.4}$$

This is more than mere notation. Since the function f is a scalar, the differential df is also a scalar; similarly, since the position $\vec{r}$ is a vector, the differential $d\vec{r}$ is also a vector. So we can interpret (12.3) as an operation which takes the vector $d\vec{r}$ and returns the scalar df. But the only way to extract a scalar from a vector is by taking the dot product with another vector. Thus $\vec{\nabla} f$, called the *gradient* of f, must be a vector. Writing

$$df = \vec{\nabla} f \cdot d\vec{r} = |\vec{\nabla} f||d\vec{r}| \cos\theta, \tag{12.5}$$

we see that when moving a fixed distance $|d\vec{r}|$ from some point, the amount by which f changes is determined by the direction one chooses to go. The rate of increase in any direction $\hat{u}$ is given by the so-called *directional derivative* $\vec{\nabla} f \cdot \hat{u}$. In other words, the directional derivative is the component of the gradient in a given direction. Since $|\cos\theta| \leq 1$, the largest change in f occurs for $\theta = 0$, *along the direction of* $\vec{\nabla} f$. So *the gradient of a function points in the direction of maximum increase of the function*. The maximal *rate* of increase is just $|\vec{\nabla} f|$.

Thus we discover that any differentiable scalar field f has a natural vector function associated with it: the gradient field $\vec{\nabla} f$. Physical examples abound: the elevation ϕ of terrain, the temperature distribution T in a room, the voltage V of a region of space — these are all scalar fields. Collectively such scalar fields are often called *potentials*. The gradient at every point gives the magnitude and direction of the greatest increase in potential. Thus $\vec{\nabla}\phi$ is the steepest way up the hill; $-\vec{\nabla}T$ gives the direction of heat flow — which, as you know, is always from hot to cold; similarly, the electric field $\vec{E} \equiv -\vec{\nabla}V$, so that $\vec{E}$ points in the direction of maximum *decrease* of V.[1]

(Although every differentiable scalar function has a gradient field, one of the questions addressed in this Part is whether the converse is true: can a vector field be expressed as the gradient of a scalar field? And the answer is a definite "maybe." Stay tuned.)

A good way to visualize a scalar field and its gradient is with a contour map, in which closed curves or surfaces indicate regions of constant potential — that is, regions in which $df = 0$. In other words, moving a distance $d\vec{r}$ along an "equipotential" will not produce a change in f. According to (12.5), then, equipotentials must be perpendicular to their gradients. With this observation, one can easily sketch the gradient field on a contour map or vice-versa.

Example 12.1 $\phi(x,y) = \sqrt{x^2 + y^2}$

We can think of this as the elevation of a circularly symmetric valley centered at the origin (Figure 12.1). To find the easiest climb out, we look at the gradient of the elevation,

$$\vec{\nabla}\phi = \frac{\hat{i}\,x + \hat{j}\,y}{\sqrt{x^2 + y^2}} = \hat{r}, \tag{12.6}$$

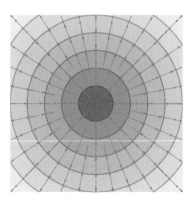

Figure 12.1 Contour lines of ϕ and field lines of $\vec{\nabla}\phi$.

a normalized field in the radial direction, independent of ϕ. Thus it makes no difference which way you choose to climb out: they're all equally steep. Note that the contours of constant ϕ are concentric circles and that the gradient field is perpendicular to them. Also, since the gradient points toward increasing potential, we easily deduce that in this contour map the darker shading indicates lower elevation.

Example 12.2 $V(x,y) = x^2 \cosh y$

The electric field associated with this voltage is

$$\vec{E} = -\vec{\nabla}V = -\hat{i}\,2x\cosh y - \hat{j}\,x^2 \sinh y. \tag{12.7}$$

---

[1] Thus $\vec{E}$ gives the direction of acceleration of a positive test charge.

As expected, the lines of $\vec{E}$ are perpendicular to the contours of constant V (Figure 12.2). So if you were only given the equipotentials, you could sketch the electric field by drawing field lines toward smaller V, taking care to ensure that they cross the equipotentials at right angles. Conversely, you could sketch the contour lines given a plot of the electric field.

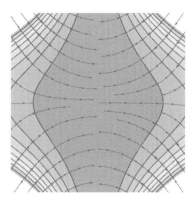

Figure 12.2 Contours of V and field lines of $\vec{E} = -\vec{\nabla} V$.

Example 12.3 **Lagrange Multiplier**

The perpendicular relationship between the gradient and equipotentials provides insight into so-called constrained optimization problems. Suppose, for example, we need to find a point P on the parabola $g(x, y) = 2x + y^2 - 4 = 0$ where the scalar field $f(x, y) = \sqrt{4x^2 + y^2}$ has the smallest value. The usual max/min approach is to solve for the points at which $\vec{\nabla} f = 0$. But that yields the origin — which, though clearly a global minimum of f, is not on the parabola. So how do we take into account the constraint $g(x, y) = 0$?

Moving infinitesimally away from an extremum does not affect the value of the field; this is why extrema are often called stationary points. But since we're constrained to move along the curve $g(x, y) = 0$, we only want points P at which $f(P)$ is stationary along the parabola. In other words, at P the curve $g(x, y) = 0$ must be tangent to an equipotential $f(x, y) = $ constant. And if the tangents of f and g are parallel, so too are their gradients. Thus rather than looking for where the gradient of f vanishes, we instead want the point(s) at which it's proportional to the gradient of g, $\vec{\nabla} f = \lambda \vec{\nabla} g$. So an extremum of f on the curve g is a solution to

$$\vec{\nabla}(f - \lambda g) = 0. \tag{12.8}$$

The parameter λ is called the *Lagrange multiplier*, the value of which can be found in the course of solving the problem. In our case, (12.8) gives the pair of equations

$$\frac{4x}{\sqrt{4x^2 + y^2}} = 2\lambda \quad \text{and} \quad \frac{y}{\sqrt{4x^2 + y^2}} = 2\lambda y. \tag{12.9}$$

One obvious solution on the parabola $y^2 = 4 - 2x$ is $x = 2, y = 0$ (and $\lambda = 1$), at which $f(x, y) = 4$. Another pair of solutions are the points $(1/4, \pm\sqrt{7/2})$, with $\lambda = 1/\sqrt{15}$. At these points, $f = \sqrt{15}/2 < 4$. On the parabola, then, the function reaches it minimum value at $(1/4, \pm\sqrt{7/2})$, as shown in Figure 12.3.

Of course we could have just solved $\frac{d}{dx}f(x, \sqrt{4 - 2x}) = 0$, but the Lagrange multiplier technique readily generalizes to higher dimensions (Problem 12.5).

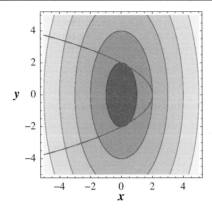

Figure 12.3 The parabola $2x + y^2 - 4 = 0$ superimposed on equipotentials of $\sqrt{4x^2 + y^2}$. Extremal points occur where g is tangent to an equipotential.

The argument we used to establish that $\vec{\nabla}f$ is a vector does not depend upon any particular differentiable f: the gradient of *any* scalar function is a vector. Indeed, "factoring out" f from (12.4),

$$\vec{\nabla}f \equiv \left(\frac{\partial}{\partial x}, \frac{\partial}{\partial y}, \frac{\partial}{\partial z} \right) f , \tag{12.10}$$

encourages us to define the gradient *operator* (often called "del") as

$$\vec{\nabla} \equiv \hat{\imath}\frac{\partial}{\partial x} + \hat{\jmath}\frac{\partial}{\partial y} + \hat{k}\frac{\partial}{\partial z} . \tag{12.11}$$

By itself, $\vec{\nabla}$ has no defined *value* at any point in space. Indeed, it is not a function but rather an *operator* which acts on some function to produce another function. On a scalar function, for instance, del acts simply: $\vec{\nabla}(f) = \vec{\nabla}f$. This, as we've just seen, is the gradient. On a *vector* function, the $\vec{\nabla}$ operator can act in two distinct ways (but see BTW 12.1):

1. Using the dot product we can construct a scalar field called the *divergence* of $\vec{A}$, denoted $\vec{\nabla} \cdot \vec{A}$. In cartesian coordinates, the components combine as one would expect,[2]

$$\vec{\nabla} \cdot \vec{A} = \partial_x A_x + \partial_y A_y + \partial_z A_z. \tag{12.12}$$

An important special case is when $\vec{A}$ itself is the gradient of a scalar, $\vec{A} = \vec{\nabla}f$:

$$\vec{\nabla} \cdot (\vec{\nabla}f) = (\vec{\nabla} \cdot \vec{\nabla})f \equiv \nabla^2 f. \tag{12.13}$$

In cartesian coordinates,

$$\nabla^2 \equiv \vec{\nabla} \cdot \vec{\nabla} = \partial_x^2 + \partial_y^2 + \partial_z^2 , \tag{12.14}$$

as you can easily work out. This scalar operator occurs so often that it has its own name: the *Laplacian*.

---

[2] In this expression we've introduced the convenient short-hand $\partial_\alpha \equiv \partial/\partial\alpha$.

2. The second way del can act on a vector uses the cross product to construct the vector field $\vec{\nabla} \times \vec{A}$, which is called the *curl* of $\vec{A}$. In cartesian coordinates, it takes the form

$$\vec{\nabla} \times \vec{A} = \sum_{ijk} \epsilon_{ijk} \hat{e}_i \partial_j A_k$$

$$= \hat{\imath}(\partial_y A_z - \partial_z A_y) + \hat{\jmath}(\partial_z A_x - \partial_x A_z) + \hat{k}(\partial_x A_y - \partial_y A_x). \tag{12.15}$$

Example 12.4 $\vec{v}_1 = \hat{\imath}xy + \hat{\jmath}y^2$

The derivatives are elementary:

$$\vec{\nabla} \cdot \vec{v}_1 = y + 2y = 3y, \tag{12.16}$$

and, since $\vec{v}_1$ is neither functionally dependent on z nor has a z-component, most of (12.15) vanishes, leaving only

$$\vec{\nabla} \times \vec{v}_1 = -x\,\hat{k}. \tag{12.17}$$

Example 12.5 $\vec{v}_2 = \frac{1}{x^2+y^2+z^2}\left(x\hat{\imath} + y\hat{\jmath} + z\hat{k}\right)$

Because each of the variables is used exactly the same way, this is actually much easier than it may first appear. Calculating

$$\partial_x\left(\frac{1}{x^2 + y^2 + z^2}\right) = \frac{-2x}{(x^2 + y^2 + z^2)^2} \tag{12.18}$$

immediately tells us that

$$\partial_y\left(\frac{1}{x^2 + y^2 + z^2}\right) = \frac{-2y}{(x^2 + y^2 + z^2)^2}$$

$$\partial_z\left(\frac{1}{x^2 + y^2 + z^2}\right) = \frac{-2z}{(x^2 + y^2 + z^2)^2}. \tag{12.19}$$

Thus

$$\vec{\nabla} \cdot \vec{v}_2 = \partial_x v_{2x} + \partial_y v_{2y} + \partial_z v_{2z}$$

$$= \frac{1}{(x^2 + y^2 + z^2)^2}\left[\left(x^2 + y^2 + z^2 - 2x^2\right) + \left(x^2 + y^2 + z^2 - 2y^2\right)\right.$$

$$\left. + \left(x^2 + y^2 + z^2 - 2z^2\right)\right]$$

$$= \frac{1}{(x^2 + y^2 + z^2)}. \tag{12.20}$$

The field $\vec{v}_2$ can be written as the gradient of scalar; you should check that $\vec{v}_2 = \vec{\nabla}f$, where

$$f = \frac{1}{2}\ln(x^2 + y^2 + z^2), \tag{12.21}$$

and then verify that $\vec{\nabla} \cdot \vec{v}_2 = \nabla^2 f$.

For the curl, start with the x-component:

$$\left(\vec{\nabla} \times \vec{v}_2\right)_x = \partial_y v_{2z} - \partial_z v_{2y}$$

$$= \frac{1}{(x^2 + y^2 + z^2)^2} \left[(-z) \cdot (2y) - (-y) \cdot (2z)\right] = 0. \tag{12.22}$$

By symmetry, the other two components of the curl will also vanish; thus $\vec{\nabla} \times \vec{v}_2 = 0$.

The divergence and curl of $\vec{v}_2$ were straightforward due to the symmetry of the field — that is, the function makes no real distinction between x, y, and z. Since neither does the divergence, the calculation of $\vec{\nabla} \cdot \vec{v}_2$ was pretty simple. Similarly, the curl is a completely *anti*symmetric operation, as should be obvious from (12.15). It's this property that colludes with the symmetry of $\vec{v}_2$ to yield a vanishing curl.

Always keep in mind the inherent symmetry of the divergence and antisymmetry of the curl; some results can be determined from these properties alone. For instance, you should (soon) be able to tell at a glance that a field like $\vec{u} = (yz, zx, xy)$ has vanishing divergence and $\vec{v} = (x^2, y^2, z^2)$ zero curl. (What about $\vec{\nabla} \times \vec{u}$ and $\vec{\nabla} \cdot \vec{v}$?)

12.2 $\vec{\nabla}$ and Vector Identities

In addition to grad, div, and curl, we can also use $\vec{\nabla}$ to form combinations such as $f\vec{\nabla}$ and $\vec{v} \cdot \vec{\nabla}$. Though perfectly legitimate expressions, they are not functions: their derivatives remain "unfulfilled," always looking for a function to differentiate. These, then, are *operator* expressions — vector and scalar, respectively. Each yields a function only when there is another function on its right to act upon. Indeed,

$$\vec{\nabla}f \neq f\vec{\nabla} \quad \text{and} \quad \vec{\nabla} \cdot \vec{v} \neq \vec{v} \cdot \vec{\nabla}.$$

Similarly, $\vec{\nabla} \times \vec{v} \neq \vec{v} \times \vec{\nabla}$. This is a fundamental algebraic distinction between vector functions and vector operators: in general they do not commute. As a result, algebraic manipulation of vector operators must be done with care. In general, you *can* manipulate $\vec{\nabla}$'s like any other vector, as long as you respect the integrity of the derivative — as we'll see in the following examples.

Example 12.6 *BAC − CAB*

Consider the vector field $\vec{\nabla} \times (\vec{\nabla} \times \vec{v})$. We'd like to try to simplify this expression using the "BAC-CAB" vector identity

$$\vec{A} \times (\vec{B} \times \vec{C}) = \vec{B}(\vec{A} \cdot \vec{C}) - \vec{C}(\vec{A} \cdot \vec{B}). \tag{12.23}$$

Identifying $\vec{A} \rightarrow \vec{\nabla}, \vec{B} \rightarrow \vec{\nabla}$, and $\vec{C} \rightarrow \vec{v}$, we get

$$\vec{\nabla} \times (\vec{\nabla} \times \vec{v}) = \vec{\nabla}(\vec{\nabla} \cdot \vec{v}) - \vec{v}(\vec{\nabla} \cdot \vec{\nabla}). \tag{12.24}$$

But this is clearly wrong: the left-hand side is a vector *function*, whereas the right-hand side is the difference of the vector function $\vec{\nabla}(\vec{\nabla} \cdot \vec{v})$ and a vector times a scalar *operator* $\vec{v}(\vec{\nabla} \cdot \vec{\nabla})$. All three terms in (12.24) are vectors, but the integrity of the derivative has been violated. Of course, for vector functions we can trivially rewrite the identity (12.23) as

$$\vec{A} \times (\vec{B} \times \vec{C}) = \vec{B}(\vec{A} \cdot \vec{C}) - (\vec{A} \cdot \vec{B})\vec{C} \tag{12.25}$$

(although "BAC-ABC" is not as mellifluous as "BAC-CAB"). Applying this to our problem respects the derivatives and so gives a satisfactory — and correct — result (as you can check by working out the components):

$$\vec{\nabla} \times (\vec{\nabla} \times \vec{v}) = \vec{\nabla}(\vec{\nabla} \cdot \vec{v}) - \vec{\nabla} \cdot \vec{\nabla}\vec{v} = \vec{\nabla}(\vec{\nabla} \cdot \vec{v}) - \nabla^2\vec{v}. \tag{12.26}$$

Example 12.7 $\vec{\nabla} \times \vec{\nabla}f$ and $\vec{\nabla} \cdot (\vec{\nabla} \times \vec{v})$

Since crossing a vector with itself gives zero, we're tempted to assert

$$\vec{\nabla} \times \vec{\nabla}f = (\vec{\nabla} \times \vec{\nabla})f \equiv 0. \tag{12.27a}$$

In this case, pairing the two $\vec{\nabla}$'s has no effect on the differentiation implied in the expression, so the conclusion should be valid. It is straightforward to verify the expression by working out the components explicitly. (We're assuming equality of mixed partials, $\partial_x\partial_y f = \partial_y\partial_x f$.)

Similarly, since $\vec{A} \cdot (\vec{A} \times \vec{B}) \equiv 0$, we're tempted to conclude

$$\vec{\nabla} \cdot (\vec{\nabla} \times \vec{v}) \equiv 0. \tag{12.27b}$$

Once again, use of the vector identity has no effect on the differentiation so the conclusion should be valid. It is straightforward to verify the expression by working out the components explicitly (again assuming equality of mixed partials).

From (12.27) we discover that both *the curl of a gradient and the divergence of a curl vanish.* Though these statements may appear trivial, they have very important mathematical and physical significance. As we'll discuss in more detail later, a vector field $\vec{E}$ which has vanishing curl can be written as the gradient of some scalar function Φ,

$$\vec{\nabla} \times \vec{E} = 0 \implies \vec{E} = \vec{\nabla}\Phi. \tag{12.28}$$

For example, the curl of an electro*static* field is identically zero; thus $\vec{E}$ can be regarded as the gradient of a scalar function called the *scalar potential*. (Physics convention defines voltage as $V \equiv -\Phi$.)

Similarly, a vector field $\vec{B}$ has zero divergence can be written as the curl of some vector field $\vec{A}$,

$$\vec{\nabla} \cdot \vec{B} = 0 \implies \vec{B} = \vec{\nabla} \times \vec{A}. \tag{12.29}$$

For instance, the divergence of the magnetic field is always zero; thus $\vec{B}$ can be generated from a vector function $\vec{A}$ called the *vector potential*.

As a derivative operator, $\vec{\nabla}$ obeys the familiar product rule,

$$\vec{\nabla}(fg) = (\vec{\nabla}f)g + f(\vec{\nabla}g). \tag{12.30}$$

Indeed, just as $\vec{\nabla}$ obeys vector identities modified to account for the integrity of the derivative, so too $\vec{\nabla}$ obeys the usual rules of calculus provided its vector nature is respected. So, for example, given the product of a scalar and a vector field $f\vec{A}$, we find

$$\begin{aligned}
\vec{\nabla} \cdot (f\vec{A}) &= \partial_x(fA_x) + \partial_y(fA_y) + \partial_z(fA_z) \\
&= (\partial_x f)A_x + f(\partial_x A_x) + (\partial_y f)A_y + f(\partial_y A_y) + (\partial_z f)A_z + f(\partial_z A_z) \\
&= (\vec{\nabla}f) \cdot \vec{A} + f(\vec{\nabla} \cdot \vec{A}).
\end{aligned} \tag{12.31}$$

Although we derived this using cartesian coordinates, the result holds in any coordinate system. In fact, we could have easily guessed this result by applying the product rule in a way that preserves the scalar property of the divergence: the right-hand side of (12.31) is the only way to use the product rule *and* produce a scalar. In this way, we can correctly surmise the result of other gradient identities in a similar fashion. For instance,

$$\vec{\nabla} \times (\vec{A}f) = f\,(\vec{\nabla} \times \vec{A}) + (\vec{\nabla}f) \times \vec{A} = (\vec{\nabla} \times \vec{A})f \; - \vec{A} \times (\vec{\nabla}f), \tag{12.32}$$

where the minus sign is due to the antisymmetry of the cross product. Sometimes this minus sign is unexpected — for example, in

$$\vec{\nabla} \cdot (\vec{A} \times \vec{B}) = \vec{B} \cdot (\vec{\nabla} \times \vec{A}) - \vec{A} \cdot (\vec{\nabla} \times \vec{B}), \tag{12.33}$$

which is the only way get a scalar-valued identity that also respects the antisymmetry of the cross product, $\vec{\nabla} \cdot (\vec{A} \times \vec{B}) = -\vec{\nabla} \cdot (\vec{B} \times \vec{A})$.

Example 12.8 Self-consistency

Let's demonstrate that (12.32) and (12.33) are consistent with the identities of Example 12.7. With $\vec{A} \to \vec{\nabla}$ and recasting (12.32) as $\vec{\nabla} \times (\vec{\nabla}f)$ we get

$$\vec{\nabla} \times (\vec{\nabla}f) = \vec{\nabla} \times \vec{\nabla}f - \vec{\nabla} \times \vec{\nabla}f = 0, \tag{12.34}$$

which is (12.27a). Similarly for (12.33),

$$\vec{\nabla} \cdot (\vec{\nabla} \times \vec{B}) = (\vec{\nabla} \times \vec{\nabla}) \cdot \vec{B} - \vec{\nabla} \cdot (\vec{\nabla} \times \vec{B}) = -\vec{\nabla} \cdot (\vec{\nabla} \times \vec{B}). \tag{12.35}$$

Since only zero equals its own negative, we recover the identity of (12.27b).

Example 12.9 $\vec{\nabla}(\vec{A} \cdot \vec{B})$

Perhaps the most surprising-at-first-glance identity is the gradient of a dot product. A common (and erroneous) guess is often something like

$$(\vec{\nabla} \cdot \vec{A})\vec{B} + \vec{A}(\vec{\nabla} \cdot \vec{B}),$$

but a straightforward component-by-component rendering shows that this isn't correct. We could instead write

$$\vec{\nabla}(\vec{A} \cdot \vec{B}) = \vec{\nabla}_A(\vec{A} \cdot \vec{B}) + \vec{\nabla}_B(\vec{A} \cdot \vec{B}), \tag{12.36}$$

where we have introduced the subscripts to emphasize which vector field is being differentiated. But this expression is awkward at best.

We can find a useful identity by clever application of BAC-CAB. Consider $\vec{A} \times (\vec{\nabla} \times \vec{B})$. Noting that the gradient operates *only* on $\vec{B}$, we find

$$\vec{A} \times (\vec{\nabla} \times \vec{B}) = \vec{\nabla}_B(\vec{A} \cdot \vec{B}) - (\vec{A} \cdot \vec{\nabla})\vec{B}, \tag{12.37}$$

where we use the subscript only when needed to clarify any ambiguity. A cumbersome expression, to be sure. Of course, we could just as easily have written

$$\vec{B} \times (\vec{\nabla} \times \vec{A}) = \vec{\nabla}_A(\vec{A} \cdot \vec{B}) - (\vec{B} \cdot \vec{\nabla})\vec{A}, \tag{12.38}$$

which is equally awkward. However, from (12.36) we see that the expression we want can be had from the sum of (12.37) and (12.38):

$$\vec{\nabla}(\vec{A} \cdot \vec{B}) = (\vec{A} \cdot \vec{\nabla})\vec{B} + (\vec{B} \cdot \vec{\nabla})\vec{A} + \vec{A} \times (\vec{\nabla} \times \vec{B}) + \vec{B} \times (\vec{\nabla} \times \vec{A}), \qquad (12.39)$$

where now the awkward $\vec{\nabla}_A$ and $\vec{\nabla}_B$ are no longer necessary.

A compilation of these and other common vector derivative identities are collected in Table 12.1.

Table 12.1 Vector derivative identities

For scalar fields f and g and vector fields $\vec{A}$ and $\vec{B}$:

$$\vec{\nabla}(fg) = f\,\vec{\nabla}g + g\,\vec{\nabla}f$$

$$\vec{\nabla} \cdot (f\vec{A}) = f(\vec{\nabla} \cdot \vec{A}) + \vec{\nabla}f \cdot \vec{A}$$

$$\vec{\nabla} \times (f\vec{A}) = f(\vec{\nabla} \times \vec{A}) + \vec{\nabla}f \times \vec{A}$$

$$\vec{\nabla} \times \vec{\nabla}f = 0$$

$$\vec{\nabla} \cdot (\vec{\nabla} \times \vec{A}) = 0$$

$$\vec{\nabla} \cdot (\vec{A} \times \vec{B}) = (\vec{\nabla} \times \vec{A}) \cdot \vec{B} - \vec{A} \cdot (\vec{\nabla} \times \vec{B})$$

$$\vec{\nabla} \times (\vec{A} \times \vec{B}) = (\vec{B} \cdot \vec{\nabla})\vec{A} - \vec{B}(\vec{\nabla} \cdot \vec{A}) + \vec{A}(\vec{\nabla} \cdot \vec{B}) - (\vec{A} \cdot \vec{\nabla})\vec{B}$$

$$\vec{\nabla} \times (\vec{\nabla} \times \vec{A}) = \vec{\nabla}(\vec{\nabla} \cdot \vec{A}) - \nabla^2\vec{A}$$

$$\vec{\nabla}(\vec{A} \cdot \vec{B}) = \vec{A} \times (\vec{\nabla} \times \vec{B}) + \vec{B} \times (\vec{\nabla} \times \vec{A}) + (\vec{A} \cdot \vec{\nabla})\vec{B} + (\vec{B} \cdot \vec{\nabla})\vec{A}$$

12.3 Different Coordinate Systems

Just as vector identities like (12.30)–(12.33) must be valid in any coordinate system, so too derivative operations like grad and div must be independent of coordinate system. For instance, if the curl of a field is found in cartesian coordinates to be $-y\hat{i}+x\hat{j}$, then calculating the curl in cylindrical coordinates must give $r\hat{\phi}$. Similarly, if a curl or divergence vanishes in one coordinate system, it must vanish in all other systems.

Calculating ∇^2, $\vec{\nabla}\cdot$ or $\vec{\nabla}\times$ in other coordinates, however, is not always so straightforward. A simple illustration can be seen using the field $\vec{r} = (x, y, z)$. The divergence is easily computed in cartesian coordinates:

$$\vec{\nabla} \cdot \vec{r} = (\partial_x, \partial_y, \partial_z) \cdot (x, y, z) = 3. \qquad (12.40)$$

In fact, it should be clear that the divergence of position is just the dimensionality of the space. A naive attempt in spherical coordinates, however, suggests

$$\vec{\nabla} \cdot \vec{r} = (\partial_r, \partial_\theta, \partial_\phi) \cdot r\hat{r} = 1, \qquad (12.41)$$

which is obviously wrong. We can isolate the error using the identity (12.31),

$$\vec{\nabla} \cdot \vec{r} = \vec{\nabla} \cdot (r\hat{r}) = (\vec{\nabla}r) \cdot \hat{r} + r(\vec{\nabla} \cdot \hat{r}). \qquad (12.42)$$

Since $\vec{\nabla}r = \hat{r}$, the first term gives the incomplete result of (12.41); to agree with (12.40), then, it must be that $\vec{\nabla} \cdot \hat{r} = 2/r$. So the issue once again is that $\hat{r}$ is not a constant vector.

As discussed earlier, Euclidean systems $\{u_1, u_2, u_3\}$ for which the lines of constant u_i may be curved — and hence have spatially varying $\hat{e}_i$ — are called curvilinear coordinates. To calculate in these non-cartesian systems, note that the components of $\vec{\nabla}$ must have consistent units of inverse distance. So the gradient must be of the form

$$\vec{\nabla} = \hat{e}_1 \frac{\partial}{\partial s_1} + \hat{e}_2 \frac{\partial}{\partial s_2} + \hat{e}_3 \frac{\partial}{\partial s_3}, \qquad (12.43)$$

where $\partial/\partial s_i \equiv \hat{e}_i \cdot \vec{\nabla}$ is the directional derivative and $ds_i \equiv \hat{e}_i \cdot d\vec{r}$ is the scalar displacement in the ith direction. For cartesian coordinates, of course, this just reproduces (12.11). But for cylindrical and spherical, (12.43) solves our problem. Consider the expression back in (2.22) for $d\vec{r}$ in cylindrical coordinates:

$$d\vec{r} = \hat{\rho}\, d\rho + \hat{\phi}\, \rho\, d\phi + \hat{k}\, dz. \qquad (12.44)$$

The displacement along, say, $\hat{\phi}$ is

$$ds_2 = \hat{\phi} \cdot d\vec{r} = \rho\, d\phi \qquad \Longleftrightarrow \qquad \frac{\partial}{\partial s_2} = \frac{1}{\rho} \frac{\partial}{\partial \phi}. \qquad (12.45)$$

Repeating for the other two coordinates results in

$$\vec{\nabla} f = (\partial_\rho f)\, \hat{\rho} + \frac{1}{\rho}(\partial_\phi f)\, \hat{\phi} + (\partial_z f)\, \hat{z}. \qquad (12.46)$$

Similarly, (2.25) in spherical coordinates

$$d\vec{r} = \hat{r}\, dr + \hat{\theta}\, r\, d\theta + \hat{\phi}\, r \sin\theta\, d\phi \qquad (12.47)$$

leads to

$$\vec{\nabla} f = (\partial_r f)\, \hat{r} + \frac{1}{r}(\partial_\theta f)\, \hat{\theta} + \frac{1}{r \sin\theta}(\partial_\phi f)\, \hat{\phi}. \qquad (12.48)$$

Thus we discover that, just as the components of a vector differ in different coordinate systems, so too do the components of the gradient operator. These different expressions for $\vec{\nabla}$ mean that the Laplacian, divergence, and curl take different forms as well. For curvilinear coordinates with scale factors h_1, h_2, and h_3 (Section 2.2), the gradient, lapacian, divergence, and curl are given by

$$\vec{\nabla} = \sum_i \frac{\hat{e}_i}{h_i} \frac{\partial}{\partial u_i} \qquad (12.49a)$$

$$\nabla^2 f = \frac{1}{h_1 h_2 h_3} \left[\partial_1 \left(\frac{h_2 h_3}{h_1} \partial_1 f \right) + \partial_2 \left(\frac{h_1 h_3}{h_2} \partial_2 f \right) + \partial_3 \left(\frac{h_1 h_2}{h_3} \partial_3 f \right) \right] \qquad (12.49b)$$

$$\vec{\nabla} \cdot \vec{A} = \frac{1}{h_1 h_2 h_3} \left[\partial_1 (h_2 h_3 A_1) + \partial_2 (h_3 h_1 A_2) + \partial_3 (h_1 h_2 A_3) \right] \qquad (12.49c)$$

$$\vec{\nabla} \times \vec{A} = \frac{1}{h_1 h_2 h_3} \begin{vmatrix} h_1 \hat{e}_1 & h_2 \hat{e}_2 & h_3 \hat{e}_3 \\ \partial/\partial u_1 & \partial/\partial u_2 & \partial/\partial u_3 \\ h_1 A_1 & h_2 A_2 & h_3 A_3 \end{vmatrix}. \qquad (12.49d)$$

We derive these results in Appendix A; their explicit form in cylindrical and spherical coordinates are given below:

- Cylindrical coordinates (ρ, ϕ, z):

$$\vec{\nabla} f = (\partial_\rho f)\, \hat{\rho} + \frac{1}{\rho}(\partial_\phi f)\, \hat{\phi} + (\partial_z f)\, \hat{z} \tag{12.50a}$$

$$\nabla^2 f = \frac{1}{\rho} \partial_\rho (\rho\, \partial_\rho f) + \frac{1}{\rho^2} \partial_\phi^2 f + \partial_z^2 f \tag{12.50b}$$

$$\vec{\nabla} \cdot \vec{A} = \frac{1}{\rho} \partial_\rho (\rho A_\rho) + \frac{1}{\rho} \partial_\phi A_\phi + \partial_z A_z \tag{12.50c}$$

$$\vec{\nabla} \times \vec{A} = \left[\frac{1}{\rho} \partial_\phi A_z - \partial_z A_\phi \right] \hat{\rho} + \left[\partial_z A_\rho - \partial_\rho A_z \right] \hat{\phi} + \frac{1}{\rho} \left[\partial_\rho (\rho A_\phi) - \partial_\phi A_\rho \right] \hat{z}. \tag{12.50d}$$

- Spherical coordinates (r, θ, ϕ):

$$\vec{\nabla} f = (\partial_r f)\, \hat{r} + \frac{1}{r}(\partial_\theta f)\, \hat{\theta} + \frac{1}{r \sin\theta}(\partial_\phi f)\, \hat{\phi} \tag{12.51a}$$

$$\nabla^2 f = \frac{1}{r^2} \partial_r (r^2 \partial_r f) + \frac{1}{r^2 \sin\theta} \partial_\theta (\sin\theta\, \partial_\theta f) + \frac{1}{r^2 \sin^2\theta} \partial_\phi^2 f \tag{12.51b}$$

$$\vec{\nabla} \cdot \vec{A} = \frac{1}{r^2} \partial_r (r^2 A_r) + \frac{1}{r \sin\theta} \partial_\theta (\sin\theta\, A_\theta) + \frac{1}{r \sin\theta} \partial_\phi A_\phi \tag{12.51c}$$

$$\vec{\nabla} \times \vec{A} = \frac{1}{r \sin\theta} \left[\partial_\theta (\sin\theta\, A_\phi) - \partial_\phi A_\theta \right] \hat{r}$$

$$+ \frac{1}{r} \left[\frac{1}{\sin\theta} \partial_\phi A_r - \partial_r (r A_\phi) \right] \hat{\theta} + \frac{1}{r} \left[\partial_r (r A_\theta) - \partial_\theta A_r \right] \hat{\phi}. \tag{12.51d}$$

In the spirit of "Things You Just Gotta' Know," you should have the cartesian expressions for $\vec{\nabla}$, $\vec{\nabla}\cdot$, $\vec{\nabla}\times$, and ∇^2 on your fingertips; the others will become familiar with use. And though much messier than in cartesian coordinates, their basic structure conforms with expectations:

1. The units are consistent throughout: all terms in grad, div, and curl have units of inverse distance (r or ρ), and all terms in the Laplacian have units of inverse distance-squared.
2. The gradient is a vector, with the u derivative associated with the $\hat{u}$-direction.
3. The divergence looks like a dot product in that it is the sum of terms, each of which is a u derivative of the $\hat{u}$-component. Similarly, the Laplacian is a sum of terms, each of which has two u derivatives.
4. The curl is completely antisymmetric: the u-component is made up of v and w derivatives of the w- and v-components, respectively.
5. Note the dissimilarity of the radial derivatives in the Laplacians (12.50b) and (12.51b) as well as in the divergences (12.50c) and (12.51c). This is easily understood: in spherical coordinates r extends out in three-dimensional space, whereas in cylindrical coordinates ρ is restricted to the plane.

It's worth reiterating that the fields obtained after applying these various derivatives must be the same no matter which coordinate system is used. So one has the freedom to work in any system. Often, problems with high symmetry can be simplified by choosing to work in a coordinate system which takes advantage of the symmetry. This is true despite the fact that non-cartesian unit vectors themselves make different contributions to div, grad, and curl.

Example 12.10 $\vec{v}_1 = \hat{i}xy + \hat{j}y^2$

This is the field of Example 12.4. Expressed in cylindrical coordinates, we find that it points purely in the radial direction,

$$\vec{v}_1 = r^2 \sin\phi\,\hat{r}. \tag{12.52}$$

As a result only one term survives in (12.50c),

$$\vec{\nabla}\cdot\vec{v}_1 = \frac{1}{r}\partial_r(r^3\sin\phi) = 3r\sin\phi. \tag{12.53}$$

Since $y = r\sin\phi$, this agrees with Example 12.4 — as it must. As for the curl, since v_1 has only an r-component and no z dependence, (12.50d) reduces to

$$\vec{\nabla}\times\vec{v}_1 = -\frac{1}{r}\partial_\phi(r^2\sin\phi)\,\hat{k} = -r\cos\phi\,\hat{k}. \tag{12.54}$$

Since $x = r\cos\phi$, this agrees with Example 12.4 — as it must.

Example 12.11 $\vec{v}_2 = \frac{1}{x^2+y^2+z^2}\left(x\hat{i} + y\hat{j} + z\hat{k}\right)$

This is the field from Example 12.5. Expressed in spherical coordinates, we find

$$\vec{v}_2 = \frac{\vec{r}}{r^2} = \frac{\hat{r}}{r}. \tag{12.55}$$

This field has *spherical symmetry:* not only does the field point purely in the radial direction, it depends functionally only on the distance r from the origin, not the angles θ and ϕ around the origin.[3] As a result only one term survives in (12.51c),

$$\vec{\nabla}\cdot\vec{v}_2 = \frac{1}{r^2}\partial_r(r) = \frac{1}{r^2}, \tag{12.56}$$

in agreement with Example 12.5. You should check that $\vec{v}_2 = \vec{\nabla}f$, where

$$f = \ln r, \tag{12.57}$$

and then verify that $\vec{\nabla}\cdot\vec{v}_2 = \nabla^2 f$. As for the curl, the spherical symmetry greatly simplifies: nowhere in (12.51d) is there an r derivative of the r-component. (Recall note 3, above.) Thus

$$\vec{\nabla}\times\vec{v}_2 = 0. \tag{12.58}$$

Example 12.12 **Laplacian of a Vector**

Since $\nabla^2 = \vec{\nabla}\cdot\vec{\nabla}$ is a scalar operator, the Laplacian of a vector is still a vector. So, for instance, in cartesian coordinates

$$\nabla^2\vec{v} \equiv \nabla^2(v_x\hat{i}) + \nabla^2(v_y\hat{j}) + \nabla^2(v_z\hat{k})$$
$$= \nabla^2(v_x)\hat{i} + \nabla^2(v_y)\hat{j} + \nabla^2(v_z)\hat{k}, \tag{12.59}$$

[3] Do you see how this symmetry was manifest in Example 12.5?

where application of the product rule is trivial. Alas, not so for non-cartesian coordinate systems. Though one can certainly derive expressions analogous to (12.50b) and (12.51b), they're messy and not terribly transparent. For example, we could work out

$$\nabla^2 \vec{v} = \vec{\nabla} \cdot (\vec{\nabla}\vec{v}), \tag{12.60}$$

but $\vec{\nabla}\vec{v}$ is a two-indexed object (specifically, a second-rank tensor), one index each for the vector components of $\vec{\nabla}$ and $\vec{v}$ (BTW 12.1). When the Laplacian of a vector is called for, it is better instead to use Example 12.6 and write

$$\nabla^2 \vec{v} = \vec{\nabla}(\vec{\nabla} \cdot \vec{v}) - \vec{\nabla} \times (\vec{\nabla} \times \vec{v}). \tag{12.61}$$

As a vector identity, this expression holds in any coordinate system; moreover, the divergence and curl are usually easier to work with than the tensor $\vec{\nabla}\vec{v}$. Consider, for example, a dipole field in spherical coordinates, $\vec{E} = \frac{1}{r^3}(2\cos\theta\,\hat{r} + \sin\theta\,\hat{\theta})$. Using (12.51c) and (12.51d) we find that both the divergence and curl of $\vec{E}$ vanish — and hence, by (12.61), $\nabla^2 \vec{E} = 0$.

12.4 Understanding ∇^2, $\vec{\nabla}\cdot$, and $\vec{\nabla}\times$

One dimension is so much simpler, having only one derivative operator, and so one measure of change: d/dx. A single application gives a function's slope, a second its concavity. Derivatives in higher dimensions, however, offer a richer array of possibilities; there are simply more ways for a field to vary. As we've seen, a scalar field Φ varies with position — but the change is direction dependent, as specified by the three directional derivatives $\hat{e}_i \cdot \vec{\nabla}\Phi$. A vector field $\vec{F}$ has even more ways to vary since changes in both magnitude *and* direction can depend on direction. Indeed, taking the gradient of each component yields a total of nine measures $\partial_i F_j$ of a field's variation. As we'll see, divergence and curl allow us to collect these quantities in a physically sensible way.

12.4.1 The Laplacian

Arguably, the most important differential equation to know by sight is the one describing simple harmonic motion,

$$\frac{d^2}{dt^2}f(t) = -\omega^2 f(t).$$

In physics and engineering, the variable t typically indicates that we're analyzing the time evolution of a system; then ω is the familiar angular frequency, usually expressed in inverse-seconds, s^{-1} (or equivalently, radians/s). Mathematically, of course, this is irrelevant. And anyway, oscillatory behavior can occur in space as well — though then the equation is usually written

$$\frac{d^2}{dx^2}f(x) = -k^2 f(x), \tag{12.62}$$

where the wave number k is the "frequency in space" (often in inverse-meters, m^{-1}). To write a differential equation for oscillation in three-dimensional space, we need to generalize (12.62) in a way that treats all three spatial variables on an equal footing. This is easily accomplished:

$$\left(\frac{\partial^2}{\partial x^2} + \frac{\partial^2}{\partial y^2} + \frac{\partial^2}{\partial z^2}\right)f(x,y,z) = -(k_x^2 + k_y^2 + k_z^2)f(x,y,z).$$

or simply

$$\nabla^2 f(x, y, z) = -k^2 f(x, y, z).$$ (12.63)

Since $\nabla^2 = \vec{\nabla} \cdot \vec{\nabla}$ is a scalar, $k^2 \equiv \vec{k} \cdot \vec{k}$ must be one as well. Note that if f happens to be a function of only one variable, say $f(x, y, z) \equiv f(x)$, (12.62) trivially re-emerges.

This, then, is the simplest interpretation of the Laplacian: it is the natural generalization of d^2/dx^2 to higher dimensions. In other words, $\nabla^2 f$ is essentially a measure of the concavity of the function. Not surprisingly, though, things are more complicated in higher dimensions. In one dimension, a point P at which both $df/dx = 0$ and $d^2 f/dx^2 > 0$ is a local minimum: $f(P)$ is less than all surrounding values. (For a local maximum, of course, $d^2 f/dx^2 < 0$, and $f(P)$ is greater than surrounding values.) But now consider a point Q in *two* dimensions at which both $\vec{\nabla} f = 0$ and $\nabla^2 f > 0$. If both $\partial_x^2 f$ and $\partial_y^2 f$ are positive, then Q is a local minimum. But it's also possible that $\partial_x^2 f > 0$ while $\partial_y^2 f < 0$. In this case, Q is not a local minimum in the usual sense — though it remains true that $f(Q)$ is less than the *average* of surrounding values.

12.4.2 The Divergence

Let's compare a field $\vec{F}$ at neighboring points both on and off a given streamline. To facilitate this, we'll choose axes for which the field at our reference point P is entirely along x; the tangent to the streamline is thus $\vec{F}(P) = F_x(P)\hat{\imath}$. Then at a distance dx along the streamline from P,

$$F_x(P + \hat{\imath}\, dx) = F_x(P) + \left.\frac{\partial F_x}{\partial x}\right|_P dx.$$ (12.64)

Since we're moving along a tangent to the streamline, this gives the change in the magnitude of $\vec{F}$. For a velocity field of water, positive $\partial_x F_x$ results in more water per unit time flowing out from P than in. Thus $\partial_x F_x$ measures the *surge* in the field as a function of position, which manifests as the strengthening of the field along a streamline.

Although the concept of surge doesn't apply across streamlines, at a perpendicular distance dy from P we have

$$F_y(P + \hat{\jmath}\, dy) = \left.\frac{\partial F_y}{\partial y}\right|_P dy,$$ (12.65)

since $F_y(P) = 0$. So the field lines at these infinitesimally separated points are not parallel: for $\partial_y F_y > 0$, they appear to *spread out* from one another.

Now these changes in flow — both surge and spread — must be the same no matter how we orient our axes. Surge measures change in the water's speed, which is a scalar; and the relative orientation of vectors is given by their dot product, which is also a scalar. Thus for an *arbitrary* orientation of axes, surge and spread are given not separately by $\partial_x F_x|_P$ and $\partial_y F_y|_P$, but rather merged into the scalar derivative $\vec{\nabla} \cdot \vec{F}|_P$. In other words, $\vec{\nabla} \cdot \vec{F}$ summarizes how the field *diverges* from a point, whether as surge or spread. Of course, a field can either diverge from or converge to a given point — but that's just a question of overall sign.

A common intuitive aid is to picture the density of sawdust floating on the water. If the field flow at P creates a change in density, then the divergence there is non-zero — positive or negative depending on whether the density increases or decreases.

Example 12.13 Surge I

The divergence is not simply a measure of the change in a field's magnitude. Consider the constant-magnitude fields $\hat{\imath} + \hat{\jmath}$ and $\hat{r}$ (Figure 12.4).

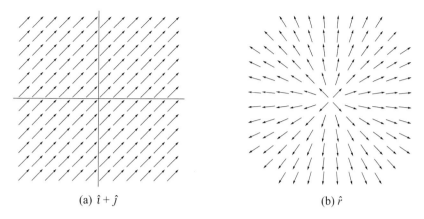

(a) $\hat{\imath} + \hat{\jmath}$ (b) $\hat{r}$

Figure 12.4 Two constant-magnitude fields: (a) $\hat{\imath} + \hat{\jmath}$ and (b) $\hat{r}$.

The first is constant in both magnitude and direction, so all its derivatives are zero. But $\hat{r}$, despite being constant in magnitude, spreads out. Floating sawdust thins as it is spreads, with the rate of spread decreasing further from the origin. And straightforward calculation confirms this: $\vec{\nabla} \cdot \hat{r} = 2/r$.

Example 12.14 Surge II

Though algebraically the fields $x\hat{\imath}$ and $y\hat{\imath}$ are similar, their behavior is very different (Figure 12.5). The surge emerging from every point in $x\hat{\imath}$ is the same, $\vec{\nabla} \cdot (x\hat{\imath}) = 1$. At every point, even along the y-axis, there is the same positive, outward divergence. This surge is responsible for the growth of the field along its streamlines.

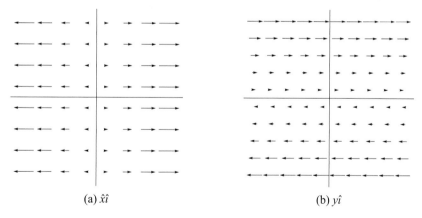

(a) $x\hat{\imath}$ (b) $y\hat{\imath}$

Figure 12.5 Two horizontal fields: (a) $x\hat{\imath}$ and (b) $y\hat{\imath}$.

The field $y\hat{\imath}$ is another matter altogether: it has vanishing divergence. Notwithstanding the change in field strength with y, the density of sawdust does not vary: there is neither surge along streamlines, nor spread between them.

Example 12.15 **Divergent Views!**

Although the concepts of surge and spread are fairly intuitive, the divergence does not distinguish between them. For instance, both $x\hat{i}$ and $y\hat{j}$ have divergence 1, all coming from surge. Since the divergence is linear, the composite field $\vec{v} = x\hat{i} + y\hat{j}$ has constant divergence $\vec{\nabla}\cdot\vec{v} = 2$. Together these non-parallel surges produce spread, as is clear by looking at the same field from a polar coordinate perspective, $\vec{v} = \rho\hat{\rho}$. But divergence is a scalar quantity — a number assigned to each point in space without any directional bias whatsoever.

Similarly, flowing water and sawdust are all well-and-good for developing intuition, but at best these metaphors only yield the sign of $\vec{\nabla}\cdot\vec{F}$ — and even then are no substitute for actually calculating the divergence. Consider, for example, the field $y\hat{i} + x\hat{j}$ (Figure 12.6a). The thinning of sawdust (positive surge) is evident as you follow along a streamline, but so too is the convergence (negative spread) of these lines. There's no reliable way to determine by sight which of these competing effects dominates. In fact, calculating the divergence yields $\vec{\nabla}\cdot\vec{F} = 0$: the positive surge exactly compensates the negative spread everywhere. Another example of ambiguous field lines is $\hat{r}/r^n$ (Figure 12.6b). The spread is obvious; indeed, to the naked eye the field looks much the same for any $n > 0$. But whereas the spread is positive independent of n, the decreasing magnitude means that the surge is negative. The divergence thus depends on n. Sawdust won't help you here. Direct calculation shows that everywhere except the origin (where the field is not defined) the divergence is negative for $n > 2$, positive for $n < 2$, and zero for $n = 2$. (We'll return to the special case $n = 2$ in Example 15.5.)

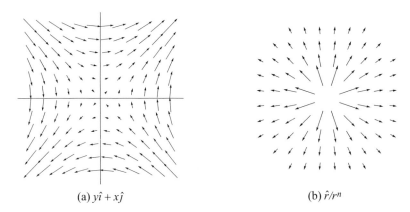

(a) $y\hat{i} + x\hat{j}$ (b) $\hat{r}/r^n$

Figure 12.6 Divergent divergences: (a) $y\hat{i} + x\hat{j}$ and (b) $\hat{r}/r^n$.

Even more insidious in this example is the possibility that $\hat{r}/r^n$ represents a field in two rather than three dimensions. (To be completely consistent, we should perhaps write the planar version as $\hat{\rho}/\rho^2$ — but it's very common to use r for both polar and spherical coordinates.) There is, in essence, less room to spread out in two dimensions — so a positive divergence in three-space may be negative in two.

Water has constant density, so it can only be expanding from a point P if there is a water spigot — a source — at P; similarly, an inward compression indicates a drain — a sink. In other words, surge and spread at P are due to the injection or removal of water at that point.[4] Fields like $r\hat{r}$ (Figure 12.7a)

---

[4] For electric fields, the sources and sinks are electric charges: the divergence is positive at a positive charge, negative at a negative charge. For gravity, mass is always positive — but since it's an attractive force, the divergence of gravitational fields is negative.

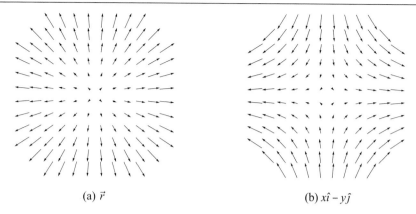

(a) $\vec{r}$ (b) $x\hat{\imath} - y\hat{\jmath}$

Figure 12.7 (a) $\vec{r}$ has uniform source distribution; (b) the solenoidal field $x\hat{\imath} - y\hat{\jmath}$.

with constant divergence require a uniform distribution of sources everywhere in space. But where field lines are straight and uniform like $\hat{\imath} + \hat{\jmath}$ — or, more subtly, a field like $x\hat{\imath} - y\hat{\jmath}$ (Figure 12.7b) in which the surge and spread just balance — there are neither sources nor sinks and the divergence is identically zero. Since these fields have vanishing divergence everywhere, they can describe constant-density flows such as water and are therefore labeled *incompressible*. In an electromagnetic context, fields satisfying $\vec{\nabla} \cdot \vec{F} = 0$ are more commonly referred to as *solenoidal* or simply "divergence free." Finally, although divergence is only defined for vector fields, recall that every differentiable scalar field Φ has a natural vector field associated with it, $\vec{F} = \vec{\nabla}\Phi$. So one should be able to express the divergence of this field in terms of Φ. And indeed, the divergence of the gradient is the Laplacian, $\vec{\nabla} \cdot \vec{F} = \vec{\nabla} \cdot \vec{\nabla}\Phi \equiv \nabla^2\Phi$.

12.4.3 The Curl

Whereas the divergence is solely concerned with the change in the ith component of $\vec{F}$ with x_i, the curl involves the six partials of the form $\partial_i F_j$, $i \neq j$. Rather than an expansion, $\partial_i F_j - \partial_j F_i$ measures the degree with which F_j morphs into F_i. That sort of continuous transformation resembles a rotation — and in fact, $\vec{\nabla} \times \vec{F}$ evaluated at a point P is a measure of the swirl of the vector field at P.

Imagine water rotating with constant angular velocity $\vec{\omega}$. The velocity of the water at $\vec{r}$ is $\vec{v} = \vec{\omega} \times \vec{r}$; a simple calculation reveals

$$\vec{\nabla} \times \vec{v} = 2\vec{\omega}. \tag{12.66}$$

So we can understand the curl as twice[5] the angular speed ω of the water, oriented along the axis of rotation — something easily measured by placing a propeller into the water and watching it spin. The flow-of-water metaphor allows us to interpret $\frac{1}{2}(\vec{\nabla} \times \vec{F}) \cdot \hat{n}$ as the angular speed of a small propeller placed at a point P with its rotation axis along $\hat{n}$. The direction of the curl, then, is along the axis around which the observed ω is greatest. Fields with vanishing curl are said to be *irrotational*.

---

[5] Twice?! See BTW 12.1.

Example 12.16 **Swirl I:** $x\hat{\imath}$ **vs.** $y\hat{\imath}$

Look back at the fields in Figure 12.5. It's not hard to see that a small propeller placed at any point in $x\hat{\imath}$ would not spin. And to be sure, $\vec{\nabla} \times (x\hat{\imath}) = 0$ — as is true of any field of the form $f(x_j)\hat{e}_j$. This may seem obvious, if only because parallel streamlines seem incongruous with any sense of swirl. On the other hand, despite its parallel streamlines, $y\hat{\imath}$ clearly has a uniform, non-zero curl pointing into the page (that is, a clockwise rotation), $\vec{\nabla} \times (y\hat{\imath}) = -\hat{k}$.

Example 12.17 **Swirl II: Significant Signs**

With a curl of $2\hat{k}$, $\vec{F}(x,y) = (x-y)\hat{\imath} + (x+y)\hat{\jmath}$ is a velocity field with unit angular speed $\omega = 1$. Indeed, in Figure 12.8 the counter-clockwise swirl is immediately obvious. As is the behavior of a small propeller at the origin. Further from the origin propeller behavior may be less clear — but as we saw in the previous example, even parallel streamlines can have a curl.

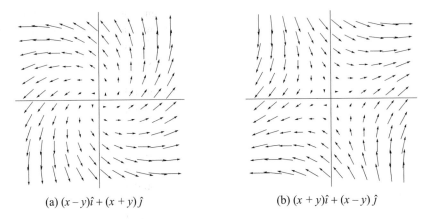

(a) $(x-y)\hat{\imath} + (x+y)\hat{\jmath}$ (b) $(x+y)\hat{\imath} + (x-y)\hat{\jmath}$

Figure 12.8 What's in a sign?

But what a difference a sign makes! Consider the effect of the substitution $y \to -y$. Near the x-axis, $\vec{F}(x,y)$ and $\vec{F}(x,-y)$ differ little. The effect of the sign flip, though, is readily apparent along the y-axis. And the repercussions are clearest at the origin: a small propeller will not rotate in $\vec{F}(x,-y)$. Indeed, the curl is zero everywhere. Though a propeller anchored at arbitrary P may be buffeted about, it will not spin.

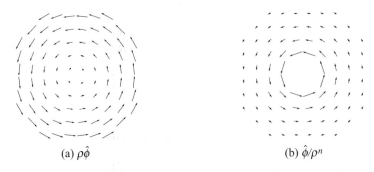

(a) $\rho\hat{\phi}$ (b) $\hat{\phi}/\rho^n$

Figure 12.9 (a) The uniform swirl of $\rho\hat{\phi}$; (b) the irrotational field $\hat{\phi}/\rho^n$.

Example 12.18   **Cautionary Curls!**

Flowing water and propellers are no substitute for actually calculating the curl. With a little insight, however, sometimes we can expose the swirl of a field more clearly. For instance, since terms like $x\hat{i}$ and $y\hat{j}$ make no contribution to the curl, the curl of $(x+y)\hat{i} + (x-y)\hat{j}$ is the same as that of $y\hat{i} + x\hat{j}$. So too, $\vec{\nabla}\times[(x-y)\hat{i} + (x+y)\hat{j}] = \vec{\nabla}\times[-y\hat{i} + x\hat{j}]$. And though the swirl of $-y\hat{i} + x\hat{j}$ may not be algebraically obvious in cartesian coordinates, in polar coordinates the field is $\rho\hat{\phi}$, which has an obvious counter-clockwise swirl pattern (Figure 12.9a). But caution is warranted, since an extended swirl pattern is not a guarantee of a non-zero curl. Consider $\hat{\phi}/\rho^n$ (Figure 12.9b). To the naked eye the field looks much the same for any $n > 0$. But whereas the overall counter-clockwise rotation is independent of n, the decreasing magnitude of the field diminishes the effect. Imagining little propellers won't help you here. Direct calculation shows that everywhere except the origin (where the field is not defined) the curl is negative for $n > 1$, positive for $n < 1$, and zero for $n = 1$.

BTW 12.1 Shear Neglect

In Section 12.1, we blithely stated that divergence and curl are the only ways $\vec{\nabla}$ can act on a vector field $\vec{v}$. But straightforward counting shows that something is missing: taking the gradient of a vector field in three dimensions results in nine independent quantities, $\partial_i v_j$. As a scalar, the divergence is one; as a vector, the curl three more. So somewhere we've neglected five degrees of freedom.

It's not too hard to figure out where to find them. Since the curl is the difference between $\partial_i v_j$ and $\partial_j v_i$, the missing information is given by the sum. Indeed, $\partial_i v_j$ can be divided into symmetric and antisymmetric parts,

$$\partial_i v_j = \frac{1}{2}\left(\partial_i v_j - \partial_j v_i\right) + \frac{1}{2}\left(\partial_i v_j + \partial_j v_i\right), \tag{12.67}$$

or using "double arrow" notation to indicate two-index arrays,

$$\partial_i v_j \rightarrow \vec{\nabla}\vec{v} = \overleftrightarrow{\phi} + \overleftrightarrow{\varepsilon}. \tag{12.68}$$

(As we'll discuss in Chapter 29, these are second-rank tensors.) The antisymmetric piece $\overleftrightarrow{\phi}$ is a matrix representation of the curl,

$$\phi_{ij} = \frac{1}{2}\sum_k \epsilon_{ijk}(\vec{\nabla}\times\vec{v})_k$$

$$\longrightarrow \frac{1}{2}\begin{pmatrix} 0 & (\vec{\nabla}\times\vec{v})_3 & -(\vec{\nabla}\times\vec{v})_2 \\ -(\vec{\nabla}\times\vec{v})_3 & 0 & (\vec{\nabla}\times\vec{v})_1 \\ (\vec{\nabla}\times\vec{v})_2 & -(\vec{\nabla}\times\vec{v})_1 & 0 \end{pmatrix}. \tag{12.69}$$

This describes the swirl of the field — and, incidentally, compensates for the factor of 2 in (12.66). The symmetric piece $\overleftrightarrow{\varepsilon}$ contains the neglected information

$$\overleftrightarrow{\varepsilon} \longrightarrow \frac{1}{2}\begin{pmatrix} 2\partial_1 v_1 & \partial_1 v_2 + \partial_2 v_1 & \partial_1 v_3 + \partial_3 v_1 \\ \partial_1 v_2 + \partial_2 v_1 & 2\partial_2 v_2 & \partial_2 v_3 + \partial_3 v_2 \\ \partial_1 v_3 + \partial_3 v_1 & \partial_2 v_3 + \partial_3 v_2 & 2\partial_3 v_3 \end{pmatrix}. \tag{12.70}$$

Not all of $\overleftrightarrow{\varepsilon}$ is new; its trace is $\vec{\nabla}\cdot\vec{v}$. If we extract this from $\overleftrightarrow{\varepsilon}$,

$$\varepsilon_{ij} \equiv \frac{1}{3}(\vec{\nabla}\cdot\vec{v})\,\delta_{ij} + \tau_{ij}, \tag{12.71}$$

what we're left with is

$$\overleftrightarrow{\tau} \longrightarrow \begin{pmatrix} \partial_1 v_1 - \frac{1}{3}\vec{\nabla}\cdot\vec{v} & \frac{1}{2}(\partial_1 v_2 + \partial_2 v_1) & \frac{1}{2}(\partial_1 v_3 + \partial_3 v_1) \\ \frac{1}{2}(\partial_1 v_2 + \partial_2 v_1) & \partial_2 v_2 - \frac{1}{3}\vec{\nabla}\cdot\vec{v} & \frac{1}{2}(\partial_2 v_3 + \partial_3 v_2) \\ \frac{1}{2}(\partial_1 v_3 + \partial_3 v_1) & \frac{1}{2}(\partial_2 v_3 + \partial_3 v_2) & \partial_3 v_3 - \frac{1}{3}\vec{\nabla}\cdot\vec{v} \end{pmatrix}. \tag{12.72}$$

Now a symmetric 3×3 matrix has six independent elements, but being traceless, $\overleftrightarrow{\tau}$ has only five free parameters — precisely the pieces we're missing. All together then,

$$(\vec{\nabla}\vec{v})_{ij} = \partial_i v_j = \frac{1}{3}\vec{\nabla}\cdot\vec{v}\,\delta_{ij} + \frac{1}{2}\sum_k \epsilon_{ijk}(\vec{\nabla}\times\vec{v})_k + \frac{1}{2}\left(\partial_i v_j + \partial_j v_i - \frac{2}{3}\vec{\nabla}\cdot\vec{v}\,\delta_{ij}\right). \tag{12.73}$$

This so-called "irreducible" decomposition of the nine partial derivatives $\partial_i v_j$ succinctly summarizes the spatial changes of a parcel of fluid. It could, of course, expand or contract; that's the divergence. It could rotate; that's the curl. The only other independent possibility is a constant volume, symmetric deformation; this five-parameter transformation is called *shear*. All spatial changes in a field can be expressed as a combination of these irreducible alternatives. (See Example 15.7.)

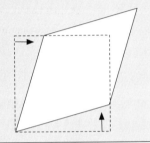

Just as a differentiable scalar function f has associated with it the vector field $\vec{\nabla}f$, so too a differentiable vector function $\vec{F}$ has a natural scalar and vector field, $\vec{\nabla}\cdot\vec{F}$ and $\vec{\nabla}\times\vec{F}$. And like the gradient, it's important to remember that divergence and curl are *local* measures, reflecting the change in density of small distributions of sawdust and the rate of spinning of small propellers around a fixed point. They are not the same as *extended* measures which characterize the overall flux and circulation of a field. Just as with the relationship between local charge density ρ and the total extended charge Q, the connections relating divergence to flux and curl to circulation require summing their local contributions over an extended region. And that means integration. So before we can establish the div/flux, curl/circulation relationships, we first need to discuss the integration of vector fields.

Problems

12.1 Calculate the gradient of the following scalar fields in both cartesian and cylindrical coordinates. Verify that the results are independent of the choice of coordinates.

(a) ρ (b) y (c) $z\rho^2$ (d) $\rho^2 \tan\phi$

12.2 Compute the gradient and Laplacian of the following scalar fields:

(a) $r\sin\theta\cos\phi$ (b) $\ln\rho^2$ (c) $x\cos(y)$ (d) $x(y^2 - 1)$.

12.3 Compute the divergence and curl of the following vector fields:

(a) $\vec{r}$ (b) $\hat{r}/r$ (c) $\hat{\theta}/r^2$ (d) $\rho z\hat{\phi}$.

12.4 Superimpose equipotential contours on the field lines found in Problem 11.1.

12.5 Consider the scalar function $f(\vec{v}) = \sum A_{ij} v_i v_j$, where $\vec{v}$ is an n-dimensional column vector and A is an $n \times n$ symmetric matrix, $A_{ij} = A_{ji}$. With the constraint $\vec{v}\cdot\vec{v} = 1$, we can think of the elements of $\vec{v}$ as the coordinates of a point on the surface of a sphere in n-dimensional space. Show that on the surface of the sphere the stationary points of f are given by vectors $\vec{v}$ satisfying $A\vec{v} = \lambda\vec{v}$, where λ is the Lagrange multiplier.

12.6 Find the curl of $\vec{B} = \frac{-y\hat{\imath}+x\hat{\jmath}}{x^2+y^2}$ in both cartesian and cylindrical coordinates.

12.7 Take the divergence of the gradient of an arbitrary radial function $f(r)$ in spherical coordinates to find an expression for the Laplacian. Verify that the result is in accord with (12.51b).

12.8 Derive the expressions in (12.50) and (12.51) from the general forms in (12.49).

12.9 Find the condition on scalar functions $f(\vec{r})$, $g(\vec{r})$ under which the Laplacian obeys the product rule, $\nabla^2 (fg) = f\nabla^2 g + g\nabla^2 f$.

12.10 Calculate the derivative of the field $\phi(\vec{r}) = x^2 + y^2 + z^2$ along the direction $\vec{v} = \hat{\imath} - 2\hat{\jmath} + \hat{k}$, and evaluate it at $P = (2, 0, -1)$.

12.11 The temperature of a system is described by the function $T(x,y,z) = x^2 y - z^2$. In what direction, and at what rate, does heat flow at point $P = (1, 2, 1)$?

12.12 An origin-centered sphere of radius 2 and the hyperbolic paraboloid $x^2 - y^2 - 2z = 1$ intersect along a curve passing through the point $P = (\sqrt{3}, 0, 1)$. What is the angle between the normals to these surfaces at P?

12.13 A plane can be identified by specifying its normal $\vec{n}$. Use this to find the planes *tangent* to the following surfaces at the given points, expressing your results in the standard form for the equation of a plane, $\alpha x + \beta y + \gamma z = k$. [Hint: What's $\vec{n}$ in terms of α, β, γ?]

(a) $\phi = x^2 + y^2 + z^2 = 6$ at the point $(2, -1, 1)$
(b) $\phi = x^2 - 5y + 4z^2 = 0$ at the point $(3, 5, -2)$

12.14 Use (12.49) to verify the form of the divergence and curl in spherical coordinates.

12.15 **Inspecting for Surge and Swirl**
Use the metaphors of sawdust and propellers to assess the divergence and curl of the fields $\vec{E}$ and $\vec{B}$ depicted in Figure 12.10. Confirm your assessments by calculating the divergences and curls directly.

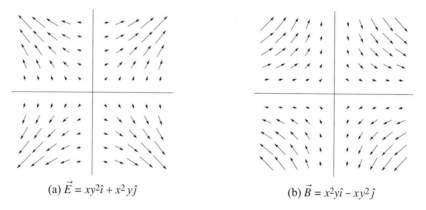

(a) $\vec{E} = xy^2\hat{\imath} + x^2 y\hat{\jmath}$ (b) $\vec{B} = x^2 y\hat{\imath} - xy^2\hat{\jmath}$

Figure 12.10 Inspecting for surge and swirl.

12.16 Calculate the divergence in both two and three dimensions of $\hat{r}/r^n$. For what values of n do the divergences flip sign? Why does the dimension of the space make a difference?

12.17 Referring to Problem 12.16, for what value of n does the divergence of $\hat{r}/r^n$ flip sign in four Euclidean dimensions? Approach this question as follows:

(a) Prediction: What does experience lead you to *expect* n to be?

(b) Though one might prefer to express r in hyperspherical coordinates (2.41), the form of the divergence in (12.49c) is not directly applicable in four dimensions. Instead, use the expressions for divergence in cylindrical and spherical coordinates to make an educated attempt at calculating $\vec{\nabla} \cdot \hat{r}/r^n$ for hyperspherical r. What value of n flips the sign?

(c) Calculate the divergence in four-dimensional cartesian coordinates and solve for n. (If you did not get this value in both parts (a) and (b), go back and try to figure out where you went wrong.)

12.18 Find the divergence and curl of the unit vector $\hat{e}_i$ in both cylindrical and spherical coordinates.

12.19 Use index notation and the equality of mixed partials to simplify each of the following expressions for derivatives of the scalar function $\phi(\vec{r})$ and vector function $\vec{A}(\vec{r})$:

(a) $\vec{\nabla} \cdot (\phi \vec{A})$
(b) $\vec{\nabla} \times (\phi \vec{A})$
(c) $\vec{\nabla} \cdot (\vec{\nabla} \times \vec{A})$
(d) $\vec{\nabla} \times \vec{\nabla}\phi$.

12.20 Use (12.27) and the $\vec{\nabla} \times (f\vec{A})$ identity in Table 12.1 to show that $\vec{\nabla}f \times \vec{\nabla}g$ can be written as a curl (and so is therefore solenoidal).

12.21 Use (12.73) to verify the vector identity of (12.61).

13 Interlude: Irrotational and Incompressible

In Example 12.7 we found that the curl of a gradient and the divergence of a curl both vanish. Let's use these identities to briefly explore vector fields in the plane.

Consider an irrotational field $\vec{E} = (E_x, E_y)$. As we saw in (12.28), $\vec{\nabla} \times \vec{E} = 0$ implies the existence of a scalar potential $\Phi(x, y)$ such that $\vec{E} = \vec{\nabla}\Phi$,

$$E_x = \frac{\partial \Phi}{\partial x} \qquad E_y = \frac{\partial \Phi}{\partial y}. \tag{13.1}$$

Now consider an incompressible field $\vec{B} = (B_x, B_y)$. The condition $\vec{\nabla} \cdot \vec{B} = 0$ reveals the existence of a vector field $\vec{A} = \vec{A}(x, y)$ such that $\vec{B} = \vec{\nabla} \times \vec{A}$. As a function of x and y, however, only A_z contributes to the curl. So we introduce a scalar field $\Psi(x, y)$ with $\vec{A} = (0, 0, \Psi)$ such that

$$B_x = \frac{\partial \Psi}{\partial y} \qquad B_y = -\frac{\partial \Psi}{\partial x}. \tag{13.2}$$

For a vector field in the plane which is *both* irrotational and incompressible, these scalar fields must satisfy

$$\frac{\partial \Phi}{\partial x} = \frac{\partial \Psi}{\partial y} \qquad \frac{\partial \Phi}{\partial y} = -\frac{\partial \Psi}{\partial x}, \tag{13.3a}$$

or in polar coordinates,

$$\frac{\partial \Phi}{\partial r} = \frac{1}{r}\frac{\partial \Psi}{\partial \phi} \qquad \frac{\partial \Psi}{\partial r} = -\frac{1}{r}\frac{\partial \Phi}{\partial r}. \tag{13.3b}$$

As discussed in Section 12.1, curves along which Φ is constant are perpendicular to its gradient. What about curves of constant Ψ? Well, if the vector field is also incompressible, then with (13.3a) we discover that $\vec{\nabla}\Phi$ and $\vec{\nabla}\Psi$ are orthogonal,

$$\begin{aligned}
\vec{\nabla}\Phi \cdot \vec{\nabla}\Psi &= \frac{\partial \Phi}{\partial x}\frac{\partial \Psi}{\partial x} + \frac{\partial \Phi}{\partial y}\frac{\partial \Psi}{\partial y} \\
&= \frac{\partial \Psi}{\partial y}\frac{\partial \Psi}{\partial x} - \frac{\partial \Psi}{\partial x}\frac{\partial \Psi}{\partial y} = 0.
\end{aligned} \tag{13.4}$$

Therefore the field $\vec{\nabla}\Psi$ must be tangent to curves of constant Φ, and likewise $\vec{\nabla}\Phi$ is tangent to curves of constant Ψ. This implies that curves of constant Φ are everywhere perpendicular to curves of constant Ψ. And since $\Phi = $ constant determines the family of equipotentials, curves of constant Ψ are streamlines. So the scalar field Ψ is often called the *stream function*.

Example 13.1 Electrostatic Fields

Electrostatic fields are irrotational; in regions without source charges, they are also incompressible. Consider, for instance, the field between the parallel plates of an ideal capacitor with surface charge density σ. The potential Φ, the negative of the voltage, is a function only of x,

$$\Phi = \sigma x/\epsilon_0, \tag{13.5}$$

where we've set $\Phi(0, y) = 0$. Consequently, vertical lines of constant x are equipotentials. Using this expression for Φ, the stream function can be found from (13.3a),

$$\Psi = \sigma y/\epsilon_0. \tag{13.6}$$

So the streamlines representing the $\vec{E}$ field are horizontal lines of constant y.

More interesting is the electric field in a plane perpendicular to two infinite parallel charged wires at $x = \pm a$ with opposite linear charge densities $\pm \lambda$. The potential is the sum of the potentials of each wire individually,

$$\Phi(P) = \frac{\lambda}{2\pi\epsilon_0}(\ln r_2 - \ln r_1) = \frac{\lambda}{4\pi\epsilon_0} \ln\left[\frac{(x+a)^2 + y^2}{(x-a)^2 + y^2}\right], \tag{13.7}$$

with zero potential on the y-axis, $\Phi(0, y) = 0$. From (13.3a), the stream function — representing the negative of the electric field — can be found (Problem 13.1):

$$\Psi(P) = -\frac{\lambda}{2\pi\epsilon_0}\left[\tan^{-1}\left(\frac{y}{x-a}\right) - \tan^{-1}\left(\frac{y}{x+a}\right)\right] = -\frac{\lambda}{2\pi\epsilon_0}(\phi_1 - \phi_2), \tag{13.8}$$

for $-\pi < \phi_{1,2} \le \pi$. The equipotentials and streamlines are shown in Figure 13.1.

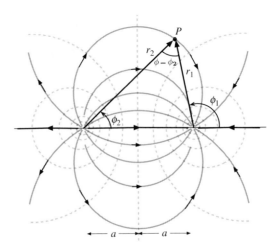

Figure 13.1 Streamlines (solid) and equipotentials (dotted) of opposite line charges (out of page).

Example 13.2 Vortex Circulation

Consider a straight wire of radius a along the z-axis. If the wire carries a steady current I, then the magnetic field outside the wire is (Example 14.12)

$$\vec{B} = \frac{\mu_0 I}{2\pi r} \hat{\phi}, \qquad r > a, \tag{13.9}$$

whose streamlines are concentric circles in the xy-plane (Figure 13.2). These circles are described by the stream function Ψ, which can be found from (13.2) in polar coordinates,

$$B_\phi = \frac{\partial \Psi}{\partial r} = \frac{\mu_0 I}{2\pi r} \qquad \longrightarrow \qquad \Psi(r, \phi) = \frac{\mu_0 I}{2\pi} \ln(r/a), \tag{13.10}$$

with the constant of integration chosen so that $\Psi(a, \phi) = 0$. Then (13.3a), also in polar coordinates, gives the potential Φ,

$$B_\phi = -\frac{1}{r} \frac{\partial \Phi}{\partial \phi} \qquad \longrightarrow \qquad \Phi(r, \phi) = -\frac{\mu_0 I}{2\pi} \phi. \tag{13.11}$$

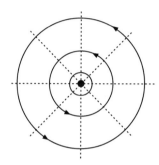

Figure 13.2 Streamlines (solid) and equipotentials (dotted) of a vortex with source out of page.

It's not difficult to see that Ψ and Φ are consistent with $B_r = 0$, and their gradients are perpendicular.[1] In the language of fluids, this system describes a whirlpool or vortex at the origin of strength $\Gamma \equiv \mu_0 I$.

A remarkable mathematical observation is that both Φ and Ψ can be combined into a single *complex* function of $z = x + iy = re^{i\phi}$ (not the z-direction in space). That's because by the usual rule of logarithms,

$$\ln z = \ln \left(re^{i\phi} \right) = \ln r + i\phi. \tag{13.12}$$

With this, we can define (the overall factor of i is conventional)

$$\Omega(z) \equiv \frac{i\Gamma}{2\pi} \ln(z/a) = \frac{i\Gamma}{2\pi} \ln(r/a) - \frac{\Gamma}{2\pi} \phi = i\Psi(x, y) + \Phi(x, y). \tag{13.13}$$

We'll return to this intriguing observation in Section 20.2.

---

[1] It may concern you that $\vec{B}$ doesn't look irrotational; we'll address this in Chapters 15 and 17.

The similarity of (13.1) and (13.2) imply that while Φ is the potential for the irrotational field (E_x, E_y), Ψ is the potential for irrotational $(-B_y, B_x)$. Thus if the fields are also incompressible, either scalar can be considered a potential and the other a stream function. Indeed, (13.3a) can be used to decouple Φ and Ψ into separate, identical second-order equations,

$$\nabla^2 \Phi = 0 \qquad \nabla^2 \Psi = 0. \qquad (13.14)$$

Of course, irrotational and incompressible fields occur in three dimensions as well; what makes $\mathbb{R}^2$ examples special is (13.2). Beyond two dimensions, streamlines cannot be described by a simple scalar. Potentials, however, continue to obey the *Laplace equation*:

$$\nabla^2 \Phi = 0. \qquad (13.15)$$

Since $\nabla^2 \Phi = \vec{\nabla} \cdot \vec{\nabla} \Phi$ and the curl of a gradient vanishes, solutions of Laplace's equation (called *harmonic functions*) describe irrotational incompressible fields. For example, a time-independent electric field is always irrotational; in a region of space without electric charge, it is also incompressible. Thus Laplace's equation describes all of electrostatics — with different configurations distinguished by conditions on Φ at the boundaries of the region. Of course, should a region *have* charge the electric field is not incompressible. If the charge distribution is given by a density function $\rho(\vec{r})$, the Laplace equation generalizes to the *Poisson equation*:

$$\nabla^2 \Phi = -\rho/\epsilon_0. \qquad (13.16)$$

We'll have much to say about these fundamental differential equations in Part VI.

Problems

13.1 Derive the stream function in (13.8).

13.2 Consider a brook or stream flowing around a rock; what are the streamlines as the water skirts around the barrier? We'll model the rock as a disk of radius a, and consider irrotational and incompressible rightward flow with uniform velocity $v_0 \hat{\imath}$ both far upstream and downstream from the rock. Notice that the boundary is itself a streamline, so that the normal component of the flow vanishes at $r = a$.

(a) Show the boundary conditions at $r = a$ and ∞ are satisfied by the velocity field

$$V_r(r, \phi) = v_0 \left(1 - \frac{a^2}{r^2} \right) \cos \phi \qquad V_\phi(r, \phi) = -v_0 \left(1 + \frac{a^2}{r^2} \right) \sin \phi.$$

(b) Find the potential Φ for this field.
(c) Find the stream function Ψ.
(d) Verify that both Ψ and Φ are harmonic functions.
(e) Make a sketch of the streamlines and equipotentials for $r > a$.

14 Integrating Scalar and Vector Fields

Integration is summation. The expression $\int_{x_1}^{x_2} f(x)\,dx$ describes a sum along the unique interval defined by the endpoints x_1 and x_2. But in higher dimensions, there are an infinite number of ways to connect endpoints. An expression like $\int_{P_1}^{P_2} f(x,y)\,dx$ asks us to sum $f(x,y)$ from $P_1 = (x_1, y_1)$ to $P_2 = (x_2, y_2)$, but it does not tell us along what path to compute it! The endpoints alone are not sufficient to determine the sum. To fully specify the integral, then, one must choose a path through space, $y = g(x)$. This reduces the number of independent variables, allowing $\int f(x, g(x))\,dx$ to be evaluated. Similarly, an integral like $\int f(x, y, z)\,dxdy$ requires the specification of a two-dimensional surface $z = g(x,y)$ on which to compute the integral. Identifying a path or surface is essential to the complete framing of these integrals — and without which, the calculation instructions are ambiguous.

It should come as no surprise then that the choice of path can affect the value of an integral. For instance, the work done by kinetic friction scales with the length of the path — even between the same end points. On the other hand, we know that when calculating the work done by gravity a path need *not* be specified. A sum of this sort, well-defined without the specification of a path, is exceptional — evidence of some underlying characteristic of the system. So what makes gravity so special? And what other systems have this remarkable property? More generally, how can we determine whether specifying a path (or a surface) is needed for any given integral? The answers to these questions transform an integral from a mere instruction to calculate into a statement about the fields being integrated.

Although our focus will be the integration of vector functions along curves and surfaces, we begin with a discussion and review of line and surface integrals of scalar functions.

14.1 Line Integrals

Integrating a function of more than one real variable over a one-dimensional region is fundamentally a one-dimensional integral. The result is called a *line integral*.

A line integral is a sum over a curve C.[1] Letting $d\ell$ denote the infinitesimal element along C (other common choices are ds and dr), its general form is

$$\int_C f(\vec{r})\,d\ell. \tag{14.1}$$

A line integral sums the contributions of f only on C; we generally do not care what values f may take elsewhere. So, for instance, if $f(\vec{r})$ is the charge density on a wire C, the integral gives the total charge on the wire.

---

[1] Despite the name "line integral," C need not be a straight segment.

Though the integration region itself is a one-dimensional entity, the curve may snake through higher dimensions; the equations which define C act to constrain the integrand to a function of only one independent variable. For example, if a path in two-dimensional space is, say, parallel with the x-axis — so that C is defined by the equation $y = y_0$ — the line integral simply reduces to the usual one-dimensional integral in x with the variable y replaced by the constant y_0,

$$\int_C f(x, y)\,d\ell = \int_{x_1}^{x_2} f(x, y_0)\,dx. \tag{14.2}$$

More generally, let $y = g(x)$ specify how C twists through space. Using

$$(d\ell)^2 = (dx)^2 + (dy)^2 \tag{14.3}$$

we can write

$$d\ell = \sqrt{(dx)^2 + (dy)^2} = dx\sqrt{1 + (dy/dx)^2}, \tag{14.4}$$

and the line integral becomes

$$\int_C f(x, y)\,d\ell = \int_{x_1}^{x_2} f(x, g(x))\,\sqrt{1 + (dg/dx)^2}\,dx, \tag{14.5}$$

which reduces to (14.2) for $g(x) = y_0$. In effect, we have projected the curve C onto a segment along x, thereby casting the problem into a more familiar form.

Incidentally, the parentheses in (14.3) are not normally used. Instead, one writes

$$d\ell^2 = dx^2 + dy^2, \tag{14.6}$$

with the implicit understanding that $dx^2 \equiv (dx)^2$, not $2x\,dx$. Though a bit sloppy, this notation is standard. Note too that the decomposition in (14.6) tacitly assumes the Pythagorean theorem; its applicability is the hallmark of a *Euclidean space*.

Example 14.1 $\quad f(\vec{r}) = \sqrt{\tfrac{1}{4}x^2 + y^2}, \quad C_1 : y(x) = \tfrac{1}{2}x^2, \ 0 < x < 1$

Applying (14.5),

$$\int_{C_1} f(\vec{r})\,d\ell = \int_0^1 \sqrt{\frac{x^2}{4} + \frac{x^4}{4}}\,\sqrt{1 + (dy/dx)^2}\,dx$$

$$= \frac{1}{2}\int_0^1 x\sqrt{1+x^2}\,\sqrt{1+x^2}\,dx = \frac{1}{2}\int_0^1 x(1+x^2)\,dx = \frac{3}{8}. \tag{14.7}$$

With minor adjustment to (14.5) we can also integrate in y:

$$\int_{C_1} f(\vec{r})\,d\ell = \int_0^{\frac{1}{2}} \sqrt{\frac{1}{2}y + y^2}\,\sqrt{1 + (dx/dy)^2}\,dy$$

$$= \int_0^{\frac{1}{2}} \sqrt{\frac{1}{2}y + y^2}\sqrt{1 + \frac{1}{2y}}\,dy = \int_0^{\frac{1}{2}} \left(y + \frac{1}{2}\right)dy = \frac{3}{8}. \tag{14.8}$$

Example 14.2 $f(\vec{r}) = \sqrt{\frac{1}{4}x^2 + y^2}, \quad C_2 : y(x) = x, \; 0 < x < 1$

Note that in the last example we didn't change the path, only the independent variable used to describe it. Changing the actual path, however, *can* affect the integral. For example, imagine the same function from $x = 0$ to 1 along the curve $C_2 : y(x) = x$,

$$\int_{C_2} f(\vec{r})\,d\ell = \int_0^1 \sqrt{\frac{1}{4}x^2 + x^2}\sqrt{1 + (dy/dx)^2}\,dx = \sqrt{\frac{5}{2}}\int_0^1 x\,dx = \sqrt{\frac{5}{8}}. \tag{14.9}$$

So this integral is clearly *path-dependent*.

The same result, of course, can be had in polar coordinates — and the calculation is even easier. In terms of r and θ, the Euclidean path element is

$$d\ell^2 = dr^2 + r^2 d\theta^2. \tag{14.10}$$

(As in (14.6), we forego parentheses.) Now the condition $y = x$ means the path C_2 is fixed at $\theta = \pi/4$. Thus $d\theta = 0$ and $d\ell = dr$:

$$\int_{C_2} f(\vec{r})\,d\ell = \int_0^{\sqrt{2}} \sqrt{\frac{1}{4}r^2\cos^2\theta + r^2\sin^2\theta}\,dr = \sqrt{\frac{5}{8}}\int_0^{\sqrt{2}} r\,dr = \sqrt{\frac{5}{8}}. \tag{14.11}$$

Example 14.3 Circles

Integrating $\int_C d\ell$ along some C gives the length of the path. If that path is along a circle centered at the origin, then $dr = 0$ and the integral is trivial. For instance, if the circle has radius a and we integrate from $\theta = 0$ to π, we get

$$\int_C d\ell = \int_C \sqrt{dr^2 + r^2 d\theta^2} = \int_0^\pi a\,d\theta = \pi a, \tag{14.12}$$

which, of course, is just half the circumference of the circle. What if the circle is not centered at the origin? Consider $r = 2a\cos\theta$, which is a circle of radius a centered at $x = a, y = 0$. As a little algebra will verify, along this circle the half-circumference is

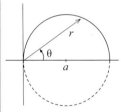

$$\int_C d\ell = \int_0^{2a} \sqrt{1 + r^2(d\theta/dr)^2}\,dr = 2a\int_0^{2a} \frac{dr}{\sqrt{1 - r^2/4a^2}} = \pi a. \tag{14.13}$$

Equivalently, we can do the integration in θ — which is even easier once the limits on θ are verified:

$$\int_C d\ell = \int_0^{\pi/2} \sqrt{r^2 + (dr/d\theta)^2}\,d\theta = 2a\int_0^{\pi/2} d\theta = \pi a. \tag{14.14}$$

Example 14.4 The Shortest Distance Between Two Points

Allowing for paths to snake through three dimensions trivially generalizes (14.6) to

$$d\ell = \sqrt{dx^2 + dy^2 + dz^2}. \tag{14.15}$$

Which path connecting two points P_1 and P_2 is the shortest? Let's choose coordinates so that both P_1 and P_2 lie on the x-axis. With functions $y = y(x)$ and $z = z(x)$ specifying C, the integral can be evaluated in x,

$$\int_C d\ell = \int_{P_1}^{P_2} dx \sqrt{1 + (dy/dx)^2 + (dz/dx)^2}.$$

Since $(dy/dx)^2 \geq 0$ and $(dz/dx)^2 \geq 0$, the integral is clearly minimized when both dy/dx and dz/dx vanish everywhere along the path. So the shortest distance between two points is a straight line. (At least in Euclidean space.)

Thus far, we have only considered paths expressed in the form $y = g(x)$ — that is, as explicit curves in space. But it's frequently more convenient to describe a path using separate functions for each dimension, say $x(t)$ and $y(t)$. The common parameter t is what unites these functions into a single path; in physics and engineering, that parameter is usually time. For instance, a particle launched from the origin with speed v_0 at an angle θ_0 near the surface of Earth has the familiar solution

$$x(t) = v_0 \cos \theta_0\, t \qquad y(t) = v_0 \sin \theta_0\, t - \frac{1}{2} g t^2. \tag{14.16a}$$

Eliminating the common parameter t in a two-dimensional problem like this is straightforward, revealing the particle's parabolic trajectory,

$$y(x) = \tan \theta_0\, x - \frac{1}{2} \frac{g}{v_0^2 \cos \theta_0^2}\, x^2. \tag{14.16b}$$

But what may be straightforward in two dimensions can be intractable for paths threading through higher dimensions. So to describe paths in higher-dimensional space we use separate functions for each dimension, which together form a vector-valued function $\vec{c}\,(t)$. The particle trajectory $\vec{c}(t) \equiv (x(t), y(t))$ in (14.16a) is a two-dimensional illustration with parameter t. We saw another example of this in Chapter 7, where orbits were characterized by the function $r(\phi)$ in (7.1), or via Kepler's equation (7.30) by the parameterized pair of functions $r(t)$ and $\phi(t)$. In general, a path C wending through n-dimensional space can be represented by a function which maps a single parameter into n dimensions. The specification of such a function is called a *parametrization* of the path C; the set of n functions are called *parametric equations*.

Though it may be obvious, it's important to emphasize that these parametric equations are functions not of space, but of a single independent parameter. Time is a common choice; arc length along the path is another. In relativity, where t is no longer an external parameter but rather a full-fledged coordinate, curves in spacetime are often parametrized by the invariant proper time τ.

Integrating along a parametrized curve should look familiar. For the simple case of a function in one dimension, $c(t) \equiv x(t)$ and $d\ell \equiv dx = \dot{x}\, dt$, where the dot indicates a t derivative. Thus

$$\int_{x_1}^{x_2} f(x)\, dx = \int_{t(x_1)}^{t(x_2)} f(x(t))\, \dot{x}(t)\, dt. \tag{14.17}$$

So (14.17) is little more than a standard change of integration variables. Similarly, a two-dimensional parametrization $\vec{c}\,(t) = (x(t),\, y(t))$ has path element

$$d\ell = \sqrt{dx^2 + dy^2} = \sqrt{\left(\frac{dx}{dt}\right)^2 + \left(\frac{dy}{dt}\right)^2}\, dt \equiv |\dot{\vec{c}}\,(t)|\, dt. \tag{14.18}$$

The vector $\dot{\vec{c}}(t) = (\dot{x}(t), \dot{y}(t))$ is the tangent vector to the curve at t; $|\dot{\vec{c}}(t)|$ can be construed as the speed along the path. The line integral of $f(x, y)$ can then be rendered

$$\int_C f(\vec{r}) \, d\ell = \int_{t_1}^{t_2} f(x(t), y(t)) \, |\dot{\vec{c}}(t)| \, dt. \tag{14.19}$$

The generalization to paths snaking through higher dimensions is straightforward.

Any given path may have many different parametrizations — but since they all represent the *same* curve C, an integral over C is independent of the choice of parametrization $\vec{c}(t)$.

Example 14.5 **The Circle**

A circular path, $x^2 + y^2 = a^2$, can be parametrized by

$$\vec{c}(t) = (a \cos \omega t, a \sin \omega t), \quad 0 \le t < 2\pi/\omega, \tag{14.20}$$

so that $|\dot{\vec{c}}| = a\omega$ for some constant ω. Strictly speaking, different ω's give different parametrizations. But integrating $d\ell$ all around the circle yields a path length independent of ω,

$$\oint_C d\ell = \int_0^{2\pi/\omega} a\omega \, dt = 2\pi a. \tag{14.21}$$

Integrating $f(x, y) = xy$ along, say, the semicircle spanning the first and fourth quadrants gives

$$\int_C f(x, y) \, d\ell = \frac{\omega a^3}{2} \int_{-\pi/2\omega}^{+\pi/2\omega} \sin(2\omega t) \, dt = \frac{a^3}{2}. \tag{14.22}$$

Again, independent of ω. A different parametrization of the same circle is

$$\vec{\gamma}(t) = (a \cos t^2, a \sin t^2), \quad 0 \le t < \sqrt{2\pi}. \tag{14.23}$$

Though the geometry of the parametrizations is the same, $\vec{c}$ and $\vec{\gamma}$ differ in their speed around the circle: $|\dot{\vec{c}}|$ is constant, whereas $|\dot{\vec{\gamma}}| = 2at$ grows linearly with t. In some sense, this varying speed allows an integral to traverse the same path C in the allotted time, and thereby return a parametrization-independent integral. Indeed, calculating the circumference using $\gamma(t)$ yields

$$\oint_C d\ell = \int_0^{\sqrt{2\pi}} 2at \, dt = 2\pi a. \tag{14.24}$$

As it should. Integrating $f(x, y) = xy$ is more difficult using γ, but you should convince yourself that the result is the same as before. Indeed, ease of integration is an important consideration when choosing a parametrization.

Example 14.6 **The Helix**

Now we need a three-dimensional parametrization. Introducing a constant b, we can use

$$\vec{c}(t) = (a \cos t, a \sin t, bt), \tag{14.25}$$

where the first two elements trace out a circle of radius a, and the third lifts the curve out of the plane with pitch $2\pi b$. Integrating once around the helix gives the length of one complete turn:

$$L = \int_C d\ell = \int_C \sqrt{dx^2 + dy^2 + dz^2} = \int_0^{2\pi} \sqrt{\left(\frac{dx}{dt}\right)^2 + \left(\frac{dy}{dt}\right)^2 + \left(\frac{dz}{dt}\right)^2}\, dt$$

$$= \int_0^{2\pi} \sqrt{a^2 + b^2}\, dt = 2\pi\sqrt{a^2 + b^2}. \tag{14.26}$$

This is, of course, completely equivalent to using (14.19) directly.

Example 14.7 **The Cycloid**

Consider a wheel of radius a rolling without slipping. As it turns through an angle $\theta = \omega t$, a spot P on the rim traces out a *cycloid* with coordinates

$$x(t) = a\theta - a\sin\theta = a(\omega t - \sin\omega t)$$
$$y(t) = a - a\cos\theta = a(1 - \cos\omega t), \tag{14.27}$$

where we've taken P to be at the origin at $t = 0$, and θ increases clockwise from the bottom of the circle (Figure 14.1). The most prominent features of this cycloid are the cusps that recur at intervals of $\omega t = 2\pi$. These cusps manifest when the wheel's forward speed $v = a\omega$ is offset by the spot's reverse speed $a\omega$ at the bottom of the wheel, bringing P momentarily to rest.

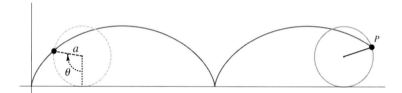

Figure 14.1 A cycloid.

Integrating over one period, the length of one hump of the cycloid is

$$L = \int_C d\ell = \int_C \sqrt{dx^2 + dy^2}$$
$$= a\omega \int_0^{2\pi/\omega} \sqrt{(1 - \cos\omega t)^2 + \sin^2\omega t}\, dt$$
$$= a\omega \int_0^{2\pi/\omega} \sqrt{2(1 - \cos\omega t)}\, dt$$
$$= 2a\omega \int_0^{2\pi/\omega} \sin(\omega t/2)\, dt = 8a. \tag{14.28}$$

Note that all of this only works if the parametrization is one-to-one — that is, if every value of the parameter t between t_1 and t_2 maps to a unique value of $\vec{c}(t)$. Such so-called *simple curves* do not not intersect or cross over themselves.

14.2 Surface Integrals

Integrating a function of more than two variables over a two-dimensional region is fundamentally a two-dimensional integral. The result is a *surface integral*.

A surface integral is a sum over a surface S. Letting dA denote the infinitesimal area element on S (da, $d\sigma$ and dS are also common choices), its general form is

$$\int_S f(\vec{r})\, dA.$$

A surface integral sums the contributions of f only on S; we generally do not care what values f may take elsewhere. Though the integration region itself is a two-dimensional entity, the surface may curve in higher dimensions; the equations which define S act to constrain the integrand to a function of only two independent variables. For example, if a surface in three-dimensional space is, say, parallel with the xy-plane – so that S is defined by the equation $z = z_0$ — the surface integral simply reduces to the usual double integral with z replaced by the constant z_0,

$$\int_C f(x, y, z)\, dA = \int_{x_1}^{x_2} \int_{y_1}^{y_2} f(x, y, z_0)\, dx\, dy. \tag{14.29}$$

More generally, let $z = g(x, y)$ specify how S curves in space. As with line integrals, this constraint is used to evaluate the integral by projecting the surface onto a plane, thereby turning the problem into a normal double integral. This is perhaps best appreciated with an example.

Example 14.8 **Mass of a Hemispherical Surface**

The integral to be performed is

$$M = \int_S \sigma(\vec{r})\, dA, \tag{14.30}$$

where the hemisphere S is given by the constraint $z = g(x, y) = \sqrt{R^2 - x^2 - y^2}$. Imagine dividing this surface into little rectangles; the area dA of each rectangle is related to the area dA' projected onto the xy-plane by $dA' = \cos\alpha\, dA$. (This is the same cosine factor that appears in Problem 2.14.) Then we can recast the integral as a sum over the disk D,

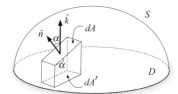

$$M = \int_D \sigma(x, y, g(x, y))\, \frac{dA'}{\cos\alpha} = \int_D \frac{\sigma(x, y, g(x, y))}{\hat{n}\cdot\hat{k}}\, dx\, dy, \tag{14.31}$$

with $\hat{n}$ the unit normal to S. Since $\hat{n} = \hat{n}(\vec{r})$ — that is, $\cos\alpha$ varies over the surface — this can be a formidable integral. But for a sphere, we know that $\hat{n} \equiv \hat{r}$ and $\hat{r}\cdot\hat{k} = \cos\alpha = z/R$. Thus

$$M = \int_D \frac{\sigma(x, y, g(x, y))}{z/R}\, dx\, dy = R \int_D \frac{\sigma(x, y, g(x, y))}{\sqrt{R^2 - x^2 - y^2}}\, dx\, dy. \tag{14.32}$$

We now have only to compute a double integral over the disk. For instance, if $\sigma = k\,(x^2 + y^2)/R^2$ for some constant k, you can easily use polar coordinates to show $M = 4\pi k R^2/3$. (If you want to work harder, stay in cartesian coordinates and use the substitution $y = \sqrt{R^2 - x^2}\sin\theta$.)

Now any surface S can be broken into infinitesimal rectangles dA and projected down into the plane — so (14.31) is actually a general result. But our ability to compute the integral in that example depended on the fact that S was spherical, so that $\hat{n} = \hat{r}$. What do you do when integrating over non-spherical surfaces?

Let's represent $z = g(x, y)$ as an equipotential surface $\phi(x, y, z) \equiv z - g(x, y) = 0$. Since its gradient is perpendicular to the surface, we can construct the unit normal to the surface[2]

$$\hat{n} \equiv \frac{\vec{\nabla}\phi}{|\vec{\nabla}\phi|} = \frac{\hat{k} - \vec{\nabla}g}{\sqrt{1 + |\vec{\nabla}g|^2}}, \tag{14.33}$$

where $\vec{\nabla}g = \hat{i}\,\partial_x g(x, y) + \hat{j}\,\partial_y g(x, y)$. Using this in (14.31) gives the general formula for projecting a surface integral over S to a double integral over D (not necessarily a disk),

$$\int_S f(\vec{r})\,dA = \int_D f(x, y, g(x, y))\,\sqrt{1 + |\vec{\nabla}g|^2}\,dx\,dy, \tag{14.34}$$

where S is given by the constraint $z = g(x, y)$. (You should verify that (14.34) reproduces the projection used in Example 14.8.) Note the similarity with (14.5) for line integrals; generalizing beyond two-dimensional surfaces should be clear.

Example 14.9 **The Hyperboloid**

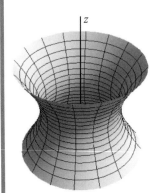

A hyperboloid is a surface described by

$$\frac{x^2}{a^2} + \frac{y^2}{b^2} - \frac{z^2}{c^2} = 1. \tag{14.35}$$

For simplicity, we'll take $a = b = c$ so that

$$z = g(x, y) = \sqrt{x^2 + y^2 - a^2}. \tag{14.36}$$

Then the surface area is given by

$$\int_S dA = \int_D \sqrt{1 + |\vec{\nabla}g|^2}\,dx\,dy$$
$$= \int_D \sqrt{\frac{2(x^2 + y^2) - a^2}{x^2 + y^2 - a^2}}\,dx\,dy$$
$$= \int_D \sqrt{\frac{2r^2 - a^2}{r^2 - a^2}}\,r\,dr\,d\phi, \tag{14.37}$$

where in the last step, we took advantage of the region's circular cross-section to convert to polar coordinates. Now unlike a sphere, the hyperboloid is not a closed surface, so we need to restrict ourselves to a finite z interval — say, $0 < z < a$. Then the region of integration D is an annulus of inner radius a and outer radius $\sqrt{2}\,a$. Cleaning things up a bit by letting $r^2/a^2 = t^2 + 1$ yields

$$\int_S dA = 2\pi \int_a^{\sqrt{2}\,a} \sqrt{\frac{2r^2 - a^2}{r^2 - a^2}}\,r\,dr = 2\pi a^2 \int_0^1 \sqrt{2t^2 + 1}\,dt. \tag{14.38}$$

We evaluated this integral back in Problem 8.6. For our hyperboloid, it yields a surface area of about $2.5\pi a^2$.

---

[2] You might reasonably object that $g(x, y) - z$ describes the same surface as ϕ, but results in $-\hat{n}$. Indeed, the sign of $\hat{n}$ must be specified.

Thus far, we have only considered surfaces in the form $z = g(x, y)$ — that is, as explicit functions over the plane. One shortcoming of this approach is that for closed surfaces the sum needs to be split in two. This is why we only considered a hemisphere in Example 14.8; we'd have to repeat the calculation for $z < 0$ to include the other half of the sphere (though symmetry guarantees that the answer will be the same). More generally, $z = g(x, y)$ must be single-valued.

So why not just integrate over the surface directly, using coordinates defined *on* the surface? Indeed, as with curves, surfaces can be described parametrically — though this time we'll need a *pair* of parameters u and v in order to characterize a two-dimensional region. Then a surface S can be represented by a set of three functions $x(u, v)$, $y(u, v)$, and $z(u, v)$ which automatically satisfy $z = g(x, y)$. Thus, rather than use three constrained coordinates (x, y, z), the integral becomes a manifestly two-dimensional affair in u and v.

Example 14.10 Mass of a Spherical surface

The simplest parametrization for a sphere is to take u and v to be the conventional spherical coordinates θ, ϕ; then the familiar transformations $x = x(\theta, \phi)$, $y = y(\theta, \phi)$, $z = z(\theta, \phi)$ between spherical and cartesian coordinates are the corresponding parametric equations. For example, let's revisit the surface integral of Example 14.8,

$$M = \int_S \sigma(\vec{r})\, dA, \tag{14.39}$$

with $\sigma = k\,(x^2 + y^2)/R^2$. Evaluating this integral directly on the surface is much easier than projecting down to the xy-plane:

$$
\begin{aligned}
M &= \frac{k}{R^2} \int_S (x^2 + y^2)\, dA \\
&= \frac{k}{R^2} \int_S \left(R^2 \sin^2\theta\right) R^2\, d\Omega \\
&= 2\pi k R^2 \int_0^\pi \sin^3\theta\, d\theta \\
&= 2\pi k R^2 \int_{-1}^1 (1 - \cos^2\theta)\, d(\cos\theta) = \frac{8\pi k R^2}{3}.
\end{aligned}
\tag{14.40}
$$

Note that we were able to integrate over the entire sphere without having to divide the integral into separate $z > 0$ and $z < 0$ patches.

Example 14.11 The Parametrized Hyperboloid

Despite the apparent ease of the previous example, the No Free Lunch Theorem looms large. Specifically, to integrate we needed to use $dA = R^2 d\Omega$. While this may be well-known for a sphere, less-familiar surfaces may require some effort before an integral can be evaluated.

Consider oblate spheroidal coordinates $\{u, \theta, \phi\}$,

$$x = \alpha \cosh u \sin\theta \cos\phi \quad y = \alpha \cosh u \sin\theta \sin\phi \quad z = \alpha \sinh u \cos\theta, \tag{14.41}$$

where $0 \le u < \infty$, $0 \le \theta < \pi$ and $0 \le \phi < 2\pi$. Since

$$\frac{x^2 + y^2}{\alpha^2 \sin^2\theta} - \frac{z^2}{\alpha^2 \cos^2\theta} = 1, \tag{14.42}$$

surfaces of constant θ are the hyperboloids

$$x = (\alpha \sin \theta) \cosh u \cos \phi \quad y = (\alpha \sin \theta) \cosh u \sin \phi \quad z = (\alpha \cos \theta) \sinh u. \tag{14.43}$$

With $\theta = \pi/4$, $\alpha = \sqrt{2}a$, this is a parametrization of the hyperboloid in Example 14.9.

As we learned back in Problem 2.17 (see also Appendix A), the scale factors in oblate spheroidal coordinates are

$$h_u = h_\theta = \alpha\sqrt{\sinh^2 u + \cos^2 \theta} = a\sqrt{2 \sinh^2 u + 1}$$
$$h_\phi = \alpha \cosh u \sin \theta = a \cosh u. \tag{14.44}$$

So the surface area of our parametrized hyperboloid surface is just

$$\int_S dA = \int_S h_u h_\phi \, du \, d\phi$$
$$= a^2 \int_S \sqrt{2 \sinh^2 u + 1} \ \cosh u \, du \, d\phi. \tag{14.45}$$

A final variable change to $t = \sinh u$ yields

$$\int_S dA = 2\pi a^2 \int \sqrt{2t^2 + 1} \ dt. \tag{14.46}$$

Confining ourselves to the same finite interval $0 < z < a$ considered in Example 14.9 gives us precisely the integral of (14.38). As it must.

BTW 14.1 Tangent Vectors

There is a nice geometrical way to understand the parametrization of a surface integral. Rather than a path $\vec{c}(t) = (x(t), y(t), z(t))$, we now have a surface

$$\vec{\Sigma}(u, v) = (x(u, v), \ y(u, v), \ z(u, v)). \tag{14.47}$$

Instead of a single tangent vector $\dot{\vec{c}} = \partial \vec{c}/\partial t$, a surface has (in general) two distinct tangent vectors,

$$\partial_u \vec{\Sigma} \equiv \frac{\partial \vec{\Sigma}(u, v)}{\partial u}, \qquad \partial_v \vec{\Sigma} \equiv \frac{\partial \vec{\Sigma}(u, v)}{\partial v},$$

which together define a surface's *tangent plane*. The simplest way to characterize this plane is with a vector normal to it. We can construct a normal by taking the cross product of the tangent vectors,

$$\vec{n}(u, v) \equiv \partial_u \vec{\Sigma} \times \partial_v \vec{\Sigma}. \tag{14.48}$$

Then by analogy with (14.18) and (14.19) for paths, we propose the area element

$$dA = |\vec{n}(u, v)| \, du dv \tag{14.49}$$

and the surface integral

$$\int_S f(\vec{r}) \, dA = \int_D f(x(u, v), \ y(u, v), \ z(u, v)) \ |\vec{n}(u, v)| \, du \, dv. \tag{14.50}$$

These expressions have some reassuring geometry to support it. Just as $\vec{c}$ specifies an orientation for paths threading through space, so too $\vec{n}$ specifies the orientation of the surfaces in the larger embedding space. Like $|\dot{\vec{c}}(t)|$, the magnitude of $\vec{n}$ gives in some sense the two-dimensional "rate" with which the parametrization covers the surface.

Note that in (14.49) and (14.50), $|\vec{n}|$ plays the role the Jacobian $|J|$ does for volume integrals. In fact, $|\vec{n}|$ can be written like a Jacobian: direct manipulation of the cross product in (14.48) leads to

$$|\vec{n}| = \sqrt{\left|\frac{\partial(x,y)}{\partial(u,v)}\right|^2 + \left|\frac{\partial(y,z)}{\partial(u,v)}\right|^2 + \left|\frac{\partial(z,x)}{\partial(u,v)}\right|^2}, \qquad (14.51)$$

where $\left|\frac{\partial(a,b)}{\partial(\alpha,\beta)}\right|$ is the Jacobian of the transformation $\{a,b\} \rightarrow \{\alpha,\beta\}$. You should verify that for a surface of the form $z = g(x,y)$, (14.51) and (14.50) reduce to (14.33) and (14.34). As another check, consider a sphere of radius a parametrized by standard spherical coordinates $\theta \equiv u$, $\phi \equiv v$,

$$x(\theta,\phi) = a\sin\theta\cos\phi \quad y(\theta,\phi) = a\sin\theta\sin\phi \quad z(\theta,\phi) = a\cos\phi.$$

The Jacobian determinants are easily found:

$$\left|\frac{\partial(x,y)}{\partial(\theta,\phi)}\right| = a^2\sin\theta\cos\theta \quad \left|\frac{\partial(y,z)}{\partial(\theta,\phi)}\right| = a^2\sin^2\theta\cos\phi \quad \left|\frac{\partial(z,x)}{\partial(\theta,\phi)}\right| = a^2\sin^2\theta\sin\phi$$

and (14.49) gives

$$dA = a^2\sin\theta\,d\theta d\phi.$$

A similar manipulation re-discovers the area element in (14.45) for hyperboloids.

BTW 14.2 Minimal Surfaces

In Example 14.4 we showed that the shortest distance between two fixed points is a straight line. Stepping up one dimension, we now ask about the surface of minimum area with a given fixed boundary — say, the shape of a soap film supported in a wire frame. For a flat frame, the soap film is flat; but what of a frame which is not so simple? What surface, for instance, does a soap film form when supported by a helical wire?

From (14.34), the surface area is given by the integral

$$A[g] \equiv \int_S dA = \int_D \sqrt{1 + |\vec{\nabla}g|^2}\,dxdy, \qquad (14.52)$$

where S is defined by $z = g(x,y)$. We've used square brackets to indicate that $A[g]$ depends not so much on the independent variables x and y (which, after all, are integrated away to find A) as it does on the *entirety* of the function $g(x,y)$. Quantities such as $A[g]$ are often called *functionals*. The problem, then, is to find the function g which minimizes the value of the functional $A[g]$ for a given boundary C. Clearly, this problem is not as simple as the familiar max-min problem of introductory calculus; instead, we use a procedure known as the *calculus of variations*.

Consider an arbitrary function $h(x,y)$ which vanishes on C, so that both g and $g + h$ have the same value on the boundary. Now, if $A[g]$ is a minimum with the given boundary conditions, then $A[g + \epsilon h]$, with ϵ an arbitrary parameter, must be bigger. More to the point, as an extremum $A[g]$ must satisfy

$$\frac{d}{d\epsilon}A[g + \epsilon h]\bigg|_{\epsilon=0} = 0. \qquad (14.53)$$

Applying this condition to (14.52) yields

$$\int_D \frac{\vec{\nabla}g \cdot \vec{\nabla}h}{\sqrt{1 + |\vec{\nabla}g|^2}}\, dxdy = 0. \tag{14.54}$$

Integrating by parts — and recalling that h vanishes on C — yields

$$\int_D \vec{\nabla} \cdot \left[\frac{\vec{\nabla}g}{\sqrt{1 + |\vec{\nabla}g|^2}} \right] h\, dxdy = 0. \tag{14.55}$$

Now the only way (14.55) can hold for an *arbitrary* function $h(x, y)$ is if the rest of the integrand vanishes identically,

$$\vec{\nabla} \cdot \left[\frac{\vec{\nabla}g}{\sqrt{1 + |\vec{\nabla}g|^2}} \right] = 0. \tag{14.56}$$

The minimal surface is the function $z = g(x, y)$, which solves this differential equation for a given boundary C. You should verify that in one dimension, the minimal "surface" is a straight line. Solutions in higher dimensions are less simple, but for surfaces which do not deviate greatly from flatness — that is, which have small $|\vec{\nabla}g|$ — (14.56) reduces to Laplace's equation, $\nabla^2 g = 0$, which is simpler. As for the helix, the minimal surface is called a *helicoid*.

14.3 Circulation

So far, we have only considered scalar fields. More possibilities arise when integrating vector fields. For instance, a vector field can be integrated component by component,

$$\int_C \vec{F}(\vec{r})\, d\ell = \left(\int_C F_x(\vec{r})\, d\ell, \int_C F_y(\vec{r})\, d\ell, \int_C F_z(\vec{r})\, d\ell \right).$$

The result, of course, is a vector. Nor is it really anything new: each component is essentially an integral of the form (14.19). But a vector field has an orientation with respect to the curve C, and this leads to new types of line integrals. Of particular importance is

$$\int_C \vec{F} \cdot d\vec{\ell}, \tag{14.57}$$

with $d\vec{\ell}$ the tangential line element along C.[3] In a cartesian basis $d\vec{\ell} = (dx, dy, dz)$, in cylindrical $d\vec{\ell} = (d\rho, \rho d\phi, dz)$, and in spherical $d\vec{\ell} = (dr, r d\theta, r \sin\theta d\phi)$. The scalar-valued integral, of course, is independent of this choice.

The integral of (14.57) measures the total contribution of the vector field *along* the curve. Recall that when $\vec{F}$ is a force, the line integral gives the work done on a particle as it moves from P_1 to P_2 — which physically manifests as an overall change in the particle's kinetic energy. This is the Work–Energy theorem. If instead the vector is an electric field $\vec{E}$, (14.57) gives the voltage difference between P_1 and P_2,

---

[3] Another possibility is the vector-valued $\int_C \vec{F} \times d\vec{\ell}$ — but $\int_C \vec{F} \cdot d\vec{\ell}$ is far more common.

$$\Delta V = -\int_{C}^{P_2}_{P_1} \vec{E} \cdot d\vec{\ell}. \tag{14.58}$$

So an ideal voltmeter directly measures the line integral of $\vec{E}$ through the meter.

When the line integral (14.57) is over a *closed* curve, the result is called the *circulation* of the field,

$$\Psi \equiv \oint_{C} \vec{F} \cdot d\vec{\ell}. \qquad \text{[Circulation]} \tag{14.59}$$

By summing over all contributions of $\vec{F}$ along and against a closed path C, circulation quantifies a vector field's net *swirl*.

You might reasonably presume that the circulation always vanishes — and indeed, that is true for many fields. But there are cases, such as for magnetic fields, which can have *non-zero* circulation. As we'll see, such fields must differ in some fundamental way from those with vanishing circulation.

Example 14.12 **Circulation and Symmetry**

Curiously, knowledge of a field's circulation — vanishing or not — can sometimes be used to solve for the field itself. This may sound absurd: a circulation is a definite integral over a curve; how can the integrand be extracted from this? And anyway, specifying the circulation is but one condition for a three-component vector field, each of which takes values over an infinite number of points. One scalar condition is hardly enough information to solve for the field. Nonetheless, it is sometimes possible to use the circulation to find the field. How? One word: symmetry.

Perhaps the most famous circulation appears in Ampère's law,

$$\oint_{C} \vec{B} \cdot d\vec{\ell} = \mu_0 I, \tag{14.60}$$

where $\vec{B}$ is the magnetic field and I a steady current flowing through a surface S bounded by the curve C. This has the status of a physical law because it gives the circulation in terms of I for *any* static magnetic field. Even so, Ampère's law can only be used to *solve* for $\vec{B}$ in the relatively few situations of high symmetry.

As an example, consider a uniformly distributed, steady current I_0 flowing down a long cylindrical wire of radius a. Cylindrical symmetry implies that $\vec{B}$ must form closed circles centered on the wire; moreover, the magnitude of the field can only depend on the distance from the cylinder's axis. Thus $\vec{B} = B(r)\hat{\phi}$. A little thought should convince you that the cylindrical symmetry of a very long wire leaves no alternatives. So even before we start to calculate, we know quite a bit about the field.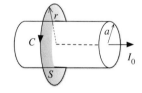

We now use this insight to *choose* a curve C coinciding with a circle of $\vec{B}$. With this choice, not only is $\vec{B}$ always along $d\vec{\ell}$, but its magnitude is constant everywhere on C. Thus the circulation integral is non-zero,

$$\oint_{C} \vec{B} \cdot d\vec{\ell} = \oint_{C} B \, d\ell = B \oint_{C} d\ell = B \cdot 2\pi r, \tag{14.61}$$

where r is the radius of the amperian loop C. Note that this result holds regardless of whether C is inside or outside the wire.

By contrast, the right-hand side of Ampère's law *does* depend on whether we're inside or outside the wire. Inside the wire ($r < a$), the current through S only extends out to the radius of C. A uniformly distributed current has constant current density within the wire — that is, constant current per area $J \equiv I_0/\pi a^2$ — so the right-hand side of Ampère gives

$$I(r < a) = J\pi r^2 = I_0\, \frac{r^2}{a^2}. \tag{14.62}$$

Bringing both sides of Ampère's law together yields

$$\vec{B}(r < a) = \frac{\mu_0 I_0}{2\pi}\, \frac{r}{a^2}\, \hat{\phi}. \tag{14.63a}$$

Outside ($r > a$), the current through S is the total in the wire, $I = I_0$. Thus

$$\vec{B}(r > a) = \frac{\mu_0 I_0}{2\pi r}\, \hat{\phi}. \tag{14.63b}$$

Note that since we could have picked *any* C coincident with a circle of $\vec{B}$, this procedure has actually yielded the function $B = B(r)$. The magnetic field grows linearly with distance, reaches a maximum at the surface of the wire, and then falls off linearly in $1/r$.

The key to this is the high symmetry, from which the precise contours of the field can be determined; since only the magnitude of the field remains to be found, the circulation condition is algebraically sufficient. To use this condition, one must identify curves upon which the field is constant and either parallel or perpendicular to $d\vec{\ell}$. Along such "amperian curves" the field may be pulled out of the circulation integral as in (14.61) — leaving only a simple integral over geometry.

As a physical statement, the right-hand side of Ampère's law gives the circulation of a magnetic field without the need to integrate. But one *can* evaluate a circulation integral directly by parametrizing the curve. As we saw in (14.18), $\dot{\vec{c}}$ is the tangent to the curve, allowing us to write $d\vec{\ell} = \dot{\vec{c}}(t)\, dt$ and expressing the line integral as

$$\int_C \vec{F} \cdot d\vec{\ell} = \int_{t_1}^{t_2} (\vec{F} \cdot \dot{\vec{c}})\, dt = \int_{t_1}^{t_2} F_{\|}\, |\dot{\vec{c}}|\, dt, \tag{14.64}$$

where $F_{\|}$ is the component of $\vec{F}$ parallel to the curve. The integral is now in the form of (14.19) with $f \equiv F_{\|}$, and can be evaluated as with any parametrized scalar integral.

Another way to evaluate these integrals is by expanding the dot product, as demonstrated in the following examples.

Example 14.13   **Work done by $\vec{B} = x^2 y\hat{\imath} - xy^2\,\hat{\jmath}$ from $P_1 = (0,0)$ to $P_2 = (1,1)$**

The work is given by

$$\int_C \vec{B} \cdot d\vec{\ell} = \int_C B_x\, dx + B_y\, dy.$$

We'll calculate this along three different paths.
C_1 : Along this path, the work is

$$\int_{C_1} \vec{B} \cdot d\vec{\ell} = \int_{C_1} x^2 y\, dx - xy^2\, dy$$

$$= \int_0^1 x^2 y\big|_{y=0}\, dx - \int_0^1 xy^2\big|_{x=1}\, dy = -\frac{1}{3}. \tag{14.65}$$

C_2 : Along this path $x = y$ and $dx = dy$, so

$$\int_{C_2} \vec{B} \cdot d\vec{\ell} = \int_{C_2} (x^2 y - xy^2)\big|_{y=x} \, dx = 0. \tag{14.66}$$

C_3 : Along the quarter circle, $x = 1 - \cos\theta$, $y = \sin\theta$, $d\vec{\ell} = (\sin\theta, \cos\theta)\, d\theta$. Then

$$\int_{C_3} \vec{B} \cdot d\vec{\ell} = \int_0^{\pi/2} \left[(1 - \cos\theta)^2 \sin\theta\right] \sin\theta d\theta - \int_0^{\pi/2} \left[(1 - \cos\theta)\sin^2\theta\right] \cos\theta d\theta$$

$$= \frac{3\pi}{8} - 1 \approx 0.18. \tag{14.67}$$

It should come as no surprise that the result integral depends on the chosen path.

Example 14.14 **Work done by $\vec{E} = xy^2\hat{\imath} + x^2 y\hat{\jmath}$ from P_1 to P_2**

Once again, we'll evaluate the work done along the three different paths:
C_1 :

$$\int_{C_1} \vec{E} \cdot d\vec{\ell} = \int_0^1 xy^2\big|_{y=0}\, dx + \int_0^1 x^2 y\big|_{x=1}\, dy = \frac{1}{2}. \tag{14.68}$$

C_2 :

$$\int_{C_2} \vec{E} \cdot d\vec{\ell} = \int_{C_2} (xy^2 + x^2 y)\big|_{y=x}\, dx = 2\int_0^1 x^3\, dx = \frac{1}{2}. \tag{14.69}$$

C_3 : This one's a little messier than its cousin in Example 14.13, but still quite doable:

$$\int_{C_3} \vec{E} \cdot d\vec{\ell} = \int_0^{\pi/2} \left[(1 - \cos\theta)\sin^3\theta + (1 - \cos\theta)^2 \sin\theta\cos\theta\right] d\theta = \frac{1}{2}. \tag{14.70}$$

It's curious that two fields as similar as $\vec{B}$ and $\vec{E}$ should give such disparate results. The work done by $\vec{B}$ is clearly path *dependent* in that the result of the line integral depends upon the curve chosen. As we discussed earlier, this is what we'd expect: identifying only the endpoints of C as it threads through space is generally insufficient to specify an integration path. By contrast, the work done by $\vec{E}$ appears to be path *independent* — the endpoints alone *do* seem to be sufficient! What is it about $\vec{E}$ that accounts for this startling result? Is there a simple characteristic feature or test that will allow us to identify a path-independent field?

Let's try something a little different: integrate from point 1 to point 2 along a path C_1, and then back to 1 along another path C_2 (Figure 14.2). Taken together, these two paths form a *closed* curve $C = C_1 + C_2$; the resulting integral is the circulation, (14.59). Now if a field is path independent, then its line integral from 1 to 2 along C_1 must be the *negative* of that from 2 back to 1 along C_2 — even if the two routes are vastly different. So the circulation around $C = C_1 + C_2$ must vanish. True path independence mandates that this construction holds for *any* path between 1 and 2. Thus the circulation of a path-independent field $\vec{F}$ must vanish for arbitrary C,

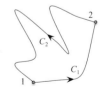

Figure 14.2 The closed curve $C = C_1 + C_2$.

$$\oint_C \vec{F} \cdot d\vec{\ell} = 0 \iff \vec{F} \text{ path independent.} \tag{14.71}$$

If the circulation vanishes for *every* closed curve C, then the integral from 1 to 2 must be the same for *any* path connecting these points. That's path independence.

Note that unlike integrating along an open curve, there is no need to specify integration limits for a circulation — which leaves the direction around C ambiguous. Unless otherwise specified, the usual convention is to integrate counterclockwise around C with the enclosed surface always on your left (as indicated in Figures 14.2). Thus a positive circulation signals an overall counter-clockwise flow of the field, a negative circulation an overall clockwise flow.

From a practical standpoint, path independence allows us the freedom to choose *any* path we wish when evaluating an integral between given endpoints. It should go without saying that one usually chooses the path which renders the integral as simple as possible. So, for instance, in Example 14.14 the integral is easiest to evaluate along C_1. And if you know that $\vec{E}$ is path-independent, then you're guaranteed that this "simplest answer" is the only answer!

Example 14.15 **Circulation of $\vec{B}$ and $\vec{E}$**

If we integrate $\vec{B}$ along C_1 and then back along, say, C_2, we get

$$\Psi = \oint_C \vec{B} \cdot d\vec{\ell} = \int_{C_1} \vec{B} \cdot d\vec{\ell} - \int_{C_2} \vec{B} \cdot d\vec{\ell},$$

where the minus sign comes from integrating *backwards* along C_2. From Example 14.13, we find

$$\oint_{C=C_1-C_2} \vec{B} \cdot d\vec{\ell} = -\frac{1}{3} - 0 = -\frac{1}{3}.$$

If instead, we integrate back along C_3 we get

$$\oint_{C_1-C_3} \vec{B} \cdot d\vec{\ell} = -\frac{1}{3} - \left(\frac{3\pi}{8} - 1\right) = \frac{2}{3} - \frac{3\pi}{8}.$$

So the circulation of $\vec{B}$ is path dependent. By contrast, the circulation of $\vec{E}$ around any combination of the three paths clearly vanishes.

Let's try to understand these results by comparing the field lines of $\vec{E}$ and $\vec{B}$. Once again, imagine that a field represents the flow of water. Suppose that a frictionless wire, bent in the shape of some closed curve C, rests on the water. Strung on the wire are beads which are free to slide along it. How do the beads respond to the water flow? Let's construct C to be everywhere either parallel or perpendicular

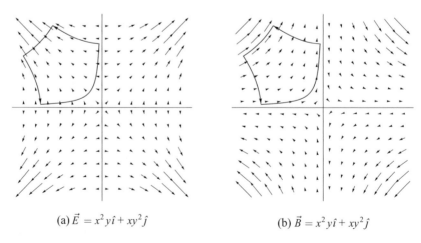

(a) $\vec{E} = x^2 y\hat{\imath} + xy^2\hat{\jmath}$ (b) $\vec{B} = x^2 y\hat{\imath} + xy^2\hat{\jmath}$

Figure 14.3 The motion of a bead on a frictionless wire.

to the field, as shown in Figure 14.3; the similarity of $\vec{E}$ and $\vec{B}$ allows us to use the same curve for both fields (Problem 14.17). Two segments of the closed wire in $\vec{E}$ are everywhere perpendicular to the field, and so do not affect on the beads. They make no contribution to the circulation. The other two segments, however, make compensating clockwise and counter-clockwise contributions. So $\vec{E}$ has zero circulation along this wire — and in fact, along any closed curve one could construct. Note that the same wire in $\vec{B}$ also has two segments which are perpendicular to the field. The net effect on the beads due to the other two, however, is less clear. Along the top left segment the field is strong, but the path short; the bottom segment is long, but the field relatively weak. Although these contributions have opposite effects on the beads, it's not at all obvious from Figure 14.3b whether they precisely cancel. In fact, we know from Example 14.15 that they don't. Since we know the circulation of $\vec{B}$ is negative, then it must be that the clockwise contribution along the shorter segment dominates the counter-clockwise contribution of the longer one.

Example 14.16 $\oint \vec{r} \cdot d\vec{\ell}$

As a purely radial field, $\vec{r}$ has no sense of circulation, and so should be path independent. To support this expectation, integrate along a circle of radius a centered at the origin. In polar coordinates, the tangent to the circle is $d\vec{\ell} = \hat{\phi} \, a \, d\phi$, and so $\vec{r} \cdot d\vec{\ell} \equiv 0$. So indeed, the circulation around this curve vanishes. But perhaps the circle is special — so let's try integrating along a square of side 2ℓ. Along segments a and c, $d\vec{\ell} = \hat{j} \, dy$, while along b and d $d\vec{\ell} = \hat{i} \, dx$. Using $\vec{r} = x\hat{i} + y\hat{j}$ gives

$$\oint_C \vec{r} \cdot d\vec{\ell} = \int_a \vec{r} \cdot d\vec{\ell} + \int_b \vec{r} \cdot d\vec{\ell} + \int_c \vec{r} \cdot d\vec{\ell} + \int_d \vec{r} \cdot d\vec{\ell}$$

$$= \int_{-\ell}^{\ell} y \, dy + \int_{-\ell}^{\ell} x \, dx + \int_{\ell}^{-\ell} y \, dy + \int_{\ell}^{-\ell} x \, dx = 0. \tag{14.72}$$

The presentation of Example 14.16 is very suggestive, but not definitive. The next thing we might try is a path that does not encircle the origin. But testing only a handful of paths is no proof — and examining an infinity of paths can be time consuming.

There's a simpler way. If the integral between two points does not depend upon the choice of path, then the only thing it *can* depend on are the endpoints themselves. So a path-independent integral must be given by the values of some function Φ at these points — specifically the difference

$$\int_1^2 \vec{F} \cdot d\vec{\ell} = \Phi(2) - \Phi(1), \tag{14.73a}$$

since the path leaves point 1 and arrives at point 2. Hence a path-independent vector field is naturally associated with a scalar function. This sounds suspiciously similar to an earlier observation: any continuous scalar field Φ is naturally associated with the vector field $\vec{\nabla}\Phi$. Moreover, the fundamental theorem of calculus applied to a gradient field looks a lot like (14.73a),

$$\int_1^2 \vec{\nabla}\Phi \cdot d\vec{\ell} = \Phi(2) - \Phi(1). \tag{14.73b}$$

Even though $\vec{F}$ and $\vec{\nabla}\Phi$ both integrate to the same difference, in order to conclude they're the same field, the integrals must hold for *any* path connecting points 1 and 2. But that's precisely the path independence that (14.73a) describes! Thus any path-independent field $\vec{F}$ has not merely an associated scalar function Φ, but in fact a scalar potential, $\vec{F} = \vec{\nabla}\Phi$.

Example 14.17   It's Got Potential!

We've claimed that $\vec{E} = xy^2\hat{i} + x^2y\hat{j}$ is path independent, but we have yet to actually prove it. To do that, we can try to find a scalar potential Φ such that $\vec{\nabla}\Phi = \vec{E}$. So we need to solve the simultaneous component equations. For E_x we find

$$\frac{\partial\Phi}{\partial x} = xy^2 \implies \Phi(x,y) = \frac{1}{2}x^2y^2 + f(y), \tag{14.74a}$$

where $f(y)$ is some unknown function of y. Similarly,

$$\frac{\partial\Phi}{\partial y} = x^2y \implies \Phi(x,y) = \frac{1}{2}x^2y^2 + g(x), \tag{14.74b}$$

where this time we have the unknown function $g(x)$. These results can easily be made consistent by setting $f(y) = g(x) = k$, a constant. Thus $\Phi = \frac{1}{2}x^2y^2 + k$ — as can be trivially verified by evaluating its gradient.

What about a field like $\vec{B} = x^2y\hat{i} - xy^2\hat{j}$ which is *not* path independent? Can't we use the same procedure to find a potential for $\vec{B}$? Let's see:

$$\frac{\partial\Phi}{\partial x} = x^2y \implies \Phi(x,y) = \frac{1}{3}x^3y + f(y) \tag{14.75a}$$

$$\frac{\partial\Phi}{\partial y} = -xy^2 \implies \Phi(x,y) = -\frac{1}{3}xy^3 + g(x). \tag{14.75b}$$

No choice of the one-variable functions f and g can make these two expressions consistent. There is no function Φ whose gradient yields $\vec{B}$.

Recall that the curl of a gradient is identically zero — so a vector field which has a potential must satisy $\vec{\nabla} \times \vec{F} = \vec{\nabla} \times \vec{\nabla}\Phi = 0$. In other words, *irrotational fields are path independent.*[4] Thus we have three equivalent mathematical statements for the path independence of a field $\vec{F}$:

$$\oint_C \vec{F} \cdot d\vec{\ell} = 0, \text{ for all } C \tag{14.76a}$$

$$\vec{F} = \vec{\nabla}\Phi \tag{14.76b}$$

$$\vec{\nabla} \times \vec{F} = 0. \tag{14.76c}$$

Path independence, (14.76a), implies the existence of a scalar function as specified by (14.76b); (14.76c) follows automatically.

The first expression in (14.76) is the most intuitive statement of path independence, the second the most profound. But by far the simplest *test* of path independence is to calculate the field's curl. For instance, you can use the curl test to easily verify that both the position $\vec{r}$ and the field $\vec{E} = xy^2\hat{i} + x^2y\hat{j}$ are path independent — and that $\vec{B} = x^2y\hat{i} - xy^2\hat{j}$ is not.

---

[4] An important caveat to this statement is addressed in Chapter 17.

Example 14.18 **Path-Independent Forces**

The simplest example of a path-independent field is a constant — for example, gravity near the surface of the Earth, $m\vec{g}$, or a uniform electrostatic field such as $\vec{E} = E_0 \hat{j}$. Actually, all gravitational and electrostatic forces are path independent. So are spring forces. Friction and electrodynamic forces, however, are path dependent.

BTW 14.3 Conservative Fields

Using the curl to test for path independence of vector fields involves only derivatives in space. Similarly, the integral of (14.76a) is a sum only over the spatial characteristics of the path. One might conclude, then, that the concept of path independence need not be concerned about such things as when or how fast one treks from point to point.

But consider friction. Though it may be physically intuitive that this is a path-dependent force, as a constant force μmg (with μ the coefficient of friction) it appears to pass the curl test! But not so fast. Unlike, say, gravity near the surface of Earth which is constant both in magnitude *and* direction, friction has an unusual vector nature: it always points opposite the direction of motion $\hat{v} \equiv \vec{v}/|\vec{v}|$. As a result $\vec{F} \cdot d\vec{\ell} \sim \hat{v} \cdot \vec{v}\,dt = v\,dt = d\ell$, and the work done by friction depends on the length of the path. So much for path independence. Similarly, drag forces, whether proportional to v or v^2, depend upon the motion along the path and are also path dependent. Then there's the magnetic force on a moving electric charge, $\vec{F} = q\vec{v} \times \vec{B}$. In this case, $\vec{F} \cdot d\vec{\ell} \equiv 0$ no matter the path — so the circulation always vanishes simply because it never does any work at all! Path independence and vanishing curl are not equivalent in these cases. Indeed, the argument leading to (14.73a) considers only the route through space; the very idea of a potential Φ which depends on $\vec{v}$ makes no sense in that context. Time-dependent fields are also problematic: what sense does it make to single out fields which may be path independent one moment but not another?

To account for these possibilities, the very concept of "path" would have to be generalized to include such considerations as how or when the path is traversed: fast or slow, yesterday or tomorrow, etc. Thus time- and velocity-dependent forces are excluded from our discussion; we will consider only fields which are functions of position. Irrotational fields which are functions only of position are called *conservative*.

By far the most important practical significance of a path-independent field is the existence of the scalar potential Φ,

$$\Phi(\vec{r}) \equiv \int_{\vec{a}}^{\vec{r}} \vec{F} \cdot d\vec{\ell}, \tag{14.77}$$

where $\Phi(\vec{a}) = 0$. Physically, the interpretation of Φ is often related to energy. If $\vec{F}$ is a force, then Φ is the negative of the potential energy U — hence the well-known relationship $\vec{F} = -\vec{\nabla}U$. If $\vec{F}$ is a gravitational or electrostatic field, then Φ is the (negative) energy per unit mass or unit charge. In the case of gravity, this is simply known as the "gravitational potential"; for electricity, it's the "electric potential" or voltage.

Mathematically, working with the scalar field Φ is often easier than working with a vector $\vec{F}$. Moreover, once the functional form of Φ is known, all line integrals of $\vec{F}$ between points P_1 and P_2 reduce to computing $\Phi(P_2) - \Phi(P_1)$. Thus two potentials which differ by a constant, $\Phi' = \Phi + \alpha$, are physically indistinguishable, $\vec{\nabla}\Phi = \vec{\nabla}\Phi'$. Any choice of α is one of convenience, equivalent to the choice of the reference point $\vec{a}$ in (14.77).

Example 14.19 **Gravitational Field of a Spherical Shell**

A point mass M has gravitational field

$$\vec{g} = -GM\hat{r}/r^2. \tag{14.78}$$

By convention, the gravitational potential is the negative of the line integral, so

$$\Phi(r) \equiv -\int_{\infty}^{r} \vec{g} \cdot d\vec{\ell} = +GM \int_{\infty}^{r} \frac{dr'}{r'^2} = -\frac{GM}{r}, \tag{14.79}$$

where we've chosen to measure with respect to infinity, since $1/r$ goes to zero there anyway. Note that it was not necessary to choose a path in order to evaluate this integral.

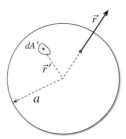

Figure 14.4 Spherical mass shell.

For more complicated mass distributions with volume density $\rho(\vec{r})$, we can let $\vec{g} \rightarrow d\vec{g}$ and $M \rightarrow dM = \rho dV$ in (14.78), and integrate to find $\vec{g}$. But why go to all the trouble of evaluating this vector-valued integral, when the same substitutions in (14.79) yield a scalar integral? In other words, it's simpler to work out Φ and use it to obtain $\vec{g} \equiv -\vec{\nabla}\Phi$. Consider the famous and historically important case of a spherically symmetric distribution of mass — in particular, a spherical shell of mass M and radius a with uniform mass distribution $\sigma = M/4\pi a^2$ (Figure 14.4). An infinitesimal area dA' has mass $dm = \sigma \, dA'$, generating a field at $\vec{r}$ given by (14.78),

$$d\vec{g} = -G \frac{(\vec{r} - \vec{r}')}{|\vec{r} - \vec{r}'|^3} \sigma \, dA'. \tag{14.80}$$

Note that $\vec{r}$ is the point at which the field $\vec{g}$ is evaluated, whereas $\vec{r}'$ is the position of the source mass $\sigma \, dA'$ which contributes to $d\vec{g}$. We can then add up all the source mass to find the total field at $\vec{r}$,

$$\vec{g}(\vec{r}) = -G \int_S \frac{(\vec{r} - \vec{r}') \sigma \, dA'}{|\vec{r} - \vec{r}'|^3}, \tag{14.81}$$

where the integral is over the entire spherical surface. This vector-valued integral must generally be decomposed it into separate component integrals. That's way too much work (although in this example the spherical symmetry certainly helps). It's far easier to use the right-hand side of (14.79) and the principle of superposition to set up the corresponding scalar integral for Φ,

$$\Phi(\vec{r}) = -G \int_S \frac{\sigma \, dA'}{|\vec{r} - \vec{r}'|}. \tag{14.82}$$

Recognizing that $|r'| = a$, the law of cosines gives $|\vec{r} - \vec{r}'| = \sqrt{r^2 + a^2 - 2ra\cos\theta}$. Then, with $dA = a^2 \, d\Omega = 2\pi a^2 \, d(\cos\theta)$, we find

$$\Phi(\vec{r}) = -2\pi G \sigma a^2 \int_{-1}^{1} \frac{d(\cos\theta)}{\sqrt{r^2 + a^2 - 2ra\cos\theta}}$$

$$= -\frac{2\pi G \sigma a}{r}\Big[(r+a) - |r-a|\Big]. \tag{14.83}$$

In order to proceed, we need to consider two possibilities: for $\vec{r}$ inside the sphere we have $|r-a| = a - r > 0$, so that

$$\Phi(r < a) = -4\pi G \sigma a = -\frac{GM}{a}, \tag{14.84a}$$

a constant. We thus recover the familiar result that the gravitational field inside a spherically symmetric shell vanishes. On the other hand, outside the sphere $|r-a| = r - a > 0$, so

$$\Phi(r > a) = -\frac{4\pi G \sigma a^2}{r} = -\frac{GM}{r}. \tag{14.84b}$$

Taking the (negative) gradient yields the same field as for a point mass, (14.78). Thus the famous result that the gravitational field outside a spherically symmetric distribution of mass is equivalent to that of a point of the same total mass located at the center. Though we derived this for a spherical shell, it's obviously true as well for a spherically symmetric solid ball, which can be viewed as a collection of nested spherical shells. The calculation is also relevant to electrostatics: direct comparison requires only the substitution $-GM \longrightarrow Q/4\pi\epsilon_0$, where Q is the total charge.

Incidentally, notice that the two expressions (14.84) agree at $r = a$. In other words, Φ is a continuous function — an important consideration, given that Φ must be differentiable if we're to find $\vec{g}$.

Example 14.20 Thermodynamic State Variables

Consider an ideal gas in equilibrium. It has well-defined values of quantities such as absolute temperature T, pressure p, volume V, and internal energy U. These are all examples of what are known as thermodynamic state variables: their values fully characterize the thermodynamic state of the system. For simple systems such as a classical ideal gas, only two variables are actually required, with others determined by, say, the equation of state $pV = nRT$ and the relation between internal energy and temperature, $U = C_V T$. (Here, n is the number of moles of gas, C_V is the heat capacity of the gas at constant volume, and $R = 8.314$ J-K^{-1}mol^{-1} is the gas constant.) So the space of states is two-dimensional — the most familiar choice leading to the famed p–V diagram.

Each point in this space corresponds to definite values of all macroscopic state variables. For example, if a bottle of gas starts with pressure p_1 and volume V_1, and ends at p_2 and V_2, the gas will go from temperature T_1 to T_2 independent of the path traversed through state space (Figure 14.5). Whether we change V while holding p fixed followed by changing p with fixed V or vice-versa — or for that matter change both simultaneouly — has no effect on the final temperature. Thermodynamic state variables are path-independent.

Strictly speaking, state variables are only defined for systems in equilibrium. Thus these integrals are only defined for processes slow enough to keep the system in equilibrium at all times. Only for such *quasi-static* processes do paths C even exist in the space of states. On the other hand, as long as the initial and final states are equilibrium states, we can *calculate* the change in a state variable by choosing any path C through the space of states we wish — even if the actual laboratory process left the space.

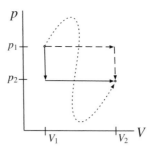

Figure 14.5 Three thermodynamic processes between (p_1, V_1) and (p_2, V_2).

Of course, there are thermodynamic quantities which *are* process dependent — the two most common being work W and heat Q. A system neither *contains* work the way it contains energy, nor does it *have* a work the way it has a temperature. So W cannot be a state variable. Similarly, despite common parlance — as well as the unfortunate historical designation of C_V as "heat capacity" — a system does not contain heat. Both heat and work represent *transfers* of energy, and are related by the First Law of Thermodynamics,

$$\dbar Q = dU + \dbar W, \tag{14.85}$$

where $\dbar Q$ is the heat transferred into the system, and $\dbar W$ the work done by system on its surroundings. The $\dbar$ notation indicates that neither $\dbar Q$ nor $\dbar W$ are exact differentials — in other words, their integrals yield path-dependent measures.

This distinction between state and non-state variables is easily demonstrated for a classical ideal gas. Plugging $dU = C_V dT$ and $\dbar W = p dV$ into the First Law we get

$$\dbar Q = C_V dT + p dV. \tag{14.86}$$

So the heat transferred during a process from an initial state with T_1 and V_1 to a final state with T_2 and V_2 is

$$Q = \int_{T_1}^{T_2} C_V dT + \int_{V_1}^{V_2} p \, dV$$

$$= C_V (T_2 - T_1) + \int_{V_1}^{V_2} \frac{nRT}{V} dV. \tag{14.87}$$

The T integral is trivial since the heat capacity is constant for a classical ideal gas. The work integral $\int p \, dV$, however, cannot be evaluated without knowing how either pressure or temperature changes with volume. For example, a constant-pressure process will clearly yield work $p \, \Delta V$, whereas a constant-temperature process yields the completely different $nRT \ln (V_2/V_1)$. Heat and work are very much path dependent, and hence not state variables.

Ultimately it is because such path-dependent quantities exist that distinguishing different processes is important; common ones include isothermal (constant T), isobaric (constant p), and adiabatic ($Q=0$). Much of classical thermodynamics is concerned with different ways of combining various processes in order to extract useful work with minimal energy wasted as heat.

Example 14.21 **Discovering Entropy**

For some quantity S to qualify as a thermodynamic state variable, its value can only depend on the state of the system, not on any process which brought the system to that state. Thus any change in S can depend only on its initial and final values. In short, S must be path independent,

$$\oint_C dS = \oint_C \vec{\nabla} S \cdot d\vec{\ell} = 0, \tag{14.88}$$

where C is any closed curve in state space.

We saw in the previous example how Q is not a state variable. But if we divide it by temperature, the path-dependent work integral becomes path *in*dependent. Picking up from (14.86):

$$\frac{d\!\!\!/Q}{T} = \frac{1}{T} C_V \, dT + \frac{1}{T} p \, dV$$

$$= C_V \frac{dT}{T} + \frac{1}{T} \frac{nRT}{V} dV = d \left[C_V \left(\ln T \right) + nR \left(\ln V \right) \right]. \tag{14.89}$$

So the quantity $dS \equiv d\!\!\!/Q/T$ is an exact differential; for any process, for any path through state space, the change

$$\Delta S = \int_1^2 \frac{d\!\!\!/Q}{T} = C_V \ln(T_2/T_1) + nR \ln(V_2/V_1). \tag{14.90}$$

depends only on the initial and final values of temperature and volume. In particular,

$$\oint_C \frac{d\!\!\!/Q}{T} = 0. \tag{14.91}$$

The macroscopic state variable S is called the *entropy* of the system. (An adiabatic process, $Q = 0$, is thus also isentropic.) Although a physical interpretation of entropy is not forthcoming from this definition, a statistical mechanics approach shows that entropy is a measure of the number of microstates Ω — that is, the number of different position and momentum configurations of gas molecules — which correspond to a single, given macroscopic thermodynamic state. Specifically, $S = k \log \Omega$, where k is Boltzmann's constant.

BTW 14.4 Conservation of Energy

Just as change in velocity $\vec{v}$ is evidence of a force, so too change in kinetic energy T is evidence of work. (Despite possible confusion with temperature, it's conventional to use T for kinetic energy — probably due to its alphabetical proximity to potential energy U.) The work done by a force $\vec{F} = m\vec{a}$ in accelerating a particle from point 1 to point 2 is related to ΔT by the Work–Energy theorem,

$$W \equiv \int_1^2 \vec{F} \cdot d\vec{\ell} = \int_1^2 m\dot{\vec{v}} \cdot \vec{v} \, dt$$

$$= \int_1^2 \frac{d}{dt} \left(\frac{1}{2} mv^2 \right) dt = \frac{1}{2} mv_2^2 - \frac{1}{2} mv_1^2 = \Delta T. \tag{14.92}$$

Note that this does *not* imply that all forces are path independent, since the right-hand side is the difference in a scalar function of speed, not position. But for conservative forces, the work done can *also* be expressed in terms of $\Phi(\vec{r})$,

$$\int_1^2 \vec{F} \cdot d\vec{\ell} = \Phi(2) - \Phi(1),$$

regardless of the path from 1 to 2. No such simplification occurs for non-conservative forces; if they act on the particle, then a path must be specified. So the total work can be simplified no further than

$$W = \Delta\Phi + W_{nc}, \tag{14.93}$$

where W_{nc} denotes the work integral of any non-conservative forces along a particular path. The similarity with the First Law of Thermodynamics, (14.85), is hardly coincidental. The Work–Energy theorem, however, provides us additional insight into mechanical systems: comparing (14.92) and (14.93) gives

$$\Delta T - \Delta\Phi = W_{nc}.$$

Defining the *potential energy* $U(\vec{r}) \equiv -\Phi(\vec{r})$ so that

$$\Delta U \equiv -\Delta\Phi = -\int \vec{F} \cdot d\vec{\ell}, \tag{14.94}$$

we get

$$\Delta T + \Delta U = W_{nc}. \tag{14.95}$$

Then in terms of the total mechanical energy $E \equiv T + U$, the Work–Energy theorem becomes

$$\Delta E = W_{nc}. \tag{14.96}$$

If *only* conservative forces act, $W_{nc} = 0$ and

$$\Delta E = 0. \tag{14.97}$$

Thus conservative forces lead directly to conservation of mechanical energy.

14.4 Flux

Circulation measures the extended swirl of a vector field $\vec{E}$ around a curve. Another measure of a vector field is how strongly it surges through some region. To do this, we define the *flux* as the product of the field's magnitude and the surface area A that it crosses,[5]

$$\Phi \equiv EA.$$

Using the flow-of-water metaphor, this makes sense: the flux should grow both with the magnitude of the field and the area of the surface. But what if only one component of $\vec{E}$ crosses A? Or the extreme case of a field that merely skips along the surface, but does not actually pierce it? This should be a zero flux situation. Clearly our simple definition of Φ must be refined.

---

[5] Context should suffice to distinguish whether Φ denotes flux or scalar potential.

Because it is embedded in a higher-dimensional space, a surface is not fully identified without giving its *orientation* within the larger space. This is done by specifying the unit normal $\hat{n}$. So a better definition of flux is

$$\Phi \equiv \vec{E} \cdot \hat{n}\, A.$$

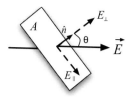

Figure 14.6 Flux of $\vec{E}$ through $\vec{A} = \hat{n}A$.

This has all the intuitive advantages of the earlier definition, but correctly gives zero for a component that skips along the surface (Figure 14.6). The dot product picks out the component $E_\perp$ that actually pierces the surface; the parallel component $E_\parallel$ resides *in* the surface and so makes no contribution to Φ.

Good. Now what about a surface that curves, so that the unit normal varies along it? In this case, we must break the surface into small flat regions, each with its own $\hat{n}$; the total flux through the surface is the sum over these regions. When the regions are infinitesimal with area da, the total flux through a surface S is the sum over all its infinitesimal regions,

$$\Phi \equiv \int_S \vec{E} \cdot \hat{n}\, da = \int_S \vec{E} \cdot d\vec{a}, \tag{14.98}$$

where we have defined the vector-valued area element $d\vec{a} \equiv \hat{n}\, da$.

In cartesian coordinates, the element da is usually rendered $dxdy$, $dydz$, or $dzdx$, depending on the surface. On a sphere of radius R, we have $da = R^2 d\Omega = R^2 \sin\theta\, d\theta d\phi$. On a cylindrical face of radius R, $da = Rd\phi dz$; on a cylinder's endcaps, $da = \rho\, d\rho d\phi$. The integral, of course, is independent of the choice of coordinates.

Note the similarities with line integrals. Just as a path has a natural direction, so too does a surface. And just as we use the vector-valued path element $d\vec{\ell}$ to define a scalar one-dimensional integral of a vector, so too we use the vector-valued area element $d\vec{a}$ to define a scalar two-dimensional integral of a vector.

When the line integral (14.98) is over a *closed* surface, it is called the *net* flux of the field,

$$\Phi_{\text{net}} \equiv \oint_S \vec{E} \cdot d\vec{a}. \qquad \text{[Net flux]} \tag{14.99}$$

Example 14.22 **Continuity — Integral Form**

The image of a quantity surging through a surface is very intuitive, and makes flux a fairly easy concept to wield. As an example, consider the conservation of electric charge. Mathematically, conservation is expressed by *continuity*: the rate at which charge Q_{in} within some volume is lost must be accounted for by the current I_{out} flowing out of the volume,

$$I_{\text{out}} = -\frac{dQ_{\text{in}}}{dt}. \tag{14.100}$$

(Since the charge inside is decreasing, we have $dQ/dt < 0$; the minus sign thus fixes the convention that outward flowing current is positive.) A quantity Q which obeys this relation is never created or destroyed: any change in Q in some region can be traced continuously as it leaves the region.

The labels "out" and "in" are cumbersome; better to incorporate these in a more mathematical fashion. So define the usual charge density ρ as

$$Q_{\text{in}} = \int_V \rho \, d\tau. \tag{14.101}$$

The movement of the charge density ρ with an average "drift" velocity $\vec{v}$ gives the *current density* $\vec{J} \equiv \rho \vec{v}$. Since ρ has units of charge per volume, $\vec{J}$ has units of current per area (measured in the plane perpendicular to $\vec{v}$). The usual current I measured by an ammeter is just the flux of $\vec{J}$ through a surface,

$$I = \int_S \vec{J} \cdot d\vec{a}. \tag{14.102}$$

Note that both Q and I are extended measures, whereas the fields $\rho(\vec{r})$ and $\vec{J}(\vec{r})$ are local quantities.

Now continuity demands an accounting of *all* charge going in or out of a region — in other words, it is concerned with the *net* flux of $\vec{J}$. This, finally, is I_{out},

$$I_{\text{out}} = \oint_S \vec{J} \cdot d\vec{a}. \tag{14.103}$$

So the continuity equation, (14.100), becomes

$$\oint_S \vec{J} \cdot d\vec{a} = -\frac{d}{dt} \int_V \rho \, d\tau, \tag{14.104}$$

where the surface S bounds the volume V. If charge suddenly appears, it didn't come out of nowhere: it had to have been carried there from somewhere.

The integral equation of continuity, (14.104), is a clear statement of a crucial property of electric charge. But evaluating the integrals can often be quite challenging. So as an instruction to calculate, (14.104) leaves much to be desired. Later we'll see how to recast this integral statement into a differential equation.

Example 14.23 **Net Flux and Gauss' Law**

Perhaps the most famous net flux appears in Gauss' law for electric fields,

$$\oint_S \vec{E} \cdot d\vec{a} = q_{encl}/\epsilon_0, \tag{14.105}$$

where q_{encl} is the total charge enclosed by S. This has the status of a physical law because it gives the net flux for any electric field. Even so, Gauss' law can only be used to *solve* for $\vec{E}$ in the relatively few situations in which the symmetry of the configuration is high enough to determine the precise contours of the field. The idea is similar to what we saw in Example 14.12 for circulation: the symmetry must be sufficient to enable us to find surfaces upon which the field is constant and either parallel or perpendicular to $d\vec{a}$. Evaluating the flux on such "Gaussian surfaces" allows the field to be pulled out of the integral — leaving only a simple integral over geometry.

As an example, consider a uniformly charged solid ball of total charge Q_0 and radius a (Figure 14.7). Now since the charge is spherically symmetric, the $\vec{E}$ field must form radial lines centered on the ball. Moreover, the magnitude of the field can only depend on the distance from the center of the sphere. So $\vec{E} = E(r)\hat{r}$. The symmetry leaves no alternatives. Thus even before we start to calculate, we know quite a bit about the field. We now use this insight to *choose* a surface S which, if concentric with the ball of charge, will be everywhere perpendicular to $\vec{E}$. With this choice, not only is $\vec{E}$ always parallel to $d\vec{a} = \hat{n} \, da$, but its magnitude is constant everywhere on S. Thus the flux integral becomes

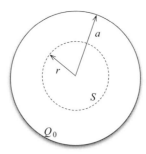

Figure 14.7 Gaussian surface S for a uniform ball of charge.

$$\oint_S \vec{E} \cdot d\vec{a} = \oint_S E \, da = E \oint_S da = E \cdot 4\pi r^2, \qquad (14.106)$$

where r is the radius of the Gaussian surface S. Note that this result holds regardless of whether S is inside or outside the ball.

By contrast, the right-hand side of Gauss' law *does* depend on whether S is inside or outside the ball. Outside ($r > a$), the charge contained in S is the total charge in the ball, $q_{encl} = Q_0$. Bringing both sides of Gauss' law together yields the familiar

$$\vec{E}(r > a) = \frac{Q_0}{4\pi \epsilon_0 r^2} \hat{r}. \qquad (14.107)$$

Inside the sphere ($r < a$), the charge enclosed only extends out to the radius of S. A uniformly distributed charge has constant density — that is, $\rho = Q_0/\frac{4}{3}\pi a^3$ — so the right-hand side of Gauss' law gives

$$q_{encl}(r < a) = \rho \frac{4}{3}\pi r^3 = Q_0 \frac{r^3}{a^3}. \qquad (14.108)$$

Thus

$$\vec{E}(r < a) = \frac{Q_0}{4\pi \epsilon_0} \frac{r}{a^3} \hat{r}. \qquad (14.109)$$

Note that since we could have picked *any* S concentric with the ball of charge, this procedure has actually yielded the function $E = E(r)$. The electric field grows linearly with distance, reaches a maximum at the surface of the sphere, and then falls off quadratically.

BTW 14.5 Gauss' Law in Different Dimensions

Unlike Coulomb's law, which is only valid for electro*statics*, Gauss' law holds in electro*dynamics* as well. The key physical implication of Gauss' law is that the net flux is the same for all surfaces enclosing the same net charge. If we take this to be a fundamental rather than a derived property of $\vec{E}$ fields, we can construct toy models in other dimensions.

Consider a point charge in a world with n spatial dimensions. Even in n dimensions, a point charge will have (hyper)spherical symmetry, so we take the surface S in Gauss' law to be the $(n-1)$-dimensional sphere enclosing q at its center. Thus the integral

$$\oint_S \vec{E} \cdot d\vec{a} = q/\epsilon_0,$$

is just the magnitude E times the area (or if you prefer, volume) of the hypersphere. As we found in Problem 8.23, that area is r^{n-1} times the area of the unit sphere S^{n-1},

$$\text{area}\left(S^{n-1}\right) = \frac{2\pi^{n/2}}{\Gamma(n/2)}.$$

So

$$E(r) = \frac{\Gamma(n/2)}{2\pi^{n/2}\epsilon_0} \frac{q}{r^{n-1}}.$$

You should verify that this correctly reproduces our $n = 3$ world. As for other worlds, we discover that in a two-dimensional universe, where the "area" of the circle S^1 grows with r, the electric field falls like $1/r$. Similarly, in a four-dimensional universe the "area" of S^3 grows like r^3, so $\vec{E}$ would fall like $1/r^3$.

As a physical statement, the right-hand side of Gauss' law gives the net flux of an electric field without having to integrate. But one *can* evaluate a flux integral of a specified function by decomposing the field along the surface — as the following examples demonstrate.

Example 14.24   Flux of $\vec{E} = xy^2 \hat{i} + x^2 y \hat{j}$

We calculate the flux through two different open surfaces, with normals chosen consistent with positive counter-clockwise circulation and the right-hand rule:

S_1: *A quarter circle of radius a in the yz-plane at $x = b$* With the unit normal along positive x, the area element is $d\vec{A} = \hat{i}\,dydz|_{x=b}$ in cartesian coordinates — though of course, the integration is best done in polar coordinates:

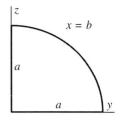

$$\int_{S_1} \vec{E} \cdot d\vec{a} = \int_{S_1} xy^2\,dy\,dz = \int_{S_1} xy^2\,r\,dr\,d\phi$$

$$= b \int_0^a r^3\,dr \int_0^{\pi/2} \cos^2\phi\,d\phi = \frac{\pi}{16}a^4 b. \qquad (14.110)$$

S_2: *A quarter cylinder of radius a, with $b < x < 2b$* Leaving the cylinder open at the bottom, there are four surfaces s_i to integrate over. The left rectangle s_1 along $z = 0$ has area element $d\vec{a}_1 = -\hat{k}\,dxdy|_{z=0}$. Similarly, the back rectangle s_2 at $y = 0$ has $d\vec{a}_2 = -\hat{j}\,dxdz|_{y=0}$. The quarter-circle cap s_3 at $x = 2b$ has area element $d\vec{a}_3 = \hat{i}\,\rho\,d\rho\,d\phi|_{x=2b}$, and on the cylindrical face s_4 we have $d\vec{a}_4 = \hat{\rho}\,a\,d\phi dx|_{\rho=a}$. The flux, then, is

$$\int_{S_2} \vec{E} \cdot d\vec{a} = \sum_i \int_{s_i} \vec{E} \cdot d\vec{a}_i$$

$$= 0 + 0 + \int_{s_3} xy^2 \rho\,d\rho d\phi|_{x=2b} + \int_{s_4} \vec{E} \cdot \hat{\rho}\,a\,d\phi dx|_{\rho=a}, \qquad (14.111)$$

where the dot product produces zero flux through s_1, and $\vec{E}$ vanishes on s_2. For the remaining integrals, we need to carefully distinguish between $y = \rho\cos\phi$ on s_3, and $y = a\cos\phi$ on s_4 — and on which we also need $\hat{\rho} = \cos\phi\,\hat{j} + \sin\phi\,\hat{k}$. All together then, we have

$$\int_{S_2} \vec{E} \cdot d\vec{a} = 2b \int_0^a \rho^3 d\rho \int_0^{\pi/2} \cos^2 \phi \, d\phi + a^2 \int_b^{2b} x^2 \, dx \int_0^{\pi/2} \cos^2 \phi \, d\phi$$

$$= \frac{\pi}{8} a^4 b + \frac{7\pi}{12} a^2 b^3 = \frac{\pi}{8} a^2 b \left(a^2 + \frac{14}{3} b^2 \right), \tag{14.112}$$

where we used results from Section 8.4.

Example 14.25 **Flux of $\vec{B} = x^2 y \hat{i} - xy^2 \hat{j}$**

Let's find the flux through the same two surfaces — plus one more for good measure:

S_1: The set-up is virtually identical as before, leading to the result

$$\int_{S_1} \vec{B} \cdot d\vec{a} = \frac{1}{3} a^3 b^2. \tag{14.113}$$

S_2: As before, the flux through the two rectangular faces vanishes, leaving

$$\int_{S_2} \vec{B} \cdot d\vec{a} = \int_{S_3} x^2 y \rho \, d\rho d\phi |_{x=2b} + \int_{S_4} \vec{B} \cdot \hat{\rho} \, a \, d\phi dx |_{\rho=a}$$

$$= 4b^2 \int_0^a \rho^2 d\rho \int_0^{\pi/2} \cos \phi \, d\phi - a^3 \int_b^{2b} x \, dx \int_0^{\pi/2} \cos^3 \phi \, d\phi$$

$$= \frac{1}{3} a^3 b^2. \tag{14.114}$$

Curious. Let's look at one more surface —

S_3: *Octant of a sphere, radius a seated in yz-plane at $x = b$* Leaving open the quarter circle in the yz-plane, there are three surfaces to integrate over. The quarter circle at $z = 0$ has $d\vec{a}_1 = -\hat{k} \, dxdy |_{z=0}$, and the one at $y = 0$ has $d\vec{a}_2 = -\hat{j} \, dxdz |_{y=0}$. Neither contributes to the flux. This leaves only the spherical surface $d\vec{a}_3 = \hat{r} \, a^2 d\Omega$. Even with the results of Section 8.4, this is messy. And tedious. But nonetheless straightforward. After much algebra, we find (Problem 14.25)

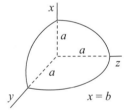

$$\int_{S_3} \vec{B} \cdot d\vec{a} = \frac{1}{3} a^3 b^2. \tag{14.115}$$

It's intriguing that two fields as similar as $\vec{E}$ and $\vec{B}$ should give such disparate results. The flux of $\vec{E}$ clearly depends upon the surface chosen. This, of course, is what we'd expect. By contrast, the flux of $\vec{B}$ is the same through all three surfaces.

If you're feeling a sense of déjà vu, it's no accident. Have a look back at Examples 14.13 and 14.14, where we examined the line integrals of the same vector fields $\vec{E}$ and $\vec{B}$. That analysis led to the concept of path independence. Recall that a field is path-independent if the line integral is the same along any curve *connecting the same points*. Thus we concluded there must be a scalar field Φ such that

$$\int_C \vec{E} \cdot d\vec{\ell} = \Phi(b) - \Phi(a), \quad \text{[Path independence]} \tag{14.116a}$$

where the integration is along any path C from a to b. We can think of the points a and b as the zero-dimensional boundary of the one-dimensional path.

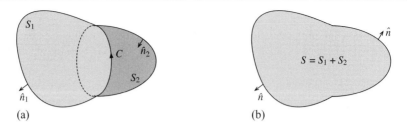

Figure 14.8 (a) Open surfaces bounded by a curve; (b) a closed surface has no bounding curve.

The similarity to the surface integrals of Example 14.25 is striking. In fact, we purposely chose the surfaces to all have the same boundary. The quarter circle in the yz-plane is the one-dimensional boundary of each of the two-dimensional open surfaces. So we propose that the flux of a "surface-independent" field must be the same through any surface *bounded by the same closed curve*. This requires the existence of a field $\vec{A}$, called the *vector potential*, such that[6]

$$\int_S \vec{B} \cdot d\vec{a} = \oint_C \vec{A} \cdot d\vec{\ell}, \qquad \text{[Surface independence]} \tag{14.116b}$$

where C is the one-dimensional boundary of a two-dimensional surface S. Why an integral on the right? Well, unlike the discrete boundary a and b in (14.116a), the boundary curve C in (14.116b) is a continuous infinity of points, so summing over them requires integration. In both cases, though, the integral of the vector field on the left can be evaluated with a lower-dimensional sum on the right. But why does $\vec{A}$ need to be a vector? Essentially for the same reason the sum over Φ in (14.116a) manifests as subtraction. There the orientation adopted in the original line integral mandates that we leave point a and arrive at point b; this orientation bequeaths the sign convention in the discrete sum over Φ. Similarly, the flux integral adopts an orientation specified by the direction of the unit normal to the surface; this orientation must be reflected in the choice of integration direction around C. Thus we must integrate using $d\vec{\ell}$ rather than $d\ell$ — and since flux is a scalar, this requires a vector $\vec{A}$ to dot into $d\vec{\ell}$. The convention is to link the direction of the unit normal $\hat{n}$ to the positive orientation around C via the right-hand rule: curl the fingers of your right hand around C, and your thumb gives $\hat{n}$, as illustrated in Figure 14.8a.

Now recall that path-independent fields have vanishing circulation,

$$\oint_C \vec{E} \cdot d\vec{\ell} = 0 \quad \forall\, C, \qquad \text{[Path independence]} \tag{14.117a}$$

which, since a closed curve has no endpoints, is consistent with (14.116a). A similar statement holds for surface independence. Indeed, just as two paths C_1 and C_2 connecting the same endpoints can be combined into a single closed path C, two surfaces bounded by the same curve can combine to form a closed surface $S = S_1 + S_2$. And just as the orientation of one of the C_i must be reversed, so too with one of the S_i. A closed surface, however, has no bounding curve — so the right-hand rule doesn't apply! On the other hand, a closed surface divides space into separate and distinct "inside" and "outside" regions, and the unit normal is used to differentiate between them. The standard convention is to take positive $\hat{n}$ as the *outward* pointing normal. In this way, a positive net flux indicates an overall outward flow, a negative net flux an inward flow. (This is the convention we used for the current I in Example 14.22.) So when combining S_1 and S_2 into a single closed surface, one of the two normals must flip direction

---

[6] Context should suffice to distinguish whether $\vec{A}$ denotes area or vector potential.

in order to provide a consistent outward convention all around S. For instance, taking $\hat{n} = \hat{n}_1 = -\hat{n}_2$ flips the sign of the flux through S_2 (Figure 14.8b). Combined into a single closed surface S, the two constituent fluxes through S_1 and S_2 *subtract*. Thus for a surface-independent field, the net flux through the closed surface must vanish:

$$\oint_S \vec{B} \cdot d\vec{a} = 0 \quad \forall\, S. \qquad \text{[Surface independence]} \tag{14.117b}$$

This is consistent with (14.116b): just as a closed curve has no endpoints, so too a closed surface doesn't have a bounding curve.

Example 14.26 Net Flux of $\vec{E} = xy^2\hat{\imath} + x^2y\hat{\jmath}$

Let's calculate the flux through two similar closed surfaces:

S_1: *A cube of side 2a centered at the origin* We break the closed integral into a sum of six open integrals, one for each cube face. Along the top and bottom faces $d\vec{a} = \pm\hat{k}dxdy$, giving $\vec{E}\cdot d\vec{a}_{\text{top/bot}} \equiv 0$. Along the front and back, $d\vec{a} = \pm\hat{\imath}dydz$, so $\vec{E} \cdot d\vec{a}_{\text{front}}\,|_{x=a} = \vec{E} \cdot d\vec{a}_{\text{back}}\,|_{x=-a} = ay^2dydz$. Thus

$$\int_{\text{front+back}} \vec{E} \cdot d\vec{a} = 2a \int_{-a}^{a} y^2dy \int_{-a}^{a} dz = \frac{8}{3}a^5. \tag{14.118a}$$

Finally, $d\vec{a}_{\text{left/right}} = \mp\hat{\jmath}dxdz$, so $\vec{E} \cdot d\vec{a}_{\text{left}}\,|_{y=-a} = +\vec{E}\cdot d\vec{a}_{\text{right}}\,|_{y=a} = ax^2dxdz$, so that we again get

$$\int_{\text{left+right}} \vec{E} \cdot d\vec{a} = 2a \int_{-a}^{a} x^2dx \int_{-a}^{a} dz = \frac{8}{3}a^5. \tag{14.118b}$$

Adding these all up gives

$$\oint_{S_1} \vec{E} \cdot d\vec{a} = \frac{16}{3}a^5. \tag{14.119}$$

S_2: *A cube of side 2a centered at (a,a,a)* Only the integration limits change — but it makes a difference:

$$\oint_{S_2} \vec{E} \cdot d\vec{a} = \int_{\text{front+right}} \vec{E} \cdot d\vec{a}$$

$$= 2a \int_0^{2a} y^2dy \int_0^{2a} dz + 2a \int_0^{2a} x^2dx \int_0^{2a} dz = \frac{64}{3}a^5. \tag{14.120}$$

Either way the net flux is non-zero, so the field in not surface independent.

Example 14.27 Net Flux of $\vec{B} = x^2y\hat{\imath} - xy^2\hat{\jmath}$

Let's calculate the flux through the same two closed surfaces. Once again the top and bottom faces again contribute nothing to the net flux. For the origin-centered cube, the remaining dot products produce odd integrands over the symmetric interval from $-a$ to a. So

$$\oint_{S_1} \vec{B} \cdot d\vec{a} = 0. \tag{14.121}$$

For the cube centered at (a,a,a), however, the intervals are no longer symmetric. Since the back face is now at $x = 0$, it makes no contribution; similarly for the left face at $y = 0$. That leaves the front and right faces at $x = 2a$ and $y = 2a$:

$$\oint_{S_2} \vec{B} \cdot d\vec{a} = \int_{\text{front+right}} \vec{B} \cdot d\vec{a}$$

$$= (2a)^2 \int_0^{2a} y \, dy \int_0^{2a} dz - (2a)^2 \int_0^{2a} x \, dx \int_0^{2a} dz = 0. \qquad (14.122)$$

As is to be expected of a surface-independent field.

Let's try to understand the results of Examples 14.26 and 14.27 by comparing the field lines of $\vec{E}$ and $\vec{B}$ (Figure 14.3). Returning to the water analogy, the flux $\oint_S \vec{F} \cdot d\vec{a}$ measures the (net) amount of water per unit time leaving the region bounded by S. For $\vec{E}$, the rate of surging water increases with distance from the coordinate axes. No closed surface could have compensating positive and negative flux contributions resulting in zero net flux. So somewhere in $\vec{E}$ there must be a source of water. By contrast, the waters of $\vec{B}$ seem to maintain constant flow along its streamlines. A little thought should convince you that *any* closed surface in $\vec{B}$ has a net flux of zero: what flows in, flows out.

This is all well-and-good — but examining the field lines is a fairly difficult way to determine if the net flux vanishes for all surfaces. And integrating over an infinity of closed surfaces as required by (14.117b) can be time consuming. What we need is a succinct mathematical statement or condition to identify a surface-independent field.

Let's continue to argue by analogy with path independent fields. Path independence equates line integrals of $\vec{F}$ along different curves connecting the same endpoints, leading to the introduction of the scalar potential Φ for path-independent fields. As we saw in (14.73), the fundamental theorem of calculus tells us that $\vec{F} \equiv \vec{\nabla}\Phi$. Similarly, surface independence equates flux integrals of $\vec{F}$ over different surfaces sharing the same boundary curve, which — as we saw in (14.116b) — leads to the introduction of the vector potential $\vec{A}$ for surface-independent fields. It too is related to $\vec{F}$ by a derivative, though clearly not via a simple gradient. In fact, the only derivative operator which takes a vector field and returns another vector field is the curl, $\vec{F} \equiv \vec{\nabla} \times \vec{A}$. This relationship between $\vec{F}$ and $\vec{A}$ elevates the definition of surface independence, (14.116b), into a fundamental result known as *Stokes' theorem*

$$\int_S (\vec{\nabla} \times \vec{A}) \cdot d\vec{a} = \oint_C \vec{A} \cdot d\vec{\ell}. \qquad (14.123)$$

We'll discuss this in more detail (and with less hand-waving) later; for now, let's see what we can do with it.[7]

Example 14.28   It's Got (Vector) Potential!

To find the vector potential for $\vec{B} = x^2 y \hat{i} - xy^2 \hat{j}$ requires solving its defining relation

$$\vec{B} = \vec{\nabla} \times \vec{A} \iff \begin{cases} x^2 y = \partial_y A_z - \partial_z A_y \\ -xy^2 = \partial_z A_x - \partial_x A_z \\ 0 = \partial_x A_y - \partial_y A_x \end{cases} . \qquad (14.124)$$

---

[7] It's worth noting that, unlike path independence, the phrase "surface independence" is rarely used, largely because it has no association with a fundamental physical quantity like energy. Instead, one usually just cites Stokes' theorem.

Without additional constraints (say, the value of $\vec{A}$ on some boundary), these equations do not fully specify $\vec{A}$. But we can trivially satisfy the z-component of the curl by *choosing* $A_x = A_y = 0$; this leads to

$$x^2 y = \partial_y A_z \longrightarrow A_z = \tfrac{1}{2}x^2 y^2 + \alpha(x,z)$$

$$xy^2 = \partial_x A_z \longrightarrow A_z = \tfrac{1}{2}x^2 y^2 + \beta(y,z). \tag{14.125}$$

Although the functions α and β are arbitrary, consistency requires both expressions in (14.125) to lead to the same $\vec{A}$. In our example, this can only happen if α and β are the same function of z, $\alpha(x,z) = \beta(y,z) \equiv \gamma(z)$. Thus one possible vector potential is

$$\vec{A}_1 = \left(\frac{1}{2}x^2 y^2 + \gamma(z) \right) \hat{k}. \tag{14.126}$$

As a quick check, taking the curl reproduces the correct $\vec{B}$ for arbirtrary $\gamma(z)$.

Setting as many components as possible to zero may yield the simplest $\vec{A}$ — but by no means the only one. If instead of setting $A_x = A_y = 0$ we were to choose, say, $A_z = 0$, we'd get

$$x^2 y = -\partial_z A_y \longrightarrow A_y = -x^2 yz + \alpha(x,y)$$

$$-xy^2 = \partial_z A_x \longrightarrow A_x = -xy^2 z + \beta(x,y). \tag{14.127}$$

Thus we find a second acceptable vector potential,

$$\vec{A}_2 = (-xy^2 z + \beta)\hat{\imath} + (-x^2 yz + \alpha)\hat{\jmath}. \tag{14.128}$$

Needless to say, α and β here are not the same as in (14.125) — in fact, the z-component of the curl constrains them to obey

$$\partial_x A_y - \partial_y A_x = 0 \longrightarrow \partial_x \alpha = \partial_y \beta. \tag{14.129}$$

With this condition, we can easily confirm that $\vec{\nabla} \times \vec{A}_2 = \vec{B}$.

Note that *no* choice of α, β, and γ in (14.126) and (14.128) can transform $\vec{A}_1$ into $\vec{A}_2$. Thus despite leading to the same field $\vec{B}$, these vector potentials are distinct functions.

The same approach applied to $\vec{E} = xy^2\hat{\imath} + x^2 y\hat{\jmath}$ fails to find a consistent vector potential. So $\vec{E}$ is not a surface-independent field.

If all we want to know is whether a field is surface independent, we don't actually have to *find* $\vec{A}$ — just determine that it exists. In fact, since we know that the divergence of a curl vanishes, all we need to do is calculate $\vec{\nabla} \cdot \vec{F}$. If it's zero, then we can be assured that $\vec{A}$ exists and the field $\vec{F}$ is surface independent. In other words, *surface-independent fields are solenoidal*. And indeed, $\vec{\nabla} \cdot \vec{B} = 0$, whereas $\vec{\nabla} \cdot \vec{E} \neq 0$. Moreover, as we discussed in Section 12.4.2, a vanishing divergence signals that there are no points at which the field lines begin or end. So no wonder the net flux of B vanishes: without sources or sinks, any field lines entering any closed surface must also exit the surface. Not so $\vec{E}$. Since $\vec{\nabla} \cdot \vec{E} = x^2 + y^2$, there must be sources distributed everywhere in the plane (except at the origin). As a result, different surfaces will enclose different quantities of this source, so the net flux depends on the closed surface chosen — as we saw in Example 14.24.

Thus we have three succinct mathematical tests for the surface independence of a vector field:

$$\oint_S \vec{F} \cdot d\vec{a} = 0 \qquad (14.130a)$$

$$\vec{F} = \vec{\nabla} \times \vec{A} \qquad (14.130b)$$

$$\vec{\nabla} \cdot \vec{F} = 0. \qquad (14.130c)$$

Surface independence, (14.130a), implies the existence of a vector function as specified by (14.130b); (14.130c) follows automatically since (as we've seen) the divergence of a curl is identically zero.

The first expression in (14.130) is the most intuitive statement of surface independence, the second the most profound. But by far the simplest *test* of surface independence is to calculate the field's divergence.

Surface independence allows us the freedom to choose *any* open surface we wish when evaluating an integral — only the boundary of the surface must be specified. It should go without saying that one usually chooses the surface which renders the integral as simple as possible.

Example 14.29   **Flux of $\vec{B}$ Through a Hemisphere Centered at $(a, a, 0)$**

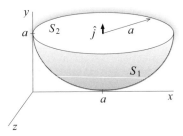

Figure 14.9 Hemisphere of radius a, centered at $x = y = a, z = 0$.

We need to calculate the flux through the "bowl" S_1 (Figure 14.9),

$$\Phi_1 = \int_{S_1} \vec{B} \cdot d\vec{a}.$$

Though we can translate $x \to x - a$, $y \to y - a$ and convert to spherical coordinates, it's messy. Fortunately, since $\vec{\nabla} \cdot \vec{B} = 0$, we know that any *closed* surface in $\vec{B}$ has zero net flux — and the hemisphere S_1 plus the disk S_2 together form a closed surface. Thus

$$\Phi_{\text{net}} = \oint_{S_1 + S_2} \vec{B} \cdot d\vec{a} = \int_{S_1} \vec{B} \cdot d\vec{a} + \int_{S_2} \vec{B} \cdot d\vec{a} = \Phi_1 + \Phi_2 = 0, \qquad (14.131)$$

where we've used outward-pointing normals for both integrals. So the flux we want is just the negative of a more-easily calculated Φ_2. On S_2, $\hat{n} = +\hat{j}\,dxdz = \hat{j}\rho\,d\rho d\phi$, where $\rho^2 = (x - a)^2 + z^2$ and $y = a$. This integral is not hard:

$$\Phi_1 = -\Phi_2 = -\int_{S_2} \vec{B} \cdot d\vec{a} = +\int_{S_2} (x-a)a^2 \, dxdz$$

$$= +a^3 \int_0^a \rho \, d\rho \int_0^{2\pi} d\phi \, (\cos\phi - 1) = -\pi a^5. \tag{14.132}$$

You should examine the field lines of $\vec{B}$ to verify that the flux through S_1 is in fact negative.

Problems

14.1 Evaluate $\int (x^2 + y^2) \, dx$ from $P_1 = (0,0)$ to $P_2 = (1,1)$ along:

(a) C_1: straight segments from $P_1 \to (1,0) \to P_2$
(b) C_2: a single straight-line path from P_1 to P_2
(c) C_3: the parabola $y = x^2$.

14.2 For a circle C of radius a, what do the following two integrals give? Be sure to explain your reasoning.

(a) $\oint_C d\ell$ (b) $\oint_C d\vec{\ell}$

14.3 Integrate $d\ell$ around a circle of radius a using:

(a) cartesian coordinates   (b) polar coordinates.

14.4 In Example 14.3, we explicitly calculated $\oint_C d\ell$ for the circle $r = 2a\cos\theta$, which has radius a and is centered at $x = a, y = 0$. Now calculate $\oint_C d\vec{\ell}$ for this circle. (Don't just state the answer.)

14.5 Consider a particle in the xy-plane moving in the circle $(\cos t^2, \sin t^2)$, with $t \geq 0$. (This is the parametrization of (14.23) with radius $a = 1$.)

(a) What is the particle's velocity as a function of t? Looking toward the origin along the positive z-axis, in which direction is the particle moving around the circle?
(b) From what point P should the particle be released in order to hit a target at $Q = (2,0)$?
(c) What is the earliest time t the particle will be at this point? What is the particle's velocity at that time?
(d) At what time τ and with what velocity should the particle be released if it were moving in the unit circle $(\cos t, \sin t)$ instead? (This is the parametrization of (14.20).)

14.6 Integrate $f(x,y,z) = xyz$ around one turn of the helix $x(t) = a\cos\omega t$, $y(t) = a\sin\omega t$, $z(t) = bt$.

14.7 For a sphere S of radius R, what do the following two integrals give? Be sure to explain your reasoning.

(a) $\oint_S dA$ (b) $\oint_S d\vec{A}$

14.8 The shape made by a free-hanging flexible chain, $y(x) = a\cosh(x/a)$, is called a *catenary*. (See Problem 6.30.) If the chain is suspended from fixed points at $x = \pm b$, find b in terms of both a and the length of the chain L.

14.9 The accuracy of a pendulum clock depends on it being isochronous: the pendulum must swing harmonically from an initial angular displacement θ_0 to $-\theta_0$ in a time $T/2 = \pi/\omega$, independent of θ_0. Amplitude-independent oscillation is given by $\ddot{s} = -\omega^2 s$, where s is the arc length and $\omega^2 = g/\ell$ for a pendulum of length ℓ. But since a hanging pendulum at some angle θ feels a gravitational acceleration $\ddot{s} = g\sin\theta$, the circular arc of a conventional pendulum clock is only approximately simple harmonic (see Problem 6.42).

So an ideal pendulum clock with $\theta = \omega t$ should follow a trajectory defined by $s = -\ell\sin\theta$; the fully isochronous trajectory which emerges is called a *tautochrone*. Solve for $d\vec{s} = (x(\theta), y(\theta))$ to show that the pendulum's tautochrone is a cycloid (Example 14.7). [Huygens built a clock in which the pendulum was constrained to follow a cycloid, but the friction undermined the design.]

14.10 Consider a simple curve C given by a function $y = f(z)$, and imagine rotating it around the z-axis to form a *surface of revolution*. If we visualize this figure as a stack of circles of radius $f(z)$, the area can be found by adding up their circumferences, $\int_S dA = \int_C 2\pi f(z)d\ell$. Use this to find the surface area of the following:

(a) The ellipsoid generated by the curve $y^2/a^2 + z^2/c^2 = 1$. For simplicity, take $c = a/2$. (Express your answer as a number times πa^2.)
(b) The hyperboloid generated by the curve $y^2/a^2 - z^2/c^2 = 1$, with $0 < z < c$. For simplicity, take $a = c$.

14.11 Evaluate $\int_S \vec{r} \cdot d\vec{A}$, over a hemisphere $z \geq 0$ of radius R.

14.12 Evaluate the surface integral $\int (x^2 + y^2 + z^2)\,dx\,dy$ on the following surfaces:

(a) S_1: the unit disk $x^2 + y^2 \leq 1, z = 0$
(b) S_2: the hemisphere $x^2 + y^2 + z^2 = 1, z \geq 0$
(c) S_3: the paraboloid $z = x^2 + y^2, 0 \leq z \leq 1$.

14.13 Derive (14.37) and (14.38).

14.14 An ellipsoid is a surface described by

$$\frac{x^2}{a^2} + \frac{y^2}{b^2} + \frac{z^2}{c^2} = 1.$$

For $a = b = c$, this is a sphere. We'll consider instead the case of an oblate spheroid, which has $a = b < c$; for simplicity, we'll take $a = 2c$. Find the surface area of this spheroid two ways:

(a) Project the surface down into a sum in the xy-plane using (14.34).
(b) Integrate over the surface using the parametrization of Problem 2.17.

Your result should agree with what you found in Problem 14.10.

14.15 Show that a scalar field Φ is completely specified by $\vec{\nabla}\Phi = \vec{E}$ if its value at a single point P is given and $E(\vec{r})$ is a known vector field.

14.16 Use the extremum condition (14.53) to derive (14.55).

14.17 It was asserted in the discussion following Example 14.15 that the similarity between the fields $\vec{E} = xy^2\hat{\imath} + x^2y\hat{\jmath}$ and $\vec{B} = x^2y\hat{\imath} - xy^2\hat{\jmath}$ allows the construction of a single closed curve C which is everywhere either parallel or perpendicular to the fields. Verify this claim.

14.18 Which of the following fields are path independent; surface independent; both; or neither?

(a) $y\hat{\imath} - x\hat{\jmath}$ (b) $\hat{r}$ (c) $-\hat{\imath} \sin x \cosh y + \hat{\jmath} \cos x \sinh y$
(d) $xy^2\hat{\imath} - x^2y\hat{\jmath}$ (e) $\rho z\hat{\phi}$

14.19 The integral of an irrotational field $\vec{F}$ along the path from points $a \to c \to d$ is 5, along $b \to d$ is -2 and along $a \to c \to e$ is 4. What's the integral along $d \to b \to a$? $e \to c \to d \to b$?

14.20 Consider two paths: C_1 is the line segment $y = -x + a$, and C_2 is the origin-centered quarter circle of radius a in the first quadrant. Calculate the following integrals connecting the points $(x, y) = (0, a)$ and $(a, 0)$ along each of these paths.

(a) $\int_C \vec{E} \cdot d\vec{\ell}$, where $\vec{E} = xy^2\hat{\imath} + x^2y\hat{\jmath}$

(b) $\int_C \vec{B} \cdot d\vec{\ell}$, where $\vec{B} = x^2 y \hat{\imath} - xy^2 \hat{\jmath}$

Are your results reasonable? Explain.

14.21 Calculate the following circulations:

(a) The work done by $x\hat{\imath} - y\hat{\jmath}$ around an origin-centered square of side 2, sides parallel to the axes.
(b) $\vec{F} = (x + y)\hat{\imath} + 3xy\hat{\jmath} + (2xz + z)\hat{k}$ around the circle $x^2 + y^2 = a^2$.
(c) $\vec{F} = -y\hat{\imath} + x\hat{\jmath} = \rho\hat{\phi}$ around a circle of radius R.
(d) $\vec{F} = r\cos^2\theta \sin\phi \, \hat{r} + r\sin^3\theta \cos\phi \, \hat{\theta} + r\sin\theta \sin^2\phi \, \hat{\phi}$ around the circle $\theta = \alpha$ lying on the surface of a sphere of radius R. Does your result make sense in the $\alpha \to \pi$ limit? Explain.

14.22 Calculate the following fluxes:

(a) $x^3\hat{\imath} + y^3\hat{\jmath} + z^3\hat{k}$ through an origin-centered cube, side a, faces parallel to the coordinate planes.
(b) $x\hat{\imath} + y\hat{\jmath} + z\hat{k}$ through the hemisphere $z = \sqrt{a^2 - x^2 - y^2}$, $z > 0$.
(c) $x\hat{\imath} + y\hat{\jmath} + z\hat{k}$ through the cylinder $x^2 + y^2 = a^2$, $0 \le z \le 2$.
(d) $r^2\hat{r}$ through a sphere of radius R.
(e) $\hat{\imath} + \hat{\jmath} + z(x^2 + y^2)^2\hat{k}$ through the cylinder $x^2 + y^2 = a^2$, $|z| \le 1$.

14.23 Find the flux of $\vec{B} = x^2 y\hat{\imath} - xy^2\hat{\jmath}$ through a spherical octant of radius a, with $|z| < a$, as shown in the figure.

14.24 Find the flux through a square in the yz-plane $x = a$, with $0 < y < a$ and $0 < z < a$, for the following fields:

(a) $\vec{E} = xy^2\hat{\imath} + x^2 y\hat{\jmath}$
(b) $\vec{B} = x^2 y\hat{\imath} - xy^2\hat{\jmath}$.

14.25 Verify the flux in (14.115) for the spherical octant S_3 of Example 14.25.

14.26 Determine whether $\vec{E} = xy^2\hat{\imath} + x^2 y\hat{\jmath}$ is surface independent either by finding a vector potential $\vec{A}$ with $\vec{E} = \vec{\nabla} \times \vec{A}$, or showing that no such $\vec{A}$ exists.

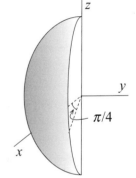

14.27 Show that the two vector potentials derived in Example 14.28 differ by a divergence $\vec{\nabla}\lambda$. Find λ. For simplicity, take $\alpha = \beta = \gamma = 0$ in both (14.126) and (14.128).

14.28 **Exploiting Symmetry**

(a) Consider a solid non-conducting sphere of radius a with charge density proportional to r, $\rho = \alpha r$. Find the electric field $\vec{E}$ both inside and outside the sphere as a function of $\vec{r}$. Express your answer in terms of the total charge Q on the ball.
(b) Consider a long cylindrical wire of radius a with a steady current density proportional to r, $\vec{J} = \alpha r\hat{k}$. Find the magnetic field $\vec{B}$ both inside and outside the cylinder as a function of $\vec{r}$. Express your answer in terms of the total current I in the wire.

14.29 Use the fact that electrostatic fields are irrotational to demonstrate that a parallel plate capacitor's field cannot be uniform (as it's often portrayed), but rather must have a fringing field at the edges of the plates.

14.30 Consider a long solenoid of radius R, n turns per length, carrying a steady current I. Let the solenoid axis be along z, and take as given that the magnetic field is (i) in the $\hat{k}$-direction, and (ii) must go to zero far from the solenoid. Use Ampère's law (14.60) to show that inside, the magnetic field $\vec{B} = \mu_0 n I\hat{k}$, but vanishes everywhere outside.

14.31   Faraday's law states that a time-dependent magnetic flux induces an electric field whose circulation is given by the rate of change of the magnetic flux,

$$\oint_C \vec{E} \cdot d\vec{\ell} = -\frac{d}{dt} \int_S \vec{B} \cdot d\vec{a},$$

where C is the bounding curve of the surface S. Taking advantage of symmetry, use Faraday's law to find the induced electric field (magnitude and direction) inside and outside the solenoid of Problem 14.30 with time-dependent current $I(t)$.

15 The Theorems of Gauss and Stokes

The complementary relationship between differentiation and integration is laid down by the fundamental theorem of calculus,

$$\int_a^b \frac{df}{dx}\, dx = f(b) - f(a). \tag{15.1}$$

Along a one-dimensional path C threading through three-dimensional space from $\vec{r}_1$ to $\vec{r}_2$, the theorem becomes

$$\int_C \vec{\nabla} f \cdot d\vec{\ell} = f(\vec{r}_2) - f(\vec{r}_1). \tag{15.2}$$

The fundamental theorem simply states that the net change of a continuous function between $\vec{r}_1$ and $\vec{r}_2$ is given by the cumulative effect of infinitesimal changes in f over a curve C connecting the points. We now want to expand the theorem's purview beyond one-dimensional integrals.

In our discussions of derivatives and integrals of vector fields, we focused particular attention on four measures: the divergence and the curl, the flux and the circulation. As the names suggest, the divergence and flux give measures of the surging of a vector field, the curl and circulation the swirl of a vector field. But though related, they are not identical: divergence is not flux, nor is curl circulation. The divergence and curl are themselves fields (scalar and vector, respectively), assigning definite values to every point in space — the expansion rate of sawdust or the spin rate of a propeller. They are *local* measures. By contrast, the flux and circulation are both (scalar) quantities defined over *extended* regions — a surface S and a curve C, respectively. The correct relationship entails the accumulation (integration) over extended regions of local, infinitesimal changes (derivatives) — in essence, the fundamental theorem of calculus.

15.1 The Divergence Theorem

To bridge the divide between local and extended measures of surging flow, let's examine the flux in the limit of infinitesimal area — in other words, let's try to localize the flux. Consider a field $\vec{E}$ and an infinitesimal, rectangular block of dimension dx, dy, dz (Figure 15.1). Because the block is infinitesimal, the net outward flux can be rendered as

$$d\Phi_{\text{net}} = \sum_{\text{side } i} \vec{E} \cdot d\vec{a}_i.$$

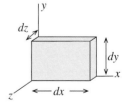

Figure 15.1 Infinitesimal rectangular block.

We tackle the surfaces in pairs. For the left and right faces in the yz-plane, evaluating $\vec{E}$ at the faces' corners on the x-axis[1] we have

$$\vec{E}(0,0,0) \cdot d\vec{a}_{\text{left}} + \vec{E}(dx,0,0) \cdot d\vec{a}_{\text{right}} = -E_x(0,0,0)\, dydz + E_x(dx,0,0)\, dydz, \qquad (15.3)$$

where the minus sign arises from the leftward-pointing unit normal, $d\vec{a}_{\text{left}} = -\hat{\imath}\, dydz$. Thus the x contribution to $d\Phi_{\text{net}}$ is

$$d\phi_1 = [E_x(dx,0,0) - E_x(0,0,0)]\, dydz = \left(\frac{\partial E_x}{\partial x} dx\right) dydz. \qquad (15.4)$$

Clearly, analogous expressions must hold for y and z:

$$d\phi_2 = \left[E_y(0,dy,0) - E_y(0,0,0)\right] dzdx = \left(\frac{\partial E_y}{\partial y} dy\right) dzdx$$

$$d\phi_3 = \left[E_z(0,0,dz) - E_z(0,0,0)\right] dxdy = \left(\frac{\partial E_z}{\partial z} dz\right) dxdy.$$

Thus

$$d\Phi_{\text{net}} = \left(\frac{\partial E_x}{\partial x} dx\right) dydz + \left(\frac{\partial E_y}{\partial y} dy\right) dzdx + \left(\frac{\partial E_z}{\partial z} dz\right) dxdy = \vec{\nabla} \cdot \vec{E}\, d\tau, \qquad (15.5)$$

where the volume element $d\tau \equiv dx\, dy\, dz$. So the infinitesimal flux is proportional to the enclosed divergence. (You should verify that the units match on each side.) Though we worked this out using cartesian coordinates, the result holds generally.

Example 15.1   Divergence from Flux

As a concrete example of the divergence emerging in the limit of vanishing volume, consider the simple case $\vec{E} = \vec{r}$. Then, for a sphere of radius R centered at the origin,

$$\vec{\nabla} \cdot \vec{E} \equiv \frac{d\Phi_{\text{net}}}{d\tau} = \lim_{\Delta S, \Delta V \to 0} \frac{\oint_{\Delta S} \vec{E} \cdot d\vec{a}}{\Delta V} = \lim_{R \to 0} \frac{R \cdot 4\pi R^2}{\frac{4}{3}\pi R^3} = 3. \qquad (15.6)$$

In practice, of course, it's easier to calculate the divergence by taking derivatives.

To complete the link from the divergence at a point to the flux over an extended region, let's assemble a collection of infinitesimal blocks into a *finite* structure of volume V bounded by a surface S (Figure 15.2). Summing all the infinitesimal fluxes $d\Phi_i$ surging through all the infinitesimal faces of all the infinitesimal boxes contained in V would seem to be a cumbersome way to compute the net flux through S. But note that fluxes through adjacent *interior* faces cancel because of their oppositely oriented normals. So the accumulation of infinitesimal fluxes is simply the net flux through the exterior surface of the entire edifice,

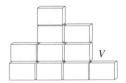

Figure 15.2 Assembly of a finite volume V.

$$\int_V \vec{\nabla} \cdot \vec{E}\, d\tau = \oint_S \vec{E} \cdot d\vec{a}. \qquad (15.7)$$

[1] We do this for simplicity; evaluating $\vec{E}$ at non-zero y and z will not affect the results.

In other words, the production rate of water within a volume can be measured either by calculating the number and size of the spigots and drains within, or by measuring the net amount of fluid which exits through the boundary. This result, known as *Gauss' theorem* or the *Divergence theorem*,[2] quantifies the relationship between flux and divergence. The divergence is the *flux density*: $\vec{\nabla} \cdot \vec{E}$ at a point P is the flux of $\vec{E}$ per unit volume at P.

This result bears repeating: the extended measure we call flux was constructed (or if you prefer, *defined*) rather intuitively as an integral over a surface S. What the divergence theorem tells us, perhaps less intuitively, is that the net flux can also be computed with a volume integral. Expressing the flux as a volume integral, $\Phi_{net} = \int_V \vec{\nabla} \cdot \vec{E} \, d\tau$, allows the identification of $\vec{\nabla} \cdot \vec{E}$ as the flux density. A useful and familiar touchstone is electric charge: like flux, a region's total charge is also an extended measure which can be found from the volume integral over the local charge density, $Q = \int_V \rho \, d\tau$. For electric fields, Gauss' law $\Phi_{net} = Q/\epsilon_0$ says that flux *is* charge — and thus the divergence is charge density, $\vec{\nabla} \cdot \vec{E} = \rho/\epsilon_0$.

Note that if the flux density vanishes everywhere, then flux must be zero for any closed surface. This, then, verifies the connection between (14.130a) and (14.130c): solenoidal fields are surface independent.

The standard expression of the divergence theorem is given in (15.7). However, there are other, vector-valued versions which are also useful — and which you're asked to derive in Problem 15.11:

$$\int_V \vec{\nabla}\Phi \, d\tau = \oint_S \Phi \, \hat{n} \, da \equiv \oint_S \Phi \, d\vec{a} \tag{15.8a}$$

$$\int_V \vec{\nabla} \times \vec{E} \, d\tau = \oint_S \hat{n} \times \vec{E} \, da \equiv \oint_S d\vec{a} \times \vec{E} \tag{15.8b}$$

Example 15.2 $\vec{E} = xy^2 \, \hat{\imath} + x^2 y \, \hat{\jmath}$

We'll demonstrate the divergence theorem using the same closed surfaces considered in Example 14.26.

S_1: *A cube of side 2a centered at the origin* We found before that $\Phi_{net} = \frac{16}{3} a^5$. To verify the divergence theorem, we integrate the divergence $\vec{\nabla} \cdot \vec{E} = x^2 + y^2$ over the volume of the cube:

$$\int_V \vec{\nabla} \cdot \vec{E} \, d\tau = \int dx \, dy \, dz \, (x^2 + y^2) = 2a \int_{-a}^{a} dx \int_{-a}^{a} (x^2 + y^2) \, dy$$

$$= 4a^2 \int_{-a}^{a} \left(x^2 + \frac{1}{3}a^2 \right) dx = 4a^2 \left(\frac{2}{3}a^3 + \frac{2}{3}a^3 \right) = \frac{16}{3}a^5 . \checkmark$$

S_2: *A cube of side 2a centered at (a, a, a)* This time, we found $\Phi_{net} = \frac{64}{3} a^5$. Let's see:

$$\int_V \vec{\nabla} \cdot \vec{E} \, d\tau = 2a \int_0^{2a} dx \int_0^{2a} (x^2 + y^2) \, dy = \frac{64}{3}a^5 . \checkmark$$

---

[2] "Divergence theorem" is often preferred to prevent confusion with Gauss' law for electric fields.

Example 15.3 $\vec{E} = r^\alpha \hat{r}$ **in a Sphere of Radius** R

Because the field is spherically symmetric, the flux through an origin-centered sphere of radius R is straightforward,

$$\oint_S \vec{E} \cdot d\vec{a} = \oint_S r^\alpha R^2 d\Omega = 4\pi R^{\alpha+2}. \tag{15.9}$$

With only an $\hat{r}$-component, the divergence in spherical coordinates is easy,

$$\vec{\nabla} \cdot \vec{E} = \frac{1}{r^2} \frac{\partial}{\partial r} \left(r^2 E_r \right) = \frac{1}{r^2} \frac{d}{dr} \left(r^{\alpha+2} \right) = (\alpha + 2) r^{\alpha-1}, \tag{15.10}$$

so that

$$\int_V \vec{\nabla} \cdot \vec{E}\, d\tau = (\alpha + 2) \int_0^R r^{\alpha-1} r^2 dr d\Omega = 4\pi R^{\alpha+2}. \tag{15.11}$$

We see from (15.11) that even when the volume V is extended to all space, the integral still converges provided E falls at least as fast $1/r^2$ — i.e., $\alpha \leq -2$. On the other hand, the field is not defined at the origin for $\alpha < 0$. This is a non-issue for the flux integral (15.9), which is unaffected by what happens at the origin. Not so the volume integral. Indeed, the divergence theorem requires fields and their first derivatives be continuous — so we would expect the theorem to fail for negative α when V includes the origin. Nonetheless, a glance at the integrand in (15.11) shows that $-2 < \alpha < 0$ works, thanks to the r^2 factor lurking in $d\tau$ — a bequest of three-dimensional space. But what about $\alpha \leq -2$? We address this in the next two Examples.

Example 15.4 $\vec{E} = r^\alpha \hat{r}$ **Between Concentric Spheres**

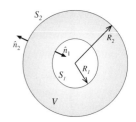

We can avoid the singularity at the origin by considering a volume V contained between concentric spheres with radii $R_2 > R_1$ — that is, a spherical shell of thickness $R_2 - R_1$. As long as $R_1 \neq 0$, the field is well-behaved throughout V. The boundary $S = S_1 + S_2$ of V, though disjointed, divides space into two distinct and unambiguous regions. One cannot pass from inside V to outside without crossing S somewhere. Thus S_1 and S_2 together form a closed surface and the divergence theorem still applies. (Physically, we can consider V as the superposition of two balls of radius R_1 and R_2 with opposite orientations.)

We did most of the work in Example 15.3. The net flux through S is

$$\oint_S \vec{E} \cdot d\vec{a} = \oint_{S_2} \vec{E} \cdot d\vec{a} + \oint_{S_1} \vec{E} \cdot d\vec{a} = 4\pi \left(R_2^{\alpha+2} - R_1^{\alpha+2} \right). \tag{15.12}$$

The minus sign arises because the unit normal on S_1, being properly oriented *out* of V, points toward smaller r — that is, $\hat{n}_1 = -\hat{r}$. As for the volume integral of the divergence, (15.11) becomes

$$\int_V \vec{\nabla} \cdot \vec{E}\, d\tau = (\alpha + 2) \int_{R_1}^{R_2} r^{\alpha-1} r^2 dr d\Omega = 4\pi \left(R_2^{\alpha+2} - R_1^{\alpha+2} \right). \tag{15.13}$$

As expected, the $R_1 \to 0$ limit is only possible for $\alpha \geq -2$. But non-zero R_1 excludes the origin from the integration, and in this region the divergence theorem holds even for $\alpha < -2$.

Example 15.5 The Case of the Missing Divergence

We can use the previous two examples to investigate the physically important case $\alpha = -2$. For a $1/r^2$ field, the right-hand side of (15.10) unambiguously gives vanishing divergence, so the volume integral of (15.11) should vanish identically — after all, there can be no flux without a flux density. But not so fast! The calculation in (15.9) demonstrates that the flux of $\vec{E} = \hat{r}/r^2$ through a sphere is indisputably 4π. This implies that $\vec{\nabla} \cdot \vec{E} \neq 0$ *some*where. So where's the missing divergence?

Carefully examining (15.10), it seems there is no way to avoid a vanishing divergence. Nor is having $\vec{\nabla} \cdot \vec{E} = 0$ so unreasonable. Consider a closed surface S of constant solid angle, with four radial sides and two concentric spherical caps (Figure 15.3). Since $\vec{E}$ has only a $\hat{r}$-component, the flux integral receives no contribution from the four side faces. And since E is constant along each of the spherical caps, their flux contributions are $\int \vec{E} \cdot d\vec{a} = +EA\big|_{\text{outer}}$ at the outer face, and $-EA\big|_{\text{inner}}$ at the inner face. But though E falls with distance-squared, A *grows* with r^2. As a result, the product EA is the *same* across each face, and so the net flux vanishes. In essence, for $\hat{r}/r^2$ the increasing separation exactly balances the diminishing surge, yielding a vanishing divergence.

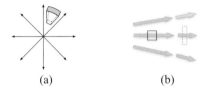

(a)   (b)

Figure 15.3 $\vec{E} = \hat{r}/r^2$: (a) the flux vanishes through surface excluding the origin; (b) diminishing surge compensates for increasing spread.

All of which is correct — up to a point. Examination of the field lines suggests that point to be the origin; a source must reside there. So somewhere our symbol-pushing has gone astray. And now that we've identified the origin as the culprit, we can more easily locate the source of the confusion: for $\alpha = -2$, the factor of $1/r^2$ in the expression for the divergence must render (15.10) invalid. But if we can't use (15.10), how then do we understand the source mathematically? Let's exclude the origin by returning to the shell of Example 15.4. Since $\vec{\nabla} \cdot (\hat{r}/r^2) = 0$ everywhere in V, the divergence theorem tells us that the net flux through the boundary $S = S_1 + S_2$ vanishes,

$$\int_V \vec{\nabla} \cdot \left(\frac{\hat{r}}{r^2}\right) d\tau = 0 = \oint_S \frac{\hat{r} \cdot \hat{n}}{r^2} \, da$$

$$= \oint_{S_1} \frac{\hat{r} \cdot \hat{n}_1}{r^2} \, da + \oint_{S_2} \frac{\hat{r} \cdot \hat{n}_2}{r^2} \, da. \tag{15.14}$$

Thus

$$\oint_{S_2} \frac{\hat{r} \cdot \hat{n}_2}{r^2} \, da = -\oint_{S_1} \frac{\hat{r} \cdot \hat{n}_1}{r^2} \, da. \tag{15.15}$$

But since we constructed the shell so that $\hat{n}_2 = -\hat{n}_1 = \hat{r}$, we have

$$\oint_{S_2} \frac{1}{r^2} \, da = +\oint_{S_1} \frac{1}{r^2} \, da = \oint_{S_1} \frac{R_1^2 \, d\Omega}{R_1^2} = 4\pi. \tag{15.16}$$

So the net flux through S_2 is both non-vanishing and independent of the size and shape of S_2. In particular, expanding S_2 all the way out to infinity has no effect. No source out there. But the flux through S_2 is also independent of the size and shape of S_1 — and in particular, shrinking S_1 all the way down to the origin does not affect the flux through S_2 at all. Thus the entire contribution to the flux, the missing divergence, must come from exactly one point, the origin. As we surmised.

If $\vec{\nabla} \cdot \vec{E}$ vanishes everywhere except at this one point, a *finite* net flux $\int_V \vec{\nabla} \cdot \vec{E}\, d\tau$ can only be had if the divergence there is infinite. This unusual behavior should sound familiar: the divergence is a delta function! Specifically,

$$\vec{\nabla} \cdot (\hat{r}/r^2) = 4\pi\, \delta(\vec{r}), \tag{15.17}$$

as you'll verify in Problem 15.25. So when V includes the origin, we get

$$\Phi_{\text{net}} = \int_V \vec{\nabla} \cdot (\hat{r}/r^2)\, d\tau = \int_V 4\pi\, \delta(\vec{r})\, d\tau = 4\pi. \tag{15.18}$$

Agreement with (15.9) for $\alpha = -2$ resolves the mystery.

Note that (15.17) and (15.18) are consistent with our interpretation of $q\, \delta(\vec{r})$ as the density of a point charge: applied to the electric field $\vec{E} = \frac{q}{4\pi\epsilon_0} \frac{\hat{r}}{r^2}$, we find the familiar result $\Phi_{\text{net}} = q/\epsilon_0$. In fact, Example 15.5 is a derivation of *Gauss' law*:

$$\oint_S \frac{\hat{r} \cdot \hat{n}}{r^2}\, da = \begin{cases} 4\pi, & \text{origin} \in V \\ 0, & \text{origin} \notin V \end{cases}. \tag{15.19}$$

Since all electrostatic fields can be constructed by superposition of individual fields, all physical electric fields obey Gauss' law.

Without a vanishing net flux, the field $\hat{r}/r^2$ is not formally surface independent. It is, however, right on the edge. By this I mean that despite having a non-zero divergence, the net flux does not depend upon the surface area of S, or even the details of its shape. What does matter is what's inside. And in fact, the right-hand side of Gauss' law betrays no details of the closed surface used to calculate flux; Φ_{net} is determined only by the enclosed net charge — independent of the area and orientation of the closed surface. Thus a cube and an ellipsoid enclosing the *same* total electric charge yield the same net flux.

BTW 15.1 Gauss' Law for Gravity and Magnetism

As we've seen, the net flux of $\vec{F}(\vec{r}) = \alpha \hat{r}/r^2$ through a sphere of radius R centered at the origin is

$$\Phi_{\text{net}} = \oint_S \vec{F} \cdot d\vec{a} = \alpha \oint_S \frac{R^2\, d\Omega}{R^2} = 4\pi\alpha, \tag{15.20}$$

independent of R. In electrostatics, $\vec{F}$ is the electric field and Coulomb's law tells us $\alpha = q/4\pi\epsilon_0$, where q is the electric charge of the source at the origin. This is Gauss' law for electricity. In gravity, $\vec{F}$ is the gravitational field and Newton's law gives $\alpha = -GM$, where M is the mass of the source at the origin. Thus we find Gauss' law for gravity. Note that unlike electricity, gravity has only positive sources — though the minus sign in the law essentially renders them as sinks.

What about magnetism? A magnetic field $\vec{B}$ obeys

$$\Phi_{\text{net}} = \oint_S \vec{B} \cdot d\vec{a} = 0, \tag{15.21}$$

for *any* closed surface S. So there are no magnetic charges to act as source or sink. Indeed, there are no magnetic field configurations which resemble electrostatic or gravitational fields of point charges or masses. (15.21) is the mathematical statement that isolated magnetic charges — "monopoles" — have never been found. Magnetic field lines must always form closed loops.

15.2 Stokes' Theorem

As we did with flux and divergence, we can bridge the divide between local and extended measures of swirling flow by examining the circulation in the limit of an infinitesimal path — that is, we'll try to localize the circulation. Consider a field $\vec{B}$ and an infinitesimal, rectangular curve of dimension dx, dy (Figure 15.4). Because the curve is infinitesimal, the counter-clockwise circulation $d\Psi$ can be rendered as

$$d\Psi = \sum_{\text{segment } i} \vec{B} \cdot d\vec{\ell}_i.$$

Figure 15.4 Infinitesimal rectangular loop.

We tackle the four segments in pairs. For the bottom and top segments along x we have (evaluating $\vec{B}$ at the segment's end on the y-axis)

$$\vec{B}(0,0,0) \cdot d\vec{\ell}_{\text{bottom}} + \vec{B}(0,dy,0) \cdot d\vec{\ell}_{\text{top}} = B_x(0,0,0)\,dx - B_x(0,dy,0)\,dx, \qquad (15.22)$$

where the minus sign arises from integrating right to left along the top element. Thus the x contribution to $d\Psi$ is

$$d\psi_1 = \left[-B_x(0,dy,0) + B_x(0,0,0)\right] dx = -\left(\frac{\partial B_x}{\partial y}dy\right) dx. \qquad (15.23)$$

Clearly, what holds for the top and bottom is also valid for the right and left,

$$d\psi_2 = \left[B_y(dx,0,0) - B_y(0,0,0)\right] dy = +\left(\frac{\partial B_y}{\partial x}dx\right) dy. $$

Thus

$$d\Psi = \left(\frac{\partial B_y}{\partial x}dx\right) dy - \left(\frac{\partial B_x}{\partial y}dy\right) dx = \left(\frac{\partial B_y}{\partial x} - \frac{\partial B_x}{\partial y}\right) da, \qquad (15.24)$$

where the area element $dA \equiv dx\,dy$. Note the emergence on the right-hand side of the z-component of the curl. Though we worked this out in cartesian coordinates, the result holds generally.

Example 15.6 Curl from Circulation

As a concrete example of the curl emerging in the limit of vanishing area, consider the field $\vec{B} = rz\hat{\phi}$ in cylindrical coordinates. The z-component of the curl can be found by integrating along a circle C_1 of radius a centered at the origin at a height z above the xy-plane, with direction given by the right-hand rule:

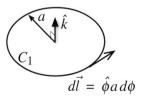

$$d\vec{l} = \hat{\phi}a\,d\phi$$

$$(\vec{\nabla} \times \vec{B})_z = \lim_{\Delta A \to 0} \frac{\oint_{C_1} \vec{B} \cdot d\vec{\ell}}{\Delta A} = \lim_{a \to 0} \frac{2\pi za^2}{\pi a^2} = 2z. \qquad (15.25)$$

This is easily verified by computing $(\vec{\nabla} \times \vec{B})_z$ directly.

The $\hat{r}$-component of the curl can be found similarly by integrating along a rectangular curve C_2 of dimensions $a\phi_0$ and z_0 on the surface of a cylinder of radius a (so $d\vec{\ell} = \pm\hat{k}\,dz$ along two segments, $\pm\hat{\phi}\,a\,d\phi$ along the other two),

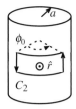
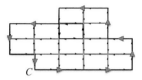

$$(\vec{\nabla} \times \vec{B}) \cdot \hat{r}\big|_{r=a} = \lim_{\Delta A \to 0} \frac{\oint_{C_2} \vec{B} \cdot d\vec{\ell}}{\Delta A} = \lim_{z_0, \phi_0 \to 0} \frac{-a^2\phi_0 z_0}{a\phi_0 z_0} = -a. \qquad (15.26)$$

The negative arises from the counter-clockwise route around C_2, as determined from the right-hand rule. In practice, of course, it's easier to calculate the curl by taking derivatives.

To complete the link from the curl at a point to the circulation over an extended region, let's assemble a collection of infinitesimal curves into a *finite* surface S bounded by a curve C (Figure 15.5). Summing all the infinitesimal circulations $d\psi_i$ swirling around all the infinitesimal loops contained within a finite S would seem to be a cumbersome way to compute the net circulation around C. But note that the contribution along adjacent *interior* segments will cancel because of their opposite directions of integration. Thus the accumulation of infinitesimal circulations throughout a surface S is the circulation around the surface's *perimeter* C,

Figure 15.5 Assembly of a finite surface S.

$$\oint_C \vec{B} \cdot d\vec{\ell} = \oint_C B_x dx + B_y dy = \int_S \left(\frac{\partial B_y}{\partial x} - \frac{\partial B_x}{\partial y} \right) dx\,dy. \qquad (15.27)$$

This is often referred to as *Green's theorem in the plane*, frequently written as

$$\int_S \left(\frac{\partial V}{\partial x} - \frac{\partial U}{\partial y} \right) dx\,dy = \oint_C U\,dx + V\,dy \qquad (15.28)$$

for continuous functions U and V.

Example 15.7 $\partial_i v_j$ **in the Plane**

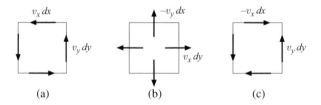

Figure 15.6 (a) Circulation; (b) flux; and (c) shear.

Green's theorem lends physical interpretations for the combinations of the four derivatives $\partial_i v_j$ of a vector field $\vec{v}$ in $\mathbb{R}^2$ shown in Figure 15.6. For instance, (15.28) with $U = v_x$ and $V = v_y$ reproduces (15.27),

$$\int_S \left(\frac{\partial v_y}{\partial x} - \frac{\partial v_y}{\partial y} \right) dx\,dy = \oint_C v_x dx + v_y dy = \oint_C \vec{v} \cdot d\vec{\ell}_1, \qquad (15.29)$$

where $d\vec{\ell}_1 = (dx, dy)$. The line integral on the right sums the contributions of $\vec{v}$ along the flow, giving the circulation of the field. But for $U = -v_y$ and $V = v_x$, Green's theorem reveals

$$\int_S \left(\frac{\partial v_x}{\partial x} + \frac{\partial v_y}{\partial y} \right) dx\, dy = \oint_C -v_y dx + v_x dy = \oint_C \vec{v} \cdot d\vec{\ell}_2, \tag{15.30}$$

with $d\vec{\ell}_2 = (dy, -dx)$. The line integral in this case sums the contributions of $\vec{v}$ perpendicular to the flow, and hence together measure the net flux of $\vec{v}$ across the curve C. (See Problem 15.18.) Finally, for $U = -v_x$ and $V = v_y$,

$$\int_S \left(\frac{\partial v_y}{\partial x} + \frac{\partial v_x}{\partial y} \right) dx\, dy = \oint_C -v_x dx + v_y dy = \oint_C \vec{v} \cdot d\vec{\ell}_3, \tag{15.31}$$

where $d\vec{\ell}_3 = (-dx, dy)$. This time the line integrals measure the contributions of $\vec{v}$ along y but against x. This is the *shear* of the field. (See BTW 12.1.)

Viewed from $\mathbb{R}^3$, the integrand on the right-hand side of (15.27) is the z-component of the curl. You might reasonably ask why that component is favored. Simple: it was already singled out by choosing an S that lies flat in the xy-plane, so that $d\vec{a} = \hat{k}\, dxdy$. In fact, this theorem-in-the-plane can be re-expressed as

$$\oint_C \vec{B} \cdot d\vec{\ell} = \int_S (\vec{\nabla} \times \vec{B}) \cdot d\vec{a}, \tag{15.32}$$

so that it holds even for closed curves which do not lie in a plane. Thus we see that *the circulation is the flux of the curl*. This result, known as *Stokes' theorem*, quantifies relationship between circulation and curl. The curl is the *circulation density*; more precisely, $(\vec{\nabla} \times \vec{B}) \cdot \hat{n}$ at P gives the circulation of $\vec{B}$ per unit area at P on a surface with unit normal $\hat{n}$.

This result bears repeating: the extended measure we call circulation was constructed — or if you prefer, *defined* — as an integral over a curve C. This seemed fairly intuitive. What Stokes' theorem tells us, rather unintuitively, is that the very same circulation can be computed with a surface integral. Unlike the divergence theorem, which identifies the scalar-valued divergence as a volume density, Stokes' theorem reveals the vector-valued curl as an *oriented* surface density. Its flux, through a surface S bounded by C, is the circulation around C.

Recall that a surface-independent field has zero divergence. But the vector field whose flux is being evaluated on the right-hand side of Stokes' theorem, (15.32), is a curl — and $\vec{\nabla} \cdot (\vec{\nabla} \times \vec{B}) \equiv 0$. Thus any field of the form $\vec{F} = \vec{\nabla} \times \vec{B}$ is surface independent: Stokes' theorem holds for *any* surface bounded by a given curve C.

The standard expression of Stokes' theorem is given in (15.32). However, there are other, vector-valued versions which are also useful — and which you're asked to derive in Problem 15.12:

$$\oint_C \Phi\, d\vec{\ell} = \int_S \hat{n} \times \vec{\nabla}\Phi\, da \equiv \int_S d\vec{a} \times \vec{\nabla}\Phi \tag{15.33a}$$

$$\oint_C d\vec{\ell} \times \vec{E} = \int_S (\hat{n} \times \vec{\nabla}) \times \vec{E}\, da \equiv \int_S (d\vec{a} \times \vec{\nabla}) \times \vec{E}. \tag{15.33b}$$

Example 15.8 $\vec{B} = x^2 y \hat{i} - xy^2 \hat{j}$

We'll demonstrate Stokes' theorem with the field of Examples 14.13 and 14.15.

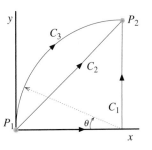

$C_2 - C_1$: We found before that the work along C_1 is $-1/3$ while along C_2 it's 0. Thus along the closed path $C_2 - C_1$ we have $\Psi = +1/3$. To verify Stoke's theorem, we must integrate the curl of $\vec{B}$ over the enclosed triangular surface. Going from P_1 to P_2 along C_2 and back along C_1 gives the circulation a clockwise orientation, so $d\vec{a} = -\hat{k}dx\,dy$. Then

$$\int_S (\vec{\nabla} \times \vec{B}) \cdot d\vec{a} = \int_S \left[-(x^2 + y^2)\hat{k} \right] \cdot (-\hat{k})\,dx\,dy$$

$$= + \int_0^1 dx \int_0^x dy(x^2 + y^2) = \frac{4}{3} \int_0^1 x^3 = \frac{1}{3} \cdot \checkmark$$

$C_1 - C_3$: In cylindrical coordinates, $x = 1 - r\cos\phi$, $y = r\sin\phi$, and a quick calculation shows the curl to be $-(r^2 - 2r\cos\phi + 1)\hat{k}$. The circulation is counter-clockwise, so $d\vec{a} = +\hat{k}\,r\,dr\,d\phi$. Then integrating over the quarter-circle gives

$$\int_S (\vec{\nabla} \times \vec{B}) \cdot d\vec{a} = - \int_0^1 r dr \int_0^{\pi/2} (r^2 - 2r\cos\phi + 1)\,d\phi = \frac{2}{3} - \frac{3\pi}{8},$$

which agrees with the sum of the line integrals computed earlier.

Example 15.9   **Straight But Not Uniform**

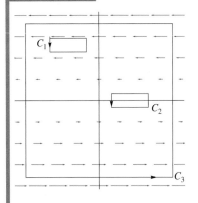

A field constant in both magnitude and direction cannot turn a propeller, so such a field is irrotational. Indeed, as a constant field the derivatives in the curl trivially vanish. But though it may seem counter-intuitive, having straight field lines alone is not a sufficient condition for irrotational fields. Consider the field $\vec{B} = -y\hat{i}$ which is straight but not uniform. (This describes, say, the magnetic field within a uniform current density in the $\hat{j}$-direction.) At first glance, the straight field lines might lead one to expect a vanishing curl; straightforward evaluation, however, yields $\vec{\nabla} \times \vec{B} = \hat{k}$. Moreover, there's obviously a non-zero circulation around a large curve like C_3, so by Stokes' theorem $\vec{B}$ must have a non-vanishing curl within this loop. Even without calculation, we can see that the circulations around C_1 and C_2 are also both non-zero. In fact, the propeller test reveals a non-zero curl everywhere out of the page. Evaluating circulations around small loops, we find that in the limit of vanishing area enclosed by a rectangular loop of dimensions Δx and Δy (Problem 15.20),

$$(\vec{\nabla} \times \vec{B}) \cdot \hat{k} = \lim_{\Delta A \to 0} \frac{\oint_C \vec{B} \cdot d\vec{\ell}}{\Delta A} = \lim_{\Delta A \to 0} \frac{\Delta x \Delta y}{\Delta A} = 1. \tag{15.34}$$

The curl is a non-zero constant.

Example 15.10 $\vec{B} = r^{\alpha}\,\hat{\phi}$ Around a Circle of Radius R

Because the field is axially symmetric, let's evaluate the circulation around an origin-centered circle of radius R,

$$\oint_C \vec{B} \cdot d\vec{\ell} = \oint_C r^{\alpha} R d\phi = R^{\alpha+1} \int_0^{2\pi} d\phi = 2\pi R^{\alpha+1}. \tag{15.35}$$

Since the field only has a $\hat{\phi}$-component, the curl in cylindrical coordinates is straightforward,

$$\vec{\nabla} \times \vec{B} = \frac{1}{r}\frac{\partial}{\partial r}\left(rB_{\phi}\right)\hat{z} = \frac{1}{r}\frac{d}{dr}\left(r^{\alpha+1}\right)\hat{z} = (\alpha+1)\,r^{\alpha-1}\,\hat{z}, \tag{15.36}$$

so that, over the circular disk bounded by C (with $d\vec{a} = r\,dr\,d\phi\,\hat{z}$),

$$\int_S (\vec{\nabla} \times \vec{B}) \cdot d\vec{a} = (\alpha+1)\int r^{\alpha-1}\,r\,dr\,d\phi = 2\pi R^{\alpha+1}. \tag{15.37}$$

This agrees with the circulation of (15.35).

Example 15.11 $\vec{B} = r^{\alpha}\hat{\phi}$ Between Concentric Circles

Similar to what we did in Example 15.4, we can avoid the singularity at the origin for $\alpha < 0$ by considering a surface S contained between concentric circles. Imagine drawing a single curve C that skirts around the origin, as shown; in the limit as $d \to 0$, C dissolves into two distinct circles C_1 and C_2 with radii $R_2 > R_1$. As long as $R_1 > 0$, the field is well-behaved throughout S. Though disjointed, the two curves together divide space into two distinct and unambiguous regions; one cannot leave S without crossing one of the circles somewhere. In this sense, C_1 and C_2 together form a closed curve and Stokes' theorem still applies — though as inherited from their construction, we must traverse C_1 and C_2 in opposite directions in order to keep the bounded surface on the left.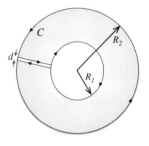

We did most of the work in Example 15.10. The circulation around C is

$$\oint_C \vec{B} \cdot d\vec{\ell} = \oint_{C_2} \vec{B} \cdot d\vec{\ell} - \oint_{C_1} \vec{B} \cdot d\vec{\ell}$$

$$= \oint r^{\alpha} R_2\,d\phi - \oint r^{\alpha} R_1\,d\phi$$

$$= 2\pi\left(R_2^{\alpha+1} - R_1^{\alpha+1}\right). \tag{15.38}$$

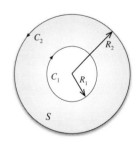

As for the flux of the curl,

$$\int_S (\vec{\nabla} \times \vec{B}) \cdot d\vec{a} = (\alpha+1)\int_S r^{\alpha-1}\,\hat{k} \cdot \hat{k}\,r\,dr\,d\phi$$

$$= 2\pi(\alpha+1)\int_{R_1}^{R_2} r^{\alpha}\,dr = 2\pi\left(R_2^{\alpha+1} - R_1^{\alpha+1}\right). \tag{15.39}$$

The $R_1 \to 0$ limit is only possible for $\alpha \geq -1$. But non-zero R_1 excludes the origin from the integration, and in this region Stokes' theorem holds even for $\alpha < -1$.

Example 15.12 **The Case of the Missing Curl**

Both (15.38) and (15.39) vanish for $\alpha = -1$. A closer look, though, invites doubt. First of all, from (15.35) we see that the circulation of $\vec{B} = \hat{\phi}/r$ around a circle is indisputably 2π. So we know what the answer must be — and it's *not* zero. Moreover, (15.37) *must* give zero because the curl vanishes identically for $\alpha = -1$, as you can easily see from (15.36). In other words, since there's *no* curl to integrate, (15.39) vanishes. But it must be that $\vec{\nabla} \times \vec{B} \neq 0$ *some*where, since we know the integral should return 2π. So where's the missing curl?

If you're experiencing déjà vu, that's because the mystery is virtually identical to that of Example 15.5. In fact, doing little more than substituting flux and divergence with circulation and curl, we come to a conclusion analogous to (15.17),

$$\vec{\nabla} \times (\hat{\phi}/r) = 2\pi\,\delta(\vec{r})\hat{k}. \tag{15.40}$$

Note that since the polar coordinate $\vec{r} = (x, y)$ lies in the plane, $\delta(\vec{r})$ in (15.40) is a two-dimensional delta function. So when S encloses the origin, we get

$$\int_S \left(\vec{\nabla} \times \vec{B}\right) \cdot d\vec{a} = \int_S 2\pi\,\delta(\vec{r})\,da = 2\pi, \tag{15.41}$$

consistent with (15.37).

Example 15.13 **Which Is Easiest?**

Consider the constant field $\vec{B} = -\hat{k}$ and the task of determining its flux through a hemisphere of radius R centered at the origin. Orienting our axes so that the base of the hemisphere lies in the xy-plane, this is fairly straightforward. In spherical coordinates, $\hat{k} = \hat{r}\cos\theta - \hat{\theta}\sin\theta$, so that

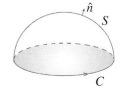

$$\int_{\text{hemisphere}} \vec{B} \cdot d\vec{a} = -\int (\hat{r}\cos\theta - \hat{\theta}\sin\theta) \cdot \hat{r}\,R^2\,d\Omega$$

$$= -2\pi R^2 \int_0^1 \cos\theta\,d(\cos\theta) = -\pi R^2. \tag{15.42}$$

Not too bad, really. But note that since $\vec{\nabla} \cdot \vec{B} = 0$, it can be expressed as the curl of the vector potential $\vec{A} = y\hat{\imath}$, so Stokes' theorem lets us evaluate the flux of $\vec{B}$ as the circulation of $\vec{A}$ around the bounding circle C in the xy-plane. The calculation isn't hard. First, since $d\vec{\ell} = \hat{\phi}\,R\,d\phi$, we only need $A_\phi = -y\sin\phi$. Second, since we're integrating in the $z = 0$ plane, θ is fixed at $\pi/2$ so that $y = R\sin\theta\sin\phi = R\sin\phi$ (in other words, we're really using polar coordinates). Thus

$$\oint_C \vec{A} \cdot d\vec{\ell} = -\oint_C y\sin\phi\,R\,d\phi$$

$$= -R^2 \int_0^{2\pi} \sin^2\phi\,d\phi = -\pi R^2. \tag{15.43}$$

Nice. But just knowing that $\vec{B}$ can be written as a curl, $\vec{B} = \vec{\nabla} \times \vec{A}$, is enough to conclude that $\vec{B}$ is surface independent. So we don't actually need $\vec{A}$: the flux of $\vec{B}$ through the hemisphere is equivalent to its flux through *any* surface bounded by C — say, the disk in the xy-plane,

$$\int_{\text{disk}} \vec{B} \cdot d\vec{a} = -\int_{\text{disk}} \hat{k} \cdot \hat{k}\, r\, dr\, d\phi = -\pi R^2. \tag{15.44}$$

Of course, we've chosen a rather modest field to demonstrate the flexibility permitted by Stokes' theorem — so which of these three integrals is simplest is, perhaps, a matter of taste. But for more flashy fields, the moral should be clear: a little forethought can pay off handsomely.

Example 15.14 **Is There a Word for "Non-Uniqueness"?**

Implicit in (15.43) is that there is only one vector potential, $\vec{A} = y\hat{\imath}$, for which $\vec{B} = \vec{\nabla} \times \vec{A} = -\hat{k}$. But this is clearly false! As a derivative, it's obvious that the curl is the same whether applied to $\vec{A}$ or $\vec{A}$ + any constant. But we can go further. The difference between vector fields $\vec{A}'$ and $\vec{A}$ with the same curl is irrotational:

$$B = \vec{\nabla} \times \vec{A}' = \vec{\nabla} \times \vec{A} \implies \vec{\nabla} \times (\vec{A}' - \vec{A}) = 0. \tag{15.45}$$

Thus all $\vec{A}$'s corresponding to the same $\vec{B}$ must differ by the gradient of a scalar,

$$\vec{A}' - \vec{A} \equiv \vec{\nabla}\lambda, \tag{15.46}$$

where λ is any differentiable scalar function. So there are in effect an *infinite* number of vector functions which correspond to the same $\vec{B}$! This is often summarized by saying that the vector potential is unique "up to the gradient of a scalar." In physics, one often says that $\vec{A}$ and $\vec{A} + \vec{\nabla}\lambda$ are "gauge equivalent."

As a result of the non-uniqueness of the vector potential, Stokes's theorem allows for even more flexibility. For instance, instead of using $\vec{A} = y\hat{\imath}$ as in Example 15.13, we could choose $\lambda(x) = x$ and compute the circulation of $\vec{A}' = \vec{A} + \vec{\nabla}x$; then

$$\oint_C \vec{A}' \cdot d\vec{\ell} = \oint_C \vec{A} \cdot d\vec{\ell} + \oint_C \hat{\imath} \cdot \hat{\phi}\, R\, d\phi$$

$$= -\pi R^2 + R \int_0^{2\pi} \sin\phi\, d\phi = -\pi R^2, \tag{15.47}$$

where we used (15.43).

Fine — but so what? Well, though gauge-equivalent $\vec{A}$'s have the same curl, there is *no* restriction on their divergence. In fact, from (15.46) we have

$$\vec{\nabla} \cdot \vec{A}' = \vec{\nabla} \cdot \vec{A} + \nabla^2\lambda. \tag{15.48}$$

So an astute choice of λ can give $\vec{A}'$ virtually any divergence we'd like! For instance, back in Example 14.28 we found a vector potential $\vec{A}_2 = -xy^2z\hat{\imath} - x^2yz\hat{\jmath}$ for which $\vec{B} = \vec{\nabla} \times \vec{A}_2 = x^2y\hat{\imath} - xy^2\hat{\jmath}$. But what if we'd prefer to work with a vector potential whose divergence vanishes? In Problem 15.23 you're asked to show that the choice $\lambda = \frac{1}{2}x^2y^2z$ yields a field

$$\vec{A}' \equiv \vec{A}_2 + \vec{\nabla}\lambda = \frac{1}{2}x^2y^2\hat{k}, \tag{15.49}$$

which is both divergenceless and whose curl is $\vec{B}$. (This is $\vec{A}_1$ of Example 14.28.)

15.3 The Fundamental Theorem of Calculus — Revisited

The theorems of Gauss and Stokes are both incarnations of the fundamental theorem of calculus. In each, the cumulation of infinitesimal changes within the region of integration equals the net change of the field over the *boundary* of the region. For instance, in (15.2),

$$f(P_2) - f(P_1) = \int_C \vec{\nabla} f \cdot d\vec{\ell}, \tag{15.50}$$

we have the accumulation of $\vec{\nabla} f \cdot d\vec{\ell}$ and the zero-dimensional boundary (endpoints) P_1 and P_2 of the curve C. Similarly (15.7),

$$\oint_S \vec{E} \cdot d\vec{a} = \int_V \vec{\nabla} \cdot \vec{E} \, d\tau, \tag{15.51}$$

declares the accumulation of $\vec{\nabla} \cdot \vec{E} \, d\tau$ equal to the evaluation of $\vec{E}$ over the two-dimensional boundary S of the volume V. Likewise in (15.32),

$$\oint_C \vec{B} \cdot d\vec{\ell} = \int_S (\vec{\nabla} \times \vec{B}) \cdot d\vec{a}, \tag{15.52}$$

we have $(\vec{\nabla} \times \vec{B}) \cdot d\vec{a}$ and the one-dimensional boundary C of the surface S.

Note that in each case, the orientations are self-consistent. In (15.32), the direction around C and the direction of the normal $\hat{n}$ on the open surface S are connected by the right-hand rule; in (15.7), the volume dictates an outward-pointing $\hat{n}$ on the closed boundary S; and the minus sign in (15.2) reflects the direction of integration along C — that is, from P_1 to P_2.

Example 15.15 **Integration by Parts**

Integration by parts, recall, is in essence the inverse of the product rule,

$$\frac{d}{dx}\left[f(x)g(x)\right] = \frac{df}{dx}\,g + f\,\frac{dg}{dx}, \tag{15.53a}$$

so that

$$\int_a^b f(x)\,\frac{dg}{dx}\,dx = f(x)g(x)\Big|_a^b - \int_a^b \frac{df}{dx}\,g(x)\,dx, \tag{15.53b}$$

where we have used the fundamental theorem,

$$\int_a^b \frac{d}{dx}\left[f(x)g(x)\right]dx = f(x)g(x)\Big|_a^b.$$

But the fundamental theorem holds in higher dimensions — and so too, then, does integration by parts. Thus

$$\vec{\nabla} \cdot (f\,\vec{E}) = f\,(\vec{\nabla} \cdot \vec{E}) + (\vec{\nabla} f) \cdot \vec{E} \tag{15.54a}$$

leads to

$$\int_V f\,(\vec{\nabla} \cdot \vec{E})\,d\tau = \oint_S f\,\vec{E} \cdot d\vec{A} - \int_V (\vec{\nabla} f) \cdot \vec{E}\,d\tau. \tag{15.54b}$$

Similarly,

$$\vec{\nabla} \times (f\,\vec{B}) = f\,(\vec{\nabla} \times \vec{B}) + (\vec{\nabla}f) \times \vec{B} \tag{15.55a}$$

implies

$$\int_S f\,(\vec{\nabla} \times \vec{B}) \cdot d\vec{a} = \oint_c f\,\vec{B} \cdot d\vec{\ell} - \int_S [(\vec{\nabla}f) \times \vec{B}] \cdot d\vec{a}. \tag{15.55b}$$

Note that in each of these integral equations, the first term on the right has been integrated once and is evaluated on the boundary of the integration region; as a result, these are often called "surface" or "boundary" terms.

BTW 15.2 The Boundary of the Boundary

By abstracting the notation a bit, we can state the fundamental theorem in a unified manner, applicable in any dimension. Let's denote the field of interest by f and its infinitesimal change by df. Similarly, we denote any integration region, regardless of its dimension, by Ω and its boundary by $\partial\Omega$. Then the fundamental theorem of calculus, in any of its incarnations, may be rendered simply as

$$\int_{\partial\Omega} f = \int_{\Omega} df. \tag{15.56}$$

Just as $\partial\Omega$ is the boundary of Ω, the boundary's boundary is denoted by $\partial\partial\Omega \equiv \partial^2\Omega$. Then applying the fundamental theorem twice gives

$$\int_{\Omega} d^2 f = \int_{\partial\Omega} df = \int_{\partial^2\Omega} f \equiv 0.$$

Why zero? Because $\partial^2\Omega$ doesn't exist: the boundary of a closed region has no boundary! Consider a gradient field $\vec{E} \equiv \vec{\nabla}\Phi$. Then

$$\int_S (\vec{\nabla} \times \vec{E}) \cdot d\vec{a} = \int_S (\vec{\nabla} \times \vec{\nabla}\Phi) \cdot d\vec{a} = \oint_C \vec{\nabla}\Phi \cdot d\vec{\ell}, \tag{15.57}$$

by the fundamental theorem as it appears in Stokes' theorem. Applying the fundamental theorem again — this time as in (15.2) — gives

$$\oint_C \vec{\nabla}\Phi \cdot d\vec{\ell} = \Phi(P_2) - \Phi(P_1) \equiv 0.$$

This last expression is identically zero because $P_2 = P_1$ — or, more precisely, there are no endpoints: *a closed curve has no bounding points.*

Similarly, for a vector field $\vec{B} \equiv \vec{\nabla} \times \vec{A}$,

$$\int_V \vec{\nabla} \cdot \vec{B}\, d\tau = \int_V \vec{\nabla} \cdot (\vec{\nabla} \times \vec{A})\, d\tau = \oint_S (\vec{\nabla} \times \vec{A}) \cdot d\vec{a}, \tag{15.58}$$

by the fundamental theorem — in this case, the divergence theorem. Applying the fundamental theorem again, this time in the form of Stokes' theorem, gives

$$\oint_S (\vec{\nabla} \times \vec{A}) \cdot d\vec{a} = \oint_C \vec{A} \cdot d\vec{\ell} \equiv 0,$$

since as the boundary of the boundary, C doesn't exist: *a closed surface has no bounding curve.*

If this analysis holds for arbitrary volumes, surfaces, and curves, then the integrands in the middle integrals of (15.57) and (15.58) must vanish identically,

$$\vec{\nabla} \times \vec{\nabla}\Phi \equiv 0 \qquad \vec{\nabla} \cdot (\vec{\nabla} \times \vec{A}) \equiv 0. \tag{15.59}$$

So the fundamental theorem provides geometric insight into these fundamental identities (which we first encountered in Example 12.7). The only caveat is that Stokes' and Gauss' theorems must be applicable — a fascinating discussion in its own right which we address in the Coda.

15.4 The Helmholtz Theorem

At this point, you're likely wondering why all the attention on divergence and curl. Aren't we more interested in a vector function $\vec{F}$ itself, rather than merely how it behaves? For example, we can in principle always use Coulomb's law (9.33) to find an electrostatic field — but for most charge distributions $\rho(\vec{r})$, this can be a formidable undertaking. So recognizing that the divergence and curl of a field are differential operations, let's ask the question more directly: can either be integrated to find the field? Can both together? Given a set of boundary conditions, just how many differential equations are needed to specify a vector field $\vec{F}$?

To answer this question, start with

$$\vec{F}(\vec{r}) = \int_V \vec{F}(\vec{r}\,') \, \delta(\vec{r} - \vec{r}\,') \, d^3 r', \tag{15.60}$$

where the volume V includes the point $\vec{r}$. The identity $\nabla^2 \left(\frac{1}{|\vec{r}-\vec{r}\,'|}\right) = -4\pi \delta(\vec{r} - \vec{r}\,')$ (Problem 15.26) then allows for the following formal manipulation:

$$\vec{F}(\vec{r}) = -\frac{1}{4\pi} \int_V \vec{F}(\vec{r}\,') \, \nabla^2 \left[\frac{1}{|\vec{r} - \vec{r}\,'|}\right] d^3 r'$$

$$= -\frac{1}{4\pi} \nabla^2 \int_V \frac{\vec{F}(\vec{r}\,')}{|\vec{r} - \vec{r}\,'|} d^3 r', \tag{15.61}$$

where, as a derivative with respect to $\vec{r}$, the ∇^2 can be pulled outside the $\vec{r}\,'$ integral. Since the integral yields a vector-valued function of $\vec{r}$, we can use $\nabla^2 \vec{v} = \vec{\nabla}(\vec{\nabla} \cdot \vec{v}) - \vec{\nabla} \times (\vec{\nabla} \times \vec{v})$ from (12.26) to find

$$\vec{F}(\vec{r}) = -\vec{\nabla} \left(\frac{1}{4\pi} \vec{\nabla} \cdot \int_V \frac{\vec{F}(\vec{r}\,')}{|\vec{r} - \vec{r}\,'|} d^3 r'\right) + \vec{\nabla} \times \left(\frac{1}{4\pi} \vec{\nabla} \times \int_V \frac{\vec{F}(\vec{r}\,')}{|\vec{r} - \vec{r}\,'|} d^3 r'\right)$$

$$\equiv -\vec{\nabla}\Phi + \vec{\nabla} \times \vec{A}. \tag{15.62}$$

Assuming these integrals converge, knowledge of the potentials allows the complete construction of the field. So let's focus on Φ and $\vec{A}$.

Start with the scalar piece:

$$\Phi(\vec{r}) \equiv \frac{1}{4\pi} \vec{\nabla} \cdot \int_V \frac{\vec{F}(\vec{r}\,')}{|\vec{r} - \vec{r}\,'|} d^3 r' = \frac{1}{4\pi} \int_V \vec{F}(\vec{r}\,') \cdot \vec{\nabla} \left(\frac{1}{|\vec{r} - \vec{r}\,'|}\right) d^3 r'$$

$$= -\frac{1}{4\pi} \int_V \vec{F}(\vec{r}\,') \cdot \vec{\nabla}' \left(\frac{1}{|\vec{r} - \vec{r}\,'|}\right) d^3 r', \tag{15.63}$$

where we've moved the r derivative $\vec{\nabla}$ back into the integral, and then exchanged it for the r' derivative $\vec{\nabla}'$ using the rather clever trick[3]

$$\vec{\nabla}\frac{1}{|\vec{r}-\vec{r}'|} = -\vec{\nabla}'\frac{1}{|\vec{r}-\vec{r}'|}. \tag{15.64}$$

With an $\vec{r}'$ derivative now within the $\vec{r}'$ integral, we can integrate by parts. Using (15.54) gives

$$\Phi(\vec{r}) = -\frac{1}{4\pi}\oint_S \frac{\vec{F}(\vec{r}')\cdot d\vec{a}'}{|\vec{r}-\vec{r}'|} + \frac{1}{4\pi}\int_V \frac{\vec{\nabla}'\cdot\vec{F}(\vec{r}')}{|\vec{r}-\vec{r}'|}\,d^3r'. \tag{15.65a}$$

A virtually identical manipulation of $\vec{A}$ in (15.62) yields

$$\vec{A}(\vec{r}) = -\frac{1}{4\pi}\oint_S \frac{d\vec{a}'\times\vec{F}(\vec{r}')}{|\vec{r}-\vec{r}'|} + \frac{1}{4\pi}\int_V \frac{\vec{\nabla}'\times\vec{F}(\vec{r}')}{|\vec{r}-\vec{r}'|}\,d^3r'. \tag{15.65b}$$

Assuming they converge, these expressions answer our question, enshrined as the *Helmholtz theorem*: in a simply connected[4] region of space V, a differentiable vector field $\vec{F}$ is *uniquely* specified by giving its divergence and curl everywhere in V, together with its value on the boundary S.

As an example, let's take V to be all of space, and consider a static field $\vec{F}(\vec{r})$ which goes to zero at infinity, with divergence and curl

$$\vec{\nabla}\cdot\vec{F} = \rho(\vec{r}) \tag{15.66a}$$

$$\vec{\nabla}\times\vec{F} = \vec{J}(\vec{r}). \tag{15.66b}$$

In order for the integrals in (15.65) to converge, the sources $\rho(\vec{r})$ and $\vec{J}(\vec{r})$ must be localized in space — which is to say, they must vanish as $r \to \infty$. This makes physical sense, since there cannot be any charge or current out at infinity. To determine how quickly they must go to zero, just count dimensions: volume integrals $d^3r \sim r^2 dr$ bring with them three powers of r. So to be convergent as $r \to \infty$, the integrands must fall *faster* than $1/r^3$. (Note that falling only as fast as r^{-3} will give a logarithmically divergent integral.) Thus from (15.65) we see that the sources $\rho(\vec{r})$ and $\vec{J}(\vec{r})$ must fall off faster than $1/r^2$ for large r. Since this is more than enough to ensure that the surface integrals in (15.65) vanish, we have

$$\Phi(\vec{r}) = \frac{1}{4\pi}\int \frac{\rho(\vec{r}')}{|\vec{r}-\vec{r}'|}\,d^3r' \tag{15.67a}$$

and

$$\vec{A}(\vec{r}) = \frac{1}{4\pi}\int \frac{\vec{J}(\vec{r}')}{|\vec{r}-\vec{r}'|}\,d^3r'. \tag{15.67b}$$

So ρ and $\vec{J}$ allow the construction of the potentials Φ and $\vec{A}$, which in turn — thanks to the decomposition of (15.62) — give the field $\vec{F}$. So indeed, a field can be determined from its divergence and curl.

But does this construction give a unique $\vec{F}$? Let's assume there are two fields $\vec{F}$ and $\vec{G} \equiv \vec{F} + \vec{\Lambda}$ which have the same divergence and curl. This requires that $\vec{\Lambda}$ be both solenoidal *and* irrotational,

$$\vec{\nabla}\cdot\vec{\Lambda} = 0, \qquad \vec{\nabla}\times\vec{\Lambda} = 0. \tag{15.68}$$

---

[3] Calling this a "trick" is a bit disingenuous, given how often it is invoked.
[4] Simply connected regions are discussed in Chapter 17.

We also require that both $\vec{F}$ and $\vec{G}$ have the same flux — that is, they both have the same condition on a bounding surface, $\vec{F} \cdot \hat{n}|_S = \vec{G} \cdot \hat{n}|_S$. Therefore the normal component of $\vec{\Lambda}$ must vanish on the surface

$$\Lambda_n|_S = \vec{\Lambda} \cdot \hat{n}|_S = 0. \tag{15.69}$$

(For V all of space, S is the surface at infinity.)

Now since $\vec{\Lambda}$ is irrotational, we know there must be a scalar field λ such that $\vec{\nabla}\lambda = \vec{\Lambda}$; being also solenoidal, λ must then satisfy Laplace's equation,

$$\nabla^2 \lambda = 0. \tag{15.70}$$

To incorporate the boundary condition, let's integrate $\lambda \nabla^2 \lambda$ by parts,

$$\int_V \lambda \, \nabla^2 \lambda \, d\tau = \oint_S \lambda \, \vec{\nabla}\lambda \cdot d\vec{a} - \int_V (\vec{\nabla}\lambda) \cdot (\vec{\nabla}\lambda) \, d\tau$$

$$= \oint_S \lambda \, \vec{\Lambda} \cdot d\vec{a} - \int_V \Lambda^2 \, d\tau. \tag{15.71}$$

The left-hand side of (15.71) vanishes, courtesy of (15.70). The surface integral on the right-hand side requires the normal component of $\vec{\Lambda}$ on S — which, as we just saw, vanishes. This leaves us with

$$\int_V (\vec{\nabla}\lambda) \cdot (\vec{\nabla}\lambda) \, d\tau = \int_V \vec{\Lambda} \cdot \vec{\Lambda} \, d\tau = 0. \tag{15.72}$$

This integrand is "positive semi-definite" — that is, $\vec{\Lambda} \cdot \vec{\Lambda} = \Lambda^2 \geq 0$. The only way this can integrate to zero is if $\vec{\Lambda}$ vanishes everywhere. Thus $\vec{F} \equiv \vec{G}$, and the solution is unique.

Problems

15.1 Repeat Problem 14.11 for the open hemisphere using either the divergence or Stokes' theorem.

15.2 Use Stokes' theorem to recalculate the circulations in Problem 14.21, using the simplest surfaces bounded by the specified curves.

15.3 Use the divergence theorem to recalculate the fluxes in Problem 14.22.

15.4 Verify the divergence theorem over a sphere of radius R for the fields:

(a) $\vec{F}_1 = r \sin \theta \, \hat{r}$ (b) $\vec{F}_2 = r \sin \theta \, \hat{\theta}$ (c) $\vec{F}_3 = r \sin \theta \, \hat{\phi}$.

15.5 In a universe in which the electric field of a point charge is $\frac{q\hat{r}}{4\pi r^3}$, show that the flux of the field depends upon the shape and size of the surface.

15.6 Calculate the flux of the field $\vec{F} = x\hat{i} + y\hat{j}$ through the paraboloid $x^2 + y^2 + z = 4, z > 0$ in two ways:

(a) Directly, using (14.33) and (14.34). (Use the outward pointing normal.)
(b) By using the divergence theorem. (Be sure to verify its applicability.)

15.7 For $\vec{F}$ irrotational, find the net flux of $\vec{r} \times \vec{F}$ over a sphere of radius R.

15.8 Show that the statements of path independence in (14.76) are equivalent.

15.9 Solenoidal fields $\vec{B}$ are surface independent — that is, for two surfaces S_1 and S_2 with a common boundary,

$$\int_{S_1} \vec{B} \cdot d\vec{a} = \int_{S_2} \vec{B} \cdot d\vec{a}.$$

Verify this (a) using Stokes' theorem and (b) using the divergence theorem.

15.10 Use the divergence and/or Stokes' theorems, as well as the identities in Table 12.1, to confirm the following realizations of integration by parts:

(a) $\int_V \Phi \vec{\nabla} \cdot \vec{E} \, d\tau = \oint_S \Phi \vec{E} \cdot d\vec{A} - \int_V \vec{E} \cdot \vec{\nabla} \Phi \, d\tau$

(b) $\int_V \Phi \nabla^2 \Phi \, d\tau = \oint_S \Phi \vec{\nabla} \Phi \cdot d\vec{A} - \int_V (\vec{\nabla} \Phi) \cdot (\vec{\nabla} \Phi) \, d\tau$

(c) $\int_V \vec{B} \cdot (\vec{\nabla} \times \vec{A}) \, d\tau = \int_V \vec{A} \cdot (\vec{\nabla} \times \vec{B}) d\tau + \oint_S (\vec{A} \times \vec{B}) \cdot d\vec{a}$

(d) $\int_S \Phi (\vec{\nabla} \times \vec{B}) \cdot d\vec{a} = \int_S (\vec{B} \times \vec{\nabla} \Phi) \cdot d\vec{a} + \oint_C \Phi \vec{B} \cdot d\vec{\ell}.$

15.11 Verify the following alternate forms of the divergence theorem:

(a) $\int_V \vec{\nabla} \Phi \, d\tau = \oint_S \Phi \, d\vec{a}$ (b) $\int_V (\vec{\nabla} \times \vec{E}) \, d\tau = \oint_S (d\vec{a} \times \vec{E}).$

[Hint: In (a), apply the divergence theorem to $\vec{F} \equiv \vec{v} \Phi$, for $\vec{v}$ a constant vector field. Use a similar approach in (b). You'll find the identities in Table 12.1 to be helpful.]

15.12 Verify the following alternate forms of Stokes' theorem:

(a) $\oint_C \Phi \, d\vec{\ell} = \int_S (d\vec{a} \times \vec{\nabla} \Phi)$ (b) $\oint_C (d\vec{\ell} \times \vec{E}) = \int_S [(d\vec{a} \times \vec{\nabla}) \times \vec{E}].$

[Hint: Use the approach of the previous problem.]

15.13 Use results from Problems 15.11–15.12 to redo Problems 14.2b and 14.7b.

15.14 Evaluate $\oint_S z \, d\vec{a}$ over the unit sphere both directly and by using the divergence theorem.

15.15 Find two fields — one each in cartesian and spherical coordinates — whose flux through a sphere of radius a yields the sphere's volume. What relevant property do these fields have in common?

15.16 Find the net flux of $\vec{E} = x^3 \hat{\imath} + y^3 \hat{\jmath} + z^3 \hat{k}$ through a origin-centered sphere of radius R.

15.17 Find the net flux of $\vec{E} = \hat{\imath} + \hat{\jmath} + z(x^2 + y^2)^2 \hat{k}$ through the cylindrical surface $x^2 + y^2 = 1, 0 \le z \le 1$.

15.18 In two-dimensions, the divergence theorem plausibly becomes

$$\int_S \vec{\nabla} \cdot \vec{E} \, da = \oint_C \vec{E} \cdot \hat{n} \, d\ell,$$

where $\hat{n}$ is the outward-pointing normal to the curve C. Use Green's theorem in the plane (15.28) to confirm this conjecture.

15.19 For scalar functions f and g, derive:

(a) Green's first identity: $\int_V (f \nabla^2 g + \vec{\nabla} f \cdot \vec{\nabla} g) d\tau = \oint_S f \vec{\nabla} g \cdot d\vec{a}$

(b) Green's second identity: $\int_V (f \nabla^2 g - g \nabla^2 f) d\tau = \oint_S (f \vec{\nabla} g - g \vec{\nabla} f) \cdot d\vec{a}.$

15.20 Verify (15.34).

15.21 Use the relation $\vec{\nabla} \cdot \vec{A} = d\Phi_{net}/dV = \lim_{V \to 0} \left[\frac{1}{V} \oint_S \vec{A} \cdot d\vec{a} \right]$ to derive:

(a) the divergence in cylindrical coordinates (12.50c) by considering an infinitesimal region of sides dr, $rd\phi$, dz

(b) the divergence in spherical coordinates (12.51c) by considering an infinitesimal region of sides dr, $rd\theta$, $r \sin \theta d\phi$.

15.22 Show that two fields which differ by the gradient of a scalar, $\vec{A}' = \vec{A} + \vec{\nabla} \lambda$, have the same circulation.

15.23 Use (15.48) to find a λ for which $\vec{\nabla} \cdot \vec{A}' = 0$ — and thus confirm (15.49). Check that $\vec{\nabla} \times \vec{A}' = \vec{\nabla} \times \vec{A}$.

15.24 Adapt Example 15.5, replacing flux with circulation and divergence with curl, to arrive at (15.40) for the two-dimensional delta function.

15.25 Verify $\vec{\nabla} \cdot (\hat{r}/r^2) = 4\pi \delta^{(3)}(\vec{r})$ and $\vec{\nabla} \times (\hat{\phi}/r) = 2\pi \delta^{(2)}(\vec{r})\hat{k}$ by integrating each with an arbitrary function $f(\vec{r})$. [Hint: Use integration by parts.]

15.26 Use the divergence of the gradient to show that

$$\nabla^2 \left(\frac{1}{r}\right) = -4\pi \delta(\vec{r}), \tag{15.73}$$

using (a) vector identities and (b) the divergence theorem.

15.27 Use the two-dimensional divergence theorem, Problem 15.18, to show that in the plane

$$\nabla^2 \ln r = 2\pi \delta(\vec{r}), \tag{15.74}$$

where $\delta(\vec{r}) \equiv \delta^{(2)}(\vec{r})$ is the two-dimensional delta function.

15.28 Emend the calculation in Problem 4.12 to find the correct identity,

$$\frac{\partial}{\partial x_i} \frac{\partial}{\partial x_j} \left(\frac{1}{r}\right) = -\frac{4\pi}{3} \delta_{ij} \delta(\vec{r}) + \frac{3x_i x_j - \delta_{ij} r^2}{r^5}. \tag{15.75}$$

15.29 The electrostatic energy of a charge distribution contained within a volume V is (see Problem 4.13)

$$U = \frac{1}{2} \int_V \rho(\vec{r}) \Phi(\vec{r}) \, d\tau,$$

for charge density ρ and $\Phi(\vec{r})$ the potential at $\vec{r}$ of the charge configuration. Show that this can be expressed in terms of the electric field $\vec{E}$ as

$$U = \frac{\epsilon_0}{2} \left[\int_V E^2 \, d\tau + \oint_S \Phi \vec{E} \cdot d\vec{a}\right].$$

Since the charge is confined within a finite volume, the size of the integration region V can be increased without affecting the value of U. What does the expression for U become if the integration volume expands to include all of space?

15.30 Consider the *Yukawa potential* $V(r) = \frac{q}{4\pi\epsilon_0} \frac{e^{-\mu r}}{r}$, for some constant μ. Note that in the $\mu \to 0$ limit, V becomes the familiar Coulomb potential of a charge q at the origin.

(a) Find the electric field $\vec{E}$ for this potential.
(b) Calculate the divergence of $\vec{E}$ from the formula in (12.51c).
(c) By Gauss' law, the flux of $\vec{E}$ gives the total charge Q contained within the surface (divided by ϵ_0); the divergence theorem allows Q to also be calculated from $\vec{\nabla} \cdot \vec{E}$. Show that for a sphere of radius R, these two integrals *do not* give the same result.
(d) Explain how the physics of the $\mu \to 0$ and $R \to \infty$ limits can identify which of the integrals is correct. What does the $R \to 0$ limit tell you about the source of the discrepancy?
(e) Recalculate $\vec{\nabla} \cdot \vec{E}$ using product rule of (12.31) to isolate $\hat{r}/r^2$, and verify that this resolves the discrepancy.

15.31 Consider the Yukawa potential $V(r)$ of Problem 15.30.

(a) Use the identity of (12.51b) to find $\nabla^2 V$. Argue that the $\mu \to 0$ limit demonstrates this result is incorrect — or at least incomplete.
(b) Show that $\nabla^2 V = \frac{q}{4\pi\epsilon_0} \left[\frac{1}{r} \nabla^2 (e^{-\mu r}) + 2\vec{\nabla}(e^{-\mu r}) \cdot \vec{\nabla}(\frac{1}{r}) + e^{-\mu r} \nabla^2 (\frac{1}{r})\right]$. Use this to find the full expression for $\nabla^2 V$.

15.32 One of the fundamental properties of the cross product is that $\vec{A} \times \vec{B}$ is perpendicular to both $\vec{A}$ and $\vec{B}$. Is this also true if $\vec{A} \equiv \vec{\nabla}$? In other words, is the curl of a vector perpendicular to that vector?

(a) Is the curl of $\vec{B}_0 = \frac{1}{3}r^2\hat{\phi}$ (cylindrical coordinates) perpendicular to $\vec{B}_0$?

(b) Let $\vec{B}_1 \equiv \vec{B}_0 + \hat{k}$. Is $\vec{\nabla} \times \vec{B}_1$ perpendicular to $\vec{B}_1$? What about $\vec{B}_2 \equiv \vec{B}_0 + z\hat{r} + r\hat{k}$? What common property of the added pieces accounts for these results?

(c) Consider the field $\vec{B}_3 \equiv \vec{B}_0 + z\hat{r} + \vec{\nabla}\Lambda$. Find a function Λ which renders $\vec{B}_3$ perpendicular to its curl.

15.33 Use Faraday's law, $\oint_C \vec{E} \cdot d\vec{\ell} = -\frac{d}{dt}\int_S \vec{B} \cdot d\vec{a}$, in the following:

(a) Show that time-dependent $\vec{E}$ fields cannot be written as the gradient of a scalar Φ.

(b) Find an expression for $\vec{E}$ in terms of the vector potential $\vec{A}$, where $\vec{B} = \vec{\nabla} \times \vec{A}$.

(c) Find the most general expression for $\vec{E}$ in terms of both scalar and vector potentials. What field combination is conservative?

15.34 For uniform circular motion, the velocity is given by $\vec{v} = \vec{\omega} \times \vec{r}$, where $\vec{\omega}$ is the constant angular velocity. Find the circulation of $\vec{v}$ in terms of $\vec{\omega}$ around the unit circle in the plane centered at the origin.

15.35 Use the flux of $\vec{B}$ to find the vector potential $\vec{A}$ both outside and inside a long solenoid of radius a, n turns per length, carrying a steady current I. For simplicity, assume that the direction of $\vec{A}$ is the same as the current, $\hat{\phi}$; this is essentially the gauge choice $\vec{\nabla} \cdot \vec{A} = 0$ — as you can use your result to verify.

15.36 Use (15.8b) to verify the vector potential in (15.65b).

15.37 Coulomb's law can be expressed as an integral over a charge distribution ρ,

$$\vec{E}(\vec{r}) = \frac{1}{4\pi\epsilon_0} \int_V \rho(\vec{r}') \frac{(\vec{r} - \vec{r}')}{|\vec{r} - \vec{r}'|^3} d^3r',$$

where the sum is cast as an integral over the primed coordinates.

(a) Show that Coulomb's law implies $\vec{\nabla} \cdot \vec{E} = \rho/\epsilon_0$, which is the differential form of Gauss' law (see Chapter 16).

(b) The converse doesn't work: one cannot derive Coulomb's law from Gauss' law, since a field is not specified by its divergence alone. We also need that $\vec{E}$ is irrotational — that is, that $\vec{E}$ can be derived from a scalar potential Φ. Write the differential form of Gauss' law in terms of Φ, and show that Coulomb's law can be confirmed from this expression.

16 Mostly Maxwell

The calculus of vector fields appears in the description of phenomena as diverse as fluid dynamics and Newtonian gravity. And of course, it is also of fundamental importance in the theory of electromagnetism, given by Maxwell's equations for electric and magnetic fields $\vec{E}$ and $\vec{B}$,

$$\oint_S \vec{E} \cdot d\vec{a} = Q_{\text{encl}}/\epsilon_0 \tag{16.1a}$$

$$\oint_S \vec{B} \cdot d\vec{a} = 0 \tag{16.1b}$$

$$\oint_C \vec{E} \cdot d\vec{\ell} = -\frac{d}{dt} \int_S \vec{B} \cdot d\vec{a} \tag{16.1c}$$

$$\oint_C \vec{B} \cdot d\vec{\ell} = \mu_0 I_{\text{encl}} + \mu_0 \epsilon_0 \frac{d}{dt} \int_S \vec{E} \cdot d\vec{a}, \tag{16.1d}$$

where Q_{encl} is the charge contained within the closed surface S, and I_{encl} is the current passing through the closed curve C. With a flux and circulation for each field, Maxwell's equations go far beyond the earlier laws of Coulomb and Biot-Savart. Indeed, Maxwell's equations represent a shift in focus from what the functions $\vec{E}(\vec{r})$ and $\vec{B}(\vec{r})$ are in any *particular* situation to how they behave in *all* situations. As such, electromagnetism provides an excellent vehicle with which to leverage a knowledge of physics in order to enhance a command of mathematics.

16.1 Integrating Maxwell

We saw back in Example 14.22 that the mathematical expression of charge conservation is given by the continuity relation

$$\oint_S \vec{J} \cdot d\vec{a} = -\frac{d}{dt} \int_V \rho \, d\tau, \tag{16.2}$$

for charge density ρ and current density $\vec{J}$. Continuity asserts that any charge which disappears from within a volume V must have passed through the bounding surface S. But are the known properties of electric and magnetic fields in concert with this? If Maxwell's equations are to describe electromagnetic phenomena, there cannot be any inconsistency between them and the conservation of charge. Specifically, can charge conservation be *derived* from Maxwell's equations?

As the only one of Maxwell's equation to use $\vec{J}$, let's begin by examining Ampère's law (16.1d). First consider time-independent fields, so that

$$\oint_C \vec{B} \cdot d\vec{\ell} = \mu_0 \int_S \vec{J} \cdot d\vec{a}. \tag{16.3}$$

Assuming that both $\vec{B}$ and $\vec{J}$ are continuous, we can shrink the bounding curve C down to a point so that S becomes a closed surface. But with C reduced to a single point, the circulation integral vanishes; thus time-independent fields have a surface independent $\vec{J}$,

$$\oint_S \vec{J} \cdot d\vec{a} = 0. \tag{16.4}$$

According to (16.2), this mandates that $Q_{\text{encl}} = \int_V \rho \, d\tau$ remains constant — exactly as it should be for time-independent fields. So Maxwell's equations are certainly compatible with charge conservation when Q is not a function of time.

But in the real world, charges and currents and fields can all vary with time. Maxwell understood that the equations of electromagnetism must be fully consistent with continuity even when $Q = Q(t)$. How can we be sure that the *total* charge remains constant? To discuss time dependence, Maxwell introduced his so-called *displacement current*, $I_{\text{disp}} \equiv \epsilon_0 \frac{d}{dt} \int_S \vec{E} \cdot d\vec{a}$; this is the term in Ampère's law which we neglected in (16.3). Evaluating the modified law for a closed surface, the left-hand side of (16.1d) vanishes as before to give

$$0 = \mu_0 \oint_S \vec{J} \cdot d\vec{a} + \mu_0 \epsilon_0 \frac{d}{dt} \oint_S \vec{E} \cdot d\vec{a}. \tag{16.5}$$

But we know the net flux of the electric field from Gauss' law, (16.1a). Thus we obtain the continuity equation,

$$\oint_S \vec{J} \cdot d\vec{a} = -\frac{dQ}{dt}. \tag{16.6}$$

What about the total charge in the universe? In that case, the surface S is at infinity. Since real physical quantities must vanish at infinity, ρ and $\vec{J}$ must fall off with distance fast enough at large distance so that all integrals of ρ and $\vec{J}$ are well-defined. And if $\vec{J}$ vanishes way out there, so will its flux. (Indeed, where would it flux *to*?) Thus the net charge of the universe is constant.

BTW 16.1 Continuity and Conservation

Imagine a quantity of charge q at point A which suddenly vanishes, simultaneously reappearing at point B. At all times the total charge is q. Since there is no current from A to carry the charge to B, this is not continuity — but the simple scenario is consistent with conservation of charge.

In fact it's too simple. One of the fundamental precepts of special relativity is that simultaneity is not absolute. What one observer regards as two simultaneous events, others will not. In particular, one group of observers will see q appear at B *before* it vanished at A, so there's a period of time in which they would see a total charge of $2q$. Another class of observers will see q appear at B *after* it vanished at A; for these viewers, the total charge would go to zero between the events. Clearly, neither scenario is compatible with the concept of conservation. For that, we must have continuity: charge that vanishes from some region must have fluxed out of that region. And since charge necessarily moves at less then the speed of light, all observers agree on the order of events. Whether charge, mass, energy, or even momentum, virtually all conservation laws in physics are expressed with a continuity equation.

Conservation of charge is one implication of Maxwell's equations — though since charge was known to be conserved, this was a requirement of any realistic theory. But Maxwell's introduction of the displacement current also led him to a revolutionary prediction. Consider that in the absence of any sources — that is, without any charge or current — Maxwell's equations (16.1) become

$$\oint_S \vec{E} \cdot d\vec{a} = 0 \qquad \oint_c \vec{E} \cdot d\vec{\ell} = -\frac{d}{dt}\int_S \vec{B} \cdot d\vec{a}$$

$$\oint_S \vec{B} \cdot d\vec{a} = 0 \qquad \oint_c \vec{B} \cdot d\vec{\ell} = \mu_0\epsilon_0 \frac{d}{dt}\int_S \vec{E} \cdot d\vec{a}. \tag{16.7}$$

We know that $\vec{E}$ and $\vec{B}$ can exist in empty space — in fact, the field concept was motivated by the problem of action at a distance. Without the displacement current, however, there are only time-independent solutions. What about *with* Maxwell's modification? Can the right-hand side of Faraday's and Ampère's laws act as sources, allowing for time-dependent fields?

Let's simplify the problem by considering $\vec{E}$ and $\vec{B}$ to be functions of only one spatial variable; then we'll answer the question by solving Maxwell's equations in empty space for the component functions $\vec{E}(x,t) = (E_x, E_y, E_z)$ and $\vec{B}(x,t) = (B_x, B_y, B_z)$.

First, we apply Gauss' law on a rectangular surface aligned with the axes, as shown in Figure 16.1. The advantage of this surface is that the dot products in the flux pick out the individual components of $\vec{E}$, allowing us to analyze them separately. So, for instance,

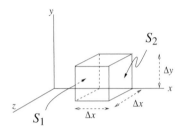

Figure 16.1 A rectangular surface.

$$\oint_S \vec{E} \cdot d\vec{a} = 0 \longrightarrow \int_{S_2} \vec{E} \cdot d\vec{a} = -\int_{S_1} \vec{E} \cdot d\vec{a}. \tag{16.8}$$

But since

$$\int_{S_2} \vec{E} \cdot d\vec{a} = E_x(x + \Delta x)\,\Delta y \Delta z \qquad \int_{S_1} \vec{E} \cdot d\vec{a} = -E_x(x)\,\Delta y \Delta z \tag{16.9}$$

we conclude E_x is independent of x. (Note the collusion between the x-component of the field, which arises from the definition of flux, and its x-dependence.) So $E_x(x)$ is constant — even as $x \to \infty$, where the field must go to zero. Thus we conclude $E_x \equiv 0$. We can then use the freedom to orient the other two axes so that $\vec{E}$ points in the y-direction, leaving us with only one unknown function, $\vec{E} = E_y(x,t)\hat{\jmath}$.

In empty space $\vec{E}$ and $\vec{B}$ satisfy the same flux law, so an identical argument gives $\vec{B} = \vec{B}(x,t)$ with $B_x = 0$. But since we've already chosen to orient our axes so that $\vec{E}$ points along $\hat{\jmath}$, the most we can say about the magnetic field is that it lies in the yz-plane, $\vec{B} = B_y(x,t)\hat{\jmath} + B_z(x,t)\hat{k}$. The flux conditions have allowed us to reduce the unknown functions from six to three, E_y, B_y, and B_z.

We next turn to the circulation conditions, and consider a rectangular path in the xy-plane (Figure 16.2). With

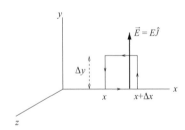

Figure 16.2 A rectangular loop in the xy-plane.

$$\oint_C \vec{E} \cdot d\vec{\ell} = [E_y(x + \Delta x) - E_y(x)]\,\Delta y \tag{16.10}$$

and

$$\frac{d}{dt}\int_S \vec{B}\cdot d\vec{a} = \frac{\partial B_z}{\partial t}\,\Delta x \Delta y,$$ (16.11)

Faraday's law (16.1c) gives

$$E_y(x+\Delta x) - E_y(x) = -\frac{\partial B_z}{\partial t}\,\Delta x.$$ (16.12)

Dividing both since by Δx and taking the limit $\Delta x \to 0$ gives

$$\frac{\partial E_y}{\partial x} = -\frac{\partial B_z}{\partial t}\,.$$ (16.13)

Since $\vec{E}$ is in the $\hat{\jmath}$-direction, the same argument applied to a rectangle in the xz-plane yields

$$\frac{\partial B_y}{\partial t} = 0\,.$$ (16.14)

Now let's apply Ampère's law to a rectangle in the xz-plane (Figure 16.3):

$$\oint_C \vec{B}\cdot d\vec{\ell} = \left[B_z(x) - B_z(x+\Delta x)\right]\Delta z$$ (16.15)

and

$$\frac{d}{dt}\int_S \vec{E}\cdot d\vec{a} = \frac{\partial E_y}{\partial t}\,\Delta x \Delta z,$$ (16.16)

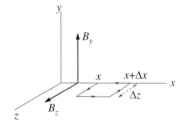

Figure 16.3 A rectangular loop in the xz-plane.

giving

$$B_z(x+\Delta x) - B_z(x) = -\mu_0\epsilon_0\frac{\partial E_y}{\partial t}\,\Delta x.$$ (16.17)

Thus

$$\frac{\partial B_z}{\partial x} = -\mu_0\epsilon_0\frac{\partial E_y}{\partial t}\,.$$ (16.18)

The same argument applied to a rectangle in the xy-plane yields

$$\frac{\partial B_y}{\partial x} = 0\,.$$ (16.19)

Equations (16.14) and (16.19) tell us that B_y depends on neither space nor time. As a constant, it has no dynamic consequences. So we drop B_y altogether, leaving $\vec{B}$ completely in the z-direction. Thus we find that *in free space, $\vec{E}$ and $\vec{B}$ are mutually perpendicular.*

Combining equations (16.13) and (16.18), we finally obtain

$$\frac{\partial^2 E_y}{\partial x^2} = \mu_0\epsilon_0\frac{\partial^2 E_y}{\partial t^2}$$ (16.20a)

$$\frac{\partial^2 B_z}{\partial x^2} = \mu_0\epsilon_0\frac{\partial^2 B_z}{\partial t^2}\,.$$ (16.20b)

The equation $\partial_x^2 f = \frac{1}{v^2}\partial_t^2 f$ describes a wave $f(x,t)$ with speed v. Thus we learn that $\vec{E}$ and $\vec{B}$ are mutually dependent waves. How fast, Maxwell wondered, do they move? Calculating, he found

$v = 1/\sqrt{\mu_0 \epsilon_0} = 3 \times 10^8$ m/s — the speed of light. Maxwell concluded that light itself is an electromagnetic wave. In a single stroke, he demonstrated that the then-separate disciplines of electricity, magnetism, and optics were actually different aspects of a unified physics of electromagnetism. Maxwell proposed the displacement current on purely theoretical grounds; experimental verification of its radical implications did not come for more than 20 years when, in 1887, Heinrich Hertz produced electromagnetic waves in the lab.

16.2 From Integrals to Derivatives

The preceding derivations are wonderful exercises in integral vector calculus. But Maxwell's equations in integral form are generally cumbersome, and often quite limiting. As we saw in Examples 14.12 and 14.23, the laws of Ampère and Gauss can only used to *solve* for the fields in cases of extreme symmetry. Fortunately, there's a better way.

We start with the following, seemingly innocuous, question: does the equality of two integrals necessarily imply the equality of the integrands? The answer is easy to appreciate in one dimension. Maybe two functions $f(x)$ and $g(x)$ with the same domain only agree between a and b, or perhaps they simply have the same area-under-the-curve between these limits. Either case would give $\int_a^b f(x)dx = \int_a^b g(x)dx$ — but it would be incorrect to conclude from this that $f(x) = g(x)$. Only if the integrals are equal for *any* choice of a and b can we assert the equivalence of f and g. Similarly, for higher-dimensional integrals over a region Ω (whether curve, surface, or volume), only if the integrals are the same for an *arbitrary* open region Ω can one conclude that the integrands are equal.

Now consider Gauss' law,

$$\oint_S \vec{E} \cdot d\vec{a} = Q_{\text{encl}}/\epsilon_0.$$

The divergence theorem allows us to convert this fairly intuitive but impractical integral equation into an easier-to-manipulate differential equation. First, the flux can be rendered

$$\oint_S \vec{E} \cdot d\vec{a} = \int_V \vec{\nabla} \cdot \vec{E} \, d\tau.$$

Then recall that $Q_{\text{encl}} = \int_V \rho \, d\tau$, where ρ is the charge density and V is the volume bounded by S used to calculate the flux. Gauss' law therefore can be rewritten in terms of integrals over V,

$$\int_V \vec{\nabla} \cdot \vec{E} \, d\tau = \frac{1}{\epsilon_0} \int_V \rho \, d\tau. \tag{16.21}$$

Since this holds for *any* surface S — and therefore, any V — the integrands must be equal:

$$\vec{\nabla} \cdot \vec{E} = \rho/\epsilon_0. \tag{16.22}$$

This is the differential form of Gauss' law — and it should come as no surprise: the integral form tells us that electric flux is charge, so the density of electric flux is electric charge density.

Applied to the flux law for magnetism, the same procedure that led to (16.22) gives

$$\oint_S \vec{B} \cdot d\vec{a} = 0 \iff \int_V \vec{\nabla} \cdot \vec{B} \, d\tau = 0, \tag{16.23}$$

so that

$$\vec{\nabla} \cdot \vec{B} = 0. \tag{16.24}$$

Now consider Faraday's law,

$$\oint_C \vec{E} \cdot d\vec{\ell} = -\frac{d}{dt} \int_S \vec{B} \cdot d\vec{a},$$

where S is any surface bounded by the curve C. (We know it's *any* surface because $\vec{\nabla} \cdot \vec{B} = 0$, so the field is surface independent.) By Stokes' theorem

$$\oint_C \vec{E} \cdot d\vec{\ell} = \int_S (\vec{\nabla} \times \vec{E}) \cdot d\vec{a}, \tag{16.25}$$

so that

$$\int_S (\vec{\nabla} \times \vec{E}) \cdot d\vec{a} = -\frac{d}{dt} \int_S \vec{B} \cdot d\vec{a} = - \int_S \frac{\partial \vec{B}}{\partial t} \cdot d\vec{a} . \tag{16.26}$$

Since this must hold for any surface,

$$\vec{\nabla} \times \vec{E} = -\frac{\partial \vec{B}}{\partial t}. \tag{16.27}$$

The same basic manipulation renders Ampère's law in the differential form

$$\vec{\nabla} \times \vec{B} = \mu_0 \vec{J} + \frac{1}{c^2} \frac{\partial \vec{E}}{\partial t}, \tag{16.28}$$

where $\mu_0 \epsilon_0 = 1/c^2$.

With the differential form of Maxwell's equations in hand, the derivation of charge conservation is almost trivial. Taking the divergence of Ampère's law, we get

$$\vec{\nabla} \cdot (\vec{\nabla} \times \vec{B}) = \mu_0 \vec{\nabla} \cdot \vec{J} + \mu_0 \epsilon_0 \vec{\nabla} \cdot \frac{\partial \vec{E}}{\partial t}. \tag{16.29}$$

Of course, the divergence of a curl is always zero; thus

$$\vec{\nabla} \cdot \vec{J} = -\epsilon_0 \frac{\partial}{\partial t} \vec{\nabla} \cdot \vec{E}, \tag{16.30}$$

or, invoking Gauss' law,

$$\vec{\nabla} \cdot \vec{J} = -\frac{\partial \rho}{\partial t}. \tag{16.31}$$

This differential equation is called the *continuity equation*; it is, of course, the differential form of the earlier continuity expression (16.2) — as direct application of the divergence theorem verifies (Problem 16.1).

A similar procedure yields the wave equation as well. Working in empty space — so $\rho = 0$ and $\vec{J} = 0$ — and taking the curl of Faraday's law yields

$$\vec{\nabla} \times (\vec{\nabla} \times \vec{E}) = -\vec{\nabla} \times \frac{\partial \vec{B}}{\partial t}, \tag{16.32}$$

or

$$-\nabla^2 \vec{E} + \vec{\nabla}(\vec{\nabla} \cdot \vec{E}) = -\frac{\partial}{\partial t}(\vec{\nabla} \times \vec{B}). \tag{16.33}$$

Since ρ vanishes in empty space, $\vec{\nabla} \cdot \vec{E} = 0$. Similarly, $\vec{J} = 0$, so by Ampère's law $\vec{\nabla} \times \vec{B}$ only depends on the time derivative of $\vec{E}$. Thus

$$-\nabla^2 \vec{E} = -\frac{\partial}{\partial t}\left(\mu_0 \epsilon_0 \frac{\partial \vec{E}}{\partial t}\right) \implies \nabla^2 \vec{E} = \mu_0 \epsilon_0 \frac{\partial^2 \vec{E}}{\partial t^2}. \tag{16.34}$$

This is the wave equation in three dimensions. A virtually identical calculation, beginning with the curl of Ampère's law, yields the wave equation for $\vec{B}$.

16.3 The Potentials

Removing all time dependence, Maxwell's equations decouple into separate physical phenomena, cleverly known as electrostatics and magnetostatics,

$$\vec{\nabla} \cdot \vec{E} = \rho/\epsilon_0 \qquad \vec{\nabla} \times \vec{E} = 0 \tag{16.35a}$$

$$\vec{\nabla} \cdot \vec{B} = 0 \qquad \vec{\nabla} \times \vec{B} = \mu_0 \vec{J}. \tag{16.35b}$$

Two are homogeneous equation, and two are inhomogeneous with sources ρ and $\vec{J}$.

Observing that an electrostatic field is irrotational invites us to define a scalar potential Φ,

$$\vec{E} = -\vec{\nabla}\Phi. \tag{16.36}$$

We have thus simplified our lives by replacing a vector field with a scalar field. Indeed, plugging this back into Gauss' law gives

$$\vec{\nabla} \cdot (-\vec{\nabla}\Phi) = \rho/\epsilon_0 \implies \nabla^2 \Phi = -\rho/\epsilon_0. \tag{16.37}$$

So we have one (second-order) differential equation governing all of electrostatics. This is Poisson's equation, which we first encountered in (13.16); the special case $\rho = 0$ yields Laplace's equation, (13.15).

Can we similarly recast (16.35b) in terms of potentials? Well, since the divergence of a curl is identically zero, we can define a vector potential $\vec{A}$ via

$$\vec{B} = \vec{\nabla} \times \vec{A}. \tag{16.38}$$

There's certainly nothing wrong with this — but it's not clear that it's very useful, since we're merely substituting one vector field for another. Pressing on though, let's take the curl and invoke Ampère's law,

$$\vec{\nabla} \times (\vec{\nabla} \times \vec{A}) = \mu_0 \vec{J} \implies \nabla^2 \vec{A} - \vec{\nabla}(\vec{\nabla} \cdot \vec{A}) = -\mu_0 \vec{J}, \tag{16.39}$$

where we used the curl-of-a-curl identity from (12.26). Note that if not for the $\vec{\nabla} \cdot \vec{A}$ term, we would again have Poisson's equation. Still, working with Φ and $\vec{A}$ ensures that the Maxwell equations for $\vec{\nabla} \cdot \vec{B}$ and the $\vec{\nabla} \times \vec{E}$ are automatically satisfied. Thus we need only solve (16.37) and (16.39) for a given ρ and $\vec{J}$.

What happens when we consider the full, time-dependent theory of Table 16.1? Since $\vec{B}$ is still divergenceless, the motivation for $\vec{A}$ is unchanged. But since $\vec{\nabla} \times \vec{E}$ is now given by Faraday's law, $\vec{E}$ is no longer irrotational: it would appear that there is no longer a scalar potential Φ. But the introduction of $\vec{A}$ yields a surprise:

$$0 = \vec{\nabla} \times \vec{E} + \partial \vec{B}/\partial t = \vec{\nabla} \times \vec{E} + \frac{\partial}{\partial t}(\vec{\nabla} \times \vec{A}) = \vec{\nabla} \times (\vec{E} + \partial \vec{A}/\partial t). \tag{16.40}$$

Table 16.1 Maxwell's equations

Gauss' law	$\oint_S \vec{E} \cdot d\vec{a} = Q_{encl}/\epsilon_0$	$\vec{\nabla} \cdot \vec{E} = \rho/\epsilon_0$
Gauss' law for magnetism	$\oint_S \vec{B} \cdot d\vec{a} = 0$	$\vec{\nabla} \cdot \vec{B} = 0$
Faraday's law	$\oint_C \vec{E} \cdot d\vec{\ell} = -\frac{d}{dt} \int_S \vec{B} \cdot d\vec{a}$	$\vec{\nabla} \times \vec{E} = -\partial\vec{B}/\partial t$
Ampère's law	$\oint_C \vec{B} \cdot d\vec{\ell} = \mu_0 I_{encl} + \frac{1}{c^2}\frac{d}{dt}\int_S \vec{E} \cdot d\vec{a}$	$\vec{\nabla} \times \vec{B} = \mu_0\vec{J} + \frac{1}{c^2}\partial\vec{E}/\partial t$

In other words, the *combination* $\vec{E} + \partial\vec{A}/\partial t$ is irrotational! So instead of (16.36), a time-dependent electric field is given by

$$\vec{E} = -\vec{\nabla}\Phi - \frac{\partial\vec{A}}{\partial t}. \tag{16.41}$$

To express the full, time-dependent Maxwell's equations in potential form, we repeat the steps which led to (16.37) and (16.39). Taking the divergence of (16.41) and using Gauss' law gives

$$\nabla^2\Phi + \frac{\partial}{\partial t}(\vec{\nabla} \cdot \vec{A}) = -\rho/\epsilon_0. \tag{16.42a}$$

Similarly, taking the curl of $B = \vec{\nabla} \times \vec{A}$ and using Ampère's law — and (12.26) for the curl of the curl — yields

$$\nabla^2\vec{A} - \frac{1}{c^2}\frac{\partial^2\vec{A}}{\partial t^2} - \vec{\nabla}\left(\vec{\nabla} \cdot \vec{A} + \frac{1}{c^2}\frac{\partial\Phi}{\partial t}\right) = -\mu_0\vec{J}. \tag{16.42b}$$

Although neither (16.42a) nor (16.42b) is as simple as Poisson's equation (16.37), Φ and $\vec{A}$ still satisfy the homogeneous Maxwell equations — leaving only these inhomogeneous Maxwell equations for the potentials.

BTW 16.2 Gauge Invariance

We used (16.38) and (16.41) to express Maxwell's equations in terms of the potentials Φ and $\vec{A}$ rather than $\vec{E}$ and $\vec{B}$. Alas, (16.42a) and (16.42b) are not very pretty. But there's one property of Φ and $\vec{A}$ we haven't yet exploited: they're not unique. Making the simultaneous changes

$$\vec{A} \longrightarrow \vec{A}' = \vec{A} + \vec{\nabla}\lambda \tag{16.43a}$$

$$\Phi \longrightarrow \Phi' = \Phi - \frac{\partial\lambda}{\partial t} \tag{16.43b}$$

for any function $\lambda(x, t)$ will also leave Maxwell's equations unaffected. We saw (16.43a) earlier in Example 15.14; it's just the statement that $\vec{A}$ is unique only up to the gradient of a scalar. Though it has no effect on the magnetic field $\vec{B}$, it does affect $\vec{E}$ via (16.41) — but this is easily patched up by (16.43b). Together, (16.43) constitute a *gauge transformation*; that they leave Maxwell's equations unchanged is known as *gauge invariance*.

Although the curl of $\vec{A}$ is unaffected by a gauge transformation, (16.43) will alter its divergence. In fact, as we saw in (15.48), these transformations give us the freedom to assign the divergence some convenient value. This is called "choosing" or "fixing" the gauge. For example, setting $\vec{\nabla} \cdot \vec{A} = 0$ — a condition known as the *Coulomb gauge* — simplifies (16.39) to a Poisson equation.

Similarly, the equations for Φ and $\vec{A}$ greatly simplify if the divergence of $\vec{A}$ satisfies the *Lorenz gauge* condition (named for the nineteenth-century Danish physicist Ludvig Lorenz)

$$\vec{\nabla} \cdot \vec{A} + \frac{1}{c^2} \frac{\partial \Phi}{\partial t} = 0. \tag{16.44}$$

Given potentials Φ' and $\vec{A}'$, (16.43) shows that physically equivalent potentials Φ and $\vec{A}$ satisfying the Lorenz gauge condition exist if there is a function $\lambda(x,t)$ such that

$$0 = \vec{\nabla} \cdot \vec{A} + \frac{1}{c^2} \frac{\partial \Phi}{\partial t} = \vec{\nabla} \cdot (\vec{A}' - \vec{\nabla}\lambda) + \frac{1}{c^2} \frac{\partial}{\partial t} \left(\Phi' + \frac{\partial \lambda}{\partial t} \right), \tag{16.45}$$

or

$$\nabla^2 \lambda - \frac{1}{c^2} \frac{\partial^2 \lambda}{\partial t^2} = F(x,t), \qquad F(x,t) = \vec{\nabla} \cdot \vec{A}' + \frac{1}{c^2} \frac{\partial \Phi'}{\partial t}. \tag{16.46}$$

This is the inhomogeneous wave equation with source $F(x,t)$. In practice, we don't need to solve for λ; all we require is the assurance that such a λ exists. So since the wave equation can be solved for well-behaved F, the Lorenz gauge condition can be met. Then choosing to work in the Lorenz gauge simply means imposing (16.44) as an auxiliary condition. This decouples (16.42a) and (16.42b) to yield separate wave equations for the potentials

$$\nabla^2 \Phi - \frac{1}{c^2} \frac{\partial^2 \Phi}{\partial t^2} = -\rho/\epsilon_0 \tag{16.47a}$$

$$\nabla^2 \vec{A} - \frac{1}{c^2} \frac{\partial^2 \vec{A}}{\partial t^2} = -\mu_0 \vec{J}. \tag{16.47b}$$

The choice of gauge generally rests of the type of electromagnetic phenomena being investigated; different gauges simplify the equations for Φ and $\vec{A}$ in a fashion appropriate to the physics being investigated. But gauge invariance assures us that they lead to the same Maxwell's equation — so the physics is independent of the choice of gauge.

Problems

16.1   Derive the differential form of the continuity equation from the integral statement of continuity, (16.2).

16.2   Heat flows from hotter to colder — that is, in the direction of $-\vec{\nabla}T$, where $T = T(\vec{r},t)$ is the temperature distribution of an object at time t. How *quickly* heat flows, however, will depend on inherent properties of the body: the mass density ρ, the specific heat c, and the thermal conductivity k — all assumed to be constant. In particular, $u = c\rho T$ is the energy density (energy/volume), which flows with energy-current density $\vec{J} = -k\vec{\nabla}T$. Use these to express conservation of energy as a continuity equation and derive the *heat equation*

$$\frac{\partial T}{\partial t} = \alpha \nabla^2 T,$$

where $\alpha \equiv k/c\rho$ is the object's *diffusivity*.

16.3   For a quantum mechanical wavefunction $\psi(\vec{r},t)$ of a particle of mass m, the probability of finding the particle within d^3r of $\vec{r}$ is $|\psi(\vec{r},t)|^2 d^3r$. So $\rho = |\psi|^2$ is the probability density. For this to be a viable interpretation, however, probability must be conserved. Show that the continuity equation (16.31) can be derived from the Schrödinger equation $\left(-\frac{\hbar^2}{2m}\nabla^2 + V(\vec{r}) \right)\psi = i\hbar\frac{\partial \psi}{\partial t}$ for complex ψ and real potential V.

16.4   The integral form of Ampère's law holds for any surface S bounded by the curve C. But surface independence only occurs for solenoidal (divergence-free) fields. Verify that $\vec{J} + \epsilon_0 \partial \vec{E}/\partial t$ is in fact surface independent.

16.5   Show that magnetic fields in empty space obey the wave equation.

16.6   **Electromagnetic Energy Flow**

   (a) Use the Lorentz force $\vec{F} = q(\vec{E} + \vec{v} \times \vec{B})$ to show that $\vec{J} \cdot \vec{E}$ is the power per volume delivered by the fields to particles with charge density ρ and current density $\vec{J} = \rho \vec{v}$.
   (b) Dotting Faraday's law with $\vec{E}$ and Ampère's with $\vec{B}$, use appropriate vector and derivative identities to arrive at *Poynting's theorem*

$$-\partial_t \left[\frac{\epsilon_0}{2} |\vec{E}|^2 + \frac{1}{2\mu_0} |\vec{B}|^2 \right] = \vec{J} \cdot \vec{E} + \vec{\nabla} \cdot \vec{S}, \tag{16.48}$$

   where $\vec{S} \equiv \frac{1}{\mu_0} (\vec{E} \times \vec{B})$ is called the Poynting vector.
   (c) Recast Poynting's theorem as a differential continuity equation. (Recall that power density is energy/volume/time — that is, the time derivative of energy density u.) What physical interpretation for the Poynting vector emerges from this equation?
   (d) Convert Poynting's theorem into an integral continuity equation. If the charges are confined within a volume V, what must be fluxing out of V?

16.7   **Magnetic Monopoles and Maxwell**
   The equation $\vec{\nabla} \cdot \vec{B} = 0$ is the mathematical assertion that magnetic monopoles do not exist (BTW 15.1). Ultimately, this equation reflects the lack of empirical evidence for isolated magnetic charges; to date, there have been no definitive sightings of an isolated north or south pole. Like an isolated electric charge q, the field of a magnetic monopole g at the origin would have the coulombic form

$$\vec{B}(\vec{r}) = \frac{\mu_0}{4\pi} \frac{g}{r^2} \hat{r}.$$

   How would Maxwell's equations need to be adapted if the existence of a monopole were confirmed? To answer this, let ρ_m and $\vec{J}_m$ denote magnetic charge density and magnetic current density, respectively. Then argue that magnetic field behavior should closely resemble the behavior of electric fields — in particular, the conservation of magnetic charge.

17 Coda: Simply Connected Regions

Let's briefly review how we arrived at the correspondence in (14.76) between path independent and irrotational fields. A field $\vec{E}(\vec{r})$ is path independent if its line integral from point P to point Q does not depend on the path taken. Thus there must exist a scalar function Φ such that

$$\int_P^Q \vec{E} \cdot d\vec{\ell} = \Phi(Q) - \Phi(P),\qquad(17.1)$$

which in turn implies vanishing circulation for any closed path so that $P \equiv Q$,

$$\oint_C \vec{E} \cdot d\vec{\ell} = 0.\qquad(17.2)$$

Citing the fundamental theorem of calculus, (17.1) leads to

$$\vec{E} = \vec{\nabla}\Phi,\qquad(17.3)$$

the cardinal feature of a conservative field. So a path-independent field $\vec{E}(\vec{r})$ is conservative, and a conservative field is path independent. The connection to irrotational fields comes from (17.3), which implies

$$\vec{\nabla} \times \vec{E} = 0.\qquad(17.4)$$

So a conservative field is also irrotational. Fine. But what about the converse — is an irrotational field always conservative? Does vanishing curl necessarily imply the existence of a potential Φ? Well, if the field is irrotational then the flux of the curl should certainly vanish,

$$\vec{\nabla} \times \vec{E} = 0 \implies \int_S (\vec{\nabla} \times \vec{E}) \cdot d\vec{a} = 0.\qquad(17.5)$$

From here, Stokes' theorem should lead directly to (17.2), and thence to (17.3). This would seem to imply that irrotational fields are also conservative.

But even simple abstract arguments collapse when confronted with a concrete counter example. Consider the magnetic field of an infinite wire carrying current along the z-axis (Example 14.12),

$$\vec{B} = \frac{\hat{\phi}}{r}.\qquad(17.6)$$

Since $\vec{\nabla} \times \vec{B} = 0$, the field is irrotational. On the other hand, integrating counterclockwise around a circle centered on the origin gives

$$\oint_C \vec{B} \cdot d\vec{\ell} = \int_0^{2\pi} \frac{\hat{\phi}}{r} \cdot \hat{\phi}\, r\, d\phi = \int_0^{2\pi} d\phi = 2\pi,\qquad(17.7)$$

independent of r. None of this is new; we saw it back in (15.35)–(15.36). Indeed, a glance at the field lines shows that path *dependence* is expected; it's just a simple, straightforward non-zero circulation. But this seems to be in clear conflict with $\vec{\nabla} \times \vec{B} = 0$. Something's not right.

17.1 No Holes Barred?

Looking at $\vec{B}$'s field lines — concentric counter-clockwise circles centered at the origin — may lead you to question the calculation of its curl. Wouldn't the flow cause a propeller to circulate around the origin? And doesn't that require that the curl *not* vanish everywhere along the path? No: remember that circulation is an extended measure, whereas the curl is a local one. So we must test for curl using infinitesimal propellers — and though it may not be obvious from the field lines, the vanishing curl tells us that nowhere along any of the field lines will one turn. To be sure, a non-zero circulation will *carry* a propeller around a field loop, but there is no curl anywhere along the path to make it spin as well. The only spot at which an infinitesimal propeller would spin is the origin.

Consider how this strange state of affairs appears to observers far from the origin. Every measurement they make in a *local* region shows unequivocally that the curl vanishes. Such observers have good reason to think the field is conservative. But were these observers to undertake a roundtrip expedition which happens to circle the origin — even along a path which stays far from the origin — they would be shocked to measure a non-zero circulation!

So what's going on? Recall that Stokes' theorem requires the surface S in (17.5) to be bounded by the closed curve C of (17.2). But for a curve in the plane which encircles the origin, any such surface would have to pass through the origin — and neither the field nor its curl is defined there. *So there is no way to choose a bound surface upon which the curl is defined everywhere!* Stokes' theorem doesn't hold if a perfectly acceptable curve doesn't bound a single acceptable surface. In such a case, the integral in (17.5) is ill-defined, and our putative proof that irrotational fields are conservative falls apart.

Let's try a topological approach. Consider the circulation of a vector field (not necessarily path independent), beginning and ending at a point P. Now imagine shrinking the loop down to just P itself, thereby generating a continuous family of circulation integrals through P (Figure 17.1). Such a shrinking-loop procedure allows us to identify a surface composed of the infinity of loops bounded by the original closed path. If the circulation is continuous along the entire continuum of shrinking paths, the curl is well-defined everywhere on the surface they sweep out. Clearly there are as many such surfaces as there are ways to shrink the path.

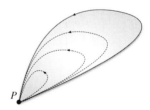

Figure 17.1 Shrinking a loop.

Of course, if the circulation is independent of the choice of closed path, then shrinking the original loop does not affect the value of the circulation at all. But that by itself doesn't require the circulation to be zero. Rather, the additional ability to reduce the loop down to a *single point* demonstrates that the integral vanishes trivially — and thus the entire infinite family of circulations also vanishes. Only in such a case are vanishing curl and path independence equivalent. The problem with $\hat{\phi}/r$ is that a path which encloses the origin cannot be shrunk to a point, because the field is not defined at the origin. True, shrinking the loop leaves the circulation unchanged — but the inability to shrink the loop all the way down to a single point means there's nothing to demand the circulation be zero. Even though any closed curve around the origin yields a perfectly good circulation, there's simply not enough space in $\mathbb{R}^2$ to accommodate a surface which excludes the origin. So there is no Φ such that $\vec{B} = \vec{\nabla}\Phi$.

You might wonder if these difficulties with $\hat{\phi}/r$ stem from restricting our attention to two dimensions. Perhaps the whole issue would disappear by considering the field in cylindrical coordinates? This would allow a surface to pass above or below the origin and thus avoid the whole problem.

But in $\mathbb{R}^3$, $\hat{\phi}/r$ is undefined at an infinite number of points: the entire z-axis! Any path which encircles the z-axis necessarily bounds a surface which is pierced by the axis. So the result is the same: the field is irrotational, but not conservative.

A region of space in which every closed curve can be shrunk continuously to a point is said to be *simply connected*. Although $\mathbb{R}^2$ is a simply connected space, remove one point from it and the remaining space is no longer simply connected. This is the source of the curiosity that is $\hat{\phi}/r$, defined not in $\mathbb{R}^2$ but rather in $\mathbb{R}^2 - \{0\}$.[1] Similarly, $\mathbb{R}^3 - \{z\text{-axis}\}$ is not simply connected. Simply put, *an irrotational field is conservative **only** in a simply connected region.*

17.2 A Real Physical Effect

Consider a solenoid, radius R, placed in the center of a circuit, its axis perpendicular to the plane of the page (Figure 17.2). Assume the solenoid, resistors, wires, and voltmeters are all ideal. Note that there is no battery anywhere.

If the current in the solenoid changes, so too do $\vec{B}$ and the magnetic flux $\Phi \equiv \int_S \vec{B} \cdot d\vec{a}$.[2] The key physical law here is Faraday's: a changing magnetic flux in the solenoid induces an emf $\mathcal{E} = -d\Phi/dt$. In terms of the underlying fields,

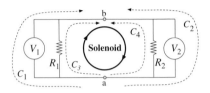

Figure 17.2 A circuit with non-trivial topology.

$$\oint_C \vec{E} \cdot d\vec{\ell} = -\frac{d}{dt} \int_S \vec{B} \cdot d\vec{a}. \tag{17.8}$$

We'll consider an increasing clockwise solenoid current; this gives $\partial \vec{B}/\partial t$ into the page, inducing a counter-clockwise electric field. For simplicity, let the solenoid current vary linearly with time so that the flux is also linear in time, $\Phi = -\alpha t$ for positive α.

The question we ask is simple: what do the voltmeters read? Well, they measure the line integral of $\vec{E}$ through the meters,

$$V_1 = -\int_{C_1}^b \vec{E} \cdot d\vec{\ell}, \qquad V_2 = -\int_{C_2}^b \vec{E} \cdot d\vec{\ell}, \tag{17.9}$$

with paths C_1 and C_2 as shown in Figure 17.2. Observing that the meters are connected (with the same polarity) between the same two points a and b, we would normally conclude $V_1 = V_2$. As you can readily show (Problem 17.1), in the region *outside* the solenoid the induced electric field is irrotational, $\vec{\nabla} \times \vec{E} = 0$ — and so the integrals of (17.9) are path independent. Right?

Wrong. Even without solving for $\vec{E}$, Faraday's law (17.8) tells us that for the counter-clockwise closed loop $C_2 - C_1$, the induced emf is

$$\mathcal{E} = \oint_{C_2 - C_1} \vec{E} \cdot d\vec{\ell} = -\frac{d}{dt} \int_S \vec{B} \cdot d\vec{a} = +\alpha, \tag{17.10}$$

where the positive sign emerges because we've chosen to integrate *along* the direction of the induced $\vec{E}$ field. (The counter-clockwise orientation of $C_2 - C_1$ gives $d\vec{a}$ out of the page, whereas $\partial \vec{B}/\partial t$ is into the page.) If we looped the other way we'd find $-\alpha$. More generally, for any closed loop C around the solenoid,

---

[1] Mathematicians often denote this as $\mathbb{R}^2 \backslash \{0\}$.
[2] Note that Φ is flux, not scalar potential.

$$\oint_C \vec{E} \cdot d\vec{\ell} = \pm n\alpha, \tag{17.11}$$

where n is the number of times C circles the solenoid. So clearly the field is *not* conservative.

What about the circulation of $\vec{E}$ around closed paths which do not encircle the solenoid, such as $C_1 - C_3$ or $C_2 - C_4$? Like the loop $C_1 - C_2$, $\vec{B}$ is zero everywhere along these paths; but unlike $C_1 - C_2$, $\vec{B}$ also vanishes *everywhere* in the surface bounded by the curves. As expected, these circulations have $n = 0$ — which permits the assertions

$$V_1 = -\int_{C_1^a}^b \vec{E} \cdot d\vec{\ell} = -\int_{C_3^a}^b \vec{E} \cdot d\vec{\ell}$$

$$V_2 = -\int_{C_2^a}^b \vec{E} \cdot d\vec{\ell} = -\int_{C_4^a}^b \vec{E} \cdot d\vec{\ell}. \tag{17.12}$$

So the reading on each meter is the voltage change across its parallel resistor.

By now it should be clear what's going on. A loop like $C_1 - C_2$, along which $B = 0$, cannot be shrunk to a point without leaving the irrotational region where $\vec{\nabla} \times \vec{E} = 0$; the area outside the solenoid is not simply connected. So irrotational does *not* imply conservative, and the integrals in (17.9) are path *dep*endent. Since $C_1 \neq C_2$, then, it's not unreasonable that $V_1 \neq V_2$.[3] Taken together (17.9) and (17.10) give $\mathcal{E} = V_2 - V_1 = \alpha$.

In fact, since the meters are connected with the same polarity, one of them must be measuring against the current. Thus V_1 and V_2 cannot even have the same sign. The counter-clockwise current through the resistors is

$$I = +\frac{V_1}{R_1} = -\frac{V_2}{R_2}. \tag{17.13}$$

Of course, $\mathcal{E} = I(R_1 + R_2)$; then from (17.10) we get $I = \alpha/(R_1 + R_2)$ and

$$V_1 = +\frac{\alpha R_1}{R_1 + R_2} \qquad V_2 = -\frac{\alpha R_2}{R_1 + R_2}. \tag{17.14}$$

Even if $R_1 = R_2$ the readings still have opposite signs.

Note that we were able to answer the question without ever solving for $\vec{E}$ (although you should be able to guess it's functional form). All we needed were its general properties — and an appreciation for the nuances of a multiply connected region. If you think this analysis is too arcane to be truly physical, see Romer (1982), where actual physical measurements of the voltages V_1 and V_2 are presented.

BTW 17.1 Winding Number and Homotopy

One remarkable feature made evident in (17.7) is that the circulation of $\hat{\phi}/r$ is insensitive to the details of the path C: the integral is 2π for *any* path which winds once around the origin. By the same token, any path which does not enclose the origin at all has vanishing circulation. In fact, we can group all closed paths into separate *equivalence classes* based on their circulation; each class is made up of closed loops with the same *winding number*,

$$n \equiv \frac{1}{2\pi} \oint_C \frac{\hat{\phi}}{r} \cdot d\vec{\ell} = \frac{1}{2\pi} \oint_C d\phi, \tag{17.15}$$

---

[3] Were the current in the solenoid constant, there would be no change in Φ, no $\partial \vec{B}/\partial t$, and no induced $\vec{E}$ — resulting in $V_1 = V_2 = 0$. Electro*static* fields are irrotational and conservative; induced, *dynamic*, $\vec{E}$ fields are not.

which is just the number of times the closed curve C encircles the origin. The sign of n reflects the orientation: n is positive for counter-clockwise circulations, negative for clockwise.

All paths with the same winding number yield the same circulation, independent of the shape of the loop. Moreover, all loops with the same winding number can be continuously deformed into one another. The proof is remarkably simple: for a continuous field, any continuous deformation of the loop must yield a continuous change in the circulation. But from (17.15), we see that n is an integer — and integers cannot change *continuously*! Thus all loops which can be continuously deformed one into another must necessarily all have the same winding number. They belong to the same *homotopy* class and are said to be *homotopic*.

Let's try to picture this with a simple example. Consider the closed curves C_0 and C_1 on $\mathbb{R}^2 - \{0\}$. Starting from any point on either curve and tracing a complete loop, it's clear that whereas ϕ_1 covers the entire range from 0 and 2π, ϕ_0 does not. As a result, C_0 can be continuously deformed into any other $n = 0$ closed loop without ever encountering the origin — including reducing it to a single point. A moment's thought should convince you that were C_0 to be deformed across the origin, ϕ_0 would start to cover the entire range from 0 to 2π and the loop would suddenly acquire winding number 1. On

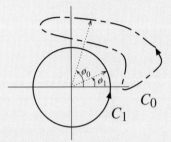

$\mathbb{R}^2 - \{0\}$, however, no continuous deformation across the origin is possible. Similarly, C_1 can be continuously deformed into any other closed loop in which ϕ_1 runs once from 0 to 2π — but it cannot be deformed into C_0 without having the loop sweep across the origin.

17.3 Single-Valued

So what makes $\hat{\phi}/r$ so special? For comparison, consider $\hat{r}/r^2$ in $\mathbb{R}^3$. As a conservative field, the integrand of the circulation must be an *exact* differential,

$$\frac{\hat{r} \cdot d\vec{\ell}}{r^2} = \frac{dr}{r^2} = -d\left[\frac{1}{r}\right]. \tag{17.16}$$

This, of course, just establishes $\Phi(r) = -1/r$ as the potential function for $\hat{r}/r^2$.[4] And the existence of a scalar potential is perhaps the clearest expression of a conservative vector field.

OK, so what about $\hat{\phi}/r$ then? A glance back at (17.7) shows that it too can be recast as an exact differential,

$$\frac{\hat{\phi} \cdot d\vec{\ell}}{r} = d\phi, \tag{17.17}$$

or, in cartesian coordinates,

$$\frac{\hat{\phi} \cdot d\vec{\ell}}{r} = \frac{-y\hat{i} + x\hat{j}}{x^2 + y^2} \cdot d\vec{\ell} = \frac{-y\,dx + x\,dy}{x^2 + y^2} = d\left[\arctan\left(\frac{y}{x}\right)\right]. \tag{17.18}$$

But wait — this would imply that $\hat{\phi}/r$ *does* have a potential and is therefore conservative. But we know that's not so! What's going on?

---

[4] We're taking the vector field to be $\vec{\nabla}\Phi$ rather than $-\vec{\nabla}\Phi$.

Let's go back to basics. We define the potential $\Phi(x)$ of a vector field $\vec{F}$ by

$$\Phi(x) = \Phi(P) + \int_P^x \vec{F} \cdot d\vec{\ell}, \qquad (17.19)$$

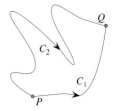

where $\Phi(P)$ is some fixed reference value. Nothing new here. But along two different paths, C_1 and C_2, from P to Q (Figure 17.3) we get

$$\Phi_1(Q) \equiv \Phi(P) + \int_{C_1} \vec{F} \cdot d\vec{\ell}, \quad \Phi_2(Q) \equiv \Phi(P) + \int_{C_2} \vec{F} \cdot d\vec{\ell}. \qquad (17.20)$$

Figure 17.3 Two paths from P to Q.

This raises the disturbing possibility that the value of the potential at any given point may not be unique. But as we've repeatedly seen, having a potential guarantees that the line integral of the vector field between any two points is independent of the path between them. In fact, each one — uniqueness and path independence — guarantees the other. So either way, the difference in the potentials along the two paths must vanish,

$$\Phi_1(Q) - \Phi_2(Q) = \oint_C \vec{F} \cdot d\vec{\ell} = 0. \qquad (17.21)$$

In other words, to be a proper potential and for $\vec{F}$ to truly be conservative, $\Phi(x)$ must be a *single-valued* function. And the arctangent in (17.18) is *not* single-valued.

A little thought reveals that as long as a closed path doesn't enclose the origin, there is no problem: all such closed loops can be continuously shrunk down to a point, so the circulation of $\hat{\phi}/r$ vanishes. And as we saw in BTW 17.1, for all such loops the function $\phi = \arctan(y/x)$ remains single-valued. In this sense, we might call the field $\hat{\phi}/r$ "quasi"-conservative. But it is not truly conservative, since any path which *does* enclose the origin runs afoul of the multi-valued nature of $\arctan(y/x)$, with the angle ϕ increasing by 2π every time we loop around the origin. So the circulation around such a loop does not vanish.

What this example demonstrates is a general property of differential equations. Continuity ensures that the equation $\vec{F} = \vec{\nabla}\Phi$ at some point P has solution Φ everywhere in a local neighborhood of P. But there is no guarantee that this local solution can be extended globally. For the field $\hat{r}/r^2$ defined on the simply connected region $\mathbb{R}^3 - \{0\}$, we find a potential which is defined both locally and globally. But $\hat{\phi}/r$ lives in the multiply connected region $\mathbb{R}^2 - \{0\}$, and its putative potential $\phi = \arctan(y/x)$ is only valid locally.

Is this common? So far, our only example of an irrotational non-conservative field has been $r^\alpha \hat{\phi}$ with $\alpha = -1$. As we've seen in (15.35) and (15.36), for $\alpha > -1$ neither the curl nor the circulation has a singularity. And while it is true that for $\alpha < -1$ the space will not be simply connected, since the field is not irrotational, the question of path independence is irrelevant. In some sense, $\alpha = -1$ is right on the border — which naturally leads to the following question: what other fields in $\mathbb{R}^2 - \{0\}$, besides $\hat{\phi}/r$, are irrotational but non-conservative?

First, let's agree that simple multiples of $\vec{B} = \hat{\phi}/r$ don't count. Even so, any irrotational non-conservative field must still be very similar to $\vec{B}$. For instance, a field $\vec{F}$ in $\mathbb{R}^2$ which is irrotational everywhere but the origin will certainly have zero circulation for any loop which *doesn't* enclose the origin. Similarly, Green's theorem (15.27) shows that the circulation also vanishes in a region which has the origin carved out — for instance, in-between the concentric circles in Example 15.11:

$$\oint_{C_2 - C_1} \vec{F} \cdot d\vec{\ell} = \oint_{C_2} \vec{F} \cdot d\vec{\ell} - \oint_{C_1} \vec{F} \cdot d\vec{\ell} = 0. \qquad (17.22)$$

This implies that the circulation of $\vec{F}$ has the same value κ for both C_1 and C_2 — and by extension, for *any* closed curve surrounding the origin. Since $\vec{F}$ is non-conservative, it must be that $\kappa \neq 0$.

We found exactly this behavior for $\vec{B}$, where the circulation came out to be 2π.[5] But this means that, even though we've stipulated that $\vec{F}$ and $\vec{B}$ are not proportional, their circulations around the origin *are*,

$$\oint_C \vec{F} \cdot d\vec{\ell} = \frac{\kappa}{2\pi} \oint_C \vec{B} \cdot d\vec{\ell}. \tag{17.23}$$

Remember that $\vec{F}$ and $\vec{B}$ each has vanishing circulation for paths which do not enclose the origin; together with (17.23) then, we discover that the field $\vec{F} - \frac{\kappa}{2\pi}\vec{B}$ is a fully path independent, conservative field,

$$\oint_C \left(\vec{F} - \frac{\kappa}{2\pi}\vec{B}\right) \cdot d\vec{\ell} = 0 \quad \forall C \iff \vec{F} - \frac{\kappa}{2\pi}\vec{B} \equiv \vec{\nabla}\Phi. \tag{17.24}$$

Thus *any* field $\vec{F}$ which is irrotational and nonconservative in $\mathbb{R}^2 - \{0\}$ must be some multiple of $\vec{B}$ *plus a gradient*. So $\vec{B} = \hat{\phi}/r$ is unique only up to a gradient. A mathematician would probably say that the irrotational, non-conservative fields in $\mathbb{R}^2 - \{0\}$ are all "cohomologous."

BTW 17.2 de Rham Cohomology

We were led to consider $\mathbb{R}^2 - \{0\}$ because $\vec{B}$ has a singularity at the excluded point. And we discovered that with one hole in the space, we have one irrotational, non-conservative field — and therefore, one cohomology class of fields. But the uniqueness of $(\hat{\phi}/r = -y\hat{\imath} + x\hat{\jmath})/r^2$ holds only for $\mathbb{R}^2 - \{0\}$; what about a space with lots of holes?

Consider $\mathbb{R}^2 - \{P\}$, where $\{P\}$ is some set of N points in the plane. We can construct a field on this space using a superposition of copies of $\vec{B}$, each copy with its singularity moved to one the P's,

$$\vec{B}_i = \frac{-(y - y_i)\hat{\imath} + (x - x_i)\hat{\jmath}}{(x - x_i)^2 + (y - y_i)^2}, \tag{17.25}$$

where $P_i = (x_i, y_i)$. Before, with only one singularity, there was only one class of simple closed curve with non-zero circulation — those which enclosed the origin. All curves in this class can be continuously deformed one into another; thus we concluded that a given irrotational, non-conservative field $\vec{F}$ has the same circulation κ for any origin-encircling curve. This led directly to (17.23) and (17.24). Now in $\mathbb{R}^2 - \{P\}$, there are N excluded points, so there are N classes of mutually deformable curves which enclose each P_i, each with its own circulation κ_i. A little thought should convince you that any closed, oriented curve enclosing *more* than one of the P_i can be constructed as a linear combination of those N basic types. Thus there are N cohomology classes of irrotational, non-conservative fields — that is, N different basic types of superpositions of the $\vec{B}_i$ which are unique up to a gradient. A theorem due to de Rham formalizes this result: the number of (cohomology) classes of irrotational, non-conservative fields on some space is equal to the number of holes in the space.

Of course, de Rham's theorem applies more generally than to $\mathbb{R}^2$. A simple example is $\mathbb{R}^3$, where the N holes in $\mathbb{R}^2$ can be extended into N lines — or, more generally, into N non-intersecting curves. In fact, de Rham's theorem in $\mathbb{R}^3$ makes an appearance in the 2001 movie *A Beautiful Mind*.

---

[5] For simplicity, we only consider so-called "simple closed curves" which do not cross themselves. A general closed curve will have circulation $n\kappa$, where n is the winding number.

Early in the movie, there's a classroom scene in which John Nash throws the textbook in the trash. He then writes the following problem on the board for the students to solve:

$$V = \{\vec{F}: \mathbb{R}^3 \backslash X \to \mathbb{R}^3 \quad \text{so} \quad \vec{\nabla} \times \vec{F} = 0\}$$
$$W = \{\vec{F} = \vec{\nabla}g\}$$
$$\dim(V/W) = ?$$

In English: V is the set of irrotational fields on $\mathbb{R}^3 - \{X\}$, W the set of conservative fields on $\mathbb{R}^3$. How many cohomology classes of irrotational, non-conservative fields exist for a given set $\{X\}$? By de Rham's theorem, when X is a single point in $\mathbb{R}^3$, the answer is zero. But if X is the z-axis, there is a single cohomology class; excluding N lines yields N cohomology classes.

   Though de Rham's theorem may seem esoteric, it is really a remarkable result. The curl, remember, is a *local* operation, depending only on values of the field near a given point. By contrast, the number of holes in a space is very much a non-local characteristic. Moreover, the number of holes is unaffected by stretching or otherwise deforming the space (short, of course, of tearing apart or gluing together parts of the space); invariant properties of this sort determine the *topology* of the space. So de Rham's theorem provides a link between local derivatives on a space and its global topology.

Problems

17.1   The induced electric field for the circuit configuration in Section 17.2 is

$$\vec{E} = \hat{\phi}\,\frac{\mu_0 n}{2}\frac{dI}{dt}\begin{cases} r, & r < R \\ \frac{R^2}{r}, & r > R \end{cases},$$

counter-clockwise since magnetic flux increases into the page (Problem 14.31).

(a) Show that $\vec{E}$ is irrotational only outside the solenoid.
(b) Use $\vec{E}$ to directly verify (17.10). Is it necessary to confine the integration to the rectangular segments of the circuit?

17.2   Magnetic Monopoles

Problem 16.7 discusses how to adapt Maxwell's equations should magnetic monopoles ever be found. Here we consider the implications of such a discovery.

(a) Like an isolated electric charge q, the field of a magnetic monopole g at the origin would have the form

$$\vec{B}(\vec{r}) = \frac{\mu_0}{4\pi}\frac{g}{r^2}\,\hat{r}.$$

Calculate the magnetic flux through a surface surrounding the monopole.
(b) Since the divergence doesn't vanish at the origin, there is no vector potential $\vec{A}$ for which $\vec{B} = \vec{\nabla} \times \vec{A}$ holds there. But in fact, there isn't even a vector potential defined over all $r \neq 0$. To verify this claim, consider a circle C at constant $\theta = \theta_0$. Let's assume a monopole $\vec{A}$ *is* well-defined, so that Stokes' theorem holds. Use this to calculate $\oint_C \vec{A} \cdot d\vec{\ell}$. [What surface bounded by C is the best choice?] Show that as $\theta_0 \to \pi$, the flux calculated in part (a) emerges. Then argue that this cannot be consistent with the circulation of $\vec{A}$ in this limit. In other words, the space is not simply connected.

(c) Although there is no $\vec{A}$ defined over all $r \neq 0$, we can find $\vec{A}$'s on more restricted domains. Show that the following choices,

$$\vec{A}_{\pm} = \pm \frac{\mu_0 g}{4\pi r} \frac{1 \mp \cos\theta}{\sin\theta} \hat{\phi},$$

are both monopole potentials, but undefined on the negative and positive z-axis ($\theta = \pi$ and 0), respectively. These singular regions, known as *Dirac strings*, can be envisioned physically as an infinitely-thin, ideal solenoid along the z-axis carrying magnetic flux out from the monopole.

(d) Show that $\vec{A}_{\pm}$ are gauge equivalent (see Example 15.14). In other words, show that the difference $\vec{A}_+ - \vec{A}_-$ has scalar potential $\Lambda = \frac{\mu_0 g}{2\pi}\phi$. Is Λ defined on the z-axis? Explain.

(e) We can avoid the singularities in $\vec{A}_{\pm}$ by constructing the piecewise field

$$\vec{A} = \begin{cases} \vec{A}_+, & \theta \leq \pi/2 \\ \vec{A}_-, & \theta \geq \pi/2. \end{cases}$$

In other words, $\vec{A}$ equals $\vec{A}_+$ in the northern hemisphere, A_- in the southern. Integrating around the equator, where both $\vec{A}_{\pm}$ apply, show that even with this construction the non-vanishing magnetic flux through a sphere centered on the monopole can only emerge if Λ in part (d) is not single-valued.

(f) Since different choices for $\vec{A}$ can relocate the string, it must be an unobservable, unphysical artifact. This is not a problem for classical electromagnetism, where potentials are mere mathematical aids for calculating physical $\vec{E}$ and $\vec{B}$ fields. In quantum mechanics, however, the potentials appear in the Schrödinger equation. In a justly famous paper, Dirac (1931) showed that the string could be rendered unobservable if the phase

$$e^{-iq\Lambda/\hbar} = e^{-i\mu_0 gq\phi/2\pi\hbar}$$

is single-valued. Show that this condition predicts that if monopoles exist, electric charge q must be quantized — that is, all electrically charged particles must come in integer multiples of a fundamental value. [For insight into the physical significance of the electromagnetic potentials, look up the Aharonov–Bohm effect.]

Part III

Calculus in the Complex Plane

The most exciting rhythms seem unexpected and complex . . .

W. H. Auden

18 Prelude: Path Independence in the Complex Plane

A complex number z represents a vector in the plane, with its cartesian form $x + iy$ construed as a decomposition over "unit vectors" 1 and i. So one might reasonably expect that problems in $\mathbb{R}^2$ can be formulated in the complex plane $\mathbb{C}$ — as we did, for example, in Section 7.2. The complex formulation of Kepler orbits presented there, however, was largely confined to algebra and time derivatives; there was no examination or exploration of the spatial behavior of functions in the complex plane.

One would certainly expect $\mathbb{R}^2$ phenomena such as flux and circulation to manifest in $\mathbb{C}$. But a function $w(z)$ in the complex plane depends on the combination $z = x + iy$ rather than x and y separately; as a result, there are continuous fields in $\mathbb{R}^2$ without corresponding continuous functions in $\mathbb{C}$. In this Prelude, we'll explore the implications of this distinction — beginning by analyzing conditions for the continuity of a function in the complex plane.

The study of functions in the complex plane is a vast subject, worthy of entire books devoted to it; excellent examples include Churchill and Brown (1984); Mathews (1988); Needham (1997); Saff and Snider (2003); and Wunsch (1983).

18.1 Analytic Functions

Consider the integral of a field $w(z)$ around a closed curve,

$$\oint_C w(z)\, dz. \tag{18.1}$$

In the complex plane, C is generally referred to as a "contour," and integrals along C are *contour integrals*. But using $dz = dx + i\, dy$, we can recast (18.1) as a familiar circulation integral in $\mathbb{R}^2$,

$$\oint_C w(z)\, dz \equiv \oint_C \vec{A} \cdot d\vec{\ell}, \tag{18.2}$$

where

$$A_x = w(x + iy) \qquad A_y = iw(x + iy). \tag{18.3}$$

Then if the field is continuous and the space simply connected, we can use Stoke's theorem (14.123) to get[1]

$$\oint_C \vec{A} \cdot d\vec{\ell} = \int_S \left(\vec{\nabla} \times \vec{A} \right) \cdot d\vec{a}. \tag{18.4}$$

---

[1] We're confined to the plane, so technically we're invoking Green's theorem, (15.27).

So $\vec{A}$ is a path independent field if

$$(\vec{\nabla} \times \vec{A}) \cdot \hat{n} = \partial_x A_y - \partial_y A_x = i\partial_x w(x + iy) - \partial_y w(x + iy) = 0. \tag{18.5}$$

In terms of w's real and imaginary parts, $w(z) = U(x, y) + iV(x, y)$ for real-valued functions U and V, this implies (Problem 18.2)

$$\frac{\partial U}{\partial x} = \frac{\partial V}{\partial y}, \qquad \frac{\partial V}{\partial x} = -\frac{\partial U}{\partial y}. \tag{18.6}$$

These conditions on U and V, which hold everywhere in S, are called the **Cauchy–Riemann equations**. Looking back at Chapter 13 reveals that they are equivalent to (13.3a) — allowing an interpretation of U and V as potential and stream functions of an irrotational, incompressible field in $\mathbb{R}^2$ (Problem 18.3).

That path independence requires conditions on derivatives of field components is nothing new; what is different is that these condition underly the *existence* of dw/dz. That's because unlike in one dimension where continuity requires $w(x) \rightarrow w(x_0)$ when approached from both the left and right, in the complex plane $w(z)$ must approach a unique $w(z_0)$ from all directions. Indeed, a little thought shows that for path independence, it could hardly be otherwise. So if a complex function $w(z)$ has continuous partial derivatives satisfying the Cauchy–Riemann equations in a neighborhood of a point z_0, it has a unique complex derivative dw/dz at z_0 (Problem 18.6). Such a function is said to be *analytic* at P; mathematicians often use the term "regular" or "holomorphic." (Path independence is not the only way to characterize analytic functions — see Table 18.1.)

Thus we discover that analytic functions in the complex plane are to contour integrals what irrotational vector fields in $\mathbb{R}^2$ are to line integrals – a result that is enshrined in Cauchy's integral theorem:

If $w(z)$ is analytic everywhere in a simply connected region S in the complex plane, then the integral of w along any closed contour C contained in S vanishes,

$$\oint_C w(z) \, dz = 0. \tag{18.7}$$

Table 18.1 Equivalent statements of analyticity

A complex function $w(z)$ is analytic in a region R if and only if any of the following hold:
- $w(z)$ obeys the Cauchy–Riemann conditions (18.6)
- the derivative $w'(z) = dw/dz$ is unique at every point in R
- $w(z)$ is independent of z^*, $dw/dz^* = 0$
- w can be expanded in a Taylor series in z, $w(z) = \sum_{n=0}^{\infty} c_n(z - a)^n$ for all $a \in R$
- $w(z)$ is a path independent field everywhere within R, $\oint_C w(z) \, dz = 0.$[‡]

[‡]The converse requires R to be simply connected; that's Cauchy's integral theorem.

Example 18.1 **Analytic Analysis**

The function $w(z) = z$ has $U = x$ and $V = y$, and so trivially satisfies the Cauchy–Riemann equations. On the other hand, neither $\Re e(z)$ nor $\Im m(z)$ is separately analytic. Similarly, $z^* = x - iy$ is not an analytic function:

$$\frac{\partial U}{\partial x} = 1 \neq \frac{\partial V}{\partial y} = -1. \tag{18.8}$$

The implication is that analytic functions cannot depend on z^*. If this seems surprising, remember that analytic $w(z)$ depends solely on $x+iy$, and so cannot also depend on the independent combination $x - iy$. As a result, continuous function $w(x,y)$ in $\mathbb{R}^2$ cannot always be represented by an analytic function $w(z)$ in the complex plane.

What about z^2? Once U and V have been identified,

$$z^2 = (x^2 - y^2) + 2ixy \equiv U + iV, \tag{18.9}$$

the partials are straightforward:

$$\frac{\partial U}{\partial x} = 2x = \frac{\partial V}{\partial y} \qquad \frac{\partial U}{\partial y} = -2y = -\frac{\partial V}{\partial x}. \tag{18.10}$$

So z^2 is analytic. Similarly,

$$w(z) = e^z = e^x e^{iy} = e^x \cos y + i e^x \sin y, \tag{18.11}$$

is also analytic:

$$\frac{\partial U}{\partial x} = e^x \cos y = \frac{\partial V}{\partial y} \qquad \frac{\partial U}{\partial y} = -e^x \sin y = -\frac{\partial V}{\partial x}. \tag{18.12}$$

For the function $1/z$, it's easier to use polar coordinates, in which the Cauchy–Riemann equations quite reasonably become (Problem 18.11)

$$\frac{\partial U}{\partial r} = \frac{1}{r}\frac{\partial V}{\partial \phi} \qquad \frac{\partial V}{\partial r} = -\frac{1}{r}\frac{\partial U}{\partial \phi}, \tag{18.13}$$

with $U = U(r,\phi)$, and $V = V(r,\phi)$. Then

$$\frac{1}{z} = \frac{1}{r}e^{-i\phi} = \frac{1}{r}\cos\phi - \frac{i}{r}\sin\phi \tag{18.14}$$

and

$$\frac{\partial U}{\partial r} = -\frac{1}{r^2}\cos\phi = \frac{1}{r}\frac{\partial V}{\partial \phi} \qquad \frac{\partial V}{\partial r} = +\frac{1}{r^2}\sin\phi = -\frac{1}{r}\frac{\partial U}{\partial \phi}. \tag{18.15}$$

Thus $1/z$ is analytic everywhere *except* at the origin. (Somehow then, it must have a veiled dependence on z^* — see Problem 18.13.) Functions like this with isolated singularities are called "meromorphic."

So a function can fail to be analytic at a point; however, it cannot be analytic at an isolated spot in an *otherwise* non-analytic region, since that could hardly guarantee path independence.[2] But functions can be analytic in separate isolated regions, or over the entire finite complex plane — in which case $w(z)$ is cleverly said to be an *entire* function.

---

[2] The oft-used phrase "analytic at P" means w is analytic at P and in a small region around P.

Example 18.2 The Point at Infinity

Thus far, we have only been considering points in the *finite* plane. That's because formally, neither $\mathbb{R}^2$ nor the complex plane includes infinity. This is fairly intuitive, since in the $r \to \infty$ limit, any sense of direction derived from the ratio of planar components becomes meaningless. Thus infinity is effectively a *single point*. The space which results from appending this point the complex plane is called the *extended complex plane*, which we'll denote $\mathbb{C} + \{\infty\}$.

Formally, infinity can be defined as the inversion of the origin, $1/0 = \infty$ and $1/\infty = 0$. So to determine if a function $w(z)$ is analytic at infinity, one defines $\zeta = 1/z$ and considers the analyticity of $\tilde{w}(\zeta) \equiv w(1/z)$ at the origin. For instance, $w(z) = 1/z$ corresponds to $\tilde{w}(\zeta) = \zeta$; since ζ is analytic at the origin, so too is $1/z$ at infinity. On the other hand, though a polynomial $w(z) = \sum_n a_n z^n$ is analytic in $\mathbb{C}$, it is not analytic at infinity.

The identification of infinity as a single point has topological consequences: unlike on the real line where $\pm\infty$ are distinct, if the entire circumference of an infinite circle becomes identified with the single point $z = \infty$, the infinite plane becomes "compactified." If you imagine gathering the edges of the plane together to a single point, then it becomes clear that $\mathbb{C} + \{\infty\}$ is topologically equivalent to a sphere. The correspondence between plane and sphere is accomplished by *stereographic projection*. Consider a sphere centered at the origin (Figure 18.1). A line extending from any point z in the complex plane to the north pole will intersect the unit sphere in only one spot, Z. For $|z| > 1$, that point lies in the sphere's northern hemisphere. As z moves further from the sphere, Z moves to higher in latitude; in the limit, all points at infinity in the plane are mapped to the north pole. Moving in the other direction, points with $|z| < 1$ map to unique points on the southern hemisphere, with the origin $z = 0$ mapping to the south pole. Points $|z| = 1$ on the sphere's equator map to themselves. You should convince yourself that circles in the complex plane centered at the origin map to circles of latitude on the sphere, and that lines through the origin extending to infinity map to circles of longitude.

This representation of the extended complex plane is known as the *Riemann sphere*. Mathematically, it's a mapping from a point $z = x + iy$ in the plane to a point Z on the sphere with $\mathbb{R}^3$ coordinates (ξ_1, ξ_2, ξ_3) (Problem 18.17),

$$\xi_1 = \frac{2\,\Re e\, z}{|z|^2 + 1} \qquad \xi_2 = \frac{2\,\Im m\, z}{|z|^2 + 1} \qquad \xi_3 = \frac{|z|^2 - 1}{|z|^2 + 1}, \tag{18.16}$$

so that the north pole is at $\vec{\xi}_N = (0,0,1)$ and the south pole at $\vec{\xi}_S = (0,0,-1)$, corresponding to $|z| \to \infty$ and 0, respectively. Note that $\sum_i \xi_i^2 = 1$. Inverting gives

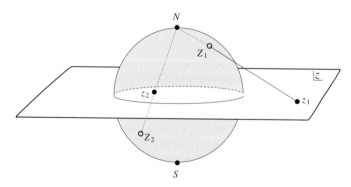

Figure 18.1 The Riemann sphere: points with $|z| > 1$ map to points Z in the northern hemisphere, points with $|z| < 1$ map to points in the southern hemisphere.

$$x = \frac{\xi_1}{1 - \xi_3} \qquad y = \frac{\xi_2}{1 - \xi_3}, \tag{18.17}$$

or in terms of the spherical coordinates of the point $Z = (\xi_1, \xi_2, \xi_3)$ on the sphere,

$$z = x + iy = \cot(\theta/2)\, e^{i\phi}. \tag{18.18}$$

This validates the observation that curves of constant ϕ on the sphere map into radial lines in the complex plane, while taking curves of constant θ into concentric circles. Conversely, all circles and lines in the complex plane map to circles on the Riemann sphere. This is true even for circles which do not enclose origin, and lines which do not pass through the origin (see Needham, 1997).

18.2 Cauchy's Integral Formula

It's hard to overstate the importance of Cauchy's integral theorem (18.7). In essence, a function's differentiability in a region S of the complex plane is all that's required for all closed contour integrals of w in S to vanish. Given how general this result is, it's perhaps no surprise that the most important applications of the Cauchy theorem are its exceptions: what happens when integrating a function which is *not* analytic throughout some region? What if there's a singularity? Consider for example the function $w(z) = c/z$, for some complex number c. Since $z = re^{i\phi}$, integrating around a circle of radius R gives $dz = iRe^{i\phi}\,d\phi$ — and thus

$$\oint_C w(z)\,dz = c \int_0^{2\pi} \left(\frac{e^{-i\phi}}{R} \right) iRe^{i\phi}\,d\phi = ic \int_0^{2\pi} d\phi = 2\pi ic. \tag{18.19}$$

A violation of the Cauchy theorem? Not at all: as we saw in Example 18.1, $1/z$ has a single, isolated singularity at $z = 0$, so it's not analytic everywhere within the contour. And although we evaluated (18.19) along a circle, the result holds for any origin-encircling contour. That's because according to Cauchy's theorem, *any continuous deformation in the shape of the contour has no effect on the integral* — that is, as long as the modification stays within an analytic region and doesn't cause C to pass over a singularity. In fact, since $1/z$ satisfies the Cauchy–Riemann equations everywhere *else* in the complex plane, its integral around any contour which does *not* enclose the origin will vanish. Moreover, a contour which winds around the origin n times yields

$$\oint_C \frac{c}{z}\,dz = 2\pi inc, \tag{18.20}$$

where n is positive when integrating counter-clockwise, negative when traversing the contour in a clockwise fashion.

If all this sounds familiar, it's because the region of analyticity for $1/z$ is not simply connected. Just as $\hat{\phi}/r$ has non-zero circulation around the origin despite a vanishing curl everywhere else, $1/z$ has non-zero contour integral around $z = 0$ even though it's analytic everywhere along C. And like $\hat{\phi}/r$ in $\mathbb{R}^2$ (Section 17.3), in the complex plane this property is effectively unique to $1/z$.

But what makes this behavior unique is unclear; after all, the analytic domain of $1/z^n$ for integer $n \neq 1$ is also not simply connected — and yet repeating the procedure in (18.19) shows that (Problem 18.15)

$$\oint_C \frac{dz}{z^n} = 0, \; n \neq 1. \tag{18.21}$$

Despite its outward appearance, this cannot be an application of Cauchy's theorem since $1/z^n$ doesn't fulfill the simply connected criterion. Rather, (18.21) emerges from

$$z^{-n} = \frac{d}{dz}\left[\frac{z^{-n+1}}{-n+1}\right], \tag{18.22}$$

since a closed contour begins and end at the same point. So as long as C doesn't pass *through* the origin, z^{-n} is effectively path independent. In fact, this holds for all integer n, positive and negative — *except* $n = 1$. In that case,

$$\frac{1}{z} = \frac{d}{dz}\left(\ln z\right). \tag{18.23}$$

For complex $z = re^{i\varphi}$, $\ln z$ can be rendered

$$\ln z = \ln r + i\varphi. \tag{18.24}$$

Thus after one complete circuit along a closed contour C, the logarithm picks up an additional phase of 2π and *does not* return to its original value — and so

$$\oint_C \frac{c}{z}\,dz = c\oint_C d\left(\ln z\right) = c\left[(\varphi + 2\pi i) - (\varphi)\right] = 2\pi ic, \tag{18.25}$$

which is (18.19). The essential distinction with (18.21)–(18.22) is that z^{-n} is a *single-valued* function, whereas the complex logarithm is *multivalued*. We discussed multivalued functions in an $\mathbb{R}^2$ context in Section 17.3, where we noted the multivalued nature of $\arctan(y/x)$, (17.18). As we can now appreciate, this angle is none other than the argument φ of $\ln z$ (Problem 18.16). We'll discuss the behavior and fundamental role of the complex logarithm in Section 19.4.

Ultimately, all of this implies that if a series expansion of a function includes negative integer powers (known as a *Laurent series*, Section 19.2), only the coefficient of the $1/z$ term, *evaluated at the singularity*, survives a contour integral surrounding the origin. As a result, this *residue* of a function is all that's needed to determine the integral — an observation which leads directly to powerful tools for evaluating integrals known as the *calculus of residues* (Chapter 21). In particular:

For w analytic in a simply connected region, a closed contour winding once around some point z in the complex plane yields

$$\oint_C \frac{w(\zeta)}{\zeta - z}\,d\zeta = 2\pi iw(z), \tag{18.26}$$

where $w(z)$ is the residue at the singularity $\zeta = z$.

This remarkable result, called **Cauchy's integral formula**, shows that $1/(\zeta - z)$ essentially acts as a delta function. And by sampling all the points z *within* the contour for residues, the integral allows for the construction of $w(z)$ using only its values ζ on C. Moreover, derivatives d/dz of this formula reveal that derivatives of w can be calculated by integration,

$$\frac{d^n w}{dz^n} = \frac{n!}{2\pi i}\oint_C \frac{w(\zeta)}{(\zeta - z)^{n+1}}\,d\zeta, \tag{18.27}$$

for C any contour in a simply connected analytic region of w.

Example 18.3 **Liouville's Theorem**

Consider a circle C of radius R centered at a point z — that is, at the singularity of the integrand in (18.27). Note that this is not a singularity of w, which is presumed analytic. Substituting $\zeta = z + Re^{i\varphi}$ and $d\zeta = iRe^{i\varphi}d\varphi$ into (18.27) with $n = 1$ we can construct w',

$$w'(z) = \frac{1}{2\pi R} \int_0^{2\pi} w(z + Re^{i\varphi}) e^{-i\varphi} d\varphi. \tag{18.28}$$

Now if $w(z)$ is a bounded function — that is, $|w(z)| \leq M$ for all z in a given region — then

$$|w'(z)| \leq \frac{1}{2\pi R} \int_0^{2\pi} |w(z + Re^{i\varphi})| \, d\varphi \leq \frac{M}{R}, \tag{18.29}$$

for all R within the analytic region. Most notably, if $w(z)$ is analytic in the extended complex plane, the inequality has to hold even as $R \to \infty$. So $|w'(z)| = 0$ for all z. Thus we have *Liouville's theorem*: a bounded, entire function $w(z)$ is a constant. In other words, w is trivial. By contrast, a polynomial is certainly not constant despite being an entire function — but then, polynomials are not bounded. Indeed, as pointed out in Example 18.2, polynomials are not analytic at infinity.

Perhaps the most important consequence of Liouville's theorem — as we already inferred from Cauchy's integral theorem — is that *non*-trivial behavior must emerge from a function's singularities in the extended complex plane.

Cauchy's integral formula (18.26) leads both to powerful insights such as Liouville's theorem, as well as practical techniques for evaluating integrals. Indeed, learning to appreciate and wield this formula is a principal goal of the coming chapters. But first we need to discuss the special challenges and opportunities complex functions present. Afterwards, we'll return to Cauchy and explore practical applications of his theorem and formula.

Problems

18.1 Determine whether the following functions are analytic — and if so, in what region of the extended complex plane:

(a) z^n for integer n
(b) $\sin z$, $\cos z$, $\tan z$
(c) $|z|$
(d) $\frac{z-i}{z+1}$
(e) $\frac{z^2+1}{z}$
(f) $P_n(z)/Q_m(z)$, for nth- and mth-order polynomials P_n, Q_m with no common roots.
(g) $x^2 + y^2$
(h) e^z
(i) e^{-iy}
(j) $\ln z$.

18.2 Consider a complex function $w(z) = U(x, y) + iV(x, y)$, where U and V are real. Show that the path-independent condition (18.5) leads directly to the Cauchy–Riemann equations (18.6).

18.3 (a) With $w(z) = U(x, y) + iV(x, y)$, show that

$$\int_C w(z) \, dz = \int_C \vec{E} \cdot d\vec{\ell} + i \int_C \vec{E} \cdot d\vec{n},$$

where the $\mathbb{R}^2$ field $\vec{E} \equiv (U, -V)$, and $d\vec{n} = (dy, -dx)$ is normal to C, $d\vec{\ell} \cdot d\vec{n} = 0$.

(b) Use Green's theorem in the plane (15.27) and the planar divergence theorem in Problem 15.18 to show that a closed contour integral vanishes if $\vec{E}$ is both irrotational and incompressible, $\vec{\nabla} \times \vec{E} = 0$, $\vec{\nabla} \cdot \vec{E} = 0$.

18.4 Show that for an analytic function $w(z) = U(x,y) + iV(x,y)$, curves of constant U and V are perpendicular to one another — that is, (U, V) is an orthogonal system.

18.5 Show that the real and imaginary parts of an analytic function $w(z) = U+iV$ both satisfy Laplace's equation, $\nabla^2 U = \nabla^2 V = 0$.

18.6 The complex derivative is defined as you would expect,

$$w'(z) = \frac{dw}{dz} = \lim_{\Delta z \to 0} \frac{w(z + \Delta z) - w(z)}{\Delta z}.$$

(a) Express $w'(z)$ in terms of $U(x,y)$ and $V(x,y)$, where $w = U + iV$.
(b) If w is analytic, then approaching a point P in the complex plane must be path independent. Thus the derivative at P must be the same whether coming in along x or y. Do this to show that a unique derivative only results if the Cauchy–Riemann equations are satisfied.

18.7 Use the approach in Problem 18.6 to show that $w(z) = z^*$ is not differentiable in the complex plane.

18.8 Show that for analytic functions $f(z)$ and $g(z)$ in some region R:

(a) the product $h(z) = f(z)g(z)$ is analytic in R
(b) $1/f(z)$ is analytic in R except at the zeros of f.

18.9 Show that analytic functions obey familiar rules of calculus:

(a) powers: $\frac{d}{dz} z^n = nz^{n-1}$
(b) products: $\frac{d}{dz} \left[f(z)g(z) \right] = f'(z)g(z) + f(z)g'(z)$
(c) quotients: $\frac{d}{dz} \left[\frac{f(z)}{g(z)} \right] = \frac{f'(z)g(z) - f(z)g'(z)}{g(z)^2}$, $g(z) \neq 0$
(d) the chain rule: $\frac{d}{dz} \left[f(g(z)) \right] = \frac{df}{dg} \frac{dg}{dz}$.

18.10 Use the results of Problem 18.7 to show that $|z|^2$ is differentiable only at the origin.

18.11 Derive the Cauchy–Riemann equations in polar coordinates, (18.8).

18.12 (a) Find expressions for the derivatives d/dz and d/dz^* in cartesian coordinates, ensuring that $\frac{d}{dz} z = \frac{d}{dz^*} z^* = 1$, and $\frac{d}{dz} z^* = \frac{d}{dz^*} z = 0$. Corroborate your results by taking derivatives of z^2.
(b) Demonstrate that an analytic function $w(z)$ is independent of z^*.

18.13 Despite its simple z dependence, since $w(z) = 1/z$ is not analytic at the origin — so it must have a veiled dependence on z^*.

(a) Express the two-dimensional Laplacian in terms of the derivatives d/dz and d/dz^* found in Problem 18.12. Use this to reformulate (15.74) (Problem 15.27) in the complex plane. Then show that $\frac{d(1/z)}{dz^*} \neq 0$ at the origin.
(b) From this, use (18.3)–(18.4) to verify that $\oint_c dz/z = 2\pi i$.

18.14 Show that $1/z^m$ is analytic everywhere but the origin.

18.15 Show that the closed integral of $1/z^n$ for integer n vanishes for all $n \neq 1$.

18.16 Show that

$$d \ln z = d \ln r + i\, d \arg z = \left(\frac{x\, dx + y\, dy}{x^2 + y^2} \right) + i \left(\frac{-y\, dx + x\, dy}{x^2 + y^2} \right),$$

where $\arg z = \arctan(y/x)$.

18.17   **Stereographic Projection**

Derive (18.16)–(18.18) using the parametric expression of a line connecting two points, $f(t) = tP + (1-t)N$, where $f(0) = N$ is the north pole $(0,0,1)$ and $f(1) = P$ is a point $Z = (\xi_1, \xi_2, \xi_3)$ on the sphere. (Start by finding t of the point of $f(t)$ in the xy-plane.)

18.18   Use (18.16)–(18.19) to verify the following, and draw sketches depicting each:

(a) Circles of radius r in the complex plane centered at the origin map to circles of latitude on the Riemann sphere — with $r > 1$ into the northern hemisphere, with $r < 1$ into the northern hemisphere.

(b) Lines through the origin in z map to circles of longitude on the Riemann sphere.

(c) z^* corresponds to a reflection of the Riemann sphere through a vertical plane containing the real axis.

(d) Inversion $1/z$ produces a π rotation of the Riemann sphere around the real axis.

(e) Points z and z' such that $z' = -1/z^*$ map to antipodal points on the Riemann sphere.

18.19   Consider a line L in the extended complex plane and its image circle C on the Riemann sphere. Draw sketches exhibiting the following statements:

(a) The image circle C passes through the north pole.

(b) The tangent T to C at the north pole is parallel to L.

(c) The angle between two intersecting lines L_1 and L_2 in the plane is the same as that between their image circles C_1 and C_2.

19 Series, Singularities, and Branches

At the risk of stating the obvious, a mathematical representation of the physical world must be able to describe change. But as implied both by Liouville's theorem and Cauchy's integral theorem, complex functions which are analytic everywhere are inadequate to the task. A description of non-trivial dynamic behavior in the complex plane thus requires functions which are not entire — that is, functions which have non-analytic points. So we begin our exploration of functions in the complex plane by surveying the different types of singularities.

19.1 Taylor Series and Analytic Continuation

The integral expression for derivatives (18.27) indicates that analytic functions are infinitely differentiable. This suggests that Taylor series on the real line can be extended to the complex plane. To verify this, let a be some point in an analytic region of $w(z)$. Then integrating along a closed contour surrounding a, Cauchy's integral formula (18.26) can be rendered as

$$
\begin{aligned}
w(z) &= \frac{1}{2\pi i} \oint_C \frac{w(\zeta)\,d\zeta}{\zeta - z} = \frac{1}{2\pi i} \oint_C \frac{w(\zeta)\,d\zeta}{(\zeta - a) - (z - a)} \\
&= \frac{1}{2\pi i} \oint_C \frac{w(\zeta)\,d\zeta}{(\zeta - a)\left[1 - \frac{z-a}{\zeta-a}\right]}.
\end{aligned}
\tag{19.1}
$$

Now since ζ represents points on the contour, and since C encloses a, we know that $\left|\frac{z-a}{\zeta-a}\right| < 1$ for all z within C. This allows us to replace the bracketed expression in the denominator with a geometric series, and then integrate term-by-term:

$$
\begin{aligned}
w(z) &= \frac{1}{2\pi i} \oint_C \frac{w(\zeta)\,d\zeta}{(\zeta - a)\left[1 - \frac{z-a}{\zeta-a}\right]} \\
&= \frac{1}{2\pi i} \oint_C \frac{w(\zeta)}{(\zeta - a)}\left[\sum_{n=0}^{\infty}\left(\frac{z-a}{\zeta-a}\right)^n\right]d\zeta \\
&= \sum_{n=0}^{\infty}\left(\frac{1}{2\pi i}\oint_C \frac{w(\zeta)\,d\zeta}{(\zeta - a)^{n+1}}\right)(z-a)^n \equiv \sum_{n=0}^{\infty} c_n\,(z-a)^n,
\end{aligned}
\tag{19.2a}
$$

where, using (18.27),

$$
c_n \equiv \frac{1}{2\pi i}\oint_C \frac{w(\zeta)\,d\zeta}{(\zeta - a)^{n+1}} = \frac{1}{n!}\left.\frac{d^n w}{dz^n}\right|_{z=a},
\tag{19.2b}
$$

verifying that we have a Taylor series of $w(z)$ centered at a in the complex plane. If this power series in z reduces to a known function on the real axis, then it provides a smooth, continuous extension — the *analytic continuation* — of the real function into the complex plane. For instance, since

$$1 + z + \frac{1}{2!}z^2 + \frac{1}{3!}z^3 + \cdots \tag{19.3}$$

is equal to e^x on the real axis, this series gives its analytic continuation over the entire complex plane. By all rights (as well as convenience), the complex function is denoted e^z. Similarly $\sin z$, $\cos z$ and many other common functions are the analytic continuations of their real counterparts. As entire functions, these examples are both straightforward and intuitive. Perhaps less intuitive is the Riemann zeta function, which in Example 6.2 we defined as the series

$$\zeta(s) \equiv \sum_{k=1}^{\infty} \frac{1}{k^s}, \tag{19.4}$$

which is convergent for $s > 1$. But we also saw in (8.79) that

$$\zeta(s) = \frac{1}{\Gamma(s)} \int_0^\infty \frac{x^{s-1}}{e^x - 1} dx, \tag{19.5}$$

which converges for complex s with real part greater than 1. Since values of this integral agree with those of the series in their common domain, (19.5) is the analytic continuation of (19.4) from real to complex s with $\Re e[s] > 1$.

But what about functions with singularities? Note that there are two basic conditions underlying (19.2): all points z must be enclosed by C in order for the geometric series to converge, and C must be in an analytic region of $w(z)$ in order for the Cauchy formula to apply. But within these bounds, the size and shape of the contour does not matter. If we expand the contour outwards, nothing changes in (19.2) unless a singularity is encountered. Then the largest distance between $z = a$ and a point on the contour, $R \equiv |\zeta - a|_{\max}$, defines a disk within which the geometric series, and hence the power series, converges. Beyond R, the power series diverges and the region under consideration ceases to be analytic. Thus we discover the important rule-of-thumb we encountered earlier in BTW 6.4: the radius of convergence of a power series is the distance R from the center to the nearest singularity (including branch points, Section 19.3) in the complex plane.

So how then can one analytically continue a power series which is only convergent within a finite disk? Consider the three series

$$w_1(z) = \sum_{n=0}^{\infty} z^n = 1 + z + z^2 + \cdots,$$

$$w_2(z) = \frac{1}{2} \sum_{n=0}^{\infty} \left(\frac{z+1}{2}\right)^n = \frac{1}{2}\left[1 + \frac{z+1}{2} + \left(\frac{z+1}{2}\right)^2 + \cdots\right], \tag{19.6}$$

$$w_3(z) = \frac{1}{1+i} \sum_{n=0}^{\infty} \left(\frac{z-i}{1-i}\right)^n = \frac{1}{1-i}\left[1 + \left(\frac{z-i}{1-i}\right) + \left(\frac{z-i}{1-i}\right)^2 + \cdots\right].$$

Inspection is sufficient to determine that these series are centered respectively at $z = 0, -1$, and i, with radii of convergence $1, 2$, and $\sqrt{2}$ (Figure 19.1). They each must have at least one singularity on each of their convergence circles. A closer look reveals that all three circles pass through $z = 1$ — and that they're all divergent there. And indeed, it's not difficult to convince yourself (Mathematica might help) that all three return the same values where their disks of convergence overlap. Thus all three are analytic continuations of one another, and together represent a complex function $w(z)$ on the union of their disks. So whereas the series w_1 is not applicable at, say, $z = -1$, we know $w(-1) = w_2(-1) = 1/2$. Similarly, $w(1 + i) = w_3(1 + i) = i$, and $w(-1 + i) = w_2(-1 + i) = w_3(-1 + i) = \frac{2}{5}(2 + i)$.

To be fair, this example is somewhat disingenuous: each of the w_i's is a geometric series which sums to $1/(1 - z)$ within their respective disks of convergence. Each then is a representation of $1/(1 - z)$ within its disk. But w_1 clearly reduces to the real function $1/(1 - x)$ for $|x| < 1$, and this allows us to identify $1/(1 - z)$ as the analytic continuation of $1/(1-x)$ everywhere in the complex plane — with the exception, of course, at $z = 1$. Note how truly remarkable this is: from the values of an analytic function in a single region, one can extend it analytically throughout the complex plane.

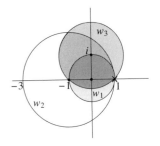

Figure 19.1 Visualizing analytic continuation.

19.2 Laurent Series

What would happen to the derivation in (19.2) if $\left|\frac{z-a}{\zeta-a}\right| > 1$? Of course the series as constructed would not converge and the derivation would collapse. But wait — why can't we have a series which converges everywhere *outside* C? Presumably that would entail an expansion in *reciprocal* integer powers of $(z - a)$. This is indeed what emerges: expanding (19.1) in a geometric series for $\left|\frac{\zeta-a}{z-a}\right| < 1$ gives

$$w(z) = -\frac{1}{2\pi i} \oint_C \frac{w(\zeta)\,d\zeta}{(z - a)\left[1 - \frac{\zeta-a}{z-a}\right]}$$

$$= -\frac{1}{2\pi i} \oint_C \frac{w(\zeta)}{(z - a)} \left[\sum_{n=0}^{\infty} \left(\frac{\zeta - a}{z - a}\right)^n\right] d\zeta$$

$$= -\sum_{n=0}^{\infty} \left(\frac{1}{2\pi i} \oint_C w(\zeta)\,(\zeta - a)^n\,d\zeta\right) \frac{1}{(z - a)^{n+1}}$$

$$= \sum_{n=1}^{\infty} \frac{b_n}{(z - a)^n}, \tag{19.7a}$$

where in the last line we let $n \to n - 1$, giving the expansion coefficients the form

$$b_n \equiv -\frac{1}{2\pi i} \oint_C w(\zeta)\,(\zeta - a)^{n-1}\,d\zeta, \quad n = 1, 2, \ldots. \tag{19.7b}$$

Although we stipulated that a sits in an analytic region of w, (19.7a) shows that if $w(z)$ has a convergent expansion in reciprocal powers around a, then a is necessarily a singular point of that series expansion. This creates no difficulties, of course, since by construction the series only represents w outside C, $|z - a| > |\zeta - a|$. Unlike Taylor series then, there are no derivative formulas for b_n analogous to (19.2b).

So from Cauchy's integral formula we've found two power series, a Taylor series which converges within a disk $r < |\zeta - a|$, and a reciprocal series which converges within a "punctured" disk $r > |\zeta - a|$. As complementary expansions, it may seem that they would never be pertinent to the same region. But Cauchy's integral theorem allows us to deform a contour without change or consequence — as long as we don't try to move it through any non-analytic points or regions. In particular, any closed contour which doesn't surround a can be deformed into the shape shown in Figure 19.2a. In the limit as $d \to 0$, the integrals along the two adjacent segments cancel, leaving integrals over concentric circles

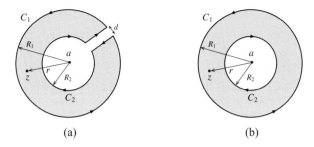

Figure 19.2 Constructing a contour to exclude the singularity at a.

C_1 and C_2 (Figure 19.2b). If this feels familiar, look back at Example 15.11. As was the case there, the combined curve $C_1 - C_2$ forms a boundary for region between the circles; the sign accounts for the negative direction along C_2, consistent with the convention that $\oint_C$ denotes a counter-clockwise path. Thus for any z between C_1 and C_2,

$$w(z) = \frac{1}{2\pi i} \oint_{C_1-C_2} \frac{w(\zeta)\,d\zeta}{\zeta - z} = \frac{1}{2\pi i} \oint_{C_1} \frac{w(\zeta)\,d\zeta}{\zeta - z} - \frac{1}{2\pi i} \oint_{C_2} \frac{w(\zeta)\,d\zeta}{\zeta - z}. \tag{19.8}$$

Points on C_1 are further from a than z is — and so the analysis of (19.2) applies. But for the integral over C_2, (19.7) applies. Both geometric series which emerge are valid in the region between the circles; together they define an *annulus* of convergence. Thus any contour C within the annulus which winds once around a within this annulus leads to a *Laurent series*

$$w(z) = \sum_{n=-\infty}^{\infty} c_n(z-a)^n \equiv c_0 + \sum_{n=1}^{\infty} \left[c_{-n}(z-a)^{-n} + c_n(z-a)^n \right], \tag{19.9}$$

where

$$c_n = \frac{1}{2\pi i} \oint_C \frac{w(\zeta)\,d\zeta}{(\zeta-a)^{n+1}}. \tag{19.10}$$

Comparison with (19.7b) shows $c_{-n} = -b_n$ for $n > 0$ — which accounts for the minus sign in (19.8). Note that if a is within C but is *not* a singularity of $w(z)$, then by Cauchy's integral theorem the c_n vanish for negative n. And with all the $c_{n<0} \equiv 0$, the Laurent expansion reduces to a Taylor series.

A Laurent series expansion (19.9) has two distinct pieces: the $n \geq 0$ "regular" or positive-power part, and a "singular" or "principal" part in negative powers, $n < 0$. The residue is c_{-1}; extracting this "principal part of the principal part" is a key element for evaluating integrals using Cauchy's integral formula (Chapter 21).

Example 19.1

Consider the Laurent expansion

$$G(t,z) \equiv \exp\left[\frac{t}{2}\left(z - \frac{1}{z} \right) \right] = \sum_{n=-\infty}^{\infty} J_n(t)\, z^n, \tag{19.11}$$

with coefficients given by (19.10),

$$J_n(t) = \frac{1}{2\pi i} \oint_C \frac{\exp\left[\frac{t}{2}\left(z - \frac{1}{z} \right) \right]}{z^{n+1}}\, dz. \tag{19.12}$$

If we choose C to be the unit circle, then $z = e^{i\varphi}$, $dz = ie^{i\varphi}d\varphi$ and the integral becomes

$$J_n(t) = \frac{1}{2\pi i}\int_0^{2\pi}\frac{\exp\left[\frac{t}{2}\left(e^{i\varphi}-e^{-i\varphi}\right)\right]}{e^{i(n+1)\varphi}}ie^{i\varphi}d\varphi = \frac{1}{2\pi}\int_0^{2\pi}e^{i(t\sin\varphi-n\varphi)}d\varphi\,. \tag{19.13}$$

As we'll see later, this is an integral representation of *Bessel functions* J_n.

Example 19.2 Annulus Analysis

Consider the function

$$w(z) = \frac{1}{z^2+z-2}. \tag{19.14}$$

Due to the singularities at $z = 1$ and -2, a single power series representation is not obtainable; rather, a different series for each region between singularities is required. We need not evaluate the integral in (19.10) (but see Problem 21.13); for this example, we can simply decompose $w(z)$ into partial fractions,

$$w(z) = \frac{1}{(z-1)(z+2)} = \frac{1}{3}\left(\frac{1}{z-1}-\frac{1}{z+2}\right), \tag{19.15}$$

and expand each piece in a geometric series around the origin — as appropriate for the region, of course. For the innermost region $|z| < 1$, each partial fraction expansion is straightforward:

$$w_1(z) = \frac{1}{3}\sum_{n=0}^{\infty}\left[-z^n-\frac{1}{2}\left(\frac{-z}{2}\right)^n\right] = -\frac{1}{3}\sum_{n=0}^{\infty}\left[1+(-1)^n\left(\frac{1}{2}\right)^{n+1}\right]z^n. \tag{19.16a}$$

For $1 < |z| < 2$, the first partial fraction must be expanded in powers of $1/z$,

$$w_2(z) = \frac{1}{3}\sum_{n=0}^{\infty}\left[\frac{1}{z^{n+1}}-\frac{1}{2}\left(\frac{-z}{2}\right)^n\right] = \frac{1}{3}\sum_{n=0}^{\infty}\left[\frac{1}{z^{n+1}}+\left(\frac{-1}{2}\right)^{n+1}z^n\right]. \tag{19.16b}$$

In the outermost region $|z| > 2$, the second fraction must also be expanded in inverse powers $1/z$ — specifically, in powers of $2/z$:

$$w_3(z) = \frac{1}{3}\sum_{n=0}^{\infty}\left[\frac{1}{z^{n+1}}-\frac{1}{z}\left(\frac{2}{-z}\right)^n\right] = \frac{1}{3}\sum_{n=0}^{\infty}\left[1-(-2)^n\right]\frac{1}{z^{n+1}}. \tag{19.16c}$$

Only sewn together piecewise do we obtain a complete series representation for $w(z)$.

A Taylor expansion's positive powers allow categorization of the zeros of a function in the complex plane: If an analytic function $w(z)$ can be written in the form

$$w(z) = (z-a)^n g(z), \qquad g(a) \neq 0, \tag{19.17}$$

for integer $n > 0$, then w is said to have a nth-order zero at $z = a$. When $n = 1$, $w(z)$ has a *simple zero* at a.

Similarly, a Laurent series allows a characterization of the different types of singularities and their distinct behavior. First, notice that Laurent series are defined in an open interval $R_1 < |z - a| < R_2$, excluding R_1 and R_2. So even if a Laurent series remains convergent for arbitrarily small non-zero R_1, the inner circle can still have an infinitesimal radius $\epsilon > 0$ around a — effectively isolating a from the analytic region. Hence a is an *isolated singularity*. Isolated singularities come in three varieties:

- **Removable Singularity**: If a function remains bounded no matter how small the neighborhood around a, then it must be that all c_{-n} are zero. In other words, the Laurent series reduces to a Taylor expansion. In this case, a is called a *removable singularity*. A simple example, familiar from introductory calculus, is

$$\frac{\sin(z-a)}{z-a} = \sum_{n=0}^{\infty} (-1)^n \frac{(z-a)^{2n}}{(2n+1)!}. \tag{19.18}$$

  Though this function is formally singular as $z = a$, the series expansion gives $w(a)$ the limiting value 1.

- **Pole**: If not all the c_{-n} are zero, but instead the Laurent expansion has a largest $n > 0$ such that $c_{-n} \neq 0$, then a is called a *pole* of order n. For $n = 1$, a is called a *simple pole*. In a neighborhood of a, the function thus has the form

$$w(z) = \frac{g(z)}{(z-a)^n}, \quad g \text{ analytic}, \ g(a) \neq 0. \tag{19.19}$$

  So an nth-order pole of w is an nth-order zero of $1/w$. For instance, since $w(z) = z$ has a simple zero at the origin, $1/z$ has a simple pole there. Unlike removable singularities in which the function remains bounded, in the neighborhood of an nth-order pole $w(z)$ behaves like $(z-a)^{-n}$. For example,

$$\frac{\sin(z-a)}{(z-a)^3} = \frac{1}{(z-a)^2} - \frac{1}{3!} + \frac{1}{5!}(z-a)^2 - \cdots \tag{19.20}$$

  has a second-order pole around $z = a$. But one need not work out the full Laurent series: the smallest positive n for which $(z-a)^n w(z)$ is finite as $z \to a$ reveals an nth-order pole. Thus the singularities of the function in Example 19.2 are simple poles at $z = 1$ and -2. Less trivially, the smallest n with finite

$$\lim_{z \to a} (z-a)^n \frac{\sin(z-a)}{(z-a)^3} = \lim_{z \to a} \frac{\sin(z-a)}{(z-a)^{3-n}} \tag{19.21}$$

  is $n = 2$.

- **Essential Singularity**: If a Laurent expansion has *no* highest power of $1/(z-a)^n$ — that is, if there's an integer $n > 0$ beyond which all $c_{-n} = 0$ — then there is no finite power n for which $(z-a)^n w(z)$ is finite. So unlike poles, there is no positive integer n for which $(z-a)^n w(z)$ is differentiable at a. In this case, a is called an *essential singularity*. If w has an essential singularity at a, then so does $1/w$.

  The remarkable thing about essential singularities is that their behavior is path dependent. Consider

$$w(z) = e^{1/z} = \sum_{n=0}^{\infty} \frac{1}{n! \, z^n} = 1 + \frac{1}{z} + \frac{1}{2! \, z^2} + \frac{1}{3! \, z^3} + \cdots, \tag{19.22}$$

which has an essential singularity at $z = 0$. But $e^{1/z}$ only diverges as $z \to 0$ along the positive real axis. Taking z to zero along the negative real axis gives $e^{1/z} \to 0$, whereas along the imaginary axis the function is oscillatory. Clearly such a function is *not* analytic at the origin.

Singularities can also occur at infinity, which maps into a zero of $\tilde{w}(\zeta) \equiv w(1/z)$. In particular, if $w(z)$ has an nth-order pole at infinity, then $\tilde{w}(\zeta)$ is analytic with an nth order zero at $\zeta = 0$. But if $w(z)$ has an essential singularity at ∞, then $\tilde{w}(\zeta)$ has an essential singularity at $\zeta = 0$ — and vice-versa, as (19.22) illustrates. More subtle is the function $\left[\sin(1/z)\right]^{-1}$, which has essential singularities at zero and infinity — but also along the real axis at $z = 1/n\pi$ for $n = \pm1, \pm 2, \ldots$. As $|n|$ gets larger and larger, these singularities get packed closer and closer. So ultimately, the singularity at $z = 0$ is not isolated.

Singularities characterized by divergence or lack of differentiability are not so different in kind from those of functions of real variables (though see BTW 19.1). But in the complex plane there is an additional type of singularity: as one traverses along an arbitrary small, closed contour around a *branch point*, a function w will not return to its previous value. Such *multivalued functions* are not analytic at a, and hence branch points are often classified as singularities. In the next section, we'll examine the challenges and implications of multivalued functions.

BTW 19.1 Picard's Theorem

An appreciation of the remarkable behavior of functions with essential singularities can be acquired by examining $|e^{1/z}| = |k|$ for constant k. The contours are circles (of infinite radius for $k = 1$), all of which appear to converge at the singularity at the origin. The smaller $|z|$, the more rapidly $e^{1/z}$ must vary as one moves around the origin. Similar plots for $\left[\sin(1/z)\right]^{-1}$, with a non-isolated singularity at $z = 0$, are even more pronounced (and pretty: check out `ComplexPlot` in Mathematica). To gain some appreciation for why this is so, write $e^{1/z} = k$ as

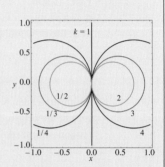

$$1/z = \ln k = \ln\left(|k|e^{i\varphi}\right) = \ln\left(|k|e^{i(\varphi+2\pi n)}\right)$$
$$= \ln|k| + i\varphi + 2\pi in, \quad n = 0, \pm 1, \pm 2, \ldots. \tag{19.23}$$

We can examine any neighborhood $|z| < \epsilon$ of the singularity simply by choosing n to be sufficiently large. Thus within a circle of radius ϵ, no matter how small, $e^{1/z}$ assumes a given value of k. But this must be true for all $k \neq 0$. This truly extraordinary behavior is encapsulated by *Picard's theorem*, which states that in the neighborhood of its essential singularity, a function assumes every complex value infinitely many times — except possibly for one particular value. (In this example, $e^{1/z}$ is never zero.)

19.3 Multivalued Functions

Unlike a real function $w(x)$ which maps values along one axis to those along another, a complex function $w(z)$ is a mapping from complex plane $z = x + iy$ to complex plane $w = u + iv$. Thus rather than the familiar y vs. x graphical layout, complex functions are depicted using separate z- and w-planes.

Perhaps the simplest non-trivial example is

$$w(z) = z^2 = (x^2 - y^2) + i(2xy) = r^2 e^{2i\varphi}. \tag{19.24}$$

As shown in Figure 19.3, arcs in z map to arcs in w — but the "rate" is twice as fast in the w-plane, as a π sweep in z maps to a full 2π sweep through w. So just as for real numbers, z^2 is two-to-one, with both $re^{i\varphi}$ and $re^{i(\varphi+\pi)}$ mapping to $r^2 e^{2i\varphi}$. Distinct from x^2, however, is the inherent periodicity of the complex plane: z^2 returns to its original value after *any* closed 2π-sweep through z. That is, as an

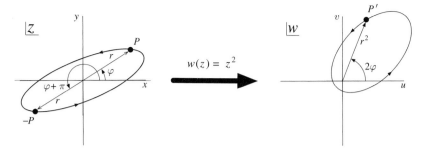

Figure 19.3 $w(z) = z^2$: the w-plane is covered twice for a single sweep through z.

analytic function everywhere in the z-plane, the choice of path is irrelevant: any contour starting and ending at P will map to a single endpoint P' in w.

The same, however, is *not* true of the inverse function

$$w(z) = z^{1/2} = \sqrt{r}\, e^{i\varphi/2}. \tag{19.25}$$

This is an example of a mutlivalued function — specifically, one-to-two: each z value maps to two points in w, one with $\arg(w) = \varphi/2$, the other $\pi + \varphi/2$. Such behavior is not so different from the real function $y = \sqrt{x}$, which is the special case of (19.25) with φ restricted to 0 and 2π. The plot of $\sqrt{x}$ in Figure 19.4 reveals two distinct *branches*, $\pm\sqrt{x}$, clearly showing that it fails the so-called "vertical line test" required of a function. Indeed, two nearby points x and $x + \epsilon$ can map to points y that are far apart. Not surprisingly, such ambiguous discontinuity poses serious difficulties for calculus — difficulties which can be sidestepped by choosing to limit the map to the positive branch only. This renders $\sqrt{x}$ a continuous and differentiable function for $x > 0$.

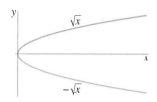

Figure 19.4 The two branches of $\sqrt{x}$.

But what happens when we lift the restriction on φ to allow all continuous values in (19.25)? As shown in Figure 19.5, one full sweep around the z-plane along C_1 from a to b covers only the right half-plane in w; a second sweep over z is required to include w's left half-plane. So $\sqrt{z}$ suffers from the same problem as $\sqrt{x}$: even if points a and c are infinitesimally close in z, they map to separate distinct points a' and c' in w. This shows that $\sqrt{z}$ has two branches, one containing a', b', the other c', d'.

With the image points in w so clearly path dependent, $z^{1/2}$ cannot be analytic; there must be a singularity. And it's not hard to locate: using the Cauchy–Riemann equations in polar coordinates (18.8) reveals that $\sqrt{z}$ is analytic everywhere *except* at the origin,

$$\frac{\partial u}{\partial r} = \frac{1}{2\sqrt{r}}\cos(\phi/2) = \frac{1}{r}\frac{\partial v}{\partial \phi} \qquad \frac{\partial v}{\partial r} = \frac{1}{2\sqrt{r}}\sin(\phi/2) = -\frac{1}{r}\frac{\partial u}{\partial \phi}. \tag{19.26}$$

Since circling the origin smoothly takes w from one branch to another, this type of singularity is called a *branch point*.

Just as the real square root has a divergent derivative at $x = 0$, so too the complex square root is not analytic at $z = 0$; just as $\sqrt{x}$ maps the positive x-axis to distinct branches along the positive and negative y-axis, $\sqrt{z}$ maps to distinct half-planes of w. However, whereas moving from one branch to another of $\sqrt{x}$ requires passing through the origin, the inherent periodicity of the complex plane allows passage between branches of $\sqrt{z}$ merely by circling the branch point, effectively blurring precisely where one branch ends and the other begins.

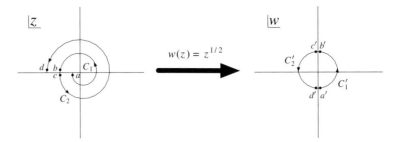

Figure 19.5 $w(z) = z^{1/2}$: two sweeps through z are required to cover the w-plane once.

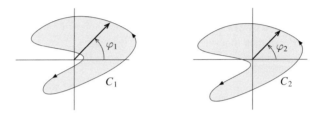

Figure 19.6 Contours in z: by including the origin, the contour C_2 yields non-unique values of $\sqrt{z}$.

So merely excluding the origin is not enough to specify a branch. Have a look at the two contours in the z-plane shown in Figure 19.6. In one full circuit around C_1 — or indeed any closed contour contained within the region defined by C_1 — the values of the phase angle φ_1 lie within $\pm\pi$. So for this region of the complex z-plane, $\sqrt{z}$ is uniquely and continuously defined within a single branch. The contour C_2, however, includes the origin, resulting in φ_2 sweeping through a full 2π range before returning to its starting point — where the function, now on the second branch, yields the negative of its starting value. But the starting point is arbitrary.

In order to enlarge the analyticity of the region delineated by C_1 to the entire complex plane, we need $\sqrt{z}$ to be a single-valued function — and to obtain that, we must forbid any contours from enclosing the origin. One imagines doing this by using scissors to cut the plane along an arbitrary curve from the branch point out to infinity, physically preventing contours from crossing. Such scissoring removes the blurring between branches by creating a *branch cut* which contours are not permitted to cross, thereby restricting φ to a single 2π interval.

Example 19.3 Choice Cuts

A common choice is to cut the z-plane along the entire negative real axis. In this way, we obtain a single-valued, analytic function on the entire complex plane — other than at the branch point singularity at the origin. The simplest range of phase consistent with the cut is $-\pi < \varphi \leq \pi$, commonly designated as the function's *principal branch* on which the function returns its *principal values*. On this branch $\sqrt{z}$ has a positive real part — and so provides an analytic continuation into the complex plane of the principal (positive) branch of $\sqrt{x}$. The second branch has $\Re e(\sqrt{z}) < 0$, with its own copy of the cut plane designated by $\pi < \varphi \leq 3\pi$. Note that the half-line $x < 0$ is formally included within each branch; it just cannot be crossed. And although analyticity requires that we work exclusively within a single branch, either one will do. But the cut does not come without cost: it imposes an artificial discontinuity in what would otherwise be a

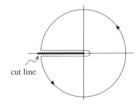

continuous passage from one branch to another. This can be seen by comparing values immediately above and below the cut. On the principal branch with φ between $\pm\pi$, approaching the cut from both sides gives in the limit

$$\text{above: } \varphi = +\pi \longrightarrow \sqrt{z} = \sqrt{r}e^{i\pi/2} = +i\sqrt{r}$$
$$\text{below: } \varphi = -\pi \longrightarrow \sqrt{z} = \sqrt{r}e^{-i\pi/2} = -i\sqrt{r}. \tag{19.27}$$

The sign flip reflects the continuous variation in the function as φ is swept continuously between $\pm\pi$. Possibly more subtle is the following:

$$\sqrt{z_1}\sqrt{z_2} = \left(r_1 e^{i\varphi_1}\right)^{1/2}\left(r_2 e^{i\varphi_2}\right)^{1/2} = \sqrt{r_1 r_2}\, e^{i(\varphi_1+\varphi_2)/2}. \tag{19.28a}$$

Simple, right? But if $|\varphi_1 + \varphi_2| > \pi$, the product crosses the branch cut. So the cut-imposed discontinuity requires $\varphi_1 + \varphi_2 \to \varphi_1 + \varphi_2 \pm 2\pi$. Thus

$$\sqrt{z_1 z_2} = \left[r_1 r_2\, e^{i(\varphi_1+\varphi_2)}\right]^{1/2} = \begin{cases} \sqrt{r_1 r_2}\, e^{i(\varphi_1+\varphi_2)/2}, & |\varphi_1 + \varphi_2| \leq \pi \\ -\sqrt{r_1 r_2}\, e^{i(\varphi_1+\varphi_2)/2}, & |\varphi_1 + \varphi_2| > \pi. \end{cases} \tag{19.28b}$$

As a result, the need to remain within a single branch of $\sqrt{z_1 z_2}$ means that the square root of a product is not always equal to the product of the square roots.

Example 19.4   Staying Within the Branch

The phase of the complex numbers $z_1 = 2e^{i3\pi/4}$ and $z_2 = e^{i\pi/2}$ places them in the principal branch, and each has a straightforward square root,

$$\sqrt{z_1} = \sqrt{2}e^{i3\pi/8} \qquad \sqrt{z_2} = e^{i\pi/4}. \tag{19.29}$$

So

$$\sqrt{z_1}\sqrt{z_2} = \sqrt{2}e^{i3\pi/8}e^{i\pi/4} = \sqrt{2}\left(\cos 5\pi/8 + i\sin 5\pi/8\right). \tag{19.30}$$

The product $z_1 z_2$, however, has phase $5\pi/4$ — a value which places the number $z_1 z_2$ across the cut and into the second branch. In the principal branch, an angle of $\pi/4$ below the negative real axis is $-3\pi/4 = 5\pi/4 - 2\pi$. Thus to secure an analytic square root function, we must use this value:

$$\sqrt{z_1 z_2} = \sqrt{2}e^{-i3\pi/4} = \sqrt{2}e^{-i\pi+i5\pi/8} = -\sqrt{2}e^{i5\pi/8} = -\sqrt{z_1}\sqrt{z_2}, \tag{19.31a}$$

or equivalently in terms of sin and cos,

$$\sqrt{z_1 z_2} = \sqrt{2}\left(\cos 3\pi/8 - i\sin 3\pi/8\right) = -\sqrt{2}\left(\cos 5\pi/8 + i\sin 5\pi/8\right). \tag{19.31b}$$

The negative real axis is certainly not the only possible cut. For instance, cutting along the positive real axis yields a branch with $0 \leq \varphi < 2\pi$ (including 0 rather than 2π is both natural and convenient). In fact, there are an infinite number of possible cuts, and they need not be straight (Figure 19.7). It's branch *points*, not branch *cuts*, which are inherent characteristics of a function, and are the only points all branches of a function share. In all cases though, the same consequences of discontinuity result.

Incidentally, branch points need not be finite; a closed contour can also be construed as surrounding $z = \infty$ — most easily visualized as encircling the north pole on the a Riemann sphere.

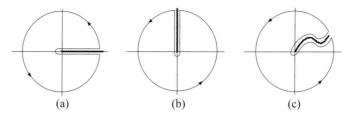

Figure 19.7 Branch cuts of $\sqrt{z}$: (a) $0 < \varphi \le 2\pi$; (b) $-3\pi/2 < \varphi \le \pi/2$; (c) r-dependent 2π range.

A straightforward way to analyze the point at infinity is to define $\zeta \equiv 1/z$ as we did earlier, and examine the behavior of $\tilde{w}(\zeta) \equiv w(1/z)$ around $\zeta = 0$. For $\sqrt{z}$, this reveals

$$\tilde{w}(\zeta) = \frac{1}{\zeta^{1/2}} = \frac{1}{\sqrt{r}} e^{-i\varphi/2}, \qquad (19.32)$$

which has multivalued behavior around the origin. So $z = \infty$ is a branch point of $\sqrt{z}$. The cuts in Figure 19.7 can therefore be regarded as linking branch points.

Example 19.5 $w(z) = \sqrt{(z-a)(z-b)}$

To ascertain whether a function has a branch point at some z_0, it's important that the contour be sufficiently close to z_0 in order not to inadvertently enclose other branch points. Consider, for instance, $w(z) = \sqrt{(z-a)(z-b)}$. (Assume for simplicity that a and b are real, with $a < b$.) From our discussion of $\sqrt{z}$, we expect both a and b to be the only finite branch points. But let's check. Introducing two sets of polar coordinates

$$r_1 e^{i\varphi_1} \equiv z - a \qquad r_2 e^{i\varphi_2} \equiv z - b, \qquad (19.33)$$

yields

$$w(z) = \sqrt{r_1 r_2}\, e^{i\varphi_1/2} e^{i\varphi_2/2}. \qquad (19.34)$$

Closed contours around *either* a or b are double-valued. But for a contour enclosing *both* a and b, each contributes a π phase around the closed path, for a total phase shift in w of 2π. So such a path poses no challenge to the analyticity of w. Moreover, since the point at infinity is enclosed by the *same* contour (though with opposite orientation), $z = \infty$ cannot be a branch point of w — as we can quickly verify:

$$\tilde{w}(\zeta) = w(1/\zeta) = \sqrt{(1/\zeta - a)(1/\zeta - b)} = \frac{1}{\zeta}\sqrt{(1 - a\zeta)(1 - b\zeta)}. \qquad (19.35)$$

Though $\tilde{w}$ diverges as $\zeta \to 0$, $1/\zeta$ is nonetheless single-valued in this limit. So unlike $\sqrt{z}$, w does not have a branch point at infinity.

All of this raises an additional challenge. We need a branch cut with disallows closed paths around a and b individually, but permits those which encircle both. Perhaps the most straightforward approach is to adapt the principal branch of $\sqrt{z}$: for the factor $\sqrt{z-a}$, impose a cut along the real axis for all $x < a$, and for $\sqrt{z-b}$ we choose a cut along the real axis for all $x < b$. This may seem a bit awkward — however, recall from (19.27) that the square root has a sign-flip discontinuity across a cut. Thus where these two cuts overlap along $x < a$, the two discontinuities *each* contribute a sign discontinuity and thus cancel. What's left then is the branch cut of Figure 19.8a linking the

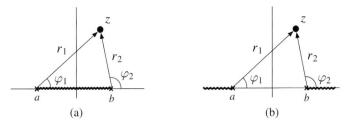

Figure 19.8 Branch cuts for $\sqrt{(z-a)(z-b)}$.

two branch points a and b. By making the two cuts along the negative real axis, both phases have the range of $\sqrt{z}$'s principal branch, $-\pi < \varphi_{1,2} \leq \pi$.

Rather than cutting to the left of the branch points, we could make the two cuts to the right along $x > a$ and $x > b$; cancellation along their common region $x > b$ again results in the cut of Figure 19.8a. Although the final cut is the same as before, the phases are now in the range $0 \leq \varphi_{1,2} < 2\pi$. Thus cutting left or right results in distinct domains of φ rather than different branches of the same cut — though since the cut is the same, the distinction is largely immaterial. Truly distinct is a third choice for the branch cuts, a hybrid of the previous two: along the real axis for all $x < a$ and $x > b$, with arguments $-\pi < \varphi_1 \leq \pi$ and $0 \leq \varphi_2 < 2\pi$ (Figure 19.8b). Even though infinity itself is not a branch point, this ostensible two-segment cut can be interpreted as a single cut from a to b passing through the north pole of the Riemann sphere. Though the phases may differ, what all these choices have in common is a 2π discontinuity across the cut — which for the square root results in a sign flip, $\left(e^{i2\pi}\right)^{1/2} = -1$.

Example 19.6 $w(z) = \sqrt{z-a} + \sqrt{z-b}$

Using (19.33) reveals that each square root has two branches, $\pm\sqrt{r_1}\,e^{i\varphi_1/2}$ and $\pm\sqrt{r_2}\,e^{i\varphi_2/2}$. For closed contours which encircle only one branch point, the *other* square root is single-valued. Thus all together $w(z)$ has four distinct branches. Unlike $\sqrt{(z-a)(z-b)}$, however, infinity is also a branch point,

$$\tilde{w}(\zeta) = \sqrt{1/\zeta - a} + \sqrt{1/\zeta - b} = \frac{1}{\sqrt{\zeta}}\left(\sqrt{1-a\zeta} + \sqrt{1-b\zeta}\right), \tag{19.36}$$

as indicated by the overall factor of $\zeta^{-1/2}$. The cut of Figure 19.8a doesn't include the branch point at infinity, and so won't work. Figure 19.8b, however, restricts this $\sqrt{z-a} + \sqrt{z-b}$ to a single branch.

We can generalize from $\sqrt{z}$ to multivalued $z^{1/m}$ for integer m. Just as $z^{1/2}$ maps z into two separate halves of the w-plane, $z^{1/m}$ similarly maps into pie slices — e.g, thirds, fourths, and fifths for $m = 3,4,5$, respectively. With m branches, m circuits around the origin are required to return to its starting value. Labeling the first branch for a given cut as the analytic function $w_1(z)$, the values in the other branches are distinguished by the m roots of unity, $w_{n+1}(z) = e^{i2\pi n/m}w_1(z)$, $n = 1$ to $m-1$. Figure 19.9 shows the structure for $w(z) = z^{1/3}$.

In fact, any function of the form $(z-a)^{n/m}$ for relatively prime integers n, m (i.e., with no common divisor) must pass through all m branches around $z = a$ to return to its starting value. One says that a is a branch point of order $(m-1)$. For instance, $(z-a)^{1/3}$, has three branches and hence requires three circuits around a, making it a second-order branch. Similarly, $(z-a)^{12/5}$ has five branches, so a is a fourth-order branch point. For $m = 1$, a is a zeroth-order branch point — which is to say, no branch point at all.

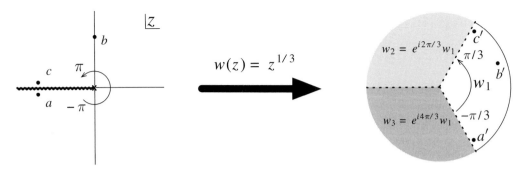

Figure 19.9 The map $w(z) = \sqrt{z}$ with cut along the negative real axis.

Example 19.7 **Deforming the Contour**

Cuts can often be leveraged to evaluate integrals. Consider the integral for real k

$$I = \int_{-\infty}^{\infty} \frac{x\,e^{ikx}}{\sqrt{x^2 + a^2}}\,dx. \tag{19.37}$$

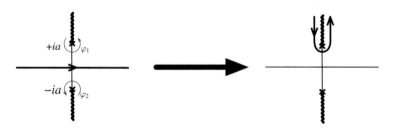

Figure 19.10 Deforming the contour along the real axis to wrap around the upper branch cut.

The behavior of the integrand is delicate for large x where the phase e^{ikx} oscillates wildly with a coefficient of order 1. But Cauchy's theorem allows us to deform the contour as long as it doesn't encounter any singularities. We can take advantage of this prerogative, analytically continuing the integrand into the complex plane by replacing x with z; this requires that we choose a cut corresponding to that in Figure 19.8b. For $k > 0$ we then wrap around the upper cut where $z \to iy$ yields a convergent, exponentially damped integral (Figure 19.10); for $k < 0$, we'd wrap the contour around the lower cut. Since the extended complex plane has a single point at ∞, the limits of integration still meet at the north pole of the Riemann sphere. Ordinarily an integral which hugs a line like this would vanish. But across a cut, the square root has a sign-flip discontinuity. Thus (Problem 19.17)

$$I = \int_{i\infty}^{ia} \frac{z\,e^{ikz}\,dz}{-i\sqrt{z^2 + a^2}} + \int_{ia}^{i\infty} \frac{z\,e^{ikz}\,dz}{i\sqrt{z^2 + a^2}}$$

$$= 2\int_{ia}^{i\infty} \frac{z\,e^{ikz}}{i\sqrt{z^2 + a^2}}\,dz. \tag{19.38}$$

Then using $z = iy$ gives the exponentially damped integral

$$I = 2 \int_a^\infty \frac{y e^{-ky}}{\sqrt{y^2 - a^2}} \, dy = 2a \, K_1(ka) \sim e^{-ka} \text{ for large } a, \tag{19.39}$$

where $K_1(x)$ is a *modified Bessel function* (see Appendix C).

19.4 The Complex Logarithm

The analytic continuation of the logarithm,

$$\ln z = \ln \left(r e^{i\varphi} \right) = \ln r + i\varphi, \tag{19.40}$$

maps the magnitude and phase of z to a single-valued real part $\ln |z|$, and a multivalued imaginary part $\arg(z) = \varphi$. Since φ increases by 2π for every complete sweep around the origin, $z = 0$ is a branch point. As for infinity, the variable change $\zeta = 1/z$ gives

$$\tilde{w}(\zeta) = \ln(1/\zeta) = -\ln(\zeta), \tag{19.41}$$

showing that $z = \infty$ is also a branch point. Making a cut from the origin out to infinity along the negative real axis gives a branch which includes real $\ln x$ for $x > 0$; this is the principal branch, $-\pi < \varphi \le \pi$.

Perhaps the most remarkable feature of the complex logarithm is that $\ln z$ *never* returns to its starting value as z circles the origin. Every circuit increases or decreases the argument by 2π. Algebraically, this is equivalent to inserting n factors of 1 into the argument:

$$\ln z = \ln \left(z \, e^{i2\pi n} \right) = \ln r + i \left(\varphi + 2\pi n \right). \tag{19.42}$$

Thus the complex logarithm has an infinite number of branches, with the principal branch given by $n = 0$. So rather than separating the w-plane into pie slices like $z^{1/m}$, $\ln z$ maps the z-plane into rectangular regions in w of height 2π. As shown in Figure 19.11, without a branch cut a circular

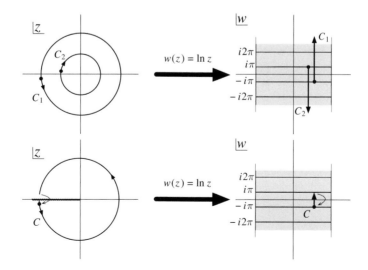

Figure 19.11 Mapping $\ln z$: continuous without cut (top); discontinuous at cut (bottom).

contour C_1 continues upward at constant r, gaining an altitude of 2π for every sweep through z. Similarly, C_2 moves downward by 2π for every one of its clockwise sweeps. However, with a branch cut these mappings are discontinuous.

It's useful to denote the principal branch of $\ln z$ as $\mathrm{Ln}\, z$ (with an upper-case "L"), so that

$$\mathrm{Ln}\, z = \mathrm{Ln}\,|z| + i\,\mathrm{Arg}\, z, \tag{19.43}$$

where $\mathrm{Arg}(z)$ (upper-case "A") is the principal argument of z, $-\pi < \mathrm{Arg}(z) \le \pi$. Then $\ln z$ can be expressed as

$$\ln z = \mathrm{Ln}\, z + i2\pi n = \mathrm{Ln}\,|z| + i\,\mathrm{Arg}(z) + i2\pi n, \quad n = \pm 1, \pm 2, \ldots . \tag{19.44}$$

Since the positive real axis has $\varphi = 0$ on the principal branch, the usual real logarithm function coincides with the principal logarithm, $\ln x = \mathrm{Ln}\, x$.

Example 19.8 $\ln z$ vs. $\mathrm{Ln}\, z$

Unlike $\ln z$, the principal value $\mathrm{Ln}\, z$ is discontinuous whenever a contour crosses the negative real axis, where $\mathrm{Arg}(z)$ abruptly changes by 2π. As a result of this cut-imposed discontinuity, identities such as $\ln z_1 z_2 = \ln z_1 + \ln z_2$ do not always hold for $\mathrm{Ln}\, z_1 z_2$. (We saw similar behavior for the square root in (19.28).)

Consider $z_1 = 1 + i$ and $z_2 = i$, both taken on the principal branch. Their arguments of course add,

$$\arg(z_1 z_2) = \arg(z_1) + \arg(z_2) = \pi/4 + \pi/2 = 3\pi/4, \tag{19.45}$$

showing that the product $z_1 z_2$ is also on the principal branch. Thus

$$\mathrm{Arg}(z_1) + \mathrm{Arg}(z_2) = \arg(z_1) + \arg(z_2), \tag{19.46}$$

and the Ln of the product is the sum of the Ln's.

By contrast, if $z_1 = z_2 = -1$, both taken on the principal branch, then

$$\arg(z_1 z_2) = \arg(z_1) + \arg(z_2) = 2\pi, \tag{19.47}$$

which places the product $\ln(z_1 z_2)$ on the second branch. But

$$\mathrm{Arg}(z_1 z_2) = \mathrm{Arg}(+1) = 0, \tag{19.48}$$

and so $\mathrm{Ln}(z_1 z_2) \ne \mathrm{Ln}\, z_1 + \mathrm{Ln}\, z_2$. Thus the identity fails. Well, sort of. It certainly fails numerically; it succeeds, however, in restricting all results to the principal branch. And therein lies its utility.

Example 19.9 **Inverse Trig Functions**

The complex continuation of trig functions is defined using their exponential representations — starting, not surprisingly, with

$$\cos z = \frac{1}{2}\left(e^{iz} + e^{-iz}\right) \qquad \sin z = \frac{1}{2i}\left(e^{iz} - e^{-iz}\right). \tag{19.49}$$

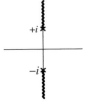

Inverse trig functions can thus be related to the logarithm. Perhaps the simplest is $\tan^{-1} z$, which we construct by taking the ratio of sine to cosine,

$$z = \tan w = -i \frac{e^{iw} - e^{-iw}}{e^{iw} + e^{-iw}}, \tag{19.50}$$

which after after a bit algebra yields $e^{2iw} = \frac{i-z}{i+z}$. Then

$$w = \tan^{-1} z = \frac{1}{2i} \ln \left(\frac{i-z}{i+z} \right). \tag{19.51}$$

Thus each branch of $\ln z$ corresponds to a branch of $\tan^{-1} z$. Since the logarithm has branch points at 0 and ∞, the inverse tangent has branch points when $(i-z)/(i+z)$ is 0 or ∞, which is at $z = \pm i$. Note that infinity is not a branch point of $\tan^{-1} z$. Then the simplest branch cuts connect $\pm i$ along the imaginary axis — either by passing through the origin, or passing through infinity with cuts from i to $i\infty$ and from $-i\infty$ to $-i$.

We'll choose between these two by linking the principal value of $\tan^{-1}$ with that of $\ln$. Specifically, Ln's branch cut along the negative real axis corresponds to negative real $u \equiv (i-z)/(i+z)$. Inspection is sufficient to verify that u is only real for imaginary z, and only negative for $|z| > 1$. Thus the principal value logarithm requires a cut in the z-plane from i to $i\infty$ and $-i\infty$ to $-i$. Indeed, since this cut doesn't cross the real axis, $\tan^{-1} z$ is the continuation of the real function $\tan^{-1} x$ — making this cut the natural choice. We can then define the principal branch of real $\tan^{-1} x$, which we'll designate as $\mathrm{Arctan}(x)$:

$$\mathrm{Arctan}\, x = \frac{1}{2i} \mathrm{Ln} \left(\frac{i-x}{i+x} \right) = \frac{1}{2i} \left[\mathrm{Ln} \left| \frac{i-x}{i+x} \right| + i\, \mathrm{Arg} \left(\frac{i-x}{i+x} \right) \right]$$

$$= \frac{1}{2} \mathrm{Arg} \left(\frac{i-x}{i+x} \right). \tag{19.52}$$

Since $-\pi < \mathrm{Arg} \leq \pi$, the principal values of real $\tan^{-1}(x)$ quite sensibly lie between the divergent values $\pm \pi/2$,

$$-\pi/2 < \mathrm{Arctan}\, x \leq \pi/2. \tag{19.53}$$

Similar manipulations reveal that other inverse trig functions are logarithms as well, which can be leveraged in the same fashion to designate their principal branches (Problem 19.28).

Inverse trig functions are not the only example of tying the principal value of the logarithm to that of other functions. The complex power z^{α} is defined as

$$z^{\alpha} = e^{\alpha \ln z} = e^{\alpha(\ln r + i\varphi + i2\pi n)} = r^{\alpha}\, e^{i\alpha(\varphi + 2\pi n)}, \tag{19.54}$$

which extends the familiar real expression $x^{\alpha} = e^{\alpha \ln(x)}$ to the complex plane. For non-integer α this is a multivalued function whose principal value is determined using Ln rather than ln,

$$z^{\alpha} = e^{\alpha \mathrm{Ln} z} = e^{\alpha(\mathrm{Ln}\, r + i\, \mathrm{Arg}(z))} = r^{\alpha}\, e^{i\alpha\, \mathrm{Arg}(z)}. \quad \text{[Principal value]} \tag{19.55}$$

This places the branch cut along the negative real axis, so that z^{α} is the analytic continuation of real x^{α}. If $\alpha = 1/m$ with integer m,

$$z^{1/m} = e^{\frac{1}{m} \mathrm{Ln} z} = e^{\frac{1}{m}[\mathrm{Ln}\, r + i\, \mathrm{Arg}(z)]} = r^{1/m}\, e^{i\, \mathrm{Arg}(z)/m}, \tag{19.56}$$

and Ln z bequeaths to $z^{1/m}$ a principal argument $\mathrm{Arg}(z)/m$.

For real rational $\alpha < 1$ such as $1/m$, a contour encircling the origin multiple times passes through each branch before returning to the principle branch; this behavior classifies $z = 0$ as an *algebraic* branch point. However, if α is real and irrational, z^α *never* returns to its original value; its branch points are said to be *logarithmic*.

BTW 19.2 Riemann Surfaces

Although branch cuts allow multivalued functions to be rendered analytic, they do so at the expense of introducing an artificial discontinuity in an otherwise continuous function. Another way to manage the problem is to reconsider the domain not as the multiply covered z-plane, but as a geometric structure which is covered exactly once. The classic image is that of a parking garage, each level of which is one of the branches of the function, connected by "ramps" at the branch cut (Figure 19.12). For example, $\sqrt{z}$ maps $z = e^{-i\pi}$ into $w = e^{-i\pi/2} = -i$; as z winds counter-clockwise around the branch point at the origin, arrival at $e^{+i\pi}$ yields $w = e^{i\pi/2} = i$. Then, rather than a branch cut along the negative real axis inducing a discontinuity back to $e^{-i\pi/2}$, the function instead takes the ramp up to the next level, where φ ranges from π to 3π and $\sqrt{z}$ goes from $e^{i\pi/2} = i$ back to $e^{i3\pi/2} = -i$. After that, the function smoothly and continuously takes the ramp back down to the first level. So the parking garage effectively doubles the area of the complex plane, rendering $\sqrt{z}$ as an analytic 1-1 map. No double value, no discontinuity, no fuss.

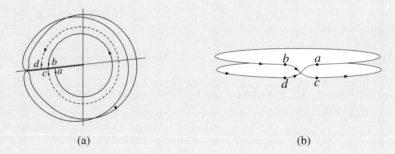

(a)   (b)

Figure 19.12 Riemann surface for $\sqrt{z}$: (a) two sheets on top of one another; (b) view along the cut line. Points a and c of Figure 19.5 are now on separate sheets.

A parking garage constructed in this way is called a *Riemann surface*; the levels are *Riemann sheets* whose only common elements are the branch points. Different functions have different Riemann surfaces. For instance, $\sqrt{z}$ has two sheets, $z^{1/3}$ three; in both cases, the surface winds back on itself so that the top level connects smoothly back to the bottom. The extreme case in the Riemann surface for $\ln z$, which has an infinite number of sheets labeled by $n = 0, \pm 1, \pm 2, \ldots$. At the opposite extreme, a single-valued function such as $w(z) = z$ has no branch points, and so has but one sheet.

The image of parking levels connected by ramps is very useful — but it's worth remembering that since branch cuts are not intrinsic to a function, the ramps can in effect be anywhere. A more accurate depiction are the two-dimensional surfaces shown in Figure 19.13, in which the value of the function $w(z)$ rises out of the complex plane. There are no branch cuts in these figures — all the sheets are instead smoothly and continuously connected, with neighboring points in z mapped to neighboring point in w. Thus one can start anywhere on the surface and designate it as sheet n; from there, going $\pm 2\pi$ around the branch point takes one to sheet $n \pm 1$.

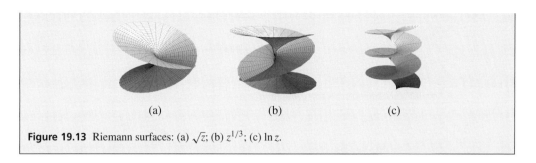

Figure 19.13 Riemann surfaces: (a) $\sqrt{z}$; (b) $z^{1/3}$; (c) $\ln z$.

Problems

19.1 Identify the type and location of singularities for each of the following, and find the residue (if applicable):

(a) $\frac{1}{z}\left(e^z - 1\right)$

(b) $\frac{\cos z}{z^4}$

(c) $\frac{\sinh z}{z^4}$

(d) $2z^2 \sin(1/z)$

(e) $\frac{\sin z - z}{z^2}$.

19.2 Find the poles of $\frac{e^z}{\sin z}$.

19.3 Show that if $Q(z)$ has a simple zero at $z - a$, then $Q'(a) = Q(z)/(z-a)\big|_{z=a}$.

19.4 Identify the residues of $\frac{1}{z^2+z-2}$ in the expansions of (19.16).

19.5 Consider $w(z) = \frac{1}{z(z-1)^2}$.

(a) Locate the poles and identify their order.
(b) Use the binomial expansion to construct a Laurent expansion around each pole, identifying their circles of convergence.
(c) Find the residue of each pole.

19.6 Since a function's Laurent coefficients are unique for a given analytic region, we need not always wrestle with the integral in (19.10) to obtain an expansion for the region beyond these singularities. Construct a geometric series centered at $z = -i$ for

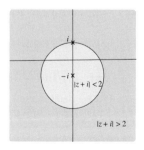

$$w(z) = \frac{1}{z^2 + 1} = \frac{1}{(z+i)(z-i)}$$

to obtain the Laurent series which:

(a) converges in the disk $|z + i| < 2$
(b) converges on the punctured disk $|z + i| > 2$.

19.7   A function $w(z)$ has a power series representation centered at a with radius of convergence R,

$$w_a(z) = \sum_{n=0}^{\infty} c_n(z-a)^n. \tag{19.57}$$

A second convergent series representation

$$w_b(z) = \sum_{n=0}^{\infty} \tilde{c}_n(z-b)^n \tag{19.58}$$

can be constructed to analytically continue w_1 beyond R, as long as b is within w_a's disk of convergence. Show that

$$\tilde{c}_n = \frac{1}{n!}\sum_{m=n}^{\infty} \frac{m!}{(m-n)!}\,c_m(b-a)^m = \sum_{m=n}^{\infty} \binom{m}{n}c_m(b-a)^m.$$

19.8   Show that the series

$$w(z) = \sum_{n=0}^{\infty}(n+1)(z+1)^n$$

can be analytically continued beyond its disk of convergence $|z+1| < 1$ to the entire punctured plane $z \neq 0$. [Hint: Integrate to get a geometric series.]

19.9   What happens to the Laurent expansion (19.7) if a is *not* a singularity of $w(z)$?

19.10   (a) Make a sketch showing how $w = u + iv = z^2$ maps the xy-grid in the first quadrant of the complex z-plane into the uv-plane. How does the figure in w change when z's second quadrant is included?
(b) Show that the image curves $v(u)$ of perpendicular lines $x = c_1, y = c_2$ with $c_i \neq 0$, also meet at right angles.

19.11   Make a sketch analogous to that in Figure 19.9 for the map $w(z) = z^{1/4}$.

19.12   Find the value of the function $w(z) = (z-1)^{1/3}$ at $z = 1 + i$ when:

(a) $w(z) > 0$ and real for $x > 1$
(b) $w(z) < 0$ and real along the negative real axis
(c) $\Re e[w(z)] = 0$ for $\Re e[z] = 1, \Im m[z] > 0$.

Be sure to specify the chosen branch cut in each case.

19.13   Using closed contours in the plane, demonstrate that $\pm i$ are the only branch points of the function $w(z) = \sqrt{z^2 + 1}$. Show how branch cuts yield a single-valued function.

19.14   For the two cuts in Figure 19.8, compare the phase of $w(z) = \sqrt{(z-a)(z-b)}$ at points a small angle $\epsilon/2$ above and below the real axis to verify that w is continuous everywhere except across the cuts.

19.15   In the construction of the cut in Figure 19.8a, we claimed that overlapping cuts cancel (see Problem 19.14). For what other fractional powers other than 1/2 would this cancellation occur?

19.16 Find the branch points and the simplest branch cuts for:

(a) $[(z-a)(z-b)]^{1/3}$
(b) $(z-a)^\alpha (z-b)^\beta$.

19.17 For the discontinuity across the upper cut in Figure 19.10, use the polar representation of (19.33) to show that $\sqrt{z^2+a^2}$ acquires a factor of $+i$ to the right of the cut, $-i$ to the left.

19.18 **Branch Points at Infinity**

(a) Show that $w(z) = \sqrt{(z-a_1)(z-a_2)\cdots(z-a_n)}$ has a branch point at infinity only for odd n.
(b) Show that $w(z) = \sqrt{z-a_1} + \sqrt{z-a_2} + \cdots + \sqrt{z-a_n}$ has a branch point at infinity for all n.

19.19 Find all the values of:

(a) $\ln(1+i)$ (b) $\ln(-1+i)$ (c) $\ln(-1-i\sqrt{3})$.

19.20 Find all the values of $\ln e^z$ for $z = 2 - \frac{5}{2}\pi i$. What's the principal value? Is there a value for which $\ln e^z = z$?

19.21 For $\ln z_1$ and $\ln z_2$ on branches n_1 and n_2, respectively, show that

$$\ln(z_1 z_2) = \ln z_1 + \ln z_2$$

and

$$\ln(z_1/z_2) = \ln z_1 - \ln z_2.$$

19.22 Determine and explain the validity (or lack thereof) of the identities

$$\mathrm{Ln}(z_1 z_2) = \mathrm{Ln}\, z_1 + \mathrm{Ln}\, z_2 \qquad \mathrm{Ln}(z_1/z_2) = \mathrm{Ln}\, z_1 - \mathrm{Ln}\, z_2$$

for (a) $z_1 = -\sqrt{3}+i$, $z_2 = 1+i$; and (b) $z_1 = i$, $z_2 = -1$.

19.23 For $z = 1 + i$, find the largest positive integer n for which $\mathrm{Ln}\, z^n = n\,\mathrm{Ln}\, z$. Does $-n$ also work?

19.24 Consider the "derivation" $\sqrt{z} = \sqrt{e^{i2\pi} z} = -\sqrt{z}$. Though each step may look reasonable, the result is clearly wrong. What's the source of the error?

19.25 Show that in general, $(z_1 z_2)^{1/m} \neq z_1^{1/m} z_2^{1/m}$.

19.26 Find the principal branch of the function $w(z) = z^{a+ib}$ for real a and b. What kind of branch point is $z = 0$? Is it a branch point for all α? Explain.

19.27 In Example 19.9, we chose a two-segment cut which connects the branch points $\pm i$ by way of infinity. Show that choosing the cut instead to connect the branch points along a straight line segment through the origin leads to principal values of $\tan^{-1} z$ between 0 and π.

19.28 (a) Show that

$$\sin^{-1} z = -i\ln\left[iz + (1-z^2)^{1/2}\right] \qquad (19.59)$$

and

$$\cos^{-1} z = -i\ln\left[z + (z^2-1)^{1/2}\right]. \qquad (19.60)$$

Note that argument of the logarithms lie on the unit circle.
(b) Construct the principal branch function $\mathrm{Sin}^{-1} z$ which analytically continues the real function $\sin^{-1} x$. What range does real $\mathrm{Sin}^{-1} x$ have?

19.29 Use (19.55) to find the principal values for each of the following:

(a) $\sqrt{a^2 - z^2}$, analytic within the circle $|z| = a$ (Figure 19.8b)

(b) $\sqrt{z^2 - a^2}$, analytic outside the circle $|z| = a$ (Figure 19.8a).

19.30 Use Mathematica's `Plot3D` to make separate plots of the magnitude `Abs[w(z)]` and the principal argument `Arg[w(z)]` for the following complex functions and interpret the results.

(a) z (b) $1/z$ (c) z^3 (d) $\sqrt{z}$

(e) $\sqrt{z+1}\sqrt{z-1}$ (f) $\ln(z^3)$ (g) $e^{1/z}$.

[Alternatively, use

$$\texttt{ComplexPlot3D[w[z],\{z, z_{min}, z_{max}\}, PlotLegends} \rightarrow \texttt{Automatic]}$$

which creates a plot of $|w(z)|$ in the complex plane with corners at z_{min} and z_{max}, colored by $\text{Arg}(z)$ — see documentation for details.]

20 Interlude: Conformal Mapping

20.1 Visualizing Maps

Consider the analytic function

$$w(z) = z^2 = x^2 - y^2 + i\,2xy = u(x,y) + iv(x,y). \tag{20.1}$$

Figure 20.1 shows how the xy cartesian grid maps into parabolas in u and v. The plot seems to imply that the right angles formed between lines of constant x and y map into right angles between their images in the w-plane. This can be confirmed by showing that where the curves intersect, their slopes are negative-reciprocals of one another (Problem 19.10).

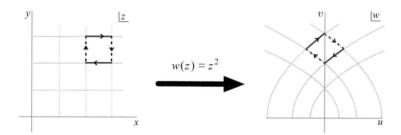

Figure 20.1 The map $w(z) = z^2$.

Deeper insight can be derived by comparing the line elements in both w and z,

$$ds^2 = du^2 + dv^2 = \left(\frac{\partial u}{\partial x}\,dx + \frac{\partial u}{\partial y}\,dy\right)^2 + \left(\frac{\partial v}{\partial x}\,dx + \frac{\partial v}{\partial y}\,dy\right)^2. \tag{20.2}$$

Since $w = z^2$ is analytic, we can use the Cauchy–Riemann equations to show

$$ds^2 = J(x,y)(dx^2 + dy^2), \tag{20.3}$$

where J is the Jacobian of the transformation (Section 2.2),

$$J = \left(\frac{\partial u}{\partial x}\right)^2 + \left(\frac{\partial v}{\partial x}\right)^2 = \left(\frac{\partial u}{\partial y}\right)^2 + \left(\frac{\partial v}{\partial y}\right)^2 = 4(x^2 + y^2). \tag{20.4}$$

With dx^2 and dy^2 having identical scale factors, (20.3) establishes that at every $w_0 = w(z_0)$ for which $J(z_0) \neq 0$, the line elements in the w- and z-planes are proportional. Thus the distance ratio which determines the angle between two directions in z must equal the angle between the corresponding directions in w. So not only do the right angles in the z-plane shown in Figure 20.1 map to right angles in w, *all* angles between arbitrary curves in z — including their orientation — are preserved at all points where the Jacobian is non-zero. Any deformation produced by an analytical map between regions in z and w is constrained by this feature.

Analytic functions such as z^2 which preserve angles and their orientation are said to be *conformal*. And it's not hard to see why they behave this way: any analytic function $w(z)$ can be expanded in a Taylor series around a point z_0,

$$w(z) = w(z_0) + w'(z_0)(z - z_0) + \cdots, \tag{20.5}$$

which for small $\xi \equiv z - z_0$ can be truncated to first order. Then

$$w(z) - w(z_0) = w'(z_0)\xi = |w'(z_0)|e^{i\alpha_0}\,\xi, \tag{20.6}$$

where $\alpha_0 \equiv \arg w'(z_0)$. So the change in w is a rotation of ξ by α_0 and a magnification by $|w'(z_0)|$. Both actions are independent of ξ's direction in the z-plane. In particular, the arguments of any two small segments emerging from $z = z_0$ will both be augmented by α_0, and hence the angle *between* them is unchanged by $w(z)$. A conformal map is therefore an analytic function $w(z)$ such that $w'(z_0) \neq 0$. In fact, $|w'(z_0)|^2 = J(z_0)$ (Problem 20.2), so the stipulation of non-zero $w'(z_0)$ is the complex expression of a non-vanishing Jacobian. This condition guarantees that w maps a neighborhood of z_0 one-to-one into a neighborhood of w_0 — and hence can be inverted. Thus not only does a conformal w take lines of constant x and y into perpendicular curves $u(a,y)$ and $v(x,b)$, so too w^{-1} maps lines of constant u and v to perpendicular curves in z.

But what about the origin? As indicated in Figure 20.1, $w = z^2$ maps the entire first quadrant in z to the upper half-plane (UHP) in w. Yet despite the fact that the x- and y-axes are perpendicular, their respective images are the positive and negative u-axis, which meet at $180°$. But then $w'(0) = 0$, so $w = z^2$ is not conformal where they meet. Points z_0 at which $w'(z_0) = 0$ are called *critical points*. Adapting the line of reasoning in (20.5)–(20.6), one can show that at a critical point, the angle between smooth intersecting curves is increased by a factor of k, where k is the order of the first non-vanishing derivative, $w^{(k)}(z_0) \neq 0$ (Problem 20.3). So for $w = z^2$, the $90°$ angle between the axes in z is doubled between their images in w.

More than just another remarkable property of analytic functions, conformal maps have significant practical applications. Perhaps most notable is that Laplace's equation $\nabla^2\Phi = 0$ is unchanged under a one-to-one analytic map. This feature allows the conformal mapping of a Laplace boundary value problem into a simpler geometric configuration (Section 20.2). For instance, an electrostatics problem with the voltage specified on parabolic boundaries as shown in Figure 20.1 can be solved in a rectangular region in the z-plane and then mapped back into w. Effective use of such techniques requires knowing how various analytic functions map between different geometric regions; Table 20.1 depicts a few common conformal maps. The following examples explore some of them.

Example 20.1 **Magnification, Rotation, and Translation**

The simplest example is a linear map

$$w_L(z) = az + b, \tag{20.7}$$

for complex constants a and b. Since $w'_L(z) = a$, it's trivially conformal for all non-zero a. (Indeed, the constant map $w(z) = b$ is *not* conformal.)

Table 20.1 Some common conformal maps

$w(z)$	z-plane	w-plane
z^2		
$1/z$		
$1/z$		
$\exp(z)$		
$\exp(z)$		
$z + 1/z$		
$\dfrac{z-1}{z+1}$		
$\mathrm{Ln}\left[i\dfrac{1-z}{1+z} \right]$		

The effects of w_L are familiar: $a = |a|e^{i\alpha}$ both magnifies and rotates z, which is then uniformly translated by b. So a region in z clearly maps to the same shape in w — for instance, lines map to lines, circles to circles, rectangles to rectangles. Figure 20.2 shows a piecemeal construction of a map which translates, rotates, and reduces by 2 a square region in z into an arguably simpler one in w.

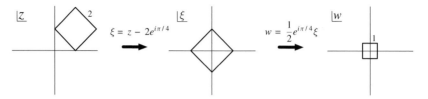

Figure 20.2 Constructing the map $w_L(z) = \frac{1}{2}e^{i\pi/4}\,z - 1$.

Example 20.2 **Inversion in the Circle**

The function $w_I(z) = 1/z$ is an analytic one-to-one map on the extended complex plane which sends points $z = |z|e^{i\alpha}$ to

$$w_I = \frac{1}{z} = \frac{1}{|z|}e^{-i\alpha}. \tag{20.8}$$

Geometrically, the flip in the sign of the phase is a reflection of z across the real axis. The reciprocal of the magnitude takes a point $|z| > 1$ outside the unit circle into the interior and vice-versa, with $z = 0$ and $z = \infty$ mapped into one another. The magnitude of points with $|z| = 1$ are unchanged: inversion maps the unit circle into itself.

In fact, all circles regardless of radius are mapped by inversion to circles — including straight lines construed as circles with infinite radius. Since stereographic projection is conformal (BTW 20.1), this is most readily appreciated by applying inversion to images in the Riemann sphere, and then stereographically projecting back into the complex plane — see Problem 20.4.

Example 20.3 **Möbius Transformations**

A map which incorporates the previous two examples is a *Möbius transformation*, which is defined as the fractional linear form

$$w_m(z) = \frac{az + b}{cz + d}, \qquad ad - bc \neq 0, \tag{20.9}$$

for complex constants a, b, c, d. Note that either $a = c = 0$ or $b = d = 0$ would result in a constant, and hence a non-conformal, transformation. Indeed, since

$$\frac{dw_m}{dz} = \frac{ad - bc}{(cz + d)^2}, \tag{20.10}$$

the condition $ad - bc \neq 0$ ensures that w_m is conformal.

Möbius transformations have several interesting properties. For instance, successive Möbius transformations also constitute a Möbius transformation. Similarly, the inverse transformation also has fractional linear form,

$$w_m^{-1}(z) = z(w) = \frac{dw - b}{-cw + a}, \tag{20.11}$$

which is also conformal for $ad - bc \neq 0$. Moreover, since

$$w_m(\infty) = a/c \qquad w_m^{-1}(a/c) = \infty$$
$$w_m^{-1}(\infty) = -d/c \qquad w_m(-d/c) = \infty, \tag{20.12}$$

the extended complex plane is included. In fact, Möbius transformations encompass all conformal maps (translation, magnification, rotation, and inversion) from the Riemann sphere to itself. Mathematicians call this an *automorphism*.

As an automorphism, Möbius transformations are composable — that is, for Möbius transformations $\xi(z)$ and $\chi(z)$, the map $\xi(\chi(z))$ is also a Möbius transformation. This underlies the intuitive construction

$$w_m(z) = \frac{a}{c} + \frac{bc - ad}{c} \frac{1}{cz + d}, \qquad c \neq 0, \tag{20.13}$$

expressing Möbius transformations as the composition of $w_L(z) = cz + d$, followed by inversion w_I, then by a different w_L. And since both w_L and w_I send circles into circles, we discover a Möbius transformation must also map circles into circles — including lines-as-circles in the extended complex plane passing through $z = -d/c$.

One immediate consequence is that since only three points are needed to specify a circle (for colinear points, that circle is a line), a Möbius transformation can be constructed from just three points z_1, z_2, z_3 and their images w_1, w_2, w_3. This can be accomplished by equating the *cross ratios* (Problem 20.7)

$$\frac{(w - w_1)(w_2 - w_3)}{(w - w_3)(w_2 - w_1)} = \frac{(z - z_1)(z_2 - z_3)}{(z - z_3)(z_2 - z_1)}, \tag{20.14}$$

where $w \equiv w_m(z)$. You should verify that $z = z_i$ corresponds to $w = w_i$. In fact, (20.14) implicitly defines the unique fractional linear transformation for which $w(z_i) = w_i$.

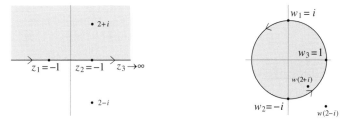

Figure 20.3 Möbius map $w(z) = \frac{z-i}{z+i}$ takes the UHP in z into the unit disk in w.

As an example, let's construct a conformal map from the real axis $y = 0$ to the unit circle $|w| = 1$. Since conformal transformations preserve orientation, we should consider the order in which we designate the points. If we assign the three $z_i = x_i$ in increasing order with $z_3 = x_3 = \infty$, then the corresponding image points w_i should be taken counterclockwise — that is, in the conventional positive sense. We'll take the three points to be

$$z_1 = -1 \quad z_2 = 1 \quad z_3 = \infty$$
$$w_1 = i \quad w_2 = -i \quad w_3 = 1. \tag{20.15}$$

Then the terms in z_3 on the right in (20.14) algebraically cancel, leaving

$$\frac{(w-i)(-i-1)}{(w-1)(-i-i)} = \frac{(z+1)}{(1+1)},\tag{20.16}$$

which reduces to the Möbus transformation (Figure 20.3)

$$w(z) = \frac{z-i}{z+i}.\tag{20.17}$$

Consistent with customary conventions, the circle's positive counter-clockwise orientation encloses the unit disk $|z| < 1$ (placing the punctured disk $|z| > 1$ outside); the same convention for the infinite-radius "circle" in z identifies the upper half-plane as "inside." Comparing, say, the locations of $w(z)$ for $z = 2 \pm i$ supports this identification. Thus the Möbus transformation of (20.17) maps the entire upper half z-plane to the unit disk in w.

This example serves as motivation to engineer more intriguing cases — say, the rectangular ribbon in the z-plane which (20.17) maps into the region between circles (Figure 20.4). Beyond the inherent aesthetic appeal, examples such as these have practical relevance. For instance, imagine that the circles in Figure 20.4 represent a non-concentric cylindrical capacitor. As we'll see, under a conformal transformation Laplace's equation $\nabla^2 \Phi = 0$ is the same in both z and w. So rather than solving for the potential and electric field in a circular geometry, we can use the Möbius transformation to map this configuration into a simpler rectangular geometry, solve there for Φ, and then map back. We'll do precisely this in Section 20.2.

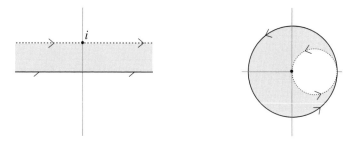

Figure 20.4 Möbius map $\frac{z-i}{z+i}$ of a ribbon in the UHP in z into the region between circles in w.

Example 20.4 **$\exp z$ and $\ln z$**

Inspection reveals that the conformal map

$$w(z) = e^z = e^x e^{iy} \equiv \rho\, e^{i\beta}, \qquad |z| < \infty\tag{20.18}$$

takes horizontal lines of constant y into radial lines of constant angle β, and vertical lines of constant x into circles of radius ρ (Figure 20.5). Notice that while the x-axis maps into the positive u-axis, the horizontal line $y = \pi$ is sent to the negative u-axis; all horizontal lines in z between $y = 0$ and π map to rays in w with angles $0 < \beta < \pi$. Similarly, the y-axis is sent to the unit circle; vertical lines $x < 0$ map to concentric circles with radii $\rho < 1$, while those with $x > 0$ are sent to circles with $\rho > 1$. Either way you slice it, the ribbon in z between $y = 0$ and π maps to the entire upper half-plane in w. Extending the ribbon's width to 2π maps one-to-one into the entire w-plane (minus the origin). For a width greater than 2π, $w(z) = e^z$ is not one-to-one, and so the map is no longer conformal. (Incidentally, notational purists would insist that e^z is multivalued, whereas $\exp(z)$

denotes the principal branch defined by the Taylor expansion $\sum z^n/n!$. In practice, however, both are used to denote the single-valued function.)

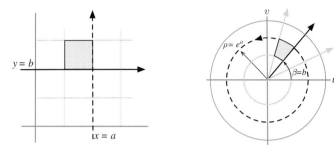

Figure 20.5 The exponential map $w(z) = e^z$.

Given the action of e^z, the inverse map $w = \ln z$ must take circles and radial lines into vertical and horizontal lines, respectively. But as a multivalued function, the domain must be restricted to lie within a 2π range,

$$w(z) = \ln z, \qquad \varphi_0 < \arg z \leq \varphi_0 + 2\pi. \qquad (20.19)$$

The branch is specified by the value of φ_0, which designates the location of a radial cut line the branch; $\varphi_0 = -\pi$ gives the principal branch, $w(z) = \operatorname{Ln} z$.

<hr>

Example 20.5 $\sin z$, $\cos z$, and $\tan z$

The analytic function $w(z) = \sin z$ is conformal between the zeros of $w'(z)$ — most simply, for $|\Re e(z)| < \pi/2$. We can visualize this map using a familiar trig identity,

$$\sin z = \sin(x + iy) = \sin x \cosh y + i \cos x \sinh y$$
$$\equiv u(x,y) + iv(x,y). \qquad (20.20)$$

For vertical lines $x = a$ in the z-plane this reduces to

$$u(a,y) = \sin a \cosh y \qquad v(a,y) = \cos a \sinh y. \qquad (20.21)$$

It should come as little surprise, given the hyperbolic trig functions, that together $u(a,y)$ and $v(a,y)$ describe hyperbolas in the w-plane (dotted curves in Figure 20.6),

$$\frac{u^2}{\sin^2 a} - \frac{v^2}{\cos^2 a} = 1, \qquad (20.22)$$

with vertices at $(\pm \sin a, 0)$.

Similarly, horizontal lines $y = b$ in the z-plane give

$$u(x,b) = \sin x \cosh b \qquad v(x,b) = \cos x \sinh b, \qquad (20.23)$$

which is an ellipse with vertices at $(\pm \cosh b, 0)$ and $(0, \pm \sinh b)$,

$$\frac{u^2}{\cosh^2 b} + \frac{v^2}{\sinh^2 b} = 1. \qquad (20.24)$$

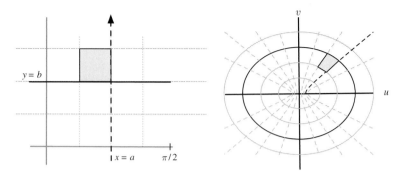

Figure 20.6 The conformal map $w(z) = \sin z$, $|\Re e(z)| < \pi/2$.

The map $w(z) = \cos z$ is conformal for $\Re e(z)$ between 0 and π. But since $\cos z = \sin(z + \pi/2)$, it's just a composition of translation $\zeta = z + \pi/2$, followed by $w = \sin \zeta$.

More interesting is $w(z) = \tan z$, which is one-to-one for $|\Re e(z)| < \pi/2$. Using familiar identities, we can write

$$\tan z = \frac{\sin z}{\cos z} = i \left(\frac{1 - e^{i2z}}{1 + e^{i2z}} \right), \tag{20.25}$$

which shows that $\tan z$ is a composition first of the magnification and rotation $\eta = 2iz$, followed by the exponential map $\zeta = e^{\eta}$ and then a Möbius transformation.

Consider a ribbon in the z-plane with $\Re e(z)$ between $\pm \pi/4$ (Figure 20.7). The first map, $\eta = 2iz$, transforms the vertical ribbon into a horizontal one with $|\Im m(\eta)| < \pi/2$. Then $\zeta = e^{\eta}$ maps this ribbon into the right half ζ-plane. Finally,

$$w(\zeta) = i \frac{1 - \zeta}{1 + \zeta} \tag{20.26}$$

sends this right half-plane into the unit circle in w.

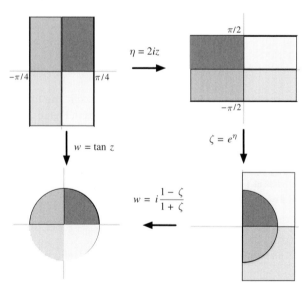

Figure 20.7 (De)constructing $w(z) = \tan z$, $|\Re e(z)| < \pi/4$.

BTW 20.1 The Mercator Projection

A sphere is not a plane. One is finite, unbounded, and curved; the other is infinite and flat. Although compactifying $\mathbb{R}^2$ by appending the point at infinity makes it *topologically* equivalent to S^2, it does not endow it with a curved geometry. Any map of the sphere to the plane will unavoidably produce distortions. A common image highlighting the distinction is of an orange peel removed intact after making a longitudinal cut from pole to pole: There is no way to flatten the peel on a table without causing it to tear.

The image of rips in the flattened peel may remind you of some wall maps of the earth. Indeed, the contrast between sphere and plane underlies one of the more confounding challenges of early maritime navigation: the difficulty of plotting a course on a map. Because charts of the time failed to account for the earth's curvature, a straight line on a map did not correspond to a constant compass heading. This necessitated difficult and error-prone course corrections mid-voyage. What was needed was a flat map on which navigators could plot a straight-line course between ports corresponding to a constant compass heading (called a rhumb line). In other words, they needed a mapping from the sphere to the plane which leaves angles unchanged — a conformal map.

In 1569, Gerardus Mercator met the challenge. On his map, a line of fixed compass heading φ relative to the north star is represented by a straight line making the same angle φ from the vertical. Since Mercator never published his derivation, the methods he used to produce the map are not entirely clear. But although he lived more than two centuries before Cauchy and Riemann, the justly famous *Mercator projection* can be reproduced using analytic functions.

As we saw with the Riemann sphere, stereographic projection maps spherical coordinates θ, ϕ onto the extended complex plane according to (18.18),

$$z = \cot(\theta/2)\, e^{i\phi}, \tag{20.27}$$

with the north pole mapped to infinity, the south pole to the origin (Figure 18.1). The simplest way to demonstrate that stereographic projection is conformal is by comparing line elements. For a sphere of radius a,

$$
\begin{aligned}
ds^2 = dx^2 + dy^2 &= a^2\, dz\, dz^* \\
&= a^2 \left| -\frac{1}{2}\csc^2(\theta/2)\, e^{i\phi}\, d\theta + i\cot(\theta/2)\, e^{i\phi}\, d\phi \right|^2 \\
&= \frac{1}{4\sin^4(\theta/2)}\, a^2 \left(d\theta^2 + \sin^2\theta\, d\phi^2 \right).
\end{aligned}
\tag{20.28}
$$

So similar to the case in (20.2)–(20.3), the common scale factor $4\sin^4(\theta/2)$ means that the ratio dy/dx is unaffected by (20.27). Stereographic projection is conformal.

As a conformal map, curves of constant ϕ and constant θ on the sphere — lines of longitude and latitude, respectively — must have images in z which remain at $90°$ to one another. And as we've previously noted, longitudes on the Riemann sphere map to radial lines in z, latitudes to concentric circles. This does not look like a conventional map of the world — but taking the logarithm of z,

$$\ln z = \ln\left[\cot(\theta/2)\right] + i\phi, \tag{20.29}$$

conformally sends the radial lines into horizontal lines, and the concentric circles into vertical lines. This is not quite Mercator's projection either, which has lines of longitude running vertically with the ϕ-axis along the equator. A $\pi/2$ rotation easily accommodates this:

$$w(z) = e^{-i\pi/2} \ln z$$
$$= -i \ln \left[\cot(\theta/2)\right] + \phi = i \ln \left[\tan(\theta/2)\right] + \phi. \tag{20.30}$$

The final piece of the derivation accounts for convention: the spherical coordinate $\theta \in [0, \pi]$ is actually the colatitude; geographer's latitude is $\lambda = \pi/2 - \theta$, running from $-\pi/2$ at the south pole up to $\pi/2$ at the north. Then if the physical paper map has width L, we rescale $w \to \frac{L}{2\pi} w$ so that the horizontal axis covers the entire 2π range of longitude, defining the map's axes as

$$u = \frac{L\phi}{2\pi} \qquad\qquad v = \frac{L}{2\pi} \ln \left[\tan\left(\lambda/2 + \pi/4\right)\right]. \tag{20.31}$$

Note that the top and bottom edges have $\lambda = \pm\pi/2$, mapping the north and south poles to $\pm\infty$, respectively; the equator is at $\lambda = v = 0$.

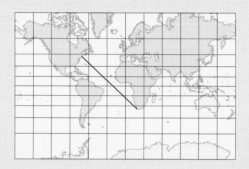

Figure 20.8 The Mercator projection, with rhumb line between Boston and Cape Town.

A flat map with axes calibrated according to (20.31) yields the Mercator projection (Figure 20.8). But though useful for compass navigation, this projection necessarily distorts other features. For instance, though a rhumb line is straight on the map, it's usually not the shortest distance between two points. The most notable distortion, however, is the depiction of relative area. Greenland is not larger than South America, nor is Antarctica larger than Africa. The source of the distortion can be traced back to the lattitude-dependent scale $\sin^4(\theta/2)$ in (20.28). In terms of the physical area element $dA = a^2 \sin\theta \, d\theta d\phi = a^2 \cos\lambda \, d\lambda d\phi$ on a sphere of radius a, an area $du \, dv$ on Mercator's map is

$$du \, dv = \left| \frac{\partial(u,v)}{\partial(\lambda,\phi)} \right| d\lambda \, d\phi$$
$$= \frac{L^2}{4\pi^2} \frac{1}{\cos\lambda} d\lambda \, d\phi = \frac{L^2}{16\pi^2 a^2} \left[\frac{1 + e^{4\pi v/L}}{e^{2\pi v/L}} \right]^2 dA. \tag{20.32}$$

So the Mercator projection magnifies the physical area element $dA = a^2 d\Omega^2$ on the sphere by a dimensionless factor that grows without bound for large $|v|$. This causes Greenland and Antarctica at high and low latitude to appear larger than their true extents on the globe.

20.2 The Complex Potential

Consider an analytic function

$$\Omega(z) = \Phi(x,y) + i\Psi(x,y). \tag{20.33}$$

Presuming that Φ and Ψ have continuous partial second derivatives, we can decouple the first-order partial differential Cauchy–Riemann equations

$$\frac{\partial\Phi}{\partial x} = \frac{\partial\Psi}{\partial y} \qquad \frac{\partial\Phi}{\partial y} = -\frac{\partial\Psi}{\partial x} \tag{20.34}$$

to get separate second-order ordinary differential equations:

$$\frac{\partial^2\Phi}{\partial x^2} = \frac{\partial}{\partial x}\frac{\partial\Psi}{\partial y} = \frac{\partial}{\partial y}\frac{\partial\Psi}{\partial x} = -\frac{\partial^2\Phi}{\partial y^2}, \tag{20.35}$$

and similarly for $\partial^2\Psi/\partial x^2$. Thus we find that Φ and Ψ are each harmonic functions — that is, they are solutions of the Laplace equation

$$\nabla^2\Phi = 0 \qquad \nabla^2\Psi = 0. \tag{20.36}$$

The converse is also true: given a real harmonic function $\Phi(x, y)$ in a simply connected region, there is another real harmonic function $\Psi(x, y)$ which together with Φ compose the analytic complex function $\Omega(z)$. The proof is straightforward. If Ω is to be analytic, Φ and Ψ must satisfy the Cauchy–Riemann equations — so given Φ, we define

$$\Psi(x, y) = \int \frac{\partial\Psi}{\partial y}\,dy + \frac{\partial\Psi}{\partial x}\,dx = \int \frac{\partial\Phi}{\partial x}\,dy - \frac{\partial\Phi}{\partial y}\,dx. \tag{20.37}$$

But for Ψ to be a proper single-valued real function, the integral in (20.37) must be path independent (Section 17.3). Recalling Green's theorem in the plane (15.27), we can close the path in a simply connected region and recast it as a surface integral,

$$\oint_C \frac{\partial\Phi}{\partial x}\,dy - \frac{\partial\Phi}{\partial y}\,dx = \int_S \left[\frac{\partial}{\partial x}\left(\frac{\partial\Phi}{\partial x}\right) - \frac{\partial}{\partial y}\left(-\frac{\partial\Phi}{\partial y}\right) \right] dx\,dy, \tag{20.38}$$

which vanishes since we stipulated that Φ is harmonic. So the integral in (20.37) is in fact path independent. Thus one can construct a complex analytic function $\Omega(z) = \Phi + i\Psi$ from a real harmonic function. The pair Φ and Ψ are called *harmonic conjugates*, and $\Omega = \Phi + i\Psi$ is the *complex potential*.

If this sounds familiar, take a look back at Chapter 13, where Φ is interpreted as the potential and Ψ the stream function of an irrotational and incompressible vector field. There we demonstrated that the curves of constant Φ are perpendicular to those of constant Ψ, and that both fields are solutions to Laplace's equation. Consider for instance electrostatics in the plane with vanishing charge density: curves of constant Ψ represent the electric field, while those of constant Φ are equipotentials. The attraction of Ω in (20.33) is that it bundles both together in a single complex function. And since Φ and Ψ are harmonic conjugates, $\Omega'(z)$ can be expressed entirely in terms of either (Problem 18.12):

$$\frac{d\Omega}{dz} = \frac{\partial\Phi}{\partial x} + i\frac{\partial\Psi}{\partial x} = \frac{\partial\Phi}{\partial x} - i\frac{\partial\Phi}{\partial y} = \frac{\partial\Psi}{\partial y} + i\frac{\partial\Psi}{\partial x} \equiv \mathcal{E}^*, \tag{20.39}$$

where $\mathcal{E}$ is the complex representation of the electric field $\vec{E}$,

$$\mathcal{E} = \frac{d\Omega}{dz}^* = \frac{\partial\Phi}{\partial x} + i\frac{\partial\Phi}{\partial y} = E_x + iE_y, \tag{20.40}$$

with magnitude

$$E = \left| \frac{d\Omega}{dz} \right|. \tag{20.41}$$

So the physics is completely determined by the complex potential.

But electrostatics is not the only physical framework for the complex potential. The heat equation discussed in Problem 16.2 reduces to Laplace's equation for a steady-state temperature distribution $T(x,y)$. So this system, too, can be described by a harmonic potential, with Φ the temperature field $T(x,y)$. Curves of constant Φ are isotherms, and the orthogonal curves of constant Ψ are streamlines of time-independent thermal gradients. The vector field describing heat flow is given by $\Omega'(z)$.

A third system — the archetype often used as the metaphor to describe the others — is incompressible fluid flow in the plane. Here, curves of constant Ψ are the streamlines of flow, curves of constant Φ are velocity equipotentials. The fluid's two-dimensional velocity field is given by (20.40), and its speed by (20.41). The physics in all these systems is completely determined by the complex potential. As a result, solving for $\Omega(z)$ with specified boundary conditions in any one of these physical frameworks gives the corresponding solution with the same boundary conditions in the others.

One might object that although recasting Φ in terms of the complex Ω has aesthetic appeal, we've merely substituted solving for Φ with solving for Ω. But there are real benefits to reformulating these systems in terms of a complex analytic function. In particular, a conformal map $w(z)$ transforms $\Omega(z)$ as

$$\Omega(x,y) = \Omega(x(u,v), y(u,v)) \longrightarrow \tilde{\Omega}(u,v), \tag{20.42}$$

which, as a composite function of analytic functions, is analytic. But if it's analytic, then its real part $\tilde{\Phi}(u,v)$ must be harmonic in the w-plane. In other words, *harmonic functions remain harmonic under conformal transformation,*

$$\nabla^2 \Phi(x,y) = 0 \quad \underset{w^{-1}}{\overset{w}{\rightleftharpoons}} \quad \nabla^2 \tilde{\Phi}(u,v) = 0. \tag{20.43}$$

The statement applies as well to the harmonic conjugate Ψ. And as harmonic conjugates, curves of constant Φ and Ψ must be orthogonal in both planes — which of course is consistent with a conformal map.

Reframing a potential problem $\nabla^2 \Phi = 0$ in $\mathbb{R}^2$ in terms of a complex analytic function Ω allows one to leverage this feature. Specifically, with a clever choice of conformal $w(z)$, challenging boundary conditions can often be mapped into a region where the corresponding boundary conditions are simpler. Moreover, once the solution with given boundary conditions is known in one complex plane, it is effectively known in all conformally related complex planes.

Example 20.6 A Cylindrical Capacitor

Consider a cylindrical capacitor with non-concentric plates meeting at the insulated point $u = 1$, $v = 0$, as shown on the right in Figure 20.4. The larger cylinder, with radius 1, is grounded; the smaller, radius 1/2, is held at voltage V_0. What is the electric field between the plates?

The solution we seek is $\tilde{\Phi}(w)$, such that

$$\nabla^2 \tilde{\Phi}(u,v) = 0. \tag{20.44}$$

Rather than solve this in the non-trivial circular geometry, under the mapping $w = \frac{z-i}{z+i}$ in (20.17), it's conformally equivalent to parallel plates with spacing d — and for which the solution is well known,

$$\Phi(x,y) = V_0 y/d = V_0 y, \tag{20.45}$$

where we've taken $d = 1$. As an analytic function, the complex potential must be a function of z; since its real part Φ is proportional to y, the imaginary part must be

$$\Psi(x,y) = -V_0 x, \tag{20.46}$$

so that $\Omega(z) = -iV_0 z$. (This is not unexpected, since the electric field $E = V_0/d$ has streamlines of constant Ψ.) To find $\tilde{\Omega}(w)$ using (20.42) we take our chosen conformal map,

$$w(z) = \frac{z - i}{z + i} = u(x, y) + iv(x, y), \tag{20.47}$$

and, after a bit of algebra, find the inverted functions

$$x(u, v) = \frac{-2v}{(1 - u)^2 + v^2} \qquad y(u, v) = \frac{1 - u^2 - v^2}{(1 - u)^2 + v^2}. \tag{20.48}$$

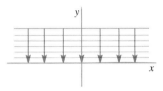 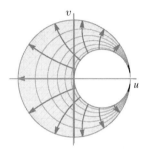

Figure 20.9 Conformal equivalence of parallel plate and cylindrical capacitors.

The potential between the cylindrical capacitor plates in w is therefore

$$\tilde{\Phi}(u, v) = \Phi(x(u, v), y(u, v)) = V_0 \frac{1 - u^2 - v^2}{(1 - u)^2 + v^2}, \tag{20.49}$$

which is a harmonic function corresponding to a grounded outer cylinder and the inner at potential V_0. The stream function is

$$\tilde{\Psi}(u, v) = \Psi(x(u, v), y(u, v)) = V_0 \frac{2v}{(1 - u)^2 + v^2}. \tag{20.50}$$

The tangents to curves of constant $\tilde{\Psi}$ are the electric fields lines $\vec{\nabla}\tilde{\Phi}$ (Figure 20.9). The magnitude of the electric field is given by (20.41) (with the roles of z and w reversed),

$$E = |\tilde{\Omega}'(w)| = \frac{2V_0}{(1 - u)^2 + v^2}. \tag{20.51}$$

The very same analysis applies to cylinders held at fixed temperatures $T = 0$ and $T = T_0$ in the absence of heat sources or sinks. So the same conformal analysis yields isotherms $\tilde{\Phi}$ and streamlines of heat flow, $\tilde{\Psi}$.

Example 20.7 **Fluid Flow Around a Rock**

Consider a brook or stream flowing around a rock; what are the streamlines as the water skirts around the barrier? We'll model the rock as a disk of radius a, and consider irrotational and incompressible rightward flow with uniform speed α both far upstream and downstream from the rock. The symmetry of the problem allows us to limit attention to the UHP in z, with the region

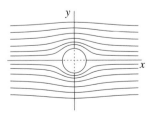

boundary made up of the x-axis together with the perimeter of the rock. Notice that the boundary is itself a streamline.

Consulting Table 20.1, we find that the conformal map

$$w(z) = z + a^2/z \qquad (20.52)$$

takes the region around and above the barrier in $z = x + iy$ into the UHP of $w = u + iv$. Note that the perimeter of the rock $r = a$ plus the two semi-infinite segments $|x| > a$ map to the u-axis — just as one requires of a flow boundary (Problem 20.14). So (20.52) sends the streamlines in z into uniform flow in w. But streamlines of uniform horizontal flow in w are simply lines of constant v — so the stream function $\tilde{\Psi}(u, v) = \beta v$ for some constant β. Similar to the previous example, this mandates that the complex potential be

$$\tilde{\Omega}(w) = \tilde{\Phi}(u, v) + i\tilde{\Psi}(u, v) = \beta(u + iv) = \beta\, w. \qquad (20.53)$$

Appeal to (20.41) shows that $|\tilde{\Omega}'(w)| = \beta$ is the uniform speed of the fluid.

Mapping back into the z-plane gives

$$\Omega(z) = \tilde{\Omega}(w(z)) = \beta\,(z + a^2/z). \qquad (20.54)$$

Notice that at z grows, the flow becomes as uniform in z as it is in w — which makes sense, since far from the rock the geometry is the same in both complex planes. Moreover, since the fluid has uniform speed α for large z, we find $\beta = \alpha$.

From the complex potential, we get the velocity field V,

$$V = \frac{d\Omega}{dz}^{*} = \alpha(1 - a^2/z^2)^{*}, \qquad (20.55)$$

which in polar coordinates on the $\{\hat{r}, \hat{\phi}\}$ basis is

$$V_r = \alpha\left(1 - \frac{a^2}{r^2}\right)\cos\phi \qquad V_\phi = -\alpha\left(1 + \frac{a^2}{r^2}\right)\sin\phi. \qquad (20.56)$$

So, as expected, the flow is entirely radial along the x-axis ($\phi = 0$), it's tangential along the surface of the rock ($r = a$), and at large r the velocity properly reduces to constant speed α in the x-direction ($\hat{\imath} = \hat{r}\cos\phi - \hat{\phi}\sin\phi$). Note as well that $V_r(a, \phi) = 0$, since fluid cannot flow into or out of the rock.

The velocity potential and stream function can be directly extracted from Ω:

$$\Phi = \Re e(\Omega) = \alpha\left(r + \frac{a^2}{r}\right)\cos\phi$$

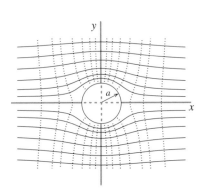

$$(20.57)$$

$$\Psi = \Im m(\Omega) = \alpha\left(r - \frac{a^2}{r}\right)\sin\phi.$$

You should confirm both that the gradient of Φ gives the velocity field in (20.56), and that the Cauchy–Riemann relations (18.8) are satisfied.

Example 20.8 Including Swirl

Back in Example 13.2 we identified the complex potential

$$\Omega_{\text{vortex}}(z) = \frac{i\Gamma}{2\pi} \ln(z/a) = -\frac{\Gamma}{2\pi}\phi + \frac{i\Gamma}{2\pi} \ln(r/a) \equiv \Phi(x,y) + i\Psi(x,y), \tag{20.58}$$

describing a vortex of strength Γ — an irrotational circulation. (If that sounds contradictory, notice the vortex domain is not simply connected.) Adding a $\ln(r/a)$ term to the stream function in (20.57) results in more realistic flow patterns, such as those depicted in Figure 20.10. One new feature which emerges is the appearance of *stagnation points* where the fluid's speed goes to zero. These occur at spots on the surface of the rock where the tangential velocity vanishes,

$$V_\phi(r=1,\phi) = \frac{\partial \Psi}{\partial r}\Big|_{r=a} = \frac{\partial}{\partial r}\left[\alpha\left(r - \frac{a^2}{r}\right)\sin\phi + \frac{\Gamma}{2\pi}\ln(r/a)\right]_{r=a} = 0. \tag{20.59}$$

For $0 < \Gamma < 4\pi\alpha a$, this is satisfied at $\phi = \sin^{-1}(-\Gamma/4\pi a\alpha)$ — so there are two stagnation points, one each in the third and fourth quadrants. As $\Gamma \to 4\pi a\alpha$, the points merge into a single "dead spot" at $\phi = -\pi/2$. Beyond that, the stagnation point moves off the surface of the rock.

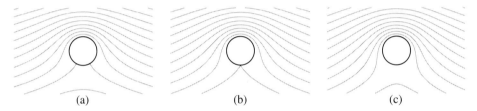

(a)   (b)   (c)

Figure 20.10 Streamlines $\Psi = \alpha\left(r - \frac{a^2}{r}\right)\sin\phi + \frac{\Gamma}{2\pi}\ln(\frac{r}{a})$ for $\frac{\Gamma}{4\pi a\alpha} =$ (a) 1.75, (b) 2.0, (c) 2.25.

BTW 20.2 The Joukowski Airfoil

The map $w(z) = z + a^2/z$ takes the circle $|z| = a$ into the segment $|u| \le 2a$ on the real axis in the w-plane. For $|z| = r > a$, however,

$$w = z + \frac{a^2}{z} = re^{i\phi} + \frac{a^2}{r}e^{-i\phi}$$

$$= \left(r + \frac{a^2}{r}\right)\cos\phi + i\left(r - \frac{a^2}{r}\right)\sin\phi \equiv u + iv \tag{20.60}$$

is an ellipse,

$$\frac{u^2}{\left(r + a^2/r\right)^2} + \frac{v^2}{\left(r - a^2/r\right)^2} = 1. \tag{20.61}$$

But not always: $w(z)$ fails to be conformal at its critical points,

$$\frac{dw}{dz} = 1 - \frac{a^2}{z^2} = 0 \quad \longrightarrow \quad z = \pm a. \tag{20.62}$$

So an off-center circle with $r > a$ which passes through one of these points will cause one end of the ellipse to degenerate into a cusp. What emerges is an airfoil; the map which produces it is called a Joukowski transform.

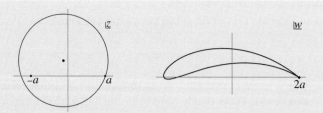

Figure 20.11 Joukoski map of a circle with $r > a$ passing through critical point $z = a$.

The Joukowski transform is a key tool in the design and analysis of airfoils. Take a uniform flow as in Example 20.7, include vortex terms, and map to configurations such as those in Figure 20.10. Moving the circle so that it passes through a critical point, the Joukowski transform reveals the flow around an airfoil (Figure 20.11).

Example 20.9 Fluid Flow Around a Log

Let's replace the rock with a thin rectangular barrier — say, a stationary log of length 1 — transverse to the flow. As in Example 20.7, we consider uniform flow with rightward speed α far upstream and downstream from the barrier. We know roughly the shape the streamlines must take; to determine their explicit functional form, we treat the log as a cut in the z-plane. Placing the log along the y-axis between ± 1, the domain of the conformal transformation

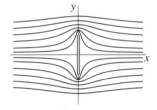

$$w(z) = \sqrt{z^2 + 1} \tag{20.63}$$

is the entire z-plane minus the segment between $\pm i$ — though symmetry allows us to restrict focus to the UHP. That (20.63) is a good choice is best illustrated by considering the effect of the inverse transformation $z(w) = \sqrt{w^2 - 1}$ (Figure 20.12). Uniform streamlines in the w UHP are mapped by $w^2 - 1$ into nested parabolas in the ξ-plane, excluding $\Re e(\xi) > -1$; these are in turn mapped to the UHP of z, minus the imaginary axis y between 0 and i.

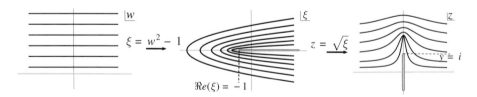

Figure 20.12 The action of $z(w) = \sqrt{w^2 - 1}$ on uniform streamlines in the w UHP.

The beauty of this map is its simplicity: since the streamlines in w are uniform, the complex potential $\tilde{\Omega}$ is the same as in (20.53) — and therefore

$$\Omega(z) = \tilde{\Omega}(w(z)) = \alpha \sqrt{z^2 + 1}. \tag{20.64}$$

The velocity field

$$V = \frac{d\Omega}{dz}^* = \alpha \left[\frac{z}{\sqrt{z^2 + 1}} \right]^* \tag{20.65}$$

correctly reduces to α for large $|z|$. So our solution of Laplace's equation obeys the required boundary condition at infinity.

The square root makes it harder to cleanly extract the velocity potential and stream function as we did before; instead we can map the streamlines in w directly into z. The streamlines in w are given by $v = v_0$, each line distinguished by the value of v_0. So in $z = x + iy$,

$$w = u + iv_0 = \sqrt{z^2 + 1} \implies x^2 - y^2 + 1 = u^2 - v_0^2, \quad xy = v_0 u. \tag{20.66}$$

Eliminating u leads to

$$(x^2 + v_0^2)(y^2 - v_0^2) = v_0^2. \tag{20.67}$$

A little investigation reveals that $v_0 = 0$ corresponds to $y = 0$ — so the boundary along x is a streamline. In addition, as x grows without bound for finite $v_0 \neq 0$, y goes to v_0 — so, far from the log, uniform horizontal flow emerges.

Problems

20.1 Where is the map $w(z) = \sin z$ conformal? By what angle are points $z = \pi, i\pi$, and $\pi/2 + i$ rotated under this map? By what factors are distances in the neighborhood of these points magnified?

20.2 Show that $|w'(z)|^2$ is the Jacobian of the transformation $J = \left| \frac{\partial(u, v)}{\partial(x, y)} \right|$.

20.3 An analytic function $w(z)$ has a critical point at z_0 if the first derivative vanishes there, $w'(z_0) = 0$. Show that at a critical point, w increases the angle between smooth intersecting curves by a factor of k, where k is the order of the first non-vanishing derivative.

20.4 Use the Riemann sphere to argue that inversion $w(z) = 1/z$ maps circles and lines in z into circles in w.

20.5 Sketch the map $w(z) = \sqrt{z}$ from the UHP of z into w, clearly indicating curves of constant u and v.

20.6 For Möbius transformations $w_m(z)$:

(a) Confirm algebraically that, as claimed, $w_m(z)$ requires only three parameters.
(b) Show that the identity map $w(z) = z$ is a Möbius transformation.
(c) Verify that (20.11) is the inverse map w_m^{-1}.
(d) Show that the composition of two Möbius transformations is also a Möbius transformation.

20.7 Show that the cross ratio $\frac{(z-z_1)(z_2-z_3)}{(z-z_3)(z_2-z_1)}$ is a Möbius transformation for $z_1 \neq z_3$.

20.8 Construct a conformal map $w(z) = (az - b)/(cz - d), ad \neq bc$, taking the right half-plane in z into the unit disk $|w| < 1$:

(a) Argue that the y-axis must map to the circle $|w| = 1$, and use this to conclude that the ratio a/c must be a pure phase, $e^{i\alpha}$. Why is this phase largely irrelevant?
(b) Examining the map for finite y, show that $d/c = -(b/a)^*$.
(c) Choose a value z_0 which maps to the center to the circle, $w(z_0) = 0$. Plot some sample points, verifying the final map.

20.9 Construct a conformal map which takes the circle $|z - 1| < 0$ to the half-plane $\Re e(w) > 0$, with $w(0) = \infty$.

20.10 Verify both exponential maps e^z shown in Table 20.1.

20.11 Consider $w(z) = \text{Ln}\left(\frac{z-1}{z+1}\right)$.

    (a) Is $w(z)$ conformal?
    (b) Where is the cut line?
    (c) Show that $w(z)$ maps the UHP in z into the ribbon $0 < v < \pi$ in the w-plane, with $|x| < 1$ mapping into the top border of the ribbon at $v = \pi$ and $|x| > 1$ into the bottom, $v = 0$. Make a sketch which could be included in Table 20.1.

20.12 Use maps from Table 20.1 to construct a conformal transformation from the ribbon $0 < y < \pi$ in the z-plane into the disk $|w| < 1$. Draw a diagram outlining the steps, identifying the parts in w corresponding to the top and bottom of the ribbon.

20.13 Use maps from Table 20.1 to construct a conformal transformation from the first quadrant in z to the punctured plane $|w| > 1$.

20.14 The flow boundary in Example 20.7 is the perimeter of the rock $r = 1$ plus the semi-infinite segments $|x| > 1$. Show that the map $w(z) = z + 1/z$ takes this boundary to the real axis in w.

20.15 Verify the electric field (20.51) for the cylindrical capacitor by taking the z derivative of the complex potential $\Omega(z)$.

20.16 In introductory physics, the fringing field of a parallel plate capacitor is generally neglected (but see Problem 14.29); ignoring edge effects is tantamount to considering an infinite capacitor. But if we can map the infinite capacitor into semi-infinite plates using a conformal map, the correct field can be found.

    (a) Show that the $w(z) = u + iv = z + e^z$ maps a capacitor with plates at $y = \pm\pi$ into parallel plates $u < -1$, $v = \pm\pi$. What happens to the line $y = 0$?
    (b) Solve for the parametric curves $(u(x,y), v(x,y))$ which correspond to equipotentials and electric field lines in the z-plane. Use Mathematica's `ParametricPlot` command to plot a selection in the w-plane.

20.17 Two semi-infinite metal plates held at potential V_0 meet at right angles at $x = y = 0$. Find the electric field and equipotentials for $x, y \geq 0$.

20.18 Since steady-state temperature problems are governed by Laplace's equation $\nabla^2 T = 0$, conformal mapping techniques apply. As an example, consider a flat semi-infinite plate which fills the UHP in z with boundary conditions

$$T(x,0) = \begin{cases} T_0, & |x| < 1 \\ 0, & |x| > 1 \end{cases} \qquad T(x, y \to \infty) \to 0.$$

Use the conformal map $w(z)$ in Problem 20.11 to take $T(x,y)$ to $\tilde{T}(u,v)$.

    (a) What are the boundary conditions for $\tilde{T}(u,v)$?
    (b) Solve Laplace's equation for $\tilde{T}$.
    (c) Use w to solve for $T(x,y)$. Verify that the solution satisfies the boundary conditions.

21 The Calculus of Residues

21.1 The Residue Theorem

A function $f(z)$ with an nth-order pole at a can be written in the form[1]

$$f(z) = \frac{g(z)}{(z-a)^n}, \qquad g(a) \neq 0, \tag{21.1}$$

for integer $n > 0$ and some analytic function $g(z)$. This is perhaps obvious, given the definition of a pole — but it's also tantalizing, given its similarity to the integrand in Cauchy's formulas, (18.26)–(18.27). So let's integrate (21.1) around a closed contour in an analytic region of g. We'll first take advantage of g being analytic to expand it in a Taylor series around a,

$$g(z) = \sum_{m=0}^{} \frac{1}{m!} g^{(m)}(a)(z-a)^m. \tag{21.2}$$

Then

$$\oint_C f(z)\,dz = \oint_C \frac{g(z)}{(z-a)^n}\,dz = \sum_{m=0}^{} \frac{g^{(m)}(a)}{m!} \oint_C (z-a)^{m-n}\,dz. \tag{21.3}$$

Note that since the order of the pole n must be greater than the lowest values of m, the sum necessarily includes negative powers of $(z-a)$ — in other words, it's a Laurent series of f. Choosing C to be the circle $z - a = Re^{i\varphi}$ gives

$$\oint_C f(z)\,dz = \sum_{m=0}^{} \frac{g^{(m)}(a)}{m!} R^{m-n+1} \int_0^{2\pi} e^{i(m-n+1)\varphi}\, i\,d\varphi$$

$$= 2\pi i \sum_{m=0}^{} \frac{g^{(m)}(a)}{m!} R^{m-n+1} \delta_{m-n,-1} = 2\pi i \frac{g^{(n-1)}(a)}{(n-1)!}. \tag{21.4}$$

As anticipated from Cauchy's formula, the only contribution from the Laurent series of f is its residue, the coefficient of $(z-a)^{-1}$ evaluated at the singularity,

$$\operatorname{Res}[f(z), a] = \frac{g^{(n-1)}(a)}{(n-1)!} = \frac{1}{(n-1)!} \lim_{z \to a} \frac{d^{n-1}}{dz^{n-1}} \left[(z-a)^n f(z) \right]. \tag{21.5}$$

---

[1] This should be compared to the definition of an nth-order zero, (19.17).

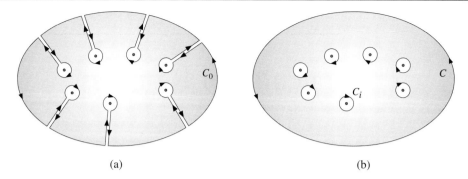

Figure 21.1 (a) Contour C_0 encloses no poles; (b) $C_0 = C + \sum_i C_i$.

Thus

$$\oint_C f(z)\, dz = 2\pi i \operatorname{Res}[f(z), a]. \tag{21.6}$$

What if $f(z)$ has several isolated singularities? We can draw a closed contour C_0 that skirts around each of the singular points (Figure 21.1). Excluding the singularities this way makes the integral around C_0 vanish, by Cauchy's theorem. And the integrals over the segments leading to and from each singularity cancel one another as the width of each "pipe" goes to zero. Thus

$$0 = \oint_{C_0} f(z)\, dz = \oint_C f(z)\, dz - \sum_i \oint_{C_i} f(z)\, dz, \tag{21.7}$$

where the minus sign accounts for the clockwise orientation of the C_i. But from (21.6), each of these small loops around a singularity contributes a factor of $2\pi i$ times its residue. Thus we discover the famous *residue theorem*,

$$\oint_C f(z)\, dz = 2\pi i \sum_{a \in C} \operatorname{Res}[f(z), a], \tag{21.8}$$

where the sum is over all the singularities a enclosed by the contour. One can view Cauchy's theorem (18.7) as a special case of this result.

The residue theorem makes the astonishing assertion that the integral of $f(z)$ around a closed contour is completely determined by the singularities of f *within* the contour. (Or maybe not so astonishing: recall Examples 15.11 and 15.12.) As a result, the evaluation of many definite integrals in the complex plane can be accomplished by a relatively simple calculation of residues (Table 21.1). For an nth-order pole, the residue can be extracted using (21.5). If the function has a simple pole ($n = 1$), this reduces to the simple

$$\operatorname{Res}[f(z), a] = \lim_{z \to a} \left[(z - a) f(z) \right]. \tag{21.9}$$

So, for example, since

$$\lim_{z \to n\pi} \left[(z - n\pi) \cot z \right] = (-1)^n \lim_{z \to n\pi} \frac{(z - n\pi)}{\sin z} = (-1)^n \frac{1}{\cos n\pi} = 1 \tag{21.10}$$

Table 21.1 Finding Res$[f(z), a]$

nth-order pole	$\frac{1}{(n-1)!} \left(\frac{d}{dz}\right)^{n-1} \left[(z-a)^n f(z)\right]_{z=a}$
Simple pole ($n=1$)	$\lim_{z \to a} \left[(z-a)f(z)\right]$
$f(z) = P/Q$, $Q(a)$ a simple zero	$P(z)/Q'(z)\big|_{z=a}$
Pole at infinity	$\lim_{z \to 0} \left[\frac{-1}{z^2} f\left(\frac{1}{z}\right)\right]$
Pole at infinity, $\lim_{|z| \to \infty} f(z) = 0$	$\lim_{|z| \to \infty} \left[-zf(z)\right]$
Laurent series	c_{-1}

is non-zero and finite, $\cot z$ has simple poles at $z = n\pi$ for integer n. And indeed, the first term in the Laurent series of $\cot z$ is $1/z$. A special case of a simple pole is a function with the form $f(z) = P(z)/Q(z)$ for analytic P and Q. A simple pole at $z = a$ implies that Q has a simple zero there, and therefore $Q'(a) = Q(z)/(z-a)\big|_{z=a}$. Then (21.9) produces the handy prescription

$$\text{Res}[f(z), a] = \frac{P(a)}{Q'(a)}, \quad P(a), Q'(a) \neq 0. \tag{21.11}$$

So, for instance,

$$\text{Res}[z/\sinh z, in\pi] = \frac{z}{d(\sinh z)/dz}\bigg|_{in\pi} = (-1)^n in\pi, \quad n \neq 0, \tag{21.12}$$

as a Laurent expansion would confirm.

For a function with an essential singularity, a Kronecker delta still emerges in the integration of $(z-a)^{m-n}$, just as occurred in (21.3). So despite not having a largest n, (21.6) still holds. Without a largest n, however, the residue-extraction formula (21.5) does not apply. Finding the residue of an essential singularity usually requires calculating the Laurent series.

Example 21.1 $f(z) = \frac{(z-1)(z-2)}{z(z+1)(3-z)}$

This function has simple poles at $z = 0, -1$, and 3, with residues

$$\text{Res}[f(z), 0] = zf(z)\big|_{z=0} = \frac{2}{3}$$

$$\text{Res}[f(z), -1] = (z+1)f(z)\big|_{z=-1} = -\frac{3}{2} \tag{21.13}$$

$$\text{Res}[f(z), 3] = (z-3)f(z)\big|_{z=3} = -\frac{1}{6}.$$

So integrating around, say, a circle of radius 2 gives

$$\oint_{|z|=2} f(z)\, dz = 2\pi i \left(\frac{2}{3} - \frac{3}{2}\right) = -i\frac{5\pi}{3}, \tag{21.14}$$

excluding the residue at $z = 3$, since that pole is outside the contour.

Example 21.2 $f(z) = 1/(z^2 \sinh z)$

Although the sinh has only one zero along the real axis ($z = 0$), it has simple zeros $in\pi$ for integer n all along the imaginary axis. So different integration contours will yield different results. The unit circle centered at the origin, for instance, encloses only the third-order pole at the origin. Its residue can be obtained from (21.5) — which, though a bit tedious, is

$$\text{Res}[f(z), 0] = \frac{1}{(3-1)!} \frac{d^{(3-1)}}{dz^{(3-1)}} \left[z^3 f(z) \right]_{z=0}$$

$$= \frac{1}{2} \frac{d^2}{dz^2} \left(\frac{z}{\sinh z} \right) \Big|_{z=0} = -\frac{1}{6}. \tag{21.15}$$

So the integral around the unit circle is

$$\oint_{|z|=1} \frac{dz}{z^2 \sinh z} = 2\pi i \, \text{Res}[f(z), 0] = -i\pi/3. \tag{21.16}$$

Because of our familiarity with the sinh, it's arguably easier to find the residue by expanding in a Laurent series around $z = 0$,

$$\frac{1}{z^2 \sinh z} = \frac{1}{z^2} \frac{1}{z + z^3/3! + z^5/5! + \cdots} = \frac{1}{z^3} \frac{1}{1 + z^2/3! + z^4/5! + \cdots}$$

$$= \frac{1}{z^3} \left(1 - z^2/3! + \mathcal{O}(z^4) \right) = \frac{1}{z^3} - \frac{1}{6z} + \cdots, \tag{21.17}$$

where in the second line we've used the binomial expansion. As expected, the residue appears as the coefficient of $1/z$. Of course, this series around $z = 0$ is only convergent in the punctured disk $0 < |z| < \pi$ — that is, out to the first singularity. So it only gives the residue for a circle of radius less than π. A circle of radius 4 includes both the pole at the origin, as well as the simple poles at $\pm i\pi$. And since $1/z^2$ is analytic for $z \neq 0$, these poles come from a function of the form P/Q with $Q = \sinh z$ — so their residues are easily found using (21.11):

$$\text{Res}[f(z), \pm i\pi] = \lim_{z \to i\pi} \frac{1}{z^2} \frac{1}{\cosh z} = \frac{1}{\pi^2}. \tag{21.18}$$

Then

$$\oint_{|z|=4} \frac{dz}{z^2 \sinh z} = 2\pi i \, (\text{Res}[f(z), 0] + 2\text{Res}[f(z), i\pi]) = -i\pi \left(\frac{1}{3} - \frac{4}{\pi^2} \right). \tag{21.19}$$

Example 21.3 $\int_{-\infty}^{\infty} dx/(x^2 + 1)$

This integral of course is along the real axis, not a closed contour in the complex plane. But if we construct a closed semicircle from an arc C_r of radius r in the upper half-plane and a base along the real axis, in the limit as its radius goes to infinity the arc C_r makes no contribution — leaving us with the integral we want,

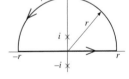

$$\oint_C f(z)\, dz = \int_{C_r} f(z)\, dz + \int_{-r}^{r} f(x)\, dx$$

$$\xrightarrow{r \to \infty} \int_{-\infty}^{\infty} \frac{dx}{x^2 + 1}. \tag{21.20}$$

This approach allows us to utilize the residue theorem. Of $f(z)$'s two simple poles at $\pm i$, the contour encloses only the one at $+i$; its residue is given by (21.9),

$$\text{Res}[f(z), i] = \lim_{z \to i}(z - i)f(z) = \lim_{z \to i}\frac{1}{z + i} = \frac{1}{2i}. \tag{21.21}$$

Therefore

$$\int_{-\infty}^{\infty}\frac{dx}{x^2 + 1} = 2\pi\,i\,\text{Res}[f(z), i] = \pi. \tag{21.22}$$

You might ask: could we not choose to close the contour in the lower half-plane? Indeed we could. To do this, we need the residue at $z = -i$,

$$\text{Res}[f(z), -i] = \lim_{z \to -i}(z + i)f(z) = \lim_{z \to -i}\frac{1}{z - i} = -\frac{1}{2i}. \tag{21.23}$$

Then, since the direction around C is now *clockwise*, the integral gives $-2\pi i$ times the residue — and so yields the same result as before. As it must.

Of course, since $\int dx/(x^2+1) = \tan^{-1}x$, you may be wondering why we used the residue theorem; clearly it wasn't necessary. But if you didn't recognize the integrand, then residues produce the result more easily. We discuss contour construction in more detail in Section 21.3, where we'll come to appreciate that application of the residue theorem is often far-and-away the quickest and easiest approach for the evaluation of certain integrals along the real axis.

Example 21.4 Residue at Infinity

For a function with a finite number of isolated singularities, there will always be a contour large enough to enclose all of them; an integral around them will give $2\pi i$ times the sum of all these residues. But what about the point at infinity? Although a function could in principle have residue there, what relevance does it have if no contour can enclose it? Actually we can enclose it easily, since any closed contour can be construed as encircling the north pole on the Riemann sphere. Then the very same C that surrounds all the finite poles also encloses infinity — though with the opposite orientation. So C can be viewed either as enclosing all the finite poles, or the single pole at infinity. Since the sum of enclosed residues must give the same result either way, the opposite orientation implies that the residue at infinity is the negative of the total residue from the isolated finite-z poles — that is,

$$\text{Res}[f(z), \infty] + \sum_i \text{Res}[f(z), a_i] = 0. \tag{21.24}$$

To verify this, we'll calculate the residue at infinity directly. Given that residues emerge from the poles of an integrand, what matters is not the pole structure of $f(z)$ but rather of $f(z)dz$ — a distinction which becomes important under a change of variables. Recalling that residue at $z = \infty$ corresponds to residue at $\zeta = 1/z = 0$,

$$f(\zeta)\,d\zeta = -\frac{1}{z^2}f\left(\frac{1}{z}\right)dz. \tag{21.25}$$

Thus the residue of $f(z)$ at infinity is the residue of $-\frac{1}{z^2}f\left(\frac{1}{z}\right)$ at the origin,

$$\text{Res}[f(z), \infty] = \text{Res}\left[-\frac{1}{z^2}f\left(\frac{1}{z}\right), 0\right]. \tag{21.26}$$

For instance, the residue at infinity of the function in Example 21.1 is

$$\text{Res}[f(z), \infty] = \text{Res}\left[-\frac{1}{z^2}\frac{(1/z - 1)(1/z - 2)}{1/z(1/z + 1)(3 - 1/z)}, 0\right]$$

$$= -\text{Res}\left[\frac{(z - 1)(2z - 1)}{z(z + 1)(3z - 1)}, 0\right]$$

$$= -\frac{(z - 1)(2z - 1)}{(z + 1)(3z - 1)}\bigg|_{z=0} = 1, \tag{21.27}$$

which is the negative of the sum of the residues from the three finite poles. The same observation similarly holds in Example 21.3. It is not true of a function like $1/\sin z$, however, which has isolated poles extending out to infinity.

If the function goes to zero at infinity, the residue there can be obtained more easily. That's because such a function can only have inverse powers in its Laurent expansion — so the highest-order term in the expansion goes like $1/z$. Thus we can find the pole at infinity directly using the simple prescription

$$\text{Res}[f(z), \infty] = -\lim_{|z| \to \infty} zf(z), \qquad \lim_{|z| \to \infty} f(z) = 0, \tag{21.28}$$

where the negative appears because a conventional counter-clockwise contour encircles infinity with the opposite orientation. This formula nimbly reproduces the result in (21.27).

Example 21.5 Residues and Partial Fractions

Residues can be used in a partial fraction decomposition. Unlike in Section 8.10, where we considered both linear and quadratic factors, here we need only be concerned with linear ones (see BTW 8.4). Consider the rational function

$$f(z) = \frac{P(z)}{Q(z)}, \tag{21.29}$$

where Q is a polynomial of degree n, and P a polynomial of degree no higher than $n - 1$. If Q has only first- and second-order zeros a_k and b_k, then f has the partial fraction decomposition

$$f(z) = \sum_{k=1}^{n} \frac{A_k}{z - a_k} + \frac{B_k}{z - b_k} + \frac{C_k}{(z - b_k)^2}. \tag{21.30a}$$

Rather than algebraically solve for the expansion coefficents as we did in Section 8.10, we can use residues. Specifically, the coefficents of the simple poles are

$$A_k = \text{Res}[f(z), a_k] \qquad B_k = \text{Res}[f(z), b_k], \tag{21.30b}$$

and of the double poles

$$C_k = \text{Res}[(z - b_k)f(z), b_k]. \tag{21.30c}$$

For instance, the partial fraction decomposition of the function in Example 21.1 is

$$\frac{(z - 1)(z - 2)}{z(z + 1)(3 - z)} = \frac{2}{3z} - \frac{1}{6(z - 3)} - \frac{3}{2(z + 1)}, \tag{21.31}$$

as you can directly verify. Similarly,

$$\frac{(z-1)(z-2)}{z(z+1)^2(3-z)} = \frac{2}{3z} - \frac{1}{24(z-3)} - \frac{5}{8(z+1)} - \frac{3}{2(z+1)^2}, \tag{21.32}$$

where the respective coefficients of $1/(z+1)$ and $1/(z+1)^2$ are

$$\text{Res}[f(z), -1] = -\frac{5}{8} \quad \text{and} \quad \text{Res}[(z+1)f(z), -1] = -\frac{3}{2}. \tag{21.33}$$

The residue theorem is a remarkable result. And as we've seen, it can also be remarkably useful. In fact, there exists a body of techniques which leverage the theorem to evaluate integrals. These techniques, which together constitute the *calculus of residues*, are best elucidated by example.

21.2 Integrating Around a Circle

Perhaps the most straightforward application of residue calculus is to angular integrals of the form

$$\int_0^{2\pi} f(\sin n\varphi, \cos m\varphi) \, d\varphi. \tag{21.34}$$

This is ostensibly a real integral over a real domain — the unit circle. But it can be reformulated in the complex plane by using

$$z = e^{i\varphi} \qquad d\varphi = \frac{dz}{iz}, \tag{21.35a}$$

and

$$\sin n\varphi = \frac{1}{2i}\left(z^n - 1/z^n\right) \qquad \cos m\varphi = \frac{1}{2}\left(z^m + 1/z^m\right), \tag{21.35b}$$

to obtain the contour integral

$$\oint_{|z|=1} f\left(\frac{1}{2i}(z^n - 1/z^n), \frac{1}{2}(z^m + 1/z^m)\right) \frac{dz}{iz}. \tag{21.36}$$

Example 21.6 $\int_0^{2\pi} \sin^2 \varphi \, d\varphi$

As our first example, let's start with an integral we know well. Making the substitutions of (21.35) converts the angular integral into a contour integral along the unit circle in the complex plane,

$$\int_0^{2\pi} \sin^2 \varphi \, d\varphi = -\frac{1}{4} \oint_{|z|=1} \left(z - \frac{1}{z}\right)^2 \frac{dz}{iz}$$

$$= \frac{i}{4} \oint_{|z|=1} \frac{1}{z^3}\left(z^4 - 2z^2 + 1\right) dz$$

$$= \frac{i}{4} \cdot 2\pi i \, \text{Res}[(z^4 - 2z^2 + 1)/z^3, 0]. \tag{21.37}$$

The third-order pole at $z = 0$ has residue

$$\text{Res}[(z^4 - 2z^2 + 1)/z^3, 0] = \frac{1}{2!}\frac{d^2}{dz^2}\left(z^4 - 2z^2 + 1\right)\Big|_{z=0} = -2, \tag{21.38}$$

and so the familiar result π emerges.

Example 21.7 $\int_0^{2\pi} \cos(2\varphi)/(5 - 4\sin\varphi)\, d\varphi$

$$\int_0^{2\pi} \frac{\cos(2\varphi)}{5 - 4\sin\varphi}\, d\varphi = \oint_{|z|=1} \frac{\frac{1}{2}\left(z^2 + \frac{1}{z^2}\right)}{5 - \frac{4}{2i}\left(z - \frac{1}{z}\right)} \frac{dz}{iz}$$

$$= -\oint_{|z|=1} \frac{z^4 + 1}{2z^2(2z - i)(z - 2i)}\, dz. \tag{21.39}$$

The integrand has a double pole at $z = 0$, as well as simple poles at $z = i/2$ and $z = 2i$. Since only the first two of these lie within the unit circle, their residues are the only contributions to the integral:

$$\text{Res}[f(z), 0] = -\frac{d}{dz}\left[\frac{z^4 + 1}{2z^2(2z - i)(z - 2i)}\right]_0 = -\frac{5i}{8} \tag{21.40}$$

and

$$\text{Res}[f(z), i/2] = -\frac{z^4 + 1}{2z^2(z - 2i)}\bigg|_{i/2} = +\frac{17i}{24}. \tag{21.41}$$

Therefore

$$\int_0^{2\pi} \frac{\cos(2\varphi)}{5 - 4\sin\varphi}\, d\varphi = 2\pi i\left(-\frac{5i}{8} + \frac{17i}{24}\right) = -\frac{\pi}{6}. \tag{21.42}$$

Example 21.8 $\int_0^{2\pi} d\varphi/(a + b\sin\varphi)$

The family of angular integrals of this form for real a, b with $a > |b|$ can be determined by residues. Using (21.35) as before,

$$\int_0^{2\pi} \frac{d\varphi}{a + b\sin\varphi} = \oint_{|z|=1} \frac{1}{a + \frac{b}{2i}\left(z - \frac{1}{z}\right)} \frac{dz}{iz}$$

$$= \frac{2}{b}\oint_{|z|=1} \frac{dz}{z^2 + 2i(a/b)z - 1}. \tag{21.43}$$

The integrand $f(z)$ has poles at[2]

$$z_{1,2} = -i\left[\frac{a}{b} \pm \sqrt{\frac{a^2}{b^2} - 1}\right] = -i\left[\frac{a}{b} \pm \frac{\sqrt{a^2 - b^2}}{|b|}\right]. \tag{21.44}$$

The product $z_1 z_2 = -1$ implies that if one root is in the unit circle $|z| < 1$, the other is necessarily outside with $|z| > 1$. And only the pole inside the circle contributes to the integral. For $b > 0$, z_2 is inside, for $b < 0$, it's z_1. Either way,

$$\int_0^{2\pi} \frac{d\varphi}{a + b\sin\varphi} = 2\pi i \frac{2}{b}\text{Res}[f(z_{\text{in}})] = \frac{2\pi}{\sqrt{a^2 - b^2}}. \tag{21.45}$$

---

[2] Since this quadratic does not have real coefficients, its roots are not complex conjugates.

21.3 Integrals Along the Real Axis

Although it may seem counterintuitive, residues can provide a vehicle for evaluating integrals along the real axis. Of course, without a closed contour the residue theorem doesn't apply — so the challenge is to find a way to close the contour in the z-plane which allows extraction of the desired integral along x.

21.3.1 Semicircular Contours

The most common circumstance is that of Example 21.3: as an integral between $\pm\infty$, we were able to construct a closed contour by appending a semicircular arc of infinite radius in the upper half-plane (UHP). Technically, this is a limiting procedure:

$$\oint_C f(z)\,dz = \lim_{r\to\infty}\left[\int_{-r}^{r} f(x)\,dx + \int_{C_r} f(z)\,dz\right], \tag{21.46}$$

where C_r is the semicircle of radius r. Then if $f(z)$ goes to zero quickly enough as $r \to \infty$, the semicircle adds nothing to the integral itself, and so

$$\int_{-\infty}^{\infty} f(x)\,dx = \oint_C f(z)\,dz = 2\pi i \sum_i \text{Res}[f(z),a_i], \tag{21.47}$$

as we saw in (21.20). This procedure works whenever there is no contribution to the contour integral from the semicircle, which requires

$$|zf(z)| \to 0 \qquad \text{as} \qquad |z| \to \infty. \tag{21.48}$$

For a ratio of polynomials $f(z) = P(z)/Q(z)$, this condition is satisfied when the order of Q is at least two more than that of P.

Example 21.9 $\int_{-\infty}^{\infty} \frac{2x+1}{x^4+5x+4}\,dx$

Since $x^4 + 5x + 4$ is three orders greater than the numerator, appending an arc in the UHP to the real axis gives a contour integral with the same value,

$$\int_{-\infty}^{\infty} \frac{2x+1}{x^4+5x+4}\,dx = \oint_C \frac{2z+1}{z^4+5z+4}\,dz, \tag{21.49}$$

allowing the desired integral to be evaluated with residues. The poles are the zeros of the denominator,

$$z^4 + 5z + 4 = (z^2+4)(z^2+1) = (z+2i)(z-2i)(z+i)(z-i) = 0, \tag{21.50}$$

so that the integrand has four simple poles at $\pm i$, $\pm 2i$. Only the two in the UHP contribute to the contour integral; they have residues

$$\text{Res}\left[\frac{2z+1}{z^4+5z+4}, i\right] = (z-i)\frac{2z+1}{z^4+5z+4}\bigg|_i$$

$$= \frac{2z+1}{(z+i)(z^2+4)}\bigg|_i = \frac{2i+1}{6i} = \frac{1}{3} - \frac{i}{6}, \tag{21.51a}$$

and

$$\text{Res}\left[\frac{2z+1}{z^4+5z+4}, 2i\right] = (z-2i)\frac{2z+1}{z^4+5z+4}\bigg|_{2i}$$

$$= \frac{2z+1}{(z+2i)(z^2+1)}\bigg|_{2i} = \frac{4i+1}{-12i} = -\frac{1}{3} + \frac{i}{12}. \tag{21.51b}$$

Thus

$$\int_{-\infty}^{\infty} \frac{2x+1}{x^4+5x+4}\, dx = 2\pi i \sum \text{residues} = \frac{\pi}{6}. \tag{21.52}$$

Were we to append a semicircle in the lower half-plane (LHP) the contribution along it would still vanish, so the residue theorem should give the same result. Of course, now we'll want the residues from poles in the LHP, which for this integrand are the complex conjugates of those in the UHP. So the sum of the LHP residues is the negative of the sum of the UHP residues. But in order to be consistent with the direction of the x integral, the integration must sweep over the LHP semicircle *clockwise*; this gives the contour integral an overall sign. Thus both semicircles return the same result. As they must.

Example 21.10 $\int_{-\infty}^{\infty} \frac{1}{x^3-i}\, dx$

Though the complex integrand implies a complex result, the integral itself is along the real axis. Closing the contour with a semicircle gives

$$\int_{-\infty}^{\infty} \frac{1}{x^3-i}\, dx = \oint_C \frac{1}{z^3-i}\, dz. \tag{21.53}$$

The integrand has poles at the three roots of i, $-i$, $e^{i\pi/6}$, and $e^{i5\pi/6}$. Since the LHP has only one pole, choosing our semicircle in the LHP is (marginally) easier. We could use (21.5) with $n = 2$ — but it's easier to use (21.11)

$$\text{Res}\left[\frac{1}{z^3-i}, -i\right] = \frac{1}{3z^2}\bigg|_{-i} = -\frac{1}{3}. \tag{21.54}$$

Then, remembering to include an overall minus sign due the clockwise sweep over the contour, we get

$$\int_{-\infty}^{\infty} \frac{1}{x^3-i}\, dx = \oint_C \frac{1}{z^3-i}\, dz = -2\pi i \left(-\frac{1}{3}\right) = \frac{2i\pi}{3}. \tag{21.55}$$

In addition to rational functions of polynomials, integrals of the form

$$\int_{-\infty}^{\infty} g(x)\, e^{ikx}\, dx, \qquad k \text{ real} \tag{21.56}$$

are particularly important (as we'll come to appreciate in Chapter 36). If $k > 0$, then extending the integral along a semicircle in the upper half-plane yields a decaying exponential:

$$\left|\int_{C_r} g(x)\, e^{ikz}\, dz\right| \leq \int_{C_r} |g(z)\, e^{ikz}\, dz|$$

$$= \int_0^\pi |g(re^{i\varphi})|\, e^{-kr\sin\varphi}\, r\, d\varphi. \tag{21.57}$$

Since $\sin \varphi \geq 0$ for φ between 0 and π, the contribution in the UHP of C_r vanishes in the $r \to \infty$ limit as long as

$$|g(z)| \to 0 \qquad \text{as} \qquad |z| \to \infty. \tag{21.58}$$

This condition is known as *Jordan's lemma*. Although for $k < 0$ a semicircle in the UHP diverges, we can close the contour and evaluate the integral in the LHP instead. Either way, as in (21.46)–(21.47), this construction gives

$$\int_{-\infty}^{\infty} g(x) \, e^{ikx} \, dx = \oint_C g(z) \, e^{ikz} \, dz. \tag{21.59}$$

Example 21.11 $I = \int_{-\infty}^{\infty} \frac{\cos kx}{x^2 + 4} \, dx$

Rather than dealing with $\cos kx$ directly, it's simpler to evaluate

$$\int_{-\infty}^{\infty} \frac{e^{ikx}}{x^2 + 4} \, dx, \tag{21.60}$$

and take the real part at the end — although since $\sin kx$ is odd and $x^2 + 4$ even, this integral should be entirely real anyway. For $k > 0$ we can close the contour with a semicircle in the UHP to get

$$I = \int_{-\infty}^{\infty} \frac{e^{ikx}}{x^2 + 4} \, dx = \oint_C \frac{e^{ikz}}{z^2 + 4} \, dz. \tag{21.61}$$

The poles are at $\pm 2i$ with residues

$$I = \text{Res}\left[\frac{e^{ikz}}{z^2 + 4}, \pm 2i\right] = \left.\frac{e^{ikz}}{z \pm 2i}\right|_{\pm 2i} = \frac{e^{\mp 2k}}{\pm 4i} = \mp \frac{i}{4} e^{\mp 2k}. \tag{21.62}$$

In the UHP, only the residue at $2i$ contributes, and we get

$$I = \int_{-\infty}^{\infty} \frac{e^{ikx}}{x^2 + 4} \, dx = 2\pi i \left(-\frac{i}{4}\right) e^{-2k} = +\frac{\pi}{2} e^{-2k}, \tag{21.63}$$

which, as expected, is real. Since the original integral doesn't specify the sign of k, however, we must consider $k < 0$ as well; this requires a circle in the LHP. But between the positive residue and the negative sweep over the contour, we would achieve the same exponentially decaying result. In other words,

$$\int_{-\infty}^{\infty} \frac{\cos kx}{x^2 + 4} \, dx = \frac{\pi}{2} e^{-2|k|}. \tag{21.64}$$

(Since $\cos kx$ is even, a dependence only on the magnitude of k was to be expected.)

Example 21.12 $\int_{-\infty}^{\infty} \frac{x \sin kx}{x^2 + 4} \, dx$

Similar to the approach of the previous example, we take $\sin kx \to e^{ikx}$ and $k > 0$, we close the contour in the LHP,

$$\int_{-\infty}^{\infty} \frac{x e^{ikx}}{x^2 + 4} \, dx = \oint_C \frac{z e^{ikz}}{z^2 + 4} \, dz. \tag{21.65}$$

The poles are the same as before; the residue at $2i$ is

$$\text{Res}\left[\frac{ze^{ikz}}{z^2+4}, 2i\right] = \frac{1}{2}e^{-2k}. \tag{21.66}$$

Thus

$$\oint_C \frac{ze^{ikz}}{z^2+4}\,dz = i\pi\,e^{-2k}. \tag{21.67}$$

Taking the imaginary part,

$$\int_{-\infty}^{\infty} \frac{x\sin kx}{x^2+4}\,dx = \pi\,e^{-2k}. \tag{21.68}$$

As an exercise, think about what the result is for $k < 0$.

Example 21.13 $\int_{-\infty}^{\infty} \frac{\sin kx}{x-ia}\,dx, a > 0$

Since the integrand is complex, we can't use $\sin kx \to e^{ikx}$ and take the imaginary part at the end. Instead, we express the integral as

$$\int_{-\infty}^{\infty} \frac{\sin kx}{x-ia}\,dx = \frac{1}{2i}\int_{-\infty}^{\infty}\frac{e^{ikx}-e^{-ikx}}{x-ia}\,dx. \tag{21.69}$$

For $k > 0$ the first term requires a semicircle in the UHP, whereas the second term requires one in the LHP. However, since the integrand has only a simple pole is at $-ia$, the UHP integration is zero. Thus we find (remembering the sign for a clockwise orientation)

$$\frac{1}{2i}\oint_C \frac{e^{ikz}-e^{-ikz}}{z-ia}\,dz = -2\pi i\cdot\left(-\frac{1}{2i}e^{-ka}\right) = +\pi\,e^{-ka}. \tag{21.70}$$

Note that the limit $a \to 0$ recovers the result we derived earlier in Example 8.16,

$$\int_0^{\infty} \frac{\sin kx}{x}\,dx = \frac{\pi}{2}. \tag{21.71}$$

21.3.2 Rectangular and Wedge Contours

As we've seen, using the residue theorem to evaluate integrals along the real axis requires the construction of closed a contour in the complex plane. Although semicircles are the most common, rectangles can occasionally be used when a semicircle simply won't do. As we'll see in the coming examples, sometimes not even rectangles — though a little cleverness may suffice to construct a contour from which the desired integral can be extracted.

Example 21.14 $\int_{-\infty}^{\infty} \frac{e^{bx}}{e^x+1}\,dx, 0 < b < 1$

Though restricting b to fall between 0 and 1 is enough to guarantee convergence, extending the integral to the complex plane presents a new challenge: the exponential e^z is periodic parallel to the imaginary axis, and thus has an infinite number of singularities at

$$z_n = (2n+1)i\pi, \qquad n = 0, \pm 1, \pm 2, \ldots. \tag{21.72}$$

Indeed, closing the contour with the conventional semicircle in either the UHP or LHP leads to divergence issues. One way to deal with the integrand is to give b a small imaginary term, $b \to b + i\epsilon$, evaluate the now-convergent residue sum, and then take the $\epsilon \to 0$ limit (Problem 21.16).

Alternatively, we can choose a different closed contour altogether — one which takes advantage of the periodicity; the rectangle in the UHP, as shown, does just this. Because $0 < b < 1$, the two vertical segments make no contribution in the $a \to \infty$ limit. The segment along the real axis gives the desired integral; the contribution of the parallel segment at $2\pi i$ is

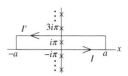

$$I' = \int_{\infty}^{-\infty} \frac{e^{b(x+2\pi i)}}{e^{x+2\pi i} + 1}\, dx = -e^{2\pi i b} \int_{-\infty}^{\infty} \frac{e^{bx}}{e^x + 1}\, dx = -e^{2\pi i b}\, I. \tag{21.73}$$

Thus the contour integral around the rectangle is

$$\oint_C \frac{e^{bz}}{e^z + 1}\, dz = \left(1 - e^{2\pi i b}\right) I. \tag{21.74}$$

The only enclosed pole is a simple one at $i\pi$, with residue given by (21.11),

$$\mathrm{Res}\left[\frac{e^{bz}}{e^z + 1}, i\pi\right] = \left.\frac{e^{bz}}{e^z}\right|_{i\pi} = -e^{i\pi b}. \tag{21.75}$$

Thus

$$I = \frac{1}{1 - e^{2\pi i b}} \cdot -2\pi i e^{i\pi b} = \frac{\pi}{\sin \pi b}. \tag{21.76}$$

Example 21.15 $\int_0^\infty \frac{dx}{1+x^3}$

Were this an even function, we could rewrite it as half the integral between $\pm\infty$ and close the contour with a semicircle in either the lower or upper HP. But the integrand is not even, and so appending an infinite semicircle is not an option. We need to find another route from the origin to $|z| = \infty$, one along which we know how its contribution is related to the desired integral. And in fact, the cube root of unity provides two such equivalent routes along radial lines at $\varphi = \pm 2\pi/3$. So we can create a closed contour by integrating along the real axis from the origin to infinity, counterclockwise along a circular arc for $120°$, and then radially inward from $|z| = \infty$ back the origin,

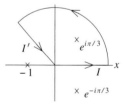

$$\oint_C \frac{dz}{1 + z^3} = I + I', \tag{21.77}$$

where, with $dz = e^{i\varphi} dr$ along a radial line,

$$I' = \int_{\infty}^{0} \frac{e^{2i\pi/3} dr}{1 + r^3} = -e^{2i\pi/3} I. \tag{21.78}$$

Thus

$$\oint_C \frac{dz}{1 + z^3} = \left(1 - e^{2i\pi/3}\right) I. \tag{21.79}$$

The enclosed simple pole at $e^{i\pi/3}$ has residue

$$\text{Res}\left[\frac{1}{1+z^3}, e^{i\pi/3}\right] = \frac{1}{3z^2}\bigg|_{e^{i\pi/3}} = \frac{1}{3}e^{-2i\pi/3}, \tag{21.80}$$

so that the desired integral is

$$I = \frac{1}{1-e^{2i\pi/3}}\, 2\pi i \cdot \frac{1}{3}e^{-2i\pi/3} = \frac{2\pi}{3\sqrt{3}}. \tag{21.81}$$

Example 21.16 $\int_0^\infty \cos x^2\, dx, \int_0^\infty \sin x^2\, dx$

Given that these integrands have no poles, they may not appear to be fodder for residue calculus. But the residue theorem still applies when the residue sum is zero — it's called Cauchy's theorem. So, recognizing that $\cos x^2$ and $\sin x^2$ are the real and imaginary parts of e^{ix^2}, let's continue the integral into the complex plane

$$\int_0^\infty e^{ix^2}\, dx \to \oint_C e^{iz^2}\, dz \tag{21.82}$$

and try to construct a contour from which we can extract the integral along the real axis. To do this, note that along a radial line $z = re^{i\pi/4}$, the integrand becomes a real Gaussian. So if we integrate out from the origin to infinity along the real axis, then make a 45° circular arc at infinity, and return to the origin along this radial line, we get

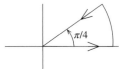

$$\oint_C e^{iz^2}\, dz = \int_0^\infty e^{ix^2}\, dx + 0 + \int_\infty^0 e^{i(re^{i\pi/4})^2}\, e^{i\pi/4}\, dr. \tag{21.83}$$

(Do you see why the circular arc doesn't contribute?) Since the contour encloses no poles, the integral must vanish — allowing straightforward extraction of the integral we want,

$$\int_0^\infty e^{ix^2}\, dx = +e^{i\pi/4}\int_0^\infty e^{-r^2}\, dr = \frac{1+i}{\sqrt{2}}\frac{\sqrt{\pi}}{2}, \tag{21.84}$$

where we've used (8.64),

$$\int_0^\infty e^{-ax^2}\, dx = \frac{1}{2}\sqrt{\frac{\pi}{a}}. \tag{21.85}$$

Thus we find that the real and imaginary parts are the same,

$$\int_0^\infty \cos x^2\, dx = \int_0^\infty \sin x^2\, dx = \frac{1}{2}\sqrt{\frac{\pi}{2}}. \tag{21.86}$$

Given the similarity between sine and cosine, this is not a great surprise. And incidentally, since $\frac{1}{\sqrt{2}}(1 \pm i) = \sqrt{\pm i}$, (21.84) shows that Gaussian integrals like (21.85) can be extended to complex exponents,

$$\int_0^\infty e^{\pm iax^2}\, dx = \frac{1}{2}\sqrt{\frac{\pm i\pi}{a}}. \tag{21.87}$$

21.4 Integration with Branch Cuts

In the calculus of residues, a branch cut should be viewed not as a bug, but a feature — another vehicle and aid in the construction of a closed contour.

Example 21.17 $\int_0^\infty \frac{\sqrt{x}}{1+x^3}\,dx$

We choose to cut along the positive real axis, restricting φ to lie between 0 and 2π. Then we can take advantage of this cut by constructing the "keyhole" contour shown. The integral around the closed path can be expressed in terms of four distinct pieces,

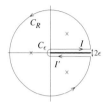

$$\oint_C \frac{\sqrt{z}}{1+z^3}\,dz = \int_0^\infty + \int_{C_R} + \int_\infty^0 + \int_{C_\epsilon} \frac{\sqrt{z}\,dz}{1+z^3}. \tag{21.88}$$

The first and third integrals are parallel to the positive real axis, a distance 2ϵ apart with the cut line between them. In the $\epsilon \to 0$ limit they reduce to integrals along x, but still differ by the discontinuity in φ: above the cut $\varphi = 0$, but below $\varphi = 2\pi$. So the cut prevents the integrals from canceling out, despite what you might otherwise expect. In fact,

$$I' = \int_\infty^0 \frac{\sqrt{z}}{1+z^3}\,dz = \int_\infty^0 \frac{\sqrt{e^{i2\pi}x}}{1+(e^{i2\pi}x)^3}\,dx = \int_\infty^0 \frac{-\sqrt{x}}{1+x^3}\,dx = +I. \tag{21.89}$$

So rather than cancel, the two integrals along x compound. The circular arcs have radii ϵ and R, and in the appropriate limits make no contribution to the contour integral because

$$\lim_{|z|\to 0} |zf(z)| = \lim_{|z|\to 0} z^{3/2} = 0 \qquad \lim_{|z|\to\infty} |zf(z)| = \lim_{|z|\to\infty} z^{-3/2} = 0. \tag{21.90}$$

All together then,

$$2I = \oint_C \frac{\sqrt{z}}{1+z^3}\,dz = 2\pi i \sum \text{residues}. \tag{21.91}$$

The pole at $e^{i\pi/3}$ has residue

$$\text{Res}\big[f(z), e^{i\pi/3}\big] = \frac{\sqrt{z}}{3z^2}\bigg|_{e^{i\pi/3}} = -\frac{i}{3}. \tag{21.92}$$

Similarly, the residues at $e^{i\pi}$ and $e^{i5\pi/3}$ have residues $i/3$ and $-i/3$, respectively. Thus

$$I = \frac{1}{2} \cdot 2\pi i \left(-\frac{i}{3}\right) = \frac{\pi}{3}. \tag{21.93}$$

Note that had we expressed the pole below the cut as $e^{-i\pi/3}$ rather than $e^{i5\pi/3}$, we'd get the wrong result: an angle $-\pi/3$ is incompatible with the cut. Working entirely on a different branch, however, the arguments of all the poles are augmented by an integer multiple of 2π, leading to the same result for I.

Example 21.18 $\int_0^\infty \frac{dx}{1+x^3}$

We evaluated this integral in Example 21.15 using a rectangular contour which took advantage of the function's periodicity along the imaginary axis. Here we use a different technique and use the logarithm to *introduce* a cut. Consider

$$\int_0^\infty \frac{\ln x}{1+x^3} \, dx \tag{21.94}$$

with a cut along the positive real axis. Along the keyhole contour of the previous example, the circular arcs C_R and C_ϵ make no contribution in the appropriate limits, leaving

$$\oint_C \frac{\ln z}{1+z^3} \, dz = \int_0^\infty \frac{\ln x}{1+x^3} \, dx + \int_\infty^0 \frac{\ln x + 2\pi i}{1+x^3} \, dx, \tag{21.95}$$

where since the first integral is above the cut, $z \to e^{i0}x$, but on second below the cut $z \to e^{i2\pi}x$. The integrals of the logarithms sum to zero, leaving

$$I = \int_0^\infty \frac{dx}{1+x^3} = -\frac{1}{2\pi i} \oint_C \frac{\ln z}{1+z^3} \, dz = -\sum \text{Res}\left[\frac{\ln z}{1+z^3}\right]. \tag{21.96}$$

The poles are the same as in the previous example, with residues

$$\text{Res}\left[\frac{\ln z}{1+z^3}, e^{i\pi/3}\right] = \frac{\ln z}{3z^2}\bigg|_{e^{i\pi/3}} = \frac{i\pi}{9} e^{-2i\pi/3}. \tag{21.97}$$

Similarly, the residues at $e^{i\pi}$ and $e^{i5\pi/3}$ have residues $i\pi/3$ and $(i5\pi/9) e^{-10i\pi/3} = (i5\pi/9) e^{+2i\pi/3}$, respectively. Thus

$$I = \int_0^\infty \frac{dx}{1+x^3} = -i\pi \left(\frac{1}{9} e^{-2i\pi/3} + \frac{1}{3} + \frac{5}{9} e^{2i\pi/3}\right) = \frac{2\pi}{3\sqrt{3}}, \tag{21.98}$$

just as we found in (21.81).

Example 21.19   The Dog Bone Contour

Consider the integral

$$I = \int_0^1 \frac{\sqrt{1-x^2}}{(x^2+a^2)} \, dx, \, a > 0. \tag{21.99}$$

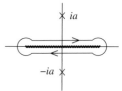

We'll cut the complex z-plane between ± 1, and use the "dog bone" contour shown. In the usual limit, the integrals around the semicircles vanish. This leaves only the segments parallel to the cut to contribute to the contour integral,

$$\oint_C \frac{\sqrt{1-z^2}}{(z^2+a^2)} \, dz = \int_{-1}^1 \frac{\sqrt{1-x^2}}{(x^2+a^2)} \, dx + \int_1^{-1} \frac{-\sqrt{1-x^2}}{(x^2+a^2)} \, dx$$

$$= 2\int_{-1}^1 \frac{\sqrt{1-x^2}}{(x^2+a^2)} \, dx = 4I, \tag{21.100}$$

where we've remembered that the square root below the cut is negative. (In Figure 19.8a, immediately above the cut $\phi_1 = 0, \phi_2 = \pi$, whereas below $\phi_1 = 0, \phi_2 = -\pi$. There's a 2π phase jump across the cut; this gives the square root its sign flip.)

The interior of the contour is not analytic, so the residue theorem doesn't apply. However, if we construe the dog bone as encircling the north pole of the Riemann sphere, then I can be found from the residues outside C. The square root's sign flip across the cut gives the residues at $\pm ia$ the same sign; direct calculation gives

$$\text{Res}\left[f(z), ia\right] = \frac{\sqrt{a^2 + 1}}{2ia}. \tag{21.101}$$

There's also residue at infinity; since $f(z) \to 0$ for large $|z|$, we use the formula

$$\text{Res}\left[f(z), \infty\right] = -\lim_{|z| \to \infty} zf(z) = -i. \tag{21.102}$$

And finally, the contour encircles the cut clockwise — so it surrounds the residues counterclockwise. Thus the contribution from infinity flips sign. All together then,

$$4I = 2\pi i \left(2 \cdot \frac{\sqrt{a^2 + 1}}{2ia} + i\right) \quad \Longrightarrow \quad I = \frac{\pi}{2a}\left(\sqrt{a^2 + 1} - a\right). \tag{21.103}$$

21.5 Integrals with Poles on the Contour

In all the residue calculus examples thus far, we have assiduously avoided poles on the contour of integration. Since such integrals are ill-defined, any meaning to be extracted from them requires taking appropriate limits as the integral approaches the pole — and that limiting procedure becomes fundamental to any significance attached to the result.

As discussed in BTW 8.1, one way to deal with an integrand which is discontinuous at an isolated point x_0 is to introduce a small $\epsilon > 0$ and take the symmetric limit

$$PV \int_a^b f(x)\, dx \equiv \lim_{\epsilon \to 0}\left[\int_a^{x_0 - \epsilon} f(x)\, dx + \int_{x_0 + \epsilon}^b f(x)\, dx\right], \tag{21.104}$$

where "PV" stands for the Cauchy principal value.[3] Though this symmetric limit is a reasonable way to assign a value to an ambiguous integral, when x_0 is a simple pole it's often easier to extend the integral into the complex plane and use residue calculus.

To do this of course requires constructing a closed contour C which overlaps with the integration interval on the real axis — usually between $\pm\infty$. Indeed, we generally close the contour with the familiar large semicircle whose contribution vanishes as its radius goes to infinity. The only novel feature comes from the need to close the 2ϵ gap in (21.104); this is achieved by diverting around x_0 with a semicircle of radius ϵ which can either include or exclude x_0 within C (Figure 21.2). Then any non-vanishing contributions to the contour integral of $f(z)$ comes from the principal value along the real axis plus the contribution from this small semicircle,

---

[3] The Cauchy principal value is distinct from values a function takes on its principal branch.

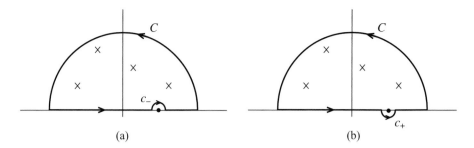

Figure 21.2 Contour C diverting around a singularity: (a) c_- excludes x_0; (b) c_+ includes x_0.

$$\oint_C f(z)\,dz = PV \int_{-\infty}^{\infty} f(x)\,dx + \lim_{\epsilon \to 0} \int_{c_\pm} f(z)\,dz$$

$$= PV \int_{-\infty}^{\infty} f(x)\,dx + \lim_{\epsilon \to 0} \int_{c_\pm} \frac{\left[(z-x_0)f(z)\right]}{z-x_0}\,dz, \tag{21.105}$$

where we've highlighted the simple pole in the $c_\pm$ integral. In the limit, the numerator is the residue of f at x_0. Then with $z - x_0 = \epsilon\,e^{i\varphi}$ in the denominator and $dz = i\epsilon\,e^{i\varphi}\,d\varphi$ on the semicircle,

$$\oint_C f(z)\,dz = PV \int_{-\infty}^{\infty} f(x)\,dx + \operatorname{Res}\left[f(z), x_0\right] \int_{c_\pm} i\,d\varphi$$

$$= PV \int_{-\infty}^{\infty} f(x)\,dx \pm i\pi \operatorname{Res}\left[f(z), x_0\right], \tag{21.106}$$

where the sign is determined by the opposite orientations of $c_\pm$. The residue theorem sets this all equal to $2\pi i$ times the sum of the residues enclosed by C, allowing the principal value to be completely determined by residue sums,

$$PV \int_{-\infty}^{\infty} f(x)\,dx = 2\pi i \sum_{z_i \text{ in } C} \operatorname{Res}\left[f(z), z_i\right] \mp i\pi \operatorname{Res}\left[f(z), x_0\right], \tag{21.107}$$

where the upper sign is for c_+, the lower for c_-. For c_-, any poles in the interior of C do not include x_0, and the separate term contributes half its residue to the principal value; for c_+, the interior poles include x_0, half of which is then subtracted off by the additional term. The choice of $c_\pm$ then is irrelevant — either way

$$PV \int_{-\infty}^{\infty} f(x)\,dx = 2\pi i \left[\sum_{z_i \text{ in } C} \operatorname{Res}\left[f(z), z_i\right] + \frac{1}{2} \sum_{z_i \text{ on } C} \operatorname{Res}\left[f(z), z_i\right] \right]. \tag{21.108}$$

So poles on the contour contribute half of their residues to the integral's principal value; they are effectively half in and half out.

Incidentally, you might have expected that integrating along either one of the skirting contours $c_\pm$ would itself yield the principal value; after all, in the $\epsilon \to 0$ limit the contours pass right through the pole. But (21.104) defined the principal value as a symmetric limit, and choosing to skirt either above or below the pole breaks that symmetry. Instead, the principal value *averages* the contributions of these two choices, restoring the symmetry.

Example 21.20 $PV \int_{-\infty}^{\infty} \frac{e^{ikx}}{x-a} dx$

For $k > 0$, Jordan's lemma (21.58) requires we close the contour with a semicircle in the UHP. Having only a simple pole on the contour, (21.108) reduces to

$$PV \int_{-\infty}^{\infty} \frac{e^{ikx}}{x-a} dx = \pi i \operatorname{Res}\left[\frac{e^{ikx}}{x-a}, a\right] = \pi i e^{ika}, \qquad k > 0. \qquad (21.109)$$

If $k < 0$, the contour must by closed with a semicircle in the LHP, which (being clockwise) flips the sign of the contour integral. Thus for all real $k \neq 0$,

$$PV \int_{-\infty}^{\infty} \frac{e^{ikx}}{x-a} dx = \operatorname{sgn}(k)\, \pi i e^{ika}. \qquad (21.110)$$

Taking real and imaginary parts, we learn

$$PV \int_{-\infty}^{\infty} \frac{\cos kx}{x-a} dx = -\pi \sin(ka), \qquad (21.111a)$$

and

$$PV \int_{-\infty}^{\infty} \frac{\sin kx}{x-a} dx = \pi \cos(ka). \qquad (21.111b)$$

The special case $a = 0$ yields the real integrals

$$PV \int_{-\infty}^{\infty} \frac{\cos kx}{x} dx = 0 \qquad (21.112a)$$

(which makes sense, since the integrand is odd), and

$$PV \int_{-\infty}^{\infty} \frac{\sin kx}{x} dx = \pi. \qquad (21.112b)$$

This should be compared to the effort required in Example 8.16 and Problem 8.7.

Example 21.21 $PV \int_{-\infty}^{\infty} \frac{x\,dx}{x^3-a^3}$

The integrand

$$f(z) = \frac{z}{z^3 - a^3} = \frac{z}{(z-z_0)(z-z_+)(z-z_-)}, \qquad (21.113)$$

has three simple poles at $z_\pm = a\, e^{\pm i2\pi/3}$ and $z_0 = a$, one each in the UHP and LHP, the third on the integration contour. Redirecting the contour around the pole at $x = a$ and completing the contour with a semicircle in the UHP, (21.108) yields

$$PV \int_{-\infty}^{\infty} \frac{x\,dx}{x^3 - a^3} = 2\pi i \left[\operatorname{Res}\left[f(z), z_+\right] + \frac{1}{2} \operatorname{Res}\left[f(z), z_0\right] \right]$$

$$= 2\pi i \left[\frac{z}{(z-a)(z - a e^{-i2\pi/3})}\bigg|_{a\,e^{i2\pi/3}} + \frac{1}{2} \frac{z}{z^2 + az + a^2}\bigg|_{a} \right]$$

$$= 2\pi i \left[\frac{1}{a}\frac{e^{i2\pi/3}}{(e^{i2\pi/3} - 1)(e^{i2\pi/3} - e^{-i2\pi/3})} + \frac{1}{6a} \right]$$

$$= 2\pi i \left[\frac{1}{a} \frac{e^{i\pi/3}}{(2i \sin \pi/3)(2i \sin 2\pi/3)} + \frac{1}{6a} \right]$$

$$= \frac{2\pi i}{a} \left[\frac{\frac{1}{2}(1 + i\sqrt{3})}{-3} + \frac{1}{6} \right] = \frac{\pi}{a\sqrt{3}}. \tag{21.114}$$

Example 21.22 $PV \int_0^\infty \frac{\ln x \, dx}{x^2 + a^2}, \, a > 0$

In this example, the troublesome spot on the contour is the logarithm's branch point at $x = 0$. With the conventional cut along the negative real axis, we construct a closed contour with a large semicircle C_R in the UHP, and a small semicircle c_ϵ which diverts around the origin. Since $|zf(z)| \to 0$ both as $|z|$ goes to 0 and to ∞, the respective contributions of c_ϵ and C_R each vanish. Then in these limits

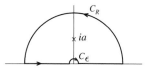

$$\oint_C \frac{\ln z}{z^2 + a^2} \, dz = \int_{-\infty}^0 \frac{\ln(e^{i\pi} x)}{x^2 + a^2} \, dx + \int_0^\infty \frac{\ln(e^{i0} x)}{x^2 + a^2} \, dx$$

$$= PV \int_{-\infty}^\infty \frac{\ln x}{x^2 + a^2} \, dx + i\pi \int_0^\infty \frac{dx}{x^2 + a^2}$$

$$= 2PV \int_0^\infty \frac{\ln x}{x^2 + a^2} \, dx + \frac{i\pi^2}{2a}, \tag{21.115}$$

where we've recognized the second integral as an inverse tangent. Setting the whole thing equal to $2\pi i$ times the residue from the simple pole at $z = ia$,

$$2\pi i \operatorname{Res} \left[\frac{\ln z}{z^2 + a^2}, ia \right] = 2\pi i \frac{\ln z}{z + ia} \Big|_{ia} = 2\pi i \frac{\ln(ia)}{2ia} = \frac{\pi}{a} \ln a + \frac{i\pi^2}{2a}, \tag{21.116}$$

yields

$$PV \int_0^\infty \frac{\ln x}{x^2 + a^2} \, dx = \frac{\pi}{2a} \ln a. \tag{21.117}$$

Rather than skirting around a pole on the contour, one could equivalently lower or lift the entire semicircular contour by an infinitesimal ϵ (Figure 21.3), sum the residues, and then send $\epsilon \to 0$. The simplest way to implement this is not to move the contour off the pole, but rather move the pole off the contour. So instead of the integral

$$\int_{-\infty}^\infty \frac{g(x)}{x} \, dx, \tag{21.118a}$$

we can lower or raise the pole by adding $\pm i\epsilon$ in the denominator and evaluate

$$\int_{-\infty}^\infty \frac{g(x)}{x \pm i\epsilon} \, dx. \tag{21.118b}$$

One advantage of this $i\epsilon$ prescription is that the contour integral can be evaluated using the usual $2\pi i$ contributions from residues — including the pole at $\pm i\epsilon$. This leads to a simpler version of (21.108): the integration identity (Problem 21.24)

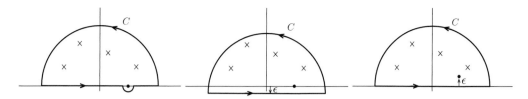

Figure 21.3 Equivalent ways to include a pole on the contour in the $\epsilon \to 0$ limit.

$$\frac{1}{x \pm i\epsilon} = PV \frac{1}{x} \mp i\pi \, \delta(x), \tag{21.119}$$

with the entire expression to be applied to $g(x)$ under an integral. For example,

$$\int_{-\infty}^{\infty} \frac{\cos x}{x} \, dx \tag{21.120}$$

has $g(x) = \cos x$, so the identity asserts that

$$\oint_C \frac{\cos z}{z \pm i\epsilon} \, dz = PV \int_{-\infty}^{\infty} \frac{\cos x}{x} \, dx \mp i\pi \int_{-\infty}^{\infty} \cos x \, \delta(x) \, dx = \mp i\pi, \tag{21.121}$$

where the principal value is zero since the integrand is odd. Direct evaluation of the left-hand side verifies this result (Problem 21.23). (This calculation should be compared with Example 21.20 in the $a \to 0$ limit.)

Mathematically there's no compelling basis for choosing between $\pm i\epsilon$; as long as the contour can be closed and the portion along the real axis extracted, either choice yields a value for the original ambiguous integral. As accustomed as we are to definite integrals having definite values, this ambiguity may seem baffling. Indeed, how could such an integral have any physical relevance? But recall that differential equations also yield ambiguous results in the absence of initial or boundary conditions; such physical criteria can also inform the decision to raise or lower a pole by $i\epsilon$. But we'll postpone this discussion to the Coda.

21.6 Series Sums with Residues

The residue theorem converts a contour integral into a straightforward discrete sum over residues. One might wonder about the reverse: taking a difficult infinite sum and converting it into a tractable contour integral.

To see how this can work, suppose we wish to evaluate the infinite sum

$$S = \sum_{n=-\infty}^{\infty} f(n), \tag{21.122}$$

where the function $f(z)$ is analytic for z real, and for which $|z^2 f(z)| \to 0$ as $|z| \to \infty$. We then introduce the auxiliary function

$$g(z) = \pi \, \cot \pi z = \pi \, \frac{\cos \pi z}{\sin \pi z}, \tag{21.123}$$

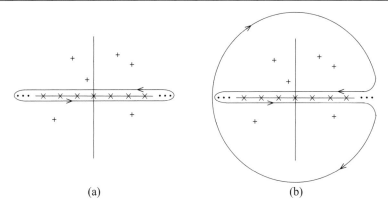

Figure 21.4 Integrating $f(z)g(z)$. The $+$'s and $\times$'s indicate poles of f and g, respectively.

which has an infinity of simple poles at real integer $z = n$. The integral of the product $f(z)g(z)$ around the contour of Figure 21.4a is determined by the poles of the product fg,

$$\frac{1}{2\pi i} \oint_C f(z)g(z)\,dz = \sum_{n=-\infty}^{\infty} \mathrm{Res}\left[f(z)\,\pi \cot \pi z, n \right]$$

$$= \sum_{n=-\infty}^{\infty} f(z)\,\frac{\pi \cos \pi z}{d\,(\sin \pi z)\,/dz}\bigg|_n = \sum_{n=-\infty}^{\infty} f(n). \tag{21.124}$$

Equivalently, since $f(z)$ falls at least as fast as $1/|z|^2$, integrating around the contour of Figure 21.4b gives a sum over the poles z_i of f,

$$\frac{1}{2\pi i} \oint_C f(z)g(z)\,dz = -\sum_i \mathrm{Res}\left[\pi \cot(\pi z_i) f(z), z_i \right], \tag{21.125}$$

where the sign comes from the contour's clockwise orientation. The equivalence of these residue sums provides a formula for the series (21.122),

$$\sum_{n=-\infty}^{\infty} f(n) = -\pi \sum_i \mathrm{Res}\left[\cot(\pi z_i) f(z), z_i \right]. \tag{21.126}$$

The same technique with the auxiliary function

$$g(z) = \frac{\pi}{\sin \pi z} = \pi \csc \pi z \tag{21.127}$$

leads to an analogous expression for alternating sums,

$$\sum_{n=-\infty}^{\infty} (-1)^n f(n) = -\pi \sum_i \mathrm{Res}\left[\csc(\pi z_i) f(z), z_i \right]. \tag{21.128}$$

This method is often referred to as a *Sommerfeld–Watson transformation*.

Example 21.23 $S = \sum_{n=0}^{\infty} \frac{1}{n^2+a^2}$

The function

$$f(z) = \frac{1}{z^2 + a^2} \tag{21.129}$$

has simple poles at $\pm ia$, so the residues needed are

$$\text{Res}\left[\frac{\pi \cot \pi z}{z^2 + a^2}, \pm ia\right] = \pi \left.\frac{\cot \pi z}{2z}\right|_{\pm ia} = \pi \frac{\cot(\pm i\pi a)}{\pm 2ia} = -\frac{\pi}{2a} \coth(\pi a). \tag{21.130}$$

Then (21.126) gives

$$\sum_{n=-\infty}^{\infty} \frac{1}{n^2 + a^2} = +\frac{\pi}{a} \coth(\pi a) \tag{21.131}$$

or

$$S \equiv \sum_{n=0}^{\infty} \frac{1}{n^2 + a^2} = \frac{1}{2a^2} + \frac{1}{2} \sum_{n=-\infty}^{\infty} \frac{1}{n^2 + a^2}$$

$$= \frac{1}{2a^2} \left[1 + \pi a \coth(\pi a)\right]. \tag{21.132}$$

Example 21.24 $S = \sum_{n=0}^{\infty} \frac{(-1)^n}{n^2+a^2}$, **real $a > 0$**

This time we need the residues

$$\pi \, \text{Res}\left[\frac{\csc \pi z}{z^2 + a^2}, \pm ia\right] = \pi \left.\frac{\csc \pi z}{2z}\right|_{\pm ia} = \pi \frac{\csc(\pm i\pi a)}{\pm 2ia} = -\frac{\pi}{2a} \frac{1}{\sinh(\pi a)}. \tag{21.133}$$

Then (21.128) gives

$$S = \sum_{n=0}^{\infty} \frac{(-1)^n}{n^2 + a^2} = \frac{1}{2a^2} + \frac{1}{2} \sum_{n=-\infty}^{\infty} \frac{(-1)^n}{n^2 + a^2}$$

$$= \frac{1}{2a^2} \left[1 + \frac{\pi a}{\sinh(\pi a)}\right]. \tag{21.134}$$

BTW 21.1 Dispersion Relations

An electromagnetic wavefunction with frequency ω propagating in vacuum along x has the form (neglecting polarization)

$$E(x,t) = E_0 e^{i(kx-\omega t)}, \qquad \omega/k = c. \tag{21.135}$$

When propagating in a medium such as air or glass, two physically distinct phenomena arise: refraction and attenuation. The first is determined by the index of refraction n, which modifies the frequency/wavenumber relationship to become $\omega/k = c/n$, $n > 1$. As the colors emerging from a prism reveal, n is a real function of ω, so this is a non-trivial modification. Attenuation is similarly characterized by a real function $\alpha = \alpha(\omega)$ parameterizing the exponential damping of the wave.

Including both effects in (21.135) gives

$$E(x,t) = E_0 e^{i\omega[n(\omega)x/c - t]} e^{-\alpha(\omega)x}. \tag{21.136}$$

The mathematical similarity of the two phenomena suggest that both can be described by replacing n with a single complex function $\mathcal{N}$,

$$\mathcal{N}(\omega) = n(\omega) + i\frac{c}{\omega}\alpha(\omega). \tag{21.137}$$

This has more than mere aesthetic appeal: if $\mathcal{N}$ is analytic, then measuring one of the phenomena allows calculation of the other.

To see how this works, consider a continuous function $f(x)$ for all x. If its analytic continuation in the complex plane falls fast enough in the UHP, we can use Cauchy's integral formula (18.26) with the contour shown to write

$$2\pi i f(x) = \oint_C \frac{f(\zeta)}{\zeta - x} \, d\zeta$$

$$= PV \int_{-\infty}^{\infty} \frac{f(\zeta)}{\zeta - x} \, d\zeta + i\pi f(x), \tag{21.138}$$

where the last term comes from the semicircle skirting under the pole at $\zeta = x$. (See Problem 21.24.) Thus we discover

$$\Re e[f(x)] = \frac{1}{\pi} PV \int_{-\infty}^{\infty} \frac{\Im m\left[f(\zeta)\right]}{\zeta - x} \, d\zeta$$

$$\Im m[f(x)] = -\frac{1}{\pi} PV \int_{-\infty}^{\infty} \frac{\Re e\left[f(\zeta)\right]}{\zeta - x} \, d\zeta. \tag{21.139}$$

Two real functions $g(x) = \Re e[f(x)]$ and $h(x) = \Im m[f(x)]$ related in this way are said to be *Hilbert transforms* of one another. In physics, (21.139) are usually referred to as the *Kramers–Kronig relations* — or simply as *dispersion relations*, since Kramers and Kronig did this work in the context of optical dispersion.

Returning now to the electromagnetic wave (21.136), we'd expect to identify $f(x)$ as the complex index of refraction $\mathcal{N}(\omega)$ in (21.137). But physics tell us that a medium becomes increasingly transparent as ω grows, so that $n(\omega)$ goes to 1 in the $\omega \to \infty$ limit. This precludes closing the contour as we did to derive (21.139). However, the dispersion relation for the real part of $\mathcal{N}$ can be constructed from $n(\omega) - 1$, which can be shown to fall as $1/\omega^2$. Moreover, since both n and α must be independent of the sign ω, both must be even functions of frequency. From these considerations, (21.139) become (Problem 21.34):

$$n(\omega) = 1 + \frac{2c}{\pi} PV \int_0^{\infty} \frac{\alpha(\omega')}{\omega'^2 - \omega^2} \, d\omega'$$

$$\alpha(\omega) = -\frac{2\omega^2}{\pi c} PV \int_0^{\infty} \frac{n(\omega') - 1}{\omega'^2 - \omega^2} \, d\omega'. \tag{21.140}$$

So refraction and absorption are in fact analytically related physical phenomena.

Problems

21.1 Find all the residues of the following functions:

(a) $\frac{1}{(z-1)(z+2)}$

(b) $\frac{e^{2iz}}{(z-1)(z+2)^3}$

(c) $\frac{e^z}{\sin z}$

(d) $\exp[1/z]$

(e) $\frac{z^2 e^z}{e^{2z}+1}$.

21.2 The contour used to evaluate the integral in Example 21.1 can be construed as surrounding two finite poles, or one finite and another at infinity with opposite orientation. Show that this view leads to the same result.

21.3 By expanding the integration contour in Example 21.2 out to the circle at infinity, compute the sum $\sum_{n=0}^{\infty}(-1)^{n+1}/n^2$.

21.4 Verify the residue in (21.102) by considering the $\zeta \equiv 1/z \to 0$ limit.

21.5 Table 21.1 gives a simple formula for finding the residue of a function at infinity if $f(z) \to 0$ as $|z| \to \infty$. If instead $f(z)$ goes to a non-zero constant, show that the residue at infinity is $\lim_{|z|\to\infty} z^2 f'(z)$.

21.6 Use residues to obtain the partial fraction decomposition of $f(z) = \frac{4-2z}{(z^2+1)(z-1)^2}$ (see Example 8.21).

21.7 Use a partial fraction decomposition to derive $\pi \cot(\pi z) = \frac{1}{z} + 2z \sum_{n=1}^{\infty} \frac{1}{z^2-n^2}$. [If your decomposition is not yielding the correct result, check to see whether your infinite series is convergent. By inserting a $1/n$ term, you should be able to recast it as a convergent series.]

21.8 Evaluate the integral in Example 21.10 with a contour in the UHP.

21.9 Complete the calculation in Example 21.8 to obtain (21.45).

21.10 Find $\oint_C e^{1/z^2} dz$ for arbitrary finite C.

21.11 Integrate $f(z) = \frac{z}{(z^2-1)(z^2+1)^2}$ over:

(a) the circle $|z| < 1/2$
(b) the circle $|z| < 2$
(c) the circle $|z - i| < 1$
(d) the ellipse centered at the origin, semi-major axis 2, semi-minor axis 1/2.

21.12 Evaluate the following integrals using calculus of residues:

(a) $\int_{-\infty}^{\infty} \frac{x^2}{x^4+1} dx$

(b) $\int_{-\infty}^{\infty} \frac{1}{(x^2+1)^2(x^2+4)} dx$

(c) $\int_{-\infty}^{\infty} \frac{x^2}{x^4+x^2+1} dx$

(d) $\int_{0}^{\infty} \frac{\cos \pi x}{x^2+1} dx$.

21.13 Use the integral (19.10) to verify the Laurent series coefficients in (19.16).

21.14 Repeat Problem 21.13 by regarding the contours as "enclosing" poles on the outside with opposite orientation.

21.15 Show that the rectangular contour of Example 21.14 can enclose any finite number of poles and still yield (21.76).

21.16 Evaluate the integral of Example 21.14 by introducing a small imaginary term, $b \to b + i\epsilon$ for $\epsilon > 0$, and closing the contour with an infinite semicircle in the UHP. Show that (21.76) emerges in the $\epsilon \to 0$ limit.

21.17 Evaluate $\int_0^\infty \frac{x\,dx}{x^4+1}$ by finding an appropriate contour in the complex plane.

21.18 Manipulate the integral of the Gaussian form e^{-z^2} around a rectangular contour of height a to find $\int_0^\infty \cos(2ax)e^{-x^2}\,dx$.

21.19 Use the complementary contour to the one chosen in Example 21.15 to evaluate the integral.

21.20 Choose an appropriate contour in the complex plane to evaluate the following integrals, using principal values as necessary:

(a) $\int_{-\infty}^\infty \frac{x}{x^3-8}\,dx$

(b) $\int_{-\infty}^\infty \frac{1}{x(x-1)(x-2)}\,dx$

(c) $\int_0^\infty \frac{\sin^2 x}{x^2}\,dx$

(d) $\int_{-\infty}^\infty \frac{x\cos x}{x^2+3x+2}\,dx$.

21.21 Evaluate the integral $\int_0^1 \frac{1}{(x^2+a^2)\sqrt{1-x^2}}\,dx$ for positive a.

21.22 In Example 21.17, we chose to cut along the positive real axis in order to aid the evaluation of the integral. But in principle, any cut should do. Show that the same result obtains with a cut along the negative real axis.

21.23 Evaluate both $\oint_C \frac{\cos z}{z \pm i\epsilon}\,dz$ and $\oint_C \frac{\sin z}{z \pm i\epsilon}\,dz$ by closing contours with semicircles at infinity.

21.24 For $g(z)$ analytic, show that

$$\lim_{\epsilon \to 0} \int_{-\infty}^\infty \frac{g(x)\,dx}{x - (x_0 \pm i\epsilon)} = PV \int_{-\infty}^\infty \frac{g(x)\,dx}{x - x_0} \pm i\pi g(x_0),$$

where g goes to zero quickly enough to close the contour with a large semicircle in the UHP or LHP in the usual fashion. [Hint: Consider separately the real and imaginary parts of $\frac{1}{x-(x_0\pm i\epsilon)}$.]

21.25 Use the technique of Section 21.6 to find the numerical values of:

(a) $\sum_{n=-\infty}^\infty \frac{1}{1+n^2}$ (b) $\sum_{n=1}^\infty \frac{(-1)^n}{n^2}$

(c) $\sum_{n=1}^\infty (-1)^{n+1}\frac{n\sin(nx)}{n^2+a^2}$, for $0 \le x < \pi$, and with real $a \ne 0$.

21.26 Adapting the technique of Section 21.6 to take account of divergences at $n = 0$, find the numerical values of $\zeta(2) = \sum_{n=1}^\infty \frac{1}{n^2}$ and $\zeta(4) = \sum_{n=1}^\infty \frac{1}{n^4}$. Verify your result for $n = 2$ by taking the $a \to 0$ limit of (21.132).

21.27 Starting with the expression between beta and gamma functions in Problem 8.20, use a keyhole contour to verify the Gamma reflection formula

$$\Gamma(a)\Gamma(1-a) = \frac{\pi}{\sin \pi a}, \qquad 0 < a < 1.$$

This identity can be analytically continued to the entire complex plane for non-integer a. [Note that the formula quickly and easily gives $\Gamma(\frac{1}{2}) = \sqrt{\pi}$.]

21.28 Show that

$$\frac{i}{2\pi} \int_{-\infty}^\infty \frac{e^{-i\omega t}}{\omega + i\epsilon}\,d\omega$$

is an integral representation of the step function $\Theta(t)$.

21.29 Suppose the only things known about a function $f(z)$ is that it has a simple pole at $z = a$ with residue α, a branch cut discontinuity across the real axis given by

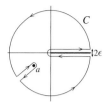

$$f(x + i\epsilon) - f(x - i\epsilon) = 2\pi i g(x), \qquad x \geq 0$$

and that f goes to zero fast enough for large $|z|$ to close a contour with a circle at infinity. With the keyhole contour shown, use this information and Cauchy's integral formula (18.26) to derive

$$f(z) = \frac{\alpha}{z - a} + \int_0^\infty \frac{g(\zeta)}{\zeta - z} \, d\zeta.$$

21.30 From the results of Problem 21.29, construct the function $f(z)$ which has a simple pole at $z = a = -1$ with residue $\alpha = 1/2$, and a branch cut discontinuity across the real axis of $2\pi i/(1 + x^2)$. Use the method of Example 21.18 to evaluate the integral.

21.31 The Argument Principle I

Consider a function $w(z)$ which is analytic everywhere on a contour C, and whose non-analytic points within C are all isolated poles. Take C to be a simple curve — that is, one which does not cross over itself.

(a) For $z = a$ an nth-order zero of $w(z)$, use (19.17) to show that $w'(z)/w(z)$ has a simple pole at a with residue n.

(b) For $z = a$ a pth-order pole of $w(z)$, use (19.19) to show that $w'(z)/w(z)$ has a simple pole at a with residue $-p$.

(c) Apply the calculus of residues to obtain the *argument principle*,

$$\frac{1}{2\pi i} \oint_C \frac{w'(z)}{w(z)} \, dz = N - P, \tag{21.141}$$

where $N = \sum_i n_i$ and $P = \sum_i p_i$ are the respective sums of the orders of all the zeros and poles of $f(z)$ within C. [In the literature, N and P are often referred to as the "number" of zeros and poles of w — which can be misleading, since their order is included.]

21.32 For $w(z) = \frac{5z^3(z+1)(z-4i)e^{-3z}}{(1-z)^3(z+3)^2(2z-10)^2}$, use the (21.141) to evaluate $\oint_C w'(z)/w(z) \, dz$ around the circles:

(a) $|z| = 1/2$ (b) $|z| = 2$ (c) $|z| = 9/2$ (d) $|z| = 6$.

21.33 The Argument Principle II

Since the argument principle stipulates that $w(z)$ has no zeros along C, the image contour C' never passes through $w = 0$. Thus $\frac{w'(z)}{w(z)} = \frac{d}{dz} \ln w(z)$ is analytic (within a branch).

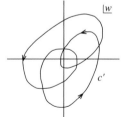

(a) Show that the phase change $\Delta_C \arg w(z)$ in one full sweep around C is

$$i\Delta_C \arg w(z) = \oint_C \frac{w'(z)}{w(z)} \, dz. \tag{21.142}$$

Relate this to the *winding number*, the net number of times $w(z)$ winds around the origin $w = 0$.

(b) A contour C in the z-plane encloses a domain D. A function w which is analytic both on C and in D maps C to the contour C' shown. If $w(z)$ is non-vanishing on C, how many zeros does $w(z)$ have within D?

21.34 Derive the dispersion relations (21.140) from (21.139).

21.35   The Stationary Phase Approximation

Consider an integral of the form

$$I(\lambda) = \int_a^b f(x)\, e^{i\lambda\varphi(x)}\, dx$$

for real $\varphi(x)$. For large $\lambda \gg 1$, even small changes in $\varphi(x)$ lead to rapid oscillation of the exponential which, if $f(x)$ varies relatively slowly, average out to zero. So the only substantial contributions to the integral occur where the phase changes most slowly — that is, in the vicinity of an extremum of $\varphi(x)$. The *stationary phase approximation* leverages this insight, replacing $\varphi(x)$ in the integrand with its Taylor expansion around an extremum x_0 which lies between a and b,

$$\varphi(x) \approx \varphi(x_0) + \frac{1}{2!}(x - x_0)^2 \varphi''(x_0).$$

Using this in the integrand approximates I as a complex Gaussian. (This is similar to the method of steepest descent explored in Problem 8.32.)

(a) Show that this procedure leads to

$$I(\lambda) \approx f(x_0)\, e^{i\lambda\varphi(x_0)} \sqrt{\frac{2\pi}{\lambda|\varphi''(x_0)|}}\, e^{is\pi/4}, \qquad (21.143)$$

where we've used (21.87), and $s = \pm 1$ is the sign of $\varphi''(x_0)$.

(b) Find an approximate large-t expression for $\int_{-\infty}^{\infty} \frac{1}{1+x^2}\, e^{itx(1-x)}dx$.

(c) Show that for large x, the Bessel function

$$J_n(x) \equiv \frac{1}{\pi} \int_0^{\pi} \cos(x \sin t - nt)\, dt$$

has the asymptotic form $J_n(x) \longrightarrow \sqrt{\frac{2}{\pi x}} \cos\left(x - \frac{n\pi}{2} - \frac{\pi}{4}\right)$.

22 Coda: Analyticity and Causality

22.1 Acting on Impulse

Consider a harmonic oscillator with natural frequency ω_0 and damping force proportional to velocity, $F_d = -bv = -2m\beta v$, where $\beta = b/2m$ has units of frequency. Then if the displacement from equilibrium is $u(t)$, Newton's second law is

$$\ddot{u} + 2\beta\dot{u} + \omega_0^2 u = 0. \tag{22.1}$$

This differential equation describes not only a mass m on a spring with $k = m\omega_0^2$, but many other oscillating systems as well. For instance, (22.1) emerges directly from Kirchhoff's voltage law for LRC circuits, with u representing charge Q, $2\beta = R/L$, and $\omega_0^2 = 1/LC$.

As discussed in Example 3.4, (22.1) has solution

$$\begin{aligned} u(t) &= e^{-\beta t}\left(Ae^{i\Omega t} + Be^{-i\Omega t}\right) \\ &= e^{-\beta t}\left(a\cos\Omega t + b\sin\Omega t\right), \end{aligned} \tag{22.2a}$$

where $\Omega^2 = \omega_0^2 - \beta^2$. This expression encompasses two types of behavior — underdamped $\beta < \omega_0$ (Ω real) and overdamped $\beta > \omega_0$ (Ω imaginary). A third, called critically damped, has $\beta = \omega_0$ ($\Omega = 0$) with solution

$$u(t) = e^{-\beta t}\left(At + B\right). \tag{22.2b}$$

These three cases summarize the decaying behavior of an unforced oscillator. But if we add a cosine driving force to (22.1) as in Example 3.5,

$$f(t) = f_0 \cos\omega t = f_0 \,\Re e\!\left[e^{i\omega t}\right], \tag{22.3}$$

then after the transients in (22.2) decay away, what's left is the complex steady-state solution found in (3.34),

$$\mathcal{U}_\omega(t) = \frac{f_0\, e^{i\omega t}}{\left(\omega_0^2 - \omega^2\right) + 2i\beta\omega}, \tag{22.4}$$

where the denominator emerges from inserting $C(\omega)e^{i\omega t}$ into the differential equation. The real solution is $u = \Re e\,\mathcal{U}_\omega$ — or equivalently, the average of $\mathcal{U}_\omega(t)$ and $\mathcal{U}_\omega^*(t) = \mathcal{U}_{-\omega}(t)$. This is a useful observation, because if the applied force $f(t)$ is a sum over a number of discrete source terms $f_\omega(t) \sim e^{i\omega t}$, each with its own value of amplitude and frequency, then each has its own solution pair $\mathcal{U}_{\pm\omega}(t)$ of the form (22.4). The full solution $u(t)$ is then a sum over these $\mathcal{U}_\omega$.

And not just for discrete sources: in the extreme limit of a continuous infinity of driving forces over a continuous range of frequencies ω, the sum of the $\mathcal{U}_\omega$'s can be expressed as the integral

$$u(t) = \int_{-\infty}^{\infty} \frac{\tilde{f}(\omega)\, e^{i\omega t}}{(\omega_0^2 - \omega^2) + 2i\beta\omega}\, d\omega, \tag{22.5}$$

where real $u(t)$ requires $\tilde{f}(\omega)$ to be an even function of frequency (as was $\mathcal{U}_\omega + \mathcal{U}_{-\omega}$). Inserting $u(t)$ into the differential equation verifies that it is a solution — provided that the sum $f(t)$ of all the applied driving forces can be expressed as the integral

$$f(t) = \int_{-\infty}^{\infty} \tilde{f}(\omega)\, e^{i\omega t}\, d\omega. \tag{22.6}$$

In Chapter 36 we'll find that $\tilde{f}(\omega)$ is the Fourier transform of $f(t)$.

Different choices for $f(t)$ in (22.6), which need not be periodic, correspond to different oscillator solutions $u(t)$. Let's investigate the case in which the forcing function is an impulse — a single hammer blow at time t_0. We know from the discussion in Example 9.1 that this can be represented mathematically using the form of the delta function in (9.6),

$$f(t) = f_0\, \delta[\omega_0(t - t_0)] = \frac{\tilde{f}_0}{2\pi} \int_{-\infty}^{\infty} e^{i\omega(t - t_0)}\, d\omega, \tag{22.7}$$

where $\tilde{f}_0 \equiv f_0/\omega_0$. Comparison with (22.6) gives $\tilde{f}(\omega) = \tilde{f}_0\, e^{-i\omega t_0}/2\pi$, showing that each frequency contributes equally to the impulse. Inserting this into (22.5) gives the response to a hammer blow at time t_0,

$$G(t) \equiv u_{\text{impulse}}(t) = \frac{\tilde{f}_0}{2\pi} \int_{-\infty}^{\infty} \frac{e^{i\omega(t - t_0)}\, d\omega}{(\omega_0^2 - \omega^2) + 2i\beta\omega}, \tag{22.8}$$

where we use $G(t)$ to denote the special case of an impulse solution. The integrand has two simple poles in the UHP (Figure 22.1) at

$$\omega_\pm = i\beta \pm \sqrt{\omega_0^2 - \beta^2} = i\beta \pm \Omega. \tag{22.9}$$

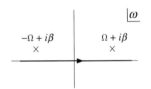

Figure 22.1 Pole structure of the damped oscillator.

For $t > t_0$, we invoke Jordan's lemma (21.58) to close the contour with a semicircle in the UHP enclosing these poles; this gives

$$G(t) = \tilde{f}_0\, \frac{e^{-\beta(t - t_0)}}{2i\Omega}\left[e^{i\Omega(t - t_0)} - e^{-i\Omega(t - t_0)}\right]$$

$$= \frac{f_0}{\Omega} \sin[\Omega(t - t_0)]\, e^{-\beta(t - t_0)}, \quad t > t_0. \tag{22.10}$$

Although (22.10) tells us the oscillator's response to the impulse for $t > t_0$, it doesn't say anything about the oscillator at earlier times. But that behavior too can be extracted from (22.8): for $t < t_0$ the integral must be evaluated by closing the contour with a semicircle in the LHP. And since there are no poles there, Cauchy's theorem gives $G(t < t_0) = 0$. We can combine this with (22.10) into a single expression for all t,

$$G(t) = \frac{f_0}{\Omega} \sin[\Omega(t - t_0)]\, e^{-\beta(t - t_0)}\, \Theta(t - t_0). \tag{22.11}$$

where

$$\Theta(t - t_0) = \begin{cases} 0, & t < t_0 \\ 1, & t > t_0 \end{cases} \tag{22.12}$$

is the Heaviside step function. The function $G(t)$ describes an initially quiescent oscillator responding to a impulse at $t = t_0$. Once the impulse occurs, the system exhibits one of the transient damped oscillations in (22.2).

There are several insights — both physical and mathematical — which illuminate the role played by the poles:

- The imaginary and real parts of the poles $\omega_\pm$ give the damping β and the parameter Ω, respectively. Moreover, the three types of oscillation which occur after (and as a result of) the impulse can be distinguished by the behavior of the poles as functions of β. For underdamped motion $\beta < \omega_0$, the magnitude $|\omega_\pm| = \omega_0$ is constant: varying the damping simply moves the poles around a circle in the complex plane, without any essential change to the behavior of the oscillator. The behavior *does* change for $\beta = \omega_0$, the critical value at which the poles converge at a single point $i\beta$ on the imaginary axis; this is critical damping. For $\beta > \omega_0$, the poles separate along the imaginary axis, corresponding to non-oscillatory overdamped motion.
- The pole structure of the integral in (22.8) also encodes the resonance characteristics of the oscillator. As we found in (3.36), at the resonant frequency

$$\omega_{res}^2 = \omega_0^2 - 2\beta^2, \tag{22.13}$$

even a driving force with small amplitude f_0 creates a large-amplitude response in the oscillator. Clearly resonance only occurs for small damping $\beta < \omega_0/\sqrt{2}$ — but for $\beta \ll \omega_0$, the real part of the pole $|\Re e(\omega_\pm)| \approx \omega_0 \approx \omega_{res}$, accurate up to second-order in β/ω_0. Since $\Im m(\omega_\pm) = \beta$, the imaginary part is obviously more sensitive to the damping, but it too can be directly related to resonance — at least for weak damping. To see how, first note that for $\omega \approx \omega_{res}$, small β implies small $\omega^2 - \omega_0^2$, so that

$$\omega_0^2 - \omega^2 = (\omega_0 + \omega)(\omega_0 - \omega) \approx 2\omega_0(\omega_0 - \omega), \qquad \beta \ll \omega_0. \tag{22.14}$$

Then the ratio of the power absorbed by the oscillator at some $\omega \approx \omega_0$ relative to that at resonance is, using (22.4),

$$\left| \frac{\mathcal{U}(\omega)}{\mathcal{U}(\omega_0)} \right|^2 = \frac{4\beta^2 \omega_0^2}{\left(\omega_0^2 - \omega^2\right)^2 + 4\beta^2 \omega_0^2}$$

$$\approx \frac{4\beta^2 \omega_0^2}{4\omega_0^2 (\omega_0 - \omega)^2 + 4\beta^2 \omega_0^2} = \frac{\beta^2}{(\omega_0 - \omega)^2 + \beta^2}. \tag{22.15}$$

The *resonance width* Γ is defined as the range of frequencies centered at ω_0 for which the power is at least half that at resonance — the so-called "full width at half maximum" (Figure 22.2),

$$\left| \frac{\mathcal{U}(\omega_0 + \Gamma/2)}{\mathcal{U}(\omega_0)} \right|^2 = \frac{\beta^2}{(\Gamma/2)^2 + \beta^2} \equiv 1/2. \tag{22.16}$$

Thus $\Gamma = 2\beta$. So the smaller the damping, the narrower the width and the larger the response — and hence the higher

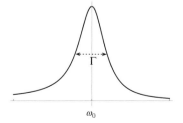

Figure 22.2 Resonance curve.

"quality" of the resonance. In the mathematical limit of zero damping, an infinite resonance response would occur precisely at $\omega = \omega_0$ with zero width.

- For $\beta \ll \omega_0$, the damping in (22.8) is mathematically equivalent to the $-i\epsilon$ prescription moving the resonance pole at $\omega = \omega_0$ off the real axis and into the UHP. This is what ultimately leads to the *causal* structure of (22.11), in which the oscillator lies quiescent until the impulse occurs at $t = t_0$ — "causing" the subsequent oscillation. The impulse response (22.11) is called the *retarded* solution G_R, because oscillation is delayed until after the impulse occurs. The cause precedes the effect. Thus the $-i\epsilon$ prescription is tantamount to fixing the oscillator's initial conditions.

- Physically, there's no choice with regard to moving poles: since the damping $\beta > 0$, the pole is necessarily in the UHP and G_R results. But mathematically, we could choose to move the pole below the real axis using negative β — or equivalently, adding $+i\epsilon$. This "anti-damping" represents a velocity-dependent infusion (rather than removal) of energy into the oscillator. In this case we'd need to close the contour in the LHP, resulting in the so-called *advanced* solution

$$G_A(t) = -\frac{f_0}{\Omega} \sin\left[\Omega(t - t_0)\right] e^{+\beta(t-t_0)} \Theta(t_0 - t), \qquad (22.17)$$

describing oscillation for $t < t_0$ until the impulse precisely brings it to rest at t_0. So the choice of $+i\epsilon$ corresponds to different initial conditions. Strictly speaking, G_A is also a causal solution — though since we're usually only interested in the signal created by an impulse, G_A is often characterized as "acausal" or "anti-causal."

- The principal value of the integral (22.8), with $\beta = 0$, is given by (21.119) as the average of the retarded and advanced solutions,

$$PV\left[G(t)\right]_{\beta=0} = \frac{1}{2}\left[G_R(t) + G_A(t)\right]. \qquad (22.18)$$

22.2 Waves on a String

Whereas an undamped oscillator is harmonic only in time, a wave is harmonic in both space and time: rather than a solution to $d^2u/dt^2 = -\omega^2 u$, a wave satisfies the wave equation $\partial_t^2 u = c^2 \nabla^2 u$, where c is the wave speed. For the one-dimensional case of a wave on a string, the solution has complex form

$$\mathcal{U}(x,t) = Ce^{i(kx-\omega t)}, \qquad (22.19)$$

with frequencies k and ω colluding to determine the wave speed, $\omega = ck$. As we did for the oscillator, we now add a source term to the wave equation

$$\frac{\partial^2 u}{\partial x^2} - \frac{1}{c^2}\frac{\partial^2 u}{\partial t^2} = f(x,t), \qquad (22.20)$$

and construct a forcing function $f(x,t)$ from a continuum of both k's and ω's. We'll again consider a unit impulse — now in both space and time,

$$f(x,t) = f_0\,\delta(x - x_0)\,\delta(t - t_0) = f_0 \int_{-\infty}^{\infty} \frac{dk}{2\pi}\, e^{ik(x-x_0)} \int_{-\infty}^{\infty} \frac{d\omega}{2\pi}\, e^{-i\omega(t-t_0)}. \qquad (22.21)$$

To reduce clutter, we'll set $f_0 = 1$, and take $x_0 = t_0 = 0$ so that the impulse occurs at the origins of both position and time. Then the steps corresponding to the derivation of (22.8) here lead to

$$G(x,t) = \int_{-\infty}^{\infty} \frac{dk}{2\pi}\, e^{ikx} \int_{-\infty}^{\infty} \frac{d\omega}{2\pi}\, \frac{e^{-i\omega t}}{\omega^2/c^2 - k^2}. \qquad (22.22)$$

as can be verified by inserting G into (22.20). As in (22.4), the denominator can be found by inserting the form (22.19) into the differential equation.

To finish the calculation requires evaluating these integrals. We'll start with the ω integral

$$I = \frac{c^2}{2\pi} \int_{-\infty}^{\infty} \frac{e^{-i\omega t}}{\omega^2 - c^2 k^2} \, d\omega. \tag{22.23}$$

Unlike the oscillator, here the simple poles lie on the real axis at $\pm ck$. So this time we have a choice for how to implement the $i\epsilon$ prescription. If there is no wave before the impulse strikes, our $i\epsilon$ choice must ensure the integral vanishes for $t < 0$. Since negative t requires the contour be closed in the UHP, we don't want the poles there — so we move them into the LHP as shown in Figure 22.3, and evaluate

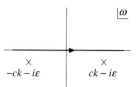

Figure 22.3 Pole structure of an undamped wave.

$$I = \frac{c^2}{2\pi} \int_{-\infty}^{\infty} \frac{e^{-i\omega t}}{(\omega - ck + i\epsilon)(\omega + ck + i\epsilon)} \, d\omega. \tag{22.24}$$

With the poles now at $\omega_{\pm} = \pm ck - i\epsilon$, closing the contour in the UHP gives $I = 0$. The effect of the impulse is thus to be seen only for $t > 0$; closing the contour in the LHP gives (Problem 22.5)

$$I(t > 0) = -\frac{c}{k} \sin(ckt) \, e^{-\epsilon t} \xrightarrow[\epsilon \to 0]{} -\frac{c}{k} \sin(ckt). \tag{22.25}$$

Comparison with the oscillator solution (22.10) shows that deforming the contour to pass above the poles like this is tantamount to adding an infinitesimal damping $\epsilon \equiv \beta$ to the system.

But we're not finished. Inserting I into (22.22), we find for $t > 0$

$$G(x, t > 0) = -\frac{c}{2\pi} \int_{-\infty}^{\infty} \frac{e^{ictk} - e^{-ictk}}{2ik} e^{ikx} \, dk$$

$$= -\frac{c}{4} \left[\text{sgn}(ct + x) + \text{sgn}(ct - x) \right] = -\frac{c}{2} \Theta(ct - |x|), \tag{22.26}$$

where we've used (21.110) with $a \to 0$ (see also Problem 8.7),

$$\int_{-\infty}^{\infty} \frac{e^{\pm i\gamma\xi}}{\xi} \, d\xi = \pm i\pi \, \text{sgn}(\gamma). \tag{22.27}$$

The step function reveals that no effect can occur at x until enough time has elapsed for a signal to get there. We can express G for all t using a second step function,

$$G_R(x, t) = -\frac{c}{2} \Theta(ct - |x|) \, \Theta(t). \tag{22.28}$$

So the impulse at $t = 0$ takes a quiescent string and creates a rectangular pulse propagating along the string in both directions from the site of the hammer blow at $x_0 = 0$. Since sufficient time must pass after the impulse for the effect to travel a distance $|x|$, this is a retarded solution: the effect of the hammer blow is delayed (i.e., retarded) as the effect of the impulse propagates along the string.

The retarded solution emerged from the deliberate choice to move the poles into the LHP. But moving the poles off the real axis is a mathematical technique for extracting a convergent result from an ambiguous integral; where we move the poles is up to us. If instead we push the poles into the UHP, the perhaps-unsurprising result is the advanced solution

$$G_A(x, t) = -\frac{c}{2} \Theta(-ct - |x|) \, \Theta(-t), \tag{22.29}$$

in which a hammer blow at $x = 0, t = 0$ precisely brings an incoming wave to rest.

22.3 The Klein–Gordon Propagator

The wavefunction $\psi(x,t)$ of a free particle of mass m satisfies the one-dimensional Schrödinger equation

$$i\hbar\frac{\partial\psi}{\partial t} + \frac{\hbar^2}{2m}\frac{\partial^2\psi}{\partial x^2} = 0. \tag{22.30}$$

Unlike classical physics, quantum mechanics is inherently complex — so there's no requirement that ψ be real. And unlike classical fields, a wavefunction describes the location of a discrete particle as it moves about; technically, $|\psi(x,t)|^2 dx$ is the probability for the particle to be found within dx of x at time t. Since energy and momentum are associated with operators

$$E \to +i\hbar\,\partial/\partial t \qquad p \to -i\hbar\,\partial/\partial x, \tag{22.31}$$

the Schrödinger equation is the quantum mechanical version of the energy–momentum relation $E = p^2/2m$.

The response to an impulse of momentum and energy at the origin at $t = 0$ is the solution to

$$i\hbar\frac{\partial G}{\partial t} + \frac{\hbar^2}{2m}\frac{\partial^2 G}{\partial x^2} = i\hbar\,\delta(x)\delta(t), \tag{22.32}$$

including a conventional i in front of the delta functions, and an $\hbar$ to ensure consistent units. The response to the impulse is the probability of the arrival ("propagation") of a particle of mass m at some position at a particular time. For this reason, the impulse solution is often referred to as the particle's *propagator*. Different ways of moving the poles off the real axis result in different types of propagators. An analysis similar to that in the previous two examples shows that there is only one pole, leading to the retarded Schrödinger propagator (Problem 22.6)

$$G_R(x,t) = \Theta(t)\sqrt{\frac{m}{2\pi i\hbar t}}\,e^{imx^2/2\hbar t}. \tag{22.33}$$

Unlike (22.28), this solution is non-zero for all $t > 0$ at all x. In other words, there is a non-zero probability to find the particle at any position, *no matter how little time has elapsed since the impulse*. But causality requires a finite propagation time between cause and effect, transmitted at a speed no greater than that of light in vacuum. So this solution is physically problematic.

But the Schrödinger equation is based upon $E = p^2/2m$, and so is intended for non-relativistic use. Indeed, with one derivative in t and two in x, it does not treat time and space on the same footing, as would be expected of a relativistic expression. One might anticipate a better result with the relativistic energy–momentum relation

$$E^2 - p^2 c^2 = m^2 c^4. \tag{22.34}$$

Making the same operator identifications as in (22.31) yields an equation with two derivatives in both space and time,

$$-\frac{1}{c^2}\frac{\partial^2\phi}{\partial t^2} + \frac{\partial^2\phi}{\partial x^2} = \frac{m^2 c^2}{\hbar^2}\phi. \tag{22.35}$$

This is the *Klein–Gordon* equation, which describes spin-zero relativistic particles such as pions and alpha particles (helium-4 nuclei). The wavefunction

$$\phi(x,t) = Ce^{i(kx-\omega t)}, \tag{22.36}$$

is a solution of the Klein–Gordon equation as long as

$$\omega^2/c^2 - k^2 = m^2 c^2/\hbar^2. \tag{22.37}$$

Comparison with (22.34) yields the famous de Broglie relations $E = \hbar\omega$ and $p = \hbar k$.

To see how causality fares for a Klein–Gordon wavefunction, we'll solve for its response to a unit impulse acting at the origin at $t = 0$,

$$\left[-\frac{\partial^2}{\partial t^2} + \frac{\partial^2}{\partial x^2} - m^2 \right] G(x,t) = i\,\delta(x)\,\delta(t), \tag{22.38}$$

where for simplicity we've set $\hbar = c = 1$.[1] Considerations similar to those in the previous examples lead to the Klein–Gordon propagator

$$G(x,t) = i \int_{-\infty}^{\infty} \frac{dp}{2\pi} e^{ipx} \int_{-\infty}^{\infty} \frac{dE}{2\pi} \frac{e^{-iEt}}{E^2 - p^2 - m^2}, \tag{22.39}$$

as can be verified by direct substitution into (22.38). As a relativistic system, particles can materialize and disappear — "creation" and "annihilation" in the common jargon. So whereas (22.26) describes the creation and propagation of a wave on a string, the retarded solution here will describe the creation and propagation of a particle of mass m after the impulse.

Unlike the Schrödinger solution with its single pole, the Klein–Gordon propagator has two — one with positive energy and one with negative energy. There are thus four distinct contours to choose from. Since $t < 0$ requires that the contour be closed in the UHP, the retarded solution emerges by choosing a contour which passes just above the poles — or equivalently, by moving the poles so that they sit below the real axis at $E_\pm = \pm\sqrt{p^2 + m^2} - i\epsilon$,

$$\int_{-\infty}^{\infty} \frac{dE}{2\pi} \frac{e^{-iEt}}{(E + i\epsilon)^2 - p^2 - m^2}. \tag{22.40}$$

With no poles in the UHP, the $t < 0$ integral vanishes. For $t > 0$, the contour must be closed in the LHP, so that in the $\epsilon \to 0$ limit

$$\int_{-\infty}^{\infty} \frac{dE}{2\pi} \frac{e^{-iEt}}{(E + i\epsilon)^2 - p^2 - m^2} = -\frac{1}{2\pi} 2\pi i \left(\frac{e^{-i\sqrt{p^2+m^2}\,t}}{2\sqrt{p^2+m^2}} + \frac{e^{i\sqrt{p^2+m^2}\,t}}{-2\sqrt{p^2+m^2}} \right)$$

$$= \frac{-i}{2E_p} \left(e^{-iE_p t} - e^{iE_p t} \right), \tag{22.41}$$

where $E_p \equiv \sqrt{p^2 + m^2} > 0$ is an even function of p, and the overall sign is due to the clockwise traversal of the contour. Plugging this back into (22.39) gives

$$G_R(x,t) = \Theta(t) \int_{-\infty}^{\infty} \frac{dp}{2\pi} \frac{e^{ipx}}{2E_p} \left(e^{-iE_p t} - e^{iE_p t} \right)$$

$$= \frac{\Theta(t)}{4\pi} \int_{-\infty}^{\infty} \frac{1}{E_p} \left[e^{-i(E_p t - px)} - e^{i(E_p t - px)} \right] dp, \tag{22.42}$$

where we let $p \to -p$ in the second term. Despite the rather agreeable appearance of the remaining integral, insights from both relativity and quantum mechanics are required to properly account for the

---

[1] Aside from being positive, the numerical values of these constants have no mathematical import. So the conventional choice in particle physics is to choose a system of units in which $\hbar = c = 1$.

poles. The astounding conclusion is that the existence of antimatter is necessary in order to maintain causality as a cardinal physical principle — see Peskin and Schroeder (1995).

Fundamental to the analysis in this Coda has been the implicit assumption that a linear combination of solutions is also a solution — in other words, that superposition holds. It was first invoked below (22.4), where we blithely stated that the sum of solutions $\mathcal{U}_\omega(t)$ and $\mathcal{U}_{-\omega}(t)$ yields a solution for a $\cos \omega t$ source term. It was taken for granted in (22.5), whose integrand represents a continuous infinity of solutions. It was similarly presumed that sources with different ω's could be combined as in (22.6) — an explicit example of which is (22.7) to describe a delta function source. In fact, superposition is fundamental to the motivation for examining impulse response in the first place. That's because the response to any physical source $f(t)$ can be constructed by superposition of impulse responses G of the system (Chapter 42). Indeed, linearity plays a crucial role throughout physics; clearly worthy of much more than implicit assumption, it's the subject of Part IV.

Problems

22.1 Show that the frequency decomposition in (22.5) requires $f(t)$ and $\tilde{f}(\omega)$ to be related by the integral (22.6).

22.2 Find the impulse response G for a critically damped oscillator, $\beta = \omega_0$.

22.3 Verify that (22.10) is a solution of the differential equation (22.1).

22.4 Verify that both (22.22) and (22.39) are solutions of their respective differential equations, (22.20)–(22.21) and (22.38).

22.5 Beginning with (22.24), fill in the steps leading to the retarded and advanced propagators in (22.28) and (22.29).

22.6 Derive the retarded solution (22.33) to the Schrödinger equation.

22.7 **Residues and Physics**
If the source term in the wave equation (22.20) has a single frequency, there is no reason to integrate over ω — reducing the impulse solution (22.22) to

$$G(x) = \int_{-\infty}^{\infty} \frac{dk}{2\pi} \frac{e^{ikx}}{\omega^2/c^2 - k^2}.$$

It's convenient to rename the integration variable q, reserving k for the ratio ω/c; then in three dimensions we get

$$G(\vec{r}) = \int \frac{d^3 q}{(2\pi)^3} \frac{e^{i\vec{q}\cdot\vec{r}}}{k^2 - q^2}, \qquad k = \omega/c. \tag{22.43}$$

(a) Show that $G(\vec{r})$ is the impulse solution to the *Helmholtz equation*,

$$\left(\nabla^2 + k^2\right) G(\vec{r}) = \delta(\vec{r}).$$

(b) Evaluate the angular integrals in (22.43). [You may wish to review Example 8.15.]
(c) For the final q integral, there are four choices for how to handle the poles — and hence four distinct impulse solutions. Find them.
(d) Given that the time dependence is $e^{-i\omega t}$, interpret the solutions corresponding to the different $i\epsilon$ prescriptions physically.

Part IV

Linear Algebra

Why hurry over beautiful things? Why not linger and enjoy them?

Clara Schumann

23 Prelude: Superposition

We saw in Chapter 16 that Maxwell predicted the existence of electromagnetic waves. And as is true of virtually all wave phenomena, E&M waves obey the *principle of superposition*: combining any number of waves produces another wave. This is why waves running into one another combine as they cross, continuing on their merry ways unscathed. By contrast, particle trajectories are altered by collisions — assuming they even survive the impact. So superposition is not universal.

What mathematical feature is responsible for superposition? Let's address this physically by considering electric fields. Intuitively, we know that if charge density ρ_1 is the source of field $\vec{E}_1$, and ρ_2 of field $\vec{E}_2$, then the total charge $\rho = \rho_1 + \rho_2$ produces the field $\vec{E} = \vec{E}_1 + \vec{E}_2$. In terms of Gauss' law, it matters not whether we use $\vec{E}$ to find the total charge, or use the individual fields to calculate ρ_1 and ρ_2 separately,

$$\rho = \vec{\nabla} \cdot \vec{E} = \vec{\nabla} \cdot (\vec{E}_1 + \vec{E}_2) = \vec{\nabla} \cdot \vec{E}_1 + \vec{\nabla} \cdot \vec{E}_2 = \rho_1 + \rho_2. \tag{23.1}$$

It's this property which underlies superposition. More generally, for superposition to hold, an operator $\mathcal{L}$ acting on f and g must be *linear*,

$$\mathcal{L}(af + bg) = a\,\mathcal{L}(f) + b\,\mathcal{L}(g), \tag{23.2}$$

for arbitrary scalars a and b. Linear operators allow one to construct a linear combination of f and g either before or after applying $\mathcal{L}$. Divergence, then, is a linear differential operator; so too are the other operators in Maxwell's equations (curl, time derivative). Similarly, flux and circulation are linear integral operators, as is Coulomb's law, which maps charge q to field $\vec{E}$. If they weren't, Maxwell would be in conflict with the reality of superposition.

Superposition is often regarded as obvious. But it shouldn't be, and it isn't.

Example 23.1 Is it Linear?

1. A one-dimensional function of the form $f(x) = mx + b$ is just a straight line — so it must be linear, right? Although we call $f(x)$ a linear function, as an operator acting on x it does not abide by the definition (23.2): on the one hand,

$$f(c_1x_1 + c_2x_2) = m(c_1x_1 + c_2x_2) + b \tag{23.3a}$$

whereas

$$c_1f(x_1) + c_2f(x_2) = c_1(mx_1 + b) + c_2(mx_2 + b). \tag{23.3b}$$

Only if the line passes through the origin — i.e., for $b = 0$ — is f actually linear. (The general form $f(x) = mx + b$ is called an *affine* transformation.)

2. In finite dimension space, a linear operator can always be expressed as a matrix,

$$\begin{pmatrix} y_1 \\ y_2 \\ y_3 \end{pmatrix} = \begin{pmatrix} m_{11} & m_{12} & m_{13} \\ m_{21} & m_{22} & m_{23} \\ m_{31} & m_{32} & m_{33} \end{pmatrix} \begin{pmatrix} x_1 \\ x_2 \\ x_3 \end{pmatrix} \quad \Longleftrightarrow \quad y_i = \sum_j M_{ij} x_j. \tag{23.4}$$

This is the generalization of the one-dimensional $y = mx$; here too, adding a non-vanishing $\vec{b}$ is disallowed.

3. If you find yourself bothered by the non-linearity of $\vec{y} = M\vec{x} + \vec{b}$, note that there is a way to express this linearly — but at the cost of scaling up in dimensionality. Specifically, we tack on an extra component to all the vectors, and augment the matrix M with an additional row and column,

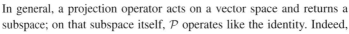

$$\vec{y} = M\vec{x} + \vec{b} \longrightarrow \begin{pmatrix} \vdots \\ \vec{y} \\ \vdots \\ 1 \end{pmatrix} = \left(\begin{array}{ccc|c} \cdots & \cdots & \cdots & \vdots \\ \cdots & M & \cdots & \vec{b} \\ \cdots & \cdots & \cdots & \vdots \\ \hline 0 & \cdots & 0 & 1 \end{array} \right) \begin{pmatrix} \vdots \\ \vec{x} \\ \vdots \\ 1 \end{pmatrix}, \tag{23.5}$$

as you can easily verify. So the transformation taking an n-dimensional vector $\vec{x}$ into $M\vec{x} + \vec{b}$ is actually a linear operation in $n + 1$ dimensions.

4. Any vector $\vec{V}$ in $\mathbb{R}^3$ can be linearly projected down to $\mathbb{R}^2$ simply by setting one of its components to zero. In matrix form, this *projection operator* can be written

$$\mathcal{P} = \begin{pmatrix} 1 & 0 & 0 \\ 0 & 1 & 0 \\ 0 & 0 & 0 \end{pmatrix}. \tag{23.6}$$

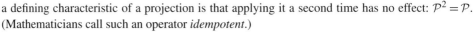

In general, a projection operator acts on a vector space and returns a subspace; on that subspace itself, $\mathcal{P}$ operates like the identity. Indeed, a defining characteristic of a projection is that applying it a second time has no effect: $\mathcal{P}^2 = \mathcal{P}$. (Mathematicians call such an operator *idempotent*.)

5. A linear operator C can map an $\mathbb{R}^3$ vector into a 2×2 complex matrix,

$$C(\vec{v}) \equiv \sum_{i=1}^{3} v_i \sigma_i = \begin{pmatrix} v_3 & v_1 - iv_2 \\ v_1 + iv_2 & -v_3 \end{pmatrix}, \tag{23.7}$$

where the σ_i are known as the *Pauli matrices*

$$\sigma_1 = \begin{pmatrix} 0 & 1 \\ 1 & 0 \end{pmatrix} \quad \sigma_2 = \begin{pmatrix} 0 & -i \\ i & 0 \end{pmatrix} \quad \sigma_3 = \begin{pmatrix} 1 & 0 \\ 0 & -1 \end{pmatrix}. \tag{23.8}$$

In particular, the position vector $\vec{r} = (x, y, z)$ becomes

$$C(\vec{r}) = \begin{pmatrix} z & x - iy \\ x + iy & -z \end{pmatrix}. \tag{23.9}$$

As we'll see, this *quaternion* representation of a vector can be used to describe rotations in $\mathbb{R}^3$.

Even with a linear operator, (23.2) only makes sense if f and g are superposable in the first place. One cannot, for instance, add mass and charge. So the divergence may be a linear operator, but it's the physical fact that electric charges add that allows the fields to add. In mathematics, superposable objects are generically known as *vectors*. This may seem curious: what about magnitude? direction? components? As we'll see, these concepts require additional structure. The most basic mathematical definition of a vector requires little more than the principle of superposition. In this primitive sense, vectors are not necessarily arrows: not only are objects like electric and magnetic fields vectors, so too are mass and charge. By contrast, in physics and engineering a vector is usually defined more narrowly in terms of the object's behavior under rotations. There's insight to be gained by appreciating how this narrowing comes about — so that's how we'll frame our discussion.

We start with an outline of the essential properties of vectors; much of the discussion is little more than an informal review of the rudiments of linear algebra. As the discussion progresses, we'll leverage our familiarity with arrows in $\mathbb{R}^n$ to develop ideas such as magnitude and direction. Along the way, we'll show that these properties can describe a host of mathematical objects that would otherwise seem to bear little or no resemblance to arrows at all. Indeed, one of the goals is to broaden the family of vectors to include matrices and, ultimately, functions — a result which leads directly into Part V.

Problems

23.1 Show which of the following operations are linear:

(a) $\hat{M}[\vec{v}] = (av_1 + bv_2, cv_1 + dv_2)$ where $\vec{v} = (v_1, v_2)$ is an $\mathbb{R}^2$ vector

(b) $\hat{T}[\vec{v}] = a\vec{v} + \vec{b}$

(c) $\hat{S}[x, y] = (x + y)^2$

(d) $\hat{N}[x, y] = |x| + |y|$

(e) $\hat{C}[\vec{A}] = \vec{\nabla} \times \vec{A}$

(f) $\hat{D}[f] = df/dx$

(g) $\hat{D}^2[f]$

(h) $(\hat{D}^2 + x^2)[f]$

(i) $\hat{D}^2[f] + \sin^2[f]$

(j) $\hat{K}[f] = \int_{P_1}^{P_2} K(x, t) f(t) \, dt$ for a given function $K(x, t)$.

23.2 Are solutions $\psi(x, t)$ to the Schrödinger equation $-\frac{\hbar^2}{2m} \frac{\partial^2 \psi}{\partial x^2} + V(x)\psi = i\hbar \frac{\partial \psi}{\partial t}$ superposable? Explain.

23.3 Use (23.2) to show that linear operators are *homogeneous* — that is, they map the zero vector to the zero vector.

24 Vector Space

Mathematics often takes a methodical, bottom-up approach. The basic program is to assemble a minimal set of assumptions or postulates necessary to produce an interesting, non-trivial mathematical result or structure. As these structures are explored and understood, additional assumptions may be appended, and the resulting superstructure studied and probed.

Take vectors. Somewhere along the line you learned that a vector is a mathematical object with both magnitude and direction. An arrow in space. And much of the time this is precisely what is meant. But mathematicians don't start there; for them, vectors are much more primitive objects that do little more than add and subtract to make new vectors. Only by incorporating additional mathematical structure does the familiar magnitude-and-direction arrow emerge.

Deciding what structure to add is, to some degree, a matter of taste and cleverness — but for the physicist or engineer, it usually comes down to empirical reality: what mathematical edifice provides a good description of the physical world? Of course, discovering the correct approach ultimately rests on the fundamental role experiment plays in science; only under its guidance can a credible mathematical structure of the physical world emerge. But as data are recorded and analyzed, mathematical rigor and consistency provide greater understanding and physical insight than could be deduced from experiment alone.

24.1 Vector Essentials

One of the first things everyone learns is that arrows can be added ("head-to-tail") to give another arrow. Actually, by definition, adding vectors *must* result in another vector; superposition is a defining characteristic of all vectors. One says that the vector space is *closed* under addition. It doesn't matter what order the vectors are added *(commutative)*,

$$\vec{A} + \vec{B} = \vec{B} + \vec{A} \tag{24.1}$$

nor how they are grouped *(associative)*

$$\vec{A} + (\vec{B} + \vec{C}) = (\vec{A} + \vec{B}) + \vec{C}. \tag{24.2}$$

Vectors can also be subtracted by defining the *inverse* vector $-\vec{A}$; since the sum of vectors produces another vector, this means that in some sense $0 \equiv \vec{A} - \vec{A}$ is also a vector. But this is probably so familiar as to be taken for granted.

The definition of an inverse can be viewed as a special case of another often-taken-for-granted property of vectors: any arrow $\vec{A}$ can be "scaled" by, well, a scalar a to obtain the vector $a\vec{A}$.

(The inverse vector is the special case $a = -1$.) It doesn't matter how many vectors are being scaled at one time *(distributive)*,

$$a(\vec{A} + \vec{B}) = a\vec{A} + a\vec{B}, \tag{24.3}$$

nor how scalings are grouped *(associative)*,

$$a(b\vec{A}) = (ab)\vec{A}. \tag{24.4}$$

At its most basic mathematical level, this completes the definition of a vector space: a collection of linear objects which can be added, subtracted, and scaled to yield more vectors. Though it may not seem like much, this modest set of properties provides sufficient scaffolding to support a rich body of concepts and relationships.

Example 24.1 Restricting Arrows

Clearly arrows in $\mathbb{R}^3$ form a vector space. Fine. But what about restrictions on this set? Well, the subset of $\mathbb{R}^3$ vectors with at least one vanishing component does not form a vector space because the collection is not closed: adding or subtracting any two of these vectors will not necessarily return a vector with a vanishing component. Including the restriction that vectors all have the *same* component vanish, however, does result in a vector space — the subspace $\mathbb{R}^2$.

So does this imply that any plane in $\mathbb{R}^3$ is a vector space? Arrows in $\mathbb{R}^3$ with tails at the origin and whose tips define a plane do not form a vector space unless the plane includes the origin. Without it, the space is not closed. Similarly, a line is a one-dimensional vector subspace only if it passes through the origin.

Example 24.2 Four-Vectors in Spacetime

Relativity recognizes the speed of light c as a universal constant of nature. This invites us to look for quantities whose units differ only by a factor of velocity, and consider the possibility that they share a common structure. The paradigmatic example is time $x_0 \equiv ct$ and position $\vec{x} = (x_1, x_2, x_3)$; together they comprise the components of a single spacetime *four-vector* $(x_0, \vec{x})$.[1] As we'll see in Chapter 29, the distinguishing feature of a four-vector $(v_0, \vec{v})$ is that, like the position $(x_0, \vec{x})$, its components in different inertial frames are related by the Lorentz transformations (1.7)–(1.8). Examples include the energy and momentum four-vector $(p_0, \vec{p})$, where $p_0 = E/c$; charge and current densities $(J_0, \vec{J})$, with $J_0 = \rho c$; and electromagnetic scalar and vector potentials $(A_0, \vec{A})$, $A_0 = \Phi/c$. In what sense these vectors "point" is unclear — and generally speaking, not relevant. But they are vectors nonetheless.

Incidentally, in SI units electric and magnetic fields also differ by a factor of c — but with a total of six components they cannot combine to form a four-vector. In Chapter 29 we'll see that together $\vec{E}$ and $\vec{B}$ form a second-rank tensor.

A few words on terminology and notation. The scalars $\{a, b, c, \ldots\}$ are often called the *field* over which the vector space is defined. So if the field is the set of all real numbers, one has a *real vector space*; if it is complex, one has a *complex vector space*. By contrast, the vectors themselves are neither real nor complex. In fact, the A beneath the arrow in $\vec{A}$ does not assume any value in the conventional sense; it is merely a label to distinguish vectors one from another.

---

[1] In physics parlance, the term "four-vector" almost always refers to a vector in spacetime.

24.2 Basis Basics

Under what circumstances can a pair of vectors be summed to zero, $a\vec{A} + b\vec{B} = 0$? Aside from $a = b = 0$, this can only be accomplished if $\vec{A}$ and $\vec{B}$ are scalar multiples of one another,

$$a\vec{A} + b\vec{B} = 0 \rightsquigarrow \vec{B} = -(a/b)\vec{A}. \tag{24.5}$$

For the usual head-to-tail addition of arrows, this is an expected result: to sum to zero, $\vec{A}$ and $\vec{B}$ must be parallel. But be careful; though picturing arrows in space may aid intuition, thus far we have not actually defined a vector's direction — so for now the word "parallel" only denotes vectors which differ by a scalar multiple. Two *non*-parallel vectors can only sum to zero if $a = b = 0$ in (24.5); no scalar multiple of $\vec{A}$ will yield $\vec{B}$. In this sense, they are independent vectors.

More generally, consider a set of n vectors and the relation

$$c_1\vec{A}_1 + c_2\vec{A}_2 + \cdots + c_n\vec{A}_n = 0. \tag{24.6}$$

If the only solution to this equation is the trivial case $c_i = 0$ for all i, the vectors $\vec{A}_i$ are said to be *linearly independent*. If it is possible to express even one of the n vectors as some linear combination of the other $n - 1$, the $\vec{A}$'s are linearly *dependent*.

This may sound abstract, but it's easy to picture: a set of arrows is linearly dependent if they can sum, head-to-tail, to zero — because if they can, then at least one of the arrows can be written as a sum of the others. Otherwise, they're independent. So any two non-parallel vectors are linearly independent. Now consider a third non-parallel vector. Viewed as arrows in the plane, the three necessarily form a linearly dependent set, since we can always scale them (with scalars, of course) to form a closed triangle. In other words, one can always combine any two to get the third. So the *maximal* set of linear independent vectors in two dimensions is two. In three-dimesional space, we can find linearly independent sets containing as many as three vectors; in n-dimensions, the maximal set has n vectors.

Clearly there's an underlying connection between the size of these maximal sets and the dimensionality of the space. In fact, the dimension of a vector space is *defined* to be the maximum number of linearly independent vectors the space allows.

Now wielding the phrase "a maximal set of linearly independent vectors" quickly becomes tiresome; instead, we call such a maximal set of vectors a *basis*. The name may suggest a basic, fundamental construct. And it is. But "basis" is actually meant to connote how such a set can be used to "support" any vector in the space — specifically, *an arbitrary vector $\vec{A}$ can always be written as a linear combination of basis vectors* $\{\vec{e}_i\}$,

$$\vec{A} = \sum_{i=1}^{n} c_i \vec{e}_i. \tag{24.7}$$

The c_i are called the *expansion coefficients* or the *components* of the vector on the given basis. Since every vector in the space can be generated this way, the basis vectors are said to *span* the space.[2] But how do we know (24.7) holds for *any* $\vec{A}$ in the space? Simple: if a vector $\vec{A}$ exists in n dimensions for which this were *not* possible, then together it and the $\vec{e}_i$'s would form an independent set — implying that by itself $\{\vec{e}_i\}$ is not maximal, and so there must be (at least) $n + 1$ linearly independent vectors. But this contradicts the premise of an n-dimensional space!

---

[2] A *spanning set* in n dimensions has at minimum n independent vectors. If it has more it cannot also be a basis, since not all the vectors will be linearly independent. See Example 24.8.

Example 24.3

Any arrow in the plane can be decomposed on the basis $\{\hat{e}_1, \hat{e}_2\} \equiv \{\hat{\imath}, \hat{\jmath}\}$, i.e., $\vec{V} = c_1 \hat{\imath} + c_2 \hat{\jmath}$. But a vector which points out of the plane cannot: we need an additional vector which is linearly independent of the other two — say, $\hat{k}$ — in order to expand any vector in $\mathbb{R}^3$. The pair $\{\hat{\imath}, \hat{\jmath}\}$ instead forms a basis in $\mathbb{R}^2$.

By the way, you should convince yourself that the vectors $\hat{\imath}$ and $\hat{\imath} + \hat{\jmath}$ also form a perfectly good basis in $\mathbb{R}^2$. If you nonetheless find yourself uncomfortable with this, then you're probably taking for granted additional structure. Stay tuned.

Example 24.4

An arrow in $\mathbb{R}^2$ can also be expanded on the polar basis $\{\hat{\rho}, \hat{\phi}\}$. The vector is the same, but the components depend on the choice of basis. Similarly, $\{\hat{\imath}, \hat{\jmath}, \hat{k}\}$ and $\{\hat{r}, \hat{\theta}, \hat{\phi}\}$ are bases in $\mathbb{R}^3$. As a real vector space, the components are real — whence the "$\mathbb{R}$" in $\mathbb{R}^n$. The generalization of $\mathbb{R}^n$ with complex components is called $\mathbb{C}^n$.

With the introduction of the concepts of basis and components, we return to the operations of vector addition and scalar multiplication: to add vectors, simply add the components on a given basis. So, for $\vec{A}$ and $\vec{B}$ with components a_i and b_i on the basis $\{\vec{e}_i\}$,

$$\vec{A} = \sum_i a_i \vec{e}_i, \quad \vec{B} = \sum_i b_i \vec{e}_i, \tag{24.8}$$

we have

$$\vec{A} + \vec{B} = \sum_i (a_i + b_i) \vec{e}_i. \tag{24.9a}$$

To multiply a vector by a scalar, simply multiply each component,

$$c\vec{A} = \sum_i (c\, a_i) \vec{e}_i. \tag{24.9b}$$

These expressions reproduce the familiar head-to-tail addition and scaling rules of arrows.

Indeed, the fundamental axioms outlined in (24.1)–(24.4), as well as the concept of a basis which springs from them, are easily framed in the paradigmatic context of arrows. Position, momentum, electric field — these are all familiar arrows in $\mathbb{R}^3$ whose directions have physical import.[3] But the same axioms also allow one to identify other, very non-arrow-like, vector spaces.

Example 24.5 The Standard Basis

The set of $n \times 1$ arrays

$$\begin{pmatrix} a_1 \\ a_2 \\ \vdots \\ a_n \end{pmatrix} \tag{24.10}$$

---

[3] In some sense, these are inequivalent *copies* of $\mathbb{R}^3$, since one cannot, e.g., add momentum and electric fields. This observation looms large in Chapter 29.

clearly obeys all the requisite rules to form the familiar space of column vectors. Rather than abstract items $\vec{A}$ and $\vec{B}$, this is a concrete example of a vector space for which the properties (24.8)–(24.9) are familiar. The simplest basis is the so-called *standard basis*

$$\hat{e}_i \rightarrow \begin{pmatrix} 0 \\ \vdots \\ 1 \\ \vdots \\ 0 \end{pmatrix}, \tag{24.11}$$

where the single non-zero component appears in the ith position.[4] In other words, $\left(\hat{e}_i \right)_\ell = \delta_{i\ell}$, where i denotes one of the n basis vectors, and ℓ designates one of its n components. This also lends itself to a natural ordering of the basis vectors: the first basis vector $\hat{e}_1$ has 1 in the first position, $\hat{e}_2$ has 1 in the second, etc. The most familiar example is undoubtedly $\mathbb{R}^3$,

$$\hat{e}_x \equiv \hat{i} \rightarrow \begin{pmatrix} 1 \\ 0 \\ 0 \end{pmatrix}, \quad \hat{e}_y \equiv \hat{j} \rightarrow \begin{pmatrix} 0 \\ 1 \\ 0 \end{pmatrix}, \quad \hat{e}_z \equiv \hat{k} \rightarrow \begin{pmatrix} 0 \\ 0 \\ 1 \end{pmatrix}, \tag{24.12}$$

so that the abstract expansion $\vec{A} = a_1 \hat{e}_x + a_2 \hat{e}_y + a_3 \hat{e}_z$ can be written concretely on the standard basis as

$$\begin{pmatrix} a_1 \\ a_2 \\ a_3 \end{pmatrix} = a_1 \begin{pmatrix} 1 \\ 0 \\ 0 \end{pmatrix} + a_2 \begin{pmatrix} 0 \\ 1 \\ 0 \end{pmatrix} + a_3 \begin{pmatrix} 0 \\ 0 \\ 1 \end{pmatrix}. \tag{24.13}$$

Notice, however, that despite notation like $\vec{A}$, nowhere in a simple column of numbers do the notions of magnitude and direction appear.

Example 24.6   A Non-Standard Basis

The defining feature of the standard basis is manifest in (24.13): the elements a_i of the column representing $\vec{A}$ are the expansion coefficients on the basis. But other bases may actually be more convenient, either mathematically or physically, than the standard basis. For example, the three vectors

$$\vec{\epsilon}_1 \rightarrow \begin{pmatrix} 1 \\ 1 \\ 1 \end{pmatrix}, \quad \vec{\epsilon}_2 \rightarrow \begin{pmatrix} 1 \\ -2 \\ 1 \end{pmatrix}, \quad \vec{\epsilon}_3 \rightarrow \begin{pmatrix} 1 \\ 0 \\ -1 \end{pmatrix} \tag{24.14}$$

also form a perfectly acceptable basis. Their order is arbitrary, though the "1-2-3" assignment provides a simple (and right-handed) convention. Note that the elements of these column vectors are their components on the standard basis — for instance, $\vec{\epsilon}_2 = \hat{e}_1 - 2\hat{e}_2 + \hat{e}_3$. But expanded on the $\{\vec{\epsilon}_i\}$ basis, an arbitrary column vector has components

---

[4] Even though we have not yet formally defined vector magnitude, we nonetheless use the conventional "hat" notation to indicate that $\hat{e}_i$ is a unit vector.

$$
\begin{pmatrix} a_1 \\ a_2 \\ a_3 \end{pmatrix} = \frac{a_1 + a_2 + a_3}{3} \begin{pmatrix} 1 \\ 1 \\ 1 \end{pmatrix} + \frac{a_1 - 2a_2 + a_3}{6} \begin{pmatrix} 1 \\ -2 \\ 1 \end{pmatrix} + \frac{a_1 - a_3}{2} \begin{pmatrix} 1 \\ 0 \\ -1 \end{pmatrix}, \tag{24.15a}
$$

or, more abstractly,

$$
\vec{A} = \left(\frac{a_1 + a_2 + a_3}{3} \right) \vec{\epsilon}_1 + \left(\frac{a_1 - 2a_2 + a_3}{6} \right) \vec{\epsilon}_2 + \left(\frac{a_1 - a_3}{2} \right) \vec{\epsilon}_3, \tag{24.15b}
$$

as can be verified with a little algebra.

Of course, $1 \times n$ rows also constitute a vector space. Superficially, the two are related by transposing one into the other – but since rows and columns cannot be added together, they form distinct vector spaces. A more formal and meaningful relationship requires additional structure (Chapter 25).

Example 24.7 **Is It a Basis?**

A straightforward way to determine whether a set of n, n-component vectors forms a basis is to use the set to construct an $n \times n$ matrix with the vectors as its columns. Recall that the determinant vanishes only if any column can be expressed as the sum of multiples of other columns. But expressing one column as a linear combination of others is just a statement of linear dependence. Thus, *if the determinant of the matrix is non-zero, the columns form a linearly independent set of vectors.* For example, consider the vectors

$$
\vec{A} = \hat{e}_1 + \hat{e}_3 \qquad \vec{B} = 2\hat{e}_1 - \hat{e}_2 + \hat{e}_3
$$
$$
\vec{C} = \hat{e}_2 - \hat{e}_3 \qquad \vec{D} = \hat{e}_2 + \hat{e}_3 \qquad \vec{E} = \hat{e}_1 - \hat{e}_2. \tag{24.16}
$$

Among these five there are $\binom{5}{3} = 10$ sets of three.[5] We can construct determinants using the standard basis. For instance, the determinant of the triplet $\{\vec{A}, \vec{B}, \vec{C}\}$ is

$$
\begin{vmatrix} 1 & 2 & 0 \\ 0 & -1 & 1 \\ 1 & 1 & -1 \end{vmatrix} = 2, \tag{24.17}
$$

so these three form a basis. On the other hand, the set $\{\vec{B}, \vec{D}, \vec{E}\}$ gives

$$
\begin{vmatrix} 2 & 0 & 1 \\ -1 & 1 & -1 \\ 1 & 1 & 0 \end{vmatrix} = 0, \tag{24.18}
$$

and so does not constitute a basis. And indeed, $\vec{B} - \vec{D} = 2\vec{E}$. Of the remaining eight triplets, only $\{\vec{A}, \vec{B}, \vec{D}\}$, $\{\vec{A}, \vec{B}, \vec{E}\}$, and $\{\vec{A}, \vec{D}, \vec{E}\}$ are not bases.

---

[5] Here, $\binom{5}{3}$ is a binomial coefficient, not a column vector.

Example 24.8 Less than a Basis?

A collection of n-component vectors, of course, cannot constitute a basis of n dimensions if the set contains fewer than n vectors. For instance, $\vec{A}$ and $\vec{B}$ in (24.16) clearly do not span three dimensions. But they are linearly independent, and so are sufficient to define a plane. In other words, they span a two-dimensional *subspace* of a larger three-dimensional space. Similarly, consider the vectors

$$\begin{pmatrix} 2 \\ 1 \\ -1 \\ 0 \end{pmatrix} \quad \begin{pmatrix} -4 \\ -1 \\ 3 \\ 2 \end{pmatrix} \quad \begin{pmatrix} 2 \\ 0 \\ -2 \\ 0 \end{pmatrix} \quad \begin{pmatrix} -2 \\ 0 \\ 2 \\ 2 \end{pmatrix}. \tag{24.19}$$

The determinant of the 4×4 matrix they form is zero, so they do not form a basis of four-dimensional space. On the other hand, inspection reveals that any *three* are linearly independent, so they span a restricted vector subspace of four-component columns — specifically, a three-dimensional "hyperplane."

Example 24.9 A Matrix Basis

Though we may be accustomed to regarding column or row vectors as n-dimensional arrows representing a force or a vector field, in and of themselves they're just matrices. Indeed, the set of $n \times n$ matrices also obeys all the required properties, and so forms a vector space. Consider the case $n = 2$. The four matrices

$$\begin{pmatrix} 1 & 0 \\ 0 & 0 \end{pmatrix}, \begin{pmatrix} 0 & 1 \\ 0 & 0 \end{pmatrix}, \begin{pmatrix} 0 & 0 \\ 1 & 0 \end{pmatrix}, \begin{pmatrix} 0 & 0 \\ 0 & 1 \end{pmatrix}, \tag{24.20}$$

are linearly independent: clearly no linear combination of three of these vectors can give the fourth. Moreover, the space of 2×2 matrices is four dimensional since any arbitrary 2×2 matrix can be written as a linear combination of these four,

$$\begin{pmatrix} a & b \\ c & d \end{pmatrix} = a \begin{pmatrix} 1 & 0 \\ 0 & 0 \end{pmatrix} + b \begin{pmatrix} 0 & 1 \\ 0 & 0 \end{pmatrix} + c \begin{pmatrix} 0 & 0 \\ 1 & 0 \end{pmatrix} + d \begin{pmatrix} 0 & 0 \\ 0 & 1 \end{pmatrix}. \tag{24.21}$$

Another basis in this space is

$$\sigma_0 = \begin{pmatrix} 1 & 0 \\ 0 & 1 \end{pmatrix}, \sigma_1 = \begin{pmatrix} 0 & 1 \\ 1 & 0 \end{pmatrix}, \sigma_2 = \begin{pmatrix} 0 & -i \\ i & 0 \end{pmatrix}, \sigma_3 = \begin{pmatrix} 1 & 0 \\ 0 & -1 \end{pmatrix}. \tag{24.22}$$

Expanded on this basis, an arbitrary 2×2 matrix is

$$\begin{pmatrix} a & b \\ c & d \end{pmatrix} = \frac{(a+d)}{2} \sigma_0 + \frac{(b+c)}{2} \sigma_1 + \frac{i(b-c)}{2} \sigma_2 + \frac{(a-d)}{2} \sigma_3, \tag{24.23}$$

as is easily verified. Comparing with (24.21), we see that the vector $\begin{pmatrix} a & b \\ c & d \end{pmatrix}$ is the same as before; only the components are basis dependent. (σ_0, of course, is just the identity matrix; the three $\sigma_{i \neq 0}$ are the Pauli matrices.)

Example 24.10 Scalar Functions as Vectors

We're familiar with a vector field $\vec{E}(\vec{r})$ having component functions $E_i(\vec{r})$ on a given basis. But restricted classes of scalar functions often form their own vector spaces. Perhaps the simplest possibility is polynomials of degree n,

$$P_n(x) = a_n x^n + a_{n-1} x^{n-1} + \cdots + a_2 x^2 + a_1 x + x_0. \tag{24.24}$$

Do these functions actually form a vector space? Well, linear combinations of polynomials will always return another polynomial. But although two polynomials of the same degree can never combine to form a polynomial of *higher* degree, two degree-n polynomials could form a polynomial of degree $n - 1$ or less. So for fixed n, the P_n's do not form a vector space. But if we relax our criterion to consider all polynomials of degree *n or less*, we'll have an $(n + 1)$-dimensional vector space with a basis given by $n + 1$ representative polynomials, $\{P_0, P_1, \ldots, P_{n-1}, P_n\}$. One possible basis is the set $\{x^m\}$, for $m = 0, 1, \ldots, n$.

What about non-polynomials — that is, functions whose power series expansions do not terminate at some highest power? The set of periodic functions, $f(t) = f(t + T)$ for some specified period T, is a particularly important example. Clearly these functions obey the strictures of distribution and associativity. Moreover, the set is closed: any linear combination of periodic functions with a given T will also be periodic with period T. We have a vector space. But finding a basis is more difficult. Intuitively we would expect sine and cosine to be candidates for a basis — and their opposite parity should convince you that they are in fact linearly independent functions. Indeed, with some restrictions a periodic function with period T can be expressed as a linear combination of basis functions $\sin(2\pi nt/T)$ and $\cos(2\pi nt/T)$ for integer n,

$$f(t) = \sum_n a_n \cos(2\pi nt/T) + b_n \sin(2\pi nt/T). \tag{24.25}$$

This is a *Fourier series* expansion of f. The frequencies $\omega_n \equiv 2\pi n/T$ are harmonics of the fundamental frequency $2\pi/T$; the vector components a_n and b_n are the relative contributions these frequencies make to $f(t)$. Note that there are an infinite number of basis functions, so this is an *infinite*-dimensional space. We'll have more to say about functions-as-vectors in Part V.

The rules laid out in (24.9) require that $\vec{A}$ and $\vec{B}$ be expanded on the same basis. The reason is not hard to see: vector components are basis dependent. Compare, for instance, (24.13) with (24.15), which provide distinct representations of the same vector $\vec{A}$; similarly for (24.21) and (24.23). Of the infinite number of bases one might choose, in the end it doesn't really matter. Still, some choices may be preferred — either because they simplify a calculation, or because they provide insight into the underlying system. Thus it's important to be able to change bases when needed.

A change of basis is almost always a linear operation, and as such is expressible as a matrix equation,

$$\vec{\varepsilon}_i = \sum_j R_{ij} \vec{e}_j. \tag{24.26}$$

The most common example is when R is a rotation – for instance, in $\mathbb{R}^2$

$$\begin{pmatrix} \vec{\varepsilon}_1 \\ \vec{\varepsilon}_2 \end{pmatrix} = \begin{pmatrix} \cos\varphi & \sin\varphi \\ -\sin\varphi & \cos\varphi \end{pmatrix} \begin{pmatrix} \vec{e}_1 \\ \vec{e}_2 \end{pmatrix} \tag{24.27}$$

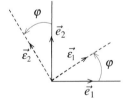

rotates the basis counterclockwise by φ (Figure 24.1). If $\{\vec{e}_i\}$ is the standard basis, then the new basis vectors are

$$\vec{\varepsilon}_1(\varphi) \to \begin{pmatrix} \cos\varphi \\ \sin\varphi \end{pmatrix} \qquad \vec{\varepsilon}_2 \to \begin{pmatrix} -\sin\varphi \\ \cos\varphi \end{pmatrix}. \tag{24.28}$$

Figure 24.1 Rotating between bases.

In this way, $R(\varphi)$ generates an infinity of bases, one for every value of $0 \le \varphi < 2\pi$.

BTW 24.1 Not All Arrows Are Vectors

As we've seen, not all vectors are arrows; conversely, not all arrows are vectors. We can, for example, represent the act of rotating an object in $\mathbb{R}^3$ with an arrow $\vec{\varphi}$ whose direction is given by the axis of rotation, and length by the angle of rotation $\varphi = |\vec{\varphi}|$. Restricting that angle to be between $-\pi$ and π, the entire set of these arrows can be depicted as a solid ball of radius π, with antipodal points on the surface identified as the same rotation. The identity is the origin. This one-to-one correspondence between rotations and points in the ball provides a visual representation of the space of $\mathbb{R}^3$ rotations.

But are these arrows actually vectors? Consider two distinct $90°$ rotations: one around the x-axis, to which we assign the arrow $\vec{\alpha} \rightarrow \hat{i}\pi/2$, and one around the y-axis, which we denote $\vec{\beta} \rightarrow \hat{j}\pi/2$. The net effect of applying two successive rotations to an object is equivalent to a single rotation by some angle φ around an axis $\hat{n}$. Clearly $\vec{\varphi}$ depends on $\vec{\alpha}$ and $\vec{\beta}$ — but is $\vec{\varphi} \equiv \vec{\alpha} + \vec{\beta}$? Well, if $\vec{\alpha}$ and $\vec{\beta}$ are vectors, Pythagoras suggests that the result of these successive $\pi/2$ rotations around perpendicular axes should be equivalent to a single rotation by $\varphi = \pi/\sqrt{2}$. It's not hard to convince oneself that this is not correct; as we'll prove later, the net rotation has $\varphi = 2\pi/3$.

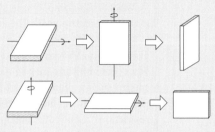

The axis $\hat{n}$ of this composite rotation holds even greater surprise. Not only does $\hat{n}$ not lie in the xy-plane, the axis depends upon the *order* of the rotations. *Rotations do not commute!* But vector addition must be commutative, so if $\vec{\alpha}$ and $\vec{\beta}$ were vectors, the order of rotation should not matter. The lesson is clear: though we can assign arrows to rotations, these arrows are not vectors. (We'll see later that $\vec{\alpha}$ followed by $\vec{\beta}$ is equivalent to a single rotation about the axis $(-1, -1, 1)$, whereas $(1, 1, 1)$ is the rotation axis for $\vec{\beta}$ and then $\vec{\alpha}$.)

The non-commutativity of rotations around different axes is quite striking the first time you see it. But it's worth observing that the difference between the final orientations becomes less pronounced as the angle of rotation decreases. In fact, *infinitesimal* rotations do commute. Infinitesimal rotations also add. Thus even though $\vec{\varphi}$ is not a vector, $d\vec{\varphi}$ is — and this allows us to define the angular velocity $\vec{\omega} \equiv d\vec{\varphi}/dt$ as a vector.

24.3 Kets and Reps

As we've seen, vectors come in many different forms: columns, rows, matrices, polynomials — even many non-polynomial functions such as sine and cosine. Given this diversity, it's worthwhile to adopt a uniform notation. Although we're accustomed to denoting vectors as $\vec{v}$ — that is, with a label capped by an arrow — it's too closely identified with $\mathbb{R}^n$. In an effort to distinguish universal attributes of vectors from properties specific to particular concrete examples, we introduce "arrow-neutral" notation that applies to any vector. So rather than denoting a vector as a label topped by an arrow, we'll place the label inside the symbol $|\ \rangle$. Instead of $\vec{v}$, then, we write $|v\rangle$ (read "ket vee").

Note that the symbol v inside the ket $|v\rangle$ does not assume any value in the conventional sense; it is merely a label used to distinguish vectors one from another. (This is no different than the v in $\vec{v}$.) For example, we will generally denote a basis vector with an index, say $|e_i\rangle$ or even just $|i\rangle$, where i runs from 1 to n, the dimension n of the space. The entire set of basis vectors is denoted $\{|i\rangle\}$. Recasting (24.8)–(24.9) in this notation, we express an arbitrary vector $|v\rangle$ on a given basis as

$$|v\rangle = \sum_{i=1}^{n} v_i|i\rangle = v_1|1\rangle + v_2|2\rangle + \cdots + v_n|n\rangle. \tag{24.29}$$

With this, addition of vectors $|v\rangle$ and $|w\rangle$ is achieved as before by expanding on a common basis and adding the components:

$$|v\rangle = \sum_{i} v_i|i\rangle \,, \ |w\rangle = \sum_{i} w_i|i\rangle \implies |v\rangle + |w\rangle = \sum_{i}(v_i + w_i)|i\rangle. \tag{24.30}$$

Similarly, to multiply a vector by a scalar c, simply multiply each component:

$$c|v\rangle = \sum_{i} cv_i|i\rangle. \tag{24.31}$$

There are many simplifications this notation allows. For instance, since the expression within the ket symbol is only a label, one can denote the vector $|v\rangle + |w\rangle$ as $|v + w\rangle$. Similarly, the vector $c|v\rangle$ can be written $|cv\rangle$.[6] This convention gives a consistent rendering of the inverse vector $-|v\rangle$:

$$|v\rangle + |-v\rangle = |v - v\rangle = |0\rangle \implies |-v\rangle \equiv -|v\rangle. \tag{24.32}$$

Similarly,

$$0|v\rangle + |w\rangle = |0v + w\rangle = |w\rangle \implies 0|v\rangle \equiv |0\rangle. \tag{24.33}$$

For notational convenience, we'll usually write the zero vector simply as 0, reserving the symbol $|0\rangle$ to denote a non-vanishing vector for which the zero label is appropriate (as for instance, in Example 24.11).

Ket notation emphasizes the vector characteristics common to what may be superficially dissimilar objects. From a vector perspective, objects such as $n \times n$ matrices and polynomials of order $\leq n^2 - 1$ are indistinguishable from rows or columns of n^2 numbers. Each provides a concrete *faithful representation* of the abstract n^2-dimensional vector space.

Example 24.11 Vector Space in Four Dimensions

Consider a four-dimensional basis $\{|0\rangle, |1\rangle, |2\rangle, |3\rangle\}$, and the superposition

$$|P\rangle = 2|0\rangle - |1\rangle + 3|2\rangle + |3\rangle. \tag{24.34}$$

This sum can be left in the abstract, or if desired evaluated in any faithful representation. For instance, the set of polynomials of order $m \leq 3$ is a four-dimensional vector space. With integer powers of x representing the basis vectors,

$$|0\rangle \to x^0 \qquad |1\rangle \to x^1 \qquad |2\rangle \to x^2 \qquad |3\rangle \to x^3 \,, \tag{24.35}$$

the expansion (24.34) is the polynomial

$$P(x) = x^3 + 3x^2 - x + 2. \tag{24.36}$$

But if we take the Pauli matrices of (24.22) as the basis representation,

$$|0\rangle \to \sigma_0 \qquad |1\rangle \to \sigma_1 \qquad |2\rangle \to \sigma_2 \qquad |3\rangle \to \sigma_3 \,, \tag{24.37}$$

---

[6] In arrow notation, the closest parallels would be $\vec{v} + \vec{w} = \overrightarrow{v + w}$ and $c\vec{v} = \overrightarrow{cv}$. Awkward.

the very same expansion gives the matrix

$$|P\rangle \rightarrow \begin{pmatrix} 3 & -1-3i \\ -1+3i & 1 \end{pmatrix}. \tag{24.38}$$

Of course, (24.34) can also be expressed as the column vector

$$|P\rangle \rightarrow \begin{pmatrix} 2 \\ -1 \\ 3 \\ 1 \end{pmatrix}, \tag{24.39}$$

where now the basis is

$$|0\rangle \rightarrow \begin{pmatrix} 1 \\ 0 \\ 0 \\ 0 \end{pmatrix} \quad |1\rangle \rightarrow \begin{pmatrix} 0 \\ 1 \\ 0 \\ 0 \end{pmatrix} \quad |2\rangle \rightarrow \begin{pmatrix} 0 \\ 0 \\ 1 \\ 0 \end{pmatrix} \quad |3\rangle \rightarrow \begin{pmatrix} 0 \\ 0 \\ 0 \\ 1 \end{pmatrix}. \tag{24.40}$$

In each case, the same scalar coefficients 2, −1, 3, 1 determine the vector sum. Each basis is a concrete representations of an abstract four-dimensional vector space.

You may be wondering why (24.34) uses an equals (=), while the other expressions have an arrow (→). The reason goes to the heart of what it means to be a representation. Though perhaps a bit overly fastidious, the arrow is meant to reflect the difference between a vector in the abstract, and any concrete representation of it. The column in (24.39) does not *equal* the vector $|P\rangle$; rather it serves as a concrete *representation* of it. The vector's plenipotentiary for that basis, if you will. By contrast, (24.34) is a full-fledged equality amongst kets.

The use of kets focuses attention on the underlying vector nature of a space, independent of representation. But to be honest, we could accomplish this almost as easily with conventional arrow notation. The real advantage of kets is revealed once vector multiplication is introduced — which we turn to next (Chapter 25).

BTW 24.2 Vector Space

The formal definition of a vector space considers vectors $\{|u\rangle, |v\rangle, |w\rangle, \ldots\}$ and scalars $\{a, b, c, \ldots\}$. These elements combine under two operations, vector addition and scalar multiplication. Vector addition is any combination rule obeying:

1. *Closure:* The sum of any two vectors is a vector, $|v\rangle + |w\rangle = |u\rangle$.
2. *Commutativity:* $|v\rangle + |w\rangle = |w\rangle + |v\rangle$.
3. *Associativity:* $|u\rangle + (|v\rangle + |w\rangle) = (|u\rangle + |v\rangle) + |w\rangle$.
4. *Zero Vector:* There is a *null vector* $|0\rangle$ such that $|v\rangle + |0\rangle = |v\rangle$.
5. *Inverse:* For every vector $|v\rangle$ there is a vector $|-v\rangle$ such that $|v\rangle + |-v\rangle = |0\rangle$.

Scalar multiplication is any combination rule which is:

1. *Closed:* The product of a scalar with a vector is another vector, $a|v\rangle = |w\rangle$.
2. *Distributive:* $a(|v\rangle + |w\rangle) = a|v\rangle + a|w\rangle$
$$(a+b)|v\rangle = a|v\rangle + b|v\rangle.$$
3. *Associative:* $a(b|v\rangle) = (ab)|v\rangle$.

Problems

24.1 Which of the following are vector spaces?

(a) The set of scalars.
(b) The set of two-dimensional columns whose first element is smaller than the second.
(c) The set of n-dimensional columns with integer-valued elements.
(d) The set of n-dimensional columns whose elements sum to zero.
(e) The set of $n \times n$ antisymmetric matrices combining under addition, $C = A + B$.
(f) The set of $n \times n$ matrices combining under multiplication rather than addition, $C = AB$.
(g) The set of polynomials with real coefficients.
(h) The set of periodic functions with $f(0) = 1$.
(i) The set of periodic functions with $f(0) = 0$.
(j) The set of functions with $cf(a) + df'(a) = 0$ for fixed c and d.

24.2 For real vector spaces V and W, we define the elements of the *product space* $U \equiv V \times W$ as the pair of vectors (v, w), with

$$a(v_1, w_1) + b(v_2, w_2) = (av_1 + bv_2, aw_1 + bw_2).$$

Is U a vector space?

24.3 Which of the following are linearly independent sets?

(a) $\begin{pmatrix} 1 \\ 0 \end{pmatrix}, \begin{pmatrix} 1 \\ 1 \end{pmatrix}$

(b) $\begin{pmatrix} 1 \\ 0 \end{pmatrix}, \begin{pmatrix} 1 \\ 1 \end{pmatrix}, \begin{pmatrix} 0 \\ 1 \end{pmatrix}$

(c) $\begin{pmatrix} 1 \\ 1 \\ 1 \end{pmatrix}, \begin{pmatrix} 1 \\ 0 \\ -1 \end{pmatrix}, \begin{pmatrix} 1 \\ -2 \\ 1 \end{pmatrix}$

(d) $\begin{pmatrix} 2 \\ 1 \\ -1 \\ 0 \end{pmatrix}, \begin{pmatrix} 2 \\ 0 \\ -2 \\ 0 \end{pmatrix}, \begin{pmatrix} -2 \\ 0 \\ 2 \\ 2 \end{pmatrix}$

(e) x, x^2, x^3

24.4 Is a complex number z linearly independent of its complex conjugate? Explain.

24.5 Use the determinant method of Example 24.7 to verify that the vectors of Example 24.6 constitute a basis in three-dimensional space.

24.6 Though a vector can be expanded on multiple bases, a vector's components on a *given* basis are unique. Prove this by assuming it's *not* true, and show that this leads to a contradiction. ("Proof by Contradiction.")

24.7 Use the standard basis $\{|\hat{e}_i\rangle\}$ of (24.11) to represent the following ket expressions for arbitrary $|v\rangle$, $|w\rangle$ as column vectors:

(a) $|v\rangle = a_1|\hat{e}_1\rangle + a_2|\hat{e}_2\rangle + a_3|\hat{e}_3\rangle$ and $|w\rangle = b_1|\hat{e}_1\rangle + b_2|\hat{e}_2\rangle + b_3|\hat{e}_3\rangle$
(b) $|v\rangle + |w\rangle = |v + w\rangle$
(c) $|w\rangle = k|v\rangle = |kv\rangle$
(d) $|v + w\rangle$ for $|v\rangle$ and $|w\rangle$ inverses of one another.

24.8 For the vector space of primary colors red, green, and blue (RGB), find a matrix representation for the basis $\{|R\rangle, |G\rangle, |B\rangle\}$. What vector represents white? black? orange?

24.9 Look up the painting by René Magritte called *La Trahison des Images* ("The Treachery of Images"). Discuss its relevance to our discussion of representations. [You can find images of the painting on line — or better, visit the Los Angeles County Museum of Art (LACMA).]

25 The Inner Product

Up until now, we have not defined multiplication of vectors — nor have we endowed vectors with either a magnitude or a sense of direction. Though none of these is a necessary feature of a vector space, one piece of additional structure is sufficient to define all three: an *inner product*.

An inner product is an operation which takes two vectors and returns a scalar; as such, it is often called the *scalar product*. In some sense, the inner product provides the intersection between the heretofore distinct mathematical sets of vectors and scalars. We will denote the inner product of two vectors $|A\rangle$ and $|B\rangle$ — whether columns, matrices, or functions — by $\langle A|B\rangle$.[1] A vector space with an inner product is called an *inner product space*.

Now an inner product is not just any scalar combination of two vectors; to qualify, $\langle A|B\rangle$ must fulfill a minimal set of criteria. We'll postpone their formal statement for now in favor of a more intuitive approach.

25.1 The Adjoint

Let's start with the obvious question: why have we chosen to denote the inner product as $\langle A|B\rangle$ rather than, say $|A\rangle \cdot |B\rangle$? To answer this, consider $\mathbb{R}^n$, where $\langle A|B\rangle \equiv \vec{A} \cdot \vec{B}$. Trying to calculate the dot product using the familiar column representation (in two dimensions, for simplicity),

$$\vec{A} \rightarrow \begin{pmatrix} a_1 \\ a_2 \end{pmatrix} \qquad \vec{B} \rightarrow \begin{pmatrix} b_1 \\ b_2 \end{pmatrix}, \tag{25.1}$$

immediately runs into a new problem. The power and efficacy of using column vectors is that they enable all abstract addition and scaling operations to be carried out using the concrete rules of matrix algebra. But such matrix manipulations can only combine two column vectors into another column vector, whereas the inner product returns a scalar! The solution to this may seem so obvious as to be not worth mentioning: we first need to *transpose* one of the vectors before the rules of matrix algebra take over,

$$\langle A|B\rangle = \begin{pmatrix} a_1 & a_2 \end{pmatrix} \begin{pmatrix} b_1 \\ b_2 \end{pmatrix} = \sum_i a_i b_i. \tag{25.2}$$

But this *is* worth mentioning because columns and rows belong to different vector spaces. Even so, clearly the two spaces are related. Since successive transpositions take a vector from one space to the other and back again, the spaces are said to be *dual* to one another.

---

[1] Other common notations are $\langle A, B\rangle$ and (A, B).

From (25.2) we can appreciate that the first slot in $\langle A|B\rangle$ treats its vector differently — a feature not manifest in notation like $|A\rangle \cdot |B\rangle$. Although we've illustrated this with a matrix representation, the rule applies generally: an inner product must be calculated with one vector and one dual vector. Conventional dot product notation does not make this distinction. (One could presumably write $\vec{A}^T \cdot \vec{B}$, but since the $\mathbb{R}^n$ dot product is symmetric, $\vec{A} \cdot \vec{B} = \vec{B} \cdot \vec{A}$, explicitly distinguishing the two slots in the inner product is unnecessary.)

But consider vectors in $\mathbb{C}^n$. One of the requirements of an inner product is that it must satisfy the condition

$$\|\vec{A}\|^2 \equiv \langle A|A\rangle \geq 0 \tag{25.3}$$

for every vector in the space (zero only if the vector itself is zero). $\|\vec{A}\|$ is called the magnitude or *norm* of $\vec{A}$.[2] You're likely most familiar with the norm as the length of the vector — which only makes sense if we require $\|\vec{A}\|$ to be both real and positive. The dot product on $\mathbb{R}^n$ clearly guarantees this,

$$\langle A|A\rangle = \begin{pmatrix} a_1 & a_2 \end{pmatrix} \begin{pmatrix} a_1 \\ a_2 \end{pmatrix} = \sum_i a_i^2 \geq 0. \tag{25.4}$$

$\mathbb{C}^n$, however, has complex a_i — so the guarantee fails. Clearly we need a more generalized formulation of $\langle A|A\rangle$ which reduces to (25.4) for real space.

The situation may sound familiar: given any complex number z, there's also no guarantee that z^2 is positive — or even real. In order for the magnitude $|z|$ to properly represent a length, one introduces the complex conjugate, such that $|z|^2 \equiv z^*z \geq 0$. The analog for vectors is called the *Hermitian conjugate* or *adjoint*, which we denote with the symbol † (read: "dagger"),

$$\|\vec{A}\|^2 \equiv \langle A|A\rangle \equiv \vec{A}^\dagger \cdot \vec{A} \geq 0. \tag{25.5}$$

A glance at (25.4) should suffice to convince you that for the length to be real and positive, the dagger must be both the transpose *and* the complex conjugate,

$$\vec{A}^\dagger \equiv \vec{A}^{T*}. \tag{25.6}$$

In other words, *the dual of a column vector in $\mathbb{C}^n$ is the row vector given by its transpose–complex-conjugate.*

Now if the adjoint is required for the norm, it must be fundamental to the inner product itself: rather than (25.2), the correct general expression is

$$\langle A|B\rangle = \vec{A}^\dagger \cdot \vec{B} \equiv \begin{pmatrix} a_1 & a_2 \end{pmatrix}^* \begin{pmatrix} b_1 \\ b_2 \end{pmatrix} = \sum_i a_i^* b_i. \tag{25.7}$$

This comes with a cost, however: to ensure a norm which is real and positive, in a complex space we must sacrifice the symmetry of the inner product. The most we can expect is[3]

$$\langle A|B\rangle = \langle B|A\rangle^*, \tag{25.8}$$

which is consistent with having a real norm when $|A\rangle \equiv |B\rangle$. In (25.7), this means

$$\begin{pmatrix} a_1 & a_2 \end{pmatrix}^* \begin{pmatrix} b_1 \\ b_2 \end{pmatrix} = \left[\begin{pmatrix} b_1 & b_2 \end{pmatrix}^* \begin{pmatrix} a_1 \\ a_2 \end{pmatrix} \right]^*, \tag{25.9}$$

---

[2] We use the double-bar notation $\|\vec{A}\|$ for vectors, reserving the familiar $|a|$ for scalars.
[3] The bracket $\langle A|B\rangle$ is a complex number, not a vector — hence the use of $*$ rather than †.

which is readily generalized to n dimensions,

$$\langle A|B \rangle = \sum_i a_i^* b_i = \left(\sum_i b_i^* a_i \right)^* = \langle B|A \rangle^*, \tag{25.10}$$

So there is a fundamental difference between the two slots in the inner product, beyond mere transposition. Indeed, the "conjugate symmetry" expressed by (25.8) is one of the defining characteristics of any inner product.

Example 25.1

Consider the vectors $|v\rangle \rightarrow \begin{pmatrix} -1 \\ 2 \end{pmatrix}$ and $|w\rangle \rightarrow \begin{pmatrix} 1 \\ -i \end{pmatrix}$. Then

$$\langle v|w \rangle = \begin{pmatrix} -1 & 2 \end{pmatrix}^* \begin{pmatrix} 1 \\ -i \end{pmatrix} = -1 - 2i, \tag{25.11a}$$

whereas

$$\langle w|v \rangle = \begin{pmatrix} 1 & -i \end{pmatrix}^* \begin{pmatrix} -1 \\ 2 \end{pmatrix} = -1 + 2i. \tag{25.11b}$$

As expected, $\langle v|w \rangle = \langle w|v \rangle^*$. Meanwhile, the lengths of the vectors are both real and positive,

$$\|\vec{v}\|^2 = \langle v|v \rangle = \begin{pmatrix} -1 & 2 \end{pmatrix}^* \begin{pmatrix} -1 \\ 2 \end{pmatrix} = 5, \tag{25.12a}$$

and

$$\|\vec{w}\|^2 = \langle w|w \rangle = \begin{pmatrix} 1 & -i \end{pmatrix}^* \begin{pmatrix} 1 \\ -i \end{pmatrix} = 2. \tag{25.12b}$$

Since the inner product is a scalar, it must be basis independent — so the use of an equal sign ($=$) is appropriate.

Complex numbers z live in the vector space $\mathbb{C}^1$, making the transpose superfluous; thus $z^\dagger \equiv z^*$. The adjoint of a matrix, on the other hand, is its transpose–complex-conjugate,

$$M_{ij}^\dagger \equiv M_{ji}^*, \tag{25.13}$$

generalizing the dual relation between column and row vectors. Note the hierarchy: generalizing the concept of absolute value from real to complex numbers requires the introduction of the complex conjugate; generalizing from numbers to matrices requires the introduction of the transpose. Complex matrices require both. The matrix analog of $z = z^*$ for real numbers, then, is $M^\dagger = M$; such matrices are said to be *self-adjoint* or *Hermitian*.

Example 25.2   **The Adjoint of the Product**

You're asked in Problem 25.6 to show that the adjoint of a product of operators is the reverse product of the adjoints,

$$(ST)^\dagger = T^\dagger S^\dagger. \tag{25.14}$$

Here, we'll just demonstrate it for matrices $S = \begin{pmatrix} i & -1 \\ 1 & i \end{pmatrix}$ and $T = \begin{pmatrix} 1 & 2i \\ -i & 1 \end{pmatrix}$:

$$(ST)^\dagger = \begin{pmatrix} 2i & -3 \\ 2 & 3i \end{pmatrix}^\dagger = \begin{pmatrix} -2i & 2 \\ -3 & -3i \end{pmatrix}, \tag{25.15a}$$

which is the same as the reverse product of the adjoints,

$$T^\dagger S^\dagger = \begin{pmatrix} 1 & i \\ -2i & 1 \end{pmatrix} \begin{pmatrix} -i & 1 \\ -1 & -i \end{pmatrix} = \begin{pmatrix} -2i & 2 \\ -3 & -3i \end{pmatrix}. \tag{25.15b}$$

Example 25.3 **Square Matrices**

Inner products can be defined on spaces other than $\mathbb{C}^n$. The trace of a square matrix, recall, is the sum of its diagonal elements, $\text{Tr}M = \sum_i M_{ii}$. In a vector space of matrices, the trace can serve as an inner product,

$$\langle A|B\rangle = \text{Tr}(A^\dagger B), \tag{25.16}$$

where $A^\dagger_{ij} \equiv A^*_{ji}$. Let's check:

$$\langle A|B\rangle = \text{Tr}\left(A^\dagger B\right) = \sum_i \left(A^\dagger B\right)_{ii} = \sum_{ij} A^\dagger_{ij} B_{ji}$$

$$= \sum_{ij} A^*_{ji} B_{ji} = \sum_{ij} B_{ji} A^*_{ji} = \sum_{ij} \left(B^\dagger_{ij} A_{ji}\right)^*$$

$$= \sum_i \left(B^\dagger A\right)^*_{ii} = \text{Tr}\left(B^\dagger A\right)^* = \langle B|A\rangle^*, \tag{25.17}$$

which verifies $\langle A|B\rangle = \langle B|A\rangle^*$. As for magnitude

$$\|A\|^2 = \text{Tr}(A^\dagger A) = \sum_{ij} A^*_{ji} A_{ji} = \sum_{ij} |A_{ji}|^2 \geq 0, \tag{25.18}$$

so the norm is indeed real and positive.

Despite the fact that our discussion has focused on rows and columns, it's important to appreciate that the adjoint $\dagger$ is a necessary operation for *every* inner product space — though for non-matrix spaces such as polynomials of degree $\leq n$, taking the transpose would seem to make little sense. In Part V, we'll discuss how to apply the adjoint to these vector spaces as well.

Dirac Notation: Bras and Kets

Despite the diversity of objects which have inner products — e.g., column vectors, matrices, and polynomials — all inner products share a common structure, independent of the representation. As a result, it's sensible to introduce notation which focuses on the commonalities of inner product spaces. We already denote a generic vector as a ket, $|v\rangle$; now that we've introduced the inner product, we need a way to distinguish a vector from its adjoint. We could just use $|v\rangle^\dagger$. Instead, we follow Dirac and "pull apart" the inner product, defining the dagger of the ket by the symbol $\langle v|$ (read: "bra vee"),

$$\langle v| \equiv |v\rangle^{\dagger}. \tag{25.19}$$

Then the inner product of two vectors $|v\rangle$ and $|w\rangle$ is just the meeting of bra and ket, $\langle v|w\rangle$. (Vertical lines are included only as needed for legibility — so $\langle v||w\rangle \equiv \langle v|w\rangle$.)

Like rows and columns, bras and kets are duals of one another: to every $|v\rangle$ is a corresponding $\langle v|$. As with kets, the v inside $\langle v|$ is merely a label to identify the vector. So

$$|v + w\rangle \equiv |v\rangle + |w\rangle \Longleftrightarrow \langle v + w| \equiv \langle v| + \langle w|. \tag{25.20}$$

Note, however, that since we've defined $|cv\rangle \equiv c|v\rangle$,

$$\langle cv| \equiv |cv\rangle^{\dagger} = \left(c|v\rangle\right)^{\dagger} = \langle v|c^{*}. \tag{25.21}$$

In other words,

$$|cv\rangle \equiv c|v\rangle \Longleftrightarrow \langle cv| \equiv \langle v|c^{*} = c^{*}\langle v|. \tag{25.22}$$

The bra-ket terminology is due to Dirac, and derives (as you may suspect) from the "bracket" $\langle w|v\rangle$. But pulling apart the bracket to create two separate entities is no joke. It allows us to manipulate bras and kets with the usual rules of algebra, with the special feature that whenever a bra comes upon a ket from the left, an inner product automatically forms. In this way, Dirac notation streamlines many otherwise clumsy calculations by exploiting the general properties of inner products without being distracted by the details of its computation.

Bras and kets also provide a straightforward expression for the adjoint $\mathcal{L}^{\dagger}$ of a linear operator. As a matter of notation, the adjoint appears by generalizing (25.22) from scalars to operators,

$$|\mathcal{L}v\rangle \equiv \mathcal{L}|v\rangle \Longleftrightarrow \langle \mathcal{L}v| \equiv \langle v|\mathcal{L}^{\dagger}. \tag{25.23}$$

(Note that unlike a complex number which can sit on either side of a bra, the adjoint operator $\mathcal{L}^{\dagger}$ in (25.23) must sit to the right of $\langle v|$ — essentially an application of (25.14).) The expression $\langle v|\mathcal{L}|w\rangle$ can thus be interpreted as the inner product either of $|v\rangle$ with $\mathcal{L}|w\rangle$, or equivalently of $\mathcal{L}^{\dagger}|v\rangle$ with $|w\rangle$. This leads directly to an inner product-based definition for the adjoint: $\mathcal{L}^{\dagger}$ is the operator such that

$$\langle v|\mathcal{L}w\rangle = \langle v|\mathcal{L}|w\rangle = \langle \mathcal{L}^{\dagger}v|w\rangle, \tag{25.24a}$$

or, invoking conjugate symmetry $\langle v|w\rangle^{*} = \langle w|v\rangle$,

$$\langle v|\mathcal{L}|w\rangle^{*} = \langle w|\mathcal{L}^{\dagger}|v\rangle. \tag{25.24b}$$

A Hermitian operator is the special case $\mathcal{L}^{\dagger} = \mathcal{L}$. These expressions may appear overly formal — especially since for matrices (25.24b) is equivalent to the transpose–complex conjugate recipe. But expressing $\mathcal{L}^{\dagger}$ in terms of bras and kets establishes a basis-independent definition which holds for any inner product space — including, as we'll see in Chapter 40, a vector space of functions for which $\mathcal{L}$ is a differential operator rather than a matrix.[4]

---

[4] Bras are best viewed as a vector space of *operators* which map a vector $|v\rangle$ into a complex number. But if the notation leaves you feeling adrift, for now just think of bras and kets as row and column vectors, and $\mathcal{L}$ as a matrix.

Example 25.4 **Adjoint Confirmation**

We can readily confirm (25.24) using the vectors $|v\rangle$ and $|w\rangle$ from Example 25.1 and the matrix S of Example 25.2:

$$\langle v|S|w\rangle^* = \langle v|Sw\rangle^* = \left[\begin{pmatrix} -1 & 2 \end{pmatrix}^* \begin{pmatrix} i & -1 \\ 1 & i \end{pmatrix} \begin{pmatrix} 1 \\ -i \end{pmatrix} \right]^* = 4 + 2i \tag{25.25a}$$

$$\langle w|S^\dagger|v\rangle = \langle w|S^\dagger v\rangle = \begin{pmatrix} 1 & -i \end{pmatrix}^* \begin{pmatrix} -i & 1 \\ -1 & -i \end{pmatrix} \begin{pmatrix} -1 \\ 2 \end{pmatrix} = 4 + 2i \tag{25.25b}$$

25.2 The Schwarz Inequality

Dirac notation allows for a straightforward definition of an inner product space:

An inner product space is a vector space endowed with an additional operation which takes two vectors $|v\rangle$ and $|w\rangle$ and returns a scalar $\langle v|w\rangle$, subject to the following criteria:

1. Positive definite norm: $\langle v|v\rangle \geq 0,$ $\qquad\qquad\qquad\qquad\qquad\qquad$ (25.26a)

$$\text{where } \langle v|v\rangle = 0 \Longleftrightarrow |v\rangle \equiv 0$$

2. Conjugate symmetry: $\langle v|w\rangle = \langle w|v\rangle^*$ $\qquad\qquad\qquad\qquad$ (25.26b)

3. Linearity: $\qquad\qquad \langle u| \big(a|v\rangle + b|w\rangle \big) = a\langle u|v\rangle + b\langle u|w\rangle.$ $\qquad$ (25.26c)

From these defining properties, the familiar geometric interpretation of the inner product as a *projection* follows from little more than the requirement $\langle C|C\rangle \geq 0$. Consider two arbitrary vectors $|A\rangle, |B\rangle$, and express their complex inner product in polar form,

$$\langle A|B\rangle = e^{i\varphi}|\langle A|B\rangle|. \tag{25.27}$$

Let's also define $|C\rangle$ as

$$|C\rangle \equiv se^{i\varphi}|A\rangle + |B\rangle, \tag{25.28}$$

for some arbitrary real scalar s. Then, using $\langle C| = se^{-i\varphi}\langle A| + \langle B|$, we find

$$0 \leq \langle C|C\rangle = s^2\langle A|A\rangle + se^{-i\varphi}\langle A|B\rangle + se^{i\varphi}\langle B|A\rangle + \langle B|B\rangle$$
$$= s^2\langle A|A\rangle + 2s\,\Re e\big[e^{-i\varphi}\langle A|B\rangle\big] + \langle B|B\rangle$$
$$= s^2\langle A|A\rangle + 2s\,|\langle A|B\rangle| + \langle B|B\rangle, \tag{25.29}$$

where we've used (25.27) in the last line. Although $\langle C|C\rangle$ depends on the real parameter s, it and the other inner products are scalars. So (25.29) describes a parabola $f(s) = as^2 + bs + c \geq 0$ with positive coefficients a, b, c. Its vertex, found by solving $f'(s) = 0$, is at $s = -b/2a$; this is where the minimum of f occurs. Therefore

$$f_{\min}(s) \geq 0 \quad \Longrightarrow \quad c \geq b^2/4a. \tag{25.30}$$

(In other words, the discriminant $b^2 - 4ac$ in the quadratic formula must be less than or equal to zero.) Thus the fundamental requirement $\langle C|C \rangle \geq 0$ reveals that *any* inner product $\langle A|B \rangle$ must obey

$$|\langle A|B \rangle|^2 \leq \langle A|A \rangle \langle B|B \rangle. \tag{25.31}$$

This is known as the *Schwarz inequality*.[5] For arrows in $\mathbb{R}^n$, it becomes

$$|\vec{A} \cdot \vec{B}| \leq \|\vec{A}\| \, \|\vec{B}\|, \tag{25.32}$$

which tells us that the number $\vec{A} \cdot \vec{B}$ lies somewhere between $\pm\|\vec{A}\| \, \|\vec{B}\|$. A simple and intuitive way to parametrize this constraint is to introduce an angle θ such that

$$\vec{A} \cdot \vec{B} = \|\vec{A}\| \, \|\vec{B}\| \cos\theta, \qquad 0 \leq \theta \leq \pi. \tag{25.33}$$

The interpretation is straightforward: if $\vec{B} = \alpha\vec{A}$, then $\vec{A} \cdot \vec{B} = \alpha\|\vec{A}\|^2 = \|\vec{A}\| \, \|\vec{B}\|$ giving $\cos\theta = 1$. Recall that earlier we defined two proportional vectors to be parallel; $\theta = 0$, then, provides a geometric picture of parallel vectors. Similarly, $\theta = \pi$ denotes "anti-parallel" vectors. On the other hand, the inner product vanishes for $\theta = \pi/2$: this is the source of the synonymy of "orthogonal" and "perpendicular." So the introduction of the inner product allows for the concepts of both vector magnitude *and* relative orientation between vectors.

Example 25.5 **Developing a Sense of Direction**

Consider arbitrary non-zero vectors $\vec{A}$ and $\vec{B}$ in $\mathbb{R}^2$, and define $\vec{C} = \vec{A} - \vec{B}$. Then

$$C^2 = (\vec{A} - \vec{B}) \cdot (\vec{A} - \vec{B}) = A^2 + B^2 - 2\vec{A} \cdot \vec{B}, \tag{25.34}$$

where $C^2 \equiv \|\vec{C}\|^2$. But $\vec{A}$, $\vec{B}$, and $\vec{C}$ form a triangle, which obeys the law of cosines

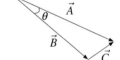

$$C^2 = A^2 + B^2 - 2AB\cos\theta, \tag{25.35}$$

where A, B, and C refer to the lengths of the sides of the triangle, and θ is the angle between $\vec{A}$ and $\vec{B}$. Comparing (25.34) and (25.35), we (re)discover (25.33).

Example 25.6 **The Triangle Inequality**

The Schwarz inequality allows us to put an upper limit on the norm of $\vec{A} + \vec{B}$:

$$\begin{aligned}
\|\vec{A} + \vec{B}\|^2 &= A^2 + B^2 + 2\vec{A} \cdot \vec{B} \\
&= A^2 + B^2 + 2AB\cos\theta \\
&\leq A^2 + B^2 + 2AB, \quad \text{since } \cos\theta \leq 1 \\
&= \left(\|\vec{A}\| + \|\vec{B}\| \right)^2.
\end{aligned} \tag{25.36}$$

Thus

$$\|\vec{A} + \vec{B}\| \leq \|\vec{A}\| + \|\vec{B}\|. \tag{25.37}$$

---

[5] My professor once quipped that the hardest part of (25.31) is spelling "Schwarz" correctly!

This is called the *triangle inequality*, which simply says that whenever two arrows are added, the length of the resulting vector cannot be greater than the sum of the lengths of the original two. (The equality only holds when $\vec{A}$ and $\vec{B}$ are parallel — which is nicely intuitive.) Similarly,

$$\|\vec{A} + \vec{B}\|^2 \geq A^2 + B^2 - 2AB, \quad \text{since } \cos\theta \geq -1$$
$$= \left(\|\vec{A}\| - \|\vec{B}\| \right)^2. \tag{25.38}$$

All together, then, we have

$$\|\vec{A}\| - \|\vec{B}\| \leq \|\vec{A} + \vec{B}\| \leq \|\vec{A}\| + \|\vec{B}\|, \tag{25.39}$$

where we have presumed $\|\vec{A}\| > \|\vec{B}\|$. Note that since this follows directly from the Schwarz inequality, the triangle inequality — and the intuition that comes with it — holds for any vector space with an inner product, not just $\mathbb{R}^n$.

Example 25.7 Inequality Insight

For $\mathbb{C}^n$ vectors $\vec{A} \rightarrow \begin{pmatrix} a_1 \\ \vdots \\ a_n \end{pmatrix}$ and $\vec{B} \rightarrow \begin{pmatrix} b_1 \\ \vdots \\ b_n \end{pmatrix}$, the Schwarz inequality becomes

$$\left| \sum_i a_i^* b_i \right|^2 \leq \left(\sum_i |a_i|^2 \right) \left(\sum_j |b_j|^2 \right). \tag{25.40}$$

It's not hard to see why this is so. The sums on the right are over the *magnitudes* of complex numbers; no relative orientation information is included. On the left, however, the sum is over complex numbers; their differing phases will result in some partial cancellations and thus yield a smaller result. Only when the vectors $\vec{A}$ and $\vec{B}$ are parallel will the two sides be equal. Similarly, for the triangle inequality,

$$\left[\sum_i \left| (a_i + b_i) \right|^2 \right]^{1/2} \leq \left[\sum_i |a_i|^2 \right]^{1/2} + \left[\sum_j |b_j|^2 \right]^{1/2}. \tag{25.41}$$

Simply put: the magnitude of the sum is smaller than the sum of the magnitudes.

It's the Schwarz inequality which underlies the familiar geometric picture of the relative orientation θ between two vectors. Parametrizing the inequality with $\cos\theta$ in particular leads to our intuitive understanding of $\langle A|B \rangle$ as a *projection* or *overlap* between two vectors. Indeed, in $\mathbb{R}^n$ where $\vec{A} \cdot \vec{B} = AB\cos\theta$, we know that $A\cos\theta$ is the component of $\vec{A}$ along $\vec{B}$, and $B\cos\theta$ is the component of $\vec{B}$ along $\vec{A}$. Note that orthogonal vectors have no overlap at all.

This geometric interpretation of the inner product in $\mathbb{R}^n$ is strong motivation to generalize it to other spaces: for any vector space with an inner product, one can define an angle θ in accord with (25.31) such that

$$\cos\theta \equiv \frac{|\langle A|B \rangle|}{\sqrt{\langle A|A \rangle \langle B|B \rangle}}. \tag{25.42}$$

Since $\langle A|B \rangle$ is a complex number, (25.42) must retain the magnitude $|\langle A|B \rangle|$ in (25.31); this restricts θ to lie between 0 and $\pi/2$. More significant is that in a complex space $\langle A|B \rangle \neq \langle B|A \rangle$, so they cannot describe precisely the same projection. The proper geometric interpretation of $\langle A|B \rangle$ is read from *right to left*:

$$\langle A|B \rangle \text{ is the projection of } |B \rangle \text{ onto } |A \rangle, \tag{25.43a}$$

whereas

$$\langle B|A \rangle \text{ is the projection of } |A \rangle \text{ onto } |B \rangle. \tag{25.43b}$$

The geometric picture provided by the Schwarz inequality encapsulates fundamental, *basis-independent* relationships amongst vectors in an inner product space. This is most-easily appreciated by noticing that the derivation of the Schwarz inequality never used a basis. One implication is that, despite appearances, $\vec{A} \cdot \vec{B}$ must be the same number independent of basis. We already knew this of course — but the Schwarz inequality makes this manifest by allowing us to write $\vec{A} \cdot \vec{B} = AB \cos\theta$, which clearly makes no reference to a basis; all that's needed is a measure of their relative orientation. In other words, $\sum_i A_i B_i$ is a basis-specific calculation of a basis-independent quantity. So magnitude and relative orientation are inherent properties of vectors in an inner product space.

25.3 Orthonormality

Since the inner product is independent of basis, we can choose any basis to work in — and a shrewd choice can greatly simplify calculation and analysis.

The inner products of basis vectors are particularly relevant. For vectors $|A \rangle$ and $|B \rangle$ with components a_i and b_i on a basis $\{|e_i \rangle\}$,

$$|A \rangle = \sum_i a_i |e_i \rangle \qquad |B \rangle = \sum_j b_j |e_j \rangle, \tag{25.44}$$

we have, using $\langle A| = \left(\sum_i a_i |e_i \rangle \right)^\dagger = \sum_i \langle e_i| a_i^*$,

$$\langle A|B \rangle = \left(\sum_i \langle e_i| a_i^* \right) \left(\sum_j b_j |e_j \rangle \right) = \sum_{ij} a_i^* b_j \langle e_i|e_j \rangle. \tag{25.45}$$

(Notice how nicely bra-ket notation works in this expression: the bra $\langle e_i|$ "passes through" the scalars a_i^* and b_j until it hits the ket $|e_j \rangle$. And with the coming-together of bra and ket, an inner product bracket naturally emerges.) From (25.45) we learn that knowledge of the inner products $\langle e_i|e_j \rangle$ between all pairs of basis vectors allows us to determine the inner product of *any* pair of vectors in the space. All we need is the matrix $g_{ij} \equiv \langle e_i|e_j \rangle$ of complex numbers. The simplest (and very often best) choice of basis is a set $\{|\hat{e}_i \rangle\}$ whose vectors all have unit length

$$\langle \hat{e}_i|\hat{e}_i \rangle = 1 \longrightarrow \text{"normalized"} \tag{25.46a}$$

and do not overlap,

$$\langle \hat{e}_i|\hat{e}_{j \neq i} \rangle = 0 \longrightarrow \text{"orthogonal"} \tag{25.46b}$$

— or simply

$$\langle \hat{e}_i|\hat{e}_j \rangle = \delta_{ij} \Longrightarrow \textbf{\textit{"orthonormal"}}. \tag{25.47}$$

In this situation, the Kronecker delta collapses one of the sums in (25.45) to give the familiar

$$\langle A|B \rangle = \sum_{ij} a_i^* b_j \, \langle \hat{e}_i | \hat{e}_j \rangle = \sum_i a_i^* b_i. \tag{25.48}$$

Example 25.8 **A Real Rotated Orthonormal Basis**

If a basis $\{|\hat{e}_1\rangle, |\hat{e}_2\rangle\}$ is orthonormal, then so too is the rotated basis $\{|\hat{\varepsilon}_1\rangle, |\hat{\varepsilon}_2\rangle\}$ of (24.27):

$$\langle \hat{\varepsilon}_1 | \hat{\varepsilon}_1 \rangle = \cos^2 \varphi \, \langle \hat{e}_1 | \hat{e}_1 \rangle + 2 \sin \varphi \cos \varphi \, \langle \hat{e}_1 | \hat{e}_2 \rangle + \sin^2 \varphi \, \langle \hat{e}_2 | \hat{e}_2 \rangle = 1$$

$$\langle \hat{\varepsilon}_2 | \hat{\varepsilon}_2 \rangle = \sin^2 \varphi \, \langle \hat{e}_1 | \hat{e}_1 \rangle - 2 \sin \varphi \cos \varphi \, \langle \hat{e}_1 | \hat{e}_2 \rangle + \cos^2 \varphi \, \langle \hat{e}_2 | \hat{e}_2 \rangle = 1$$

$$\langle \hat{\varepsilon}_1 | \hat{\varepsilon}_2 \rangle = -\cos \varphi \sin \varphi \, \langle \hat{e}_1 | \hat{e}_1 \rangle + (\cos^2 \varphi - \sin^2 \varphi) \, \langle \hat{e}_1 | \hat{e}_2 \rangle + \sin \varphi \cos \varphi \, \langle \hat{e}_2 | \hat{e}_2 \rangle = 0.$$

By contrast, the basis $\{|\hat{e}_1\rangle + |\hat{e}_2\rangle, -|\hat{e}_1\rangle + |\hat{e}_2\rangle\}$ is orthogonal but not orthonormal, whereas the basis $\{\frac{1}{\sqrt{2}}(|\hat{e}_1\rangle + |\hat{e}_2\rangle), |\hat{e}_2\rangle\}$ is normalized but not orthogonal.

Example 25.9 **Projecting Out Vector Components**

Consider a vector expanded on an arbitrary orthonormal basis

$$|v\rangle = v_1 |\hat{\varepsilon}_1\rangle + v_2 |\hat{\varepsilon}_2\rangle.$$

Projecting this onto the basis vector $|\hat{\varepsilon}_1\rangle$ demonstrates the efficacy of orthonormality:

$$\langle \hat{\varepsilon}_1 | v \rangle = v_1 \langle \hat{\varepsilon}_1 | \hat{\varepsilon}_1 \rangle + v_2 \langle \hat{\varepsilon}_1 | \hat{\varepsilon}_2 \rangle = v_1.$$

For instance, on the rotated orthonormal basis of (24.28) with $\varphi = \pi/4$,

$$|\hat{\varepsilon}_{1,2}\rangle \equiv \frac{1}{\sqrt{2}} \left(\pm |\hat{e}_1\rangle + |\hat{e}_2\rangle \right) \to \frac{1}{\sqrt{2}} \begin{pmatrix} \pm 1 \\ 1 \end{pmatrix}, \tag{25.49}$$

the vectors $|v\rangle$ and $|w\rangle$ from Example 25.1 have components

$$v_1 \equiv \langle \hat{\varepsilon}_1 | v \rangle = \frac{1}{\sqrt{2}} \begin{pmatrix} 1 & 1 \end{pmatrix}^* \begin{pmatrix} -1 \\ 2 \end{pmatrix} = \frac{1}{\sqrt{2}} \tag{25.50a}$$

$$v_2 \equiv \langle \hat{\varepsilon}_2 | v \rangle = \frac{1}{\sqrt{2}} \begin{pmatrix} -1 & 1 \end{pmatrix}^* \begin{pmatrix} -1 \\ 2 \end{pmatrix} = \frac{3}{\sqrt{2}} \tag{25.50b}$$

and

$$w_1 \equiv \langle \hat{\varepsilon}_1 | w \rangle = \frac{1}{\sqrt{2}} \begin{pmatrix} 1 & 1 \end{pmatrix}^* \begin{pmatrix} 1 \\ -i \end{pmatrix} = \frac{1 - i}{\sqrt{2}} \tag{25.51a}$$

$$w_2 \equiv \langle \hat{\varepsilon}_2 | w \rangle = \frac{1}{\sqrt{2}} \begin{pmatrix} -1 & 1 \end{pmatrix}^* \begin{pmatrix} 1 \\ -i \end{pmatrix} = \frac{-1 - i}{\sqrt{2}}. \tag{25.51b}$$

(Because the inner product is complex, we must read from right-to-left to correctly identify a component — e.g., $\langle \hat{\varepsilon}_2 | w \rangle$ is the projection of $|w\rangle$ onto $|\hat{\varepsilon}_2\rangle$.) All together, (25.50)–(25.51) give

$$|v\rangle = -|\hat{e}_1\rangle + 2|\hat{e}_2\rangle = \frac{1}{\sqrt{2}} |\hat{\varepsilon}_1\rangle + \frac{3}{\sqrt{2}} |\hat{\varepsilon}_2\rangle \tag{25.52a}$$

and

$$|w\rangle = |\hat{e}_1\rangle - i|\hat{e}_2\rangle = \frac{1-i}{\sqrt{2}}\,|\hat{\varepsilon}_1\rangle + \frac{-(1+i)}{\sqrt{2}}\,|\hat{\varepsilon}_2\rangle, \qquad (25.52b)$$

as is easily verified by summing the column vectors directly.

Example 25.10 **A Basis in $\mathbb{C}^2$**

It may not be obvious at first, but the basis

$$|\hat{e}_I\rangle \to \frac{1}{\sqrt{5}}\begin{pmatrix} 1 \\ 2i \end{pmatrix} \quad |\hat{e}_{II}\rangle \to \frac{1}{\sqrt{5}}\begin{pmatrix} 2 \\ -i \end{pmatrix}, \qquad (25.53)$$

is orthonormal:

$$\langle \hat{e}_I|\hat{e}_I\rangle = \frac{1}{5}\begin{pmatrix} 1 & 2i \end{pmatrix}^*\begin{pmatrix} 1 \\ 2i \end{pmatrix} = 1 \quad \langle \hat{e}_{II}|\hat{e}_{II}\rangle = \frac{1}{5}\begin{pmatrix} 2 & -i \end{pmatrix}^*\begin{pmatrix} 2 \\ -i \end{pmatrix} = 1, \qquad (25.54)$$

and

$$\langle \hat{e}_I|\hat{e}_{II}\rangle = \frac{1}{5}\begin{pmatrix} 1 & 2i \end{pmatrix}^*\begin{pmatrix} 2 \\ -i \end{pmatrix} = 0. \qquad (25.55)$$

The components of a vector $|A\rangle \to \begin{pmatrix} a_1 \\ a_2 \end{pmatrix}$ on this basis are

$$a_I \equiv \langle \hat{e}_I|A\rangle = \frac{1}{\sqrt{5}}\begin{pmatrix} 1 & 2i \end{pmatrix}^*\begin{pmatrix} a_1 \\ a_2 \end{pmatrix} = \frac{a_1 - 2ia_2}{\sqrt{5}} \qquad (25.56)$$

and

$$a_{II} \equiv \langle \hat{e}_{II}|A\rangle = \frac{1}{\sqrt{5}}\begin{pmatrix} 2 & -i \end{pmatrix}^*\begin{pmatrix} a_1 \\ a_2 \end{pmatrix} = \frac{2a_1 + ia_2}{\sqrt{5}}, \qquad (25.57)$$

where $a_{1,2}$ are the components on the standard basis. Thus

$$|A\rangle = \left(\frac{a_1 - 2ia_2}{\sqrt{5}}\right)|\hat{e}_I\rangle + \left(\frac{2a_1 + ia_2}{\sqrt{5}}\right)|\hat{e}_{II}\rangle. \qquad (25.58)$$

You should verify that $|a_I|^2 + |a_{II}|^2 = |a_1|^2 + |a_2|^2$.

Of course, any basis can be used to expand a vector. Vector sums can be carried out by adding components on a common basis, and vector scaling simply by multiplying each component by the same scalar. So then, what unique features does orthogonality offer — and in particular, why might one prefer an orthonormal basis?

The answer has everything to do with the inner product — in both interpretation and execution. Being orthogonal, the basis vectors have no overlap, so the projection of a vector $|A\rangle$ onto any basis vector $|\hat{e}_i\rangle$ is *separate and distinct* from projections along the other $|\hat{e}_{j\neq i}\rangle$. And being normalized, the basis vectors contribute nothing to the *magnitude* of these components. So an expansion like (25.58) cleanly distinguishes between the "pieces" (components) of a vector, and the mutually orthogonal "directions" (unit basis vectors).

In short, it's only on an orthonormal basis that the inner product is the simple sum of the product of the vectors' components, as in (25.48). And as we just saw, those components are easily projected out: the jth component of $|A\rangle = \sum_i a_i |\hat{e}_i\rangle$ is the projection of $|A\rangle$ in the jth "direction" —

$$\langle \hat{e}_j | A \rangle = \langle \hat{e}_j | \left(\sum_i a_i |\hat{e}_i\rangle \right) = \sum_i a_i \langle \hat{e}_j | \hat{e}_i \rangle = \sum_i a_i \delta_{ij} = a_j. \qquad (25.59)$$

This is especially transparent in the standard basis,

$$|\hat{e}_1\rangle \rightarrow \begin{pmatrix} 1 \\ 0 \end{pmatrix} \quad |\hat{e}_2\rangle \rightarrow \begin{pmatrix} 0 \\ 1 \end{pmatrix}, \qquad (25.60)$$

where a vector $|A\rangle = a_1 |\hat{e}_1\rangle + a_2 |\hat{e}_2\rangle$ has components

$$\langle \hat{e}_1 | A \rangle = \begin{pmatrix} 1 & 0 \end{pmatrix} \begin{pmatrix} a_1 \\ a_2 \end{pmatrix} = a_1 \qquad \langle \hat{e}_2 | A \rangle = \begin{pmatrix} 0 & 1 \end{pmatrix} \begin{pmatrix} a_1 \\ a_2 \end{pmatrix} = a_2.$$

On a *different* orthonormal basis $\{|\hat{e}_j'\rangle\}$, the very same vector has different components,

$$a_j' \equiv \langle \hat{e}_j' | A \rangle = \langle \hat{e}_j' | \sum_i a_i |\hat{e}_i\rangle = \sum_i a_i \langle \hat{e}_j' | \hat{e}_i \rangle. \qquad (25.61)$$

So to find the components on a new basis, all we need are the inner products between the two bases. The elements of the matrix $\langle \hat{e}_j' | \hat{e}_i \rangle$ are often referred to as the *direction cosines* between bases. For the bases of (24.28), that matrix is the rotation matrix $R(\varphi)$.

Example 25.11 **Expanding a Basis Vector I**

A basis vector, of course, is a vector like any other. So consider what happens when $|A\rangle$ is itself one of the orthonormal basis vectors — that is, $|A\rangle \equiv |\hat{e}_\ell\rangle$ for some fixed ℓ. Then since all the $|\hat{e}_i\rangle$ are orthonormal, all their inner products with $|A\rangle$ vanish except for $i = \ell$, and the usual

$$a_i \equiv \langle \hat{e}_i | A \rangle \rightarrow \begin{pmatrix} a_1 \\ a_2 \\ \vdots \\ a_n \end{pmatrix}, \qquad (25.62a)$$

becomes

$$(\hat{e}_\ell)_i = \langle \hat{e}_i | \hat{e}_\ell \rangle = \delta_{i\ell}. \qquad (25.62b)$$

Thus the ith component of the ℓth basis vector, expanded on the basis $\{\hat{e}_i\}$, is $\delta_{i\ell}$. So only the ℓth component is non-vanishing,

$$|A\rangle \equiv |\hat{e}_\ell\rangle \rightarrow \begin{pmatrix} 0 \\ \vdots \\ 1 \\ \vdots \\ 0 \end{pmatrix}, \qquad (25.63)$$

where the 1 appears in the ℓth position. The columns representing the other $n - 1$ basis vectors will all have a zero in their ℓth positions.

Example 25.12   **A Non-Standard Basis**

Consider an $\mathbb{R}^3$ basis $\{|\epsilon_i\rangle\}$ represented by the column vectors in Example 24.6,

$$|\epsilon_1\rangle \rightarrow \begin{pmatrix} 1 \\ 1 \\ 1 \end{pmatrix} \quad |\epsilon_2\rangle \rightarrow \begin{pmatrix} 1 \\ -2 \\ 1 \end{pmatrix} \quad |\epsilon_3\rangle \rightarrow \begin{pmatrix} 1 \\ 0 \\ -1 \end{pmatrix}. \tag{25.64}$$

It's straightforward to check that they're linearly independent — and so form a basis in $\mathbb{R}^3$. Moreover, since $\langle\epsilon_1|\epsilon_2\rangle = \langle\epsilon_2|\epsilon_3\rangle = \langle\epsilon_3|\epsilon_1\rangle = 0$, they form an orthogonal basis. But they're not orthonormal:

$$\langle\epsilon_1|\epsilon_1\rangle = 3, \quad \langle\epsilon_2|\epsilon_2\rangle = 6, \quad \langle\epsilon_3|\epsilon_3\rangle = 2. \tag{25.65}$$

We can easily *normalize* the vectors, though, to obtain

$$|\hat{\epsilon}_1\rangle \rightarrow \frac{1}{\sqrt{3}}\begin{pmatrix} 1 \\ 1 \\ 1 \end{pmatrix} \quad |\hat{\epsilon}_2\rangle \rightarrow \frac{1}{\sqrt{6}}\begin{pmatrix} 1 \\ -2 \\ 1 \end{pmatrix}, \quad |\hat{\epsilon}_3\rangle \rightarrow \frac{1}{\sqrt{2}}\begin{pmatrix} 1 \\ 0 \\ -1 \end{pmatrix}, \tag{25.66}$$

where the coefficients are called the *normalization constants*. On this orthonormal basis, an arbitrary vector has $\hat{\epsilon}_1$-component

$$\langle\hat{\epsilon}_1|A\rangle = \frac{1}{\sqrt{3}}\begin{pmatrix} 1 & 1 & 1 \end{pmatrix}\begin{pmatrix} a_1 \\ a_2 \\ a_3 \end{pmatrix} = \frac{1}{\sqrt{3}}(a_1 + a_2 + a_3). \tag{25.67a}$$

Similarly,

$$\langle\hat{\epsilon}_2|A\rangle = \frac{1}{\sqrt{6}}(a_1 - 2a_2 + a_3) \quad \text{and} \quad \langle\hat{\epsilon}_3|A\rangle = \frac{1}{\sqrt{2}}(a_1 - a_3). \tag{25.67b}$$

Thus

$$|A\rangle = \left(\frac{a_1 + a_2 + a_3}{\sqrt{3}}\right)|\hat{\epsilon}_1\rangle + \left(\frac{a_1 - 2a_2 + a_3}{\sqrt{6}}\right)|\hat{\epsilon}_2\rangle + \left(\frac{a_1 - a_3}{\sqrt{2}}\right)|\hat{\epsilon}_3\rangle, \tag{25.68}$$

or, in the column vector representation,

$$\begin{pmatrix} a_1 \\ a_2 \\ a_3 \end{pmatrix} = \frac{a_1 + a_2 + a_3}{3}\begin{pmatrix} 1 \\ 1 \\ 1 \end{pmatrix} + \frac{a_1 - 2a_2 + a_3}{6}\begin{pmatrix} 1 \\ -2 \\ 1 \end{pmatrix} + \frac{a_1 - a_3}{2}\begin{pmatrix} 1 \\ 0 \\ -1 \end{pmatrix}. \tag{25.69}$$

Notice that the normalization constants appear both in the basis vectors, (25.66), as well as the vector components, (25.67). Thus it is the *squared* normalization constants which appear in the expansion on an orthogonal basis.

Example 25.13   **The Pauli Basis**

An orthogonal basis for 2×2 matrices can be constructed from the three Pauli matrices of Example 24.9,

$$\sigma_1 = \begin{pmatrix} 0 & 1 \\ 1 & 0 \end{pmatrix} \quad \sigma_2 = \begin{pmatrix} 0 & -i \\ i & 0 \end{pmatrix} \quad \sigma_3 = \begin{pmatrix} 1 & 0 \\ 0 & -1 \end{pmatrix}. \tag{25.70}$$

With the inner product introduced in Example 25.3, this basis is orthogonal,

$$\text{Tr}(\sigma_i^\dagger \sigma_j) = 2\delta_{ij}. \tag{25.71}$$

The σ_i can be rendered orthonormal by dividing each by $\sqrt{2}$ — though it's more common to rescale the inner product as

$$\langle A|B \rangle \equiv \frac{1}{2}\text{Tr}(A^\dagger B). \tag{25.72}$$

With this, the σ_i form a complete orthonormal basis for the three-dimensional space of 2×2 traceless matrices.

You may recall from Example 24.9 that all 2×2 complex matrices can be expanded on the basis of the Pauli matrices and the identity $\mathbb{1}$. In fact, under the inner product of (25.72), $\mathbb{1}$ provides a fourth orthonormal basis vector. Defining $\sigma_0 \equiv \mathbb{1}$ for notational convenience, any 2×2 matrix A can be expanded as

$$A = \sum_{\alpha=0}^{3} a_\alpha \sigma_\alpha = \frac{1}{2}\sum_{\alpha=0}^{3}\text{Tr}[\sigma_\alpha^\dagger A]\sigma_\alpha$$

$$= \frac{1}{2}\text{Tr}(A)\mathbb{1} + \frac{1}{2}\sum_{i=1}^{3}\text{Tr}(\sigma_i A)\,\sigma_i, \tag{25.73}$$

where the components $a_\alpha \equiv \langle \sigma_\alpha | A \rangle$.

One of the more intriguing applications of the σ_i is as a representation for $\mathbb{R}^3$. Just as an $\mathbb{R}^3$ vector $\vec{v}$ can be expanded on the standard basis of columns (24.12), so too it can be expanded on the basis of Pauli matrices,

$$V \equiv \vec{v} \cdot \vec{\sigma} = \sum_i v_i \sigma_i = \begin{pmatrix} v_z & v_x - iv_y \\ v_x + iv_y & -v_z \end{pmatrix}, \tag{25.74}$$

where projection gives $v_i = \langle \sigma_i | V \rangle = \frac{1}{2}\text{Tr}(\sigma_i^\dagger V) = \frac{1}{2}\text{Tr}(\sigma_i V)$, since the σ_i are Hermitian, $\sigma_i = \sigma_i^\dagger$. Indeed, with Hermitian σ and real v_i, the matrix V is Hermitian. So any $\mathbb{R}^3$ vector $\vec{v}$ can be represented by a 2×2 traceless, Hermitian matrix V.

25.4 Building a Better Basis: Gram–Schmidt

It's only with the introduction of an inner product that the concept of an orthonormal basis emerges. Until then, not only is there no notion of a vector's length, the closest we got to any sense of direction was defining parallel vectors as being scalar multiples of one another. Indeed, without an inner product there is no definition — let alone quantification — of an angle between vectors. The inner product not only provides such an angle, but reveals a special angle at which the inner product vanishes: $\varphi = \pi/2$. That alone deserves its own title, orthogonal; any other (non-parallel) vectors are simply "oblique." Moreover, the inner product provides an intuitive, straightforward method to transform an arbitrary set of linearly independent (oblique) vectors into an orthonormal set. This is called Gram–Schmidt orthogonalization.

Consider the linear independent vectors $\{|\phi_1\rangle, |\phi_2\rangle, \ldots, |\phi_n\rangle\}$. Start by normalizing the first basis vector,[6]

---

[6] The order is arbitrary; the subscript in $|\phi_i\rangle$ simply labels the order chosen.

$$|\hat{e}_1\rangle \equiv \frac{|\phi_1\rangle}{\sqrt{\langle\phi_1|\phi_1\rangle}}, \tag{25.75}$$

where $\sqrt{\langle\phi_1|\phi_1\rangle} \equiv \||\phi_1\rangle\|$ is the norm of $|\phi_1\rangle$. The unit vector $|\hat{e}_1\rangle$ is the first member of the new orthonormal set. Now take a second vector $|\phi_2\rangle$ and *remove its component along* $|\hat{e}_1\rangle$,

$$|e_2\rangle \equiv |\phi_2\rangle - \langle\hat{e}_1|\phi_2\rangle\,|\hat{e}_1\rangle. \tag{25.76a}$$

Remember that $\langle\hat{e}_1|\phi_2\rangle$ is a number, so (25.76a) is just vector addition, $|e_2\rangle \equiv |\phi_2\rangle - c|\hat{e}_1\rangle$, with the explicit identification of c as the projection of $|\phi_2\rangle$ along $|\hat{e}_1\rangle$. Normalize the result,

$$|\hat{e}_2\rangle \equiv \frac{|e_2\rangle}{\sqrt{\langle e_2|e_2\rangle}}, \tag{25.76b}$$

and we have our second orthonormal vector. Similarly, subtracting out the $|\hat{e}_1\rangle$ and $|\hat{e}_2\rangle$ "directions" from $|\phi_3\rangle$ and normalizing,

$$|e_3\rangle \equiv |\phi_3\rangle - \langle\hat{e}_1|\phi_3\rangle\,|\hat{e}_1\rangle - \langle\hat{e}_2|\phi_3\rangle\,|\hat{e}_2\rangle \quad \longrightarrow \quad |\hat{e}_3\rangle \equiv \frac{|e_3\rangle}{\sqrt{\langle e_3|e_3\rangle}}, \tag{25.77}$$

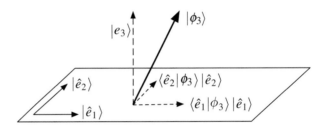

Figure 25.1 Gram–Schmidt orthogonalization.

yields the third, as depicted in Figure 25.1. The pattern should be clear: the mth orthonormal vector is constructed by subtracting off all its components which "point" in one of the previously established $m-1$ directions,

$$|e_m\rangle \equiv |\phi_m\rangle - \sum_{\ell=1}^{m-1}\langle\hat{e}_\ell|\phi_m\rangle\,|\hat{e}_\ell\rangle \quad \longrightarrow \quad |\hat{e}_m\rangle \equiv \frac{|e_m\rangle}{\sqrt{\langle e_m|e_m\rangle}}. \tag{25.78}$$

As long as the $|\phi_m\rangle$ are linearly independent, at least one of the inner products in (25.78) will be non-zero, yielding a new orthogonal vector.

Example 25.14 **Gram–Schmidt in $\mathbb{C}^3$**

The three vectors

$$|\phi_1\rangle \rightarrow \begin{pmatrix} 1 \\ -i \\ 1 \end{pmatrix} \qquad |\phi_2\rangle \rightarrow \begin{pmatrix} i \\ 0 \\ i \end{pmatrix} \qquad |\phi_3\rangle \rightarrow \begin{pmatrix} 1 \\ i \\ -1 \end{pmatrix} \tag{25.79}$$

are linearly independent, but clearly not orthogonal. Normalizing $|\phi_1\rangle$ gives

$$|\hat{e}_1\rangle \equiv \frac{|\phi_1\rangle}{\||\phi_1\rangle\|} \rightarrow \frac{1}{\sqrt{3}}\begin{pmatrix} 1 \\ -i \\ 1 \end{pmatrix}. \tag{25.80}$$

To get the second orthogonal direction, subtract the $|\hat{e}_1\rangle$ component from $|\phi_2\rangle$,

$$|e_2\rangle \equiv |\phi_2\rangle - \langle \hat{e}_1|\phi_2\rangle\, |\hat{e}_1\rangle$$

$$\rightarrow \begin{pmatrix} i \\ 0 \\ i \end{pmatrix} - \left[\frac{1}{\sqrt{3}} \begin{pmatrix} 1 & -i & 1 \end{pmatrix}^* \begin{pmatrix} i \\ 0 \\ i \end{pmatrix} \right] \cdot \frac{1}{\sqrt{3}} \begin{pmatrix} 1 \\ -i \\ 1 \end{pmatrix}$$

$$= \begin{pmatrix} i \\ 0 \\ i \end{pmatrix} - \frac{2i}{3} \begin{pmatrix} 1 \\ -i \\ 1 \end{pmatrix} = \frac{1}{3} \begin{pmatrix} i \\ -2 \\ i \end{pmatrix}. \tag{25.81}$$

Then

$$|\hat{e}_2\rangle \equiv \frac{|e_2\rangle}{\||e_2\rangle\|} \rightarrow \frac{1}{\sqrt{6}} \begin{pmatrix} i \\ -2 \\ i \end{pmatrix}. \tag{25.82}$$

Finally, subtracting the $|\hat{e}_1\rangle$ and $|\hat{e}_2\rangle$ components from $|\phi_3\rangle$,

$$|e_3\rangle \equiv |\phi_3\rangle - \langle \hat{e}_1|\phi_3\rangle\, |\hat{e}_1\rangle - \langle \hat{e}_2|\phi_3\rangle\, |\hat{e}_2\rangle$$

$$\rightarrow \begin{pmatrix} 1 \\ i \\ -1 \end{pmatrix} - \left[\frac{1}{\sqrt{3}} \begin{pmatrix} 1 & -i & 1 \end{pmatrix}^* \begin{pmatrix} 1 \\ i \\ -1 \end{pmatrix} \right] \cdot \frac{1}{\sqrt{3}} \begin{pmatrix} 1 \\ -i \\ 1 \end{pmatrix}$$

$$- \left[\frac{1}{\sqrt{6}} \begin{pmatrix} i & -2 & i \end{pmatrix}^* \begin{pmatrix} 1 \\ i \\ -1 \end{pmatrix} \right] \cdot \frac{1}{\sqrt{6}} \begin{pmatrix} i \\ -2 \\ i \end{pmatrix}$$

$$= \begin{pmatrix} 1 \\ i \\ -1 \end{pmatrix} + \frac{1}{3} \begin{pmatrix} 1 \\ -i \\ 1 \end{pmatrix} + \frac{i}{3} \begin{pmatrix} i \\ -2 \\ i \end{pmatrix} = \begin{pmatrix} 1 \\ 0 \\ -1 \end{pmatrix}. \tag{25.83}$$

gives $|\hat{e}_3\rangle$,

$$|\hat{e}_3\rangle \equiv \frac{|e_3\rangle}{\||e_3\rangle\|} \rightarrow \frac{1}{\sqrt{2}} \begin{pmatrix} 1 \\ 0 \\ -1 \end{pmatrix}. \tag{25.84}$$

We thus obtain the orthonormal basis

$$|\hat{e}_1\rangle \rightarrow \frac{1}{\sqrt{3}} \begin{pmatrix} 1 \\ -i \\ 1 \end{pmatrix} \qquad |\hat{e}_2\rangle \rightarrow \frac{1}{\sqrt{6}} \begin{pmatrix} i \\ -2 \\ i \end{pmatrix} \qquad |\hat{e}_3\rangle \rightarrow \frac{1}{\sqrt{2}} \begin{pmatrix} 1 \\ 0 \\ -1 \end{pmatrix}. \tag{25.85}$$

Taking the vectors of the original non-orthogonal basis in a different order will likely result in a distinct orthonormal basis. Indeed, had we orthogonalized them in reverse order, we would have obtained

$$|\hat{e}_3'\rangle \rightarrow \frac{1}{\sqrt{3}} \begin{pmatrix} 1 \\ i \\ -1 \end{pmatrix} \qquad |\hat{e}_2'\rangle \rightarrow \frac{i}{\sqrt{2}} \begin{pmatrix} 1 \\ 0 \\ 1 \end{pmatrix} \qquad |\hat{e}_1'\rangle \rightarrow \frac{1}{\sqrt{6}} \begin{pmatrix} 1 \\ -2i \\ -1 \end{pmatrix}. \tag{25.86}$$

Example 25.15 **Gram–Schmidt with 2 × 2 Matrices**

The matrices

$$|M_1\rangle \rightarrow \begin{pmatrix} 1 & 1 \\ 1 & 1 \end{pmatrix} \qquad |M_2\rangle \rightarrow \begin{pmatrix} 0 & i \\ i & 0 \end{pmatrix}$$

$$|M_3\rangle \rightarrow \begin{pmatrix} 1 & 0 \\ 0 & i \end{pmatrix} \qquad |M_4\rangle \rightarrow \begin{pmatrix} 0 & -1 \\ 1 & 0 \end{pmatrix} \tag{25.87}$$

are linearly independent, and thus form a basis for 2×2 complex matrices. They are not, however, orthonormal under the inner product

$$\langle M_i | M_j \rangle \equiv \frac{1}{2} \mathrm{Tr}\left(M_i^\dagger M_j \right) \neq \delta_{ij}, \tag{25.88}$$

as you can readily verify. Normalizing $|M_1\rangle$ gives

$$|\hat{e}_1\rangle \equiv \frac{|M_1\rangle}{\||M_1\rangle\|} \rightarrow \frac{1}{\sqrt{2}} \begin{pmatrix} 1 & 1 \\ 1 & 1 \end{pmatrix}. \tag{25.89}$$

Subtracting the projection of $|M_2\rangle$ along $|\hat{e}_1\rangle$ yields

$$|e_2\rangle \equiv |M_2\rangle - \langle \hat{e}_1 | M_2 \rangle\, |\hat{e}_1\rangle$$

$$\rightarrow \frac{i}{2} \begin{pmatrix} -1 & 1 \\ 1 & -1 \end{pmatrix}, \tag{25.90}$$

which, when normalized, gives

$$|\hat{e}_2\rangle \equiv \frac{|e_2\rangle}{\||e_2\rangle\|} \rightarrow \frac{i}{\sqrt{2}} \begin{pmatrix} -1 & 1 \\ 1 & -1 \end{pmatrix}. \tag{25.91}$$

Continuing this way, we find

$$|\hat{e}_3\rangle \rightarrow \frac{1-i}{\sqrt{2}} \begin{pmatrix} 1 & 0 \\ 0 & -1 \end{pmatrix}. \tag{25.92}$$

and

$$|\hat{e}_4\rangle \equiv |M_4\rangle \rightarrow \begin{pmatrix} 0 & -1 \\ 1 & 0 \end{pmatrix}. \tag{25.93}$$

25.5 Completeness

One of the most important features of a basis is its ability to "support" any vector — that is to say, any vector in the space can be expanded on any given basis. One says that the basis *spans* the space; this capacity is often called *completeness*. And just as (25.47) provides a simple statement of orthonormality, we'd like to have a succinct statement of completeness. Such a statement can be used as a test: a set of vectors which does *not* satisfy this completeness condition, even if linearly independent, cannot be a basis for the space.

Consider a vector $|A\rangle$ and an arbitrary orthonormal basis $\{|e_\ell\rangle\}$. Using (25.59),

$$|A\rangle = \sum_\ell a_\ell |\hat{e}_\ell\rangle = \langle \hat{e}_1|A\rangle |\hat{e}_1\rangle + \langle \hat{e}_2|A\rangle |\hat{e}_2\rangle + \cdots + \langle \hat{e}_n|A\rangle |\hat{e}_n\rangle$$

$$= |\hat{e}_1\rangle\langle \hat{e}_1|A\rangle + |\hat{e}_2\rangle\langle \hat{e}_2|A\rangle + \cdots + |\hat{e}_n\rangle\langle \hat{e}_n|A\rangle, \tag{25.94}$$

where we've moved the number $a_\ell \equiv \langle \hat{e}_\ell|A\rangle$ to the other side of $|\hat{e}_\ell\rangle$ — which, though a trivial maneuver, suggests that we can factor out the $|A\rangle$ on the far right:

$$|A\rangle = \Big[|\hat{e}_1\rangle\langle \hat{e}_1| + |\hat{e}_2\rangle\langle \hat{e}_2| + \cdots + |\hat{e}_n\rangle\langle \hat{e}_n| \Big]|A\rangle. \tag{25.95}$$

Each term in the sum $\sum_i |\hat{e}_i\rangle\langle \hat{e}_i|$ acts on $|A\rangle$ to project out the ith component. But since (25.94) must hold for *arbitrary* $|A\rangle$, it must be that

$$\sum_\ell |\hat{e}_\ell\rangle\langle \hat{e}_\ell| = |\hat{e}_1\rangle\langle \hat{e}_1| + |\hat{e}_2\rangle\langle \hat{e}_2| + \cdots + |\hat{e}_n\rangle\langle \hat{e}_n| = \mathbb{1}. \tag{25.96}$$

This identity, called the *completeness relation*, reinforces the intuitive picture of a vector as the sum of its projections $\mathcal{P}_\ell \equiv |\hat{e}_\ell\rangle\langle \hat{e}_\ell|$ along each basis direction. Indeed, for a complete basis, there *are* no other directions! Thus $\mathcal{P} \equiv \sum_\ell \mathcal{P}_\ell = \mathbb{1}$ is a projection onto the entire space. Note that this picture requires the basis to be orthonormal, because only then will we have a proper projection obeying $\mathcal{P}^2 = \mathcal{P}$:

$$\mathcal{P}^2 = \left(\sum_\ell |\hat{e}_\ell\rangle\langle \hat{e}_\ell| \right)\left(\sum_m |\hat{e}_m\rangle\langle \hat{e}_m| \right)$$

$$= \sum_{\ell m} |\hat{e}_\ell\rangle\langle \hat{e}_\ell|\hat{e}_m\rangle\langle \hat{e}_m| = \sum_{\ell m} |\hat{e}_\ell\rangle\delta_{\ell m}\langle \hat{e}_m| = \sum_\ell |\hat{e}_\ell\rangle\langle \hat{e}_\ell|. \tag{25.97}$$

So orthonormality ensures the sum over projections gets it all — no more, no less. This insight, implicit in the Gram–Schmidt procedure, is revealed by recasting the general formula (25.78) as

$$|e_m\rangle = \left[\mathbb{1} - \sum_{\ell=1}^{m-1} |\hat{e}_\ell\rangle\langle \hat{e}_\ell| \right]|\phi_m\rangle \equiv \mathcal{P}_\perp^{(m)}|\phi_m\rangle, \tag{25.98}$$

where $P_\perp^{(m)}$ is the projection onto the subspace *orthogonal* to the $m-1$ previously-constructed basis vectors. Once all n vectors have been constructed, the basis is complete and $P_\perp^{n+1}$ vanishes. There is no $n+1$st independent vector.

Example 25.16 **Inside the Outer Product $|A\rangle\langle B|$**

In a column vector representation, the inner product is

$$\langle A|B\rangle = \begin{pmatrix} \cdots & a_i & \cdots \end{pmatrix}^* \begin{pmatrix} \vdots \\ b_i \\ \vdots \end{pmatrix} = \sum_i a_i^* b_i, \tag{25.99}$$

which yields a complex number. However, the "outer product" $|A\rangle\langle B|$ is an array of products of the components,

$$|A\rangle\langle B| \rightarrow a_i b_j^*. \tag{25.100}$$

This can be suggestively rendered using columns and rows as

$$|A\rangle\langle B| \to \begin{pmatrix} \vdots \\ a_i \\ \vdots \end{pmatrix} \begin{pmatrix} \cdots & b_j & \cdots \end{pmatrix}^* = \begin{pmatrix} \ddots & \vdots & \iddots \\ \cdots & a_i b_j^* & \cdots \\ \iddots & \vdots & \ddots \end{pmatrix}. \qquad (25.101)$$

In this form it's clear the outer product is an *operator*, which acts on a vector and returns another vector,

$$|A\rangle\langle B|\big[|C\rangle\big] = |A\rangle\langle B|C\rangle \to a_i \sum_j b_j^* c_j. \qquad (25.102)$$

Example 25.17   **Completeness in $\mathbb{R}^3$**

In the standard basis, using only the familiar rules of matrix multiplication, the inner products are

$$\langle \hat{e}_1|\hat{e}_1\rangle = \langle \hat{e}_2|\hat{e}_2\rangle = \langle \hat{e}_3|\hat{e}_3\rangle$$

$$= \begin{pmatrix} 1 & 0 & 0 \end{pmatrix}\begin{pmatrix} 1 \\ 0 \\ 0 \end{pmatrix} = \begin{pmatrix} 0 & 1 & 0 \end{pmatrix}\begin{pmatrix} 0 \\ 1 \\ 0 \end{pmatrix} = \begin{pmatrix} 0 & 0 & 1 \end{pmatrix}\begin{pmatrix} 0 \\ 0 \\ 1 \end{pmatrix} = 1, \qquad (25.103a)$$

$$\langle \hat{e}_1|\hat{e}_2\rangle = \langle \hat{e}_2|\hat{e}_3\rangle = \langle \hat{e}_3|\hat{e}_1\rangle$$

$$= \begin{pmatrix} 1 & 0 & 0 \end{pmatrix}\begin{pmatrix} 0 \\ 1 \\ 0 \end{pmatrix} = \begin{pmatrix} 0 & 1 & 0 \end{pmatrix}\begin{pmatrix} 0 \\ 0 \\ 1 \end{pmatrix} = \begin{pmatrix} 0 & 0 & 1 \end{pmatrix}\begin{pmatrix} 1 \\ 0 \\ 0 \end{pmatrix} = 0. \qquad (25.103b)$$

So the representation faithfully renders orthonormality.

The completeness sum over a column vector basis manifests as

$$\sum_\ell |\hat{e}_\ell\rangle\langle\hat{e}_\ell| \to \sum_\ell \left(\hat{e}_\ell\right)_i \left(\hat{e}_\ell\right)_j^*, \qquad (25.104)$$

each term having the form of (25.101). In the standard basis, this gives

$$|\hat{e}_1\rangle\langle\hat{e}_1| + |\hat{e}_2\rangle\langle\hat{e}_2| + |\hat{e}_3\rangle\langle\hat{e}_3|$$

$$\to \begin{pmatrix} 1 \\ 0 \\ 0 \end{pmatrix}\begin{pmatrix} 1 & 0 & 0 \end{pmatrix} + \begin{pmatrix} 0 \\ 1 \\ 0 \end{pmatrix}\begin{pmatrix} 0 & 1 & 0 \end{pmatrix} + \begin{pmatrix} 0 \\ 0 \\ 1 \end{pmatrix}\begin{pmatrix} 0 & 0 & 1 \end{pmatrix}$$

$$\qquad (25.105)$$

$$= \begin{pmatrix} 1 & 0 & 0 \\ 0 & 0 & 0 \\ 0 & 0 & 0 \end{pmatrix} + \begin{pmatrix} 0 & 0 & 0 \\ 0 & 1 & 0 \\ 0 & 0 & 0 \end{pmatrix} + \begin{pmatrix} 0 & 0 & 0 \\ 0 & 0 & 0 \\ 0 & 0 & 1 \end{pmatrix} = \mathbb{1}.$$

So the basis is complete. Indeed, any two of the terms $|\hat{e}_\ell\rangle\langle\hat{e}_\ell|$ in (25.105) do not sum to the identity, so they are *not* complete and therefore cannot be a basis of $\mathbb{R}^3$ (though they are complete in $\mathbb{R}^2$).

Example 25.18 **Being $\mathbb{1}$ with the Basis**

As we saw in Example 24.6, the $\{\vec{\epsilon}_i\}$ basis of $\mathbb{R}^3$ is not normalized. And summing over the projections onto these non-normalized vectors does *not* give the identity:

$$\sum_{i=1}^{3} |\epsilon_i\rangle\langle\epsilon_i| = \begin{pmatrix} 1 \\ 1 \\ 1 \end{pmatrix} \begin{pmatrix} 1 & 1 & 1 \end{pmatrix} + \begin{pmatrix} 1 \\ -2 \\ 1 \end{pmatrix} \begin{pmatrix} 1 & -2 & 1 \end{pmatrix} + \begin{pmatrix} 1 \\ 0 \\ -1 \end{pmatrix} \begin{pmatrix} 1 & 0 & -1 \end{pmatrix}$$

$$= \begin{pmatrix} 1 & 1 & 1 \\ 1 & 1 & 1 \\ 1 & 1 & 1 \end{pmatrix} + \begin{pmatrix} 1 & -2 & 1 \\ -2 & 4 & -2 \\ 1 & -2 & 1 \end{pmatrix} + \begin{pmatrix} 1 & 0 & -1 \\ 0 & 0 & 0 \\ -1 & 0 & 1 \end{pmatrix} \neq \mathbb{1}. \qquad (25.106)$$

Normalizing the basis, however, divides each matrix in (25.106) by the square of the corresponding normalization constants in (25.65) — 3, 6, and 2, respectively — and the completeness relation $\sum_{i=1}^{3} |\hat{\epsilon}_i\rangle\langle\hat{\epsilon}_i| = \mathbb{1}$ emerges in full.

Example 25.19 **Rotating Completeness**

If a basis $\{\hat{e}_1, \hat{e}_2\}$ in $\mathbb{R}^2$ is complete, then so too is the rotated basis $\{\vec{\varepsilon}_1, \vec{\varepsilon}_2\}$ of (24.27):

$$\hat{\varepsilon}_1 \hat{\varepsilon}_1^\dagger + \hat{\varepsilon}_2 \hat{\varepsilon}_2^\dagger = \left(\hat{e}_1 \cos\varphi + \hat{e}_2 \sin\varphi\right)\left(\hat{e}_1 \cos\varphi + \hat{e}_2 \sin\varphi\right)^\dagger$$

$$+ \left(-\hat{e}_1 \sin\varphi + \hat{e}_2 \cos\varphi\right)\left(-\hat{e}_1 \sin\varphi + \hat{e}_2 \cos\varphi\right)^\dagger$$

$$= \hat{e}_1 \hat{e}_1^\dagger \cos^2\varphi + \hat{e}_2 \hat{e}_2^\dagger \sin^2\varphi + \left(\hat{e}_1 \hat{e}_2^\dagger + \hat{e}_2 \hat{e}_1^\dagger\right) \sin\varphi\cos\varphi$$

$$+ \hat{e}_1 \hat{e}_1^\dagger \sin^2\varphi + \hat{e}_2 \hat{e}_2^\dagger \cos^2\varphi - \left(\hat{e}_1 \hat{e}_2^\dagger + \hat{e}_2 \hat{e}_1^\dagger\right) \sin\varphi\cos\varphi$$

$$= \hat{e}_1 \hat{e}_1^\dagger + \hat{e}_2 \hat{e}_2^\dagger = \mathbb{1}. \qquad (25.107)$$

So rotating a complete orthonormal basis yields a complete orthonormal basis.

It's well worth emphasizing that the order of bra and ket is crucial: the inner product is a sum over the product vector components to produce a scalar, whereas the completeness relation is a sum over the set of basis vectors and yields an operator. Simply put, $\langle A|B\rangle \neq |B\rangle\langle A|$. And the difference looms large in the following:

Any basis $\{|\hat{e}_\ell\rangle\}$ that obeys

$$\langle\hat{e}_\ell|\hat{e}_m\rangle = \delta_{\ell m} \quad \text{and} \quad \sum_\ell |\hat{e}_\ell\rangle\langle\hat{e}_\ell| = \mathbb{1} \qquad (25.108)$$

is called a ***complete orthonormal basis***.

Now you might reasonably object that the expression "complete basis" is redundant. After all, a basis is a minimal spanning set by definition — so isn't an orthonormal basis *always* complete? For a finite-dimensional basis this is certainly true; for an *infinite*-dimensional space, however, things are more subtle. What does it even mean to be a "minimal" infinite set? How can we be sure we've projected along *every* direction? And how do we even know if the sum over these projections *converges*? The mathematical statement of completeness in (25.108) is ultimately the guarantor that a putative basis does indeed span a space of infinite dimensionality. This means not only that infinite sums $\sum_i c_i|\hat{e}_i\rangle$ converge, but that they converge to vectors within the space. In other words, just as for finite dimensions, an infinite-dimensional vector space must be closed. As a result, the completeness statement in (25.108) is often referred to as the *closure relation*. We'll have much more to say about the challenges of infinite-dimensional inner product spaces in Part V.

But whether the space is finite or infinite, the completeness relation is a common and useful tool for projecting vectors onto a given basis — with little more effort than it takes to multiply by $\mathbb{1}$:

$$|A\rangle = \mathbb{1}|A\rangle = \left(\sum_\ell |\hat{e}_\ell\rangle\langle\hat{e}_\ell|\right)|A\rangle$$

$$= \sum_\ell |\hat{e}_\ell\rangle\langle\hat{e}_\ell|A\rangle = \sum_\ell \langle\hat{e}_\ell|A\rangle|\hat{e}_\ell\rangle$$

$$\equiv \sum_\ell a_\ell|\hat{e}_\ell\rangle. \tag{25.109}$$

This is just the reverse of the derivation of (25.96). But a *change* of basis from $\{|\hat{e}_\ell\rangle\}$ to $\{|\hat{e}_\ell'\rangle\}$ can also be accomplished this way:

$$|A\rangle = \sum_\ell a_\ell|\hat{e}_\ell\rangle = \sum_\ell a_\ell\mathbb{1}|\hat{e}_\ell\rangle$$

$$= \sum_\ell a_\ell\left(\sum_m |\hat{e}_m'\rangle\langle\hat{e}_m'|\right)|\hat{e}_\ell\rangle$$

$$= \sum_{\ell m} \langle\hat{e}_m'|\hat{e}_\ell\rangle\, a_\ell|\hat{e}_m'\rangle \equiv \sum_m a_m'|\hat{e}_m'\rangle. \tag{25.110}$$

So the new components are given by matrix multiplication

$$a_m' = \sum_\ell \langle\hat{e}_m'|\hat{e}_\ell\rangle\, a_\ell. \tag{25.111}$$

This is just a rederivation of (25.61).

Example 25.20 **Expanding a Basis Vector II**

Let's expand the basis vectors $\hat{e}_{1,2}$ on the basis $\hat{\epsilon}_\ell$ basis of Example 25.9 — this time by using completeness rather then the column representation:

$$\hat{e}_1 = \left(\sum_\ell \hat{\epsilon}_\ell\hat{\epsilon}_\ell^\dagger\right)\cdot\hat{e}_1$$

$$= \hat{\epsilon}_1(\hat{\epsilon}_1^\dagger\cdot\hat{e}_1) + \hat{\epsilon}_2(\hat{\epsilon}_2^\dagger\cdot\hat{e}_1) = \frac{1}{\sqrt{2}}(\hat{\epsilon}_1 - \hat{\epsilon}_2), \tag{25.112}$$

where we've used the representation-independent relation $\hat{\epsilon}_{1,2} = \frac{1}{\sqrt{2}}(\pm\hat{e}_1 + \hat{e}_2)$. Since this is an example in $\mathbb{R}^n$, we've used dot product notation. But the deployment of the dot and the dagger can confuse; Dirac notation is cleaner:

$$|\hat{e}_1\rangle = \left(\sum_\ell |\hat{\epsilon}_\ell\rangle\langle\hat{\epsilon}_\ell|\right)|\hat{e}_1\rangle$$

$$= |\hat{\epsilon}_1\rangle\langle\hat{\epsilon}_1|\hat{e}_1\rangle + |\hat{\epsilon}_2\rangle\langle\hat{\epsilon}_2|\hat{e}_1\rangle = \frac{1}{\sqrt{2}}\left(|\hat{\epsilon}_1\rangle - |\hat{\epsilon}_2\rangle\right). \tag{25.113}$$

Similarly,

$$|\hat{e}_2\rangle = \left(\sum_\ell |\hat{\epsilon}_\ell\rangle\langle\hat{\epsilon}_\ell|\right)|\hat{e}_y\rangle$$

$$= |\hat{\epsilon}_1\rangle\langle\hat{\epsilon}_1|\hat{e}_2\rangle + |\hat{\epsilon}_2\rangle\langle\hat{\epsilon}_2|\hat{e}_2\rangle = \frac{1}{\sqrt{2}}\left(|\hat{\epsilon}_1\rangle + |\hat{\epsilon}_2\rangle\right). \qquad (25.114)$$

In a mere two dimensions this may not seem like the best way to invert (25.49), but in higher-dimensional spaces insertion of the completeness relation is a crucial tool for mapping between bases.

<hr>

Example 25.21 **Spin-1/2**

Quantum mechanics describes a spin-1/2 particle as a system with only two orientations, spin up and spin down. The general state of the system $|\psi\rangle$ is a linear combination of these two possibilities,

$$|\psi\rangle = a|\hat{e}_\uparrow\rangle + b|\hat{e}_\downarrow\rangle, \qquad (25.115)$$

where the magnitude-squared of a coefficient is the probability that a measurement finds that spin state,

$$\mathscr{P}(\uparrow) = |a|^2 \qquad \mathscr{P}(\downarrow) = |b|^2. \qquad (25.116)$$

Thus the projection of $|\psi\rangle$ along a "possible outcome" (i.e., basis) vector determines the probability of measuring that outcome. In this example there are only two possible outcomes, so these probabilities must sum to 100%, $|a|^2 + |b|^2 = 1$. Thus $|\psi\rangle$ must be normalized, $\langle\psi|\psi\rangle = 1$.

So $|\psi\rangle$ is a vector in a two-dimensional "state space," and can be represented by a two-dimensional column vector in $\mathbb{C}^2$,

$$|\psi\rangle \rightarrow \begin{pmatrix} a \\ b \end{pmatrix} = a\begin{pmatrix} 1 \\ 0 \end{pmatrix} + b\begin{pmatrix} 0 \\ 1 \end{pmatrix}. \qquad (25.117)$$

This is, of course, the standard basis — and in the usual quantum mechanical convention, represents spin up and down of $\pm\hbar/2$ along the z-axis. But that convention by no means implies that the spin *cannot* be measured along another axis — spin up and down along a different axis simply define a different basis. The state space is also spanned, for instance, by

$$|\hat{e}_+\rangle \rightarrow \frac{1}{\sqrt{2}}\begin{pmatrix} 1 \\ 1 \end{pmatrix} \qquad |\hat{e}_-\rangle \rightarrow \frac{1}{\sqrt{2}}\begin{pmatrix} 1 \\ -1 \end{pmatrix}. \qquad (25.118)$$

One can show that with the usual quantum conventions, $|\hat{e}_\pm\rangle$ represent spin up and spin down along the x-axis. In order to find the probabilities that a measurement along *this* axis yields up or down, just insert a complete set of states,

$$|\psi\rangle = |\hat{e}_+\rangle\langle\hat{e}_+|\psi\rangle + |\hat{e}_-\rangle\langle\hat{e}_-|\psi\rangle$$

$$= \frac{1}{\sqrt{2}}\left[(a+b)\,|\hat{e}_+\rangle + (a-b)\,|\hat{e}_-\rangle\right]. \qquad (25.119)$$

It's still the same normalized state,

$$\langle \psi | \psi \rangle = \frac{1}{2}|a+b|^2 + \frac{1}{2}|a-b|^2 = |a|^2 + |b|^2 = 1, \qquad (25.120)$$

but now the probabilities for finding spin up and down along the chosen axes are manifest, $\mathscr{P}(\pm) = \frac{1}{2}|a\pm b|^2$. Thus we have the stunning result that even a particle which is in a pure spin-up state along the z-axis, (25.115) with $b \equiv 0$, has equal probabilities of being measured to have spin up or down along x.

25.6 Matrix Representation of Operators

Consider an linear operator $\mathcal{M}$ such that

$$|W\rangle = \mathcal{M}|V\rangle, \qquad (25.121)$$

for vectors $|V\rangle$ and $|W\rangle$. We can project this onto a basis $\{\hat{e}_i\}$ to find the relationship between the components,

$$w_i' \equiv \langle \hat{e}_i | W \rangle = \langle \hat{e}_i | \mathcal{M} | V \rangle$$
$$= \sum_j \langle \hat{e}_i | \mathcal{M} | \hat{e}_j \rangle \langle \hat{e}_j | V \rangle \equiv \sum_j M_{ij} v_j, \qquad (25.122)$$

where in the second line we inserted the identity in the form of the completeness relation in (25.108). In terms of column vectors, (25.122) is

$$\begin{pmatrix} w_1' \\ w_2' \\ \vdots \\ w_n' \end{pmatrix} = \begin{pmatrix} M_{11} & M_{12} & \cdots & M_{1n} \\ M_{21} & M_{22} & \cdots & M_{2n} \\ \vdots & & & \vdots \\ M_{n1} & M_{n2} & \cdots & M_{nn} \end{pmatrix} \begin{pmatrix} v_1 \\ v_2 \\ \vdots \\ v_n \end{pmatrix}. \qquad (25.123)$$

Thus when vectors are represented as n-dimensional column vectors, operators become $n \times n$ matrices mapping column vectors into other column vectors — as expected of linear operators in a finite-dimensional vector space. But whereas the components of a vector $|V\rangle$ can be obtained with a single projection $v_i = \langle \hat{e}_i | V \rangle$ onto the orthonormal $\{|\hat{e}_i\rangle\}$ basis, (25.122) reveals that the ijth *matrix element* of $\mathcal{M}$ requires *two* projections,

$$M_{ij} \equiv \langle \hat{e}_i | \mathcal{M} | \hat{e}_j \rangle, \qquad (25.124)$$

one for the rows, one for the columns. To get some feeling for this, recall that $\langle \hat{e}_i | \mathcal{M} \rightarrow \hat{e}_i^\dagger \cdot \mathcal{M}$ yields a row of numbers — each one the projection of a column of M on the basis vector $\hat{e}_i$. The inner product of this new row with the column $\hat{e}_j$ yields the complex number M_{ij}. This is particularly straightforward using the standard basis. For instance, in $\mathbb{R}^2$

$$\langle \hat{e}_2 | \mathcal{M} | \hat{e}_1 \rangle = \begin{pmatrix} 0 & 1 \end{pmatrix}^* \begin{pmatrix} M_{11} & M_{12} \\ M_{21} & M_{22} \end{pmatrix} \begin{pmatrix} 1 \\ 0 \end{pmatrix} = M_{21}, \qquad (25.125)$$

as it should. Thus just as the components of a vector are basis-dependent, so too are the matrix elements of $\mathcal{M}$. In (25.125), M_{21} is the 21-element of $\mathcal{M}$ on the standard basis. To find the matrix elements $M_{ab}' \equiv \langle \hat{\epsilon}_a | \mathcal{M} | \hat{\epsilon}_b \rangle$ on a different basis $\{|\hat{\epsilon}_a\rangle\}$ for $a = 1, 2$, we can use completeness — but we need to deploy it twice in order to change the basis:

$$\mathcal{M} \xrightarrow[\hat{\epsilon} \text{ basis}]{} M'_{ab} \equiv \langle \hat{\epsilon}_a | \mathcal{M} | \hat{\epsilon}_b \rangle = \langle \hat{\epsilon}_a | \mathbb{1} \mathcal{M} \mathbb{1} | \hat{\epsilon}_b \rangle$$

$$= \langle \hat{\epsilon}_a | \left(\sum_i | \hat{e}_i \rangle \langle \hat{e}_i | \right) \mathcal{M} \left(\sum_j | \hat{e}_j \rangle \langle \hat{e}_j | \right) | \hat{\epsilon}_b \rangle$$

$$= \sum_{ij} \langle \hat{\epsilon}_a | \hat{e}_i \rangle \langle \hat{e}_i | \mathcal{M} | \hat{e}_j \rangle \langle \hat{e}_j | \hat{\epsilon}_b \rangle$$

$$\equiv \sum_{ij} a_i^* M_{ij} b_j, \tag{25.126}$$

where $a_i \equiv \langle \hat{e}_i | \hat{\epsilon}_a \rangle$, $b_i \equiv \langle \hat{e}_i | \hat{\epsilon}_b \rangle$ are the components of $| \hat{\epsilon}_a \rangle$ and $| \hat{\epsilon}_b \rangle$ on the $| \hat{e}_i \rangle$ basis.

Example 25.22 Basis of Operations

Consider the operator $\mathcal{M}$ with standard basis elements $\langle \hat{e}_1 | \mathcal{M} | \hat{e}_1 \rangle = \langle \hat{e}_2 | \mathcal{M} | \hat{e}_2 \rangle = 1$ and $\langle \hat{e}_1 | \mathcal{M} | \hat{e}_2 \rangle = \langle \hat{e}_2 | \mathcal{M} | \hat{e}_1 \rangle = 2$, that is,

$$\mathcal{M} \xrightarrow[\text{standard basis}]{} M = \begin{pmatrix} \langle \hat{e}_1 | \mathcal{M} | \hat{e}_1 \rangle & \langle \hat{e}_1 | \mathcal{M} | \hat{e}_2 \rangle \\ \langle \hat{e}_2 | \mathcal{M} | \hat{e}_1 \rangle & \langle \hat{e}_2 | \mathcal{M} | \hat{e}_2 \rangle \end{pmatrix} = \begin{pmatrix} 1 & 2 \\ 2 & 1 \end{pmatrix}. \tag{25.127}$$

Then on the orthonormal basis

$$| \hat{\epsilon}_1 \rangle \to \frac{1}{\sqrt{2}} \begin{pmatrix} 1 \\ 1 \end{pmatrix} \qquad | \hat{\epsilon}_2 \rangle \to \frac{1}{\sqrt{2}} \begin{pmatrix} 1 \\ -1 \end{pmatrix}, \tag{25.128}$$

$\mathcal{M}$ has matrix elements

$$M'_{11} \equiv \langle \hat{\epsilon}_1 | \mathcal{M} | \hat{\epsilon}_1 \rangle = \frac{1}{\sqrt{2}} \begin{pmatrix} 1 & 1 \end{pmatrix}^* \begin{pmatrix} 1 & 2 \\ 2 & 1 \end{pmatrix} \frac{1}{\sqrt{2}} \begin{pmatrix} 1 \\ 1 \end{pmatrix} = 3 \tag{25.129a}$$

and

$$M'_{12} \equiv \langle \hat{\epsilon}_1 | \mathcal{M} | \hat{\epsilon}_2 \rangle = \frac{1}{\sqrt{2}} \begin{pmatrix} 1 & 1 \end{pmatrix}^* \begin{pmatrix} 1 & 2 \\ 2 & 1 \end{pmatrix} \frac{1}{\sqrt{2}} \begin{pmatrix} 1 \\ -1 \end{pmatrix} = 0. \tag{25.129b}$$

Similarly, $M'_{22} = -1$ and $M'_{21} = 0$. So on this basis,

$$\mathcal{M} \xrightarrow[\hat{\epsilon} \text{ basis}]{} M' = \begin{pmatrix} 3 & 0 \\ 0 & -1 \end{pmatrix}. \tag{25.130}$$

On the orthonormal basis

$$| \hat{\varepsilon}_1 \rangle \to \frac{1}{\sqrt{2}} \begin{pmatrix} 1 \\ i \end{pmatrix} \qquad | \hat{\varepsilon}_2 \rangle \to \frac{1}{\sqrt{2}} \begin{pmatrix} 1 \\ -i \end{pmatrix}, \tag{25.131}$$

we find

$$M''_{11} \equiv \langle \hat{\varepsilon}_1 | \mathcal{M} | \hat{\varepsilon}_1 \rangle = \frac{1}{\sqrt{2}} \begin{pmatrix} 1 & i \end{pmatrix}^* \begin{pmatrix} 1 & 2 \\ 2 & 1 \end{pmatrix} \frac{1}{\sqrt{2}} \begin{pmatrix} 1 \\ i \end{pmatrix} = 1 \tag{25.132a}$$

and

$$M''_{12} \equiv \langle \hat{\varepsilon}_1 | \mathcal{M} | \hat{\varepsilon}_2 \rangle = \frac{1}{\sqrt{2}} \begin{pmatrix} 1 & i \end{pmatrix}^* \begin{pmatrix} 1 & 2 \\ 2 & 1 \end{pmatrix} \frac{1}{\sqrt{2}} \begin{pmatrix} 1 \\ -i \end{pmatrix} = -2i. \tag{25.132b}$$

You should verify that $M''_{22} = M''_{11}$ and $M''_{21} = M''^*_{12}$ — giving the matrix

$$\mathcal{M} \xrightarrow[\hat{\varepsilon} \text{ basis}]{} M'' = \begin{pmatrix} 1 & -2i \\ 2i & 1 \end{pmatrix}. \tag{25.133}$$

These are all equivalent representations of the same operator $\mathcal{M}$; only its matrix elements are basis dependent.

Incidentally, inspection of these matrix representations reveals all three to be Hermitian, $M_{ij} = M^*_{ji}$ — consistent with the assertion of (25.24) that hermiticity $\mathcal{M} = \mathcal{M}^\dagger$ is a basis-independent attribute of an operator.

There is little doubt that orthonormal bases are generally preferred; the challenge is how to choose amongst the infinite possibilities. Even with the Gram–Schmidt procedure, not only can different linearly independent sets $\{|\phi_i\rangle\}$ yield distinct orthonormal bases, just choosing the vectors of a set in a different order can lead to a different result. With so many possible bases, how does one decide which to use? Ultimately, the choice is arbitrary — though not surprisingly, some bases are more convenient than others. The simplicity of the standard basis is certainly appealing. But frequently the best choice is dictated by the operators, which often have preferred bases that both simplify the mathematics, and provide greater physical insight.

What might such a preferred basis be? As (25.130) shows, at least some operators have a diagonal matrix representation. Diagonal matrices are much simpler to work with, so oftentimes the best choice of orthonormal basis is the one on which the operator is diagonal (for instance, the $|\hat{\varepsilon}\rangle$ basis in Example 25.22). For Hermitian operators, such a basis always exists. We'll discover how to find these diagonalizing bases in Chapters 26 and 27.

Problems

25.1   Given the following vectors:

$$|1\rangle \to \begin{pmatrix} 1 \\ i \end{pmatrix} \quad |2\rangle \to \begin{pmatrix} -i \\ 2i \end{pmatrix} \quad |3\rangle \to \begin{pmatrix} e^{i\varphi} \\ -1 \end{pmatrix} \quad |4\rangle \to \begin{pmatrix} 1 \\ -2i \\ 1 \end{pmatrix} \quad |5\rangle \to \begin{pmatrix} i \\ 1 \\ i \end{pmatrix}.$$

(a) Show that each vector has positive norm. Then normalize each.
(b) Verify that the pairs of the same dimensionality satisfy the requirement of conjugate symmetry, $\langle A|B\rangle = \langle B|A\rangle^*$.

25.2   A proper inner product must have $\langle A|A\rangle \geq 0$. Using only $|cA\rangle = c|A\rangle$ and conjugate symmetry $\langle A|B\rangle = \langle B|A\rangle^*$, verify that even a vector $|B\rangle \equiv |iA\rangle$ must have $\langle B|B\rangle \geq 0$. Show that this leads to the result $\langle cA| = c^*\langle A|$.

25.3   (a) Prove that vectors in an inner product space obey the parallelogram law: $\langle A|A\rangle + \langle B|B\rangle = \frac{1}{2}\langle A + B|A + B\rangle + \frac{1}{2}\langle A - B|A - B\rangle$. Thus the sum of the square of the sides = the sum of square of the diagonals.
(b) Prove the Pythagorean theorem for orthogonal vectors $|A\rangle$ and $|B\rangle$, $\langle A|A\rangle + \langle B|B\rangle = \langle A + B|A + B\rangle$.

25.4   For vectors $|v\rangle \to \begin{pmatrix} 1 \\ -i \end{pmatrix}$, $|w\rangle \to \begin{pmatrix} -1 \\ 1 \end{pmatrix}$, and the matrix operator $\mathcal{M} \to \begin{pmatrix} 1 & i \\ 2 & 1 \end{pmatrix}$:

(a) Show that $|u\rangle = |\mathcal{M}v\rangle = \mathcal{M}|v\rangle$ and $\langle u| = \langle \mathcal{M}v| = \langle v|\mathcal{M}^\dagger$ are adjoints of one another.
(b) Verify (25.24).

25.5   Use bra-ket manipulation to verify the equivalent definitions of adjoint in (25.24).

25.6   **Is This a Dagger Which I See Before Me?**
Using only the definition of the adjoint operator $\mathcal{T}^\dagger$,

$$\langle v|\mathcal{T}^\dagger|w\rangle = \langle w|\mathcal{T}|v\rangle^*,$$

(a) Show that $\mathcal{T}^{\dagger\dagger} = \mathcal{T}$. [The identity $z^{**} = z$ for complex numbers is a special case.]
(b) Given a second operator $\mathcal{S}$, show that the adjoint of the product is the reverse-product of the adjoints,

$$(\mathcal{S}\mathcal{T})^\dagger = \mathcal{T}^\dagger \mathcal{S}^\dagger. \tag{25.134}$$

In fact, do it twice: first using a third vector $|u\rangle = \mathcal{T}|v\rangle$, and a second time using the $\mathcal{T}|v\rangle \equiv |\mathcal{T}v\rangle$.

25.7   Find the angle between the following vector pairs:

(a) $|v\rangle \rightarrow \begin{pmatrix} 4 \\ -3 \end{pmatrix}$, $|w\rangle \rightarrow \begin{pmatrix} 1 \\ 1 \end{pmatrix}$

(b) $|v\rangle \rightarrow \begin{pmatrix} 1-i \\ i \\ -2i \end{pmatrix}$, $|w\rangle \rightarrow \begin{pmatrix} 2i \\ 1 \\ 2+i \end{pmatrix}$

(c) $|v\rangle \rightarrow \begin{pmatrix} 1 & i \\ -i & 2 \end{pmatrix}$, $|w\rangle \rightarrow \begin{pmatrix} i & 1+i \\ -i & 1 \end{pmatrix}$.

25.8   Demonstrate that each of the following bases is orthonormal:

(a) $|1\rangle \rightarrow \begin{pmatrix} 1 \\ 0 \end{pmatrix}$, $|2\rangle \rightarrow \begin{pmatrix} 0 \\ 1 \end{pmatrix}$

(b) $|+\rangle \rightarrow \frac{1}{\sqrt{2}}\begin{pmatrix} 1 \\ i \end{pmatrix}$, $|-\rangle \rightarrow \frac{1}{\sqrt{2}}\begin{pmatrix} 1 \\ -i \end{pmatrix}$

(c) $|\uparrow\rangle \rightarrow \frac{1}{\sqrt{2}}\begin{pmatrix} 1 \\ 1 \end{pmatrix}$, $|\downarrow\rangle \rightarrow \frac{1}{\sqrt{2}}\begin{pmatrix} 1 \\ -1 \end{pmatrix}$

(d) $|I\rangle \rightarrow \frac{1}{3}\begin{pmatrix} i-\sqrt{3} \\ 1+2i \end{pmatrix}$, $|II\rangle \rightarrow \frac{1}{3}\begin{pmatrix} 1-2i \\ i+\sqrt{3} \end{pmatrix}$.

25.9   Use projections to find the components of the vectors

$$|A\rangle \rightarrow \begin{pmatrix} 1-i \\ 2+2i \end{pmatrix} \qquad |B\rangle \rightarrow \begin{pmatrix} 1+i \\ 2i \end{pmatrix}$$

on the bases in Problem 25.8 — i.e., find $a_{1,2}$ such that $|A\rangle = a_1|1\rangle + a_2|2\rangle$, and similarly the pairs $a_\pm$, $a_{\uparrow,\downarrow}$, and $a_{I,II}$ for the other three bases. Calculate the norm squared of $|A\rangle$ and $|B\rangle$ in each basis.

25.10   Use the trace inner product (25.72) of Example 25.13 to verify the vector decomposition of (24.23).

25.11   Show how the Schwarz inequality and the directional derivative df/du lends meaning to the gradient vector $\vec{\nabla}f$.

25.12   Show that the triangle inequality (25.37) leads to (25.41).

25.13   Use bra-ket notation to normalize $|A\rangle = |\hat{e}_1\rangle + |\hat{e}_2\rangle + |\hat{e}_3\rangle$.

25.14   Rederive the components found in (25.50) and (25.51) of Example 25.9 using only the representation-independent relation of (25.49),

$$\hat{\epsilon}_{1,2} = \frac{1}{\sqrt{2}}\left(\pm\hat{e}_1 + \hat{e}_2\right),$$

together with the knowledge that $\{\hat{e}_1, \hat{e}_2\}$ is an orthonormal basis. No column or row vectors allowed!

25.15 For an inner product space with orthonormal basis $\{|\hat{e}_i\rangle\}$, prove that the norm $\||A\rangle - \sum_i a_i|\hat{e}_i\rangle\|$ is minimized for $a_i = \langle\hat{e}_i|A\rangle$. For simplicity, assume the vector space is real.

25.16 Verify that $|\hat{e}_1\rangle, |\hat{e}_2\rangle$, and $|\hat{e}_3\rangle$ of (25.75)–(25.77) are orthonormal.

25.17 Following the Gram–Schmidt approach in Example 25.15, derive the vectors $|\hat{e}_3\rangle$ and $|\hat{e}_4\rangle$ in (25.92)–(25.93).

25.18 **Expansion on an Orthogonal Basis**
Though orthonormal bases are often preferred, it's not uncommon to use orthogonal bases instead. Consider the expansion of a vector $|v\rangle = \sum_m c_m|\phi_m\rangle$ on a basis with orthogonality relation $\langle\phi_n|\phi_m\rangle = k_m\delta_{mn}$. (An orthonormal basis obtains when $k_m \equiv 1$ for all m.)

(a) Show that the expansion coefficients are given by

$$c_n = \frac{1}{k_n}\langle\phi_n|v\rangle. \tag{25.135}$$

(b) Show that the completeness relation must be

$$\sum_m \frac{1}{k_m}|\phi_m\rangle\langle\phi_m| = \mathbb{1}. \tag{25.136}$$

25.19 By inserting a complete set of basis vectors $|\hat{e}_i\rangle$, show that the definition of adjoint in (25.24) is the transpose–complex-conjugate of the matrix $M_{ij} = \langle\hat{e}_i|\mathcal{M}|\hat{e}_j\rangle$.

25.20 Use bras and kets, together with the completeness relation for arbitrary orthonormal bases $\{|\hat{e}_i\rangle\}$ and $\{|\hat{\epsilon}_a\rangle\}$, to show that the trace of an operator A is independent of basis.

25.21 An operator $\mathcal{P}$ is a projection if $\mathcal{P}^2 = \mathcal{P}$.

(a) Give an intuitive explanation why this must be so.
(b) Under what condition is the operator $|\hat{e}_i\rangle\langle\hat{e}_j|$ an orthogonal projection — that is, it projects onto orthogonal subspaces?
(c) Show that the operator $\mathcal{P}_\perp^{(m)}$ defined in (25.98) is an orthogonal projection.
(d) Without using the explicit form of $\mathcal{P}_\perp^{(m)}$, verify that $|A\rangle \equiv \mathcal{P}_\perp^{(m)}|V\rangle$ is orthogonal to $|B\rangle \equiv |V\rangle - \mathcal{P}_\perp^{(m)}|V\rangle$.

25.22 The notation $\vec{a} \cdot \vec{\sigma}$ is a common and convenient short hand for $\sum_i a_i\sigma_i$ — but is it an inner product? Explain.

25.23 An arbitrary 2×2 complex matrix can be expanded on the Pauli basis $\{\sigma_0, \sigma_1, \sigma_2, \sigma_3\}$ as $M = \sum_{\alpha=0}^{3} c_\alpha\sigma_\alpha$, where $\sigma_0 \equiv \mathbb{1}$. Show that this leads to the completeness relation

$$\sum_{\alpha=0}^{3} \sigma_\alpha\sigma_\alpha^\dagger = \sum_{\alpha=0}^{3} (\sigma_\alpha)_{ij}(\sigma_\alpha)_{k\ell} = 2\delta_{i\ell}\delta_{jk}.$$

25.24 **Pauli Algebra**

(a) Verify that the Pauli matrices σ_i are all traceless and Hermitian.
(b) Show that

$$\sigma_\ell\sigma_m = \mathbb{1}\delta_{\ell m} + i\sum_n \epsilon_{\ell mn}\sigma_n. \tag{25.137}$$

Use this to prove that $\mathrm{Tr}\left(\sigma_i\sigma_j\right) = 2\delta_{ij}$. [The identity (25.137) is often written without the sum, since for given ℓ and m on the left, only one value of n makes a non-zero contribution.]
(c) Verify that for vectors $\vec{a}$ and $\vec{b}$,

$$(\vec{a} \cdot \vec{\sigma})(\vec{b} \cdot \vec{\sigma}) = \vec{a} \cdot \vec{b}\mathbb{1} + i\vec{\sigma} \cdot (\vec{a} \times \vec{b}). \tag{25.138}$$

(d) Show that the matrix representation (25.74) of a vector respects the $\mathbb{R}^3$ inner product — that is, $\langle A|B\rangle = \frac{1}{2}\text{Tr}(A^\dagger B) = \vec{a}\cdot\vec{b}$.

25.25 Use the Gram–Schmidt procedure to verify (25.86), thus demonstrating that different orderings of the original linearly independent set can yield different orthonormal bases.

25.26 Verify that the following vectors are linearly independent,

$$|\phi_1\rangle \to \begin{pmatrix} 1 \\ 1 \\ 1 \\ 1 \end{pmatrix} \qquad |\phi_2\rangle \to \begin{pmatrix} 1 \\ -1 \\ 1 \\ -1 \end{pmatrix} \qquad |\phi_3\rangle \to \begin{pmatrix} 1 \\ -i \\ 1 \\ i \end{pmatrix} \qquad |\phi_4\rangle \to \begin{pmatrix} 1 \\ 1 \\ i \\ i \end{pmatrix},$$

and then use the Gram–Schmidt procedure to construct an orthonormal basis $\{|\hat{e}_i\rangle\}$ from them.

25.27 A spin-1/2 particle is in the normalized state $|\psi\rangle \to \begin{pmatrix} a \\ b \end{pmatrix}$. (See Example 25.21.) One can show that the basis vectors

$$|+\rangle \to \frac{1}{\sqrt{2}}\begin{pmatrix} 1 \\ i \end{pmatrix} \qquad |-\rangle \to \frac{1}{\sqrt{2}}\begin{pmatrix} 1 \\ -i \end{pmatrix}$$

respectively describe spin up and down along the y-axis. What are the probabilities that measurements on the state $|\psi\rangle$ return either spin up or down along y? What is the sum of these two probabilities?

25.28 As discussed in Example 25.21, a spin-1/2 particle is described in quantum mechanics by a two-dimensional state space. We'll work in the orthonormal basis $\{|\uparrow\rangle, |\downarrow\rangle\}$ — which, by general convention, are the states spin up and spin down along the z-axis. [In Example 25.21 these were labeled $|\hat{e}_{\uparrow,\downarrow}\rangle$ — but really, why carry the extra $\hat{e}$'s around?] The three so-called spin operators are

$$S_x = \frac{\hbar}{2}\left(|\uparrow\rangle\langle\downarrow| + |\downarrow\rangle\langle\uparrow|\right)$$
$$S_y = \frac{i\hbar}{2}\left(|\downarrow\rangle\langle\uparrow| - |\uparrow\rangle\langle\downarrow|\right)$$
$$S_z = \frac{\hbar}{2}\left(|\uparrow\rangle\langle\uparrow| - |\downarrow\rangle\langle\downarrow|\right),$$

where $\hbar$ is Planck's constant. Show that

$$S_i S_j = \frac{\hbar^2}{4}\delta_{ij} + i\frac{\hbar}{2}\epsilon_{ijk}S_k. \qquad (25.139)$$

Find the matrix representations of these operators on the standard basis $|\uparrow\rangle \to \begin{pmatrix} 1 \\ 0 \end{pmatrix}, |\downarrow\rangle \to \begin{pmatrix} 0 \\ 1 \end{pmatrix}$. Verify that they obey (25.139).

25.29 Consider a two-dimensional orthonormal basis $|1\rangle, |2\rangle$ and an operator $\mathcal{T}$ such that

$$\mathcal{T}|1\rangle = |2\rangle \qquad \mathcal{T}|2\rangle = |1\rangle.$$

(a) Express $\mathcal{T}$ as a sum of outer products.
(b) Find three *distinct* matrix representations for $\mathcal{T}$.

25.30 Find matrices S and T representing the operators $\mathcal{S}$ and $\mathcal{T}$ in the orthonormal basis $\{|\hat{e}_i\rangle\}$, $i = 1$ to 3, such that

$$\mathcal{S}|\hat{e}_1\rangle = |\hat{e}_1\rangle \quad \mathcal{S}|\hat{e}_2\rangle = 0 \quad \mathcal{S}|\hat{e}_3\rangle = -|\hat{e}_3\rangle$$
$$\mathcal{T}|\hat{e}_1\rangle = |\hat{e}_3\rangle \quad \mathcal{T}|\hat{e}_2\rangle = |\hat{e}_2\rangle \quad \mathcal{T}|\hat{e}_3\rangle = |\hat{e}_1\rangle.$$

Express both as a sum of outer products. What is their matrix representations in the standard basis?

25.31 Follow the steps in (25.126) to find the matrix M'' in (25.133) directly from M' in (25.130).

26 Interlude: Rotations

After many pages of fairly formal discussion of vectors, matrices, and inner products, we turn (pun intended) to something a bit more physical: rotations. A rotation is a linear transformation which leaves one point, the origin, fixed. Of course, all linear transformations necessarily map the zero vector into itself (Problem 23.3) — but the entire class of linear transformations is too large for our purposes. For instance, we must be able to rotate back and forth between $|v\rangle$ and $|v'\rangle = \mathcal{R}|v\rangle$; this requires a rotation operator $\mathcal{R}$ to be invertible, and therefore represented by a square matrix R. A rotation must also leave vectors' lengths and relative orientations unchanged — so we'll only consider transformations $\mathcal{R}$ which preserve the inner product. This means that rotations not only take $|v\rangle$ into $|v'\rangle$, but also map entire orthonormal bases into one another.

But why an entire chapter devoted to rotations? Well, there's the profound observation that the physical world appears to be fundamentally rotationally invariant, implying that coordinate frames which differ only by rotation are physically indistinguishable. Thus the laws of physics themselves must be expressible in an inherently rotationally – invariant way (Chapter 29).

More broadly, $\mathbb{R}^n$ rotations provide an intuitive illustration of continuous symmetries appearing all over physics — including rotations in complex space, translations in space and time, and Lorentz transformations. Mathematically, these are all examples of *groups*. The technical criteria defining a group are given in BTW 26.2 — but as we'll see, a continuous group of transformations can be understood intuitively as generalized rotations. As as result, rotations serve as a powerful physical paradigm for many of the linear operators you're likely to encounter. So command of rotations and rotational symmetry provides insight into other continuous symmetries. (Lorentz transformations, for instance, can be viewed as rotations in spacetime.) Moreover, the ability to wield rotations allows one to formulate a problem in one basis, and then rotate the system to analyze it in a second, perhaps more physically-transparent basis. Instances include not just simplifying equations of motion, but also moving to a particle's rest frame (Example 26.15), as well as the extremely important phenomenon of normal modes (Chapter 28).

After discussing the equivalence of rotating a vector versus rotating the basis, we consider rotations in the plane for real and complex Euclidean space. Once the fundamentals have been established, we move to the more subtle case of rotations in three-space.

26.1 Active and Passive Transformations

Consider a vector $\vec{v} = (v_1, v_2)$ in the plane. Rotating $\vec{v}$ counter-clockwise through an angle φ results in a new vector $\vec{v}'$. The polar coordinates of both vectors are easily found from Figure 26.1:

$$
\begin{aligned}
v_1 &= v \cos \alpha & v_1' &= v \cos \beta \\
v_2 &= v \sin \alpha & v_2' &= v \sin \beta,
\end{aligned}
\tag{26.1}
$$

where $|\vec{v}| = |\vec{v}'| = v$. Since $\beta = \alpha + \varphi$, a familiar trig identity yields

$$v_1' = v_1 \cos\varphi - v_2 \sin\varphi \qquad (26.2a)$$
$$v_2' = v_1 \sin\varphi + v_2 \cos\varphi. \qquad (26.2b)$$

So a rotation mixes up the components v_i to produce new components v_i',

$$v_i' = \sum_j R_{ij} v_j, \qquad (26.3)$$

where R_{ij} is the *rotation matrix*,

$$R(\varphi) \equiv \begin{pmatrix} \cos\varphi & -\sin\varphi \\ \sin\varphi & \cos\varphi \end{pmatrix}. \qquad (26.4)$$

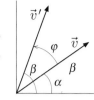

Figure 26.1
Active rotation.

Note that $\varphi > 0$ gives a counter-clockwise rotation; $R(-\varphi)$, then, is a clockwise rotation. Positive and negative rotations are obviously *inverses* of one another. Furthermore, since $\cos\varphi$ is an even function and $\sin\varphi$ an odd function, a negative rotation is just the *transpose* of $R(\varphi)$. In other words,

$$R^{-1}(\varphi) = R(-\varphi) = R^T(\varphi). \qquad (26.5)$$

As we'll see, $R^{-1} = R^T$ is a defining property of rotations in $\mathbb{R}^n$.

Example 26.1 **Maxwell's Equations and Duality**

Not all rotations are concerned with physical space. Consider the close relation between electric and magnetic fields in the absence of sources ρ and $\vec{J}$:

$$\vec{\nabla} \cdot \vec{E} = 0 \qquad \vec{\nabla} \times \vec{E} = -\partial_t \vec{B}$$
$$\vec{\nabla} \cdot \vec{B} = 0 \qquad \vec{\nabla} \times \vec{B} = \mu_0 \epsilon_0 \partial_t \vec{E}. \qquad (26.6)$$

Let's examine the effect of the following rotation,

$$\vec{E}' = \vec{E} \cos\varphi - c\vec{B} \sin\varphi$$
$$c\vec{B}' = \vec{E} \sin\varphi + c\vec{B} \cos\varphi. \qquad (26.7)$$

This rotation mixes electric and magnetic fields into one another; it occurs not in physical space, but rather in the more abstract space of fields.[1] Thus the derivatives in (26.6) do not act on φ. The remarkable thing is that (26.7) actually leaves Maxwell's equations invariant. In other words, if $\vec{E}$ and $\vec{B}$ are solutions of Maxwell's equations, so are $\vec{E}'$ and $\vec{B}'$. This is called a *duality transformation*, and the invariance of Maxwell's equations under (26.7) is known as the duality property of electromagnetic fields. Of course, we cannot actually perform a duality transformation in the laboratory, but the symmetry of Maxwell's equations under this purely mathematical rotation still tells us something about real electric and magnetic fields. In particular, a duality transformation with $\varphi = \pi/2$ essentially interchanges $\vec{E}$ and $\vec{B}$. Thus for any field configuration $(\vec{E}, c\vec{B})$ which may turn up in the lab, duality mandates that there exists a way to produce the same configuration with $\vec{E}$ and $c\vec{B}$ replaced by $c\vec{B}'$ and $\vec{E}'$, respectively. For instance, the electric field between the plates of an ideal parallel-plate capacitor are straight with constant magnitude, while the $\vec{B}$ field is zero. Thus it must be possible to generate a magnetic field which is also constant in magnitude and points in a straight line, but with $\vec{E} = 0$. The electromagnetic fields of an ideal solenoid have just these features.

---

[1] The factor of c is included to give the field-space components $(\vec{E}, c\vec{B})$ the same SI units.

Thus far have we have made no mention of the basis; indeed, the presumption of (26.3) is that the basis is the same for both $\vec{v}$ and $\vec{v}\,'$. We can achieve the same result, however, by leaving $\vec{v}$ fixed and instead rotate the axes *clockwise* by the same amount φ,

$$\begin{pmatrix} |\hat{e}_1'\rangle \\ |\hat{e}_2'\rangle \end{pmatrix} = \begin{pmatrix} \cos\varphi & -\sin\varphi \\ \sin\varphi & \cos\varphi \end{pmatrix} \begin{pmatrix} |\hat{e}_1\rangle \\ |\hat{e}_2\rangle \end{pmatrix}. \tag{26.8}$$

This is fairly intuitive — and a straightforward exercise verifies that the components of $\vec{v}$ on the primed basis are the *same* as the components of $\vec{v}\,'$ on the unprimed basis (Problem 26.1).

But wait, isn't the rotation matrix in (26.8) a *counter*-clockwise rotation? A quick way to check that (26.3), (26.4), and (26.8) are consistent is to look for the sign of the sine: negative in the top right corner means that (26.3) yields $v_1' < v_1$ — so the vector is being rotated counter-clockwise. But that same sign gives $|\hat{e}_1'\rangle$ as the vector *difference* of $|\hat{e}_1\rangle$ and $|\hat{e}_2\rangle$ in (26.8) — so it's being rotated clockwise (Figure 26.2). In the first case, the rotation mixes up the *components* of $\vec{v}$; in the second it's the *vectors* $|\hat{e}_i\rangle$ which are being mixed. In both cases though, the new vector components are given by

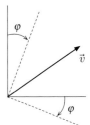

Figure 26.2
Passive rotation.

$$v_i' = \langle \hat{e}_i | \mathcal{R} | v \rangle. \tag{26.9}$$

Associativity provides us the option of applying $\mathcal{R}$ on $|v\rangle$ and then projecting onto the fixed basis, or first letting $\mathcal{R}$ operate to the left and afterwards take the inner product with fixed $|v\rangle$. To first rotate $|v\rangle$, insert a complete set of vectors:

$$v_i' = \langle \hat{e}_i | \mathcal{R} | v \rangle = \sum_k \langle \hat{e}_i | \mathcal{R} | \hat{e}_k \rangle \langle \hat{e}_k | v \rangle = \sum_k R_{ik} v_k, \tag{26.10}$$

where $R_{ik} \equiv \langle \hat{e}_i | \mathcal{R} | \hat{e}_k \rangle$ is the matrix representation of $\mathcal{R}$ in the unprimed basis. Thus we get (26.3). Note that we inserted a sum over basis vectors, but the result is a sum over vector components. To compare this with $\mathcal{R}$ acting first on the basis, we'll use (25.24b) — noting that in real space $v_i = v_i^*$ and $\dagger \to T$:

$$v_i' = \langle \hat{e}_i | \mathcal{R} | v \rangle = \langle v | \mathcal{R}^T | \hat{e}_i \rangle = \sum_k \langle v | \hat{e}_k \rangle \langle \hat{e}_k | R^T | \hat{e}_i \rangle$$

$$= \sum_k \langle v | \hat{e}_k \rangle R_{ki}^T = \sum_k \langle v | R_{ik} | \hat{e}_k \rangle \equiv \langle v | \hat{e}_i' \rangle. \tag{26.11}$$

Thus

$$|\hat{e}_i'\rangle = \sum_k R_{ik} | \hat{e}_k \rangle. \tag{26.12}$$

So the matrix in (26.4) which rotates a vector counter-clockwise is also the operator which rotates a basis clockwise. In the first case the rotation mixes components, in the second it mixes basis vectors.

An expression of the form $|v'\rangle = \mathcal{R} |v\rangle$ is said to be an *active* transformation because it maps a vector $|v\rangle$ to a *new* vector $|v'\rangle$ on the *same basis*,

$$|v\rangle = \sum_i v_i | \hat{e}_i \rangle \xrightarrow{\mathcal{R}} |v'\rangle = \sum_i v_i' | \hat{e}_i \rangle. \qquad \text{[Active]} \tag{26.13}$$

The v_i' are components of the new vector $|v'\rangle$; there is no primed basis.

A *passive* transformation effects a change of basis from $\{|\hat{e}_i\rangle\} \xrightarrow{\mathcal{R}} \{|\hat{e}_i'\rangle\}$ such that

$$|v\rangle = \sum_i v_i |\hat{e}_i\rangle = \sum_i v_i' |\hat{e}_i'\rangle. \qquad \text{[Passive]} \qquad (26.14)$$

In this picture, the components of the vector change from $v_i \to v_i'$ *not* because $|v\rangle$ is transformed, but rather so that it stays fixed in space as the basis rotates "beneath" it. There is no transformed vector $|v'\rangle$.

Example 26.2 **From Active to Passive**

Under an active rotation $|v'\rangle = \mathcal{R}|v\rangle \longrightarrow \begin{pmatrix} v_1' \\ v_2' \end{pmatrix}$, where the components of $|v'\rangle$ are expressed on a fixed basis,

$$\begin{pmatrix} v_1' \\ v_2' \end{pmatrix} = (v_1 \cos\varphi - v_2 \sin\varphi) \begin{pmatrix} 1 \\ 0 \end{pmatrix} + (v_1 \sin\varphi + v_2 \cos\varphi) \begin{pmatrix} 0 \\ 1 \end{pmatrix}. \qquad (26.15)$$

In the passive picture, there is no $|v'\rangle$. Instead, $|v\rangle$ has components v_i' on the new basis

$$|\hat{e}_1'\rangle \longrightarrow \begin{pmatrix} \cos\varphi \\ -\sin\varphi \end{pmatrix} \qquad |\hat{e}_2'\rangle \longrightarrow \begin{pmatrix} \sin\varphi \\ \cos\varphi \end{pmatrix}. \qquad (26.16)$$

Thus

$$|v\rangle \to \begin{pmatrix} v_1 \\ v_2 \end{pmatrix} = v_1 \begin{pmatrix} 1 \\ 0 \end{pmatrix} + v_2 \begin{pmatrix} 0 \\ 1 \end{pmatrix}$$

$$= v_1' \begin{pmatrix} \cos\varphi \\ -\sin\varphi \end{pmatrix} + v_2' \begin{pmatrix} \sin\varphi \\ \cos\varphi \end{pmatrix}$$

$$= (v_1 \cos\varphi - v_2 \sin\varphi) \begin{pmatrix} \cos\varphi \\ -\sin\varphi \end{pmatrix} + (v_1 \sin\varphi + v_2 \cos\varphi) \begin{pmatrix} \sin\varphi \\ \cos\varphi \end{pmatrix}, \qquad (26.17)$$

as is easily confirmed. The components in (26.15) and (26.17) are the same.

Example 26.3 **Sum of Components vs. Sum over Vectors**

Consider the standard basis vectors $|\hat{e}_1\rangle \to \begin{pmatrix} 1 \\ 0 \end{pmatrix}$ and $|\hat{e}_2\rangle \to \begin{pmatrix} 0 \\ 1 \end{pmatrix}$, as well as the primed basis vectors

$$|\hat{e}_1'\rangle \longrightarrow \begin{pmatrix} \cos\varphi \\ \sin\varphi \end{pmatrix} = \begin{pmatrix} \cos\varphi & -\sin\varphi \\ \sin\varphi & \cos\varphi \end{pmatrix} \begin{pmatrix} 1 \\ 0 \end{pmatrix}$$

$$= \cos\varphi \begin{pmatrix} 1 \\ 0 \end{pmatrix} + \sin\varphi \begin{pmatrix} 0 \\ 1 \end{pmatrix} \qquad (26.18a)$$

and

$$|\hat{e}_2'\rangle \longrightarrow = \begin{pmatrix} -\sin\varphi \\ \cos\varphi \end{pmatrix} = \begin{pmatrix} \cos\varphi & -\sin\varphi \\ \sin\varphi & \cos\varphi \end{pmatrix} \begin{pmatrix} 0 \\ 1 \end{pmatrix}$$

$$= -\sin\varphi \begin{pmatrix} 1 \\ 0 \end{pmatrix} + \cos\varphi \begin{pmatrix} 0 \\ 1 \end{pmatrix}. \qquad (26.18b)$$

The first equality in each of these expressions is an active counter-clockwise rotation applied *individually* to each standard basis vector. A sum over components. The second equality is the expansion of the primed vectors on the original basis. A sum over vectors.

Note that each $|\hat{e}_i\rangle$ was individually rotated counter-clockwise. Assembling the basis into a column vector of vectors, we find that the basis transforms as

$$
\begin{pmatrix} |\hat{e}_1'\rangle \\ |\hat{e}_2'\rangle \end{pmatrix} = \begin{pmatrix} \cos\varphi & \sin\varphi \\ -\sin\varphi & \cos\varphi \end{pmatrix} \begin{pmatrix} |\hat{e}_1\rangle \\ |\hat{e}_2\rangle \end{pmatrix}, \tag{26.19}
$$

which is a positive, counter-clockwise change of basis.

As straightforward and intuitive as the active/passive relationship is, Example 26.3 touches on a common source of confusion. By convention, counter-clockwise rotations are taken to be positive. So (26.19) is a positive passive rotation of the basis as a whole, corresponding to the positive active rotations of each basis vector individually in (26.18). Often, the matrix in (26.19) is referred to as a passive rotation matrix, distinct from the active matrix (which is its transpose). All too frequently this creates more confusion than clarity. It's better to establish one convention for the rotation matrix, and let the parity of sine and cosine take care of the signs. Then all we need is this simple rule: a positive rotation of an arbitrary vector $|v\rangle$ is equivalent to a negative rotation of the basis as a whole.

Physically, an active transformation affects the object of study, while a passive transformation acts on the device used to measure it. Mathematically speaking, if a problem is framed in terms of two bases, then a passive viewpoint applies. For example, increasing an object's momentum $\vec{p} \to \vec{p}'$ is active; a Lorentz transformation to instead describe it in a different frame of reference is passive. The difference is largely one of perspective; ultimately, how one interprets (26.9) is a matter of taste and cognitive convenience. Since the v_i' are the same either way, we'll use both active and passive language, as appropriate for the circumstances.

BTW 26.1 Beyond Rotations

The active and passive pictures apply to transformations beyond the concrete example of rotations. But a passive approach only makes sense for transformations which take vectors back into the same vector space, $\mathcal{T} : V \to V$. Indeed, one of the strengths of the active approach is that it remains sensible for mapping into a *different* vector space. This works even when the dimensions of the spaces differ. In such a case, T_{ij} is not a square matrix and so is not *invertible* — that is, the inverse matrix T^{-1} does not exist. But an active perspective still applies. A good example is the projection operator (23.6): since it maps $\mathbb{R}^3$ to $\mathbb{R}^2$, $\mathcal{P}$ can be expressed as the 2×3 matrix

$$
\mathcal{P} = \begin{pmatrix} 1 & 0 & 0 \\ 0 & 1 & 0 \end{pmatrix}, \tag{26.20}
$$

so that

$$
|w\rangle = \mathcal{P}|v\rangle \longrightarrow \begin{pmatrix} w_1 \\ w_2 \end{pmatrix} = \begin{pmatrix} 1 & 0 & 0 \\ 0 & 1 & 0 \end{pmatrix} \begin{pmatrix} v_1 \\ v_2 \\ v_3 \end{pmatrix} = \begin{pmatrix} v_1 \\ v_2 \end{pmatrix}. \tag{26.21}
$$

The non-existence of the inverse $\mathcal{P}^{-1}$ is easy to understand: a projection results in the loss of information. In fact, there are an infinite number of $|v\rangle$'s which give the same $|w\rangle$, so $\mathcal{P}^{-1}|w\rangle$ cannot return a unique $|v\rangle$.

26.2 What Makes a Rotation a Rotation?

If you think of a vector as a mathematical representation of a physical object, then an operator is a mathematical representation of a physical action. As such, most of the properties of rotations can be gleaned just from study of the rotation matrix. Thus our focus will be not on the vectors, but rather on the rotation operators themselves. And many of its properties are quite intuitive. For instance, successive rotations must also be a rotation (the space of rotation operators is closed under multiplication). Thus any single rotation by $\theta = \varphi_1 + \varphi_2$,

$$R(\theta) = \begin{pmatrix} \cos(\varphi_1 + \varphi_2) & -\sin(\varphi_1 + \varphi_2) \\ \sin(\varphi_1 + \varphi_2) & \cos(\varphi_1 + \varphi_2) \end{pmatrix}, \tag{26.22}$$

must be equivalent to successive rotations by φ_1 and φ_2,

$$\begin{aligned} R(\varphi_1 + \varphi_2) &= R(\varphi_2)R(\varphi_1) \\ &= \begin{pmatrix} \cos\varphi_2 & -\sin\varphi_2 \\ \sin\varphi_2 & \cos\varphi_2 \end{pmatrix} \begin{pmatrix} \cos\varphi_1 & -\sin\varphi_1 \\ \sin\varphi_1 & \cos\varphi_1 \end{pmatrix} \\ &= \begin{pmatrix} \cos\varphi_1 \cos\varphi_2 - \sin\varphi_1 \sin\varphi_2 & -\sin\varphi_1 \cos\varphi_2 - \cos\varphi_1 \sin\varphi_2 \\ \sin\varphi_1 \cos\varphi_2 + \cos\varphi_1 \sin\varphi_2 & \cos\varphi_1 \cos\varphi_2 - \sin\varphi_1 \sin\varphi_2 \end{pmatrix}. \end{aligned} \tag{26.23}$$

Comparing (26.23) and (26.22) reveals the familiar trigonometric identities for $\sin(\varphi_1 + \varphi_2)$ and $\cos(\varphi_1 + \varphi_2)$. Since the right-most operator is the first to act on a vector, successive transformations are always denoted from right to left. Thus (26.23) is a rotation by φ_1 followed by a second rotation by φ_2.[2]

Another interesting feature of rotations is the observation that no rotation at all is, mathematically speaking, a perfectly good rotation,

$$R(\varphi=0) = \begin{pmatrix} 1 & 0 \\ 0 & 1 \end{pmatrix} \equiv \mathbb{1}. \tag{26.24}$$

Moreover, any rotation by φ can be easily undone by a second rotation through $-\varphi$; the net result is equivalent to the identity,

$$R(-\varphi)R(\varphi) = R(0) = \mathbb{1} \implies R^{-1}(\varphi) = R(-\varphi). \tag{26.25}$$

This is clearly consistent with (26.22).

Intuitively we know that the properties outlined in (26.23)–(26.25) hold not only for rotations in $\mathbb{R}^2$ but for rotations in $\mathbb{R}^n$ generally. They also hold for Lorentz transformations in spacetime. Indeed, there are many continuous transformations which obey these same conditions; as a result much of your intuition regarding rotations can be transferred directly to them. Such collections of rotation-like transformations are called *groups* (BTW 26.2).

Rotations are a ubiquitous feature of any inner product space. But it cannot be that rotations in $\mathbb{R}^n$, rotations in $\mathbb{C}^n$, and Lorentz transformations all use the same rotation matrix. We've thus far neglected one important property of rotations, the one which ultimately distinguishes the space in which the rotation occurs.

---

[2] This may seem overly fussy, since $R(\varphi_2)R(\varphi_1) = R(\varphi_1)R(\varphi_2)$. But this is only true of rotations in the plane; as we'll see, in higher dimensions rotations around different axes do *not* commute.

26.2.1 Rotations in Real Euclidean Space

Any vector must have some scalar quantity associated with it. As a Euclidean space, the scalar in $\mathbb{R}^n$ is the familiar pythagorean length $\|\vec{v}\| \equiv \sqrt{\sum_i v_i^2}$. Indeed, to be a rotation in $\mathbb{R}^n$, R_{ij} should change only a vector's orientation, not its length. So under a rotation,

$$\|\vec{v}\|^2 = \sum_i v_i v_i = \sum_i v_i' v_i'$$

$$= \sum_i \left(\sum_k R_{ik} v_k \right) \left(\sum_\ell R_{i\ell} v_\ell \right) \qquad (26.26)$$

$$= \sum_{ik\ell} R_{ik} R_{i\ell} v_k v_\ell = \sum_{k\ell} \left(\sum_i R_{ki}^T R_{i\ell} \right) v_k v_\ell.$$

So if the magnitude of the vector is invariant, we must have

$$\sum_i R_{ki}^T R_{i\ell} = \delta_{k\ell} \iff R^T R = \mathbb{1}, \qquad (26.27)$$

or simply

$$R^T = R^{-1}. \qquad (26.28)$$

Any real $n \times n$ matrix which obeys (26.28) is said to be *orthogonal*, since its columns and rows form an orthogonal (actually, orthonormal) basis of $\mathbb{R}^n$.

We could have derived (26.28) just as easily by insisting that the inner product of two different vectors remains unchanged under a rotation. Let's use bra-ket notation this time: given vectors $|v\rangle$ and $|w\rangle$ and $|v'\rangle = \mathcal{R}|v\rangle$ and $|w'\rangle = \mathcal{R}|w\rangle$, we have

$$\langle v'|w'\rangle = \langle \mathcal{R}v|\mathcal{R}w\rangle = \langle v|\mathcal{R}^T \mathcal{R}|w\rangle = \langle v|w\rangle. \qquad (26.29)$$

So an orthogonal transformation preserves both the norms and the relative orientation of vectors. In $\mathbb{R}^2$, orthogonal matrices compose a group called $O(2)$ ("oh-two"); in $\mathbb{R}^n$, the group is $O(n)$. The "O" stands for orthogonal; the "n" indicates that the scalar invariant is the n-dimensional pythagorean quadratic $\sum_{i=1}^n v_i^2$.

Incidentally, as the rotation angle goes to zero, rotation matrices must reduce to the identity. This is easily verified for the paradigmatic rotation of (26.4). Moreover, varying the rotation angle does not alter R's determinant — so reduction to the identity $R(\varphi \to 0) = \mathbb{1}$ is only true of orthogonal matrices with determinant +1. There are so-called "proper" rotations. $O(2)$ matrices with determinant +1 form a group by itself — the "special" subgroup called $SO(2)$ ("ess-oh-two").[3]

> ### Example 26.4   Don't Calculate the Inverse!
>
> Beginners often try to verify whether a matrix is orthogonal the hard way by calculating the inverse and comparing it with the transpose. *Do not do this!* One can often verify the orthogonality of a matrix by simply inspecting its column vectors. For instance, the rotation
>
> $$R = \begin{pmatrix} \cos\theta & -\sin\theta \\ \sin\theta & \cos\theta \end{pmatrix} \qquad (26.30)$$

---

[3] In Section 26.3 we'll see that all orthogonal matrices have determinant ± 1. Those with -1, deemed "improper rotations," do not form a group — see BTW 26.2.

clearly has orthonormal columns. Similarly, inspection reveals that the 4×4 matrix

$$A = \begin{pmatrix} 1 & 1 & 1 & 1 \\ -1 & 1 & -1 & 1 \\ 1 & -1 & -1 & 1 \\ 1 & 1 & -1 & -1 \end{pmatrix} \qquad (26.31)$$

has orthogonal columns; dividing A by 2 renders them orthonormal. So $\frac{1}{2}A$ is an orthogonal transformation in $\mathbb{R}^4$.

Alas, inspection for orthonormal columns can often become tedious. But rather than ask whether the inverse is the transpose, ask instead whether the transpose is the inverse. Take, for instance,

$$B = \begin{pmatrix} 1+\sqrt{3} & -1+\sqrt{3} & 1 \\ -1 & 1+\sqrt{3} & -1+\sqrt{3} \\ 1-\sqrt{3} & -1 & 1+\sqrt{3} \end{pmatrix}. \qquad (26.32)$$

A simple calculation gives

$$B^T B = 9\,\mathbb{1}. \qquad (26.33)$$

Thus $\frac{1}{3}B$ is orthogonal.

Note that because of all the off-diagonal zeros that must emerge, testing a matrix M for orthogonality can be pretty quick; in particular, if M is *not* orthogonal, the answer will often become clear long before you finish working out $M^T M$. Moreover, the product $M^T M$ is itself a symmetric matrix — which means that for an $n \times n$ matrix, only $n(n+1)/2$ independent multiplications are required.

Example 26.5 Real Rotations and Complex Numbers

Multiplying two complex numbers has a straightforward geometric interpretation. Multiplying $z_1 = r_1 e^{i\theta_1}$ by $z_2 = r_2 e^{i\theta_2}$ *rotates* z_1 by θ_2 and *magnifies* it by r_2, yielding the new number $Z = z_2 z_1 = r_2 r_1 e^{i(\theta_2 + \theta_1)}$. But as a rotation in the plane, it can also be represented as real 2×2 matrix,

$$z = a + ib \rightarrow \begin{pmatrix} a & -b \\ b & a \end{pmatrix}. \qquad (26.34)$$

With this assignment, addition and multiplication of complex numbers correspond directly to addition and multiplication of 2×2 matrices: given $z_1 = a_1 + ib_1$ and $z_2 = a_2 + ib_2$,

$$z_1 + z_2 \rightarrow \begin{pmatrix} a_1 & -b_1 \\ b_1 & a_1 \end{pmatrix} + \begin{pmatrix} a_2 & -b_2 \\ b_2 & a_2 \end{pmatrix} = \begin{pmatrix} a_1 + a_2 & -b_1 - b_2 \\ b_1 + b_2 & a_1 + a_2 \end{pmatrix}, \qquad (26.35)$$

which is the form of (26.34) for the sum of two complex numbers. Similarly,

$$z_1 z_2 \rightarrow \begin{pmatrix} a_1 & -b_1 \\ b_1 & a_1 \end{pmatrix} \cdot \begin{pmatrix} a_2 & -b_2 \\ b_2 & a_2 \end{pmatrix} = \begin{pmatrix} a_1 a_2 - b_1 b_2 & -a_1 b_2 - b_1 a_2 \\ a_1 b_2 + b_1 a_2 & a_1 a_2 - b_1 b_2 \end{pmatrix}, \qquad (26.36)$$

which is (26.34) for their product. This shows that the matrix of (26.34) is a faithful representation of a complex number. To be a rotation, though, the matrix representing a pure phase must be orthogonal. This is most easily displayed using the polar form, $a = r\cos\theta$ and $b = r\sin\theta$. Then we can factor the r out of the matrix in (26.34), leaving a 2×2 real matrix with determinant $+1$,

$$z \to r \begin{pmatrix} \cos\theta & -\sin\theta \\ \sin\theta & \cos\theta \end{pmatrix}. \tag{26.37}$$

Comparing with (26.30), we see clearly that multiplication by a complex number is equivalent to magnification by r together with rotation in the plane.

26.2.2 Rotations in Complex Euclidean Space

So proper rotations in $\mathbb{R}^n$ are represented by orthogonal matrices $R^T = R^{-1}$ with $\det R = 1$. What about rotations in complex space $\mathbb{C}^n$?

   We denote the complex analog of an orthogonal transformation R by U, and retrace the steps that led to (26.28). The derivation is virtually identical, requiring only that we swap out transpose for adjoint, $T \to \dagger$. The result is that in order for the norm of a vector in $\mathbb{C}^n$ to be constant, the rotation matrix U must satisfy

$$U^\dagger = U^{-1}. \tag{26.38}$$

A matrix which obeys (26.38) is called *unitary*; its columns (or rows) form an orthonormal set of complex vectors. Note that real, orthogonal matrices are also unitary, but unitary matrices are not necessarily orthogonal. The only other notable difference between unitary and orthogonal matrices is the determinant. As you'll demonstrate in Problem 26.8, whereas the determinant of an orthogonal matrix is restricted to ± 1, a unitary matrix has $\det U = e^{i\alpha}$ for some real parameter α. (One often says that the determinant can be anywhere on the unit circle: magnitude 1, polar angle α.) "Proper" unitary matrices, which reduce to the identity, have $\det U = +1$ — that is, $\alpha = 0$.

   You might reasonably ask under what circumstances one would ever encounter a $\mathbb{C}^n$ vector — after all, laboratory measurements always return real values! Nonetheless, many physical phenomena are described by complex vectors. In particular, physical states in quantum mechanics are vectors in complex space. For instance, electron spin is described by a $\mathbb{C}^2$ vector ξ. The standard basis $\begin{pmatrix} 1 \\ 0 \end{pmatrix}$ and $\begin{pmatrix} 0 \\ 1 \end{pmatrix}$ represents spin up and down, respectively — with a typical spin state given by $\begin{pmatrix} c_1 \\ c_2 \end{pmatrix}$ for complex $c_{1,2}$. So to describe, say, a rotating electron, we need to know how a physical $\mathbb{R}^3$ rotation manifests on the $\mathbb{C}^2$ quantum state vectors.

Example 26.6   Rotation in $\mathbb{C}^2$

How many parameters are necessary to specify a rotation in $\mathbb{C}^2$? Well, to start, a complex 2×2 matrix

$$U = \begin{pmatrix} a & b \\ c & d \end{pmatrix} \tag{26.39}$$

has 8 free parameters (since each of the elements is complex). But a complex rotation must be unitary,

$$U^\dagger U = \mathbb{1} \iff \begin{cases} aa^* + cc^* = 1 \\ bb^* + dd^* = 1 \\ a^*b + c^*d = 0, \end{cases} \tag{26.40}$$

which, since the first two equations are real, imposes a total of 4 (rather than 6) constraints. We also want a proper rotation with determinant +1. But unitary matrices all have $\det U = e^{i\alpha}$, so choosing $\alpha = 0$ imposes one additional condition,

$$\det U = 1 \iff ad - cb = 1. \tag{26.41}$$

Thus we have a total of 5 constraints, leaving only 3 parameters to specify a rotation. We can neatly summarize all this by writing

$$U = \begin{pmatrix} a & -b \\ b^* & a^* \end{pmatrix}, \quad |a|^2 + |b|^2 = 1, \tag{26.42}$$

which nicely reduces to (26.34) in $\mathbb{R}^2$. Let's check the action of U on a complex vector $\vec{\xi} = \begin{pmatrix} \xi_1 \\ \xi_2 \end{pmatrix}$:

$$\vec{\xi}' \equiv U\vec{\xi} = \begin{pmatrix} a & -b \\ b^* & a^* \end{pmatrix} \begin{pmatrix} \xi_1 \\ \xi_2 \end{pmatrix} = \begin{pmatrix} a\xi_1 - b\xi_2 \\ b^*\xi_1 + a^*\xi_2 \end{pmatrix}. \tag{26.43}$$

Then

$$\langle \xi | \xi \rangle = \vec{\xi}'^\dagger \cdot \vec{\xi}'$$

$$= |a\xi_1 - b\xi_2|^2 + |b^*\xi_1 + a^*\xi_2|^2$$

$$= \left(|a|^2 + |b|^2 \right)\left(|\xi_1|^2 + |\xi_2|^2 \right)$$

$$= |\xi_1|^2 + |\xi_2|^2 = \|\vec{\xi}\|^2. \tag{26.44}$$

So U is in fact a unitary rotation operator in $\mathbb{C}^2$. Unitary matrices which preserve the length of a two-dimensional complex vector compose a group called $U(2)$ ("you-two"). The subset of $U(2)$ matrices with determinant +1 form the "special" subgroup $SU(2)$ ("ess-you-two").

Example 26.7 $U(1)$ and $SO(2)$

For real φ, a complex number $z = e^{i\varphi}$ is unitary, since $z^*z = 1$; pure phases form the group $U(1)$. We saw in Example 26.5 that this one-dimensional group describes rotations in the plane: rotations in one-dimensional complex space are equivalent to rotations in two-dimensional real space. Thus a pure phase can be represented as a proper, orthogonal 2×2 matrix,

$$e^{i\varphi} \quad \longleftrightarrow \quad \begin{pmatrix} \cos\varphi & -\sin\varphi \\ \sin\varphi & \cos\varphi \end{pmatrix}. \tag{26.45}$$

One says that $U(1)$ and $SO(2)$ are *isomorphic*.

Example 26.8 **Malice Aforethought**

The strength of the parametrization in (26.42) is its simple use of two complex variables. There are, of course, other ways to write out an $SU(2)$ matrix. For instance, as we saw in Example 25.13, any complex 2×2 matrix can be written as a linear combination of the identity and the three Pauli matrices,

$$\sigma_1 = \begin{pmatrix} 0 & 1 \\ 1 & 0 \end{pmatrix} \quad \sigma_2 = \begin{pmatrix} 0 & -i \\ i & 0 \end{pmatrix} \quad \sigma_3 = \begin{pmatrix} 1 & 0 \\ 0 & -1 \end{pmatrix}. \tag{26.46}$$

So let's introduce the four parameters n_0, n_1, n_2, n_3 and write

$$U = n_0 \mathbb{1} + i\vec{n} \cdot \vec{\sigma}, \tag{26.47}$$

where[4] $\vec{n} \cdot \vec{\sigma} \equiv \sum_{i=1}^{3} n_i \sigma_i$. Expanded like this, U is often called a *quaternion*; in matrix form,

$$U = \begin{pmatrix} n_0 + in_3 & i(n_1 - in_2) \\ i(n_1 + in_2) & n_0 - in_3 \end{pmatrix}. \tag{26.48}$$

For real $\vec{n}$, this has the form of (26.42); indeed,

$$UU^\dagger = \begin{pmatrix} n_0^2 + n_1^2 + n_2^2 + n_3^2 & 0 \\ 0 & n_0^2 + n_1^2 + n_2^2 + n_3^2 \end{pmatrix}. \tag{26.49}$$

Thus to be unitary, we need only impose $\det U = n_0^2 + n_1^2 + n_2^2 + n_3^2 = 1$. As expected, a $\mathbb{C}^2$ rotation is fully specified with only three independent parameters.

It's common and convenient to introduce a parameterizing angle φ such that

$$n_0 \equiv \cos(\varphi/2) \qquad \vec{n} \equiv \hat{n} \sin(\varphi/2), \tag{26.50}$$

so that $n_0^2 + \vec{n} \cdot \vec{n} = 1$ is automatic. Then (26.47) becomes

$$U(\varphi) = \mathbb{1} \cos(\varphi/2) + i\hat{n} \cdot \vec{\sigma} \sin(\varphi/2). \tag{26.51}$$

The use of the half-angle $\varphi/2$ is an example of "malice aforethought" — which is a classy way to say that we've introduced it now for later convenience. As we'll see, (26.51) is the representation in $\mathbb{C}^2$ of a rotation around $\hat{n}$ by φ in $\mathbb{R}^3$. So if we rotate an electron, the corresponding quantum spin state transforms with $U(\varphi)$.

Incidentally, note that (26.51) looks a lot like our old friend $\cos + i\sin$ — a resemblance which leads us to posit that U can be written as

$$U(\varphi) = e^{i\varphi \hat{n} \cdot \vec{\sigma}/2}. \tag{26.52}$$

Recall that the function of a matrix is defined by its Taylor expansion. Here this entails evaluating powers of Pauli matrices, but the discovery that $\sigma_i^2 = \mathbb{1}$ renders demonstrating the equivalence of (26.51) and (26.52) a straightforward exercise (Problem 26.25). And the exponential form is very convenient. For instance, since the Pauli matrices are Hermitian, $\sigma_i^\dagger = \sigma_i$,

$$U^\dagger = e^{-i\varphi \hat{n} \cdot \vec{\sigma}^\dagger/2} = e^{-i\varphi \hat{n} \cdot \vec{\sigma}/2} = U^{-1}, \tag{26.53}$$

so in this paramterization unitarity is essentially automatic. Similarly, showing that U is a proper rotation is also simple — at least, once we invoke the remarkable identity (Problem 6.16)

$$\det[e^M] = e^{\text{Tr}[M]}. \tag{26.54}$$

Then, since the σ_i are all traceless,

$$\det\left[e^{i\varphi \hat{n} \cdot \vec{\sigma}/2}\right] = e^{i(\varphi/2)\text{Tr}[\hat{n} \cdot \vec{\sigma}]} = +1. \tag{26.55}$$

The relative simplicity of (26.53) and (26.55) is one reason we chose to expand on the basis of Pauli matrices. In fact, a unitary matrix can always be written in the form $U = \exp[iH]$, where H is a Hermitian matrix; for $\det U = +1$, H must be traceless.

---

[4] The use of $\vec{\sigma}$ is a notational shortcut: the σ_i's are not components of a vector in $\mathbb{R}^3$.

26.2.3 Spacetime Rotations

One can define rotations in any inner product space — including Minkowski space. We saw in (1.7) that the Lorentz transformations can be written as

$$x_0' = \gamma(x_0 - \beta x_1)$$
$$x_1' = \gamma(x_1 - \beta x_0),$$

(26.56)

where $\beta \equiv V/c$ and $\gamma = (1 - \beta^2)^{-1/2}$. In this form, the transformation is manifestly symmetric between space and time. We can write the transformation in terms of a "rotation matrix" Λ (*Lambda* for "Lorentz")

$$\Lambda = \begin{pmatrix} \gamma & -\gamma\beta \\ -\gamma\beta & \gamma \end{pmatrix},$$

(26.57)

which, acting on the column vector $X \to \begin{pmatrix} x_0 \\ x_1 \end{pmatrix}$, puts (26.56) into matrix form,

$$X' = \Lambda X.$$

(26.58)

What about the scalar associated with this vector? If spacetime were equivalent to $\mathbb{R}^2$, a Lorentz transformation would leave $x_0^2 + x_1^2$ invariant. Instead, the scalar "length" of the vector is the invariant interval

$$s^2 \equiv x_0^2 - x_1^2.$$

(26.59)

The invariance of the interval means that the coordinates x_0' and x_1' emerging from a Lorentz transformation must lie somewhere on a hyperbola specified by the fixed value of s^2; precisely where on the hyperbola depends on the relative speed β of the two inertial frames (Figure 26.3). This is very different from rotations in $\mathbb{R}^2$, which constrain vectors to remain on a circle, the precise location depending on the angle of rotation. Even so, the two are sufficiently similar that we can algebraically re-parameterize Λ to look like a rotation. Assigning $\sin\varphi = -\gamma\beta$ and $\cos\varphi = \gamma$, however, is not compatible with $\gamma = (1 - \beta^2)^{1/2}$ — nor does such a parametrization give $|\det \Lambda| = 1$. However, as we saw in BTW 3.1, one can take the angle of rotation $\Phi \equiv i\varphi$ to be *imaginary*, so that $\gamma\beta = \sinh\Phi$ and $\gamma = \cosh\Phi$. This allows us to write our transformation matrix as

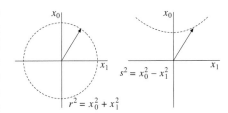

Figure 26.3 Circular vs. hyperbolic invariants.

$$\Lambda \equiv \begin{pmatrix} \cosh\Phi & -\sinh\Phi \\ -\sinh\Phi & \cosh\Phi \end{pmatrix}.$$

(26.60)

Since $\cosh^2 - \sinh^2 = 1$, this is a proper transformation.

Example 26.9 **Rapidity**

Consider two inertial frames with relative speed V along their common x-axis. A particle moving with velocity $u' = dx'/dt'$ in the moving frame is related to the velocity $u = dx/dt$ measured in the "lab" frame by the relativistic velocity addition formula

$$u = \frac{u' + V}{1 + u'V/c^2},$$

(26.61)

which can be derived by taking derivatives of (26.56). So contrary to intuition, velocity is not simply additive. Indeed, inspection reveals that if both u' and V are less than c, then so too is u. (For example, $u' = \frac{1}{2}c$ and $V = \frac{3}{4}c$ yields $u = \frac{10}{11}c$, not $\frac{5}{4}c$.) Thus no material particle can be observed going faster than light. Similarly, if either u' or V equals c, we find $u = c$: the speed of light is the same in all frames — as required (Example 1.3).

Thinking of a Lorentz transformation as a rotation in hyperbolic space lends clarity to this counter-intuitive velocity addition. A rotation by φ_1 followed by a rotation by φ_2 (about the same axis) is equivalent to a single rotation by $\varphi_1 + \varphi_2$. But with spacetime coordinate axes t and x, the velocity u is not the angle but the slope of the spacetime position vector — and slopes don't add. In $\mathbb{R}^2$ the slope m is given by $\tan \varphi$; for hyperbolic space the slope β can be written in terms of an angle Φ,

$$\tanh \Phi \equiv V/c \equiv \beta. \tag{26.62}$$

Called the *rapidity*, Φ is precisely the Lorentz rotation angle in (26.60). If we similarly introduce rapidities ϕ and ϕ',

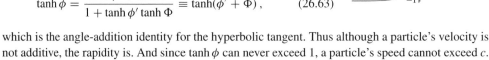

$$\tanh \phi \equiv u/c \quad \tanh \phi' \equiv u'/c,$$

the velocity addition formula (26.61) becomes

$$\tanh \phi = \frac{\tanh \phi' + \tanh \Phi}{1 + \tanh \phi' \tanh \Phi} \equiv \tanh(\phi' + \Phi), \tag{26.63}$$

which is the angle-addition identity for the hyperbolic tangent. Thus although a particle's velocity is not additive, the rapidity is. And since $\tanh \phi$ can never exceed 1, a particle's speed cannot exceed c. Moreover, as u'/c runs from -1 to 1, the rapidity ϕ' goes from $-\infty$ to $+\infty$. So something already moving at c has $\phi' = \infty$; adding this to any other rapidity will still yield $\phi = \infty$.

Underlying the parameterization (26.60) is the fact that Λ is *not* an orthogonal transformation: although the inverse of $\Lambda(\beta)$ is $\Lambda(-\beta)$ (as expected of a rotation), it is not equal to its transpose,

$$\Lambda^{-1}(\beta) = \Lambda(-\beta) \neq \Lambda^T(\beta). \tag{26.64}$$

So just what is the defining characteristic of a Lorentz transformation? To answer this, we introduce the matrix

$$\eta = \begin{pmatrix} 1 & 0 \\ 0 & -1 \end{pmatrix}, \tag{26.65}$$

called the *metric* (Section 30.2). We can use η to keep track of the relative sign in the invariant interval by rendering (26.59) as the matrix equation

$$s^2 = X^T \eta X. \tag{26.66}$$

Then using (26.58) and the invariance of s^2, we learn what makes a Lorentz transformation special:

$$s^2 = X'^T \eta X' = \left(X^T \Lambda^T\right) \eta \left(\Lambda X\right) \equiv X^T \eta X \tag{26.67}$$

so that

$$\Lambda^T \eta \Lambda = \eta, \tag{26.68}$$

as is easily verified with (26.60). Just as $R^T R = \mathbb{1}$ is the fundamental condition for rotations in $\mathbb{R}^n$, (26.68) is the defining relation of rotations in spacetime. In fact, its derivation is virtually identical to the steps leading to (26.28) — the only difference being the substitution of the metric η in place of the identity $\mathbb{1}$. [Since Lorentz transformations preserve the difference rather than the sum of squares, the Λ's are not $SO(2)$ matrices. The relative sign in the invariant interval is indicated by denoting the group as $SO(1,1)$.]

BTW 26.2 What Is a Group?

$U(1)$, $SO(2)$, $SU(2)$, $SO(1,1)$, ... it can quickly become a little confusing. But recognition of their fundamental commonality — specifically, that all are *groups* — helps to translate this potpourri of ciphers into a meaningful classification of transformations.

Formally, a set G with some mathematical operation $\circ$ defined amongst its elements $a, b, c, \ldots$ forms a *group* if:

1. the combination of any two group elements is also a group element,

$$c = a{\circ}b, \quad a,b \in G \Longrightarrow c \in G, \tag{26.69a}$$

2. there exists an identity element $\mathbb{1} \in G$, such that

$$\mathbb{1}{\circ}a = a{\circ}\mathbb{1} = a, \quad \forall\, a \in G, \tag{26.69b}$$

3. every element in the group has an inverse which is also in the group,

$$a \in G \Longrightarrow a^{-1} \in G, \quad \text{such that } a{\circ}a^{-1} = a^{-1}{\circ}\,a = \mathbb{1}, \tag{26.69c}$$

4. the operation $\circ$ is associative,

$$a{\circ}(b{\circ}c) = (a{\circ}b){\circ}c, \quad \forall\, a,b,c \in G. \tag{26.69d}$$

Note that these conditions are precisely those of rotation matrices, where $\circ$ is regular matrix multiplication. So rotations form a group — the *rotation group*. Since rotations are continuous functions of φ, it's a continuous group. The *Lorentz group* of Lorentz transformations in special relativity is also continuous. But groups can also be discrete. For example, with $\circ$ simple addition and the zero as the identity, the integers form a group. So too are permutations applied to a collection of objects. Although discrete groups satisfy the criteria of (26.69), they do not behave like rotations. Nonetheless, the appearance of a group structure in a physical system is usually evidence of underlying symmetry. The duality symmetry between electric and magnetic fields in Example 26.1 is one illustration. As another, consider the discrete group of $\pi/2$ rotations in the plane. A system which remains unchanged after application of any of this group's four distinct elements must have the symmetry of a square. On the other hand, a two-dimensional system which is invariant under *any* rotation in the plane has an underlying continuous circular symmetry.

Some groups have subgroups. For instance, the group of orthogonal transformations $O(2)$ has within it the subgroup $SO(2)$ of matrices with determinant $+1$. As a group all by itself, any product of these so-called proper rotations is itself a proper rotation. By contrast, orthogonal matrices with determinant -1 do not include the identity as one of their own, and so do not form a group.

26.3 Improper Orthogonal Matrices: Reflections

Thus far we've restricted our attention in real Euclidean space to proper rotations with det $= +1$. But orthogonal matrices with determinant -1 also preserve the lengths and relative orientations of vectors. We want to explore in more detail what earns them the moniker "improper."

First, let's verify the claim that $R^T R = \mathbb{1}$ alone doesn't fully characterize the rotation matrix. Taking the determinant of $R^T R$ gives

$$1 = \det(R^T R) = \det(R^T)\det(R) = (\det R)^2,$$

to find

$$\det R = \pm 1. \tag{26.70}$$

These two possibilities are more distinct than you might initially realize. For one thing, the set of orthogonal matrices with determinant -1 does not include the identity (and so by itself does not form a group). In $\mathbb{R}^2$, the "closest" we can get to the identity is a matrix of the form

$$B_0 = \begin{pmatrix} 1 & 0 \\ 0 & -1 \end{pmatrix}. \tag{26.71}$$

Applied to a vector $\vec{v}$, we find

$$B_0 \vec{v} = \begin{pmatrix} 1 & 0 \\ 0 & -1 \end{pmatrix} \begin{pmatrix} v_x \\ v_y \end{pmatrix} = \begin{pmatrix} v_x \\ -v_y \end{pmatrix}. \tag{26.72}$$

This is not a rotation, but rather a *reflection* of $\vec{v}$ across the x-axis — as if a mirror lies along the x-axis. Similarly, $-B_0$ reflects across y. Vector magnitude is clearly unchanged under reflections; their "impropriety" doesn't undermine their status as orthogonal transformations.

Of course one can also map the component $v_y \to -v_y$ with a proper rotation by $\varphi = 2\tan^{-1}(v_y/v_x)$. What makes (26.72) distinct is that B_0 applies to *arbitrary* $\vec{v}$, whereas the value of φ depends on the specific vector being rotated. Moreover, two successive reflections across the same axis yield the identity, $B_0^2 = \mathbb{1}$. This is certainly not the general case for rotations.

Reflections across the coordinate axes are easy enough — but what about across an arbitrary axis $\hat{n}$ through the origin? With $\hat{n} \cdot \hat{\imath} = \cos\varphi$, we *could* work this out trigonometrically using the slope $\tan\varphi$ of the axis to calculate the position of the image of an arbitrary point. This is tedious. Easier — as well as far more powerful and intuitive — is to use rotations. First, use R^{-1} to actively rotate vectors clockwise by φ; this aligns the reflection axis $\hat{n}$ with the x-axis. In this orientation, the matrix B_0 gives the correct reflection. After reflecting, rotate the reflected vector back to the original orientation to get the final result. Assembling this into a single operation $B(\varphi)$ — and remembering that matrix multiplication should be read from right to left — yields the composite operator

$$B(\varphi) = R(\varphi)B_0 R^{-1}(\varphi) = \begin{pmatrix} \cos\varphi & -\sin\varphi \\ \sin\varphi & \cos\varphi \end{pmatrix} \begin{pmatrix} 1 & 0 \\ 0 & -1 \end{pmatrix} \begin{pmatrix} \cos\varphi & \sin\varphi \\ -\sin\varphi & \cos\varphi \end{pmatrix}$$

$$= \begin{pmatrix} \cos\varphi & -\sin\varphi \\ \sin\varphi & \cos\varphi \end{pmatrix} \begin{pmatrix} \cos\varphi & \sin\varphi \\ \sin\varphi & -\cos\varphi \end{pmatrix}$$

$$= \begin{pmatrix} \cos 2\varphi & \sin 2\varphi \\ \sin 2\varphi & -\cos 2\varphi \end{pmatrix}, \tag{26.73}$$

where we've used familiar trig identities in the last step. $B(\varphi)$ is a reflection across an axis at an angle φ with the x-axis. Since this is valid for arbitrary $\vec{v}$, the matrix multiplication of (26.73) is a general result. As a quick consistency check, note that $B(0) = B_0$ and $B(\pi/2) = -B_0$. Moreover, as the product of orthogonal matrices, B is also orthogonal. And clearly $\det B = -1$, as expected of the product of an odd number of determinant -1 matrices. But the product of an even number of reflections has determinant $+1$, and hence is a proper rotation. So although

$$B(\pi/2)B(0) = \begin{pmatrix} -1 & 0 \\ 0 & -1 \end{pmatrix} = -\mathbb{1} \tag{26.74}$$

reflects across both coordinate axes, it's actually a proper rotation by π. In general, any improper rotation B' can be written as the product of a proper rotation R and a reflection B,

$$B' = RB. \tag{26.75}$$

Example 26.10 **Reflection Decomposition**

With determinant -1, the orthogonal matrix

$$T = \begin{pmatrix} 3/5 & 4/5 \\ 4/5 & -3/5 \end{pmatrix} \tag{26.76}$$

is an improper rotation. Using (26.73), we find $\tan\varphi = 1/2$, so this is a reflection across the line $y = x/2$. But we can also factor T as in (26.75) to get

$$T = R(\varphi)B_0 = \begin{pmatrix} 3/5 & -4/5 \\ 4/5 & 3/5 \end{pmatrix}\begin{pmatrix} 1 & 0 \\ 0 & -1 \end{pmatrix}, \tag{26.77}$$

which shows that T is equivalent to reflection across the x-axis followed by a rotation through $\cos^{-1}(3/5) = 53°$.

Example 26.11 **On Further Reflection**

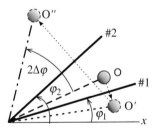

Figure 26.4 Mirror #1 forms image O' of object O; mirror #2 takes image O' into image O''. This double reflection is equivalent to a $2\Delta\varphi = 2(\varphi_2 - \varphi_1)$ rotation mapping $O \rightarrow O''$.

Consider two mirrors oriented at angles φ_1 and φ_2 with the x-axis (Figure 26.4). The image O'' in the second mirror of an image O' formed by the first is described by the transformation

$$B(\varphi_2)B(\varphi_1) = \begin{pmatrix} \cos 2\varphi_2 & \sin 2\varphi_2 \\ \sin 2\varphi_2 & -\cos 2\varphi_2 \end{pmatrix}\begin{pmatrix} \cos 2\varphi_1 & \sin 2\varphi_1 \\ \sin 2\varphi_1 & -\cos 2\varphi_1 \end{pmatrix}$$

$$= \begin{pmatrix} \cos 2\varphi_2 \cos 2\varphi_1 + \sin 2\varphi_2 \sin 2\varphi_1 & \cos 2\varphi_2 \sin 2\varphi_1 - \sin 2\varphi_2 \cos 2\varphi_1 \\ \sin 2\varphi_2 \cos 2\varphi_1 - \cos 2\varphi_2 \sin 2\varphi_1 & \sin 2\varphi_2 \sin 2\varphi_1 + \cos 2\varphi_2 \cos 2\varphi_1 \end{pmatrix}$$

$$= \begin{pmatrix} \cos 2(\varphi_2 - \varphi_1) & -\sin 2(\varphi_2 - \varphi_1) \\ \sin 2(\varphi_2 - \varphi_1) & \cos 2(\varphi_2 - \varphi_1) \end{pmatrix}$$

$$= R(2\Delta\varphi). \tag{26.78}$$

So two reflections through mirrors which meet at an angle $\Delta\varphi = \varphi_2 - \varphi_1$ are together equivalent to a rotation by $2\Delta\varphi$. This is easily demonstrated by holding a mirror at $45°$ to a wall mirror. Looking into the wall mirror, you'll see that it shows your image in the hand mirror to be rotated $90°$.[5] (Remember this next time you use one of those segmented mirrors when buying clothes.) The equivalence of an even number of successive reflections to a single rotation holds in three dimensions as well, with the rotation axis given by the line of intersection between the planes of the first and last mirrors.

In $\mathbb{R}^3$, a transformation such as $(x, y, z) \to (x, -y, -z)$ is not actually a reflection, since the matrix operator has determinant $+1$. In fact, it's not hard to visualize that it's a π rotation around the x-axis. On the other hand, the determinant -1 matrices

$$B_x = \begin{pmatrix} -1 & 0 & 0 \\ 0 & 1 & 0 \\ 0 & 0 & 1 \end{pmatrix} \quad B_y = \begin{pmatrix} 1 & 0 & 0 \\ 0 & -1 & 0 \\ 0 & 0 & 1 \end{pmatrix} \quad B_z = \begin{pmatrix} 1 & 0 & 0 \\ 0 & 1 & 0 \\ 0 & 0 & -1 \end{pmatrix} \tag{26.79}$$

all represent true reflections through mirrors in the yz-, zx-, and xy-planes, respectively. (The subscripts give the directions normal to these planes.) Odd products of these matrices map a right-handed coordinate system into a left-handed one — something no proper rotation can do.

Whereas the product of any two of the reflections in (26.79) is a proper rotation, the product of all three is a reflection called the *parity* operator,

$$P \equiv \begin{pmatrix} -1 & 0 & 0 \\ 0 & -1 & 0 \\ 0 & 0 & -1 \end{pmatrix}. \tag{26.80}$$

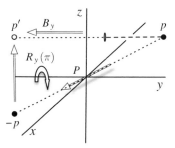

Figure 26.5 A reflection B_y through the xz plane is equivalent to the combined operations of inversion $P : p \to -p$ and a π rotation around the y axis.

P is a reflection through the origin — i.e., an inversion, $(x, y, z) \to (-x, -y, -z)$. As illustrated in Figure 26.5, any planar reflection can be decomposed as an inversion times a π rotation around the normal to the plane. Indeed, since P affects all coordinates equally, the simplest version of the decomposition (26.75) is as the product of a rotation and inversion,

$$B = PR = RP. \tag{26.81}$$

And since $P = -\mathbb{1}$, the required rotation is just $R = -B$.

---

[5] If you're not sure you're looking at the correct image, touch the right side of your nose: the image resulting from two reflections should also be touching the right side of its nose.

BTW 26.3 Through the Looking Glass

Although common convention opts to use right-handed coordinate systems, one should expect physics to be disinterested in that choice. But perhaps you object, pointing out that quantities such as angular momentum and magnetic field which are defined using cross products clearly depend on the handedness of the system. Though this is undeniably true, the choice of right- or left-hand rule does not produce any observable effects since directly measured quantities have an *even* number of cross products. For instance, the presence of a magnetic field is revealed by the force it exerts on a moving charge. But that force is $\vec{F} = q\vec{v} \times \vec{B}$ — requiring one cross product to find $\vec{B}$ and a second to find $\vec{F}$. Similarly, a gyroscope's $\vec{L}$ is needed to determine the direction of precession — but precession also requires a torque, which itself entails a cross product. In both cases, right- and left-hand rules predict identical observations. Since right- and left-handed systems are mirror images of one another, one reasonably concludes that a universe through the looking glass behaves precisely the same as ours: no measurement can tell us which universe is being observed. One says that physics is mirror-symmetric — or, more technically, that physics is invariant under parity.

The notion of mirror symmetry extends to quantum mechanics, where a quantum state is assigned a parity, plus or minus, which describes the effect of inversion on it. Parity invariance requires that replacing all states with their mirror images yields the same physics. Which everyone thought to be the case until 1956, when Lee and Yang proposed that the weak nuclear force, responsible for radioactive decay, is *not* invariant under inversion. The experiment to look for such parity violation was performed in 1957 by Wu and her collaborators, who measured the electron emerging from the beta decay of ^{60}Co nuclei. Cooling the nuclei to about 1 K

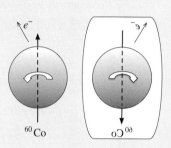

in order to align the nuclear spins with a magnetic field, it was found that the electrons emerged preferentially aligned with the spin of the Co nuclei. Were the same experiment performed in a mirror-image universe, then, electrons would be emitted *anti*-aligned with the spin of the nucleus. So Wu's results clearly distinguish left and right! (What experimental result would have supported parity conservation?) Thus was parity found to be violated in weak interactions — with the earth-shattering conclusion that mirror symmetry is *not* reflected in the physical world!

26.4 Rotations in $\mathbb{R}^3$

Rotations in the plane are straightforward, not least because only a single parameter is needed to specify $R(\varphi)$. Step up even one dimension, however, and things become considerably more complicated. And interesting. A rotation in $\mathbb{R}^3$, for instance, requires three parameters: an angle of rotation φ plus the two spherical coordinates of the axis $\hat{n}$. But the number of parameters is just the beginning of the story.

The simplest 3×3 orthogonal matrices are obtained by embedding $\mathbb{R}^2$ rotations within them:

$$R_z(\varphi) = \begin{pmatrix} \cos\varphi & -\sin\varphi & 0 \\ \sin\varphi & \cos\varphi & 0 \\ 0 & 0 & 1 \end{pmatrix}, \tag{26.82}$$

where the subscript on R indicates the axis of rotation. Rotations around the z-axis leave the third (i.e., z) component as little more than a spectator:

$$R_z(\varphi)\vec{v} = \begin{pmatrix} \cos\varphi & -\sin\varphi & 0 \\ \sin\varphi & \cos\varphi & 0 \\ 0 & 0 & 1 \end{pmatrix} \begin{pmatrix} v_x \\ v_y \\ v_z \end{pmatrix} = \begin{pmatrix} v_x\cos\varphi - v_y\sin\varphi \\ v_x\sin\varphi + v_y\cos\varphi \\ v_z \end{pmatrix}. \tag{26.83}$$

We can similarly write down rotations about the x- and y-axes,

$$R_x(\varphi) = \begin{pmatrix} 1 & 0 & 0 \\ 0 & \cos\varphi & -\sin\varphi \\ 0 & \sin\varphi & \cos\varphi \end{pmatrix}, \quad R_y(\varphi) = \begin{pmatrix} \cos\varphi & 0 & \sin\varphi \\ 0 & 1 & 0 \\ -\sin\varphi & 0 & \cos\varphi \end{pmatrix}. \tag{26.84}$$

Positive rotations in $\mathbb{R}^3$ are right-handed — which is to say, counter-clockwise when looking down the rotation axis towards the origin. The apparent sign flip of the sines in R_y is due to this choice: positive rotations around y take z into x. In other words, to retain a right-handed convention, the three matrices are related by cyclic permutation of both columns and rows.

Example 26.12   Precession

Rotations can be dynamic by making φ a function of time. For instance, a vector $\vec{r}$ rotating around z with constant angular velocity $\vec{\omega} = \omega\hat{k}$ can be described by

$$\begin{pmatrix} x(t) \\ y(t) \\ z(t) \end{pmatrix} = \begin{pmatrix} \cos\omega t & -\sin\omega t & 0 \\ \sin\omega t & \cos\omega t & 0 \\ 0 & 0 & 1 \end{pmatrix} \begin{pmatrix} x \\ y \\ z \end{pmatrix}, \tag{26.85}$$

where $\begin{pmatrix} x \\ y \\ z \end{pmatrix}$ is the position at $t = 0$. For small times $\delta t \ll 1/\omega$,

$$\begin{pmatrix} x(t) \\ y(t) \\ z(t) \end{pmatrix} = \left[\mathbb{1} + \begin{pmatrix} 0 & -\omega\delta t & 0 \\ \omega\delta t & 0 & 0 \\ 0 & 0 & 0 \end{pmatrix} \right] \begin{pmatrix} x \\ y \\ z \end{pmatrix}. \tag{26.86}$$

The change $\delta\vec{r} = \vec{r}(t) - \vec{r}(0)$ is thus

$$\begin{pmatrix} \delta x \\ \delta y \\ \delta z \end{pmatrix} = \begin{pmatrix} 0 & -\omega\delta t & 0 \\ \omega\delta t & 0 & 0 \\ 0 & 0 & 0 \end{pmatrix} \begin{pmatrix} x \\ y \\ z \end{pmatrix}. \tag{26.87}$$

Defining the small vector-valued angle $\delta\vec{\varphi} \equiv \vec{\omega}\delta t$, this is equivalent to the cross product

$$\delta\vec{r} = \delta\vec{\varphi} \times \vec{r}. \tag{26.88}$$

Though we motivated this from R_z, as a vector equation it's valid for infinitesimal rotations around any axis. And in the $\delta t \to 0$ limit, we recover the familiar

$$\vec{v} = \frac{d\vec{r}}{dt} = \vec{\omega} \times \vec{r}, \tag{26.89}$$

which holds for any constant $\vec{\omega}$.

BTW 26.4   Rotations in $\mathbb{R}^n$

How many angles are required to completely specify a rotation in $\mathbb{R}^n$? Well, since the rotation matrix is $n \times n$, one might guess that no fewer than n^2 parameters are needed. By this logic, of course, rotations in the plane would require four angles, whereas we know only one is needed.

The problem is that we don't want just *any* $n \times n$ matrix, we want an $n \times n$ *orthogonal* matrix. The requirement that columns of an orthogonal matrix are all normalized imposes n constraints; that they must also be mutually orthogonal, another $\frac{1}{2}n(n-1)$ constraints. Thus

$$\text{\# independent parameters} = n^2 - n - \frac{n(n-1)}{2} = \frac{1}{2}n(n-1). \tag{26.90}$$

This is equivalent to the binomial coefficient $\binom{n}{2}$, which is just the number of independent orthogonal planes in $\mathbb{R}^n$. So we can think of the number of required parameters as the dimensionality of a basis of planes. In one dimension there are no planes, so the concept of rotation doesn't exist. In two dimensions, there is only one independent plane; a single angle of rotation suffices to specify a rotation. In three dimensions, there are three independent planes, and so three rotation angles are needed. Four-dimensional space requires $\binom{4}{2} = 6$ angles.

It's worth emphasizing how special — or if you prefer, atypical — rotations in $\mathbb{R}^3$ are. Specifically, only for $n = 3$ is the number of parameters equal to the dimensionality of the space. One consequence is that only in three dimensions is angular velocity $\vec{\omega}$ a vector. In two dimensions, ω is just a single number, the angular speed. And the six independent ω's needed in four dimensions are too many to be assembled into an $\mathbb{R}^4$ vector. Moreover, only for $n = 3$ is the cross product of two vectors also a vector. So (26.89) is specific to $\mathbb{R}^3$.

The matrices of (26.82) and (26.84) are clearly orthogonal. But the ease of embedding $\mathbb{R}^2$ rotations into $\mathbb{R}^3$ distracts from what is perhaps the most significant difference between rotations in $\mathbb{R}^3$ and rotations in the plane. Consider rotating a vector around the y-axis,

$$\vec{v}' = R_y(\alpha)\vec{v}, \tag{26.91}$$

followed by a rotation around x,

$$\vec{v}'' = R_x(\beta)\vec{v}' = R_x(\beta)R_y(\alpha)\vec{v}. \tag{26.92}$$

Applying the rotations in reverse order gives instead

$$\vec{w}'' = R_y(\alpha)\vec{w}' = R_y(\alpha)R_x(\beta)\vec{v}. \tag{26.93}$$

Big deal? Well yes, it is: since the matrices *do not commute*,

$$R_x R_y \neq R_y R_x, \tag{26.94}$$

the final vectors are not the same, $\vec{w}'' \neq \vec{v}''$. (Rotations in $\mathbb{R}^3$ are said to be *non-abelian*. Rotations in the plane, which commute, are *abelian*.)

Example 26.13   **That Does Not Commute!**

Let's look at the extreme case of successive $90°$ rotations about perpendicular axes. First rotating around y and then around x yields

$$R_x(\pi/2)R_y(\pi/2) = \begin{pmatrix} 1 & 0 & 0 \\ 0 & 0 & -1 \\ 0 & 1 & 0 \end{pmatrix} \begin{pmatrix} 0 & 0 & 1 \\ 0 & 1 & 0 \\ -1 & 0 & 0 \end{pmatrix} = \begin{pmatrix} 0 & 0 & 1 \\ 1 & 0 & 0 \\ 0 & 1 & 0 \end{pmatrix}. \tag{26.95}$$

This is equivalent to a single rotation by 120° around the axis

$$\hat{n} = \frac{1}{\sqrt{3}} \begin{pmatrix} 1 \\ 1 \\ 1 \end{pmatrix}, \tag{26.96}$$

as you can confirm by showing that $\left(\begin{smallmatrix} 0 & 0 & 1 \\ 1 & 0 & 0 \\ 0 & 1 & 0 \end{smallmatrix} \right)^3 = \mathbb{1}$, and that $\hat{n}$ is unaffected by the transformation, $R\hat{n} = \hat{n}$. By contrast, rotating in the reverse order gives

$$R_y(\pi/2)R_x(\pi/2) = \begin{pmatrix} 0 & 0 & 1 \\ 0 & 1 & 0 \\ -1 & 0 & 0 \end{pmatrix} \begin{pmatrix} 1 & 0 & 0 \\ 0 & 0 & -1 \\ 0 & 1 & 0 \end{pmatrix} = \begin{pmatrix} 0 & 1 & 0 \\ 0 & 0 & -1 \\ -1 & 0 & 0 \end{pmatrix}, \tag{26.97}$$

which is equivalent to a single rotation by 120° around the axis

$$\hat{m} = \frac{1}{\sqrt{3}} \begin{pmatrix} 1 \\ 1 \\ -1 \end{pmatrix}. \tag{26.98}$$

Clearly, the order matters. The axes differ by a single rotation of $\varphi = \cos^{-1}\left(\hat{n} \cdot \hat{m}\right)$, around an axis $\hat{q}$ parallel to $\hat{n} \times \hat{m}$. As you'll verify in Problem 26.20, the rotation R_q which takes $\hat{n}$ into $\hat{m}$ is

$$R_q = \frac{1}{3} \begin{pmatrix} 2 & -1 & 2 \\ -1 & 2 & 2 \\ -2 & -2 & 1 \end{pmatrix}, \tag{26.99}$$

which is a $\varphi = 70.5°$ rotation around

$$\hat{q} = \frac{1}{\sqrt{2}} \begin{pmatrix} -1 \\ 1 \\ 0 \end{pmatrix}. \tag{26.100}$$

Example 26.14 Boosts and Rotations

The lack of commutativity also occurs for spacetime transformations. Although (26.57) was only concerned with Lorentz transformations in "1+1" dimensional spacetime, in 3+1 dimensions the Lorentz boost matrix

$$\Lambda_{tx} = \begin{pmatrix} \gamma & -\gamma\beta & 0 & 0 \\ -\gamma\beta & \gamma & 0 & 0 \\ 0 & 0 & 1 & 0 \\ 0 & 0 & 0 & 1 \end{pmatrix} \tag{26.101}$$

acts on the four-vector $(x_0, x_1, \vec{x}_\perp)$, where $\vec{x}_\perp = (x_2, x_3)$ represents the directions transverse to the frames' relative motion. Like (26.82) and (26.84), this is the embedding of rotations in the plane into a higher-dimensional space — in this case, embedding spacetime rotations mixing t and x components into the larger Minkowski space. But there's no reason we can't mix t and y or t and z,

$$\Lambda_{ty} = \begin{pmatrix} \gamma & 0 & -\gamma\beta & 0 \\ 0 & 1 & 0 & 0 \\ -\gamma\beta & 0 & \gamma & 0 \\ 0 & 0 & 0 & 1 \end{pmatrix} \qquad \Lambda_{tz} = \begin{pmatrix} \gamma & 0 & 0 & -\gamma\beta \\ 0 & 1 & 0 & 0 \\ 0 & 0 & 1 & 0 \\ -\gamma\beta & 0 & 0 & \gamma \end{pmatrix}. \tag{26.102}$$

Lorentz transformations such as these, which mix a four-vector's space and time components, are called *boosts*. The direction of boosts is given by the direction of the frames' relative velocity; for general $\vec{\beta}$, the Lorentz matrix becomes[6]

$$\Lambda = \begin{pmatrix} \gamma & -\beta_x\gamma & -\beta_y\gamma & -\beta_z\gamma \\ -\beta_x\gamma & 1 + \frac{\gamma-1}{\beta^2}\beta_x^2 & \frac{\gamma-1}{\beta^2}\beta_x\beta_y & \frac{\gamma-1}{\beta^2}\beta_x\beta_z \\ -\beta_y\gamma & \frac{\gamma-1}{\beta^2}\beta_y\beta_x & 1 + \frac{\gamma-1}{\beta^2}\beta_y^2 & \frac{\gamma-1}{\beta^2}\beta_y\beta_z \\ -\beta_z\gamma & \frac{\gamma-1}{\beta^2}\beta_z\beta_x & \frac{\gamma-1}{\beta^2}\beta_z\beta_y & 1 + \frac{\gamma-1}{\beta^2}\beta_z^2 \end{pmatrix}, \tag{26.103}$$

which reduces to (26.101) for $\beta_y = \beta_z = 0$. Any combination of boosts along a common spatial axis is equivalent to a single boost along that axis. But like rotations in $\mathbb{R}^3$, boosts along different directions do not commute.

By expanding beyond one spatial dimension, we pick up another feature: boosts no longer encompass all possible spacetime transformations. Strictly speaking, the full Lorentz group should include *all* operations which leave the four-vector norm $v_0^2 - \vec{v} \cdot \vec{v}$ unchanged. In addition to boosts, any orthogonal rotation of the *spatial* components of a four-vector will have no affect on the norm. So a general Lorentz transformation can be written as the product of rotations and boosts. Together, boosts and $\mathbb{R}^3$ rotations form the *Lorentz group*.

Example 26.15 **Thomas Precession**

Although boosts along a common spatial axis commute, products of boosts along different axes cannot always be written in the form (26.103) as a single boost. In fact, boosts alone can give rise to rotations (as the group structure combining boosts and rotations implies). It's not too hard to see this physically: a boost along x_1 is a rotation in the x_0x_1-plane, whereas a boost along x_2 rotates x_0 and x_2. So the combination of the two boosts mixes x_1 and x_2.

Consider an electron in a circular orbit. Although moving with constant speed, the electron's centripetal acceleration v^2/r disallows the use of a single inertial rest frame. In fact, it's not obvious that special relativity even applies to accelerating frames. We get around this using a very common stratagem: imagine an infinity of inertial frames whose relative velocities differ infinitesimally from each other. At any moment in time, one of these frames can serve as the instantaneous rest frame of an accelerating particle. We wish to compare the instantaneous rest frames of the electron at time t and an infinitesimal time later $t + \delta t$. Set up the coordinates so that at time t, the electron moves in the lab frame S with speed v in the x_1-direction; the instantaneous rest frame, S', moves with velocity $v\,\hat{e}_1$ with respect S. A time δt later, the electron has experienced

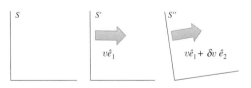

---

[6] An analogous expression for spatial rotations around an arbitrary $\hat{n}$ is derived in Section B.1.

an infinitesimal acceleration in the x_2-direction; the instantaneous rest frame, S'', now moves with velocity $(v\,\hat{e}_1 + \delta v\,\hat{e}_2)$ with respect to S. (Recall that velocity and acceleration are perpendicular in centripetal motion.) Assuming $\delta v \ll v$, we want to construct the Lorentz transformation mapping S' into S''. The transformation from the lab to frame S' is given by $|x'\rangle = \Lambda_1(\beta)|x\rangle$,

$$
\begin{pmatrix} x'_0 \\ x'_1 \\ x'_2 \\ x'_3 \end{pmatrix} = \begin{pmatrix} \gamma & -\beta\gamma & & \\ -\beta\gamma & \gamma & & \\ & & 1 & \\ & & & 1 \end{pmatrix} \begin{pmatrix} x_0 \\ x_1 \\ x_2 \\ x_3 \end{pmatrix}. \tag{26.104}
$$

Similarly, the transformation from the lab to S'' is given by $|x''\rangle = \Lambda_2(\delta\beta)|x\rangle$,

$$
\begin{pmatrix} x''_0 \\ x''_1 \\ x''_2 \\ x''_3 \end{pmatrix} = \begin{pmatrix} \gamma & -\beta\gamma & -\gamma\delta\beta & 0 \\ -\beta\gamma & \gamma & \frac{\gamma-1}{\beta}\delta\beta & 0 \\ -\gamma\delta\beta & \frac{\gamma-1}{\beta}\delta\beta & 1 & 0 \\ 0 & 0 & 0 & 1 \end{pmatrix} \begin{pmatrix} x_0 \\ x_1 \\ x_2 \\ x_3 \end{pmatrix}, \tag{26.105}
$$

where we have neglected terms past first order in $\delta\beta$. Inverting $|x'\rangle = \Lambda_1|x\rangle$ gives $|x\rangle = \Lambda_1^{-1}|x'\rangle$, so that

$$
|x''\rangle = \Lambda_{\text{tot}}|x\rangle = \Lambda_2(\delta\beta)\Lambda_1^{-1}(\beta)|x'\rangle. \tag{26.106}
$$

Then the Lorentz transformation between frames S' and S'' is

$$
\Lambda_{\text{tot}} \equiv \Lambda_2(\delta\beta)\Lambda_1^{-1}(\beta). \tag{26.107}
$$

Finding the inverse $\Lambda_1^{-1}(\beta)$ is not difficult: a Lorentz transformation is a type of rotation, so it must be that $\Lambda_1^{-1}(\beta) = \Lambda_1(-\beta)$. The transformation Λ_{tot} is thus

$$
\Lambda_{\text{tot}} = \begin{pmatrix} \gamma & -\beta\gamma & -\gamma\delta\beta & 0 \\ -\beta\gamma & \gamma & \frac{\gamma-1}{\beta}\delta\beta & 0 \\ -\gamma\delta\beta & \frac{\gamma-1}{\beta}\delta\beta & 1 & 0 \\ 0 & 0 & 0 & 1 \end{pmatrix} \begin{pmatrix} \gamma & +\beta\gamma & & \\ +\beta\gamma & \gamma & & \\ & & 1 & \\ & & & 1 \end{pmatrix}
$$

$$
= \left(\begin{array}{cc|cc} 1 & 0 & -\gamma\delta\beta & 0 \\ 0 & 1 & \frac{\gamma-1}{\beta}\delta\beta & 0 \\ \hline -\gamma\delta\beta & -\frac{\gamma-1}{\beta}\delta\beta & 1 & 0 \\ 0 & 0 & 0 & 1 \end{array} \right) \equiv \mathbb{1} + \Delta L + \Delta\Omega, \tag{26.108}
$$

where

$$
\Delta L \equiv \left(\begin{array}{c|ccc} 0 & 0 & -\gamma\delta\beta & 0 \\ 0 & & & \\ -\gamma\delta\beta & & & \\ 0 & & & \end{array} \right), \quad \Delta\Omega \equiv \left(\begin{array}{c|ccc} 0 & & & \\ \hline & 0 & \frac{\gamma-1}{\beta}\delta\beta & \\ & -\frac{\gamma-1}{\beta}\delta\beta & 0 & \\ & & & 0 \end{array} \right). \tag{26.109}
$$

Since we're neglecting terms greater than order $\delta\beta$, we can write Λ_{tot} as

$$\Lambda_{\text{tot}} = \mathbb{1} + \Delta L + \Delta\Omega = (\mathbb{1} + \Delta L)(\mathbb{1} + \Delta\Omega)$$

$$= \begin{pmatrix} 1 & 0 & -\gamma\delta\beta & 0 \\ 0 & 1 & & \\ -\gamma\delta\beta & & 1 & \\ 0 & & & 1 \end{pmatrix} \begin{pmatrix} 1 & & & \\ & 1 & \frac{\gamma-1}{\beta}\delta\beta & \\ & -\frac{\gamma-1}{\beta}\delta\beta & 1 & \\ & & & 1 \end{pmatrix}$$

$$= \Lambda_2 \left(\gamma\delta\beta\right) R_3 \left(\frac{1-\gamma}{\beta}\delta\beta\right). \tag{26.110}$$

Λ_2 is a boost in the x_2-direction with infinitesimal velocity $\gamma\delta\beta$; R_3 is a rotation within S' through an angle $\delta\epsilon = (1-\gamma)\delta\beta/\beta$ about the x_3'-axis.

All of this means that an object executing circular motion at a relativistic speed will precess, i.e., in addition to its circular motion, it will also undergo rotation with angular speed

$$\Omega = \frac{\delta\epsilon}{\delta t} = \frac{1-\gamma}{v}\frac{\delta v}{\delta t} = \frac{1-\gamma}{v}\frac{v^2}{r} = (1-\gamma)\omega_0, \tag{26.111}$$

where ω_0 is the angular speed of the object's orbital motion. Thus an electron revolving around a nucleus will have its spin axis pointing in a different direction after each orbit; such precession of the spin axis yields observable effects on the spectral lines of some atoms. This result, called *Thomas precession*, is a purely relativistic effect: the precession frequency $\Omega = (1-\gamma)\omega_0 \to 0$ for $\beta \ll 1$.

26.5 Rotating Operators: Similarity Transformations

It's the lack of commutativity that gives rotations in $n \geq 3$ dimensions their character. By contrast, rotations in the plane always commute; this is equivalent to consecutive $\mathbb{R}^3$ rotations around the *same* axis. But rotations about different axes cannot be fundamentally different from one another — after all, one observer's z-axis can be another's x-axis. And since any two axes are related by a rotation, so too must the rotation operators themselves. To relate rotations around one axis to those around another, we need to learn how to rotate a rotation matrix.

The issue actually goes beyond rotation matrices. Consider, for instance, the moment of inertia I (Section 4.4), which as an operator maps the angular velocity $\vec{\omega}$ to angular momentum $\vec{L}$,

$$\vec{L} = I\vec{\omega}. \tag{26.112}$$

If we apply a symmetry operator S to the entire system (which could be a rotation), both $\vec{L}$ and $\vec{\omega}$ transform as $\vec{L}' = S\vec{L}$ and $\vec{\omega}' = S\vec{\omega}$. But if vectors change under S, then so too must the operator I — in other words, there should be a matrix I' such that $\vec{L}' \equiv I'\vec{\omega}'$. We can try to find I' by applying S to both sides of (26.112),

$$\vec{L}' = S\vec{L} = SI\vec{\omega}. \tag{26.113}$$

Alas, since in general S and I will not commute, $\vec{\omega}'$ does not emerge on the right — without which, we cannot isolate and identify I'. We can, however, circumvent the conflict between these operators by inserting a factor of $\mathbb{1} = S^{-1}S$, as follows:

$$\vec{L}' = SI\mathbb{1}\vec{\omega} = SI(S^{-1}S)\vec{\omega} = (SIS^{-1})S\vec{\omega} \equiv (SIS^{-1})\vec{\omega}', \tag{26.114}$$

where we have used the associativity of matrix multiplication. Thus we identify

$$I' \equiv S I S^{-1}. \tag{26.115}$$

This expression, called a *similarity transformation*, applies to matrix operators generally. Matrices T and $T' \equiv S T S^{-1}$ are said to be similar because, though their elements are not identical, they represent the same abstract linear operator $\mathcal{T}$. In particular, they have the same trace,

$$\text{Tr}(T') = \text{Tr}(S \, T \, S^{-1}) = \text{Tr}(S^{-1} S \, T) = \text{Tr}(T), \tag{26.116}$$

and the same determinant,

$$\det(T') = \det(S \, T \, S^{-1}) = \det(S)\det(T)\det(S^{-1}) = \det(T). \tag{26.117}$$

As we'll see, they also have the same eigenvalues.

We can think of S as a generalized rotation which effects a change of basis by rotating the axes. When S *is* an actual rotation, a similarity transformation becomes the unitary transformation $T' = UTU^{\dagger}$ in complex space, and in real space the orthogonal transformation $T' = RTR^{T}$.

If (26.115) looks familiar, it may be because a similarity transformation is precisely what we applied to the basic reflection B_0 in (26.73) to obtain $B(\varphi)$. The underlying idea is not merely similar — it's identical. We can speak of similarity transformations actively rotating operators, or consider the transformation from T to T' passively as the same operator expanded on a different basis.[7]

Example 26.16 **Rotating Rotations**

For a right-handed system, a positive $90°$ rotation around the z-axis takes $x \to y$ and $y \to -x$. Thus the rotation operators R_x and R_y should transform accordingly. Let's check. From (26.82) we have

$$R_z(\pi/2) = \begin{pmatrix} 0 & -1 & 0 \\ 1 & 0 & 0 \\ 0 & 0 & 1 \end{pmatrix}. \tag{26.118}$$

Then applying a similarity transformation to $T \equiv R_x(\alpha)$ in (26.84) gives

$$R'_x(\alpha) = R_z\,(\pi/2)\,R_x(\alpha)\,R_z^T(\pi/2)$$

$$= \begin{pmatrix} 0 & -1 & 0 \\ 1 & 0 & 0 \\ 0 & 0 & 1 \end{pmatrix} \begin{pmatrix} 1 & 0 & 0 \\ 0 & \cos\alpha & -\sin\alpha \\ 0 & \sin\alpha & \cos\alpha \end{pmatrix} \begin{pmatrix} 0 & 1 & 0 \\ -1 & 0 & 0 \\ 0 & 0 & 1 \end{pmatrix}$$

$$= \begin{pmatrix} \cos\alpha & 0 & \sin\alpha \\ 0 & 1 & 0 \\ -\sin\alpha & 0 & \cos\alpha \end{pmatrix} = R_y(\alpha), \tag{26.119}$$

---

[7] Similarity transformations are often given as $S^{-1}TS$. The difference is largely a question of who's S and who's S^{-1}.

and R_y transforms into

$$R_y'(\alpha) = R_z\,(\pi/2)\,R_y(\alpha)\,R_z^T(\pi/2)$$

$$= \begin{pmatrix} 0 & -1 & 0 \\ 1 & 0 & 0 \\ 0 & 0 & 1 \end{pmatrix} \begin{pmatrix} \cos\alpha & 0 & \sin\alpha \\ 0 & 1 & 0 \\ -\sin\alpha & 0 & \cos\alpha \end{pmatrix} \begin{pmatrix} 0 & 1 & 0 \\ -1 & 0 & 0 \\ 0 & 0 & 1 \end{pmatrix}$$

$$= \begin{pmatrix} 1 & 0 & 0 \\ 0 & \cos\alpha & \sin\alpha \\ 0 & -\sin\alpha & \cos\alpha \end{pmatrix} = R_x(-\alpha) \equiv R_{-x}(\alpha), \qquad (26.120)$$

as expected.

Example 26.17 **A Different Perspective**

It may be helpful to look at similarity transformations from the perspective of the vector being rotated (Figure 26.6). Consider $\vec{v}_2 = R_y(\alpha)\vec{v}_1$. Reading (26.119) from right to left, first $\vec{v}_1$ is rotated using $R_z^T(\pi/2) = R_z(-\pi/2)$. Because this rotation is through $90°$, the resulting vector $\vec{w}_1$ has the same orientation with respect to the x-axis that $\vec{v}_1$ has with respect to y. Next is the rotation by α around x, which gives a vector $\vec{w}_2$ having the same x-component as $\vec{w}_1$. Finally reversing the original z rotation takes $\vec{w}_2$ to $\vec{v}_2$, whose orientation with respect to y is identical to the orientation of $\vec{w}_2$ with respect to x. This is exactly the same result obtained by directly rotating $\vec{v}_1$ around y by α.

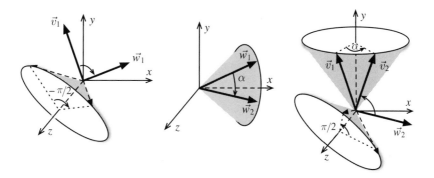

Figure 26.6 Active view of a similarity transformation.

Rotations around a coordinate axis in $\mathbb{R}^3$ take the simple forms shown in (26.82) and (26.84). For rotations around other axes, the matrices $R_{\hat{n}}(\theta)$ are less simple – but, as in Example 26.16, they can always be written as a similarity transformation of another rotation. Intuitively, the rotation S which maps one axis into another, $\hat{n} = S\hat{m}$, should be the same matrix that transforms a rotation operator around $\hat{m}$ into a rotation around $\hat{n}$ by similarity — and indeed,

$$R_{S\hat{m}}(\varphi) = SR_{\hat{m}}(\varphi)S^{-1}. \qquad (26.121)$$

So a $\pi/2$ rotation operator which takes $\hat{x}$ into $\hat{y}$ is the same operator which takes the $R_x(\alpha)$ to $R_y(\alpha)$ — just as we saw in Example 26.16.

The bottom line is that all rotations by a given angle φ are essentially equivalent; regardless of how convoluted a rotation matrix may appear, it is only a similarity transformation away from one of the canonical rotations about a coordinate axis. This observation allows one to determine both the rotation angle and the axis of rotation from inspection of any proper 3×3 orthogonal matrix. The angle is derived using the invariance of the trace under similarity transformations: (26.82) and (26.84) imply that for any rotation R_n,[8]

$$\cos \varphi = \frac{1}{2} [\text{Tr}(R_n) - 1].$$ (26.122)

Moreover, the axis of rotation $\hat{n}$ is just the direction left unscathed by a rotation,

$$R_n \hat{n} = \hat{n}.$$ (26.123)

It's trivial to see that this holds for R_x, R_y, and R_z.

Example 26.18

Consider the matrix

$$A = \frac{1}{2} \begin{pmatrix} -1 & 1 & \sqrt{2} \\ -1 & 1 & -\sqrt{2} \\ -\sqrt{2} & -\sqrt{2} & 0 \end{pmatrix}.$$ (26.124)

Although it doesn't look like any of the rotation matrices of (26.82) and (26.84), it's definitely a proper rotation because

$$A^T A = \mathbb{1},$$ (26.125)

and $\det A = +1$. The axis of rotation satisfies $A\hat{n} - \hat{n}\mathbb{1} = 0$; elimination reveals[9]

$$\hat{n} = \pm \frac{1}{\sqrt{3}} \begin{pmatrix} 0 \\ \sqrt{2} \\ -1 \end{pmatrix}.$$ (26.126)

The angle of rotation around $\hat{n}$ is

$$\theta = \cos^{-1} \left[\frac{\text{Tr}(A) - 1}{2} \right] = \frac{2\pi}{3}.$$ (26.127)

And indeed, $A^3 \equiv \mathbb{1}$ — as expected of a $120°$ rotation.

Example 26.19   A First Encounter with Diagonalization

Just as one often tries to choose axes in order to simplify a vector's components, so too for matrix operators. And very often the simplest form is as a diagonal matrix. Consider for instance transforming the matrix

$$T = \begin{pmatrix} 3 & 0 & -1 \\ 0 & -2 & 0 \\ -1 & 0 & 3 \end{pmatrix}$$ (26.128)

---

[8] To avoid notation creep, we omit the hat and write R_n rather than $R_{\hat{n}}$.
[9] In Chapter 27, we'll treat this as an eigenvalue problem.

using

$$S = \begin{pmatrix} 1 & 0 & 1 \\ 1 & 0 & -1 \\ 0 & \sqrt{2} & 0 \end{pmatrix} \tag{26.129}$$

to get the similar matrix T'. Though calculating S^{-1} can be tedious, notice that the columns of S are orthogonal; multiplying by $1/2$ renders them orthonormal. So it must be that $S^{-1} = S^T/2$. With this, straightforward matrix multiplication gives

$$T' = STS^{-1} = \begin{pmatrix} 2 & 0 & 0 \\ 0 & 4 & 0 \\ 0 & 0 & -2 \end{pmatrix}. \tag{26.130}$$

You should verify that $S/\sqrt{2}$ is a rotation by $\varphi = 98.4°$ around $\hat{n} = (1 + \sqrt{2}, 1, 1)$. Note that, as expected, both T and T' have the same trace and determinant. In Chapter 27 we'll learn how to determine whether a given matrix even *has* a diagonal form, and if so how to construct the rotation S which diagonalizes it.

Given an orthogonal matrix, it's straightforward to determine the axis $\hat{n}$ and rotation angle φ. But the more important challenge is the reverse: to construct a rotation operator for a given rotation angle around a specified axis. Appendix B explores three of the most common approaches.

26.6 Generating Rotations

Rotations in $\mathbb{R}^3$ do not commute. As we saw in Example 26.13 for the extreme case of $90°$ rotations, the ramifications of this can be significant. But we also know that rotation by $\varphi = 0$ is a perfectly good (and proper) rotation, $R(0) = \mathbb{1}$ — and of course $\mathbb{1}$ commutes with all other rotations. This raises the interesting question of how non-commutativity kicks in — abruptly for all angles $\epsilon > 0$, no matter how small? Or little by little as the angle increases to finite values?

Since the angle of rotation is a continuous parameter, a proper rotation by an infinitesimal angle ϵ should differ infinitesimally from the identity. Indeed, rotations form a continuous group of transformations — so we should be able to use some calculus on the matrices themselves. This in turns informs our choice of the allowed range of the rotation angle: to ensure continuity in order to analyze infinitesimal rotations near $\mathbb{1}$, we will take $-\pi < \varphi \leq \pi$.

Consider rotations by an infinitesimal angle ϵ. Expanding (26.82) through second order gives

$$R_z(\epsilon) = \begin{pmatrix} 1 - \frac{\epsilon^2}{2} & -\epsilon & 0 \\ \epsilon & 1 - \frac{\epsilon^2}{2} & 0 \\ 0 & 0 & 1 \end{pmatrix}, \tag{26.131}$$

with similar results for $R_x(\epsilon)$ and $R_y(\epsilon)$,

$$R_x(\epsilon) = \begin{pmatrix} 1 & 0 & 0 \\ 0 & 1 - \frac{\epsilon^2}{2} & -\epsilon \\ 0 & \epsilon & 1 - \frac{\epsilon^2}{2} \end{pmatrix} \quad R_y(\epsilon) = \begin{pmatrix} 1 - \frac{\epsilon^2}{2} & 0 & \epsilon \\ 0 & 1 & 0 \\ -\epsilon & 0 & 1 - \frac{\epsilon^2}{2} \end{pmatrix}. \tag{26.132}$$

Then a rotation about y followed by a rotation around x gives

$$R_x(\epsilon)R_y(\epsilon) = \begin{pmatrix} 1 - \frac{\epsilon^2}{2} & 0 & \epsilon \\ \epsilon^2 & 1 - \frac{\epsilon^2}{2} & -\epsilon \\ -\epsilon & \epsilon & 1 - \epsilon^2 \end{pmatrix}, \tag{26.133a}$$

whereas rotating in reverse order we find

$$R_y(\epsilon)R_x(\epsilon) = \begin{pmatrix} 1 - \frac{\epsilon^2}{2} & \epsilon^2 & \epsilon \\ 0 & 1 - \frac{\epsilon^2}{2} & -\epsilon \\ -\epsilon & \epsilon & 1 - \epsilon^2 \end{pmatrix}, \tag{26.133b}$$

where we have only kept terms through ϵ^2. We define the *commutator* of two operator A and B as

$$[A, B] \equiv AB - BA \tag{26.134}$$

so that commuting operators have $[A, B] = 0$. Calculating the commutator of our infinitesimal rotations gives

$$\begin{aligned} [R_x(\epsilon), R_y(\epsilon)] &\equiv R_x(\epsilon)R_y(\epsilon) - R_y(\epsilon)R_x(\epsilon) \\ &= \begin{pmatrix} 0 & -\epsilon^2 & 0 \\ \epsilon^2 & 0 & 0 \\ 0 & 0 & 0 \end{pmatrix} = R_z(\epsilon^2) - \mathbb{1}. \end{aligned} \tag{26.135}$$

Clearly the rotations don't commute. But as $\epsilon \to 0$, the rotation $R_z(\epsilon^2)$ on the right-hand side goes to $\mathbb{1}$ faster than each of the individual rotations $R_{x,y}(\epsilon)$ on the left. So the commutator $[R_x, R_y]$ goes to zero faster than the individual rotations. In other words, $[R_x, R_y] = 0$ to first order in ϵ. Thus unlike finite rotations, *infinitesimal* rotations commute. (This verifies the claim made in BTW 24.1.)

To explore the deviation from commutativity, we introduce the Hermitian matrix L_3 as the coefficient of the first-order term,

$$R_z(\epsilon) = \begin{pmatrix} 1 & -\epsilon & 0 \\ \epsilon & 1 & 0 \\ 0 & 0 & 1 \end{pmatrix} \equiv \mathbb{1} - i\epsilon L_3, \tag{26.136}$$

so that

$$L_3 \equiv \begin{pmatrix} 0 & -i & 0 \\ i & 0 & 0 \\ 0 & 0 & 0 \end{pmatrix}. \tag{26.137a}$$

Infinitesimal rotations about x and y similarly lead to

$$L_1 \equiv \begin{pmatrix} 0 & 0 & 0 \\ 0 & 0 & -i \\ 0 & i & 0 \end{pmatrix} \qquad L_2 \equiv \begin{pmatrix} 0 & 0 & i \\ 0 & 0 & 0 \\ -i & 0 & 0 \end{pmatrix}. \tag{26.137b}$$

We can then write an infinitesimal rotation about an axis $\hat{n}$ as

$$R_n(\epsilon) = \mathbb{1} - i\epsilon\,\hat{n} \cdot \vec{L}, \tag{26.138}$$

where $\vec{L} \equiv (L_1, L_2, L_3)$ and $\hat{n} \cdot \vec{L} = \sum_i \hat{n}_i L_i$. These matrices — which encode how R_n deviates from the identity — are called the *generators* of rotations.

Bestowing the L_i with the honorific "generator" may seem overly formal; after all, how often does one actually *need* an infinitesimal rotation? But in fact, we can use them to construct ("generate") finite rotations. To see how, let's use (26.136) to express a $\varphi + d\varphi$ rotation around z as

$$R_z(\varphi + d\varphi) = R_z(d\varphi)R_z(\varphi) = (\mathbb{1} - id\varphi\, L_3)R_z(\varphi). \tag{26.139}$$

So the infinitesimal matrix-valued change dR_z is

$$dR_z \equiv R_z(\varphi + d\varphi) - R_z(\varphi) = -id\varphi L_3 R_z(\varphi). \tag{26.140}$$

Integrating, and using $R_z(0) = \mathbb{1}$, we get

$$R_z(\varphi) = e^{-i\varphi L_3}, \tag{26.141a}$$

where, recall, the exponential of a matrix is defined by its Taylor series (6.32). In fact, Taylor expansion reveals this exponential form to be equivalent to the canonical matrix R_z of (26.82). The same, of course, holds for the other coordinate axes as well:

$$R_x(\varphi) = e^{-i\varphi L_1} \qquad R_y(\varphi) = e^{-i\varphi L_2}. \tag{26.141b}$$

Generalizing to an arbitrary axis,

$$R_n(\varphi) = e^{-i\varphi\, \hat{n}\cdot\vec{L}} = e^{-i\vec{\varphi}\cdot\vec{L}}. \tag{26.142}$$

Even without expanding, we can verify that (26.142) is a proper $\mathbb{R}^3$ rotation by using the antisymmetry $L_i^T = -L_i$ of the generators,

$$\begin{aligned} R_n(\varphi)R_n^T(\varphi) &= \left(e^{-i\vec{\varphi}\cdot\vec{L}}\right)\left(e^{-i\vec{\varphi}\cdot\vec{L}}\right)^T \\ &= e^{-i\vec{\varphi}\cdot\vec{L}}\, e^{-i\vec{\varphi}\cdot\vec{L}^T} = e^{-i\vec{\varphi}\cdot\vec{L}}\, e^{+i\vec{\varphi}\cdot\vec{L}} = \mathbb{1}. \end{aligned} \tag{26.143}$$

So R_n is orthogonal. (But see Problem 6.43.) And because the L_i are traceless, (26.54) guarantees $\det R = +1$.

[You may be wondering why we pulled out a factor of i when defining the generators in (26.136)–(26.138). After all, with or without the factor of i, R must be orthogonal (that is, unitary and real). Indeed, in classical mechanics one often works with real, antisymmetric generators $N_k \equiv -iL_k$ (see Section B.1). But Hermitian generators are more convenient for quantum mechanics, which uses manifestly unitary rotation operators,

$$\begin{aligned} R_n(\varphi)R_n^\dagger(\varphi) &= \left(e^{-i\vec{\varphi}\cdot\vec{L}}\right)\left(e^{-i\vec{\varphi}\cdot\vec{L}}\right)^\dagger \\ &= e^{-i\vec{\varphi}\cdot\vec{L}}\, e^{+i\vec{\varphi}\cdot\vec{L}^\dagger} = e^{-i\vec{\varphi}\cdot\vec{L}}\, e^{+i\vec{\varphi}\cdot\vec{L}} = \mathbb{1}. \end{aligned} \tag{26.144}$$

This is also reminiscent of complex numbers $z = e^{-i\varphi}$ — which are themselves unitary rotations for real φ.]

BTW 26.5 Continuous, Connected, and Compact

Formally speaking, a group is continuous if its elements are distinguished by one or more continuously varying parameters α_i, such that small changes in the α_i produce a small change in the group element. This is clearly true of rotations angles and rotation matrices. The dimension of the group is just the number of parameters required to specify elements of the group. With only the single angle φ, rotations in the plane are one dimensional; rotations is $\mathbb{R}^3$ require three parameters (an angle and a rotation axis) and so form a three-dimensional group.

Two other properties distinguish continuous groups. If any element can evolve into every other element by a continuous variation of the parameters, the group is said to be *connected*. Since all proper rotations can be continuously reduced to the identity, they can clearly be continuously connected to one another. So proper rotations form a *connected* space. Even though improper rotations do not form a group, they still constitute a connected space. But proper and improper rotations together do not constitute a connected space: a determinant 1 rotation cannot continuously evolve into a determinant −1 reflection.

Rotations are also *compact* since the full range of rotation angle(s) maps continuously to the full range of rotation operators. Technically, every infinite sequence of angles corresponds to an infinite sequence of rotations whose limit is also a group element. Lorentz transformations, by contrast, are not compact since boosts from frame to frame are distinguished by the open interval $0 \leq v/c < 1$. Though a sequence of boosts can have the limit $v/c \to 1$, no boost — that is, no element of the Lorentz group — has $v/c = 1$.

If rotations do not commute, then neither can the generators. Straightforward calculation using (26.137) reveals (Problem 26.34)[10]

$$[L_i, L_j] = i\epsilon_{ijk}L_k. \tag{26.145}$$

Formally, the commutation relation amongst the generators is known as a *Lie algebra* ("lee"). Continuous symmetry groups such as rotations are called Lie groups.

Mathematically, the commutator has much in common with the cross product. For instance, the cyclic symmetry of (26.145) expresses the right-handed convention enshrined in the canonical rotation matrices R_x, R_y, and R_z of (26.82) and (26.84). Indeed, since $(L_k)_{\ell m} = -i\epsilon_{k\ell m}$, (26.145) is really just our old friend the epsilon identity of (4.25) (Problem 26.35). And the antisymmetry of ϵ_{ijk} guarantees that the commutator vanishes for $i = j$, reflecting the fact that successive rotations about the same axis commute.

All of which is nice — but what advantages do non-commutative L_i have over non-commutative R_n? For one thing, rotation matrices are angle-dependent, with their largest elements for $\varphi = 90°$ and vanishing as $\varphi \to 0$. Similarly, any product of R's depends both on axis and angle. By contrast, (26.145) is *independent* of both axis and rotation angle — and therefore directly reflects the source of non-commutativity. Moreover, the commutator of rotations is *not* a rotation. (Remember, it's the *product* of R's that yields another R, not their sum. There is no $R = 0$ rotation.) On the other hand, the generators form a vector space, with the L_i's themselves providing a basis. This makes them much easier to work with, justifying the shift of focus from rotations R_n to generators L_i. But as we'll see in the Example 26.20, the underlying physics is perhaps even more compelling.

Example 26.20 Generators and Angular Momentum

Thus far, we've focused almost exclusively on the rotation of vectors. But what happens to a scalar function $f(\vec{r})$ under rotation? When $\vec{r}$ goes to $\vec{r}' = R\vec{r}$, what happens to f? Since it's not a column vector, its rotation is obviously not given by a matrix. Indeed, we expect any change in f to be given by a derivative.

---

[10] Technically, there should be a sum over k — but for given i and j, the ϵ_{ijk} renders it superfluous.

Let's denote the effect a rotation of the coordinates has on f as a new function f' of the rotated coordinates, $\mathcal{R}f(\vec{r}) \equiv f'(R\vec{r})$. Now as a scalar, the numerical value of f at every point is unchanged by a rotation. For instance, if the far corner of the room has a given temperature before the coordinate axes are rotated, then it still has that same temperature afterward. In other words, the new function at the new coordinates must equal the old function at the old coordinates, $f'(R\vec{r}) = f(\vec{r})$ — or, since this must hold between any two positions related by rotation,

$$f'(\vec{r}) = f(R^{-1}\vec{r}). \tag{26.146}$$

Now consider an infinitesimal rotation by $\delta\vec{\varphi}$. From (26.88), $R^{-1}\vec{r} = \vec{r} - \delta\vec{\varphi} \times \vec{r}$ (the minus sign comes from using R^{-1}). Taylor expanding to first order,

$$f'(\vec{r}) = f(R^{-1}\vec{r}) = f(\vec{r}) - (\delta\vec{\varphi} \times \vec{r}) \cdot \vec{\nabla}f$$

$$= f(\vec{r}) - \delta\vec{\varphi} \cdot (\vec{r} \times \vec{\nabla}f) = \left[1 - \delta\vec{\varphi} \cdot (\vec{r} \times \vec{\nabla})\right]f(\vec{r}). \tag{26.147}$$

Thus the effect of a rotation on a scalar function is given by

$$R(\delta\varphi) = 1 - \delta\vec{\varphi} \cdot (\vec{r} \times \vec{\nabla}). \tag{26.148}$$

Comparison with (26.138) identifies the generators

$$\vec{L} \equiv -i\vec{r} \times \vec{\nabla}. \tag{26.149}$$

As expected, this is a differential rather than a matrix operator. Even so, careful application of the product rule verifies that the three components $-i(\vec{r} \times \vec{\nabla})_i$ obey the commutation relation of (26.145) (Problem 26.34c). And quantum mechanics provides a physical interpretation: recalling that the momentum operator is $\vec{p} = -i\hbar\vec{\nabla}$, we find[11]

$$\hbar\vec{L} = \vec{r} \times \vec{p}. \tag{26.150}$$

So the generators of rotation are the components of angular momentum! Because it is based on $\vec{r} \times \vec{p}$, the operator in (26.149) describes the quantum mechanical manifestation of classical angular momentum (which is why the generators are denoted with the letter L).

It's worth noting that this *orbital* angular momentum is distinct from *spin* angular momentum. Indeed, the spin generators S_i are matrix, rather than differential, operators — and so have no connection to classical $\vec{r} \times \vec{p}$. For spin-$\frac{1}{2}$, the generators are proportional to the Pauli matrices, $S_i = \sigma_i/2$ — see (26.52). Direct calculation easily verifies that they obey the same commutator (26.145) as the L_i,

$$\left[\frac{\sigma_i}{2}, \frac{\sigma_j}{2}\right] = i\epsilon_{ijk}\frac{\sigma_k}{2}. \tag{26.151}$$

As Hermitian operators satisfying this commutator, the S_i generate unitary rotations in two-dimensional complex space — and hence have all the physical characteristics and hallmarks of angular momentum.

---

[11] The operators $\vec{L}$ are dimensionless; the units of angular momentum come from the factor of $\hbar$.

Problems

26.1   Use rotations in the plane to validate the expression (26.14) for passive transformations with constant vector $|v\rangle$.

26.2   Starting with the active rotation in the plane, $|v'\rangle = v_1'|\hat{e}_1\rangle + v_2'|\hat{e}_2\rangle$, substitute the expressions for the v_i' and collect terms to define a primed basis. Is this basis a clockwise or counter-clockwise rotation of the unprimed basis? Explain.

26.3   Show that the rotation matrix R relating orthonormal bases, $\hat{\epsilon}_i = \sum_j R_{ij}\hat{e}_j$, is $R_{ij} = \hat{\epsilon}_i \cdot \hat{e}_j$. What rotation connects $\hat{e}_1 \to \frac{1}{\sqrt{2}}\begin{pmatrix} 1 \\ 1 \end{pmatrix}$, $\hat{e}_2 \to \frac{1}{\sqrt{2}}\begin{pmatrix} 1 \\ -1 \end{pmatrix}$ with $\hat{\epsilon}_1 \to \begin{pmatrix} 0.6 \\ 0.8 \end{pmatrix}$, $\hat{\epsilon}_2 \to \begin{pmatrix} 0.8 \\ -0.6 \end{pmatrix}$? What is the rotation angle?

26.4   Express m identical rotations around $\hat{n}$ as a single rotation operation. Use this result to derive trig identities for $\cos 3\varphi$ and $\sin 3\varphi$.

26.5   Determine whether the following matrices are orthogonal, unitary, and/or Hermitian — or none of the above:

(a) $A = \frac{1}{\sqrt{6}}\begin{pmatrix} 1 & -\sqrt{3} & \sqrt{2} \\ -2 & 0 & \sqrt{2} \\ 1 & \sqrt{3} & \sqrt{2} \end{pmatrix}$

(b) $B = \frac{1}{2\sqrt{2}}\begin{pmatrix} 2 & 2i \\ -2i & 2 \end{pmatrix}$

(c) $C = \begin{pmatrix} i & 1 & i \\ 2 & i & 0 \\ i & 1 & -i \end{pmatrix}$

(d) $D = \exp\left[i \begin{pmatrix} 0 & -i \\ i & 0 \end{pmatrix} \right]$

(e) $E = \begin{pmatrix} 0 & -i \\ i & 0 \end{pmatrix}$.

26.6   How many independent parameters are required to specify the following matrices in n dimensions?

(a) Symmetric $S = S^T$
(b) Antisymmetric $A = -A^T$
(c) Orthogonal matrix $O^T = O^{-1}$
(d) Unitary matrix $U^\dagger = U^{-1}$
(e) Hermitian matrix $H^\dagger = H$

26.7   Use 2×2 orthogonal matrices to verify $i^2 = -1$ and $i^3 = -i$.

26.8   Show that the determinant of a unitary matrix is a pure phase, $\det U = e^{i\alpha}$.

26.9   The *spherical basis* $\{\hat{e}_+, \hat{e}_-, \hat{e}_0\}$,

$$\hat{e}_\pm = \mp\frac{1}{\sqrt{2}}\left(\hat{e}_x \pm i\hat{e}_y\right), \qquad\qquad \hat{e}_0 = \hat{e}_z,$$

is a complex representation of vectors in $\mathbb{R}^3$. Construct the matrix T that maps the cartesian basis into the spherical basis. What type of transformation is this? Write out the inner product $\vec{A} \cdot \vec{B}$ on this basis. [The spherical basis is used extensively in the quantum mechanical description of angular momentum.]

26.10   Electromagnetic Duality
Problem 16.7 shows that magnetic monopoles essentially put electric and magnetic field on an equal footing. We'll explore this symmetry in more detail using the duality transformation of Example 26.1.

(a) Confirm that (26.7) leaves the Maxwell equations (26.6) invariant.

(b) Show that duality symmetry is spoiled in the presence of electric sources.

(c) If monopoles exist, Maxwell's equations become (see Problem 16.7)

$$\vec{\nabla} \cdot \vec{E} = \rho_e/\epsilon_0 \qquad \vec{\nabla} \times \vec{E} + \partial_t \vec{B} = -\mu_0 \vec{J}_m$$
$$\vec{\nabla} \cdot \vec{B} = \mu_0 \rho_m \qquad \vec{\nabla} \times \vec{B} - \mu_0 \epsilon_0 \partial_t \vec{E} = \mu_0 \vec{J}_e.$$

Here ρ_e, $\vec{J}_e$ denote electric charge and current density, while ρ_m, $\vec{J}_m$ are magnetic charge and current density. Show that duality symmetry is restored if electric charge q and magnetic charge g (divided by c) transform with the same rotation as in (26.7). [Discovery of monopoles would essentially reduce the distinction between electric and magnetic phenomena to a conventional choice of φ.]

(d) Show that the force felt by a charge q and monopole g moving with velocity $\vec{v}$ through electric and magnetic fields

$$\vec{F} = q \left(\vec{E} + \vec{v} \times \vec{B} \right) + g \left(\vec{B} - \vec{v} \times \vec{E}/c^2 \right)$$

is duality invariant. [Discovery of monopoles would essentially reduce the distinction between electric and magnetic phenomena to a conventional choice of φ.]

26.11 Generalize the matrix derivation leading to (26.28) for rotation in $\mathbb{R}^n$ to show that rotation matrices in complex space must be unitary, (26.38).

26.12 One can envision doing a proper rotation in $\mathbb{R}^n$ by choosing an axis in n-space, and rotating around it in an $(n-1)$-dimensional hyperplane. Similarly, to do a proper rotation in $n-1$ dimensions, choose an axis and rotate around it in the $(n-2)$-dimensional hyperplane. Use this recursive argument to determine the total number of parameters needed to specify a proper rotation in $\mathbb{R}^n$.

26.13 Groups and vectors have very similar criteria. Is a group also a vector space? Explain.

26.14 What directions do reflections in $\mathbb{R}^2$ leave invariant? We'll answer this using the geometric approach outlined above (26.73):

(a) What are the invariant directions of B_0 in (26.71)? Interpret these directions, and the action of B_0 on them, geometrically. (No calculation necessary!)

(b) Rotate the directions found in (a) to find the invariant directions of $B(\varphi)$. Verify they are correct by direction application of $B(\varphi)$.

26.15 Verify that the reflections of (26.79), as well as the parity inversion (26.80), change a systems's handedness.

26.16 Lest you fear that (26.88) allows $|\vec{r}|$ to vary — violating what we expect of rotations — take the inner product of both sides with $\vec{r}$ to conclude that all's well.

26.17 Use (26.88) to verify that (i) infinitesimal rotations by $\delta\vec{\varphi}_1 = \delta\varphi_1 \hat{n}_1$ and $\delta\vec{\varphi}_2 = \delta\varphi_2 \hat{n}_2$ commute, and (ii) the result of two successive infinitesimal rotations is a single rotation by $\delta\vec{\varphi}_1 + \delta\vec{\varphi}_2$. In other words, infinitesimal rotations add vectorially. (See BTW 24.1.)

26.18 Find the angle, axis, and/or reflection axis for each of the following:

(a) $A = \frac{1}{4} \begin{pmatrix} 3 & \sqrt{6} & -1 \\ -\sqrt{6} & 2 & -\sqrt{6} \\ -1 & \sqrt{6} & 3 \end{pmatrix}$

(b) $B = \begin{pmatrix} 0 & 1 \\ 1 & 0 \end{pmatrix}$

(c) $C = \frac{1}{2} \begin{pmatrix} \sqrt{3} & 0 & 1 \\ 0 & -2 & 0 \\ -1 & 0 & \sqrt{3} \end{pmatrix}$

(d) $D = -\frac{1}{\sqrt{2}} \begin{pmatrix} \sqrt{2} & 0 & 0 \\ 0 & 1 & 1 \\ 0 & 1 & -1 \end{pmatrix}$.

26.19 Rotate $\vec{v} = (1,1,0)$ in succession by $90°$ around x and y. Show that the order of these successive rotations matter. What's the angle between the different final vectors?

26.20 Verify that R_q of (26.99) is a $\varphi = 70.5°$ rotation around the axis

$$\hat{q} = \frac{1}{\sqrt{2}} \begin{pmatrix} -1 \\ 1 \\ 0 \end{pmatrix}.$$

26.21 Use the canonical rotations of (26.82) and (26.84) and similarity transformations to construct the matrix for rotation by α around $\hat{n} = \frac{1}{\sqrt{2}}(0,1,1)$.

26.22 Use index notation to compare the active rotation of a vector $\vec{v}$ and a matrix M by an orthogonal matrix R. How would you presume a three-index object T_{ijk} transforms? [We'll have much to say about this in Chapter 29.]

26.23 A *dyad* is a two-index object formed by placing two vectors side-by-side to yield the matrix $M_{ij} = u_i v_j$. Since a rotation acts on each vector in the usual way, a dyad transforms as

$$M'_{k\ell} = \left(\sum_k R_{ki} u_i \right) \left(\sum_\ell R_{\ell j} v_j \right) = \sum_{k\ell} R_{ki} R_{\ell j} u_i v_j = \sum_{k\ell} R_{ki} R_{\ell j} M_{ij},$$

which is a similarity transformation of M. The same rotation can be calculated in a vector representation by incorporating $\vec{u}$ and $\vec{v}$ into a single column $w_a = (uv)_a$, where the index a runs from 1 to n^2 in one-to-one correspondence with every permutation of the pair $\{ij\}$. Rotating $\vec{w}$ then manifests as $w'_a = \sum_b T_{ab} w_b$, where the $n^2 \times n^2$ matrix T_{ab} is the product of elements $R_{ki} R_{\ell j}$ in which a and b are in the same one-to-one correspondence with permutations of $\{k\ell\}$ and $\{ij\}$, respectively.

(a) Write out the column representation of $\vec{w}$ in terms of u_i and v_j for $n = 2$.
(b) Write out the matrix T_{ab} for rotation in the plane $R(\varphi)$.
(c) Verify that $\sum_b T_{ab} w_b$ produces the same rotation as the similarity transformation RMR^T.

26.24 Proper rotations in $\mathbb{R}^3$ it can be elegantly characterized as linear transformations $\mathcal{R}$ which preserve both the dot product,

$$\mathcal{R}\vec{v} \cdot \mathcal{R}\vec{w} = \vec{v} \cdot \vec{w} \tag{26.152a}$$

and the cross product,

$$\mathcal{R}\vec{v} \times \mathcal{R}\vec{w} = \mathcal{R}(\vec{v} \times \vec{w}). \tag{26.152b}$$

The first preserves the norms and relative orientation of vectors, the second their handedness (Biedenharn and Louck, 1981). Use index notation to verify these statements.

26.25 Verify (see Example 26.8)

$$U(\varphi) \equiv e^{i\varphi \, \hat{n} \cdot \vec{\sigma}/2} = \mathbb{1} \cos(\varphi/2) + i\hat{n} \cdot \vec{\sigma} \sin(\varphi/2).$$

[Remember that all the Pauli matrices have $\sigma_i^2 = \mathbb{1}$. You also may find (25.138) from Problem 25.24 helpful.]

26.26 Show that Lorentz transformations leave the interval $s^2 = x_0^2 - x_1^2$ invariant by verifying (26.68), $\Lambda^T \eta \Lambda = \eta$.

26.27 Show that (26.141) do in fact produce the canonical $\mathbb{R}^3$ rotations of (26.82) and (26.84).

26.28 Show that a general rotation cannot be expressed as a symmetric matrix.

26.29 Use the first-order expression in (26.138) to show that the unitarity of rotations, $R_z^\dagger R_z = \mathbb{1}$, requires the generators to be Hermitian.

26.30 Use the first-order expression in (26.138) to decompose a finite rotation around z by φ as a succession of N rotations by small φ/N. Show that as N gets bigger (so that each individual rotation becomes smaller and (26.136) more reliable) the exponential form (26.141a) emerges.

26.31 Verify the Jacobi identity for the commutator of matrices,

$$[A,[B,C]] + [B,[C,A]] + [C,[A,B]] = 0.$$

26.32 What is the generator of rotation for a vector in $\mathbb{R}^2$? (See Example 26.5.)

26.33 **Generators**

(a) Show that a proper rotation in $\mathbb{R}^n$ can be expressed in the form $R = e^A$, where the generator A is a real antisymmetric matrix. Use A's antisymmetry to determine the number of parameters required to specify a rotation in $\mathbb{R}^n$. (You may find (26.54) helpful.)

(b) Show that a unitary matrix can be expressed in the form $U = e^{iH}$, where the generator H is a Hermitian matrix, $H = H^\dagger$. Use this result to determine the number of parameters required to specify a rotation in $\mathbb{C}^n$. How many are required if $\det U = 1$?

26.34 Angular momentum in quantum mechanics is described by a three-element family of operators J_i satisfying the commutation relation of (26.145),

$$[J_i, J_j] = i\epsilon_{ijk}J_k.$$

Whether the J operators are matrices or differential operators, if they satisfy the commutation relation, they play the same basic role generating rotations. Show that each of the following are generators of angular momentum:

(a) The matrices of (26.137).

(b) Spin $S_k = \sigma_k/2$, where σ_i are the Pauli matrices.

(c) Orbital angular momentum $L_k = -i(\vec{r} \times \vec{\nabla})_k = -i\sum_{\ell n} \epsilon_{k\ell n}r_\ell\partial_n$. (Be sure to use the product rule.) This angular momentum corresponds to the classical expression $\hbar\vec{L} = \vec{r} \times \vec{p}$.

26.35 (a) Show that the L matrices of (26.137) can be expressed in terms of the Levi-Civita epsilon, $(L_k)_{\ell m} = -i\epsilon_{k\ell m}$.

(b) Verify, using (26.138) and index notation, that the commutator in (26.145) is equivalent to the familiar identity $\sum_i \epsilon_{ijk}\epsilon_{i\ell m} = \delta_{j\ell}\delta_{km} - \delta_{jm}\delta_{k\ell}$.

26.36 Show that the commutation relation (26.145) can be expressed in terms of the cross product, $\vec{L} \times \vec{L} = i\vec{L}$. [Note that for a vector $\vec{L} = (L_1, L_2, L_3)$ with non-commuting components, the cross product $\vec{L} \times \vec{L} \neq 0$.]

26.37 (a) Verify the commutator identity $[A, BC] = B[A, C] + [A, B]C$ for arbitrary operators $A, B,$ and C. [Note the similarity to "BAC-CAB."]

(b) For three operators J_i obeying $[J_i, J_j] = i\epsilon_{ijk}J_k$, use the commutator identity to show that the J_i commute with $J^2 = J_1^2 + J_2^2 + J_3^2$.

(c) Use matrix multiplication to verify this for the L_i of (26.137).

26.38 [Adapted from Namias (1977)] Expressing angular momentum's cartesian components $L_k = -i(\vec{r} \times \vec{\nabla})_k = -i\sum_{\ell n} \epsilon_{k\ell n}r_\ell\partial_n$ in spherical coordinates can require tedious use of the chain rule and scale factors (see almost any quantum mechanics text). Let's try a simpler approach:

(a) On general grounds, write

$$L_x = -i\left(y\,\partial_z - z\,\partial_y\right) = a\,\partial_r + b\,\partial_\theta + c\,\partial_\phi,$$

where a, b, and c are functions of (r, θ, ϕ). To find them, apply L_x first to $(x^2 + y^2 + z^2)^{1/2} = r$, then to $x = r \sin\theta \cos\phi$ and $z = r \cos\theta$. You should find

$$L_x = +i \left(\sin\phi \, \partial_\theta + \cot\theta \cos\phi \, \partial_\phi \right).$$

Apply the same approach to L_y and L_z, to find

$$L_y = -i \left(\cos\phi \, \partial_\theta - \cot\theta \sin\phi \, \partial_\phi \right) \quad \text{and} \quad L_z = -i\partial_\phi.$$

(b) We can use these results to find $L^2 = L_x^2 + L_y^2 + L_z^2$ in spherical coordinates — but noting that none of the L_i depend on r, we can write

$$L^2 = a \, \partial_\theta^2 + b \, \partial_\phi^2 + c \, \partial_\theta \partial_\phi + d \, \partial_\theta + e \, \partial_\phi.$$

This time operate on $x + iy = r \sin\theta e^{i\phi}$, $z(x + iy) = r^2 \sin\theta \cos\theta e^{i\phi}$, and $z = r \cos\theta$ to show that

$$L^2 = - \left[\frac{1}{\sin\theta} \, \partial_\theta (\sin\theta \, \partial_\theta) + \frac{1}{\sin^2\theta} \, \partial_\phi^2 \right].$$

[Hint: Use the fact that L^2 is real, so that all the imaginary parts must sum to zero.] How does this compare with the Laplacian ∇^2 in spherical coordinates?

27 The Eigenvalue Problem

As we've seen, linear systems allow physical relationships to be communicated using mathematical expressions valid in any basis. This naturally invites the question: is there a *preferred* basis, one which highlights certain features and directions? Very often, the answer is yes. Take, for example, rotations in $\mathbb{R}^3$. The rotation axis $\hat{n}$ associated with a rotation R is given by the intuitive expression

$$R\hat{n} = \hat{n}. \tag{27.1}$$

Only vectors proportional to $\hat{n}$ are left unchanged by the rotation — so choosing $\hat{n}$ as a basis vector may be advantageous. And Euler's rotation theorem (BTW 27.1) guarantees that $R\hat{n} = \hat{n}$ has a non-trivial solution $\hat{n}$ for proper rotations in $\mathbb{R}^3$.

And what of improper rotations? Well, for reflections in $\mathbb{R}^2$, that invariant direction must be the line of reflection $\vec{u}$ (the mirror, if you will), so that any vector $\vec{w}$ *orthogonal* to $\vec{u}$ is reflected into $-\vec{w}$. In this case, then, we should recast (27.1) as

$$R\vec{v} = \lambda\vec{v}, \tag{27.2}$$

with *two* solutions: $\lambda_1 = +1, \vec{v}_1 = \vec{u}$ (the mirror), and $\lambda_2 = -1, \vec{v}_2 = \vec{w}$ (orthogonal to the mirror).

There are similar physical interpretations for other linear transformations. For instance, one can always find axes in which a complicated moment of inertia has only non-zero elements along its diagonal. In other words, I has special directions $\vec{\omega}$ such that angular momentum $\vec{L}$ is parallel to angular velocity,

$$L_i = \sum I_{ij}\omega_j = \lambda\omega_j. \tag{27.3}$$

Working with these *principal axes* and *principal moments of inertia* λ can greatly simplify the analysis of general rigid-body rotations.

These examples, and countless more, reflect inherent properties of their operators. They all have in common the *eigenvalue equation*

$$A|v\rangle = \lambda|v\rangle, \tag{27.4}$$

for an operator A and some scalar λ. (We use ket notation to emphasize that we need not be in $\mathbb{R}^n$.) Note that only A is specified, so the solution requires finding both the *eigenvectors* $|v\rangle$'s and their associated *eigenvalues* λ's. One clear advantage of working with eigenvectors is that an operator A can be replaced by scalar multiplication λ. More importantly, the λ's and $|v\rangle$'s are characteristic (in German, "eigen") of the operator; as we'll see, knowledge of an operator's eigenvalues and eigenvectors is tantamount to knowledge of the operator itself.

27.1 Solving the Eigenvalue Equation

Finding the eigenvalues and eigenvectors of an operator means finding solutions of

$$(A - \lambda \mathbb{1}) |v\rangle = 0. \tag{27.5}$$

We're tempted, of course, to just multiply both sides by the inverse operator $(A - \lambda \mathbb{1})^{-1}$. But this procedure only returns $|v\rangle = 0$ — a solution to be sure, but clearly not characteristic of the operator. The only way to obtain a *non-trivial* solution is if this procedure is invalid. And the procedure is only invalidated if the inverse operator does not exist. For matrices, this implies

$$\det(A - \lambda \mathbb{1}) = 0. \tag{27.6}$$

This is called the *secular determinant*, which leads to the *secular* or *characteristic* equation of the matrix. The roots of this polynomial are the eigenvalues of A.

Example 27.1

Let's start with the symmetric 2×2 matrix

$$A = \begin{pmatrix} 5 & 2\sqrt{2} \\ 2\sqrt{2} & 3 \end{pmatrix}. \tag{27.7}$$

Calculating the secular determinant, we find

$$\begin{vmatrix} 5 - \lambda & 2\sqrt{2} \\ 2\sqrt{2} & 3 - \lambda \end{vmatrix} = (5 - \lambda)(3 - \lambda) - 8 = 0, \tag{27.8}$$

or

$$\lambda^2 - 8\lambda + 7 = 0. \tag{27.9}$$

This is called the secular polynomial; its solutions are the eigenvalues — in this example, $\lambda_1 = 7$, $\lambda_2 = 1$. (The order is arbitrary.) The set of eigenvalues is called the *spectrum* of the operator.

So we're halfway there. All that's left is to find the eigenvectors corresponding to each of the eigenvalues. We do this using the eigenvalue equation itself,

$$\begin{pmatrix} 5 & 2\sqrt{2} \\ 2\sqrt{2} & 3 \end{pmatrix} \begin{pmatrix} a \\ b \end{pmatrix} = \lambda \begin{pmatrix} a \\ b \end{pmatrix}, \tag{27.10}$$

and solve for a, b. For instance, for $\lambda_1 = 7$ we have

$$\left. \begin{matrix} -2a + 2\sqrt{2}\,b = 0 \\ 2\sqrt{2}\,a - 4b = 0 \end{matrix} \right\} \longrightarrow a = \sqrt{2}\,b. \tag{27.11}$$

Note that these two equations are linearly *de*pendent, so the best we can do is solve for the ratio a/b. Choosing $b = 1$ yields the eigenvector

$$|v_1\rangle \rightarrow \begin{pmatrix} \sqrt{2} \\ 1 \end{pmatrix}. \tag{27.12}$$

Similarly, for $\lambda_2 = 1$, we find

$$\left. \begin{array}{l} 4a + 2\sqrt{2}b = 0 \\ 2\sqrt{2}a + 2b = 0 \end{array} \right\} \longrightarrow b = -\sqrt{2}\,a. \tag{27.13}$$

Taking $a = 1$ yields the eigenvector

$$|v_2\rangle \rightarrow \begin{pmatrix} 1 \\ -\sqrt{2} \end{pmatrix}. \tag{27.14}$$

Substituting $|v_1\rangle$ and $|v_2\rangle$ into the original eigenvalue equation verifies these solutions.

Example 27.2

The matrix

$$B = \begin{pmatrix} -1 & -1 & 5 \\ -1 & 5 & -1 \\ 5 & -1 & -1 \end{pmatrix} \tag{27.15}$$

has the secular determinant

$$\begin{vmatrix} -1-\lambda & -1 & 5 \\ -1 & 5-\lambda & -1 \\ 5 & -1 & -1-\lambda \end{vmatrix} = (-1-\lambda)\,[(5-\lambda)(-1-\lambda) - 1]$$

$$+ [(1+\lambda) + 5] + 5\,[1 - 5(5-\lambda)] = 0. \tag{27.16}$$

Cleaning this up a bit leads to the charateristic equation

$$\lambda^3 - 3\lambda^2 - 36\lambda + 108 = (\lambda - 3)(\lambda^2 - 36) = 0, \tag{27.17}$$

with solutions $\lambda = 3, \pm 6$. These eigenvalues allow us to find the eigenvectors $\begin{pmatrix} a \\ b \\ c \end{pmatrix}$ from the eigenvalue equation,

$$\begin{pmatrix} -1 & -1 & 5 \\ -1 & 5 & -1 \\ 5 & -1 & -1 \end{pmatrix} \begin{pmatrix} a \\ b \\ c \end{pmatrix} = \lambda \begin{pmatrix} a \\ b \\ c \end{pmatrix}. \tag{27.18}$$

For $\lambda_1 = 3$, this gives

$$-4a - b + 5c = 0$$

$$-a + 2b - c = 0$$

$$5a - b - 4c = 0. \tag{27.19}$$

Three equations, three unknowns — no problem, right? Well, only if the equations are linearly independent. Which, since they sum to zero, they're clearly not. So with only two independent conditions, the most we can deduce from this system is $a = b = c$; making the simplest choice yields the eigenvector

$$|v_1\rangle \rightarrow \begin{pmatrix} 1 \\ 1 \\ 1 \end{pmatrix}. \tag{27.20}$$

As for $\lambda_{2,3} = \pm 6$, a similar analysis yields the eigenvectors

$$|v_2\rangle \rightarrow \begin{pmatrix} 1 \\ -2 \\ 1 \end{pmatrix}, \qquad |v_3\rangle \rightarrow \begin{pmatrix} 1 \\ 0 \\ -1 \end{pmatrix}. \tag{27.21}$$

As before, straightforward insertion of these into the original eigenvalue equation verifies the solutions.

From these examples, we can already start to glean and appreciate certain general features. In both cases the equations for the eigenvectors are not linearly independent. Indeed, an n-dimensional system will only have $n-1$ conditions. This is due to the nature of the eigenvalue equation itself: if $|v\rangle$ solves (27.4), then so too does any multiple $k|v\rangle$ with the same eigenvalue λ. In other words, the eigenvalue equation only specifies a special *direction* in space; there is no constraint on the overall length of an eigenvector. It's very common to use this freedom to express the vector as simply as possible. For instance, we rendered $|v_2\rangle$ in (27.21) in the form $\begin{pmatrix} 1 \\ -2 \\ 1 \end{pmatrix}$ rather than, say, $\begin{pmatrix} 1/2 \\ -1 \\ 1/2 \end{pmatrix}$. It can also be convenient, given that only their directions are fixed, to normalize eigenvectors to unit length. For Example 27.1 these are

$$|\hat{v}_1\rangle \rightarrow \frac{1}{\sqrt{3}} \begin{pmatrix} \sqrt{2} \\ 1 \end{pmatrix} \quad |\hat{v}_2\rangle \rightarrow \frac{1}{\sqrt{3}} \begin{pmatrix} 1 \\ -\sqrt{2} \end{pmatrix}, \tag{27.22}$$

and for Example 27.2,

$$|\hat{v}_1\rangle \rightarrow \frac{1}{\sqrt{3}} \begin{pmatrix} 1 \\ 1 \\ 1 \end{pmatrix} \quad |\hat{v}_2\rangle \rightarrow \frac{1}{\sqrt{6}} \begin{pmatrix} 1 \\ -2 \\ 1 \end{pmatrix} \quad |\hat{v}_3\rangle \rightarrow \frac{1}{\sqrt{2}} \begin{pmatrix} 1 \\ 0 \\ -1 \end{pmatrix}. \tag{27.23}$$

Notice that (27.22) and (27.23) are orthonormal bases of $\mathbb{R}^2$ and $\mathbb{R}^3$, respectively.

It's no coincidence that the number of eigenvalues is equal to the dimensionality of the original matrix. We can easily trace this back to the secular determinant, which for an $n \times n$ matrix yields an nth-order polynomial in λ; the fundamental theorem of algebra then assures us of n solutions. But though the theorem guarantees n roots to an nth-order polynomial, it does not claim that all these eigenvalues are real. Or distinct. Nor does it tell us anything about the eigenvectors. Let's step through a few examples illustrating some of the different possibilities.

Example 27.3 Real and Complex Matrices

Consider the real matrix

$$C_1 = \begin{pmatrix} 1 & 2 \\ -2 & 1 \end{pmatrix}. \tag{27.24}$$

The characteristic equation

$$\lambda^2 - 2\lambda + 5 = 0 \tag{27.25}$$

gives the spectrum $\lambda_{1,2} = 1 \pm 2i$. So complex eigenvalues can arise even for real matrices. Moreover, the eigenvectors form a basis of $\mathbb{C}^2$,

$$|\hat{v}_1\rangle \to \frac{1}{\sqrt{2}} \begin{pmatrix} 1 \\ i \end{pmatrix} \qquad |\hat{v}_2\rangle \to \frac{1}{\sqrt{2}} \begin{pmatrix} 1 \\ -i \end{pmatrix}. \tag{27.26}$$

On the other hand, the almost-identical matrix

$$C_2 = \begin{pmatrix} 1 & 2i \\ -2i & 1 \end{pmatrix} \tag{27.27}$$

has the same eigenvectors but *real* eigenvalues -1 and 3, whereas the real antisymmetric matrix

$$C_3 = \begin{pmatrix} 0 & 2 \\ -2 & 0 \end{pmatrix} \tag{27.28}$$

has purely imaginary eigenvalues $\pm 2i$.

Now consider the complex-valued matrix

$$C_4 = \begin{pmatrix} 4 & -\sqrt{2} & i\sqrt{6} \\ -\sqrt{2} & 5 & i\sqrt{3} \\ i\sqrt{6} & i\sqrt{3} & 3 \end{pmatrix}. \tag{27.29}$$

The algebra is a bit more tedious, but it's not hard to confirm that the secular determinant leads to

$$\lambda^3 - 12\lambda^2 + 54\lambda - 108 = (\lambda - 6)\left(\lambda^2 - 6\lambda + 18\right) = 0. \tag{27.30}$$

Recall that roots of a polynomial with real coefficients occur in complex-conjugate pairs. For an odd-order polynomial, then, one of the roots must be its *own* complex conjugate — in other words, real. Indeed, solving (27.30) yields the eigenvalues $\lambda_1 = 6$ and $\lambda_{2,3} = 3 \pm 3i$. The eigenvector components, however, are real:

$$|\hat{v}_1\rangle \to \frac{1}{3} \begin{pmatrix} -\sqrt{3} \\ \sqrt{6} \\ 0 \end{pmatrix} \qquad |\hat{v}_2\rangle \to \frac{1}{\sqrt{6}} \begin{pmatrix} \sqrt{2} \\ 1 \\ \sqrt{3} \end{pmatrix} \qquad |\hat{v}_3\rangle \to \frac{1}{\sqrt{6}} \begin{pmatrix} -\sqrt{2} \\ -1 \\ \sqrt{3} \end{pmatrix}. \tag{27.31}$$

Example 27.4 A Degenerate System

The 3×3 matrix

$$D = \begin{pmatrix} 3 & -8 & -2 \\ -8 & -9 & -4 \\ -2 & -4 & 6 \end{pmatrix} \tag{27.32}$$

presents a different kind of challenge. The characteristic equation

$$(\lambda + 14)(\lambda - 7)^2 = 0 \tag{27.33}$$

appears to have only two solutions $\lambda_1 = -14$ and $\lambda_2 = 7$. But as a double root, 7 counts as two eigenvalues, $\lambda_2 = \lambda_3 = 7$. Repeated eigenvalues are said to be *degenerate*; in this system, $\lambda = 7$ has degree of degeneracy 2 — or, more simply, is "doubly degenerate."

The first eigenvector is straightforward enough,

$$|v_1\rangle \to \begin{pmatrix} 2 \\ 4 \\ 1 \end{pmatrix}. \tag{27.34}$$

The degenerate part of the spectrum, though, requires a bit of circumspection. With $\lambda = 7$, the eigenvalue equation yields the system of equations

$$4a + 8b + 2c = 0$$
$$8a + 16b + 4c = 0$$
$$2a + 4b + c = 0. \tag{27.35}$$

As expected, they are not linearly independent. But in this case, the system is equivalent to only a single condition, $c = -2(a + 2b)$. So rather than one overall normalization choice at our disposal, we have two free parameters. In other words, the "degenerate eigensubspace" is two-dimensional. We can, if we wish, arbitrarily take $c = 0$ to get (up to normalization, reflecting the second free parameter)

$$|v_{c=0}\rangle \rightarrow \begin{pmatrix} -2 \\ 1 \\ 0 \end{pmatrix}, \tag{27.36a}$$

or we could let $a = 0$ or $b = 0$ to find

$$|v_{a=0}\rangle \rightarrow \begin{pmatrix} 0 \\ 1 \\ -4 \end{pmatrix} \qquad |v_{b=0}\rangle \rightarrow \begin{pmatrix} 1 \\ 0 \\ -2 \end{pmatrix}. \tag{27.36b}$$

All three are eigenvectors with $\lambda = 7$, but they are not linearly independent — which is hardly a surprise given that it's a two-dimensional subspace. Indeed, any two degenerate eigenvectors $|v_1\rangle$ and $|v_2\rangle$ can be linearly combined to form another degenerate eigenvector $|u\rangle \equiv \alpha|v_1\rangle + \beta|v_2\rangle$:

$$A|u\rangle = A\big(\alpha|v_1\rangle + \beta|v_2\rangle\big) = \alpha A|v_1\rangle + \beta A|v_2\rangle$$
$$= \alpha\lambda|v_1\rangle + \beta\lambda|v_2\rangle = \lambda\big(\alpha|v_1\rangle + \beta|v_2\rangle\big) = \lambda|u\rangle. \tag{27.37}$$

With so much freedom, what criteria should we invoke to choose $|v_2\rangle$ and $|v_3\rangle$? Well, notice that any vector with $c = -2(a + 2b)$ — that is, the entire degenerate eigensubspace — is orthogonal to $|v_1\rangle$. So why not choose the degenerate eigenvectors to be orthogonal to one another as well? Any such pair will do — for instance

$$|v_2\rangle \rightarrow \begin{pmatrix} 1 \\ 0 \\ -2 \end{pmatrix} \qquad |v_3\rangle \rightarrow \begin{pmatrix} 8 \\ -5 \\ 4 \end{pmatrix}. \tag{27.38}$$

Then the trio of eigenvectors $(|v_1\rangle, |v_2\rangle, |v_3\rangle)$ forms an orthogonal basis of $\mathbb{R}^3$.

Example 27.5 Non-Orthogonal Eigenvectors

The previous examples all yielded orthogonal eigenvectors. In the non-degenerate cases this was automatic, whereas with a degenerate spectrum we had the freedom to choose mutually orthogonal eigenvectors. But not all matrices have orthogonal eigenvectors. Consider

$$E = \begin{pmatrix} -2 & -4 & -3 \\ 3 & 5 & 3 \\ -2 & -2 & -1 \end{pmatrix}. \tag{27.39}$$

The usual techniques yield eigenvalues $1, -1, +2$ and eigenvectors

$$\begin{pmatrix} 1 \\ 0 \\ -1 \end{pmatrix} \quad \begin{pmatrix} 1 \\ -1 \\ 1 \end{pmatrix} \quad \begin{pmatrix} 1 \\ -1 \\ 0 \end{pmatrix}. \tag{27.40}$$

Though linearly independent — and hence, a basis — they are not all orthogonal. And without a degenerate subspace, we do not have the freedom to choose them to be orthogonal.

Example 27.6 **Common Eigenvectors**

Eigenvalues — even degenerate ones — must correspond to distinct eigenvectors in order to form an eigenbasis.[1] However, *different* matrices can have the same eigenvectors. The matrix

$$F = \begin{pmatrix} 4 & 5 & 1 \\ -1 & -2 & -1 \\ -3 & -3 & 0 \end{pmatrix} \tag{27.41}$$

has non-degenerate spectrum $3, 0, -1$, but with same eigenvectors as the matrix E of Example 27.5. These matrices have something else in common: they commute, $EF = FE$. As we'll see, this is no mere coincidence.

Example 27.7 **A Defective Matrix**

Not all matrices have an eigenbasis. For instance,

$$G = \begin{pmatrix} 1 & -1 & 1 \\ 1 & 3 & -1 \\ 1 & 1 & -1 \end{pmatrix} \tag{27.42}$$

has $\lambda_1 = -1$ and doubly degenerate $\lambda_{2,3} = 2$. The first eigenvector is straightforward,

$$|v_1\rangle \to \begin{pmatrix} -1 \\ 1 \\ 3 \end{pmatrix}. \tag{27.43}$$

The degenerate sector, though, is strange. We expect two linearly independent conditions amongst the eigenvalue equation's three relations,

$$a - b + c = 2a$$
$$a + 3b - c = 2b$$
$$a + b - c = 2c. \tag{27.44}$$

But the only non-trivial solution has $c = 0$ and $a = -b$. Thus there's only *one* linearly independent eigenvector in the two-dimensional degenerate subspace,

---

[1] Remember: the zero vector is not characteristic of an operator, and hence is not an eigenvector.

$$|v_2\rangle \rightarrow \begin{pmatrix} 1 \\ -1 \\ 0 \end{pmatrix}. \tag{27.45}$$

Matrices like G whose eigenvectors do not form a basis are said to be *defective*. Since eigenvectors associated with *distinct* eigenvalues are linearly independent, a defective matrix always has a degeneracy in its spectrum.

BTW 27.1 Euler's Rotation Theorem

The relationship of a rotation to its axis is most-clearly expressed as an eigenvalue problem with $\lambda = 1$,

$$R\hat{n} = \hat{n}. \tag{27.46}$$

The interpretation of R as a rotation requires assurance that a solution $\hat{n}$ in $\mathbb{R}^3$ exists. This assurance is provided by Euler's rotation theorem.

We could establish the theorem by solving the eigenvalue equation for, say, the general $\mathbb{R}^3$ Euler rotation $R(\alpha, \beta, \gamma)$ in (B.18) of Appendix B. Far more pleasant is to start with the condition of a vanishing secular determinant,

$$\det(R - \mathbb{1}) = 0. \tag{27.47}$$

The most elegant way to demonstrate that something vanishes is to show that it's equal to its negative. With that goal in mind, use $R^T = R^{-1}$ to verify

$$R - \mathbb{1} = R(\mathbb{1} - R^{-1}) = R(\mathbb{1} - R^T) = R(\mathbb{1} - R)^T. \tag{27.48}$$

Then

$$\det(R - \mathbb{1}) = \det[R(\mathbb{1} - R)^T] = \det(R)\det(\mathbb{1} - R)$$
$$= \det(R)\det(-\mathbb{1})\det(R - \mathbb{1}). \tag{27.49}$$

So the theorem is proved as long as $\det(R)\det(-\mathbb{1}) = -1$. But proper rotations have $\det(R) = +1$, and in three dimensions $\det(-\mathbb{1}) = -1$. Done. (And what of the other two eigenvectors of proper $\mathbb{R}^3$ rotations? See Example 27.10.) It's the orthogonality of R, the essence of what makes a rotation matrix a rotation matrix, which makes it work. Notice, however, that the theorem only holds in *odd*-dimensional spaces. Indeed, there are no rotation axes in $\mathbb{R}^2$. But the theorem does show that a *reflection* in $\mathbb{R}^2$ has an invariant direction — the mirror itself.

27.2 Normal Matrices

As we've seen, eigenvalues can be real or complex, degenerate or non-degenerate. Eigenvectors can form either orthogonal or skew bases — or on occasion fail to form a basis at all. In short, we're all over the place.

But remember, eigensystems reflect the characteristics of the operator. So we should focus our attention on *classes* of matrices. Given the importance of orthonormal vectors, we'll limit almost all our attention to matrices with an orthogonal eigenbasis. Collectively, these are called *normal matrices N* (possibly to contrast with "defective" matrices, such as the one in Example 27.7 — but see Chapter 28).

Normal matrices can be identified as those which commute with their adjoints, $[N, N^\dagger] = 0$. You'll validate this definition in Problem 27.41; for now, let's see some examples.

27.2.1 Hermitian Operators

By definition, Hermitian operators have $H = H^\dagger$ — so they are certainly normal,

$$[H, H^\dagger] = [H, H] = 0. \tag{27.50}$$

For eigenvector $|v\rangle$ and corresponding eigenvalue λ, the eigenvalue equation allows us to write (Problem 27.11)

$$\begin{aligned} \lambda \langle v|v\rangle = \langle v|\lambda|v\rangle &= \langle v|H|v\rangle \\ &= \langle v|H^\dagger|v\rangle^* = \langle v|H|v\rangle^* = \langle v|\lambda|v\rangle^* = \lambda^* \langle v|v\rangle, \end{aligned} \tag{27.51}$$

where in the third step we used the definition of operator adjoint in (25.24b). Thus we learn that *the eigenvalues of Hermitian operators are real*. As we see from Example 27.5, the converse is not generally true. If all the eigenvalues of a *normal* matrix are real, however, then it's Hermitian. Stated formally: a normal matrix has real eigenvalues if and only if ("iff") the matrix is Hermitian.

Now consider two eigenvectors $|v_{1,2}\rangle$ with eigenvalues $\lambda_{1,2}$. Since H is Hermitian and the λ's real, we use (25.24a) to get

$$\lambda_2 \langle v_1|v_2\rangle = \langle v_1|Hv_2\rangle = \langle Hv_1|v_2\rangle = \lambda_1 \langle v_1|v_2\rangle, \tag{27.52a}$$

so that

$$(\lambda_1 - \lambda_2) \langle v_1|v_2\rangle = 0. \tag{27.52b}$$

Thus for non-degenerate eigenvalues, the eigenvectors of Hermitian operators are orthogonal. And the eigenvectors of any degenerate eigenvalues can always be *chosen* to be orthogonal (using Gram–Schmidt, for instance). So *the eigenvectors of an $n \times n$ Hermitian matrix form an orthogonal basis in $\mathbb{C}^n$*.

Take a moment to appreciate how remarkable all this is: only when a complex matrix is Hermitian does its action on a complex eigenvector always produce a real eigenvalue. Moreover, any $\mathbb{C}^n$ vector can be expanded on a Hermitian operator's eigenbasis — that is, each term in the expansion is an eigenvector of the operator.

The matrices of Examples 27.1, 27.2, 27.4, as well as C_2 of Example 27.3, are all Hermitian, and all have real eigenvalues and eigenvectors. The first three are real symmetric matrices, an important subclass of Hermitian. Indeed, the special case of real symmetric matrices abound in physics and engineering; as Hermitian matrices, they come with their own characteristic eigenbasis in $\mathbb{R}^n$.

Example 27.8 Linear Image of a Sphere

Consider a vector $\vec{v}$ in $\mathbb{R}^n$, with

$$\|\vec{v}\|^2 = \sum_i v_i^2 = 1. \tag{27.53}$$

Such a vector describes a unit sphere in n-dimensions. How does the sphere change if we transform with an $n \times n$ matrix,[2]

$$\vec{u} = B^{-1}\vec{v}. \tag{27.54}$$

---

[2] We are not assuming B is orthogonal, so we wrote the transformation as $\vec{u} = B^{-1}\vec{v}$ rather than $\vec{u} = B\vec{v}$ to avoid unwieldy inverse-transpose notation in (27.55).

In other words, we want to find the geometric figure described by $\vec{u}$ subject to

$$1 = \|\vec{v}\|^2 = \|B\vec{u}\|^2 = \vec{u}^T B^T \cdot B\vec{u} = \vec{u}^T \cdot B^T B\vec{u}. \tag{27.55}$$

As a linear operator, B^{-1} necessarily maps the quadratic form (27.53) in v_i into a quadratic form (27.55) in u_i. Having only required B to be invertible, however, it may seem impossible to say anything more about the form of the image. But though B's properties are unspecified, the matrix $A \equiv B^T B$ is real and symmetric — *so it has an orthonormal eigenbasis*. And with its eigenbasis $\{\hat{a}_i\}$ comes eigenvalues λ_i.

Expanding $\vec{u}$ on A's eigenbasis,

$$\vec{u} = \sum_i u_i \hat{a}_i, \tag{27.56}$$

gives

$$1 = \vec{u}^T \cdot A\vec{u} = \sum_{ij} \left(u_i \hat{a}_i^T \right) \cdot A \left(u_j \hat{a}_j \right) = \sum_{ij} u_i u_j \lambda_j \delta_{ij} = \sum_i \lambda_i u_i^2, \tag{27.57}$$

where we've used both $A\hat{a}_j = \lambda_j \hat{a}_j$ and the orthonormality of the eigenbasis. So the quadratic form described by

$$\sum_i \lambda_i u_i^2 = 1, \tag{27.58}$$

is determined by the eigenvalues. Of course without specifying B, we cannot solve for the λ's. Still, since

$$0 \leq \|B\hat{a}_i\|^2 = \hat{a}_i^T B^T B\hat{a}_i = \hat{a}_i^T A\hat{a}_i = \lambda_i, \tag{27.59}$$

we know that all the eigenvalues are greater than or equal to zero. In fact, since an invertible B implies an invertible A, a zero eigenvalue is disallowed (Problem 27.34). So all the λ_i's must be positive. Were they all equal to 1, (27.58) would still describe a sphere. In general, however, the linear image of a sphere describes an ellipsoid with semi-principal axes $\lambda_i^{-1/2}$; the eigenvectors give the directions of the principal axes.

Though the λ_i are the eigenvalues of A, we can glean from (27.59) that (the absolute value of) the eigenvalues of B are $\sqrt{\lambda_i}$ — which means that those of B^{-1} are $1/\sqrt{\lambda_i}$. Thus the lengths of the ellipsoid's semi-principal axes are the magnitudes of the eigenvalues of the original transformation B^{-1} from $\vec{v}$ to $\vec{u}$, (27.54).

Example 27.9 **Principal Axes**

As we saw in Section 4.4, the moment of inertia tensor is an inherent property of a rigid body, independent of its motion and choice of coordinates. So even though rotational motion is complicated, perhaps there are special axes $\hat{n}$ around which the angular momentum $\vec{L}$ is parallel to the angular velocity $\vec{\omega} = \omega\hat{n}$. And that's where the eigenvalue problem comes in: as a real symmetric matrix, we are assured that there are *always* three such axes — the eigenvectors of

$$\begin{pmatrix} L_1 \\ L_2 \\ L_3 \end{pmatrix} = \begin{pmatrix} I_{11} & I_{12} & I_{13} \\ I_{21} & I_{22} & I_{23} \\ I_{31} & I_{32} & I_{33} \end{pmatrix} \begin{pmatrix} \omega_1 \\ \omega_2 \\ \omega_3 \end{pmatrix} = I_0 \begin{pmatrix} \omega_1 \\ \omega_2 \\ \omega_3 \end{pmatrix}. \tag{27.60}$$

The normalized eigenvectors $\hat{n}$ give the directions of the body's *principal axes*; the three eigenvalues I_0 are the *principal moments of inertia* (Hand and Finch, 1998; Taylor, 2005). The existence of these eigensolutions means that every rigid body has three axes around which rotation is fairly simple — specifically, for which $\vec{L}$ and $\vec{\omega}$ are parallel. It's only around these axes that an object can rotate at constant speed, torque free. But even if the rotation is about some other axis, the basis of eigenvectors still simplifies the analysis. To appreciate this, let's return to the dumbbell of Example 4.8. The elements of the moment of inertia tensor are given in (4.61) and (4.62). Although tedious to solve for directly, it's not hard to confirm eigenvalues $I_{0_1} = I_{0_2} = 2mr^2$ and $I_{0_3} = 0$, with eigenvectors

$$\hat{n}_1 = \begin{pmatrix} -\sin\phi \\ \cos\phi \\ 0 \end{pmatrix} \quad \hat{n}_2 = \begin{pmatrix} -\cos\theta\cos\phi \\ -\cos\theta\sin\phi \\ \sin\theta \end{pmatrix} \quad \hat{n}_3 = \begin{pmatrix} \sin\theta\cos\phi \\ \sin\theta\sin\phi \\ \cos\theta \end{pmatrix}. \tag{27.61}$$

As expected, they form an orthonormal basis of $\mathbb{R}^3$. Since $\hat{n}_3 \equiv \hat{r}$ is in the direction along the dumbbell, $\hat{n}_{1,2}$ must span the plane perpendicular to the dumbbell. That these latter two directions are physically indistinguishable is attested to by the degeneracy of their eigenvalues. In other words, the moment of inertia for rotation around either $\hat{n}_1$ or $\hat{n}_2$ is the same. The zero eigenvalue I_{0_3} reflects our use of point masses — which, lacking any spatial extent, have no moment of inertia around the axis $\hat{n}_3$ connecting them.

Choosing $\hat{n}_1$ and $\hat{n}_2$ as the basis for the degenerate subspace places the angular momentum $\vec{L}$ in (4.59) to lie entirely along $\hat{n}_2$ (see the figure in Example 4.8). So as the dumbbell rotates around the z-axis, $\vec{L}$ precesses together with $\hat{n}_2$ around $\vec{\omega} = \omega\hat{k}$. Even for constant ω, the change in the direction of $\vec{L}$ cannot be sustained without an applied torque — a clear indication that the rotation is not around a principal axis. However, for the special case $\theta = \pi/2$, the dumbbell lies in the xy-plane and $\hat{n}_2$ aligns with $\hat{k}$. Since $\vec{L}$ is now parallel to $\vec{\omega}$, there is no precession and a torque is not required to sustain the motion. (For $\theta = 0$ the dumbbell is along $\hat{k} = \hat{n}_3$, and we again have constant $\vec{L}$ — though since $I_{0_3} = 0$ that constant value is zero.)

So rotation around principal axes yields the simplest description of the motion. But even when the rotation is not around one of the principal axes, it can always be expressed on the basis of principal axes. For our dumbbell with $\vec{\omega} = \omega\hat{k}$,

$$\begin{aligned} \vec{\omega} &= \hat{e}_1 \left(\hat{n}_1 \cdot \vec{\omega} \right) + \hat{n}_2 \left(\hat{n}_2 \cdot \vec{\omega} \right) + \hat{n}_3 \left(\hat{n}_3 \cdot \vec{\omega} \right) \\ &= \omega \left(\sin\theta\,\hat{n}_2 + \cos\theta\,\hat{n}_3 \right), \end{aligned} \tag{27.62}$$

in agreement with (4.63). So the general motion can be seen as a combination of rotations around principal axes $\hat{n}_2$ and $\hat{n}_3$.

In quantum mechanics, physical quantities such as energy or angular momentum (commonly called *observables*) are represented by Hermitian operators, and physical states by vectors $|\psi\rangle$. The equation $H|\phi_j\rangle = \lambda_j|\phi_j\rangle$ says that measuring the observable H on one of its *eigenstates* $|\phi_j\rangle$ yields the result λ_j. Even in a non-eigenstate, a measurement can only produce one of H's eigenvalues; the guarantee of real λ's ensures a physically acceptable measurement result. It's a discrete eigenvalue spectrum that puts the "quantum" in quantum mechanics.

BTW 27.2 Quantum Measurement and Expectation

For non-eigenstates, the hermiticity of H guarantees that a state $|\psi\rangle$ can be expanded as a superposition of H's eigenstates $|\phi_i\rangle$,

$$|\psi\rangle = \sum_i c_i|\phi_i\rangle. \tag{27.63}$$

One of the postulates of quantum mechanics is that a measurement *collapses* this expansion down to the eigenstate corresponding to the measured eigenvalue,

$$|\psi\rangle \xrightarrow[\text{of } H]{\text{measurement}} \lambda_j |\phi_j\rangle. \tag{27.64}$$

So the act of measurement alters the state from the superposition $|\psi\rangle$ to the eigenstate $|\phi_j\rangle$. Only when $|\psi\rangle$ is itself an eigenstate does a measurement have no effect on the system.

Quantum mechanics also assigns physical meaning to the expansion coefficients $c_j = \langle\phi_j|\psi\rangle$: the *probability* that a measurement of H yields the result λ_j is $|c_j|^2$. This interpretation requires that $|\psi\rangle$ be normalized, so that $\langle\psi|\psi\rangle = \sum_i |c_i|^2 = 1$. The simplest illustration is the case of spin-1/2 in Example 25.21.

Now if $|\psi\rangle$ is an eigenstate $|\phi_j\rangle$, all the $c_{i\neq j}$ vanish leaving only $|c_j| = 1$. But for the general superposition state, measurement of H can yield any of its eigenvalues. Over many identical measurements on the state $|\psi\rangle$, the *average* or *expectation* value $\langle H\rangle$ is given by the probability-weighted sum of its eigenvalues

$$\langle H\rangle = \sum_i |c_i|^2 \lambda_i = \sum_i |\langle\phi_i|\psi\rangle|^2 \lambda_i$$
$$= \sum_i \langle\psi|\phi_i\rangle\langle\phi_i|\psi\rangle\lambda_i = \sum_i \langle\psi|H|\phi_i\rangle\langle\phi_i|\psi\rangle$$
$$= \langle\psi|H|\psi\rangle, \tag{27.65}$$

where we've used the completeness of the eigenbasis, $\sum_i |\phi_i\rangle\langle\phi_i| = \mathbb{1}$. The average deviation from the mean is given by the standard deviation σ_H,

$$\sigma_H^2 \equiv \langle\,(H - \langle H\rangle\,)^2\rangle = \langle H^2\rangle - \langle H\rangle^2 \geq 0. \tag{27.66}$$

The standard deviation quantifies the inherent uncertainty ΔH in the superposition $|\psi\rangle$. For $|\psi\rangle \equiv |\phi_j\rangle$ an eigenstate of H, the expectation value is just λ_j and the uncertainty vanishes.

Anti-Hermitian matrices, $Q^\dagger = -Q$, are closely related to Hermitian matrices. In fact, they are simply i times a Hermitian matrix, $Q = iH$. So the eigenvectors of anti-Hermitian operators form a basis with purely *imaginary* eigenvalues. Real antisymmetric matrices are special cases — and indeed, matrix C_3 in (27.28) has imaginary eigenvalues.

Just as a real square matrix can always be written as the sum of a symmetric and antisymmetric matrix, so too any complex square matrix is the sum of a Hermitian and anti-Hermitian matrix, $M = H_1 + iH_2$.

27.2.2 Unitary Operators

Let U be a unitary operator, so that $U^\dagger U = \mathbb{1}$. Since

$$[U, U^\dagger] = UU^\dagger - U^\dagger U = \mathbb{1} - \mathbb{1} = 0, \tag{27.67}$$

it's clearly normal. For eigenvector $|v\rangle$ with eigenvalue λ we have

$$|\lambda|^2 \langle v|v\rangle = \langle v|\lambda^*\lambda|v\rangle = \langle v|U^\dagger U|v\rangle = \langle v|v\rangle, \tag{27.68}$$

demonstrating that *the eigenvalues of unitary operators have magnitude one* — which is to say, pure phases $\lambda = e^{i\alpha}$ lying on the unit circle. Using this result in an argument very similar to that of (27.52) shows that *the eigenvectors of unitary operators are orthogonal* (Problem 27.14).

Example 27.10 **Keepin' It Real**

Recall that when restricted to real space, unitary matrices reduce to orthogonal rotations R with $R^T R = \mathbb{1}$. Even so, the eigenvalues are still pure phases with unit magnitude. But since R is real, the eigenvalues and the elements of the eigenvectors occur in complex-conjugate pairs,

$$R\hat{q} = \lambda\hat{q} \quad\longleftrightarrow\quad R\hat{q}^* = \lambda^*\hat{q}^*. \tag{27.69}$$

But in an *odd*-dimensional space like $\mathbb{R}^3$, the fundamental theorem of algebra guarantees that a rotation matrix R has at least one real eigenvalue and hence one real eigenvector $\hat{n}$. So the other two eigenvalues are complex conjugates $\lambda_{2,3} = e^{\pm i\varphi}$ corresponding to complex eigenvectors $\hat{q}_2$ and $\hat{q}_3 = \hat{q}_2^*$. How are we to interpret these in the context of rotations in real space?

Let's first consider this physically. A rotation matrix in $\mathbb{R}^3$ picks out one special direction in space — the axis $\hat{n}$. There is no other distinctive direction. But rather than specify the rotation by its axis, one can instead identify the *plane* of rotation. Though we often define a plane by its unit normal $\hat{n}$, we can also use any two linearly independent vectors orthogonal to $\hat{n}$. This is what $\hat{q}_{2,3}$ — which as eigenvectors of a normal operator are orthogonal both to $\hat{n}$ and to each other — describe.

Just as the real and imaginary parts of a complex number $z = a + ib$ give two real numbers, so too the real and imaginary parts of a $\mathbb{C}^2$ vector reveal two $\mathbb{R}^2$ vectors,

$$\hat{q}_2' \equiv \frac{1}{i\sqrt{2}}\left(\hat{q}_2 - \hat{q}_3\right) \qquad \hat{q}_3' \equiv \frac{1}{\sqrt{2}}\left(\hat{q}_2 + \hat{q}_3\right), \tag{27.70}$$

where the $\sqrt{2}$ is for normalization. This is a unitary transformation within the two-dimensional eigensubspace. But as a linear combinations of *non*-degenerate eigenvectors, $\hat{q}_{2,3}'$ are not eigenvectors of R. How do they behave under our $\mathbb{R}^3$ rotation? Well, with $\lambda_{2,3} = e^{\pm i\varphi}$:

$$\begin{aligned} R\hat{q}_2' &= \frac{1}{i\sqrt{2}}\left(R\hat{q}_2 - R\hat{q}_3\right) = \frac{1}{i\sqrt{2}}\left(\lambda_2\hat{q}_2 - \lambda_3\hat{q}_3\right) \\ &= \frac{1}{i\sqrt{2}}\left[(\cos\varphi + i\sin\varphi)\,\hat{q}_2 - (\cos\varphi - i\sin\varphi)\,\hat{q}_3\right] \\ &= \cos\varphi\,\hat{q}_2' - \sin\varphi\,\hat{q}_3', \end{aligned} \tag{27.71a}$$

and

$$\begin{aligned} R\hat{q}_3' &= \frac{1}{\sqrt{2}}\left(R\hat{q}_2 + R\hat{q}_3\right) = \frac{1}{\sqrt{2}}\left(\lambda_2\hat{q}_2 + \lambda_3\hat{q}_3\right) \\ &= \frac{1}{\sqrt{2}}\left[(\cos\varphi + i\sin\varphi)\,\hat{q}_2 + (\cos\varphi - i\sin\varphi)\,\hat{q}_3\right] \\ &= \sin\varphi\,\hat{q}_2' + \cos\varphi\,\hat{q}_3'. \end{aligned} \tag{27.71b}$$

This, obviously, is a planar rotation, with φ the phase of the complex eigenvalues; $\hat{q}_2'$ and $\hat{q}_3'$ specify the plane of rotation.

Notice that the rotation angle comes not from the real eigenvalue, but rather from the complex eigenvalues, $e^{i\varphi} + e^{-i\varphi} = 2\cos\varphi$. Moreover, we might expect that values of φ for which these eigenvalues are real have particular physical import. And indeed, they are only real for $\varphi = 0$ and π, corresponding to $R = \pm 1$ — the only rotations in the plane which map vectors into scalar multiples of themselves. So ironically, although the real eigenvector $\hat{n}$ is easily interpreted, it's the complex eigensubspace which fully characterizes the rotation.

Example 27.11 Stationary States

The eigenbases of unitary operators are fundamental to quantum mechanics. Time evolution, for instance, is governed by an operator $U(t)$ which takes a state $|\psi(0)\rangle$ into $|\psi(t)\rangle$,

$$|\psi(t)\rangle = U(t)|\psi(0)\rangle. \tag{27.72}$$

In order to preserve probability, $|\psi\rangle$ must remain normalized for all t,

$$1 = \langle\psi(t)|\psi(t)\rangle = \langle\psi(0)|U^\dagger(t)U(t)|\psi(0)\rangle = \langle\psi(0)|\psi(0)\rangle. \tag{27.73}$$

Thus time evolution is governed by a unitary $U(t)$. For instance, the state of a spin-1/2 particle $|\psi\rangle$ can be represented by a two-dimensional complex column vector, and U by a 2×2 unitary matrix — see Examples 26.6 and 26.8.

Since unitary operators are normal, $|\psi\rangle$ can be expanded on an orthonormal basis of its eigenvectors called *stationary states*,

$$|\psi(0)\rangle = \sum_n c_n|\hat{\phi}_n\rangle, \tag{27.74}$$

where since the eigenvalues of a unitary operator are pure phases,

$$U(t)|\hat{\phi}_n\rangle = e^{-i\alpha(t)}|\hat{\phi}_n\rangle. \tag{27.75}$$

This leads to the natural question: what physical characteristics determine an eigenstate of time evolution? As we've just seen, mathematically only the phase of a stationary state changes in time. Furthermore, just as for orthogonal rotations $R(\varphi) = R(\varphi - \varphi_0)R(\varphi_0)$, unitary time evolution obeys $U(t) = U(t - t_0)U(t_0)$. Thus the time-dependent phase angle α must have the form $\alpha(t) = \omega t$ for some frequency ω. But frequency is energy in quantum mechanics, $E = \hbar\omega$. Thus time evolution U and its eigenstates must be intimately connected to the energy operator H called the *Hamiltonian*. Putting all this together, it must be that

$$U(t) = e^{-iHt/\hbar}, \tag{27.76}$$

for time-independent Hermitian H such that $H|\hat{\phi}_n\rangle = E|\hat{\phi}_n\rangle$. (See Problem 26.33. In fact, the Hamiltonian is the *generator* of time evolution — compare Section 26.6, (26.139)–(26.141).) Then (27.75) becomes

$$\begin{aligned}
U(t)|\hat{\phi}_n\rangle &= e^{-iHt/\hbar}|\hat{\phi}_n\rangle \\
&= \left(1 - iHt/\hbar - \frac{1}{2!}H^2t^2/\hbar^2 + \frac{i}{3!}H^3t^3/\hbar^3 + \cdots\right)|\hat{\phi}_n\rangle \\
&= \left(1 - iEt/\hbar - \frac{1}{2!}E^2t^2/\hbar^2 + \frac{i}{3!}E^3t^3/\hbar^3 + \cdots\right)|\hat{\phi}_n\rangle = e^{-i\omega t}|\hat{\phi}_n\rangle.
\end{aligned} \tag{27.77}$$

So $|\hat{\phi}_n\rangle$ is an eigenstate of both H and $U(t)$, with eigenvalues E and $e^{-i\omega t}$, respectively. A stationary state is thus a state of definite energy E. The energy eigenvalue equation $H|\hat{\phi}_n\rangle = E|\hat{\phi}_n\rangle$ is the *time-independent Schrödinger equation*.

27.3 Diagonalization

Normal matrices are those with an orthogonal eigenbasis; the most common examples are listed in Table 27.1. So when presented with a linear system governed by a normal matrix, it's only natural to choose the basis defined by the operator itself. We saw some of the advantages that come from this choice in Examples 27.9 and 27.11. But the benefits and insights run deeper still.

Table 27.1 Common normal matrices

	Matrix	Eigenvalues	Eigenbasis
Hermitian	$H^\dagger = H$	Real	$\mathbb{C}^n$
Anti-Hermitian	$Q^\dagger = -Q$	Imaginary	$\mathbb{C}^n$
Real symmetric	$S^T = S$	Real	$\mathbb{R}^n$
Real antisymmetric	$A^T = -A$	Imaginary	$\mathbb{R}^n$
Unitary	$U^\dagger = U^{-1}$	Pure phase	$\mathbb{C}^n$
Orthogonal	$O^T = O^{-1}$	Pure phase	$\mathbb{R}^n$

27.3.1 Why Be Normal?

Given a normalized eigenvector $|\hat{v}\rangle$ of a normal matrix T, we can easily project out its eigenvalue,

$$T|\hat{v}\rangle = \lambda|\hat{v}\rangle \quad \Longrightarrow \quad \lambda = \langle \hat{v}|T|\hat{v}\rangle. \tag{27.78}$$

As characterisitc attributes of the operator, eigenvalues must be scalars — so this projection must hold even when rotated. To verify this claim for an arbitrary rotation U, we insert $UU^\dagger = \mathbb{1}$ twice to recast (27.78) in terms of the rotated vectors $|\hat{v}'\rangle = U^\dagger|\hat{v}\rangle$,

$$\lambda = \langle \hat{v}|(UU^\dagger)T(UU^\dagger)|\hat{v}\rangle = \langle \hat{v}|U(U^\dagger TU)U^\dagger|\hat{v}\rangle \equiv \langle \hat{v}'|T'|\hat{v}'\rangle, \tag{27.79}$$

where $T' = U^\dagger TU$. The expression (27.79) can be interpreted as complementary rotations of both the eigenvector and the operator such that λ remains the same.

We can, in any coordinate basis, project out all the λ's piecemeal using (27.78), one eigenvalue and eigenvector at a time. It's more elegant, however, to project out the eigenvalues all at once by constructing the specific unitary matrix whose columns are formed by the complete orthonormal basis of its eigenvectors,

$$U = \left(\begin{pmatrix} \uparrow \\ \hat{v}_1 \\ \downarrow \end{pmatrix} \begin{pmatrix} \uparrow \\ \hat{v}_2 \\ \downarrow \end{pmatrix} \cdots \begin{pmatrix} \uparrow \\ \hat{v}_n \\ \downarrow \end{pmatrix} \right). \tag{27.80}$$

It's not hard to appreciate the efficacy of this choice: the action of T on each column of U is just its eigenvalue times that same column,

$$TU = \left(\begin{pmatrix} \uparrow \\ \lambda_1\hat{v}_1 \\ \downarrow \end{pmatrix} \begin{pmatrix} \uparrow \\ \lambda_2\hat{v}_2 \\ \downarrow \end{pmatrix} \cdots \begin{pmatrix} \uparrow \\ \lambda_n\hat{v}_n \\ \downarrow \end{pmatrix} \right) \equiv U\Lambda, \tag{27.81}$$

where Λ is the diagonal matrix of eigenvalues,

$$\Lambda = \begin{pmatrix} \lambda_1 & 0 & \cdots & 0 \\ 0 & \lambda_2 & \cdots & 0 \\ \vdots & \vdots & \ddots & \vdots \\ 0 & 0 & \cdots & \lambda_n \end{pmatrix}. \tag{27.82}$$

So TU encapsulates the entire eigensystem (Problem 27.15),

$$TU = U\Lambda \quad \Longrightarrow \quad \Lambda = U^\dagger TU. \tag{27.83}$$

$U^\dagger TU$ is a unitary transformation induced by the vector rotation $|\hat{v}'\rangle = U^\dagger|\hat{v}\rangle$. Thus the matrices Λ and T are different matrix representations of the same abstract operator $\mathcal{T}$. The unitary transformation $\Lambda = U^\dagger TU$ *diagonalizes* the matrix T. On the original coordinate basis, $\mathcal{T} \to T$ with eigenvectors $\hat{v}_1, \hat{v}_2, \ldots, \hat{v}_n$. But on the diagonal basis, $\mathcal{T} \to \Lambda$ with trivial standard-basis eigenvectors (as is always the case for a diagonal matrix).

Example 27.12

As we found in Example 27.2, the matrix

$$B = \begin{pmatrix} -1 & -1 & 5 \\ -1 & 5 & -1 \\ 5 & -1 & -1 \end{pmatrix} \tag{27.84}$$

has normalized eigenvectors

$$|\hat{v}_1\rangle \to \frac{1}{\sqrt{3}}\begin{pmatrix} 1 \\ 1 \\ 1 \end{pmatrix} \qquad |\hat{v}_2\rangle \to \frac{1}{\sqrt{6}}\begin{pmatrix} 1 \\ -2 \\ 1 \end{pmatrix} \qquad |\hat{v}_3\rangle \to \frac{1}{\sqrt{2}}\begin{pmatrix} 1 \\ 0 \\ -1 \end{pmatrix}. \tag{27.85}$$

Assembling them into a unitary matrix

$$U = \begin{pmatrix} \frac{1}{\sqrt{3}} & \frac{1}{\sqrt{6}} & \frac{1}{\sqrt{2}} \\ \frac{1}{\sqrt{3}} & -\frac{2}{\sqrt{6}} & 0 \\ \frac{1}{\sqrt{3}} & \frac{1}{\sqrt{6}} & -\frac{1}{\sqrt{2}} \end{pmatrix}, \tag{27.86}$$

we can rotate B into diagonal form:

$$U^\dagger BU = \begin{pmatrix} 3 & 0 & 0 \\ 0 & 6 & 0 \\ 0 & 0 & -6 \end{pmatrix}. \tag{27.87}$$

Note that the diagonal elements are the eigenvalues of B. Diagonalization corresponds to a rotation $U^\dagger$ which renders the eigenvectors on the standard basis.

$$U^\dagger|\hat{v}_1\rangle \to \begin{pmatrix} 1 \\ 0 \\ 0 \end{pmatrix} \qquad U^\dagger|\hat{v}_2\rangle \to \begin{pmatrix} 0 \\ 1 \\ 0 \end{pmatrix} \qquad U^\dagger|\hat{v}_3\rangle \to \begin{pmatrix} 0 \\ 0 \\ 1 \end{pmatrix}. \tag{27.88}$$

Although (27.88) expresses active rotation, diagonalization is perhaps best construed as a passive transformation which orients the coordinate axes along the eigenvectors of the system. In this view, only the matrix representation of the eigenvectors has changed; their directions in space have not.

Example 27.13 **Matrix Powers**

Since functions of matrices are defined by their Taylor expansions, one is not-infrequently required to consider powers of matrices. Unfortunately, there is often little recourse beyond tiresome successive multiplications. But a diagonalizable matrix significantly diminishes the required effort. Multiple insertions of $\mathbb{1} = UU^\dagger$ and use of (27.83) gives

$$T^m = U(U^\dagger T U)(U^\dagger T U) \cdots (U^\dagger T U)U^\dagger = U\Lambda^m U^\dagger. \tag{27.89}$$

Thus the problem is largely reduced to calculating powers of a diagonal matrix,

$$\Lambda^m = \begin{pmatrix} \lambda_1^m & 0 & \cdots & 0 \\ 0 & \lambda_2^m & \cdots & 0 \\ \vdots & \vdots & \ddots & \vdots \\ 0 & 0 & \cdots & \lambda_n^m \end{pmatrix}. \tag{27.90}$$

For the matrix of (26.128) and say, $m = 4$, this procedure gives

$$T^4 = \frac{1}{2}\begin{pmatrix} 1 & 1 & 0 \\ 0 & 0 & \sqrt{2} \\ 1 & -1 & 0 \end{pmatrix}\begin{pmatrix} 2^4 & 0 & 0 \\ 0 & 4^4 & 0 \\ 0 & 0 & (-2)^4 \end{pmatrix}\begin{pmatrix} 1 & 0 & 1 \\ 1 & 0 & -1 \\ 0 & \sqrt{2} & 0 \end{pmatrix}$$

$$= \begin{pmatrix} 136 & 0 & -120 \\ 0 & 16 & 0 \\ -120 & 0 & 136 \end{pmatrix}, \tag{27.91}$$

which you can readily (if tediously) verify by calculating T^4 directly. Enjoy.

We showed earlier that similarity transformations preserve both the trace and the determinant. To expand on this result, consider an arbitrary operator S. Multiplying the eigenvalue equation $T|v\rangle = \lambda|v\rangle$ from the left by S^{-1}, two insertions of $\mathbb{1} = SS^{-1}$ and with $S^{-1}|v\rangle = |v'\rangle$, we find

$$S^{-1}\big[T(SS^{-1})|v\rangle = \lambda(SS^{-1})|v\rangle\big] \quad \Longrightarrow \quad T'|v'\rangle = \lambda|v'\rangle. \tag{27.92}$$

So similar matrices have *identical* eigenvalues. Diagonalization reveals an important link between these two results: *the trace of a matrix is the sum of its eigenvalues, and the determinant their product,*

$$\operatorname{Tr} T = \sum_i \lambda_i \qquad \det T = \prod_i \lambda_i. \tag{27.93}$$

You can verify this for every one of the examples in Section 27.1.

Though we claimed that S is arbitrary, that's not quite so: it must be invertible. Which means that its columns must be linearly independent. Which means that for a diagonalizing matrix, the eigenvectors must form a basis. In fact, the two are inseparable: a matrix is diagonalizable if and only if it has an eigenbasis.[3]

Since in general an eigenbasis need not be orthonormal, diagonalization is not confined to normal matrices. However, though any matrix whose eigenvectors form a basis can be diagonalized, only normal matrices have orthonormal eigenbases and hence are *unitarily* diagonalizable — which is to say, $S^{-1} = S^\dagger$. This is an important distinction, because it means we can diagonalize a system while

---

[3] So the defective matrix of Example 27.7 is not diagonalizable. Nonetheless, (27.93) holds for all square matrices — including defective ones.

simultaneously maintaining the magnitudes and relative orientations of all the vectors in the space. Moreover, the ability to write $T = U\Lambda U^\dagger$ means that a normal matrix can be considered as a sequence of elementary operations: a rotation $U^\dagger$, then a simple scaling by λ's, followed by a return rotation U — just as we saw in Example 26.16.

Example 27.14 **Sequential Action**

Consider the matrix

$$M = \frac{1}{2}\begin{pmatrix} 1 & 1 \\ 1 & 1 \end{pmatrix}. \tag{27.94}$$

It's clearly neither a rotation nor a reflection. But as a real symmetric matrix it is normal, so a unitary diagonalization provides an interpretation of M as a sequence of actions. Its eigensystem is

$$\lambda_1 = 1, \ |\hat{v}_1\rangle \to \frac{1}{\sqrt{2}}\begin{pmatrix} 1 \\ 1 \end{pmatrix} \qquad \lambda_2 = 0, \ |\hat{v}_2\rangle \to \frac{1}{\sqrt{2}}\begin{pmatrix} -1 \\ 1 \end{pmatrix}, \tag{27.95}$$

so that

$$U = \frac{1}{\sqrt{2}}\begin{pmatrix} 1 & -1 \\ 1 & 1 \end{pmatrix} \qquad \Lambda = \begin{pmatrix} 1 & 0 \\ 0 & 0 \end{pmatrix}. \tag{27.96}$$

Thus $M = U\Lambda U^T$ can be understood as a $-\pi/4$ rotation, followed by a projection onto x, followed by a $+\pi/4$ rotation.

Example 27.15 **Building a Normal Matrix**

Let's construct a real 2×2 matrix with given eigenvalues. First, arrange the eigenvalues in a diagonal matrix

$$\Lambda = \begin{pmatrix} \lambda_1 & 0 \\ 0 & \lambda_2 \end{pmatrix}. \tag{27.97}$$

Then use the rotation matrix

$$R = \begin{pmatrix} \cos\varphi & -\sin\varphi \\ \sin\varphi & \cos\varphi \end{pmatrix}, \tag{27.98}$$

to apply the similarity transformation,[4]

$$T = R\Lambda R^T = \begin{pmatrix} \lambda_1 \cos^2\varphi + \lambda_2 \sin^2\varphi & (\lambda_1 - \lambda_2)\sin\varphi\cos\varphi \\ (\lambda_1 - \lambda_2)\sin\varphi\cos\varphi & \lambda_1 \sin^2\varphi + \lambda_2 \cos^2\varphi \end{pmatrix}. \tag{27.99}$$

The easiest way to obtain the eigenvectors is to apply R to the standard basis,

$$|v_1\rangle \to R\begin{pmatrix} 1 \\ 0 \end{pmatrix} = \begin{pmatrix} \cos\varphi \\ \sin\varphi \end{pmatrix} \qquad |v_2\rangle \to R\begin{pmatrix} 0 \\ 1 \end{pmatrix} = \begin{pmatrix} -\sin\varphi \\ \cos\varphi \end{pmatrix}. \tag{27.100}$$

Indeed, $T|v_i\rangle = (R\Lambda R^T)(R|e_i\rangle) = R\Lambda|e_i\rangle = \lambda_i R|e_i\rangle = \lambda_i|v_i\rangle$. So it all works.

---

[4] One could, of course, rotate in the opposite sense by letting $R \longleftrightarrow R^T$.

Example 27.16 Spectral Decomposition

Since eigenvalues are unchanged under similarity transformations, we can talk about the eigenvalues, trace, and determinant of the abstract operator $\mathcal{A}$, independent of the basis for a given matrix representation A. The eigenvectors $|\hat{v}_i\rangle$ are also invariant, in the sense that they pick out fixed directions in space. How we *represent* those directions in a column vector, however, depends upon their orientation with respect to the chosen coordinate axes (the standard basis). When the eigen directions align with the coordinate axes $(\hat{e}_i)_j = \delta_{ij}$, $\mathcal{A}$ is given by the diagonal matrix Λ.

And just as we can expand a vector $|a\rangle$ on an arbitrary orthonormal basis,

$$|a\rangle = \sum_i |\hat{v}_i\rangle\langle\hat{v}_i|a\rangle = \sum_i c_i|\hat{v}_i\rangle, \tag{27.101}$$

so too we can expand an operator $\mathcal{A}$,

$$\mathcal{A} = \sum_{ij} |\hat{v}_i\rangle\langle\hat{v}_i|\mathcal{A}|\hat{v}_j\rangle\langle\hat{v}_j| = \sum_{ij} A_{ij}|\hat{v}_i\rangle\langle\hat{v}_j|, \tag{27.102}$$

as we saw in Section 25.6. But if $\mathcal{A}$ is a normal operator and $\{|\hat{v}\rangle\}$ its eigenbasis, then we get the much simpler *spectral decomposition* of an operator,

$$\mathcal{A} = \sum_{ij} |\hat{v}_i\rangle\langle\hat{v}_i|\mathcal{A}|\hat{v}_j\rangle\langle\hat{v}_j| = \sum_{ij} \lambda_j|\hat{v}_i\rangle\langle\hat{v}_i|\hat{v}_j\rangle\langle\hat{v}_j| = \sum_j \lambda_j|\hat{v}_j\rangle\langle\hat{v}_j|, \tag{27.103}$$

which is a sum over orthogonal projections, each weighted by the eigenvalue. Applied to the standard matrix basis, this is

$$\lambda_1 \begin{pmatrix} 1 \\ 0 \end{pmatrix} \begin{pmatrix} 1 & 0 \end{pmatrix} + \lambda_2 \begin{pmatrix} 0 \\ 1 \end{pmatrix} \begin{pmatrix} 0 & 1 \end{pmatrix} = \begin{pmatrix} \lambda_1 & 0 \\ 0 & \lambda_2 \end{pmatrix}. \tag{27.104}$$

On the basis of (27.100) we get the same result as mutliplying out $R\Lambda R^T$,

$$\lambda_1 \begin{pmatrix} \cos\varphi \\ \sin\varphi \end{pmatrix} \begin{pmatrix} \cos\varphi & \sin\varphi \end{pmatrix} + \lambda_2 \begin{pmatrix} -\sin\varphi \\ \cos\varphi \end{pmatrix} \begin{pmatrix} -\sin\varphi & \cos\varphi \end{pmatrix}$$

$$= \begin{pmatrix} \lambda_1\cos^2\varphi + \lambda_2\sin^2\varphi & (\lambda_1 - \lambda_2)\sin\varphi\cos\varphi \\ (\lambda_1 - \lambda_2)\sin\varphi\cos\varphi & \lambda_1\sin^2\varphi + \lambda_2\cos^2\varphi \end{pmatrix}. \tag{27.105}$$

Example 27.17 The Hessian

To second order, the Taylor expansion of a scalar function $f(\vec{r}) = f(x,y)$ is (Example 6.17)

$$f(x + dx, y + dy) - f(x,y) = \frac{\partial f}{\partial x}dx + \frac{\partial f}{\partial y}dy$$

$$+ \frac{1}{2}\left(\frac{\partial^2 f}{\partial x^2}dx^2 + 2\frac{\partial^2 f}{\partial x\partial y}dxdy + \frac{\partial f}{\partial y^2}dy^2\right) + \cdots$$

$$= \vec{\nabla}f \cdot d\vec{r} + \frac{1}{2}d\vec{r} \cdot \overset{\leftrightarrow}{H} \cdot d\vec{r}, \tag{27.106}$$

where the symmetric matrix of second derivatives $(f_{ij} \equiv \partial_i\partial_j f)$,

$$H(x,y) = \begin{pmatrix} f_{xx} & f_{xy} \\ f_{yx} & f_{yy} \end{pmatrix}, \tag{27.107}$$

is called the *Hessian*. The first derivatives $\vec{\nabla}f$ vanish at an equilibrium point, so it's the quadratic form $dr^T H dr$ which determines the type of equilibrium. If the point is a minimum, then moving in any direction must increase f; thus the Hessian has positive definite eigenvalues. Similarly, both eigenvalues are negative if the point is a maximum. At a saddle point, f increases along one eigendirection, decreases along the other — so the λ's have opposite sign. Thus the Hessian's eigenspectrum gives the generalization to higher dimensions of the familiar second-derivative concavity test.

Example 27.18   Gaussian Integrals in $\mathbb{R}^n$

Recall that convergence of the real Gaussian integral $\int_{-\infty}^{\infty} dx \exp\left(-\frac{1}{2}ax^2\right) = \sqrt{2\pi/a}$ requires a to be positive definite, $a > 0$. We want to generalize this to a real n-component vector $\vec{x}$, with n different integration variables x_i, and an exponent which can include not just quadratic terms like x_i^2 but *bilinear* cross terms of the form $x_i x_j$ — all with different coefficients. Happily, we can arrange these coefficients of the various terms in a symmetric matrix A, allowing the exponent to be written $\frac{1}{2}\sum_{ij} A_{ij} x_i x_j$. The n-dimensional integral, then, becomes

$$ I = \int d^n x \, e^{-\frac{1}{2} x^T A x}. \tag{27.108} $$

Convergence requires the scalar function $Q(\vec{x}) = x^T A x$ be positive definite for all non-zero vectors $\vec{x}$. But what does this tell us about A?

As a real symmetric matrix, A can be diagonalized with a rotation R, $D = R^T A R$ (D for diagonal). Making the change of variables $y = R^T x$ gives

$$ I = \int |J| \, d^n y \, e^{-\frac{1}{2} y^T (R^T A R) y} = \int d^n y \, e^{-\frac{1}{2} y^T D y}, \tag{27.109} $$

where as an orthogonal transformation, the Jacobian $|J| = |R| = 1$. Since D is diagonal, we get $y^T D y = \sum_{ij} y_i \left(\lambda_i \delta_{ij}\right) y_j = \sum_i \lambda_i y_i^2$. Thus

$$ I = \int d^n y \, e^{-\frac{1}{2} \sum_i \lambda_i y_i^2} = \prod_i \int dy \, e^{-\frac{1}{2}\lambda_i y_i^2}. \tag{27.110} $$

So convergence requires all the eigenvalues be positive definite. And with that proviso, we can finish the calculation:

$$ I = \int d^n x \, e^{-\frac{1}{2} x^T A x} = \prod_i \sqrt{\frac{2\pi}{\lambda_i}} = \sqrt{\frac{(2\pi)^n}{\det A}}. \tag{27.111} $$

Though diagonalization allowed us to derive this result, only the determinant of A is actually required in the answer.

So how is having only positive λ_i's related to the original condition $x^T A x > 0$? Well, expanding an arbitrary $\vec{x} = \sum_i c_i \hat{u}_i$ on the eigenbasis $\{\hat{u}_i\}$ of A reveals

$$ Q(\vec{x}) = x^T A x = \sum_{ij} c_i c_j \, \hat{u}_i^T A \, \hat{u}_j = \sum_{ij} c_i c_j \lambda_j \hat{u}_i^T \hat{u}_j $$

$$ = \sum_{ij} c_i c_j \lambda_j \delta_{ij} = \sum_i \lambda_i c_i^2 > 0. \tag{27.112} $$

Thus having all positive eigenvalues is equivalent to $x^T A x > 0$ for all non-zero vectors $\vec{x}$. Normal matrices with eigenvalues $\lambda_i \geq 0$ are positive; those like A with $\lambda_i > 0$ are positive definite.

27.3.2 Simultaneous Diagonalization

Choosing a good basis is crucial to an analysis of linear systems. One of the benefits of diagonalizable matrices is that they come with their own preferred basis in which complicated matrix relations often reduce to multiplication by eigenvalues. But what if the system has more than one operator? Which basis should one choose?

Consider a diagonalizable operator A with an eigenvector $|v\rangle$ such that $A|v\rangle = a|v\rangle$. Given a second normal operator B, what can we say about $B|v\rangle$? Well:

$$A\left(B|v\rangle\right) = \left(BA + [A,B]\right)|v\rangle = \left(aB + [A,B]\right)|v\rangle, \qquad (27.113)$$

since as a number, a commutes with operators. Now *if* A and B commute — that is, $[A,B] = 0$ – we get

$$A\left(B|v\rangle\right) = a\left(B|v\rangle\right). \qquad (27.114)$$

So whatever other properties B may have, we know it takes an eigenvector of A to an eigenvector $|u\rangle \equiv B|v\rangle$ of A. If a is a non-degenerate eigenvalue, $|u\rangle$ must be the *same* as $|v\rangle$ — or at least parallel to it. In other words, it must be that $B|v\rangle \equiv b|v\rangle$.

Thus we learn that *commuting diagonalizable matrices have a common eigenbasis*. We need not decide whose basis to use: they're identical. So *commuting diagonalizable matrices can be simultaneously diagonalized*.

Example 27.19 **Commuting Matrices**

When A has a degeneracy, the situation is more subtle — but the freedom to choose eigenvectors in the degenerate subspace of A allows us to find $|v\rangle$'s which are also eigenvectors of B. Consider, for instance commuting matrices

$$A = \frac{1}{6}\begin{pmatrix} 7 & -2 & 1 \\ -2 & 10 & -2 \\ 1 & -2 & 7 \end{pmatrix} \qquad B = \frac{1}{3}\begin{pmatrix} 2 & -1 & -7 \\ -1 & -4 & -1 \\ -7 & -1 & 2 \end{pmatrix}, \qquad (27.115)$$

with eigenvalues

$$a_1 = 2, \; a_2 = a_3 = 1 \qquad b_1 = -1, \; b_2 = -2, \; b_3 = 3. \qquad (27.116)$$

As real symmetric matrices, the eigenvectors of each form a basis in $\mathbb{R}^3$. For B we find

$$|w_1\rangle \to \begin{pmatrix} 1 \\ -2 \\ 1 \end{pmatrix} \qquad |w_2\rangle \to \begin{pmatrix} 1 \\ 1 \\ 1 \end{pmatrix} \qquad |w_3\rangle \to \begin{pmatrix} 1 \\ 0 \\ -1 \end{pmatrix}, \qquad (27.117)$$

which are orthogonal, as expected. On the other hand, the degeneracy in A's spectrum allows some freedom choosing the eigenvectors of A,

$$|v_1\rangle \to \begin{pmatrix} 1 \\ -2 \\ 1 \end{pmatrix} \qquad |v_{2,3}\rangle \to \begin{pmatrix} 2x \\ x+z \\ 2z \end{pmatrix}. \qquad (27.118)$$

As commuting matrices, A and B automatically share the eigenvector $|v_1\rangle = |w_1\rangle$ in the non-degenerate subspace. And though there's lots of freedom in the choice of degenerate eigenvectors $|v_{2,3}\rangle$, note that $x = z = 1/2$ yields $|w_2\rangle$, while $x = -z = 1/2$ results in $|w_3\rangle$.

Lack of commutativity, of course, is a hallmark of matrix multiplication — with both mathematical and physical implications. Recall, for instance, that finite rotations in $\mathbb{R}^3$ do not commute (Section 26.4). As diagonalizable matrices, we then know they do not share eigenvectors; physically this means that the rotations are necessarily around different axes.

In quantum mechanics, a measurement is represented mathematically as the action of a Hermitian operator on a state $|\psi\rangle$. The operator's eigenvalues give the spectrum of possible measurement results (BTW 27.2); its eigenvectors are states with a definite value of that observable. Since a second, commuting observable shares those eigenvectors, they are states of definite values of both observables. A *complete set of commuting observables* (CSCO) is a maximal set of Hermitian operators which all commute with one another, and which together have a single common eigenbasis whose individual vectors have definite values of each of the observables. As a maximal set, they thus completely characterize the quantum state.

An important commuting example are the operators for energy and angular momentum. The energy is given by the Hamiltonian H, whose eigenvalue equation $H|\psi\rangle = E|\psi\rangle$ describes states $|\psi\rangle$ of definite energy E — and, as such, are often labeled $|E\rangle$. As we saw in Example 27.11, for time-independent H these eigenstates are stationary states of constant energy. When angular momentum $\vec{J}$ is also conserved, it will commute with H and so have common eigenvectors. But recalling Problem 26.34, the components obey the commutation relation

$$[J_i, J_j] = i\hbar\, \epsilon_{ijk} J_k. \tag{27.119}$$

(To give J_i units of angular momentum, we've scaled $\vec{J} \to \hbar\vec{J}$ in (26.145).) So each component J_i has its own distinct eigenbasis. As a result, only one component can be included in our CSCO. The usual convention chooses J_z, with eigenvalue equation written

$$J_z|m\rangle = \hbar m|m\rangle, \tag{27.120}$$

where for historical reasons, the dimensionless m is called the *magnetic quantum number*. And as you demonstrated in Problem 26.37, although the individual components don't commute, each component *does* commute with the scalar J^2; its eigenvalue equation is usually expressed as

$$J^2|j\rangle = \hbar^2 j(j+1)|j\rangle \tag{27.121}$$

with quantum number j. Taken together, the three operators H, J^2, and J_z form a CSCO with eigenbasis $|E, j, m\rangle$.

Example 27.20

Consider a quantum mechanical system with angular momentum operator

$$J^2 = \hbar^2 \begin{pmatrix} 2 & 0 & 0 \\ 0 & 2 & 0 \\ 0 & 0 & 2 \end{pmatrix}. \tag{27.122}$$

This matrix is not merely diagonal, it's proportional to the identity; its three identical eigenvalues correspond to the triply degenerate quantum number $j = 1$. From a physics standpoint, this degeneracy means that a measurement of angular momentum alone is insufficient to distinguish the three physical states. Indeed, any three orthogonal vectors can serve as the eigenbasis. So we need a second, commuting hermitian operator to differentiate between states. Since J^2 is proportional to the identity, any 3×3 matrix will do — say, the matrix

$$J_z = \frac{\hbar}{6} \begin{pmatrix} 1 & -2 & 1 \\ -2 & 4 & -2 \\ 1 & -2 & 1 \end{pmatrix}. \tag{27.123}$$

If this were to represent the z-component of angular momentum, then its eigenvalues correspond to quantum numbers $m_1 = m_2 = 0$, $m_3 = 1$. (An actual J_z operator would have $m = 0, \pm 1$.) Straightforward calculation yields the non-degenerate eigenvector

$$|j, m_3\rangle \rightarrow \begin{pmatrix} 1 \\ -2 \\ 1 \end{pmatrix}, \tag{27.124}$$

and a degenerate pair of the form

$$|j, m_{1,2}\rangle \rightarrow \begin{pmatrix} b \\ \frac{1}{2}(b+d) \\ d \end{pmatrix}. \tag{27.125}$$

We include both j and m in the ket labeling since each is a simultaneous eigenvector of both J^2 and J_z. Only $|j, m_3\rangle$, however, is uniquely determined by these measurements; the other two states remain indistinguishable. To lift this residual double degeneracy requires an additional measurement, which means we need a third commuting hermitian operator. So we turn to the Hamiltonian — say, the matrix

$$H = \frac{2\hbar\omega}{3} \begin{pmatrix} 2 & -1 & -1 \\ -1 & 2 & -1 \\ -1 & -1 & 2 \end{pmatrix} \tag{27.126}$$

which has energy eigenvalues $E_1 = 0$ and $E_2 = E_3 = 2\hbar\omega$, and eigenvectors

$$|E_1\rangle \rightarrow \begin{pmatrix} 1 \\ 1 \\ 1 \end{pmatrix} \qquad |E_{2,3}\rangle \rightarrow \begin{pmatrix} a \\ -(a+c) \\ c \end{pmatrix}. \tag{27.127}$$

Because H commutes with both J^2 (trivially) and J_z (less trivially), all three can share an eigenbasis — the degeneracies providing the freedom to choose values of the parameters to accomplish this. The choice $b = d = 1$ aligns $|j, m_{1,2}\rangle$ with $|E_1\rangle$, whereas $a = b = -c = -d = 1$ aligns it with $|E_{2,3}\rangle$. Similarly, $a = c = 1$ orients $|E_{2,3}\rangle$ with $|j, m_3\rangle$. Thus we obtain three orthogonal eigenvectors common to the operators H, J^2, and J_z, each with a distinct triplet of measured values:

$$|E_1, j, m_1\rangle \rightarrow \begin{pmatrix} 1 \\ 1 \\ 1 \end{pmatrix} \qquad |E_2, j, m_1\rangle \rightarrow \begin{pmatrix} 1 \\ 0 \\ -1 \end{pmatrix} \qquad |E_2, j, m_3\rangle \rightarrow \begin{pmatrix} 1 \\ -2 \\ 1 \end{pmatrix}. \tag{27.128}$$

BTW 27.3 The Uncertainty Principle

As we discussed in BTW 27.2, numerous measurements on identically prepared quantum mechanical states will not produce the same result — leading to the expressions for expectation value and uncertainty of (27.65)–(27.66). Only when $|\psi\rangle$ is an eigenstate of the measured observable does the uncertainty vanish. But what of non-commuting Hermitian operators A and B? Since they do not share eigenvectors, it's impossible for *both* observables to have zero uncertainty. So then the question is just how small the uncertainty can be.

We can answer this using the Schwarz inequality. For notational cleanliness, define the shifted operator $A' = A - \langle A\rangle$. (Recall that $\langle A\rangle$ is the weighted average of eigenvalues — so it's a scalar.) The standard deviation σ_A quantifies the inherent uncertainty ΔA of A in $|\psi\rangle$. This is given by the inner product implicit in the expectation value (27.66),

$$\sigma_A^2 = \langle A'^2\rangle \equiv \langle\alpha|\alpha\rangle \qquad \sigma_B^2 = \langle B'^2\rangle \equiv \langle\beta|\beta\rangle, \tag{27.129}$$

where $|\alpha\rangle \equiv A'|\psi\rangle$ and $|\beta\rangle \equiv B'|\psi\rangle$. By the Schwarz inequality,

$$\langle\alpha|\alpha\rangle\langle\beta|\beta\rangle \geq |\langle\alpha|\beta\rangle|^2, \tag{27.130}$$

we have

$$\sigma_A^2\sigma_B^2 \geq |\langle A'B'\rangle|^2. \tag{27.131}$$

Now we can algebraically separate $\langle A'B'\rangle$ into the sum

$$A'B' = \frac{1}{2}[A',B'] + \frac{1}{2}\{A',B'\}, \tag{27.132}$$

where $\{A',B'\} \equiv A'B' + B'A'$ is often called the *anti*-commutator. You should convince yourself that (27.132) is a separation of $A'B'$ into Hermitian and anti-Hermitian parts. Since the eigenvalues of Hermitian operators are real and those of anti-Hermitian operators imaginary, this separation allows the expectation value $\langle A'B'\rangle$ to be broken up into real and imaginary parts. Then the right-hand side of (27.131) becomes

$$\begin{aligned}
|\langle A'B'\rangle|^2 &= \frac{1}{4}|\langle[A',B']\rangle|^2 + \frac{1}{4}|\langle\{A',B'\}\rangle|^2 \\
&\geq \frac{1}{4}|\langle[A',B']\rangle|^2 \\
&= \frac{1}{4}|\langle[A,B]\rangle|^2,
\end{aligned} \tag{27.133}$$

since $[A',B'] = [A,B]$. Plugging this back into (27.131) gives

$$\sigma_A\sigma_B \geq \frac{1}{2}|\langle[A,B]\rangle|. \tag{27.134}$$

That's a lot of hardware on the right-hand side: the absolute value of the expectation value of the commutator. Perhaps it's easier to think of it as the magnitude of the mean value of $C = [A,B]$. Ultimately, it's C which determines just how precisely two observables can be simultaneously specified. For example, when trying to minimize the uncertainties in both the x- and y-components of angular momentum, the best one can do is limited by (27.119),

$$\Delta J_x\Delta J_y \geq \frac{\hbar}{2}|\langle J_z\rangle|. \tag{27.135}$$

> More famous is the case of position and momentum; these operators satisfy $[X, P] = i\hbar$ (Problem 36.18), resulting in
>
> $$\Delta x \Delta p \geq \frac{\hbar}{2}, \tag{27.136}$$
>
> which you may recognize as the celebrated Heisenberg uncertainty principle.

27.4 The Generalized Eigenvalue Problem

Introducing a non-trivial operator B on the right-hand side of the eigenvalue equation gives the so-called "generalized" eigenvalue problem,

$$A\vec{v} = \lambda B \vec{v}. \tag{27.137}$$

As long as B is invertible we can multiply (27.137) from the left by B^{-1} to get the familiar eigenvalue equation for $C = B^{-1}A$. But this has two drawbacks. First, it requires the calculation of a matrix inverse, a not-always-pleasant task; second, even if A and B are Hermitian, C is usually not (Problem 27.30). So we lose many of the advantages of hermiticity — in particular, the eigenvectors are not necessarily orthogonal. We can solve the generalized eigenvalue problem, however, merely by replacing $\mathbb{1}$ in the regular eigenvalue problem with B. This not only yields the same $\vec{v}$'s and λ's as C, but reestablishes a form of orthogonality.

Example 27.21

To find non-trivial solutions of (27.137), we must solve

$$\det(A - \lambda B) = 0, \tag{27.138}$$

which is the usual secular determinant with $\mathbb{1} \to B$. For Hermitian matrices

$$A = \begin{pmatrix} 4 & 1 \\ 1 & -2 \end{pmatrix} \quad \text{and} \quad B = \begin{pmatrix} 5 & 3 \\ 3 & 2 \end{pmatrix}, \tag{27.139}$$

straightforward calculation reveals solutions

$$\lambda_1 = -9, \quad \vec{v}_1 = \begin{pmatrix} 4 \\ -7 \end{pmatrix} \quad \text{and} \quad \lambda_2 = 1, \quad \vec{v}_2 = \begin{pmatrix} 2 \\ -1 \end{pmatrix}. \tag{27.140}$$

As anticipated, $\vec{v}_1$ and $\vec{v}_2$ are not orthogonal. Note, however, that

$$v_1^T B v_2 = 0. \tag{27.141}$$

This generalizes the usual orthogonality of eigenvectors, with $\mathbb{1}$ replaced by B.

It is often convenient to recast (27.141) as a more-conventional inner product by absorbing $B^{1/2}$ into each of the vectors, where (Problem 27.36)

$$B^{\frac{1}{2}} = \begin{pmatrix} 2 & 1 \\ 1 & 1 \end{pmatrix}, \tag{27.142}$$

with $B = B^{\frac{1}{2}} B^{\frac{1}{2}}$. Then in terms of the vectors

$$\vec{u}_1 \equiv B^{\frac{1}{2}} \vec{v}_1 = \begin{pmatrix} 1 \\ -3 \end{pmatrix} \qquad \vec{u}_2 \equiv B^{\frac{1}{2}} \vec{v}_2 = \begin{pmatrix} 3 \\ 1 \end{pmatrix}, \tag{27.143}$$

the orthogonality statement in (27.141) takes the familiar form

$$\vec{u}_1 \cdot \vec{u}_2 = 0. \tag{27.144}$$

In fact, multiplying the original generalized eigenvalue equation (27.137) by $B^{-1/2}$ together with a clever insertion of $\mathbb{1} = B^{-\frac{1}{2}} B^{+\frac{1}{2}}$ gives

$$B^{-\frac{1}{2}} \left[A\left(B^{-\frac{1}{2}} B^{\frac{1}{2}} \right) \vec{v} = \lambda B \vec{v} \right] \tag{27.145}$$

or

$$A'\vec{u} = \lambda \vec{u}, \qquad A' = B^{-\frac{1}{2}} A B^{-\frac{1}{2}}. \tag{27.146}$$

$B = \mathbb{1}$ gives the conventional eigenvalue problem; but for the $B^{\frac{1}{2}}$ in (27.142),

$$A' = \begin{pmatrix} 0 & 3 \\ 3 & -8 \end{pmatrix}, \tag{27.147}$$

which has eigenvalues $\lambda_1 = -9$, $\lambda_2 = 1$, and eigenvectors $\vec{u}_{1,2}$ of (27.143). Note that A' is Hermitian — in fact, $B^{-\frac{1}{2}} A B^{-\frac{1}{2}}$ is Hermitian for any Hermitian A and B (Problem 27.38).

Problems

27.1   Find the eigenvalues and eigenvectors of the following matrices:

(a) $A = \begin{pmatrix} 1 & 2 \\ 2 & 1 \end{pmatrix}$

(b) $B = \begin{pmatrix} 1 & 2 \\ -2 & 1 \end{pmatrix}$

(c) $C = \begin{pmatrix} 1 & 2i \\ 2i & 1 \end{pmatrix}$

(d) $D = \begin{pmatrix} 1 & 2i \\ -2i & 1 \end{pmatrix}$.

What common property do the matrices with real eigenvalues have?

27.2   Show that, for a two-dimensional system, the trace and determinant are sufficient to find the eigenvalues directly. Use this method to find the eigenvalues of $A = \begin{pmatrix} 1 & 2 \\ 2 & 1 \end{pmatrix}$.

27.3   Show that merely changing the sign of two of the imaginary elements in the matrix of (27.29),

$$H = \begin{pmatrix} 4 & -\sqrt{2} & i\sqrt{6} \\ -\sqrt{2} & 5 & i\sqrt{3} \\ -i\sqrt{6} & -i\sqrt{3} & 3 \end{pmatrix},$$

results in real eigenvalues. Why is this not unexpected?

27.4 Find the eigenvalues and orthogonal eigenvectors of $S = \frac{1}{2} \begin{pmatrix} 1 & 2 & -1 \\ 2 & -2 & 2 \\ -1 & 2 & 1 \end{pmatrix}$.

27.5 Find the eigenvalues and eigenvectors of $M = \begin{pmatrix} 1 & 2 & 1 \\ 2 & 1 & 2 \\ 1 & 1 & 1 \end{pmatrix}$. Are the eigenvectors orthogonal? Explain.

27.6 Show that although the matrices E and F of Examples 27.5 and 27.6 have different eigenvalues, they share a common eigenbasis. Verify that these matrices commute, $[E, F] = 0$.

27.7 Compare and contrast the eigensystems of rotations

$$R = \begin{pmatrix} \cos \varphi & -\sin \varphi \\ \sin \varphi & \cos \varphi \end{pmatrix}$$

and reflections

$$S = \begin{pmatrix} \cos \varphi & \sin \varphi \\ \sin \varphi & -\cos \varphi \end{pmatrix}$$

in $\mathbb{R}^2$. How does Euler's rotation theorem (BTW 27.1) bear on these results?

27.8 Find the eigenvalues and eigenvectors of an $\mathbb{R}^2$ rotation. Show how the complex parts of the eigensystem can be used to find both the plane and angle of rotation. (See Example 27.10.)

27.9 **The Fredholm Alternative**
Consider the inhomogenous equation

$$(A - \lambda \mathbb{1}) |v\rangle = |\rho\rangle \neq 0,$$

where λ is not necessarily an eigenvalue of A. The *Fredholm alternative* delineates the only two possible situations:

(a) Show that if $(A - \lambda \mathbb{1})^{-1}$ exists, there is a unique solution $|v\rangle = \sum_n a_n |\hat{\phi}_n\rangle$ with

$$a_n = \frac{\langle \hat{\phi}_n | \rho \rangle}{\lambda_n - \lambda},$$

where $|\hat{\phi}_n\rangle$ is an eigenvector of A with eigenvalue λ_n. Recast the expansion for $|v\rangle$ in terms of a projection operator G with depends only on properties of A. What happens if λ is an eigenvalue?

(b) If $(A - \lambda \mathbb{1})^{-1}$ does not exist, show that there is still a solution $|v\rangle$ as long as $|\rho\rangle$ does not have a component along any of the eigenvectors, $\langle \hat{\phi}_n | \rho \rangle = 0$.

27.10 Solve the eigenvalue problem for the canonical rotation R_z (26.82), and interpret the complex eigenvectors and eigenvalues in light of Example 27.10.

27.11 **Hermitian Operators**

(a) Explain each step in (27.51).
(b) Explain why real eigenvalues are crucial to the argument of (27.52).

27.12 Consider an anti-Hermitian matrix $Q = -Q^\dagger$.

(a) Show that the eigenvalues of Q are purely imaginary.
(b) Show that the non-degenerate eigenvectors of Q are orthogonal.

27.13 In (27.68) we proved that the eigenvalues of unitary operators are pure phases. Here's another approach:

(a) Show that an eigenvector of an operator M is also an eigenvector of M^{-1}. What's its eigenvalue?
(b) Use $\det(A^\dagger) = (\det A)^*$ to show that the eigenvalues of M and $M^\dagger$ are complex conjugates of one another. (The eigenvectors are the same only if M is a normal matrix.)
(c) Show that eigenvalues of unitary operators are pure phases.

27.14 Eigenvalues of a unitary operator are pure phases. Use this to show that non-degenerate eigenvectors of a unitary operator are orthogonal.

27.15 Denoting the ith component of the ℓth eigenvector as $v_i^{(\ell)}$, use index notation to verify $TU = U\Lambda$ from (27.83). (See Problem 4.3.)

27.16 Explicitly diagonalize matrix D of Example 27.3,

$$D = \begin{pmatrix} 4 & -\sqrt{2} & i\sqrt{6} \\ -\sqrt{2} & 5 & i\sqrt{3} \\ i\sqrt{6} & i\sqrt{3} & 3 \end{pmatrix}.$$

27.17 Show that on the basis of eigenvectors in (27.61), the moment of inertia given in (4.61) and (4.62) for the rotating dumbbell is diagonal.

27.18 Verify the steps in (27.65) leading to the expression $\langle H \rangle = \langle \psi | H | \psi \rangle$ for the expectation (average) value of an observable H in a superposition $|\psi\rangle = \sum_i c_i |\phi_i\rangle$ of H eigenstates.

27.19 Show that normal matrices with all real eigenvalues must be Hermitian.

27.20 Consider the set of vectors

$$|1\rangle \to \frac{1}{\sqrt{3}} \begin{pmatrix} -1 \\ \sqrt{2} \\ 0 \end{pmatrix} \qquad |2\rangle \to \frac{1}{\sqrt{6}} \begin{pmatrix} \sqrt{2} \\ 1 \\ \sqrt{3} \end{pmatrix} \qquad |3\rangle \to \frac{1}{\sqrt{6}} \begin{pmatrix} -\sqrt{2} \\ -1 \\ \sqrt{3} \end{pmatrix}.$$

(a) Verify that they form a complete orthonormal basis.
(b) Weight each basis projection in the completeness relation by the numbers $\lambda_1 = 6, \lambda_2 = 3 + 3i, \lambda_3 = 3 - 3i$. Verify that the resulting matrix M has eigenvectors $|i\rangle$ with eigenvalues λ_i. This "weighted completeness" sum is the spectral decomposition of M (Example 27.16).

27.21 Find the principal axes and principal moments of inertia for rotation around one corner of the cube in Problem 4.16. [If you haven't already done that problem, you'll have to do part (b) now.]

27.22 Use (27.103) to construct the matrix M with eigensystem

$$\lambda_1 = 2, \hat{v}_1 = \frac{1}{\sqrt{6}} \begin{pmatrix} 1 \\ 2 \\ 1 \end{pmatrix} \qquad \lambda_2 = -2, \hat{v}_2 = \frac{1}{\sqrt{2}} \begin{pmatrix} 1 \\ 0 \\ -1 \end{pmatrix} \qquad \lambda_3 = 1, \hat{v}_3 = \frac{1}{\sqrt{3}} \begin{pmatrix} 1 \\ -1 \\ 1 \end{pmatrix}.$$

27.23 Find the extrema of the surface

$$z = 2x^2 + y^2 - 2xy^2, \tag{27.148}$$

and use the Hessian to classify them. Verify your results visually using Mathematica's `ContourPlot3D` to plot the surface.

27.24 Evaluate the integral (27.108) for the 2×2 matrix A in Example 27.1.

27.25 The Gaussian integral in Example 27.18 presumes that A is a symmetric matrix. How would the analysis be affected if A were not symmetric?

27.26 For any normal matrix M with eigenvalues λ_i, show that:

(a) $\text{Tr}(M) = \sum_i \lambda_i$
(b) $\det(M) = \Pi_i \lambda_i$.

27.27 Consider matrices

$$M_1 = \begin{pmatrix} 0 & a & 0 \\ b & 0 & c \\ 0 & d & 0 \end{pmatrix} \qquad M_2 = \begin{pmatrix} 0 & a & b \\ 0 & 0 & c \\ d & 0 & 0 \end{pmatrix}$$

with real, positive elements and non-degenerate eigenspectra. What can you determine about the eigenvalues without actually solving for them?

27.28 DetExp=ExpTr

(a) For matrix $M = e^A$ and unitary transformation U, show that

$$U^\dagger M U = \exp[U^\dagger A U]. \tag{27.149}$$

(b) Show that for A a normal matrix,

$$\det[\exp(A)] = \exp[\mathrm{Tr}(A)]. \tag{27.150}$$

(c) Use Mathematica's `MatrixExp` command to verify both identities for the matrix A of (26.128).

27.29 Show that an operator T is diagonalizable if and only if $\langle Tv|Tw \rangle = \langle T^\dagger v|T^\dagger w \rangle$.

27.30 (a) Show that the inverse of a Hermitian matrix is Hermitian.
(b) Show that the product of unitary matrices U_1, U_2 is also unitary, but that the product of Hermitian matrices H_1, H_2 is not generally Hermitian. Find the condition for which $H_1 H_2$ is Hermitian.

27.31 Show that two commuting Hermitian matrices can be diagonalized by the same unitary transformation.

27.32 Find the condition for which the product of Hermitian matrices A and B shares an eigenbasis with both A and B.

27.33 Show that the commutator of two Hermitian operators is anti-Hermitian.

27.34 Show that an invertible operator cannot have vanishing eigenvalues.

27.35 The Fibonnacci sequence $1, 1, 2, 3, 5, 8, 13, 21, 34, \ldots$ is defined by the recursive formula $f_n = f_{n-1} + f_{n-2}$ for $n \geq 2$ with $f_0 = 0$ and $f_1 = 1$. Using the procedure outlined in Example 27.13, we can obtain an expression that yields f_n directly. First, find the matrix F which takes the column $v_n = \begin{pmatrix} f_n \\ f_{n-1} \end{pmatrix}$ and returns $v_{n+1} = \begin{pmatrix} f_{n+1} \\ f_n \end{pmatrix}$ so that F^n maps v_1 into v_{n+1}. Then diagonalize F to find a formula for f_n in terms of its eigenvalues. Show that in the limit as $n \to \infty$, the ratio between consecutive terms f_{n+1}/f_n converges to the *golden ratio* $\phi = \frac{1}{2}(1 + \sqrt{5})$. [Express the eigenvector components in terms of the eigenvalues — and don't normalize them: diagonalize using the inverse rather than the transpose.]

27.36 Matrix Square Root

(a) Show that defining the square root $D^{1/2}$ of a diagonal matrix D is a straightforward exercise in matrix multiplication.
(b) Describe a procedure for finding the square root of a diagonalizable matrix M.
(c) Find the square roots of $M_1 = \begin{pmatrix} 1 & 2i \\ -2i & 1 \end{pmatrix}$ and $M_2 = \begin{pmatrix} 3 & 0 & -1 \\ 0 & 1 & 0 \\ -1 & 0 & 3 \end{pmatrix}$. Verify your results.

27.37 Given a complex function $f(z)$, define

$$f(M) = \sum_j f(\lambda_j) |\hat{v}_j\rangle \langle \hat{v}_j|,$$

where $M = \sum_j \lambda_j |\hat{v}_j\rangle \langle \hat{v}_j|$ is the spectral decomposition of a normal matrix. Use this to find both the square root $M^{1/2}$ and the exponential $\exp(M)$ of the matrices in Problem 27.36.

27.38 For Hermitian A and B, show that $B^{-\frac{1}{2}} A B^{-\frac{1}{2}}$ is Hermitian.

27.39 Following the approach of Example 27.14, work out an interpretation of $M = \frac{1}{2} \begin{pmatrix} -1 & \sqrt{3} \\ \sqrt{3} & 1 \end{pmatrix}$ as a sequence of three operations.

27.40 Show that both real symmetric and real antisymmetric matrices are diagonalizable.

27.41 Any matrix can be written as the sum of a Hermitian and an anti-Hermitian matrix, $M = H + Q$ — or, perhaps more suggestively,

$$M = H_1 + iH_2,$$

where both H_1 and H_2 are Hermitian. We've already shown that the eigenvectors of a Hermitian matrix H form an orthonormal basis in the space — and hence H is diagonalizable. Show that an operator M is diagonalizable as long as H_1 and H_2 share an eigenbasis.

27.42 **The Generalized Eigenvalue Problem**

(a) Solve the generalized eigenvalue problem (27.138) for Hermitian operators

$$A = \begin{pmatrix} 6 & 3 & 3\sqrt{3} \\ 3 & 9/2 & \sqrt{3} \\ 3\sqrt{3} & \sqrt{3} & 5/2 \end{pmatrix} \qquad B = \frac{1}{4} \begin{pmatrix} 2 & 1 & \sqrt{3} \\ 1 & 7/2 & 0 \\ \sqrt{3} & 0 & 4 \end{pmatrix}.$$

(b) Verify that the three eigenvectors $\vec{v}_i$ satisfy the generalized orthogonality relation (27.141).
(c) After verifying that

$$B^{1/2} = \frac{1}{\sqrt{8}} \begin{pmatrix} \sqrt{3} & 0 & 1 \\ 0 & -\sqrt{3} & 2 \\ 1 & 2 & \sqrt{3} \end{pmatrix},$$

show that the $\vec{u}_i = B^{1/2} \vec{v}_i$ are orthogonal.

In Problems 27.43–27.46, use Mathematica commands such as `Eigensystem` *to find eigenvalues and eigenvectors, and* `MatrixExp[M]` *for the exponential of a matrix.*

27.43 Back in Section 26.6 we found that the Hermitian matrices

$$L_1 = \begin{pmatrix} 0 & 0 & 0 \\ 0 & 0 & -i \\ 0 & i & 0 \end{pmatrix} \quad L_2 = \begin{pmatrix} 0 & 0 & i \\ 0 & 0 & 0 \\ -i & 0 & 0 \end{pmatrix} \quad L_3 = \begin{pmatrix} 0 & -i & 0 \\ i & 0 & 0 \\ 0 & 0 & 0 \end{pmatrix}$$

generate rotations in $\mathbb{R}^3$. A fundamental property of these generators is the commutation relation

$$[L_i, L_j] = i\epsilon_{ijk} L_k,$$

implying that these matrices cannot be simultaneously diagonalized. Still, one would expect it to be advantageous to choose a basis in which one of the three is diagonal. This is what's done in quantum mechanics, where the convention is to work in the eigenbasis of L_3, whose eigenvalues 0, ± 1 yield the quantized values of the z-component of angular momentum in units of $\hbar$. Problems 27.43–27.46 explore the form rotations take on this basis.

(a) Construct the normalized eigenbasis of L_3, ordering the eigenvectors from highest to lowest eigenvalue. Leave their phases unspecified — that is, each eigenvector should have an overall factor $e^{i\varphi_\ell}$, $\ell = 1, 2, 3$. [Note that Mathematica gives the eigenvectors as rows, not columns.]

(b) Transform each of the generators into this basis; label the new matrices J_1, J_2, and J_3, respectively. As an error check, verify that they are Hermitian and obey the same commutation relation as before.

(c) Choose phase angles φ_ℓ between $\pm\pi$ so that, like Pauli matrices $\sigma_{1,2}$, non-zero elements of J_1 are real and positive, and those of J_2 are imaginary. (You should have only two independent constraints on three phase angles — and yet this suffices. Why?) Verify that in this phase convention, L_3's eigenbasis is the complex spherical basis of Problem 26.9.

27.44   From Euler to Wigner

(a) As discussed in Section B.2, an arbitrary rotation can be decomposed into successive coordinate-axis rotations using Euler angles α, β, γ as in (B.17),

$$R(\alpha, \beta, \gamma) = R_z(\alpha)R_y(\beta)R_z(\gamma).$$

Show that if the individual R's are generated according to (26.141) using the L matrices in Problem 27.43, then $R(\alpha, \beta, \gamma)$ is the Euler matrix of (B.18) on the standard basis.

(b) On the complex spherical basis of of Problem 2.5, the same Euler rotation is represented instead by the unitary matrix

$$D(\alpha, \beta, \gamma) = U_z(\alpha)U_y(\beta)U_z(\gamma),$$

where the individual U's are generated by the J_1, J_2, and J_3 matrices found in Problem 27.43. Show that this produces the matrix

$$D(\alpha, \beta, \gamma) \equiv \begin{pmatrix} e^{-i(\alpha+\gamma)}\cos^2\frac{\beta}{2} & -\frac{1}{\sqrt{2}}e^{-i\alpha}\sin\beta & e^{-i(\alpha-\gamma)}\sin^2\frac{\beta}{2} \\ \frac{1}{\sqrt{2}}e^{-i\gamma}\sin\beta & \cos\beta & -\frac{1}{\sqrt{2}}e^{i\gamma}\sin\beta \\ e^{i(\alpha-\gamma)}\sin^2\frac{\beta}{2} & \frac{1}{\sqrt{2}}e^{i\alpha}\sin\beta & e^{i(\alpha+\gamma)}\cos^2\frac{\beta}{2} \end{pmatrix}.$$

This unitary representation of the orthogonal $R(\alpha, \beta, \gamma)$ is called a Wigner D matrix (*Darstellung* being the German word for "representation").

(c) Show that D can be derived directly from R by similarity transformation (cf. (27.149) in Problem 27.27).

27.45   Block Diagonal Form and Irreducibility

In Problem 26.23 you were asked to find a vector representation for the similarity transformation of a 2×2 matrix. Not surprisingly, this quickly becomes tedious beyond two dimensions; fortunately, the Mathematica commands `Flatten`, `ArrayFlatten`, and `TensorProduct` can help.

(a) As a warm-up, use Mathematica to redo Problem 26.23 for rotation R in the plane. First define the dyad $M_{ij} = u_i v_j$, and then use `w=Flatten[M]` to flatten this 2×2 matrix into a four-dimensional column vector. The rotation matrix T in this representation is given by the *tensor product* $R \otimes R$ — in Mathematica, `TensorProduct[R,R]`. (You can also use the escape shortcut `R<esc>t*<esc>R` to get R⊗R.) As you'll see, Mathematica keeps track of all the array subgroups; this can be flattened into a true 4×4 matrix using `T = ArrayFlatten[R⊗R]`. Finally, use `Simplify` to compare the results of $T\vec{w}$ with the similarity transformation RMR^T. (For R^T, use `R<esc>tr<esc>`.)

(b) Now suitably emboldened, let's explore rotations of a tensor product in $\mathbb{R}^3$. Start with the matrix generators L_1, L_2, L_3 from Problem 27.43, and verify that $L^2 = L_1^2 + L_2^2 + L_3^2$ is proportional to the identity. (Note that Mathematica distinguishes between `L.L` and `L^2`.) Then use `MatrixExp` to get the familiar rotation matrices $R_x(\alpha)$, $R_y(\beta)$, and $R_z(\gamma)$.

(c) Use `ArrayFlatten[R⊗R]` to find the 9×9 matrices $T_x(\alpha)$, $T_y(\beta)$, and $T_z(\gamma)$ describing rotations in the nine-dimensional vector representation. Recalling the definition of generator in (26.136), expand each T to first order using `Series` to extract the three generators K_1, K_2, K_3. (The 9×9 identity matrix is `IdentityMatrix[9]`.) Verify that they are Hermitian and obey the same commutation relation as the L_i's.

(d) As in Problem 27.43, we want to diagonalize K_3. Use `Eigensystem` to confirm that K_3 is highly degenerate — so there's a lot more flexibility beyond choosing a phase convention. Show that $K^2 \equiv \sum_i K_i^2$, although not diagonal, commutes with K_3. Thus we can try to lift the degeneracies of K_3 by finding its simultaneous eigenvectors with K^2. To do this, use `Eigensystem[a Ksq + b K3]`, where parameters a and b allow one to distinguish the eigenvalues of each operator. Verify that each of the nine eigenvalue pairs (k^2, k_3) is unique.

(e) To construct the diagonalizing rotation matrix U, arrange the eigenvectors into sets of common K^2 eigenvalue, ordering them from lowest to highest k^2; order the eigenvectors within each of these subsets from highest to lowest k_3 (as we did in Problem 27.43 for J_3). Leave their phases unspecified — that is, each eigenvector should have an overall factor $e^{i\varphi_\ell}$, $\ell = 1$ to 9.

(f) Transform each of the generators K_i into this eigenbasis, labeling the new matrices J_1, J_2, and J_3. Show that the phase choices $\varphi_2 = \varphi_7 = \pi$, $\varphi_6 = \varphi_8 = \pi/2$, with the rest zero, render the non-zero elements of J_1 real and positive, and those of J_2 imaginary. You should find not only diagonal J_3 (and J^2), but also *block diagonal* $J_{1,2}$ — specifically, each is composed of three isolated matrices of dimension 1, 3, and 5 arrayed down the diagonal. (The 1×1 matrices should be trivial, whereas the 3×3 should be identical to matrices $J_{2,3}$ found in Problem 27.43.)

(g) Finally, use `MatrixExp` to construct the unitary rotation (see Problem 27.44),

$$D(\alpha, \beta, \gamma) = e^{-i\alpha J_3} e^{-i\beta J_2} e^{-i\gamma J_3}.$$

The result is a block diagonal rotation operator, each block distinguished by the eigenvalue of J^2. This is an example of the general result: rotations can be decomposed into *irreducible* pieces, distinguished by the eigenvalue of J^2, which only transform amongst themselves.

27.46 The main challenge in Problem 27.45 is finding the generators for the tensor product space — a need predicated on using the left-hand side of (27.149) in Problem 27.27, where M is one of the rotations $\exp[-iJ\varphi]$. But surely the right-hand side of (27.149) simpler, since one can begin with matrices $K = U^\dagger L U$ which are already expressed on the desired basis? Indeed, since the right-hand side requires not a tensor product of rotations but rather a tensor sum over generators, it's not unreasonable to guess that the generators in the tensor product space are given by $J_i \equiv K_i \otimes \mathbb{1} + \mathbb{1} \otimes K_i$. Validate this by repeating the derivation of $D(\alpha, \beta, \gamma)$ using the right-hand side of (27.149). (You'll still need to find the simultaneous eigenvectors of J_3 and J^2 — but all the phases can be set to zero.)

28 Coda: Normal Modes

We've discussed vectors and matrices, linear independence and inner products, rotations and reflections, eigenvectors and eigenvalues. Together, they comprise a mathematical construct at once both beautiful and captivating. Perhaps most compelling to the physicist and engineer, though, are the insights it imparts to physical systems.

Consider the two-mass, three-spring configuration of Figure 28.1. Imagine pulling m_1 to the left and releasing it from rest. The presence of the middle spring immediately causes the block to lose energy and amplitude. But as it slows, the second block begins to oscillate with increasing amplitude. Absent friction, m_1's loss is m_2's gain. Eventually, the second block begins to slow as the first regains speed. Thus neither mass has a single, well-defined frequency.

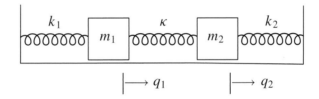

Figure 28.1 Coupled two-mass system.

Solving for the position functions $q_i(t)$ would seem a daunting task. Additional masses and springs make the system even more complicated — and intimidating. But as we'll see, the whole is greater than the sum of the parts.

28.1 Decoupling Oscillators

Considering only so-called "nearest neighbor" interactions, Newton's second law yields the equations of motion ($\dot{q} \equiv dq/dt$)

$$m_1\ddot{q}_1 = -k_1 q_1 - \kappa(q_1 - q_2)$$
$$m_2\ddot{q}_2 = -k_2 q_2 - \kappa(q_2 - q_1), \tag{28.1}$$

where $q_{1,2}$ are the longitudinal displacements of the masses from their equilibrium positions. The source of the system's non-trivial behavior — both physically and mathematically — is the coupling of the middle spring. Were the middle spring removed, we'd have two uncoupled differential equations for $q_1(t)$ and $q_2(t)$,

$$m_1\ddot{q}_1 = -k_1 q_1 \qquad m_2\ddot{q}_2 = -k_2 q_2, \tag{28.2}$$

describing two independent harmonic oscillators with frequencies $\sqrt{k_1/m_1}$ and $\sqrt{k_2/m_2}$.

Since the middle spring couples the masses into a single unified physical system, let's try to cast the differential equations in a unified mathematical form. We begin by writing the displacements $q_1(t), q_2(t)$ as components of a two-dimensional column vector on the standard basis,

$$\begin{pmatrix} q_1 \\ q_2 \end{pmatrix} = q_1 \begin{pmatrix} 1 \\ 0 \end{pmatrix} + q_2 \begin{pmatrix} 0 \\ 1 \end{pmatrix}, \tag{28.3}$$

or

$$|Q(t)\rangle = q_1(t)|q_1\rangle + q_2(t)|q_2\rangle, \tag{28.4}$$

where the pair

$$|q_1\rangle \longrightarrow \begin{pmatrix} 1 \\ 0 \end{pmatrix} \qquad |q_2\rangle \longrightarrow \begin{pmatrix} 0 \\ 1 \end{pmatrix} \tag{28.5}$$

forms an orthonormal basis. For obvious reasons, let's call it the position basis. By defining the matrix operators

$$M \longrightarrow \begin{pmatrix} m_1 & 0 \\ 0 & m_2 \end{pmatrix} \qquad K \longrightarrow \begin{pmatrix} k_1 + \kappa & -\kappa \\ -\kappa & k_2 + \kappa \end{pmatrix}, \tag{28.6}$$

the original coupled differential equations (28.1) can be written

$$M|\ddot{Q}\rangle = -K|Q\rangle, \tag{28.7}$$

where the sign of K was chosen so that (28.7) mirrors Hooke's law.

The joint position vector $|Q\rangle$ together with the usual rules of matrix multiplication fulfill all the axioms of a real inner product space. Since the configuration of the blocks is given by $|Q(t)\rangle$, this vector space is often referred to as "configuration space." But don't confuse the functions $q_i(t)$ with the kets $|q_i\rangle$. The solutions $q_{1,2}(t)$ we seek are the *components* of the configuration space vector $|Q\rangle$ on the position basis $\{|q_1\rangle, |q_2\rangle\}$. Note too that we're placing all the time dependence in the components, not the basis vectors.

All this may seem to be little more than a shift in notation, merely exchanging two coupled differential equations for one matrix equation. It's certainly not obvious from Figure 28.1 that q_1 and q_2 are components of a physical vector. But in fact, we're doing something very incisive here: by identifying $|Q\rangle$ as a vector in an inner product space, we can apply a new set of mathematical tools to the problem. These in turn will lead to deeper physical insight.

An old and venerable approach is to guess a solution and see if it works. In this case, the unified form of (28.7) invites us to focus not on the positions of the masses individually, but rather to look for *collective* motions. In particular, one can imagine that although the blocks do not generally oscillate with a fixed frequency, there are certain motions in which the spring forces are distributed in such a way that the masses oscillate in concert with a single common frequency. Denoting this state of motion by $|\Phi\rangle$, we make the physically motivated *ansatz*[1]

$$|Q(t)\rangle = A \cos(\omega t + \delta)|\Phi\rangle. \tag{28.8}$$

---

[1] Postulating the mathematical form of a solution is called "making an ansatz" (pl. ansätze); it's really just a fancy way of saying "educated guess." But a good ansatz can greatly simplify a problem — as we'll see. Conversely, should the subsequent analysis fail to yield a solution, the ansatz is likely wrong. (That will not be the case here.)

To see what this describes, let ϕ_1 and ϕ_2 be the components on the position basis,

$$|\Phi\rangle \rightarrow \begin{pmatrix} \phi_1 \\ \phi_2 \end{pmatrix}. \tag{28.9}$$

Then the masses in the collective state $|\Phi\rangle$ oscillate with frequency ω and relative amplitudes $\phi_{1,2}$,

$$q_{1,2}(t) = \langle q_{1,2}|Q(t)\rangle = A\cos(\omega t + \delta)\langle q_{1,2}|\Phi\rangle = A\cos(\omega t + \delta)\,\phi_{1,2}. \tag{28.10}$$

The ansatz is only valid if it leads to consistent expressions for ω and $\phi_{1,2}$.

That, however, is easier than it may first appear. Though the ansatz may have been physically motivated, its mathematical incentive is equally compelling. Specifically, it renders the derivatives of $|Q(t)\rangle$ as

$$|\ddot{Q}\rangle = -\omega^2|Q\rangle, \tag{28.11}$$

thus converting our system of *differential* equations (28.7) into the time-independent *algebraic* equation

$$K|\Phi\rangle = +\omega^2 M|\Phi\rangle. \tag{28.12}$$

This is the generalized eigenvalue equation with $\lambda = \omega^2$. Thus we expect not one, but actually two solutions $|\Phi_{1,2}\rangle$ with frequencies $\omega_{1,2}$. The collective motions described by $|\Phi_{1,2}\rangle$ are known as the eigenmodes — or, more commonly, the *normal modes* of the system (Hand and Finch, 1998; Taylor, 2005). These are the special motions singled out by the ansatz; they are certainly not the most general solutions. Taken together, however, they provide a second complete orthogonal basis upon which to decompose $|Q\rangle$. So in addition to (28.4), we now have

$$|Q\rangle = \zeta_1(t)|\Phi_1\rangle + \zeta_2(t)|\Phi_2\rangle. \tag{28.13}$$

Thus the *general configuration of the blocks is a linear combination of the normal modes.* But unlike the still-unknown and complicated $q_i(t)$, the time dependence of the so-called "normal coordinates" $\zeta_{1,2}(t)$ comes directly from the ansatz,

$$\zeta_i(t) = A_i\cos(\omega_i t + \delta_i), \tag{28.14}$$

where A_i is the amplitude, and δ_i the phase of the ith mode.

Example 28.1 Equal m's, Equal k's

For simplicity, let's use identical masses and springs, $m_1 = m_2 \equiv m$ and $\kappa = k_1 = k_2 \equiv k$, reducing (28.6) to

$$M \longrightarrow \begin{pmatrix} m & 0 \\ 0 & m \end{pmatrix}, \quad K \longrightarrow \begin{pmatrix} 2k & -k \\ -k & 2k \end{pmatrix}. \tag{28.15}$$

Then solving (28.12) yields eigenfrequencies ω^2 and eigenvectors $|\Phi\rangle$,

$$\omega_1^2 = \frac{k}{m} \qquad |\Phi_1\rangle \longrightarrow \begin{pmatrix} 1 \\ 1 \end{pmatrix} \tag{28.16a}$$

$$\omega_2^2 = \frac{3k}{m} \qquad |\Phi_2\rangle \longrightarrow \begin{pmatrix} 1 \\ -1 \end{pmatrix}. \tag{28.16b}$$

The eigenvectors characterize just the sort of collective motion we anticipated. The first represents symmetric, equal-amplitude motion of the masses. Since the coupling spring is never stretched or compressed, each mass oscillates with its uncoupled frequency $\sqrt{k/m}$. By contrast, eigenvector $|\Phi_2\rangle$ describes anti-symmetric motion in which the masses oscillate $180°$ out of phase. Because the middle spring exerts equal and opposite force on the blocks, their amplitudes are the same — though they oscillate at a higher frequency $\omega_2 > \omega_1$. Note that the solution with the higher symmetry has the lower frequency.

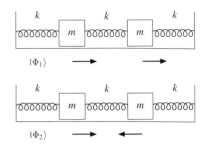

The positions $q_{1,2}(t)$ of each block individually can be read off from the matrix representation of (28.13),

$$|Q\rangle \rightarrow \begin{pmatrix} q_1 \\ q_2 \end{pmatrix} = A_1 \cos(\omega_1 t + \delta_1) \begin{pmatrix} 1 \\ 1 \end{pmatrix} + A_2 \cos(\omega_2 t + \delta_2) \begin{pmatrix} 1 \\ -1 \end{pmatrix}. \qquad (28.17)$$

More formally, the $q_i(t)$ are found by by projecting $|Q\rangle$'s normal mode expansion onto the position basis,

$$q_1(t) = \langle q_1|Q(t)\rangle$$
$$= \zeta_1(t)\langle q_1|\Phi_1\rangle + \zeta_2(t)\langle q_1|\Phi_2\rangle$$
$$= A_1 \cos(\omega_1 t + \delta_1) + A_2 \cos(\omega_2 t + \delta_2), \qquad (28.18a)$$

and

$$q_2(t) = \langle q_2|Q(t)\rangle$$
$$= \zeta_1(t)\langle q_2|\Phi_1\rangle + \zeta_2(t)\langle q_2|\Phi_2\rangle$$
$$= A_1 \cos(\omega_1 t + \delta_1) - A_2 \cos(\omega_2 t + \delta_2), \qquad (28.18b)$$

where in this example, $\langle q_i|\Phi_j\rangle = \pm 1$.

The solutions $q_{1,2}(t)$ vindicate our intuition: neither mass oscillates with a single frequency. Rather, their motions are each governed by *two* frequencies; the non-trivial motion of each block is due to interference between the normal modes as their contributions pass in and out of phase with one another over time.

The four constants $A_{1,2}$ and $\delta_{1,2}$ are determined not by the eigenvalue problem, but rather from the system's initial conditions required by the original differential equation — typically specified as values of $q_{1,2}(0)$ and $\dot{q}_{1,2}(0)$. A common example is to hold the masses at non-zero positions and release them from rest — that is, $q_1(0) = a$, $q_2(0) = b$, and $\dot{q}_{1,2}(0) = 0$. Since we chose to write the solutions as cosines, both $\dot{q}_{1,2}(t)$ are sines; releasing the blocks from rest thus gives $\delta_{1,2} = 0$. This in turn leads to the simple expression $q_{1,2}(0) = A_1 \pm A_2$ for the initial positions. Putting it all together yields the full solution (Problem 28.3)

$$q_{1,2}(t) = A_1 \cos(\omega_1 t) \pm A_2 \cos(\omega_2 t)$$
$$= \frac{a}{2}\left[\cos(\omega_1 t) \pm \cos(\omega_2 t)\right] + \frac{b}{2}\left[\cos(\omega_1 t) \mp \cos(\omega_2 t)\right]. \qquad (28.19)$$

Individual normal modes are easily excited by choosing initial conditions consistent with the eigenvectors. The simplest way is taking either $a = b$ or $a = -b$ and releasing from rest. A glance at (28.19) reveals that with these initial positions, $q_1 = \pm q_2$ for all t. So unlike $|q_{1,2}\rangle$, normal modes are states which do not evolve into one another over time. In other words, although the positions of the masses are time dependent, the *form* of the collective motion is not; once set into an eigenmode, the system remains in that eigenmode. For this reason, the normal modes are often said to be "stationary." (Normal modes of definite ω are direct analogs of the quantum mechanical stationary states discussed in Example 27.11.)

Of the infinity of bases we could choose, the position basis (28.5) is the most physically intuitive, but the eigenbasis (28.9) is the best choice for conducting the analysis. The key to that analysis is the existence of single-frequency stationary collective motions — the normal modes. The ansatz that led to them is the assertion that a solution exists with relative amplitudes ϕ_n all having a common time dependence, $|Q\rangle = \zeta(t)|\Phi\rangle$. Indeed, plugging this into the original differential equation (28.7), and using (28.11) and (28.12), yields

$$M\ddot{\zeta}|\Phi\rangle = -K\zeta|\Phi\rangle = -\omega^2 M\zeta|\Phi\rangle \tag{28.20}$$

or

$$\ddot{\zeta} = -\omega^2\zeta. \tag{28.21}$$

So the normal coordinates ζ_i are actually *decoupled* oscillators — though unlike the q_i, the ζ_i are decoupled even for non-zero κ. This is why they have such simple time dependence. In other words, the components ζ_i of $|Q\rangle$ on the eigenbasis represent *independent harmonic oscillators* with frequencies given by the eigenvalues. Although they don't describe the positions of real physical oscillators, linear combinations yield the general, multi-frequency solutions $q_i(t)$. So even though the time dependence of the q_i's is complicated, they are ultimately governed by the characteristic frequencies of the system. Solving for these is usually the first order of business. And that's an eigenvalue problem.

   Indeed, solving for the normal modes is essentially an exercise in diagonalization. And it's not hard to appreciate that the off-diagonal elements of the K matrix in (28.6) are the mathematical hallmarks of coupling. In fact, it's the *absence* of off-diagonal elements in the normal mode basis which makes the fictitious oscillators ζ_i independent. But the coupling didn't *create* normal modes, it merely *rotated* them. Without the middle spring, the system would already be diagonal with two single-frequency stationary states: in state $|q_1\rangle$, the first block oscillates with frequency $\sqrt{k_1/m_1}$ while the second block remains at rest at equilibrium; in $|q_2\rangle$ the second block oscillates with frequency $\sqrt{k_2/m_2}$ while the first remains at rest. In this case, the normal modes coincide with the position basis, so that normal coordinates and position coordinates are the same. The middle spring essentially rotates the normal modes from the positions basis $|q_{1,2}\rangle$ into the eigenbasis $|\Phi_{1,2}\rangle$ of the coupled system. With equal masses and springs, the net effect of the coupling is to lift the degenerate frequencies of the original system into the non-degenerate eigenfrequencies $\omega_{1,2}$.

Example 28.2 Transverse Oscillations

The previous example considered longitudinal motion of the masses; what about transverse motion? Specifically, let's consider two masses fixed on a string, equally spaced between two walls (Figure 28.2).

   The equations of motion are

$$m_1\ddot{q}_1 = -T_1 \sin\alpha - \tau \sin\beta$$
$$m_2\ddot{q}_2 = \tau \sin\beta - T_2 \sin\gamma, \tag{28.22}$$

where the q_i are now the transverse displacements from equilibrium. For small displacements, we have

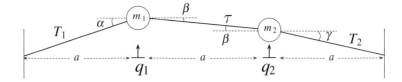

Figure 28.2 Transverse motion of masses on a string.

$$\alpha \approx q_1/a \qquad \beta \approx (q_1 - q_2)/a \qquad \gamma \approx q_2/a. \tag{28.23}$$

In the previous example, this is equivalent to the (unstated) restriction that the displacements not exceed the elastic limit of the springs. In both cases, it keeps the system linear — here, by allowing us to linearize (28.22) using the small-angle approximation $\sin\theta \approx \theta$ to get

$$\begin{pmatrix} m_1 & 0 \\ 0 & m_2 \end{pmatrix} \begin{pmatrix} \ddot{q}_1 \\ \ddot{q}_2 \end{pmatrix} = -\frac{1}{a} \begin{pmatrix} T_1 + \tau & -\tau \\ -\tau & T_2 + \tau \end{pmatrix} \begin{pmatrix} q_1 \\ q_2 \end{pmatrix}. \tag{28.24}$$

This is precisely the same mathematical relationship in (28.6) and (28.7), with $ka \rightarrow T$ and $\kappa a \rightarrow \tau$. Thus the entire normal mode analysis there applies here as well. Moreover, with only transverse motion, the horizontal components of tension must all balance; in other words, in the small-angle approximation $T_1 = T_2 = \tau$. So for identical masses, there is a symmetric mode and an antisymmetric mode with frequencies $\sqrt{T/am}$ and $\sqrt{3T/am}$, respectively. Just as in Example 28.1.

Example 28.3 The Double Pendulum

At first blush the double pendulum may appear very different from a spring-and-mass system. For starters, the coupling can't be switched off: there's no analog to removing the coupling spring by letting $\kappa \rightarrow 0$. But when the pendulums are confined to small oscillations in the plane, the one-dimensional motion of the two masses yields a two-dimensional linear system. And indeed, the two angles ϑ_1 and ϑ_2 are sufficient to locate the masses; these generalized coordinates will be the elements of our two-dimensional position vector $|\Phi\rangle$. So this is a system with two normal modes. It's reasonable to anticipate one mode with the masses swinging in phase, and the other with them swinging out of phase at a higher eigenfrequency. Since the second pendulum pivots on the first, though, it should also have a larger amplitude — which means this time

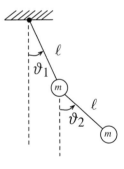

we do not expect the two modes to be precisely the symmetric and antisymmetric ones of (28.16).

One can show that in the small-angle approximation, the equations of motion are

$$\left. \begin{array}{l} 2\ddot{\vartheta}_1 + \ddot{\vartheta}_2 = -\frac{2g}{\ell}\vartheta_1 \\ \ddot{\vartheta}_1 + \ddot{\vartheta}_2 = -\frac{g}{\ell}\vartheta_2, \end{array} \right\} \implies M|\ddot{Q}\rangle = -K|Q\rangle, \tag{28.25}$$

where the configuration this time is given by $|Q\rangle \rightarrow \begin{pmatrix} \vartheta_1 \\ \vartheta_2 \end{pmatrix}$, and

$$M \rightarrow \begin{pmatrix} 2 & 1 \\ 1 & 1 \end{pmatrix} \qquad K \rightarrow \frac{g}{\ell}\begin{pmatrix} 2 & 0 \\ 0 & 1 \end{pmatrix}. \tag{28.26}$$

Invoking the ansatz and solving the general eigenvalue problem yields

$$\omega_{1,2}^2 = (2 \pm \sqrt{2})\frac{g}{\ell} \qquad |\Phi_{1,2}\rangle \longrightarrow \begin{pmatrix} 1 \\ \mp\sqrt{2} \end{pmatrix}. \tag{28.27}$$

Note that

$$M^{-1}K = \frac{g}{\ell}\begin{pmatrix} 2 & -1 \\ -2 & 2 \end{pmatrix} \tag{28.28}$$

is not a normal matrix, so the modes are not naively orthogonal. They do, however, satisfy the general orthogonality (27.141), $\langle\Phi_1|M|\Phi_2\rangle = 0$.

Perhaps the most remarkable feature of normal modes is their ubiquity and universality. Whether springs, tensions, gravity, LC circuits, molecular forces — even the identity of neutrinos — all linear oscillatory systems have normal modes with natural (or *resonant*) frequencies. Though the physical systems may be very different, the underlying mathematics is the same. In all cases, the general motion can be decomposed into a linear superposition of its normal modes $|\Phi_\ell\rangle$,

$$|Q(t)\rangle = \sum_\ell A_\ell \cos(\omega_\ell t + \delta_\ell)|\Phi_\ell\rangle. \tag{28.29}$$

Thus coupled linear oscillators provide a dynamic physical manifestation for the abstract concept of a basis.

28.2 Higher Dimensions

The number of normal modes is equal to the dimensionality of configuration space, which grows with the number of oscillators. It also grows with the types of motion considered. Thus far, we've only considered one-dimensional motion. In general, however, the mass-and- spring systems can also move both up and down as well as in and out; each dimension contributes an additional normal mode per mass. So N masses moving in three dimensions have $3N$ normal modes. For simplicity we'll stick with one-dimensional motion, and increase the dimensionality of the vector space by adding mass–spring elements.

Example 28.4 **Three-Mass System**

Whereas the previous examples each have two normal modes, a three-mass system has three. Let's again take all the masses and springs to be identical. Then the equations of motion are

$$m\ddot{q}_1 = -kq_1 - k(q_1 - q_2)$$
$$m\ddot{q}_2 = -k(q_2 - q_1) - k(q_2 - q_3)$$
$$m\ddot{q}_3 = -k(q_3 - q_2) - kq_3. \tag{28.30}$$

Invoking the ansatz (28.8), we find in the three-dimensional configuration space the eigenvalue equation

$$k\begin{pmatrix} 2 & -1 & 0 \\ -1 & 2 & -1 \\ 0 & -1 & 2 \end{pmatrix}\begin{pmatrix} \phi_1 \\ \phi_2 \\ \phi_3 \end{pmatrix} = m\omega^2\begin{pmatrix} \phi_1 \\ \phi_2 \\ \phi_3 \end{pmatrix}. \tag{28.31}$$

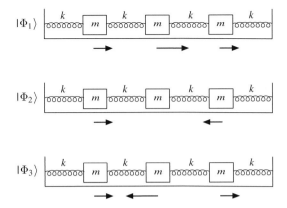

Figure 28.3 Normal modes of the three-mass system.

Solving reveals solutions[2]

$$\omega_1^2 = (2 - \sqrt{2})\frac{k}{m} \qquad |\Phi_1\rangle \longrightarrow \begin{pmatrix} 1 \\ \sqrt{2} \\ 1 \end{pmatrix} \tag{28.32a}$$

$$\omega_2^2 = \frac{2k}{m} \qquad |\Phi_2\rangle \longrightarrow \begin{pmatrix} 1 \\ 0 \\ -1 \end{pmatrix} \tag{28.32b}$$

$$\omega_3^2 = (2 + \sqrt{2})\frac{k}{m} \qquad |\Phi_3\rangle \longrightarrow \begin{pmatrix} 1 \\ -\sqrt{2} \\ 1 \end{pmatrix}. \tag{28.32c}$$

As we saw in the two-mass systems, the mode with the greatest symmetry has the lowest frequency. In modes 1 and 3 (Figure 28.3), the outer masses oscillate in phase while the middle block moves with larger amplitude — in phase for $|\Phi_1\rangle$, out of phase for $|\Phi_2\rangle$. By contrast, in mode 2 the central mass remains at rest while the outer two oscillate asymmetrically around it. Notice that there's no analog of the symmetric mode of (28.16a) in which all three masses are in phase with the same amplitude. The reason is easy to understand: when all three blocks have the same displacement from equilibrium, the force on the middle block vanishes — and no force, no acceleration. So the state $q_1 = q_2 = q_3$ is not stationary.

Example 28.5   **Zero Modes**

Chemical bonds between atoms are often modeled as springs; a molecule's vibrations can thus be decomposed as a linear combination of its normal modes. But a molecule is not usually fixed between walls. How does this affect the analysis? Well, if we detach the masses from the walls, the first and fourth springs of Example 28.4 are rendered irrelevant. We thus drop the solitary kq_1 and kq_3 terms from (28.30), leaving us with (after inserting the ansatz)

$$k\begin{pmatrix} 1 & -1 & 0 \\ -1 & 2 & -1 \\ 0 & -1 & 1 \end{pmatrix}\begin{pmatrix} \phi_1 \\ \phi_2 \\ \phi_3 \end{pmatrix} = m\omega^2 \begin{pmatrix} \phi_1 \\ \phi_2 \\ \phi_3 \end{pmatrix}, \tag{28.33}$$

---

[2] The determinant gives a daunting *sixth*-order equation in ω; better to solve the cubic in ω^2.

and eigenmodes

$$\omega_1^2 = 0 \qquad |\Phi_1\rangle \longrightarrow \begin{pmatrix} 1 \\ 1 \\ 1 \end{pmatrix} \tag{28.34a}$$

$$\omega_2^2 = \frac{k}{m} \qquad |\Phi_2\rangle \longrightarrow \begin{pmatrix} 1 \\ 0 \\ -1 \end{pmatrix} \tag{28.34b}$$

$$\omega_3^2 = \frac{3k}{m} \qquad |\Phi_3\rangle \longrightarrow \begin{pmatrix} 1 \\ -2 \\ 1 \end{pmatrix}. \tag{28.34c}$$

Modes 2 and 3 here are similar to those of (28.32), with minor but not insignificant changes in frequency and relative amplitude. In mode 2, for instance, the outer masses are each connected by a single spring to a motionless central mass; thus each should oscillate with squared-frequency k/m. By contrast, in Example 28.4 the outer masses are each connected to *two* springs, thereby doubling the effective spring constant.

Mode 1 is the symmetric mode absent in the previous example. Here it corresponds to the net translation of the entire system, without any stretching of the springs. This isn't oscillation at all — or, if you prefer, it's zero-frequency oscillation. Hence it's called a *zero mode*, a direct result of the springs not being anchored to the wall. With no anchor, the system is invariant under translations.

The molecule is similarly unchanged by rotation, so there are rotational zero modes as well. In general, zero modes reflect underlying translational and rotational symmetries of the system as a whole — one zero mode for each symmetry. Since these modes do not reflect *internal* characteristics of the molecule, they are often disregarded.

One should always begin a coupled oscillation problem by first trying to intuit the normal modes of a system. Consider, for instance, adding another mass and spring to the system of Example 28.4. With four oscillators moving in one dimension, there are four normal modes (Figure 28.4). If the middle two masses move together in phase with identical amplitudes, the central spring is never stretched or compressed; together, the pair acts like the central block of Example 28.4. So we expect modes 1 and 3 of (28.32) to be essentially the same as modes 1 and 3 here. Similarly, since the center of mass

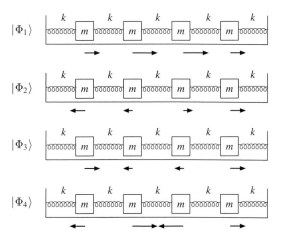

Figure 28.4 Normal modes for the four-mass system.

remains fixed in modes 2 and 4, they are akin to the antisymmetric mode in Example 28.1. You should convince yourself — without calculation — that $\omega_1 < \omega_2 < \omega_3 < \omega_4$.

Using intuition to figure out normal modes is a fine pastime; even so, calculation is required to find the eigenfrequencies and amplitudes. Of course, only the secular determinant stands in your way, but for most matrices larger than 3×3 its direct evaluation is tedious — if not insufferable. As the dimensionality grows, numerical and computational software can certainly be your best friend.

Example 28.6   A Bracelet of Masses and Springs

Oftentimes a system has symmetry that can be exploited which not only simplifies a calculation, but deepens insight into the physics. As an example, let's confine a chain of four identical beads and springs to a frictionless circular wire to form a bracelet.[3] Take a moment to predict the normal modes.

The bracelet is not identical to a linear chain of masses and springs: disconnecting the first and fourth masses from the wall and attaching them to one another introduces off-diagonal couplings between them. You should verify that the matrix $M^{-1}K$ becomes

$$T \equiv M^{-1}K = \frac{k}{m} \begin{pmatrix} 2 & -1 & 0 & -1 \\ -1 & 2 & -1 & 0 \\ 0 & -1 & 2 & -1 \\ -1 & 0 & -1 & 2 \end{pmatrix}, \tag{28.35}$$

where the zeros indicate that only adjacent masses in the bracelet are coupled. We want to find the four normal modes $|\Phi_\ell\rangle$ and their eigenfrequecies ω_ℓ.

Now as 4×4 matrices go, this one's not so awful — but with a little physical insight, we can make the calculation easier. That's because the bracelet possesses symmetry the linear chain does not have: it is indistinguishable from the same bracelet rotated by an integer multiple of $90°$. The representation of a $\pi/2$ rotation in configuration space (where T lives) is given by multiples of the discrete shift operation

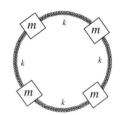

$$S = \begin{pmatrix} 0 & 1 & 0 & 0 \\ 0 & 0 & 1 & 0 \\ 0 & 0 & 0 & 1 \\ 1 & 0 & 0 & 0 \end{pmatrix}, \tag{28.36}$$

which takes $q_n \to q_{n+1}$ (modulo 4).

A quick glance at the columns verifies that S is orthogonal (it does, after all, represent an $\mathbb{R}^2$ rotation). So it's diagonalizable. Moreover, since S leaves the bracelet invariant, it doesn't matter whether it acts before or after T. So even without calculation we know $[S, T] = 0$ — and therefore S and T share a common eigenbasis. This is a useful insight, since the eigenvectors of S are easier to calculate directly. In particular, four successive shifts must produce no change at all, $S^4 = \mathbb{1}$. So we don't even need to set up the secular determinant: we already know that the eigenvalues s must satisfy $s^4 = 1$; that is, they are the fourth roots of unity[4]

$$s^4 = 1 \implies s_\ell = e^{(2\pi i\ell)/4} = e^{i\pi\ell/2}, \qquad \ell = 0, 1, 2, 3. \tag{28.37a}$$

The eigenvectors $|s_\ell\rangle$ can then be found to have components $\langle q_n|s_\ell\rangle = e^{i\pi\ell n/2}$, $n = 0$ to 3 for each of the four eigenvectors,

---

[3] Examples 28.6 and 28.7 are adapted from Georgi (1993).
[4] S is not Hermitian, so complex eigenvalues are allowed.

$$|s_0\rangle \rightarrow \begin{pmatrix} 1 \\ 1 \\ 1 \\ 1 \end{pmatrix} \quad |s_1\rangle \rightarrow \begin{pmatrix} 1 \\ i \\ -1 \\ -i \end{pmatrix} \quad |s_2\rangle \rightarrow \begin{pmatrix} 1 \\ -1 \\ 1 \\ -1 \end{pmatrix} \quad |s_3\rangle \rightarrow \begin{pmatrix} 1 \\ -i \\ -1 \\ i \end{pmatrix}. \tag{28.37b}$$

Each $|s_\ell\rangle$ has a phase difference of $\Delta\varphi = \ell\pi/2$ between adjacent masses, consistent with the bracelet's periodicity. It's not hard to see that $|s_{0,2}\rangle$ are normal modes, with adjacent masses having a relative phase of 0 or π. By contrast, neighboring masses in $|s_{1,3}\rangle$ are $\pm\pi/2$ out of phase, and so describe traveling waves moving in opposite directions. Notice that their eigenvalues are complex conjugates, $s_3 = s_1^*$.

Now, because S and T commute, the $|s_\ell\rangle$ must also be eigenvectors of T — which is not only easy to verify, but allows direct evaluation of the normal mode frequencies

$$\omega_0^2 = 0 \qquad \omega_1^2 = 2k/m \qquad \omega_2^2 = 4k/m \qquad \omega_3^2 = 2k/m. \tag{28.38}$$

Indeed, since we already have the eigenvectors, there's no need to solve the secular determinant for T; this is the benefit of having a symmetry operator.

The bracelet's zero mode $|\Phi_0\rangle \equiv |s_0\rangle$ describes rotation of the entire system without oscillation; mode 2 is $|\Phi_2\rangle \equiv |s_2\rangle$, the analog of the linear chain's fourth mode. We have not yet fully solved the normal mode problem, however: the traveling waves $|s_1\rangle$ and $|s_3\rangle$, though eigenvectors of T, are not normal modes (as you might expect, since they are complex). But since they form a two-dimensional degenerate eigensubspace of T, we can take linear combinations of them to find the normal modes. In fact, just as the superposition of oppositely directed traveling waves on a fixed string yield standing waves, so too for normal modes:

$$|\Phi_1\rangle = \frac{|s_1\rangle + |s_3\rangle}{2} \rightarrow \begin{pmatrix} 1 \\ 0 \\ -1 \\ 0 \end{pmatrix} \quad |\Phi_3\rangle = \frac{|s_1\rangle - |s_3\rangle}{2i} \rightarrow \begin{pmatrix} 0 \\ 1 \\ 0 \\ -1 \end{pmatrix}, \tag{28.39}$$

as you can verify directly. Incidentally, since $|s_1\rangle$ and $|s_3\rangle$ are not degenerate in S, these linear combinations are not eigenvectors of S; rather, $S|\Phi_{1,3}\rangle = \mp|\Phi_{3,1}\rangle$ — which makes sense since these modes are $\pi/2$ rotations of one another.

Generalizing to a bracelet with N beads, the eigenvalues s of S are the N roots of unity

$$s_\ell = e^{2\pi i\ell/N}, \tag{28.40a}$$

and the nth component of their eigenvectors $|s_\ell\rangle$ on the position basis is

$$\langle q_n|s_\ell\rangle = e^{2\pi i\ell n/N}, \tag{28.40b}$$

where, as in (28.37), $\ell = 0$ to $N-1$ labels the mode, and $n = 0$ to $N-1$ the elements of the column vectors for each ℓ. The T matrix is the $N \times N$ version of (28.35):

$$T \equiv \frac{k}{m} \begin{pmatrix} 2 & -1 & 0 & 0 & \cdots & -1 \\ -1 & 2 & -1 & 0 & \cdots & 0 \\ 0 & -1 & 2 & -1 & \cdots & 0 \\ 0 & 0 & -1 & 2 & \cdots & 0 \\ \vdots & \vdots & \vdots & \vdots & \ddots & -1 \\ -1 & 0 & 0 & 0 & -1 & 2 \end{pmatrix}. \tag{28.41}$$

Because we've limited ourselves to nearest-neighbor interactions, the nth bead only "knows about" masses $n \pm 1$. Thus we can find the eigenfrequencies merely by considering one bead and its two neighbors, focusing on a 3×3 submatrix within T. With ϕ_n the nth component of a T eigenvector on the position basis, the eigenvalue equation connecting ϕ_n with $\phi_{n\pm1}$ is

$$\omega^2 \phi_n = \frac{k}{m} \left(-\phi_{n+1} + 2\phi_n - \phi_{n-1} \right). \tag{28.42}$$

Since T and S share an eigenbasis, we can then use $\phi_n = \langle q_n | \phi_\ell \rangle$ and (28.40b) to obtain

$$\omega_\ell^2 = \frac{k}{m} \left(-e^{2\pi i\ell/N} + 2 - e^{-2\pi i\ell/N} \right)$$

$$= \frac{2k}{m} (1 - \cos 2\pi \ell/N) = \frac{4k}{m} \sin^2(\pi \ell/N). \tag{28.43}$$

(As a quick check, note that this reduces to (28.38) for $N = 4$.) The periodicity of $\sin^2(\pi \ell/N)$ ensures that, as before, the eigenvectors $|s_\ell\rangle$ with complex-conjugate elements have degenerate normal frequencies ω_ℓ. And as in (28.39), the real and imaginary parts of these eigenvectors give the normal modes of the N-bracelet (Problem 28.17).

Example 28.7   The N-Chain

Having solved the N-bracelet, we turn to finding a solution for a linear chain of identical masses and springs for arbitrary N. That would seem a daunting task, given that as the number of masses increases, so does the difficulty of the calculation. Ironically, however, an *infinite* chain of masses and springs is easier to solve than the finite chain (at least for $N > 3$ or so). And we can then use its basis to construct the solution for finite N.

Symmetry is the key. An infinite chain is invariant under the shift $q_n \rightarrow q_{n+1}$, so there's a symmetry operator S with 1's running down the off-diagonal, and zeros everywhere else. Granted, the prospect of calculating the secular determinant for an infinite-dimensional matrix may seem unattractive — but in fact we've already solved it! The symmetry of the chain is the same as that of the bracelet; only the periodicity condition is absent. So the eigenvalues and eigenvectors of S should be the same as (28.40), without the $2\pi i\ell/N$ in the exponent. Replacing $2\pi \ell/N$ by some constant α yields eigenvalues

$$s_\alpha = e^{i\alpha}, \tag{28.44}$$

where we take α to be real. (Indeed, the infinite matrix S is unitary.) Since $S|s_\alpha\rangle = e^{i\alpha}|s_\alpha\rangle$, the eigenvectors can be represented as infinite column vectors with adjacent masses shifted in phase by $e^{\pm i\alpha}$,

$$|s_\alpha\rangle \rightarrow \begin{pmatrix} \vdots \\ e^{-2i\alpha} \\ e^{-i\alpha} \\ 1 \\ e^{+i\alpha} \\ e^{+2i\alpha} \\ \vdots \end{pmatrix}, \tag{28.45}$$

which, as you'll surely appreciate, we prefer not to write out in its entirety. Far easier is simply to denote the individual amplitudes as

$$\langle q_n | s_\alpha \rangle = e^{in\alpha}, \qquad (28.46)$$

with integer $-\infty < n < \infty$. Now for this approach to work, S must commute with $T \equiv M^{-1}K$. Although they do not commute for the finite chain, the action of S — moving each mass over one spot — has no effect on an *infinite* chain. So it must be that $[S, T] = 0$ for the infinite chain (Problem 28.18). Therefore the eigenvectors (28.45) are also eigenvectors of T,

$$T|s_\alpha\rangle = \omega_\alpha^2 |s_\alpha\rangle. \qquad (28.47)$$

A calculation virtually identical to (28.42)–(28.43) gives

$$\omega_\alpha^2 = \frac{k}{m} \left(-e^{i\alpha} + 2 - e^{-i\alpha} \right) = \frac{2k}{m} \left(1 - \cos\alpha \right). \qquad (28.48)$$

The case $\alpha = 0$ corresponds to $\phi_n = 1$ for all n, which is the zero mode (whatever that means for an infinitely long chain). All the other eigenvectors are doubly degenerate in ω^2, with $\pm\alpha$ returning the same eigenfrequency. So, as with the bracelet, the normal mode $|\Phi_\alpha\rangle$ associated with frequency ω_α is given by real combinations of $|s_\alpha\rangle$ and $|s_{-\alpha}\rangle$.

So much for the infinite chain; what about the finite case with N masses? We can reproduce this on the infinite chain by artificially imposing constraints $q_0 = q_{N+1} = 0$. Physically, this is equivalent to taking the masses at those positions to infinity. Since the $|s_{\pm\alpha}\rangle$ form a complete basis, the finite-N normal modes will emerge from real linear combinations consistent with these boundary conditions.

Examining (28.45) and (28.46) shows that we can only achieve $q_0 = 0$ by subtraction — that is, the correct finite-N mode must have the form

$$|\Phi_\alpha\rangle \equiv |s_\alpha\rangle - |s_{-\alpha}\rangle. \qquad (28.49)$$

Though this properly holds position $n = 0$ fixed, at $n = N + 1$ this yields

$$q_{N+1} = \langle q_{N+1}|\Phi_\alpha\rangle = e^{i\alpha(N+1)} - e^{-i\alpha(N+1)} = 2i\sin[\alpha(N+1)]. \qquad (28.50)$$

So $q_{N+1} = 0$ can only be satisfied if α takes on equally spaced discrete values,

$$\alpha = \frac{\ell\pi}{N+1}, \quad \ell = 1, 2, 3, \dots, N. \qquad (28.51)$$

Until now, we've had no restriction on α other than it being real; it's the imposition of boundary conditions that results in the "quantization" condition (28.51).

A finite chain of N identical masses and springs thus has normal modes

$$|\Phi_\ell\rangle \rightarrow \begin{pmatrix} \sin(\frac{\ell\pi}{N+1}) \\ \sin(\frac{2\ell\pi}{N+1}) \\ \vdots \\ \sin(\frac{N\ell\pi}{N+1}) \end{pmatrix}, \qquad (28.52a)$$

with eigenfrequencies

$$\omega_\ell^2 = \frac{2k}{m} \left[1 - \cos\left(\frac{\ell\pi}{N+1}\right) \right] = \frac{4k}{m} \sin^2\left(\frac{\ell\pi}{2(N+1)}\right). \qquad (28.52b)$$

These expressions easily reproduce our earlier results for $N = 2$ and 3 (Problem 28.16). The normal modes for a four-mass chain with positions 0 and 5 held fixed, Figure 28.4, are reproduced in Figure 28.5 with the curve $\sin(x\ell\pi/5)$ superimposed for each mode.[5] As N grows, so too does the length of the chain; the masses remain equidistant along the chain, with relative amplitudes given by

$$\phi_n^{(\ell)} = \langle q_n|\Phi_\ell\rangle = \sin\left(\frac{n\ell\pi}{N+1}\right), \tag{28.53}$$

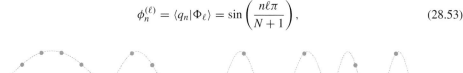

Figure 28.5 Normal modes $\ell = 1$ to 4 of a four-mass linear chain ($N = 4$).

where, as before, the superscript in $\phi_n^{(\ell)}$ denotes the eigenvector and the subscript its component on the position basis. Verifying the orthogonality of the eigenmodes is straightforward for relatively small N; for general N, the demonstration is a bit lengthier. Using (28.52a):

$$\langle \Phi_\ell|\Phi_m\rangle = \sum_{k=1}^{N} \sin\left(k\frac{\ell\pi}{N+1}\right)\sin\left(k\frac{m\pi}{N+1}\right) = \sum_{k=1}^{N+1} \sin\left(k\frac{\ell\pi}{N+1}\right)\sin\left(k\frac{m\pi}{N+1}\right)$$

$$= \frac{1}{2}\sum_{k=1}^{N+1}\left\{\cos\left[\frac{k\pi}{N+1}(\ell-m)\right] - \cos\left[\frac{k\pi}{N+1}(\ell+m)\right]\right\}$$

$$= \frac{N+1}{2}\delta_{\ell m}, \tag{28.54}$$

where we used the results of Problem 6.8. (Since $\sin n\pi = 0$, including $k = N + 1$ adds nothing — but it simplifies the evaluation of the sum.)

All we've really done in (28.52a) is expand the normal modes of the finite chain on the complete basis of the infinite chain. But you may be puzzled, since the infinite chain has an infinite number of basis vectors, whereas the finite chain exists in an N-dimensional space. So there should be infinite variations of the N finite-chain modes, distinguished by their form *outside* the region 1 through N. Except there aren't: (28.52) show that the modes of the finite chain simply repeat themselves, periodically, all along the infinite chain. In other words, there are only N distinct modes consistent with the conditions $q_0 = q_{N+1} = 0$. Thus the infinity of modes are reduced to merely the N modes of the finite chain.

So finally we can say that the most general motion of a linear chain of N identical, equally spaced masses is given by a sum over the orthogonal basis of normal modes

$$q_n(t) = \sum_{\ell=1}^{N} A_\ell \langle q_n|\Phi_\ell\rangle \cos(\omega_\ell t + \delta_\ell)$$

$$= \sum_{\ell=1}^{N} A_\ell \sin\left(\frac{n\ell\pi}{N+1}\right)\cos(\omega_\ell t + \delta_\ell), \tag{28.55}$$

with the A_ℓ and δ_ℓ determined by initial conditions.

---

[5] Though Figure 28.5 represents the amplitudes of longitudinal oscillations, as we saw in Example 28.2, the same applies to transverse motion as well.

Looking at Figure 28.5, you might wonder what happens as more and more masses are added between fixed endpoints 0 and 5. We address this question in Part V.

Problems

28.1 Show that for the case of Example 28.1 with identical springs and mass, the eigenmodes can be found just by adding and subtracting the original coupled equations (28.1). (In principle, one can always use elementary algebra to decouple a system of linear equations. But while the algebra itself may be elementary, a more complicated system entails more elaborate manipulations — which might not be so bad if we knew what linear combinations to take! The secular determinant, though often tedious, provides a standard approach to finding the right combinations.)

28.2 Show that the basis of normal modes in Example 28.1 diagonalizes the system.

28.3 Starting from (28.17) or (28.18), verify (28.19).

28.4 To better appreciate the role of the coupling, find the normal mode frequencies for the system in Example 28.1 when the middle spring has a force constant $\kappa \neq k$.

28.5 **Beats**

Consider the system in Example 28.1 when the middle spring has force constant $\kappa \neq k$.

(a) Find the positions of the masses as a function of time when released from rest with $q_1(0) = 1, q_2(0) = 0$.
(b) For weak coupling, $\kappa \ll k$, the normal mode frequencies are similar (see Problem 28.4). As a result, their sum and difference become more meaningful. Reexpress the solution in (a) in terms of $\omega_{av} = \frac{1}{2}(\omega_1 + \omega_2)$ and $\Delta\omega = \frac{1}{2}(\omega_1 - \omega_2)$; simplify using familiar trig identities. Use Mathematica's Plot command to display $q_1(t)$ and $q_2(t)$ with $\omega_1 = 2\pi$ and $\omega_2 = 1.1\omega_1$. Interpret physically.

28.6 Verify the solution to the double pendulum, (28.27).

28.7 Two masses are suspended vertically, m_1 from the ceiling with a spring of stiffness k_1, and m_2 from the bottom of m_1 with a spring with stiffness k_2. Find the normal modes of the system for $m_1 = m_2 = m$ and $k_1 = 3k, k_2 = 2k$. (Consider vertical motion only.)

28.8 Solving the eigenvector problem merely gives the general solution; a complete solution requires initial conditions to nail down the unknown constants. Write out the general solution using the eigensystem found in Example 28.4, assuming the blocks are released from rest with initial positions a, b, and c, respectively. Verify that (i) $a = b/\sqrt{2} = c$, (ii) $a = -c, b = 0$, and (iii) $a = -b/\sqrt{2} = c$ each yields normal modes with a single eigenfrequency.

28.9 A two-dimensional coupled system has normal modes $|1\rangle \rightarrow \begin{pmatrix} 1 \\ -2 \end{pmatrix}$ and $|2\rangle \rightarrow \begin{pmatrix} 3 \\ 2 \end{pmatrix}$, with frequencies $\omega_1 = \pi \text{ s}^{-1}$ and $\omega_2 = 2\pi \text{ s}^{-1}$. What are the positions of the two oscillators at time $t = 1$ s if at $t = 0$:

(a) the system is released from rest with $q_1(0) = 4$ cm and $q_2(0) = 8$ cm?
(b) the oscillators are at their equilibrium positions with $\dot{q}_1(0) = 4$ cm/s and $\dot{q}_2(0) = 8$ cm/s?

28.10 If carbon monoxide (CO) has a natural vibrational frequency of 6.42×10^{13} Hz, find the effective spring constant in N/m.

28.11 Find the normal modes and frequencies for the three-mass system of Example 28.4 with outer masses $m_1 = m_3 = m$ and inner mass $m_2 = M$.

28.12 Three identical masses m are connected by two springs $k_1 = k$ and $k_2 = nk$ with unknown n. The vibrational modes are

$$\begin{pmatrix} 3 + \sqrt{3} \\ -3 + \sqrt{3} \\ -2\sqrt{3} \end{pmatrix} \qquad \begin{pmatrix} 3 - \sqrt{3} \\ -3 - \sqrt{3} \\ 2\sqrt{3} \end{pmatrix}.$$

The normal mode frequencies are $(3 \pm \sqrt{3})$, 0 (in units of k/m). Use the spectral decomposition (27.103) to determine which frequency goes with which mode, and find n.

28.13 Sketch the three vibrational normal modes of a water molecule.

28.14 The voltage drop across a capacitor is Q/C, and across an inductor is $L dI/dt$, where $Q(t)$ is the charge on the capacitor, and the current $I = dQ/dt$.

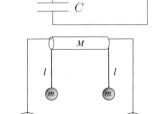

(a) Find the natural frequency ω_0 of an LC circuit.
(b) Consider two LC circuits arranged so that the magnetic flux through each inductor contributes flux to the other, as shown. The *mutual inductance* M contributes additional voltage drops $M dI_1/dt$ to circuit 2 and $M dI_2/dt$ to circuit 1. Find the normal modes and eigenfrequencies.
(c) Write out the general expression for the currents in each circuit.

28.15 Two identical pendulums of mass m and length ℓ are suspended from the ends of a drinking straw of mass M threaded on a frictionless horizontal wire. What are the normal modes for small oscillations in the plane of the frame? [Hint: Very little calculation is actually needed.] What are the eigenfrequencies?

28.16 From the N-chain eigenmodes of (28.52):

(a) Verify the eigenfrequencies and eigenmodes of the two- and three-mass chains, (28.16) and (28.34) respectively. [Rather than using a calculator, you may find the half-angle formula for $\sin(x/2)$ empowering.]
(b) Find the modes for the four-mass chain depicted in Figures 28.4 and 28.5, and verify the eigenfrequency hierarchy asserted in the accompanying discussion. [You may find Problem 5.6 helpful.]

28.17 Find the normal modes for a bracelet of six beads with the same mass.

28.18 Verify that in Example 28.7, $[S, T] \neq 0$ for finite N, but commutes for the infinite chain. [Hint: Use index notation.]

28.19 The change induced in eigenvalues and normal modes by a coupling spring κ depends on the relative size of κ to the other k's in system. To explore this for linear coupled systems generally, consider a Hermitian matrix

$$H = \begin{pmatrix} \omega_1 & A \\ A^* & \omega_2 \end{pmatrix} = \begin{pmatrix} \omega_{av} + \Delta\omega & A \\ A^* & \omega_{av} - \Delta\omega \end{pmatrix},$$

with $\omega_{av} = \frac{1}{2}(\omega_1 + \omega_2)$ and $\Delta\omega = \frac{1}{2}(\omega_1 - \omega_2)$. Denote the unperturbed (i.e., $A = 0$) eigenvectors as $|1\rangle$ and $|2\rangle$, with frequencies ω_1 and ω_2, respectively.

(a) Find the eigenfrequencies $\omega_\pm$, and show that the normal mode basis $|+\rangle$, $|-\rangle$ is related to the standard basis by the unitary rotation

$$U = \begin{pmatrix} \cos\frac{\theta}{2}\, e^{-i\varphi} & \sin\frac{\theta}{2}\, e^{i\varphi} \\ -\sin\frac{\theta}{2}\, e^{-i\varphi} & \cos\frac{\theta}{2}\, e^{i\varphi} \end{pmatrix},$$

where $\tan \theta = |A|/\Delta \omega$, $0 \le \theta < \pi$, and $A = |A|e^{-i\varphi}$.

(b) The matrix U shows that the normal modes are only sensitive to the difference in the unperturbed eigenvalues — making $|A|/\Delta \omega$, or equivalently θ, the key parameter. Find $\omega_\pm$ and $|\pm\rangle$ in limit of both the weak coupling $|A| \ll |\Delta \omega|$ ($\theta \sim 0$), and strong coupling $|A| \gg |\Delta \omega|$ ($\theta \sim \pi/2$).

(c) Make a sketch of $\omega_\pm - \omega_{\mathrm{av}}$ vs. $\Delta \omega$; highlight and discuss the various features.

(d) If the system starts in state $|1\rangle$ at time $t = 0$, how long before it has evolved completely into state $|2\rangle$ in the strong coupling limit? What about for weak coupling? [It's easier to use the exponential form of the ansatz, $e^{i(\omega t + \delta)}$.]

28.20 Quantum Mechanical Resonance

The molecular hydrogen ion H_2^+ has two protons and one electron. The symmetry of the system makes it equally likely to find the electron around either proton; symmetry also dictates that these two possibilities, states $|1\rangle$ and $|2\rangle$, have the same energy E_0. Quantum mechanically, however, there is a non-zero probability for the electron to tunnel through the 13.6 eV barrier and find itself around the other proton — that is, to transition between states $|1\rangle$ and $|2\rangle$. The probability for this to occur increases as the distance between the protons decreases. We can model this phenomenologically using the Hamiltonian H of Problem 28.19 with $\omega_1 = \omega_2 \equiv E_0/\hbar$. (Recall that the Hamiltonian governs time evolution, Example 27.11.) Use the results of that problem to argue that a bound state of the molecule exists. Since this one-electron bond requires oscillation between the states $|1\rangle$ and $|2\rangle$, it's called a quantum resonance.

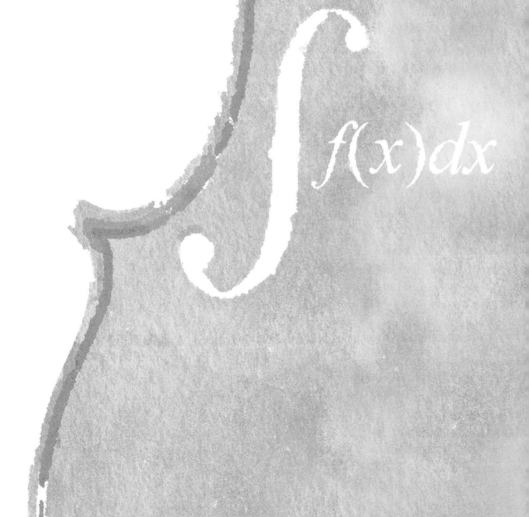

Entr'acte: Tensors

The music is not in the notes, but in the silence in between.

Wolfgang Amadeus Mozart

29 Cartesian Tensors

29.1 The Principle of Relativity

According to our definitions, both velocity and angular velocity are vectors. As are momentum and angular momentum, force and torque, electric fields and magnetic fields. Moreover, they are all vectors in $\mathbb{R}^3$ with the familiar Euclidean dot product. But though we may be inclined to regard them as belonging to a single inner product space, they must in fact live in distinct $\mathbb{R}^3$'s — after all, it makes no sense to add an electric field and a velocity. Rather than belonging to the same space, the common feature of these vectors is their dimensionality and inner product. One says that they belong to the same *equivalence class* of vector spaces defined by these characteristics. This is what allows us to indicate and decompose all these quantities on the same free-body diagram.

For instance, consider that universal standard of introductory mechanics, a mass on an inclined plane (Figure 29.1). The system is, of course, governed by Newton's second law — which, as a vector equation, requires the decomposition of forces (gravity, normal, friction) into mutually perpendicular components in order to solve for the motion. There are an infinite number of choices for this decomposition. We could, for instance, use the x-, y- axes with gravity pointing along $-\hat{\jmath}$. Since the motion is inherently one-dimensional, however, decomposing along the x'-, y'- axes shown puts the motion entirely along $+\hat{\imath}'$ and thus leads to the simplest calculation. This only works because position, velocity, acceleration and force belong to the same equivalence class, and so can share a common basis.

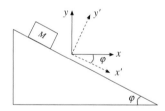

Figure 29.1 Which basis is best?

Of course, the choice of that common basis is ultimately irrelevant: though the vector components and the simplicity of ensuing algebra may depend on the axes, physical results do not. In this sense, *vectors possess an objective physical reality not shared by their components.* Vector equations like $\vec{F} = m\vec{a}$ reflect this reality by maintaining their validity for any choice of the axes: even though a passive rotation transforms the component equations $F_i = ma_i$ into a different set $F_{i'} = ma_{i'}$, as members of the same equivalence class the components of $\vec{F}$ and $\vec{a}$ change *in precisely the same fashion*, depending only on the properties and characteristics of the rotation matrix R:

$$\begin{aligned} F_x &= ma_x \\ F_y &= ma_y \end{aligned} \quad \underset{R^{-1}}{\overset{R}{\rightleftarrows}} \quad \begin{aligned} F_{x'} &= ma_{x'} \\ F_{y'} &= ma_{y'}. \end{aligned} \tag{29.1}$$

The physical equivalence of the two sets of equations relies not simply on being able to use the same basis for both vectors, but on the ability to use the identical rotation to change to another basis. This is the appeal of vector equations: coordinate-independent statements that hold for *any* orientation of coordinate frame.

There's deep physical insight lurking here. One says that $\vec{F} = m\vec{a}$ is a *covariant* expression — which is a technical way of saying that any observer, regardless of basis, can use the exact same vector equation to describe a system.[1] Indeed, in order to make meaningful statements about the physical world, one must distinguish between those elements which reflect physically significant aspects of the system, and those which merely betray an observer's orientation or bias. This observation is fundamental to the Principle of Relativity, which asserts that *the laws of physics are the same for all observers*. Put simply, it must be possible to express a physical law without relying on or referring to any observer's perspective, frame of reference, or choice of coordinates. It must be covariant.

The Principle of Relativity applies far beyond rotations in $\mathbb{R}^3$ to include a host of coordinate transformations. For example, translation-covariant equations are independent of the choice of origin; Lorentz symmetry underlies the arbitrary choice of inertial frame; and covariance under general spacetime transformations yields the fundamental equivalence of all observers in general relativity. In all cases, a mathematical relationship which purports to be a law of nature must be expressible in coordinate-independent form. Thus laws of nature must be expressible in terms of scalars and vectors — or more generally, *tensors*.

Tensors are often superficially distinguished by the number of indices required to specify their components. Scalars have none, vectors one. But there are also quantities with more indices. Since these are less familiar, we begin our tensor discussion by showing how two-index objects naturally emerge in concrete physical examples: stress and strain. In fact, tensor analysis originated from the study of stress and strain within continuous bodies — hence the word "tensor." We'll limit the presentation for now to cartesian bases; in the next chapter, we'll expand the discussion to include non-cartesian tensors.

29.2 Stress and Strain

Consider the forces within a continuous body. Stress is defined as the internal force per unit area exerted at an imaginary surface within the body. So stress times area yields the force, each component of which is given by

$$dF_i = \sum_j \sigma_{ij} da_j, \qquad (29.2)$$

where σ_{ij} is called the *stress tensor* (Figure 29.2). Why is stress described by a two-index object? After all, since both force $\vec{F}$ and area $d\vec{a}$ are vectors, one might initially conclude that stress must be a scalar — hardly something in need of a multi-index description. A little thought, however, reveals that there are two *different* types of stress. The stress component normal to the surface is familiar as *pressure*, whereas its tangential component gives rise to *shear* which tries to tear the body apart. A single number is inadequate to describe this situation. Indeed, using two vector indices makes sense — one for the force component, the other the surface orientation. So σ_{ij} describes the ith component of the stress (force/area) acting on a surface with a normal in the jth direction: the diagonal components give the pressure (normal stress), the off-diagonal components the shear.

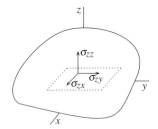

Figure 29.2 Internal stresses within a body.

---

[1] An *invariant quantity* has the same value on all coordinate frames; a *covariant expression* has the same form in all coordinate frames.

The word "component" to describe the elements of the stress matrix is not incidental. The individual elements of σ_{ij} clearly depend on the choice of axes. Just as with the components of a vector, the numbers σ_{ij} are basis-dependent quantities whose values change in a prescribed way under rotation — as we'll see, in a covariant fashion which respects and preserves the vector nature of both $\vec{F}$ and $d\vec{a}$. The elements σ_{ij} are thus components of the basis-independent tensor $\overset{\leftrightarrow}{\sigma}$.

Whereas stress summarizes the internal forces at work in a body, strain describes the deformations of a body due to external forces — deformations which induce additional stress in order to restore equilibrium. So where there's strain, there's stress. Consider a rod of length L and cross section A, strained to a length $L + \Delta L$. For small ΔL, the relationship between induced stress $\sigma = F/A$ normal to the surface and the strain $\varepsilon \equiv \Delta L/L$ is given by Hooke's law,

$$F = k\Delta L \Longrightarrow \sigma = Y\varepsilon, \qquad (29.3)$$

where $Y = kL/A$ is Young's (or the elastic) modulus of the material. Since ΔL can be negative, (29.3) describes both *tensile* and *compressive* stress and strain.

But if stress is a tensor, then so too is strain. Moving beyond one dimension, we define a vector field $\vec{u}(\vec{r})$ which gives the magnitude and direction of a body's displacement at position $\vec{r}$. To analyze displacement due to internal stress, we need to focus not on the overall displacement of an object — the so-called *rigid body motion* — but rather on the relative displacements at different points within the body. These deformations result in changes to the body's shape. In the one-dimensional case, this required comparing L with $L + \Delta L$; in three dimensions, we Taylor expand components of $\vec{u}$ (see Figure 29.3),

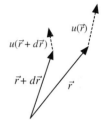

$$u_i(\vec{r} + d\vec{r}) = u_i(\vec{r}) + \vec{\nabla}u_i \cdot d\vec{r} + \cdots, \qquad (29.4)$$

Figure 29.3
Displacement u due to internal stress.

where we presume $|\vec{\nabla}u_i| \ll 1$ so that first order is sufficient. We can put this into tensor notation by placing the vectors $\vec{\nabla}$ and $\vec{u}$ side by side:

$$\vec{u}(\vec{r} + d\vec{r}) = \vec{u}(\vec{r}) + \sum_j \partial_j \vec{u}\, dr_j = \vec{u}(\vec{r}) + \vec{\nabla}\vec{u} \cdot d\vec{r}. \qquad (29.5)$$

In matrix form,

$$\vec{\nabla}\vec{u} = \begin{pmatrix} \frac{\partial u_x}{\partial x} & \frac{\partial u_x}{\partial y} & \frac{\partial u_x}{\partial z} \\ \frac{\partial u_y}{\partial x} & \frac{\partial u_y}{\partial y} & \frac{\partial u_y}{\partial z} \\ \frac{\partial u_z}{\partial x} & \frac{\partial u_z}{\partial y} & \frac{\partial u_z}{\partial z} \end{pmatrix}. \qquad (29.6)$$

Clearly this is where the strain resides. We start to delve into it by breaking $\vec{\nabla}\vec{u}$ up into symmetric and antisymmetric parts,

$$\vec{\nabla}\vec{u} \equiv \overset{\leftrightarrow}{\varepsilon} + \overset{\leftrightarrow}{\phi}, \qquad (29.7)$$

where six of the nine elements of $\vec{\nabla}\vec{u}$ are contained in the symmetric tensor

$$\varepsilon_{ij} = \frac{1}{2}\left(\partial_i u_j + \partial_j u_i\right), \qquad (29.8a)$$

while the other three are housed in the antisymmetric

$$\phi_{ij} = \frac{1}{2}\left(\partial_i u_j - \partial_j u_i\right). \qquad (29.8b)$$

This separation into symmetric and antisymmetric parts is always possible, and — as you'll show in Problem 29.11 — the symmetry of a tensor is unchanged under rotation. In other words, a symmetric/antisymmetric decomposition is covariant.

In terms of these new tensors, (29.5) becomes

$$\vec{u}(\vec{r} + d\vec{r}) = \vec{u}(\vec{r}) + \overset{\leftrightarrow}{\varepsilon} \cdot d\vec{r} + \overset{\leftrightarrow}{\phi} \cdot d\vec{r}. \tag{29.9}$$

Let's turn to $\overset{\leftrightarrow}{\phi}$ first. If we follow the example set out in (B.3), the antisymmetric tensor $\overset{\leftrightarrow}{\phi}$ can be repackaged as an $\mathbb{R}^3$ vector $\vec{\varphi}$, courtesy of the Levi-Civita epsilon,

$$\phi_{ij} \equiv -\sum_k \epsilon_{ijk} \varphi_k. \tag{29.10}$$

Rewriting the contribution of $\overset{\leftrightarrow}{\phi}$ in terms of $\vec{\varphi}$ reveals

$$\sum_j \phi_{ij} dr_j = -\sum_{jk} \epsilon_{ijk} \varphi_k dr_j = +\sum_{jk} \epsilon_{ikj} \varphi_k dr_j, \tag{29.11}$$

or

$$d\vec{u} = \vec{\varphi} \times d\vec{r}, \tag{29.12}$$

which describes a rotation of $d\vec{r}$ by φ around $\hat{\varphi}$. Rather than a deformation, this is a form of rigid body motion.

Deformations, then, must be described by the *strain tensor* $\overset{\leftrightarrow}{\varepsilon}$. Hooke's law (29.3) connecting stress and strain becomes the tensor relation (Problem 29.14)

$$\sigma_{ij} = \sum_{k\ell} Y_{ijk\ell} \varepsilon_{k\ell}, \tag{29.13}$$

where $Y_{ijk\ell}$ is often called the elasticity tensor. Just as for the spring constant k in one dimension, $Y_{ijk\ell}$ quantifies the stiffness of the material in different directions.

29.3 The Equivalence Class of Rotations

As arguably the most primitive of physical vectors, we take the position $\vec{r}$ as the defining prototype for the equivalence class of $\mathbb{R}^n$ vectors. Any real, n-component object $\vec{v}$ which transforms under rotations in the same manner as $\vec{r}$,

$$v_{i'} = \sum_j R_{i'j} v_j, \tag{29.14}$$

belongs to the equivalence class. As we did in (29.1), we place the primes on the indices rather than on the vector to indicate that $\vec{v}$ is the same physical object before and after rotation (Schutz, 2009). In other words, we take the passive perspective in which the effect of R is to rotate the basis, relating the components v_i of $\vec{v}$ on the old basis with its components $v_{i'}$ on the new. (Since this notational change allows (29.14) to be written as, say, $v_{i'} = \sum_i R_{i'i} v_i$, it has the ancillary advantage of increasing the number of parameters available to use as indices.)

In common physics and engineering parlance, any one-index object which behaves like $\vec{r}$ under rotation *is* a vector. Since $\vec{v}$ has a single index, it is also referred to as a tensor of rank-one. Scalars, which have no index and are unchanged by a rotation, are zero-rank tensors. The simplest way to construct a rank-two tensor is to take two rank-one tensors and set them side-by-side — just as we did in (29.5) — to form a *dyad*,

$$\overset{\leftrightarrow}{T} = \left(\sum_i v_i \hat{e}_i \right) \left(\sum_j w_j \hat{e}_j \right) \equiv \sum_{ij} v_i w_j \, \hat{e}_i \otimes \hat{e}_j. \tag{29.15}$$

The basis for this construct is not a set a vectors, but rather a *tensor product* of vectors $\hat{e}_i \otimes \hat{e}_j$. And the ability to expand $\overset{\leftrightarrow}{T}$ on a basis highlights a fundamental distinction: an array of numbers has *elements*, but a tensor has *components*.

<hr>

Example 29.1 **The Tensor Product**

The tensor product $\hat{e}_i \otimes \hat{e}_j$ is neither an inner product nor a cross product; rather, it's a mapping of two one-index vectors into a two-index object. It may look more familiar to you in bra-ket notation, $\hat{e}_i \otimes \hat{e}_j \longleftrightarrow |\hat{e}_i\rangle \langle \hat{e}_j|$, or as the matrix $\hat{e}_i \otimes \hat{e}_j \rightarrow \hat{e}_i \hat{e}_j^T$. So just as $\hat{e}_i$ is a basis vector whose standard matrix form is a column with a single non-zero component in the ith position, e.g.,

$$\hat{e}_3 \rightarrow \begin{pmatrix} 0 \\ 0 \\ 1 \end{pmatrix}, \tag{29.16}$$

similarly, $\hat{e}_i \otimes \hat{e}_j$ is a basis tensor whose standard matrix representation has a single non-zero component in the ijth position, e.g.,

$$\hat{e}_2 \otimes \hat{e}_3 \rightarrow \begin{pmatrix} 0 & 0 & 0 \\ 0 & 0 & 1 \\ 0 & 0 & 0 \end{pmatrix}. \tag{29.17}$$

This makes clear that the tensor product is not commutative, $\hat{e}_i \otimes \hat{e}_j \neq \hat{e}_j \otimes \hat{e}_i$. So a tensor product is an *ordered* pair of directions in space.

<hr>

Manipulating cartesian tensors components follows the familiar rules of index algebra. For instance, the sum of tensors is given by the sum of their components,

$$\overset{\leftrightarrow}{T} + \overset{\leftrightarrow}{U} = \left(T_{ij} + U_{ij} \right) \hat{e}_i \otimes \hat{e}_j. \tag{29.18}$$

Similarly, the product of a tensor and a vector is

$$\begin{aligned} \overset{\leftrightarrow}{T} \cdot \vec{v} &= \left(\sum_{ij} T_{ij} \, \hat{e}_i \otimes \hat{e}_j \right) \cdot \left(\sum_k v_k \hat{e}_k \right) \\ &= \sum_{ijk} T_{ij} v_k \, \hat{e}_i \otimes \left(\hat{e}_j \cdot \hat{e}_k \right) \\ &= \sum_{ijk} T_{ij} v_k \, \hat{e}_i \delta_{jk} = \sum_{ij} T_{ij} v_j \hat{e}_i \,. \end{aligned} \tag{29.19}$$

Thus $\overset{\leftrightarrow}{T}$ operates on vector $\vec{v}$ to produce another vector $\vec{w}$,

$$w_i = \sum_j T_{ij} v_j. \tag{29.20}$$

This is sometimes called "right" contraction of T_{ij} to distinguish it from "left" contraction $\vec{v} \cdot \overset{\leftrightarrow}{T}$, or $\sum_j v_i T_{ij}$. Either way, it's just matrix multiplication of components. The underlying index manipulations apply to tensors of higher rank as well.

On the cartesian tensor product basis, the dyad $\overset{\leftrightarrow}{T}$ has components

$$T_{ij} \equiv v_i w_j. \tag{29.21}$$

Because a dyad is just two vectors side-by-side, under a rotation it transforms as

$$T_{i'j'} = v_{i'} w_{j'} = \left(\sum_k R_{i'k} v_k \right) \left(\sum_\ell R_{j'\ell} w_\ell \right)$$
$$= \sum_{k\ell} R_{i'k} R_{j'\ell} v_k w_\ell = \sum_{k\ell} R_{i'k} R_{j'\ell} T_{k\ell}. \tag{29.22}$$

Although not all second-rank tensors can be written as a dyad, by definition all rank-two tensors transform in this way,

$$T_{i'j'} = \sum_{k\ell} R_{i'k} R_{j'\ell} T_{k\ell}, \tag{29.23}$$

with each vector index of T_{ij} contracting with its own rotation matrix. A little manipulation reveals

$$T_{i'j'} = \sum_{k\ell} R_{i'k} R_{j'\ell} T_{k\ell} = \sum_{k\ell} R_{i'k} R^T_{\ell j'} T_{k\ell} = \sum_{k\ell} R_{i'k} T_{k\ell} R^T_{\ell j'}. \tag{29.24}$$

So second-rank tensors in $\mathbb{R}^n$ transform by orthogonal similarity transformation.

Though both $\overset{\leftrightarrow}{T}$ and T_{ij} are often referred to as tensors, as we've seen the matrix elements T_{ij} are actually the components of $\overset{\leftrightarrow}{T}$ on a basis. And just as the components of a vector depend upon the coordinate system, so too the components of a tensor. But since they're all related by rotations, measuring the components in one frame allows their determination in *any* frame.

Example 29.2   The Moment of Inertia Tensor

Using (4.57) and noting that r^2 is a scalar,

$$I_{i'j'} = \sum_{k,\ell} R_{i'k} R_{j'\ell} I_{k\ell}$$

$$= \sum_{k,\ell} R_{i'k} R_{j'\ell} \left[\int \left(r^2 \delta_{k\ell} - r_k r_\ell \right) \rho(\vec{r}) \, d\tau \right]$$

$$= \int \left[\sum_{k,\ell} R_{i'k} R_{j'\ell} \delta_{k\ell} \, r^2 - \left(\sum_k R_{i'k} r_k \right) \left(\sum_\ell R_{j'\ell} r_\ell \right) \right] \rho \, d\tau$$

$$= \int \left[\delta_{i'j'} \, r^2 - r_{i'} r_{j'} \right] \rho \, d\tau, \tag{29.25}$$

which, obviously, is exactly the same expression written in primed coordinates. So the moment of inertia is indeed a tensor, as claimed. Note that a sum of tensors of the same rank is also a tensor. So since we already know $r_i r_j$ is a second-rank tensor (a dyad, in fact) and r^2 a scalar, simply demonstrating that δ_{ij} is a second-rank tensor is itself sufficient to establish that I_{ij} is as well. (In fact, as we'll see in Section 29.5, δ_{ij} is an *invariant tensor*.)

As a matrix equation, $L_i = \sum_j I_{ij}\omega_j$ says that I_{ij} maps the contribution of the jth component of $\vec{\omega}$ to the ith component of $\vec{L}$. And due to the behavior of vectors and tensors under rotations, this linking of components of $\vec{\omega}$ is covariant,

$$\vec{L} = \overleftrightarrow{I}\,\vec{\omega}, \tag{29.26}$$

valid for any coordinate frame — including (and especially) on its diagonal basis. It's the tensorial nature of $\overleftrightarrow{I}$ that promotes it from a mere collection of numbers into a quantity with an objective physical reality not shared by its components.

Broadening (29.23) in a natural way, a rank-three tensor is defined as a three-index object which transforms as

$$A_{i'j'k'} = \sum_{\ell mn} R_{i'\ell}R_{j'm}R_{k'n}A_{\ell mn}. \tag{29.27}$$

The same is true for higher ranks: a tensor of rank r transforms with r rotation matrices — one for each vector index,

$$A_{i'j'k'\ell'\cdots} = \sum_{abcd\cdots} R_{i'a}R_{j'b}R_{k'c}R_{\ell'd}\cdots A_{abcd\cdots}. \tag{29.28}$$

Without question, the prospect of performing by hand a rotation on a high-rank tensor is not terribly inviting; happily, such opportunities are rare. But the larger impact of (29.28) is not as an instruction to calculate; rather, by establishing an equivalence class all transforming with the same R, it provides a means to implement covariance and comply with the Principle of Relativity.

Example 29.3 **Indices Do Not a Tensor Make**

Recording the temperature at regularly spaced locations on a three-dimensional cartesian grid produces an array of numbers T_{ijk}. But temperature is not in any way fundamentally connected to position; any mixture of these measurements has no fundamental meaning. So despite superficial appearances, the array T_{ijk} is not a tensor.

Similarly, as we saw back in Example 23.1, $\vec{y} = M\vec{x} + \vec{b}$ in N dimensions can be expressed as a linear transformation in $N + 1$ dimensions by adding a column and row to M,

$$\vec{y} = M\vec{x} + \vec{b} \longrightarrow \begin{pmatrix} \vdots \\ \vec{y} \\ \vdots \\ 1 \end{pmatrix} = \begin{pmatrix} \cdots & \cdots & \cdots & \vdots \\ \cdots & M & \cdots & \vec{b} \\ \cdots & \cdots & \cdots & \vdots \\ \hline 0 & \cdots & 0 & 1 \end{pmatrix} \begin{pmatrix} \vdots \\ \vec{x} \\ \vdots \\ 1 \end{pmatrix}. \tag{29.29}$$

Letting $\tilde{M}$ denote this augmented matrix, we can write this in index notation as

$$\tilde{y}_i = \sum_j \tilde{M}_{ij}\tilde{x}_j. \tag{29.30}$$

Despite the indices, this is *not* a tensor equation: rotations in $\mathbb{R}^{n+1}$ would improperly mix the augmented elements with those of the N-dimensional space.

Table 29.1 Tensors in N dimensions

	Components	Rank	Elements
Scalar S	S	0	1
Vector $\vec{V}$	V_i	1	N
Tensor $\overleftrightarrow{T}$	T_{ij}	2	N^2

Be careful not to confuse or conflate tensor rank with the dimensionality of the underlying space. A tensor's rank is specified by the *number* of indices; the dimension of the space is the *range* of values an index can take. Thus a tensor of rank r in N dimensions has N^r components (Table 29.1). A rank zero has one element in any N; it's a scalar. A vector can be construed as a one-dimensional array of N components. Similarly, a rank-two tensor can be thought of as a vector of vectors. A third-rank tensor is a vector whose elements are second-rank tensors, $A_{ijk} = (A_{ij})_k$; you can picture this as a three-dimensional matrix, a collection of ij-planes standing parallel with one another along the third k direction coming out of the page. This progression quickly becomes too much to picture; fortunately the most common tensors have rank $r \leq 2$. This is why, since $r = 0$ and 1 tensors have their own special names, in common parlance the word "tensor" most-often connotes a second-rank tensor.

The rank of a tensor field can be changed by taking derivatives. In particular, since $\vec{\nabla} \longrightarrow \frac{\partial}{\partial x_i}$ transforms as a vector (Problem 29.12), it changes the rank r of the field by 1. For instance, a scalar ϕ is mapped to the gradient $\partial_i \phi$ ($r = 0 \rightarrow 1$). A vector $\vec{v}$ can be mapped in two ways: to a scalar by the contraction of indices, $\sum_i \partial_i v_i$ ($1 \rightarrow 0$), or to a second-rank tensor with a dyadic product $T_{ij} = \partial_i v_j$ ($1 \rightarrow 2$). In fact, a Taylor expansion of a function $f(\vec{r})$ is essentially a tensor sum (see Example 6.17),

$$f(\vec{r} + \vec{a}) = \left(1 + \sum_i a_i \frac{\partial}{\partial x_i} + \frac{1}{2!} \sum_{ij} a_i a_j \frac{\partial}{\partial x_i} \frac{\partial}{\partial x_j} + \cdots \right) f(\vec{r}). \tag{29.31}$$

The zeroth-order term is a scalar, the first-order term introduces the vector $\partial_i f$, and the second-order term a second-rank symmetric tensor $\partial_i \partial_j f$. So analysis of a system around equilibrium naturally focuses on a second-rank tensor. This observation underlies the relevance of the rank-two tensors of stress and strain.

Example 29.4 **The Power of Covariance**

Though (29.31) is a standard result of multivariable calculus, let's explore it in the context of rotational covariance. Start with the one-dimensional Taylor series,

$$f(x + a) = \left(1 + a \frac{d}{dx} + \frac{a^2}{2!} \frac{d^2}{dx^2} + \cdots \right) f(x) = \sum_{n=0}^{\infty} \frac{1}{n!} \left(a \frac{d}{dx}\right)^n f(x). \tag{29.32}$$

This is the expansion we'd get in higher dimensions if $\vec{a}$ were aligned with the x-axis, $\vec{a} = a\hat{\imath}$. Indeed, we can write the operator suggestively as

$$a \frac{d}{dx} = (a\hat{\imath}) \cdot \left(\hat{\imath} \frac{d}{dx}\right), \tag{29.33}$$

acting on $f(x)$. But the dot product is a covariant expression; were we to rotate the basis, the operator would become $\vec{a} \cdot \vec{\nabla}$, giving the Taylor expansion the form

$$f(\vec{r} + \vec{a}) = \sum_{n=0}^{\infty} \frac{1}{n!} \left(\vec{a} \cdot \vec{\nabla} \right)^n f(\vec{r}), \tag{29.34}$$

which is (29.31). In other words, if we can find a covariant expression which we know holds in one frame, then — by virtue of it being a tensor expression — it must hold in all frames.

29.4 Tensors and Pseudotensors

The equivalence class definition of a tensor (29.28) pertains to general orthogonal transformations, without distinction between proper and improper rotations. In $\mathbb{R}^3$, recall that improper rotations can be expressed as a product of a proper rotation and the inversion or parity operator $P_{ij} = -\delta_{ij}$, (26.80). This allows discussion of the differences between proper and improper rotations to focus only on parity. Now since under parity our paradigmatic vector changes sign, $\vec{r} \xrightarrow{P} -\vec{r}$, all vectors in the equivalence class should *also* flip sign under inversion. Take velocity, for example: throw an object at a mirror and its image comes back toward you, so $\vec{v} \xrightarrow{P} -\vec{v}$. On the other hand, imagine a rotating top with its angular velocity $\vec{\omega}$ pointing into the mirror: since the top's image rotates the *same* way, $\vec{\omega}$ is not reversed under reflection, $\vec{\omega} \xrightarrow{P} +\vec{\omega}$. Thus, despite its behavior under rotations, $\vec{\omega}$ does not actually belong to the equivalence class. Vectors such as velocity which transform like $\vec{r}$ are called *polar* vectors; those like $\vec{\omega}$ which do not flip sign under inversion form a second equivalence class called *axial* vectors or *pseudovectors*. Both kinds occur regularly in physics: position, momentum, and electric fields are all polar vectors; angular velocity, angular momentum and magnetic fields are axial vectors.

Notice that these $\mathbb{R}^3$ axial vectors are defined by cross products of polar vectors. Although $\vec{A} \times \vec{B}$ is a one-index object which behaves as a vector under proper rotations, under parity

$$\vec{A} \times \vec{B} \xrightarrow{P} (-\vec{A}) \times (-\vec{B}) = +\vec{A} \times \vec{B}. \tag{29.35}$$

Even so, the dot product of two axial vectors still yields a scalar. On the other hand, the dot product of a polar vector $\vec{C}$ with an axial vector

$$s \equiv \vec{C} \cdot (\vec{A} \times \vec{B}) \tag{29.36}$$

is a *pseudoscalar*, since $s \to -s$ under parity. Magnetic flux $\vec{B} \cdot \vec{a}$ is a notable example.

What we're actually seeing here are the transformation properties of the ϵ_{ijk} hiding within a cross product. As a rank-three tensor, it *should* transform with three R's,

$$\sum_{\ell mn} R_{i\ell} R_{jm} R_{kn} \epsilon_{\ell mn}. \tag{29.37}$$

For $ijk = 123$ and its cyclic permutations, this is just the determinant of R; anti-cyclic permutations, however, give $-\det R$. All together then,

$$\sum_{\ell mn} R_{i\ell} R_{jm} R_{kn} \epsilon_{\ell mn} = \epsilon_{ijk} \det(R), \tag{29.38}$$

just as we found back in (4.42). Thus we discover what's going on: ϵ_{ijk} is only a tensor for proper rotations, $\det R = +1$. So it transforms not as in (29.27), but rather as a *pseudotensor*

$$\epsilon_{i'j'k'} = \det(R) \sum_{abc} R_{i'a} R_{j'b} R_{k'c} \epsilon_{abc}, \tag{29.39}$$

which subsumes both proper and improper rotations. More generally, a pseudotensor is an indexed object with transformation rule

$$A_{i'j'k'\ell'\ldots} = \det(R) \sum_{abcde\cdots} R_{i'a} R_{j'b} R_{k'c} R_{\ell'd} \cdots A_{abcd\cdots}. \tag{29.40}$$

Under parity, tensors of rank r transform as $(-1)^r$, one sign for each vector index; pseudotensors, however, transform as $(-1)^{r+1}$. So ϵ_{ijk} is a rank-three pseudotensor.

29.5 Tensor Invariants and Invariant Tensors

Though the components of a tensor are frame dependent, certain scalar combinations of components remain the same after a rotation. For vectors there's only one such invariant, the magnitude $v^2 = \sum_i v_i v_i$. For tensors of higher rank, we can find invariants by similarly summing away indices to produce scalars. We can contract the two indices of a second-rank tensor to get

$$\sum_{ij} T_{ij} \delta_{ij} = \sum_i T_{ii} = \mathrm{Tr}(T). \tag{29.41a}$$

But with more indices, there are more ways to produce scalars. For example,

$$\sum_{ijk} T_{ij} T_{jk} \delta_{ik} = \sum_{ij} T_{ij} T_{ji} = \mathrm{Tr}(T^2). \tag{29.41b}$$

In three dimensions, we can also use ϵ_{ijk} to form

$$\sum_{ijk\ell mn} \epsilon_{ijk} \epsilon_{\ell mn} T_{i\ell} T_{jm} T_{kn} = \frac{1}{3!} \det(T). \tag{29.41c}$$

These invariants should be familiar from our discussion of similarity transformations. But you might wonder: if $\mathrm{Tr}(T)$ and $\mathrm{Tr}(T^2)$ are invariant, then surely so is $\mathrm{Tr}(T^3)$? Which it is — but it turns out that a second-rank tensor in N dimensions has only N *independent* invariants. So although $\mathrm{Tr}(T^3)$ is invariant, it's not independent of the three given in (29.41).

You might reasonably argue that the scalar invariants in (29.41) reflect not properties of T, but rather properties of combinations of T with δ_{ij} and ϵ_{ijk}. But consider the effect of rotations on the Kronecker δ:

$$\sum_{ij} R_{k'i} R_{\ell'j} \delta_{ij} = \sum_i R_{k'i} R_{\ell'i} = \sum_i R_{k'i} R_{i\ell'}^T = \delta_{k'\ell'}. \tag{29.42a}$$

In other words, the array of numbers δ is unchanged. This is even more transparent in matrix notation,

$$\mathbb{1}' = R\,\mathbb{1}\,R^T = \mathbb{1}, \tag{29.42b}$$

which merely says that the identity is invariant under rotations. So there's no reason to ever use notation like δ' or $\mathbb{1}'$. Tensors such as δ_{ij} which have precisely the same elements in all coordinate systems are called *invariant* or *isotropic* tensors. Similarly, (29.39) shows that if we restrict ourselves to proper rotations, ϵ_{ijk} is also an invariant tensor.

It's the invariance of δ_{ij} and ϵ_{ijk} which renders the familiar index notation expressions for trace and determinant covariant. Thus the scalars of (29.41) are indeed properties of $\overset{\leftrightarrow}{T}$ alone.

Note well the distinction between a scalar invariant and an invariant tensor. A scalar is a *single* element unaffected by rotation, whereas in N-dimensional space, δ has N^2 elements, and ϵ has N^3 (most of which happen to be zero). There are no invariant tensors of rank one, and all invariant tensors of rank two are proportional to the identity. In three dimensions, invariant tensors of rank three are proportional to ϵ_{ijk}, whereas those of rank four have the form

$$T_{ijkl} = \alpha\,\delta_{ij}\,\delta_{kl} + \beta\,\delta_{ik}\,\delta_{jl} + \gamma\,\delta_{il}\,\delta_{jk}, \tag{29.43}$$

for constants α, β, γ.

Example 29.5

A function of invariant tensors, such as the expression $\epsilon_{mik}\epsilon_{mj\ell}$, is also invariant. In fact, as a fourth-rank tensor, the familiar epsilon identity

$$\sum_m \epsilon_{mik}\epsilon_{mj\ell} = \delta_{ij}\,\delta_{kl} - \delta_{il}\,\delta_{jk} \tag{29.44}$$

is just a special case of (29.43) with $\alpha = -\gamma = 1$, and $\beta = 0$. These values are a clear reflection of the Levi-Civita symbol's antisymmetry.

Example 29.6 Tensor Duals

In (29.10), we repackaged a vector $\vec{\varphi}$ as an antisymmetric tensor $\overset{\leftrightarrow}{\phi}$, claiming that φ_k and $\phi_{ij} = -\sum_k \epsilon_{ijk}\varphi_k$ are interchangeable representations of the underlying rotation. This assertion is founded on the invariance of the Levi-Civita epsilon.

In fact, any $\mathbb{R}^3$ vector $\vec{B}$ can be rewritten as a second-rank antisymmetric tensor $\overset{\leftrightarrow}{T}$,

$$T_{ij} = \sum_k \epsilon_{ijk} B_k \longrightarrow \begin{pmatrix} 0 & B_3 & -B_2 \\ -B_3 & 0 & B_1 \\ B_2 & -B_1 & 0 \end{pmatrix}. \tag{29.45}$$

Inverting gives (Problem 29.25)

$$B_i = \frac{1}{2}\sum_{jk} \epsilon_{ijk} T_{jk}. \tag{29.46}$$

Since ϵ_{ijk} is an invariant tensor, these are covariant expressions under proper rotations. As a result, $\overset{\leftrightarrow}{T}$ is an equivalent way to package the components of B; the two are said to be "duals" of one another. The cross product is an excellent example:

$$(\vec{A} \times \vec{B})_i = \sum_{jk} \epsilon_{ijk} A_j B_k = \sum_j A_j \left(\sum_k \epsilon_{ijk} B_k \right)$$

$$\equiv \sum_j T_{ij} A_j. \tag{29.47}$$

So in the context of tensor analysis, the cross product of two vectors is actually the result of a second-rank tensor acting on a vector.

BTW 29.1 Irreducible Tensors

Applying a rotation to a tensor mixes up its components. But just how "mixed" are the new components? Does every one of the elements mix with every other? Well, sure — after all, that's what generally happens with the elements of a vector under rotation. But for higher-rank tensors there are *combinations* of components that mix just amongst themselves.

The key is to use the invariant tensors δ_{ij} and ϵ_{ijk} to separate out these combinations. For instance, recall that any second-rank tensor can be divided into symmetric and antisymmetric pieces, $T_{ij} = S_{ij} + A_{ij}$, where

$$S_{ij} = \frac{1}{2}\left(T_{ij} + T_{ji}\right) = \frac{1}{2}\sum_{\ell m}\left(\delta_{i\ell}\delta_{jm} + \delta_{im}\delta_{j\ell}\right) T_{\ell m}, \tag{29.48}$$

and

$$A_{ij} = \frac{1}{2}\left(T_{ij} - T_{ji}\right)$$
$$= \frac{1}{2}\sum_{\ell m}\left(\delta_{i\ell}\delta_{jm} - \delta_{im}\delta_{j\ell}\right) T_{\ell m} = \frac{1}{2}\sum_{k\ell m}\epsilon_{kij}\epsilon_{k\ell m}T_{\ell m}. \tag{29.49}$$

That these can be isolated using invariant tensors means that they are separately covariant pieces of T. Similarly, since $\text{Tr}(T) = \sum_{k\ell} T_{k\ell}\delta_{k\ell}$ is a scalar, this combination of tensor components is unchanged under rotation. All together, then, we can decompose T_{ij} as

$$T_{ij} = A_{ij} + S_{ij}$$
$$= \frac{1}{3}\text{Tr}(T)\delta_{ij} + \frac{1}{2}\left(T_{ij} - T_{ji}\right) + \frac{1}{2}\left(T_{ij} + T_{ji} - \frac{2}{3}\text{Tr}(T)\delta_{ij}\right). \tag{29.50}$$

Thus any second-rank tensor can be expanded as the sum of a term proportional to the identity, an antisymmetric tensor, and a traceless symmetric tensor. The importance of this decomposition is that each of these individual pieces transform *within themselves*: the subspaces of traceless symmetric tensors, antisymmetric tensors, and $\mathbb{1}\times$trace do not mix with one another under rotations. We have decomposed T_{ij} into its *irreducible* components — in other words, further tensor reduction is not possible. All of which implies that each piece represents different physical aspects of the system. For the stress tensor σ_{ij}, the scalar piece is the pressure, the traceless symmetric piece the shear; as a symmetric tensor, it doesn't have an antisymmetric piece. Strain too is symmetric — and we used this property to isolate it in (29.7)–(29.8) by identifying and discarding antisymmetric rigid rotation.

Note that all nine elements of T_{ij} have been accounted for: the trace is zero rank and has one component, the antisymmetric tensor is equivalent to a first-rank dual with three elements, and the rank-two symmetric traceless tensor has five elements. We've fully decomposed a second-rank tensor into a collection of tensors of rank ≤ 2. The invariant tensors δ_{ij} and ϵ_{ijk} ensure that the sum (29.50) is over tensors of the same rank, as required of tensor sums. The decomposition for higher rank r works similarly as a sum of tensor with ranks $j = 0$ to r, each with $2j+1$ elements,

$$T^{(r)} = \sum_{j=0}^{r} T^{(j)}. \tag{29.51}$$

This decomposition is exactly what happens with angular momentum in quantum mechanics: the states decompose into irreducible representations j ("irreps") of rotations, each with $2j+1$ elements (Biedenharn and Louck, 1981; Sakurai and Napolitano, 2021). That angular momentum is the generator of rotations (Section 26.6) is hardly coincidental.

Problems

29.1 Explain which of the following are tensors:

(a) a spreadsheet of student grades
(b) a rotation matrix $R_{\hat{n}}(\varphi)$
(c) the average temperature in the room
(d) a two-dimensional array χ which maps an applied electric field $\vec{E}$ into a crystal's polarization $\vec{P}$.

29.2 A square ceramic tile is divided into an $n \times n$ cartesian grid. The temperature of each cell is measured, and the data arrayed in an $n \times n$ matrix T_{ij}. Is T_{ij} a tensor? Explain.

29.3 Is the component of a vector a scalar? Explain.

29.4 For the inclined plane example of Section 29.1, use rotations to determine the force equation in the unprimed frame from the primed frame. Show that the block has the same speed $v(t)$ in both bases. (Assume acceleration from rest.)

29.5 For vectors $\vec{v}$, $\vec{w}$ and second-rank tensors $\overset{\leftrightarrow}{T}$, $\overset{\leftrightarrow}{U}$, explain what type of tensor each of the following is:

(a) $\vec{w} \cdot \overset{\leftrightarrow}{T}$
(b) $\vec{w} \cdot \overset{\leftrightarrow}{T} \cdot \vec{v}$
(c) $\overset{\leftrightarrow}{T} \cdot \overset{\leftrightarrow}{U}$
(d) $\overset{\leftrightarrow}{T}\vec{v}$ (no dot product)
(e) $\overset{\leftrightarrow}{T}\overset{\leftrightarrow}{U}$ (no dot product).

29.6 Explain which of the following are tensors and which, pseudotensors:

(a) torque
(b) angular velocity ω
(c) the precession $\vec{\Omega}$ of a gyroscope
(d) an electric field
(e) a magnetic field
(f) the Lorentz force $q\vec{E} + \vec{v} \times \vec{B}$
(g) $\vec{r} \times (\vec{E} \times \vec{B})$
(h) $\vec{r} \cdot (\vec{E} \times \vec{B})$.

29.7 How many components does a general fourth-rank tensor A_{ijkl} have in two dimensions?

29.8 Given tensors $\overset{\leftrightarrow}{S}$ and $\overset{\leftrightarrow}{T}$ and scalars a and b, show that $\overset{\leftrightarrow}{W} = a\overset{\leftrightarrow}{S} + b\overset{\leftrightarrow}{T}$ is also a tensor.

29.9 Show that the trace of a second-rank tensor is a scalar.

29.10 Consider a rank-two tensor in two dimensions, T_{ij}. Use the invariance of ϵ_{ij}, together with the elements of the proper rotation matrix, to explicitly verify that $\sum_{ij} \epsilon_{ij} T_{ij}$ is a scalar.

29.11 (a) Show that a tensor T_{ij} can always be written as $T_{ij} = S_{ij} + A_{ij}$, where $S_{ij} = S_{ji}$ and $A_{ij} = -A_{ji}$.
(b) Show that a tensor which is symmetric or antisymmetric in one frame has that symmetry in all frames. Do it twice: once using index notation, and once using matrix notation.

29.12 We saw in Section 12.1 that the gradient operator $\vec{\nabla} \longleftrightarrow \frac{\partial}{\partial x_i}$ is a vector. We want to see how this manifests in the framework of cartesian tensors.

(a) Given that we know the divergence of a vector is a scalar, show by direct calculation that the gradient transforms as a vector in $\mathbb{R}^2$.
(b) Use index notation to abstract and generalize this to cartesian coordinates in n-dimensions. Where in your calculation must you assume that the rotation matrix R is not a function of position x_i?

(c) Rather than use our knowledge of the divergence, use the chain rule on the operator $\frac{\partial}{\partial x_i}$ to show that it is, indeed, a vector.

29.13 We asserted that since both force and area are known vectors, (29.2) can be a covariant statement only if stress σ_{ij} is a second-rank tensor. Use the tensor definition (29.28) to verify this claim.

29.14 Use the known tensorial ranks of stress and strain to show that the elasticity tensor of (29.7) has rank four.

29.15 Consider the dyadic form $\vec{\nabla}\vec{v}$. Show that:

(a) $\vec{\nabla} \cdot \vec{v} = \mathrm{Tr}(\vec{\nabla}\vec{v})$
(b) $\vec{\nabla} \times \vec{v}$ is twice the dual of $\vec{\nabla}\vec{v}$.

29.16 Use the divergence theorem to show that the force density on a body is the divergence of the stress tensor,

$$\vec{f} \equiv \vec{\nabla} \cdot \overset{\leftrightarrow}{\sigma}. \tag{29.52}$$

29.17 Archimedes' Principle
A body of volume V in a uniform gravitational field $-g\hat{k}$ is submerged (partially or totally) in a fluid of density ρ. Starting from (29.2) for the force on an area element of the body, derive Archimedes' Principle: the total buoyant force on the body exerted by the fluid is equal to the weight of the displaced fluid. (Think physically about what the force density (29.52) must be.)

29.18 In Problem 16.6 we looked at energy conservation in the context of electromagnetism; here we'll consider momentum. This requires more care, because the conserved quantity is a vector rather than a scalar. Since $\vec{F} = d\vec{p}/dt$, we start with the Lorentz force density on charges, $\vec{f} = \rho(\vec{E} + \vec{v} \times \vec{B})$.

(a) Use Maxwell's equations and the vector identity for $\vec{\nabla}(\vec{A} \cdot \vec{B})$ to verify that the Lorentz force density is

$$\vec{f} = \vec{\nabla} \cdot \overset{\leftrightarrow}{\sigma} - \frac{1}{c^2}\frac{\partial \vec{S}}{\partial t},$$

where $\vec{S}$ is the Poynting vector (Problem 16.6), and

$$\sigma_{ij} = \sigma_{ji} = \epsilon_0\left(E_iE_j - \frac{1}{2}\delta_{ij}|\vec{E}|^2\right) + \frac{1}{\mu_0}\left(B_iB_j - \frac{1}{2}\delta_{ij}|\vec{B}|^2\right) \tag{29.53}$$

is the *Maxwell stress tensor*. (See also BTW 12.1.)

(b) Show that the total force on charges contained within a volume V can be expressed in differential form as a continuity equation

$$\vec{\nabla} \cdot (-\overset{\leftrightarrow}{\sigma}) = -\frac{\partial}{\partial t}\left(\vec{\mathcal{P}}_{\text{charge}} + \vec{\mathcal{P}}_{\text{fields}}\right), \tag{29.54}$$

where

$$\vec{p} = \int_V \vec{\mathcal{P}}_{\text{charge}}\, d\tau \quad \text{and} \quad \vec{\mathcal{P}}_{\text{fields}} = \mu_0\epsilon_0\vec{S}.$$

Interpret (29.53) physically.

29.19 The upper plate of a parallel plate capacitor at $z = d/2$ has surface charge density α, while the lower plate at $z = -d/2$ has charge density $-\alpha$. [We use α rather than the traditional σ to avoid confusion with the stress tensor $\overset{\leftrightarrow}{\sigma}$.]

(a) Find all the components of the Maxwell stress tensor $\overset{\leftrightarrow}{\sigma}$ (Problem 29.18).

(b) Find the force per unit area of the top plate.

(c) Find the momentum per area per time crossing the xy-plane between the plates.

29.20 The standard demonstration that the stress tensor is symmetric examines the equilibrium-restoring torques generated by strain within the body — and that, in equlilbrium, the angular momentum is constant. Assume the stresses act on a closed surface of a small volume.

(a) Use $dF_i = \sum_j \sigma_{ij} da_j$ to write an expression for the torque $d\tau_i$ around some origin.

(b) Use the divergence theorem to show that the torque volume density — equivalently, the change in angular momentum per unit volume — is

$$\sum_{jk} \epsilon_{ijk}\, r_j f_k + \sum_{kj} \epsilon_{ijk} \sigma_{kj},$$

where $\vec{f}$ is the force volume density (29.52) given in Problem 29.16. (You may find also (15.8) to be helpful.) The first term is the angular momentum generated by acceleration a distance $\vec{r}$ from the origin; the second is the contribution from internal stress.

(c) The choice of origin is arbitrary — but since the first term is not relevant to stress and strain, the intelligent choice is $\vec{r} = 0$. Show that this leads to the conclusion $\sigma_{ij} = \sigma_{ji}$.

29.21 As an alternative derivation of the strain tensor, consider neighboring positions $\vec{r}$ and $\vec{r} + d\vec{r}$ within an object that are deformed to $\vec{r} + \vec{u}(\vec{r})$ and $\vec{r} + d\vec{r} + \vec{u}(\vec{r} + d\vec{r})$. These strained positions are separated by a distance Δ. Show that to lowest order in $\partial u/\partial x$,

$$\Delta^2 = \sum_{ij} \left(\delta_{ij} + 2\varepsilon_{ij} \right) dx_i dx_j,$$

where

$$\varepsilon_{ij} = \varepsilon_{ji} \equiv \frac{1}{2} \left(\frac{\partial u_i}{\partial x_j} + \frac{\partial u_j}{\partial x_i} \right)$$

is the strain tensor. Show that by including the $\frac{1}{2}$ in the definition, the deformation scales as ε. What happened to the rigid rotation in (29.9)?

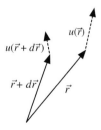

29.22 (a) Show that a rank-one tensor has only one invariant in any number of dimensions.

(b) Find all the *linear* invariants of a third-rank tensor T_{ijk}.

29.23 Show that the individual pieces in (29.50) are irreducible — that is, that under rotation they transform separately from one another.

29.24 (a) Verify that the cross product transforms as a vector under rotation.

(b) Show that the cross product $\vec{A} \times \vec{B}$ can be expressed as an antisymmetric second-rank tensor $\overset{\leftrightarrow}{C}$. Broaden this result to higher dimensions to explain why only for $N = 3$ is the cross product of two vectors a vector.

29.25   Let $T_{ij} = \sum_k \epsilon_{ijk} B_k$ be duals of one another.

(a) Show that $B_i = \frac{1}{2} \sum_{jk} \epsilon_{ijk} T_{jk}$. Are both B and T pseudotensors? Explicitly verify this for the matrix of (29.45).

(b) Show that if T_{ij} is not antisymmetric, only its antisymmetric part contributes to $\vec{B}$.

(c) Show that the dual of a rotated vector is equivalent to rotating the dual tensor.

30 Beyond Cartesian

In physics, the cognitive leap from scalars to vectors involves assigning a direction to a measurement or quantity. When you consider that tensors merely require assigning two or more directions, it may seem that the subject is much ado about nothing. From an intuition standpoint, one might argue this is largely true. But from a technical perspective, consider that we've only discussed cartesian tensors in $\mathbb{R}^n$ — and as we've seen many times, things are often simplest in cartesian coordinates. Moreover, the Principle of Relativity disallows working only in cartesian coordinates: if physical law is to be expressed in a coordinate-free fashion, tensors must appertain both to curvilinear coordinates with their space-dependent basis vectors, as well as to physically-relevant non-Euclidean spaces in which Pythagoras no longer reigns. This is the challenge addressed in this chapter.

The distinctive simplifying features of cartesian coordinates are their constant basis vectors and trivial scale factors (see Section 2.2). These underlie their signature virtue: the ability to manipulate vector components without the need to keep track of basis vectors — for instance, $\vec{v} \cdot \vec{w} = \sum_i v_i w_i$. This won't work in spherical coordinates, since $\vec{v} \cdot \vec{w} \neq v_r w_r + v_\theta w_\theta + v_\phi w_\phi$. But the relative simplicity of cartesian systems is compelling — and we can replicate it for non-cartesian systems by accommodating some of their distinct features. We start with a presentation and analysis of a non-cartesian coordinate system in $\mathbb{R}^2$ which displays these new features. Then we'll see how they can be incorporated with minor modifications into our usual index notation.

30.1 A Sheared System

The components c_i of a vector $\vec{A} = \sum_i c_i \hat{e}_i$ on a basis $\{\hat{e}_i\}$ can be equivalently visualized as parts of $\vec{A}$ which lie *along* the ith direction, as well as orthogonal projections of $\vec{A}$ *onto* the ith basis vector. But in fact, these interpretations — "along" and "projection onto" — are equivalent only for orthogonal bases.

Consider a sheared coordinate system in which normalized basis vectors $\hat{e}_u, \hat{e}_v$ make angles α and β with the cartesian axes, so that $\alpha + \beta + \varphi = \pi/2$ as shown in Figure 30.1. Though not orthogonal, the $\hat{e}_i$ are linearly independent — as they must be to form a basis; representing them on the standard cartesian column basis, they are

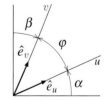

Figure 30.1 The sheared basis $\{\hat{e}_u, \hat{e}_v\}$.

$$\hat{e}_u \to \begin{pmatrix} \cos\alpha \\ \sin\alpha \end{pmatrix} \qquad \hat{e}_v \to \begin{pmatrix} \sin\beta \\ \cos\beta \end{pmatrix}. \qquad (30.1)$$

The orthogonal projections of an arbitrary vector $\vec{A}$ onto this basis are determined in familiar fashion,

$$A_u \equiv \hat{e}_u \cdot \vec{A} \qquad A_v \equiv \hat{e}_v \cdot \vec{A}. \qquad (30.2)$$

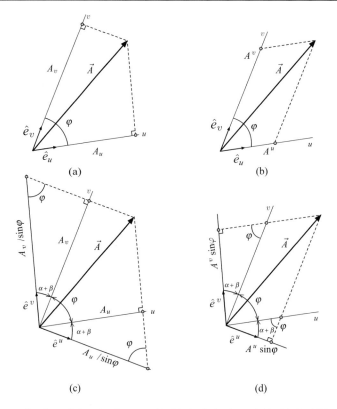

Figure 30.2 Orthogonal and parallel projections on: (a), (b) the $\hat{e}_a$ basis; and (c), (d) the $\hat{e}^a$ basis.

But as revealed in Figure 30.2a, $\vec{A} \neq A_u\hat{e}_u + A_v\hat{e}_v$. What is needed is the decomposition of $\vec{A}$ *along* the $\vec{e}_i$, as shown in Figure 30.2b. This parallel projection supports the vector expansion

$$\vec{A} = A^u\hat{e}_u + A^v\hat{e}_v, \tag{30.3}$$

where superscripts are used to distinguish these components from those in (30.2). Still, as linearly independent components, A_u and A_v can also be used to reconstruct $\vec{A}$. To see how, symmetrically extend the orthogonal projections in Figure 30.2a as shown in Figure 30.2c. This allows identification of the "dual" basis $\{\hat{e}^u, \hat{e}^v\}$. Then

$$\vec{A} = \frac{1}{\sin\varphi}\left(A_u\hat{e}^u + A_v\hat{e}^v\right) = A_u\vec{e}^u + A_v\vec{e}^v, \tag{30.4}$$

where we've absorbed the factor of $\sin\varphi$ to define the unnormalized basis vectors $\vec{e}^a \equiv \hat{e}^a/\sin\varphi$. This makes (30.4) resemble (30.3). By extending the projections in Figure 30.2b as shown in Figure 30.2d, the $\vec{e}^a$ basis can be identified by orthogonal projection as well.

Although there are an infinite number of bases, $\{\hat{e}_a\}$ and $\{\vec{e}^a\}$ are intimately related to one another. Just as $\hat{e}_u, \hat{e}_v$ are inclined inward by α, β from the cartesian, the $\vec{e}^u$ and $\vec{e}^v$ axes are inclined outward by β, α. So even though neither is an orthogonal basis, they are orthogonal to one another,

$$\hat{e}_u \cdot \vec{e}^v = \vec{e}^u \cdot \hat{e}_v = 0. \tag{30.5a}$$

Moreover, despite the $\vec{e}^a = \hat{e}^a/\sin\varphi$ not being normalized,

$$\hat{e}_u \cdot \vec{e}^u = \hat{e}_v \cdot \vec{e}^v = 1, \tag{30.5b}$$

since $\cos(\alpha + \beta) = \sin\varphi$. Using (30.1) we find

$$\vec{e}^{\,u} \rightarrow \frac{1}{\cos(\alpha+\beta)}\begin{pmatrix}\cos\beta\\-\sin\beta\end{pmatrix} \qquad \vec{e}^{\,v} \rightarrow \frac{1}{\cos(\alpha+\beta)}\begin{pmatrix}-\sin\alpha\\\cos\alpha\end{pmatrix}. \tag{30.6}$$

These bases are complementary, one neither more fundamental than the other. Complementary as well are A^a and A_a, which are the components *along* $\hat{e}_a$ and $\vec{e}^{\,a}$, respectively,

$$\vec{A} = A^u\hat{e}_u + A^v\hat{e}_v = A_u\vec{e}^{\,u} + A_v\vec{e}^{\,v}, \tag{30.7}$$

as well as orthogonal projections *onto* the other basis,

$$A_a = \hat{e}_a \cdot \vec{A} \qquad\qquad A^a = \vec{e}^{\,a} \cdot \vec{A}. \tag{30.8}$$

The relationship between these components can be found using (30.5)–(30.8),

$$A_u \equiv \vec{e}_u \cdot \vec{A} = A^u + A^v \cos\varphi$$

$$A_v \equiv \vec{e}_v \cdot \vec{A} = A^u \cos\varphi + A^v, \tag{30.9}$$

as a little trigonometry confirms. For $\alpha = \beta = 0, \varphi = \pi/2$, they're all the same: $A^i = A_i, \hat{e}_i = \vec{e}^{\,i}$.

30.2 The Metric

The existence of dual bases is a general feature of non-cartesian systems. But whereas the relationship between the sheared bases in Section 30.1 can be visualized with the aid of a little trignometry, in others the relationship is not so transparent. What we need then is a covariant index notation which can be applied to all bases in a fairly straightforward manner.

We start with the basic notion that the scalar distance between infinitesimally-separated points ds is a system-independent measure. We want to retain the simple cartesian form $d\vec{s} = \sum_i \hat{e}_i dx_i$ — but having recognized the existence of dual bases, also adopt the subscript/superscript indices introduced in the sheared system. So for arbitrary coordinates $(u^1, u^2, u^3, \ldots)$, we assert a cartesian-esque line element

$$d\vec{s} = \sum_a \vec{e}_a\, du^a. \tag{30.10}$$

(Here we use indices $a, b, c, \ldots$ in order to reserve $i, j, k, \ldots$ to denote cartesian bases and components.) This form dictates that any non-trivial scale factors must be hidden in the basis vectors — just as we did in (30.4). We can gather them neatly together into a single entity by calculating the distance-squared between neighboring points,

$$ds^2 = d\vec{s} \cdot d\vec{s} = \left(\sum_a \vec{e}_a\, du^a\right) \cdot \left(\sum_b \vec{e}_b\, du^b\right)$$

$$= \sum_{ab} (\vec{e}_a \cdot \vec{e}_b)\, du^a\, du^b \equiv \sum_{ab} g_{ab}\, du^a\, du^b. \tag{30.11}$$

The elements of the symmetric matrix

$$g_{ab} = g_{ba} \equiv \vec{e}_a \cdot \vec{e}_b \tag{30.12}$$

are the components of the tensor $\overleftrightarrow{g}$, called the *metric*. We know it's a tensor because it maps the two vectors du^a and du^b into a scalar, which can only be done by contracting them with a second-rank downstairs tensor. It's called the metric because it determines the *measure* of distance ds between neighboring points.

Comparison of (30.11) with the cartesian expression $ds^2 = dx^2 + dy^2 + dz^2$ reveals that the cartesian metric is just the identity

$$g_{ab} \longrightarrow \delta_{ij} = \begin{pmatrix} 1 & & \\ & 1 & \\ & & 1 \end{pmatrix}. \tag{30.13}$$

It hardly seems worth the effort — and is why no distinction is made between upstairs and downstairs cartesian indices. But the very same distance increment in spherical coordinates is

$$ds^2 = dr^2 + r^2\, d\theta^2 + r^2\, \sin^2\theta\, d\phi^2, \tag{30.14}$$

so that g_{ab} is the diagonal matrix

$$g_{ab} \longrightarrow \begin{pmatrix} 1 & & \\ & r^2 & \\ & & r^2 \sin^2\theta \end{pmatrix}. \tag{30.15}$$

The metric is where the scale factors reside; using it will allow us to achieve our goal of calculating using vector components without the explicit use of basis vectors.

Clearly, the precise form of g_{ab} depends on the coordinates. Nor is it necessarily diagonal. And as we've just seen, it need not be a matrix of constants; in fact, the metric is usually a function of position $g_{ab}(\vec{r})$. This means that despite being a symmetric tensor, in general it can only be diagonalized at a single point. The $\mathbb{R}^n$ metric is the notable exception in that it can be rendered diagonal everywhere. Moreover, as we'll see in Section 30.5, the same change of variables which maps cartesian into spherical coordinates also takes (30.13) into (30.15). They are different representations of the same tensor $\overleftrightarrow{g}$.

Example 30.1 The Minkowski Metric

The Minkowski spacetime metric is conventionally denoted $\eta_{\mu\nu}$, with spacetime coordinates denoted with Greek indices running from 0 to 3. The relevant spacetime distance is called the invariant interval

$$ds^2 = c^2 dt^2 - d\vec{r}^2 \equiv \sum_{\mu\nu} \eta_{\mu\nu}\, dx^\mu dx^\nu, \tag{30.16}$$

where $x^0 = ct$. In cartesian spatial coordinates,

$$\eta_{\mu\nu} \equiv \begin{pmatrix} 1 & & & \\ & -1 & & \\ & & -1 & \\ & & & -1 \end{pmatrix}. \tag{30.17}$$

Unlike the $\mathbb{R}^n$ metric, $\eta_{\mu\nu}$ cannot be transformed into the identity. Its *signature* — the difference between the number of its positive and negative elements — is a defining attribute of Minkowski space. The metric encodes this essential feature.

Example 30.2 The Shear Metric

We can determine the metric for the sheared basis from Figure 30.1,

$$g_{ab} \rightarrow \begin{pmatrix} \vec{e}_u \cdot \vec{e}_u & \vec{e}_u \cdot \vec{e}_v \\ \vec{e}_v \cdot \vec{e}_u & \vec{e}_v \cdot \vec{e}_v \end{pmatrix} = \begin{pmatrix} 1 & \cos\varphi \\ \cos\varphi & 1 \end{pmatrix}, \tag{30.18}$$

where $\cos\varphi = \sin(\alpha + \beta)$. Expressed in terms of a line element,

$$ds^2 = \sum_{ab} g_{ab}\, du^a du^b = (du^2 + 2\cos\varphi\, du\, dv + dv^2), \tag{30.19}$$

which is the law of cosines. (Note that φ is a parameter of the basis, not a coordinate.) The non-zero off-diagonal elements, or equivalently the non-vanishing $du\, dv$ term, indicates that the coordinates are not orthogonal; the factor of 2 in the cross term reflects the symmetry of the metric, $g_{ab} = g_{ba}$. And though the metric may not look Euclidean, it is. After all, we're still in $\mathbb{R}^2$.

The norm of a vector follows the template in (30.11):

$$\|\vec{A}\|^2 = \left(\sum_a A^a \hat{e}_a\right) \cdot \left(\sum_b A^b \hat{e}_b\right) = \sum_{ab} g_{ab} A^a A^b. \tag{30.20}$$

Applied to the sheared system, (30.9) correctly returns the law of cosines,

$$\|\vec{A}\|^2 = \sum_{ab} g_{ab} A^a A^b = g_{uu} A^u A^u + g_{uv} A^u A^v + g_{vu} A^v A^u + g_{vv} A^v A^v$$

$$= \left(A^u\right)^2 + 2\cos\varphi A^u A^v + \left(A^v\right)^2. \tag{30.21}$$

30.3 Upstairs, Downstairs

The expression (30.20) for the norm generalizes to the scalar product of two vectors

$$\vec{A} \cdot \vec{B} = \sum_{ab} g_{ab} A^a B^b. \tag{30.22}$$

This suggests a way to further streamline the notation. If we define

$$A_b = \sum_a g_{ab} A^a \qquad\qquad B_a = \sum_b g_{ab} B^b \tag{30.23}$$

with subscript indices, then the inner product can be written simply as

$$\vec{A} \cdot \vec{B} = \sum_a A^a B_a = \sum_b A_b B^b. \tag{30.24}$$

In other words, (30.22) can equally be interpreted as the metric g_{ab} first acting to lower the index of B^b before contracting with A^a or vice-versa. We can use (30.9) to verify this for the norm in the sheared system:

$$\vec{A} \cdot \vec{A} = \sum_a A^a A_a = A^u \left(A^u + A^v \cos\varphi\right) + A^v \left(A^u \cos\varphi + A^v\right)$$

$$= \left(A^u\right)^2 + 2\cos\varphi A^u A^v + \left(A^v\right)^2. \tag{30.25}$$

The beauty of (30.24) is how closely it resembles index notation for cartesian systems; the only cost is the introduction of an "upstairs/downstairs" distinction in the placement of indices. A fundamental feature of this notation is that to achieve a covariant result, index sums must always be over one subscript and one superscript. The scalar $\vec{A} \cdot \vec{B}$ is the simplest covariant example, with no net indices remaining after the sum. In an expression with net indices, they must match in elevation on both sides of an equation. So an expression like

$$A_a = \sum_b M^{ab} B_b \qquad \text{[incorrect]} \qquad (30.26)$$

is invalid, since the leftover a on the right is upstairs, whereas the left side requires a downstairs index. On the other hand,

$$A_a = \sum_b M_{ab} B^b \qquad (30.27)$$

is acceptable — the dual nature of the bases and components compensating to ensure a covariant result. A summation requires one vector component of each type — though it's immaterial which is which. To see this, we adapt (30.23) to define $M_{ac} = \sum_b g_{bc} M_a{}^b$ and $B_b = \sum_c g_{bc} B^c$; then

$$A_a = \sum_c M_{ac} B^c = \sum_b M_a{}^b B_b, \qquad (30.28)$$

where the index spacing on M distinguishes the row (left) index from the column (right) index. A summation without balanced upstairs/downstairs indices is almost certainly wrong.

Although there's no way to avoid all those pesky summations, we can tidy things up a bit. If you peruse our use of index notation going back to Chapter 4, you'll notice that summations are always over an index which appears twice in a product. In fact, one never has an index appearing more than twice, and a single index is rarely summed over. The *Einstein summation convention* leverages these observations to dispense with the summation sign Σ altogether. Thus (30.27) becomes

$$A_a = M_{ab} B^b, \qquad (30.29)$$

in which the sum is *implied* by the repeated dummy index — always with one upstairs, the other downstairs. We'll use this convention for the remainder of the chapter.

As a first example, let's calculate the inverse of the metric. Of course, g_{ab} must be invertible, since one has to be able to go back and forth between upstairs and downstairs bases and components. So in keeping with our modified index notation, there must be a matrix g^{ab} such that

$$A^a = g^{ab} A_b. \qquad \text{[Implied sum over } b] \qquad (30.30)$$

It's not hard to verify that g_{ab} and g^{ab} are inverses of one another:

$$\begin{aligned} \vec{A} \cdot \vec{B} = A^a B_a &= \left(g^{ab} A_b \right) \left(g_{ac} B^c \right) \\ &= \left(g^{ab} g_{ac} \right) A_b B^c = A_b B^b, \end{aligned} \qquad (30.31)$$

and so as expected

$$g^{ab} g_{ac} = g_b{}^c \equiv \delta_b^c. \qquad (30.32)$$

Referring back to (30.12) indicates this to be the metric-tensor version of the inter-basis orthonormality we saw in (30.5),

$$g_b^c = \vec{e}_b \cdot \vec{e}^{\,c} = \delta_b^c, \qquad (30.33)$$

implying that the metric also raises and lowers the index on basis vectors (Problem 30.5):

$$\vec{e}_a \equiv g_{ab}\vec{e}^b \qquad\qquad \vec{e}^a = g^{ab}\vec{e}_b. \tag{30.34}$$

Example 30.3 **Upstairs/Downstairs in Spherical Coordinates**

Using the template (30.10), the downstairs basis in spherical coordinates can be read off from the line element

$$d\vec{s} = \hat{r}\,dr + \hat{\theta}\,rd\theta + \hat{\phi}\,r\sin\theta\,d\phi \tag{30.35}$$

to get

$$\vec{e}_r = \hat{r} \qquad \vec{e}_\theta = r\hat{\theta} \qquad \vec{e}_\phi = r\sin\theta\,\hat{\phi}. \tag{30.36}$$

We can also extract the metric directly from $d\vec{s}\cdot d\vec{s}$ to rediscover (30.15),

$$g_{ab} \rightarrow \begin{pmatrix} 1 & & \\ & r^2 & \\ & & r^2\sin^2\theta \end{pmatrix}. \tag{30.37}$$

The usual convention leaves upstairs vector components A^a free of scale factors in order that they most closely resemble cartesian components. Then using the metric, the downstairs components can be found directly:

$$A_r = g_{ra}A^a = g_{rr}A^r + g_{r\theta}A^\theta + g_{r\phi}A^\phi = A^r$$
$$A_\theta = g_{\theta a}A^a = g_{\theta r}A^r + g_{\theta\theta}A^\theta + g_{\theta\phi}A^\phi = r^2 A^\theta$$
$$A_\phi = g_{\phi a}A^a = g_{\phi r}A^r + g_{\phi\theta}A^\theta + g_{\phi\phi}A^\phi = r^2\sin^2\theta A^\phi. \tag{30.38}$$

The inverse of g_{ab} is

$$g^{ab} = \begin{pmatrix} 1 & & \\ & 1/r^2 & \\ & & 1/r^2\sin^2\theta \end{pmatrix}, \tag{30.39}$$

which with $\vec{e}^a = g^{ab}\vec{e}_b$ gives the upstairs basis:

$$\vec{e}^r = g^{ra}\vec{e}_a = \vec{e}_r = \hat{r}$$
$$\vec{e}^\theta = g^{\theta a}\vec{e}_a = \vec{e}_r = \hat{\theta}/r$$
$$\vec{e}^\phi = g^{ra}\vec{e}_a = \vec{e}_r = \hat{\phi}/r\sin\theta. \tag{30.40}$$

For spherical and other orthogonal coordinates, the bases $\vec{e}_a$ and $\vec{e}^a$ are related by the metric to their more-familiar orthonormal versions — which for clarity, we'll denote here as $\hat{\epsilon}_a$ (independent of elevation),

$$\vec{e}_a = \sqrt{g_{aa}}\,\hat{\epsilon}_a \qquad\qquad \vec{e}^a = \hat{\epsilon}_a/\sqrt{g_{aa}}. \qquad \text{(no implied sum)} \tag{30.41}$$

The reciprocal distribution of scale factors amongst bases and components allows straightforward verification of $\vec{A} = A^a\vec{e}_a = A_a\vec{e}^a$.

Index elevation is notation; what fundamentally distinguishes the bases and their components is how they transform. And having enlarged our purview beyond cartesian bases, the space of transformations will no longer be confined to orthogonal rotations. Invoking our new index protocols, let's express the transformation of components and bases as

$$A^{b'} = T^{b'}_a A^a \qquad \vec{e}_{b'} = \vec{e}_a S^a_{b'}. \qquad (30.42)$$

Associativity easily establishes the relationship between S and T:

$$\vec{A} = \vec{e}_{b'} A^{b'} = \left(\vec{e}_a S^a_{b'} \right) \left(T^{b'}_c A^c \right) = \left(S^a_{b'} T^{b'}_c \right) \vec{e}_a A^c = \vec{e}_a A^a = \vec{A}, \qquad (30.43)$$

demonstrating that S and T are inverses. For orthogonal transformations, this is just the familiar $T^{-1} = T^T$. Indeed, for orthogonal transformations T is a rotation matrix; between inertial frames, T is the matrix Λ of Lorentz transformations, (26.103). (For the sheared system, see Problem 30.7.) We'll address the question of how to subsume these into a generally applicable expression for T in Section 30.5.

Example 30.4 **Constructing Coordinate Transformations**

We can find the transformation between coordinate systems by projecting both vector expansions

$$\vec{A} = \sum_{a'} \vec{e}_{a'} A^{a'} = \sum_a \vec{e}_a A^a \qquad (30.44)$$

onto the $\vec{e}^{\,b'}$ basis,

$$\vec{e}^{\,b'} \cdot \vec{A} = \sum_{a'} \left(\vec{e}^{\,b'} \cdot \vec{e}_{a'} \right) A^{a'} = \sum_{a'} \left(\vec{e}^{\,b'} \cdot \vec{e}_a \right) A^a. \qquad (30.45)$$

Invoking the mutual orthogonality expressed by (30.33), this reduces to

$$A^{b'} = \sum_a \left(\vec{e}^{\,b'} \cdot \vec{e}_a \right) A^a. \qquad (30.46)$$

Comparison with (30.42) identifies $T^{b'}_a$ as the matrix of inner products amongst basis vectors — which is fundamentally the same as the matrix in (25.61). For the transformation from sheared to cartesian components, (30.46) becomes

$$\begin{pmatrix} A^x \\ A^y \end{pmatrix} = \begin{pmatrix} \vec{e}^{\,x} \cdot \vec{e}_u & \vec{e}^{\,x} \cdot \vec{e}_v \\ \vec{e}^{\,y} \cdot \vec{e}_u & \vec{e}^{\,y} \cdot \vec{e}_v \end{pmatrix} \begin{pmatrix} A^u \\ A^v \end{pmatrix}. \qquad (30.47a)$$

Reading directly from Figure 30.1 yields (recall that indices $i, j, k, \ldots$ denote cartesian components)

$$T^i_a = \begin{pmatrix} \cos \alpha & \sin \beta \\ \sin \alpha & \cos \beta \end{pmatrix}. \qquad (30.47b)$$

Don't conflate the distinct roles of a tensor transformation *between* systems or frames and the action of the metric *within* a frame. A transformation T between frames (such as a rotation or a Lorentz transformation) maps between prime and unprimed. By contrast, the metric establishes the dual relationship within a frame. In fact, the ability to raise and lower indices is arguably the most common role the metric plays in a calculation.

Incidentally, as (30.42) and (30.43) show — and as we saw in Section 26.1 — an upstairs vector component varies in the opposite, or "contrary" fashion as the downstairs basis. So upstairs indices are said to be *contravariant*. Similarly, A_a varies with, or "co" the downstairs basis, and so downstairs indices are said to be *covariant*.[1] In the same way, $\vec{e}^a$ is often called a contravariant basis vector, while A_a is referred to as a covariant vector (though strictly speaking it's a covariant component of a vector.) These names are historical — and to a large degree irrelevant. Simply by adhering to the rule that summation over a dummy index requires one upstairs and one downstairs, we seldom need worry which is which.

Although the new upstairs/downstairs notation takes some getting used to, its utility comes from being able to largely forget about the differences between A_a, A^a and $\vec{e}_a$, $\vec{e}^a$. The transformation rule defining a tensor is given by adapting (29.28) to the new notation,

$$A^{a'b'c'd'\cdots} = T^{a'}_a T^{b'}_b T^{c'}_c T^{d'}_d \cdots A^{abcd\cdots},\tag{30.48}$$

with an implied sum over every pair of repeated indices — always with one upstairs, the other downstairs. Analogous expressions hold for the transformation of downstairs and mixed-elevation tensors.

The beauty of a tensorial basis such as $\{\vec{e}_a\}$ is that index algebra proceeds almost as if it we were using a cartesian basis. Just as we can work with cartesian vector components A^i without explicit use of basis vectors, so too in, say, spherical — as long as we maintain the upstairs/downstairs distinction and the metric that relates them. In particular, we can take any cartesian expression amongst tensor components and, by imposing the up/down summation rule and inserting factors of g_{ab}, obtain an expression valid in any system in $\mathbb{R}^n$. For instance,

$$\vec{A} \cdot \vec{A} = A^i A_i = \delta_{ij} A^i A^j = g_{ab} A^a A^b \,.\tag{30.49}$$

The ability to write a general equation valid in all frames and all coordinates is the mathematical expression of the Principle of Relativity.

BTW 30.1 Metric Spaces

A *metric space* is a set of objects $P = \{u, v, w, \ldots\}$ endowed with a real-valued function $d(u, v)$ satisfying the following criteria:

$$1.\ d(u,v) \geq 0,\ \text{where } d(u,v) = 0 \Longleftrightarrow u = v \tag{30.50a}$$

$$2.\ d(u,v) = d(v,u) \tag{30.50b}$$

$$3.\ d(u,w) \leq d(u,v) + d(v,w).\quad \text{(Triangle inequality)} \tag{30.50c}$$

As with an inner product, d maps two objects into a number. But although a metric space is not necessarily an inner product space (or even a vector space), the converse holds: an inner product space is also a metric space with

$$d(u,v) = \sqrt{g_{ab}(u-v)^a (u-v)^b},\tag{30.51}$$

where g_{ab} is the metric. Whereas inner products $\langle u|v \rangle$ describe the relative orientation of vectors, $d(u,v) \equiv \sqrt{\langle u - v|u - v \rangle}$ defines the distance between vectors. The most familiar example is the distance between infinitesimally separated positions in $\mathbb{R}^3$ given long ago by Pythagoras: $\|(\vec{r} + d\vec{r}) - \vec{r}\|^2 = \delta_{ij}dx^i\, dx^j$.

Incidentally, since the spacetime interval between different vectors is not positive definite, the Minkowski metric $\eta_{\mu\nu}$ violates (30.50a). Technically, $\eta_{\mu\nu}$ is an "indefinite" or a pseudo-Euclidean metric.

---

[1] This use of "covariant" is distinct from its sense in the Principle of Relativity.

30.4 Lorentz Tensors

In Minkowski spacetime, indexed objects transform not by orthogonal rotations, but rather with the larger group of Lorentz transformations of (26.103) — often given in the abbreviated form

$$\Lambda^{\mu'}_{\ \nu} \longrightarrow \left(\begin{array}{c|c} \gamma & -\vec{\beta}\gamma \\ \hline -\vec{\beta}\gamma & \mathbb{1} + \frac{\gamma-1}{\beta^2}\vec{\beta}\vec{\beta} \end{array} \right), \tag{30.52}$$

where $\mathbb{1}$ is the 3×3 unit matrix and $\vec{\beta}\vec{\beta}$ a 3×3 dyad. Since the Lorentz matrix Λ is not the same as orthogonal rotations in $\mathbb{R}^4$, this is a distinct equivalence class of tensors — encapsulated by the Lorentz metric (30.17). The relative simplicity of this cartesian metric results in a relatively simple relation between upstairs and downstairs vectors,

$$A^\mu = (A^0, \vec{A}) \qquad \text{but} \qquad A_\mu = (A^0, -\vec{A}). \tag{30.53}$$

In this way, the relative sign between space and time — the hallmark of spacetime — is automatically included by our tensor notation.

Example 30.5 **Lorentz Scalars and Vectors**

Remarkably, a great deal of physics can be derived with little more than consideration of spacetime's tensorial structure. The simplest cases are the contraction of all indices — such as

$$A^\mu A_\mu = A_0^2 - \vec{A} \cdot \vec{A}, \tag{30.54}$$

which is often called a *Lorentz scalar* to distinguish it from scalars in $\mathbb{R}^n$. As a scalar, it must be the same in all frames. The most fundamental example is the interval ds between two events in space and time, with

$$ds^2 = dx^\mu dx_\mu = c^2 dt^2 - d\vec{x}^2. \tag{30.55}$$

As a Lorentz scalar, all frames agree on the value of ds^2 — which is why it's called the *invariant* interval. So, for example, given two events with $ds^2 > 0$ (called a *timelike* interval), there exists a frame in which the events occur at the same place, $d\vec{x}^2 = 0$. Then $ds^2 = c^2 dt^2 \equiv c^2 d\tau^2$, where the *proper time* $d\tau$ is the time measured between the events in this frame. But since observers in any inertial frame measure the same ds^2, those for whom the events do not occur at the same place must necessarily measure larger time intervals dt in order to get the same ds^2; this is the essence of *time dilation*.

   We can use the proper time to construct physically relevant Lorentz vectors (see Example 24.2). Consider for instance that, as the ratio of a vector and a scalar, the velocity $\vec{u} = d\vec{x}/dt$ is a vector in $\mathbb{R}^3$. We can similarly define the so-called four-velocity U^μ by differentiating the four-vector x^μ by the Lorentz scalar τ. (We can't use dx^μ/dt since $x_0 = ct$ is a four-vector *component* — and dividing vector components will not yield a vector.) After verifying $dt/d\tau = 1/\sqrt{1 - u^2/c^2} \equiv \gamma$ from (30.55), the spacetime components of the four-velocity can be shown to be

$$U^0 \equiv \frac{dx^0}{d\tau} = \frac{dx^0}{dt}\frac{dt}{d\tau} = \gamma c \qquad \vec{U} \equiv \frac{d\vec{x}}{d\tau} = \frac{d\vec{x}}{dt}\frac{dt}{d\tau} = \gamma \vec{u}, \tag{30.56}$$

whose magnitude is

$$U^\mu U_\mu = (U^0)^2 - \vec{U}^2 = \gamma^2(c^2 - \vec{u}^2) = \gamma^2 c^2\left(1 - \frac{\vec{u}^2}{c^2}\right) = c^2, \tag{30.57}$$

which, as a frame-independent universal constant, is clearly a Lorentz scalar. Note that in the non-relativistic limit $u/c \ll 1$, $\vec{U}$ reduces to the non-relativistic $\vec{u}$.

Motivated by the non-relativistic expression for momentum, we consider the relativistic form $\vec{p} = \gamma m\vec{u}$, and define the four-momentum

$$P^\mu = mU^\mu = (\gamma mc, \gamma m\vec{u}) = (E/c, \vec{p}), \tag{30.58}$$

where $E = \gamma mc^2$ is the energy (Example 6.19). The associated scalar invariant is

$$P_\mu P^\mu = \eta_{\mu\nu}P^\mu P^\nu = (E/c)^2 - \vec{p} \cdot \vec{p} = \gamma^2 m^2 c^2\left(1 - u^2/c^2\right) = m^2 c^2. \tag{30.59}$$

In other words, the familiar relation $E^2 - p^2 c^2 = m^2 c^4$ is nothing more than a constant c^2 times the magnitude-squared of the momentum four-vector. From this we discover that mass is a frame-independent Lorentz scalar. (For a particle at rest, (30.59) becomes the World's Most Famous Equation.)

Example 30.6 The Electromagnetic Field Tensor

As a further example of the physical insight tensors provide, consider the Lorentz force due to electromagnetic fields,

$$\frac{d\vec{p}}{dt} = q(\vec{E} + \vec{u} \times \vec{B}). \tag{30.60}$$

We know from the study of electromagnetism that this expression is valid in all inertial frames, despite the fact that different frames observe different fields. Whereas Alice standing still next to a uniform line charges sees only an electric field, Bob walking along the line with constant velocity will measure both electric and magnetic fields. Yet both can use (30.60) to calculate the force on a test charge in each frame.

In other words, (30.60) is a Lorentz covariant statement. So it must be possible to make it manifestly covariant by writing it in tensor notation. The first step is to replace the non-relativistic time t with the invariant proper time τ defined in (30.55). Using $dt/d\tau = \gamma$ and (30.56),

$$\frac{d\vec{p}}{d\tau} = \frac{dt}{d\tau}\frac{d\vec{p}}{dt} = \gamma q(\vec{E} + \vec{u} \times \vec{B})$$

$$= \frac{q}{m}\left(P^0\vec{E}/c + \vec{P} \times \vec{B}\right), \tag{30.61}$$

where we've written the right-hand side using the four-momentum, (30.58). Now the left-hand side is the spatial portion of $dP^\mu/d\tau$, so (30.61) must be the spatial part of a four-vector expression of the form

$$\frac{dP^\mu}{d\tau} = \frac{q}{m}\left(\text{some mix of } P^\mu, \vec{E} \text{ and } \vec{B}\right). \tag{30.62}$$

Somehow the right-hand side must be a four-vector. The only way for this to work covariantly is if on the right, P^μ is being contracted with a second-rank tensor constructed from $\vec{E}$ and $\vec{B}$. So using little more than the rules of index algebra, we conclude that there must be a tensor F (for "field" not "force") such that

$$\frac{dP^\mu}{d\tau} = \frac{q}{m}F^\mu{}_\nu P^\nu = \frac{q}{m}F^{\mu\nu}P_\nu. \tag{30.63}$$

The second-rank *electromagnetic field tensor* $F^{\mu\nu}$ contains in one structure the electric and magnetic fields as observed in any inertial frame of reference. From the expression for the Lorentz force, we can deduce the components of F (Problem 30.11),

$$F^{\mu\nu} = \begin{pmatrix} 0 & -E_x/c & -E_y/c & -E_z/c \\ E_x/c & 0 & -B_z & B_y \\ E_y/c & B_z & 0 & -B_x \\ E_z/c & -B_y & B_x & 0 \end{pmatrix}. \tag{30.64}$$

The rows and columns are labeled from 0 to 3. So, for instance, $F^{02} = -E_y/c$ and $F^{13} = -\epsilon^{132}B_2 = +B_y$.

As an antisymmetric tensor in four dimensions, $F^{\mu\nu}$ has only six independent elements — exactly the number of components required to specify $\vec{E}$ and $\vec{B}$. Moreover, the transformation of a second-rank tensor mixes these six components with one another under a change of inertial frame; it's this behavior that underlies the observations of Alice and Bob (Problem 30.12).

Further insight comes from F's two tensor invariants. Direct calculation shows (Problem 30.13)

$$F_{\mu\nu}F^{\nu\mu} = 2(\vec{E}^2/c^2 - B^2) \tag{30.65a}$$

and

$$\epsilon^{\mu\nu\alpha\beta}F_{\alpha\beta}F_{\mu\nu} = \frac{4}{c}(\vec{E} \cdot \vec{B}), \tag{30.65b}$$

where $\epsilon^{\mu\nu\rho\sigma}$ is the invariant antisymmetric tensor

$$\epsilon^{\mu\nu\rho\sigma} = \begin{cases} -1, & (\mu\nu\rho\sigma) \text{ even permutations of } 0123 \\ +1, & (\mu\nu\rho\sigma) \text{ odd permutations of } 0123 \\ 0, & \text{else.} \end{cases} \tag{30.66}$$

(Note that the equivalence of even and cyclic permutations does not hold beyond three dimensions.) We immediately discover two things about electric and magnetic fields. First, although the fields may differ in different frames, (30.65a) reveals that if one field dominates in one frame, it dominates in all frames. In particular, if there exists a frame with $\vec{E} \neq 0$ and $\vec{B} = 0$, then there is *no* frame in which $\vec{E} = 0$ and $\vec{B} \neq 0$. The second invariant, (30.65b), shows that if $\vec{E}$ and $\vec{B}$ are perpendicular in one frame, they are perpendicular in all frames — unless one field vanishes (consistent with (30.65a), of course). These are non-trivial properties of electromagnetic fields, discovered using little more than the Principle of Relativity.

30.5 General Covariance

In Chapter 29 we used rotations to define the equivalence class of tensors. But as we've seen, rotations alone are insufficient to fully implement the Principle of Relativity. This is true even in $\mathbb{R}^n$, as the non-orthogonal sheared system makes clear. Moreover, as a rank-two tensor the metric $\overleftrightarrow{g}$ should transform with two copies of the rotation operator. But as Example 30.3 demonstrated, the metric need not be a matrix of constants; it can be — and usually is — a function of position. A rotation cannot accommodate this. We need more general tensor transformation matrices which are also functions of position, reducing to rotations where appropriate.

Rotations also have the physical shortcoming that they are limited to frames with a common origin. If a system is translation invariant, for instance, the choice of origin can have no physical significance — and tensors must be able to accommodate this. Enlarging tensor transformations to include translations, however, means the position vector $\vec{r}$, which is origin dependent, is no longer a tensor. On the other hand, the vector *difference* between positions remains a physically relevant, origin-independent quantity. So while we can no longer use $\vec{r} = \left(u^1, \ldots, u^N\right)$ as our paradigmatic tensor, $d\vec{r}$ fits the bill. Better still, elementary calculus tells us what the new transformation law should be. Under a general coordinate change from unprimed u^a to primed $u^{a'}$, the chain rule gives

$$du^{a'} = \frac{\partial u^{a'}}{\partial u^b} du^b, \tag{30.67}$$

with an implied sum over b. You should convince yourself that this includes as special cases both rotations and Lorentz transformations between cartesian coordinates. Moreover, (30.67) is linear in du^b, with $\partial u^{a'}/\partial u^b$ familiar as the Jacobian matrix of the transformation. For instance, a change from cartesian to polar coordinates — even if the two have different origins — is given by the transformation matrix

$$T^{a'}_{\ b} \equiv \frac{\partial u^{a'}}{\partial u^b} \longrightarrow \frac{\partial(r, \phi)}{\partial(x, y)} = \begin{pmatrix} \partial r/\partial x & \partial r/\partial y \\ \partial\phi/\partial x & \partial\phi/\partial y \end{pmatrix} = \begin{pmatrix} x/r & y/r \\ -y/r^2 & x/r^2 \end{pmatrix}. \tag{30.68}$$

Then (30.67) correctly yields

$$dr = \frac{x}{r} dx + \frac{y}{r} dy \qquad d\phi = -\frac{x}{r^2} dx + \frac{y}{r^2} dy. \tag{30.69}$$

But shifting focus from u^a to du^a represents a more significant change than simply allowing for translation invariance. Constrained only by requirements of differentiability and invertibility,[2] (30.67) allows for *any* arbitrary change of coordinates. In physical terms, this is the recognition that coordinates are mere support structure used to describe the underlying physical reality — and as an objective reality, physical law cannot and should not reflect the scaffolding. In relativity, this implies that covariance of physical law is not restricted to the special case of inertial frames and Lorentz rotations, but applies more generally to arbitrary non-inertial frames and general coordinate transformations. This idea is often called *general covariance*, and is one of the underpinnings of general relativity (Weinberg, 1972; Schutz, 2009).

Implementation of general covariance follows from the declaration that a single-index object belongs to the equivalence class of vectors if it transforms using the same Jacobian matrix as du^a,

$$A^{a'} = \frac{\partial u^{a'}}{\partial u^b} A^b. \tag{30.70}$$

Notice though that (30.70) is only consistent with the upstairs/downstairs summation convention if $\partial/\partial u^b$ is actually a *downstairs* vector operator. In fact, the chain rule shows that the gradient transforms as

$$\frac{\partial}{\partial u^{a'}} = \frac{\partial u^b}{\partial u^{a'}} \frac{\partial}{\partial u^b}, \tag{30.71a}$$

---

[2] In techno-speak, "diffeomorphisms on a smooth manifold."

or simply

$$\partial_{a'} \equiv \frac{\partial u^b}{\partial u^{a'}} \partial_b. \tag{30.71b}$$

The gradient of a scalar $\partial_a \Phi$ thus transforms as a downstairs vector; with this as paradigm, we define all downstairs vectors as single-index objects A_a which transform like the gradient,

$$A_{a'} = \frac{\partial u^b}{\partial u^{a'}} A_b. \tag{30.72}$$

The invariance of the inner product is all we need to verify this:

$$A^{a'} A_{a'} = \left(\frac{\partial u^{a'}}{\partial u^b} A^b \right) \left(\frac{\partial u^c}{\partial u^{a'}} A_c \right)$$

$$= \left(\frac{\partial u^{a'}}{\partial u^b} \frac{\partial u^c}{\partial u^{a'}} \right) A^b A_c = \delta_b^c A^b A_c = A^c A_c. \tag{30.73}$$

Higher-rank tensors — upstairs, downstairs, or mixed — follow suit: as in (30.48), one transformation matrix $\partial u^{a'}/\partial u^b$, or its inverse $\partial u^b/\partial u^{a'}$, for each index (see Table 30.1).

Table 30.1 Tensor transformations (summation convention)

Scalar	$\Phi = \Phi'$
Vector	$A^{a'} = \frac{\partial u^{a'}}{\partial u^c} A^c$
	$A_{a'} = \frac{\partial u^c}{\partial u^{a'}} A_c$
Tensor	$T^{a'b'} = \frac{\partial u^{a'}}{\partial u^c} \frac{\partial u^{b'}}{\partial u^d} T^{cd}$
	$T_{a'b'} = \frac{\partial u^c}{\partial u^{a'}} \frac{\partial u^d}{\partial u^{b'}} T_{cd}$
	$T^{a'}{}_{b'} = \frac{\partial u^{a'}}{\partial u^c} \frac{\partial u^d}{\partial u^{b'}} T^c{}_d$

Example 30.7 Tensor Transformation of the Metric

As a second-rank downstairs tensor field, the metric transforms as

$$g_{a'b'} = \frac{\partial u^a}{\partial u^{a'}} \frac{\partial u^b}{\partial u^{b'}} g_{ab}. \tag{30.74}$$

Since the cartesian metric is the Kronecker delta, transformation from cartesian into other coordinates is straightforward. For polar coordinates, the metric is

$$g_{rr} = \frac{\partial x^i}{\partial r} \frac{\partial x^j}{\partial r} \delta_{ij} = \left(\frac{\partial x}{\partial r} \right)^2 + \left(\frac{\partial y}{\partial r} \right)^2 = 1$$

$$g_{\phi\phi} = \frac{\partial x^i}{\partial \phi} \frac{\partial x^j}{\partial \phi} \delta_{ij} = \left(\frac{\partial x}{\partial \phi} \right)^2 + \left(\frac{\partial y}{\partial \phi} \right)^2 = r^2 \tag{30.75}$$

$$g_{r\phi} = g_{\phi r} = \frac{\partial x^i}{\partial r} \frac{\partial x^j}{\partial \phi} \delta_{ij} = \frac{\partial x}{\partial r} \frac{\partial x}{\partial \phi} + \frac{\partial y}{\partial r} \frac{\partial y}{\partial \phi} = 0 \, .$$

Example 30.8 **The Downstairs Basis in Spherical Coordinates**

The transformation (30.72) of a downstairs vector component must also apply to downstairs basis vectors,

$$\vec{e}_{a'} = \frac{\partial u^b}{\partial u^{a'}} \vec{e}_b. \tag{30.76}$$

Flipping all the way back to Section 2.1 (see also Problem 6.36) reminds us that the basis vectors are the gradient of position,

$$\vec{e}_b \equiv \frac{\partial \vec{r}}{\partial u^b}. \tag{30.77}$$

This means that any downstairs $\mathbb{R}^n$ basis can be found by tensor transformation of the cartesian basis,

$$\vec{e}_{a'} = \frac{\partial \vec{r}}{\partial u^{a'}} = \frac{\partial x^j}{\partial u^{a'}} \frac{\partial \vec{r}}{\partial x^j} = \frac{\partial x^j}{\partial u^{a'}} \hat{x}_j. \tag{30.78}$$

For spherical coordinates $u^1 = r, u^2 = \theta, u^3 = \phi$ in $\mathbb{R}^3$, this gives

$$\begin{pmatrix} \vec{e}_r \\ \vec{e}_\theta \\ \vec{e}_\phi \end{pmatrix} \equiv \begin{pmatrix} \vec{e}_1 \\ \vec{e}_2 \\ \vec{e}_3 \end{pmatrix} = \begin{pmatrix} \partial x^1/\partial u^1 & \partial x^2/\partial u^1 & \partial x^3/\partial u^1 \\ \partial x^1/\partial u^2 & \partial x^2/\partial u^2 & \partial x^3/\partial u^2 \\ \partial x^1/\partial u^3 & \partial x^2/\partial u^3 & \partial x^3/\partial u^3 \end{pmatrix} \begin{pmatrix} \hat{x}_1 \\ \hat{x}_2 \\ \hat{x}_3 \end{pmatrix}$$

$$= \begin{pmatrix} \sin\theta\cos\phi & \sin\theta\sin\phi & \cos\theta \\ r\cos\theta\cos\phi & r\cos\theta\sin\phi & -r\sin\theta \\ -r\sin\theta\sin\phi & r\sin\theta\cos\phi & 0 \end{pmatrix} \begin{pmatrix} \hat{x}_1 \\ \hat{x}_2 \\ \hat{x}_3 \end{pmatrix}$$

$$= \begin{pmatrix} \hat{r} \\ r\hat{\theta} \\ r\sin\theta\hat{\phi} \end{pmatrix}, \tag{30.79}$$

where we've used (2.19) in the last line.

30.6 Tensor Calculus

So far, we've been largely concerned with tensor algebra. What about tensor calculus? In particular, is the integral or derivative of a tensor also a tensor?

30.6.1 Tensor Integration

Reviewing Section 14.1 on line integrals, you'll realize that the unstated common thread through most of that presentation is the metric. This should not be surprising, given the metric's role in determining distance. And in fact, from (30.11) we see that the finite path length along a curve C is

$$\int_C ds = \int_C \sqrt{|g_{ab} \, du^a du^b|}. \tag{30.80}$$

(The use of absolute values allows us to include non-Euclidean metrics such as $\eta_{\alpha\beta}$.) So there's really nothing new here.

As we well know, the volume element $d\tau = du^1 du^2 \cdots du^n$ transforms with a factor of the Jacobian determinant,

$$d\tau = |J|\, d\tau' = \det\left(\frac{\partial u}{\partial u'}\right) d\tau'. \tag{30.81}$$

This shows that, despite expectations, $d\tau$ is not a scalar. (It's actually a *tensor density* — see Problem 30.18.) This means that a volume integral $\int d\tau$ is not a covariant expression

To construct a proper scalar volume element, we need somehow to account for the Jacobian. Surely the metric, the arbiter of distance, can help. Indeed, under a tensor transformation the determinant $g \equiv \det g_{ab}$ transforms as

$$g = \det g_{ab} = \det\left(\frac{\partial u^{c'}}{\partial u^a}\frac{\partial u^{d'}}{\partial u^b} g_{c'd'}\right) = \det\left(\frac{\partial u'}{\partial u}\right)\det\left(\frac{\partial u'}{\partial u}\right)g'. \tag{30.82}$$

So like $d\tau$, g is not a scalar; unlike $d\tau$, g transforms with not one, but two factors of the Jacobian. Using $\det\left(\partial u/\partial u'\right) = 1/\det\left(\partial u'/\partial u\right)$, we can leverage this distinction to construct the scalar quantity

$$\sqrt{|g'|}\, d\tau' = \sqrt{|g|}\, d\tau, \tag{30.83}$$

where again the absolute value allows for non-Euclidean metrics. Since orthogonal systems have diagonal metrics, the determinant g is simply the product of the scale factors — so (30.83) gives the familiar volume element. The volume integral $\int \sqrt{|g|}\, d\tau$ is a covariant expression.

Of course, we can construct a scalar in this way using the determinant of any second-rank tensor. The significance of g comes from the known form of the volume element in cartesian coordinates, $d\tau = \sqrt{\det(\mathbb{1})}\, d\tau$. The appearance of $\mathbb{1}$ is neither arbitrary nor irrelevant; in these coordinates, the identity *is* the metric. And that's the point: as a valid tensor expression in one coordinate frame, $\sqrt{|g|}\, d\tau$ must describe the same quantity in all frames. Thus (30.83) is the correct invariant volume element. The same fundamental argument also holds for the covariance of (30.80).

30.6.2 Tensor Derivatives

Tensor differentiation is more subtle than integration. Recall that the paradigmatic downstairs vector is the gradient, (30.71). As a result, the gradient of a scalar is a rank-one (downstairs) tensor. But what about the derivative of a vector, $\partial_b A^a \equiv \partial A^a/\partial u^b$? The indices give it a tentative identification as a second-rank mixed tensor, with one upstairs and one downstairs index. But tensors are not identified by indices; tensors are defined by their transformations. So let's see: under a change of coordinates, the chain and product rules give

$$\partial_{b'} A^{a'} = \partial_{b'}\left(\frac{\partial u^{a'}}{\partial u^c} A^c\right)$$

$$= \frac{\partial u^d}{\partial u^{b'}}\frac{\partial}{\partial u^d}\left(\frac{\partial u^{a'}}{\partial u^c} A^c\right)$$

$$= \partial_d A^c \frac{\partial u^d}{\partial u^{b'}}\frac{\partial u^{a'}}{\partial u^c} + A^c \frac{\partial^2 u^{a'}}{\partial u^c \partial u^d}\frac{\partial u^d}{\partial u^{b'}}. \tag{30.84}$$

Now look closely at what we have: on the left, there's what appears to be a second-rank tensor. And indeed, the first term on the right is precisely the transformation required of a mixed rank-two tensor. However, the second term does not belong in such a transformation; it emerges because the transformation matrix $\partial u^{a'}/\partial u^c$ can be a function of position. Though $\partial^2 u^{a'}/\partial u^c \partial u^d$ vanishes for rotations, in general the derivative of a tensor is not a tensor.

The problem is that a vector component implicitly depends on the basis. Thus any change in A^a will reflect not only a difference in $\vec{A}$, but also include effects of a changing basis. (Not for the first time do we come to appreciate the simplicity of the position-independent cartesian basis.) As we did back in Chapter 2, we need to consider not $\partial_b A^a$, but rather

$$\frac{\partial}{\partial u^b}(A^a \vec{e}_a) = \frac{\partial A^a}{\partial u^b}\vec{e}_a + A^a\frac{\partial \vec{e}_a}{\partial u^b}. \tag{30.85}$$

The vector $\partial \vec{e}_a/\partial u^b$ can be expanded on the basis $\vec{e}_a$ itself; we'll write this as

$$\frac{\partial \vec{e}_a}{\partial u^b} \equiv \Gamma^c_{ab}\,\vec{e}_c, \tag{30.86}$$

where the *Christoffel symbol* Γ^c_{ab} is the cth component of $\partial \vec{e}_a/\partial u^b$ on the basis. As we'll see in a moment, the introduction of Christoffel symbols will allow us to continue using index notation without the need for explicit reference to basis vectors.

Example 30.9 Γ^c_{ab} in Polar and Spherical Coordinates

Before scrutinizing (30.85), let's pause to examine the Christoffel symbol. In n dimensions, it's an array of n^3 elements. Since $\vec{e}_a = \partial\vec{r}/\partial u^a$ (Section 2.1), we get $\partial_b \vec{e}_a = \partial^2\vec{r}/\partial u^a \partial u^b = \partial_a \vec{e}_b$. So Γ is symmetric in its lower two indices,

$$\Gamma^c_{ab} = \Gamma^c_{ba}. \tag{30.87}$$

This leaves only $n^2(n+1)/2$ independent elements. In $\mathbb{R}^n$, all of these elements vanish for the position-independent cartesian basis; the same, however, is not true in curvilinear coordinates. To find Γ we start by expanding the $\vec{e}_a$ on the cartesian basis $\hat{x}_j$,

$$\vec{e}_a = \frac{\partial\vec{r}}{\partial u^a} = \frac{\partial x^j}{\partial u^a}\frac{\partial\vec{r}}{\partial x^j} = \frac{\partial x^j}{\partial u^a}\hat{x}_j, \tag{30.88}$$

which is the tensor transformation (30.76). For instance, in $\mathbb{R}^2$ polar coordinates this gives

$$\vec{e}_r = \cos\phi\,\hat{i} + \sin\phi\,\hat{j} = \hat{r} \qquad \vec{e}_\phi = -r\sin\phi\,\hat{i} + r\cos\phi\,\hat{j} = r\hat{\phi}, \tag{30.89}$$

which is consistent with (2.3)–(2.5). Taking derivatives of these basis vectors with respect to $u^1 = r, u^2 = \phi$ and comparing with (30.86) identifies the six independent elements:

$$\Gamma^r_{rr} = 0 \qquad \Gamma^\phi_{rr} = 0 \qquad \Gamma^r_{\phi\phi} = -r \qquad \Gamma^\phi_{\phi\phi} = 0 \tag{30.90a}$$

and

$$\Gamma^\phi_{r\phi} = \Gamma^\phi_{\phi r} = 1/r \qquad \Gamma^r_{r\phi} = \Gamma^r_{\phi r} = 0. \tag{30.90b}$$

We can then reproduce (2.9) as follows:

$$\begin{aligned}
\frac{d\vec{r}}{dt} = \frac{d}{dt}(u^r\vec{e}_r) &= \frac{du^r}{dt}\vec{e}_r + u^r\left(\frac{du^j}{dt}\frac{\partial\vec{e}_r}{\partial u^j}\right) \\
&= \dot{r}\vec{e}_r + u^r\dot{u}^j\Gamma^k_{rj}\vec{e}_k \\
&= \dot{r}\vec{e}_r + r\dot{\phi}\,(1/r)\,\vec{e}_\phi \\
&= \dot{r}\,\hat{r} + r\dot{\phi}\,\hat{\phi},
\end{aligned} \tag{30.91}$$

where $\Gamma^r_{r\phi}$ is the only non-zero contribution to the sum over k in the second line, and in the last line we've used $\vec{e}_r \equiv \hat{r}$ and $\vec{e}_\phi \equiv r\hat{\phi}$. A similar calculation (Problem 30.25) reveals that the 18 independent Γ's in spherical coordinates all vanish except

$$\Gamma^r_{\theta\theta} = -r \qquad \Gamma^r_{\phi\phi} = -r\sin^2\theta \qquad \Gamma^\theta_{\phi\phi} = -\sin\theta\cos\theta$$

$$\Gamma^\theta_{r\theta} = \Gamma^\theta_{\theta r} = \frac{1}{r} \qquad \Gamma^\phi_{r\phi} = \Gamma^\phi_{\phi r} = \frac{1}{r} \qquad \Gamma^\phi_{\theta\phi} = \Gamma^\phi_{\phi\theta} = \cot\theta. \tag{30.92}$$

Then, using (30.79),

$$\frac{d\vec{r}}{dt} = \frac{d}{dt}(u^r\vec{e}_r) = \dot{r}\vec{e}_r + u^r\dot{u}^j\Gamma^k_{rj}\vec{e}_k$$

$$= \dot{r}\vec{e}_r + \dot{\theta}\,\vec{e}_\theta + \dot{\phi}\,\vec{e}_\phi = \dot{r}\hat{r} + r\dot{\theta}\,\hat{\theta} + r\sin\theta\,\dot{\phi}\,\hat{\phi}. \tag{30.93}$$

So how does the introduction of the Christoffel symbol remedy the non-tensorial derivative of (30.84)? Using (30.86), we can relabel some of the dummy indices to recast (30.85) as

$$\partial_b(A^a\vec{e}_a) = \left(\partial_b A^a + A^c\,\Gamma^a_{bc}\right)\vec{e}_a. \tag{30.94}$$

Because of the implied sum over a, this is the expansion of $\partial_b(A^a\vec{e}_a)$ on the basis; the expression in parentheses on the right-hand side then is its mixed, two-indexed component on the basis,

$$D_bA^a \equiv \frac{\partial A^a}{\partial u^b} + A^c\,\Gamma^a_{bc}. \tag{30.95}$$

The first term on the right represents our original, unsuccessful attempt to find the tensor derivative of a vector; the second term compensates for the change in the basis to actually make it all work. Thus we identify D_b as a vector operator which acts on a vector A^a to produce a second-rank tensor. Problem 30.27 asks you to formally verify that D is a tensor operator, even though separately neither ∂ nor Γ is. (So despite appearances, Γ^a_{bc} is not a tensor — hence its designation as a "symbol.")

The same basic argument shows that derivatives of higher-rank tensors require a Christoffel symbol for each index — for example,

$$D_cT^{ab} = \frac{\partial T^{ab}}{\partial u^c} + T^{db}\,\Gamma^a_{cd} + T^{ad}\,\Gamma^b_{cd} \tag{30.96}$$

is a third-rank tensor. A tensor with downstairs indices is treated essentially the same, though in addition to appropriate index balancing, the sign in front of the Christoffel symbol flips — for instance (Problem 30.29),

$$D_cT_{ab} = \frac{\partial T_{ab}}{\partial u^c} - T_{db}\,\Gamma^d_{ac} - T_{ad}\,\Gamma^d_{bc}. \tag{30.97}$$

A scalar function has no index, so D reduces to the ordinary derivative,

$$D_a\Phi = \partial_a\Phi. \tag{30.98}$$

The operator D defined in (30.95)–(30.98) is known as the *covariant derivative*.[3] The name is not intended as a description of a downstairs index, but rather to invoke the more meaningful sense of tensors retaining their form under general coordinate transformations. Indeed, combining D_a with the principal of general covariance is a powerful tool. Consider for instance an expression in cartesian

---

[3] A common notation uses a comma for the regular derivative and a semi-colon for the covariant derivative:
$A_{a,b} \equiv \partial_b A_a$, and $A_{a;b} \equiv A_{a,b} + A^c\Gamma^a_{bc}$.

coordinates. Since $\Gamma^i_{jk} \equiv 0$ on the cartesian basis, the expression is identical using either ∂_j or D_j. But choosing to write it in terms of D_j renders the expression "manifestly" covariant: if this tensor formulation is correct in cartesian coordinates, it remains valid in any coordinates. In other words, cartesian expressions can be rendered generally covariant by replacing derivatives with covariant derivatives.

Example 30.10 The Covariant Derivative of the Metric

In general, a space's metric is a function of position. What sets $\mathbb{R}^n$ apart is that coordinates exist (cartesian) in which the metric is constant *throughout* the space, $g_{ij} = \delta_{ij}$. Thus $\partial_k g_{ij} = 0$ at every point. In other coordinates the spatial variation of the basis must be taken into account by using the Christoffel symbols and (30.97). But, by general covariance, if $\partial_k g_{ij} = 0$ everywhere in cartesian, then $D_c g_{ab} = 0$ in all coordinates,

$$D_c g_{ab} = \frac{\partial g_{ab}}{\partial u^c} - g_{db}\,\Gamma^d_{ac} - g_{ad}\,\Gamma^d_{bc} = 0. \tag{30.99}$$

So even though in both cylindrical and spherical coordinates the $\mathbb{R}^3$ metric is a non-trivial function of position, $D_k g_{ij} = 0$ in those coordinates (Problem 30.30).

From (30.99), we can find the relationship between Christoffel symbols and derivatives of the metric (Problem 30.24),

$$\Gamma^c_{ab} = \frac{1}{2} g^{cd} \left(\partial_a g_{bd} + \partial_b g_{da} - \partial_d g_{ab}\right). \tag{30.100}$$

(Note the cyclic permutation of the downstairs indices on the right-hand side.) Since (30.12) directly relates the metric to the basis vectors, it's not unexpected to discover that Γ depends on derivatives of the metric.

Example 30.11 Covariant Motion

For cartesian coordinates in the plane, the acceleration $\vec{a}$ can be expressed via the chain rule as

$$\vec{a} = \frac{d\vec{V}}{dt} = \left(\frac{d\vec{r}}{dt} \cdot \vec{\nabla}\right)\vec{V} = \left(\vec{V} \cdot \vec{\nabla}\right)\vec{V}, \tag{30.101a}$$

or in component form

$$a^i = \frac{dV^i}{dt} = V^k \frac{\partial V^i}{\partial x^k}, \tag{30.101b}$$

where $V^i = dx^i/dt$ is the velocity. Though correct on the cartesian basis, as written it's not a covariant tensor expression. We remedy this by replacing the derivative ∂ with the covariant derivative D:

$$a^b = \frac{DV^b}{dt} \equiv V^c D_c V^b$$

$$= V^c \left(\frac{\partial V^b}{\partial x^c} + V^d \Gamma^b_{cd}\right) = \frac{dV^b}{dt} + \Gamma^b_{cd} V^c V^d. \tag{30.102}$$

When indices b, c, d represent cartesian coordinates, the Christoffel symbol vanishes and (30.101) re-emerges. Since this tensor expression is correct in one set of coordinates (cartesian), by general covariance it must be correct in all coordinates. For instance, using (30.90) we find in polar coordinates,

$$a^r = \dot{V}^r + V^j V^k \Gamma^r_{jk} = \ddot{r} - r\dot{\phi}^2$$

$$a^\phi = \dot{V}^\phi + V^j V^k \Gamma^\phi_{jk} = \ddot{\phi} + V^r \dot{\phi}\frac{1}{r} + V^\phi \dot{r}\frac{1}{r} = \ddot{\phi} + \frac{2}{r}\dot{r}\dot{\phi}. \tag{30.103}$$

These are components of acceleration on the basis $\vec{e}_r = \hat{r}$ and $\vec{e}_\phi = r\hat{\phi}$ — identical to the results of (2.12). For an inverse-square central force $\vec{F} = -k\hat{r}/r^2$, this provides an alternative derivation of the equations of motion in Problem 2.13a.

30.7 Geodesics, Curvature, and Tangent Planes

One type of motion of particular physical interest is that of a free particle — that is, motion in the absence of external forces. This can be described tensorially by setting acceleration to zero in (30.102),

$$\frac{DV^b}{dt} = \frac{dV^b}{dt} + \Gamma^b_{cd}V^c V^d = 0. \tag{30.104a}$$

This is the *geodesic equation*, also frequently expressed in terms of position,

$$\frac{d^2 u^b}{dt^2} + \Gamma^b_{cd}\frac{du^c}{dt}\frac{du^d}{dt} = 0. \tag{30.104b}$$

The solution $u^b(t)$, called a *geodesic*, describes the trajectory of a free particle. Valid in any coordinate system, the geodesic equation is the covariant statement of Newton's first law for force-free motion (a.k.a. Galileo's law of inertia). A geodesic, then, must be a straight line — though we'll need to consider just what this means. Still, we can verify the plane meaning in polar coordinates by setting the expressions in (30.103) to zero to get

$$\ddot{r} - r\dot{\phi}^2 = 0 \qquad\qquad \ddot{\phi} + \frac{2}{r}\dot{r}\dot{\phi} = 0, \tag{30.105}$$

which describe a constant-velocity, straight-line trajectory $r = a + bt$, $\phi = \phi_0$.

BTW 30.2 Geodesics Are Extreme

A particle moving along a curve C covers a distance given by (30.80),

$$S = \int_C ds = \int_C \sqrt{|g_{ab}\,du^a du^b|} = \int_{t_1}^{t_2} \left|g_{ab}\frac{du^a}{dt}\frac{du^b}{dt}\right|^{1/2} dt, \tag{30.106}$$

where the functions $u^a(t)$ define C between t_1 and t_2. It's natural to ask when this path length is an extremum — that is, for what trajectory $u(t)$ does any variation δu give $\delta S = 0$? If we vary C from $u(t)$ to $u(t) + \delta u(t)$, the change in distance δS can be found by treating δ as a derivative operator and using the product rule,

$$\delta S = \frac{1}{2}\int_{t_1}^{t_2}\left|g_{ab}\frac{du^a}{dt}\frac{du^b}{dt}\right|^{-1/2}$$

$$\times \left[\left(\frac{\partial g_{ab}}{\partial u^c}\delta u^c\right)\frac{du^a}{dt}\frac{du^b}{dt} + 2g_{ab}\frac{d(\delta u^a)}{dt}\frac{du^b}{dt}\right]dt. \tag{30.107}$$

This can be cleaned up a bit using $ds/dt = \sqrt{\left| g_{ab} \frac{du^a}{dt} \frac{du^b}{dt} \right|}$ from (30.106),

$$\delta S = \frac{1}{2} \int_{t_1}^{t_2} \frac{dt}{ds} \left[\partial_c g_{ab} \delta u^c \frac{du^a}{dt} \frac{du^b}{dt} + 2 g_{ab} \frac{d(\delta u^a)}{dt} \frac{du^b}{dt} \right] dt$$

$$= \int_{s_1}^{s_2} \left[\frac{1}{2} \partial_c g_{ab} \delta u^c \frac{du^a}{ds} \frac{du^b}{ds} + g_{ab} \frac{d(\delta u^a)}{ds} \frac{du^b}{ds} \right] ds . \tag{30.108}$$

Now let's stipulate that the new path $u + \delta u$ has the same endpoints as the original path — in other words, we take $\delta u(s_1) = \delta u(s_2) = 0$. Then if we integrate the second term by parts, not only can we extract an overall factor of δu from the square brackets (with some dummy-index relabeling), we can do so while discarding the surface term; this yields the expression

$$\delta S = \int_{s_1}^{s_2} \left[\frac{1}{2} \partial_c g_{ab} \frac{du^a}{ds} \frac{du^b}{ds} - \partial_a g_{bc} \frac{du^a}{ds} \frac{du^b}{ds} - g_{ac} \frac{d^2 u^a}{ds^2} \right] \delta u^c \, ds . \tag{30.109}$$

This may still look like a mess, but due to the vanishing covariant derivative of g_{ab}, (30.99), each partial derivative of the metric can be replaced with Christoffel symbols. After some rearrangement, this yields the relatively simple expression

$$\delta S = - \int_{s_1}^{s_2} \left[\frac{d^2 u^c}{ds^2} + \Gamma_{ab}^c \frac{du^a}{ds} \frac{dx^b}{ds} \right] g_{cd} \, \delta u^d \, ds. \tag{30.110}$$

Since δu is completely arbitrary, the path $u(s)$ that solves the differential equation

$$\frac{d^2 u^c}{ds^2} + \Gamma_{ab}^c \frac{du^a}{ds} \frac{dx^b}{ds} = 0 \tag{30.111}$$

must be the trajectory for with $\delta S = 0$ This is the geodesic equation (30.104) — the covariant statement of Newton's first law. Thus we discover that the trajectory of a free particle has extremal path length between given points.

[This derivation is an example of the calculus of variations (see also BTW 14.2). In the language of Lagrangian mechanics, (30.111) is the Euler–Lagrange equation whose solution extremizes the action S with Lagrangian $\sqrt{|g_{ab} \, du^a du^b|}$.]

You may be forgiven for thinking the geodesic equation is a lot of effort for something as straightforward as a line. But we introduced Christoffel symbols to accommodate position-dependent bases, so they apply not only in $\mathbb{R}^n$, but to other spaces as well. Such spaces, which are continuous and differentiable everywhere, are called *differentiable manifolds*. Clearly $\mathbb{R}^n$ fits the bill; so too does the curved spherical surface S^2. A counter example is the cone, which is not differentiable at its tip.

A particularly intuitive way to regard differentiable manifolds is as collections of tangent spaces. For example, imagine planes tangent at various points in $\mathbb{R}^2$ and S^2. In $\mathbb{R}^2$, all the planes are indistinguishable copies of $\mathbb{R}^2$ itself; it has zero curvature. By contrast, the tangent planes of S^2 are distinct — both from the sphere and from one another. But as a differentiable manifold, these tangent planes must be in $\mathbb{R}^2$: and indeed, within each plane in a small-enough region around the tangent point P, the surface of a sphere is a flat Euclidean plane (Problem 30.21).[4] Cartesian coordinates can be used

---

[4] My favorite "proof" that the sphere is a differentiable manifold is historical: for thousands of years, people thought Earth was flat!

within each of its tangent planes, where the Christoffel symbols all vanish. It's only as one ventures along the sphere beyond a small neighborhood around P that the Γ's begin to deviate from zero. In this way, the Γ's reflect the curvature of the manifold; they are the "connections" between tangent planes.

Returning then to the issue of geodesics, what does it mean for a path to be "straight" on a curved manifold? The usual answer is that it's the shortest distance between two points. Which is fine as far as it goes — but BTW 30.2 notwithstanding, this is not manifest in the geodesic equation. Indeed, (30.104b) depends only on the local (initial) data provided by $u^b(t)$ and its first derivative, not the trajectory's endpoints. So how can you determine if every step along your path is straight?

Mathematically, the answer must be encoded in the geodesic equation, whose solution depends not merely on initial data, but also on Christoffel symbols — which, as (30.100) shows, are direct reflections of the space's metric. So "straight" must somehow depend on the space itself. For example, on the surface of a sphere, geodesics are segments of great circles connecting any two nearby points (Problem 30.28); these segments are also the shortest path between these points.[5] If we were to focus on a tiny piece of a great circle through an arbitrary point P on S^2, we'd find that it lies along a straight line in the plane tangent to the sphere at P. So a great circle can be construed as the succession of infinitesimal pieces of straight Euclidean lines. To local, two-dimensional observers confined to the surface of the sphere, a similarly confined free particle whizzing by will appear to take a straight line path; to an observer able to measure a finite stretch of the trajectory, however, the free particle will appear to accelerate! The implications of this for gravity are profound — see BTW 30.3.

BTW 30.3 The Principle of Equivalence

Galileo's apocryphal experiment dropping different weights from the Tower of Pisa demonstrated that the acceleration of gravity is independent of an object's mass. This is often passed over as a trivial fact since, by Newton's second law, the applied force mg is set equal to ma, and the m's cancel. But it's actually more subtle than that. Just as an applied electric field imparts a force qE on a charge, a gravitational field acts with a force mg on a mass. In this expression, mass behaves as a "charge" — commonly referred to as an object's gravitational mass m_g. By contrast, the ma of Newton's second law quantifies a particle's resistance to *any* applied force, and so is called the inertial mass m_i. Since these masses play very different physical roles, there's no reason they should be equal. And this has observable consequences: if the gravitational-to-inertial mass ratio depends on an object's composition, different objects would have different accelerations $a = (m_g/m_i)g$ in the same gravitational field. Yet increasingly precise measurements over centuries have shown that $m_g = m_i$. This implies that all particles with the same initial velocity — regardless of their composition or mass — have the same trajectory in a gravitational field. It's as if space itself establishes the path, rather than the interaction between an object and a gravitational field.

But as we know, there *are* special paths determined only by the properties of the space: any two nearby points can be connected by a unique geodesic of extremal length. Since the geodesic equation (30.111) is the tensor statement of Newton's first law, these are paths of free particles. So the equality of gravitational and inertial mass can be explained by regarding objects subject only to gravity as free particles — that is, by viewing gravity not as a force, but as arising from some characteristic which determines a space's geodesics. The deviation of geodesics from the expected Euclidean straight line motion would then be indicative of a gravitational force.

This invites speculation that gravitational orbits may be geodesics in curved three-dimensional space. But a unique orbit requires specification of both initial direction and speed, whereas a geodesic requires only an initial direction to define a trajectory — so this idea can quickly be

---

[5] The same two points are also connected by the longer, complementary segment of the same great circle — but that's a topological rather than a geometrical feature of S^2.

> discarded. Notice, however, that an initial direction in *spacetime* includes the initial velocity. Perhaps, then, we should expand our view to include both space and time, hypothesizing that gravitational orbits are geodesics in four-dimensional spacetime. This line of thought leads to general relativity.

Problems

30.1 Of the following, which are legal index expressions? Explain why the others are incorrect expressions. (Assume the summation convention.)

(a) $B_i C^i$
(b) $A_i = B_{ij} C^j$
(c) $A^j = B_{ji} C^i$
(d) A_{ii}
(e) A^k_k
(f) $T_{ij} = A_i B_j C_k C^k$
(g) $A^j_i = B_{ijk} C^k$
(h) $C_i = T_{jk} S^k_i U^{jk}$

30.2 Use (30.70) and (30.72) to show that δ^a_b is an invariant tensor, but δ_{ab} is not.

30.3 The contraction of dummy indices is only allowed when one is upstairs, the other downstairs. Use (30.70) and (30.72) to verify this by demonstrating that $A^j B_j$ is invariant under coordinate transformations while $A^j B^j$ is not. And as long as you're in the neighborhood, show that $A^j B_j = A_j B^j$.

30.4 Use basis completeness $\vec{e}^{\,b} \vec{e}_b = \mathbb{1}$ to verify that upstairs and downstairs bases are mutually orthonormal.

30.5 Confirm the mapping between the upstairs and downstairs bases in (30.34) for sheared coordinates.

30.6 (a) Find the downstairs basis $\{\vec{e}_r, \vec{e}_\phi, \vec{e}_z\}$ for cylindrical coordinates.
(b) Use the results of (a) to find the upstair basis $\{\vec{e}^{\,r}, \vec{e}^{\,\phi}, \vec{e}^{\,z}\}$ in terms of the conventional orthonormal $\{\hat{r}, \hat{\phi}, \hat{k}\}$ basis.
(c) For an arbitrary vector $\vec{A}$, find the relationship between components A_r, A_ϕ, A_z and A^r, A^ϕ, A^z.
(d) Verify that $A^i \vec{e}_i = A_i \vec{e}^{\,i}$.

30.7 The transformation between primed and unprimed systems is given by $A^{a'} = T^{a'}_{\ b} A^b$, such that $\vec{A} = \vec{e}_{a'} A^{a'} = \vec{e}_b A^b$.

(a) Use the mutual orthogonality of upstairs and downstairs bases within a system to find an expression of the transformation matrix $T^{a'}_{\ b}$ in terms of the basis vectors.
(b) Find the transformation matrix from sheared to cartesian systems $T^i_{\ b}$, where $i = x, y$ and $b = u, v$.
(c) Find the position $A^u = A^v = 1$ in the cartesian frame. Verify that the length of $\vec{A}$ is the same in both frames.
(d) Show that the shear metric g_{ab} can be transformed into the cartesian metric δ_{ij}, thereby demonstrating that shear coordinates describe $\mathbb{R}^2$. How does this approach compare to (30.74)?

30.8 Show that the cartesian metric $\eta_{\mu\nu}$ of special relativity (30.17) is an invariant tensor.

30.9 As we saw in (30.71), the gradient operator $\partial_a \equiv \partial/\partial u^a$ transforms as a downstairs vector.

(a) Write out the components of the four-gradient operator ∂_μ in Minkowski space. What are the components of ∂^μ? How does this compare with (30.53)?

(b) Give a physical interpretation for the operator $\partial_\mu \partial^\mu$.

(c) Write out in component form the four-divergence, $\partial_\mu A^\mu$. Referring to (16.31), what are the physical implications of a physical field A^μ with $\partial_\mu A^\mu = 0$?

30.10 As (30.59) shows, mass is a Lorentz scalar. So, for instance, a box containing a pion has the same mass as one containing the two photons it decays into. At least it will if mass is conserved under electromagnetism. Confirm this by showing that $\frac{d}{d\tau}\left(P_\mu P^\mu\right) = 0$, using (30.63) for $dP^\mu/d\tau$. [Hint: $F^{\mu\nu} = -F^{\nu\mu}$.]

30.11 We derived (30.63) based almost entirely on tensor considerations. But is it right? Use the field tensor $F^{\mu\nu}$ in (30.64) to verify that the spatial components $dP^i/d\tau$ give the familiar Lorentz force $q(\vec{E} + \vec{v} \times \vec{B})$. Can you find a physical interpretation for the time component? [Recall that $dt = \gamma d\tau$, $\vec{p} = \gamma m \vec{v}$ and $\mathcal{E} = \gamma mc^2 = P_0 c$ — where $\gamma = 1/\sqrt{1 - \beta^2}$, $\beta = v/c$. To avoid confusion with electric field E, we use $\mathcal{E}$ to denote energy; you may also find $F^{ij} = -\epsilon^{ijk} B_k$ helpful.]

30.12 In Example 30.6 we claimed that different observers measure different electromagnetic fields from the same source. Consider a long line of charge along the z-axis with uniform charge density λ. Alice, at rest next to the line, measures $\vec{E} = \hat{\rho}\lambda/2\pi\epsilon_0\rho$, $\vec{B} = 0$. Bob, walking parallel the line with speed v, sees the charge moving relative to him — so he measures not only a different $\vec{E}$, but a non-zero $\vec{B}$ as well.

(a) Use (30.64) to express Alice's observed fields in terms of the electromagnetic field tensor $F^{\mu\nu}$.

(b) Use (30.52) to write the 4×4 Lorentz transformation matrix Λ for boosts from Alice's frame to Bob's. [Recall $\gamma = 1/\sqrt{1 - \beta^2}$, $\beta = v/c$.]

(c) Use (30.52) and the transformation rule for rank-two tensors to find the electromagnetic fields measured in Bob's frame. You can do this as an explicit index sum — but matrix multiplication is much easier.

(d) Verify that the Lorentz scalar invariants of (30.65) are indeed the same for both observers.

30.13 **Electromagnetic Invariants**

(a) Find the electromagnetic field tensor $F_{\mu\nu}$ from the upstairs form $F^{\mu\nu}$ in (30.64).

(b) Use (30.66) to show that the tensor dual $G^{\mu\nu} = \frac{1}{2}\epsilon^{\mu\nu\alpha\beta} F_{\alpha\beta}$

$$G^{\mu\nu} = \begin{pmatrix} 0 & B_x & B_y & B_z \\ -B_x & 0 & -E_z/c & E_y/c \\ -B_y & E_z/c & 0 & -E_x/c \\ -B_z & -E_y/c & E_x/c & 0 \end{pmatrix}.$$

(c) Verify the scalar invariants of (30.65).

30.14 The four-vector electromagnetic current is $J^\mu = (\rho c, \vec{J})$, where ρ is the charge density, and $\vec{J}$ the current density. We can also express it in terms of the four-velocity of (30.56) as $J^\mu = \rho_0 U^\mu$. Interpret ρ_0 by considering the Lorentz scalar associated with J^μ. What's the covariant expression of charge conservation? (See Problem 30.9.)

30.15 (a) For $F_{\mu\nu}$ the electromagnetic field tensor found in Problem 30.13, show that

$$\partial_\mu F_{\alpha\beta} + \partial_\alpha F_{\beta\mu} + \partial_\beta F_{\mu\alpha} = 0$$

is the Lorentz covariant form of the homogeneous Maxwell equations in Table 16.1.

(b) Show that $\partial_\alpha F^{\alpha\beta} = \mu_0 J^\alpha$, where J^α is the electromagnetic four-current in Problem 30.14, is the Lorentz covariant form of the inhomogeneous Maxwell equations. Verify that this expression is consistent with charge conservation.

30.16 Use (30.74) to derive the $\mathbb{R}^3$ metric in cylindrical coordinates starting from spherical.

(a) Transform directly from the metric in spherical coordinates.

(b) Use a two-step process from spherical to cartesian to cylindrical by using (30.74) twice. Despite appearing more tedious than the direct method in (a), what makes this approach simpler?

30.17   Verify that $\det\left(\partial u/\partial u'\right) = 1/\det\left(\partial u'/\partial u\right)$, where $J \equiv \det\left(\partial u/\partial u'\right)$ is the Jacobian of the transformation from u to u'.

30.18   **Tensor Densities**

Under a tensor transformation, the determinant of the metric $g \equiv \det g_{ab}$ transforms not as a tensor, but rather as a *tensor density*, (30.82)

$$g = \left|\frac{\partial u'}{\partial u}\right| \left|\frac{\partial u'}{\partial u}\right| g',$$

where $\left|\partial u'/\partial u\right| \equiv \det\left(\partial u'/\partial u\right)$. With those two factors of $\left|\partial u'/\partial u\right|$, g is said to be a tensor density of weight -2.

(a) Demonstrate, in cartesian, cylindrical, and spherical coordinates, that $\sqrt{g}\,d\tau$, where $d\tau = du^1 du^2 du^3$, is a true scalar tensor. What tensor density weight does $d\tau$ have?

(b) Is the Levi-Civita epsilon

$$\epsilon_{a_1 a_2 \dots a_n} = \begin{cases} +1, & \{a_1 a_2 \dots a_n\} \text{ an even permutation of } \{123 \dots n\} \\ -1, & \{a_1 a_2 \dots a_n\} \text{ an odd permutation of } \{123 \dots n\} \\ 0, & \text{else} \end{cases}$$

a tensor or a tensor density? If the latter, what is its weight?

30.19   Consider the simple curve $C: y = x$, parameterised as $x = t, y = t$. Use (30.80) to find the distance along C from $t = 0$ to 1:

(a) in cartesian coordinates

(b) in the shear coordinates.

30.20   Using the shear metric, integrate to find the area outlined by the u-, v-axes between $u = v = 0$ and $u = a$, $v = b$. Interpret your result. [Note that in two-dimensional space, the invariant volume element described in (30.83) is an area element.]

30.21   (a) Find the metric on the surface of the sphere of radius R in conventional spherical coordinates.

(b) Use the metric to determine the physical distance s between the north pole and a point on the circle $\theta = \theta_0$. Observers at the north pole interpret s as the radial distance from their position.

(c) Find the circumference C of the circle $\theta = \theta_0$ as measured by observers on the sphere. How do they reconcile the fact that $C \neq 2\pi s$? Draw a sketch explaining the discrepancy.

(d) In a small region tangent to an arbitrary point on the sphere, show that the space can be approximated as a Euclidean plane.

30.22   **The Schwarzschild Metric**

In a static, spherically symmetric spacetime with mass M at the origin, the solution to Einstein's equation of general relativity is called the Schwarzschild metric — which in terms of the line element is

$$ds^2 = \left(1 - \frac{r_s}{r}\right)(c\,dt)^2 - \left(1 - \frac{r_s}{r}\right)^{-1} dr^2 - r^2\left(d\theta^2 + \sin^2\theta\,d\phi^2\right),$$

where G is Newton's gravitational constant, and the quantity $r_s \equiv 2GM/c^2$ is called the *Schwarzschild radius*. Since $g_{\mu\nu} = g_{\mu\nu}(r)$, this metric is both time independent and spherically symmetric. Note that for $r \gg r_s$, $g_{\mu\nu}$ reduces to the flat metric $\eta_{\mu\nu}$ of special relativity.

(a) Show that the physical distance Δs along a radial line between $r = r_1$ and r_2 (both greater than r_s) at fixed time is *greater* than $\Delta r = |r_2 - r_1|$. Thus the coordinate r is *not* radial distance.

(b) Find the physical area of a sphere centered at the origin at an instant of time. What physical interpretation does this confer on r?

(c) Find the proper time $\Delta \tau = \Delta s/c$ measured by a clock at rest at coordinate position $r > r_s$ in terms of coordinate time Δt. Under what circumstances do these times agree?

(d) Alice, at rest at infinity, watches as Bob leaves and heads to smaller r. What does Alice say about the passage of time for Bob as he nears the so-called "horizon" at $r = r_s$?

[The Schwarzschild metric has two singularities: $r = 0$ and r_s. The first is a real physical singularity — this is where the black hole is. The second can be shown to be a reflection of Schwarzschild coordinate r: though it has physical ramifications for Bob, no physical divergence occurs there.]

30.23 Examine the form of $D_\alpha D_\beta$ on a Lorentz scalar Φ, (30.98), and argue from general covariance that $\Gamma^\mu_{\alpha\beta}$ must be symmetric in its lower two indices.

30.24 Verify the expression

$$\Gamma^c_{ab} = \frac{1}{2} g^{cd} \left(\partial_b g_{ad} + \partial_a g_{db} - \partial_d g_{ba} \right)$$

for the Christoffel symbols Γ in two ways:

(a) Take the inner product of (30.86) with a basis vector and use (30.12).
(b) Using the vanishing of the covariant derivative of the metric, (30.99).

30.25 Verify the Christoffel symbols for polar coordinates in $\mathbb{R}^2$, (30.90), and spherical coordinates in $\mathbb{R}^3$, (30.92).

30.26 Find the Christoffel symbols for the sheared system.

30.27 **Index Mania!**
The results of Example 30.9 are essentially a repackaging of familiar scale factors into a tensorial context. But again, indices do not a tensor make.

(a) Following the approach in (30.84), verify that Γ^a_{bc} is not a tensor by showing that under a change of coordinates

$$\Gamma^{a'}_{b'c'} = \frac{\partial u^{a'}}{\partial u^f} \frac{\partial^2 u^f}{\partial u^{b'} \partial u^{c'}} + \frac{\partial u^{a'}}{\partial u^f} \frac{\partial u^d}{\partial u^{b'}} \frac{\partial u^e}{\partial u^{c'}} \Gamma^f_{de}. \tag{30.112}$$

This is why it's often referred to as the Christoffel "symbol."

(b) Use $\frac{\partial u^{a'}}{\partial u^f} \frac{\partial u^f}{\partial u^{c'}} = \delta^{a'}_{c'}$ to write the inhomogeneous term in (30.112),

$$\frac{\partial u^{a'}}{\partial u^f} \frac{\partial^2 u^f}{\partial u^{b'} \partial u^{c'}} = -\frac{\partial^2 u^{a'}}{\partial u^d \partial u^f} \frac{\partial u^d}{\partial u^{b'}} \frac{\partial u^f}{\partial u^{c'}}.$$

(c) Use these results to show that the covariant derivative, (30.95), is a tensor.

30.28 **Geodesics on S^2**

(a) Show that the geodesic equations on a sphere of radius R can be written

$$\frac{d^2\theta}{dt^2} - \sin\theta \cos\theta \left(\frac{d\phi}{dt} \right)^2 = 0 \qquad \frac{d}{dt} \left(\sin^2\theta \frac{d\phi}{dt} \right) = 0.$$

(b) Symmetry allows one to place the initial position on the equator and initial velocity along $\hat{\phi}$. Find the geodesic for these initial conditions.
(c) Argue that all geodesics on S^2 are segments of great circles.

30.29 (a) For D_a to be a true derivative operator, it should obey the product rule. Show that applying the product rule to $A^a B^b$ reproduces (30.96).
(b) We asserted in (30.97) that the covariant derivative of a tensor with a vector index flips the sign in front of the Christoffel symbol. Verify this for a downstairs vector A_a by taking the covariant derivative of $A^a A_a$.

30.30 (a) Use (30.12) and (30.86) to show that the covariant derivative of the metric vanishes.

(b) Verify that $D_a g_{bc} = 0$ in both cylindrical and spherical coordinates.

(c) Show that $\partial_a g_{bc} = 0$ on the surface of a cylinder, but not on the surface of the sphere. Can you explain the apparent discrepancy in terms of curvature?

30.31 Back in (12.49) we gave the form of the gradient, divergence, Laplacian, and curl in general curvilinear coordinates. Since the metric summarizes the spatial dependence of basis vectors $\mathbb{R}^3$, it's natural to return to these expressions.

(a) The gradient of a scalar in orthogonal coordinates is $\vec{\nabla}\Phi = \vec{e}^a \partial_a \Phi$. Rewrite this in term of the conventional unit vectors $\hat{e}_a$ in (30.41), and the metric elements g_{aa} (no implied sum). Compare your result to the expression in (12.49a).

(b) Use (30.41) to write down the tensorial divergence $D_a A^a$ in terms of the $A^{\hat{a}}$ component on the orthonormal basis $\hat{e}_a$. Then use (and prove, if you're so inclined)

$$\Gamma^a_{ab} = \frac{1}{\sqrt{g}} \partial_b \sqrt{g}, \qquad g = \det g_{ab}$$

to show that $D_a A^a$ reproduces (12.49c).

(c) Use the results above to find the Laplacian, and compare with (12.49b).

(d) Use the results of Problem 30.18 to write down the tensorial expression for the curl $(\vec{\partial} \times \vec{A})^a$. Then recast $(\vec{\partial} \times \vec{A})^1$ in terms of of the components of $\vec{A}$ on the $\hat{e}_a$ basis. Compare with (12.49d).

Part V

Orthogonal Functions

Mathematics is the art of giving the same name to different things.

Henri Poincaré

31 Prelude: 1 2 3 . . . Infinity

31.1 The Continuum Limit

We found in Example 28.7 that each normal mode $|\Phi_\ell\rangle$ of the N-chain with fixed ends is a finite N-dimensional vector in which the nth mass has amplitude given by (28.53),

$$\phi_n^{(\ell)} = \langle q_n|\Phi_\ell\rangle = \sin\left(\frac{n\ell\pi}{N+1}\right), \tag{31.1}$$

where both n and ℓ run from 1 to N. For $N = 4$, Figure 28.5 suggests that by inserting additional masses, each mode ℓ will start to resemble a standing wave; a full standing wave should emerge as the number of such insertions goes to infinity. This infinity is not, however, the simple $N \to \infty$ limit of an infinitely long chain. Rather, a standing wave arises only in the *continuum limit*: instead of leaving the mass spacing a fixed and letting the length L of the chain go to infinity as masses are added on, we take $N \to \infty$ while simultaneously letting $a \to 0$, such that $L = (N+1)a$ is held fixed. As we'll see, this transforms our punctuated chain of discrete masses m and springs κ into a continuous string of uniform mass density μ and tension T; the normal modes become the standing waves that can be seen emerging in Figure 28.5.

To determine the normal modes in the continuum limit, start by defining the location of the nth mass as $x \equiv na$. As the spacing a decreases, the blocks become closer and closer until in the limit x becomes a continuous variable. Thus the argument of the sine function in (31.1) becomes

$$\frac{n\ell\pi}{N+1} = \frac{\ell\pi(na)}{(N+1)a} \longrightarrow \frac{\ell\pi x}{L}, \tag{31.2}$$

so that the discrete n-component column vector of (28.52a) manifests as a continuous x-dependent function,

$$\begin{pmatrix} \sin(\frac{\ell\pi}{N+1}) \\ \sin(\frac{2\ell\pi}{N+1}) \\ \sin(\frac{3\ell\pi}{N+1}) \\ \vdots \\ \sin(\frac{N\ell\pi}{N+1}) \end{pmatrix} \longrightarrow \sin(\ell\pi x/L). \tag{31.3}$$

In other words, in the continuum limit the normal modes $|\Phi_\ell\rangle$ are represented on the position basis not as column vectors with discrete components $\phi_n^{(\ell)}$, but as continuous functions $\phi_\ell(x)$,

$$\phi_n^{(\ell)} = \sin\left(\frac{n\ell\pi}{N+1}\right) \longrightarrow \phi_\ell(x) = \sin(\ell\pi x/L), \tag{31.4}$$

where $0 \leq x \leq L$. Since with every additional mass comes an additional normal mode, ℓ now runs from 1 to ∞ — one for each standing wave of a fixed string with length L.

What about the normal mode frequencies, ω_ℓ? From (28.52b) we have[1]

$$\omega_\ell = 2\sqrt{\frac{\kappa}{m}} \sin\left(\frac{\ell\pi}{2(N+1)}\right)$$

$$= 2\sqrt{\frac{\kappa}{m}} \sin\left(\frac{\ell\pi a}{2L}\right) \approx \sqrt{\frac{\kappa}{m}}\left(\frac{\ell\pi a}{L}\right), \tag{31.5}$$

for small a. In the $a \to 0$ limit, the mass density $m/a \to \mu$ and tension $\kappa a \to T$ emerge, so that

$$\omega_\ell \longrightarrow \sqrt{\frac{T}{\mu}}\left(\frac{\ell\pi}{L}\right) = vk_\ell, \tag{31.6}$$

where we've identified $v = \sqrt{T/\mu}$ as the speed of a wave, and defined the wave number $k_\ell \equiv \ell\pi/L$.

Just as the N-component column vectors form a basis in N dimensions, we posit that the functions $\phi_\ell(x)$, with $\ell = 1$ to ∞, form a basis in an infinite-dimensional vector space. If so, then even in the continuum limit the general motion can be expanded on a basis of normal modes, as (28.55) becomes

$$q(x,t) = \sum_{\ell=1}^{\infty} A_\ell \sin(k_\ell x) \cos(\omega_\ell t + \delta_\ell), \tag{31.7}$$

with the sum over modes running to infinity.

Remarkably, two distinct infinities have emerged in the continuum limit. As the discrete masses merge into a continuous distribution, the column vector describing the collective displacements q_n along the chain becomes a field $q(x)$ on a string. In some imprecise sense we can think of this field as an infinite-dimensional column vector with continuous, rather than discrete, components. The continuum limit also results in the number of modes going to infinity. This one is a countable infinity describing normal modes $\phi_\ell(x)$ in an infinite-dimensional vector space. Both infinities are evident in (31.7): the discrete sum over modes ℓ, the continuum of positions x.

Ultimately, the validity of (31.7) as a general solution rests on the presumption that the functions $\sin(k_\ell x)$ constitute a basis. These well-behaved functions fulfill all the criteria of a vector space, such as closure and associativity. And that's a powerful insight. Even so, finite is not infinite. When expanding on a basis in a finite vector space, we never had to confront issues of convergence; for an infinite-dimensional space that's no longer true. But how does one define convergence of a vector sum? Similarly, though in the finite case the phrase "complete basis" might be considered redundant, in a space of infinite dimension how can we be sure we don't need additional vectors to complete a putative basis?

We'll address these issues in due course. For now we'll assume we have a basis, and turn our attention first to questions of orthogonality — or more broadly, how to apply our intuition of vector magnitude and relative orientation in finite dimensions to infinite-dimensional function space. We'll use this to leverage a discussion of the eigenvalue problem when the eigenvectors are represented by functions rather than column vectors; in the process, we'll discover in what sense an infinite set of functions can constitute a basis.

---

[1] For reasons that will become clear momentarily, we use κ to denote the spring constant.

31.2 An Inner Product of Functions

To introduce the concept of orthogonality amongst functions requires an inner product. Obviously, a discrete sum over components on a particular basis will not do. But in the usual way Riemann sums lead to integrals, in the continuum limit the inner product of vectors $|\phi\rangle$ and $|\psi\rangle$ becomes

$$\langle \psi|\phi \rangle = \sum_{n=1}^{N} \psi_n^* \phi_n \rightarrow \int_a^b \psi^*(x)\phi(x)\,dx, \tag{31.8}$$

for the closed interval $a \leq x \leq b$. It's straightforward to verify that this expression, which takes two vectors (functions) and returns a scalar, satisfies the criteria (25.26) for an inner product. So a family of functions $\phi_m(x)$ which satisfies

$$\langle \phi_\ell|\phi_m \rangle = \int_a^b \phi_\ell^*(x)\phi_m(x)\,dx = k_m \delta_{\ell m} \tag{31.9}$$

is orthogonal. Then the basis expansion is simply

$$|f\rangle = \sum_m c_m|\phi_m\rangle \longrightarrow f(x) = \sum_m c_m \phi_m(x), \tag{31.10}$$

and the projection in (25.135) manifests as

$$c_n = \frac{1}{k_n}\langle \phi_n|f \rangle = \frac{1}{k_n}\int_a^b \phi_n^*(x)f(x)\,dx. \tag{31.11}$$

Of course, all of this necessitates that these integrals converge. For arbitrary functions f and g, the Schwarz inequality of (25.40),

$$\left| \int_a^b f^*(x)g(x)\,dx \right|^2 \leq \left(\int_a^b |f(x)|^2\,dx \right)\left(\int_a^b |g(x)|^2\,dx \right), \tag{31.12}$$

is meaningful only for vectors with finite norm,

$$\langle f|f \rangle = \int_a^b |f(x)|^2\,dx < \infty. \tag{31.13}$$

Functions which satisfy this criterion are said to be "square integrable" — and only for square-integrable functions is (31.8) a valid inner product. Thus full characterization of an inner product requires the specification of the integration limits a and b. Complex-valued, square-integrable functions are said to belong to the vector space $\mathbb{L}^2[a,b]$; when the limits are $\pm\infty$, it's sometimes called $\mathbb{L}^2[\mathcal{R}]$; ("ell-two of ar"). The $\mathbb{L}$ honors the French mathematician Henri Lebesgue (please don't pronounce the "s"), while the "2" refers to the exponent in (31.13).[2]

Note that, context aside, (31.12) is a standard calculus result (Problem 31.2). Indeed, computing the $\mathbb{L}^2$ inner product is merely an exercise in integration — just as computing the $\mathbb{R}^n$ dot product is an exercise in matrix multiplication. Neither procedure has an inherent relation to the abstract concept of inner product. And this distinction is a large part of the power of bra-ket notation: the details of an inner-product calculation are irrelevant to the general properties of the space. Bras and kets allow us to

---

[2] Incidentally, amongst the family of Lebesgue spaces $\mathbb{L}^p$, only $p = 2$ yields an inner product space with all its attendant intuitive properties of magnitude and relative orientation of vectors (see Reed and Simon 1980).

highlight features such as projections without getting distracted by the details of any particular inner product space. And often a basic integral need only be computed once, ready to be deployed at any time. For example, the normal modes $|\Phi_\ell\rangle$ of the fixed string are orthogonal (Problem 31.5),

$$\int_0^L \sin{(\ell\pi x/L)} \sin{(m\pi x/L)} \, dx = \frac{L}{2} \delta_{\ell m}. \tag{31.14}$$

Thus any time we need the inner product of these normal modes, we simply write

$$\langle\Phi_\ell|\Phi_m\rangle = \frac{L}{2} \delta_{\ell m}. \tag{31.15}$$

The fact that this came from an integral is largely irrelevant.

Inner product spaces like $\mathbb{R}^n$ and $\mathbb{L}^2$ are examples of what are known as *Hilbert spaces*. Examples abound, from classical mechanics to signal processing. But perhaps its most notable appearance in physics is in quantum mechanics, where the state of a system $|\psi\rangle$ is a vector in Hilbert space. The ability to normalize a state,

$$\langle\psi|\psi\rangle = 1,$$

depends on the square integrability of wavefunctions, (31.13). Without this, the standard probabilistic interpretation of wavefunctions would not be viable.

Hilbert space extends the familiar properties of physical Euclidean space to inner product spaces more generally, with virtually all our intuition intact. Though we may find ourselves a long way from the picture of arrows in space, the concepts of magnitude and relative orientation still pertain. And perhaps most crucially, a defining feature of Hilbert space is that infinite sums converge to vectors in the space. This ultimately is what justifies the normal mode expansion of (31.7).

In this unit, we'll explore examples of orthogonal functions which regularly appear in physics and engineering. Many of these are commonly defined as solutions to differential equations — but in an effort to instill insight and intuition, we focus first on their properties, most notably their orthogonality. We'll then leverage our familiarity with these functions when we discuss techniques for solving differential equations in Part VI.

BTW 31.1 To Infinity — and Beyond!

The two infinities in (31.7) — the normal modes ℓ and the position x — are not mathematically equivalent. In fact, they're not even the same size. The normal modes can be ordered and labeled, $\phi_1, \phi_2, \phi_3, \ldots$. Though we'll never get to the end, we can be certain we haven't missed any. Formally, this is known as being in one-to-one correspondence with the integers, the paradigmatic infinite ordered set. Any infinite set such as the ϕ_n which can be put in one-to-one correspondence with the integers is said to be a *countable* or *denumerable* infinity. By contrast, the positions x cannot be ordered: as a continuum of values, there is no way to go from, say, $x = 1$ to $x = 1 + \epsilon$ without skipping over $1 + \epsilon/2$ for all positive ϵ no matter how small. Such a set is called a *continuous* or *non-denumerable* infinity. Since it cannot be ordered, it is a *larger* infinity than the set of integers.

Georg Cantor used the Hebrew letter $\aleph$ ("alef") to denote a denumerable infinity as $\aleph_0$ and a non-denumerable ones as $\aleph_1$. So taking the continuum limit of a discrete chain of masses results in an $\aleph_0$ infinity of modes, each with an $\aleph_1$ infinity of positions. Incredibly, these are but two of an infinity of different types of infinite sets denoted by the so-called *transfinite* cardinal numbers $\aleph_n$. (For more, see Halmos (2017), originally published in 1960.)

Problems

31.1 How do we know that the normalization factor k_n in (31.9) must be real and positive for any orthogonal basis?

31.2 Start with the double integral $\int_a^b |\psi(x)\phi(y) - \psi(y)\phi(x)|^2 dx\,dy$ over the closed interval $a \le x \le b$ to verify the Schwarz inequality, (31.12) — no inner product spaces necessary.

31.3 Use the Schwarz inequality to show that the sum of square-integrable functions is also square integrable.

31.4 (a) Show that over a finite interval $[a, b]$, square-integrable functions are also absolutely integrable.

(b) Show that the converse isn't true for the function $\frac{1}{\sqrt{x}}$ between 0 and 1.

(c) Show that absolutely integrable functions are also square integrable if the function is bounded — that is, for functions with $\max_{x \in [a,b]} f(x) = M < \infty$.

31.5 Show that the normal modes of the fixed string, $\phi_\ell(x) = \sin(\ell\pi x/L)$ satisfy the orthogonality relation

$$\int_0^L \sin(\ell\pi x/L)\sin(m\pi x/L)\,dx = \frac{L}{2}\delta_{\ell m}.$$

[Hint: Use the trig identity for $\cos(A \pm B)$.]

31.6 Refer to Problem 6.32b to show that (28.42) in the continuum limit $\phi_n \to \phi(x)$ becomes the differential equation

$$\frac{d^2\phi}{dx^2} = -k^2\phi,$$

where $\omega/k = v = \sqrt{T/\mu}$ as in (31.6). (Be careful to distinguish the spring constant κ from the wave number k.) This is an eigenvalue equation for the operator $\mathcal{D}^2 = d^2/dx^2$; possible eigenfunction solutions are the sines of (31.4). [As we'll see, the restriction to sines and particular eigenmodes ℓ comes from boundary conditions, not the differential equation.]

32 Eponymous Polynomials

We begin our foray into orthogonal functions by considering polynomials. Polynomials, of course, are functions of the form $\sum_{\ell=0}^{n} a_\ell x^\ell$, where the a_ℓ are constants and the integer n is finite and non-negative. As we saw in Example 24.10, the space of all polynomials of degree $\leq n$ is an $(n+1)$-dimensional vector space; an obvious choice of basis is the set $\{x^\ell\}$, $\ell = 0, 1, 2, \ldots, n$. But is this an orthogonal basis? The question, of course, is meaningless without first specifying an inner product. We'll use three different inner products and the Gram–Schmidt procedure (Section 25.4) to generate three of the most common orthogonal bases of polynomials from the linearly independent set $\{x^\ell\}$.

32.1 Legendre Polynomials

The inner product[1]

$$\langle \phi_m | \phi_\ell \rangle = \int_a^b \phi_m(x)\phi_\ell(x)\,dx \tag{32.1}$$

is not fully specified without fixing the integration limits. Since our integrands consist of positive powers of x, convergence requires finite limits; the simplest choice is ± 1. Then our inner product is

$$\langle \chi_m | \chi_\ell \rangle = \int_{-1}^{1} x^m x^\ell \, dx = \begin{cases} \frac{2}{\ell+m+1}, & \ell + m \text{ even} \\ 0, & \ell + m \text{ odd,} \end{cases} \tag{32.2}$$

where we take x^ℓ to be the polynomial representation of the ket $|\chi_\ell\rangle$. Although parity considerations render the odd vectors orthogonal to the even vectors, that alone does not yield a fully orthogonal basis. To construct an orthonormal basis of polynomials $\hat{P}_\ell(x)$ using Gram–Schmidt, we begin with $\ell = 0$ and $x^0 = 1$. The normalization condition $\langle \hat{P}_0 | \hat{P}_0 \rangle = 1$ gives the first (or "zeroth") basis vector

$$|\hat{P}_0\rangle \rightarrow \hat{P}_0(x) \equiv \frac{1}{\sqrt{2}}. \tag{32.3}$$

For the next vector, we need to remove from $|\chi_1\rangle$ its component orthogonal to $|\hat{P}_0\rangle$ — but since x^0 and x^1 have opposite parity, they're already orthogonal,

$$|\hat{P}_1\rangle = c\left(|\chi_1\rangle - |\hat{P}_0\rangle \underbrace{\langle \hat{P}_0 | \chi_1 \rangle}_{0} \right) = c|\chi_1\rangle, \tag{32.4}$$

---

[1] We're only considering real polynomials, so we'll drop the complex conjugation.

which in integral form is

$$\hat{P}_1(x) = c \left[x - \frac{1}{\sqrt{2}} \left(\int_{-1}^{1} \frac{1}{\sqrt{2}} x' \, dx' \right)^{0} \right] = cx. \tag{32.5}$$

The normalization condition $\langle \hat{P}_1 | \hat{P}_1 \rangle = 1$ fixes the coefficient c to give

$$\hat{P}_1(x) = \sqrt{\frac{3}{2}} x. \tag{32.6}$$

Similarly, for $n = 2$ we need to remove the $|\hat{P}_0\rangle$ and $|\hat{P}_1\rangle$ components from $|\chi_2\rangle$ — only one of which is non-zero:

$$|\hat{P}_2\rangle = c \left(|\chi_2\rangle - |\hat{P}_0\rangle \langle \hat{P}_0 | \chi_2 \rangle - |\hat{P}_1\rangle \langle \hat{P}_1 | \chi_2 \rangle^{0} \right), \tag{32.7}$$

or

$$\hat{P}_2(x) = c \left[x^2 - \frac{1}{\sqrt{2}} \left(\int_{-1}^{1} \frac{1}{\sqrt{2}} x'^2 \, dx' \right) - \sqrt{\frac{3}{2}} x \left(\int_{-1}^{1} \sqrt{\frac{3}{2}} x'^3 \, dx' \right)^{0} \right]$$

$$= c \left(x^2 - \frac{1}{3} \right). \tag{32.8}$$

Normalizing gives

$$\hat{P}_2(x) = \sqrt{\frac{5}{2}} \frac{1}{2} (3x^2 - 1). \tag{32.9}$$

We can of course continue in this rather tedious fashion, but it's worth pausing to note an emerging pattern: each of the $\hat{P}_\ell$'s in (32.3), (32.6), and (32.9) is multiplied by a factor of $\sqrt{\frac{1}{2}(2\ell + 1)}$. It is both cleaner and conventional to absorb this factor into the polynomials themselves, thereby defining a set of orthogonal (rather than orthonormal) polynomials $P_\ell(x) = \sqrt{\frac{2}{2\ell+1}} \hat{P}_\ell(x)$ such that

$$\langle P_\ell | P_m \rangle = \int_{-1}^{1} P_\ell(x) P_m(x) \, dx = \frac{2}{2\ell + 1} \delta_{\ell m}. \tag{32.10}$$

The orthogonal set of $P_\ell(x)$'s are known as the *Legendre polynomials*, the first several of which are listed inside the back cover; the first four are plotted in Figure 32.1. The normalization choice in (32.10) gives the P_ℓ's the same normalization as the elements of the original set $\{x^\ell\}$; this choice also results in the convention $P_\ell(1) = 1$ for all ℓ.

We can use the orthogonal basis of Legendre polynomials to expand a function,

$$f(x) = \sum_{\ell=0}^{\infty} c_\ell P_\ell(x), \quad |x| \le 1. \tag{32.11}$$

The expansion coefficients c_ℓ are determined by orthogonality,

$$\langle P_\ell | f \rangle = \langle P_\ell | \left(\sum_m c_m | P_m \rangle \right)$$

$$= \sum_m c_m \langle P_\ell | P_m \rangle = \sum_m c_m \frac{2}{2m + 1} \delta_{m\ell} = \frac{2}{2\ell + 1} c_\ell. \tag{32.12}$$

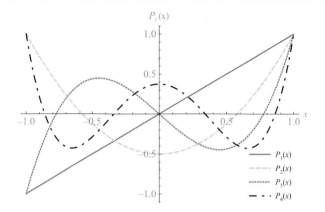

Figure 32.1 Legendre polynomials $P_\ell(x)$, $\ell = 1$ to 4.

Thus

$$c_\ell = \frac{2\ell + 1}{2} \langle P_\ell | f \rangle = \frac{2\ell + 1}{2} \int_{-1}^{1} P_\ell(x) f(x)\, dx, \tag{32.13}$$

so that

$$f(x) = \sum_{\ell=0} c_\ell P_\ell(x) = \sum_{\ell=0} \left(\frac{2\ell + 1}{2} \int_{-1}^{1} P_\ell(x') f(x')\, dx' \right) P_\ell(x). \tag{32.14}$$

You should convince yourself that, together, (32.12)–(32.14) are equivalent to inserting a complete set of states:

$$|f\rangle = \sum_{\ell=0} |\hat{P}_\ell\rangle \langle \hat{P}_\ell | f \rangle = \sum_{\ell=0} \frac{2\ell + 1}{2} |P_\ell\rangle \langle P_\ell | f \rangle. \tag{32.15}$$

Example 32.1 **Legendre Series of $f(x) = |x|$**

Applying (32.14) to $f(x) = |x|$,

$$|x| = \sum_{\ell=0} \left(\frac{2\ell + 1}{2} \int_{-1}^{1} P_\ell(x') |x'|\, dx' \right) P_\ell(x)$$

$$= \sum_{\ell \text{ even}} \left(\frac{2\ell + 1}{2} \int_{-1}^{1} P_\ell(x') |x'|\, dx' \right) P_\ell(x)$$

$$= \sum_{\ell \text{ even}} (2\ell + 1) \left(\int_{0}^{1} P_\ell(x') x'\, dx' \right) P_\ell(x). \tag{32.16}$$

The first three non-zero coefficients are easily found to be $c_0 = 1/2$, $c_2 = 5/8$, and $c_4 = -3/16$; all the odd c_ℓ's vanish. Thus

$$|x| = \frac{1}{2} P_0(x) + \frac{5}{8} P_2(x) - \frac{3}{16} P_4(x) + \cdots, \tag{32.17}$$

valid through order x^5. In general, (32.11) gives an infinite series expansion — though if $f(x)$ is itself an mth-order polynomial, its Legendre series will terminate at $\ell = m$ (Problem 32.3).

| Example 32.2 | **Electric Potential of a Spherical Shell** |

Noting that $|\cos\theta| \leq 1$, together with $0 \leq \theta \leq \pi$ in spherical coordinates, we see that a function $f(\theta)$ can be expanded on the basis of Legendre polynomials $P_\ell(\cos\theta)$. Consider for instance a non-conducting spherical shell of radius a with an electric potential $V_0(\theta)$ on its surface. The absence of any ϕ-dependence allows us to expand this surface voltage in a Legendre series

$$V_0(\theta) = \sum_{\ell=0} c_\ell P_\ell(\cos\theta), \tag{32.18}$$

with expansion coefficients given by orthogonality, (32.13):

$$c_\ell(a) = \frac{2\ell+1}{2} \int_0^\pi P_\ell(\cos\theta)V_0(\theta) \sin\theta \, d\theta. \tag{32.19}$$

This integral only gives the coefficients at $r = a$ — which by itself is not terribly useful if we already know $V_0(\theta)$. However, (32.19) would become valuable if we can leverage it to determine the coefficients $c_\ell(r)$ off the sphere and thereby obtain a Legendre expansion $V(r,\theta) = \sum c_\ell(r)P_\ell(\cos\theta)$. To do this, first convince yourself that the $c_\ell(r)$ have units of voltage, and so must be functions of the dimensionless ratio r/a. Then note that since $V(r,\theta)$ must go to zero at infinity, the $c_\ell(r)$ must as well. But for finite r too far away to resolve angular variations due to the surface potential $V_0(\theta)$, the sphere will resemble a point charge — and so V must have the familiar and simple form $1/r$, independent of θ. Since only $P_0(\cos\theta)$ is independent of θ, we conclude that at large $r \gg a$, the Legendre sum for $V(r,\theta)$ must reduce to a single term proportional to $\left(\frac{a}{r}\right) P_0(\cos\theta) = \frac{a}{r}$. The implication then is that contributions for $\ell > 0$ must have coefficients which vary with higher powers of a/r (otherwise they'd be present for very large r). And indeed, as we move to smaller r the increasingly prominent variations in θ are encoded with higher-ℓ Legendre polynomials, each accompanied by a coefficient proportional to $(a/r)^{\ell+1}$. (We'll validate this in Example 32.5, and later in (41.72).) Introducing the constants b_ℓ,

$$c_\ell(r) = b_\ell \left(\frac{a}{r}\right)^{\ell+1}, \tag{32.20}$$

so that $b_\ell \equiv c_\ell(a)$, the potential for all $r \geq a$ can be written as the expansion

$$V(r > a, \theta) = \sum_{\ell=0} b_\ell \left(\frac{a}{r}\right)^{\ell+1} P_\ell(\cos\theta), \tag{32.21}$$

which reduces to (32.18) at $r = a$. Together, (32.19)–(32.21) provide a complete expression for the potential outside the sphere. (Problem 32.5 explores the region $r < a$.) The individual terms in the series for $V(r)$ represent contributions from different averages, or *moments*, of that charge distribution. The simplest is the $\ell = 0$ term which, having no θ dependence, gives the contribution to $V(r)$ without regard for the arrangement of charge on the sphere; this is known as the monopole term. By contrast, the $\ell = 1$ term is proportional to $\cos\theta$, representing the dipole component of $V(r)$. Similarly, $\ell = 2$ is the quadrupole moment, and $\ell = 3$ the octupole (since $2^\ell = 4$ and 8, respectively). The decomposition (32.21) is called the *multipole expansion* of the potential.

32.2 Laguerre and Hermite Polynomials

Since Legendre polynomials $P_\ell(x)$ are only defined for $|x| \leq 1$, they cannot serve as a basis for functions outside this interval. In principle we could construct versions of Legendre polynomials for $|x| \leq a$ for any finite a, but over an infinite domain the divergent integrals of x^n preclude (32.1) serving as an inner product. From a calculus standpoint, though, creating a convergent integral isn't terribly difficult: simply insert a function $w(x)$ such that

$$\langle\psi|\phi\rangle = \int_a^b \psi^*(x)\phi(x)\,w(x)\,dx \tag{32.22}$$

is finite. But wait — can we simply insert a factor into the integral and declare it an inner product? No, of course not; an inner product must obey certain conditions. But a look back at the criteria listed in (25.26) shows that as long as $w(x) \geq 0$ on the domain, (32.22) is a legitimate inner product. Including such *weight functions* in the definition of an inner product allows consideration of non-square-integrable functions $\phi(x)$ as vectors, with all the usual Schwarz-based intuition this implies. Thus the question of whether functions are vectors in an inner product space requires specification of a weight function in addition to the integration limits; as a result, the space of such vectors is often denoted $\mathbb{L}^2[a,b;w]$.[2]

For the basis $\{x^n\}$, we can accommodate the interval $0 \leq x < \infty$ by introducing the weight function e^{-x} to dampen out run-away contributions for large x,

$$\langle\chi_\ell|\chi_m\rangle = \int_0^\infty x^\ell x^m e^{-x}\,dx. \tag{32.23}$$

Although this inner product doesn't allow for parity arguments, that doesn't mean the integrals are difficult. Indeed for polynomials, they're all of the form

$$\int_0^\infty x^n e^{-x}\,dx = \Gamma(n+1) = n!\,. \tag{32.24}$$

With this, we can apply the Gram–Schmidt procedure to derive the *Laguerre polynomials* $L_n(x)$, with orthogonality condition

$$\langle L_n|L_m\rangle = \int_0^\infty L_n(x)L_m(x)\,e^{-x}\,dx = \delta_{nm}. \tag{32.25}$$

Note that unlike the Legendre polynomials by convention, the L_n's are defined to have unit norm. The first few Laguerre polynomials are given inside the back cover; the first four are plotted in Figure 32.2.

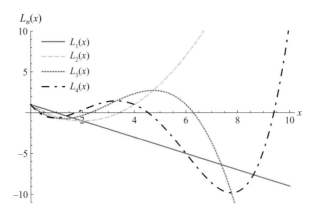

Figure 32.2 Laguerre polynomials $L_n(x)$, $n = 1$ to 4.

---

[2] Sometimes one says a function ϕ is square-integrable with weight function w — though we can always define $\tilde{\phi} \equiv \sqrt{w}\phi$ to obtain a true square-integrable function. Whether one does this is largely a matter of convention or convenience.

As with Legendre polynomials, we can use Laguerre polynomials to expand a function for all finite $x \geq 0$,

$$f(x) = \sum_{n=0} c_n L_n(x), \quad 0 \leq x < \infty, \tag{32.26}$$

where, by orthogonality (Problem 32.4),

$$c_n = \langle L_n | f \rangle = \int_0^\infty L_n(x) f(x) e^{-x} \, dx. \tag{32.27}$$

Example 32.3 **Laguerre Series of $f(x) = |x|$**

Applying (32.27) to $f(x) = |x|$,

$$|x| = \sum_{n=0} \left(\int_0^\infty L_n(x') |x'| e^{-x'} dx' \right) L_n(x)$$

$$= \sum_{n=0} \left(\int_0^\infty L_n(x') x' e^{-x'} dx' \right) L_n(x). \tag{32.28}$$

In this case, the only non-zero coefficients are $c_0 = 1$ and $c_1 = -1$, so that the Laguerre series is

$$|x| = L_0(x) - L_1(x). \tag{32.29}$$

A glance at the table inside the back cover shows this is simply $|x| = x$. But this shouldn't be a surprise, since in choosing a Laguerre expansion we've restricted the domain to $x \geq 0$, on which $f(x)$ is in fact a first-order polynomial.

If we want an orthogonal basis for all $|x| < \infty$, the weight factor e^{-x} is no help. Instead, we choose the Gaussian weight $w(x) = e^{-x^2}$ to get the inner product

$$\langle \chi_\ell | \chi_m \rangle = \int_{-\infty}^\infty x^\ell x^m e^{-x^2} \, dx. \tag{32.30}$$

Parity considerations apply, so that (see Problem 8.19)

$$\int_{-\infty}^\infty x^n e^{-x^2} \, dx = \begin{cases} \Gamma(\frac{n+1}{2}), & n \text{ even} \\ 0, & n \text{ odd} \end{cases}. \tag{32.31}$$

Applying Gram–Schmidt results in the orthogonal basis of *Hermite Polynomials* $H_n(x)$, which satisfy

$$\langle H_n | H_m \rangle = \int_{-\infty}^\infty H_n(x) H_m(x) e^{-x^2} \, dx = \sqrt{\pi} \, 2^n n! \, \delta_{nm}. \tag{32.32}$$

The first few Hermite polynomials are listed inside the back cover; the first four are plotted in Figure 32.3. Expanding a function for all finite x on a basis of Hermite polynomials takes the expected form

$$f(x) = \sum_m c_m H_m(x), \quad |x| < \infty, \tag{32.33}$$

where orthogonality gives (Problem 32.4),

$$c_m = \frac{1}{\sqrt{\pi} \, 2^m m!} \langle H_m | f \rangle = \frac{1}{\sqrt{\pi} \, 2^m m!} \int_{-\infty}^\infty H_m(x) f(x) e^{-x^2} \, dx. \tag{32.34}$$

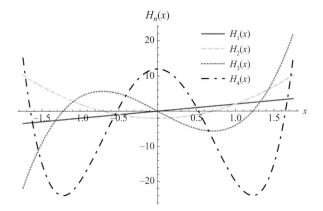

Figure 32.3 Hermite polynomials $H_n(x)$, $n = 1$ to 4.

Example 32.4 **Hermite Series of $f(x) = |x|$**

Applying (32.34) to $f(x) = |x|$,

$$|x| = \sum_m \left(\frac{1}{\sqrt{\pi}\, 2^m m!} \int_{-\infty}^{\infty} H_m(x')|x'|\, e^{-x'^2} dx' \right) H_m(x)$$

$$= \sum_{m \text{ even}} \left(\frac{2}{\sqrt{\pi}\, 2^m m!} \int_0^{\infty} H_m(x')\, x'\, e^{-x'^2} dx' \right) H_m(x). \tag{32.35}$$

Because $f(x)$ is even, only the even m's contribute; the first three non-vanishing coefficients give the fourth-order Hermite series

$$|x| = \frac{1}{\sqrt{\pi}} \left(H_0(x) + \frac{1}{2} H_2(x) - \frac{1}{96} H_4(x) + \cdots \right). \tag{32.36}$$

Comparison for all three representations of $f(x) = |x|$ through fourth order is shown in Figure 32.4. The Laguerre series is exact — at least for $x \geq 0$, where the L_n are defined. Though the other two series extend to negative x, only the Hermite polynomials are defined for all x. Including more terms in these two expansions improves the approximation — in particular, near the origin where $f'(x)$ is discontinuous.

How does expansion in orthogonal polynomials compare with Taylor series? In the case of $|x|$, the discontinuity of its derivative at the origin means that a Taylor expansion around $x = 0$ does not exist. Similarly, although the Taylor series of $1/(1 + x^2)$ exists, it's restricted by a radius of convergence of 1 (see BTW 6.4). By contrast, the Legendre series converges uniformly for $|x| \leq 1$, the Laguerre series for $x \geq 0$, and the Hermite series for all x. Limitations of Taylor series stem from a reliance on local data provided by the derivatives of the function at some chosen point. By contrast, a sum over an orthogonal basis of functions is constructed from data distilled from integrals of the function over the specified domain of the basis functions. The individual contributions can be interpreted physically in various ways, depending on the application. In Example 32.2, we saw that each P_ℓ in the Legendre series of (32.21) gives a multipole contribution to the potential $V(r)$. In quantum mechanics, a Legendre series usually represents an expansion over states with definite values of orbital angular momentum ℓ. The quantum mechanical bound states of the electron in the hydrogen atom include Laguerre polynomials L_m in their wavefunctions, with quantized energies dependent on the index m. Hermite polynomials similarly underlie states of definite energy for the quantum harmonic oscillator (see Example 39.9).

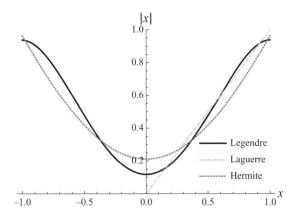

Figure 32.4 Fourth-order expansions of $|x|$.

32.3 Generating Functions

The Gram–Schmidt procedure is by no means the only pathway to orthogonal polynomials. Nor, despite being conceptually straightforward, does it tells us much about their properties beyond orthogonality. Other approaches provide methods and tools for analyzing and working with these polynomials. Let's start down these alternative paths with some physics.

Example 32.5 **Expanding the Electric Potential**

The electric potential at $\vec{r}$ due to a point charge q at position $\vec{r}'$ is given by the familiar expression $V(\vec{r}) = \frac{1}{4\pi\epsilon_0}\frac{q}{|\vec{r}-\vec{r}'|}$. We'll choose a coordinate system for which the line segment connecting these two points is along z; in other words, we take $\vec{r} = r\hat{k}$ and $\vec{r}' = r'\hat{k}$. Then the electric potential is proportional to

$$\frac{1}{|\vec{r}-\vec{r}'|} = \frac{1}{|r-r'|}.$$ (32.37)

For $r' < r$, we can express this as a binomial expansion,

$$\frac{1}{r(1-r'/r)} = \frac{1}{r}\sum_{\ell=0}^{\infty}\left(\frac{r'}{r}\right)^{\ell}, \qquad r' < r.$$ (32.38)

For $r' > r$ we'd factor out r' instead of r to obtain an equivalent expression in powers of r/r'. We can combine these possibilities in the single expression

$$\frac{1}{|r-r'|} = \frac{1}{r_>}\sum_{\ell=0}^{\infty}\left(\frac{r_<}{r_>}\right)^{\ell} = \frac{1}{r_>}\sum_{m=0}^{\infty}t^{\ell}, \qquad t \equiv \frac{r_<}{r_>},$$ (32.39)

where $r_>$ and $r_<$ are the greater and lesser of r and r', respectively.

This expression relates the distance $R = |\vec{r}-\vec{r}'| = |r-r'|$ to the ratio of the magnitudes of r and r'. So if we rotate $\vec{r}$ to an angle θ off the z-axis, the only change to the right-hand side of (32.39) would be some function of θ — most likely a different function of θ for each term in the sum, each of which must reduce to 1 for $\theta = 0$. So we introduce a family of functions $P_{\ell}(\cos\theta)$ with $P_{\ell}(1) = 1$, such that

$$\frac{1}{|\vec{r} - \vec{r}'|} = \sum_{\ell=0} \frac{r_<^\ell}{r_>^{\ell+1}} P_\ell(\cos\theta) = \frac{1}{r_>} \sum_{\ell=0} t^\ell P_\ell(\cos\theta). \tag{32.40}$$

Note that the introduction of these unknown P_ℓ's stems from completely general considerations — but since the left-hand side is a *known* function of θ, we can fully solve for them. Letting $x \equiv \cos\theta$, we use

$$|\vec{r} - \vec{r}'| = \sqrt{r^2 + r'^2 - 2rr'\cos\theta} \equiv r_> \sqrt{1 + t^2 - 2xt} \tag{32.41}$$

to get

$$G(x,t) \equiv \frac{1}{\sqrt{1 - 2xt + t^2}} = \sum_{\ell=0} t^\ell P_\ell(x). \tag{32.42}$$

The expansion on the right defines the $P_\ell(x)$'s as the Taylor coefficients of the power series expansion in t; we can take derivatives of the closed form of G on the left to extract them,

$$P_\ell(x) = \frac{1}{\ell!} \left. \frac{\partial^\ell G(x,t)}{\partial t^\ell} \right|_{t=0}. \tag{32.43}$$

$G(x,t)$ is called a *generating function* because (32.43) can be used to "generate" the unknown P_ℓ's; straightforward manipulation reveals

$$P_0 = 1 \qquad P_1 = x \qquad P_2 = \frac{1}{2}(3x^2 - 1) \qquad P_3 = \frac{1}{2}(5x^3 - 3x) \quad \cdots, \tag{32.44}$$

which are none other than the Legendre polynomials. (Indeed, the similarity of (32.40) to (32.21) is no coincidence.)

Similar to (32.43) is the *Rodrigues formula* (Problem 32.15)

$$P_\ell(x) = \frac{(-1)^\ell}{2^\ell \ell!} \left[\frac{d}{dx} \right]^\ell \left(1 - x^2\right)^\ell, \tag{32.45}$$

which can be used to find Legendre polynomials directly without the need of the auxiliary variable t. Generating functions and Rodrigues formulas for Legendre and other orthogonal polynomials are presented in Table 32.1; they are especially useful for determining properties and identities of their polynomials.

Table 32.1 Generating functions and Rodrigues formulas

Polynomial ϕ_n	c_n	$G(x,t) = \sum_n c_n \phi_n(x) t^n$ $G(x,t)$	$\phi_n = \frac{1}{h_n w(x)} \frac{d^n}{dx^n} \left[w(x) f(x)^n \right]$ h_n	$w(x)$	$f(x)$
Legendre P_n	1	$(1 - 2xt + t^2)^{-1/2}$	$(-1)^n 2^n n!$	1	$1 - x^2$
Hermite H_n	$\frac{1}{n!}$	$e^{2xt - t^2}$	$(-1)^n$	e^{-x^2}	1
Laguerre L_n	1	$\frac{e^{-xt/(1-t)}}{1-t}$	$n!$	e^{-x}	x
Chebyshev T_n	1	$\frac{1 - xt}{1 - 2xt + t^2}$	$(-1)^n 2^{n+1} \frac{\Gamma(n+1/2)}{\sqrt{\pi}}$	$(1 - x^2)^{-1/2}$	$1 - x^2$

Example 32.6 **Generating Orthogonality**

Consider the following integral of the Legendre $G(x,t)$,

$$\sum_{nm} t^n t^m \int_{-1}^{1} P_n(x) P_m(x) dx = \int_{-1}^{1} G^2(x,t)\, dx$$

$$= \int_{-1}^{1} \frac{dx}{1 - 2xt + t^2} = \frac{1}{t}\left[\ln(1+t) - \ln(1-t)\right]. \tag{32.46}$$

Using the familiar expansion $\ln(1 \pm t) = \sum_{n=1} \frac{(-1)^{n+1}}{n} (\pm t)^n$, the difference between logs leaves only even powers of t,

$$\sum_{nm} t^n t^m \int_{-1}^{1} P_n(x) P_m(x) dx = \sum_{n=0} \frac{2}{2n+1} t^{2n}. \tag{32.47}$$

Thus one of the sums on the left must collapse with a Kronecker delta, leading directly to the orthogonality condition

$$\int_{-1}^{1} P_n(x) P_m(x) dx = \frac{2}{2n+1} \delta_{nm}. \tag{32.48}$$

(Of course, we already knew this — but had our first encounter with the P_m's occurred in the context of electric potential, this would be a crucial and unexpected discovery.)

We can also use G to deduce the parity of the P_m's: inspection of its closed form reveals that $G(x,t) = G(-x, -t)$ — and so therefore

$$\sum_{m=0} t^m P_m(x) = \sum_{m=0} (-t)^m P_m(-x) \quad \Longrightarrow \quad P_m(-x) = (-1)^m P_m(x). \tag{32.49}$$

Example 32.7 **Generating Recursion**

With a little more effort, a generating function can also be used to derive recursion relations amongst polynomials. Consider for example the Hermite generating function from Table 32.1,

$$G(x,t) = e^{2xt - t^2}. \tag{32.50}$$

The partial of G with respect to x,

$$\frac{\partial}{\partial x} G(x,t) = 2t\, G(x,t), \tag{32.51}$$

can be expressed in terms of the power series expansion $G = \sum_{n=0} \frac{1}{n!} H_n(x)\, t^n$,

$$\sum_{n=0} \frac{1}{n!} H_n'(x)\, t^n = 2t \sum_{m=0} \frac{1}{m!} H_m(x)\, t^m$$

$$= 2 \sum_{m=0} \frac{1}{m!} H_m(x)\, t^{m+1}. \tag{32.52}$$

Shifting the index $m \to n-1$ allows us to collect things together in a single sum

$$\sum_{n=1} \frac{1}{n!} \left[H_n'(x) - 2nH_{n-1} \right] t^n = 0. \tag{32.53}$$

Since the coefficients of like-powers of t must vanish separately, we discover

$$H_n'(x) = 2nH_{n-1}(x), \tag{32.54}$$

which you're invited to explicitly verify for any $H_{n\geq 1}$. (If we formally set $H_{-1} \equiv 0$, it holds for $n = 0$ as well.)

What if instead we were to take a partial derivative with respect to t? Well, since

$$\frac{\partial}{\partial t} G(x,t) = 2(x-t)\, G(x,t), \tag{32.55}$$

we get

$$\sum_{n=0} \frac{1}{n!} H_n(x)\, n\, t^{n-1} = 2(x-t) \sum_{m=0} \frac{1}{m!} H_m(x)\, t^m.$$

Bringing everything to one side and shifting indices,

$$\sum_{n=1} \frac{1}{n!} \left[H_{n+1}(x) - 2x\, H_n(x) + 2n\, H_{n-1}(x) \right] t^n = 0, \tag{32.56}$$

reveals

$$H_{n+1} - 2x\, H_n + 2n\, H_{n-1} = 0. \tag{32.57}$$

Given H_0 and H_1, this three-term recursion relation provides a simple way to calculate higher-n Hermite polynomials. Analogous recursive formulas up and down the "index ladder" are given for other orthogonal polynomials in Table 32.2.

Let's take a look at the relationship between orthogonality and three-term recursion relations more generally. Consider

$$(2m+1)xP_m(x) = (m+1)P_{m+1}(x) + mP_{m-1}(x) \tag{32.58}$$

for Legendre polynomials, showing that xP_m is a linear combination of $P_{m\pm 1}$. We can understand this using orthogonality: the inner product of xP_m and P_ℓ is

$$\langle \ell|x|m \rangle = \int_{-1}^{1} P_\ell(x)\, x\, P_m(x)\, dx. \tag{32.59}$$

By parity, this is non-zero only for $\ell + m + 1$ even. Moreover, since $xP_m(x)$ is an $(m+1)$-order polynomial, the inner product vanishes for $\ell > m+1$ (see Problem 32.17). In other words, xP_m only has components along $P_{\ell \leq m+1}$. Similarly, $xP_\ell(x)$ only has components along $P_{m \leq \ell+1}$. Taken together, this shows that $\langle \ell|x|m \rangle$ is non-zero only for $\ell = m \pm 1$, as the recursion relation asserts. A similar analysis holds for Laguerre and Hermite polynomials. Thus orthogonality implies the existence of a three-term recursion relation. The converse is also true: indexed polynomials $\phi_n(x)$ which satisfy a three-term recursion relation of the form $(a_n x + b_n)\phi_n = c_n \phi_{n+1} + d_n \phi_{n-1}$ are orthogonal.

Table 32.2 Recursion relations

Polynomial ϕ_n	$(a_n x + b_n)\phi_n = c_n \phi_{n+1} + d_n \phi_{n-1}$				$\alpha(x)\phi_n' = \beta(x)\phi_n + \gamma_n \phi_{n-1}$		
	a_n	b_n	c_n	d_n	$\alpha(x)$	$\beta(x)$	γ_n
Legendre P_n	$2n+1$	0	$n+1$	n	$1-x^2$	$-nx$	n
Laguerre L_n	-1	$(2n+1)$	$n+1$	n	x	n	$-n$
Hermite H_n	2	0	1	$2n$	1	0	$2n$
Chebyshev T_n	2	0	1	1	$1-x^2$	$-nx$	n

The other set of recursive identities in Table 32.2 involve the derivatives ϕ_n'. We can combine these with the three-term recursion relations to obtain second-order differential equations. For instance, combining the Hermite polynomial identities

$$2xH_n = H_{n+1} + 2nH_{n-1} \tag{32.60}$$

and

$$H_n' = 2nH_{n-1}, \tag{32.61}$$

gives

$$H_{n+1} - 2xH_n + H_n' = 0. \tag{32.62}$$

Then, taking the derivative and using (32.61) with $n \to n+1$ yields *Hermite's equation*

$$H_n'' - 2xH_n' + 2nH_n = 0. \tag{32.63}$$

Similar manipulations produce *Legendre's equation*,

$$(1-x^2)P_n'' - 2xP_n' + n(n+1)P_n = 0, \tag{32.64}$$

and *Laguerre's equation*,

$$xL_n'' + (1-x)L_n' + nL_n = 0. \tag{32.65}$$

Somehow these differential equations must encode the orthogonality of their solutions; we'll address this question in Chapter 40.

Problems

32.1 Show that $\int_{-1}^1 P_\ell(x)\, dx \le \sqrt{2\int_{-1}^1 |P_\ell(x)|^2\, dx}$.

32.2 For $\ell = 0$ to 3, use the Gram–Schmidt procedure to generate:

(a) Legendre polynomials $P_\ell(x)$
(b) Laguerre polynomials $L_\ell(x)$.

32.3 When $f(x)$ is itself a polynomial of order m, the expansions of (32.11), (32.26), and (32.33) do not go beyond order m. (See Problem 32.17a.) Verify that the families of orthogonal polynomials listed inside the back cover give an exact expansion of $f(x) = 2x^3 - x^2 + 3x + 1$ with only coefficients c_ℓ for $\ell = 0$ to 3.

32.4 Write out the steps leading to (32.27) and (32.34) to show the role orthogonality plays.

32.5 Adapt the analysis leading to (32.21) to find $V(r < a, \theta)$, the potential inside the non-conducting spherical shell. [Note that V must be continuous at $r = a$.] Find the potential both inside and outside the sphere when its surface is held as $V(a, \theta) = V_0 \cos 3\theta$.

32.6 Show that with a simple scaling, the domain of Legendre series can be extended beyond $|x| \leq 1$ to expand a function over $[-a, a]$.

32.7 Find the first three terms in the Legendre series for $f(x) = e^{ikx}$.

32.8 Find the first three terms in the Laguerre series for $f(x) = e^{+x/2}$.

32.9 Use Gram–Schmidt orthogonalization to generate the first five Chebyshev polynomials $T_n(x)$, $n = 0$ to 4 listed inside the back cover. You may find useful the integral

$$\int_{-1}^{1} \frac{x^n}{\sqrt{1 - x^2}} \, dx = \frac{\sqrt{\pi}}{(n/2)!} \, \Gamma\left(\frac{n+1}{2}\right), \quad n \text{ even.}$$

32.10 Find the Chebyshev series for $f(x) = |x|$ through fourth-order.

32.11 Use the generating function in Table 32.1 to derive the Hermite polynomials $H_n(x)$, $n = 0, 1$, and 2.

32.12 Use the generating function in Table 32.1 to determine the parity of the Hermite polynomials.

32.13 Using the generating function in Table 32.1, derive the orthogonality relation for Laguerre polynomials.

32.14 What family of real functions does $G(x, t) = \exp\left[te^{ix}\right]$ generate?

32.15 (a) Show that the function $f_n(x) = (x^2 - 1)^n$ satisfies the differential equation

$$(1 - x^2)f_n'' + 2(n - 1)xf_n' + 2nf_n = 0.$$

(b) Apply the Leibniz rule (5.22) to this equation, revealing that $\phi^{(n)}(x) \equiv \frac{d^n}{dx^n} f_n$ is a solution to Legendre's equation (32.64).

(c) Fix the normalization of ϕ_n to match that of the Legendre polynomials, $P_n(1) = 1$, to obtain the Rodrigues formula (32.45).

32.16 Use the Rodrigues formula (32.45) to generate Legendre polynomials for $\ell = 0, 1$, and 2.

32.17 Whether the Rodrigues formula (32.45) provides a less tedious way to generate the P_ℓ's than Gram–Schmidt is, perhaps, a matter of taste. But as an expression valid for all ℓ, it *is* useful for determining general properties of the P_ℓ's. Use the Rodrigues formula and integration by parts to verify the following:

(a) $\int_{-1}^{1} x^m P_\ell(x) \, dx = 0$ for $m < \ell$. [Hint: What are the surface terms?]

(b) The orthogonalization condition $\langle P_\ell | P_m \rangle = \frac{2}{2\ell+1} \delta_{\ell m}$. You're likely to find useful the integral

$$\int_{-1}^{1} (x^2 - 1)^m \, dx = (-1)^m \, 2^{2m+1} \frac{(m!)^2}{(2m+1)!}.$$

32.18 Use the three-term recursion relations for all four eponymous polynomial ϕ_n in Table 32.2 to find ϕ_3 and ϕ_4 from $\phi_{n<3}$.

32.19 Use recursion and the generating function to show that the Legendre polynomials satisfy

$$P'_{m+1} - P'_{m-1} = (2m + 1)P_m.$$

32.20 Use the derivative recursion relation, together with the generating function, to derive Legendre's equation (32.64).

32.21 Use integration by parts and Legendre's equation (32.64) to show that the derivatives of the Legendre polynomials are orthogonal over $[-1, 1]$ with weight function $1 - x^2$,

$$\int_{-1}^{1} P'_n(x) P'_m(x)(1 - x^2) \, dx = \frac{2n(n + 1)}{2n + 1} \, \delta_{nm}, \quad n \geq 1, \tag{32.66}$$

where $P'_n(x) \equiv \frac{dP_n}{dx}$. [If you'd like to appreciate the relative ease of this approach, try deriving (32.66) using the Rodrigues formula. First use Table 32.2 to construct an identity for P'_n; after that, the argument proceeds much as in Problem 32.17 — but watch your surface terms!]

32.22 **Creation and Annihilation of Quanta**
The function $\psi_n(x) \equiv H_n(x) e^{-x^2/2}$ is square integrable with weight 1; the constant $c_n \equiv 1/\sqrt{\sqrt{\pi} \, 2^n n!}$ ensures normalization, $\langle \psi_n | \psi_n \rangle = 1$. We define the operators $\hat{a}$ and $\hat{a}^\dagger$ as[3]

$$\hat{a} \equiv \frac{1}{\sqrt{2}} \left(x + \frac{d}{dx} \right) \qquad \hat{a}^\dagger \equiv \frac{1}{\sqrt{2}} \left(x - \frac{d}{dx} \right).$$

Use the identities in Table 32.2 to verify that

$$\hat{a} \psi_n(x) = \sqrt{n} \, \psi_{n-1}(x) \qquad \hat{a}^\dagger \psi_n(x) = \sqrt{n + 1} \, \psi_{n+1}(x).$$

Because of their action on $\psi_n(x)$, $\hat{a}$ and $\hat{a}^\dagger$ are called annihilation and creation operators, respectively. In quantum mechanics, ψ_n represents a state of the harmonic oscillator with energy $E_n = \hbar\omega\left(n + \frac{1}{2}\right)$, where ω is the natural frequency of the oscillator. Thus $\hat{a}^\dagger$ maps a state with energy E_n into a state with energy E_{n+1} — one says that $\hat{a}^\dagger$ "creates" one quantum of energy. Similarly, $\hat{a}$ annihilates one quantum of energy. Use these operators to demonstrate that, while there is no upper bound on energy, the harmonic oscillator has a minimum-energy ground state.

32.23 The associated Legendre functions $P_\ell^m(x)$ are solutions to the differential equation

$$(1 - x^2) \frac{d^2 P_\ell^m}{dx^2} - 2x \frac{dP_\ell^m}{dx} + \left(\ell(\ell + 1) - \frac{m}{1 - x^2} \right) P_\ell^m(x) = 0. \tag{32.67}$$

Comparison with (32.64) shows that the $m = 0$ solutions are Legendre polynomials — that is, $P_\ell(x) \equiv P_\ell^0(x)$. The solutions for non-zero m can be generated from the Legendre polynomials. Take m derivatives of Legendre's equation (32.64) to show

$$P_\ell^m(x) = (1 - x^2)^{m/2} \frac{d^m}{dx^m} P_\ell(x) = \frac{(-1)^\ell}{2^\ell \ell!} (1 - x^2)^{m/2} \frac{d^{\ell+m}}{dx^{\ell+m}} (1 - x^2)^\ell.$$

32.24 The associated Laguerre polynomials $L_n^k(x)$ are solutions of a variant of Laguerre's equation,

$$x L_n^{k\prime\prime} + (k + 1 - x) L_n^{k\prime} + n L_n^k = 0. \tag{32.68}$$

For $k = 0$, the associated polynomials are equivalent to the Laguerre polynomials, $L_n^0(x) \equiv L_n(x)$. For integer $k \geq 0$:

(a) Verify the identity $L_n^k(x) = (-1)^k \left(\frac{d}{dx}\right)^k L_{n+k}(x)$ using Laguerre's equation (32.65). (Unlike the associated Legendre functions, the L_n^k are all nth-order polynomials in x.) Use this to find L_2^1. The phase $(-1)^k$ is convention, complementing the conventional alternating phase of the L_n's.

---

[3] The notation implies they're Hermitian conjugates; we'll validate this in Chapter 40.

(b) From the generating function for the Laguerre polynomials (Table 32.1), show that the associated Laguerre polynomials have generating function

$$G^{(k)}(x,t) \equiv (-1)^k \frac{e^{-xt/(1-t)}}{(1-t)^{k+1}}.$$

(c) Use the generating function to verify the orthogonality relation for the associated polynomials,

$$\int_0^\infty x^k e^{-x} L_n^k(x) L_m^k(x) \, dk = \frac{(k+n)!}{n!} \delta_{nm}.$$

(The weight function $x^k e^{-x}$ is derived in Problem 40.13.)

32.25 Use Gram–Schmidt orthogonalization with $w = x^k e^{-x}$ for integer $k \geq 0$ to derive the associated Laguerre polynomials for $n = 0, 1$, and 2, with normalization given in Problem 32.24c.

32.26 We applied the Gram–Schmidt procedure to the set $\{x^n\}$, $n = 0, 1, 2, \ldots$ over $[-1, 1]$ with weight $w(x) = 1$ to construct the Legendre polynomials $P_n(x)$. Use Mathematica to construct a new basis of orthonormal polynomials $\tilde{P}_n(x)$ which start with $n = 1$. (In other words, remove $x^0 = 1$ from the original basket of linear independent vectors.) Are the $\tilde{P}_n$'s complete?

33 Fourier Series

33.1 A Basis of Sines and Cosines

Not all sets of orthogonal functions are polynomials. Indeed, (31.7) asserts that a natural basis for the space of periodic functions is composed of the archetypical periodic functions sine and cosine. Over a single period they are certainly square integrable; but are they orthogonal? Consider the infinite set $\{\sin a\xi, \cos b\xi\}$ for any positive real numbers a and b — real, because otherwise the functions do not oscillate, and positive because sine and cosine have definite parity, so negative a and b contribute nothing new.

To find out if these functions are orthogonal, we take inner products by calculating integrals of $\sin a\xi \sin b\xi$, $\cos a\xi \cos b\xi$, and $\sin a\xi \cos b\xi$ over the basic 2π period. For instance,

$$
\begin{aligned}
\int_{-\pi}^{\pi} \sin a\xi \sin b\xi \; d\xi &= \frac{1}{2} \int_{-\pi}^{\pi} \left\{ \cos[(a-b)\xi] - \cos[(a+b)\xi] \right\} d\xi \\
&= \int_{0}^{\pi} \left\{ \cos[(a-b)\xi] - \cos[(a+b)\xi] \right\} d\xi \\
&= \left[\frac{\sin[(a-b)\xi]}{a-b} - \frac{\sin[(a+b)\xi]}{a+b} \right]\Big|_{0}^{\pi} \\
&= \frac{\sin[(a-b)\pi]}{a-b} - \frac{\sin[(a+b)\pi]}{a+b}.
\end{aligned}
\tag{33.1}
$$

For $\sin a\xi$ and $\sin b\xi$ to be orthogonal, this integral must vanish for all $a \neq b$. *This requires that we restrict a and b to be integers.*[1]

What about for $a = b$? Does this give a non-zero, finite result? Although the second term in (33.1) vanishes, we might worry about the first term's denominator. Writing $\Delta = a - b$, however, we recover the familiar limit

$$
\lim_{\Delta \to 0} \frac{\sin \Delta\pi}{\Delta} = \pi.
$$

(We could also calculate $\int_{-\pi}^{\pi} \sin^2 n\xi \; d\xi$ for integer n directly.) For the special case $a = b = 0$, both terms in (33.1) give π, which then subtract to zero — as it should, since in this case we're just integrating $\sin 0 = 0$. Thus we find that for integer n, $\{\sin n\xi\}$ forms an orthogonal set.

---

[1] The integral also vanishes for half-integer a and b, but this does not result in a distinct orthogonal set.

The parallel calculation for $\cos a\xi \cos b\xi$,

$$\int_{-\pi}^{\pi} \cos a\xi \cos b\xi \, d\xi = \frac{\sin[(a-b)\pi]}{a-b} + \frac{\sin[(a+b)\pi]}{a+b}, \tag{33.2}$$

similarly restricts a and b to be integers. In fact, the only substantive difference with (33.1) is the special case $a = b = 0$. Each term on the right of (33.2) gives π, as before, but now they *add* to give 2π — as it should, since in this case we're merely integrating $\cos 0 = 1$. Thus we find that $\{\cos n\xi\}$, for integer n, forms an orthogonal set.

There only remains to show that

$$\int_{-\pi}^{\pi} \sin a\xi \cos b\xi \, d\xi = 0, \tag{33.3}$$

which, as an odd integrand over a symmetric interval, is obviously true.

So we have established that for integer n, $\{\sin n\xi, \cos n\xi\}$ forms an orthogonal set of functions. Introducing the notation $\phi_n^-(\xi) \equiv \sin n\xi$ and $\phi_n^+(\xi) \equiv \cos n\xi$, where the $\pm$ indicate the function's parity, we can easily summarize our results:

$$\langle \phi_n^- | \phi_m^- \rangle = \begin{cases} \pi \, \delta_{mn}, & n \neq 0 \\ 0, & n = m = 0 \end{cases} \tag{33.4a}$$

$$\langle \phi_n^+ | \phi_m^+ \rangle = \begin{cases} \pi \, \delta_{mn}, & n \neq 0 \\ 2\pi, & n = m = 0 \end{cases} \tag{33.4b}$$

$$\langle \phi_n^- | \phi_m^+ \rangle = 0. \tag{33.4c}$$

The expansion of a periodic function $f(\xi)$ on the basis $\{\phi_n^\pm(\xi)\}$ can be written

$$f(\xi) = \sum_{n=0} a_n \phi_n^+(\xi) + \sum_{n=0} b_n \phi_n^-(\xi). \tag{33.5}$$

Applying (31.11) gives

$$a_n = \frac{1}{k_n} \langle \phi_n^+ | f \rangle = \frac{1}{\pi} \int_{-\pi}^{\pi} \cos(n\xi) f(\xi) \, d\xi \tag{33.6a}$$

$$b_n = \frac{1}{k_n} \langle \phi_n^- | f \rangle = \frac{1}{\pi} \int_{-\pi}^{\pi} \sin(n\xi) f(\xi) \, d\xi, \tag{33.6b}$$

where $k_n = \pi$ for all $n > 0$. These expressions can also apply for $n = 0$, which is trivial in the case of b_0 since $\phi_0^- \equiv 0$. But the different normalization of ϕ_0^+ affects its contribution to the expansion. We can accommodate this by inserting the extra factor of 2 by hand into (33.5), rewriting it as

$$f(\xi) = \frac{a_0}{2} + \sum_{n=1}^{\infty} (a_n \cos n\xi + b_n \sin n\xi), \tag{33.7}$$

where now the sum begins at $n = 1$. The expansion defined by (33.6) and (33.7) is known as the *Fourier series* representation of f. Note that for odd $f(\xi)$ all the a_n's vanish, whereas for even functions, all the $b_n \equiv 0$; both contribute for functions without definite parity.

Example 33.1 $f(\xi) = \xi, |\xi| < \pi$

Since ξ is an odd function, we can immediately conclude that all the $a_n = 0$. So we need only calculate the b_n from (33.6b), which gives[2]

$$b_n = \frac{1}{\pi} \int_{-\pi}^{\pi} \sin(n\xi) f(\xi) \, d\xi = \frac{1}{\pi} \int_{-\pi}^{\pi} \xi \sin(n\xi) \, d\xi$$

$$= \frac{2}{\pi} \int_{0}^{\pi} \xi \sin(n\xi) \, d\xi = (-1)^{n+1} \frac{2}{n}. \tag{33.8}$$

Thus (33.7) gives the Fourier representation

$$\xi = 2 \sum_{n=1}^{\infty} (-1)^{n+1} \frac{\sin n\xi}{n}. \tag{33.9}$$

Figure 33.1 shows the first four standing waves contributing to this sum; Figure 33.2 demonstrates how the series slowly builds towards $f(\xi) = \xi$.

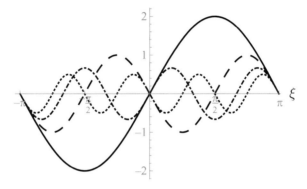

Figure 33.1 Standing waves contributing to the Fourier series for $f(\xi) = \xi$: solid, n = 1; dashed, $n = 2$; dot-dashed, $n = 3$; dotted, $n = 4$.

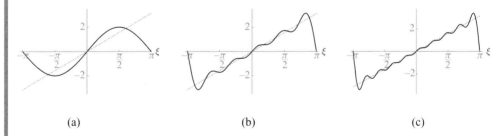

(a)   (b)   (c)

Figure 33.2 Fourier series for $f(\xi) = \xi$ for $-\pi < \xi < \pi$, through (a) $n = 1$; (b) $n = 5$; (c) $n = 10$.

---

[2] Though it can be empowering to use pen and paper, it's faster to use a symbolic manipulation system such as Mathematica or Wolfram Alpha — both to integrate and to plot up the results.

Example 33.2 **Fourier Series of** $f(\xi) = |\xi|, |\xi| < \pi$

As an even function, we can immediately conclude that all the $b_n = 0$. So we need only calculate the a_n:

$$a_n = \frac{1}{\pi} \int_{-\pi}^{\pi} \cos(n\xi) f(\xi) \, d\xi = \frac{2}{\pi} \int_{0}^{\pi} \xi \cos(n\xi) \, d\xi$$

$$= \frac{2}{\pi n^2} \left[(-1)^n - 1 \right] = \begin{cases} \frac{-4}{\pi n^2}, & n \text{ odd} \\ 0, & n \text{ even} \end{cases}, \qquad (33.10)$$

where we've used $\cos n\pi = (-1)^n$ for integer n. The n in the denominator necessitates that we compute a_0 separately:

$$a_0 = \frac{1}{\pi} \int_{-\pi}^{\pi} |\xi| \, d\xi = \frac{2}{\pi} \int_{0}^{\pi} \xi \, d\xi = \pi. \qquad (33.11)$$

Thus (33.7) gives the series

$$|\xi| = \frac{\pi}{2} - \frac{4}{\pi} \sum_{n \text{ odd}} \frac{\cos n\xi}{n^2}. \qquad (33.12)$$

This is shown through $n = 5$ in Figure 33.3, plotted between ± 1 for easier comparison with the polynomial expansions of Figure 32.4. That (33.12) is a pure cosine series reflects $|\xi|$'s even parity; that only odd n contribute to the sum is due to its "subparity": the even halves of $|\xi|$ have odd reflection symmetry around their midline (Problem 33.10).

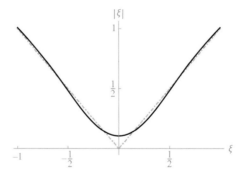

Figure 33.3 Fourier expansion of $f(\xi) = |\xi|$ through $n = 5$.

It's curious how the Fourier series for such similar functions as ξ and $|\xi|$ differ: whereas (33.9) converges quite slowly, the series of (33.12) converges rapidly — as is evident when comparing of Figures 33.2 and 33.3. The Fourier series for ξ particularly appears to have difficulty converging at the endpoints $\pm\pi$; we'll return to this curiosity in Section 34.1.

33.2 Examples and Applications

Most physical applications are not directly concerned with phase ξ, but rather with distance x or time t. So to expand a function $f(x)$ in a Fourier series, the basis functions need to be expressed in terms of x. With $\xi \equiv x\pi/L$, (33.6)–(33.7) become

$$a_n = \frac{1}{L} \int_{-L}^{L} \cos(n\pi x/L) f(x) \, dx \qquad b_n = \frac{1}{L} \int_{-L}^{L} \sin(n\pi x/L) f(x) \, dx \qquad (33.13)$$

and

$$f(x) = \frac{a_0}{2} + \sum_{n=1}^{\infty} a_n \cos(n\pi x/L) + b_n \sin(n\pi x/L), \tag{33.14}$$

giving a series with period $2L$. This is a superposition of standing waves of wavelengths $\lambda_n = 2L/n$ and frequencies $k_n \equiv 2\pi/\lambda_n = n\pi/L$, integer multiples of the so-called fundamental frequency π/L. For time rather than space the period is usually denoted T, requiring basis functions $\sin(2\pi nt/T)$ and $\cos(2\pi nt/T)$ with frequencies $\omega_n \equiv 2\pi n/T$.

> **Example 33.3** **Fourier Series of $f(x) = 1 + 4x^2 - x^3$, $-L < x < L$**
>
> Although $f(x)$ lacks definite parity, there are no contributions to the a_n from x^3, nor to the b_n from 1 or x^2. Straightforward application of (33.13)–(33.14) gives
>
> $$a_n = (-1)^n \frac{16L^2}{\pi^2 n^2} \qquad b_n = (-1)^n \frac{2L^3}{\pi^3 n^3} \left(\pi^2 n^2 - 6 \right), \tag{33.15}$$
>
> and $a_0 = 2(1 + 4L^2/3)$. Figure 33.4 plots the separate contributions of these coefficients through $n = 3$; even these truncated series are plausible representations of the parabola $1 + 4x^2$ and the cubic $-x^3$.

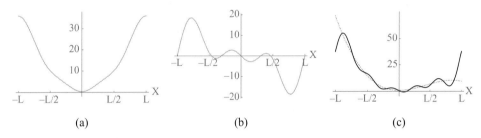

Figure 33.4 Through $n = 3$: (a) $a_n \cos(n\pi x/L)$; (b) $b_n \sin(n\pi x/L)$ series of (33.15); (c) the sum of (a) and (b) gives the Fourier expansion (solid) of $1 + 4x^2 - x^3$ (dot-dashed).

Incidentally, since there is no requirement that f be real, a_n and b_n can be complex. Nonetheless, in general parlance a *complex Fourier series* usually refers to an expansion on the basis of complex exponentials $|n\rangle \to e^{in\pi x/L}$,

$$f(x) = \sum_{n=-\infty}^{\infty} c_n e^{+in\pi x/L}. \tag{33.16}$$

For this to be an orthogonal projection of $|f\rangle$ onto basis vector $|n\rangle$ requires the orthogonality of the $\{e^{in\pi x/L}\}$ — and indeed, straightforward calculation reveals

$$\langle m|n \rangle = \int_{-L}^{L} \left(e^{im\pi x/L} \right)^* e^{in\pi x/L} dx = \int_{-L}^{L} e^{i(n-m)\pi x/L} dx = 2L\, \delta_{mn}. \tag{33.17}$$

Thus the expansion coefficients in (33.16) are given by (note the complex conjugation of the basis function)[3]

$$c_n = \frac{1}{k_n} \langle n|f \rangle = \frac{1}{2L} \int_{-L}^{L} e^{-in\pi x/L} f(x)\, dx. \tag{33.18}$$

---

[3] Normalization rule: the integral for c_n is divided by the period, for a_n and b_n by half the period.

Whether expanding in sines and cosines or in complex exponentials, the Fourier series must all be equivalent; you're asked in Problem 33.4 to verify for $n > 0$ that

$$c_0 = \frac{1}{2} a_0 \qquad c_n = \frac{1}{2}(a_n - ib_n) \qquad c_{-n} = \frac{1}{2}(a_n + ib_n). \tag{33.19}$$

Example 33.4 **Complex Fourier Representation of $f(x) = |x|$, $|x| < L$**

Even though $|x|$ is a real function, we can still use (33.16) and (33.18). The expansion coefficients are

$$c_n = \frac{1}{2L} \int_{-L}^{L} e^{-in\pi x/L} |x| \, dx = \frac{L}{\pi^2 n^2} \left[(-1)^n - 1 \right] = \begin{cases} \frac{-2L}{\pi^2 n^2}, & n \text{ odd} \\ 0, & n \text{ even} \end{cases} \tag{33.20}$$

and

$$c_0 = \frac{1}{2L} \int_{-L}^{L} |x| \, dx = \frac{L}{2}. \tag{33.21}$$

Then

$$|x| = c_0 + (c_1 e^{i\pi x/L} + c_{-1} e^{-i\pi x/L}) + (c_3 e^{3i\pi x/L} + c_{-3} e^{-3i\pi x/L}) + \cdots$$
$$= \frac{L}{2} - \frac{4L}{\pi^2} \left[\cos(\pi x/L) + \frac{1}{9} \cos(3\pi x/L) + \cdots \right], \tag{33.22}$$

which agrees with (33.12) for $L = \pi$. Of course despite the complex basis, the Fourier series of a real function must be real. Unlike this example, however, the c_n for a real function need not be real (Problem 33.6).

So Fourier analysis decomposes a function into orthogonal components, each contribution a multiple of a fundamental (i.e., lowest) frequency ω_0. If we think of $f(t)$ as representing a sound wave, then Fourier series can be construed as a collection of N students, each holding a unique tuning fork of frequency $n\omega_0$ with instructions to hit the fork with a magnitude $|c_n|$, $n = 1$ to N.[4] Each has been instructed to hit his or her fork with a magnitude $|c_n|$. (We're ignoring relative phases here — but that's just a timing issue.) As $N \to \infty$, the emerging sound will be more and more indistinguishable from $f(t)$.

Example 33.5 **Musical Instruments**

Have you ever wondered why you're able to distinguish different musical instruments with your eyes closed — even when they're playing the same note? (Imagine the tremendous loss to music if this weren't so!) The explanation is revealed by Fourier analysis: instruments playing the same fundamental frequency have the same *pitch*, but the contributions from higher multiples — the *harmonics* or *overtones* — differ from instrument to instrument. In other words, it's the spectrum of c_n's which give an instrument its characteristic sound, or *timbre*.

Consider a violin, guitar, and clarinet, all playing A above middle C (commonly used in orchestra tuning). We cannot only hear the difference, we can see it too: Figure 33.5 shows the waveforms for each instrument. Though each is playing a fundamental frequency of $f_0 = 440$ Hz, the curves are quite distinct. Using (33.18) with $\omega_0 \equiv \pi/L = 2\pi f_0$ reveals the different overtones

---

[4] The zero-frequency contribution c_0 merely fixes the average value of the series; as such, it's sometimes given the electronic-circuitry moniker "DC offset."

for each instrument, shown in Figure 33.6. Though the violin's largest contribution comes from the fundamental at 440 Hz, the total contribution of higher harmonics is comparable. The guitar demonstrates that even though the fundamental determines the note we hear, it need not be the dominant contribution. The harmonic signature can also tell us something about the instrument. For instance, a hollow tube of constant diameter open at one end — which is a fair description of a clarinet — has all the even harmonics missing from its spectrum of standing sound waves. And in fact, the harmonic spectrum of the clarinet in Figure 33.6 is dominated by the odd harmonics; this is what accounts for its "hollow" sound. By contrast, an oboe has more of a conical shape, effectively restoring the even harmonics.

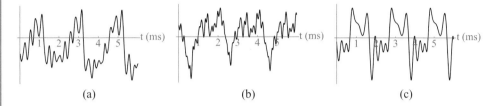

(a)   (b)   (c)

Figure 33.5 Waveforms produced by a (a) violin, (b) guitar, and (c) clarinet playing A above middle C. (Created using the `SoundNote` command in Mathematica.)

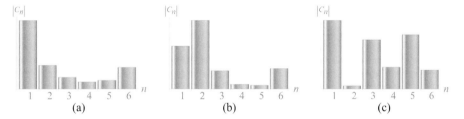

(a)   (b)   (c)

Figure 33.6 The Fourier $|c_n|$ spectra for a (a) violin, (b) guitar, and (c) clarinet playing A above middle C.

Perhaps you noticed that the waveforms in Figure 33.5 are not perfectly periodic. This can be due to several effects, including a variation in volume. The violin waveform also includes vibrato, the small, rapid change in pitch produced by the musician rocking her finger on the fingerboard. This results in a fundamental frequency which itself oscillates around 440 Hz. Plucked instruments like the guitar are affected by harmonics which decay away at different rates; these *transients* cause the harmonic spectrum to change in time.

Fourier analysis can not only be used to decompose a wave into its harmonic components, the process can also be reversed to construct a wave with whatever spectrum we'd like. Indeed, knowledge of an instrument's harmonic spectrum allows one to reconstruct a waveform which can be all-but-indistinguishable from the instrument itself. This is the principle behind electronic music synthesizers (and is a lot more practical than passing out tuning forks).

The spectrum of expansion coefficients encodes the features of a function $f(x)$. This is true for any basis of orthogonal functions (though of course the spectra are not the same). Consider, for example, Fourier and Legendre expansions of $f(x) = e^{-x}$. Looking back at Figure 32.1, we would expect low-order Legendre polynomials to pick out most of the features of f; by contrast, the oscillations of the Fourier basis should require the superposition of many standing waves to construct such a smoothly varying function. Figure 33.7 shows that both these predictions hold true. (Since Legendre series are restricted

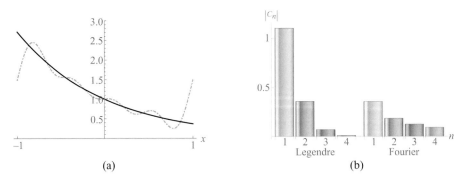

Figure 33.7 Fourier and Legendre expansions of $f(x) = e^{-x}$ through $n = 4$. (a) At this scale the Legendre series (solid) is indistinguishable from e^{-x}; (b) series spectra $|c_n|$.

to $|x| \leq 1$, for proper comparison the Fourier series shown has period 2.) Clearly the Legendre series converges much faster — so much so, that at the scale shown in Figure 33.7, an expansion through $n = 4$ is indistinguishable from e^{-x}. On the other hand, the Fourier series has significant contributions from higher harmonics. A Hermite expansion has a spectrum similar to Legendre, a Laguerre series less so — but still narrower than Fourier.[5]

A notable feature of the Fourier expansion in Figure 33.7 is the increased oscillation and overshoot ("ringing") near the endpoints $\pm L$. As we discuss in the next section, that's because the Fourier series treats $\pm L$ not as endpoints, but rather as discontinuities in the periodic repetition of e^{-x}. This effect, called *Gibbs phenomenon*, occurs because more rapid changes in $f(x)$ require higher frequency contributions to its Fourier spectrum — and the extreme limit of rapid change is a discontinuity.

One might reasonably expect that including more terms in the Fourier series reduces the ringing; as we'll see, however, that is not the case. But to appreciate this counter-intuitive result first requires a discussion of the convergence of functional expansions — which we take up in Chapter 34.

33.3 Even and Odd Extensions

It may seem curious that all these examples concern ostensibly non-periodic functions; even the musical tones of Example 33.5 have a beginning and end. How is it, then, that we can use a periodic basis? The answer can be gleaned from Figure 33.8, which plots the expansion (33.22) of $|x|$ through $n = 5$ to just beyond $\pm 2L$. Clearly, the Fourier series only reproduces $|x|$ for $|x| \leq L$, repeating these values indefinitely over every $2L$ interval. Thus (33.22) is actually an expansion of a triangle wave whose periodicity comes not from $|x|$ (which has no inherent scale), but from the basis functions used in the expansion. So even though (33.22) can represent $|x|$ within $\pm L$, strictly speaking there is no Fourier series for a function $|x|$ within $\pm L$ which vanishes for all $|x| > L$. The same, of course, holds true for any function $f(x)$ defined over a limited interval.

Real signals, however, have finite extent — say, the amount of time a horn player holds a note. This would seem to imply that Fourier series have limited utility. But usually we're only interested in the Fourier modes over that limited interval (say, the standing waves on a violin string), unconcerned about what the expansion might give elsewhere. In such cases, not only are Fourier series useful, we have flexibility in the choice of series. Suppose we want a Fourier series for an arbitrary function defined only between 0 and L. To find a Fourier representation on this limited domain requires that

---

[5] And, of course, a Laguerre series is restricted to $x \geq 0$.

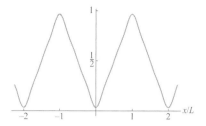

Figure 33.8 Extended view of the Fourier expansion for $|x|/L$ through $n = 5$.

we consider $f(x)$ as a segment of a fictitious periodic wave defined for all x. One approach is that of Figure 33.9a, which shows how an arbitrary $f(x)$ can be construed as one full period L of this putative wave. Since periodicity L requires fundamental frequency $2\pi/L$, we get the Fourier expansion

$$f(x) = \frac{a_0}{2} + \sum_{n=1}^{\infty} a_n \cos(2n\pi x/L) + b_n \sin(2n\pi x/L), \qquad 0 < x < L \qquad (33.23a)$$

with

$$a_n = \frac{2}{L} \int_0^L \cos(2n\pi x/L) f(x)\, dx \qquad b_n = \frac{2}{L} \int_0^L \sin(2n\pi x/L) f(x)\, dx, \qquad (33.23b)$$

where, as before, the a_n and b_n integrals are divided by half the period — in this case, $L/2$. This series reconstructs the entire fictitious wave; it's only a representation of our function $f(x)$ between 0 and L (which, recall, is all we're interested in).[6]

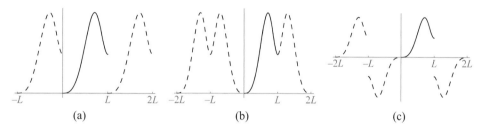

      (a)   (b)   (c)

Figure 33.9 Periodic extensions of a function beyond $x \in [0, L]$: (a) repetition, period L; (b) even extension $F^{(+)}(x)$, period $2L$; (c) odd extension $F^{(-)}(x)$, period $2L$.

This is straightforward enough, but it still requires evaluating integrals for both a_n and b_n. If instead we were to consider $f(x)$ as if it were a *half*-period of a wave $F^{(\pm)}(x)$ of period $2L$ with definite parity, we'd have only one set of coefficients. Figures 33.9b and c show even and odd extensions of $f(x)$ defined between 0 and L. For the even extension all the b_n vanish, so

$$f(x) = \frac{a_0}{2} + \sum_{n=1}^{\infty} a_n \cos(n\pi x/L), \qquad 0 < x < L, \qquad (33.24a)$$

where

$$a_n = \frac{1}{L} \int_{-L}^{L} \cos(n\pi x/L)\, F^{(+)}(x)\, dx = \frac{2}{L} \int_0^L \cos(n\pi x/L) f(x)\, dx. \qquad (33.24b)$$

[6] If imposing a period to analyze a finite interval sounds familiar, look back at the derivation of the solution to the N-chain from the infinite chain in Example 28.7.

Similarly, for the odd extension there are no a_n (including $n = 0$), leaving only

$$f(x) = \sum_{n=1}^{\infty} b_n \sin(n\pi x/L), \qquad 0 < x < L, \tag{33.25a}$$

with

$$b_n = \frac{1}{L} \int_{-L}^{L} \sin(n\pi x/L) F^{(-)}(x)\, dx = \frac{2}{L} \int_{0}^{L} \sin(n\pi x/L) f(x)\, dx. \tag{33.25b}$$

So for a function $f(x)$ with $0 \leq x \leq L$, its *Fourier cosine* series (33.24) is the Fourier series of the even extension $F^{(+)}(x)$; its *Fourier sine* series (33.25) is the Fourier series of the odd extension $F^{(-)}(x)$. Both actually follow the usual pattern of integrating over one period and dividing by half the period. But (33.24b) and (33.25b) take advantage of the fact that for both odd and even extensions, the integrals over $F^{(\pm)}(x)$ are even. So not only can we change the lower limit to 0 and multiply by 2, we can also replace $F^{(\pm)}(x)$ by $f(x)$. Thus the Fourier coefficients emerge just by integrating $f(x)$ over its defined domain; we never need bother with the full even- or odd-extended function. (Since the factor of 2 here has a different provenance than that in (33.23b), the expressions for Fourier cosine and sine series are often listed as distinct formulas — as in Table 33.1.)

Table 33.1 Fourier series

Period	Basis	Expansion coefficients	Fourier representation of $f(x)$ on $[0, L]$
L	$\cos\left(\frac{2n\pi x}{L}\right)$ $\sin\left(\frac{2n\pi x}{L}\right)$	$a_n = \frac{2}{L} \int_0^L \cos\left(\frac{2n\pi x}{L}\right) f(x)\, dx$ $b_n = \frac{2}{L} \int_0^L \sin\left(\frac{2n\pi x}{L}\right) f(x)\, dx$	$\frac{a_0}{2} + \sum_{n=1}^{\infty} a_n \cos\left(\frac{2n\pi x}{L}\right) + b_n \sin\left(\frac{2n\pi x}{L}\right)$
L	$e^{2in\pi x/L}$	$c_n = \frac{1}{L} \int_0^L e^{-2in\pi x/L} f(x)\, dx$	$\sum_{n=-\infty}^{\infty} c_n e^{2in\pi x/L}$
$2L$ even extension	$\cos\left(\frac{n\pi x}{L}\right)$	$a_n = \frac{2}{L} \int_0^L \cos\left(\frac{n\pi x}{L}\right) f(x)\, dx$	$\frac{a_0}{2} + \sum_{n=1}^{\infty} a_n \cos\left(\frac{n\pi x}{L}\right)$
$2L$ odd extension	$\sin\left(\frac{n\pi x}{L}\right)$	$b_n = \frac{2}{L} \int_0^L \sin\left(\frac{n\pi x}{L}\right) f(x)\, dx$	$\sum_{n=1}^{\infty} b_n \sin\left(\frac{n\pi x}{L}\right)$

Example 33.6 **Fourier Expansions of $f(x) = x^2$, $\quad 0 < x < L$**

Taking $f(x)$ to represent one full cycle of a function of period L produces a sawtooth wave with fundamental frequency $2\pi/L$:

$$a_n = \frac{2}{L} \int_0^L x^2 \cos\left(\frac{2n\pi x}{L}\right) dx = \frac{L^2}{n^2 \pi^2}$$

$$b_n = \frac{2}{L} \int_0^L x^2 \sin\left(\frac{2n\pi x}{L}\right) dx = -\frac{L^2}{n\pi}, \tag{33.26}$$

and $a_0 = 2L^2/3$. This period-L series is distinct from an odd extension of $f(x) = x^2$ with period $2L$. In this construction, all the a_n are zero — leaving only

$$b_n = \frac{2}{L} \int_0^L x^2 \sin\left(\frac{n\pi x}{L}\right) dx = -2L^2 \begin{cases} \frac{1}{n\pi}, & n \text{ even} \\ \frac{4-n^2\pi^2}{n^3\pi^3}, & n \text{ odd} \end{cases} . \tag{33.27}$$

An even extension of period $2L$ gives a cosine series with

$$a_n = \frac{2}{L} \int_0^L x^2 \cos\left(\frac{n\pi x}{L}\right) dx = (-1)^n \frac{4L^2}{n^2\pi^2}, \tag{33.28}$$

and $a_0 = 2L^2/3$ (as before).

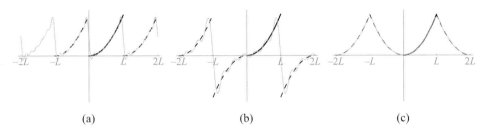

(a)   (b)   (c)

Figure 33.10 Fourier expansions of $f(x) = x^2$ between 0 and L through $n = 6$: (a) period L, (b) sine series, period $2L$, (c) cosine series, period $2L$.

All three expansions are depicted in Figure 33.10; comparison shows that the rate of convergence may play a role when choosing one series representation over another. In particular, note that the ringing evident around the discontinuities in Figures 33.10a,b is absent in Figure 33.10c, which has no discontinuities.

Example 33.7 **The Particle in a Box**

The choice of how to expand a function periodically is not always ours to make. Consider the quantum mechanics of a particle free to move about within a box of impenetrable walls. In one dimension, this is described by the potential energy

$$V(x) = \begin{cases} 0, & 0 < x < L \\ \infty, & \text{elsewhere} \end{cases}, \tag{33.29}$$

often referred to as an infinite square well. For this system, the time-independent Schrödinger equation for a particle of mass m and energy E

$$\left[-\frac{\hbar^2}{2m} \frac{d^2}{dx^2} + V(x) \right] \phi(x) = E\phi(x), \tag{33.30}$$

reduces to the familiar differential equation

$$\frac{d^2}{dx^2} \phi(x) = -k^2 \phi(x), \tag{33.31}$$

where $k^2 = 2mE/\hbar^2$. Since the particle cannot get out of the box, the wavefunction for both $x < 0$ and $x > L$ must be identically zero. Thus to ensure a continuous wavefunction, (33.31) must be solved subject to the boundary conditions

$$\phi(0) = \phi(L) = 0. \tag{33.32}$$

Though the general solution is $\phi(x) = A\cos kx + B\sin kx$, the condition at $x = 0$ requires $A = 0$. At $x = L$ we find

$$\phi(L) = B\sin(kL) = 0. \tag{33.33}$$

Taking $B = 0$ would give the trivial solution $\phi(x) = 0$; that won't work. The only other choice is a restriction on the frequency k to integer multiples of π/L,

$$k \rightarrow k_n = n\pi/L. \tag{33.34}$$

Thus the solutions are given by the family of functions

$$\phi_n(x) = \sin\left(\frac{n\pi x}{L}\right), \qquad n = 1, 2, 3, \ldots. \tag{33.35}$$

Since these solutions form an orthogonal basis, any state of the particle $\psi(x)$ can be expressed as a Fourier sine series

$$\psi(x) = \sum_n b_n \phi_n(x) = \sum_n b_n \sin\left(\frac{n\pi x}{L}\right). \tag{33.36}$$

A sine series such as (33.36) can be construed as a sum not over the complete set of functions $\phi_n(x)$, but equivalently as a sum over the single function $\sin(x/L)$,

$$\psi(x) = \sum_n b_n \sin\left(\alpha_n \frac{x}{L}\right), \tag{33.37}$$

where $\alpha_n \equiv n\pi$ are the zeros of $\sin x$. In this view, orthogonality is not amongst different basis functions $\phi_n(x)$, but rather between copies of the single function $\sin(x/L)$, as regulated by the condition at L. The same is true of a cosine series. Because of the periodicity of sine and cosine — specifically, the regular spacing between their zeros — this would seem a distinction without a difference. In Section 40.3 we'll look at orthogonal functions which are not as simple as sines and cosines, for which the distinction makes a difference.

Incidentally, notice that the ϕ_n of (33.35) are identical to the normal mode standing waves of the fixed string, (31.7). That the same behavior often appears in very different physical contexts is a truly remarkable feature of the physical world. But though both are described by standing waves, the interpretations are starkly different. For the particle in the box, the standing wave condition (33.34) results in *energy quantization*,

$$E_n = \frac{\hbar^2 k_n^2}{2m} = \frac{\hbar^2 n^2 \pi^2}{2mL^2}. \tag{33.38}$$

This shows that the ground state $\phi_1(x)$ has positive energy, which is not true of a classical particle in a box. A quantum particle never rests.

Problems

33.1   Find the Fourier series with period 2π representing the square wave

$$f(x) = \begin{cases} 0, & -\pi < x \le 0 \\ 1, & 0 < x \le \pi \end{cases}.$$

33.2   Verify the Fourier coefficients in Example 33.3.

33.3   Show that the basis $\{e^{inx}\}$ is orthogonal for integer n — without using $\cos nx + i \sin nx$.

33.4   Derive (33.19) in two ways:

(a) Starting from the Fourier series $f(x) = \frac{a_0}{2} + \sum_{n=1}^{\infty} a_n \cos nx + b_n \sin nx$.
(b) Starting from the projection $c_n = \frac{1}{2\pi} \int_{-\pi}^{\pi} e^{-inx} f(x)\, dx$.

33.5   Expand the Dirac delta function $\delta(x - a)$ in a complex Fourier series.

33.6   (a) What general condition must hold amongst the c_n of a real function? Show that this is consistent with the coefficients c_n in Example 33.4.
(b) Find the condition amongst the c_n for a function with definite parity.

33.7   Find the Fourier expansion for $f(x) = \sin x \cos x$ between $\pm\pi$. What trig identity emerges from the series?

33.8   Expand $e^{-x/L}$ for $|x| < L$ in a Fourier series with period $2L$. Plot the odd and even parts of the series separately, as well as their sum. [You may find (33.19) helpful.] Use Mathematica to make plots showing how the series better approximates $e^{-x/L}$ as the number of included terms increases.

33.9   Expand Legendre polynomials $P_2(x)$ and $P_3(x)$ in a Fourier series for $|x| < 1$ with period 2.

33.10   Make a prediction for the odd/even n contributions of the a_n and b_n for the following functions; sketches will help. Then calculate these coefficients to verify your predictions. [See the discussion of "subparity" in Example 33.4.]

(a) $f(x) = \begin{cases} 0, & |x| > \pi/2 \\ 1, & -\pi/2 < x < \pi/2 \end{cases}$

(b) $g(x) = \begin{cases} 0, & -\pi < x < 0 \\ 1, & 0 < x < \pi \end{cases}$

33.11   Find Fourier representations of $\sin(x + \pi/4)$ by treating it as:

(a) a function of period 2π
(b) a function of period π
(c) an even function of period 2π
(d) an odd function of period 2π.

Plot your results for parts (b)–(d). Why do some of these graphs exhibit ringing while others do not?

33.12   Find Fourier representations of both $\sin^2 x$ and $\cos^2 x$ for $0 < x < \pi$ by treating them as:

(a) functions of period π
(b) even functions of period 2π
(c) an odd functions of period 2π.

Plot your results for part (c). Why does one of these graphs exhibit ringing while the others do not?

33.13   Making sine even and cosine odd:

(a) expand $\sin x$, $0 < x < 2\pi$, in a Fourier cosine series
(b) expand $\cos x$, $0 < x < 2\pi$, in a Fourier sine series.

33.14   The particle in Example 33.7 was confined to a box with ends at 0 and L. But the physics can only depend upon the size of the box, not the choice of its coordinates. Show how having the box instead sit between $\pm L/2$ yields the same energies and — despite superficial appearances — the same normal modes. What are the advantages and disadvantages of $[0, L]$ versus $[-L/2, L/2]$?

33.15   The log sine function is defined as the integral $S_n \equiv \int_0^\pi [\ln(\sin x)]^n \, dx$. Calculate S_1 and S_2 using the Fourier series for $\ln(\sin x)$ in Problem 6.14.

33.16   (a) Find a Fourier series for the function $f(x) = \begin{cases} 0, & -\pi < x < 0 \\ x, & 0 < x < \pi \end{cases}$.

(b) Evaluating this Fourier series at $x = 0$ and $x = \pi$ may appear to give *two* possible values for the infinite sum $\sum\limits_{n\ \mathrm{odd}} \frac{1}{n^2}$. What are those two values — and which one is correct? Explain the discrepancy.

33.17   Use relevant portions of the series in Example 33.3 with $L = \pi$ to evaluate the following infinite sums:

(a) $\zeta(2) = \sum\limits_{n=1} \frac{1}{n^2}$

(b) $\sum\limits_{n=1} \frac{(-1)^{n+1}}{n^2}$.

34 Convergence and Completeness

Expansions in orthogonal functions share the same underlying structure: for an orthogonal basis $\{\phi_n(x)\}$ under the inner product

$$\langle \phi_n | \phi_m \rangle = \int_a^b \phi_n^*(x)\phi_m(x)w(x)dx = k_n\delta_{nm}, \tag{34.1a}$$

a square-integrable function $f(x)$ can be expanded as

$$|f\rangle = \sum_n c_n|\phi_n\rangle \longrightarrow f(x) = \sum_n c_n\,\phi_n(x), \tag{34.1b}$$

with expansion coefficients

$$c_n = \langle \phi_n | f \rangle = \frac{1}{k_n}\int_a^b \phi_n^*(x)f(x)w(x)dx. \tag{34.1c}$$

For any given function in any particular instance, some orthogonal bases may not be appropriate because they're defined on a different interval. Or perhaps one basis yields faster convergence or greater physical insight than others. Not that we always get to choose: as we saw in Example 33.7, oftentimes a particular series emerges in the natural course of finding a solution.

But as we briefly discussed in Section 25.5, none of this really works unless the basis is complete. For an infinite-dimensional space, this is a subtle issue. How can we be sure we've exhausted (34.1c) and projected along *every* "direction"? And how do we even know if the infinite sum over these projections (34.1b) converges — and if so, whether it converges to a vector within the space? There are in fact many different ways in which a function expansion can be said to converge. In this chapter, we'll explore three — the last of which will lead to a statement of completeness in an infinite-dimensional inner product space.

34.1 Pointwise and Uniform Convergence

We start with a deceptively straightforward question: is a function actually *equal* to its expansion,

$$f(x) \stackrel{?}{=} \lim_{N\to\infty} \sum_{n=0}^{N} c_n\phi_n(x), \tag{34.2}$$

at least within the radius of convergence? The answer requires that we specify a type, or mode, of convergence — which is to say, the manner in which an infinite sum of functions, defined as the limit of the partial sums,

$$S(x) = \sum_{n=0}^{\infty} c_n \phi_n(x) \equiv \lim_{N \to \infty} \sum_{n=0}^{N} c_n \phi_n(x), \tag{34.3}$$

converges. $S(x)$ can be regarded as an infinite set of infinite series, one for every value of x. The expansion is said to be *pointwise convergent* if each of these series converges; the function represented by the expansion is defined by the collection of the individual convergent values. We can then determine whether this limiting function $S(x)$ is identical to the desired $f(x)$.

Example 34.1   **A Second Look at Three Expansions**

Recall the familiar power series $\sum_{n=0}^{\infty} x^n$. The ratio test confirms that the infinite sum is absolutely convergent for all $|x| < 1$, and therefore the series is pointwise convergent on the open interval $(-1, 1)$. This geometric series is, as we know, a representation of the function $1/(1-x)$ for $|x| < 1$. Moreover, it's fairly simple to verify that integrating term by term produces a power series for $-\ln(1 - x)$; similarly, we can differentiate term by term to get a power series representation of $1/(1-x)^2$. Both of these series are valid for all $|x| < 1$.

But now consider the expansion

$$S(x) = \frac{1}{2}P_0(x) + \frac{5}{8}P_2(x) - \frac{3}{16}P_4(x) + \cdots, \tag{34.4}$$

which is the Legendre representation of $|x|$ we found in (32.17). It too is a pointwise convergent series. But as an expansion in even Legendre polynomials, the derivative $S'(x)$ is an odd function — which means $S'(0) = 0$. So although $S(x)$ represents $|x|$ between ± 1, its derivative is certainly not a faithful representation of the derivative, which is undefined at $x = 0$. An identical argument holds for the Hermite expansion of $|x|$, (32.36).

Finally, in Example 33.1 we found the Fourier series

$$\sum_{n=1}^{\infty} (-1)^{n+1} \frac{\sin nx}{n} = \sin x - \frac{1}{2}\sin 2x + \frac{1}{3}\sin 3x + \cdots \tag{34.5}$$

to be a representation of $f(x) = x$, $|x| < \pi$ (Figure 33.2). Unlike $\sum x^n$ at ± 1, this series converges at $\pm \pi$ — but *not* to $f(\pm \pi)$. Instead, at all integer multiples of π the series collapses to zero, which is the *average* of the jump at the points of discontinuity. Moreover, the derivative series, $\cos x - \cos 2x + \cos 3x + \cdots$ is divergent. So although (34.5) is pointwise convergent, it is not equivalent to $f(x) = x$.

What these examples demonstrate is that a pointwise convergent series for $f(x)$ may not always equal the function everywhere — and even if it does, it may not be possible to differentiate it to get a series for $f'(x)$. Nor for that matter is there any guarantee that it can be integrated term by term to obtain a series for $\int f(x)\, dx$. The problem is that requiring only that the infinite sums converge at every x does not impose any condition on the *relative* rates of convergence of these series. This is evident in the overshoot seen in Figure 33.10b, indicating a rate of convergence which depends on x. As a result, discontinuities can arise in the collection of series which define $S(x)$, but which don't manifest in any of the individual terms or partial sums. Indeed, the idea of independent convergence at each x seems antithetical to an orthogonal series with coefficients (34.1c) constructed from the entire interval. Surely some level of collusion amongst these sums is desirable.

A stronger, collusive condition is *uniform convergence*, which requires not simply that the series yields a finite result at every x in the interval $[a, b]$, but that the rates of convergence at each point are comparable — thereby preventing the emergence of discontinuities (see BTW 34.1). Although a uniform rate of convergence implies pointwise convergence, uniform convergence confers on a series expansion $S(x) = \sum u_n(x)$ of continuous functions $u_n(x)$ the following additional properties:

- $S(x)$ is continuous.
- The integral of the sum is the sum of the integrals:

$$\int_\alpha^x \left(\sum_{n=0}^\infty u_n(x') \right) dx' = \sum_{n=0}^\infty \left(\int_\alpha^x u_n(x') \, dx' \right), \tag{34.6a}$$

for all α and x in the interval $[a, b]$.
- If in addition, $\sum u_n'(x)$ is also uniformly convergent for continuous $u_n'(x)$, then the derivative of the sum is the sum of the derivatives:

$$\frac{d}{dx} \sum_{n=0}^\infty u_n(x) = \sum_{n=0}^\infty \frac{du_n}{dx}. \tag{34.6b}$$

Clearly, uniform convergence is something to be desired; with it, manipulation of a series representation of f is virtually equivalent to working with its closed form.

BTW 34.1 The Weierstrass M Test

The formal statements of pointwise and uniform convergence are essentially that of (6.9), applied to the sequence of partial sums $S_N(x)$: for any given number $\epsilon > 0$, no matter how small, there is always some number N_0 in a convergent sequence such that at every x in the interval $[a, b]$,

$$|S(x) - S_N(x)| < \epsilon, \tag{34.7}$$

for all $N > N_0$. In other words, if you set a precision goal of ϵ, there comes a point N_0 in a convergent series beyond which all the subsequent terms $S_{N > N_0}(x)$ are within ϵ of $S(x)$. The distinction between pointwise and uniform convergence lies in the choice of N_0. For pointwise convergence, N_0 depends not just on the precision goal, but on the location within the interval as well — that is, $N_0 = N_0(\epsilon, x)$. Thus the rate of convergence need not be the same at all points. The stricter requirement $N_0 = N_0(\epsilon)$ ensures a consistent rate of convergence over the entire interval; this is uniform convergence.

Using (34.7) to analyze a continuous infinity of infinite series is not terribly practical. The Weierstrass M test reduces the proof of uniform convergence to a condition on a single infinite series. Start with a function series (not necessarily orthogonal) $\sum_n u_n(x)$ in which each term has an absolute upper bound — the same M_n for all x in the interval,

$$|u_n(x)| \le M_n. \tag{34.8}$$

This allows us to construct the inequality chain

$$\sum_{n=0}^\infty u_n(x) \le \sum_{n=0}^\infty |u_n(x)| \le \sum_{n=0}^\infty M_n. \tag{34.9}$$

The M test is then essentially a comparison test: if $\sum_{n=0}^\infty M_n < \infty$, then the series $\sum_{n=0}^\infty u_n(x)$ is uniformly convergent on the interval. Consider, for example, the series

$$\sum_{n=1}^\infty \frac{\cos nx}{n^s}, \tag{34.10}$$

for real s. Then

$$\sum_{n=1}^{\infty} \frac{\cos nx}{n^s} \leq \sum_{n=1}^{\infty} \frac{|\cos nx|}{n^s} \leq \sum_{n=1}^{\infty} \frac{1}{n^s}. \tag{34.11}$$

Thus the series (34.10) is uniformly convergent for all $s > 1$.

Though the M test provides a sufficient rather than necessary condition for uniform convergence, it is nonetheless a powerful tool. And its proof is straightforward: using (34.8) we get

$$|S(x) - S_N(x)| \leq \sum_{n=N+1}^{\infty} |u_n(x)| \leq \sum_{n=N+1}^{\infty} M_n. \tag{34.12}$$

But if $\sum_{n=0}^{\infty} M_n$ converges, then $\lim_{N \to \infty} \sum_{n=N+1}^{\infty} M_n = 0$. Which means there must be an N_0 and precision goal ϵ which satisfies (34.7), and so the series $S(x)$ converges uniformly. (If you find this type of analysis compelling, consider taking a course on real analysis.)

Power series are always uniformly convergent, and the ability to flip the order of sum and integral or sum and derivative validates many of the manipulations in Chapters 6–8. But as we saw in Example 34.1, expansions over orthogonal functions do not necessarily have these properties — and unlike power series, sweeping general statements are difficult. However, we can say that for a function $f(x)$ on a closed interval $[a, b]$, the so-called "Dirichlet conditions" ("derishlay"),

- f must be absolutely integrable, $\int_a^b |f(x)| \, dx < \infty$,
- f can have only a finite number of jump discontinuities,
- f can have only a finite number of extrema,

are sufficient (but not necessary) to conclude that an orthogonal expansion $S(x)$ converges pointwise to $\frac{1}{2} \left[f(x_+) + f(x_-) \right]$, where $f(x_\pm) \equiv \lim_{x \to x_\pm} f(x)$. In other words, at a jump discontinuity S converges to the average of the values of f across the gap. But over any closed interval $[a, b]$ on which f is continuous and f' piecewise continuous, the series converges uniformly.

The case of Fourier series requires an additional comment. Since $f(x)$ must be periodic, jump discontinuities include those at the ends of intervals — such as we saw in Figure 33.10a,b. Thus in order for the Fourier series to be uniformly convergent, not only must f be continuous over the period, we must also require the function to have the same value at the endpoints of the interval. For example, the expansion in Figure 33.10c has $f(-L) = f(L)$, yielding a smooth function; by contrast, Figure 33.10b has $f(-L) = -f(L)$, resulting in jump discontinuities.

Also evident in Figure 33.10 is one of the more intriguing features of orthogonal series expansions: Gibbs phenomenon, the large oscillations ("ringing") that occur near a jump discontinuity. This is direct evidence that the expansions do not converge uniformly. The simplest example is perhaps that of a Fourier expansion of a step function (Problem 33.1):

$$f(x) = \begin{cases} 0, & -\pi < x \leq 0 \\ 1, & 0 < x \leq \pi \end{cases}, \tag{34.13}$$

for which $a_0 = 1$, $a_{n>1} = 0$, and $b_n = 2/n\pi$ for odd n. We already know that as $N \to \infty$, the Fourier series for this function converges to the mean value $1/2$, so one might reasonably expect the ringing near $x = 0$ to be an artifact of using partial sums rather than the full infinite series. In other words, a truncation error. But Figure 34.1 implies something unexpected: although increasing the number

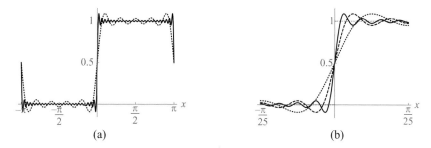

Figure 34.1 (a) Fourier series of the step function through $N = 10$ (dotted), 50 (solid); (b) Close-up of ringing near the discontinuity through $N = 50$ (dotted), 100 (dashed), 200 (solid).

of terms in the series pushes the ringing ever-closer to the point of discontinuity and thus improves overall convergence, the height of the overshoot near $x = 0$ appears never to decrease below a certain value. This is, in fact, the case; other orthogonal series expansions exhibit the same behavior.

BTW 34.2 Gibbs Overshoot

To verify the implication of Figure 34.1 and determine the size of the overshoot, we examine the partial sum of the complex Fourier series:

$$S_N(x) = \sum_{n=-N}^{N} c_n e^{inx}$$

$$= \sum_{n=-N}^{N} \left[\frac{1}{2\pi} \int_{-\pi}^{\pi} f(x') e^{-inx'} dx' \right] e^{inx}$$

$$= \frac{1}{2\pi} \int_{-\pi}^{\pi} f(x') \left[\sum_{n=-N}^{N} e^{in(x-x')} \right] dx'$$

$$\equiv \frac{1}{2\pi} \int_{-\pi}^{\pi} f(x') D_N(x - x') dx'. \qquad (34.14)$$

(Note that a finite sum of finite functions always commutes with integration and differentiation.) D_N is often called the *Dirichlet kernel*; in Problem 6.9, you found the closed form

$$D_N(y) = \frac{\sin[(N + 1/2)y]}{\sin(y/2)}. \qquad (34.15)$$

Inserting this into (34.14), the partial sum for the step function (34.13) becomes

$$S_N(x) = \frac{1}{2\pi} \int_{-\pi}^{\pi} f(x') D_N(x - x') dx'$$

$$= \frac{1}{2\pi} \int_{0}^{\pi} D_N(x - x') dx'$$

$$= \frac{1}{2\pi} \int_{0}^{\pi} \frac{\sin[(N + \frac{1}{2})(x - x')]}{\sin[\frac{1}{2}(x - x')]} dx'. \qquad (34.16)$$

With a change to integration variable $t \equiv (N + \frac{1}{2})(x - x') = \frac{1}{2}(2N + 1)(x - x')$,

$$S_N(x) = \frac{1}{2\pi} \int_{\frac{1}{2}(2N+1)(x-\pi)}^{\frac{1}{2}(2N+1)x} \frac{\sin t}{\sin [t/(2N + 1)]} \frac{2dt}{2N + 1}$$

$$= \frac{1}{\pi} \int_{\frac{1}{2}(2N+1)(x-\pi)}^{\frac{1}{2}(2N+1)x} \frac{\sin t}{t} \frac{t/(2N + 1)}{\sin [t/(2N + 1)]} dt. \tag{34.17}$$

This may look unduly complicated — but in fact, for our purposes it's a simplification: since we're interested in the convergence properties near $x = 0$, we need only concern ourselves with large N and small x. In this regime the second factor in the integrand goes to 1, allowing the partial sum to be rendered

$$S_N(x) \to \frac{1}{\pi} \int_{-\infty}^{Nx} \frac{\sin t}{t} dt$$

$$= \frac{1}{\pi} \left(\int_{-\infty}^{0} \frac{\sin t}{t} dt + \int_{0}^{Nx} \frac{\sin t}{t} dt \right) = \frac{1}{2} + \frac{1}{\pi} \mathrm{Si}(Nx), \tag{34.18}$$

where the *sine integral* function is

$$\mathrm{Si}(u) \equiv \int_{0}^{u} \frac{\sin t}{t} dt, \tag{34.19}$$

and we've used $\mathrm{Si}(\infty) = \pi/2$ (Example 8.16).

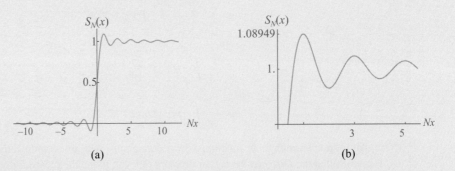

Figure 34.2 (a) Large-N Fourier series $S_N(x)$ of a step function; (b) close-up of ringing.

Figure 34.2 displays the partial sum S_N as a function of Nx. Note that as the number of terms N increases, the plots represent smaller values of x — but the profile of the function is the same. For any point arbitrarily close to the discontinuity, we can find large-enough N until S_N at that point is within any precision goal $\epsilon > 0$ set in (34.7). But though pointwise convergent, it's not uniformly convergent since the rate of convergence is clearly position dependent. Moreover, though larger N allows convergence at smaller x, this just moves the site of the overshoot closer to $x = 0$, with its value remaining a constant of about 8.94% (Problem 34.3). At no finite value of N does the overshoot vanish. The ringing around the jump is unavoidable.

34.2 Parseval's Theorem

So orthogonal function expansions are not uniformly convergent around jumps; nor does pointwise convergence allow us to remove the question mark from (34.2). What we'd like is a mode of convergence under which we can assert that a function is equivalent to its orthogonal series expansion despite jump discontinuities.

Even with ringing, we can see from Figure 34.1 that the partial sums become better overall representations of f as N increases. Nonetheless, the ringing is unavoidable. In some sense, the problem stems from trying to use a countable infinity of coefficients c_n to reconstruct a function over an uncountable infinity of points x. So rather than comparing values at each point, perhaps we should define convergence using some average measure of the difference between f and S over the entire interval from a to b.

Since our functions are vectors in an inner product space, a natural full-interval comparison should involve the inner product $\langle \phi | \psi \rangle = \int_a^b \phi^* \psi$. Indeed, we know from one of the defining criteria of inner products, (25.26a), that only the zero vector has vanishing norm. We can use this to establish the equivalence of two vectors: given $|f - S\rangle \equiv |f\rangle - |S\rangle$,

$$\langle f - S | f - S \rangle = 0 \quad \Longrightarrow \quad |f\rangle \equiv |S\rangle. \tag{34.20}$$

So for the function space $\mathbb{L}^2[a, b; w]$, if

$$\lim_{N \to \infty} \|f - S_N\|^2 = \lim_{N \to \infty} \int_a^b |f(x) - S_N(x)|^2 \, w(x) \, dx = 0, \tag{34.21}$$

then f and S are considered to be equal in the *mean-square* sense, and the infinite series $S(x)$ is said to be *mean-square* (or $\mathbb{L}^2$) *convergent* to $f(x)$. All the examples we've seen, whether uniform and merely pointwise convergent, are mean-square convergent. Unless stated otherwise, we'll use "convergence" to denote this mean-square sense.[1]

To develop intuition for mean-square convergence, let's examine how well the partial sums approximate a function $f(x)$. We use σ to denote the magnitude of the difference between vectors $|f\rangle$ and $|S_N\rangle$,

$$\sigma^2 \equiv \|f - S_N\|^2 = \langle f - S_N | f - S_N \rangle$$
$$= \langle f | f \rangle - \langle f | S_N \rangle - \langle S_N | f \rangle + \langle S_N | S_N \rangle. \tag{34.22}$$

This is often referred to as the mean-square deviation — or simply, the error. Inserting the orthogonal expansion $|S_N\rangle = \sum_n^N s_n |\hat{\phi}_n\rangle$ gives[2]

$$\sigma^2 = \langle f | f \rangle - \sum_n^N s_n \langle f | \hat{\phi}_n \rangle - \sum_n^N s_n^* \langle \hat{\phi}_n | f \rangle + \sum_{n,m}^N s_n^* s_m \langle \hat{\phi}_n | \hat{\phi}_m \rangle$$
$$= \langle f | f \rangle - \sum_n^N s_n \langle \hat{\phi}_n | f \rangle^* - \sum_n^N s_n^* \langle \hat{\phi}_n | f \rangle + \sum_n^N |s_n|^2, \tag{34.23}$$

---

[1] Recall that the Dirichlet conditions listed in the previous section require a function to be absolutely integrable, whereas mean-square convergence demands that it be square integrable. For finite limits a and b, however, a square-integrable function is also absolutely integrable — see Problem 31.4.

[2] The lower limit could be $n = 0$ or 1, depending on the choice of ϕ_n's.

where for simplicity we've taken the $|\hat{\phi}_n\rangle$ to be normalized. If we complete the square by adding and subtracting $\sum_n^N |\langle\hat{\phi}_n|f\rangle|^2$, the error becomes

$$\sigma^2 = \langle f|f\rangle - \sum_n^N |\langle\hat{\phi}_n|f\rangle|^2 + \sum_n^N |s_n - \langle\hat{\phi}_n|f\rangle|^2. \qquad (34.24)$$

Now the best approximation must have the smallest error, which can be achieved by choosing coefficients to minimize the last term. Remarkably, that choice is

$$s_n = \langle\hat{\phi}_n|f\rangle. \qquad (34.25)$$

In other words, the best mean-square approximation to $f(x)$ is given by orthogonal projections onto the basis — which, of course, is what we've been doing all along. Moreover, adding the next term to the expansion to get S_{N+1} requires only that we calculate the new coefficient s_{N+1}; there's no need to adjust or recalculate the other N coefficients. Although this "best expansion" pertains to orthogonal bases generally (Fourier, Legendre, Hermite, ...), the s_n are often referred to as "generalized Fourier coefficients."

With the choice $s_n = \langle\hat{\phi}_n|f\rangle$, the minimum error is

$$\sigma^2_{\min} = \langle f|f\rangle - \sum_n^N |\langle\hat{\phi}_n|f\rangle|^2. \qquad (34.26)$$

Since $\sigma^2 \geq 0$, this leads directly to *Bessel's inequality*,

$$\langle f|f\rangle \geq \sum_n^N |s_n|^2. \qquad (34.27)$$

The inequality is telling us that there remain orthogonal "directions" which haven't been included in the expansion. Indeed, as N goes to infinity we know the expansion converges and the mean-square error goes to zero — at which point Bessel's inequality saturates to become *Parseval's theorem*,

$$\langle f|f\rangle = \sum_n^\infty |s_n|^2. \qquad (34.28)$$

Parseval's theorem is essentially a statement of the Pythagorean theorem: the sum of the squares of the "legs" (orthogonal projections) is equal to the square of the "hypotenuse" (the vector). More fundamentally, the transition from Bessel to Parseval shows that including all orthogonal directions in an expansion gives a faithful, mean-square representation of the function. This conforms with our intuitive understanding of a basis in finite-dimensional space (but see BTW 34.3). Thus Parseval's theorem expresses the completeness of the basis $\{|\hat{\phi}_n\rangle\}$.

If the basis is orthogonal rather than orthonormal, $\langle\phi_n|\phi_m\rangle = k_n\delta_{nm}$, then just as we saw in (31.11), the coefficients (34.25) pick up a factor of k_n,

$$s_n = \frac{1}{k_n}\langle\phi_n|f\rangle, \qquad (34.29)$$

and Parseval's theorem becomes

$$\langle f|f\rangle = \sum_n^\infty k_n |s_n|^2. \qquad (34.30)$$

For Fourier series, this is (Problem 34.5)

$$\frac{1}{2\pi} \int_{-\pi}^{\pi} |f(x)|^2 \, dx = \left| \frac{a_0}{2} \right|^2 + \frac{1}{2} \sum_{n=1}^{\infty} |a_n|^2 + |b_n|^2 = \sum_{n=-\infty}^{\infty} |c_n|^2. \tag{34.31}$$

Recall that a wave's energy and intensity are proportional to the square of its amplitude. So (34.31) states that the energy is given by the sum of the energies of each harmonic. In the case of sound, the intensity is volume (loudness); thus the intensity of a musical chord is the direct sum of the volume of its individual notes. For functions of time such as current or voltage, the squared coefficients give the average power contributed at ω_n; as a result, the collection $\{|s_n|^2\}$ is known as the signal's *power spectrum*.

Example 34.2 **Parseval in Practice**

For complex Fourier series, Parseval's theorem is

$$\frac{1}{2\pi} \int_{-\pi}^{\pi} |f(x)|^2 \, dx = \sum_{n=-\infty}^{\infty} |c_n|^2. \tag{34.32}$$

This can be used to evaluate infinite sums. For instance, in Example 33.4 we found that $f(x) = |x|$ has expansion coefficients $c_n = -2L/\pi^2 n^2$ for odd n and $c_0 = L/2$. So with $L = \pi$, we can integrate directly to get

$$\frac{1}{2\pi} \int_{-\pi}^{\pi} |x|^2 \, dx = \frac{\pi^2}{3}, \tag{34.33}$$

which Parseval sets equal to

$$\sum_{n=-\infty}^{\infty} |c_n|^2 = \left(\frac{\pi}{2} \right)^2 + 2 \sum_{\text{odd } n>0} \frac{4}{\pi^2 n^4}. \tag{34.34}$$

A little algebra leads to the non-trivial result

$$\sum_{n \text{ odd}}^{\infty} \frac{1}{n^4} = \frac{\pi^4}{96}. \tag{34.35}$$

BTW 34.3 Hilbert Space

As intuitive as mean-square convergence may be, a subtle technical point lurks beneath its surface. Treating f and S_N as vectors in an inner product space led to (34.21), but mean-square convergence only makes sense if the limit $S = \lim_{N \to \infty} S_N$ is *also* in the space; after all, vector spaces must be closed. Consider the one-dimensional space of rational numbers. One can add and subtract these in any finite amount, and the result is still a rational number. But back in Example 6.7 we claimed (and you verified in Problem 33.17) that $\zeta(2) = \sum_{n=1}^{\infty} \frac{1}{n^2}$ converges to $\pi^2/6$. Thus despite the fact that every partial sum approximation to $\zeta(2)$ is rational, in the infinite limit we get an irrational number. We can use this feature to identify a larger space: the set of rational numbers, together with the limits of all convergent sequences of partial sums, yields the space of real numbers.

Similarly, we expanded the step function of (34.13) on a basis of $\sin nx$'s, all of which vanish at $x = 0$ — as does every one of the partial sums. So it's quite reasonable to presume one is confined to a space of vectors with $f(0) = 0$; this would certainly be the case in a finite-dimensional space. But in the $N \to \infty$ limit, we can end up with a function which is *non*-zero at the origin.

Formally enlarging the defining criteria of our space to include the limits of all such sequences produces a "complete inner product space" known as a *Hilbert space*. Hilbert spaces are complete in the sense that the limit of any Cauchy sequence is also in the space — see BTW 6.2, with the absolute value in (6.18) replaced by the $\mathbb{L}^2$ norm.

This may seem unduly fastidious. Indeed, finite-dimensional inner product spaces have no completeness issues, and so can automatically be classified as Hilbert spaces. But in infinite dimensions, many important results only hold if the space is complete. Specifically, Hilbert space extends the familiar properties of Euclidean $\mathbb{R}^3$ to inner products spaces generally. The power of Hilbert space is that it allows $\mathbb{R}^3$ intuition to pertain even in infinite-dimensional inner product spaces.

The connection between a complete inner product space and the Parseval statement of completeness derives from mean-square convergence, (34.21): only if the sequence of partial sums has a limit in the space does the statement $|f\rangle = \sum_n^\infty c_n|\phi_n\rangle$ makes sense. By choosing the c_n to be generalized Fourier coefficients of orthogonal projection, Parseval follows.

As an expression of completeness, Parseval's theorem (34.30) is fairly intuitive. But since completeness is a property of the basis, we'd prefer a statement that explicitly uses the $\phi_n(x)$ without reference to an arbitrary function f and its coefficients s_n. We can find such an expression by inserting the coefficients of (34.29) into S_N:

$$S_N(x) = \sum_n^N s_n\phi_n(x) = \sum_n^N \left[\frac{1}{k_n}\int_a^b \phi_n^*(x')f(x')w(x')\,dx'\right]\phi_n(x)$$

$$= \int_a^b f(x')\left[w(x')\sum_n^N \frac{1}{k_n}\phi_n^*(x')\phi_n(x)\right]dx'. \tag{34.36}$$

If the ϕ_n form a complete basis, then S_N converges to $f(x)$ as N goes to infinity. Examination of (34.36) in this limit reveals the completeness relation

$$\delta(x-x') = w(x')\sum_n^\infty \frac{1}{k_n}\phi_n^*(x')\phi_n(x). \tag{34.37}$$

This can be viewed both as a statement of completeness as well as an orthogonal series expansion of the delta function. The Fourier basis e^{inx}, having $w(x) = 1$ and $k_n = 1/2\pi$ for all n, is perhaps the simplest:[3]

$$\delta(x-x') = \frac{1}{2\pi}\sum_{n=-\infty}^\infty e^{in(x-x')}, \tag{34.38}$$

where despite the complex basis functions, the sum over $\phi_n^*(x')\phi_n(x)$ is real. Figure 34.3 shows how this sum approach a delta function; the same progression occurs for other function bases, though with differing rates of convergence.

An equivalent expression to (34.37) which better reflects the symmetry of the δ-function in x and x' is

$$\delta(x-x') = \sqrt{w(x)\,w(x')}\sum_n^\infty \frac{1}{k_n}\phi_n^*(x')\phi_n(x). \tag{34.39}$$

---

[3] Note that (34.38) is the $N \to \infty$ limit of the Dirichlet kernel D_N in BTW 34.2.

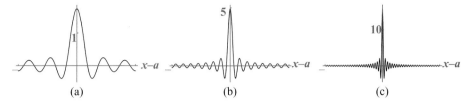

Figure 34.3 Approaching δ: Fourier completeness $\frac{1}{2\pi}\sum_{n=-N}^{N}e^{in(x-x')}$ through $N = 5, 15,$ and 50.

Then if we define the square-integrable, orthonormal functions $\hat{\phi}_n(x) = \sqrt{\frac{w(x)}{k_n}}\,\phi_n(x)$, the completeness relation becomes

$$\delta(x - x') = \sum_n \hat{\phi}_n^*(x')\hat{\phi}_n(x), \tag{34.40}$$

which should be compared to the finite-dimensional completeness relation of Section 25.5, $\mathbb{1} = \sum_i |\hat{e}_i\rangle\langle\hat{e}_i|$. In fact, $|\phi_n\rangle\langle\phi_n| \to \int \hat{\phi}_n^*(x')\hat{\phi}_n(x)dx'$ is a projection operator:

$$|\phi_n\rangle\langle\phi_n|f\rangle = \int \hat{\phi}_n^*(x')\hat{\phi}_n(x)f(x')\,dx'$$

$$= \left(\int \hat{\phi}_n^*(x')f(x')\,dx'\right)\hat{\phi}_n(x) = c_n\hat{\phi}_n(x). \tag{34.41}$$

So $\int dx'\,\hat{\phi}_n^*(x')\hat{\phi}_n(x)$ projects out the component of f "along" $\hat{\phi}_n$ and directs it back along $\hat{\phi}_n$.

Example 34.3 Inserting Completeness

Just as we can insert a complete set of states $\mathbb{1} = \sum_i |\hat{e}_i\rangle\langle\hat{e}_i|$ in a discrete sum, we can insert the identity in a continuous sum — for instance, in the inner product of two functions:

$$\langle f|g\rangle = \int_a^b f^*(x)g(x)\,w(x)dx$$

$$= \int_a^b f^*(x)\delta(x - x')g(x')\,w(x')\,dx\,dx'$$

$$= \int_a^b f^*(x)\left[w(x)\sum_n^{\infty}\frac{1}{k_n}\phi_n^*(x')\phi_n(x)\right]g(x')\,w(x')\,dx\,dx'$$

$$= \sum_n \frac{1}{k_n}\left[\int_a^b f^*(x)\phi_n(x)\,w(x)\,dx\right]\left[\int_a^b \phi_n^*(x')g(x')\,w(x')\,dx\right]. \tag{34.42}$$

These integrals should be familiar: they are the projections of $|f\rangle$ and $|g\rangle$ on the basis, $\langle f|\phi_n\rangle = k_n c_n^*$ and $\langle \phi_n|g\rangle = k_n d_n$. Thus by choosing a basis, the inner product of f and g can be expressed as

$$\langle f|g\rangle = \sum_n k_n\, c_n^* d_n. \tag{34.43}$$

For a complex Fourier series, where $k_n = 2\pi$ independent of n, this is simply

$$\frac{1}{2\pi}\int_{-\pi}^{\pi} f^*(x)g(x)\,dx = \sum_n c_n^* d_n. \tag{34.44}$$

Because (34.42) is really just the insertion of a complete set of states, it's more transparent in bra-ket notation: using the completeness relation

$$\sum_n \frac{1}{k_n} |\phi_n\rangle \langle\phi_n| = \mathbb{1}, \tag{34.45}$$

we have

$$\langle f|g\rangle = \langle f| \left[\sum_n \frac{1}{k_n} |\phi_n\rangle \langle\phi_n| \right] |g\rangle$$

$$= \sum_n \frac{1}{k_n} \langle f|\phi_n\rangle \langle\phi_n|g\rangle$$

$$= \sum_n \frac{1}{k_n} \left(k_n c_n^* \right) \left(k_n d_n \right) = \sum_n k_n c_n^* d_n. \tag{34.46}$$

In this form, it's easier to appreciate that it's not really anything new — other than that integration is required to actually calculate the components c_n and d_n.

Problems

34.1   (a) Use (34.11) to determine whether the Fourier series of x^2, $|x| < \pi$, is uniformly convergent.

(b) Calculate the derivative of the Fourier series. Is it uniformly convergent? What about the second derivative?

34.2   Find the relationship between the Fourier coefficients a_n and b_n for a function $f(x)$, and the Fourier coefficients α_n and β_n for $f'(x)$.

34.3   **Gibbs Overshoot**

(a) Use Mathematica to reproduce the plots in Figure 34.1 of the step function

$$f(x) = \begin{cases} 0, & -\pi < x \le 0 \\ 1, & 0 < x \le \pi \end{cases}.$$

(b) Show that the derivative of the Fourier series for $f(x)$ up through $n = N$ has the closed form (see Problem 6.10)

$$\frac{1}{\pi} \frac{\sin 2Nx}{\sin x}.$$

(c) The extrema of $f(x)$ give the heights of the ringing around $x = 0$; the Gibbs overshoot can be found by evaluating f at the smallest positive extremal point x_1. Show that

$$\lim_{N \to \infty} f(x_1) = \frac{1}{2} + \frac{1}{\pi} \int_0^\pi \frac{\sin t}{t} \, dt \approx 1.08949.$$

Thus no matter how many terms are included in the Fourier series for $f(x)$, the overshoot is an unavoidable 8.9%.

34.4   Rederive (34.26) for the minumum error by subjecting (34.23) to a classic max/min calculus analysis. (For simplicity, assume everything is real.)

34.5   Verify that Parseval's theorem for Fourier series is

$$\frac{1}{2\pi} \int_{-\pi}^{\pi} |f(x)|^2 \, dx = \left| \frac{a_0}{2} \right|^2 + \frac{1}{2} \sum_{n=1}^{\infty} |a_n|^2 + |b_n|^2 = \sum_{n=-\infty}^{\infty} |c_n|^2.$$

34.6 Use the Weierstrass M test (34.9) to show that a power series $S = \sum_{n=0}^{\infty} a_n x^n$ is uniformly convergent:

(a) Argue that if the power series converges for some $x = x_0 \neq 0$, then there must be an $M \geq 0$ for which $|a_n x_0^n| \leq M$ for all n.

(b) Recast S as a geometric series in x/x_0 to show that S is uniformly convergent for all $|x| < |x_0|$. What is the largest x_0 for which this holds?

34.7 Express the completeness relation (34.40) in terms of Legendre, Laguerre, and Hermite polynomials. Use Mathematica to demonstrate that they are complete bases by showing, as in Figure 34.3, that $\sum \hat{\phi}^*(x')\hat{\phi}(x)$ approaches the delta function.

34.8 Use a Fourier series for $f(x) = x^2$ with $|x| < \frac{1}{2}$ to find the numerical value of $\zeta(4) = \sum_{n=1}^{\infty} \frac{1}{n^4}$.

34.9 Use a Fourier sine series of $f(x) = 1$, $0 < x < 1$, to evaluate the sum $\sum_{n \text{ odd}} \frac{1}{n^2}$.

34.10 Use the Fourier coefficients for $f(x) = e^{iax}$ with $|x| < \pi$ to show that

$$\pi^2 \csc^2(\pi a) = \frac{\pi^2}{\sin^2(\pi a)} = \sum_{n=-\infty}^{\infty} \frac{1}{(a-n)^2}.$$

35 Interlude: Beyond the Straight and Narrow

All the examples of orthogonal expansions we've seen thus far — whether from 0 to 2π or 0 to ∞, whether between ± 1 or $\pm\infty$ — have been along a one-dimensional domain. But what about beyond the straight and narrow confines of a line? In this Interlude, we'll consider important generalizations of Fourier expansions beyond one dimension by considering how the basis functions e^{ikx} manifest in higher dimensions and in different coordinates. As we'll see, the transition from e^{ikx} to $e^{i\vec{k}\cdot\vec{r}}$ will lead to the introduction of new bases of orthogonal functions having wide applications throughout mathematical science and engineering — and which, not coincidentally, we'll encounter many times in the chapters to come. The goal of this Interlude is to develop familiarity with these functions in advance of our foray into partial differential equations, whose solutions often involve normal mode expansions on these new bases.

35.1 Fourier Series on a Rectangular Domain

To adapt Fourier series to higher dimensions, a powerful if simple criterion is that the result should reduce to a familiar expansion when limited to a single dimension. The most straightforward illustration of this is cartesian coordinates in $\mathbb{R}^2$, for which

$$e^{i\vec{k}\cdot\vec{r}} = e^{i(k_x x + k_y y)} = e^{ik_x x} e^{ik_y y}. \tag{35.1}$$

Rather than a finite string, consider a $L_1 \times L_2$ rectangular region. We can obtain Fourier series with, say, periods $2L_1$ and $2L_2$ by restricting $k_{x,y}$ to be

$$k_x \to n\pi/L_1 \qquad k_y \to m\pi/L_2, \tag{35.2}$$

for integers n, m. So a complex Fourier expansion over the rectangle looks like

$$f(x, y) = \sum_{n,m=-\infty}^{\infty} \tilde{f}_{nm} e^{i\pi\left(\frac{nx}{L_1} + \frac{my}{L_2}\right)}, \tag{35.3}$$

with coefficients given by the double integral (Problem 35.1)

$$\tilde{f}_{nm} = \frac{1}{4L_1 L_2} \int_{-L_1}^{L_1} \int_{-L_2}^{L_2} f(x, y)\, e^{-in\pi x/L_1}\, e^{-im\pi y/L_2}\, dx\, dy. \tag{35.4}$$

For a real Fourier series, (35.3) is equivalent to

$$f(x, y) = \left[\frac{a_0}{2} + \sum_{n=1}^{\infty} a_n \cos\left(\frac{n\pi x}{L_1}\right) + b_n \sin\left(\frac{n\pi x}{L_1}\right) \right]$$
$$\times \left[\frac{\tilde{a}_0}{2} + \sum_{m=1}^{\infty} \tilde{a}_m \cos\left(\frac{m\pi y}{L_2}\right) + \tilde{b}_m \sin\left(\frac{m\pi y}{L_2}\right) \right], \qquad (35.5)$$

with two independent sets of Fourier coefficients derived from similar two-dimensional projection integrals. Note that for either x or y held fixed, this reduces to a one-dimensional Fourier series. Generalizing to higher dimensions is straightforward, with one integer index associated with each dimension. So in cartesian coordinates the Fourier basis is largely unchanged; appropriate discretization of $\vec{k}$'s components allows for familiar expansions in $\sin kx$ and $\cos kx$, or $e^{\pm ikx}$.

Example 35.1 The Reciprocal Crystal Lattice[1]

Consider a crystal as a periodic array of atoms forming a regular lattice structure. Being periodic, each lattice point is related to one another by translation; repeated indefinitely, this inherent translational symmetry implies that a crystal looks the same when viewed from any lattice site. Such a structure is called a Bravais ("brah-vey") lattice.[2] The simplest example is a two-dimensional rectangular lattice as depicted in Figure 35.1a, where the so-called "primitive vectors" $\vec{a}_1 = a\hat{i}$ and $\vec{a}_2 = b\hat{j}$ define a unit cell which repeats periodically throughout the crystal. Of course, not all crystals are rectangular or two dimensional. In general, one defines a linearly independent (though not necessarily orthogonal) basis of primitive vectors $\vec{a}_1, \vec{a}_2$, and $\vec{a}_3$ between adjacent lattice points to specify a unit cell. The lattice is then mathematically constructed by translation, each new cell generated from the unit cell using a vector $\vec{R}_n$ composed of integer multiples of the $\vec{a}_i$,

$$\vec{R}_n = n_1 \vec{a}_1 + n_2 \vec{a}_2 + n_3 \vec{a}_3. \qquad (35.6)$$

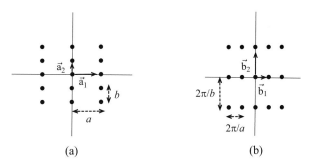

(a)   (b)

Figure 35.1 Rectangular lattice in two dimensions: (a) direct lattice (b) reciprocal lattice.

The set of points $\vec{R}_n$ gives the vector location of each lattice node $n \equiv (n_1, n_2, n_3)$; this representation of the crystal's structure is called its *direct lattice*.

Now as a periodic structure, a crystal's properties such as electron density $\rho(\vec{r})$ are themselves periodic. And being periodic, they can be expanded in a Fourier series,

---

[1] See, e.g., Kittel (2012).

[2] Each lattice site need not be occupied by a single constituent; rather, each Bravais node can represent a "basis" arrangement of atoms or molecules.

$$\rho(\vec{r}) = \sum_m \tilde{\rho}_m \, e^{i\vec{k}_m \cdot \vec{r}}, \tag{35.7}$$

where since $\rho(\vec{r})$ is real, the Fourier coefficients must satisfy $\tilde{\rho}_{-m} = \tilde{\rho}_m^*$. The periodicity of $\rho(\vec{r})$ restricts the $\vec{k}_m$ in (35.7) to those Fourier space vectors satisfying the condition $\rho(\vec{r}) = \rho(\vec{r} + \vec{R}_n)$,

$$\sum_m \tilde{\rho}_m e^{i\vec{k}_m \cdot \vec{r}} = \sum_m \tilde{\rho}_m e^{i\vec{k}_m \cdot (\vec{r} + \vec{R}_n)}, \tag{35.8}$$

or

$$e^{i\vec{k}_m \cdot \vec{R}_n} = 1, \tag{35.9}$$

for all n in the direct lattice. This implies that just as lattice positions are given by a family of vectors $\vec{R}_n$, positions in k-space can similarly be generated from a family of vectors $\vec{G}_m$,

$$\vec{G}_m = m_1 \vec{b}_1 + m_2 \vec{b}_2 + m_3 \vec{b}_3, \tag{35.10}$$

which, in order to satisfy (35.9), must have integers m_i such that

$$\vec{a}_i \cdot \vec{b}_j = 2\pi \, \delta_{ij}. \tag{35.11}$$

So the Fourier-space vectors $\vec{G}_m$ form their own Bravais lattice known as the *reciprocal lattice*. For the simplest case of a one-dimensional lattice with spacing a, the reciprocal lattice vector is $b = 2\pi/a$, and $\vec{G}_m$ reduces to the familiar $k_m = 2\pi m/a$. In three dimensions, (35.11) is satisfied by

$$\vec{b}_1 = 2\pi \frac{\vec{a}_2 \times \vec{a}_3}{V} \qquad \vec{b}_2 = 2\pi \frac{\vec{a}_3 \times \vec{a}_1}{V} \qquad \vec{b}_3 = 2\pi \frac{\vec{a}_1 \times \vec{a}_2}{V}, \tag{35.12}$$

where $V = |\vec{a}_1 \cdot (\vec{a}_2 \times \vec{a}_3)|$ is the volume of a unit cell. In two dimensions, one finds

$$\vec{b}_1 = 2\pi \frac{\vec{a}_2 \times (\vec{a}_1 \times \vec{a}_2)}{|\vec{a}_1 \times \vec{a}_2|^2} \qquad \vec{b}_2 = 2\pi \frac{\vec{a}_1 \times (\vec{a}_2 \times \vec{a}_1)}{|\vec{a}_1 \times \vec{a}_2|^2}, \tag{35.13}$$

with unit cell area $|\vec{a}_1 \times \vec{a}_2|$. For the rectangular example depicted in Figure 35.1b, these primitive vectors generate the reciprocal lattice

$$\vec{G}_m = m_1 \frac{2\pi}{a} \hat{\imath} + m_2 \frac{2\pi}{b} \hat{\jmath}, \tag{35.14}$$

for arbitrary integers m_1 and m_2.

Thus we discover that a crystal can be described by two lattices, the $\vec{R}_n$ of (35.6) in configuration space, and the $\vec{G}_m$ of (35.10) in "reciprocal space." Though distinct, (35.11) shows that the direct lattice's translational symmetry has been bequeathed to the reciprocal lattice and its primitive vectors $\vec{b}_j$. Who bequeaths to whom is moot: though the direct lattice is more intuitive, Parseval's theorem holds that the two lattices are mathematically equivalent representations of the crystal.

35.2 Expanding on a Disk

A Fourier expansion over a rectangle in cartesian coordinates requires imposing conditions on $\vec{k}$ to obtain the necessary periodicity. By contrast, polar coordinates are already periodic in one of its dimensions,

$$e^{i\vec{k} \cdot \vec{r}} = e^{ikr \cos \gamma}, \tag{35.15}$$

where $\gamma = \phi - \phi'$ is the difference between the orientations ϕ and ϕ' of $\vec{r}$ and $\vec{k}$ in the plane. And being periodic, (35.15) can be expanded as a Fourier series in γ,

$$e^{ikr\cos\gamma} = \sum_{n=-\infty}^{\infty} i^n J_n(kr)\, e^{in\gamma} = \sum_{n=-\infty}^{\infty} i^n J_n(kr)\, e^{-in\phi'} e^{in\phi}, \qquad (35.16)$$

where as we'll see in a moment, the i^n renders the J_n real. Note that the expansion coefficients J_n are actually functions of kr — in other words, "constant" only in the sense that they are independent of γ. So (35.16) can be construed as a collection of Fourier series on nested concentric circles, with the coefficient function J_n given by the usual Fourier projection integral

$$J_n(x) = \frac{1}{2\pi i^n} \int_{-\pi}^{\pi} e^{ix\cos\gamma - in\gamma}\, d\gamma. \qquad (35.17a)$$

Equivalent integral representations are (Problem 35.3)

$$J_n(x) = \frac{1}{2\pi} \int_{-\pi}^{\pi} e^{ix\sin\gamma - in\gamma}\, d\gamma, \qquad (35.17b)$$

and the manifestly real

$$J_n(x) = \frac{1}{\pi} \int_{0}^{\pi} \cos(x\sin\gamma - n\gamma)\, d\gamma. \qquad (35.17c)$$

These expressions can be taken as the definition of *Bessel functions* for integer n — the same functions we found in (19.13). They are oscillatory but not periodic, having both decaying amplitude and non-uniform spacing between zeros (Figure 35.2). Nonetheless, they have many similarities with sin and cos. For instance, as shown in Appendix C, for large x they reduce to

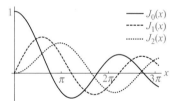

$$J_n(x) \xrightarrow{x\,\text{large}} \sqrt{\frac{2}{\pi x}} \cos\left(x - \frac{n\pi}{2} - \frac{\pi}{4}\right). \qquad (35.18a)$$

Figure 35.2 Bessel functions $J_n(x)$, $n = 0, 1, 2.$

Thus asymptotically at least, successive J_n's become $\pi/2$ out of phase with one another, with zeros regularly spaced by π. For small x they behave as

$$J_n(x) \xrightarrow{x\,\text{small}} \frac{x^n}{2^n n!}, \qquad (35.18b)$$

so that like cos and sin, $J_0(0) = 1$ and $J_{n\neq0}(0) = 0$.

The similarities Bessel functions have with sine and cosine emerge from the angular periodicity of a disk — but their utility often surfaces even when no underlying disk is manifest. Let's pause to examine two such examples.

Example 35.2 **Frequency Modulation**

Although an FM radio station nominally broadcasts at its *carrier frequency* $\omega_c/2\pi$, it does not actually transmit at that fixed frequency. In fact, no information can be sent with a single, time-independent frequency. Rather, ω_c is modulated,

$$\Psi(t) = A\cos\left(\omega_c t + \psi_m(t)\right), \qquad (35.19)$$

where ψ_m encodes the information (e.g., music) being broadcast. In other words, the actual broadcast frequency is time dependent,

$$\omega = \omega_c + d\psi_m/dt. \tag{35.20}$$

Thus the spectrum broadcast by the radio station is not a delta function at ω_c. To determine the actual spectrum, we'll assume sinusoidal modulation ψ and write the broadcast signal as

$$\Psi(t) = A \cos(\omega_c t + \beta \sin \omega_m t). \tag{35.21}$$

The *modulation index* β is thus the maximum phase deviation due to $\psi_m(t)$ — or equivalently, since $\omega = \omega_c + \beta \omega_m \cos \omega_m t$, it's the ratio of the maximum change in frequency to the modulating frequency,

$$\beta = \frac{\omega - \omega_c}{\omega_m} \equiv \frac{\Delta\omega}{\omega_m}. \tag{35.22}$$

So β is the key parameter determining the breadth of an FM radio station's broadcast spectrum, its *bandwidth*. This is verified by expanding (35.21) in a Fourier series (Problem 35.8),

$$A \cos(\omega_c t + \beta \sin \omega_m t) = \sum_{n=-\infty}^{\infty} J_n(\beta) \cos[(\omega_c + n\omega_m)t]. \tag{35.23}$$

Thus the spectrum extends indefinitely, with *sideband* frequencies $\omega_c + n\omega_m$ of amplitude $J_n(\beta)$ arrayed symmetrically around the carrier frequency. But though technically an infinite bandwidth, the behavior of the Bessel functions limits the practical extent of the spectrum. For small β, (35.18b) indicates that sidebands beyond the first few make negligible contribution, resulting in a relatively narrow bandwidth. For large β, (35.18a) shows that the contributions drop off as $1/\sqrt{\beta}$, resulting in a larger bandwidth — though still effectively finite. Indeed, as coefficients in the Fourier expansion (35.16), the J_n are subject to Parseval's theorem, (34.31), which reveals

$$\sum_{n=-\infty}^{\infty} |J_n(\beta)|^2 = 1. \tag{35.24}$$

As a convergent sum, the terms must drop off with increasing n. An inspection of Bessel functions would reveal that $|J_n(\beta)|^2$ becomes negligible for $n > \beta + 1$ — so common practice defines bandwidth using "Carson's bandwidth rule" (CBR),

$$\text{CBR} = 2\frac{\omega_m}{2\pi}(\beta + 1) = 2(\Delta f + f_m) \quad \text{(in Hz)}, \tag{35.25}$$

which includes at least 98% of the total power of the signal (Problem 35.9). For instance, standard FM radio is designed for a maximum modulating frequency of $f_m = 15$ kHz.[3] A modulation index of $\beta \approx 5.0$ corresponds to a frequency deviation of $\Delta f = 75$ kHz and thus a bandwidth of CBR = 180 kHz — which explains why FM stations are spaced 200 kHz apart. The usual FM radio frequency range of 88–108 MHz therefore has sufficient bandwidth for 100 radio stations.

Example 35.3   Kepler's Equation

In Section 7.3, we introduced three angles, or "anomalies," used to track the position of a satellite in an elliptical orbit. As described in Figure 7.1, the mean anomaly μ is the angle of a fictitious point moving in a circle of radius a with uniform angular speed, whereas the eccentric anomaly η tracks the orbit along the so-called auxiliary circle. Kepler's equation (7.30) relates the two,

---

[3] Since the human ear is sensitive up to ~20 kHz, this excludes the uppermost frequencies. But AM (*amplitude modulation*) radio cuts off at about 5 kHz, so FM yields higher fidelity.

$$\mu = \eta - \epsilon \sin \eta, \tag{35.26}$$

where $0 \leq \epsilon < 1$ is the eccentricity of the orbit. Calculating the orbit requires that Kepler's equation be inverted to give $\eta = \eta(\mu, \epsilon)$. Convergence of the Taylor series (7.33), however, is limited to eccentricities of less than about 2/3. Bessel developed an alternative approach by expanding $\epsilon \sin \eta$ in a Fourier sine series,

$$\eta(\mu) = \mu + \epsilon \sin \eta = \mu + \sum_{k=1}^{\infty} b_k \sin k\mu. \tag{35.27}$$

Using Kepler's equation and integration by parts,

$$\begin{aligned}
b_k &= \frac{2}{\pi} \int_0^\pi \epsilon \sin \eta \, \sin k\mu \, d\mu \\
&= \frac{2}{\pi} \int_0^\pi [\eta(\mu) - \mu] \sin(k\mu) \, d\mu \\
&= -\frac{2}{k\pi} [\eta(\mu) - \mu] \cos(k\mu) \Big|_0^\pi + \frac{2}{k\pi} \int_0^\pi \cos(k\mu)(d\eta/d\mu) \, d\mu.
\end{aligned} \tag{35.28}$$

As is evident in Figure 7.1, the mean and eccentric anomalies agree only at perigee and apogee — that is, $\mu = \eta$ at 0 and π, as is easily verified by (35.26). So the surface term in (35.28) vanishes, and we're left with

$$\begin{aligned}
b_k &= \frac{2}{k\pi} \int_0^\pi \cos(k\mu) \frac{d\eta}{d\mu} \, d\mu \\
&= \frac{2}{k\pi} \int_0^\pi \cos[k(\eta - \epsilon \sin \eta)] \, d\eta = \frac{2}{k} J_k(k\epsilon),
\end{aligned} \tag{35.29}$$

where we've used (35.17c). Thus

$$\eta(\mu) = \mu + 2 \sum_{k=1}^{\infty} \frac{1}{k} J_k(k\epsilon) \sin(k\mu). \tag{35.30}$$

Unlike the expansion in powers of ϵ, this series converges for all $\epsilon < 1$.

Incidentally, although Bessel did not discover the J_n (see Example 41.2), it was his work such as this on planetary orbits which led to them acquiring his name.

35.3 On a Sphere: The $Y_{\ell m}$'s

Having considered the Fourier basis on a disk, we move next to examine a ball. Given their similarities, one should anticipate an expression analogous to (35.16) — though clearly not with Bessel functions $J_n(kr)$ which depend on cylindrical r. Instead, we'll discover *spherical* Bessel functions $j_n(kr)$, with lower-case j and spherical r. Since there are now two periodic angles θ and ϕ to consider, defining the j_n with a single Fourier projection as in (35.17a) seems naive. On the other hand, $e^{i\vec{k}\cdot\vec{r}} = e^{ikr\cos\gamma}$ has only one relevant angle, so defining j_n by one-dimensional projection must be possible. To address and resolve this dilemma, we'll first consider how to expand a function on the surface of a sphere before continuing in Section 35.4 with the expansion of $e^{i\vec{k}\cdot\vec{r}}$ within a ball.

We already have separate orthogonal expansions in θ and ϕ — the Legendre polynomials $P_\ell(\cos\theta)$ for $0 \leq \theta \leq \pi$ with integer ℓ, and $e^{im\phi}$ for $0 \leq \phi < 2\pi$ with integer m. So we might expect an expansion for a square-integrable function $f(\theta, \phi)$ on the sphere to look something like

$$f(\theta,\phi) \stackrel{?}{=} \left(\sum_{\ell=0}^{\infty} c_\ell P_\ell(\cos\theta)\right)\left(\sum_{m=-\infty}^{\infty} d_m e^{im\phi}\right). \tag{35.31}$$

Holding ϕ constant to reduce this to a one-dimensional expansion reveals a Legendre series between $\cos\theta = \pm 1$. Similarly, setting $\theta = \pi/2$ gives a Fourier expansion on a circle. So far, so good. But since the radius of those Fourier circles scales like $\sin\theta$, holding θ fixed at another value results in the same formal Fourier series in ϕ despite being on a circle with smaller radius. In particular, for $\theta = 0$ there is no circle at all and no ϕ to speak of, necessitating a contribution from only $m = 0$ — but there's nothing in the Legendre expansion requiring a collapse of the Fourier series to facilitate this.

So (35.31) is at best incomplete. Though the assumption that the basis can be factored into the product of functions in θ and ϕ has aesthetic appeal, clearly these functions cannot be completely unaware of one another. In fact, we would expect the two separate functions in θ and ϕ to share an index — just as we saw in (35.16) for polar coordinates. To properly scale with $\sin\theta$, we can introduce an m-dependent factor of $\sin^m\theta = (1 - \cos^2\theta)^{m/2}$ to the Legendre polynomials; to avoid changing their overall order in $\cos\theta$, we'll take m derivatives of $P_\ell(x)$ with respect to $x = \cos\theta$. Doing this results in the correct expansion — not in terms of Legendre polynomials, but rather the *associated Legendre functions* $P_\ell^m(x)$ (see Problem 32.23),

$$P_\ell^m(x) = (1 - x^2)^{m/2}\left(\frac{d}{dx}\right)^m P_\ell(x)$$

$$= \frac{(-1)^\ell}{2^\ell \ell!}(1 - x^2)^{m/2}\left(\frac{d}{dx}\right)^{\ell+m}(1 - x^2)^\ell, \qquad m \geq 0, \tag{35.32}$$

where we've used the Rodrigues formula for Legendre polynomials, (32.45). Note that for $m = 0$, they reduce to the Legendre polynomials, $P_\ell^0(x) \equiv P_\ell(x)$.

The derivative structure in (35.32) shows that $P_\ell^m \equiv 0$ for $m > \ell$, but would allow m down to $-\ell$. Associated Legendre functions with negative m, consistent with (35.32), are given by

$$P_\ell^{-m} = (-1)^m \frac{(\ell - m)!}{(\ell + m)!} P_\ell^m, \tag{35.33}$$

thereby defining the P_ℓ^m for integer m between $\pm\ell$. All together then, for every integer $\ell \geq 0$ there are $2\ell + 1$ values of m running in integer steps from $-\ell$ to ℓ. The plots in Figure 35.3 exhibit their general form, and Table 35.1 displays the $P_\ell^m(\cos\theta)$ for $\ell \leq 3$.[4]

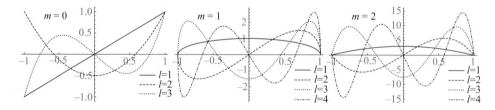

Figure 35.3 Associated Legendre functions $P_\ell^m(x)$ have $\ell - m$ nodes. Note that $P_\ell^{m\neq 0}(\pm 1) = 0$.

---

[4] P_ℓ^m is often called an associated Legendre *polynomial*, despite being a polynomial in x only for m even. But it is a "bivariate" polynomial in $\sin\theta$ and $\cos\theta$ of order m and $\ell - m$, respectively.

Table 35.1 Associated Legendre functions $P_\ell^m(\cos\theta)$

$P_0^0 = 1$	$P_1^0 = \cos\theta$	$P_2^0 = \frac{1}{2}\left(3\cos^2\theta - 1\right)$	$P_3^0 = \frac{1}{2}\left(5\cos^3\theta - 3\cos\theta\right)$
	$P_1^1 = \sin\theta$	$P_2^1 = 3\sin\theta\cos\theta$	$P_3^1 = \frac{3}{2}\sin\theta\left(5\cos^2\theta - 1\right)$
		$P_2^2 = 3\sin^2\theta$	$P_3^2 = 15\sin^2\theta\cos\theta$
			$P_3^3 = 15\sin^3\theta$

Like the Legendre polynomials, the $P_\ell^m(x)$ are orthogonal in ℓ, though with the m-dependent normalization

$$\int_{-1}^{1} P_\ell^m(x)P_{\ell'}^m(x)\,dx = \frac{2}{2\ell+1}\frac{(\ell+m)!}{(\ell-m)!}\,\delta_{\ell\ell'}. \tag{35.34}$$

This expression nicely reduces to the familiar Legendre polynomial normalization for $m = 0$; it is also applicable for both $\pm m$. In practice, however, one rarely needs to remember any of this, because the products of Legendre and Fourier basis functions which appear in the corrected version of (35.31) are typically merged into a single normalized *spherical harmonic* [5]

$$Y_{\ell m}(\theta,\phi) \equiv (-1)^m \sqrt{\frac{2l+1}{4\pi}}\sqrt{\frac{(\ell-m)!}{(\ell+m)!}}\,P_\ell^m(\cos\theta)e^{im\phi}, \tag{35.35}$$

with each ℓ having $2\ell+1$ m's. Table 35.2 gives their explicit form for $\ell \leq 3$.

Table 35.2 Spherical harmonics $Y_{\ell m}(\theta,\phi) = (-1)^m\sqrt{\frac{2l+1}{4\pi}}\sqrt{\frac{(\ell-m)!}{(\ell+m)!}}\,P_\ell^m(\cos\theta)e^{im\phi}$

$Y_{00} = \sqrt{\frac{1}{4\pi}}$	$Y_{20} = \sqrt{\frac{5}{16\pi}}\left(3\cos^2\theta - 1\right)$	$Y_{30} = \sqrt{\frac{7}{16\pi}}\left(5\cos^3\theta - 3\cos\theta\right)$
	$Y_{2\pm1} = \mp\sqrt{\frac{15}{8\pi}}\sin\theta\cos\theta\,e^{\pm i\phi}$	$Y_{3\pm1} = \mp\sqrt{\frac{21}{64\pi}}\sin\theta\left(5\cos^2\theta - 1\right)e^{\pm i\phi}$
$Y_{10} = \sqrt{\frac{3}{4\pi}}\cos\theta$	$Y_{2\pm2} = \sqrt{\frac{15}{32\pi}}\sin^2\theta\,e^{\pm 2i\phi}$	$Y_{3\pm2} = \sqrt{\frac{105}{32\pi}}\sin^2\theta\cos\theta\,e^{\pm 2i\phi}$
$Y_{1\pm1} = \mp\sqrt{\frac{3}{8\pi}}\sin\theta\,e^{\pm i\phi}$		$Y_{3\pm3} = \mp\sqrt{\frac{35}{64\pi}}\sin^3\theta\,e^{\pm 3i\phi}$

Example 35.4 **Some Useful Properties of the $Y_{\ell m}$'s**

Using (35.35) together with (35.33), one can easily verify that

$$Y_{\ell,-m}(\theta,\phi) = (-1)^m Y_{\ell m}^*(\theta,\phi), \tag{35.36}$$

which is simpler than (35.33) alone. Reflection through the origin (parity), $P : \vec{r} \to -\vec{r}$, takes $\theta \to \pi - \theta$ and $\phi \to \phi + \pi$, from which follows (Problem 35.11)

$$Y_{\ell m}(\theta,\phi) \xrightarrow{P} Y_{\ell m}(\pi - \theta, \phi + \pi) = (-1)^\ell Y_{\ell m}(\theta,\phi). \tag{35.37}$$

One says that the $Y_{\ell m}$'s have parity $(-1)^\ell$.

---

[5] Whether to include $(-1)^m$ here in the $Y_{\ell m}$'s or with the P_ℓ^m's in (35.32) is a matter of convention. The former is common in quantum mechanics; Mathematica opts for the latter.

It's important to appreciate that $m = 0$ and ϕ independence go hand-in-hand. So not only does setting $m = 0$ reduce (35.35) to

$$Y_{\ell 0}(\theta, \phi) = \sqrt{\frac{2\ell + 1}{4\pi}} \, P_\ell(\cos \theta), \tag{35.38}$$

but any system with axial symmetry — in other words, independent of ϕ — must be restricted to $m = 0$. Along the z-axis, for instance, $\theta = 0$ and ϕ is not defined, so the only non-vanishing spherical harmonics must have $m = 0$ — and indeed, due to the factors of $\sin \theta$ for $m \neq 0$,

$$Y_{\ell m}(\hat{z}) = \sqrt{\frac{2\ell + 1}{4\pi}} \, P_\ell(1)\delta_{m0} = \sqrt{\frac{2\ell + 1}{4\pi}} \, \delta_{m0}. \tag{35.39}$$

At the opposite extreme, for maximal $m = \pm\ell$ one can show that

$$Y_{\ell \pm \ell}(\theta, \phi) \sim \sin^\ell \theta \, e^{\pm i\ell\phi} = \left(\frac{x \pm iy}{r}\right)^\ell. \tag{35.40}$$

Table 35.3 summarizes these and other notable properties of spherical harmonics.

Spherical harmonics are to the surface of a sphere what the $e^{im\phi}$ are to the circumference of a circle. Just as the real and imaginary parts of $e^{\pm im\phi}$ describe normal modes on the circle, so too spherical harmonics describe standing waves on the sphere. The lowest vibrational mode $Y_{00} = 1/\sqrt{4\pi}$ corresponds to the fundamental: with no angular dependence, the entire spherical surface undulates periodically in what's often called the "breathing mode." For $\ell > 0$, the $\ell - m$ zeros of P_ℓ^m along each longitude (see Figure 35.3 and Table 35.1) merge to form $\ell - m$ nodal lines of latitude. Similarly, the zeros of $\cos m\phi$ (or $\sin m\phi$) along each latitude symmetrically divide the sphere with m nodal great circles of longitude. The standing wave patterns of $\Re e\,(Y_{\ell m})$ and $\Im m\,(Y_{\ell m})$ thus have a total of ℓ nodal lines. From this we deduce that ℓ alone determines the wavelength of the mode on the unit sphere: $\lambda \simeq 2\pi/\ell$; this is the spherical analog of Fourier wave number. So it must be that m distinguishes between $2\ell + 1$ different configurations the sphere can accommodate at that wavelength with a distance $\sim \pi/\ell$ between extrema. Thus each (ℓ, m) pair specifies a normal mode of oscillation of a flexible spherical shell. The distinct configurations for $\ell = 4$ are shown in Figure 35.4 (Mollweide projection).[6]

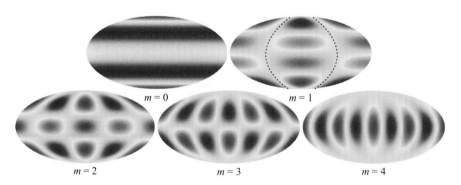

Figure 35.4 Standing waves on a sphere: spherical harmonics $\Re e\,(Y_{4m})$. Each harmonic has $\ell - m$ nodes of latitude and m nodes of longitude. The longitudinal nodal line shown for $m = 1$ is a single great circle passing through the poles.

---

[6] For reference, here's Earth in Mollweide projection:

Example 35.5 **Electron Orbitals**

Though plots like those in Figure 35.4 give a direct illustration of standing wave patterns, it's common to display nodal configurations using the magnitude of the spherical harmonics $|Y_{\ell m}|$ as in Figures 35.5 and 35.6. If these look somewhat familiar, it's because they depict electron orbitals of hydrogenic atoms such as H and singly ionized He. For quantum systems, $\ell\hbar$ denotes the particle's total orbital angular momentum, $m\hbar$ its z-component, and $|Y_{\ell m}(\theta,\phi)|^2 \, d\Omega$ gives the probability of finding a particle with these values within solid angle $d\Omega$ around θ,ϕ. This interpretation is reflected Figures 35.5 and 35.6: when m is maximal, $|Y_{\ell\ell}| \sim |\sin^\ell \theta|$ so that the electron is most likely found around $\theta = \pi/2$ — and indeed, a large z-component of angular momentum corresponds to motion in the xy-plane. Similarly, $Y_{\ell 0}$ is a function of $\cos\theta$, resulting in orbitals which are largest near $\theta = 0$ and π — just as expected of motion with no component of angular momentum along z.

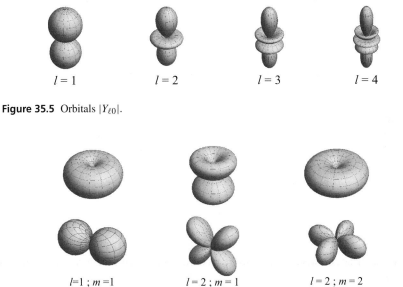

$l = 1$ $l = 2$ $l = 3$ $l = 4$

Figure 35.5 Orbitals $|Y_{\ell 0}|$.

$l=1 \,;\, m=1$ $l = 2 \,;\, m = 1$ $l = 2 \,;\, m = 2$

Figure 35.6 Orbital plots: top row, $|Y_{\ell m}|$; bottom row, $|\Re e(Y_{\ell m})|$.

As proper normal modes, the spherical harmonics should be mutually orthogonal. In fact, their orthogonality is a direct consequence of the separate orthogonality of $P_\ell^m(\cos\theta)$ and $e^{im\phi}$,

$$\int Y_{\ell'm'}^*(\theta,\phi)Y_{\ell m}(\theta,\phi)\, d\Omega = \delta_{\ell\ell'}\delta_{mm'}, \tag{35.41}$$

their unit length deriving from the product of the separate Legendre and Fourier normalizations in (35.35). The $Y_{\ell m}$'s also form a complete basis — and so we can finally answer the question we began with: a function $f(\hat{n}) \equiv f(\theta,\phi)$ can be expanded on the surface of a sphere not as in (35.31), but rather as

$$f(\hat{n}) = \sum_{\ell=0}^{\infty} \sum_{m=-\ell}^{\ell} a_{\ell m} Y_{\ell m}(\theta,\phi), \tag{35.42}$$

where the expansion coefficients are given by

$$a_{\ell m} = \int Y_{\ell m}^*(\theta,\phi)f(\theta,\phi)\, d\Omega. \tag{35.43}$$

Table 35.3 Properties of spherical harmonics

Orthonormality:	$\int Y^*_{\ell'm'}(\theta,\phi)Y_{\ell m}(\theta,\phi)\,d\Omega = \delta_{\ell\ell'}\delta_{mm'}$
Completeness:	$\sum_{\ell,m} Y^*_{\ell m}(\theta,\phi)Y_{\ell m}(\theta',\phi') = \delta(\cos\theta - \cos\theta')\,\delta(\phi - \phi')$
Complex conjugation:	$Y_{\ell,-m} = (-1)^m Y^*_{\ell m}$
Parity:	$Y_{\ell m}(-\hat{n}) = (-1)^\ell Y_{\ell m}(\hat{n})$
Addition theorem:	$\sum_{m=-\ell}^{\ell} Y^*_{\ell m}(\hat{a})Y_{\ell m}(\hat{b}) = \frac{2\ell+1}{4\pi}P_\ell(\hat{a}\cdot\hat{b})$

Differential equations:	$r^2\nabla^2 Y_{\ell m} = -\ell(\ell+1)Y_{\ell m}$	$-i\frac{\partial}{\partial\phi}Y_{\ell m} = mY_{\ell m}$
	$\pm e^{\pm i\phi}\left(\partial_\theta \pm i\cot\theta\,\partial_\phi\right)Y_{\ell m} = \sqrt{(\ell\mp m)(\ell\pm m+1)}\,Y_{\ell,m\pm 1}$	

Special cases:	$Y_{\ell 0}(\theta,\phi) = \sqrt{\frac{2\ell+1}{4\pi}}\,P_\ell(\cos\theta)$	$Y_{\ell m}(\hat{z}) = \sqrt{\frac{2\ell+1}{4\pi}}\,\delta_{m0}$
	$Y_{\ell,\pm\ell} = c_\ell\sin^\ell\theta\,e^{\pm i\ell\phi} = c_\ell\left(\frac{x\pm iy}{r}\right)^\ell,$	$c_\ell = \frac{(\mp)^\ell}{2^\ell\ell!}\sqrt{\frac{(2\ell+1)!}{4\pi}}$

Though this is a two-dimensional integral over the sphere, it's just the usual projection of a function onto the basis. Since each mode ℓ corresponds to an angular resolution $\theta \simeq \pi/\ell$, (35.42) is to be interpreted as an expansion in ever-finer angular scales and wavelengths, with m specifying one of the $2\ell + 1$ orientations at each scale.

The completeness relation can be found by using (35.43) to expand the δ-function,

$$\delta(\hat{n} - \hat{n}') = \sum_{\ell=0}^{\infty}\sum_{m=-\ell}^{\ell} a_{\ell m}Y_{\ell m}(\hat{n})$$

$$= \sum_{\ell=0}^{\infty}\sum_{m=-\ell}^{\ell}\left[\int Y^*_{\ell m}(\hat{\xi})\,\delta(\hat{\xi} - \hat{n}')\,d\Omega_\xi\right]Y_{\ell m}(\hat{n})$$

$$= \sum_{\ell=0}^{\infty}\sum_{m=-\ell}^{\ell} Y^*_{\ell m}(\hat{n}')Y_{\ell m}(\hat{n}), \tag{35.44}$$

where

$$\delta(\hat{n} - \hat{n}') = \delta(\cos\theta - \cos\theta')\delta(\phi - \phi') = \frac{1}{\sin\theta}\,\delta(\theta - \theta')\delta(\phi - \phi'). \tag{35.45}$$

Since $\delta(\hat{n} - \hat{n}')$ should only depend on the angle γ between $\hat{n}$ and $\hat{n}'$ rather than their spherical coordinates, any choice of axes should lead to the same result. So let's simplify (35.44) by choosing to orient the axes so that $\hat{n}'$ is along z,

$$\delta(\hat{n} - \hat{z}) = \sum_{\ell=0}^{\infty}\sum_{m=-\ell}^{\ell} Y^*_{\ell m}(\hat{z})Y_{\ell m}(\hat{n}). \tag{35.46}$$

With this orientation, there can be no ϕ dependence; indeed, $\delta(\hat{n} - \hat{z})$ is axially symmetric. Hence it must be that only $m = 0$ contributes to the sum:

$$\delta(\hat{n} - \hat{z}) = \sum_{\ell=0}^{\infty} Y_{\ell 0}(\hat{z})Y_{\ell 0}(\hat{n}), \tag{35.47}$$

where we've noted that $Y_{\ell 0}$ is real. This can be simplified further using $Y_{\ell 0}(\hat{n}) = \sqrt{\frac{2l+1}{4\pi}} P_\ell(\cos\theta)$ and $P_\ell(\hat{z}) = P_\ell(1) = 1$:

$$\delta(\hat{n} - \hat{z}) = \sum_{\ell=0}^{\infty} \frac{2l+1}{4\pi} P_\ell(\cos\theta) = \sum_{\ell=0}^{\infty} \frac{2l+1}{4\pi} P_\ell(\hat{n} \cdot \hat{z}). \tag{35.48}$$

Finally, since dot products are independent of the orientation of the axes, we can return to the general case: for arbitrary $\hat{n}, \hat{n}'$

$$\delta(\hat{n} - \hat{n}') = \sum_{\ell=0}^{\infty} \frac{2l+1}{4\pi} P_\ell(\cos\gamma), \qquad \cos\gamma = \hat{n} \cdot \hat{n}'. \tag{35.49}$$

Now compare this with (35.44): orthogonality in ℓ requires these two expressions to be equal term by term. Thus we find the *spherical harmonic addition theorem*,

$$P_\ell(\cos\gamma) = \frac{4\pi}{2\ell+1} \sum_{m=-\ell}^{\ell} Y_{\ell m}^*(\theta', \phi') Y_{\ell m}(\theta, \phi). \tag{35.50}$$

For $\ell = 1$, this reproduces the expression found in Problem 2.8 relating the angle between vectors $\hat{a}$ and $\hat{b}$ to their spherical coordinates,

$$\cos\gamma = \cos\theta \cos\theta' + \sin\theta \sin\theta' \cos(\phi - \phi'). \tag{35.51}$$

The addition theorem generalizes this geometrical result.

Example 35.6   **Invoking Orthogonality**

Since $\cos\gamma$ is basis independent, the addition theorem holds for any orientation of the cartesian axes. For instance, in Section 32.3 we discovered

$$\frac{1}{|\vec{r} - \vec{r}'|} = \sum_{\ell=0}^{\infty} \frac{r_<^\ell}{r_>^{\ell+1}} P_\ell(\cos\gamma), \qquad \vec{r} \cdot \vec{r}' = rr' \cos\gamma, \tag{35.52}$$

where $r_>$ and $r_<$ are the greater and lesser of r and r', respectively. The addition theorem allows us to express this in terms of the orientations of $\vec{r}$ and $\vec{r}'$ within any choice of coordinate axes,

$$\frac{1}{|\vec{r} - \vec{r}'|} = \sum_{\ell=0}^{\infty} \frac{4\pi}{2\ell+1} \frac{r_<^\ell}{r_>^{\ell+1}} \sum_{m=-\ell}^{\ell} Y_{\ell m}^*(\theta', \phi') Y_{\ell m}(\theta, \phi). \tag{35.53}$$

Despite the double sum, the ability to include the coordinate orientations of $\vec{r}$ and $\vec{r}'$ can be very useful. For example, suppose we want to calculate the average

$$\frac{1}{4\pi} \int \frac{d\Omega}{|\vec{r} - \vec{r}'|}. \tag{35.54}$$

We could certainly insert (35.51) into (35.52), orient the z-axes to lie along $\vec{r}'$, and integrate. But far simpler is to use (35.53) and evaluate

$$\int \frac{d\Omega}{|\vec{r} - \vec{r}'|} = 4\pi \sum_{\ell=0}^{\infty} \frac{1}{2\ell+1} \frac{r_<^\ell}{r_>^{\ell+1}} \sum_{m=-\ell}^{\ell} Y_{\ell m}^*(\theta', \phi') \int Y_{\ell m}(\theta, \phi) \, d\Omega. \tag{35.55}$$

Note that the ϕ integral gives zero for all $m \neq 0$; even so, integrating $Y_{\ell 0}$ for arbitrary ℓ may seem daunting. Fortunately, there's an easier way: by inserting $1 = \sqrt{4\pi}\, Y_{00}$ into the integrand, orthonormality easily reveals

$$\int Y_{\ell m}(\theta, \phi)\, d\Omega = \sqrt{4\pi} \int Y_{\ell m}(\theta, \phi) Y_{00}^* \, d\Omega = \sqrt{4\pi}\, \delta_{\ell 0} \delta_{m0}. \tag{35.56}$$

Thus

$$\frac{1}{4\pi} \int \frac{d\Omega}{|\vec{r} - \vec{r}'|} = \sum_{\ell=0}^{\infty} \frac{1}{2\ell+1} \frac{r_<^\ell}{r_>^{\ell+1}} \sum_{m=-\ell}^{\ell} Y_{\ell m}^*(\theta', \phi') \sqrt{4\pi}\, \delta_{\ell 0} \delta_{m0}$$

$$= \frac{1}{r_>} \sqrt{4\pi}\, Y_{00}^*(\theta', \phi') = \frac{1}{r_>}. \tag{35.57}$$

As a complete basis, the spherical harmonics obey a version of Parseval's theorem, (34.28) — that is, a relation amongst the magnitude-squared of a function's expansion coefficients:

$$\int |f(\hat{n})|^2 d\Omega = \sum_{\ell, m} \sum_{\ell', m'} a_{\ell m}^* a_{\ell' m'} \int Y_{\ell m}^*(\hat{n}) Y_{\ell' m'}(\hat{n})\, d\Omega$$

$$= \sum_{\ell=0}^{\infty} \left[\sum_{m=-\ell}^{\ell} |a_{\ell m}|^2 \right] \equiv \sum_{\ell=0}^{\infty} (2\ell+1) C_\ell, \tag{35.58}$$

where the C_ℓ's are the m-averaged squared coefficients,

$$C_\ell \equiv \frac{1}{2\ell+1} \sum_{m=-\ell}^{\ell} |a_{\ell m}|^2. \tag{35.59}$$

The C_ℓ's record the intensity contribution at each angular scale and wavelength, collectively known as the signal's angular power spectrum.

BTW 35.1 The Cosmic Microwave Background

A fundamental prediction of modern cosmology is the existence of isotropic, universe-pervading radiation with a blackbody spectrum. According to the standard model of cosmology, about 10^{-6} seconds after the Big Bang the early Universe was a plasma of electrons and nuclei in thermal equilibrium with a gas of photons. This hot dense plasma was opaque to the photons, effectively trapping them. But the Universe cooled as it expanded, and once the temperature dropped sufficiently to allow electrons and nuclei to combine into neutral atoms, the photons decoupled from matter and began freely streaming — all the while maintaining a blackbody spectrum. This decoupling occurred when the Universe was about 380,000 years old at a temperature of about 3000 K; some 13.8 Gyr of expansion later, the radiation has now cooled to a temperature of ~ 2.7 K. This places it in the microwave part of the electromagnetic spectrum, and so the radiation is known as the *cosmic microwave background* (CMB). Its serendipitous discovery in 1964 earned Penzias and Wilson the Nobel prize.

The CMB has an almost-perfect blackbody spectrum — in fact, the most precise ever discovered. But measurements of its intensity in a direction $\hat{n}$ reveal a blackbody temperature $T(\hat{n})$ which can differ slightly from the $\bar{T} = 2.7$ K average. Though these variations from the mean are small — the fractional deviations are on the order of 1 part in 10^5 — they are physically relevant: temperature fluctuations in different directions reflect the state of the pressure (i.e., sound) waves present in the photon gas at the time of decoupling, and therefore record perturbations in the distribution of normal matter at that time. These perturbations are the seeds which evolved into the large-scale structure we see today. Thus the CMB provides a picture of the Universe at the tender age of 380,000 years — specifically, an image of a spherical shell 13.8 billion light years in radius known as the "surface of last scattering."

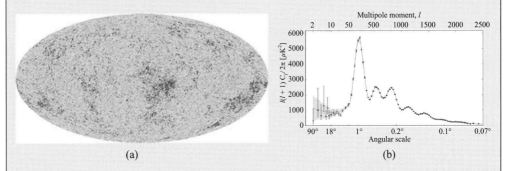

(a)   (b)

Figure 35.7 (a) CMB temperature fluctuations, showing the imprint of sound waves at the time of decoupling. (© ESA and the Planck Collaboration) (b) The power spectrum of CMB temperature fluctuations. The curve is the best fit to the standard model of cosmology. (© ESA and the Planck Collaboration)

Analysis of CMB temperature variations begins by defining the temperature fluctuation field (Figure 35.7a)

$$\frac{\delta T(\hat{n})}{\bar{T}} \equiv \frac{T(\hat{n}) - \bar{T}}{\bar{T}}. \tag{35.60}$$

Since this field is arrayed over the surface of a sphere, it's appropriate to expand it not in a Fourier series, but rather over the basis of spherical harmonics,

$$\frac{\delta T(\hat{n})}{\bar{T}} = \sum_{\ell=0}^{\infty} \sum_{m=-\ell}^{\ell} a_{\ell m} Y_{\ell m}(\hat{n}), \tag{35.61}$$

where

$$a_{\ell m} = \frac{1}{\bar{T}} \int Y_{\ell m}^*(\hat{n})\, \delta T(\hat{n})\, d\Omega. \tag{35.62}$$

The goal of theoretical models is to predict the power spectrum (35.59), and compare it with the values statistically extracted from the data.

To do this extraction, one looks for correlations in temperature fluctuations (35.61) between directions $\hat{n}$ and $\hat{n}'$,

$$C(\varphi) = \left\langle \frac{\delta T(\hat{n})}{\bar{T}} \frac{\delta T(\hat{n}')}{\bar{T}} \right\rangle, \tag{35.63}$$

where the angle brackets denote averaging over an ensemble of temperature fields. With only a dependence on the angle between $\hat{n}$ and $\hat{n}'$, we can use the spherical harmonic addition theorem to find (Problem 35.20),

$$C(\varphi) = \frac{1}{4\pi} \sum_{\ell=0}^{\infty} (2\ell+1) C_\ell P_\ell(\cos\varphi), \tag{35.64}$$

with CMB angular power spectrum

$$C_\ell = \frac{1}{2\ell+1} \sum_{m=-\ell}^{\ell} \langle |a_{\ell m}|^2 \rangle. \tag{35.65}$$

This ℓ-dependent data set is what's compared with theoretical predictions.

There's just one problem: there's only one Universe from which to take data, so we do not have an ensemble of temperature fields to average. However, we can mimic an ensemble by averaging over many different parts of space for a given φ — the idea being that if the regions are far enough apart, their fluctuations should be independent of one another. (This obviously breaks down for large φ, resulting in a limit to the statistics for small ℓ. This limitation, known as *cosmic variance*, is reflected by the error bars in Planck satellite data shown in Figure 35.7b). The various peaks in the power spectrum indicate significant hot and cold spots with angular separation π/ℓ. The largest angular scale or wavelength is limited by the size of the Universe at the time of decoupling; theoretical calculations put this fundamental mode at $\sim 1°$, which the Planck observations confirm. The higher harmonics reflect the relative content (e.g., matter, radiation, dark matter, and dark energy) of the early Universe. [See Dodelson and Schmidt (2020) for more detail.]

35.4 From Shell to Ball

Proceeding from the expansion of a function on the surface of the sphere to a full spherical coordinate expansion in $\mathbb{R}^3$ is straightforward. Using (35.42) we write

$$f(\vec{r}) = \sum_{\ell=0}^{\infty} \sum_{m=-\ell}^{\ell} a_{\ell m}(r) Y_{\ell m}(\theta,\phi), \tag{35.66}$$

which can be construed as a collection of spherical harmonic expansions on nested concentric spherical shells, with $a_{\ell m}(r)$ determined by projection onto each shell.

For the Fourier basis $f(\vec{r}) = e^{i\vec{k}\cdot\vec{r}} = e^{ikr\cos\gamma}$, the appearance of $\vec{k}$ and $\vec{r}$ allows two implementations of (35.66). First consider $\hat{k}$ to be held fixed and do a spherical harmonic expansion in $\hat{r}$,

$$e^{i\vec{k}\cdot\vec{r}} = \sum_{\ell=0}^{\infty} \sum_{m=-\ell}^{\ell} a_{\ell m}(r,\vec{k}) Y_{\ell m}(\hat{r}), \tag{35.67}$$

followed by an expansion of $a_{\ell m}(r,\vec{k})$ around $\hat{k} = (\theta_k, \phi_k)$ for fixed r,

$$e^{i\vec{k}\cdot\vec{r}} = \sum_{\ell,m} \left[\sum_{\ell'm'} \tilde{a}_{\ell m\ell'm'}(r,k) Y_{\ell'm'}(\hat{k}) \right] Y_{\ell m}(\hat{r})$$

$$= \sum_{\ell m} \sum_{\ell'm'} \tilde{a}_{\ell m\ell'm'}(r,k) Y_{\ell'm'}(\hat{k}) Y_{\ell m}(\hat{r}). \tag{35.68}$$

This much must be true for generic $f(\vec{k},\vec{r})$. But from the specific form of $e^{i\vec{k}\cdot\vec{r}}$, one can show that this quadruple sum reduces to (Problem 35.22)

$$e^{i\vec{k}\cdot\vec{r}} = \sum_{\ell=0}^{\infty} \sum_{m=-\ell}^{\ell} c_\ell(kr) Y_{\ell m}^*(\hat{k}) Y_{\ell m}(\hat{r}). \qquad (35.69)$$

Parity considerations show $c_\ell(kr)$ to be proportional to i^ℓ (Problem 35.23); then introducing the real functions $j_\ell(kr) = c_\ell(kr)/4\pi i^\ell$, the expansion becomes

$$e^{i\vec{k}\cdot\vec{r}} = 4\pi \sum_{\ell=0}^{\infty} \sum_{m=-\ell}^{\ell} i^\ell j_\ell(kr) Y_{\ell m}^*(\hat{k}) Y_{\ell m}(\hat{r}), \qquad (35.70a)$$

or, courtesy of the spherical harmonic addition theorem (35.50),

$$e^{ikr\cos\gamma} = \sum_{\ell=0}^{\infty} i^\ell(2\ell+1) j_\ell(kr) P_\ell(\cos\gamma). \qquad (35.70b)$$

These expressions should be compared with the expressions in (35.16) for the analogous expansion in polar coordinates. Just as there we introduced cylindrical Bessel functions J_n, here the expansion defines the *spherical Bessel functions j_ℓ*. Projection onto the basis of P_ℓ's yields the integral

$$j_\ell(x) = \frac{(-i)^\ell}{2} \int_0^\pi e^{ix\cos\theta} P_\ell(\cos\theta)\sin\theta\, d\theta = \frac{(-i)^\ell}{2} \int_{-1}^1 e^{ixu} P_\ell(u)\, du. \qquad (35.71)$$

Superficially, the j_ℓ's behave much like the J_n's. They are oscillatory but not periodic, having both decaying amplitude and non-uniform spacing between zeros which asymptotically goes to π — except for $j_0(x) = \sin(x)/x$, whose zeros are quite regular for all x. But j_0 is indicative of another property: unlike cylindrical Bessel functions, the j_ℓ can be expressed in terms of more elementary functions; the first few are listed in Table C.4.

Having considered planes, disks, and balls, we're now ready to compare and contrast the form of the Fourier basis function $e^{i\vec{k}\cdot\vec{r}}$ in the three most-common $\mathbb{R}^3$ coordinate systems,

$$e^{i\vec{k}\cdot\vec{r}} = e^{ik_x x} e^{ik_y y} e^{ik_z z} \qquad \text{(Cartesian)} \qquad (35.72a)$$

$$= e^{ik_z z} \sum_{n=-\infty}^{\infty} i^n J_n(k_\perp r) e^{-in\phi'} e^{in\phi} \qquad \text{(Cylindrical)} \qquad (35.72b)$$

$$= 4\pi \sum_{\ell=0}^{\infty} \sum_{m=-\ell}^{\ell} i^\ell j_\ell(kr) Y_{\ell m}^*(\hat{k}) Y_{\ell m}(\hat{r})$$

$$= \sum_{\ell=0}^{\infty} i^\ell(2\ell+1) j_\ell(kr) P_\ell(\cos\gamma), \qquad \text{(Spherical)} \qquad (35.72c)$$

where $k^2 = k_x^2 + k_y^2 + k_z^2 = k_\perp^2 + k_z^2$. These expansions employ three different orthogonal bases on which to decompose a plane wave. Not surprisingly, cartesian is the simplest, cleanly factoring into a product three independent one-dimensional phases. And though the polar and spherical expansions are less simple, they can be viewed as expressions of Huygen's principle, which construes each point on a planar wavefront as a superposition of circular or spherical waves (see Example 36.4). Further physical insight comes from quantum mechanics. Since momentum is given by the de Broglie relation $\vec{p} = \hbar\vec{k}$, plane waves describe states with fixed, definite momentum. On the other hand, spherical harmonics describe states of definite angular momentum such as electron orbitals (Example 35.5). So (35.72c) expresses a quantum state of definite momentum as a superposition of all possible angular momenta. This decomposition into individual *partial waves* $j_\ell(kr) Y_{\ell m}(\theta, \phi)$ is called a plane wave's *partial wave expansion* (Section 43.2).

Problems

35.1   (a) Verify the two-dimensional Fourier projection of (35.4) by regarding (35.3) as a one-dimensional Fourier series at fixed x.
(b) Find the projection integrals for the coefficients in (35.5).

35.2   Adapt (35.16) to exapnd (35.15) in cylindrical coordinates r, ϕ and z.

35.3   (a) Show that the all three integral representations of $J_n(x)$ in (35.17) are equivalent.
(b) Verify that these integral representations satisfy Bessel's equation, (C.1) (Appendix C) for integer $\nu = n$. Being equivalent, just show this for any one of the integrals. [Hint: Look for an opportunity to integrate by parts.]

35.4   Use any of the integral representations of (35.17) to verify the behavior of $J_n(x)$ for small x, (35.18b).

35.5   Use any of the integral representations of (35.17) to verify that

$$\frac{dJ_0}{dx} = -J_1(x),$$

which is reminiscent of cos and sin. Of course, the complementary relation $J_1' = J_0$ cannot also be valid, since then J_0 and J_1 *would* be cosine and sine; instead, show that

$$\frac{d}{dx}(xJ_1) = xJ_0.$$

35.6   Derive Bessel's equation for J_0,

$$xJ_0'' + J_0' + xJ_0 = 0,$$

from the derivative identities in Problem 35.5.

35.7   Though integral representations of J_n are useful, sometimes a series expansions is preferable. Verify that the series

$$J_n(x) = \sum_{m=0}^{\infty} \frac{(-1)^m}{m!\,(m+n)!} \left(\frac{x}{2}\right)^{2m+n}$$

solves Bessel's equation (C.1).

35.8   (a) Find the Fourier series for $e^{ix\sin\phi}$.
(b) Derive the Fourier expansions

$$\cos(x\sin\phi) = \sum_{n=-\infty}^{\infty} J_n(x)\cos n\phi \qquad \sin(x\sin\phi) = \sum_{n=-\infty}^{\infty} J_n(x)\sin n\phi. \qquad (35.73)$$

(c) Derive the Fourier series in (35.23).

35.9   Use Mathematica to show that 98% of the contribution to $\sum_{n=-\infty}^{\infty}|J_n(\beta)|^2 = 1$, (35.24), comes from $n \le \beta + 1$. (Take $1 \le \beta \le 10$.)

35.10   Use (35.32) and (35.33) to find the associated Legendre functions $P_3^m(x)$.

35.11   Verify the parity of spherical harmonics, (35.37).

35.12   Show $Y_{\ell 0}(\hat{n}) = \sqrt{\frac{2\ell+1}{4\pi}}\,P_\ell(\cos\theta)$, and $Y_{\ell m}(\hat{z}) = \sqrt{\frac{2\ell+1}{4\pi}}\,\delta_{m0}$.

35.13   As discussed above Figure 35.4, mode ℓ on the unit sphere corresponds to dimensionless wavelength $\lambda = 2\pi/\ell$. This can be made more precise by comparing the differential equation for $Y_{\ell m}$ in Table 35.3 with that of two-dimensional harmonic motion. Argue that $\lambda \approx \frac{2\pi}{\ell+1/2}$ for large ℓ. This is known as Jeans' rule.

35.14 A satellite can resolve gravitational field variations at the surface equal roughly to its altitude. For a satellite in orbit 200 km above Earth, approximately what should be the maximum ℓ when fitting the data with a superposition of spherical harmonics?

35.15 Show that the expansion (35.42)–(35.43) reduces to a Legendre series when the function has azimuthal symmetry, $f(\theta,\phi) = f(\theta)$.

35.16 Evaluate the sum $\sum_m |Y_\ell^m(\theta,\phi)|^2$.

35.17 (a) Verify that the expression (35.50) is real.
  (b) Show that

$$\int Y_{\ell m}(\hat{r}) P_\ell(\cos\gamma)\, d\Omega = \frac{4\pi}{2\ell+1} Y_{\ell m}(\hat{k}),$$

  where $\cos\gamma = \hat{r}\cdot\hat{k}$.
  (c) Use spherical harmonics to explicitly verify (35.51).

35.18 Consider a spherical shell of radius a with charge density

$$\rho(\vec{r}) = \frac{q}{2\pi a^2}\cos\theta\,\delta(r-a).$$

  Use Coulomb's law (9.33), the spherical harmonic expansion (35.53), and orthogonality to find the potential inside and outside the sphere.

35.19 Notice that the sum in (35.50) resembles the inner product of two $(2\ell+1)$-dimensional vectors with component index m.

  (a) Argue that the sum must in fact yield a scalar quantity, unaffected by rotation. Thus the value of ℓ remains constant under rotation.
  (b) Consider how $Y_{\ell m}$ is affected by change in ϕ to argue that the value of m depends on the choice of the z-axis. Hence like a vector, the individual components are changed by rotation, but its magnitude ℓ remains constant.

35.20 The *cosmological principle* asserts that on the largest scales, the universe is invariant both under translations (so that it's the same everywhere) and rotations (so that it looks the same in every direction). The universe is said to be *homogeneous* and *isotropic*, respectively.

  (a) Show that (35.63) can be written

$$C(\varphi) = \sum_{\ell m}\sum_{\ell'm'}\langle a_{\ell m}\, a_{\ell'm'}^*\rangle\, Y_{\ell m}(\hat{n}) Y_{\ell'm'}^*(\hat{n}').$$

  (b) Argue that homogeneity and isotropy imply that the ensemble average $\langle a_{\ell m}\, a_{\ell'm'}^*\rangle$ must not only be proportional to $\delta_{\ell\ell'}$ and $\delta_{mm'}$, but can only depend on ℓ — allowing us to write

$$\langle a_{\ell m}\, a_{\ell'm'}^*\rangle = \delta_{\ell\ell'}\delta_{mm'}C_\ell\,.$$

  [An ensemble of universes being unavailable, the best estimate of the angular power spectrum C_ℓ is given by (35.65), the m-average over the measured values $a_{\ell m}$.]
  (c) Use the addition theorem (35.50) to verify (35.64).

35.21 (a) Verify the eigenvalue equations

$$L_z Y_{\ell m} = m Y_{\ell m}\qquad L_z \equiv -i\partial_\phi,$$

and, using (32.67) from Problem 32.23,

$$L^2 Y_{\ell m} = \ell(\ell + 1) Y_{\ell m} \qquad L^2 \equiv -r^2 \nabla^2_{\theta\phi},$$

where $\nabla^2_{\theta\phi}$ is the angular part of the Laplacian in (12.51b).

(b) Show that

$$L^2 = L_x^2 + L_y^2 + L_z^2,$$

where

$$L_x \equiv +i \left(\sin\phi \, \partial_\theta + \cot\theta \cos\phi \, \partial_\phi \right) \qquad L_y \equiv -i \left(\cos\phi \, \partial_\theta - \cot\theta \sin\phi \, \partial_\phi \right).$$

In quantum mechanics, these operators are associated with cartesian components of angular momentum — see Example 26.20. (Including a factor of $\hbar$ gives them units of angular momentum.)

(c) The L^2 eigenspace is $2\ell+1$ degenerate; L_z lifts this degeneracy, with the $Y_{\ell m}$ their common eigenvectors. The spherical harmonics are not, however, eigenvectors of L_x and L_y. What does this imply about the commutators amongst the L_i? What about the commutator of the L_i with L^2? [We'll see in Chapter 40 that these operators are Hermitian.]

(d) The common metaphor for the degenerate eigenvectors of L^2 is that of a ladder with $2\ell + 1$ rungs, starting at $-m$ at the bottom and climbing to $+m$ at the top. We can "climb" up and down with *ladder operators*

$$L_\pm \equiv L_x \pm iL_y.$$

Write out these operators in terms of θ and ϕ. Then to see what happens at the top and bottom of the ladder, calculate $L_+ Y_{\ell\ell}$ and $L_- Y_{\ell\,-\ell}$.

(e) One can show that $L_\pm$ move up and down the degeneracy ladder as

$$L_\pm Y_{\ell m} = \sqrt{(\ell \mp m)(\ell \pm m + 1)}\, Y_{\ell\,m\pm 1}.$$

Verify this for $\ell = 2$ by starting with Y_{22} and using $L_\pm$ to find all the spherical harmonics Y_{2m}.

35.22 Noting that $e^{i\vec{k}\cdot\vec{r}}$ is a scalar, show that the quadruple sum in (35.68) reduces to the double sum of (35.69).

35.23 In (35.69):

(a) Explicitly demonstrate that either spherical harmonic can have the complex conjugate.
(b) Verify that $c_\ell(kr)$ is proportional to i^ℓ.

35.24 Derive (35.70b) from (35.70a) by choosing the z-axis to align with $\vec{k}$.

35.25 Use (35.71) to show that $j_\ell(0) = \delta_{\ell 0}$.

Fourier Transforms

A basis in $\mathbb{C}^N$ consists of a finite number of eigenvectors $\phi^{(\ell)}$ with discrete components $\phi_n^{(\ell)}$ — such as those in (31.1). In the continuum limit to $\mathbb{L}^2$, these discrete components essentially merge to produce the eigenfunctions $\phi_\ell(x)$ of (31.4). One can imagine the continuum limit "squeezing" the vector components $\phi_n^{(\ell)}$ into continuous functions $\phi_\ell(x)$ over the finite interval from a to b, even as each individual mode maintains a discrete identity. But what if the interval were infinite?

What indeed. Though many of properties still have direct analogs with matrix systems, one distinctive feature occurs when the domain $b - a$ is *not* finite: part or all of the eigenvalue spectrum can have a continuous, rather than a discrete, infinity of values. And with a continuous spectrum, the familiar eigenfunction expansion $\psi(x) = \sum_n c_n \phi_n(x)$ evolves into an integral over a continuous infinity of eigenfunctions. So rather than a discrete basis of continuous functions, we obtain continuous basis of continuous functions.

36.1 From Fourier Sum to Fourier Integral

Consider again the differential equation $d^2 f/dx^2 = -k^2 f$, this time over all space, $|x| < \infty$. The solutions for real k are still plane waves Ae^{ikx}, but in the absence of additional constraints there's no reason to expect a discrete eigenspectrum of k's. So the familiar Fourier sum over the discrete infinity of normal modes should become an integral over a continuum of plane waves e^{ikx}. To derive this integral, we'll examine Fourier series with $k_n = 2\pi n/L$ in the $L \to \infty$ limit. Since the discrete sum over n is in unit steps $\Delta n = 1$, we can write $\Delta k_n = 2\pi \Delta n/L$ to get

$$
\begin{aligned}
f(x) = \sum_{n=-\infty}^{\infty} c_n e^{i2\pi nx/L} &= \sum_{n=-\infty}^{\infty} c_n e^{i2\pi nx/L} \Delta n \\
&= \sum_{n=-\infty}^{\infty} c_n e^{ik_n x} \frac{L}{2\pi} \Delta k_n \\
&= \sum_{n=-\infty}^{\infty} \frac{L}{2\pi} \left[\frac{1}{L} \int_{-L/2}^{L/2} e^{-ik_n y} f(y)\, dy \right] e^{ik_n x} \Delta k_n,
\end{aligned}
\tag{36.1}
$$

where in the last step we used the familiar integral expression for the c_n.

Now you might reasonably object that a square-integrable $f(x)$ over all space is *not* periodic — so the Fourier series expansion we started with in (36.1) doesn't apply. Indeed, by letting $L \to \infty$, we're in effect attempting a Fourier analysis for a function $f(x)$ with *infinite* period. This may seem absurd — but let's see if it gives a convergent result. In the limit, the spacing $\Delta k_n = 2\pi/L$ between frequency eigenvalues causes the discrete spectrum of k_n's to merge into a continuous spectrum of k's. As anticipated then, $\sum_n \Delta k_n \to \int_{-\infty}^{\infty} dk$. So after canceling the diverging factors of L, we can take the $L \to \infty$ limit to get

$$f(x) = \frac{1}{2\pi} \int_{-\infty}^{\infty} \left[\int_{-\infty}^{\infty} e^{-iky} f(y)\, dy \right] e^{ikx}\, dk$$

$$\equiv \frac{1}{\sqrt{2\pi}} \int_{-\infty}^{\infty} \tilde{f}(k)\, e^{ikx}\, dk, \tag{36.2}$$

which converges for square integrable $\tilde{f}$. The functions $f(x)$ and $\tilde{f}(k)$ are said to be *Fourier transform pairs*,

$$\tilde{f}(k) \equiv \mathcal{F}[f] = \frac{1}{\sqrt{2\pi}} \int_{-\infty}^{\infty} f(x)\, e^{-ikx}\, dx \tag{36.3a}$$

$$f(x) \equiv \mathcal{F}^{-1}[\tilde{f}] = \frac{1}{\sqrt{2\pi}} \int_{-\infty}^{\infty} \tilde{f}(k)\, e^{ikx}\, dk, \tag{36.3b}$$

where $\mathcal{F}$ is the integral operator mapping $f \to \tilde{f}$. The operation integrates out all x dependence, substituting a k dependence in its place — and vice-versa for $\mathcal{F}^{-1}$. Note that the Fourier transform is a linear operator: $\mathcal{F}[af + bg] = a\tilde{f} + b\tilde{g}$.

Like the discrete set c_n of Fourier series coefficients, $\tilde{f}$ is the spectral decomposition of f; indeed, (36.3a) is the projection of f onto a continuous basis of plane waves. The familiar Fourier sum over harmonics of f thus becomes the Fourier integral (36.3b). So the image of Fourier series as a collection of students arrayed with tuning forks still applies — though now we need a continuous infinity of students and a continuous spectrum of tuning forks. Each student hits a fork with magnitude $|\tilde{f}(k)|$ (ignoring relative phases) in order to produce the aperiodic signal $f(x)$. And just as the collection of Fourier series coefficients $|c_n|^2$ are associated with the energy contribution of each discrete harmonic n, so too the function $|\tilde{f}(k)|^2$ for continuous k.

Example 36.1   A Decaying Exponential

Finding Fourier transform pairs is largely an exercise in integration. As an illustration, let's calculate $\mathcal{F}[f]$ for $f(x) = \frac{a}{2} e^{-a|x|}$, $a > 0$:

$$\tilde{f}(k) = \frac{1}{\sqrt{2\pi}} \int_{-\infty}^{\infty} \frac{a}{2} e^{-a|x|} e^{-ikx}\, dx$$

$$= \frac{a}{2\sqrt{2\pi}} \left[\int_{-\infty}^{0} e^{-(ik-a)x}\, dx + \int_{0}^{\infty} e^{-(ik+a)x}\, dx \right]$$

$$= \frac{a}{2\sqrt{2\pi}} \left[-\frac{1}{(ik-a)} + \frac{1}{(ik+a)} \right] = \frac{1}{\sqrt{2\pi}} \frac{1}{1 + k^2/a^2}. \tag{36.4}$$

The inverse transform can be calculated as outlined in Problem 8.30, verifying that $\mathcal{F}^{-1}[\tilde{f}(k)] = f(x)$; we can also use Cauchy's integral formula (18.26).

Despite some similarities, f and $\tilde{f}$ are clearly distinct functions (Figure 36.1). In particular, note that as a grows, $f(x)$ drops off more rapidly whereas $\tilde{f}(k)$ broadens and flattens. This inverse relationship between $f(x)$ and $\tilde{f}(k)$ is an essential feature of Fourier transform pairs. And it's true even in the extreme limit: as $a \to \infty$, $\tilde{f}(k)$ broadens to a constant $1/\sqrt{2\pi}$ while $f(x)$ narrows to a delta function (see inside the back cover). It works the other way as well: $\frac{2}{a} f(x)$ goes to a constant 1 as $a \to 0$, while $\frac{2}{a}\tilde{f}(k) \to \sqrt{2\pi}\, \delta(k)$. So zero-width in one space corresponds to infinite width in the other.

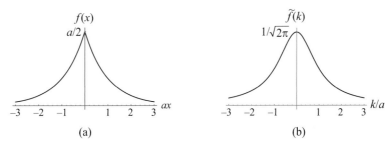

Figure 36.1 Fourier pairs: (a) $f(x) = \frac{a}{2} e^{-a|x|}$; (b) $\tilde{f}(k) = \frac{1}{\sqrt{2\pi}} (1 + k^2/a^2)^{-1}$.

Not that this is actually new: direct application of $\mathcal{F}$ and $\mathcal{F}^{-1}$ reveals the familiar representation of the delta function in (9.6),

$$\delta(x) = \frac{1}{2\pi} \int_{-\infty}^{\infty} e^{ikx}\, dk \qquad\qquad \delta(k) = \frac{1}{2\pi} \int_{-\infty}^{\infty} e^{-ikx}\, dx \,. \qquad (36.5)$$

These are superpositions of equal-magnitude contributions from the entire continuous infinity of plane waves $e^{\pm ikx}$, each weighted by the constant amount $1/2\pi$.[1]

The virtually identical forms of $\mathcal{F}$ and $\mathcal{F}^{-1}$ in (36.3) would seem to imply that f and $\tilde{f}$ have equal status, that they are essentially equivalent. This may seem an outlandish claim, given that f and $\tilde{f}$ are certainly not identical: as Table 36.1 shows, Fourier transform pairs are in almost all cases fundamentally different functions. Is it possible, say, that $\mathcal{F}$ introduces some spurious information in $\tilde{f}$ which is not contained in f?

Table 36.1 Fourier transform pairs

$f(x) = \mathcal{F}^{-1}[\tilde{f}] = \frac{1}{\sqrt{2\pi}} \int_{-\infty}^{\infty} \tilde{f}(k) e^{+ikx} dk$	$\tilde{f}(k) = \mathcal{F}[f] = \frac{1}{\sqrt{2\pi}} \int_{-\infty}^{\infty} f(x) e^{-ikx} dx$
e^{-ax}	$\sqrt{2\pi}\, \delta(k - ia)$
$e^{-a|x|}$	$\sqrt{\frac{2}{\pi}} \frac{a}{k^2 + a^2}$ $(a > 0)$
e^{iax}	$\sqrt{2\pi}\, \delta(k - a)$
$\cos ax$	$\sqrt{\frac{\pi}{2}} [\delta(k + a) + \delta(k - a)]$
$\sin ax$	$i\sqrt{\frac{\pi}{2}} [\delta(k + a) - \delta(k - a)]$
$e^{-\frac{1}{2} a^2 x^2}$	$\frac{1}{a} e^{-\frac{k^2}{2a^2}}$
$\delta(x)$	$\frac{1}{\sqrt{2\pi}}$
1	$\sqrt{2\pi}\, \delta(k)$
$\frac{1}{x^2 + a^2}$	$\sqrt{\frac{\pi}{2a^2}} e^{-|k|a}$ $(a > 0)$

[1] Perhaps you noticed and are concerned by the fact that neither a δ function nor a plane wave is square integrable. We'll address this in Section 36.4.

You may recall that we had similar concerns about a periodic function and its Fourier coefficients c_n. Yet Parseval's theorem (34.28) establishes that these are equivalent mean-square representations of the same vector $|f\rangle$. To see how this plays out for Fourier transforms, insert (36.3b) into $\int_{-\infty}^{\infty} |f(x)|^2 \, dx$ and re-order the integrations:

$$
\int_{-\infty}^{\infty} |f(x)|^2 \, dx = \int_{-\infty}^{\infty} \left(\frac{1}{\sqrt{2\pi}} \int_{-\infty}^{\infty} \tilde{f}(k') e^{ik'x} \, dk' \right)^* \left(\frac{1}{\sqrt{2\pi}} \int_{-\infty}^{\infty} \tilde{f}(k) e^{ikx} \, dk \right) dx
$$

$$
= \int_{-\infty}^{\infty} \tilde{f}^*(k') \tilde{f}(k) \left[\frac{1}{2\pi} \int_{-\infty}^{\infty} e^{i(k-k')x} \, dx \right] dk' \, dk
$$

$$
= \int_{-\infty}^{\infty} \tilde{f}^*(k') \tilde{f}(k) \, \delta(k - k') \, dk' \, dk = \int_{-\infty}^{\infty} |\tilde{f}(k)|^2 \, dk, \tag{36.6}
$$

where we've used (36.5). So a function and its Fourier transform have the same norm. One can similarly show that $\int_{-\infty}^{\infty} f(x)^* g(x) \, dx = \int_{-\infty}^{\infty} \tilde{f}(k)^* \tilde{g}(k) \, dk$. In other words, the inner product of two vectors is invariant under Fourier transform. We have a name for a mapping which leaves unchanged the relative orientation of all the vectors in the space: *unitary*. So one can think of $\mathcal{F}$ as a "rotation" between the x and k bases. Just as $\mathcal{F}[f]$ decomposes $f(x)$ into its frequency components $\tilde{f}(k)$, so too $\mathcal{F}^{-1}[\tilde{f}]$ decomposes $\tilde{f}(k)$ into its spatial components $f(x)$. Neither f nor $\tilde{f}$ is primary; rather, they are equivalent representations of the abstract Hilbert space vector $|f\rangle$. The statement

$$
\int_{-\infty}^{\infty} |f(x)|^2 dx = \int_{-\infty}^{\infty} |\tilde{f}(k)|^2 \, dk \tag{36.7}
$$

is called *Plancherel's theorem* (though it's often filed under Parseval's name).

Example 36.2 Plancherel in Practice

We can directly verify Plancherel's theorem for $f(x) = e^{-a|x|}$ with $a > 0$. First,

$$
\int_{-\infty}^{\infty} |f(x)|^2 dx = \int_{-\infty}^{\infty} e^{-2a|x|} \, dx = 2 \int_{0}^{\infty} e^{-2ax} \, dx = \frac{1}{a}. \tag{36.8}
$$

Then, using the results of Example 36.1 gives $\tilde{f}(k)$:

$$
\int_{-\infty}^{\infty} |\tilde{f}(k)|^2 dk = \frac{4}{\pi a^2} \int_{0}^{\infty} \frac{dk}{(1 + k^2/a^2)^2} = \frac{1}{a}. \tag{36.9}
$$

As a second example, consider

$$
\int_{-\infty}^{\infty} e^{-\frac{1}{2}ax^2 - bx} \, dx, \qquad a > 0. \tag{36.10}
$$

We can evaluate this by completing the square — but if a table of Fourier transforms is handy, it's even easier. That's because the integral is the inner product of two vectors with x-space representations $e^{-\frac{1}{2}ax^2}$ and e^{-bx}. Referring to Table 36.1, we can calculate the inner product in k space:

$$
\int_{-\infty}^{\infty} e^{-\frac{1}{2}ax^2} e^{-bx} \, dx = \int_{-\infty}^{\infty} \left(\frac{1}{\sqrt{a}} e^{-k^2/2a} \right) \left(\sqrt{2\pi} \, \delta(k - ib) \right) dk
$$

$$
= \sqrt{\frac{2\pi}{a}} \, e^{+b^2/2a}. \tag{36.11}
$$

Though this example is relatively simple, the moral is clear: if the calculation is challenging in x, try to do it in k!

A few comments about notation and conventions:

- Since $\mathcal{F}$ is a unitary operation, we denote Fourier transform pairs as f and $\tilde{f}$ to emphasize their status as two representations of the same abstract vector $|f\rangle$. [The notation $f(x)$, $F(k)$ is also common, as is the less-transparent $f(x)$, $g(k)$.]
- Fourier integrals are over all x or all k, so which sign in $e^{\pm ikx}$ goes with $\mathcal{F}$ and which with $\mathcal{F}^{-1}$ is irrelevant. But one must be consistent.
- In addition to mapping between x and k, $\mathcal{F}$ is also commonly used in the time-frequency domains, mapping between t and ω. Note that in both cases, the spaces have reciprocal units. In quantum mechanics, where $E = \hbar\omega$ and $p = \hbar k$, ω and k space are often referred to as energy and momentum space, respectively.
- Between $\mathcal{F}$ and $\mathcal{F}^{-1}$ there is an overall factor of $1/2\pi$ — but there are different conventions for how to distribute it. A common physics convention is to place the 2π entirely within the k-integration ("for every dk a $1/2\pi$"). In signal processing, one often removes all prefactors by substituting $\nu = \omega/2\pi$, so that frequency ν is in hertz. Here we choose the symmetric convention in which $\mathcal{F}$ and $\mathcal{F}^{-1}$ are each defined with a factor of $1/\sqrt{2\pi}$ out front. With this choice, the only distinction between operators $\mathcal{F}$ and $\mathcal{F}^{-1}$ is the sign of the exponent.[2]

Finally, the Fourier integrals in (36.3) apply in one-dimensional space. But $\mathcal{F}$ easily generalizes to n-dimensional space simply by replacing position x with $\vec{r}$ and wave number k with the wave vector $\vec{k}$ to get

$$\tilde{f}(\vec{k}) = \int \frac{d^n r}{(2\pi)^{n/2}}\, e^{-i\vec{k}\cdot\vec{r}} f(\vec{r}) \tag{36.12a}$$

$$f(\vec{r}) = \int \frac{d^n k}{(2\pi)^{n/2}}\, e^{i\vec{k}\cdot\vec{r}} \tilde{f}(\vec{k}), \tag{36.12b}$$

where the integrals are over all $\vec{r}$ and all $\vec{k}$, respectively. Note that a factor of $(2\pi)^{-1/2}$ is required for each integration dimension.

BTW 36.1 Dense Space

Is membership in $\mathbb{L}^2[\mathbb{R}]$ necessary to ensure that the Fourier integrals in (36.3) converge? A typical square integrable $f(x)$ certainly makes increasingly negligible contributions to $\tilde{f}$ near the integration limits as $L \to \infty$. But in fact, we need not require square integrability:

$$\tilde{f}(k) = \int_{-\infty}^{\infty} e^{-ikx} f(x)\, dx \le \int_{-\infty}^{\infty} |e^{-ikx} f(x)|\, dx \le \int_{-\infty}^{\infty} |f(x)|\, dx. \tag{36.13}$$

Thus as long as $f(x)$ is absolutely integrable, we're good. The same, of course, must be true of $\tilde{f}(k)$ as well.

This raises an interesting question: our functions are vectors in $\mathbb{L}^2$, which is a Hilbert space; absolutely integrable functions are in $\mathbb{L}^1$, which is not. We saw in Problem 31.4 that every bounded integrable function is *also* square integrable — but that's only over a finite region of integration;

---

[2] Mathematica's `FourierTransform` uses $\mathcal{F}[f] = \sqrt{\frac{|b|}{(2\pi)^{1-a}}} \int_{-\infty}^{\infty} f(t) e^{ib\omega t} dt$, with a and b specified by the `FourierParameters`$\to\{a, b\}$ option. The default is $\{0, 1\}$; our convention is $\{0, -1\}$.

it doesn't hold for infinite limits. So in order for the vectors-in-an-inner-product interpretation to apply, we must restrict ourselves to functions in the intersection $\mathbb{L}^1 \cap \mathbb{L}^2$. But then, what of functions in $\mathbb{L}^2$ which are *not* also in $\mathbb{L}^1$? Well as a Hilbert space, all Cauchy sequences of functions in $\mathbb{L}^2$ converge in the space (see BTWs 6.2 and 34.3). And as it happens, $\mathbb{L}^1 \cap \mathbb{L}^2$ is *dense* in $\mathbb{L}^2$ — which means that any vector in $\mathbb{L}^2$ can be expressed as the mean-square limit of a sequence of functions in $\mathbb{L}^1 \cap \mathbb{L}^2$. So we can define the Fourier transform of any function in $\mathbb{L}^2$ as the mean-square limit of a sequence of Fourier transforms of functions in $\mathbb{L}^1 \cap \mathbb{L}^2$.

36.2 Physical Insights

If a function and its Fourier transform are unitarily equivalent, how do their representations of $|f\rangle$ differ? Beyond the simple observation that x and k (or t and ω) have inverse units, how do distinct functions f and $\tilde{f}$ package and encode the same information?

Arguably the most notable feature of Fourier transform pairs is illustrated by the reciprocal relationship between the x and k spectra we encountered in Example 36.1. Let's delve into this further by exploring three situations in which this essential feature provides physical insight.

Example 36.3 **Wave Packets**

Consider a bundle of wiggles localized in an slice of space or time having length of order L, with no signal anywhere outside this region. In other words, a pulse — or, more technically, a *wave packet*. Since a wave packet is defined not just between $\pm L$ but over all space (or time), a Fourier series is not appropriate. To put it another way, a countable infinity of harmonics is unable to reconstruct a true wave pulse over all space. To do that requires the continuous infinity of k's of a Fourier transform.

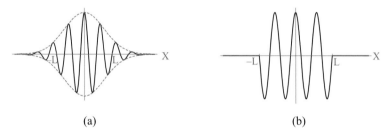

(a)   (b)

Figure 36.2 Wave pulses: (a) standard wave packet; (b) idealized wave packet, (36.14).

For simplicity, consider an idealized wave packet which looks like a cosine between $\pm L$, but is identically zero everywhere else (Figure 36.2b),

$$f(x) = \begin{cases} \cos k_0 x, & |x| < L \\ 0, & |x| > L \end{cases}. \tag{36.14}$$

Its spectral representation $\tilde{f}(k)$ is

$$\tilde{f}(k) = \mathcal{F}[f] = \frac{1}{\sqrt{2\pi}} \int_{-\infty}^{\infty} e^{-ikx} f(x)\, dx = \frac{1}{\sqrt{2\pi}} \int_{-L}^{L} e^{-ikx} \cos(k_0 x)\, dx$$

$$= \frac{1}{2\sqrt{2\pi}} \int_{-L}^{L} \left[e^{-i(k-k_0)x} + e^{-i(k+k_0)x} \right] dx. \tag{36.15}$$

Now if we were dealing with a harmonic wave extending over all space — that is, (36.14) in the $L \to \infty$ limit — we would find

$$\mathcal{F}[\cos k_0 x] = \sqrt{\frac{\pi}{2}} \left[\delta(k - k_0) + \delta(k + k_0)\right]. \tag{36.16}$$

Note that $k_0 < 0$ is the same frequency as $k_0 > 0$, just phase shifted by π. So in fact, only one frequency contributes: the sign of k_0 reflects the well-known decomposition of a standing wave as the sum of oppositely moving traveling waves with the same frequency.

    None of these results are new — but since the δ function only emerges when integrating between infinite limits, we deduce something new and fundamental for finite L: a localized pulse does not, indeed *cannot*, have a single, well-defined frequency. Though $\tilde{f}$ may be dominated by k_0, there will necessarily be other frequencies which contribute. Performing the integral in (36.15) for finite L reveals

$$\tilde{f}(k) = \frac{1}{\sqrt{2\pi}} \left[\frac{\sin[(k + k_0)L]}{k + k_0} + \frac{\sin[(k - k_0)L]}{k - k_0}\right], \tag{36.17}$$

with the spectrum dominated by frequencies $k \approx k_0$,

$$\tilde{f}(k) \approx \frac{1}{\sqrt{2\pi}} \frac{\sin[(k - k_0)L]}{k - k_0}, \quad k \approx k_0. \tag{36.18}$$

 The function $\sin(x)/x$ is called the sinc function; its square is proportional to the wave intensity I (Figure 36.3)

$$I(k) \sim \left|\frac{\sin[(k - k_0)L]}{k - k_0}\right|^2. \tag{36.19}$$

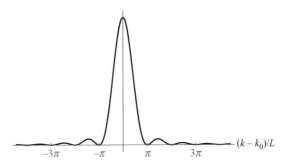

Figure 36.3 The intensity of a localized pulse $I(k) \sim \left[\frac{\sin[(k-k_0)L]}{k-k_0}\right]^2$.

Clearly the primary contribution comes from the central peak around k_0 located between the zeros of $\tilde{f}(k)$ at $|k - k_0| = \pi/L$. Thus the spectrum is dominated by k values in the range $k = k_0 \pm \Delta k$, where $\Delta k = \pi/L$. Since the pulse itself is only defined over a region $\Delta x = 2L$, we make the remarkable observation that for our idealized wave packet

$$\Delta x \Delta k = 2\pi, \tag{36.20}$$

independent of L. This naturally leads one to ask about the product $\Delta x \Delta k$ for more realistic wave packets. In particular, just how low can it go?

    The answer: not all the way to zero. Since wave packets vanish in the infinite region beyond some finite L, their spectral decomposition requires a far more delicate superposition than that of standing waves on a string. It takes a continuous infinity of k's in order to obtain both constructive interference for $|x| < L$ and totally destructive interference for $|x| > L$. And the larger the destructive region, the greater the frequency range of plane waves needed in the superposition. Thus not only is reciprocal breadth of the x and k spectra quite general, their product necessarily has a lower bound

$$\Delta x \Delta k \gtrsim 1. \tag{36.21}$$

The value of the lower bound depends upon the precise definitions of Δx and Δk; what matters is that it's non-zero.

Consider, say, a trumpet blast. The shorter the blast, the more difficult it is to determine the note played. Indeed, pulses with shorter durations Δt have greater uncertainty in the frequency $\Delta \omega = 2\pi \Delta \nu$; that uncertainty is a measure of the breadth of frequency components contained in the pulse. Since a trumpet plays in a frequency range of about 150–950 Hz, a blast needs to last more than roughly $(400 \text{ Hz})^{-1} = 2.5$ ms before $\Delta\omega/\omega$ is small enough to discern the note being played. Only an infinitely held note truly has a single frequency.[3]

The lower bound also has famous consequences in quantum mechanics, in which wave packets describe particles as localized distributions with uncertainty in position Δx and spectral breadth Δk. Since momentum $p = \hbar k$, (36.21) becomes the familiar form of the Heisenberg uncertainty principle,

$$\Delta x \Delta p \gtrsim \hbar, \tag{36.22}$$

where the lower bound represents nature's ultimate precision limit. Imagine, then, that we're able to precisely determine a particle's position x_0, giving it the quantum wavefunction $\psi(x) = \delta(x - x_0)$. That certainty in position comes at the cost of complete uncertainty in k, since as we saw in Example 36.1, $\mathcal{F}[\delta] = \frac{1}{\sqrt{2\pi}}$ — which is to say, the same constant contribution to $\psi(x)$ for all k. The complement is also true: a particle with definite momentum has a single value of k and hence a spectral breadth $\Delta k = 0$. Thus its uncertainty in position is infinite.

Example 36.4   Fraunhofer Diffraction

Diffraction is a well-known illustration of an inverse relationship: the smaller the opening, the broader the diffraction pattern. Perhaps less familiar is that diffraction is a physical example of the reciprocal relationship between a function and its Fourier transform.

According to Huygens' principle,[4] each point on the wavefront of a monochromatic plane wave of light with frequency $\omega = ck$ can be considered a source of an outgoing *spherical wave*,

$$\Psi(r,t) = \psi(r)e^{-i\omega t} = A\frac{e^{i(kr-\omega t)}}{r}, \tag{36.23}$$

where the $1/r$ accounts for the decrease of wave amplitude, and $e^{i(kr-\omega t)}$ tracks the advance of a spherical wavefront in space and time. (For simplicity, we neglect polarization.) Huygens describes wave propagation as the superposition of such wave sources at each point on a plane wavefront to produce the next wavefront. So if light encounters an opaque screen in the xy-plane with a small aperture at $z = 0$, most of these Huygen wavelets will be blocked; what then is the intensity arriving at a point P on a screen a distance $z > 0$ away?

---

[3] You can explore this phenomenon using Mathematica's `Play` command. Listen to, say, a low 300 Hz with `Play[Sin[600π t],{t,0,T}]`, and vary the pulse length T.

[4] "Hoy-gens" (g as in "game").

Because the light at the opening is coherent, any phase differences between the wavelets will be due to path differences taken to get to P. Assigning cartesian coordinates u, v to points within the plane of the aperture, the wave arriving at P is the superposition of all the spherical waves emanating from the opening,

$$\psi(P) = A \int_S \frac{e^{ikr}}{r} \, du \, dv, \tag{36.24}$$

where A is a constant, and the integral is over the surface S of the aperture. Each contributing wave is a squared-distance $r^2 = (x - u)^2 + (y - v)^2 + z^2$ from P at (x, y, z) (Figure 36.4).

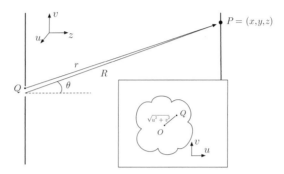

Figure 36.4 Point Q at position (u, v) within an aperture. Inset: View toward screen from behind.

For small z, the pattern of light will reveal an image of the opening; this is called "near field" or Fresnel ("fre-nel") diffraction. We will consider instead large z, for which only the position of the opening is relevant; this is called "far-field" or Fraunhofer diffraction. For an arbitrary aperture point Q with coordinates $u, v \ll z$, the distance to P can be expanded as

$$r = \left[(x - u)^2 + (y - v)^2 + z^2 \right]^{1/2}$$

$$\approx R \left[1 - \frac{2(xu + yv)}{R^2} \right]^{1/2} \approx R \left[1 - \frac{xu + yv}{R^2} \right], \tag{36.25}$$

where $R = \sqrt{x^2 + y^2 + z^2}$ is the distance from the origin to P. So for sufficiently small u, v, the field at P becomes

$$\psi(P) = A \frac{e^{ikR}}{R} \int_S e^{-ik(xu+yv)/R} \, du \, dv$$

$$\equiv A \frac{e^{ikR}}{R} \int_S e^{-i(k_x u + k_y v)} \, du \, dv, \tag{36.26}$$

where

$$\vec{k}_\perp = (k_x, k_y) = (kx/R, ky/R) = k(\sin\theta \cos\phi, \, \sin\theta \sin\phi). \tag{36.27}$$

Though (36.26) looks like a two-dimensional Fourier transform, the integration region is limited to S. So let's replace the single opening with a pattern of holes and slits — called a "mask" — described by a function $f(u, v)$ such that

$$f(u,v) = \begin{cases} 1, & \text{opening in the mask} \\ 0, & \text{else} \end{cases}. \tag{36.28}$$

Then as long of f is non-zero only for u, v small compared to z, we can replace the integral over S with an integral over the entire uv-plane to get

$$\begin{aligned}
\psi(P) &= A \frac{e^{ikR}}{R} \int_{-\infty}^{\infty} e^{-i(k_x u + k_y v)} f(u,v) \, du \, dv \\
&= A \frac{e^{ikR}}{R} \int e^{-i\vec{k}_\perp \cdot \vec{u}} f(\vec{u}) \, d^2 u \\
&= 2\pi A \frac{e^{ikR}}{R} \tilde{f}(\vec{k}_\perp),
\end{aligned} \tag{36.29}$$

with $\tilde{f}(\vec{k}_\perp)$ the two-dimensional Fourier transform of $f(\vec{u})$. So for incident light of intensity I_0, the intensity at P is

$$I \sim |\psi(P)|^2 = I_0 \frac{|\tilde{f}(\vec{k}_\perp)|^2}{R^2}. \tag{36.30}$$

Thus the Fourier transform $\tilde{f}$ of the mask function f gives its diffraction pattern on a far-away screen! The simplest mask is that of a single narrow slit,

$$f(u) = h_a(u) \equiv \begin{cases} 1, & |u| < a/2 \\ 0, & |u| > a/2 \end{cases}, \tag{36.31}$$

sometimes called the "top hat" function.[5] Having no v dependence, (36.29) reduces to the one-dimensional Fourier transform we found in Example 36.3: it's the sinc function of (36.17) and Figure 36.3 with $k_0 = 0$, $L = a/2$. So the diffraction pattern h_a produces is simply

$$|\tilde{h}(k)|^2 = \frac{1}{2\pi} \frac{\sin^2(k_x a/2)}{(k/2)^2} = \frac{a^2}{2\pi} \text{sinc}^2(k_x a/2). \tag{36.32}$$

The pattern for a finite-width rectangular slit, shown in Figure 36.5a, is described by the product of sinc functions, one each in k_x and k_y. For a circular mask

$$f(r) = \begin{cases} 1, & |r| < a \\ 0, & |r| > a \end{cases}, \tag{36.33}$$

the Fourier integral should be evaluated in polar coordinates $d^2 u \equiv r dr d\gamma$,

$$\begin{aligned}
\tilde{f}(\vec{k}_\perp) &= \int e^{-ik_\perp r \cos\gamma} f(r) \, d^2 u \\
&= \int_0^a \left(\int_0^{2\pi} e^{-ikr \sin\theta \cos\gamma} \, d\gamma \right) r \, dr,
\end{aligned} \tag{36.34}$$

where γ is the angle between $\vec{k}_\perp$ and $\vec{u}$, and we've used (36.27) to write $|\vec{k}_\perp| = k_\perp = k \sin\theta$. The Bessel function integral in (35.17a) reveals the angular integral in γ to be $2\pi J_0(kr \sin\theta)$; then

$$\tilde{f}(\vec{k}_\perp) = 2\pi \int_0^a J_0(kr \sin\theta) \, r \, dr = 2\pi a^2 \frac{J_1(ka \sin\theta)}{ka \sin\theta}, \tag{36.35}$$

---

[5] $h_a(x)$ is often denoted $\Pi_a(x)$, which looks like a top hat. In Mathematica, it's `HeavisidePi[x/a]`.

as can be verified with the identity (C.18b). Note the parallel: a rectangular opening results in a sinc function, whereas a circular opening gives what we might think of as a Bessel-sinc function — sometimes called a jinc function.[6] The resulting intensity,

$$I = 4\pi^2 a^2 I_0 \,\text{jinc}^2(ka \sin \theta), \qquad (36.36)$$

produces the concentric diffraction pattern in Figure 36.5b called the *Airy disk*.

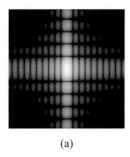

(a)

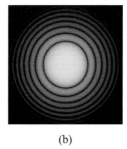

(b)

Figure 36.5 Fraunhofer diffraction patterns $|\tilde{f}|^2$: (a) rectangular slit (b) circular hole.

Example 36.5   Bragg Scattering

A crystal's structure can be determined using X-ray diffraction. Incoming X-rays interact with lattice electrons, and the scattered light from different parts interfere. As in the previous example, we consider far-field diffraction. Then the electron distribution $\rho(\vec{r})$ can be determined from the interference pattern which results, with the reciprocal lattice introduced in Example 35.1 playing a crucial role.

Consider an incoming X-ray with wave vector $\vec{k}$ scattering elastically from the crystal. The scattered wave, with wave vector $\vec{k}'$, is the superposition of contributions from individual lattice electrons; each contribution emerges from its site at $\vec{r}'$ and arrives at point P on the screen with amplitude proportional to $e^{i\vec{k}'\cdot(\vec{r}-\vec{r}')}$. It's the $\vec{r}'$-dependent path difference which leads to the diffraction pattern. The amount of scattering coming from each individual lattice electron must also be proportional to the incoming wave $e^{i\vec{k}\cdot\vec{r}'}$ incident on an electron. So the wave amplitude arriving at P, given by the integral over all scatterers, is proportional to the superposition

$$\int_V e^{i\vec{k}\cdot\vec{r}'} e^{i\vec{k}'\cdot(\vec{r}-\vec{r}')} \rho(\vec{r}')\,d^3r' = e^{i\vec{k}'\cdot\vec{r}} \int_V \rho(\vec{r}')e^{-i\vec{q}\cdot\vec{r}'}\,d^3r', \qquad (36.37)$$

where $\vec{q} = \vec{k}' - \vec{k}$ is the scattering event's *momentum transfer* ($\hbar k$ being momentum in quantum mechanics). The diffraction pattern is thus determined by the *form factor* $F(q)$, the Fourier transform of the electron distribution $\rho(\vec{r})$,

$$F(\vec{q}) = \int \rho(\vec{r})\,e^{-i\vec{q}\cdot\vec{r}'}\,d^3r', \qquad (36.38)$$

where we've omitted the irrelevant factor of $(2\pi)^{3/2}$. Since the electron distribution of a crystal is periodic, we can expand $\rho(\vec{r})$ in a Fourier series

---

[6] The name "sinc" is a contraction of the Latin *sinus cardinalis* — so designated because of its cardinal importance in signal processing. The name "jinc" is thought to be witty.

$$\rho(\vec{r}) = \sum_m \tilde{\rho}(\vec{G}_m)\, e^{i\vec{G}_m \cdot \vec{r}}, \tag{36.39}$$

where $\vec{G}_m$ is a vector in the crystal's reciprocal lattice, (35.10). Inserting this Fourier series into (36.38) reveals

$$F(\vec{q}) = \sum_m \int \rho(\vec{G}_m)\, e^{i\left(\vec{G}_m - \vec{q}\right)\cdot \vec{r}'}\, d^3r' = \sum_m \tilde{\rho}(\vec{G}_m) \int e^{i\left(\vec{G}_m - \vec{q}\right)\cdot \vec{r}'}\, d^3r'$$

$$= (2\pi)^3 \sum_m \tilde{\rho}(\vec{G}_m)\, \delta(\vec{G}_m - \vec{k}' + \vec{k}). \tag{36.40}$$

Noting that $\vec{G}_{-m} = -\vec{G}_m$, this shows that X-rays will only scatter in directions $\vec{k}'$ which satisfy

$$\vec{k}' = \vec{k} \pm \vec{G}, \tag{36.41}$$

for any reciprocal lattice vectors $\pm\vec{G}$. Thus $\vec{G} \equiv \vec{q}$ is the momentum transfer.

For elastic scattering, $|\vec{k}| = |\vec{k}'| \equiv k$ — so squaring both sides of (36.41) gives

$$k^2 = k^2 + G^2 \pm 2\vec{k}\cdot\vec{G} \quad \Longrightarrow \quad \hat{k}\cdot\hat{G} = \pm\frac{1}{2}|\vec{G}|, \tag{36.42}$$

which is *Bragg's law* for X-ray diffraction. The geometric construction in Figure 36.6 of the Bragg condition $\hat{k}\cdot\hat{G} = \pm\frac{1}{2}|\vec{G}|$ shows that the tip of the incident wave vector must lie in a plane which is the perpendicular bisector of a reciprocal lattice vector. These reciprocal-space planes are called *Bragg planes*. It terms of the angle θ which $\vec{k}$ and $\vec{k}'$ make with the Bragg plane, the Bragg condition is $|\vec{G}| = 2k\sin\theta = (4\pi/\lambda)\sin\theta$. But notice that the relation between the reciprocal and direct lattices (35.14) implies that the distance d between successive direct lattice planes satisfies $|\vec{G}_m| = 2\pi m/d$. Thus (36.42) is equivalent to the statement of the Bragg condition familiar from intro physics,

$$2d\sin\theta = m\lambda. \tag{36.43}$$

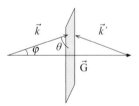

Figure 36.6 The Bragg plane bisects the reciprocal lattice vector $\vec{G} = \vec{k}' - \vec{k}$.

36.3 Complementary Spaces

Due to Plancherel's theorem (36.7) establishing the unitarity of $\mathcal{F}$, Fourier transforms can be used to map a system from one space to the other. Considerable insight can be derived — both mathematical and physical — by considering the same problem in two equivalent ways, essentially from two different perspectives. Moreover, it is often the case that a difficult analysis in one space becomes a more tractable problem in the other.

To appreciate the potential, it's important to recognize the various complementary properties of f and $\tilde{f}$; several are listed in Table 36.2. For instance, a straightforward change of variables demonstrates that the reciprocal nature discussed in the previous section is common to all Fourier transform pairs:

Table 36.2 Fourier transform symmetries

	x space	k space
Scaling	$f(ax)$	$\frac{1}{a}\tilde{f}(k/a)$
Translation $\leftrightarrow$ phase shift	$f(x-a)$	$e^{-ika}\tilde{f}(k)$
Damping $\leftrightarrow$ frequency shift	$e^{-bx}f(x)$	$\tilde{f}(k-ib)$
Real $\leftrightarrow$ reflection	$f^*(x)=f(x)$	$\tilde{f}^*(k)=\tilde{f}(-k)$
Real and even $\leftrightarrow$ real	$f(-x)=f(x)$	$\tilde{f}^*(k)=\tilde{f}(k)$
Real and odd $\leftrightarrow$ imaginary	$f(-x)=-f(x)$	$\tilde{f}^*(k)=-\tilde{f}(k)$
Derivative $\leftrightarrow$ multiplication	$\left(\frac{d}{dx}\right)^n f$	$(ik)^n\tilde{f}(k)$
Multiplication $\leftrightarrow$ derivative	$(-ix)^n f(x)$	$d^n\tilde{f}/dk^n$

$$\mathcal{F}[f(ax)] = \frac{1}{\sqrt{2\pi}}\int_{-\infty}^{\infty} f(ax)e^{-ikx}\,dx$$

$$= \frac{1}{\sqrt{2\pi}}\int_{-\infty}^{\infty} f(y)e^{-iky/a}\frac{1}{a}\,dy = \frac{1}{a}\tilde{f}(k/a). \tag{36.44}$$

So if f has characteristic width a, then $\tilde{f}$ has width $1/a$; by Plancherel's theorem, any change in breadth of one is balanced by opposite change of height in the other. This is most clearly manifest for a Gaussian, the only entry in Table 36.1 which is its own Fourier transform. In fact, for $a = 1$ the Gaussian is an eigenfunction of $\mathcal{F}$ (Problem 36.36).

Example 36.6 **The Dirac Comb**

The periodic sequence of delta functions

$$s_a(x) \equiv \sum_{n=-\infty}^{\infty} \delta(x-na) \tag{36.45}$$

is called a *Dirac comb*; it's also known as the "Shah" function, $\mathrm{III}_a(x)$.[7] One of its most intriguing and useful properties is that, like a Gaussian, its Fourier transform is *also* a Dirac comb. This is fairly intuitive: the Fourier transform of a periodic function with period a only has non-zero contributions for k an integer multiple of the fundamental frequency $2\pi/a$. Moreover, since a single delta function has equal contributions at all frequencies, each harmonic in an infinite periodic array of δ's will have infinite amplitude. It must be, then, that the contributions of the remaining, non-harmonic frequencies vanish by destructive interference. And indeed, explicit calculation reveals (Problem 36.15)

$$\tilde{s}_a(k) = \frac{\sqrt{2\pi}}{a}\sum_{m=-\infty}^{\infty} \delta\left(k-2\pi m/a\right) = \frac{\sqrt{2\pi}}{a}s_{2\pi/a}(k). \tag{36.46}$$

So the Fourier transform of a periodic train of delta functions is a periodic train of delta functions — though the inherent reciprocal relationship remains evident.

---

[7] The use of this symbol almost certainly originated as a pictograph of a comb. The name is from its similarity to the Cyrillic letter sha, Ш, which was adapted from the Hebrew letter shin, ש.

Example 36.7   The Reciprocal Lattice, Revisited

The Dirac comb is a natural tool in the description of a crystal's structure, its reciprocal properties providing a clean link between the direct and reciprocal lattices. In fact, $s(x)$ describes a one-dimensional lattice with spacing a. The direct lattice of a two-dimensional rectangular crystal array with primitive vectors $\vec{a}_1 = a\,\hat{\imath}$ and $\vec{a}_2 = b\,\hat{\jmath}$, as depicted in Figure 35.1, can be described by

$$s(x, y) = \sum_{n_1, n_2 = -\infty}^{\infty} \delta(x - n_1 a)\, \delta(y - n_2 b). \tag{36.47}$$

This naturally generalizes to Bravais lattices in any number of dimensions,

$$s(\vec{r}) = \sum_{n=-\infty}^{\infty} \delta(\vec{r} - \vec{R}_n), \tag{36.48}$$

where the sum is over lattice sites n, and $\vec{R}_n$ is the vector distance between sites.

The reciprocal lattice is the Fourier transform of the direct lattice. For a two-dimensional rectangular lattice, (36.46) gives

$$\tilde{s}(\vec{k}) = \frac{2\pi}{ab} \sum_{m_1, m_2 = -\infty}^{\infty} \delta\left(k_x - \frac{2\pi m_1}{a}\right) \delta\left(k_y - \frac{2\pi m_2}{b}\right). \tag{36.49}$$

Thus we get a reciprocal lattice with primitive vectors $\vec{b}_1 = \hat{\imath}\, 2\pi/a$ and $\vec{b}_2 = \hat{\jmath}\, 2\pi/b$ with k-space lattice sites given by $\vec{k}_m \equiv \vec{G}_m$, where

$$\vec{G}_m = m_1 \frac{2\pi}{a}\, \hat{\imath} + m_2 \frac{2\pi}{b}\, \hat{\jmath}, \tag{36.50}$$

just as we found in (35.14). A reciprocal lattice in three dimensions is then

$$\tilde{s}(\vec{k}) = \frac{(2\pi)^{3/2}}{V} \sum_m \delta(\vec{k} - \vec{G}_m). \tag{36.51a}$$

If instead we calculate the Fourier transform directly from (36.48) we find

$$\tilde{s}(\vec{k}) = \frac{1}{(2\pi)^{3/2}} \sum_n \int \delta(\vec{r} - \vec{R}_n) e^{-i\vec{k}\cdot\vec{r}}\, d\tau = \frac{1}{(2\pi)^{3/2}} \sum_n e^{-i\vec{k}\cdot\vec{R}_n}. \tag{36.51b}$$

Comparison of these two expressions for $\tilde{s}$ reveals that only when $\vec{k}$ is a reciprocal lattice vector $\vec{G}_m$ does the sum in (36.51b) contribute; all other $\vec{k}$'s must result in destructive interference.

Another complementary relation between Fourier transform pairs emerges from translation:

$$\mathcal{F}\big[f(x - a)\big] = \frac{1}{\sqrt{2\pi}} \int_{-\infty}^{\infty} f(x - a) e^{-ikx}\, dx$$

$$= \frac{1}{\sqrt{2\pi}} \int_{-\infty}^{\infty} f(y) e^{-ik(y+a)}\, dy$$

$$= e^{-ika} \frac{1}{\sqrt{2\pi}} \int_{-\infty}^{\infty} f(y) e^{-iky}\, dy = e^{-ika}\tilde{f}(k). \tag{36.52}$$

So moving a distance a along a wave of frequency k is accompanied by an advance in phase by ka — exactly as one might expect. And the symmetry between the operators $\mathcal{F}$ and $\mathcal{F}^{-1}$ ensures that this works both ways: a frequency translation $k \to k - k_0$ results in a phase shift $f(x) \to e^{ik_0 x} f(x)$. For imaginary $k_0 = ib$, this gives the relationship between damping and complex frequency.

Example 36.8 $\mathcal{F}$ Without Integrating

The Fourier transform of $e^{-\frac{1}{2}ax^2 - bx}$ can be found by direct appeal to Table 36.2,

$$\mathcal{F}\left[e^{-\frac{1}{2}ax^2 - bx}\right] = \mathcal{F}\left[e^{-bx}e^{-\frac{1}{2}ax^2}\right]$$

$$= \mathcal{F}\left[e^{-\frac{1}{2}ax^2}\right]_{k-ib} = \frac{1}{\sqrt{a}} e^{-(k-ib)^2/2a}, \tag{36.53}$$

which is both more sophisticated and much simpler than completing the square.

The Fourier basis e^{ikx} is an eigenfunction of the operator $\mathcal{D} = d/dx$. This observation underlies the effect of $\mathcal{F}$ on the derivative:

$$\mathcal{F}\left[\frac{df}{dx}\right] = \frac{1}{\sqrt{2\pi}} \int_{-\infty}^{\infty} e^{-ikx} \frac{df}{dx}\, dx$$

$$= \frac{1}{\sqrt{2\pi}} e^{-ikx} f(x)\Big|_{-\infty}^{\infty} - \frac{1}{\sqrt{2\pi}} \int_{-\infty}^{\infty} \frac{d}{dx}\left[e^{-ikx}\right] f(x)\, dx$$

$$= +ik \frac{1}{\sqrt{2\pi}} \int_{-\infty}^{\infty} e^{-ikx} f(x)\, dx$$

$$= ik\tilde{f}(k), \tag{36.54}$$

where, for physical square-integrable $f(x)$, the surface term vanishes. Thus $\mathcal{F}$ essentially "diagonalizes" the derivative operator, unitarily mapping $\mathcal{D}$ into k-space, where it becomes multiplication by k.

The Fourier transform of powers of $\mathcal{D}$ requires integrating by parts multiple times, so that

$$\mathcal{F}\left[\frac{d^n f}{dx^n}\right] = (ik)^n \tilde{f}(k). \tag{36.55}$$

In more than one dimension, the gradient brings down factors of $\vec{k}$ — for instance,

$$\mathcal{F}\left[\nabla^2 f(\vec{r})\right] = i\vec{k} \cdot i\vec{k}\, \tilde{f}(\vec{k}) \equiv -k^2 \tilde{f}(\vec{k}). \tag{36.56}$$

Of course, e^{ikx} is also an eigenfunction of d/dk. So just as $\mathcal{F}\left[df/dx\right]$ maps to multiplication by k, $\mathcal{F}^{-1}\left[d\tilde{f}/dk\right]$ maps back to multiplication by x (though they differ by a sign — see Problem 36.10).

Example 36.9 The Quantum Momentum Operator

The Fourier transform of a quantum wavefunction $\psi(x)$ yields $\tilde{\psi}(k)$. Because of the de Broglie relation $p = \hbar k$, this is often referred to as the momentum space wavefunction. In this space, the momentum operator $\mathcal{P}$ is given by the relation

$$\mathcal{P}\tilde{\psi}(k) = \hbar k\, \tilde{\psi}(k). \tag{36.57}$$

In other words, in k space the momentum operator is just multiplication by p, $\mathcal{P} \to p$. But how does $\mathcal{P}$ manifest in position space — that is, how does $\mathcal{P}$ act on $\psi(x)$? The answer is given via inverse Fourier transform:

$$\mathcal{F}^{-1}\left[\hbar k \tilde{\psi}(k)\right] = \hbar \left(-i\frac{d}{dx}\right)\psi(x).$$ (36.58)

Thus we find the position–space momentum operator

$$\mathcal{P} \to \frac{\hbar}{i}\frac{d}{dx}.$$ (36.59)

Example 36.10 Generating Translations

Taken together, the correspondence between translation and phase shift on the one hand, differentiation and multiplication on the other, provide insight into the translation operator. As we saw in Example 6.16, translation of a function from $x \to x + a$ is given by Taylor expansion,

$$f(x+a) = \sum_{n=0}^{\infty}\frac{1}{n!}\left(a\frac{d}{dx}\right)^n f(x) = e^{a\frac{d}{dx}}f(x).$$ (36.60)

Applying a Fourier transform gives

$$\mathcal{F}\left[f(x+a)\right] = \sum_{n=0}^{\infty}\frac{1}{n!}(ika)^n \tilde{f}(k) = e^{ika}\tilde{f}(k).$$ (36.61)

So the relationship uncovered in Example 36.9 between momentum and spatial derivatives hints at an intimate connection between momentum and spatial translation (Problem 36.19).

A powerful technique suggested by (36.55) uses $\mathcal{F}$ to transform a differential equation into an algebraic equation. Then solving the differential equation follows a road map through Fourier space:

$$\begin{array}{ccc}
\mathcal{L}_x f(x) = r(x) & \xrightarrow{\ \mathcal{F}\ } & \tilde{\mathcal{L}}_k \tilde{f}(k) = \tilde{r}(k) \\
 & & \downarrow \\
f(x) & \xleftarrow{\ \mathcal{F}^{-1}\ } & \tilde{f}(k) = \tilde{\mathcal{L}}_k^{-1}\tilde{r}(k)
\end{array}$$ (36.62)

where $\mathcal{L}_x$ is a differential operator in x, $\tilde{\mathcal{L}}_k$ the corresponding algebraic operator in k. Solving a differential equation then reduces to the integral $\tilde{\mathcal{L}}_k^{-1}\tilde{r}$.

Example 36.11 $f''(x) - f(x) = -e^{-|x|}$

Fourier transforming this second-order equation gives

$$-k^2\tilde{f} - \tilde{f} = -\sqrt{\frac{2}{\pi}}\frac{1}{1+k^2} \quad \Longrightarrow \quad \tilde{f}(k) = \sqrt{\frac{2}{\pi}}\frac{1}{(1+k^2)^2}.$$ (36.63)

Then, for $f(x) \to 0$ at $\pm\infty$,

$$f(x) = \sqrt{\frac{2}{\pi}}\,\mathcal{F}^{-1}\left[\frac{1}{(1+k^2)^2}\right]$$

$$= \frac{1}{\pi}\int_{-\infty}^{\infty}\frac{e^{ikx}}{(1+k^2)^2}\,dk = \frac{1}{2}\,(1+|x|)\,e^{-|x|}.$$ (36.64)

Example 36.12 Poisson Equation

Back in Section 16.3 we found that the relationship between the electric potential Φ and charge density ρ is given by Poisson's equation, $\nabla^2\Phi = -\rho/\epsilon_0$. The basic challenge of electrostatics is to solve for Φ given charge distribution ρ. Direct solution essentially involves integrating $\rho(\vec{r})$ twice — which, depending on the form of ρ, may not be so straightforward. But we can use Fourier transforms to map the equation into Fourier space,

$$(ik)^2 \tilde{\Phi}(\vec{k}) = -\tilde{\rho}(\vec{k})/\epsilon_0, \tag{36.65}$$

so that $\tilde{\mathcal{L}}_k = -k^2$. We then need only to calculate $\mathcal{F}^{-1}\big[\tilde{\rho}/k^2\big]$ to obtain $\Phi(\vec{r})$. Let's demonstrate for the simple case of a point charge at $\vec{r}'$, $\rho(\vec{r}) = q\,\delta(\vec{r}-\vec{r}')$, with the potential going to zero at infinity. Using $\mathcal{F}\big[\delta(\vec{r}-\vec{r}')\big] = (2\pi)^{-3/2}e^{-i\vec{k}\cdot\vec{r}'}$:

$$\begin{aligned}
\Phi(\vec{r}) &= \frac{1}{\epsilon_0}\mathcal{F}^{-1}\big[\tilde{\rho}/k^2\big] = \frac{q}{(2\pi)^{3/2}\epsilon_0}\frac{1}{(2\pi)^{3/2}}\int \frac{e^{i\vec{k}\cdot(\vec{r}-\vec{r}')}}{k^2}\,d^3k \\[2mm]
&= \frac{q}{8\pi^3\epsilon_0}\int \frac{e^{ik|\vec{r}-\vec{r}'|\cos\theta}}{k^2}\,k^2\,dk\,d(\cos\theta)d\phi \\[2mm]
&= \frac{q}{4\pi^2\epsilon_0}\int_0^\infty dk \int_{-1}^1 e^{ik|\vec{r}-\vec{r}'|u}du = \frac{q}{2\pi^2\epsilon_0}\int_0^\infty dk\,\frac{\sin(k|\vec{r}-\vec{r}'|)}{k|\vec{r}-\vec{r}'|} \\[2mm]
&= \frac{q}{2\pi^2\epsilon_0|\vec{r}-\vec{r}'|}\int_0^\infty d\xi\,\frac{\sin(\xi)}{\xi} = \frac{q}{4\pi\epsilon_0}\frac{1}{|\vec{r}-\vec{r}'|},
\end{aligned} \tag{36.66}$$

which is the well-known potential of a point charge.

Example 36.13 A Damped Driven Oscillator

Consider the differential equation

$$\mathcal{L}_t u = \left[\frac{d^2}{dt^2} + 2\beta\frac{d}{dt} + \omega_0^2\right]u(t) = f(t), \tag{36.67}$$

which describes the motion of a damped oscillator with natural frequency ω_0 and driving force f (see Example 3.5). A Fourier transform maps this differential equation in t into an algebraic equation in ω',[8]

$$\tilde{\mathcal{L}}_{\omega'}\tilde{u} = \big[-\omega'^2 + 2i\beta\omega' + \omega_0^2\big]\tilde{u}(\omega') = \tilde{f}(\omega'), \tag{36.68}$$

so that

$$\tilde{u}(\omega') = \frac{\tilde{f}(\omega')}{\omega_0^2 - \omega'^2 + 2i\beta\omega'}, \tag{36.69}$$

giving the solution to (36.67) as

$$u(t) = \mathcal{F}^{-1}\left[\frac{\tilde{f}(\omega')}{\omega_0^2 - \omega'^2 + 2i\beta\omega'}\right]. \tag{36.70}$$

---

[8] We use ω' rather than ω to forestall confusion in (36.71).

For instance, if $f(t) = f_0 \cos \omega t$ is a periodic driving force with frequency ω, then we can find $\tilde{f}$ in Table 36.1 to obtain the solution

$$
\begin{aligned}
u(t) &= \mathcal{F}^{-1}\left[\frac{\tilde{f}(\omega')}{\omega_0^2 - \omega'^2 + 2i\beta\omega'}\right] \\
&= f_0 \frac{1}{\sqrt{2\pi}}\int_{-\infty}^{\infty} \frac{\sqrt{\frac{\pi}{2}}\left[\delta(\omega - \omega') + \delta(\omega + \omega')\right]}{\omega_0^2 - \omega'^2 + 2i\beta\omega'}e^{i\omega't}\,d\omega' \\
&= \frac{1}{2}f_0\left[\frac{e^{i\omega t}}{\omega_0^2 - \omega^2 + 2i\beta\omega} + \frac{e^{-i\omega t}}{\omega_0^2 - \omega^2 - 2i\beta\omega}\right] \\
&= f_0 \frac{(\omega_0^2 - \omega^2)\cos \omega t + 2\beta\omega \sin \omega t}{(\omega_0^2 - \omega^2)^2 + 4\beta^2\omega^2},
\end{aligned}
\tag{36.71}
$$

as you can verify by acting on $u(t)$ with $\mathcal{L}_t$.

But wait: as a second-order ordinary differential equation, the full solution to (36.67) requires knowledge of $u(0)$ and $\dot{u}(0)$. And yet there appears to be no role for initial conditions in (36.71).[9] This is a direct result of having a purely periodic driving force, subtly implying that $f(t)$ has no actual start time. Moreover, the solution (36.71) is not unique: due to the linearity of $\mathcal{L}_t$, adding any solutions of the free oscillator $\mathcal{L}_t u = 0$ to the driven solution (36.71) still satisfies (36.67). As we first saw back in Example 3.4, solutions to $\mathcal{L}_t u = 0$ for $t > 0$ have an overall factor of $e^{-\beta t}$, and so damp away to zero. Once these "transient" solutions die out, what's left is the "steady-state" solution (36.71).

Concerns about initial conditions and uniqueness can be addressed by replacing $\tilde{f} = \mathcal{F}\left[\cos \omega' t\right]$ in (36.71) with $\mathcal{F}\left[\Theta(t)\cos \omega' t\right]$, with $\Theta(t)$ the step function of (22.12). This yields (36.71) together with a linear combination of homogeneous solutions such that $u(0) = 0$ and $\dot{u}(0)$ is discontinuous — as expected of a sudden application of a force to a previously quiescent system. We'll return to this example in Sections 36.6 and 42.4.

The Fourier transform approach to differential equations is a powerful method — and despite its formulaic procedure, is actually fairly intuitive. Consider that if the driving force is periodic with frequency ω, then surely that periodicity is reflected in the solution to (36.67). So inserting the ansatz $u = Ce^{i\omega t}$ into the differential equation gives

$$
\left(-\omega^2 + 2i\beta\omega + \omega_0^2\right)Ce^{i\omega t} = f_0 e^{i\omega t},
\tag{36.72}
$$

to quickly and easily reveal

$$
u(t) = \left[\frac{f_0}{(\omega_0^2 - \omega^2) + 2i\beta\omega}\right]e^{i\omega t}.
\tag{36.73}
$$

For $f(t) = f_0 \cos \omega t$, taking the real part gives (36.71). We can obtain this same solution by superposing two copies of $u(t)$ — one with $+\omega$, the other $-\omega$. But here's where it really gets interesting: as a linear system, we can accommodate any number of discrete source terms of the form $Ce^{i\omega t}$, each contributing its own (36.73) with its own values of C and ω; their sum gives the solution to the differential equation. (This observation was noted earlier following (22.4).) What (36.69) demonstrates is that any driving force $f(t)$ which has a Fourier transform can be construed as a continuous collection of individual

---

[9] The previous two examples included boundary conditions at infinity.

sources $\tilde{f}(\omega) = \langle \omega | f \rangle$, each a component of f on a basis of plane waves $e^{i\omega t}$. This renders the solution (36.70) to the differential equation as an integral — not a Fourier series sum over standing waves with discrete frequencies ω_n, but rather a Fourier transform integral over the basis $\{e^{i\omega t}\}$ of continuous ω.

36.4 A Basis of Plane Waves

What happens when we try to calculate the orthogonality of plane waves? In fact, we already know:

$$\int_{-\infty}^{\infty} e^{ikx} e^{-ik'x} \, dx = 2\pi \delta(k - k'). \tag{36.74}$$

Though not the Kronecker delta orthogonality we're accustomed to, conceptually the Dirac delta plays the same role. Indeed, (36.74) shows that $e^{ik'x}$ and e^{ikx} are orthogonal for all $k \neq k'$; their inner product is non-zero only for $k = k'$. Of course, that non-zero value is infinity — but then, a delta function must ultimately be deployed inside an integral. In terms of this "generalized" Dirac orthogonality relation, $\left\{ \frac{1}{\sqrt{2\pi}} e^{ikx} \right\}$ is an orthonormal basis. It's also a complete basis: rather than the discrete sum $\frac{1}{2\pi} \sum_n e^{ik_n x} e^{-ik_n x'} = \delta(x - x')$, the continuum limit gives

$$\frac{1}{2\pi} \int_{-\infty}^{\infty} e^{ikx} e^{-ikx'} \, dk = \delta(x - x'). \tag{36.75}$$

Comparison with (36.74) shows that orthogonality in x-space is completeness in k-space, and vice-versa — a reflection of the equal status of x and k representations of an abstract state $|f\rangle$. Introducing complete bases $\{|x\rangle\}$ and $\{|k\rangle\}$ for continuous x and k, the function $f(x)$ is the projection of $|f\rangle$ onto the x basis, and $\tilde{f}(k)$ its projection onto the k basis $|k\rangle$,

$$f(x) \equiv \langle x | f \rangle \qquad \tilde{f}(k) \equiv \langle k | f \rangle. \tag{36.76}$$

In fact, as the parallel expressions in Table 36.3 indicate, little here is actually new conceptually. For instance,

$$\tilde{f}(k) = \langle k | f \rangle = \int \langle k | x \rangle \langle x | f \rangle \, dx = \int \langle k | x \rangle f(x) \, dx, \tag{36.77}$$

where we've inserted the completeness of the $|x\rangle$ basis. Direct comparison with (36.3a) reveals

$$\langle k | x \rangle = \frac{1}{\sqrt{2\pi}} e^{-ikx}. \tag{36.78}$$

Thus $\frac{1}{\sqrt{2\pi}} e^{-ikx}$ is the x-space representation of $|k\rangle$. Since $\langle x | k \rangle = \langle k | x \rangle^*$,

$$\langle x | k \rangle = \frac{1}{\sqrt{2\pi}} e^{+ikx} \tag{36.79}$$

Table 36.3 Discrete versus continuous bases

	Discrete	Continuous
Expansion	$|f\rangle = \sum_n c_n |\alpha_n\rangle$	$|f\rangle = \int d\alpha \, c(\alpha) |\alpha\rangle$
Orthogonality	$\langle \alpha_m | \alpha_n \rangle = \delta_{nm}$	$\langle \alpha | \beta \rangle = \delta(\alpha - \beta)$
Completeness	$\sum_n |\alpha_n\rangle\langle\alpha_n| = \mathbb{1}$	$\int d\alpha \, |\alpha\rangle\langle\alpha| = 1$

is the k-space representation of $|x\rangle$. Thus

$$f(x) = \langle x|f\rangle = \int \langle x|k\rangle \langle k|f\rangle \, dk = \int \langle x|k\rangle \tilde{f}(k) \, dk, \tag{36.80}$$

which is just (36.3b). Moreover, orthogonality of $|x\rangle$ is consistent with completeness in $|k\rangle$ (and vice-versa):

$$\langle x|x'\rangle = \langle x| \left[\int dk \, |k\rangle \langle k| \right] |x'\rangle = \int dk \, \langle x|k\rangle \langle k|x'\rangle$$

$$= \frac{1}{2\pi} \int dk e^{ik(x-x')} = \delta(x - x'). \tag{36.81}$$

That the plane wave bases are complex conjugates of one another underlies the Fourier transform as the unitary map between equivalent representations of $|f\rangle$. Indeed, comparing (36.79) with (25.61) identifies $\langle x|k\rangle$ as a unitary "matrix" mapping between spaces.

Example 36.14   **Quantum Momentum and Position States**

Let $|p\rangle$ denote a quantum mechanical state of definite momentum $p = \hbar k$. This can be expanded on a basis of states with definite position $|x\rangle$ using the completeness relation

$$|p\rangle = \int dx \, |x\rangle \langle x|p\rangle. \tag{36.82}$$

The difference between $\langle x|k\rangle$ and $\langle x|p\rangle$ is normalization, with $2\pi \to 2\pi\hbar$ in (36.79). This is easily confirmed by projecting (36.82) onto $|p'\rangle$:

$$\langle p'|p\rangle = \int dx \, \langle p'|x\rangle \langle x|p\rangle$$

$$= \frac{1}{2\pi\hbar} \int dx \, e^{i(p-p')x/\hbar} = \frac{1}{\hbar}\delta\left[(p-p')/\hbar\right] = \delta(p - p'). \tag{36.83}$$

Physically, this says that a state $|\psi\rangle \equiv |p\rangle$ of definite momentum manifests as an infinite spike when represented in momentum space, $\tilde{\psi}(p') = \langle p'|\psi\rangle = \langle p'|p\rangle$. Its representation in position, however, is spread over all space, $\psi(x) = \langle x|\psi\rangle = \frac{1}{\sqrt{2\pi\hbar}} e^{ipx/\hbar}$. In fact, it's the oscillatory nature of $e^{ipx/\hbar}$ that puts the "wave" in wavefunction, with $\lambda = 2\pi\hbar/p = h/p$. Thus emerges the de Broglie relation.

That any of this works at all is rather remarkable — especially considering that a Fourier transform is an expansion of a square-integrable function on a basis which is not itself square integrable! In other words, the eigenfunctions $e^{\pm ikx}$ of $\frac{d^2\phi}{dx^2} = -k^2\phi$ don't belong to the Hilbert space $\mathbb{L}^2$. Nor, for that matter, can a pure plane wave e^{ikx} be physically realized. Moreover, there's nothing inherent in the operator $\mathcal{D}^2 = (d/dx)^2$ requiring its eigenvalues to be real. It is perhaps astonishing then that Fourier transform integrals nonetheless converge for real k and square-integrable f, and that decomposition of $\mathbb{L}^2$ vectors into a continuum of harmonic components still works. Since $e^{\pm ikx}$ has finite inner products with all $\mathbb{L}^2$ functions, it's as if e^{ikx} lives just beyond the boundary of $\mathbb{L}^2$, barely missing out being a member. It seems not unreasonable to append to the space $\mathbb{L}^2$ such curious non-normalizable functions. In this larger vector space, called a "rigged" Hilbert space, $\mathcal{D}^2$ can be considered Hermitian over $\pm\infty$ with real eigenvalue k and orthogonal eigenfunctions $e^{\pm ikx}$ — see de la Madrid (2005).

36.5 Convolution

Perhaps you noticed that nowhere amongst the properties listed in Table 36.2 is a formula for the Fourier transform of the product of two functions. The reason is simple, if unexpected: the Fourier transform of the product is *not* the product of the transforms,

$$\mathcal{F}[fg] \neq \mathcal{F}[f]\mathcal{F}[g], \tag{36.84}$$

as can be readily verified. The correct relationship is more intriguing:

$$\mathcal{F}[f]\mathcal{F}[g] = \left[\frac{1}{\sqrt{2\pi}}\int_{-\infty}^{\infty} f(u)\,e^{-iku}\,du\right]\left[\frac{1}{\sqrt{2\pi}}\int_{-\infty}^{\infty} g(v)\,e^{-ikv}\,dv\right]$$

$$= \frac{1}{2\pi}\int_{-\infty}^{\infty} f(u)g(v)\,e^{-ik(u+v)}\,du\,dv. \tag{36.85}$$

Substituting v with $w = u + v$,

$$\mathcal{F}[f]\mathcal{F}[g] = \frac{1}{2\pi}\int_{-\infty}^{\infty} e^{-ikw}\left[\int_{-\infty}^{\infty} f(u)g(w-u)\,du\right]dw$$

$$\equiv \frac{1}{\sqrt{2\pi}}\mathcal{F}[f*g], \tag{36.86}$$

where $f * g$ is called the *convolution* of f and g,

$$(f*g)(w) \equiv \int_{-\infty}^{\infty} f(u)g(w-u)\,du. \tag{36.87}$$

What is convolution? As a mechanical operation, convolving "slides" the functions f and g over one another in opposite directions, computing the area under their product at each value of w. The integral $f * g$ can also be regarded as the weighted sum of an input signal $f(u)$ with weight $g(w - u)$ — or vice-versa, since (as can be readily verified)

$$f*g = \int_{-\infty}^{\infty} f(u)g(w-u)\,du = \int_{-\infty}^{\infty} f(w-u)g(u)\,du = g*f. \tag{36.88}$$

Example 36.15 **There's a Hole in the Bucket, Dear Liza**[10]

Convolution emerges naturally in systems for which one rate-of-change process depends on another. For instance, imagine filling a bucket with water at a rate of $f(t)$ (in, say, liters per minute). If the bucket has a hole in the bottom, the rate at which water leaks out depends both on the size of the hole and the height of the water in the bucket at a given time. Using Bernoulli's equation for incompressible fluid flow, one can determine the amount of leakage $A_0 g(t)$, where A_0 is the quantity of water in the bucket at $t = 0$. Now in a time du, $f(u)du$ liters pour into the bucket; by time $t > u$, however, the leak reduces this quantity to $f(u)g(t-u)du$. In the meantime though, more water has flowed in — so the total amount of water in an initially empty bucket as a function of time is

---

[10] https://youtu.be/zYY6Q4nRTS4

$$A(t) = \int_0^t f(u)g(t-u)\,du, \tag{36.89}$$

which is the finite convolution $(f * g)\,(t)$. (With water flow beginning at time zero, $f(u < 0) = 0$; for times greater than t water has yet to flow in, $f(u > t) = 0$. These restrictions on f allow the integration limits can be extended to $\pm\infty$.)

The relationship (36.86) for the product of Fourier transforms is enshrined in the *convolution theorem*:

$$\mathcal{F}[f\,]\mathcal{F}[g] = \frac{1}{\sqrt{2\pi}}\,\mathcal{F}[f * g\,] \tag{36.90a}$$

$$\mathcal{F}[fg] = \frac{1}{\sqrt{2\pi}}\,\mathcal{F}[f\,] * \mathcal{F}[g\,]. \tag{36.90b}$$

The product of the Fourier transforms is the transform of the convolution; the Fourier transform of the product is the convolution of the Fourier transforms. The theorem is also valid with $\mathcal{F}$ replaced everywhere by $\mathcal{F}^{-1}$. Note that the factors of $\sqrt{2\pi}$ arise from the 2π convention (introduced after Example 36.2), so statements of the theorem often appear without them.

But how, you might ask, did convolution $f * g$ emerge in place of the product fg? To answer this, have a look back at Problem 6.20. There we proved that for power series $A(x) = \sum_{n=0}^{\infty} a_n x^n$ and $B(x) = \sum_{m=0}^{\infty} b_m x^m$, the product $C = AB = \sum_{\ell=0}^{\infty} c_\ell x^\ell$ has coefficients

$$c_\ell = (a * b)_\ell \equiv \sum_{n=0}^{\ell} a_n b_{\ell-n}. \tag{36.91}$$

The coefficient c_ℓ is the discrete convolution of the coefficients a_n and b_n. Just as we can regard a_n and b_n as components of A and B on the discrete basis $\{x^n\}$, similarly $\tilde{f}(k)$ and $\tilde{g}(k)$ are the component functions of $f(x)$ and $g(x)$ on the continuous basis $\{e^{ikx}\}$. Then just as $(a * b)_\ell$ is the coefficient of $C = AB$ on $\{x^n\}$, so too $\tilde{h}(k) = \tilde{f} * \tilde{g}$ is the component function of $h(x) = f(x)g(x)$ on $\{e^{ikx}\}$,

$$\tilde{h}(k) = \langle k|h\rangle = \langle k|fg\rangle = \frac{1}{\sqrt{2\pi}}\int dx\, e^{-ikx} f(x)g(x) = \frac{1}{\sqrt{2\pi}}\tilde{f} * \tilde{g}, \tag{36.92}$$

where we've used (36.90b).

Example 36.16   Convolving With δ

A particularly simple yet important example is convolution of a function with the Dirac delta,

$$f * \delta \equiv \left[f(x) * \delta(x)\right](w) = \int_{-\infty}^{\infty} f(u)\,\delta(w-u)\,du = f(w). \tag{36.93}$$

Thus $\delta(x)$ acts as the identity under convolution — as the convolution theorem confirms:

$$\sqrt{2\pi}\,\mathcal{F}[f(x)\,]\mathcal{F}[\delta(x)\,] = \tilde{f}(k) = \mathcal{F}[f * \delta\,]. \tag{36.94}$$

Convolving with a shifted delta function $\delta(x-a)$ shifts f,

$$\left[f(x) * \delta(x-a)\right](w) = \int_{-\infty}^{\infty} f(u)\,\delta(w-a-u)\,du = f(w-a). \tag{36.95}$$

In this case, the convolution theorem results in a phase shift,

$$\sqrt{2\pi}\,\mathcal{F}\big[f(x)\big]\mathcal{F}\big[\delta(x-a)\big]=\tilde{f}(k)\,e^{-ika}, \tag{36.96}$$

exactly as we saw (36.52).

The convolution theorem renders convolution a crucially important tool in multiple areas of mathematics, science, and engineering. And we've already seen a few applications. In Example 36.11, (36.64) is the (inverse) transform of two factors of $\tilde{h}=(1+k^2)^{-1}$. So from (36.90b), we now know that the solution f can be cast as the "autoconvolution" of h — that is, the convolution of $\mathcal{F}^{-1}\big[\tilde{h}\big]$ with itself (Problem 36.26),

$$f(y)=\frac{1}{\sqrt{2\pi}}\,\mathcal{F}^{-1}\big[\tilde{h}\big]*\mathcal{F}^{-1}\big[\tilde{h}\big]=\frac{1}{2}\int_{-\infty}^{\infty}e^{-|x|}e^{-|y-x|}\,dx. \tag{36.97}$$

More compelling perhaps is the consequence of the convolution theorem for electrostatics. From (36.65), solutions to Poisson's equation for a given charge distribution $\rho(\vec{r})$ are given by the inverse transform

$$\Phi(\vec{r})=\frac{1}{\epsilon_0}\,\mathcal{F}^{-1}\left[\frac{\tilde{\rho}(\vec{k})}{k^2}\right]=\frac{1}{(2\pi)^{3/2}\,\epsilon_0}\,\rho(\vec{r})*\mathcal{F}^{-1}\big[1/k^2\big], \tag{36.98}$$

where three factors $\sqrt{2\pi}$ arise from working in three dimensions. A calculation virtually identical to that of (36.66) shows that $\mathcal{F}^{-1}\big[1/k^2\big]=\sqrt{\frac{\pi}{2}}\frac{1}{|\vec{r}|}$. Thus

$$\Phi(\vec{r})=\frac{1}{(2\pi)^{3/2}\,\epsilon_0}\,\rho(\vec{r})*\sqrt{\frac{\pi}{2}}\frac{1}{|\vec{r}|}=\frac{1}{4\pi\epsilon_0}\int_V\frac{\rho(\vec{r}')}{|\vec{r}-\vec{r}'|}\,d^3r', \tag{36.99}$$

which is Coulomb's law, (9.33).

Example 36.17 Two-Slit Diffraction

If we take the sum of two delta functions

$$g(x)=\delta(x+d/2)+\delta(x-d/2), \tag{36.100}$$

and convolve it with the top hat function $h_a(x)$ of (36.31), we get

$$c(x)=h_a*g=h_a(x+d/2)+h_a(x-d/2), \tag{36.101}$$

which is the mask for two slits of width a separated by distance d. By itself, this may not seem much — but since the mask is expressed as a convolution, we can use the convolution theorem to easily find

$$\mathcal{F}\big[h*g\big]=\sqrt{2\pi}\,\tilde{h}(k)\tilde{g}(k)=\sqrt{2\pi}\left[\sqrt{\frac{2}{\pi}}\frac{\sin(ka/2)}{k}\right]\left[\sqrt{\frac{2}{\pi}}\cos(kd/2)\right], \tag{36.102}$$

and thence the two-slit diffraction pattern (Figure 36.7)

$$I(k)=\frac{8}{\pi}\frac{\sin^2(ka/2)}{k^2}\cos^2(kd/2). \tag{36.103}$$

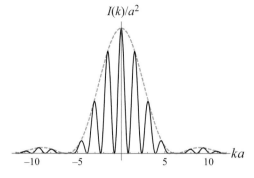

Figure 36.7 Two-slit diffraction pattern, $d = 3a$, within a sinc function envelope $[\sin(x)/x]^2$.

Example 36.18 **The Reciprocal Lattice, Once Again**

Just as convolution with the delta function sum of (36.100) gave a two-slit mask, the Dirac comb introduced in Example 36.6 can generate periodic structure. Specifically, the convolution of a function with $s_a = \sum_{n=-\infty}^{\infty} \delta(x - na)$ returns a periodic function with period a (Figure 36.8),

$$g(w) \equiv f * s_a = \sum_{n=-\infty}^{\infty} \int_{-\infty}^{\infty} f(u)\, \delta(w - na - u)\, du$$

$$= \sum_{n=-\infty}^{\infty} f(w - na) = g(w + ma), \tag{36.104}$$

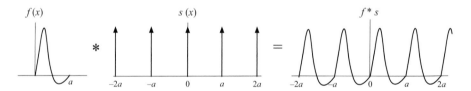

Figure 36.8 The convolution of a pulse with the Dirac comb.

since a shift in w by any integer multiple m of a has no effect on the infinite sum. So the Dirac comb can be used to construct a periodic function from any $f(x)$. For instance, suppose we want to determine the distribution of electric charge within a one-dimensional crystal. If each unit cell has charge density $\rho(x)$, then the crystal's charge distribution is the convolution of ρ with s_a,

$$\rho_a(x) = \sum_{n=-\infty}^{\infty} \int du\, \rho(u)\, \delta(x - na - u) = \rho * s_a. \tag{36.105}$$

The crystal's X-ray diffraction pattern is determined by the Fourier transform,

$$\mathcal{F}[\rho * s_a] = \sqrt{2\pi}\, \tilde{\rho}(k)\, \tilde{s}_a(k)$$

$$= \frac{2\pi}{a} \tilde{\rho}(k) \, s_{2\pi/a}(k) = \frac{2\pi}{a} \sum_{m=-\infty}^{\infty} \tilde{\rho}(k) \, \delta(k - 2\pi m/a)$$

$$= \frac{2\pi}{a} \sum_{m=-\infty}^{\infty} \tilde{\rho}(2\pi m/a) \, \delta(k - 2\pi m/a), \tag{36.106}$$

where we've used (36.46). Thus the diffraction pattern is comprised of an array of spots separated by the reciprocal of the crystal lattice spacing a, with intensity proportional to the magnitude-squared of $\tilde{\rho}$.

Example 36.19 **Linear Translation-Invariant Systems**

Measurement inevitably introduces error. A recorded signal is unavoidably contaminated by noise and distortion: photos are blurred, sounds garbled. Removing distortion to reveal the true signal is crucial — and the convolution theorem is a fundamental tool in this process.

Consider a microphone M listening to sound from a source S to create a recording R. In the process $S \xrightarrow{M} R$, one says that the output R is the response of M to the input S,

$$R = M[S]. \tag{36.107}$$

We make two assumptions about M. First that it's linear: the response to a sum of signals is the sum of the responses of each individually,

$$M[aS_1 + bS_2] = aR_1 + bR_2. \tag{36.108a}$$

We also assume that only time *intervals* are relevant — that is, the operator M is indifferent to time translation

$$R(t - T) = M[S(t - T)]. \tag{36.108b}$$

With these assumptions, M is a *linear translation-invariant* (LTI) system.[11]

The prominent feature of LTI systems is that they can be characterized by their response to a unit impulse $\delta(t)$,

$$\Delta(t) \equiv M[\delta(t)]. \tag{36.109}$$

To see why this is so, consider an arbitrary sound input $v(t)$, discretized as

$$v(t) = \sum_i v(\tau_i) \, \delta(t - \tau_i). \tag{36.110}$$

This describes $v(t)$ as a sequence of delta function impulses, each weighted with independent volume (intensity) $v(\tau_i)$. The response to this sequential input is

$$R = M\left[\sum_i v(\tau_i) \, \delta(t - \tau_i) \right] = \sum_i v(\tau_i) \, M\left[\sum_i \delta(t - \tau_i) \right]$$

$$\equiv \sum_i v(\tau_i) \, \Delta(t - \tau_i). \tag{36.111}$$

---

[11] Though T in LTI often denotes time, the analysis applies to spatially invariant systems as well.

This is the discrete convolution $v * \Delta$ — which in the limit of vanishing spacing between impulses becomes

$$R(t) = \int v(\tau)\Delta(t - \tau)\,d\tau. \tag{36.112}$$

Then the convolution theorem (36.90a) gives

$$\tilde{R}(\omega) = \sqrt{2\pi}\,\tilde{v}(\omega)\tilde{\Delta}(\omega). \tag{36.113}$$

With knowledge of Δ, the input v can be extracted from the recorded data. A simple example is two-slit diffraction, where $\tilde{R}$ is given by (36.103). In this case, Δ is the top hat function h, whose Fourier transform is the sinc function, (36.32); thus the input signal must be the source of the $\cos(kd/2)$. Taking its inverse Fourier transform reveals that we're observing two point objects of equal magnitude, a distance d apart. Of course, an ideal unit impulse is not realistic — but any inherent blurring of a signal due to measurement can be encoded in Δ by measuring the response to an impulse of known intensity. In this context, Δ is called the system's *point spread function* (PSF) — and it's by no means limited to microphones or one-dimensional diffraction. In astronomy, a telescope's PSF is a two-dimensional function $\Delta(x, y)$ defined over the area of the lens, usually deduced from observation of far-away stars. The relation (36.113) is also used to analyze the effect of applied filters, with $\tilde{\Delta}$ referred to as a *transfer function* — the name applying to the Fourier transform of Δ because filters are usually designed to affect the spectrum of the input signal (for instance, low-pass, high-pass, and band-pass filters). In this language, a two-slit diffraction mask is the application of a filter with transfer function $\mathrm{sinc}(k)$.

36.6 Laplace Transforms[12]

Fourier transforms allow one to describe a physical system in terms of position $\vec{r}$ or wave vector $\vec{k}$, at time t or with frequency ω. But working in Fourier space often has a simpler and more direct physical interpretation — particularly in quantum mechanics, where momentum and energy are $\hbar\vec{k}$ and $\hbar\omega$ respectively.

Even so, Fourier transforms do have limitations — perhaps most notably with its focus on square-integrable functions. Although the periodic driving force in Example 36.13 is not in $\mathbb{L}^2$, we were able to reach just outside the space by utilizing a Dirac delta (see the discussion at the end of Section 36.4). In general, however, functions which are not square integrable — even some which may be of physical interest, such as a constant or a linear driving force — do not have convergent Fourier integrals. This can be addressed by inserting a factor $e^{-\sigma t}$ for real $\sigma > 0$ to damp out the driving force for large t, rendering the system both more mathematically stable and physically realistic. Of course, this refinement likely exacerbates convergence as $t \to -\infty$. But unlike (36.71), any realistic driving force has to be switched on at some $t = 0$; considering only functions f which vanish for $t < 0$ effectively sets to the lower integration limit to zero, eliminating any convergence concerns for negative t. Problem 36.38 brings these modifications together, resulting in the *Laplace transform* of f,[13]

$$F(s) = \mathscr{L}[f(t)] \equiv \int_0^\infty f(t)\,e^{-st}\,dt, \tag{36.114}$$

with complex $s \equiv \sigma + i\omega$. For a function f continuous on $0 \le t < \infty$, the integral converges as long as $|f(t)|$ grows more slowly than $e^{\sigma t}$ for large t. In other words, $F(s)$ exists if $|f(t)|$ diverges no faster

---

[12] This section uses results from Chapter 21.

[13] Our conventional $\sqrt{2\pi}$ factor is omitted, to be recovered in the inverse transformation (36.118).

than $e^{\beta t}$ for some $\beta < \sigma$. Such functions are said to be of *exponential order*. So whereas convergence of a Fourier transform requires square-integrability along ω, a Laplace transform requires f to be of exponential order, with $\mathscr{L}$ convergent for complex s in the right half-plane $\Re e(s) > \beta$.

Example 36.20 **Finding $\mathscr{L}[f]$**

The function $f(t) = t^a = e^{a \ln t}$ is of exponential order with arbitrarily small $\beta > 0$. Using $u = st$ and (8.78) gives

$$\mathscr{L}[t^a] = \int_0^\infty t^a e^{-st} \, dt = \frac{1}{s^{a+1}} \int_0^\infty u^a e^{-u} \, du = \frac{\Gamma(a+1)}{s^{a+1}}, \qquad \Re e(s) > 0. \tag{36.115}$$

For integer $a = n \geq 0$, this becomes $\mathscr{L}[t^n] = \frac{n!}{s^{n+1}}$. Similarly,

$$\mathscr{L}[e^{at}] = \int_0^\infty e^{-(s-a)t} \, dt = \frac{1}{s-a}, \qquad \Re e(s) > \Re e(a). \tag{36.116}$$

Together they reveal a translation property similar to $\mathcal{F}[e^{at}f(t)]$,

$$\mathscr{L}[e^{at}t^n] = \mathscr{L}[t^n]_{s-a} = \frac{n!}{(s-a)^{n+1}}, \qquad \Re e(s) > a. \tag{36.117}$$

The Laplace transform essentially extends the Fourier transform to include complex frequencies. So the inverse transform is an integral in the complex s-plane (Figure 36.9),

$$f(t) \equiv \mathscr{L}^{-1}[F(s)] = \frac{1}{2\pi i} \int_C F(s) \, e^{st} \, ds$$

$$= \frac{1}{2\pi i} \int_{\sigma - i\infty}^{\sigma + i\infty} F(s) \, e^{st} \, ds, \tag{36.118}$$

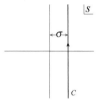

Figure 36.9 Contour for evaluating $\mathscr{L}^{-1}$.

where since $\mathscr{L}[f(t)]$ only converges in the right half-plane of s, the contour must be offset by some $\sigma > \beta$. This leaves any poles of $F(s)$ necessarily sitting to the left of C, giving (36.118) the form of what's called a *Bromwich integral*. Closing the contour with a semicircle at infinity allows the integral to be evaluated by the residue theorem (Section 21.1) — yielding zero when closed to the right, the sum of resides when closed to the left.

A list of commonly used Laplace transform pairs is given in Table 36.4.

Table 36.4 Laplace transform pairs

$f(t)$	$\mathscr{L}[f(t)]$	$\Re e(s) >$
t^n	$\frac{n!}{s^{n+1}}$	0
e^{at}	$\frac{1}{s-a}$	$\Re e(a)$
$t^n e^{at}$	$\frac{n!}{(s-a)^{n+1}}$	$\Re e(a)$
$\sin at$	$\frac{a}{s^2+a^2}$	0
$\cos at$	$\frac{s}{s^2+a^2}$	0
$\sinh at$	$\frac{a}{s^2-a^2}$	$\|a\|$
$\cosh at$	$\frac{s}{s^2-a^2}$	$\|a\|$
$\frac{1}{t}\left(e^{bt} - e^{at}\right)$	$\ln\left(\frac{s-a}{s-b}\right)$	$\Re e(b)$
$\delta(t - t_0)$	e^{-st_0}	0

Example 36.21 **The Inverse Transform of $F(s) = \frac{s}{s^2-a^2}$**

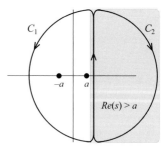

The pole structure of $F(s)$ requires the integration path in (36.118) to lie anywhere to the right of $s = a$. By Jordan's lemma (21.58), the contour can be closed with an infinite semicircle in either the left half-plane for $t > 0$ or the right half-plane for $t < 0$ without making additional contributions to the integral (Problem 36.49). In both cases, $F(s)$ is then given by the sum of the residues from any enclosed singularities. The contour C_1 encloses the poles at $\pm a$, so that for $t > 0$,

$$f(t) = \frac{1}{2\pi i} \int_{C_1} \frac{s\, e^{st}}{s^2 - a^2}\, ds = \text{Res}\left[\frac{s\, e^{st}}{s^2 - a^2}, a\right] + \text{Res}\left[\frac{s\, e^{st}}{s^2 - a^2}, -a\right]$$

$$= \frac{s\, e^{st}}{s + a}\bigg|_{s=a} + \frac{s\, e^{st}}{s - a}\bigg|_{s=-a}$$

$$= \frac{1}{2}\left[e^{at} + e^{-at}\right] = \cosh at. \tag{36.119}$$

By contrast, C_2 encloses no poles — so by Cauchy's theorem (18.7), the integral vanishes for $t < 0$. Note that, as expected, $\cosh at \sim e^{at}$ is of exponential order for large t.

Extending Fourier analysis to include complex frequencies is more natural than it may first appear. Physically, it provides straightforward inclusion of damped oscillations (as we first saw in Example 3.4). Mathematically, Laplace transforms utilize the entire eigenspace of the operator d/dt, not just the those eigenfunctions with real eigenvalues. The cost, of course, is the loss of unitarity and orthogonal bases. So Plancherel's theorem doesn't apply. But as important as unitary transformations are, any invertible linear operator can be construed as a change of basis.

Table 36.5 Properties of Laplace transform $F(s) = \mathcal{L}[f(t)]$

$f(at)$	$\frac{1}{a}F(s/a)$
$f(t - a)$	$e^{-as}F(s)$
$e^{at}f(t)$	$F(s - a)$
$tf(t)$	$-F'(s)$
$t^n f(t)$	$(-1)^n F^{(n)}(s)$
df/dt	$sF(s) - f(0)$
d^2f/dt^2	$s^2 F(s) - sf(0) - \dot{f}(0)$
$\int_0^t f(\tau)\, d\tau$	$\frac{1}{s}F(s)$
$(f * g)(t) = \int_0^t f(\tau)g(t - \tau)\, d\tau$	$F(s)G(s)$

Laplace transform pairs have complementary relationships not unlike those of Fourier pairs (compare Tables 36.2 and 36.5). Perhaps the most noteworthy difference is their mapping of the derivative operator. As expansions in eigenvectors of d/dt, both Fourier and Laplace transforms map derivatives into multiplication. But $\mathcal{L}$ also includes initial data (Problem 36.45):

$$\mathcal{L}\left[\frac{df}{dt}\right] = sF(s) - f(0), \tag{36.120a}$$

and similarly

$$\mathscr{L}\left[\frac{d^2f}{dt^2}\right] = s^2F(s) - sf(0) - \dot{f}(0). \tag{36.120b}$$

(This presumes of course that the transforms exist — that is, both $\dot{f}(t)$ and $\ddot{f}(t)$ are of exponential order.) The appearance of $f(0)$ and $\dot{f}(0)$ in these expressions makes Laplace transforms a fundamental tool for solving initial value problems.

Example 36.22 Two Masses, Three Springs

The equations of motion for the two-mass system of Example 28.1,

$$m\ddot{q}_1 = -kq_1 - k(q_1 - q_2) \qquad m\ddot{q}_2 = -kq_2 - k(q_2 - q_1), \tag{36.121}$$

after a Laplace transform become

$$s^2 Q_1(s) - sq_1(0) - \dot{q}_1(0) = -\frac{2k}{m}Q_1(s) + \frac{k}{m}Q_2(s)$$

$$s^2 Q_2(s) - sq_2(0) - \dot{q}_2(0) = -\frac{2k}{m}Q_2(s) + \frac{k}{m}Q_1(s). \tag{36.122}$$

For initial conditions $q_1(0) = q_2(0) = \dot{q}_1(0) = 0$ and $\dot{q}_2(0) = v_0$, this simplifies to

$$\left(s^2 + \frac{2k}{m}\right)Q_1(s) = \frac{k}{m}Q_2(s) \qquad \left(s^2 + \frac{2k}{m}\right)Q_2(s) = \frac{k}{m}Q_1(s) + v_0. \tag{36.123}$$

The most elegant way to decouple these algebraic equations is to add and subtract them and solve for $Q_\pm = Q_1 \pm Q_2$,

$$Q_+(s) = \frac{v_0}{s^2 + k/m} \qquad Q_-(s) = -\frac{v_0}{s^2 + 3k/m}. \tag{36.124}$$

Notice how the normal mode frequencies $\omega_1^2 = k/m$ and $\omega_2^2 = 3k/m$ emerge as poles — making them directly relevant to the inverse transform integral in the complex s-plane. Alternatively, we can consult Table 36.4. Either way, taking advantage of the linearity of Laplace transforms yields solutions (Problem 36.50)

$$q_{1,2}(t) = \frac{1}{2}v_0\left[\frac{1}{\omega_1}\sin\omega_1 t \mp \frac{1}{\omega_2}\sin\omega_2 t\right] \tag{36.125}$$

which clearly obey the initial conditions. Indeed, notice (and appreciate) how initial conditions are incorporated from the outset, rather than being applied at the end.

Example 36.23 The Damped Driven Oscillator, Revisited

The oscillator of Example 36.13 is governed by the differential equation

$$\left[\frac{d^2}{dt^2} + 2\beta\frac{d}{dt} + \omega_0^2\right]u(t) = f(t). \tag{36.126}$$

With $U(s) = \mathscr{L}[u(t)]$, $F(s) = \mathscr{L}[f(t)]$, $u_0 = u(0)$, and $v_0 = \dot{u}(0)$, its Laplace transform is

$$\left(s^2 + 2\beta s + \omega_0^2\right)U(s) - (su_0 + v_0) - 2\beta u_0 = F(s), \tag{36.127}$$

so that

$$U(s) = \frac{F(s)}{s^2 + 2\beta s + \omega_0^2} + \frac{u_0(s + 2\beta) + v_0}{s^2 + 2\beta s + \omega_0^2}. \tag{36.128}$$

Since the first term includes the driving force, it must be the steady-state solution $U_{st}(s)$ (but see Problem 36.44). The second, which depends on initial conditions, gives the transient solution $U_{tr}(s)$ not included in the Fourier transform. We can solve for $U_{tr}(t)$ by making clever use of Laplace transform tables. Start by completing the square in the denominator to get

$$U_{tr}(s) = \frac{u_0(s + \beta)}{(s + \beta)^2 + \Omega^2} + \frac{u_0\beta + v_0}{(s + \beta)^2 + \Omega^2}, \tag{36.129}$$

where $\Omega^2 = \omega_0^2 - \beta^2$. Then referring to Table 36.4, and using the translation property in Table 36.5, we find (Problem 36.53)

$$u_{tr}(t) = e^{-\beta t}\left(u_0 \cos \Omega t + \frac{1}{\Omega}(\beta u_0 + v_0)\sin \Omega t\right), \tag{36.130}$$

which can be shown to be equivalent to (3.28).

Problems

36.1 Verify that $\mathcal{F}$ and $\mathcal{F}^{-1}$ are inverse operations.

36.2 Show that the Fourier transform of a Gaussian is a Gaussian.

36.3 Find the Fourier transform of

$$f(x) = \begin{cases} 1, & |x| < a \\ 0, & |x| > a \end{cases}.$$

What happens to your answer in the limit $a \to \infty$? Discuss the implications for the general relationship between the range of x values in $f(x)$ and k values in $\tilde{f}(k)$.

36.4 Use Table 36.1 to calculate $\int_{-\infty}^{\infty} e^{-a|x|} \cos bx \, dx$, $a > 0$.

36.5 Show that $\langle f|g\rangle$ is the same when calculated in x or k space — that is,

$$\int_{-\infty}^{\infty} f^*(x)g(x)\,dx = \int_{-\infty}^{\infty} \tilde{f}^*(k)\,\tilde{g}(k)\,dk.$$

36.6 Use $\mathcal{F}$ and $\mathcal{F}^{-1}$ to verify (36.5).

36.7 Verify Plancherel's theorem (36.7) using the bra-ket notation for continuous bases in Section 36.4.

36.8 Use Plancherel's theorem to evaluate $\int_{-\infty}^{\infty} \frac{dx}{(x^2+a^2)^2}$.

36.9 Find the Fourier transforms of the following in terms of $\tilde{f} = \mathcal{F}(f)$:

(a) $e^{-bx}f(ax)$
(b) $f^*(x)$
(c) $f(x), f$ even.

36.10 Find $\mathcal{F}^{-1}\left[d\tilde{f}/dk\right]$ in terms of $f(x)$.

36.11 Using only properties listed in Tables 36.1–36.2 (no integration!), find the following:

(a) $\mathcal{F}\left[e^{-ak^2}\right]$

(b) $\mathcal{F}\left[xe^{-x^2/2}\right]$

(c) $\mathcal{F}\left[x^2 e^{-a^2x^2/2}\right]$

(d) $\mathcal{F}^{-1}\left[\frac{e^{2ik}}{1+k^2}\right]$

(e) $\mathcal{F}\left[\frac{d^2}{dx^2} - x^2\right]$.

36.12 Because the Heaviside step function of (9.27),

$$\Theta(x) = \begin{cases} 1, & x > 0 \\ 0, & x < 0, \end{cases}$$

is largely constant, its Fourier transform should be dominated by low frequencies. The problem is that evaluating $\mathcal{F}[\Theta]$ directly gives an integral with zero as its lower limit — precisely where Θ is undefined. To find $\tilde{\Theta}(k)$, take the following approach:

(a) Use the relationship between $\mathcal{F}[f]$ and $\mathcal{F}[f']$ to find an expression for $\tilde{\Theta}(k)$ valid for all $k \neq 0$. [Your result should be reminiscent of the Fourier coefficients for the square wave in Problem 33.1.]

(b) Noting that $k = 0$ represents the zero-frequency (DC) average of the step function, you can find the Fourier transform of this component directly. Show that adding this contribution to the result in (b) gives

$$\tilde{\Theta}(k) = \frac{1}{\sqrt{2\pi}} \frac{1}{ik} + \sqrt{\frac{\pi}{2}} \delta(k). \tag{36.131}$$

[The first term is the Cauchy principle value — see BTW 8.1.]

36.13 **Flipping a Switch**

Imagine a signal $\phi(t)$ turned on at $t = 0$. Using the Heaviside step function Θ, we can express the process, valid for both positive and negative t, as $\psi(t) = \Theta(t)\phi(t)$. Use (36.131) to show that

$$\tilde{\psi}(\omega) = \frac{1}{2\pi i} \int_{-\infty}^{\infty} \frac{\tilde{\phi}(\omega - \omega')}{\omega'} d\omega' + \frac{1}{2} \tilde{\phi}(\omega).$$

[If this looks familiar, see (21.119) and Problem 21.24.]

36.14 The Airy function Ai(x) is defined as a solution to Airy's differential equation $\frac{d^2}{dx^2}f(x) = xf(x)$. Use Fourier transforms to find an integral representation of Ai(x). [The conventional normalization for Ai is $\frac{1}{\sqrt{2\pi}}$ times this result.]

36.15 Consider the Dirac comb (36.45),

$$s_a(x) = \sum_{n=-\infty}^{\infty} \delta(x - na).$$

Note that this string of delta functions is itself periodic.

(a) Show that the Fourier transform $\tilde{s}_a(k)$ is the Fourier series of a single delta function. What is the series' periodicity?

(b) Use the periodicity to express $\tilde{s}_a(k)$ as an infinite sum, thereby confirming (36.46).

36.16 For a function $f(x)$ and it Fourier transform $\tilde{f}(k)$, use the Dirac comb and the invariance of the inner product to derive the *Poisson summation formula*

$$\sum_{n=-\infty}^{\infty} f(\alpha n) = \frac{\sqrt{2\pi}}{\alpha} \sum_{m=-\infty}^{\infty} \tilde{f}(2\pi m/\alpha).$$

36.17 For $f(x) = e^{-bx}, b > 0$, use the Poisson summation formula (Problem 36.16) to show (see also Example 21.23)

$$\coth(x) = x \sum_{m=-\infty}^{\infty} \frac{1}{x^2 + \pi^2 m^2} = -\frac{1}{x} + 2x \sum_{m=0}^{\infty} \frac{1}{x^2 + \pi^2 m^2}.$$

36.18 On a quantum wavefunction $\psi(x)$, the position operator is defined as $X\psi(x) = x\psi(x)$. The momentum operator (36.59) obeys $P\psi(x) = \frac{\hbar}{i} \frac{d}{dx} \psi(x)$. Show that $[X,P]\psi(x) = i\hbar\,\psi(x)$, where $[X,P] = XP - PX$. The non-commutativity of position and momentum implies that they do not have a common eigenbasis; this leads directly to the Heisenberg uncertainty principle — see BTW 27.3.

36.19 Infinitesimal translations of a quantum wavefunction $\psi(\vec{r})$ can be obtained with the operator (compare Section 26.6 for infinitesimal rotations),

$$T_n(\epsilon) = \mathbb{1} + i\epsilon\,\hat{n} \cdot \vec{p}/\hbar,$$

such that $T_n(\epsilon)\psi(\vec{r}) = \psi(\vec{r} + \epsilon\hat{n})$. Show that the generator of translations is the momentum operator $\vec{p} = -i\hbar\vec{\nabla}$. What's the operator for finite translation?

36.20 Show that the Fourier transform of the product is (proportional to) the convolution of the Fourier transforms, (36.90b):

$$\mathcal{F}[fg] = \frac{1}{\sqrt{2\pi}}\,\mathcal{F}[f] * \mathcal{F}[g]. \tag{36.132}$$

36.21 Consider the signals $f(x) = 4\sin(8\pi x)$ and $g(x) = 8 + 6\cos(2\pi x)$. Describe the effects of convolution by examining the Fourier spectrum of the product $h(x) = f(x)g(x)$.

36.22 Find the convolution of $f(x) = \sqrt{\frac{a}{2\pi}}e^{-ax^2/2}$ and $g(x) = \sqrt{\frac{b}{2\pi}}e^{-bx^2/2}$.

36.23 Show that convolution is both commutative and associative: $f * g = g * f$ and $(f * g) * h = f * (g * h)$.

36.24 Use a series of diagrams at regular time steps to display the convolution of two top hat functions, $(h_a * h_a)(t)$. What happens to the discontinuities in h_a? [This is an example of how convolution can be used to smooth a signal.]

36.25 Find the Fourier transform of the triangle function,

$$\Lambda_a(x) = \begin{cases} 1 - |x/a|, & |x| \le |a| \\ 0, & \text{else} \end{cases}$$

[You'll likely find it easier to do Problem 36.24 first.]

36.26 Show that the convolution in (36.97) gives the same result as the Fourier transform in (36.64).

36.27 Find the function $f(x)$ such that $f * f = e^{-x^2}$.

36.28 The *autocorrelation* of a function

$$\int_{-\infty}^{\infty} f^*(u)f(t + u)\,du$$

is similar to the convolution $f * f$, but without the sign flip in u. Show that the Fourier transform of the autocorrelation is the power spectrum $|\tilde{f}(k)|^2$.

36.29 **So Fix It, Dear Henry**
What becomes of $A(t)$ in (36.89) if the hole is plugged?

36.30 Use the analysis of Example 36.15 to interpret Coulomb's law (36.99),

$$\Phi(\vec{r}) = \frac{1}{4\pi\epsilon_0} \int_V \frac{\rho(\vec{r}')}{|\vec{r} - \vec{r}'|} \, d^3r'.$$

36.31 **A Low-Pass Filter**
The charge $Q(t)$ on a capacitor in an RC circuit changes at a rate $I = dQ/dt$ on a characteristic time-scale $\tau = RC$. Consider a square voltage pulse which acts for a time τ,

$$V_{in}(t) = \begin{cases} V_0, & t < \tau \\ 0, & t > \tau \end{cases}.$$

If the capacitor is uncharged at $t = 0$, it acts as a short circuit, resulting in an initial current $I_0 = V_0/R$. Show that the convolution $Q(t) = \frac{1}{R} V_{in}(t) * e^{-t/\tau}$ yields the familiar continuous charging and discharging curves. (Express your result in terms of the maximum charge deposited on the capacitor.) By examining the shape of the incoming voltage pulse and the voltage $V_{out}(t) = Q/C$ across the capacitor, explain why this circuit is called a "low-pass filter."

36.32 Find the steady-state response of a damped oscillator with constant driving force $F(t) = \alpha$.

36.33 Despite your best efforts, your data set $\{f_n(t)\}$ is noisy, with high-frequency fluctuations obscuring the smooth signal you're looking for.

(a) Explain how convolving with the top hat function (36.31) can filter out high-frequency noise $|\omega| > \omega_0$.
(b) Analysis of your signal requires taking a derivative. Should you smooth your data as in part (a) before or after taking the derivative? Explain.

36.34 Consider an infinite rod with initial temperature distribution $T(x, 0) = f(x)$. In one spatial dimension, the heat equation becomes (Problem 16.2),

$$\frac{\partial T}{\partial t} = \alpha \frac{\partial^2 T}{\partial x^2}.$$

(a) Take the Fourier transform in x to map this partial differential equation into an ordinary differential equation in t.
(b) Solve for $\tilde{T}(k, t)$ in terms of the initial condition $\tilde{f}(0)$.
(c) Use $\mathcal{F}^{-1}$ and the convolution theorem to find an integral expression for the solution $T(x, t)$.

36.35 A time-invariant linear system — that is, one which is only sensitive to time intervals — can be expressed as a convolution, (36.112). Show that the converse is also true: if a linear system can be expressed as a convolution, then it is time invariant.

36.36 **The $\mathcal{F}$ Eigensystem**
As an operator mapping functions from $\mathbb{L}^2$ to $\mathbb{L}^2$, it's natural to ask about the $\mathcal{F}$ eigensystem $\mathcal{F}[\phi] = \lambda\phi$, where eigenfunctions are identified by their functional form, not the basis x or k.

(a) Show that $\mathcal{F}^4[\phi(x)] = \phi(x)$. Thus conclude that $\lambda^4 = 1$. In what way does this reflect the unitarity of Fourier transforms?
(b) Show that $\mathcal{F}$ maps the differential operator $\mathcal{H} = -\frac{d^2}{dx^2} + x^2$ into itself — that is, the operators $\mathcal{F}$ and $\mathcal{H}$ commute. Thus eigenfunctions of $\mathcal{H}$ are also eigenfunctions of $\mathcal{F}$.
(c) The Schrödinger equation for the harmonic oscillator has solutions $\phi_n(x) = H_n(x)e^{-x^2/2}$, where $H_n(x)$ is a Hermite polynomial. Use the Rodrigues formula in Table 32.1 to show that

$$\mathcal{F}[\phi_n] = (-i)^n \phi_n.$$

Note that $\phi_0(x)$ is a Gaussian (see Problem 36.2).
(d) For $f(x)$ even, show that $f(x) + \tilde{f}(x)$ is an eigenfunction of $\mathcal{F}$ with $\lambda = 1$.

36.37 A Diffraction Grating

Consider a coherent light source of intensity I_0, normally incident on a screen of N evenly spaced slits a distance d apart.

(a) Find the far-field interference pattern $I(k)$, assuming the slits have negligible width. (You may wish to review Problem 6.9.) How does the intensity of the central peak $I(0)$ scale with the number of slits?

(b) In the continuum limit $N \to \infty$ with $a \equiv Nd$ held fixed, show that the intensity relative to that of the central peak, $I(k)/I(0)$, gives the single-slit diffraction pattern found in (36.32).

(c) Find the intensity pattern made by slits of finite width $a < d$. Use Mathematica to plot the pattern for $N = 10, d = 5, a = 1$.

36.38 From Fourier to Laplace

For arbitrary $f(t)$ of exponential order β:

(a) Extract the Laplace transform (36.114) from the Fourier transform of $g(t) = \Theta(t)f(t)e^{-\sigma t}$. What's the relationship between $\tilde{g}(\omega)$ and $F(s)$? Verify the region of convergence as the right half-plane $\Re e(s) > \beta$.

(b) Use the inverse Fourier transform to find the inverse Laplace integral (36.118). [Note that (36.118) presumes that $f(t) = 0$ for $t < 0$.]

36.39 Find the Laplace transforms of the following:

(a) $\cos \omega t$

(b) $\sin^2 \omega t$

(c) $1/\sqrt{t}$.

36.40 Use Table 36.4, convolution (Table 36.5) and/or residue calculus to find the inverse Laplace transforms of the following:

(a) $\dfrac{1}{s(s^2+4s+3)}$

(b) $\dfrac{1}{(s-1)(s^2+1)}$

(c) $\dfrac{e^{-2s}}{(s-2)^2}$

(d) $\tan^{-1}\left(\frac{a}{s}\right)$ [see Problem 3.12].

36.41 Verify the convolution theorem for Laplace transforms given in Table 36.5. Why does the convolution integral have different limits from (36.87)?

36.42 Use convolution (Table 36.5) to verify the integration rule for $\int_0^t f(\tau)d\tau$ in Table 36.4.

36.43 Fourier vs. Laplace

Factors of $\sqrt{2\pi}$ aside, discuss the relationship between the Fourier and Laplace transforms of the real function $f(t) = e^{at}$ for:

(a) $a < 0$

(b) $a > 0$

(c) $a = 0$ [see Problem 36.12].

36.44 Since only the first term in (36.128) depends on the driving force, we called it U_{st} and identified it as the steady-state solution. But this is not quite correct. We'll use Mathematica's `LaplaceTransform` and `InverseLaplaceTransform` commands to sort this out.

(a) For driving force $f(t) = f_0 \cos \omega t$, calculate the inverse Laplace transform of U_{st} in (36.128). Plot the result to reveal that it is not, strictly speaking, a steady-state solution.

(b) Use the `Apart` command to obtain a partial fraction decomposition of U_{st}. Calculate the inverse Laplace transform of each separately and plot them to discover which is the transient contribution. [Could you have determined this without calculating $\mathscr{L}^{-1}$?]

(c) Why does $U_{st}(s)$ include a transient that the Fourier solution in Example 36.13 does not?

36.45 Derive the expressions in (36.120) for the Laplace transform of df/dt and d^2f/dt^2. What's the general expression for $\mathscr{L}\left[\frac{d^n f}{dt^n}\right]$?

36.46 Use Laplace transforms to solve $\frac{d^2f}{dt^2} - 3\frac{df}{dt} + 2f = g(t)$, where:

(a) $g(t) = 0$, $f(0) = 1$, $\dot{f}(0) = 0$
(b) $g(t) = t$, $f(0) = 1$, $\dot{f}(0) = -1$
(c) $g(t) = e^{-t}$, $f(0) = 0$, $\dot{f}(0) = 1$.

36.47 Show that the Laplace transform of a periodic function $f(t) = f(t + T)$ is

$$F(s) = \frac{1}{1 - e^{-sT}} \int_0^T f(t)\, e^{-st}\, dt.$$

36.48 Verify the following limits:

(a) $\lim_{s \to 0} F(s) = 0$
(b) $\lim_{t \to 0} f(t) = \lim_{s \to \infty} sF(s)$
(c) $\lim_{t \to \infty} f(t) = \lim_{s \to 0} sF(s)$.

36.49 Adapt the argument leading to Jordan's lemma (21.58) to show that the infinite semicircles in Example 36.21 make no contribution to the inverse Laplace transform.

36.50 Verify (36.122). Then derive the solutions $q_{1,2}(t)$ for the following initial conditions:

(a) $q_1(0) = q_2(0) = \dot{q}_1(0) = 0$ and $\dot{q}_2(0) = v_0$
(b) $q_1(0) = a, q_2(0) = b$, and $\dot{q}_1(0) = \dot{q}_2(0) = 0$.

36.51 Use Laplace transforms to find the response of a quiescent oscillator to a hammer blow impulse at time $t = t_0$ — in other words, solve oscillator equation (36.126) with $f(t) = f_0\delta(t - t_0)$.

36.52 Use Laplace transforms to solve the forced undamped oscillator

$$\ddot{f} + \omega^2 f = t,$$

with initial conditions $f(0) = a, \dot{f}(0) = b$.

36.53 Derive (36.130) from (36.129), and verify that it's equivalent to (3.28).

37 Coda: Of Time Intervals and Frequency Bands

A continuous signal $f(t)$ is ultimately recorded digitally as a sequence of discrete values. Intuitively, the more measurements we take the better. An infinity of measurements being impractical, however, it's essential to explore how best to sample a signal in order to obtain a data set which faithfully represents $f(t)$ — and to do so without wasting storage space (Lyons, 2011; Proakis and Manolakis, 2006).

From our study of orthogonal functions we know that a continuous infinity of values $f(t)$ can be constructed with only a countable infinity of numbers c_n — and that finite partial sums can provide good approximations. But in all the examples we've seen, the expansion coefficients c_n are weighted averages of a known function over some interval, not direct samples of an unknown function. What we'd like is a basis on which the expansion coefficients are the discrete measurements we've made of $f(t)$, and with which a convergent expansion can be constructed.

Of course, taking measurements every τ seconds gives a sampling rate of $1/\tau$ Hz. So it's to be expected that the choice of sampling frequency should be influenced by the spectral content of $f(t)$. Hence Fourier analysis is relevant. But if we're going to focus on the sampling rate $1/\tau$ in hertz, it's convenient to replace ω with $2\pi \nu$ and redistribute the $\sqrt{2\pi}$'s to get the Fourier convention

$$\tilde{f}(\nu) = \int_{-\infty}^{\infty} f(t)\, e^{-2\pi i\nu t}\, dt \qquad f(t) = \int_{-\infty}^{\infty} \tilde{f}(\nu)\, e^{2\pi i\nu t}\, d\nu. \tag{37.1}$$

This is the form generally used in signal processing.[1] As you might imagine, it leads to a convolution theorem devoid of those pesky factors of $\sqrt{2\pi}$,

$$\mathcal{F}[fg] = \mathcal{F}[f] * \mathcal{F}[g] \qquad \mathcal{F}[f]\mathcal{F}[g] = \mathcal{F}[f * g]. \tag{37.2}$$

Indeed, given the dearth of 2π factors you might reasonably wonder why one doesn't always opt for this convention. The answer is that it's an issue of context: in physics the angular frequency ω (in s^{-1}) is the more relevant quantity, and so the reciprocal spaces t and ω are more appropriate. But in this Coda, we'll adopt the convention of (37.1). Although our focus will be on t and ν, the analysis applies to position $\vec{r}$ and wave vector $\frac{1}{2\pi}\vec{k}$ as well — and so has direct relevance to image processing: a digital image, after all, is a two-dimensional sampled array of pixels of a continuous function $f(x, y)$ designating a grayscale value.[2]

---

[1] In the notation of footnote 2 in Chapter 36, it's the convention $\{0, -2\pi\}$.
[2] Color images require a three-component red-green-blue (RGB) vector.

37.1 Sampling and Interpolation

Let's assume that the spectrum of $f(t)$ is *band-limited* — that is, $\tilde{f}(\nu)$ is non-zero only within some finite interval (*bandwidth*) $\Delta\nu = 2\mu$. For simplicity, we'll consider a centered spectrum between $\pm\mu$,

$$\tilde{f}(|\nu| > \mu) = 0, \tag{37.3}$$

so that μ is the maximum frequency in $f(t)$'s spectrum. Then the Fourier integral can be expressed over a finite interval,

$$f(t) = \int_{-\infty}^{\infty} \tilde{f}(\nu)\, e^{2\pi i \nu t}\, d\nu = \int_{-\mu}^{\mu} \tilde{f}(\nu)\, e^{2\pi i \nu t}\, d\nu. \tag{37.4}$$

Since the first goal is to determine an appropriate sampling rate, we'll let $t = n\tau$ for integer n and look for a constraint on τ:

$$f(n\tau) = \int_{-\mu}^{\mu} \tilde{f}(\nu)\, e^{2\pi i \nu n\tau}\, d\nu. \tag{37.5}$$

The collection of points $\{f(n\tau)\}$ represents our measured data set.

Since $\tilde{f}(|\nu| > \mu) \equiv 0$, we can use a Fourier series to expand $\tilde{f}$ as a function with period 2μ. In fact, the integral in (37.5) already has the basic form of a Fourier coefficient (33.18) with this periodicity,

$$\tilde{c}_n = \frac{1}{2\mu} \int_{-\mu}^{\mu} \tilde{f}(\nu)\, e^{+in\pi\nu/\mu}\, d\nu. \tag{37.6}$$

So choosing $\tau = 1/2\mu$ in (37.5) identifies the measured samples $f(n\tau)$ as Fourier coefficients.

What does this tell us? First, that a sampling rate of 2μ is sufficient to determine the Fourier coefficients of $\tilde{f}(\nu)$. But $\tilde{f}(\nu)$ is the spectral content of $f(t)$, and so $f(t)$ can be determined by inverse Fourier transform. Thus sampling at a frequency of at least 2μ, called the *Nyquist rate*, is sufficient to determine the original signal $f(t)$. This essential result of signal processing is known as the *sampling theorem*.

The factor of two in the Nyquist rate may seem surprising. But consider a finite superposition of harmonics $A\cos(2\pi\nu t + \varphi)$. For any one of the contributing modes, three independent measurements within a single period are sufficient to determine its parameters A, φ, and ν. And to have at least three equally spaced measurements within one period, the measurements must be separated by no more than half a period. Hence a sampling rate of at least twice the highest frequency μ in the superposition is required to resolve all frequencies $\nu \leq \mu$. Even if μ is not the highest frequency, it's the largest that can be resolved with a sampling rate of 2μ; higher frequencies must be sampled at a higher rate. The highest resolvable frequency is called the *Nyquist frequency*.[3] Frequencies below this are *over-sampled*, frequencies above are *under-sampled* (Figure 37.1).

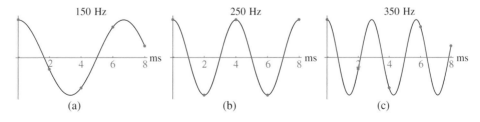

150 Hz   250 Hz   350 Hz

(a)   (b)   (c)

Figure 37.1 Sampling points at 500 Hz: (a) over-sampling; (b) critical sampling; and (c) under-sampling.

[3] It's not uncommon to conflate Nyquist *frequency* μ with Nyquist *rate* 2μ, so caution is advised.

Example 37.1 **Sampling a Triad**

The choice of sampling frequency has important implications for the recording industry. Consider the three-note musical cord C-E-G with frequencies 262 Hz, 327.5 Hz, and 393 Hz, respectively; this is called a C major triad, defined by the frequency ratios 4:5:6. Displayed in Figure 37.2 is the chord (solid), sampled at a rate of $v_s = 655$ Hz. Superimposed over the chord is each of the constituent notes (dotted), as labeled above each plot. Observing the phase of each note at the time of each sampled point shows that for the C, sampling occurs more than three times for every period; it's over-sampled. The E is being sampled at its Nyquist rate. Only the G is unresolved at this sampling frequency.

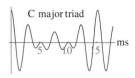

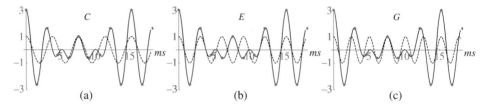

Figure 37.2 Sampling a C major triad at 655 Hz: (a) C, 262 Hz; (b) E, 327.5 Hz; (c) G, 393 Hz.

Indeed, sampling at any frequency below $2 \times 393 = 786$ Hz will distort the perceived sound of the chord. Obviously, there are musical notes and sounds of even higher frequency. Since the human ear is sensitive to a frequency range of about 20 Hz to 20 kHz, digital recordings must employ a sampling rate of at least 40 kHz.

Though the derivation of the sampling theorem reveals the relevance of the Nyquist rate 2μ, (37.6) identifies $f(n\tau)$ as coefficients in the Fourier series of $\tilde{f}(v)$, and not as expansion coefficients over a (still-to-be-identified) basis in t. We can try to discretize $f(t)$ with a train of periodic impulses — that is, by multiplication with a Dirac comb (Figure 37.3),

$$f_\tau(t) \equiv f(t)s_\tau(t) = \sum_{n=-\infty}^{\infty} f(t)\,\delta(t-n\tau) = \sum_{n=-\infty}^{\infty} f(n\tau)\,\delta(t-n\tau). \qquad (37.7)$$

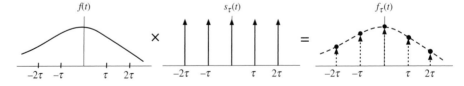

Figure 37.3 Sampling: the discretization of a continuous signal.

This however is not the continuous basis expansion we're after; rather, $f_\tau(t)$ is comprised of equally spaced discrete measurements of the original signal, each contribution an impulse weighted by $f(n\tau)$. Its Fourier transform $\tilde{f}_\tau$, however, *is* a continuous function — courtesy of the periodizing feature of the Dirac comb, (36.104):[4]

---

[4] Discretizing is essentially the converse of periodizing. In fact, the operations are Fourier transforms of one another — one multiplication by a comb, the other convolving with a comb.

$$\tilde{f}_\tau(\nu) \equiv \mathcal{F}\big[f(t)s_\tau(t)\big] = \big[\tilde{f} * \tilde{s}_\tau\big](\nu)$$

$$= \frac{1}{\tau} \int_{-\infty}^{\infty} \tilde{f}(\xi)\, s_{1/\tau}(\nu - \xi)\, d\xi$$

$$= \frac{1}{\tau} \sum_m \int_{-\infty}^{\infty} \tilde{f}(\xi)\, \delta(\nu - \xi - m/\tau)\, d\xi$$

$$= \frac{1}{\tau} \sum_m \tilde{f}(\nu - m/\tau) = \nu_s \sum_m \tilde{f}(\nu - m\nu_s), \tag{37.8}$$

where $\nu_s \equiv 1/\tau$ is the sampling frequency, and we've used (36.45)–(36.46).[5] So it appears that the discrete data set $f(n\tau)$ can be used construct $f_\tau(t)$, from which one calculates $\tilde{f}_\tau(\nu)$, whose inverse Fourier transform gives multiple copies of $\mathcal{F}^{-1}\big[\tilde{f}\big]$ — and hence ultimately reveals the original signal $f(t)$ in full. Though quite a mouthful, it nonetheless seems too easy; there must be a catch.

Indeed there is. The final expression in (37.8) is the sum of frequency-shifted copies of $\tilde{f}(\nu)$; only if these shifted copies do not overlap can $\tilde{f}(\nu)$ be extracted. This can be done when $f(t)$ has a band-limited spectrum, the condition we started with in (37.3): since the copies of $\tilde{f}(\nu)$ in $\tilde{f}_\tau(\nu)$ are separated by $\nu_s = 1/\tau$, a large sampling rate should yield separate and distinct copies. Which makes sense: a large ν_s means more measurements within any given time interval and therefore a higher-fidelity data set. The challenge is to choose τ small enough so that copies of $\tilde{f}$ don't overlap (obscuring the spectral content of f), but not so small as to render the sampling inefficient. In other words, we dare not under-sample (large τ), and we'd prefer not to over-sample (small τ).

The sampling theorem tells us that we need to separate the copies by at least 2μ. So if we set $\tau = 1/\nu_s = 1/2\mu$ in (37.8), we can isolate a single copy of $\tilde{f}$ by multiplying with a top hat function (36.31) of width 2μ,

$$h_{2\mu}(\nu)\tilde{f}_{1/2\mu}(\nu) = h_{2\mu} \cdot \big(\tilde{f} * \tilde{s}_{1/2\mu}\big) = \nu_s\, \tilde{f}(\nu). \tag{37.9}$$

In this expression $h_{2\mu}$ acts as a *low-pass filter*, allowing "through" only the selected frequency band $|\nu| < 2\mu$; all other copies of $\tilde{f}_{1/2\mu}$ are discarded. In this way, we've essentially inverted (37.5), obtaining $\tilde{f}(\nu)$ directly from the sampled values $f(n\tau)$. And from here, $f(t)$ can be found by inverse Fourier transform, using the convolution theorem twice:

$$2\mu f(t) = \mathcal{F}^{-1}\big[h_{2\mu} \cdot \big(\tilde{f} * \tilde{s}_{1/2\mu}\big)\big]$$

$$= \mathcal{F}^{-1}\big[h_{2\mu}\big] * \mathcal{F}^{-1}\big[\tilde{f} * \tilde{s}_{1/2\mu}\big]$$

$$= \mathcal{F}^{-1}\big[h_{2\mu}\big] * \big(f \cdot s_{1/2\mu}\big). \tag{37.10}$$

In consultation with Table 37.1, the remaining convolution can be expressed as a discrete convolution (Problem 37.2)

$$f(t) = \sum_{n=-\infty}^{\infty} f(n/2\mu)\, \mathrm{sinc}_\pi(2\mu t - n), \tag{37.11}$$

where $\mathrm{sinc}_\pi(x) \equiv \mathrm{sinc}(\pi x)$. So sampling at or above the Nyquist rate 2μ allows the recovery of the original signal. Of course, it's an infinite sum: even at the Nyquist rate, an infinite number of measurements $f(n/2\mu)$ are required for full reconstruction of $f(t)$. This should not come as a surprise. The significance of (37.11) instead lies in that it's an expansion of a band-limited $f(t)$ on the complete orthogonal basis $\{\mathrm{sinc}_\pi\}$ of $\mathbb{L}^2$ (Problem 37.4); this is the basis we set out to find. Thus the finite partial

---

[5] Do you see what happened to the 2π's in (36.46)?

Table 37.1 Common filters

Filter	$\tilde{f}(v) = \int f(t)\, e^{-2\pi i v t}\, dt$	$f(t) = \int \tilde{f}(v)\, e^{2\pi i v t}\, dv$
sinc	$\operatorname{sinc}_\pi(av) = \sin(\pi a v)/\pi a v$	$\frac{1}{a} h_a(t)$
top hat, rectangle	$h_a(v)^{\ddagger} = \begin{cases} 1, & \lvert v \rvert < \frac{a}{2} \\ 0, & \text{else} \end{cases}$	$a \operatorname{sinc}_\pi(at)$
triangle	$\Lambda_a(v) = \begin{cases} 1 - \lvert v/a \rvert, & \lvert v \rvert \le \lvert a \rvert \\ 0, & \text{else} \end{cases}$	$a \operatorname{sinc}_\pi^2(at)$
Dirac comb, sha	$\frac{1}{a} s_{1/a}(v) = \frac{1}{a}\sum_n \delta(v - n/a)$	$s_a(t)^{\ddagger\ddagger}$

‡Also: Π_a, rect_a ‡‡Also: Ш_a

sums $\sum_{n=-N}^{N} f(n/2\mu)\operatorname{sinc}_\pi(2\mu t - n)$ converge to $f(t)$ in the mean-square sense. This allows one to interpolate good estimates for the values of $f(t)$ between the discrete $2N+1$ samples; the larger N, the better these values. For this reason, (37.11) is called the *interpolation formula*.

Example 37.2 **On the Basis of Sincs**

The Nyquist rate for the C major triad (Example 37.1) is 786 Hz, twice that of its highest note G. Sampling at a lower rate not only returns a truncated spectrum, it gives insufficient spacing between the shifted $\tilde{f}$'s in the sum of (37.8). These, of course, are not unrelated. Figure 37.4 shows the results of the orthogonal expansion (37.11) with $N = 10$ for three different sampling rates. Only above the Nyquist rate $2\mu = 786$ Hz is the expansion applicable. Figure 37.5 (together with Figure 37.4c) exhibits the convergence of (37.11) with increasing N for $v_s = 800$ Hz.

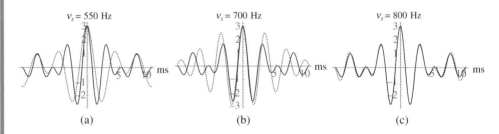

Figure 37.4 Interpolation, $N = 10$: (a), (b) below and (c) above Nyquist rate of the C major triad.

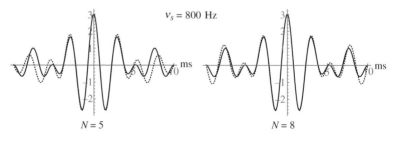

Figure 37.5 Convergence of the interpolation formula above the Nyquist rate for the C major triad.

37.2 Aliasing

Though of fundamental significance, the sampling theorem has one major weakness: it requires that $f(t)$ be band-limited. Not that there's any difficulty in principle with a band-limited signal — the C major triad, for instance (albeit with a discrete spectrum). But a musical chord is played for a finite amount of time. In fact, all physically realistic signals must have a beginning and an end. And as we've seen, a finite signal necessarily contains arbitrarily high frequency components. Thus one cannot have a signal that is time-limited (and hence physically realistic), while simultaneously band-limited and so subject to the sampling theorem.

A straightforward demonstration of the incompatibility of a band-limited signal $\tilde{f}(|\nu| > \mu) = 0$ also being time-limited starts with the identity

$$\tilde{f}(\nu) = h_{2\mu}(\nu)\,\tilde{f}(\nu). \tag{37.12}$$

Taking the inverse Fourier transform gives the identity's form in t,[6]

$$f(t) = \mathcal{F}^{-1}\big[\tilde{f}(\nu)\big] = \mathcal{F}^{-1}\big[h_{2\mu}\big] * \mathcal{F}^{-1}\big[\tilde{f}(\nu)\big] = 2\mu\,\mathrm{sinc}_\pi(2\mu t) * f(t). \tag{37.13}$$

Although its oscillations decay, the sinc function continues to have non-zero values all the way out to infinity (Figure 37.6) — and so its convolution with $f(t)$ is decidedly *not* time-limited. Given the symmetry of Fourier transforms, the converse is also true: a signal which is non-zero only over a finite interval cannot also be band-limited. This is not a trivial shortcoming. As Figure 37.4 illustrates, the interpolation formula (37.11) is only valid for sampling frequencies greater than or equal to 2μ.

Figure 37.6 $\mathrm{sinc}_\pi(ax)$.

What are the practical implications of this? Well, a band-limited signal is subject to the sampling theorem and so the sinc interpolation of (37.11) can be used to fully reconstruct the signal. Which sounds great — until you realize that since this signal cannot also be time-limited, any *finite* collection of measurements $f(n\tau)$ necessarily leads to an approximate reconstruction of $f(t)$. On the other hand, a time-limited signal cannot also be band-limited — and so the sampling theorem doesn't apply. Of course, interpolation of a band-limited signal could produce a fast-enough convergence to $f(t)$ over some finite time interval of interest. Similarly, if it happens that the spectrum is relatively small above some value μ, then the interpolation formula may still give acceptable results.

But there's another challenge, simultaneously subtle and intriguing. Since there are no realistic band-limited signals, no matter what sampling frequency one chooses there will always be frequencies above the Nyquist frequency that remain under-sampled; as we've seen, a sampling rate of ν_s can only resolve frequencies up to $\nu_s/2$. And yet frequencies above $\nu_s/2$ can still affect (some might say infect) the sampled data. Have a look back at Figure 37.1: at a sampling rate of 500 Hz, the undulations occurring between the measurements of a 350 Hz signal are completely missed. As a result, these sampled points will masquerade, or *alias*, as a 150 Hz signal (Figure 37.7a). This is one of the hazards of under-sampling. Aliasing is a straightforward consequence of Fourier decomposition on a periodic basis of sines and cosines. Specifically, sampling a sinusoidal signal $f(t)$ at a rate $\nu_s = 1/\tau$ yields measured points $f(n\tau)$ — which, as a familiar trig identity easily demonstrates, can be rendered

---

[6] If you're feeling a sense of déjà vu, see (37.9)–(37.11).

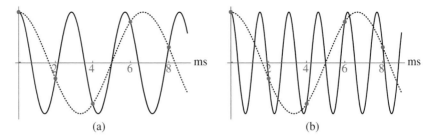

Figure 37.7 Sampling at $\nu_s = 500$ Hz: both under-sampled signals (a) $\nu_s - \nu_{\text{alias}} = 350$ Hz and (b) $\nu_s + \nu_{\text{alias}} = 650$ Hz (solid) alias as $\nu_{\text{alias}} = 150$ Hz (dotted).

$$f(n\tau) = A\cos(2\pi n\nu\tau + \varphi) = A\cos(2\pi n\nu/\nu_s + \varphi)$$

$$= A\cos\left[\frac{2\pi n}{\nu_s}(\nu - k\nu_s) + \varphi\right], \tag{37.14}$$

for any integer k. Though this may seem little more than an exercise in periodicity, it demonstrates that there is inevitably a signal of frequency $\nu > \nu_s/2$ in samples $f(n\tau)$ indistinguishable from those of a signal with $\nu_{\text{alias}} = |\nu - k\nu_s| \leq \nu_s/2$,

$$f(n\tau) = A\cos\left(\frac{2\pi n}{\nu_s}\nu + \varphi\right) = A\cos\left(\pm\frac{2\pi n}{\nu_s}\nu_{\text{alias}} + \varphi\right). \tag{37.15}$$

The first equality shows how data points can be fit to reveal signal frequency ν; the second warns that the fit need not be to the unique ν below the Nyquist frequency. The value of k implicit in (37.15) is that for which a frequency ν_{alias} is critically sampled or over-sampled at ν_s. In other words, ν_{alias} is the smallest-magnitude difference between the signal frequency ν and a multiple $k\nu_s$ of the sampling frequency. So at any given sampling rate, frequencies above $\nu_s/2$ are not merely under-sampled, they are mistaken for a contribution to $f(t)$ at the frequency $\nu_{\text{alias}} \leq \nu_s/2$. Figure 37.7 shows how sampling at 500 Hz aliases both 350 and 650 Hz signals, with $k = 1$, as 150 Hz. Notice that whereas the actual signals are under-sampled, the aliased frequency appears over-sampled. The most visually captivating example of this phenomenon is the "wagon wheel effect" seen in movies and television, in which the rotation of a spoked wheel appears slowed or even reversed — see Problems (37.7)–(37.8).

Despite the challenges imposed by time-limited signals and aliasing, the sampling theorem remains of enormous practical importance — though to wield it effectively requires the use of anti-alias filters. For instance, low-pass filters are used to pass all frequencies below some specified μ, essentially creating a band-limited signal. By choosing μ to be the desired (or expected) Nyquist frequency, sampling can proceed at the Nyquist rate 2μ. An ideal low-pass filter $h_{2\nu_c}(\nu)$, blocking all $|\nu|$ greater than a specified cut-off ν_c, acts by simple multiplication with $\tilde{f}(\nu)$ — or equivalently, as convolution of $f(t)$ with the sinc function (see Table 37.1). Unlike the top hat's strict cut-off, however, real low-pass filters pass all frequencies up to μ, but gradually attenuate frequencies above μ within an interval called its transition band; complete filtering only occurs above this. As a result, sampling at 2μ could result in aliasing of frequencies in the transition band. This can be avoided by using a sampling rate more than twice the highest transition-band frequency. For instance, even though the human ear is only sensitive up to 20 kHz, the typical sampling rate for audio recordings is 44.1 kHz rather than 40 kHz in order to accommodate a 2 kHz wide transition band.

Problems

37.1 Under the Fourier conventions of (37.1):

(a) Show that Plancherel's theorem (36.7) is unchanged.
(b) Show that the convolution theorem simplifies to (37.2).

37.2 Verify the derivation of the interpolation formula, (37.11).

37.3 **A Sinc-ing Feeling**

(a) Use Fourier transforms to show

$$\lim_{n\to\infty} n\,\mathrm{sinc}_\pi\,(nt) = \delta(t).$$

(b) Use the periodizing property of the Dirac comb (36.104) to show that

$$\sum_{n=-\infty}^{\infty} \mathrm{sinc}_\pi\,(t-n) = 1.$$

(c) Use the interpolation formula (37.11) to show that

$$\sum_{n=-\infty}^{\infty} \mathrm{sinc}_\pi\,(t-n)\,\mathrm{sinc}_\pi\,(n-a) = \mathrm{sinc}_\pi\,(t-a).$$

(d) Evaluate $\sum_{n=-\infty}^{\infty} \mathrm{sinc}_\pi^2\,(t-n)$.
(e) Evaluate $\int_{-\infty}^{\infty} \mathrm{sinc}_\pi\,(\tau-t)\,\mathrm{sinc}_\pi\,(t)\,dt$.

37.4 Solve the following using Plancherel's theorem (36.7) and the convolution theorem (37.2), consulting Tables 36.2 and 37.1 as needed. [Remember that we're using the conventions of (37.1).]

(a) Show that $\phi_n(t) = \sqrt{2\mu}\,\mathrm{sinc}_\pi\,(2\mu t - n)$ are orthonormal functions.
(b) Show that the expansion coefficients $f(n/2\mu)$ in the interpolation formula (37.11) are the projections of $f(t)$ onto the ϕ_n basis.

37.5 Use the interpolation formula (37.11) to show that an aliased signal $g(t)$ can be constructed from the sampled points $f(n/2\mu)$ of the actual signal $f(t)$.

37.6 What frequency is aliased in a 150 Hz signal sampled at 40 Hz?

37.7 The second hand on a clock goes around once a minute. Describe the apparent motion of the hand when a strobe light flashes every.

(a) 75 seconds   (b) 60 seconds   (c) 45 seconds
(d) 30 seconds   (e) 15 seconds.

37.8 Consider a typical four-blade fan, whirring with frequency ν, being filmed at the standard rate of 24 frames/s (fps).

(a) How high does ν need to be to avoid aliasing?
(b) For what range of ν will the film depict the fan rotating backward?

37.9 An industrial-scale six-blade rotor sitting at floor level turns with frequency ν_{rot}. Given that the factory's overhead fluorescent lights flicker at 120 Hz (twice the 60 Hz AC frequency), what value of ν_{rot} poses a real danger to workers? Explain.

Part VI

Differential Equations

Don't only practice your art, but force your way into its secrets.

Ludwig von Beethoven

38 Prelude: First Order First

The breadth of phenomena described by differential equations is impressive. Consider the following (very incomplete) list:

$$\text{Newton:} \quad F = m\frac{d^2x}{dt^2}$$

$$\text{Poisson:} \quad \nabla^2\Phi = \rho$$

$$\text{Schrödinger:} \quad \left(-\frac{\hbar^2}{2m}\nabla^2 + V\right)\psi = E\psi$$

$$\text{Oscillation:} \quad \frac{d^2f}{dt^2} = -\omega^2 f$$

$$\text{Waves:} \quad \nabla^2 f = \frac{1}{v^2}\frac{\partial^2 f}{\partial t^2}$$

$$\text{Diffusion:} \quad \nabla^2 f = \frac{1}{\kappa}\frac{\partial f}{\partial t}$$

$$\text{Radioactive decay:} \quad \frac{dN}{dt} = -kN$$

Whether classical or quantum mechanics, waves or fields, the physical world seems to be ruled by differential equations. So knowing how to work with them is of fundamental importance.

We begin with some basic classification vocabulary. An *ordinary differential equation* (ODE) has only total derivatives in one independent variable. In the examples listed above, Newton's second law, oscillation, and radioactive decay are all ODEs. The rest have partial derivatives in several independent variables, and so are called *partial differential equations* (PDEs).

An *n*th-order *linear* differential equation has the form $\mathcal{L}_q u(q) = r(q)$, where the differential operator

$$\mathcal{L}_q = a_n(q)\frac{d^n}{dq^n} + a_{n-1}(q)\frac{d^{n-1}}{dq^n} + \cdots + a_1(q)\frac{d}{dq} + a_0(q) \tag{38.1}$$

is independent of $u(q)$. An *inhomogeneous* equation has $r(q) \neq 0$. Poisson's equation is inhomogeneous in $\vec{r}$; Newton's second law is inhomogeneous when the applied force depends on t. The others in our list have $r = 0$ and so are *homogeneous*. A crucially important property of homogeneous linear equations is that solutions can be superposed to form other solutions:

$$\mathcal{L}(c_1 u_1 + c_2 u_2) = c_1\,\mathcal{L}(u_1) + c_2\,\mathcal{L}(u_2). \tag{38.2}$$

In addition to linearity, our foray into differential equations will rely on a confluence of concepts and techniques discussed earlier, including power series, Hermitian differential operators, projections, special functions, diagonalization, impulse solutions, orthogonal bases of eigenfunctions, and normal mode expansions.

As implied by our list of differential equations, most of physics seems to be concerned with second-order equations. For completeness, however, we begin with a review of techniques to solve first-order differential equations. Let's start with the simple case of a body in free fall,

$$m\frac{dv}{dt} = mg, \tag{38.3}$$

where we've taken the downward direction to be positive. This couldn't be simpler: if the body starts with velocity v_0 at $t = 0$, then

$$v(t) = v_0 + gt. \tag{38.4}$$

What if we include the effects of air resistance, or drag? For low speeds, the drag is proportional (and opposite) to the velocity, $F_d \equiv -bv$, where the constant $b > 0$ summarizes details of the drag force (e.g., density of air and cross-sectional area of the body). Including this in (38.3) gives

$$\frac{dv}{dt} = -\frac{b}{m}v + g. \tag{38.5}$$

Moving v and t to opposites sides of the equation,

$$\frac{dv}{bv/m - g} = -dt, \tag{38.6}$$

we can integrate to find

$$\ln(bv/m - g) = -bt/m + \text{constant} \quad \Longrightarrow \quad v(t) = ke^{-bt/m} + mg/b, \tag{38.7}$$

where $k = v_0 - mg/b$. This result reveals the phenomenon of *terminal velocity* — that is, the speed at which the drag force exactly cancels the pull of gravity so that the body ceases to accelerate:

$$v_{\text{term}} \equiv v(t \to \infty) = mg/b. \tag{38.8}$$

What makes these differential equations so tractable — (38.3) trivially, (38.5) a bit less so — is that dv and dt can be isolated on either side of the equation, allowing them to be integrated separately. In general, a first-order differential equation is said to be *separable* if it can be written in the form

$$\alpha(x)\,dx + \beta(y)\,dy = 0. \tag{38.9}$$

If so, then

$$\int \alpha(x)\,dx = -\int \beta(y)\,dy + C, \tag{38.10}$$

where the integration constant C is determined from a specified initial or boundary condition. Remarkably, a general linear first-order equation

$$\frac{dy}{dx} + q(x)y = r(x) \tag{38.11}$$

can be rendered separable by introducing an *integrating factor*. This is a function $w(x)$, which when multiplied onto (38.11) allows the left-hand side to become

$$\frac{d}{dx}(wy) = wy' + w'y.\tag{38.12}$$

For this to be in concert with (38.11), the integrating factor must satisfy $w' = qw$ — that is,

$$w(x) = \exp\left[\int q(x)\,dx\right].\tag{38.13}$$

Having thus massaged the differential equation into the form

$$\frac{d}{dx}(wy) = wr,\tag{38.14}$$

we easily identify the solution,

$$y(x) = \frac{1}{w(x)}\int w(x)r(x)\,dx.\tag{38.15}$$

Note that, as in (38.10), two integrations — one each to find $w(x)$ and $y(x)$ — are still required.

For the example of (38.5) we have $q(t) = b/m$, yielding the integrating factor

$$w(t) = e^{bt/m}.\tag{38.16}$$

Then (38.15) leads to the same solution as before,

$$v(t) = e^{-bt/m}\int g\,e^{bt/m}dt = e^{-bt/m}\left[\frac{mg}{b}e^{bt/m} + k\right] = \frac{mg}{b} + ke^{-bt/m},\tag{38.17}$$

where k is the constant of integration.

Example 38.1 The Falling Raindrop[1]

Let our falling object be a raindrop, starting from rest with initial mass m_0. As the raindrop passes through a stationary cloud of mist, inelastic collisions with mist droplets will cause the drop to increase its mass as it falls. Now with variable mass, rather than $F = ma$ we must write $F = dp/dt$, with

$$\frac{dp}{dt} = m\dot{v} + \dot{m}v.\tag{38.18}$$

Neglecting drag, the differential equation to solve is

$$m\dot{v} + \dot{m}v = mg.\tag{38.19}$$

Clearly the larger the drop's surface area, the greater its interaction with the mist. So if we assume that its mass increases at a rate proportional to its surface area, $dm/dt \sim 4\pi r^2$, then presuming a drop's density $\rho \sim m/r^3$ is constant gives

$$dm/dt = km^{2/3},\tag{38.20}$$

---

[1] Sokal (2010).

for constant k. We can use this to replace t with m as the independent variable via

$$\frac{dv}{dt} = \frac{dm}{dt}\frac{dv}{dm} \equiv km^{2/3}v',$$

(38.21)

where $v' \equiv dv/dm$. Then the differential equation for $v = v(m)$ is

$$v' + \frac{v}{m} = \frac{g}{km^{2/3}},$$

(38.22)

with initial condition $v(m_0) = 0$. Although not separable, the integrating factor[2]

$$w(m) = \exp\left[\int \frac{dm}{m}\right] = m$$

(38.23)

allows us to easily obtain

$$v(m) = \frac{1}{m}\left(\frac{g}{k}\int m^{1/3}\,dm\right)$$

$$= \frac{3g}{4k}m^{1/3} + \frac{C}{m} = \frac{3g}{4k}m^{1/3}\left[1 - \left(\frac{m_0}{m}\right)^{4/3}\right],$$

(38.24)

where the integration constant was chosen to satisfy $v(m_0) = 0$. Putting this into (38.19) gives the raindrop's acceleration:

$$\dot{v} = g - \frac{\dot{m}}{m}v = \frac{g}{4}\left[1 + 3\left(\frac{m_0}{m}\right)^{4/3}\right].$$

(38.25)

Thus the downward acceleration decreases asymptotically to $g/4$ as m grows; there is no terminal velocity.

The separable form in (38.9) is a special case of the general first-order equation

$$\alpha(x,y)\,dx + \beta(x,y)\,dy = 0.$$

(38.26)

Its resemblance to

$$d\Phi(x,y) = \frac{\partial\Phi}{\partial x}\,dx + \frac{\partial\Phi}{\partial y}\,dy$$

(38.27)

invites us to look for a scalar function Φ whose partials are α and β,

$$\frac{\partial\Phi}{\partial x} = \alpha(x,y) \quad \text{and} \quad \frac{\partial\Phi}{\partial y} = \beta(x,y).$$

(38.28)

This doesn't guarantee that Φ exists; but if it does, then the equality of mixed partials $\partial_y\partial_x\Phi = \partial_x\partial_y\Phi$ requires

$$\frac{\partial\alpha}{\partial y} = \frac{\partial\beta}{\partial x}.$$

(38.29)

---

[2] Had we expressed the physics in terms of momentum rather than speed, the left-hand side becomes $mv' + v = d(mv)/dm$ and there would have been no need for an integrating factor.

A differential equation which can be rendered as $d\Phi = 0$ is said to be *exact*. And we've seen it before: an exact differential is path independent, $\oint_C d\Phi \equiv 0$, the condition (38.29) consistent with Green's theorem in the plane, (15.28). So if (38.29) is satisfied, we can use (38.28) to find Φ. And since the differential equation is then equivalent to $d\Phi = 0$, its solution is simply $\Phi(x, y) = \text{constant}$.

Example 38.2 **The Falling Raindrop, Revisited**

Rearranging (38.22) into the form

$$m\,dv + \left(v - \frac{g}{k}m^{1/3}\right)dm = 0,\tag{38.30}$$

comparison with (38.26) identifies

$$\alpha(v, m) = m \qquad \beta(v, m) = v - \frac{g}{k}m^{1/3}.\tag{38.31}$$

Since $\partial_m\alpha = \partial_v\beta = 1$, this is an exact differential. So we can use (38.28) to find Φ:

$$\partial_v\Phi = \alpha(v, m) \quad\Longrightarrow\quad \Phi(v, m) = mv + c_1(m)$$

$$\partial_m\Phi = \beta(v, m) \quad\Longrightarrow\quad \Phi(v, m) = mv - \frac{3g}{4k}m^{4/3} + c_2(v).\tag{38.32}$$

These expressions are consistent for $c_1 = -\frac{3g}{4k}m^{4/3}$, $c_2 = 0$. Thus we find that the quantity

$$\Phi(v, m) = mv - \frac{3g}{4k}m^{4/3} = C\tag{38.33}$$

remains constant as the growing raindrop falls. The condition $v(m_0) = 0$ gives $C = -\frac{3g}{4k}m_0^{4/3}$, leading directly to the rediscovery of (38.24).

For an *in*exact differential equation, we may be able to find integrating factor w such that

$$\frac{\partial}{\partial y}(w\alpha) = \frac{\partial}{\partial x}(w\beta)\tag{38.34}$$

to make it exact. The simplest situation is when w is a function of either x or y alone; so too, then, is its derivative. For $w = w(x)$, (38.34) gives a linear first-order differential equation,

$$\frac{dw}{dx} = \frac{w}{\beta}\left[\frac{\partial\alpha}{\partial y} - \frac{\partial\beta}{\partial x}\right] \equiv p(x)w(x),\tag{38.35}$$

for non-vanishing $\beta(x, y)$. And we already know how to solve a separable equation like this using the integrating factor

$$w(x) = \exp\left[\int p(x)\,dx\right].\tag{38.36}$$

So the condition that an integrating factor $w(x)$ exists is simply that

$$p(x) \equiv \frac{1}{\beta}\left[\frac{\partial\alpha}{\partial y} - \frac{\partial\beta}{\partial x}\right]\tag{38.37}$$

must be a function of x. Similarly, an integrating factor $w = w(y)$ exists if

$$p(y) \equiv \frac{1}{\alpha} \left[\frac{\partial \beta}{\partial x} - \frac{\partial \alpha}{\partial y} \right] \tag{38.38}$$

is a function of y.

Example 38.3 The Falling Raindrop, Yet Again

Modeling mass accretion as a function only of the falling drop's surface area A is incomplete. Surely the rate of growth depends on the volume Av swept out per unit time within the cloud of mist. So to improve our physical model, we'll replace (38.20) with

$$\frac{dm}{dt} = km^{2/3}v. \tag{38.39}$$

The steps leading to (38.22) now give

$$v \, dv + \left(\frac{v^2}{m} - \frac{g}{km^{2/3}} \right) dm = 0. \tag{38.40}$$

This is neither separable nor linear. Nor is it exact: reading α and β off (38.40), we easily see that $\partial_m \alpha \neq \partial_v \beta$. However,

$$p(m) = \frac{1}{\alpha} \left[\frac{\partial \beta}{\partial v} - \frac{\partial \alpha}{\partial m} \right] = \frac{1}{v} \left[\frac{2v}{m} - 0 \right] = \frac{2}{m} \tag{38.41}$$

is only a function of m. Thus we can use the integrating factor

$$w(m) = \exp \left[\int p \, dm \right] = m^2 \tag{38.42}$$

to get an exact differential. From (38.28):

$$\partial_v \Phi = m^2 \alpha \quad \Longrightarrow \quad \Phi(v, m) = \frac{1}{2} m^2 v^2 + c_1(m)$$

$$\partial_m \Phi = m^2 \beta \quad \Longrightarrow \quad \Phi(v, m) = \frac{1}{2} m^2 v^2 - \frac{3g}{7k} m^{7/3} + c_2(v). \tag{38.43}$$

Consistency requires $c_1(m) = -\frac{3g}{7k} m^{7/3}$, $c_2(v) = 0$. So the solution to (38.40) is

$$\Phi(v, m) = \frac{1}{2} m^2 v^2 - \frac{3g}{7k} m^{7/3} = C. \tag{38.44}$$

Imposing the initial condition $v(m_0) = 0$, we get

$$\frac{1}{2} m^2 v^2 = \frac{3g}{7k} m^{7/3} \left[1 - \left(\frac{m_0}{m} \right)^{7/3} \right]. \tag{38.45}$$

From this, one can show (Problem 38.9)

$$\dot{v} = \frac{g}{7} \left[1 + 6 \left(\frac{m_0}{m} \right)^{7/3} \right], \tag{38.46}$$

which differs from the simpler model's result, (38.25).

The falling raindrop is an old and venerable example of the motion of an object with varying mass. But arguably the most well-known example is that of a rocket, whose mass decreases as consumed fuel is essentially ejected out the rear. By Newton's third law, the resulting reaction force on the rocket propels it upward.

Consider a rocket, ascending straight up through the atmosphere. If mass dm_e is ejected in time dt with velocity $-u$ relative to the rocket, then its momentum change is $-u\,dm_e$, and the force on it is $-u\,dm_e/dt$. By Newton's third law, the reaction force on the rocket is $+u\,dm_e/dt = -u\,dm/dt$, where $dm < 0$ is the rocket's mass loss. Then Newton's second law gives a first-order differential equation for the rocket's acceleration,

$$m\frac{dv}{dt} = -u\frac{dm}{dt} + F_{\text{ext}}, \tag{38.47}$$

where F_{ext} is the net external force on the rocket. The reaction force $-u\,dm/dt > 0$, which contributes as if it were an external force, is called the rocket's *thrust*.

The simplest (if unrealistic) example of (38.47) is $F_{\text{ext}} = 0$ (but see Problem 38.7). In this case, with constant u the differential equation is easily separated and solved to get the *rocket equation*,

$$v(m) = v_0 + u\ln\left(m_0/m\right), \tag{38.48}$$

where $m_0 = m_{\text{r}} + m_{\text{f}}$ is the initial combined mass of the rocket and its fuel. Note that an increase in speed depends on the ratio of initial to final mass, independent of the time it takes to burn the fuel. Moreover, the logarithm puts a severe constraint on the rocket's acceleration: even if $m_{\text{f}} = 9.9m_{\text{r}}$ (and assuming all the fuel is consumed), the rocket can only increase its speed by $u\ln(100) = 4.6u$.

Problems

38.1 Bernoulli's Equation

(a) Verify that Bernoulli's equation,

$$u' + f(x)u = g(x)u^n,$$

is non-linear for $n \neq 0, 1$.

(b) Dividing both sides by u^n, show that the variable change $v = u^{1-n}$ reduces this to a linear equation in v.

38.2 According to Newton's law of cooling, an object's temperature $T(t)$ changes at a rate proportional to the difference between T and the temperature T_{env} of the surrounding environment. The proportionality constant $b > 0$ depends on such things as an object's thermal conductivity and characteristic size. Write down a differential equation for this law. Solve the equation for constant T_{env}, and verify that $T(t)$ describes both cooling and warming of the object.

38.3 A radium-226 nucleus α-decays to radon-222 with a rate $\lambda_1 \approx 1/2300 \text{ yr}^{-1}$; radon-222 α-decays to polonium 218 at a rate of $\lambda_2 \approx 1/5.5 \text{ d}^{-1}$. [These numbers are provided for context; they are not actually needed.]

(a) Show that the change dN_2 in the number of ^{222}Rn nuclei is given by

$$dN_2 = \lambda_1 N_1 dt - \lambda_2 N_2 dt,$$

where N_1 is the number of ^{226}Ra nuclei.

(b) Find an expression for the number of radon nuclei $N_2(t)$, assuming there are none present initially. [Hint: What's $N_1(t)$?]

(c) Noting that $\lambda_2 \gg \lambda_1$, find the equilibrium balance of radium and radon nuclei in the sample.

38.4   The first-order equation

$$2x^2 u \frac{du}{dx} + xu^2 + 2 = 0$$

is an example of an "isobaric" equation for which there's a value of m (called a weight) such that u has the same dimensions as x^m. This allows the equation to be rendered separable with the substitution of the dimensionless $v(x) = u(x)/x^m$ in place of u.

(a) Confirm that the equation is isobaric by substituting $v(x)$ and determining the value of m for which the equation is separable.

(b) With this value of m, solve for $u(x)$ with $u(1) = k$. Verify the solution by substituting it back into the differential equation.

38.5   When we included the effects of air resistance on an object in free fall, we took the drag force to be proportional to velocity. But at high speed or for a sufficiently large object, atmospheric drag is quadratic in v. Find $v(t)$ for a falling object subject to quadratic drag. What is the terminal velocity?

38.6   Examples 38.1 and 38.3 examined the effects of increasing mass on a falling raindrop, neglecting atmospheric drag. Since drag should depend on the cross-sectional area of the raindrop, it will vary, like mass accretion, as $r^2 \sim m^{2/3}$. Parametrizing the drag force as $F_d = -bm^{2/3} v^\alpha$, find the acceleration of a raindrop falling from rest for:

(a) linear drag ($\alpha = 1$) with speed-independent mass accretion, (38.20)
(b) quadratic drag ($\alpha = 2$) with linear speed accretion, (38.39).

For simplicity, take $m(t = 0) = 0$. (You may find it helpful at some point to make the variable change $u = v^\alpha$.)

38.7   Consider a rocket which ejects mass at a constant rate, $m = m_0 - kt$. If the rocket takes off vertically from rest, find its speed as a function of time if the only external force is (a) gravity mg; (b) linear drag bv. Verify that both cases reduce to the rocket equation in the absence of the external force.

38.8   Verify (38.38).

38.9   Verify (38.46).

39 Second-Order ODEs

A second-order linear ordinary differential equation has the conventional form

$$\mathcal{L}_u(x) = \left[\frac{d^2}{dx^2} + p(x)\frac{d}{dx} + q(x) \right] u(x) = r(x), \tag{39.1}$$

in which the coefficient of the second derivative term is 1. Establishing this convention permits general results to be unambiguously expressed in terms of p, q and r; we'll discover several such formulas as we explore solution methods. But even after obtaining two linearly independent homogeneous solutions $u_1(x)$ and $u_2(x)$, specified boundary or initial conditions must still be satisfied. For instance, if $u_1(0) = 1, u_1'(0) = 0$ and $u_2(0) = 0, u_2'(0) = 1$, then

$$u(x) = Au_1(x) + Bu_2(x) \tag{39.2}$$

is a homogeneous solution satisfying initial conditions $u(0) = A$, $u'(0) = B$. As we'll see, a similar construction holds for solutions to the inhomogeneous equation.

39.1 Constant Coefficients

Perhaps the simplest form of a linear second-order ODE is when the coefficients p and q in (39.1) are constants. But even that relative simplicity describes a wealth of physics. We'll define $\beta \equiv p/2$ and $\omega_0^2 \equiv q$ so that the homogeneous equation takes the form

$$\frac{d^2u}{dt^2} + 2\beta\frac{du}{dt} + \omega_0^2 u = 0, \tag{39.3}$$

where for what may be obvious physical (though mathematically irrelevant) reasons, we've denoted the independent variable as t. And indeed, (39.3) is just Newton's second law for a harmonic oscillator of natural frequency ω_0 and restoring force $-m\omega_0^2 u$, subject to a damping proportional to velocity, $F_d = -bv = -2m\beta v$. Note that $\beta = b/2m$ has units of frequency; the factor of 2 is for future convenience.[1]

We'll solve (39.3) by leveraging some of the results for first-order equations. With $D \equiv d/dt$ and taking advantage of having constant coefficients, (39.3) can be written

$$\left(D^2 + 2\beta D + \omega_0^2\right)u(t) = (D - \alpha_1)(D - \alpha_2)\,u(t) = 0, \tag{39.4}$$

---

[1] Chapter 22 considered this system in the complex plane; here we explore different techniques.

where the two roots are

$$\alpha_{1,2} = -\beta \pm \sqrt{\beta^2 - \omega_0^2}. \tag{39.5}$$

This factoring of a second-order operator into the product of two first-order operators results in two first-order equations,

$$(D - \alpha_2)\, u(t) = v(t), \tag{39.6a}$$

which introduces the function $v(t)$ — which, in order to satisfy (39.4), must be the solution of

$$(D - \alpha_1)\, v(t) = 0. \tag{39.6b}$$

This first-order homogeneous equation has the simple solution

$$v(t) = Ae^{\alpha_1 t}, \tag{39.7}$$

so we only need to solve

$$(D - \alpha_2)\, u(t) = Ae^{\alpha_1 t}. \tag{39.8}$$

With the integrating factor $w(t) = e^{-\alpha_2 t}$ from (38.13), the solution (38.15) is

$$u(t) = e^{\alpha_2 t} \int e^{-\alpha_2 t} v(t)\, dt = e^{\alpha_2 t} \int Ae^{(\alpha_1 - \alpha_2)t}\, dt = \frac{Ae^{\alpha_1 t} + Be^{\alpha_2 t}}{\alpha_1 - \alpha_2}. \tag{39.9}$$

Now, if $\alpha_2 \neq \alpha_1$ — in other words, for $\beta \neq \omega_0$ — the denominator can be absorbed into the integration constants A and B to give two linearly independent solutions to our second-order equation,

$$u_{1,2}(t) = \exp\left[-\left(\beta \pm \sqrt{\beta^2 - \omega_0^2}\right) t\right]. \tag{39.10}$$

Now perhaps these don't look like oscillation — but look more closely. Physically, the motion of the oscillator should depend on the relative strengths of the damping and restoring forces. What our result is telling us is that this physical comparison is expressed mathematically by the relative sizes of β and ω_0. If in fact $\beta > \omega_0$, then

$$u(t) = e^{-\beta t}\left[Ae^{-\sqrt{\beta^2 - \omega_0^2}\, t} + Be^{+\sqrt{\beta^2 - \omega_0^2}\, t}\right], \quad \beta > \omega_0. \tag{39.11a}$$

In this case, the damping is so large as to prevent any oscillation from occurring. On the other hand, if $\omega_0 > \beta$, then (39.10) becomes

$$u(t) = e^{-\beta t}\left[Ae^{-i\sqrt{\omega_0^2 - \beta^2}\, t} + Be^{+i\sqrt{\omega_0^2 - \beta^2}\, t}\right]$$
$$= e^{-\beta t}\left[c_1 \cos\left(\sqrt{\omega_0^2 - \beta^2}\, t\right) + c_2 \sin\left(\sqrt{\omega_0^2 - \beta^2}\, t\right)\right], \quad \omega_0 > \beta. \tag{39.11b}$$

(Since $u(t)$ must be real, the appearance of complex exponentials requires complex $A = B^*$.) This is the motion we expected: oscillation with decreasing amplitude as the damping steadily removes energy from the system. And if $\beta = \omega_0$? In this instance, re-evaluating the integral in (39.9) with $\alpha_1 = \alpha_2 = -\beta$ gives

$$u(t) = (At + B)\, e^{-\beta t}, \quad \beta = \omega_0. \tag{39.11c}$$

Now the damping is just strong enough to prevent oscillation, but not so strong as to further retard motion. So $\beta = \omega_0$ results in the smallest relaxation time for the oscillator to return to equilibrium — as is easily confirmed by direct comparison of the exponential rates of decay in all three cases.

Despite their vastly different behavior, all three possibilities described by (39.11) emerge from the same law of motion, from the same differential equation. Of course, we've seen them before: these are respectively overdamped, underdamped, and critically damped oscillation of Example 3.4. Moreover, the ubiquity of oscillatory phenomena implies that second-order differential equations with constant coefficients should be common — although as (39.11a) shows, oscillation does not necessarily emerge. Nonetheless, with a little physical insight, we could have arrived at the solution faster merely by making the ansatz $u = e^{\alpha t}$. Plugging this into the differential equation yields a quadratic equation for α with roots $\alpha_{1,2}$ — and thence the general result

$$u(t) = Ae^{\alpha_1 t} + Be^{\alpha_2 t}, \qquad \alpha_1 \neq \alpha_2. \tag{39.12}$$

Complex α yields oscillatory solutions. But if $\alpha_1 = \alpha_2$, the single root must be real (BTW 8.4); as shown in (39.11c), the second independent solution is $te^{\alpha t}$.

It's worth noting a closely related ansatz for a differential equation of the form

$$c_2 x^2 \frac{d^2 u}{dx^2} + c_1 x \frac{du}{dx} + c_0 u = 0, \tag{39.13}$$

where the c_i are constant. If it's not immediately clear, take a moment to convince yourself that $u = x^\alpha$ is a wise ansatz. Two other standard methods listed in Table 39.1 can be used to reduce a second-order equation to first order; these are explored in Problems 39.10–39.11.

Table 39.1 Ansätze for $\frac{d^2 u}{dx^2} + p\frac{du}{dx} + qu = 0$

p, q constant	$u = e^{\alpha x}$	$u(x) = \begin{cases} c_1 e^{\alpha_1 x} + c_2 e^{\alpha_2 x}, & \alpha_1 \neq \alpha_2 \\ (c_1 + c_2 x)e^{\alpha x_1}, & \alpha_1 = \alpha_2 \end{cases}$
$p(x) \sim x$ q constant	$u = x^\alpha$	$u(x) = \begin{cases} c_1 x^{\alpha_1} + c_2 x^{\alpha_2}, & \alpha_1 \neq \alpha_2 \\ (c_1 + c_2 \ln x)x^{\alpha_1}, & \alpha_1 = \alpha_2 \end{cases}$
$p = p(x), q = 0$	$v = \frac{du}{dx}$	$v(x) = \exp\left[-\int p(x)\,dx\right]$
$p = p(u, u')$ $q = q(u, u')$	$v = \frac{du}{dx}, \frac{d^2 u}{dx^2} = v\frac{dv}{du}$	$v\frac{dv}{du} + pv = -qu$

39.2 The Wronskian

Superposing homogeneous solutions u_1 and u_2 to construct other homogeneous solution requires that $u_{1,2}$ be linearly independent. That's certainly true in each of the three types of damped oscillation. But, in general, how does one establish that functions are linearly independent? Finite column vectors, recall, can be assembled into a matrix; if its determinant is non-zero, then the columns form a linearly independent set. It turns out that a similar procedure works for functions as well.

For linearly independent homogeneous solutions u_1 and u_2, any other homogeneous solution can be expressed as a superposition,

$$u(x) = c_1 u_1 + c_2 u_2. \tag{39.14a}$$

Since $c_{1,2}$ are constants, we also have

$$u'(x) = c_1 u_1' + c_2 u_2'. \tag{39.14b}$$

Since the defining differential equation relates u'' to u and u', a second derivative gives nothing new. So we combine (39.14) into the matrix expression

$$\begin{pmatrix} u \\ u' \end{pmatrix} = \begin{pmatrix} u_1 & u_2 \\ u_1' & u_2' \end{pmatrix} \begin{pmatrix} c_1 \\ c_2 \end{pmatrix}. \tag{39.15}$$

In this form, it's clear that a solution — that is, a unique pair of coefficients $c_{1,2}$ — exists only if the matrix is invertible. In other words, if the *Wronskian*

$$W(x) \equiv \begin{vmatrix} u_1 & u_2 \\ u_1' & u_2' \end{vmatrix} \tag{39.16}$$

is non-zero, then u_1 and u_2 are linearly independent. Remarkably, whether or not $W(x)$ vanishes can be determined by testing only one point. The reason is that (as you'll show in Problem 39.3)

$$\frac{d}{dx} W(x) = \frac{d}{dx} \left(u_1 u_2' - u_1' u_2 \right) = -p(x) W(x), \tag{39.17}$$

where $p(x)$ is the coefficient function in (39.1). Thus

$$W(x) = W(a) \exp \left[-\int_a^x p(x') \, dx' \right]. \tag{39.18}$$

So if the Wronskian is non-zero at a point a, it's non-zero everywhere. Incidentally, notice that $W(x)$ is only defined up to an overall constant: in (39.18) by the choice of a, for (39.16) in that homogeneous solutions $u_{1,2}$ are only defined up to overall constants in the first place.

Example 39.1 The Wronskian Test of Independence

The familiar solutions for simple harmonic motion $d^2u/dx^2 + k^2 u = 0$ are certainly linearly independent:

$$W(x) \equiv \begin{vmatrix} \cos kx & \sin kx \\ -k \sin kx & k \cos kx \end{vmatrix} = k. \tag{39.19}$$

This is consistent with (39.18), which trivially gives $W(x) = $ constant. So $W(x) \neq 0$ for all x. Well, at least for $k \neq 0$ — which makes sense, given that $\sin kx \to 0$ in that case. Since actual oscillation necessarily has non-vanishing frequency, this is usually not an issue. We can, however, avoid the problem altogether by taking the solution to be $\sin(kx)/k$ instead: not only does this give $W = 1$, but the $k \to 0$ limit yields the correct solution of $d^2u/dx^2 = 0$, the line $u = ax + b$.

A similar thing occurs for the damped oscillation solution in (39.10), for which

$$W(x) \equiv e^{-2\beta t} \begin{vmatrix} e^{-\Omega t} & e^{+\Omega t} \\ -\Omega e^{-\Omega t} & \Omega e^{+\Omega t} \end{vmatrix} = 2\Omega e^{-2\beta t}, \tag{39.20}$$

where $\Omega \equiv \sqrt{\beta^2 - \omega_0^2}$. Comparison with (39.18)

$$W(t) = W(a)e^{2\beta t}, \tag{39.21}$$

gives $W(a) = 2\Omega$. So the solutions are linearly independent as long as $\beta \neq \omega_0$.

A non-trivial bonus of (39.18) is that it gives $W(x)$ without having to solve for u_1 and u_2. So as the arbiter of linear independence, the Wronskian can be used to generate a second solution from the first. The demonstration is straightforward:

$$\frac{d}{dx}\left(\frac{u_2}{u_1}\right) = \frac{u_1 u_2' - u_1' u_2}{u_1^2} = \frac{W}{u_1^2}. \tag{39.22}$$

Since the left-hand side is an exact differential, we can integrate to find

$$u_2(x) = u_1(x) \int \frac{W(x)}{u_1^2(x)} dx = u_1(x) \int \frac{\exp\left[-\int^x p(t) dt\right]}{u_1^2(x)} dx. \tag{39.23}$$

Example 39.2 **Finding u_2**

Finding the Wronskian for a differential equation starts with identifying the coefficient function $p(x)$ — and that requires referring to the standard form in (39.1). Thus Legendre's equation

$$(1 - x^2)\phi'' - 2x\phi' + \ell(\ell + 1)\phi = 0, \tag{39.24}$$

has $p(x) = -2x/(1 - x^2)$, leading to the Wronskian

$$W(x) = W(a)\exp\left[-\int \frac{-2x}{1 - x^2} dx\right] = \frac{1}{1 - x^2}, \tag{39.25}$$

where we've taken $W(a) = -1$. Though divergent at ± 1, W is nowhere zero. Using (39.23), we find the linearly independent partner solutions of the Legendre polynomials P_ℓ,

$$Q_\ell(x) = P_\ell(x) \int \frac{1}{(1 - x^2)P_\ell^2(x)} dx. \tag{39.26}$$

Even without solving for any specific ℓ, it's clear that the Q_ℓ, cleverly known as *Legendre functions of the second kind*, are not polynomials. Direct calculation for $\ell = 0$ and 1 gives

$$Q_0(x) = \frac{1}{2}\ln\left(\frac{1 + x}{1 - x}\right) \qquad Q_1(x) = \frac{1}{2}x\ln\left(\frac{1 + x}{1 - x}\right) - 1. \tag{39.27}$$

Plugging these back into (39.24) verifies that they're solutions.

And still we're not done exploring the wonders of the Wronskian. A solution $u_p(x)$ to the inhomogeneous equation

$$\mathcal{L}u_p = \left[\frac{d^2}{dx^2} + p(x)\frac{d}{dx} + q(x)\right] u_p(x) = r(x) \tag{39.28}$$

is known as a *particular solution*. Now because $u_{1,2}$ are linearly independent functions, they can be used to reproduce u_p and u_p' at some point $x = x_0$,

$$u_p(x_0) = c_1\, u_1(x_0) + c_2\, u_2(x_0)$$
$$u'_p(x_0) = c_1\, u'_1(x_0) + c_2\, u'_2(x_0). \tag{39.29}$$

This obviously cannot hold with the same constants for all x, if only because constant $c_{1,2}$ necessarily give homogeneous rather than inhomogeneous solutions. But if we imagine doing this for every point in the domain, the infinite collection of constants $c_{1,2}$ which emerge define *functions* $a_{1,2}(x)$,

$$u_p(x) = a_1(x)\, u_1(x) + a_2(x)\, u_2(x). \tag{39.30}$$

This process is ironically known as the *variation of constants* (less ironically as the *variation of parameters*). Remarkably, it leads to a particular solution without laboriously visiting every x to construct the coefficients $a_i(x)$: consistency between (39.29) and (39.30) mandates that if the derivative of u_p is to give u'_p,

$$u'_p(x) = a_1(x)\, u'_1(x) + a_2(x)\, u'_2(x). \tag{39.31}$$

then we must have

$$a'_1\, u_1(x) + a'_2\, u_2(x) = 0. \tag{39.32}$$

Clearly one condition cannot determine both a_1 and a_2 — and in any event, since we're looking for a particular solution, somehow we'll have to take $r(x)$ into account. This suggests that we turn to the differential equation itself: taking the derivative of (39.31) to get a second condition for $a_{1,2}(x)$, and using the differential equation to replace $u''_{1,2}$, you'll find (Problem 39.5)

$$a_1(x) = -\int \frac{u_2(x)r(x)}{W(x)}\,dx$$
$$a_2(x) = +\int \frac{u_1(x)r(x)}{W(x)}\,dx, \tag{39.33}$$

where once again the Wronskian emerges to play a crucial role. These integrals for $a_{1,2}(x)$, together with (39.30), yield a particular solution u_p.

Just as general homogeneous solutions to linear ODEs do not take into account boundary conditions, so too their particular solutions.[2] Though boundary conditions can be satisfied with the choice of integration constants in (39.33), it's often easier to disregard them and instead leverage linearity. This is done by noting that any linear combination of homogeneous solutions can be added to u_p,

$$u(x) = u_p(x) + c_1 u_1(x) + c_2 u_2(x), \tag{39.34}$$

to get another inhomogeneous solution,

$$\mathcal{L}u(x) = \mathcal{L}\big[u_p + c_1 u_1 + c_2 u_2\big]$$
$$= \mathcal{L}u_p + c_1 \mathcal{L}u_1 + c_2 \mathcal{L}u_2 = \mathcal{L}u_p = r. \tag{39.35}$$

With the freedom to add a *complementary solution* $u_c \equiv c_1 u_1 + c_2 u_2$, the coefficients c_1 and c_2 can be chosen to satisfy boundary conditions for u_p. Hence u_c "complements" u_p.

---

[2] "Particular solution" is also sometimes used to denote the inhomogeneous solution which does satisfy particular boundary conditions. The distinction is relevant for non-linear equations, where (39.34) is inapplicable.

Example 39.3 $d^2u/dx^2 = r(x)$, $u(0) = u(L) = 0$

The homogeneous solutions of the operator d^2u/dx^2 have the form $a + bx$. Any two linearly independent choices will do; we'll choose

$$u_1(x) = L \qquad u_2(x) = x. \tag{39.36}$$

The Wronskian is also L, so the general solution is

$$u(x) = c_1 L + c_2 x - \int^x x' r(x')\, dx' + x \int^x r(x')\, dx', \tag{39.37}$$

with $c_1 L + c_2 x$ the complementary solution. For instance, if $r = x$,

$$u(x) = c_1 L + c_2 x + \frac{1}{6} x^3. \tag{39.38}$$

The boundary conditions can be satisfied by taking $c_1 = 0$, $c_2 = -L^2/6$. Thus

$$u(x) = \frac{1}{6} x(x^2 - L^2), \tag{39.39}$$

as direct substitution into the differential equation readily confirms.

Example 39.4   **The Driven Oscillator**

An underdamped oscillator loses amplitude as energy is lost, eventually coming to rest. Energy can be pumped back in by subjecting the oscillator to an external force $F(t)$. This shows up in the differential equation as an inhomogeneous term,

$$\frac{d^2u}{dt^2} + 2\beta \frac{du}{dt} + \omega_0^2 u = F(t)/m. \tag{39.40}$$

Consider the case of a periodic driving force, $F(t) = F_0 \cos \omega t$. Although the frequency ω is independent of the oscillator's natural frequency ω_0, since there are now two frequency (or equivalently, time) scales in the problem, it's reasonable to expect their relative magnitudes to be physically relevant. To find out how, we need to solve (39.40).

We've done this before using both Fourier and Laplace transforms. This time we'll use the homogeneous solutions of (39.11b) and $W(t) = e^{-2\beta t}$ in (39.32) to construct the integrals

$$a_1(t) = -\frac{F_0}{m} \int e^{+\beta t} \sin\left(\sqrt{\omega_0^2 - \beta^2}\, t\right) \cos \omega t\, dt$$

$$a_2(t) = \frac{F_0}{m} \int e^{+\beta t} \cos\left(\sqrt{\omega_0^2 - \beta^2}\, t\right) \cos \omega t\, dt, \tag{39.41}$$

both of which are easier to evaluate as complex exponentials. Dropping overall constants other than F_0/m, (39.30) then delivers the particular solution (Problem 39.15),

$$u_p(t) = \frac{F_0}{m} \frac{\left(\omega_0^2 - \omega^2\right) \cos(\omega t) + 2\beta\omega \sin(\omega t)}{\left(\omega_0^2 - \omega^2\right)^2 + 4\beta^2 \omega^2}. \tag{39.42}$$

As expected, the amplitude grows linearly with F_0. But a dramatic response occurs for any F_0 when the driving is set at the resonant frequency $\omega^2 = \omega_0^2 - 2\beta^2$. The smaller the damping, the more dramatic the response, and the closer ω is to the oscillator's natural frequency ω_0. (The idealized case with no damping, $\beta = 0$, is unphysical — so perhaps we need not be concerned that u_p diverges at resonance in this limit. But see Problem 39.16 nonetheless.)

39.3 Series Solutions

Let's recap the results of the previous section. Confronted with a second-order linear ODE, we can calculate the Wronskian from (39.18). With $W(x)$ and one homogeneous solution, (39.23) provides a second, linearly independent homogeneous solution. Then with these two homogeneous solutions, we can find the particular solution $u_p(x)$ to the inhomogeneous solution using (39.29). To this we can add a complementary solution as in (39.34) to satisfy initial or boundary conditions. That's quite a progression — all of which hinges on the ability to find the first homogeneous solution. We turn to that now.

Whenever a solution $u(x)$ of an ordinary homogeneous differential equation is sufficiently smooth, it can be expanded in a Taylor series,

$$u(x) = \sum_{m=0}^{\infty} c_m x^m, \tag{39.43}$$

where $c_m = \frac{1}{m!} d^m u/dx^m \big|_{x=0}$. This provides a valid representation of $u(x)$ within the radius of convergence around the origin. Obviously, if we know $u(x)$ we can compute the coefficients c_m. But we don't; that's the point. So instead, we take (39.43) as an ansatz: by plugging it into the differential equation we can *solve* for the c_m. Since the power series is unique, the collection of c_m's provides a solution to the differential equation.

The ansatz, however, doesn't apply if the function cannot be written as a simple power series — for instance, if the lowest power of x is negative or non-integer. We can account for these possibilities with the modified ansatz

$$u(x) = x^{\alpha} \sum_{m=0}^{\infty} c_m x^m, \tag{39.44}$$

where α need not be a positive integer. The only condition we impose is that $c_0 \neq 0$, so that, by construction, the lowest power of the series is x^{α}. This modified ansatz produces what is known as a *generalized power series solution*; its use is often called the *Frobenius method*.

Example 39.5 Finding α: The Indicial Equation

Consider the differential equation

$$4xu'' + 2u' - u = 0. \tag{39.45}$$

Inserting the ansatz (39.44) gives

$$\sum_{m=0}^{\infty} c_m 2(m + \alpha)(2m + 2\alpha - 1)x^{m+\alpha-1} - \sum_{m=0}^{\infty} c_m x^{m+\alpha} = 0. \tag{39.46}$$

Since powers of x are linearly independent, their coefficients must vanish separately. The lowest power contributing in (39.46) is $\alpha - 1$, which gives

$$c_0 \alpha(2\alpha - 1) = 0. \tag{39.47}$$

Higher powers in (39.46) lead to the recursion relation

$$c_m = \frac{1}{2(m + \alpha)(2m + 2\alpha - 1)} c_{m-1}. \tag{39.48}$$

Now the ansatz requires $c_0 \neq 0$, so (39.47) determines the values for the index α — and hence is known as the *indicial equation*. Its two solutions, $\alpha = 0$ and $1/2$, correspond to two independent solutions of the differential equation. There is thus no condition on c_0; it's completely unconstrained by the differential equation itself, and must ultimately be determined from initial or boundary conditions.

1. For $\alpha = 1/2$, the recursion relation yields a tower of coefficients supported at its base by c_0,

$$c_m = \frac{1}{(2m + 1)2m} c_{m-1} = \frac{1}{(2m + 1)2m} \cdot \frac{1}{(2m - 1)(2m - 2)} c_{m-2}$$

$$\vdots$$

$$= \frac{c_0}{(2m + 1)!}. \tag{39.49}$$

So the first solution is (up to an overall constant)

$$u_1(x) = x^{1/2} \left(1 + \frac{1}{3!}x + \frac{1}{5!}x^2 + \frac{1}{7!}x^3 + \cdots \right). \tag{39.50}$$

2. For the second solution, $\alpha = 0$ gives another tower of coefficients:

$$c_m = \frac{1}{2m(2m - 1)} c_{m-1} = \frac{1}{2m(2m - 1)} \cdot \frac{1}{(2m - 2)(2m - 3)} c_{m-2}$$

$$\vdots$$

$$= \frac{c_0}{(2m)!}, \tag{39.51}$$

so that

$$u_2(x) = x^0 \left(1 + \frac{1}{2!}x + \frac{1}{4!}x^2 + \frac{1}{6!}x^3 + \cdots \right), \tag{39.52}$$

where including $x^0 = 1$ allows for easier comparison with (39.50).

Look closely and you'll see that $u_1(x) = \sinh\sqrt{x}$ and $u_2(x) = \cosh\sqrt{x}$; plugging either the series or closed forms back into (39.45) confirms that they're solutions. But since the differential equation is linear, its most general solution is

$$u(x) = au_1(x) + bu_2(x), \tag{39.53}$$

where the constants a and b are given by initial or boundary conditions.

Example 39.6 **Not Exactly Sine and Cosine**

This time, let's examine

$$x^2 u'' + 2xu' + x^2 u = 0, \tag{39.54}$$

which, if not for the u' term, would yield oscillatory solutions. In fact, we expect this term to act like a damping force. Let's see: using the ansatz (39.44) gives

$$\sum_{m=0} c_m(m+\alpha)(m+\alpha+1)x^{m+\alpha} + \sum_{m=0} c_m x^{m+\alpha+2} = 0, \tag{39.55}$$

where we've combined the contributions from (39.54)'s first two terms. With a little clever shifting of indices, we can recast this as

$$c_0\,\alpha(\alpha+1)x^\alpha + c_1\,(\alpha+1)(\alpha+2)x^{\alpha+1}$$
$$+ \sum_{m=2} \big[c_m\,(m+\alpha)(m+\alpha+1) + c_{m-2}\big] x^{m+\alpha} = 0. \tag{39.56}$$

Starting with the lowest power in the ansatz, x^α, (39.56) gives the indicial equation

$$c_0\,\alpha(\alpha+1) = 0, \tag{39.57a}$$

and so $\alpha = 0$ or -1. For the coefficients of $x^{\alpha+1}$ we get another quadratic in α,

$$c_1\,(\alpha+1)(\alpha+2) = 0, \tag{39.57b}$$

whereas collecting powers greater than $\alpha+1$ leads to a recursion relation,

$$c_m = -\frac{1}{(m+\alpha)(m+\alpha+1)}c_{m-2}. \tag{39.57c}$$

1. For $\alpha = 0$, (39.57b) requires $c_1 = 0$. For the rest of the coefficients, we get

$$c_m = -\frac{1}{m(m+1)}c_{m-2}. \tag{39.58}$$

This yields a single tower of even-m coefficients supported at its base by c_0,

$$c_m = -\frac{1}{m(m+1)}c_{m-2} = \frac{-1}{m(m+1)}\cdot\frac{-1}{(m-1)(m-2)}c_{m-4}$$

$$\vdots$$

$$= \frac{(-1)^{m/2}}{(m+1)!}c_0, \qquad m \text{ even.} \tag{39.59}$$

With these coefficients, the ansatz yields the even solution (up to an overall constant)

$$u_1(x) = \left(1 - \frac{1}{3!}x^2 + \frac{1}{5!}x^4 - \frac{1}{7!}x^6 + \cdots\right)$$

$$= \frac{1}{x}\left(x - \frac{1}{3!}x^3 + \frac{1}{5!}x^5 - \frac{1}{7!}x^7 + \cdots\right) = \frac{\sin x}{x}, \tag{39.60}$$

as can be verified by insertion into (39.55).
2. For $\alpha = -1$, (39.57b) gives no condition on c_1. We therefore have the freedom to *choose* a value; for simplicity, we'll take $c_1 = 0$. The recursion relation then once again yields a single tower of even-m coefficients supported by c_0,

$$c_m = -\frac{1}{m(m-1)}c_{m-2} = -\frac{1}{m(m-1)} \cdot -\frac{1}{(m-2)(m-3)}c_{m-4}$$

$$\vdots$$

$$= \frac{(-1)^{m/2}}{m!}c_0, \qquad m \text{ even.} \tag{39.61}$$

Putting these results into the ansatz with $\alpha = -1$ gives the odd solution (up to an overall constant)

$$u_2(x) = \left(\frac{1}{x} - \frac{1}{2!}x + \frac{1}{4!}x^3 - \frac{1}{6!}x^5 + \cdots\right)$$

$$= \frac{1}{x}\left(1 - \frac{1}{2!}x^2 + \frac{1}{4!}x^4 - \frac{1}{6!}x^6 + \cdots\right) = \frac{\cos x}{x}. \tag{39.62}$$

Choosing $c_1 = 0$ made finding a second solution easier. Another choice would simply lead to an admixture of u_1 and u_2 — which, though an acceptable independent second solution, would not have the advantage of definite parity.

Before diving into another example, let's take a moment to see what insight we can glean from mere inspection of a differential equation. For example, counting powers of x reveals what sort of recursion relation to expect. In (39.54), although d^2/dx^2 lowers x^m by two degrees, the factor of x^2 restores them both. Similarly for the term $x\,d/dx$. On the other hand, the x^2u term adds two degrees to the series. So the first two terms give a series in x^m, whereas the third a series in x^{m+2}. Hence the every-other-m recursion we found. By contrast, the first two terms of (39.45) give a series in x^{m-1}, the third term in x^m.

Related to this is parity: if the differential operator $\mathcal{L}_x$ has definite parity,

$$\mathcal{L}_{-x} = \pm\mathcal{L}_x, \tag{39.63}$$

then even and odd solutions of $\mathcal{L}_x[u(x)] = 0$ can be found. Although (39.54) has definite parity, (39.45) does not — so only the first has even and odd solutions.

Practical as questions of recursion and parity may be with regards to solving a differential equation, far more important is the question of whether an equation even *has* a series solution — and if it does, with what radius of convergence. Consider a second-order differential equation in the general form

$$u'' + p(x)u' + q(x)u = 0. \tag{39.64}$$

The existence of a generalized series solution hinges on the behavior of the functions p and q. A point x_0 is said to be *ordinary* if both p and q are finite there. But even if they diverge, a series solution may exist. Specifically, if $(x - x_0)p(x)$ and $(x - x_0)^2q(x)$ have convergent power series, then the larger of the two α's emerging from the ansatz will yield a generalized series solution around x_0. Another way to say this is that what matters is not whether p and q have singularities at some point x_0, but rather the rate at which they diverge there: as long as

$$\lim_{x \to x_0}(x - x_0)p(x) \qquad \text{and} \qquad \lim_{x \to x_0}(x - x_0)^2\,q(x) \tag{39.65}$$

are finite, then at least one generalized series solution exists around x_0, called a *regular* singularity. The radius of convergence of the series is equal to the distance from x_0 to the nearest singularity of either p or q. If either of the limits in (39.65) diverge, however, then x_0 is an *irregular* or *essential* singularity, and no series solution exists around that point.

Even if the singularity structure guarantees one generalized series solution, what about a *second* solution? As we saw in (39.23), u_2 can be generated from u_1 and the Wronskian,

$$u_2(x) = u_1(x) \int \frac{W(x)}{u_1^2(x)} \, dx. \tag{39.66}$$

Let's assume for the moment that in addition to the generalized series solution $u_1 = x^{\alpha_1} \sum a_m x^m$, our differential equation has a second solution $u_2 = x^{\alpha_2} \sum a_m' x^m$ with $\alpha_1 \geq \alpha_2$. Then it's straightforward to show that the Wronskian is also a generalized power series,

$$W = x^{\alpha_1 + \alpha_2 - 1} \sum_{k=0} b_k x^k. \tag{39.67}$$

Writing $u_1^{-2} = x^{-2\alpha_1} \sum_{k=0} c_k x^k$, (39.66) gives[3]

$$u_2(x) = u_1(x) \int \left(x^{\alpha_2 - \alpha_1 - 1} \sum_{k=0} d_k x^k \right) dx$$

$$= u_1(x) \int x^{-\Delta - 1} (d_0 + d_1 x + \cdots) \, dx, \tag{39.68}$$

where $\Delta \equiv \alpha_1 - \alpha_2 \geq 0$. For non-integer Δ, the assumption that u_2 is a generalized series solution is thus validated. But if $\Delta = 0$, the integral will necessarily contribute a logarithm to u_2, contradicting that assumption. (Indeed, $\alpha_1 = \alpha_2$ would render the presumed second solution proportional to u_1, and hence give $W = 0$.) Similarly, a log term emerges for integer $\Delta > 0$ — *unless* it happens that the coefficient $d_{k=\Delta} = 0$. So in these two cases, the assertion that there are two series solutions fails. However, a second series augmented by a logarithm works. Thus, if $u_1(x)$ is the generalized series solution corresponding to $\alpha_1 \geq \alpha_2$, then:

- for $\alpha_1 - \alpha_2 \neq$ integer, there is a second independent generalized series solution;
- for $\alpha_1 - \alpha_2 =$ integer (including 0), the second solution has the form

$$u_2(x) = A u_1(x) \ln |x| + x^{\alpha_2} \sum_m c_m x^m, \tag{39.69}$$

which can be used as an ansatz in the differential equation. Now A could turn out to be zero, in which case the differential equation has two series solutions — as we saw in Examples 39.6 and 39.5. But whenever $A \neq 0$, the second solution has a logarithmic divergence. This is unavoidable for $\alpha_1 = \alpha_2$, since otherwise $A = 0$ would just return the first solution u_1.

Example 39.7 **Bessel's Equation**

Bessel's differential equation is

$$x^2 u'' + x u' + (x^2 - \lambda^2) u = 0, \tag{39.70}$$

whose solutions constitute a family of solutions distinguished by the value of the constant λ. With only a regular singularity at $x = 0$, a series expansion around the origin exists which converges for all finite x.

---

[3] A power series for u_1^{-2} exists only if $c_0 \neq 0$ — which is consistent with the series ansatz.

Once again, we expect every-other-m recursion relations which, reflecting the equation's definite parity, describe even and odd solutions. In this example, however, the algebra simplifies a bit if we pull the factorial out of the coefficients — that is, rather than the ansatz (39.44), we use

$$u(x) = x^\alpha \sum_m \frac{1}{m!} c_m x^m. \tag{39.71}$$

Plugging this into Bessel's equation yields

$$\sum_{m=0}^{\infty} \frac{1}{m!} c_m [(m+\alpha)^2 - \lambda^2] x^{m+\alpha} + \sum_{m=0}^{\infty} \frac{1}{m!} c_m x^{m+\alpha+2} = 0. \tag{39.72}$$

A little index shifting gives

$$c_0 (\alpha^2 - \lambda^2) x^\alpha + c_1 [(\alpha+1)^2 - \lambda^2] x^{\alpha+1}$$
$$+ \sum_{m=2}^{\infty} \left[\frac{c_m}{m!} ((m+\alpha)^2 - \lambda^2) + \frac{c_{m-2}}{(m-2)!} \right] x^{m+\alpha} = 0. \tag{39.73}$$

The indicial equation

$$c_0 (\alpha^2 - \lambda^2) = 0 \tag{39.74a}$$

requires $\alpha = \pm\lambda$. With this, the coefficients of $x^{\alpha+1}$ in (39.73) give

$$c_1 (1 \pm 2\lambda) = 0, \tag{39.74b}$$

and the recursion relation becomes

$$c_m = -\frac{(n-1)}{(m \pm 2\lambda)} c_{m-2}. \tag{39.74c}$$

Equation (39.74b) can only be satisfied by $c_1 = 0$ — except for $\lambda = \pm 1/2$, but even then we can *choose* $c_1 = 0$. So again we have a tower of even-m coefficients supported by c_0, leading to the series solution for $\alpha = +\lambda$

$$u(x) = x^\lambda \left[1 - \frac{1}{2!} \frac{1}{2(1+\lambda)} x^2 + \frac{1}{4!} \frac{1 \cdot 3}{4(1+\lambda)(2+\lambda)} x^4 - \frac{1}{6!} \frac{1 \cdot 3 \cdot 5}{8(1+\lambda)(2+\lambda)(3+\lambda)} x^6 + \cdots \right]$$

$$= x^\lambda \left[1 - \frac{1}{(1+\lambda)} \left(\frac{x}{2}\right)^2 + \frac{1}{2!(1+\lambda)(2+\lambda)} \left(\frac{x}{2}\right)^4 - \frac{1}{3!(1+\lambda)(2+\lambda)(3+\lambda)} \left(\frac{x}{2}\right)^6 + \cdots \right]. \tag{39.75}$$

We can spruce this up by dividing by $2^\lambda \lambda!$ to get the Bessel function (Section 35.2)

$$J_\lambda(x) \equiv \frac{1}{\lambda!} \left(\frac{x}{2}\right)^\lambda - \frac{1}{(1+\lambda)!} \left(\frac{x}{2}\right)^{2+\lambda}$$

$$+ \frac{1}{2!(2+\lambda)!} \left(\frac{x}{2}\right)^{4+\lambda} - \frac{1}{3!(3+\lambda)!} \left(\frac{x}{2}\right)^{6+\lambda} + \cdots$$

$$= \sum_{m=0}^{\infty} \frac{(-1)^m}{m! \, \Gamma(m+\lambda+1)} \left(\frac{x}{2}\right)^{2m+\lambda}, \tag{39.76}$$

where we've used $\Gamma(m+\lambda+1) = (m+\lambda)!$ in order to generalize beyond positive integer values of λ. For non-integer $\alpha = \lambda \equiv \nu$ we get two linearly independent series solutions, the Bessel functions $J_{\pm\nu}$ — so the general solution to the differential equation is

$$u(x) = a\,J_\nu(x) + b\,J_{-\nu}(x), \qquad \lambda \equiv \nu \neq \text{integer}. \tag{39.77}$$

But one shouldn't declare victory too soon: for integer $\lambda = n$, $J_{\pm n}$ are *not* linearly independent. To verify this, note that $1/\Gamma(z) = 0$ for integer $z \leq 0$ (Problem 8.12c). So the first n contributions to the sum in (39.76) don't contribute to J_{-n}:

$$J_{-n}(x) = \sum_{m=0} \frac{(-1)^m}{m!\,\Gamma(m-n+1)} \left(\frac{x}{2}\right)^{2m-n}$$

$$= \sum_{m=n} \frac{(-1)^m}{m!\,\Gamma(m-n+1)} \left(\frac{x}{2}\right)^{2m-n}. \tag{39.78}$$

Shifting the index $m \to m + n$ gives

$$J_{-n}(x) = \sum_{m=0} \frac{(-1)^{m+n}}{(m+n)!\,\Gamma(m+1)} \left(\frac{x}{2}\right)^{2m+n}$$

$$= (-1)^n \sum_{m=0} \frac{(-1)^m}{\Gamma(m+n+1)\,m!} \left(\frac{x}{2}\right)^{2m+n} = (-1)^n J_n(x). \tag{39.79}$$

Thus replacing integer $\lambda \equiv n \to -n$ in (39.76) does not yield an independent solution. We can, however, generate one from the Wronskian using (39.23) — though this solution will diverge at least logarithmically at the origin.

39.4 Legendre and Hermite, Re-revisited

There are many ways to derive orthogonal polynomials: Gram–Schmidt, generating functions — and as we'll see in Chapter 40, as eigenfunctions of Hermitian differential operators. But for a polynomial solution to emerge from a generalized series ansatz, somehow the presumed infinite series must terminate in order to obtain a finite sum. Let's see how this can occur.

Example 39.8 Legendre's Equation

We can use the ansatz of (39.44) to derive the solutions to Legendre's equation

$$(1 - x^2)u'' - 2xu' + \lambda u = 0. \tag{39.80}$$

Here λ is a general, unspecified parameter; in the course of solving the differential equation, we'll discover the origin of the familiar form $\lambda = \ell(\ell + 1)$ for integer $\ell \geq 0$.

Since the equation has definite parity, we look for even and odd solutions emerging from an every-other-m recursion relation. Straightforward application of the ansatz yields the indicial equation,

$$c_0\,\alpha(\alpha - 1) = 0, \tag{39.81a}$$

a condition on c_1,

$$c_1\,\alpha(\alpha + 1) = 0, \tag{39.81b}$$

and the recursion relation

$$c_m = \frac{(m + \alpha - 1)(m + \alpha - 2) - \lambda}{(m + \alpha)(m + \alpha - 1)} c_{m-2}. \tag{39.81c}$$

The indicial equation gives $\alpha = 1$ and 0; as before, we'll take them one at a time:

1. For $\alpha = 1$, (39.81b) can only be satisfied with $c_1 = 0$. So once again, we get a tower of even c_m's supported by c_0,

$$c_m = \frac{m(m - 1) - \lambda}{(m + 1)m} c_{m-2}. \tag{39.82}$$

With these coefficients, the power series expansion gives (up to an overall constant)

$$
\begin{aligned}
u_1(x) &= x + \frac{(2 \cdot 1 - \lambda)}{3!} x^3 + \frac{(2 \cdot 1 - \lambda)(4 \cdot 3 - \lambda)}{5!} x^5 \\
&\quad + \frac{(2 \cdot 1 - \lambda)(4 \cdot 3 - \lambda)(6 \cdot 5 - \lambda)}{7!} x^7 + \cdots \\
&= x + \frac{(2 - \lambda)}{3!} x^3 + \frac{(2 - \lambda)(12 - \lambda)}{5!} x^5 \\
&\quad + \frac{(2 - \lambda)(12 - \lambda)(30 - \lambda)}{7!} x^7 + \cdots.
\end{aligned} \tag{39.83}
$$

Even though all the coefficients have even m, odd $\alpha = 1$ gives an odd solution.

2. $\alpha = 0$: This time, we get no condition on c_1 — but since we just found an odd solution, we can get an even one by choosing $c_1 = 0$. So yet again we get a tower of even c_m's,

$$c_m = \frac{(m - 1)(m - 2) - \lambda}{m(m - 1)} c_{m-2}, \tag{39.84}$$

giving solution (up to an overall constant)

$$
\begin{aligned}
u_2(x) &= 1 + \frac{(0 \cdot 1 - \lambda)}{2!} x^2 + \frac{(0 \cdot 1 - \lambda)(2 \cdot 3 - \lambda)}{4!} x^4 \\
&\quad + \frac{(0 \cdot 1 - \lambda)(2 \cdot 3 - \lambda)(4 \cdot 5 - \lambda)}{6!} x^6 + \cdots \\
&= 1 - \frac{\lambda}{2!} x^2 - \frac{\lambda(6 - \lambda)}{4!} x^4 - \frac{\lambda(6 - \lambda)(20 - \lambda)}{6!} x^6 + \cdots.
\end{aligned} \tag{39.85}
$$

Before declaring the problem solved, we should first determine the radii of convergence of these power series. Straightforward application of the ratio test reveals

$$x^2 < \lim_{m \to \infty} \left| \frac{c_{m-2}}{c_m} \right| = 1 \tag{39.86}$$

for both u_1 and u_2. So we have independent series solutions of Legendre's equation for $|x| < 1$. This is consistent with our discussion around (39.64): the generalized power series is convergent out to the first singularity of $p(x) = -2x/(1 - x^2)$. But what about *on* the radius of convergence? For that, we need to keep the leading term in $1/m$ for large m (see Problem 6.4),

$$\left| \frac{c_{m-2}}{c_m} \right| \longrightarrow \frac{1 - 3/m}{1 - 1/m} \approx 1 - \frac{2}{m}. \tag{39.87}$$

This behavior is the same as for the harmonic series $\sum_m 1/m$, which we know diverges (see Example 6.6 and BTW 6.1). So both series $u_{1,2}$ are *unbounded*, converging to ever larger values as x approaches ± 1. A solution which can be arbitrarily large within some small $\epsilon > 0$ of $x = \pm 1$ is not square-integrable — and so is physically untenable as, say, a basis in a normal mode expansion.

But a remarkable thing happens for certain choices of λ: inspection of (39.83) and (39.85) reveals that when $\lambda = \ell(\ell + 1)$ for integer $\ell \geq 0$, one of the two infinite series *terminates* and becomes a polynomial of order ℓ. So for these special values of λ, one of the two solutions $u_{1,2}$ *is* finite on the entire closed interval $|x| \leq 1$. Let's write out the first few:

$$\ell = 0 : u_2 \longrightarrow 1$$

$$\ell = 1 : u_1 \longrightarrow x$$

$$\ell = 2 : u_2 \longrightarrow 1 - 3x^2$$

$$\ell = 3 : u_1 \longrightarrow x - \frac{5}{3}x^3. \tag{39.88}$$

These should look familiar: aside from their conventional normalization, they are the Legendre polynomials $P_\ell(x)$, found here in a completely different fashion from our earlier Gram–Schmidt approach. So for every integer ℓ, we get one finite solution on $[-1, 1]$. The other series remains divergent at $x = \pm 1$; we can use the Wronskian to find the closed form for these so-called Legendre functions of the second kind, $Q_\ell(x)$ (see Example 39.2).

Example 39.9 Hermite's Equation

The time-independent Schrödinger equation for a particle of mass m and energy E in a one-dimensional simple harmonic potential of frequency ω is

$$\left[-\frac{\hbar^2}{2m}\frac{d^2}{dx^2} + \frac{1}{2}m\omega^2 x^2 \right]\psi(x) = E\psi(x). \tag{39.89a}$$

Recasting in terms of dimensionless $z \equiv \sqrt{\frac{m\omega}{\hbar}}\, x$, this takes the sleeker form

$$\frac{d^2\phi}{dz^2} + \left(\varepsilon - z^2 \right)\phi = 0, \tag{39.89b}$$

where $\varepsilon = 2E/\hbar\omega$ and $\phi(z) \equiv \psi\left(\sqrt{\frac{\hbar}{m\omega}}\, z \right)$. Now a clever ansatz can greatly facilitate finding a solution to a differential equation. In this case, inspection reveals that the solutions for large $|z|$ are $e^{\pm z^2/2}$; discarding the divergent one, we make the ansatz

$$\phi(z) = u(z)\, e^{-z^2/2}, \tag{39.90}$$

for some polynomial $u(z)$. Inserting this into (39.90) gives Hermite's equation

$$u'' - 2zu' + 2\lambda u = 0, \qquad \lambda = \frac{1}{2}(\varepsilon - 1). \tag{39.91}$$

Of course, having emerged from the full differential equation, solutions to Hermite's equation must still include the divergent form we discarded in the ansatz — but let's see where that ansatz (39.90) takes us.

Applying the now-familiar generalized series method to Hermite's equation gives

$$\sum_m c_m \, (m + \alpha) \, (m + \alpha - 1) \, z^{m+\alpha-2} - 2 \sum_m c_m \, (m + \alpha - \lambda) \, z^{m+\alpha} = 0, \tag{39.92}$$

leading to the indicial equation

$$c_0 \alpha \, (\alpha - 1) = 0, \tag{39.93a}$$

a condition on c_1,

$$c_1 \alpha \, (\alpha + 1) = 0, \tag{39.93b}$$

and the recursion relation

$$c_{m+2} \, (m + \alpha + 2) \, (m + \alpha + 1) = 2 c_m \, (m + \alpha - \lambda). \tag{39.93c}$$

The two series solutions correspond to $\alpha = 1$ and 0.

1. The $\alpha = 1$ case requires $c_1 = 0$ and yields the even-coefficient tower

$$c_m = \frac{2(m - \lambda - 1)}{m(m + 1)} \, c_{m-2}. \tag{39.94}$$

So the first solution is odd:

$$u_1(z) = z \left(1 + \frac{2(1 - \lambda)}{3!} z^2 + \frac{2^2(1 - \lambda)(3 - \lambda)}{5!} z^4 + \frac{2^3(1 - \lambda)(3 - \lambda)(5 - \lambda)}{7!} z^6 + \cdots \right)$$

$$= z + \sum_{n=1}^{\infty} 2^n \, [(1 - \lambda)(3 - \lambda) \cdots (2n - 1 - \lambda)] \, \frac{z^{2n+1}}{(2n + 1)!}. \tag{39.95}$$

2. Since Hermite's equation has definite parity, for $\alpha = 0$ we choose $c_1 = 0$ to get the coefficient tower

$$c_m = \frac{2(m - \lambda - 2)}{m(m - 1)} \, c_{m-2} \tag{39.96}$$

and the even solution

$$u_2(z) = 1 + \frac{2(-\lambda)}{2!} z^2 + \frac{2^2(-\lambda)(2 - \lambda)}{4!} z^4$$

$$+ \frac{2^3(-\lambda)(2 - \lambda)(4 - \lambda)}{6!} z^6 + \cdots$$

$$= 1 + \sum_{n=1}^{\infty} 2^n \, [(-\lambda)(2 - \lambda) \cdots (2n - \lambda)] \, \frac{z^{2n}}{(2n)!}. \tag{39.97}$$

The radius of convergence for both these series is

$$z^2 \le \lim_{m \to \infty} \left| \frac{c_{m-2}}{c_m} \right| = \lim_{m \to \infty} \frac{m}{2} = \infty, \tag{39.98}$$

so the series converges for all finite z. But the large z behavior of the series is dominated by large m — and the ratio $m/2$ in (39.98) can be shown to match the same ratio for the series expansion of

e^{+z^2}. In other words, for large z both u_1 and u_2 grow as e^{+z^2} — and the factor of $e^{-z^2/2}$ from (39.90) is not enough to tame it. But once again, there is a clever way out: if $\lambda = n$ for integer $n \geq 0$, then one of the two series terminates to become a polynomial. Let's write out the first few:

$$n = 0 : u_2 \longrightarrow 1$$

$$n = 1 : u_1 \longrightarrow z$$

$$n = 2 : u_2 \longrightarrow 1 - 2z^2$$

$$n = 3 : u_1 \longrightarrow z - \frac{2}{3}z^3. \tag{39.99}$$

Normalization aside, these are the Hermite polynomials $H_n(z)$. So the physical solutions to the Schrödinger equation (39.89) are square-integrable wavefunctions $\phi_n(z) = H_n(z)e^{-z^2/2}$. Most remarkably, integer n acquires physical significance: the ϕ_n have *quantized*, equally spaced energies $E_n = (n + \frac{1}{2})\hbar\omega$. And since they started with $n = 0$, the quantum oscillator has ground-state energy $E_0 = \frac{1}{2}\hbar\omega$ called the oscillator's *zero-point energy*. This non-vanishing minimum energy is a reflection of the uncertainty principle: a quantum particle never rests [see also (33.38)].

Our choice of examples notwithstanding, a series solution need not have a closed-form expression or designation — though we can always *assign* one. Consider the differential equation $\frac{d^2u}{dx^2} + k^2u = 0$. Of course we know the solutions are sines and cosines. But why of course? It's fair to say that this equation is trivial largely because both the equation and its solutions are so familiar. But it's a mistake to conflate trivial with familiar. [As I often say to students: don't let the "duh" be the enemy of the "wow."] So imagine you're the first person to undertake an analysis of $u'' + k^2u = 0$. Using the generalized series ansatz $u(x) = \sum_m c_m x^{m+\alpha}$ you discover odd and even solutions

$$u^-(x) = x \left[1 - \frac{1}{3!}k^2x^2 + \frac{1}{5!}k^4x^4 - \cdots \right]$$

$$u^+(x) = x^0 \left[1 - \frac{1}{2!}k^2x^2 + \frac{1}{4!}k^4x^4 - \cdots \right], \tag{39.100a}$$

or, multiplying u^- by k,

$$u_1(x) \equiv ku^-(x) = \sum_{m=0} \frac{(-1)^m}{(2m+1)!}(kx)^{2m+1}$$

$$u_2(x) \equiv u^+(x) = \sum_{m=0} \frac{(-1)^m}{(2m)!}(kx)^{2m}. \tag{39.100b}$$

A quick calculation verifies that these series are convergent for all finite x. Since your series solutions are only defined up to an overall constant, the simplest and most natural choice is to specify initial conditions $u_1(0) = 0$, $u_2(0) = 1$ and $u_1'(0) = 1$, $u_2'(0) = 0$. And with that, your functions are fully defined. Virtually anything and everything we want to know about $u_{1,2}(x)$ can be gleaned from the series in (39.100). And not merely their values at any particular value of x: their derivatives, integrals, and even algebraic relations such as $u_1^2 + u_2^2 = 1$ and their periodicity can be deduced (Problem 6.21). Once all the properties of your new functions have been determined and tabulated, there's little reason to continue to write them as infinite series; far less cumbersome would be to express them in closed form. But there is no closed form! Your functions are new to mathematics, unrelated to basic algebraic functions. So you decide to give them succinct names with which to *designate* the series expansions. Since you're writing the original paper, your choice has a high likelihood of being widely adopted. Grasping for celebrity, you name them after yourself. And so it was that Professor Sine passed into posterity.[4]

And so it is that many so-called "special functions" (e.g., Bessel, Legendre, and Hermite) are often defined as series solutions of differential equations. Designations such as $J_\nu, P_\ell,$ and H_n merely reflect conventional notation, with mathematical tables and databases serving as archives of their various properties (for instance, Abramowitz and Stegun, 1966). But rather than egotism, functions which bear someone's name are more likely to have acquired their moniker out of honor and respect for seminal contributions.

Problems

39.1 Our solution to (39.3) could be found more quickly with a little physical insight: explain the phyiscal motivation for the ansatz $u(t) = Ae^{i\xi t}$, and use it to obtain the solution (39.10).

39.2 Use the Wronskian to determine which (if any) of the following functions are linearly independent:

(a) $\sin kx, \cos kx, e^{-ikx}$
(b) $x, 1+x, x^2$
(c) x, x^2, x^3
(d) $e^{\alpha x}, xe^{\alpha x}$.

39.3 Derive (39.18) from (39.16).

39.4 Find the Wronskian for the following "named" differential equations:

(a) Bessel: $x\frac{d}{dx}\left(x\frac{du}{dx}\right) + \left(x^2 - \nu^2\right)u(x) = 0$
(b) Associated Legendre: $(1 - x^2)u'' - 2xu' + \left[\ell(\ell+1) - \frac{m}{1-x^2}\right]u = 0$
(c) Mathieu: $u'' + \left[a - 2q\cos(2x)\right]u = 0$.

39.5 Complete the derivation of (39.33).

39.6 For the so-called Sturm–Liouville operator $\frac{d}{dx}\left[p(x)\frac{d}{dx}\right] + q(x)$, show that the product $p(x)W(x)$ is a constant.

39.7 Find the general solutions of the following differential equations:

(a) $u'' + 2u' + u = e^{-x}$
(b) $u'' - u = e^x$
(c) $u'' + u = \sin x$.

39.8 Use variation of parameters to find a particular solution for

$$\frac{d^2u}{dt^2} + 2\beta\frac{du}{dt} + \omega_0^2 u = \delta(t - t').$$

39.9 A complementary solution u_c can always be added to the particular solution u_p so that (39.34) satisfies conditions $u(x_0) = a$, $u'(x_0) = b$. Show algebraically that this is true only if the Wronskian $W(x_0) \neq 0$ — in other words, only if the homogeneous solutions $u_{1,2}$ are linearly independent.

39.10 Consider the second-order equation (39.1) with $q(x) = 0$, so that there is no term in y. Clearly one solution is constant u; what about the other? Find the general solution to the following differential equations by defining $v(x) \equiv du/dx$:

(a) $xu'' - 2u' = 0$

---

[4] In reality, the word "sine" emerged from a mistranslation: it seems the Arabic *jiba* (the chord of an arc) was confused with *jaib* (bend or fold) — in Latin, *sinus*. So much for posterity.

(b) $u'' + 2xu' = 0$

(c) $xu'' - u' - u'^2 = 0$.

39.11 Consider a second-order equation for which there is no explicit appearance of x. Find the general solutions to the following differential equations by changing the independent variable from x to u using $v \equiv du/dx$ (see Table 39.1):

(a) $u'' + uu' = 0$

(b) $2uu'' - u'^2 = 0$.

39.12 Either by hand or using Mathematica's DSolve command, solve the damped oscillator for the following:

(a) $u(0) = 1, u'(0) = 0$

(b) $u(0) = 0, u'(0) = 1$.

Plot the solutions for $\beta = \omega_0/5, \omega_0$, and $5\omega_0$.

39.13 Find the solution to the damped driven oscillator with $F = F_0 \cos \omega t$ for initial conditions:

(a) $u(0) = x_0, u'(0) = 0$

(b) $u(0) = 0, u'(0) = v_0$.

39.14 Find the general solution to the damped oscillator driven by a constant force $F = F_0$. Give a physical explanation for this result.

39.15 Derive (39.42). Then use the differential equation to verify that it's an inhomogeneous solution of the damped driven oscillator.

39.16 In the absence of damping, the resonant frequency of a forced oscillator is equal to its natural frequency, $\omega = \omega_0$; without damping to dissipate energy, a driving force at this frequency causes the solution to diverge. But (39.42) gives no indication of how the oscillator's amplitude grows over time. Use the ansatz $u_p(t) = \xi(t) \cos(\omega t - \delta)$ to show that the general solution for $\beta = 0, \omega = \omega_0$ is

$$u(t) = \sqrt{c_1^2 + \left(c_2 + \frac{F_0}{2m\omega} t\right)^2} \cos(\omega t - \varphi),$$

for constant $c_{1,2}$ and time-dependent phase $\tan \varphi = \frac{1}{c_1}\left(c_2 + \frac{F_0}{2m\omega} t\right)$. Show that for large t the amplitude diverges as $t \sin \omega t$.

39.17 (a) Find two independent solutions to the homogeneous equation

$$\frac{d^2u}{dt^2} + 3\frac{du}{dt} + 2u = 0.$$

(b) Find the particular solution u_p for the source $\cos t$.

(c) Find the full solution solution to the inhomogeneous equation with initial conditions $u(0) = 1$ and $u'(0) = 0$.

39.18 Find the solution to

$$x^2 \frac{d^2u}{dx^2} - 2u = x, \quad u(0) = 1, \quad u'(0) = 1.$$

[Hint: start with the ansatz $u(x) = x^\alpha$ — see (39.13).]

39.19 **The Inverted Pendulum I**

A simple pendulum of length ℓ satisfies $\ddot{\theta} + \omega_0^2 \sin\theta = 0$, where $\omega_0^2 = g/\ell$ and θ is measured from equilibrium at $\theta = 0$. With a stick instead of a string, the pendulum can be inverted by placing the support at the bottom. Let φ measure the angular displacement from the pendulum's vertical position such that

$\varphi + \theta = \pi$. (The mass of the stick is negligible compared with the weight at its end, so that the momentum of inertia around the support is just $m\ell^2$.)

(a) Show that $\varphi = 0$ is an unstable equilibrium — that is, at the slightest perturbation from $\varphi = 0$, the inverted pendulum falls over.

(b) An external horizontal force applied at the support will contribute an angular acceleration proportional to $\ddot{x}\cos\theta = -\ddot{x}\cos\varphi$ (Phelps and Hunter, 1965). With or a sinusoidal driving force $x(t) = A\cos\omega t$, the equation of motion becomes

$$\ddot{\varphi} - \omega^2\alpha\cos\omega t\cos\varphi - \omega_0^2\sin\varphi = 0,$$

where $\alpha = A/\ell$. Show that for small φ, this force can stabilize the inverted pendulum by finding a finite solution $\varphi(t)$. (This is effectively what unicyclists do.) By contrast, show that the normal pendulum can become unstable. [Hint: Compare the equations of motion in φ and θ.]

39.20 The Inverted Pendulum II: Mathieu's Equation

Let an external vertical force be applied at the support of the inverted pendulum in Problem 39.19. From the perspective of an observer on the pendulum bob, the vertical acceleration $\ddot{y}$ acts as an addition to gravity, $g \to g + \ddot{y}$. (This is an application of the equivalence principal, see BTW 30.3.) So for a sinusoidal driving force $y(t) = A\cos\omega t$, the equation of motion is

$$\ddot{\varphi} + \left(\alpha\omega^2\cos\omega t - \omega_0^2\right)\varphi = 0, \qquad (39.101)$$

with $\alpha = A/\ell$. This is a form of *Mathieu's equation* ("mah-tyoo," with "oo" as in "book"),

$$\frac{d^2\psi}{d\eta^2} + \left[a - 2q\cos(2\eta)\right]\psi = 0,$$

with dimensionless parameters a and q. Its solutions are the non-trivial, two-parameter Mathieu functions $C(a, q, \eta)$ and $S(a, q, \eta)$.

Notice that if the applied force overwhelms gravity, $\alpha\omega^2 \gg \omega_0^2$, the pendulum oscillates around $\varphi = 0$ rather than fall over. Thus the vertical force can stabilize an inverted pendulum — though intuitively, we expect this to occur only for small amplitude relative to the length of the stick, $\alpha \ll 1$. Still, with two time scales in the problem $(\omega^{-1}, \omega_0^{-1})$, obtaining analytical solutions $\varphi(t)$ is challenging. Instead, we'll explore the pendulum's behavior using an ansatz, and then verify it numerically using Mathematica.

(a) For $\alpha\omega^2 \gg \omega_0^2$, we expect the pendulum to exhibit relatively slow oscillation around $\varphi = 0$, superposed with small oscillations of more-rapid frequency ω. So we'll take for our ansatz $\varphi(t) \approx \xi(t)(1 + \alpha\cos\omega t)$ with the expectation that $\xi(t)$ varies slowly over time scales $\sim \omega^{-1}$. Insert this into (39.101) and average over one period of the applied force to show that ξ oscillates harmonically around $\varphi = 0$ with frequency $\sqrt{\frac{1}{2}\alpha^2\omega^2 - \omega_0^2}$. Thus stability requires $\alpha\omega > \sqrt{2}\omega_0$.

(b) Use Mathematica's DSolve command to solve (39.101) for initial conditions $\varphi(0) = 0.1, \dot{\varphi}(0) = 0$. Make a series of plots of the solution from $\alpha\omega^2$ between ω_0^2 and $100\omega_0^2$, with $\omega_0^2 = 10\ \text{s}^{-2}$ and $\alpha = 0.1$. How good is our ansatz?

39.21 Show that if a linear second-order differential operator $\mathcal{L}_x$ has definite parity, then even and odd homogeneous solutions exist.

39.22 Derive the expressions in (a) (39.56), (b) (39.73), and (c) (39.92).

39.23 Verify the recursion and indicial equations for (a) Legendre's equation, Example 39.8, and (b) Hermite's equation, Example 39.9.

39.24 For a differential operator of the form $x^2D^2 + P(x)xD + Q(x)$, show that the indicial equation can be determined directly and simply from the coefficients in the convergent power series $P = \sum_{m=0} p_m x^m$ and $Q = \sum_{m=0} q_m x^m$. Verify your result for Examples 39.5, 39.6, and 39.7.

39.25   Use (39.23) for the system in Example 39.6 to find u_2 from u_1 and vice-versa.

39.26   In the discussion around (39.67) and (39.68), we found that the assumption of a second generalized series solution is not always valid. Show that for series solution u_1, the ansatz $u_2(x) = Au_1 \ln x + v(x)$, where $v(x) = x^{\alpha_2} \sum b_m x^m$, yields consistent results. For $\alpha_1 = \alpha_2$, where must the sum in the power series for $v(x)$ begin?

39.27   One cannot use (39.65) directly to classify the point $x = \infty$. Instead, use $z = 1/x$ in the canonical form of (39.64) and examine the behavior at $z = 0$. In this way, show that the point $x = \infty$ is (a) an ordinary point if $2x - x^2 p(x)$ and $x^4 q(x)$ are finite; (b) a regular singular point if $xp(x)$ and $x^2 q(x)$ are finite. Then verify that Legendre's equation has a regular singularity at infinity, while Hermite's equation has an irregular (or essential) singularity at infinity.

39.28   Find two independent series solutions to Airy's equation,

$$u'' - xu = 0.$$

[Note that the conventional Airy functions $\text{Ai}(x)$, $\text{Bi}(x)$ are linear combinations of these series solutions.]

39.29   **Laguerre, Re-revisited**
Consider Laguerre's differential equation

$$xu'' + (1 - x)u' + \lambda u = 0.$$

(a) Determine the parity and singularity structure of this equation.
(b) Explain why only one of the solutions is a generalized series.
(c) Show that the series solution u_1, though convergent, is unbounded for large x. Come up with a way to resolve this problem, and identify the solutions which emerge.

40 Interlude: The Sturm–Liouville Eigenvalue Problem

40.1 Whence Orthogonality?

Orthogonality is an inherent property of sines and cosines, though it's hard to see that it has anything to do with triangles. But as we've learned, one doesn't need triangles: sines and cosines can be defined as series solutions of

$$\frac{d^2\phi}{dx^2} = -k^2\phi. \tag{40.1}$$

Just as this equation signals that it has solutions of definite parity, surely, somehow, it encodes the orthogonality of its solutions. We saw in Section 27.2 that eigenvectors of a Hermitian matrix form an orthogonal basis. So if we can extend the notion of a Hermitian or self-adjoint operator from $\mathbb{C}^n$ to $\mathbb{L}^2$, then we can take hermiticity as both evidence of, and justification for, the existence of an orthogonal basis of eigenvectors.

There's no doubt that (40.1) looks like an eigenvalue equation for the operator $\mathcal{D}^2 = (d/dx)^2$ with eigenvalue $\lambda = -k^2$. But even though orthogonal eigenfunctions exist for real k, this alone is not sufficient to establish $\mathcal{D}^2$ as a Hermitian operator. Different conditions imposed at the domain boundaries can yield not only oscillatory solutions $\sin kx$, $\cos kx$, and $e^{\pm ikx}$, but also non-oscillatory solutions $\sinh \kappa x$, $\cosh \kappa x$, and $e^{\pm \kappa x}$ with imaginary $k = i\kappa$. (Recall that for the particle in the box, Example 33.7, real k emerged after application of boundary conditions.) This leads us to posit that hermiticity of $\mathbb{L}^2$ operators is determined by *both* the differential operator and boundary conditions. But then what types of boundary conditions are required to obtain a Hermitian differential operator?

To answer this, we first need to define just what an expression like $(d/dx)^\dagger$ means; clearly it makes no sense to take the transpose–complex-conjugate of a derivative. This presents us with an opportunity to exhibit the power of abstract notation: to avoid being anchored in a matrix mindset, we'll express matrix hermiticity in terms of bras and kets; then we'll ask how that abstract statement manifests in $\mathbb{L}^2$.

Actually, we've already done this. Back in (25.24a), we defined the adjoint of a linear operator $\mathcal{L}$ as that operator $\mathcal{L}^\dagger$ for which

$$\langle \psi | \mathcal{L}\phi \rangle = \langle \mathcal{L}^\dagger \psi | \phi \rangle. \tag{40.2}$$

A self-adjoint or Hermitian operator is one for which $\mathcal{L} = \mathcal{L}^\dagger$. We used this bra-ket-based definition in (27.51)–(27.52) to show that Hermitian operators have real eigenvalues and orthogonal eigenvectors. These properties should thus be features of Hermitian operators in any inner product space.

Applied to matrices, (40.2) results in the transpose–complex-conjugate recipe. But for operators acting on arbitrary functions ϕ and ψ in $\mathbb{L}^2$, it becomes

$$\int_a^b \psi^*(x)\mathcal{L}[\phi(x)]\,dx = \int_a^b [\mathcal{L}^\dagger \psi(x)]^*\phi(x)\,dx. \tag{40.3}$$

Let's use this to find the adjoint of $\mathcal{D} = d/dx$; the key is to use integration by parts to move the derivative from one vector to the other:

$$\langle \psi | \mathcal{D} | \phi \rangle = \int_a^b \psi^*(x) \frac{d\phi}{dx}\, dx = \psi^*(x)\phi(x)\Big|_a^b - \int_a^b \frac{d\psi^*}{dx}\phi(x)\, dx$$

$$= \psi^*(x)\phi(x)\Big|_a^b + \int_a^b \left(-\frac{d\psi}{dx}\right)^* \phi(x)\, dx. \tag{40.4}$$

If we restrict ourselves to functions which vanish on the boundary, we can identify $(d/dx)^\dagger = -d/dx$. Thus D is anti-Hermitian rather than Hermitian. So we would expect D^2, which requires two integrations-by-part, to be Hermitian — and therefore have real eigenvalues and orthogonal eigenfunctions (Problems 40.3–40.4).

This then is the answer as to why the solutions $\sin nx$ and $\cos nx$ are orthogonal over a full period: integer n ensures that the surface term vanishes, yielding a Hermitian operator with a complete basis of eigenfunctions and real eigenfrequencies. And since the differential equation (40.1) is linear, its general solution is a sum over the eigenbasis — in other words, a Fourier series.

40.2 The Sturm–Liouville Operator

Of course, we know of several more families of orthogonal functions; surely they too are eigenfunctions of a Hermitian operator. Indeed, $\mathcal{D}^2\phi = -k^2\phi$ is the simplest of second-order differential equations — so we're just getting started. A more general linear second-order differential eigenvalue equation can be written as

$$\mathcal{L}\phi = \left[\alpha(x)\frac{d^2}{dx^2} + \beta(x)\frac{d}{dx} + \gamma(x)\right]\phi(x) = \lambda\phi(x), \tag{40.5}$$

with real functions α, β, γ. Is $\mathcal{L}$ Hermitian? Moving the derivatives as we did before yields (Problem 40.7)

$$\langle \psi | \mathcal{L}\phi \rangle = \int_a^b \psi^*(x) \left(\alpha\frac{d^2}{dx^2} + \beta\frac{d}{dx} + \gamma\right)\phi(x)\, dx$$

$$= \int_a^b \left[\left(\alpha\frac{d^2}{dx^2} + (2\alpha' - \beta)\frac{d}{dx} + (\alpha'' - \beta' + \gamma)\right)\psi(x)\right]^* \phi(x)\, dx$$

$$+ \text{ surface term.} \tag{40.6}$$

Even though the operator inside the brackets doesn't look at all like $\mathcal{L}$, it does reduce to $\mathcal{L}$ if

$$\alpha'(x) = \beta(x). \tag{40.7}$$

This is trivially satisfied by (40.1), so you might reasonably presume that most operators $\mathcal{L}$ obeying (40.7) are too special to be of much general interest. But in fact (40.5) can be modified to obey this condition. The secret is to introduce the function

$$w(x) \equiv \frac{1}{\alpha(x)} \exp\left[\int^x \frac{\beta(t)}{\alpha(t)}\, dt\right]. \tag{40.8}$$

If we multiply (40.5) from the left by $w(x)$, the coefficient function α becomes

$$\alpha(x) \longrightarrow \tilde{\alpha}(x) \equiv w(x)\,\alpha(x) = \exp\left[\int^x \frac{\beta(t)}{\alpha(t)}\, dt\right], \tag{40.9}$$

and similarly $\beta \to \tilde{\beta} = w\beta$, $\gamma \to \tilde{\gamma} = w\gamma$. This may not seem like much — but the new coefficients $\tilde{\alpha}, \tilde{\beta}$ satisfy (40.7),

$$\tilde{\alpha}' = \frac{d}{dx} \exp\left[\int^x (\beta/\alpha)\,dt\right] = \frac{\beta(x)}{\alpha(x)} \exp\left[\int^x (\beta/\alpha)\,dt\right]$$

$$= w(x)\beta(x) \equiv \tilde{\beta}(x). \qquad (40.10)$$

A second-order differential operator (40.5) can always be made to satisfy (40.7) in this way. Renaming $\tilde{\alpha} \to -p$ and $\tilde{\gamma} \to q$ to conform with common custom, the resulting operator is said to be in *Sturm–Liouville* form

$$\mathcal{L} = -\frac{d}{dx}\left(p(x)\frac{d}{dx}\right) + q(x) = -p(x)\frac{d^2}{dx^2} - p'(x)\frac{d}{dx} + q(x), \qquad (40.11)$$

where the functions p, p', q, and w are all assumed real and well behaved between a and b, and the sign convention is set by taking $p > 0$. Then the *Sturm–Liouville* eigenvalue equation emerging from (40.5) is[1]

$$-\frac{d}{dx}\left(p(x)\frac{d}{dx}\right)\phi(x) + q(x)\phi(x) = \lambda w(x)\phi(x). \qquad (40.12)$$

The result is an operator $\mathcal{L} = -\frac{d}{dx}\left(p(x)\frac{d}{dx}\right) + q(x)$ with real eigenvalues, orthogonal eigenfunctions, and weight function $w(x)$ (Problem 40.4). In other words, $\mathcal{L}$ is Hermitian.

Well, almost — we still need to ensure that the surface term from (40.6) vanishes,

$$p(x)\left[\psi'^*(x)\phi(x) - \psi^*(x)\phi'(x)\right]\Big|_a^b = 0. \qquad (40.13)$$

We'll first explore the simplest case in which $p(x) = 0$ at both a and b. In fact, we can use this condition to choose a and b. That's because the zeros of p are singular points of the second-order equation (40.12) — so when they exist, it makes sense to use successive zeros to determine the boundaries a and b.

Example 40.1 Orthogonal Polynomials

Comparing (40.12) with Legendre's differential equation (32.64),

$$(1 - x^2)\phi'' - 2x\phi' + \ell(\ell + 1)\phi = 0, \qquad (40.14)$$

gives $p(x) = (1 - x^2) > 0$ between ± 1. Since $p'(x) = -2x$, Legendre's equation is already in Sturm–Liouville form with eigenvalue $\lambda = \ell(\ell + 1)$ and weight $w(x) = 1$. Then, since the surface term (40.13) vanishes for $a = -1$ and $b = 1$, the Legendre differential operator is Hermitian. So we need not solve the differential equation to know that its eigenfunctions are orthogonal and the eigenvalues real. (In Example 39.8 we verified that square-integrable solutions only exist for integer ℓ, so that the eigenvalues are $\lambda = 0, 2, 6, \ldots$.)

As a second-order differential equation, you might wonder about a second solution for integer ℓ. These are called the "Legendre functions of the second kind," usually denoted $Q_\ell(x)$ (see Example 39.2). As the name indicates, they are not polynomials — for example, $Q_0(x) = \frac{1}{2}\ln\left(\frac{1+x}{1-x}\right)$. Although square-integrable between ± 1, both Q_ℓ and Q'_ℓ diverge at $x = \pm 1$ so that, despite $p(\pm 1) = 0$, the surface term (40.13) does not vanish. And indeed, the Q_ℓ's are not mutually orthogonal.

---

[1] Although w has been absorbed into p and q, it still appears next to the eigenvalue λ.

Unlike Legendre's equation, Laguerre's differential equation (32.65),

$$x\phi'' + (1 - x)\phi' + n\phi = 0, \tag{40.15}$$

is not in Sturm–Liouville form. But hitting it from the left with

$$w(x) = \frac{1}{x} \exp\left[\int^x \frac{1-t}{t}\, dt\right] = \frac{1}{x} \exp\left[\ln(x) - x\right] = e^{-x} \tag{40.16}$$

gives

$$xe^{-x}\phi'' + (1 - x)e^{-x}\phi' + ne^{-x}\phi = 0 \tag{40.17a}$$

or

$$-\frac{d}{dx}\left(xe^{-x}\phi'\right) = ne^{-x}\phi. \tag{40.17b}$$

Comparison with the template of (40.12), we identify $p(x) = xe^{-x}$, $q(x) = 0$, and $\lambda = n$. Note that $p(x)$ is positive for all finite $x > 0$, but vanishes at both $x = 0$ and infinity — so the surface term in (40.13) vanishes. So once again, without actually solving the differential equation we know that the eigenfunctions $\phi_n(x)$ are orthogonal with weight e^{-x} over $0 \leq x < \infty$. (As with Legendre, the second Laguerre solution diverges at the origin; for $n = 0$, it's the exponential integral function $\mathrm{Ei}(x) = \int_{-x}^{\infty} e^{-t}/t\, dt$.)

Example 40.2   **Associated Legendre Functions**

An important generalization of (40.14) is the two-parameter equation

$$(1 - x^2)\phi'' - 2x\phi' + \left[\ell(\ell + 1) - \frac{m}{1 - x^2}\right]\phi = 0. \tag{40.18}$$

The square-integrable solutions on $[-1, 1]$ are called the *associated Legendre functions* $P_\ell^m(x)$ with $\ell = 0, 1, 2, \ldots$ and $-\ell \leq m \leq \ell$ (Problem 32.23 and Section 35.3). A Sturm–Liouville perspective shows that the $P_\ell^m(x)$ obey *two* orthogonality relations. One has weight function $w(x) = 1$; explicit calculation reveals

$$\int_{-1}^{1} P_\ell^m(x)P_{\ell'}^m(x)\, dx = \frac{2}{2\ell + 1}\frac{(\ell + m)!}{(\ell - m)!}\,\delta_{\ell\ell'}, \tag{40.19a}$$

as claimed in (35.34). The second has $w(x) = 1/(1 - x^2)$,

$$\int_{-1}^{1} P_\ell^m(x)P_\ell^{m'}(x)\,\frac{dx}{1 - x^2} = \frac{(\ell + m)!}{m(\ell - m)!}\,\delta_{mm'}, \quad m \neq 0. \tag{40.19b}$$

These expressions display separate orthogonality of P_ℓ^m in each of its eigenvalues: (40.19a) applies for fixed m, (40.19b) for fixed ℓ. In other words, neither is a "double-delta" relation $\delta_{\ell\ell'}\delta_{mm'}$. For instance, straightforward integration shows that P_3^1 and P_2^2 are not orthogonal with either weight. But for the case $w = 1$ and (40.19a), the phase $e^{im\phi}$ — itself a Sturm–Liouville eigenfunction — can be appended to $P_\ell^m(\cos\theta)$ to form the spherical harmonics $Y_{\ell m}$ in (35.35), orthogonal over the sphere in both ℓ and m.

Though $p(a) = p(b) = 0$ suffices to make the surface term (40.13) vanish (at least for ϕ, ψ, and their first derivatives finite on the boundary), in general we need to apply conditions on ϕ and ψ at both

a and b. These boundary conditions must not only be compatible with a vanishing surface term, but also ensure that the solutions χ form a vector space so that linear combinations of eigensolutions also solve. The homogeneous restrictions

$$c_1 \chi(a) + d_1 \chi'(a) = 0 \qquad c_2 \chi(b) + d_2 \chi'(b) = 0, \qquad (40.20a)$$

satisfy both requirements. (Note that $c = d = 0$ at either boundary is disallowed.) These conditions are said to be "separated" because they place constraints at a and b independently. Two special cases have their own names:

- homogeneous *Dirichlet* conditions: $\chi(a) = \chi(b) = 0$
- homogeneous *Neumann* conditions: $\chi'(a) = \chi'(b) = 0$.   (40.20b)

Dirichlet ("derishlay") conditions describe, say, a string held fixed at both ends, whereas a string with Neumann ("noy-mahn") conditions is free at each end. The general constraints of (40.20a) can then be framed as elastic conditions between these extremes.

Finally, if $p(a) = p(b)$ the surface term vanishes with *periodic boundary conditions*

$$\chi(a) = \chi(b) \qquad \chi'(a) = \chi'(b). \qquad (40.20c)$$

Both separated and periodic conditions yield self-adjoint operators with a complete orthogonal eigenbasis.

Example 40.3 **Going in Circles**

Amongst other things, $\phi'' = -k^2 \phi$ is the Schrödinger equation for a free particle with momentum $\hbar k$.[2] In Example 33.7, we confined the particle to a one-dimensional box by limiting ourselves to functions with $\phi(0) = \phi(L) = 0$. These Dirichlet boundary conditions satisfy (40.13), and indeed the eigenfunctions are the mutually orthogonal $\sin(n\pi x/L)$ for integer $n \geq 1$; the general solution is a Fourier sine series.

But if we change the boundary conditions to the periodic

$$\phi(0) = \phi(L) \qquad \phi'(0) = \phi'(L), \qquad (40.21)$$

then rather than confined to a one-dimensional segment of length L, the particle is in effect restricted to a circle of circumference L. Since the new boundary conditions satisfy (40.13), the solutions are orthogonal and the eigenvalues real. But now both sine and cosine are eigenfunctions, and the eigenfrequencies $k_n = 2n\pi/L$ are doubly degenerate for $n \neq 0$. The general solution is thus the full Fourier series

$$\phi(x) = \frac{a_0}{2} + \sum_{n=1}^{} a_n \cos\left(\frac{2n\pi x}{L}\right) + b_n \sin\left(\frac{2n\pi x}{L}\right), \qquad (40.22)$$

or equivalently,

$$\phi(x) = \sum_{n=-\infty}^{\infty} c_n e^{i2\pi nx/L}. \qquad (40.23)$$

---

[2] The $p > 0$ convention in (40.12) identifies $\lambda = +k^2$ as the eigenvalue of $\mathcal{L} = -\mathcal{D}^2$, whereas in physics we usually think of $\lambda = -k^2$ as the eigenvalue of $\mathcal{D}^2$. Though obviously not in conflict, confusion can arise when discussing the sign of λ.

This has real, physical implications: the quantized energies on the circle

$$E_n = \frac{p_n^2}{2m} = \frac{\hbar^2 k_n^2}{2m} = \frac{2\hbar^2 n^2 \pi^2}{mL^2} \qquad (40.24)$$

are four times that for the particle in a box, (33.38).

Example 40.4   Yes, It's Really Orthogonal

Let's look at $\phi'' = -k^2 \phi$ once again, this time with boundary conditions

$$\phi'(0) = 0 \qquad \phi'(L) = \phi(L). \qquad (40.25)$$

Though the differential equation may be familiar, these boundary conditions are not. But since they satisfy (40.20a) with $c_1 = 0$ and $c_2 = -d_2$, we know they yield a Hermitian operator. Thus the eigenvalues k^2 are real; we can't, however, be sure of their signs. So let's look at the three possibilities individually:

$k^2 = 0$: This is the simplest case, having as solution the line $\phi = \alpha x + \beta$. But it's even simpler than that, since the boundary conditions give $\alpha = \beta = 0$. So in fact there is no solution with eigenvalue zero.

$k^2 < 0$: With $k = i\kappa$, we can express the solution as $\phi = \alpha \cosh(\kappa x) + \beta \sinh(\kappa x)$. Conditions (40.25) give $\beta = 0$ and $\tanh(\kappa L) = +1/\kappa$ — which, as a transcendental equation, must be solved numerically (Figure 40.1a). Since tanh grows linearly for small argument before bending over and asymptotically going to 1, $\tanh(\kappa L)$ crosses $1/\kappa$ exactly once at some value we denote κ_1 to indicate the first (i.e., smallest) eigenvalue under the conditions of (40.25). The corresponding eigenfunction is $\phi_1(x) = \cosh \kappa_1 x$.

$k^2 > 0$: This has the familiar solution $\phi = \alpha \cos(kx) + \beta \sin(kx)$. As in the previous case, the condition at $x = 0$ gives $\beta = 0$ — though this time the transcendental equation arising from the condition at L is $\tan(kL) = -1/k$. Once again, this must be solved numerically (Figure 40.1b). But a simple plot reveals the basic features: since the tangent is periodic, there are multiple spots where its graph crosses that of $-1/k$. Thus even absent a full solution, we can conclude that there is an infinite tower of positive solutions $k_2 < k_3 < k_4 < \cdots$ with corresponding eigenfunctions $\phi_n(x) = \cos k_n x$.

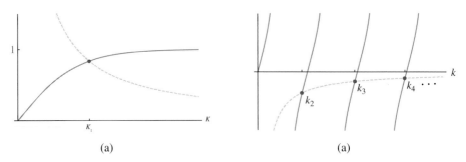

(a)   (a)

Figure 40.1 The solutions (a) κ_1 of $\tanh(\kappa L) = +1/\kappa$; (b) k_i of $\tan(kL) = -1/k$.

The full eigenfunction expansion supports the general solution

$$\phi(x) = c_1 \cosh(\kappa_1 x) + \sum_{n=2}^{\infty} c_n \cos(k_n x). \qquad (40.26)$$

Now we don't normally think of cosh as an orthogonal function. Yet as an eigensolution of a Hermitian operator, it must be orthogonal to all the other eigenfunctions. In other words, Sturm–Liouville theory requires $\int_0^L \cosh(\kappa_1 x) \cos(k_n x)\, dx = 0$ for all integer $n \geq 2$; explicit verification necessitates finding the numerical values of κ_1 and k_n (Problem 40.20).

BTW 40.1 Hermitian vs. Self-Adjoint

Our discussion in this section is motivated by what's called the *spectral theorem*, which states that the eigenvectors of a self-adjoint operator form a complete orthogonal basis. A self-adjoint operator is one which is identical to its dagger, $\mathcal{L} = \mathcal{L}^\dagger$. For matrices this is equivalent to being Hermitian; for operators acting on vectors in continuous, infinite-dimensional spaces, however, there is an important (if subtle) distinction between being Hermitian and self-adjoint.

An operator $\mathcal{L}$ is Hermitian if it satisfies

$$\langle \psi | \mathcal{L}\phi \rangle = \langle \mathcal{L}\psi | \phi \rangle. \tag{40.27}$$

This yields the transpose–complex-conjugate rule for matrices — so Hermitian matrices are also self adjoint. In $\mathbb{L}^2$, not only must the operator on either side of (40.27) have the same form, but any surface term needs to vanish. Such an operator is Hermitian. But for an operator in $\mathbb{L}^2$ to also be self-adjoint, the domains of both $\mathcal{L}$ and $\mathcal{L}^\dagger$ must be the same set of functions. Only then are $\mathcal{L}$ and $\mathcal{L}^\dagger$ truly identical; only then can one invoke the spectral theorem. So although self-adjoint operators are Hermitian, Hermitian operators are not necessarily self-adjoint. And in quantum mechanics, only self-adjoint operators with their complete orthogonal eigenbasis represent observable quantities.

We can appreciate this subtlety by examining the surface term

$$\langle \psi | \mathcal{L}\phi \rangle - \langle \mathcal{L}^\dagger \psi | \phi \rangle = p(x) \left[\psi'^*(x)\phi(x) - \psi^*(x)\phi'(x) \right] \Big|_a^b. \tag{40.28}$$

Although the domain of any operator in $\mathbb{L}^2$ includes only square integrable functions between a and b, we usually need to further restrict the space in order that the surface term vanish. Consider, for instance, the deceptively simple case of $\mathcal{D}^2$ over $0 \leq x < \infty$. For square integrable ϕ and ψ which go to zero at infinity, we have only the requirement

$$\psi'^*(0)\phi(0) - \psi^*(0)\phi'(0) = 0. \tag{40.29}$$

We can render $\mathcal{D}^2$ Hermitian for $x > 0$ by limiting its domain to $\mathbb{L}^2$ functions with $\phi(0) = \phi'(0) = 0$. This restriction, however, need not apply to ψ and ψ', which can have *any* finite values whatsoever and still satisfy (40.29). Herein lies the catch: according to (40.27), ψ is in the domain of $(\mathcal{D}^2)^\dagger$ rather than $\mathcal{D}^2$ — so the extra restriction on ϕ gives $\mathcal{D}^2$ a smaller domain than $(\mathcal{D}^2)^\dagger$. Thus $\mathcal{D}^2$ and $(\mathcal{D}^2)^\dagger$ are *not* in fact equivalent, and the spectral theorem does not apply.

There are several mathematical subtleties lurking here [see Gieres (2000) and references therein], but the crucial result is this: while Hermitian operators have real eigenvalues, only self-adjoint operators also have an orthogonal basis of eigenfunctions — a basis of normal modes. (Indeed, $\phi(0) = \phi'(0) = 0$ does not yield *any* eigenfunction solutions for $\mathcal{D}^2$.) Similarly, only self-adjoint operators can be exponentiated to get a unitary operator. These are essentially equivalent statements of the spectral theorem for self-adjoint operators.

So why is such an important distinction hiding out in a BTW? Because for most cases one encounters in physics, Hermitian operators are also self-adjoint. (Some unusual counter-examples in quantum mechanics are discussed in Problems 40.9 and 40.10.) As a result, physicists are rarely concerned with the difference, and so tend (as we do) to use the terms Hermitian and self-adjoint synonymously.

40.3 Beyond Fourier

The paradigmatic Sturm–Liouville system is that of Fourier — not merely due to the ubiquity of periodic functions, but because $\mathcal{D}^2$ is the simplest second-order differential operator. Though some properties of Fourier series are specific to sines and cosines, any Sturm–Liouville differential equation obeying either separated or periodic boundary conditions has the following features — some of which are reminiscent of Hermitian matrices:

- The eigenvectors form an orthogonal basis

$$\int_a^b \phi_m^*(x)\,\phi_n(x)\,w(x)\,dx = k_n\,\delta_{nm} \tag{40.30}$$

which is complete in the mean-square sense of (34.21). Thus any well-behaved function defined on $[a,b]$ can be expanded on the eigenbasis,

$$\psi(x) = \sum_n c_n\phi_n(x), \tag{40.31}$$

where

$$c_m = \langle \phi_m|f\rangle = \frac{1}{k_n}\int_a^b \phi_m^*(x) f(x)\, w(x)\, dx. \tag{40.32}$$

This eigenfunction expansion is sometimes called a *generalized Fourier series*. [Incidentally, aside from restrictions imposed by boundary conditions, the domain of a Sturm–Liouville operator is not all of $\mathbb{L}^2$, but only those functions ϕ which have second derivatives. Remarkably, this restricted set of functions still provides a basis for $\mathbb{L}^2$.]
- The eigenvalues λ are all real.
- For $b - a$ finite, the spectrum of eigenvalues forms a discrete, countably infinite ordered set (as we saw in Example 40.4),

$$\lambda_1 \le \lambda_2 \le \lambda_3 \le \cdots, \tag{40.33}$$

and this natural progression is used to label the eigenfunctions ϕ_n.[3] Since $\lambda_{n\to\infty}$ goes to infinity, there is no highest eigenvalue. However, the lower bound

$$\lambda_n \ge \frac{1}{k_n}\left[-p\phi_n^*\phi_n'\big|_a^b + \int_a^b q(x)|\phi_n(x)|^2\, dx\right] \tag{40.34}$$

means that there is a lowest eigenvalue (Problem 40.27). So for Sturm–Liouville operators with both $q \ge 0$ and $-p\phi_n^*\phi_n'\big|_a^b \ge 0$, all the eigenvalues are positive.

---

[3] When the lowest eigenvalue is zero, the numbering convention is often shifted to begin at $n = 0$.

- Real eigenfunctions are oscillatory — specifically, between a and b there are $n-1$ nodes α_i where $\phi_n(\alpha_i) = 0$. Oscillatory, however, does not imply periodic: unlike sine and cosine, these zeros need not be uniformly spaced. Nor must ϕ have fixed amplitude.
- For separated boundary conditions (40.20a), the eigenvalue spectrum is non-degenerate. With periodic conditions (40.20c), degeneracies can occur — though a second-order differential equation has only two independent solutions, so at most the eigenfunctions are doubly-degenerate.

Sturm–Liouville systems appear throughout physics and engineering, but they have particular resonance in quantum mechanics. Schrödinger surmised (correctly) that the theoretical difficulties posed by the observed discrete energies of hydrogen could be described by a Sturm–Liouville eigenvalue equation — and indeed, his eponymous time-independent equation,

$$-\frac{\hbar^2}{2m}\phi''(x) + V(x)\phi(x) = E\phi(x), \tag{40.35}$$

is just that: a Sturm–Liouville equation with real, discrete eigenvalues E. Furthermore, those discrete eigenvalues precisely coincide with square-integrable eigenfunctions (Problems 40.18 and 40.19). And knowing that there exists a lowest eigenvalue guarantees that quantum systems have a minimum energy ground state. For example, the superfluid phase of liquid helium flows without friction because in its ground state, it cannot lose energy to dissipative forces.

We explored in Examples 33.7 and 40.3 how different boundary conditions affect the particle in a box; in the following examples, we'll vary the shape of the well and examine some of the non-Fourier eigensolutions which result.

Example 40.5 The Infinite Circular Well

Rather than restricting a particle in a one-dimensional box of length L, consider a particle confined within a disk of radius a,

$$V(r) = \begin{cases} 0, & 0 \leq r < a \\ \infty, & r \geq a \end{cases}. \tag{40.36}$$

This potential differs from the one-dimensional square well of (33.29) only in the independent variable $r = \sqrt{x^2 + y^2}$. But in the Schrödinger equation (33.31), d^2/dx^2 must be replaced with the Laplacian ∇^2, which in polar coordinates gives

$$\frac{1}{r}\frac{\partial}{\partial r}\left(r\frac{\partial u}{\partial r}\right) + \frac{1}{r^2}\frac{\partial^2 u}{\partial \phi^2} = -k^2 u, \tag{40.37}$$

where $k^2 = 2mE/\hbar^2$ as before. We'll tackle here the simplest case of circularly symmetric solutions, for which $u(\vec{r})$ reduces to the purely radial function $R(r)$ — resulting in an effective one-dimensional equation,

$$\frac{d}{dr}\left(r\frac{dR}{dr}\right) = -k^2 r R(r) \tag{40.38}$$

This is the simplest form of Bessel's equation. As a second-order equation it has two solutions, the Bessel function J_0, and the *Neumann function* N_0, where the zero subscripts indicate the absence of angular influence in (40.38). Thus

$$R(r) = AJ_0(kr) + BN_0(kr). \tag{40.39}$$

Though oscillatory, these functions are not periodic; in fact, neither can be expressed in terms of more elementary functions. Nonetheless, (40.38) is a Sturm–Liouville system with $p = r$, $q = 0$, weight $w = r$, and eigenvalue $\lambda = +k^2$ — hence the solutions will be orthogonal if the surface term in (40.13) vanishes. Do the boundary conditions make that happen? Well, $r = 0$ is no problem, but as it represents the center rather than the edge of the well, there is no condition on R other than continuity. As shown in Appendix C, Neumann functions diverge at the origin, so we discard it by setting $B = 0$. The solution must also be continuous at $r = a$, so confining a particle inside the circular well requires $R(a) = 0$. This restricts ka to be a zero α_n of the Bessel function,

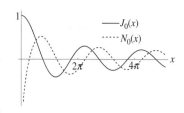

$$J_0(ka) = 0 \quad \Longrightarrow \quad k_n = \alpha_n/a. \tag{40.40}$$

A similar restriction to zeros of the eigenfunction occurs in (33.33)–(33.34) for the square well. But unlike sine and cosine, the zeros of J_0 are not integer multiples of π — the first three, for instance, are[4]

$$\alpha_1 = 2.4048 \qquad \alpha_2 = 5.5201 \qquad \alpha_3 = 8.6537. \tag{40.41}$$

Thus the eigenvalues k are quantized — though rather than the square well energies of (33.38), the circular well has energies

$$E_n = \frac{\hbar^2 k_n^2}{2m} = \frac{\hbar^2 \alpha_n^2}{2ma^2}. \tag{40.42}$$

With the application of boundary conditions, the Sturm–Liouville surface term vanishes and the eigensolutions $J_0(\alpha_n r/a)$ form an orthogonal basis. As a Sturm–Liouville system, one can readily find the orthogonality relation (Problem 40.23)

$$\int_0^a J_0(\alpha_n r/a) \, J_0(\alpha_m r/a) \, r \, dr = \frac{a^2}{2} J_0'(\alpha_n)^2 \, \delta_{mn}. \tag{40.43}$$

(Though this may look daunting, it's essentially the same expression as Fourier orthogonality — see Problem 40.23.) The general solution of the infinite circular well is a linear combination of eigensolutions,

$$\psi(r) = \sum_n c_n R_n(r) = \sum_n c_n J_0(\alpha_n r/a), \tag{40.44}$$

where the expansion coefficients are given by the projection

$$c_n = \frac{2}{a^2 J_1(\alpha_n)^2} \int_0^a J_0(\alpha_n r/a) \, \psi(r) \, r \, dr. \tag{40.45}$$

Together, (40.44)–(40.45) define a *Fourier–Bessel series*, and should be compared with the Fourier series of (33.37) — which, as was noted there, can also be construed as an expansion over zeros of a single function.

---

[4] These can be found in tables — or using the Mathematica command BesselJZero.

Figure 40.2 The first four circularly symmetric normal modes, $J_0(k_n r)$.

The first four normal modes are shown in Figure 40.2. Due to the circular symmetry, J_0 is readily apparent in the radial profiles. And it's no coincidence that these figures resemble waves on the surface of water in a cylindrical container, or the vibrations of a circular drumhead: Bessel functions are the normal modes of those systems as well. We'll have much more to say about this in Chapter 41.

Example 40.6 **Quarkonium**

As a final variation on the quantum particle in a box, let's replace the right-hand wall of the infinite well with an increasing linear potential,

$$V(x) = \begin{cases} \sigma x, & x > 0 \\ \infty, & x < 0 \end{cases}, \tag{40.46}$$

where σ is a constant with dimensions of force. This describes, for instance, a particle in a gravitational field near the surface of the Earth, $\sigma = mg$. It's also a useful model for the constant attractive force between a heavy quark–anti-quark pair in a meson: since increasing the distance x between them requires more and more energy, the quarks are confined in a bound state called *quarkonium*.[5] The picture is that of an elastic string holding the quarks together with a tension σ on the order 1 GeV/fm — meaning that a billion electron volts of energy are needed to separate the quarks by a mere 10^{-15} m!

The wavefunction describing this system is found by inserting the quarkonium potential into the Schrödinger equation (40.35):

$$\frac{d^2\phi}{dx^2} + \frac{2m}{\hbar^2}(E - \sigma x)\phi = 0. \tag{40.47}$$

This is a Sturm–Liouville system with $p = 1$, $q = -2m\sigma x/\hbar^2$, weight $w = 1$, and eigenvalue $\lambda = 2mE/\hbar^2$. For given energy E, the behavior of the solution depends on x: for small $\sigma x < E$, the wavefunction ϕ is oscillatory, whereas for large $\sigma x > E$ an exponential behavior takes over. (Contrast this with $J_n(x)$, which is oscillatory for all x.) It would be a little cleaner were this behavior shift to occur at the origin — so we recast (40.47) in terms of the variable $u \equiv \kappa(x - E/\sigma)$, where κ is a constant introduced to make u dimensionless. A little algebra then reveals the standard form of *Airy's equation* for $\xi(u) = \phi(x)$,

$$\frac{d^2\xi}{du^2} = u\,\xi, \tag{40.48}$$

with $\kappa^3 = 2m\sigma/\hbar^2$ (Problem 40.21). The solutions are the *Airy functions* Ai(u) and Bi(u); as Figure 40.3 shows, they display precisely the behavior we anticipated for positive and negative u.

---

[5] The usual quarkonium potential between heavy quarks includes a Coulomb-like attraction — but for large enough separation, only the confining portion given in (40.46) contributes.

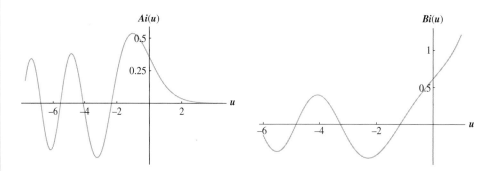

Figure 40.3 The Airy functions Ai(u) and Bi(u). (Note the different scales.)

To obtain real energy eigenvalues and orthogonal eigenfunctions requires, of course, that the by-now-familiar surface term (40.13) vanishes at both ends of the potential. Whereas Ai goes to zero at infinity, Bi blows up — so we discard it. Continuity then requires that the eigenfunction goes to zero where the infinite potential wall looms at $x = 0$:

$$\xi(u)\Big|_{x=0} = \text{Ai}\,(-\kappa E/\sigma) = 0. \tag{40.49}$$

This condition yields discrete, quantized energies

$$E_n = -\left(\frac{\sigma}{\kappa}\right)\alpha_n = -\left(\frac{\hbar^2\sigma^2}{2m}\right)^{1/3}\alpha_n > 0, \quad n = 1, 2, 3, \ldots, \tag{40.50}$$

where the $\alpha_n < 0$ are the zeros of Ai. As with Bessel functions, the zeros are not evenly distributed; the first three are[6]

$$\alpha_1 = -2.3381 \qquad \alpha_2 = -4.0880, \qquad \alpha_3 = -5.5206. \tag{40.51}$$

With the vanishing of the surface term, we obtain a complete basis of eigenfunctions $\phi_n(x) = \text{Ai}\,(\kappa x + \alpha_n)$ obeying the orthogonality relation

$$\int_0^\infty \text{Ai}\,(\kappa x + \alpha_n)\,\text{Ai}\,(\kappa x + \alpha_m)\,dx = \frac{1}{\kappa}\text{Ai}'(\alpha_n)^2\,\delta_{mn}, \tag{40.52}$$

with quantized eigenvalues $\lambda_n = -\kappa^2\alpha_n > 0$. The general solution to (40.47) is then the linear combination of eigensolutions

$$\psi(x) = \sum_n c_n\phi_n(x) = \sum_n c_n\,\text{Ai}\,(\kappa x + \alpha_n), \tag{40.53}$$

where the expansion coefficients are given by the projection

$$c_n = \frac{\kappa}{\text{Ai}'(\alpha_n)^2}\int_0^\infty \text{Ai}\,(\kappa x + \alpha_n)\,\psi(x)\,dx. \tag{40.54}$$

---

[6] These can be found in tables — or using the Mathematica command `AiryAiZero`.

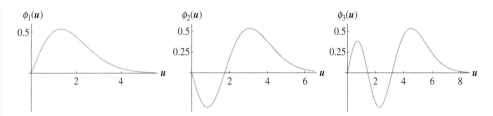

Figure 40.4 The first three quarkonium eigensolutions.

The unfamiliar shifts by α_n in the eigenfunctions $\phi_n(x) = \mathrm{Ai}\,(\kappa x + \alpha_n)$ are essentially dictated by the Sturm–Liouville criteria. Although the physical problem is defined only for $x > 0$, the oscillations of the Airy functions occur exclusively for negative arguments. Since the zeros of Ai are needed at $x = 0$, they are shifted, one by one and eigenfunction by eigenfunction, into the region $x > 0$. Moreover, in order to be orthogonal Sturm–Liouville, eigenfunctions must be oscillatory — and indeed, you can appreciate from Figure 40.4 that alternating regions of positive and negative ϕ_n are needed in order for the orthogonality integral to vanish.

Although these examples have been of the Schrödinger equation, Sturm–Liouville systems appear in both classical and quantum physics, so the occurrence of so-called "special" functions such as Legendre and Bessel is not rare. Moreover, just as whenever we happen upon the differential equation $\phi'' = -k^2\phi$ we can just write $\phi(x) = A\cos kx + B\sin kx$ without laboring to solve the differential equation, familiarity with special functions allows us to do the same — should we be fortunate (or clever) enough to recognize their defining differential equations.

Problems

40.1   What are the eigenvalues and eigenfunctions of $\mathcal{D} = d/dx$? Though it may seem counterintuitive, $\mathcal{D}$ needs a factor of i to have real eigenvalues. Show under what conditions $i\mathcal{D} = i\,d/dx$ is Hermitian for:

(a) $x \in [-\pi, \pi]$
(b) $x \in [0, \pi]$.

Verify that the eigenfunctions are orthogonal and the eigenvalues real.

40.2   Find $(id/dx)^\dagger$ if the surface term does not vanish. [Hint: Think δ function.]

40.3   Show under what conditions $\mathcal{D}^2 = \frac{d^2}{dx^2}$ is a Hermitian operator for $x \in [-\pi, \pi]$. Verify that the eigenfunctions are orthogonal and the eigenvalues real.

40.4   Use (40.3) to show quite generally that with a vanishing surface term, the resulting Hermitian operator $\mathcal{L}$ has:

(a) real eigenvalues λ
(b) orthogonal eigenfunctions with non-degenerate λ.

Do each twice: once in integral form, and once using only bra-ket notation.

40.5   (a) Show that $\mathcal{D}^2 = \frac{d^2}{dx^2}$ is a Hermitian operator with either Dirichlet or Neumann boundary conditions, (40.20b).
(b) Show, both in integral form and with bra-ket notation, that the eigenvalue spectrum of $\mathcal{D}^2$ with Neumann conditions is positive or zero.
(c) Show, both in integral form and with bra-ket notation, that the eigenvalue spectrum of $\mathcal{D}^2$ with Dirichlet conditions is strictly positive — that is, $\lambda = 0$ is not an eigenvalue.

40.6 (a) Presuming that the integrals converge, show that the position operator X, defined by $X\phi(x) = x\phi(x)$, is Hermitian for real x. Then argue that any function $f(X)$ which can be Taylor expanded is also Hermitian.

(b) Find the eigenfunction of X.

(c) So both X and $\mathcal{P} = -i\mathcal{D}$ (with appropriate boundary conditions) are Hermitian. What about the product $X\mathcal{P}$? (See Problem 27.32.)

40.7 Verify (40.6), including the surface term, (40.13).

40.8 The *spectral theorem* states that the eigenvectors of a self-adjoint operator A form a complete orthogonal basis. Thus any $\mathbb{L}^2$ vector $|\psi\rangle$ can be expanded on the eigenbasis of A, $|\psi\rangle = \sum_n c_n |a_n\rangle$, where $A|a_n\rangle = a_n|a_n\rangle$. This means that $|\psi\rangle$ itself does not have a definite value of A (unless all the c_n's but one are zero). For ψ normalized, $\langle\psi|\psi\rangle = \sum_n |c_n|^2 = 1$, quantum mechanics says that $|c_n|^2$ is the probability a measurement of A on ψ yields a_n. Show that the average or *expectation value* of A,

$$\langle A\rangle = \sum_n |c_n|^2 a_n$$

is equivalent to the basis-independent expression

$$\langle A\rangle = \langle\psi|A|\psi\rangle.$$

Problems 40.9 and 40.10 rely on BTW 40.1.

40.9 The quantum mechanical momentum operator is given by $P = -id/dx$. (For simplicity, we've set $\hbar = 1$.)

(a) Is P Hermitian for the infinite square well of (33.29)? self adjoint? What are the implications for the observability of momentum eigenstates of the square well?

(b) What about the square well energies, $P^2/2m$? What makes this different than P?

(c) Examine the surface term to find a general boundary condition which would render P self-adjoint. Give a physical interpretation.

40.10 Consider a particle of mass m in the infinite square well of (33.29), with normalized wavefunction

$$\psi(x) = \begin{cases} \sqrt{\frac{30}{L^3}}\, x\left(1 - \frac{x}{L}\right), & 0 < x < L \\ 0, & \text{else.} \end{cases}$$

As we saw in Problem 40.8, there are two equivalent ways to calculate the expectation value of a self-adjoint operator.

(a) Expand $\psi(x)$ on the basis of energy eigenfunctions $\phi_n(x) = \sqrt{\frac{2}{L}}\sin\left(\frac{n\pi x}{L}\right)$. (Feel free to use Mathematica.)

(b) Use the eigenenergies $E_n = n^2\pi^2\hbar^2/2mL^2$ and the expansion coefficients from (a) to calculate the expectation value $\langle H^2\rangle$ the first way, where $H = P^2/2m$. (You may find $\sum_{n=\text{odd}} \frac{1}{n^2} = \frac{1}{8}\pi^2$ helpful.)

(c) Now calculate the expectation value the second way, $\langle H^2\rangle = \langle\psi|H^2|\psi\rangle$.

(d) Resolve the discrepancy between these two calculations of $\langle H^2\rangle$.

40.11 Put Hermite's equation (32.62),

$$H_n'' - 2xH_n' + 2nH_n = 0, \qquad |x| < \infty,$$

into Sturm–Liouville form.

40.12 Put Chebyshev's equation

$$(1 - x^2)T_n'' - xT_n' + n^2 T_n = 0, \qquad |x| < 1,$$

into Sturm–Liouville form.

40.13 (a) Use (32.68) to find the weight function $w(x)$ for the associated Laguerre polynomials $L_m^k(x)$. In what parameter are they orthogonal?

(b) Verify that the functions $\phi_m^k(x) \equiv \sqrt{w} L_m^k(x)$ are orthogonal with $w = 1$, and that the associated Laguerre equation becomes the Sturm–Liouville

$$x\phi_m^{k\,\prime\prime} + \phi_m^{k\,\prime} - \frac{1}{4x}[x^2 - 2(k+1)x + k^2]\phi_m^k = -m\phi_m^k. \tag{40.55}$$

40.14 Show that the eigenfunctions of a Sturm–Liouville operator $\mathcal{L}$ can be chosen to be real.

40.15 Verify that both separated and periodic boundary conditions of (40.20) yield Hermitian Sturm–Liouville operators.

40.16 **The Hydrogen Atom**

Consider the Schrödinger equation for the hydrogen atom. With Coulomb potential $V(r) = -e^2/4\pi\epsilon_0 r$, the radial portion $R(r)$ of the electron wavefunction is the solution to

$$-\frac{\hbar^2}{2m_e}\frac{1}{r^2}\frac{d}{dr}\left(r^2\frac{dR}{dr}\right) + \left[-\frac{e^2}{4\pi\epsilon_0}\frac{1}{r} + \frac{\hbar^2\ell(\ell+1)}{2m_e r^2}\right]R = ER,$$

where m_e is the electron mass. The derivatives come from the radial portion of the Laplacian in spherical coordinates. The term proportional to $\ell(\ell+1)$ is the influence of the angular part, the famed centrifugal barrier $L^2/2I$ of classical physics — but here with integer $\ell \geq 0$ coming from the Legendre polynomials. Since the Schrödinger equation is a Hermitian eigenvalue equation, it's not unreasonable for our analysis to be guided by Messrs. Sturm and Liouville.

(a) The first step is to identify physical scales with which to simplify the equation. With the inverse length scale $\kappa = \sqrt{-2m_e E}/\hbar > 0$ (bound state energies E are negative), rewrite the equation for R in terms of the dimensionless variable $z \equiv \kappa r$ to find

$$\frac{d}{dz}\left(z^2\frac{dR}{dz}\right) + \left[\frac{2z}{\kappa a_0} - \ell(\ell+1)\right]R = z^2 R,$$

where the Bohr radius $a_0 = \hbar/m_e c\alpha = 0.5A$, and the fine structure constant $\alpha = e^2/4\pi\epsilon_0\hbar c = 1/137$. [The electron mass, $m_e c^2 = 0.511$ MeV.] It's best to combine the two natural length scales a_0 and $1/\kappa$ into a single dimensionless parameter — so rearrange terms to put the equation into Sturm–Liouville form with eigenvalue $\epsilon \equiv 2/\kappa a_0$.

(b) Define $\phi(z)$ in terms of R to get

$$z\phi'' + \phi' - \frac{1}{4z}[4z^2 + (2\ell+1)^2]\phi = -\epsilon\phi,$$

so that the solutions ϕ are orthogonal with weight 1 and eigenvalue ϵ.

(c) The equation is now in a form remarkably similar to (40.55) in Problem 40.13 for the associated Laguerre polynomials; show that with $x = 2z$ the equations are the same. What are the quantized energies? Then verify that, up to normalization, the radial wavefunctions are

$$R_{n\ell}(r) = \left(\frac{2r}{na_0}\right)^\ell e^{-r/na_0} L_{n-\ell-1}^{2\ell+1}(2r/na_0),$$

where integer $n \geq 1$.

40.17 **The Quantum Pendulum**

Energy eigenfunctions of the quantum plane pendulum of mass m and length ℓ are solutions to the time-independent Schrödinger equation

$$-\frac{\hbar^2}{2m\ell^2}\frac{d^2\psi}{d\theta^2} + mg\ell(1 - \cos\theta)\psi = E\psi, \qquad \psi(\theta) = \psi(\theta + 2\pi), \tag{40.56}$$

where θ is the usual angular displacement from the equilibrium position $\theta = 0$.

(a) Without solving the differential equation, show that the wavefunctions ψ form a complete orthogonal basis.

(b) Recast (40.56) into the form of Mathieu's equation

$$\frac{d^2\psi}{d\eta^2} + \left[a - 2q\cos(2\eta)\right]\psi = 0 \tag{40.57}$$

with dimensionless parameters a and q. The solutions are the even and odd Mathieu functions $C(a,q,\eta)$ and $S(a,q,\eta)$. These are highly non-trivial, usually requiring numerical rather than analytic methods. But what do they simplify to for $q = 0$?

(c) Looking at (40.57), it's not hard to appreciate that C and S are periodic only for specific "characteristic" values of a. In fact, there are ordered sets of characteristic functions, $a_n(q)$ for C and $b_n(q)$ for S, which yield periodic solutions. (As with the Fourier basis, a_n starts at $n = 0$, b_n at $n = 1$.) With these, one can construct the one-parameter periodic Mathieu functions $ce_n(q,\eta)$ and $se_n(q,\eta)$ — sometimes called "elliptic cosine" and "elliptic sine" functions. Though non-trivial, Mathematica can lessen the challenge (Trott, 1996):

$$ce_{n\_Integer?NonNegative}[q\_,\eta\_] = \text{MathieuC}[\text{MathieuCharacteristicA}[q,n],q,\eta] \tag{40.58}$$

$$se_{n\_Integer?Positive}[q\_,\eta\_] = \text{MathieuS}[\text{MathieuCharacteristicB}[q,n],q,\eta].$$

Use Mathematica's Plot3D command to make plots of ce_n and se_n for $n = 0, 1, 2$ with $0 \le \eta \le 2\pi$ and $0 \le q \le 5$.

(d) Integrate pairs of ce_n and se_n in Mathematica for various values of $q > 0$, and extrapolate the results to determine the orthogonality relations for $ce_n(q,\eta)$ and $se_n(q,\eta)$.

(e) From the connection between Mathieu's η and the pendulum's θ, inspect the plots in (c) to show that the periodicity of the pendulum restricts the characteristic values to even-indexed a_{2n} and b_{2n}. Use Mathematica to find the pendulum's five lowest energies for $q = 4$ in units of $\hbar^2/2m\ell^2$.

40.18   **The Importance of Being Square Integrable (Analytical)**

(a) Show that the only $E \le 0$ solution of the infinite square well compatible with the boundary conditions is the non-normalizable $\phi(x) = 0$. (Thus the ground state necessarily has $E > 0$.)

(b) Use considerations of slope and curvature to argue that normalizable solutions of the Schrödinger equation (40.35) must have $E > V_{\text{min}}$.

40.19   **The Importance of Being Square Integrable (Numerical)**

Perhaps the best way to appreciate the collusion between the square-integrability of Sturm–Liouville solutions and their eigenvalues is graphically. Consider the *finite* square well centered at the origin,

$$V(x) = \begin{cases} 0, & |x| < L/2 \\ V_0, & |x| > L/2 \end{cases}, \tag{40.59}$$

where the walls of the well have finite height V_0. To prepare for numerical solution, express the Schrödinger equation using the dimensionless $\xi \equiv x/L$,

$$\frac{d^2\phi}{d\xi^2} = -\frac{2mL^2}{\hbar^2}[E - V]\phi(\xi) \equiv -[\varepsilon - v(\xi)]\phi(\xi),$$

where $v(\xi)$ and ε are the potential and eigenenergy measured in units of $\hbar^2/2mL^2$. The only free parameter is the (dimensionless) wall height v_0; this we want to set high enough to ensure at least two bound states. Using the infinite well eigenenergies $\varepsilon = n^2\pi^2$ as a guide, take $v_0 = 25$.

Solve this eigenvalue problem via the so-called "shooting" method: set the initial conditions $\phi(0), \phi'(0)$, and vary ε until the value which yields a square integrable solution is found. For even solutions, take

$\phi(0) = 1$, $\phi'(0) = 0$; for odd, $\phi(0) = 0$, $\phi'(0) = 1$. (Since the eigenvalue problem does not dictate overall normalization, 1 can be replaced by any non-zero value.)

Use Mathematica's NDSolve command to solve for $\phi(\xi)$. Plotting the solutions, you should discover that for almost all values of ε, the solutions diverge at large $|\xi|$ and hence are not square integrable. By carefully varying ε, however, you should be able to zero in on acceptable solutions at specific eigenenergies. Examine plots to the find both the ground state (even) and the first excited state (odd) to five significant figures.

40.20 For the system in Example 40.4, use Mathematica to show that the ground state $\phi_0(x) = \cosh(\kappa_0 x)$ is orthogonal to the first three eigenstates $\phi_n(x) = \cos(k_n x)$.

40.21 Derive Airy's equation (40.48) from the quarkonium Schrödinger equation (40.47).

40.22 For the Sturm–Liouville operator $\mathcal{L} = -\frac{d}{dx}\left(p\frac{d}{dx}\right) + q$:

(a) Verify that

$$\int_a^b \left[\psi^* (\mathcal{L}\phi) - \phi^* (\mathcal{L}\psi)\right] dx = p \left(\psi'\phi - \psi\phi'\right) \big|_a^b. \qquad (40.60)$$

(b) For non-degenerate eigenstates ϕ_i and ϕ_j of $\mathcal{L}$ with weight w, show that

$$\int_a^b \phi_i^*(x)\phi_j(x)\, w(x)\, dx = \frac{p\left(\phi_i'\phi_j^* - \phi_i^*\phi_j'\right)\big|_a^b}{\lambda_j - \lambda_i}. \qquad (40.61)$$

Note that orthogonality only obtains when the surface term goes to zero.

40.23 Use (40.61) to verify the orthogonality relations for the following:

(a) Fourier functions $\sin(n\pi x/a)$ between 0 and a
(b) Bessel functions J_0, (40.43)
(c) Airy functions Ai, (40.52).

40.24 Problem 40.23 establishes orthogonality when the independent variable is scaled by the functions' zeros. Use (40.61) to show that these same functions are similarly orthogonal when scaled with the zeros of their derivatives.

40.25 Use Legendre's equation (32.64) to show that the derivatives of the Legendre polynomials $P_n'(x)$ must be orthogonal over $[-1, 1]$ with weight function $1 - x^2$. (You need not find the normalization k_n.)

40.26 Consider the Sturm–Liouville operator $\mathcal{L}$ such that

$$\mathcal{L}\psi(x) = f(x),$$

for some $\mathbb{L}^2$ function $f(x)$. Presuming that ψ satisfies separated or periodic boundary conditions, it can be expanded on a basis of $\mathbb{L}^2$'s eigenfunctions. Use this to show that on the orthonormal eigenbasis of ϕ_n's,

$$\psi(x) = \sum_n \frac{1}{\lambda_n}\phi_n(x) \int_a^b \phi_n^*(x')f(x')\, dx' = \int_a^b G(x,x')f(x')\, dx'$$

(presuming, of course, $\lambda_n \neq 0$), where the function

$$G(x, x') \equiv \sum_n \frac{1}{\lambda_n}\phi_n(x)\phi_n^*(x') \qquad (40.62)$$

is called the *Green's function* of $\mathcal{L}$. Comparison with the original differential equations shows that G is essentially the inverse of $\mathcal{L}$ — in bra-ket notation,

$$|\psi\rangle = \sum_n \frac{1}{\lambda_n}|\phi_n\rangle\langle\phi_n|f\rangle = G|f\rangle, \qquad \text{where } G = \sum_n \frac{1}{\lambda_n}|\phi_n\rangle\langle\phi_n|.$$

40.27 (a) Express the Rayleigh quotient introduced in Problem 4.14 in bra-ket notation.

(b) Replacing $\vec{v}$ with $\phi(x)$, and the real symmetric matrix A with the Hermitian differential operator $\mathcal{L}$, use the Rayleigh quotient to find an integral expression for the eigenvalue λ of a Sturm–Liouville eigenfunction.

(c) Confirm that each eigenvalue λ_n has the lower bound of (40.34), and therefore that Sturm–Liouville systems have a lowest eigenvalue.

40.28 (a) Apply the Sturm–Liouville operator $\mathcal{L}$ to a superposition of orthogonal eigenfunctions $\psi = \sum_n c_n \phi_n$ to find an upper bound on the lowest eigenvalue,

$$\lambda_1 \leq \frac{\langle\psi|\mathcal{L}|\psi\rangle}{\langle\psi|\psi\rangle}. \tag{40.63}$$

Under what circumstances does the equality hold? What happens if ψ is orthogonal to the ground state?

(b) Consider the centered infinite square well

$$V(x) = \begin{cases} 0, & |x| < L/2 \\ \infty, & |x| > L/2 \end{cases}.$$

Assume we do not know the solutions; instead, with a good ansatz for the ground state solution, we can use (40.63) to estimate the ground state energy. Since the ground state is even and must vanish at $\pm L/2$, a parabolic trial solution seems not unreasonable,

$$\psi(x) = A\left[1 - 4\left(\frac{x}{L}\right)^2\right],$$

where A is the normalization constant. Use this choice for ψ in (40.63) to get an upper bound of the ground state energy. Compare it to the known ground state energy.

(c) The first excited state is odd, so it's automatically orthogonal to the ground state. Thus a reasonable odd trial function should give a good estimate for the energy. Come up with such a trial function, and use it to estimate the energy. Compare it to the known energy.

40.29 We'll use the method of Problem 40.28 to numerically estimate the ground and first excited eigenenergies of the finite square well in (40.59).

(a) As a warm-up, take as trial wavefunction $\psi(\xi) = Ae^{-\xi^2}$ to find an estimate for the ground state. (You can use Mathematica if you like.) Comparison with your result in Problem 40.19 should reveal OK, but not great, agreement. We can do better.

(b) Introduce a parameter b into the trial wavefunction,

$$\psi(\xi) = Ae^{-b\xi^2},$$

and use this to find an upper bound as a function of b. Then with Mathematica's FindRoot command, find the value of b which minimizes the upper bound to get a better estimate for the ground state energy.

(c) Repeat the procedure for the first excited state (which is odd).

41 Partial Differential Equations

By itself, a single ordinary differential equation is limited in its capacity to describe physical systems. Living in three spatial dimensions as we do, ODEs are simply inadequate as descriptions of the natural world. Encompassing a wider range of phenomena requires dealing with functions of more than one independent variable — and for that, we need partial differential equations (PDEs).

There are three basic types of linear second-order PDEs; their point of departure is the Laplacian operator ∇^2. As the natural extension of the second derivative d^2/dx^2 to higher dimensions, ∇^2 has relevance for determining extrema — though as briefly discussed in Section 12.4.1, it's not as simple as in one dimension. Echoing the "concave up/concave down" rule of introductory calculus, if $\nabla^2 \Phi(\vec{r}_0) > 0$ at some point $\vec{r}_0$, then $\Phi(\vec{r}_0)$ is smaller than the *average* of Φ over neighboring points; if $\nabla^2 \Phi(\vec{r}_0) < 0$, then $\Phi(\vec{r}_0)$ is larger than the surrounding average. So one can think of the Laplacian as a measure of the difference between the value of Φ at $\vec{r}_0$ and its average value over the surface of a sphere centered at $\vec{r}_0$. The third basic type is enshrined in Laplace's equation,

$$\nabla^2 \Phi(\vec{r}) = 0, \tag{41.1}$$

for which $\Phi(\vec{r}_0)$ is equal to the surrounding average. Indeed, any extremum within the sphere could not be the average over the sphere. Perhaps the most familiar physics exemplar is electrostatics, which in a charge-free region is governed by Laplace's equation. So the voltage at an interior point is the average of the voltage at neighboring points.

As another example, consider the temperature distribution within a material. The rate at which temperature T changes at $\vec{r}_0$ is proportional to the difference in temperature with its surroundings. But if the temperature at a point is the average of its surroundings, then any heat flow in must be balanced with heat flow out, $\vec{\nabla} \cdot \vec{\nabla} T = \nabla^2 T = 0$ — and thus $T = T(\vec{r})$, independent of time. A temperature distribution satisfying Laplace's equation is said to be static or steady state.[1]

Of course, this implies that a distribution with $\nabla^2 T \neq 0$ must be time dependent, its non-zero time derivative at $\vec{r}_0$ determined by the surrounding average temperature (see Problem 16.2),

$$\nabla^2 T = \frac{1}{\alpha} \frac{\partial T}{\partial t}. \tag{41.2}$$

This PDE is called the *heat equation*. The constant α is the material's thermal diffusivity, with dimensions distance^2/time.

Unlike the heat equation, the *wave equation* has a second-order time derivative,

$$\nabla^2 \Psi = \frac{1}{v^2} \frac{\partial^2 \Psi}{\partial t^2}, \tag{41.3}$$

---

[1] Since Laplace's equation describes both this and electrostatics, static temperature can be analyzed as a DC circuit, with T and heat flow the analogs of voltage and electric current.

where v is the speed of the wave. This can be viewed as an acceleration $\partial_t^2 \Psi$ proportional to the difference of Ψ from its local average — not unlike Newton's second law for a spring displaced from equilibrium. But clearly (41.3) treats space and time on a more-equal footing than the heat equation. Consider a wave on a string. A point on the string oscillates harmonically in time with frequency ω; similarly, a snapshot of the string reveals a simple harmonic undulation in space with frequency k. But they're not independent: the collusive link is the wave speed, $v = \omega/k$.

BTW 41.1 Elliptic, Hyperbolic, and Parabolic

Consider the general linear second-order PDE in two dimensions (either both spatial, or one space and one time) in the form

$$\left[a\,\frac{\partial^2}{\partial u^2} + b\,\frac{\partial^2}{\partial u \partial v} + c\,\frac{\partial^2}{\partial v^2} + d\,\frac{\partial}{\partial u} + e\,\frac{\partial}{\partial v} + f \right] \Psi = r(u,v). \tag{41.4}$$

This PDE can be classified based on the coefficients of its second-order terms — so we set the lower-order terms aside and focus on the operator

$$\mathbb{D}^2 = a\,\partial_u^2 + b\,\partial_u\partial_v + c\,\partial_v^2 = D^T A D, \tag{41.5}$$

where D^T is the row vector (∂_u, ∂_v), and

$$A = \begin{pmatrix} a & b/2 \\ b/2 & c \end{pmatrix}. \tag{41.6}$$

The symmetry of A derives from the equivalence of mixed partials, $\partial_u\partial_v = \partial_v\partial_u$. As a symmetric matrix, it can be diagonalized with a change of variables $(u,v) \rightarrow (\xi_1, \xi_2)$ to get

$$\mathbb{D}^2 = \lambda_1 \partial_{\xi_1}^2 + \lambda_2 \partial_{\xi_2}^2. \tag{41.7}$$

By rescaling the coordinates $\xi_i \rightarrow \xi_i/\sqrt{|\lambda_i|}$, the coefficients of $\partial_{\xi_i}^2$ can scaled to be either ± 1 or 0. But we don't need to work out the details; the signs of the eigenvalues are sufficient to classify the PDE — and this we can deduce from the determinant,

$$|A| = ac - b^2/4 = \lambda_1 \lambda_2. \tag{41.8}$$

There are three possibilities:

- $b^2 < 4ac$: the eigenvalues have the same sign, $\mathbb{D}^2 = \partial_{\xi_1}^2 + \partial_{\xi_2}^2$
- $b^2 > 4ac$: the eigenvalues have opposite signs, $\mathbb{D}^2 = \partial_{\xi_1}^2 - \partial_{\xi_2}^2$
- $b^2 = 4ac$: one eigenvalue is zero, $\mathbb{D}^2 = \partial_{\xi_1}^2$.

The analysis is easily extended to higher dimensions, writing the most general form of the linear second-order operator as

$$\mathbb{D}^2 = \sum_{ij} a_{ij} \frac{\partial^2}{\partial u_i \partial u_j}, \qquad a_{ij} = a_{ji}. \tag{41.9}$$

Diagonalizing A now yields more than two eigenvalues, so there are more possibilities — but the three above are the most important:

- If all the eigenvalues have the same sign, the PDE can be rendered

$$\sum_{i\geq 1} \partial_{\xi_i}^2 \Psi + \text{lower order terms} = 0. \tag{41.10a}$$

The relative signs between its quadratic terms (the "signature" of $\mathbb{D}^2$) is the same as in the algebraic equation for an ellipse. Hence (41.10a) is called an *elliptic* PDE. The paradigmatic example is Laplace's equation

$$\nabla^2 \Psi(\vec{r}) = \left(\partial_x^2 + \partial_y^2 + \partial_z^2 \right) \Psi = 0. \tag{41.10b}$$

The Helmholtz equation (Section 41.2) is also elliptic.

- If all the eigenvalues are non-zero, with one having opposite sign from the others, then

$$-\partial_{\xi_1}^2 \Psi + \sum_{i\geq 2} \partial_{\xi_i}^2 \Psi + \text{lower order terms} = 0. \tag{41.11a}$$

Based on the signature, this is called a *hyperbolic* PDE — best illustrated by the wave equation

$$\nabla^2 \Psi - \frac{1}{c^2} \frac{\partial^2}{\partial t^2} \Psi = 0. \tag{41.11b}$$

- If one λ is zero and the rest have the same sign, the equation is *parabolic*,

$$\sum_{i\geq 2} \partial_{\xi_i}^2 \Psi + \text{lower order terms} = 0. \tag{41.12a}$$

Lacking a ∂_t^2 term, its exemplar is the heat equation,

$$\nabla^2 \Psi - \frac{1}{\alpha} \frac{\partial}{\partial t} \Psi = 0. \tag{41.12b}$$

We begin with a straightforward method to reduce a PDE in n dimensions to n individual ODEs. Not surprisingly, such reduction is not alway possible. Moreover, an equation which separates into ODEs in some coordinate systems may not separate in others. Nonetheless, *separation of variables* is often the simplest way to solve a PDE. We begin in "1+1" dimensions — that is, one space and one time. Afterwards, we'll apply the technique to the Laplacian in different $\mathbb{R}^3$ coordinate systems; it's there that we'll encounter many of the special functions introduced in Chapter 35.

41.1 Separating Space and Time

Separation of variables begins with the proposal that a PDE's solution $\Psi(x,t)$ can be cleanly factored into the product of one-dimensional functions,

$$\Psi(x,t) \equiv \psi(x)\tau(t). \tag{41.13}$$

The goal is to transform a two-variable PDE into two single-variable ODEs. Consider the wave equation in 1+1 dimensions,

$$\frac{\partial^2 \Psi}{\partial x^2} = \frac{1}{v^2} \frac{\partial^2 \Psi}{\partial t^2}, \tag{41.14}$$

with constant wave speed v. Inserting the ansatz (41.13) gives

$$\tau(t) \frac{d^2 \psi}{dx^2} = \frac{1}{v^2} \psi(x) \frac{d^2 \tau}{dt^2}, \tag{41.15}$$

where the derivatives are now total rather than partial. Dividing both sides by $\Psi = \psi \tau$ yields

$$\frac{1}{\psi(x)} \frac{d^2 \psi}{dx^2} = \frac{1}{v^2 \tau(t)} \frac{d^2 \tau}{dt^2}. \tag{41.16}$$

Everything on the left in this expression is a function only of x, whereas the right side is solely concerned with time. Since x and t are independent variables, the only way this separation can be true is if each side equals a constant — indeed, the *same* constant. Denoting this constant as $-k^2$, we get

$$\frac{d^2 \psi}{dx^2} = -k^2 \psi(x) \qquad\qquad \frac{d\tau}{dt} = -k^2 v^2 \tau(t), \tag{41.17a}$$

having solutions

$$\psi(x) = a \cos kx + b \sin kx \qquad\qquad \tau(t) = c \cos \omega_k t + d \sin \omega_k t, \tag{41.18}$$

where $\omega_k \equiv kv$. So for any given value of k, the wavefunction is

$$\Psi_k(x, t) = (a_k \cos kx + b_k \sin kx)(c_k \cos \omega_k t + d_k \sin \omega_k t), \tag{41.19}$$

or, using a common streamlined notation,

$$\Psi_k(x, t) = \begin{Bmatrix} \cos kx \\ \sin kx \end{Bmatrix} \begin{Bmatrix} \cos \omega_k t \\ \sin \omega_k t \end{Bmatrix}. \tag{41.20}$$

Electing to use $-k^2$ rather than $+k^2$ as the separation constant gave (41.17) the veneer of oscillation in both space and time — which, though physically sensible for an undamped wave, is not actually required by the separated ODEs themselves. In fact, we know virtually nothing about k since it appears nowhere in the original PDE — so it's rather mysterious that a solution would require a supplementary parameter. How do we choose a value for it?

Recall that a system described by a differential equation is not fully specified without suitable boundary and initial conditions. Only together — PDE plus conditions — can a full solution be obtained. So the form of k must be dictated by constraints imposed by boundary conditions. Still, there are generally many values of k consistent with these conditions. How do we choose which to use?

The answer is simultaneously simple and deep: as a linear differential equation, we need not pick a single value of k; instead, we sum over *all* k's to get a solution which is a linear superposition of the $\Psi_k(x, t)$. Thus we effectively sum away all k dependence, removing any concern with k not appearing in the original PDE.

Notice the convergence of several ideas. The separation constant emerging from the ansatz (41.13) serves as the connection between time and space; this manifests physically as the relation between frequency and wave number, $\omega = kv$. The boundary conditions determine the set of wavelengths $\lambda = 2\pi/k$ compatible with the system. Then linearity admits any superposition of ψ_k's to also be a solution. This is tantamount to summing over different oscillation frequencies ω. Moreover, the Sturm–Liouville form of the separated ODEs together with boundary conditions produce orthogonal eigenfunctions. Thus the general solution to the wave equation is an expansion in normal modes.

Example 41.1 **A String Fixed at Both Ends**

A string of length L with both ends fixed has boundary conditions

$$\Psi(0,t) = \Psi(L,t) = 0. \tag{41.21}$$

The condition at $x = 0$ picks out $\sin kx$ from the first bracket in (41.20) — or equivalently, requires a_k in (41.19) to be zero for all k. Then holding the string fixed at $x = L$ can only be satisfied for k with restricted values,

$$k = \frac{n\pi}{L}, \qquad n = 1, 2, 3, \ldots. \tag{41.22}$$

Thus from (41.19),

$$\Psi_k(x,t) = \sin(n\pi x/L)\left[c_k \cos \omega_k t + d_k \sin \omega_k t\right], \tag{41.23}$$

where we've absorbed b_k into c_k and d_k. Since the wave equation is second-order in time, we'll need two initial conditions,

$$\Psi(x,0) = f(x) \qquad \dot{\Psi}(x,0) = g(x), \tag{41.24}$$

where the functions f and g are arbitrary position and velocity profiles. The c_k are determined by f and the d_k by g — both via Fourier orthogonality,

$$c_k = \frac{2}{L} \int_0^L f(x) \sin(n\pi x/L) \, dx$$

$$d_k = \frac{1}{\omega_k} \frac{2}{L} \int_0^L g(x) \sin(n\pi x/L) \, dx. \tag{41.25}$$

For a plucked string starting at rest $g(x) = 0$ with an initial triangular configuration,

$$f(x) = \begin{cases} x, & 0 < x < L/2 \\ L - x, & L/2 < x < L, \end{cases} \tag{41.26}$$

we get

$$\Psi(x,t) = \frac{4L}{\pi^2} \sum_{n \text{ odd}} \frac{(-1)^{(n-1)/2}}{n^2} \sin\left(\frac{n\pi x}{L}\right) \cos\left(\frac{n\pi vt}{L}\right). \tag{41.27}$$

The $\omega_n = n\pi v/L$ are the system's normal mode frequencies; the wave speed is determined by the string tension and linear mass density, $v = \sqrt{T/\rho}$.

Example 41.2 **Hanging by a Thread**

Their familiarity aside, sines and cosines are not mere abstract functions: as Example 41.1 shows, they are easily observed as the normal modes of a horizontal string. The Bessel function J_0 can similarly be observed — as was demonstrated by Daniel Bernoulli in 1732, some 80 years before Bessel's work.

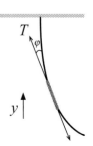

   Rather than a plucked string stretched tightly between two fixed points, consider a rope of length L and uniform linear mass density ρ suspended from the ceiling so that it can swing freely in a plane. Unlike a string fixed at both ends with constant tension, here T increases with height y as the weight of rope hanging beneath increases; if we set $y = 0$ at the free end of the rope, $T(y) = \rho g y$. Let $h(y, t)$ describe the horizontal displacement at a point y as a function of time. For small displacement, the net horizontal restoring force on an element Δy is the difference between the tensions at each end,

$$\Delta F = \frac{\partial F}{\partial y} \Delta y = \frac{\partial}{\partial y} (T \sin \varphi) \Delta y$$

$$\approx \frac{\partial}{\partial y} (T \tan \varphi) \Delta y = \frac{\partial}{\partial y} \left(T \frac{\partial h}{\partial y} \right) \Delta y, \tag{41.28}$$

with φ the small angle the element makes with the vertical. Then from Newton's second law $\Delta F = (\rho \Delta y) \ddot{h}$ we get the modified wave equation

$$\frac{\partial^2 h}{\partial t^2} = \frac{1}{\rho} \frac{\partial}{\partial y} \left(T \frac{\partial h}{\partial y} \right) = g \left(y \frac{\partial^2 h}{\partial y^2} + \frac{\partial h}{\partial y} \right). \tag{41.29}$$

This is a separable equation, revealing a sinusoidal time dependence and a horizontal displacement $u(z)$ given by (Problem 41.4)

$$z^2 \frac{d^2 u}{dz^2} + z \frac{du}{dz} + z^2 \omega^2 u = 0, \tag{41.30}$$

where $z^2 \equiv 4y/g = 4y/\omega_0^2 L$, and $\omega_0^2 = g/L$ is the natural frequency of a simple pendulum with the same length. This is Bessel's equation (39.70) with $\lambda = 0$ — and so has solution

$$u(z) = A J_0(\omega z) + B N_0(\omega z). \tag{41.31}$$

Since $N_0(z)$ diverges at $z = 0$, finite oscillations require $B = 0$. In addition, a rope fixed at the ceiling must have $u = 0$ at $y = L$ — which is $z = 2/\omega_0$. In a similar fashion to the analysis in Example 40.5, this condition restricts ω to discrete values ω_n proportional to the zeros of the Bessel function,

$$J_0 (2\omega_n/\omega_0) = 0 \quad \Longrightarrow \quad \omega_n = \frac{\omega_0}{2} \alpha_n, \tag{41.32}$$

where $J_0(\alpha_n) = 0$. The ω_n are the rope's normal mode frequencies; the first four modes are depicted in Figure 41.1.

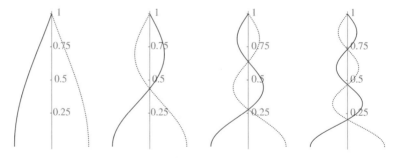

Figure 41.1 The first four normal modes of the hanging rope, $J_0(\alpha_n \sqrt{y/L})$. The nth zero is at the top where $y/L = 1$; the axis crossings occur at $\sqrt{y/L} = \alpha_k/\alpha_n$, $k = 1$ to $n-1$.

The general motion of the rope as a function of time can be described as a linear combination of eigensolutions — the Fourier–Bessel series

$$h(y,t) = \sum_n c_n J_0 \left(\alpha_n \omega_0 z/2\right) \cos(\omega_n t + \delta_n). \tag{41.33}$$

If, for instance, the rope has some initial displacement profile $h(y,0) = f(y)$, and is released from rest so that all the $\delta_n = 0$, then at time $t = 0$ the series becomes

$$h(y,0) = f(y) = f(gz^2/4) = \sum_n c_n J_0 \left(\alpha_n \omega_0 z/2\right), \tag{41.34}$$

with expansion coefficients given by the projection

$$c_n = \frac{2}{\omega_0^2 J_1(\alpha_n)^2} \int_0^{2/\omega_0} J_0 \left(\alpha_n \omega_0 z/2\right) f(gz^2/4) \, z \, dz, \tag{41.35}$$

courtesy of the Sturm–Liouville orthogonality of J_0, (40.43).

Example 41.3 **Heat Diffusion**

In one spatial dimension, the heat equation (41.2) becomes

$$\frac{\partial^2 T}{\partial x^2} = \frac{1}{\alpha} \frac{\partial T}{\partial t}. \tag{41.36}$$

The ansatz (41.13) (with $\Psi \to T$) leads to the separated ODEs (Problem 41.1)

$$\frac{d^2 \psi}{dx^2} = -k^2 \, \psi(x) \qquad \frac{d\tau}{dt} = -\alpha k^2 \, \tau(t), \tag{41.37a}$$

with solutions

$$\psi(x) = a \cos(kx) + b \sin(kx) \qquad \tau(t) = c \, e^{-\alpha k^2 t}, \tag{41.37b}$$

for some as-yet-unspecified separation constant k. For example, a solid bar of length L with its ends held at zero temperature has boundary conditions

$$T(x=0,t) = T(x=L,t) = 0. \tag{41.38}$$

Imposed on the $\psi(x)$ of (41.37b) requires both $a = 0$ and

$$k = \frac{n\pi}{L}, \qquad n = 1,2,3,\ldots, \tag{41.39}$$

just as was the case for the particle in a box, Example 33.7. We can now express the solution as the superposition

$$T(x,t) = \sum_{n=1}^{\infty} b_n \sin(n\pi x/L) \, e^{-\left(n^2 \pi^2 \alpha/L^2\right)t}. \tag{41.40}$$

Physically, the ends of the bar $x = 0$ and L act as heat sinks, so the temperature along the length of the bar decays over time at a rate which depends on the diffusivity α. For the details, we need to determine the b_n. If the bar's initial temperature profile is given by $T_0(x)$,

$$T(x,t=0) = T_0(x), \tag{41.41}$$

then the b_n are the Fourier sine series coefficients

$$b_n = \frac{2}{L} \int_0^L T_0(x) \sin (n\pi x/L) \ dx. \tag{41.42}$$

For the triangular initial profile of (41.26), this gives

$$T(x,t) = \frac{4L}{\pi^2} \sum_{n \text{ odd}} \frac{(-1)^{(n-1)/2}}{n^2} \sin \left(\frac{n\pi x}{L} \right) e^{-\alpha(n\pi/L)^2 t} \tag{41.43}$$

for the diffusion of heat along the bar as a function of time $t \geq 0$. Notice that, for $t < 0$, the solution grows unphysically without bound. This reflects the fact that being first order in the time derivative, the heat equation is not time-reversible; physical solutions occur only for $t \geq 0$. Notice too that solutions $\tau(t)$ are not orthogonal — but of course their defining ODE is not a Sturm–Liouville equation.

Example 41.4   The Wine Cellar Puzzle

Wines are traditionally aged and stored in underground vaults or caves which maintain a relatively stable temperature, despite variations in daily and seasonal temperature at surface level. To better understand this phenomenon, take the cellar to lie a large positive distance $y = L$ below ground level at $y = 0$, and model the temperature variation at the surface $T(0,t)$ as the real part of the periodic function $T_0 e^{i\omega t}$. Then for the separated ODEs of (41.37), the separation constant is

$$k = \left[\frac{1}{\alpha \tau} \frac{d\tau}{dt} \right]^{1/2} = \left(\frac{i\omega}{\alpha} \right)^{1/2} = (1 + i)\sqrt{\frac{\omega}{2\alpha}}. \tag{41.44}$$

So the spatial ODE has solution

$$\psi(y) = ae^{iky} + be^{-iky} = ae^{(i-1)y/\delta} + be^{-(i-1)y/\delta} \tag{41.45}$$

where $\delta \equiv \sqrt{2\alpha/\omega}$ has units of distance. To prevent the second term growing without bound as the depth of the cellar increases, we set $b = 0$; applying the spatial condition at $y = 0$ yields $a = T_0$. All together then, the temperature as a function of depth and time is

$$T(y,t) = \Re \left[T_0 \, e^{(i-1)y/\delta} e^{i\omega t} \right] = T_0 \, e^{-y/\delta} \cos(\omega t + y/\delta). \tag{41.46}$$

The resolution of the puzzle is immediately clear: the effects of the heat source at ground level are exponentially damped with depth y. The dimensional parameter δ is an example of a *skin depth*, a measure of the attenuation rate.

41.2 The Helmholtz Equation

Generalizing the separation of variables to higher dimensions is straightforward, starting with the observation that a PDE in n dimensions requires $n - 1$ independent separation constants. And whereas the separated time equation is unaffected by additional spatial dimensions, new features arise from the substitution of d^2/dx^2 with the Laplacian ∇^2. With the notational adjustment $\psi(x) \to \psi(\vec{r})$ in (41.13), the separated spatial equation becomes an eigenvalue equation for the Laplacian known as the *Helmholtz equation*,

$$\left(\nabla^2 + k^2 \right) \psi(\vec{r}) = 0. \tag{41.47}$$

Though a differential equation in space, the influence of the time dependence of the physical system still reverberates through the separation constant k. The special case $k \equiv 0$ is Laplace's equation, which describes systems completely divorced from time; familiar examples include electrostatics and the time-independent heat equation. Since the specific form of the Laplacian depends on the choice of coordinates, the Helmholtz equation produces different separated spatial ODEs in different coordinates system. There are only 11 $\mathbb{R}^3$ coordinate systems in which the Laplace and Helmholtz equations can be separated; the 10 non-cartesian ones are presented in Table A.1. Here we'll work out the general solutions to the Helmholtz equation in the three most common coordinate choices: cartesian, cylindrical, and spherical. Then in Section 41.3 we'll adapt these results to various boundary conditions.

Cartesian

In cartesian coordinates, the separation ansatz is

$$\psi(\vec{r}) = \chi(x)\Upsilon(y)\zeta(z). \tag{41.48}$$

Plugging this into the Helmoltz equation and dividing by ψ gives

$$\frac{1}{\chi}\frac{d^2\chi}{dx^2} = -\frac{1}{\Upsilon}\frac{d^2\Upsilon}{dy^2} - \frac{1}{\zeta}\frac{d^2\zeta}{dz^2} - k^2. \tag{41.49}$$

The left-hand side is only a function of x, whereas the right-hand side is only of y and z. As before, this can only be true if

$$\frac{1}{\chi}\frac{d^2\chi}{dx^2} = -\alpha^2 \qquad\qquad \frac{1}{\Upsilon}\frac{d^2\Upsilon}{dy^2} + \frac{1}{\zeta}\frac{d^2\zeta}{dz^2} + k^2 = \alpha^2, \tag{41.50}$$

for some separation constant α. We can then separate y and z using

$$\frac{1}{\Upsilon}\frac{d^2\Upsilon}{dy^2} = -\frac{1}{\zeta}\frac{d^2\zeta}{dz^2} - k^2 + \alpha^2, \tag{41.51}$$

requiring a second separation constant

$$\frac{1}{\Upsilon}\frac{d^2\Upsilon}{dy^2} = -\beta^2 \qquad\qquad \frac{1}{\zeta}\frac{d^2\zeta}{dz^2} = \alpha^2 + \beta^2 - k^2 \equiv -\gamma^2, \tag{41.52}$$

where we've introduced γ in order that all three separated ODEs appear equivalent,

$$\frac{1}{\chi}\frac{d^2\chi}{dx^2} = -\alpha^2 \qquad \frac{1}{\Upsilon}\frac{d^2\Upsilon}{dy^2} = -\beta^2 \qquad \frac{1}{\zeta}\frac{d^2\zeta}{dz^2} = -\gamma^2. \tag{41.53}$$

(This is reasonable: the order of separation is, after all, completely arbitrary.) Thus, using the streamlined notation introduced in (41.20),

$$\psi_{\alpha\beta\gamma}(\vec{r}) = \begin{Bmatrix} \cos\alpha x \\ \sin\alpha x \end{Bmatrix} \begin{Bmatrix} \cos\beta y \\ \sin\beta y \end{Bmatrix} \begin{Bmatrix} \cos\gamma z \\ \sin\gamma z \end{Bmatrix} = \left\{ e^{\pm i\alpha x} \right\} \left\{ e^{\pm i\beta y} \right\} \left\{ e^{\pm i\gamma z} \right\}. \tag{41.54}$$

In the absence of boundary conditions there is nothing whatsoever to distinguish one cartesian coordinate from another. Boundary conditions both impose constraints on the separation constants, and dictate the correct linear combination within each bracket. Nonetheless, because these ODEs come from the same PDE, they are not independent of one another: it must be the case that

$$\alpha^2 + \beta^2 + \gamma^2 = k^2. \tag{41.55}$$

Note that for $k^2 \geq 0$, only one of these squared constants must be positive. So solutions like $\psi_{\alpha\beta\gamma}(\vec{r}) = \cos \alpha x \, \sinh \beta y \, e^{-\gamma z}$ or $\cosh \alpha x \, e^{-i\beta y} \cos \gamma z$ in which oscillation occurs in only one or two of the variables are allowed. For $k^2 < 0$, none of the separation constants α, β, γ need be real. Since both sinh and cosh lack zeros, the Sturm–Liouville analysis of Section 40.2 implies that solutions with negative separation parameter need not form a complete orthogonal basis.[2] We'll explore this further in Section 41.3.

Cylindrical

Unlike cartesian coordinates, the Laplacian in cylindrical and spherical coordinates does not utilize the three independent variables symmetrically. Thus separating the Helmholtz equation in these systems leads to different sets of ODEs. In cylindrical coordinates, inserting the ansatz

$$\psi(\vec{r}) = R(r)\Phi(\phi)\zeta(z) \tag{41.56}$$

in the Helmholtz equation with $\nabla^2 = \frac{1}{r}\partial_r (r\partial_r) + \frac{1}{r^2}\partial_\phi^2 + \partial_z^2$ leads to

$$\frac{1}{R}\frac{1}{r}\frac{d}{dr}\left(r\frac{dR}{dr}\right) + \frac{1}{\Phi}\frac{1}{r^2}\frac{d^2\Phi}{d\phi^2} + \frac{1}{\zeta}\frac{d^2\zeta}{dz^2} = -k^2. \tag{41.57}$$

Introducing the constant γ^2, we can separate the z contribution to get

$$\frac{d^2\zeta}{dz^2} = -\gamma^2\zeta. \tag{41.58}$$

For convenience, we introduce a constant β so that the remaining coupled equation in r and ϕ is

$$\frac{1}{rR}\frac{d}{dr}\left(r\frac{dR}{dr}\right) + \frac{1}{r^2\Phi}\frac{d^2\Phi}{d\phi^2} = -k^2 + \gamma^2 \equiv -\beta^2. \tag{41.59}$$

Separation of this equation requires multiplication by r^2 to get

$$\frac{r}{R}\frac{d}{dr}\left(r\frac{dR}{dr}\right) + \beta^2 r^2 = -\frac{1}{\Phi}\frac{d^2\Phi}{d\phi^2}. \tag{41.60}$$

Setting both sides equal to a constant to m^2 yields for the radial coordinate

$$r\frac{d}{dr}\left(r\frac{dR}{dr}\right) + \left(\beta^2 r^2 - m^2\right)R = 0. \tag{41.61}$$

This is Bessel's equation (39.70), with $x = \beta r$ and $\lambda = m$. So R is a linear combination of Bessel and Neumann functions $J_m(\beta r)$ and $N_m(\beta r)$ (see Appendix C).

Finally, for ϕ there's the ubiquitous

$$\frac{d^2\Phi}{d\phi^2} = -m^2\Phi. \tag{41.62}$$

---

[2] If k is imaginary, (41.47) is often called the "modified" Helmholtz equation.

Despite its similarity to (41.58), the periodicity of ϕ requires that a rotation by 2π leaves the solution unchanged — that is, $\Phi(\phi)$ must be *single-valued*,

$$e^{\pm im(\phi+2\pi)} = e^{\pm im\phi} e^{\pm i2\pi m} \equiv e^{\pm im\phi}. \tag{41.63}$$

So m is limited to real integers. There is no such general restriction on γ.

Putting everything together, the solution to the Helmholtz equation in cylindrical coordinates is

$$\psi_{m\beta\gamma}(r,\phi,z) = \begin{Bmatrix} J_m(\beta r) \\ N_m(\beta r) \end{Bmatrix} \begin{Bmatrix} \cos m\phi \\ \sin m\phi \end{Bmatrix} \left\{ e^{\pm i\gamma z} \right\}. \tag{41.64}$$

Note how boundary conditions on a separation constant in any one of the ODEs affects the solution in another, thereby tying everything together into a PDE solution. But unlike the simple relationship linking the three cartesian separation constants in (41.55), here only two constants are similarly related, $\beta^2 + \gamma^2 = k^2$. For $k \geq 0$, either β or γ can be imaginary, producing modified Bessel functions (Appendix C) for the radial solution or exponential growth and decay along z. Moreover, any boundary-condition restrictions on β and γ impact the system's behavior in time. For the heat equation, this manifests in the decay rate of each Helmholtz eigenfunction — as we saw in (41.43). For the wave equation, each k corresponds to a specific frequency $\omega_k = kv$, so that the Helmholtz eigensolutions are the normal modes of wave vibration.

For the special case of Laplace's equation in the plane, the Bessel functions in (41.64) are replaced by powers of r,

$$\psi_m(r,\phi) = \begin{Bmatrix} r^m \\ r^{-m} \end{Bmatrix} \begin{Bmatrix} \cos m\phi \\ \sin m\phi \end{Bmatrix}, \tag{41.65}$$

as direct substitution can readily verify.

Spherical

Separating the Helmholtz equation in spherical coordinates with the ansatz

$$\psi(\vec{r}) = R(r)\Theta(\theta)\Phi(\phi) \tag{41.66}$$

leads to the ODEs (Problem 41.8)

$$\frac{d^2\Phi}{d\phi^2} = -m^2\Phi \tag{41.67a}$$

$$\frac{1}{\sin\theta} \frac{d}{d\theta}\left(\sin\theta \frac{d\Theta}{d\theta} \right) + \left(\lambda - \frac{m^2}{\sin^2\theta} \right)\Theta(\theta) = 0, \tag{41.67b}$$

and

$$\frac{d}{dr}\left(r^2 \frac{dR}{dr} \right) + \left(k^2 r^2 - \lambda \right)R(r) = 0. \tag{41.67c}$$

As in the cylindrical case, requiring single-valued solutions restricts m to integer values. As for the Θ equation, letting $x = \cos\theta$ with $\chi(x) \equiv \Theta(\theta)$ transforms it to

$$\frac{d}{dx}\left[(1 - x^2)\frac{d\chi}{dx} \right] + \left(\lambda - \frac{m^2}{1 - x^2} \right)\chi = 0, \tag{41.68}$$

which is the associated Legendre equation whose solutions are $P_\ell^m(x)$, $Q_\ell^m(x)$ with $\lambda = \ell(\ell + 1)$ for integer $\ell \geq 0$ and integer $|m| \leq \ell$ (see Example 40.2 and Section 35.3). This leaves the radial equation

$$\frac{d}{dr}\left(r^2\frac{dR}{dr}\right) + \left[k^2r^2 - \ell(\ell+1)\right]R(r) = 0, \tag{41.69}$$

which is the spherical Bessel equation with spherical Bessel and Neumann solutions $j_\ell(kr)$ and $n_\ell(kr)$ (see Appendix C).

All together, the Helmholtz solution in spherical coordinates is

$$\psi_{\ell m}(r,\theta,\phi) = \begin{Bmatrix} j_\ell(kr) \\ n_\ell(kr) \end{Bmatrix} \begin{Bmatrix} P_\ell^m(\cos\theta) \\ Q_\ell^m(\cos\theta) \end{Bmatrix} \left\{e^{\pm im\phi}\right\}. \tag{41.70}$$

Although there's no relation analogous to (41.55), it remains the case that each of the separation constants ℓ and m appears in two of the brackets. This, together with $|m| \leq \ell$, is what unites the disparate pieces into a PDE solution.

Similar to (41.65), Laplace's equation solutions swap out the Bessel functions with a linear combination of r^ℓ and $r^{-\ell-1}$,

$$\psi_{\ell m}(r,\theta,\phi) = \sum_{\ell=0}^{\infty}\sum_{m=-\ell}^{\ell}\left(A_{\ell m}r^\ell + \frac{B_{\ell m}}{r^{\ell+1}}\right)Y_{\ell m}(\theta,\phi), \tag{41.71}$$

where the spherical harmonic $Y_{\ell m}(\theta,\phi) = P_\ell^m(\cos\theta)e^{im\phi}$, and we've dropped the Q_ℓ^m to obtain finite ψ for all θ. A configuration with azimuthal symmetry has contributions only from $m = 0$, simplifying the Laplace equation solution to

$$\psi_\ell(r,\theta) = \sum_{\ell=0}^{\infty}\left(A_\ell r^\ell + \frac{B_\ell}{r^{\ell+1}}\right)P_\ell(\cos\theta). \tag{41.72}$$

This validates the argument used to find the multipole expansion in Example 32.2.

BTW 41.2   Eigenfunctions of the Laplacian

The Helmholtz equation is the eigenvalue equation for the Laplacian,

$$\nabla^2\psi(\vec{r}) = -k^2\psi(\vec{r}). \tag{41.73}$$

The simplest eigenfunctions are plane waves

$$\psi_{\vec{k}}(\vec{r}) = e^{i\vec{k}\cdot\vec{r}}, \tag{41.74}$$

which take their clearest form when separated in cartesian coordinates,

$$\nabla^2\left[e^{ik_xx}e^{ik_yy}e^{ik_zz}\right] = -k^2\left[e^{ik_xx}e^{ik_yy}e^{ik_zz}\right], \quad k^2 = k_x^2 + k_y^2 + k_z^2. \tag{41.75}$$

But plane waves are not the only eigenfunctions of the Laplacian. Reading from Table 41.1 and keeping only non-divergent functions, the eigenvalue equation in cylindrical coordinates can be written

$$\nabla^2\left[J_m(\beta r)e^{im\phi}e^{i\gamma z}\right] = -k^2\left[J_m(\beta r)e^{im\phi}e^{i\gamma z}\right], \quad k^2 = \beta^2 + \gamma^2. \tag{41.76}$$

Similarly, separating the Laplacian in spherical coordinates the eigenvalue equation manifests as

$$\nabla^2\left[j_\ell(kr)Y_{\ell m}(\theta,\phi)\right] = -k^2\left[j_\ell(kr)Y_{\ell m}(\theta,\phi)\right]. \tag{41.77}$$

Thus we identify the non-cartesian Laplacian eigenfunctions

$$\zeta_{\beta m \gamma}(r, \phi, z) = J_m(\beta r) e^{im\phi} e^{i\gamma z} \tag{41.78}$$

and

$$\xi_{k\ell m}(r, \theta, \phi) = j_\ell(kr) Y_{\ell m}(\theta, \phi). \tag{41.79}$$

Their physical interpretation derives from examining the asymptotic behavior for large r. Using (C.22)–(C.23),

$$J_m(kr) \sim \frac{1}{\sqrt{kr}} \big[e^{i(kr - m\pi/2 - \pi/4)} + e^{-i(kr - m\pi/2 - \pi/4)} \big], \tag{41.80}$$

which is a superposition of incoming and outgoing circular waves in the plane. Similarly, (C.38)–(C.39) gives

$$j_\ell(kr) \sim \frac{1}{2ikr} \big[e^{i(kr - \ell\pi/2)} - e^{-i(kr - \ell\pi/2)} \big], \tag{41.81}$$

which is the sum of incoming and outgoing spherical waves. These are physically distinct waveforms, so although $\psi_{k_x k_y k_z}$, $\zeta_{\beta m \gamma}$ and $\xi_{k\ell m}$ are all finite eigenfunctions of ∇^2, they are obviously not equivalent. Indeed, each is also an eigenfunction of other operators. The plane wave for instance is an eigenfunction of the individual cartesian components of the gradient,

$$\partial_j \psi_{\vec{k}} = i k_j \psi_{\vec{k}}. \tag{41.82}$$

By contrast, neither $\zeta_{\beta m \gamma}$ nor $\xi_{k\ell m}$ are eigenfunctions of $\vec{\nabla}$. Rather, ζ is an eigenfunction of the ordinary differential operators in (41.58), (41.61), (41.62) with eigenvalues β, m, and γ; similarly ξ satisfies the ODEs of (41.67) with eigenvalues k, ℓ, and m. With appropriate boundary conditions, each is an orthogonal basis of Sturm–Liouville eigenfunctions with real eigenvalues.

As with commuting Hermitian matrices (Section 27.3.2), commuting Hermitian differential operators share a common eigenbasis. Since all three cartesian components of $\vec{\nabla}$ commute with one another, $\partial_i \partial_j = \partial_j \partial_i$, they share the basis of plane waves. The three differential operators in the other two cases also mutually commute, defining bases of cylindrical and spherical waves. In all three instances, three mutually commuting ODE operators are required to specify the eigenfunctions of the Laplacian. Of course, with different bases of eigenfunctions spanning the same vector space, it must be possible to express one in terms of the others. We worked this out in Chapter 35; the relationships between the bases are given in (35.72).

41.3 Boundary Value Problems

Solving a differential equation has two distinct aspects. First is the calculus leading to general solutions; once achieved, the results can be saved in a table. The remaining task is a *boundary value problem* (BVP) in which boundary conditions are applied to determine the specific solution Ψ. The need for such conditions is clear not just mathematically, but physically as well. For instance, waves on a string differ depending on whether one or both ends are held fixed. Similarly, heat diffusion through a conducting rod differs depending upon whether an end is held at fixed temperature or is insulated from its surroundings.

Table 41.1 General solutions of $(\nabla^2 + k^2)\psi = 0$

	Parameters	$k^2 \neq 0$ (Helmholtz)	$k^2 = 0$ (Laplace)
Cylindrical[†]	$k^2 = \beta^2 + \gamma^2$ integer m	$\left\{ \begin{matrix} J_m(\beta r) \\ N_m(\beta r) \end{matrix} \right\} \left\{ \begin{matrix} \cos m\phi \\ \sin m\phi \end{matrix} \right\} \{ e^{\pm i\gamma z} \}$	$\left\{ \begin{matrix} J_m(\beta r) \\ N_m(\beta r) \end{matrix} \right\} \left\{ \begin{matrix} \cos m\phi \\ \sin m\phi \end{matrix} \right\} \{ e^{\pm \beta z} \}$
Spherical[‡]	integer $\ell \geq 0$ integer $|m| \leq \ell$	$\left\{ \begin{matrix} j_\ell(kr) \\ n_\ell(kr) \end{matrix} \right\} \left\{ \begin{matrix} P_\ell^m(\cos\theta) \\ Q_\ell^m(\cos\theta) \end{matrix} \right\} \{ e^{\pm im\phi} \}$	$\left\{ \begin{matrix} r^\ell \\ r^{-\ell-1} \end{matrix} \right\} \left\{ \begin{matrix} P_\ell^m(\cos\theta) \\ Q_\ell^m(\cos\theta) \end{matrix} \right\} \{ e^{\pm im\phi} \}$
Polar	integer m	$\left\{ \begin{matrix} J_m(kr) \\ N_m(kr) \end{matrix} \right\} \left\{ \begin{matrix} \cos m\phi \\ \sin m\phi \end{matrix} \right\}$	$\left\{ \begin{matrix} r^m \\ r^{-m} \end{matrix} \right\} \left\{ \begin{matrix} \cos m\phi \\ \sin m\phi \end{matrix} \right\}$

[†] For $\beta^2 < 0$, Bessel functions $J, N \longrightarrow$ modified Bessel functions I, K (Appendix C)
[‡] For $k^2 < 0$, Bessel functions $j, n \longrightarrow$ modified Bessel functions i, k (Appendix C)

There are three basic types of boundary conditions:

- Dirichlet: the value of Ψ is specified everywhere on the boundary
- Neumann: the normal derivative $\frac{\partial \Psi}{\partial n} = \vec{\nabla}\Psi \cdot \hat{n}$ is specified everywhere on the boundary
- Cauchy: both Ψ and $\vec{\nabla}\Psi \cdot \hat{n}$ are specified everywhere on the boundary.

One can also have mixed conditions in which Dirichlet conditions are applied to some portions of the boundary, Neumann conditions to the other portions. Robin conditions specify a linear combination of Ψ and its normal derivative everywhere on the boundary. When these specified values are zero, the conditions are said to be homogeneous; otherwise they are inhomogeneous.

Dirichlet and Neumann conditions can be given straightforward intuitive physical interpretations. For a vibrating string of length L, Dirichlet conditions fix the ends of the string at specified positions — say,

$$\Psi(0,t) = a \qquad \Psi(L,t) = b. \qquad (41.83)$$

For the heat equation, these same conditions fix the temperatures at the ends of a bar by placing them in contact with heat reservoirs with temperatures a and b. Heat can flow in or out, constrained only by the fixed temperatures at the ends. By contrast, an insulated bar can have no heat flow in or out; heat flux being proportional to the gradient of temperature, this gives the homogeneous Neumann condition

$$\partial_x \Psi(0,t) = \partial_x \Psi(L,t) = 0. \qquad (41.84)$$

Since the vertical tension on a vibrating string is proportional to the gradient of displacement $\partial_x \Psi(x,t)$, these same conditions describe a string with free ends.

Generalizing to boundary conditions in higher dimensions is straightforward. And although we've been discussing conditions in space, the full boundary for a PDE is that of the mathematical domain, which may be a combination of space and time. So whereas the Helmholtz equation requires spatial boundary conditions, they must be combined with boundary conditions in time in order to produce solutions to the wave and heat equations. Indeed, an ODE's initial value problem (IVP) is a type of BVP: specifying Ψ and $\dot{\Psi}$ at $t = 0$ are Cauchy conditions on the three-dimensional (hyper)surface $\{x, y, z\}|_{t=0}$ of the four-dimensional domain of space and time.

One must also distinguish between open and closed boundaries. In this context, a closed boundary is one which completely surrounds the domain — like the fixed edges of a drumhead on its perimeter. An open boundary has at least one direction extending to infinity — say, the diffusion of heat over open-ended time.

All of this leads naturally to the question of which combinations of conditions and boundaries are sufficient to produce stable, unique solutions. For instance, solutions to the heat equation such as in (41.43) are only valid for the open-boundary domain $t \geq 0$; negative t would result in a divergent solution. But consider the familiar initial value data used to solve second-order ODEs. The initial position and speed constitute a Cauchy condition — so one might expect Dirichlet and Neumann conditions, with only half the data of Cauchy, to be insufficient to yield a solution to second-order PDEs. Curiously, the opposite is true: for many PDEs, Dirichlet or Neumann conditions alone are sufficient; in these cases it turns out that, unlike ODEs, Cauchy conditions *over*-constrain the problem. Table 41.2 summarizes the conditions and boundary types which yield unique stable solutions for the different classes of PDEs.

Table 41.2 Boundary conditions for unique stable solutions

PDE	Class	Boundary condition	Boundary type
Laplace	Elliptic	Dirichlet, Neumann	Closed
Wave	Hyperbolic	Cauchy	Open
Heat	Parabolic	Dirichlet, Neumann	Open

Example 41.5 **Laplace's Equation in Cartesian Coordinates**

Consider a rectangular, U-shaped conductor, open at the top; a cross-section in the xy-plane looks like ⊔. With $k = 0$, (41.55) shows that it is impossible for all three separated cartesian equations to describe periodic solutions. Now if the width a is much smaller than its extent out of the page, the solution is effectively independent of z. Thus the problem reduces to solving the two-dimensional Laplace equation for voltage $V(x, y)$, with (41.55) simplified to

$$\alpha^2 = -\beta^2. \tag{41.85}$$

This dictates that only one of the two remaining ODEs has an oscillatory solution; the other must describe exponential growth and decay. Which is which is determined by boundary conditions. If we ground the parallel sides at $x = 0$ and a, the solution must satisfy the Dirichlet boundary conditions in x

$$V(0, y) = V(a, y) = 0. \tag{41.86a}$$

Insulating the sides from the bottom, we can hold the bottom at a constant potential V_0; this, together with the requirement that $V(x, y)$ must go to zero for large y give boundary conditions in y

$$V(x, 0) = V_0 \qquad V(x, y \to \infty) = 0. \tag{41.86b}$$

Of the two sets of conditions, only (41.86a) is consistent with oscillation — so we choose $\alpha^2 > 0$ to find

$$\chi(x) = \sin(n\pi x/a), \qquad \Upsilon(y) = e^{-n\pi y/a} \qquad n = 1, 2, 3, \ldots, \tag{41.87}$$

giving the solution's superposition the form

$$V(x,y) = \sum_n c_n \sin(n\pi x/a)e^{-n\pi y/a}.$$ (41.88)

We've yet to incorporate V_0 into our solution — and in fact, the constants c_n can be found from the remaining condition

$$V(x,0) = V_0 = \sum_n c_n \sin(n\pi x/a),$$ (41.89)

which is a Fourier sine series for the constant V_0 with expansion coefficients

$$c_n = \frac{2V_0}{a} \int_0^a \sin(n\pi x/a)\,dx = \frac{4V_0}{\pi n}, \qquad n = 1,3,5,\ldots.$$ (41.90)

Thus

$$V(x,y) = \frac{4V_0}{\pi} \sum_{n\,\text{odd}} \frac{1}{n} \sin(n\pi x/a)\, e^{-n\pi y/a}.$$ (41.91)

The boundary conditions of (41.86) make it easy to deduce in which dimension the voltage must oscillate; applying this insight early in the BVP as we did makes the subsequent algebra a bit simpler. But taking $\alpha^2 < 0$ must still produce the same correct result (Problem 41.10).

With a Neumann condition along the boundary at $y = 0$, we'd have

$$\left.\frac{\partial V}{\partial y}\right|_{y=0} = \sigma(x).$$ (41.92)

To implement this condition for constant $\sigma(x) = \sigma_0$, we merely swap (41.90) with

$$c_n = -\frac{a}{n\pi}\frac{2\sigma_0}{a} \int_0^a \sin(n\pi x/a)\,dx = -\frac{4\sigma_0 a}{\pi^2 n^2}, \qquad n = 1,3,5,\ldots.$$ (41.93)

But though the procedure is much the same, the physics described is not: the condition (41.92) specifies the normal component of the electric field $E_y = -\partial_y V$, with $-\sigma$ the surface charge density on the conductor.

Example 41.6 Occult Oscillation

Though it would appear from (41.55) with $k = 0$ that at least one dimension must be periodic, a little physical insight and reasoning reveals that to be absurd: one can always use batteries and ground wires to apply virtually any boundary conditions — including those which undermine expectations of oscillation.

Consider a rectangular conducting frame, with sides of length a and b, having two adjacent sides grounded and the other two held at a constant V_0 — say,

$$V(0,y) = 0 \qquad\qquad V(a,y) = V_0$$ (41.94a)

and

$$V(x,0) = 0 \qquad\qquad V(x,b) = V_0.$$ (41.94b)

In this problem, it's not at all obvious whether the voltage is periodic in x or y. In fact, the symmetry of the problem mandates that the behavior must be the *same* in both. But this is not consistent with (41.85) — and so it would appear that these boundary conditions are not compatible with a separation of variables solution.

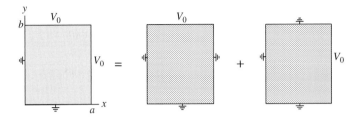

Figure 41.2 Boundary condition in superposition.

But don't be hasty: the way forward once again relies on the linearity of the differential equation and superposition. The idea is to express the physical configuration as a sum of unambiguous configurations whose boundary conditions *are* consistent with (41.85), as indicated in Figure 41.2. Then the solution to the differential equation is given by the sum of solutions whose periodic behavior is manifest. Applying the grounded boundary conditions for each gives

$$V(x,y) = \sum_{n=1}^{\infty} A_n \sin(n\pi x/a) \sinh(n\pi y/a) + B_n \sinh(n\pi x/b) \sin(n\pi y/b). \tag{41.95}$$

The remaining boundary conditions

$$V(x,b) = \sum_n A_n \sin(n\pi x/a) \sinh(n\pi b/a) = V_0$$

$$V(a,y) = \sum_n B_n \sinh(n\pi a/b) \sin(n\pi y/b) = V_0 \tag{41.96}$$

yield expansion constants

$$A_n \sinh(n\pi b/a) = \frac{2}{a} \int_0^a V_0 \sin(n\pi x/a)\, dx = \frac{4V_0}{n\pi}, \quad n \text{ odd}$$

$$B_n \sinh(n\pi a/b) = \frac{2}{b} \int_0^b V_0 \sin(n\pi y/b)\, dy = \frac{4V_0}{n\pi}, \quad n \text{ odd}. \tag{41.97}$$

All together then,

$$V(x,y) = \frac{4V_0}{\pi} \sum_{n \text{ odd}} \frac{1}{n} \left[\frac{\sin(n\pi x/a)\sinh(n\pi y/a)}{\sinh(n\pi b/a)} + \frac{\sinh(n\pi x/b)\sin(n\pi y/b)}{\sinh(n\pi a/b)} \right]. \tag{41.98}$$

This generalizes to arbitrary boundary voltages $f(y)$ and $g(x)$ by substituting these functions into the integrals of (41.97).

Example 41.7   **Laplace in Cylindrical Coordinates**

Consider a conducting cylinder of radius R centered at the origin in the xy-plane, extending a distance L upward along z. To find the voltage everywhere inside the cylinder, we need the solution to Laplace's equation in cylindrical coordinates — readily found in Table 41.1,

$$V_m(r,\phi,z) = \begin{Bmatrix} J_m(\beta r) \\ N_m(\beta r) \end{Bmatrix} \begin{Bmatrix} \cos m\phi \\ \sin m\phi \end{Bmatrix} \{e^{\pm \beta z}\}. \tag{41.99}$$

As boundary conditions, let the top and bottom faces be grounded, and the cylindrical surface held fixed at a constant voltage V_0,

$$V(r,\phi,0) = V(r,\phi,L) = 0 \qquad V(R,\phi,z) = V_0. \tag{41.100}$$

These conditions imply oscillation in the z-direction, which requires imaginary $\beta \equiv ik$. This recasts (41.99) as

$$V_m(r,\phi,z) = \begin{Bmatrix} I_m(kr) \\ K_m(kr) \end{Bmatrix} \begin{Bmatrix} \cos m\phi \\ \sin m\phi \end{Bmatrix} \begin{Bmatrix} \cos kz \\ \sin kz \end{Bmatrix}, \tag{41.101}$$

where I and K are modified Bessel functions (Appendix C). Since V must be finite everywhere, we discard K_m, which diverges at $r = 0$. We can also drop $\cos kz$, which is inconsistent with $V = 0$ at $z = 0$. Grounding the top face requires that k be proportional to the zeros of $\sin kz$,

$$k \to k_n = \frac{n\pi}{L}, \qquad n = 1,2,3,\dots. \tag{41.102}$$

Applying these constraints gives

$$V(r,\phi,z) = \sum_{\substack{m=0 \\ n=1}}^{\infty} I_m(k_n r)(a_{nm}\cos m\phi + b_{nm}\sin m\phi)\sin k_n z, \tag{41.103}$$

where the expansion coefficients a and b depend on both indices. Though modified Bessel functions are not orthogonal, "Fourier functions" $\sin k_n z$ are. Using the boundary condition at $r = R$, we multiply both sides by $\sin k_\ell z$ and integrate over z to find

$$\frac{4V_0}{n\pi} = \sum_{m=0} I_m(k_n R)(a_{nm}\cos m\phi + b_{nm}\sin m\phi), \qquad n \text{ odd}. \tag{41.104}$$

Since the problem has cylindrical symmetry (indeed, the boundary conditions place no constraint on ϕ), the solution must be independent of ϕ. Thus only $m = 0$ contributes, and the single surviving coefficient is

$$a_{n0} = \frac{4V_0}{n\pi I_0(k_n R)}. \tag{41.105}$$

Putting it all together, the solution is

$$V(r,\phi,z) = \frac{4V_0}{\pi} \sum_{n \text{ odd}} \frac{\sin(k_n z)I_0(k_n r)}{n\, I_0(k_n R)}. \tag{41.106}$$

Deriving solutions to the Helmholtz equation follows the same pattern as for Laplace, with boundary conditions determining the discrete set of eigenvalues k_n. But when $(\nabla^2 + k^2)\psi = 0$ emerges from the wave or heat equation, there remain initial conditions to apply in order to recombine the space and time solutions into the solution of the full PDE. For Helmholtz eigenfunctions $\phi_n(\vec{r}; k_n)$, this is the normal mode expansion

$$\psi(\vec{r}, t) = \sum_n c_n \phi_n(\vec{r}; k_n)\tau(t; k_n), \tag{41.107}$$

where, for diffusivity α and wave speed $v = \omega/k$,

$$\tau(t; k_n) = \begin{cases} e^{-\alpha k_n^2 t}, & \text{heat equation} \\ ae^{i\omega_n t} + be^{-i\omega_n t}, & \text{wave equation} \end{cases}. \tag{41.108}$$

The initial conditions determine the expansion coefficients by orthogonal projection; Examples 41.1 and 41.3 are $(1 + 1)$-dimensional illustrations.

Indeed, the fundamental role played by orthogonal projection underlies our focus on oscillation: Sturm–Liouville solutions are oscillatory (Section 40.3). But orthogonal eigenfunctions only emerge if the Sturm–Liouville surface term in (40.6) vanishes, and this requires homogeneous (or periodic) boundary conditions — see (40.20). Moreover, the superposition of these eigenfunctions is only justified if they form a vector space — after all, the sum of functions which satisfy, say, $\phi(0) = 1$, cannot also satisfy that condition.

The importance of this is perhaps most readily appreciated in the case of the two-dimensional rectangular frame: one dimension needs to have homogeneous boundary conditions in order to be able to satisfy the inhomogeneous condition in the other. The wave and heat equations, however, add an additional mathematical dimension to the problem. This necessitates that both spatial dimensions have Sturm–Liouville solutions so that the initial condition $\psi(x, y, 0) = f(x, y)$ can be satisfied with a double Fourier integral at the $t = 0$ "boundary." But physical configurations with inhomogeneous spatial conditions certainly occur; not only must they have solutions, they should be of the same basic form as those with homogeneous conditions. Once again, superposition is the key to finding such solutions.

Example 41.8 Inhomogeneous Boundary Conditions

Consider the temperature distribution $T(\vec{r}, t)$ of a rectangular plate with sides of length a and b, initial temperature distribution $T_0(x, y)$, and fixed temperatures along the perimeter given by

$$T(x = 0, y, t) = 0 \qquad T(x = a, y, t) = 0$$

$$T(x, y = 0, t) = 0 \qquad T(x, y = b, t) = T_b(x). \tag{41.109}$$

For non-zero T_b, these Dirichlet conditions are not homogeneous, and so do not support orthogonal normal modes in y. This is not a problem when solving only the Laplace equation; as we've seen, the Sturm–Liouville solutions in x can be used to satisfy the condition in y. But as a Helmholtz equation, we need orthogonal eigenfunctions in both x and y in order satisfy the initial condition $T(\vec{r}, 0) = T_0(x, y)$.

Let's consider this physically: whereas heat can flow across the perimeter in order to maintain fixed temperatures at the edges, there is no other heat flux into or out of the plate. So the full solution will settle into a time-independent equilibrium solution $T_{eq} = T_{eq}(x, y)$ for large t,

$$T_{eq}(\vec{r}) = \lim_{t \to \infty} T(\vec{r}, t). \tag{41.110}$$

Being independent of time, the heat equation reduces for large t to Laplace's,

$$\nabla^2 T_{eq} = \frac{1}{\alpha} \frac{\partial T_{eq}}{\partial t} \equiv 0 \tag{41.111}$$

with the boundary conditions of (41.109). And this we know how to solve. In fact, we can transcribe the solution directly from (41.98) for constant T_b:

$$T_{eq}(x, y) = \frac{4T_b}{\pi} \sum_{n \, \text{odd}} \frac{1}{n} \left[\frac{\sin(n\pi x/a) \sinh(n\pi y/a)}{\sinh(n\pi b/a)} \right]. \tag{41.112}$$

Superposition does the rest, allowing the full solution to be expressed as

$$T(x, y, t) = \tilde{T}(x, y, t) + T_{eq}(x, y), \tag{41.113}$$

with homogeneous conditions on $\tilde{T}(x, y, t)$ in both x and y. Thus $\tilde{T}(x, y, t)$ can be expanded in normal modes,

$$\tilde{T}(x, y, t) = \sum_{nm} c_{nm} \sin(n\pi x/a) \sin(m\pi y/b) \, e^{-\alpha \left[\left(\frac{n\pi}{a} \right)^2 + \left(\frac{m\pi}{b} \right)^2 \right] t}, \tag{41.114}$$

able to satisfy the initial condition by orthogonal projection (Problems 41.22–41.23),

$$c_{nm} = \frac{4}{ab} \int_0^a \int_0^b \left[T_0(x', y') - T_{eq}(x', y') \right] \sin(n\pi x'/a) \sin(m\pi y'/b) \, dx' \, dy'. \tag{41.115}$$

The function $\tilde{T}$ is the complementary solution to T_{eq} — see (39.35).

Example 41.9   Convection at the Boundary

A spherical container of radius R has initial temperatures T_0 at the center and T_1 at the surface. Due to the spherical symmetry, only the radial direction is relevant — so $T(\vec{r}, t) = T(r, t) = R(r)\tau(t)$, with $R(r)$ satisfying the Helmholtz equation in spherical coordinate r. Consultation with Table 41.1 gives

$$R(r) = j_0(kr), \tag{41.116}$$

where we discard the Neumann functions since they diverge at $r = 0$, and spherical symmetry is consistent only with $\ell = 0$. At time $t = 0$, the sphere is placed in a freezer maintaining a constant temperature T_{fr}. For $T_1 = T_{fr} = 0$, this is a homogeneous Dirichlet problem. If $T_1 = T_{fr} \neq 0$ the condition is inhomogeneous — but since constant T_1 is a solution to Laplace's equation, we can solve the homogeneous problem for $\tilde{T}(r, t) = T(r, t) - T_1$. These cases are explored in Problem 41.29. More interesting is the situation in which $T_1 \neq T_{fr}$, since then the temperature at $r = R$ cannot remain at the initial value T_1. Instead, heat transfer across the surface changes the temperature at $r = R$ at a rate proportional to its difference with T_{fr},

$$\left. \frac{dT}{dt} \right|_{r=R} = -b \left[T(R, t) - T_{fr} \right], \tag{41.117}$$

where $b > 0$ reflects properties of the sphere. This linear combination of T and its normal derivative is an example of a Robin condition; for constant T_{fr}, it's homogeneous in $\tilde{T}(r, t) = T(r, t) - T_{fr}$ (Problem 41.30).

41.4 The Drums

Many of the special functions which emerge as solutions to the Helmholtz equation were introduced back in Chapter 35. But their most physically resonant interpretation is as normal modes. This is traditionally illustrated by the vibrating modes of a two-dimensional taut membrane, such as a drumhead; we'll follow this tradition by examining normal mode solutions of the wave equation for different drum geometries.

Example 41.10 **A Rectangular Drumhead**

Consider a rectangular $L_1 \times L_2$ membrane held fixed all around its perimeter,

$$\psi(0,y) = \psi(x,0) = 0 \qquad \psi(L_1,y) = \psi(x,L_2) = 0. \qquad (41.118)$$

From (41.54), the drum's mutually orthogonal modes are the direct product of a fixed string's one-dimensional normal modes,

$$\psi_{nm}(x,y) = \sin\left(\frac{n\pi x}{L_1}\right) \sin\left(\frac{m\pi y}{L_2}\right), \qquad n,m = 1,2,3,\ldots, \qquad (41.119)$$

where, as in Example 41.1, the cosine terms drop out in order to satisfy the Dirichlet conditions at 0 and $L_{1,2}$. Each orthogonal mode (n,m) has corresponding frequency $\omega_{nm} = k_{nm}v$, with

$$k_{nm}^2 = \pi^2 \left(\frac{n^2}{L_1^2} + \frac{m^2}{L_2^2}\right). \qquad (41.120)$$

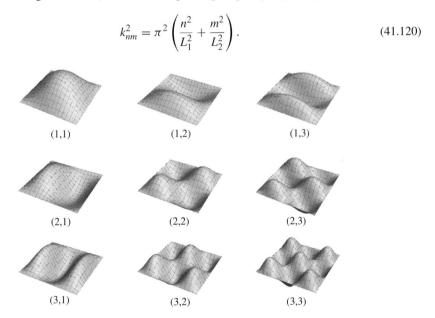

Figure 41.3 Normal modes (n,m) of a square drumhead.

Several modes of a square drumhead, $L_1 = L_2 \equiv L$, are depicted in Figure 41.3. Note that unlike a string whose extrema are uniformly spaced $\pi/k = L/n$ apart, here the quadratic relation $k^2 = k_x^2 + k_y^2$ yields extrema spaced $L/\sqrt{n^2 + m^2}$ apart.

Including the time-dependence of each mode, the full solution to the wave equation is the superposition

$$\Psi(x,y,t) = \sum_{nm} \psi_{nm}(x,y) \left(a_{nm} \cos \omega_{nm} t + b_{nm} \sin \omega_{nm} t \right), \qquad (41.121)$$

with the sets of constants a_{nm} and b_{nm} extracted from initial conditions

$$\Psi(x,y,0) = f(x,y) \qquad \dot{\Psi}(x,y,0) = g(x,y) \qquad (41.122)$$

by the projection integrals

$$a_{nm} = \frac{1}{L_1 L_2} \int_{-L_1}^{L_1} \int_{-L_2}^{L_2} f(x,y) \, \psi_{nm}(x,y) \, dx \, dy$$

$$\omega_{nm} b_{nm} = \frac{1}{L_1 L_2} \int_{-L_1}^{L_1} \int_{-L_2}^{L_2} g(x,y) \, \psi_{nm}(x,y) \, dx \, dy. \qquad (41.123)$$

Example 41.11 A Circular Drumhead

The vibrational normal modes of a circular drumhead are prescribed by the Helmholtz equation in polar coordinates. Without a z dependence, we set $\gamma = 0$ in Table 41.1 and write

$$\psi_m(r,\phi) = J_m(kr) \begin{Bmatrix} \cos m\phi \\ \sin m\phi \end{Bmatrix}, \qquad (41.124)$$

where the Neumann function is discarded because it diverges at $r = 0$. We've also sensibly chosen the ϕ eigenfunctions to be real. Since the drumhead must be fixed all around its circumference at $r = R$, this is a Dirichlet problem with

$$\psi(r = R, \phi) = 0. \qquad (41.125)$$

Thus the normal mode frequencies of vibration are determined by the zeros of the Bessel functions,

$$J_m(kR) = 0 \quad \Longrightarrow \quad k_{jm} R = \alpha_{jm}, \qquad (41.126)$$

with α_{jm} the jth zero of $J_m(x)$.[3] The drum's normal modes are therefore

$$\psi_m(r,\phi) = J_m(\alpha_{jm} r/R) \begin{Bmatrix} \cos m\phi \\ \sin m\phi \end{Bmatrix}, \quad \begin{array}{l} j = 1,2,3,\ldots \\ m = 0,1,2,\ldots, \end{array} \qquad (41.127)$$

with frequencies $\omega_{jm} = k_{jm} v$. Though these eigenfunctions are doubly degenerate in m, their nodal patterns are simple rotations of one another. The first four distinct modes are depicted in Figure 41.4.

All together, the general solution to the wave equation is given by the rather cumbersome superposition

$$\psi(r,\phi,t) = \sum_{jm} J_m(\alpha_{jm} r/R) \left(a_{jm} \cos m\phi + b_{jm} \sin m\phi \right) \cos \omega_{jm} t$$

$$+ \sum_{jm} J_m(\alpha_{jm} r/R) \left(c_{jm} \cos m\phi + d_{jm} \sin m\phi \right) \sin \omega_{jm} t. \qquad (41.128)$$

---

[3] This expands the notation of (40.40) beyond the zeros of J_0.

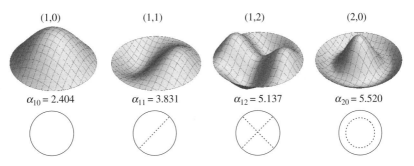

Figure 41.4 The lowest modes (j, m) of a circular drumhead; the circles show their nodal patterns.

In this form, the a_{jm} and b_{jm} are determined from initial condition $\psi(r, \phi, 0) = f(r, \phi)$, and the c_{jm} and d_{jm} from $\dot{\psi}(r, \phi, 0) = g(r, \phi)$ using both Fourier and Bessel orthogonality. For instance,

$$b_{jm} = \frac{2}{\pi R^2 J_{m+1}^2(\alpha_{jm})} \int_0^R \int_{-\pi}^{\pi} f(r, \phi) J_m(\alpha_{jm} r/R) \sin(m\phi) \, r \, dr \, d\phi. \tag{41.129}$$

It's worth comparing the modes of a circular drum with those of a rectangular one. In the latter, Dirichlet boundary conditions separately restrict the values of $k_{x,y}$; for the circular drum the restriction is only pertinent to $k = |\vec{k}|$. In both geometries, though, the k's are constrained by the zeros of their eigenfunctions, resulting to Sturm–Liouville orthogonality — for the circular drum, of Bessel functions,

$$\int_0^R J_m(\alpha_{im} r/R) J_m(\alpha_{jm} r/R) \, r \, dr = \frac{R^2}{2} J_{m+1}^2(\alpha_{jm}) \, \delta_{ij}. \tag{41.130}$$

So modes having the same m but distinct j's derive their orthogonality from that of the J_m's on the disk, whereas the orthogonality of modes with the same j but different m's derives from that of $\sin(m\phi)$ and $\cos(m\phi)$ on the circle. Since the zeros of Bessel functions have neither uniform spacing nor a simple algebraic form, numerical methods are needed to solve for them.[4] Moreover, the spacing $\sim R/\alpha_{jm}$ between a mode's extrema more closely resembles that of standing waves on a string than of a rectangular drum (a reflection of the one-dimensional radial dependence of the Bessel functions).

Remarkably, some higher modes are concentrated around the circumference, leaving the middle largely undisturbed (Figure 41.5). These are known as "whispering modes" after the Whispering Gallery inside London's St. Paul's Cathedral, an elevated circular walkway around the base of the dome. Rather than vibrations of an elastic drumhead, these oscillations are low-intensity sound waves which propagate around the circumference with little interference. So whispers can be heard all along the curved wall, far further than a whisper can be expected to reach directly.

---

[4] These can be found in tables — or using the Mathematica command `BesselJZero`.

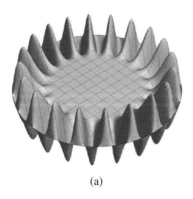

(a)   (a)

Figure 41.5 (a) A whispering mode, $J_m(\alpha_{jm}r/R)$, $m = 20, j = 1$. (b) The Whispering Gallery in St. Paul's Cathedral. (© The Chapter of St. Paul's. Photo: Graham Lacdao)

Example 41.12 **Airball**

The real Helmholtz solutions for a ball can be found directly from Table 41.1,

$$\psi_{\ell m}(r,\theta,\phi) = j_\ell(kr)P_\ell^m(\cos\theta)\begin{Bmatrix}\cos m\phi \\ \sin m\phi\end{Bmatrix}, \qquad (41.131)$$

where we've discarded the Neumann function n_ℓ which diverges at $r = 0$, as well as the Q_ℓ^m's which diverge at $\cos\theta = \pm 1$. Of course, it's difficult to envision a solid rubber ball as a drum since holding the spherical boundary at $r = R$ makes little sense in that context. A better drum analog is the pattern of sound waves generated inside a heavy sphere — say, the "ringing" a basketball makes when struck.[5] For this system, the wave equation describes the acoustic air pressure ψ within the ball, with the radial component of particle velocity vanishing at $r = R$. Since the particle velocity depends on the gradient of the pressure, the boundary condition constrains the normal derivative of the pressure,

$$\left.\frac{d\psi}{dr}\right|_{r=R} \equiv u'(r = R) = 0. \qquad (41.132)$$

So this is a Neumann problem: rather than the zeros of j_ℓ, we need those of its derivative j'_ℓ. In other words, the k in (41.131) must satisfy

$$k_{n\ell}R = \beta_{n\ell}, \qquad (41.133)$$

where $\beta_{n\ell}$ denotes the nth zero of j'_ℓ. Note that this condition satisfies boundary criteria for a Sturm–Liouville system in (40.13), so the solution results in normal modes with integers $n \geq 1$ and $\ell \geq 0$,

$$\psi_{n\ell m} = j_\ell(\beta_{n\ell}\, r/R)P_\ell^m(\cos\theta)\begin{Bmatrix}\cos m\phi \\ \sin m\phi\end{Bmatrix}, \qquad m = 0,1,2,\ldots,\ell, \qquad (41.134)$$

or, perhaps more conveniently,

$$\psi_{n\ell m} = j_\ell(\beta_{n\ell}\, r/R)Y_{\ell m}(\theta,\phi), \qquad m = -\ell, -\ell+1, \ldots, \ell-1, \ell. \qquad (41.135)$$

---

[5] We distinguish ringing from the low-frequency "thump" made when the ball hits the floor.

In either form, the frequencies $\omega_{n\ell} = k_{n\ell}v$ are independent of m, rendering the modes $2\ell + 1$-fold degenerate.

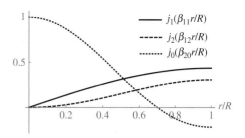

A physical feel for these modes can be developed by perusing the zeros of j'_ℓ listed in Table C.6. The first non-trivial zero and lowest normal mode frequency emerges from $\beta_{11} = 2.08$, corresponding to the triply degenerate modes $(n, \ell, m) = (1, 1, m)$. The radial piece of these modes, $j_1(k_{11}r)$, smoothly passes from a pressure node at $r = 0$ to an antinode at $r = R$. So the shapes of these modes at any given r (that is, the pressure amplitude as a function of θ and ϕ) correspond to the first orbitals shown in Figures 35.5 and 35.6. Because pressure nodes are antinodes of particle velocity, in these modes air in the ball oscillates back and forth, with maximum velocity at the pressure nodal surfaces $\theta = \pi/2$ ($m = 0$) or $\theta = 0, \pi$ ($m = \pm 1$).

The next normal mode frequency has $\beta_{12} = 3.34$, corresponding to the five-fold degenerate $(1, 2, m)$ modes. Since $j_2(k_{12}r)$ has a similar radial profile to $j_1(k_{11}r)$, these orbitals shapes are also determined by the spherical harmonics: the second orbital in Figure 35.5 depicts the $m = 0$ mode, and the second and third orbitals in 35.6 $|m| = 1, 2$. It's not until the third frequency with $\beta_{20} = 4.49$ do we get to the non-degenerate mode $(2, 0, 0)$. Since Y_{00} is constant, this mode is spherically symmetric. So it's a radial mode with pressure antinodes both at $r = 0$ and R, and an antinode of air velocity at $r/R \approx 0.7$.

The actual normal mode frequencies, of course, depend both on the speed of sound and the size of the ball. More details, together with experimental data, can be found in Russell (2010).

BTW 41.3 "Can One Hear the Shape of a Drum?"

The different frequency spectra of drums reasonably leads one to ask whether a drum's harmonic signature is tantamount to a fingerprint. In other words, can one use the spectrum to identify the drum. This is an example of an *inverse problem*: rather than solve for the spectrum of a given drum, one tries to determine the drum given a particular spectrum. This is the deceptively deep question Mark Kac (1966) posed in his seminal article "Can one hear the shape of a drum?"

Consider first a one-dimensional drum — a string fixed at both ends. The harmonics, which constitute the eigenspectrum of the one-dimensional Laplacian, are all integer multiples of the fundamental frequency $\omega_n/2\pi = f_n = nf_1$. The simple integer ratios $n/(n + 1)$ between adjacent harmonics are responsible for the pleasant quality of the sound. (It's no coincidence that $\sum 1/n$ is called the harmonic series.) These relationships between eigenfrequencies, together with the pitch of the fundamental $f_1 \sim 1/L$, are sufficient to determine the shape of the "drum" — in this case, a one-dimensional string of length L. This is largely true not just of strings and string instruments, but also of most wind instruments. Indeed, almost all non-percussive instruments in an orchestra are one-dimensional drums.

Kac of course knew this; his question concerns drums in N dimensions, $N \geq 2$. A rectangular drum, for instance, has the "pythagorean" relation amongst its eigenfrequencies of (41.120),

$$k_{nm}^2 \sim \frac{n^2}{L_1^2} + \frac{m^2}{L_2^2}. \tag{41.136}$$

The spectrum for a circular drum, which depends on the zeros of Bessel functions, doesn't even have this simple form. Neither sounds terribly musical (but see Problem 41.25) — but even so, surely this is a difference one can hear. In fact, years before Kac posed his question it was discovered that the eigenvalues λ_n of the N-dimensional Laplacian, listed in order of increasing size, scale as

$$\lambda_n \sim 4\pi^2 \left(\frac{n}{B_N V}\right)^{2/N}, \qquad n \text{ large}, \tag{41.137}$$

where B_N is the volume of the unit ball in N dimensions (see Problem 8.23). Thus one can, in effect, "hear" the N-volume of a drum — or the area A for $N = 2$. A later $N = 2$ result showed that the perimeter L could also be teased out from the eigenfrequencies,

$$\sum_{n=1}^{\infty} e^{-\lambda_n t} \sim \frac{A}{2\pi t} - \frac{L}{\sqrt{32\pi t}}, \qquad t \to 0. \tag{41.138}$$

Thus from the spectrum one can "hear" both the area and perimeter. The only case in which this reveals the *shape* of the drum, however, is when $L^2 = 4\pi A$, which holds only for circles. So one can in effect distinguish a circular drum by how it sounds.

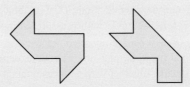

Figure 41.6 Isospectral planar drums (Gordan and Webb, 1996).

But the more general question, whether the shape of a drum corresponds to a unique spectrum, remained. At the time of his paper, Kac knew that distinct isospectral drums exist for $N = 16$, but could only speculate about the more mundane case $N = 2$. Finally, in 1991, examples in the plane were discovered of different shaped drums having the same spectrum (Figure 41.6). So as a general rule, one cannot hear the shape of a drum.

Problems

41.1   Use the separation ansatz (41.13) to show that the (1+1)-dimensional heat equation separates into the ODEs of (41.37a).

41.2   Starting with the full Schrödinger equation,

$$\left(-\frac{\hbar^2}{2m}\nabla^2 + V(\vec{r})\right)\Psi(\vec{r},t) = i\hbar\frac{\partial \Psi(\vec{r},t)}{\partial t},$$

derive the time-independent Schrödinger equation,

$$\left(-\frac{\hbar^2}{2m}\nabla^2 + V(\vec{r})\right)\phi(\vec{r}) = E\phi(\vec{r}), \tag{41.139}$$

for a particle of definite energy $E = \hbar\omega$.

41.3   What are the normal mode frequencies of the string in Example 41.1 if the end at $x=0$ is released? (See (41.84).)

41.4   Show that the modified wave equation of (41.29) separates into a familiar equation in time and Bessel's equation (41.30) in space.

41.5   In BTW 41.1 we defined and distinguished between three classes of PDEs. Which of the three is $\partial_x \partial_y \Psi = 0$? Find the variables $u(x, y)$ and $v(x, y)$ which renders this equation in the canonical form for that class.

41.6   **d'Alembert's Solution**
Physical considerations imply that wave equation solutions depend not on x and t separately, but rather on the combination $x \pm vt$ with wave speed v.

(a) Show that the general solution to the wave equation is the superposition of right- and left-moving traveling waves, $\psi(x, t) = f(x - vt) + g(x + vt)$.
(b) Apply the initial conditions $\psi(x, 0) = \alpha(x)$, $\partial_t \psi(x, 0) = \beta(x)$ to this general form to derive the solution

$$\psi(x, t) = \frac{1}{2} [\alpha(x - vt) + \alpha(x + vt)] + \frac{1}{2v} \int_{x-vt}^{x+vt} \beta(\xi) \, d\xi. \qquad (41.140)$$

(c) Show that for finite v the solution is causal: points x which can be influenced by a wave emitted from x_0 at $t = 0$ must lie within the cone $|x - x_0| \le vt$.

41.7   Recast (41.67b) in terms of the variable $x = \cos\theta$.

41.8   Verify (41.70) by separating the Helmholtz equation in spherical coordinates.

41.9   [Adapted from Verbin (2015)] Affixing a mass m to the lower end of the hanging rope in Example 41.2 adds weight mg to the tension, $T = mg + \rho g y = \rho g(\mu + y)$, where $\mu = m/\rho$.

(a) Modify (41.30) to account for the added mass.
(b) What values of z do the ends of the rope correspond to? What boundary conditions apply there?
(c) Argue physically that for large m, the normal mode frequencies should approach those of a horizontal string fixed at both ends, $\omega_n = \sqrt{\frac{T}{\rho}} \frac{n\pi}{L}$ (Example 41.1). Then argue mathematically that for large m, the asymptotic forms (C.8) and (C.13) of J_n and N_n are applicable.
(d) Apply the boundary conditions using these asymptotic forms to find an expression for the first-order correction to the normal mode frequencies of the fixed horizontal string.

41.10   In Example 41.5, we took $\alpha^2 > 0$ based on consideration of the boundary conditions. Show that had we taken $\beta^2 > 0$ instead the same solution (41.91) still emerges.

41.11   Show how the two-dimensional Laplace equation calculation in Example 41.5 can be mapped directly into the analysis and solution of a (1+1)-dimensional heat equation problem.

41.12   A rectangular conducting frame with sides a and b has all four sides held at constant potential V_0.

(a) Use uniqueness arguments (and scant calculation) to find the potential $V(x, y)$ everywhere inside the frame.
(b) Adapt Example 41.6 to express $V(x, y)$ as an infinite sum. Use Mathematica's `Plot3D` command to show it agrees with the result in (a).

41.13   A hollow cube of side a sits in the octant $x, y, z > 0$ with one corner at the origin.

(a) Find the voltage if all the faces are grounded except for $x = 0$, which is held at constant potential $V = V_0$.
(b) Repeat with the face at $z = a$ also held at $V = V_0$.

41.14   A conducting cylinder of radius R, its base centered at the origin in the xy-plane, extends a distance L upward along z. The cylindrical surface is grounded everywhere except on the top, where $V(r, \phi, L)$ is held at a constant V_0. Find the voltage everywhere inside the cylinder.

41.15 Laplacian Chess
An infinite metal chessboard fills the plane $z = 0$. Its squares have sides of length a, the black ones held at constant potential V_0 and the white ones grounded. Find the potential $V(x, y, z)$ above and below the plane. Use Mathematica's `Plot3D` to show the behavior of $V(x, y_0, z)$ for different values of y_0.

41.16 Consider a cylinder of height L and radius a with fixed voltage on the surface

$$V(a, \phi, z) = \begin{cases} V_0, & 0 < z < L/2 \\ -V_0, & L/2 < z < L \end{cases}.$$

Imagine that identical copies of this cylinder are arranged in an infinite stack along z — essentially a one-dimensional periodic "crystal lattice" of cylinders. Find the potential inside the cylinder.

41.17 The ends of a rod of length L are held at constant temperatures T_1 and T_2. Find the temperature $T(x, t)$ for an initial distribution $T(x, 0) = x$.

41.18 Find the temperature distribution along a circular ring of radius R with initial distribution $f(\theta) = \cos^2 \theta$.

41.19 Verify that the function $\tilde{T}$ defined in (41.113) has the claimed boundary conditions.

41.20 As in Examples 41.8 and 41.9, the solution $u(x)$ to a Helmholtz equation with inhomogeneous conditions can be expressed as a superposition $u = \tilde{u} + v$, where $v(x)$ satisfies the required boundary conditions, and $\tilde{u}$ obeys homogeneous conditions. Since the Laplacian is a second-order operator, the most general form of v is

$$v(x) = a\left(x^2 + c_1 x + d_1\right) + b\left(x^2 + c_2 x + d_2\right).$$

Find the functions $v(x)$ appropriate for:

(a) Dirichlet conditions, $u'(0) = a$, $u(b) = b$
(b) Neumann conditions, $u'(0) = a$, $u'(b) = b$
(c) mixed conditions, $u'(0) = a$, $u(b) = b$.

Problems 41.21–41.24 consider a rectangular plate with sides a and b and initial condition

$$T(x, y, t = 0) = \begin{cases} 0, & 0 < y < b/2 \\ T_0, & b/2 < y < b \end{cases}.$$

In each case, use Mathematica's `Plot3D` to display various space and time slices for $a = 2, b = 1$, and $T_0 = 10$.

41.21 Find the temperature distribution of the plate as a function of time if all sides are held at zero temperature.

41.22 Find the temperature distribution of the plate as a function of time when the side at $y = b$ is held at $3T_0$, the other three at $T = 0$. How does the solution change if the side at $y = b$ is held at 0 and that at $y = 0$ at $3T_0$?

41.23 Find the temperature distribution of the plate as a function of time when three sides are held at zero temperature, the fourth satisfying

$$T(a, y, t) = T_a(y) = y.$$

41.24 Find the temperature distribution of the plate as a function of time when one side is held at zero temperature, the other three satisfying

$$T(a, y, t) = y \qquad T(x, 0, t) = x^2 \qquad T(x, b, t) = 3T_0.$$

41.25 Western music is based on a 12-tone equal temperament tuning dividing the octave into 12 equal intervals, with a frequency ratio $2^{1/12}$ between semitones,

C	C#	D	D#	E	F	F#	G	G#	A	A#	B
1	1.059	1.122	1.189	1.260	1.335	1.414	1.498	1.587	1.682	1.782	1.888

A timpani (also called a kettledrum) is unusual amongst drums in that it can produce a clear pitch with quasi-harmonic overtones. As a circular drumhead with Dirichlet boundary conditions, the ideal timpani's normal mode frequencies are determined by the zeros of the Bessel function J_n. Which modes are excited depends on where the drumhead is struck.

(a) Show that if struck in the center, only the modes $(0, n)$, corresponding the eigenfunctions J_0, are excited. Compare the ratios of the zeros α_{0n} with those of the equal temperament scale to decide whether the sound produced is musically harmonic. [In fact, J_0 severely constrains the harmonic content, producing a "hollow" sound.]

(b) Professional timpanists strike the drum about 2/3 out from the center, exciting the "preferred modes" $(1, n)$. Explain why this suppresses the $(0, n)$ modes.

(c) Look up (or use Mathematica to find) the resonant frequencies of the $(1, n)$ modes for $n = 2$ to 6 and explain why these give the timpani a sense of harmonicity.

[For more on the physics of music and musical instruments, see Backus (1977) and Benade (1990).]

41.26 Elliptical Coordinates

(a) With the ansatz $u(x, y) = R(\rho)\Phi(\phi)$, show that the two-dimensional Helmholtz equation separates in elliptical coordinates with foci at $\pm c$,

$$x = c \cosh \rho \cos \phi \qquad y = c \sinh \rho \sin \phi.$$

[See Problem 2.16, Table A.1, and (12.49b).]

(b) Show that the two separated ODEs can be put in the form of Mathieu's equation (40.57),

$$\frac{d^2\Phi}{d\phi^2} + (a - 2q \cos 2\phi)\Phi(\phi) = 0,$$

and the so-called modified Mathieu equation,

$$\frac{d^2 R}{d\rho^2} - (a - 2q \cosh 2\rho)R(\rho) = 0.$$

Notice that these equations are related by $\rho = \pm i\phi$ — not unlike the Bessel and modified Bessel equations (see Appendix C). The solutions are Mathieu functions $C(a, q, \phi)$, $S(a, q, \phi)$ and modified Mathieu functions $C(a, q, i\rho)$, $S(a, q, i\rho)$.

41.27 Normal Modes of an Elliptical Drumhead [Adapted from Trott (1996)]

The normal modes of an elliptical drum are determined from boundary conditions on the solutions of the separated ODEs in Problem 41.26. Although initially a and q are independent, the condition on Φ imposes functional relationships $a = a_n(q)$; the boundary condition on R then picks out discrete values of q from these curves, corresponding to the normal mode frequencies.

(a) What value $\rho = \rho_0$ describes the boundary of an elliptical drumhead with semi-major axis 5 and semi-minor axis 3? Where are the foci $\pm c$?

(b) Imposing π- or 2π-periodicity in ϕ reduces the space of solutions $\Phi(\phi)$ to the elliptic cosine and sine functions $ce_n(q, \phi)$ and $se_n(q, \phi)$ introduced in Problem 40.17. Use (40.58) to make Mathematica plots of these for $q = 4$, $n = 0, 1, 2$. [Like $\sin n\phi$, se_n requires $n > 0$.]

(c) Since the ODEs share the parameters a and q, the periodic condition on Φ also restricts the radial solutions to linear combinations of the modified elliptic sine and cosine $ce_n(q, i\rho)$ and $-i\,se_n(q, i\rho)$. [Just as $\sinh \rho = -i \sin(i\rho)$ is real, so too $-i\,se_n(q, i\rho)$.] Make Mathematica plots of these functions evaluated at the drum perimeter $\rho = \rho_0$ for $n = 0, 1, 2$.

(d) To identify the normal modes, we need to impose the boundary condition in ρ — which requires finding the zeros of $ce_n(q, i\rho)$ and $se_n(q, i\rho)$. We can locate them with Mathematica's FindRoot command, using the plots in (c) to choose reasonable starting values. For instance, a plot of $ce_0(q, i\rho_0)$ shows the first zero to be near $q = 1.5$; with this value as the starting value of FindRoot's search,

$$\texttt{FindRoot}[ce_0, [q, i\rho 0], \{q, 1.5\}]//\texttt{Chop}$$

returns 1.73531. [Chop replaces numbers $|\epsilon| < 10^{-10}$ with 0.] Similarly,

$$\texttt{FindRoot}[se_1, [q, i\rho 0], \{q, 5\}]//\texttt{Chop}$$

gives 5.43008. Find the values of q for the first five normal modes.

(e) Use the parity of $ce_n(q, \xi)$ and $se_n(q, \xi)$ in ξ (even and odd, respectively), together with the requirement of continuity across $\rho = 0$, to argue that the drum's normal modes are given by the products $ce_n(q, \phi)ce_n(q, i\rho)$ and $se_n(q, \phi)se_n(q, i\rho)$.

(f) Make plots of the first five normal modes by having Mathematica convert from elliptical to cartesian, restricting the domain to the ellipse — for instance,

```
Plot3D[
        Evaluate[TransformedField[
        {{"Elliptic",c} → "Cartesian"},
        ce₀[1.73531, iρ]ce₀[1.73531,φ],{ρ,φ} → {x,y}]],
        {x, − c Cosh[ρ0],c Cosh[ρ0]},{y, − c Sinh[ρ0],c Sinh[ρ0]},
        PlotPoints → 50,
        RegionFunction →
```

$$\texttt{Function}[\{x, y\}, \frac{x^2}{c^2 \texttt{Cosh}[\rho 0]^2} + \frac{y^2}{c^2 \texttt{Sinh}[\rho 0]^2} < 1]$$

```
        ]
```

gives the fundamental (lowest) mode of the drum. [For a different perspective, replace Plot3D with ContourPlot.]

41.28   I once overheard a student remark that he didn't know why we bothered with Neumann functions, since they are "always tossed out" for being divergent at $r = 0$. So I asked the class to consider an annular drum — a circular membrane with a concentric circular hole and homogeneous Dirichlet conditions at both the inner and outer radii $a < b$. Though an annulus is a strange shape for a drum, it still has normal modes. [For a discussion of how such a drum sounds, see Gottlieb (1979).]

(a) Find the linear combination of J_n and N_n which satisfies the boundary conditions, identifying (but not solving for) any parameters introduced.

(b) Find the radial modes $R_m(u)$ for $m = 1$ to 3, in terms of dimensionless $u = r/a$ and the ratio $\gamma = b/a = 2$. Use Mathematica's FindRoot or Reduce to find the needed zeros for $n = 0$ and 1. [You should discover the two sets are not terribly different.] Use Plot to show these R_m satisfy the boundary conditions and Integrate to verify their orthogonality.

(c) Use Plot3D to plot the first two normal modes for both $n = 0$ and 1.

41.29   [Adapted from Fourier's 1822 treatise *Théorie Analytique de la Chaleur*.] A spherical container of radius R is filled with ice cream. Initially the ice cream has temperature $T_0 = -15°C$ at the center, increasing radially like r^2 until reaching $T_1 = 0°C$ at the surface. At time $t = 0$, the ice cream is placed in a freezer of constant temperature T_{fr}.

(a) Set up the problem mathematically for a (poor) freezer with $T_{fr} = 0°C$, including initial and boundary conditions. Is this a Dirichlet problem or a Neumann problem?

(b) Use Mathematica to solve for the temperature $T(r,t)$ as a normal mode sum. [Mathematica was unavailable to Fourier — be impressed!] Use the `Plot` command to verify the initial distribution as a function of $u = r/R$. Then use `Plot3D` to display how the distribution evolves in dimensionless time $\tau = \alpha t/R^2$. Is the final distribution what you expect?

(c) Repeat the analysis for a proper freezer with $T_{\text{fr}} = -10°\text{C}$ (see Example 41.9). You can use `FindRoot` or `Reduce` to find the needed zeros.

41.30 What if the spherical container of Problem 41.29 is an insulated thermos, so that no heat flows in or out. Again taking $T_0 = -15°\text{C}$ and $T_1 = 0°\text{C}$, use Mathematica to solve for the temperature $T(r,t)$. [It may be helpful to refer to Problem 40.24, and consider how j_0' is related to j_1.] Use the `Plot` command to verify the initial distribution as a function of $u = r/R$. Then use `Plot3D` to display how the distribution evolves in dimensionless time $\tau = \alpha t/R^2$.

42 Green's Functions

42.1 A Unit Source

Consider a linear differential operator $\mathcal{L}$ and the inhomogeneous equation

$$\mathcal{L}\psi(\vec{r}) = \rho(\vec{r}), \qquad \rho \neq 0. \tag{42.1}$$

Whereas homogeneous solutions reflect properties of $\mathcal{L}$, an inhomogeneous solution must clearly depend on the source ρ as well. It would be beneficial, both conceptually and as a practical matter, to separate these influences by deriving a general expression for $\psi(\vec{r})$ which reflects the characteristics $\mathcal{L}$ in a form that can be applied to any given ρ.

Remarkably, we can do just that by considering one specific case — that of a point source at $\vec{r}\,'$ of unit magnitude, $\delta(\vec{r} - \vec{r}\,')$. The inhomogeneous solution with this source is called the *Green's function* $G(\vec{r}, \vec{r}\,')$ of the operator,[1]

$$\mathcal{L}\, G(\vec{r}, \vec{r}\,') = \delta(\vec{r} - \vec{r}\,'). \tag{42.2}$$

In other words, $G(\vec{r}, \vec{r}\,')$ is the field at $\vec{r}$ due to a unit point source at $\vec{r}\,'$. An infinitesimal source $\rho(\vec{r}\,')\, d^3r'$ at $\vec{r}\,'$ then produces the field $G(\vec{r}, \vec{r}\,')\rho(\vec{r}\,')\, d^3r'$ at $\vec{r}$. As indicated in Figure 42.1, superposing all such infinitesimal contributions for all $\vec{r}\,'$ in V yields the full field at $\vec{r}$

$$\psi(\vec{r}) = \int_V G(\vec{r}, \vec{r}\,')\rho(\vec{r}\,')\, d^3r'. \tag{42.3}$$

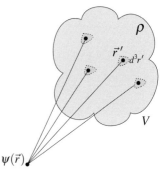

Figure 42.1 The field at $\vec{r}$ due to the superposition of infinitesimal sources $\rho(\vec{r}\,')d^3r'$.

That this is a solution of (42.1) is easily demonstrated by moving $\mathcal{L}$, which acts only on functions of $\vec{r}$, inside the $\vec{r}\,'$ integral:

$$\mathcal{L}\psi(\vec{r}) = \mathcal{L} \int_V G(\vec{r}, \vec{r}\,')\rho(\vec{r}\,')\, d^3r'$$

$$= \int_V \left[\mathcal{L}G(\vec{r}, \vec{r}\,') \right] \rho(\vec{r}\,')\, d^3r'$$

$$= \int_V \delta(\vec{r} - \vec{r}\,')\rho(\vec{r}\,')\, d^3r' = \rho(\vec{r}). \tag{42.4}$$

---

[1] Green's functions are named for George Green. Following the usual practice of named functions (e.g., Bessel functions), G would be a "Green function." This usage exists, but "Green's function" (often without an apostrophe) predominates — presumably to avoid association with color.

So the Green's function approach solves an inhomogeneous differential equation in two distinct steps: first find an operator's Green's function with given boundary conditions, and then integrate it against the source to find the inhomogeneous solution ψ with the same boundary conditions.

We actually found the Green's function for the Laplacian in Example 9.5 — in fact, it's hiding in plain sight: Coulomb's law (9.33) has precisely the form of (42.3). Thus the Green's function with a zero-at-infinity boundary condition is

$$G(\vec{r},\vec{r}') \equiv \frac{1}{4\pi} \frac{1}{|\vec{r} - \vec{r}'|}. \tag{42.5}$$

Verification that this produces the Poisson equation of electrostatics, $\nabla^2\Phi = -\rho/\epsilon_0$, follows as in (42.4) — courtesy of the familiar identity (Problem 15.26)[2]

$$\nabla^2 \frac{1}{|\vec{r} - \vec{r}'|} = -4\pi\,\delta(\vec{r} - \vec{r}'). \tag{42.6}$$

The relative ease with which we found this particular G was aided by familiarity with electrostatics; deriving the Green's functions for other differential operators is generally not so straightforward. In particular, one must accommodate the discontinuity in G demanded in (42.2) by the delta function at $\vec{r} = \vec{r}'$. The basic approach is best described with examples; we'll start with one-dimensional problems, first in bounded space and then in unbounded time. [For a rigorous exposition of Green's functions see Stakgold (1979).]

Example 42.1 **Poisson in One Dimension**

Consider the Poisson equation in one dimension,

$$\frac{d^2}{dx^2}\psi(x) = \rho(x) \tag{42.7}$$

over a finite region $0 \le x \le L$, with Dirichlet boundary conditions,

$$\psi(0) = \psi(L) = 0. \tag{42.8}$$

The Green's function is the solution to

$$\frac{d^2}{dx^2}G(x,x') = \delta(x - x'), \tag{42.9}$$

with the same boundary conditions. As inhomogeneous ODEs go, this Green's function is not particularly difficult to find — after all, everywhere other than $x = x'$, G must be a linear function $ax + b$ with constants a and b. Even so, the inhomogeneous solution is subtle. For one thing, the solutions cannot be identical on either side of x'; some response to the delta function must manifest. This implies that there are actually *four* unknown constants, two each on either side of x'. The homogeneous solution for $x < x'$ need only obey the boundary condition at $x = 0$, whereas for $x > x'$ it must satisfy only the condition at $x = L$. That takes care of two unknowns, resulting in the piecewise form

$$G(x,x') = \begin{cases} ax, & 0 < x < x' \\ b(x - L), & x' < x < L \end{cases}. \tag{42.10}$$

---

[2] A sign confusion lurks: mathematically, Poisson's equation is generally given as $\nabla^2\Phi = \rho$; in electrostatics, the sign of ρ flips — a consequence of the convention assigning the electron a negative charge.

We'll need two "interior" conditions at $x = x'$ in order to determine the remaining constants a and b; both can be found from (42.9). First, the delta function requires the derivative dG/dx to have a jump discontinuity at $x = x'$. To verify this, we isolate the effect of the delta function by integrating the differential equation over an infinitesimal interval of width 2ϵ centered around $x = x'$. One integration confirms the discontinuity in dG/dx as $\epsilon \to 0$,

$$\int_{x'-\epsilon}^{x'+\epsilon} \frac{d^2}{dx^2} G(x,x')\, dx = \frac{d}{dx} G(x,x') \Big|_{x'-\epsilon}^{x'+\epsilon} = 1, \tag{42.11}$$

where the right-hand side follows from integrating over the delta function. Thus G must have a kink at $x = x'$. On the other hand, G itself must be continuous across x', since otherwise the second derivative would yield δ' rather than δ. Applying these two interior conditions to (42.10),

$$b - a = 1 \qquad ax' = b(x' - L), \tag{42.12}$$

yields

$$a = \frac{1}{L}(x' - L) \qquad b = \frac{x'}{L}. \tag{42.13}$$

Thus the Green's function is

$$G(x,x') = \frac{1}{L}x_<(x_> - L) = \frac{1}{L}\begin{cases} x(x' - L), & x < x' \\ x'(x - L), & x' < x \end{cases}, \tag{42.14}$$

where $x_<$ and $x_>$ denote the lesser and greater of x and x', respectively.

With G in hand, we can now solve the inhomogeneous equation for a given source. For instance, $\rho(x) = x$ gives

$$\psi(x) = \int_0^L G(x,x')\rho(x')\, dx' = \frac{1}{L}\int_0^x [x'(x - L)]x'\, dx' + \frac{1}{L}\int_x^L [x(x' - L)]x'\, dx'$$
$$= \frac{1}{6}x(x^2 - L^2), \tag{42.15}$$

where the integration limits derive from the piecewise form of G. Notice that ψ obeys both the inhomogeneous differential equation as well as the boundary conditions.

Example 42.2   Duhamel's Principle

Unlike the finite domain of the previous boundary value problem, initial value problems are defined on $0 \le t < \infty$. Consider the driven oscillator

$$\left[\frac{d^2}{dt^2} + 2\beta \frac{d}{dt} + \omega_0^2\right] u(t) = f(t), \tag{42.16}$$

with initial conditions $u(0) = u_0$, $\dot{u}(0) = v_0$. An initial value problem may seem ill-suited to a Green's function with a kink at some arbitrary $t' > t$. But with the identity

$$f(t) = \int_0^t f(t')\delta(t - t')\, dt', \tag{42.17}$$

the driving force can be construed as a continuous sequence of individual impulses, each occurring at its unique time t' with magnitude $f(t')$. Each impulse creates a homogeneous response for $t > t'$.

Since by definition the response at time t to a unit impulse at t' is $G(t - t')$, the response to each applied impulse is $f(t')G(t - t')$. And as a linear system, this response adds to the behavior of the oscillator when it reaches t. The particular solution $u_p(t)$ then is the superposition of all impulse responses up to that time,[3]

$$u_p(t) = \int_0^t f(t')G(t - t')\,dt', \tag{42.18}$$

This has a clear and intuitive causal structure: there is no contribution from an impulse until it's applied. The response of the oscillator at time t depends not just on $f(t)$, but on all the earlier impulses $f(t' < t)$ as well. This construction of the full solution as a superposition of staggered initial-value solutions is often referred to as *Duhamel's principle*.

A Green's function's required kink thus comes into focus: since the initial value problem is a superposition of staggered initial value problems, G must have a kink at the instant of each impulse. For instance, satisfying the (dis)continuity conditions to bring $u(t < t') = 0$ into accord with the underdamped solution for $t > t'$ yields (Problem 42.14)

$$G(t,t') = \Theta(t - t')\frac{1}{\Omega}\sin\left[\Omega(t - t')\right]e^{-\beta(t-t')}, \quad \Omega = \sqrt{\omega_0^2 - \beta^2}, \tag{42.19}$$

just as we found in (22.11) through other methods. Since the contribution of each pulse is delayed (retarded) until time t', this is called a "retarded" Green's function. The full solution to (42.18) is

$$u(t) = u_h(t) + \int_0^t f(t')G(t - t')\,dt', \tag{42.20}$$

where u_h is a homogeneous solution satisfying the initial conditions, describing the behavior of the oscillator before application of the driving force.

BTW 42.1 Green's Function and the Variation of Constants

We obtained the solution (42.15) back in Example 39.3 using the variation of constants method — a technique which is essentially equivalent to using Green's functions. To see this, notice that placing constant lower limits on the integrals of (39.33) produces a complementary solution. So (39.34) can be recast as

$$\psi(x) = \phi_2(x)\int_a^x \frac{\phi_1(x')\rho(x')}{p(x')W(x')}\,dx' - \phi_1(x)\int_b^x \frac{\phi_2(x')\rho(x')}{p(x')W(x')}\,dx', \tag{42.21}$$

where $\phi_{1,2}$ are homogeneous solutions and W their Wronskian. The function p is that of the Sturm–Liouville operator (40.12) — so ρ/p in (42.21) is the source r of (39.33). Now if we take a and b to be the endpoints of the domain, it makes sense to flip the sign in the second integral,

$$\psi(x) = \int_a^x \frac{\phi_2(x)\phi_1(x')r(x')}{W(x')}\,dx' + \int_x^b \frac{\phi_1(x)\phi_2(x')r(x')}{W(x')}\,dx'. \tag{42.22}$$

Comparison with (42.15) allows identification of the emergent Green's function,

$$G(x,x') = \frac{1}{pW}\begin{cases}\phi_1(x')\phi_2(x), & x < x' \\ \phi_1(x)\phi_2(x'), & x' < x\end{cases}, \tag{42.23}$$

---

[3] Incidentally, notice that this is a convolution integral — see Example 36.15.

where we've implicitly noted that the product pW is constant for Sturm–Liouville operators (Problem 39.6). The final task is to choose linearly independent $\phi_{1,2}$ so that ψ obeys the boundary conditions. (We need only specify $\phi_{1,2}$ up to overall multiplicative constants, since any such factors will cancel when divided by the Wronskian.) For instance, conditions $\psi(a) = \psi(b) = 0$ are satisfied by choosing homogeneous solutions such that $\phi_1(a) = 0$ and $\phi_2(b) = 0$ — precisely what we did in (42.10).

Example 42.3 **Poisson Unbound**

Boundary value problems can also have infinite domain; unlike time, however, spatial domains can be more than one-dimensional. Consider

$$\nabla^2 G(\vec{r}, \vec{r}\,') = -\delta(\vec{r} - \vec{r}\,') \tag{42.24}$$

over all of $\mathbb{R}^3$ — so we'll use spherical coordinates with $0 \le r < \infty$. (The sign is appropriate to electrostatics — see footnote 2 in this chapter.) Since G must be finite at both $r = 0$ and ∞, homogeneous solutions on either side of $\vec{r}\,'$ can be read from Table 41.1,

$$G(\vec{r}, \vec{r}\,') = \sum_{\ell=0}^{\infty} \sum_{m=-\ell}^{\ell} \begin{cases} a_{\ell m} r^\ell \, Y_{\ell m}(\theta, \phi), & 0 < r < r' \\ b_{\ell m} \frac{1}{r^{\ell+1}} Y_{\ell m}(\theta, \phi), & r' < r < \infty \end{cases}. \tag{42.25}$$

The coefficients $a_{\ell m}$ and $b_{\ell m}$ are functions of r', to be determined by the continuity of G and the discontinuity of $\vec{\nabla} G$. The first easily reveals

$$b_{\ell m} = r'^{2\ell+1} a_{\ell m}. \tag{42.26}$$

The second condition requires accommodation of the discontinuity. As in (42.11), we can isolate the delta function singularity by integrating,

$$\int_V \nabla^2 G(\vec{r}, \vec{r}\,') \, d^3 r = -1. \tag{42.27}$$

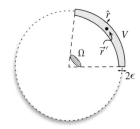

The volume of integration, of course, must enclose the source at $\vec{r}\,'$; we'll choose a 2ϵ-thick spherical cap of solid angle Ω with $d\vec{a} = \hat{r} r^2 d\Omega$. Then, courtesy of the divergence theorem, we evaluate (42.27) as a flux integral,

$$\int_V \nabla^2 G(\vec{r}, \vec{r}\,') \, d^3 r = \oint_S \vec{\nabla} G(x, x') \cdot d\vec{a} = \int \left(\frac{\partial G}{\partial r} \bigg|_{r'+\epsilon} - \frac{\partial G}{\partial r} \bigg|_{r'-\epsilon} \right) r^2 d\Omega. \tag{42.28}$$

Then (42.27) can be satisfied with

$$\frac{\partial G}{\partial r} \bigg|_{r'+\epsilon} - \frac{\partial G}{\partial r} \bigg|_{r'-\epsilon} = -\frac{1}{r'^2} \, \delta(\cos\theta - \cos\theta')\delta(\phi - \phi')$$

$$= -\frac{1}{r'^2} \sum_{\ell m} Y_{\ell m}^*(\theta', \phi') \, Y_{\ell m}(\theta, \phi), \tag{42.29}$$

using the completeness of spherical harmonics, (35.44)–(35.45). This is our sought-after discontinuity condition, the analog of (42.11). Applying it and (42.26) to (42.25) gives

$$a_{\ell m} = \frac{1}{2\ell + 1} \frac{1}{r'^{\ell+1}} Y^*_{\ell m}(\theta', \phi').$$

(42.30)

All together then,

$$G(\vec{r}, \vec{r}') = G(\vec{r}', \vec{r}) = \sum_{\ell, m} \frac{1}{2\ell + 1} Y^*_{\ell m}(\theta', \phi') Y_{\ell m}(\theta, \phi) \begin{cases} \frac{r^\ell}{r'^{\ell+1}}, & 0 < r < r' \\ \frac{r'^\ell}{r^{\ell+1}}, & 0 < r < r' \end{cases}$$

$$= \sum_{\ell, m} \frac{1}{2\ell + 1} \frac{r^\ell_<}{r^{\ell+1}_>} Y^*_{\ell m}(\theta', \phi') Y_{\ell m}(\theta, \phi)$$

$$= \frac{1}{4\pi} \sum_\ell \frac{r^\ell_<}{r^{\ell+1}_>} P_\ell(\cos \gamma), \qquad \cos \gamma = \hat{r} \cdot \hat{r}',$$

(42.31)

where we've used the spherical harmonic addition theorem (35.50). This may look familiar: comparison with (35.52) or (35.53) shows that

$$G(\vec{r}, \vec{r}') = \frac{1}{4\pi} \frac{1}{|\vec{r} - \vec{r}'|},$$

(42.32)

confirming (42.5).

Example 42.4 The Wave Equation in 2+1 Dimensions

Sometimes symmetry can simplify the derivation of a Green's function. For instance, for a monochromatic wave $\Psi(\vec{r}, t) = \psi(r)e^{-i\omega t}$ with a source $f(r)e^{-i\omega t}$, the wave equation reduces to the radial Helmholtz equation

$$\left(\nabla^2 + k^2\right)\psi = f(r), \qquad \omega/k = v.$$

(42.33)

The Green's function given by

$$\left(\nabla^2 + k^2\right)G(\vec{r}, \vec{r}') = \delta(\vec{r} - \vec{r}'),$$

(42.34)

describes a wave emitted by a unit source from $\vec{r}'$. Due to translational symmetry, we can take $\vec{r}'$ to be the origin of our coordinates, reducing the Green's function to $G(r) \equiv G(\vec{r}, 0)$. Having worked in spherical coordinates in the previous example, this time we'll consider polar coordinates, so that (42.34) becomes

$$\partial_r^2 G(r) + \frac{1}{r}\partial_r G(r) + k^2 G(r) = \delta(r).$$

(42.35)

Everywhere but at the origin, this is a homogeneous Bessel equation with general solution

$$\psi(r) = a J_0(kr) + b N_0(kr),$$

(42.36)

where the ϕ-independence of (42.35) requires $m = 0$. Although there are clear similarities to the circular drumhead of Example 41.11, the boundary conditions are very different. In this case, we need a solution describing a wave emanating from the origin which goes to zero at infinity. Not only do both J_0 and N_0 fall off for large r, but with a delta function at the origin there is no reason to throw out the Neumann function.

A comparison to waves on a string can be helpful. The familiar standing wave solutions for fixed ends can be portrayed as the interference of a traveling wave $e^{i(kx-\omega t)}$ with its reflection $e^{i(kx+\omega t)}$ off the boundary. Without the boundary, an infinite string would support only the traveling wave, and the general solution is more-intuitively expressed not as $a\cos kx + b\sin kx$, but rather as $c\,e^{ikx} + d\,e^{-ikx}$. Similarly, the solution in (42.36) is better expressed on the basis of *Hankel functions* (Appendix C),

$$G(r) = cH_0^{(1)}(kr) + dH_0^{(2)}(kr), \qquad (42.37)$$

where

$$H_n^{(1,2)}(kr) = J_n(kr) \pm iN_n(kr). \qquad (42.38)$$

A physical picture can be extracted by examining their asymptotic form in (C.23),

$$H_0^{(1,2)}(kr) \xrightarrow{kr\text{ large}} \sqrt{\frac{2}{\pi kr}}\, e^{\pm i(kr - \pi/4)}, \qquad (42.39)$$

which describes radial waves in the plane with intensity falling as $1/r$. Appending the source's time-dependence $e^{-i\omega t}$ identifies $H_0^{(1)}$ as an outgoing wave, $H_0^{(2)}$ an ingoing wave. So the solution we want requires $d = 0$ in (42.37), leaving

$$G(r) = cH_0^{(1)}(kr). \qquad (42.40)$$

To find c, we use the discontinuity imposed by the delta function — similar to what we did in Example 42.3. The result is (Problem 42.7)

$$G(r) = -\frac{i}{4}\, H_0^{(1)}(kr). \qquad (42.41)$$

Because this depends only on the distance $r = |\vec{r}|$ from the source, we can easily reinstate arbitrary source position $\vec{r}\,'$ to get

$$G(\vec{r},\vec{r}\,') = -\frac{i}{4}\, H_0^{(1)}(k|r - r'|). \qquad (42.42)$$

Manifest in each of the solutions (42.14), (42.23), (42.31), and (42.42) is symmetry between $\vec{r}$ and $\vec{r}\,'$,

$$G(\vec{r},\vec{r}\,') = G(\vec{r}\,',\vec{r}). \qquad (42.43a)$$

This in fact is a basic property of Green's functions for Hermitian $\mathcal{L}$.[4] Thus for Sturm–Liouville equations, not only is the Green's function the response at $\vec{r}$ due to a unit source at $\vec{r}\,'$, it's also the response at $\vec{r}\,'$ to a source at $\vec{r}$ (Problem 42.18). In addition, since the unit source depends only on the difference $\vec{r} - \vec{r}\,'$, the Green's function for a differential operator $\mathcal{L}$ with constant coefficients should have the same dependence — that is, G should be invariant under translations

$$G(\vec{r},\vec{r}\,') = G(\vec{r} - \vec{r}\,') = G(|\vec{r} - \vec{r}\,'|), \qquad (42.43b)$$

where the last equality follows from the additional assumption of isotropy. Both (42.32) and (42.42) clearly exhibit the symmetries of (42.43b) — though not (42.14). The failure of translation invariance in that instance is due not to the operator, but rather to its domain. The finite string is defined on the

---

[4] For non-Hermitian operators, this generalizes to $G(\vec{r},\vec{r}\,') = G^*(\vec{r}\,',\vec{r})$.

bounded domain between 0 and L, which spoils translational symmetry; Examples 42.3 and 42.4, on the other hand, are defined on unbounded $\mathbb{R}^3$.

A Green's function on an unbounded domain is said to be the operator's *fundamental solution*. With finite boundary conditions, an operator's Green's function can be written as

$$G_{\text{BC}}(\vec{r}, \vec{r}') = G(\vec{r}, \vec{r}') + F(\vec{r}, \vec{r}'), \tag{42.44}$$

where $F(\vec{r}, \vec{r}')$ is a solution of the homogeneous equation $\mathcal{L}F = 0$ and the subscript BC denotes the finite boundary condition. In other words, F is the complementary homogeneous solution to the inhomogeneous solution G.

Example 42.5 **Image Charge inside a Conducting Sphere**

Consider a charge distribution $\rho(\vec{r}')$ outside a grounded conducting sphere of radius a. The conductor's electrons will respond to ρ by rearranging themselves to make the surface an equipotential; this induced surface charge distribution contributes to the total potential outside the sphere. To find it, we need to solve Poisson's equation for $r > a$ with homogeneous Dirichlet conditions at $r = a$.

The fundamental Green's function with a unit source outside the sphere at $|\vec{r}'| > a$ is given by (42.31). Adding to this the contribution from the induced charge on the sphere — that is, the general homogeneous (i.e., Laplace) solution — gives the Dirichlet Green's function

$$4\pi G_D(\vec{r}, \vec{r}') = -\frac{1}{|\vec{r} - \vec{r}'|} + F(\vec{r}, \vec{r}') \tag{42.45}$$

$$= -\sum_{\ell, m} \frac{1}{2\ell + 1} \frac{r_<^\ell}{r_>^{\ell+1}} Y_{\ell m}^*(\theta', \phi') Y_{\ell m}(\theta, \phi) + \sum_{\ell, m} \frac{B_{\ell m}(\vec{r}')}{r^{\ell+1}} Y_{\ell m}(\theta, \phi), \tag{42.46}$$

where we've used (41.71) with $A_{\ell m} = 0$ since G_D must remain finite as r extends out to infinity. To evaluate it on the sphere's surface, we identify $r_< = a$ and $r_> = r'$. Then the Dirichlet condition for our grounded sphere becomes

$$0 = \sum_{\ell, m} \left[-\frac{1}{2\ell + 1} \frac{a^\ell}{r'^{\ell+1}} Y_{\ell m}^*(\theta', \phi') + \frac{B_{\ell m}(\vec{r}')}{a^{\ell+1}} \right] Y_{\ell m}(\theta, \phi). \tag{42.47}$$

Since the spherical harmonics are complete, the bracketed term must vanish separately for each ℓ, m pair — and so

$$B_{\ell m} = \frac{1}{2\ell + 1} \frac{a^{2\ell+1}}{r'^{\ell+1}} Y_{\ell m}^*(\theta', \phi'). \tag{42.48}$$

The contribution of the sphere's induced charge distribution is thus

$$F(\vec{r}, \vec{r}') = \sum_{\ell, m} \frac{1}{2\ell + 1} \frac{a^{2\ell+1}}{(rr')^{\ell+1}} Y_{\ell m}^*(\theta', \phi') \, Y_{\ell m}(\theta, \phi)$$

$$= \frac{a}{r'} \sum_{\ell, m} \frac{1}{2\ell + 1} \frac{(a^2/r')^\ell}{r^{\ell+1}} Y_{\ell m}^*(\theta', \phi') \, Y_{\ell m}(\theta, \phi) = \frac{a/r'}{|\vec{r} - \frac{a^2}{r'^2}\vec{r}'|}, \tag{42.49}$$

where we've used (42.31)–(42.32) in the last step. So F describes the potential of a fictitious point charge with relative magnitude $q' = a/r'$ at position $\vec{r}_0 \equiv a^2 \vec{r}'/r'^2$ inside the sphere. Fictitious, because there is no actual charge for $r < a$; rather, together with the charge q outside the sphere, q' produces the same zero equipotential at $r = a$ as does the surface charge distribution induced on the sphere by the unit source at $\vec{r}'$. In some sense, q' is a reflection of the charge outside, and is therefore known as the "image charge." And since Dirichlet conditions give unique solutions, either configuration — the induced charge or the image charge — produces the same potential for $r \geq a$.

The physical interpretation of F as arising from a fictitious charge distribution underlies the *method of images*. Rather than formally solving for F, image charges are "placed" in the region external to the real charge distribution so that the total field, due to both real and image charges, satisfies the required boundary conditions. If such an image configuration can be found, the analysis greatly simplifies — both mathematically and physically. In the present example, the image charge leads directly to the closed form for the Dirichlet Green's function

$$
\begin{aligned}
4\pi G_D(\vec{r}, \vec{r}') &= -\frac{1}{|\vec{r} - \vec{r}'|} + \frac{a/r'}{|\vec{r} - \vec{r}_0|} \\
&= -\frac{1}{\sqrt{r^2 + r'^2 - 2rr' \cos\gamma}} + \frac{1}{\sqrt{(rr'/a)^2 + a^2 - 2rr' \cos\gamma}},
\end{aligned}
\tag{42.50}
$$

with γ the angle between $\vec{r}$ and $\vec{r}'$. Not only does this give $G_D|_{r=a} = 0$, but the symmetry $G_D(\vec{r}, \vec{r}') = G_D(\vec{r}', \vec{r})$ is manifest. Integrating G_D against ρ as in (42.3) yields the electric potential Φ in the presence of the conducting sphere — whose (real) induced charge is readily found from the normal derivative $\partial\Phi/\partial n|_S = -\sigma$.

BTW 42.2 Dirichlet, Neumann, and the Fundamental Solution

When the domain is all space, the Green's function is the so-called fundamental solution. But as we've seen, smaller domains give different Green's functions. Consider scalar fields Φ and G, and Green's second identity (Problem 15.19b)

$$
\int_V \left(\Phi \nabla^2 G - G \nabla^2 \Phi \right) d^3r = \oint_S \left(\Phi \vec{\nabla} G - G \vec{\nabla} \Phi \right) \cdot d\vec{a}.
\tag{42.51}
$$

For the familiar case of a Poisson field Φ and its fundamental Green's solution,

$$
\nabla^2 \Phi(\vec{r}) = -\rho(\vec{r}) \qquad G(\vec{r}, \vec{r}') = \frac{1}{4\pi |\vec{r} - \vec{r}'|},
\tag{42.52}
$$

this can be unpacked using (42.26) to get

$$
\Phi(\vec{r}) = \int_V G(\vec{r}, \vec{r}') \rho(\vec{r}') d^3r' + \oint_S \left[G(\vec{r}, \vec{r}') \vec{\nabla}' \Phi(\vec{r}') - \Phi(\vec{r}') \vec{\nabla}' G(\vec{r}, \vec{r}') \right] \cdot d\vec{a}',
\tag{42.53}
$$

where $\vec{\nabla}'$ denotes differentiation with respect to $\vec{r}'$. Though this expression would seem to be in conflict with (42.3), note that as the volume expands to include all of space, the surface S is at infinity where the fields go to zero — and Coulomb's law emerges. Still, for finite boundaries (42.53) is problematic, since it would seem to require specification of *both* the field Φ and its normal derivative $\vec{\nabla}\Phi \cdot \hat{n} = \partial\Phi/\partial n$ everywhere on S. In other words, both Dirichlet *and* Neumann conditions are demanded over the entire boundary — despite the fact that this would over constrain the problem (see Table 41.2).

The problem is that a Green's function must obey the same boundary conditions as the solution being sought, and we're trying to generate a solution Φ with boundary conditions inconsistent with the fundamental solution in (42.52). But we can make them consistent by taking advantage of the freedom expressed in (42.44). Notice that the integrand of the first surface integral in (42.53) is the product of G with $\partial\Phi/\partial n$, while the second is the product of Φ with $\partial G/\partial n$. If *both* Φ and G

obey the same type of boundary condition on S, then the trouble with over-constraining the system vanishes. So, for instance, if we add a Laplace solution F to get a Green's function which obeys homogeneous Dirichlet conditions on S, the Green's function which results,

$$\nabla^2 G_D(\vec{r}, \vec{r}') = \delta(\vec{r} - \vec{r}'), \qquad G_D(\vec{r}, \vec{r}') = 0 \text{ on } S, \tag{42.54}$$

causes the first surface integral in (42.53) to vanish. Then (42.53) becomes the Dirichlet solution

$$\Phi_D(\vec{r}) = \int_V G(\vec{r}, \vec{r}') \rho(\vec{r}') \, d^3 r' - \oint_S \Phi(\vec{r}') \vec{\nabla}' G(\vec{r}, \vec{r}') \cdot d\vec{a}'. \tag{42.55}$$

This applies to Example 42.5, with a homogeneous condition on Φ.

Neumann conditions are a bit more subtle. Though it would appear that a homogenous Neumann Green's function would render (42.53) a Neumann solution for Φ, $\partial G / \partial n|_S = 0$ is actually inconsistent with the fundamental solution (Problem 42.10). But with

$$\nabla^2 G_N(\vec{r}, \vec{r}') = \delta(\vec{r} - \vec{r}'), \qquad \frac{\partial G_N}{\partial n} = -\frac{1}{A} \text{ on } S, \tag{42.56}$$

where A is the surface area of S, Neumann conditions on Φ are sufficient to obtain a unique solution with Neumann conditions of Φ,

$$\Phi_N(\vec{r}) = \langle \Phi \rangle_S + \int_V G(\vec{r}, \vec{r}') \rho(\vec{r}') \, d^3 r' + \oint_S G_N(\vec{r}, \vec{r}') \vec{\nabla}' \Phi(\vec{r}') \cdot d\vec{a}', \tag{42.57}$$

where $\langle \Phi \rangle_S$ is the average of Φ over the surface.

42.2 The Eigenfunction Expansion

It's not always easy or possible to find a closed form for a Green's function. But for Sturm–Liouville systems, G can be expressed as an expansion in eigenfunctions. Consider a Hermitian operator $\mathcal{L}$ and the homogeneous equation[5]

$$(\mathcal{L} - \lambda) |\phi\rangle = 0. \tag{42.58}$$

Recall that for matrix operators, the condition for a non-trivial solution is that $\mathcal{L} - \lambda \mathbb{1}$ not have an inverse. A similar thing holds for differential operators, where as we've seen the eigenvalues are determined from boundary conditions rather than a matrix determinant. By contrast, the inhomogeneous equation

$$(\mathcal{L} - \lambda) |\psi\rangle = |\rho\rangle \tag{42.59}$$

only has a solution if $\mathcal{L} - \lambda \mathbb{1}$ *does* has an inverse. As we'll soon see, the inverse of a differential operator $\mathcal{L} - \lambda$ is the Green's function.

Clearly λ in (42.59) cannot be an eigenvalue of $\mathcal{L}$ since then $\mathcal{L} - \lambda \mathbb{1}$ doesn't have an inverse. Nonetheless, for a Hermitian operator the homogeneous solutions form a complete orthonormal basis, allowing $\mathcal{L}$ to be fully characterized by its eigensystem (see (27.103) and BTW 40.1). In other words, solutions to (42.58) for non-eigenvalue λ must be reflected in those of (42.59) — a clear echo of the

---

[5] Note that Example 42.1 does not have this Helmholtz form, since $\lambda = 0$ is not an eigenvalue of $\mathcal{L} = d^2/dx^2$ with Dirichlet boundary conditions.

argument made regarding (42.1). The challenge is to discover how the homogeneous normal modes of a Sturm–Liouville differential equation collude to produce an inhomogeneous solution.

Any homogeneous solution $|\Phi\rangle$ can be expanded on $\mathcal{L}$'s eigenbasis,

$$|\Phi\rangle = \sum_n c_n |\hat{\phi}_n\rangle, \tag{42.60}$$

for which the λ appearing in (42.58) must be one of the eigenvalues λ_n:

$$(\mathcal{L} - \lambda)\,|\Phi\rangle = (\mathcal{L} - \lambda) \left(\sum_n c_n |\hat{\phi}_n\rangle \right)$$

$$= \sum_n c_n\,(\lambda_n - \lambda)\,|\hat{\phi}_n\rangle = 0 \implies \lambda = \lambda_n. \tag{42.61}$$

Though this doesn't work for an inhomogeneous solution $|\psi\rangle$, both it and the source $|\rho\rangle$ can nonetheless be expanded on the basis of $|\hat{\phi}_n\rangle$'s,

$$|\psi\rangle = \sum_n a_n |\hat{\phi}_n\rangle \qquad\qquad |\rho\rangle = \sum_n b_n |\hat{\phi}_n\rangle. \tag{42.62}$$

To find $|\psi\rangle$, we need the a_n; this can be accomplished by inserting these expansions into (42.59),

$$\sum_n [a_n\,(\lambda_n - \lambda) - b_n]\,|\hat{\phi}_n\rangle = 0 \implies a_n = \frac{b_n}{\lambda_n - \lambda} = \frac{\langle \hat{\phi}_n | \rho \rangle}{\lambda_n - \lambda}. \tag{42.63}$$

Thus

$$|\psi\rangle = \sum_n \frac{\langle \hat{\phi}_n | \rho \rangle}{\lambda_n - \lambda} |\hat{\phi}_n\rangle = \left(\sum_n \frac{|\hat{\phi}_n\rangle \langle \hat{\phi}_n|}{\lambda_n - \lambda} \right) |\rho\rangle \equiv G|\rho\rangle, \tag{42.64}$$

where the operator

$$G = \sum_n \frac{|\hat{\phi}_n\rangle \langle \hat{\phi}_n|}{\lambda_n - \lambda} \tag{42.65}$$

is a weighted projection onto the basis of homogeneous solutions. And since the $|\hat{\phi}_n\rangle$ are eigenvectors of a Hermitian $\mathcal{L}$, we can use completeness to reveal

$$(\mathcal{L} - \lambda)\,G = \sum_n (\lambda_n - \lambda) \frac{|\hat{\phi}_n\rangle \langle \hat{\phi}_n|}{\lambda_n - \lambda} = \sum_n |\hat{\phi}_n\rangle \langle \hat{\phi}_n| = \mathbb{1}. \tag{42.66}$$

So the operator G is the inverse of the operator $\mathcal{L} - \lambda$.

To transition from abstract bras and kets to functions, one formally projects (42.66) onto the position basis, $\phi_n(\vec{r}) = \langle \vec{r} | \phi_n \rangle$. Or one can simply rely on the expression of completeness for an orthonormal basis of eigenfunctions in (34.40),

$$\sum_n \hat{\phi}_n(\vec{r}) \hat{\phi}_n^*(\vec{r}') = \delta(\vec{r} - \vec{r}'), \tag{42.67}$$

to conclude that the projection operator G in (42.65) is represented on the position basis by the function

$$G(\vec{r}, \vec{r}') = \sum_n \frac{\hat{\phi}_n(\vec{r}) \hat{\phi}_n^*(\vec{r}')}{\lambda_n - \lambda}, \tag{42.68}$$

which must satisfy

$$(\mathcal{L} - \lambda)\, G(\vec{r}, \vec{r}') = \delta(\vec{r} - \vec{r}'). \tag{42.69}$$

So $G(\vec{r}, \vec{r}')$ is a Green's function. The one-dimensional case in Example 42.1 has $\lambda = 0$; its eigenfunctions and eigenvalues of d^2/dx^2 consistent with the boundary conditions are[6]

$$\hat{\phi}_n(x) = \sqrt{\frac{2}{L}}\, \sin(n\pi x/L), \qquad \lambda_n = -n^2 \pi^2/L^2, \quad n = 1, 2, 3, \ldots. \tag{42.70}$$

Then according to (42.68),

$$G(x, x') = -\frac{2L}{\pi^2} \sum_n \frac{1}{n^2} \sin(n\pi x'/L) \sin(n\pi x/L), \tag{42.71}$$

which is the Fourier expansion, in either x or x', of the solution in (42.14).

Example 42.6 A Driven String

A string subject to a periodic driving force $\cos \omega t$ acting at position x' on the string satisfies the inhomogeneous (1+1)-dimensional wave equation

$$\frac{\partial^2 \Psi}{\partial x^2} - \frac{1}{v^2} \frac{\partial^2 \Psi}{\partial t^2} = \delta(x - x') \cos \omega t. \tag{42.72}$$

In the absence of damping, the periodicity of the driving force induces the same harmonic time dependence in Ψ,

$$\Psi(x, t) = \psi(x) \cos \omega t, \tag{42.73}$$

transforming the wave equation into the Helmholtz equation

$$\left(\frac{d^2}{dx^2} + k^2 \right) \psi(x) = \delta(x - x'). \tag{42.74}$$

Since $k^2 = \omega^2/v^2$ is determined by the driving force, it is *not* an eigenvalue; rather, it is the $-\lambda$ of (42.59) with $\mathcal{L} = d^2/dx^2$. For a fixed string with Dirichlet conditions

$$\psi(0) = \psi(L) = 0, \tag{42.75}$$

the eigenfunction expansion (42.68) is

$$G_D(x, x') = -\frac{2}{L} \sum_n \frac{\sin(n\pi x/L) \sin(n\pi x'/L)}{n^2 \pi^2/L^2 - k^2}. \tag{42.76}$$

Note that this properly reduces to (42.71) in the $k \to 0$ limit.

Though an eigenfunction expansion of a Green's function is generally not as useful as a closed expression — nor is the requisite jump discontinuity always manifest — it does clearly exhibit an important caveat: if the λ in (42.59) is an eigenvalue of $\mathcal{L}$, then (42.65) blows up and no solution exists. For instance, the Green's function in (42.76) clearly diverges for $k = n\pi/L$; physically, this is a manifestation of resonance. Mathematically, either $\mathcal{L} - \lambda$ in (42.59) with specified boundary conditions

---

[6] Negative λ_n arises from the sign convention for the eigenvalue equation, $\mathcal{L}|\phi\rangle = \lambda|\phi\rangle$.

has an inverse or it doesn't; either there's a homogeneous solution or an inhomogeneous solution — but not both. Of course, if G is expressed as an eigenfunction expansion, one can always just skip over any such n's; the result is called the *modified Green's function* $\tilde{G}$,

$$\tilde{G}(\vec{r}, \vec{r}') = \sum_{n \notin \Omega} \frac{\hat{\phi}_n(\vec{r})\hat{\phi}_n^*(\vec{r}')}{\lambda_n - \lambda}, \tag{42.77}$$

where Ω is the subspace of degenerate eigenfunctions $\hat{\phi}_n$ for which $\lambda_n = \lambda$. This procedure may seem rather ad hoc — nor is it immediately obvious that $\tilde{G}$ is a useful quantity, since unlike G it does not span $\mathcal{L}$'s eigenspace. Indeed, $\tilde{G}$ satisfies not the defining equation (42.69), but rather

$$(\mathcal{L} - \lambda)\,\tilde{G}(\vec{r}, \vec{r}') = \delta(\vec{r} - \vec{r}') - \sum_{n \in \Omega} \hat{\phi}_n(\vec{r})\hat{\phi}_n^*(\vec{r}'). \tag{42.78}$$

What makes this useful is that the eigensubspace Ω can be characterized not just by λ, but by the source ρ as well. To discover the relationship, insert $|\psi\rangle = \sum_n a_n |\hat{\phi}_n\rangle$ into (42.59) and project onto $|\hat{\phi}_m\rangle$:

$$\langle\hat{\phi}_m|\rho\rangle = \sum_n a_n \langle\hat{\phi}_m|\mathcal{L} - \lambda|\hat{\phi}_n\rangle$$

$$= \sum_n a_n(\lambda_n - \lambda)\langle\hat{\phi}_m|\hat{\phi}_n\rangle = a_m(\lambda_m - \lambda). \tag{42.79}$$

Now if $a_m = 0$, then the inhomogeneous solution $|\psi\rangle$ has no component along $|\hat{\phi}_m\rangle$; in other words, the source does not excite the mth eigenmode, $\langle\hat{\phi}_m|\rho\rangle = 0$. So there is no inherent conflict which would disallow an inhomogeneous solution. Note that this is a characteristic of the source, not the operator $\mathcal{L} - \lambda$.

However, (42.79) can also be satisfied by $\lambda = \lambda_m$ corresponding to homogeneous solution $|\hat{\phi}_m\rangle$, in which case $(\mathcal{L} - \lambda)^{-1}$ cannot span the full eigenspace of $\mathcal{L}$ as in (42.65). In this instance, (42.79) shows that unless $\langle\hat{\phi}_m|\rho\rangle = 0$, there is no solution: the sum in (42.68) blows up. But if the source is orthogonal to $\hat{\phi}_m$, then once again this mode doesn't contribute to the sum and a solution exists (see Problem 27.9).

Since the collection of $|\hat{\phi}_m\rangle$'s which must be orthogonal to $|\rho\rangle$ is the same as the subspace Ω of degenerate eigenfunctions in (42.78), we have a practical way to characterize Ω,

$$\langle\hat{\phi}_m|\rho\rangle = \int_0^L \hat{\phi}_m^*(x)\rho(x)\,dx = 0, \qquad |\phi_m\rangle \in \Omega. \tag{42.80}$$

Most notably for a Green's function, the unit source $\delta(x - x')$ does *not* obey this condition, so the usual G cannot generate an inhomogeneous solution if $\lambda = \lambda_m$. But as (42.78) implies, a modified Green's function $\tilde{G}$ can be constructed not from the unit source, but rather from

$$\rho(\vec{r}) = \delta(\vec{r} - \vec{r}') - \sum_{n \in \Omega} \hat{\phi}_n(\vec{r})\hat{\phi}_n^*(\vec{r}'). \tag{42.81}$$

With this modified impulse, the defining orthogonality condition for Ω is satisfied,

$$\int_0^L \left[\delta(x - x') - \sum_{n \in \Omega} \hat{\phi}_n(x)\hat{\phi}_n^*(x') \right] \hat{\phi}_m(x')\,dx' = 0. \tag{42.82}$$

Example 42.7 **A Modified Green's Function**

Consider (42.59) with $\mathcal{L} = d^2/dx^2$ and $\lambda = 0$,

$$\frac{d^2 T}{dx^2} = q(x), \tag{42.83}$$

with Neumann conditions,

$$T'(0) = T'(L) = 0. \tag{42.84}$$

Physically, this one-dimensional Poisson equation can describe the steady-state temperature distribution of a rod of length L with ends insulated so that no heat flows through them. The normalized eigenfunctions with these boundary conditions are

$$\hat{\phi}_n(x) = \sqrt{\frac{\varepsilon_n}{L}} \cos(n\pi x/L), \qquad \varepsilon_0 = 1, \quad \varepsilon_{n>0} = 2, \tag{42.85}$$

with $\lambda_n = -(n\pi/L)^2$. Since $\lambda = 0$, the only problematic eigenfunction is $\hat{\phi}_0 = 1/\sqrt{L}$. Thus a solution only exists if

$$\langle \hat{\phi}_0 | q \rangle = \frac{1}{\sqrt{L}} \int_0^L q(x)\, dx = 0, \tag{42.86}$$

and $\tilde{G}$ satisfies

$$\frac{d^2 \tilde{G}}{dx^2} = \delta(x - x') - \frac{1}{L}. \tag{42.87}$$

A physical interpretation helps clarify what's happening. With a time-independent source and insulated ends, the temperature increases without limit over time, leading to an unphysical divergence. The orthogonality condition (42.80) requires that this heat be removed, as indicated by the uniformly distributed sink on the right-hand side of (42.87). Only then does a solution exist.

Solving for $\tilde{G}$ follows as in Example 42.1 (Problem 42.31). In the process, you'll discover that the discontinuity condition on $\partial_x \tilde{G}$ is automatically satisfied — leaving $\tilde{G}$ one independent constraint short of specifying the final constant. The auxiliary condition

$$\langle \hat{\phi}_0 | \tilde{G} \rangle = \int_0^L \hat{\phi}_0(x)\, \tilde{G}(x, x')\, dx = 0 \tag{42.88}$$

is commonly chosen to fully specify the solution. With this, the modified Green's function is

$$\tilde{G}_N(x, x') = -\frac{1}{2L} \left(x^2 + x'^2 \right) + x_> - \frac{L}{3}. \tag{42.89}$$

The general solution to (42.83) is then the sum of the homogeneous and particular solutions,

$$T(x) = C\hat{\phi}_0(x) + \int_0^L \tilde{G}_N(x, x')\, q(x')\, dx', \tag{42.90}$$

assuming, of course, the source q satisfies (42.87):

$$\frac{d^2T}{dx^2} = C\frac{d^2\hat{\phi}_0}{dx^2} + \int_0^L \frac{d^2\tilde{G}_N(x,x')}{dx^2} q(x')\,dx'$$

$$= 0 + \int_0^L \left[\delta(x-x') - \frac{1}{L}\right] q(x')\,dx' = q(x). \tag{42.91}$$

BTW 42.3 The Complex λ-Plane

The derivation of (42.64) was predicated on the presumption that the eigenfunctions form a complete orthonormal basis. A remarkable proof can be developed by considering the pole structure of the Green's function in the complex λ-plane. Considered as an analytic function of λ (Chapter 18), the Green's function

$$G(\vec{r},\vec{r}';\lambda) = -\sum_n \frac{\hat{\phi}_n(\vec{r})\hat{\phi}_n^*(\vec{r}')}{\lambda - \lambda_n} \tag{42.92}$$

has simple poles at the eigenvalues λ_n of $\mathcal{L}$, all of them on the real axis. The residue at each pole is the eigenfunction product $\phi_n(\vec{r})\phi_n^*(\vec{r}')$; the goal is to show that the sum over these residues is a delta function. Formally, that sum emerges from the contour integral

$$\oint_C G(\vec{r},\vec{r}';\lambda)\,d\lambda = -2\pi i\sum_n \hat{\phi}_n(\vec{r})\hat{\phi}_n^*(\vec{r}'), \tag{42.93}$$

where the contour encloses all the poles in the finite plane. Since $G(\vec{r},\vec{r}';\lambda)$ only varies as $1/\lambda$, the integral (and hence the residue sum) does not vanish even for contours extending to infinity. What does vanish, though, is the integral of G/λ,

$$\oint_C G(\vec{r},\vec{r}';\lambda)\,\frac{d\lambda}{\lambda} = 0. \tag{42.94}$$

Operating with $\mathcal{L}$ and invoking (42.69),

$$0 = \mathcal{L}\oint_C G(\vec{r},\vec{r}';\lambda)\,\frac{d\lambda}{\lambda} = \oint_C \left[\mathcal{L}G(\vec{r},\vec{r}';\lambda)\right]\frac{d\lambda}{\lambda}$$

$$= \oint_C \left[\delta(\vec{r}-\vec{r}') + \lambda G\right]\frac{d\lambda}{\lambda} \tag{42.95}$$

$$= 2\pi i\,\delta(\vec{r}-\vec{r}') + \oint_C G(\vec{r},\vec{r}';\lambda)\,d\lambda,$$

where we've used (18.25) in the last step. Thus

$$\oint_C G(\vec{r},\vec{r}';\lambda)\,d\lambda = -2\pi i\delta(\vec{r}-\vec{r}'). \tag{42.96}$$

Comparison with (42.93) establishes the completeness of the $\hat{\phi}_n$ basis.

42.3 Going Green in Space and Time

Up until now, we've considered initial value problems for Green's function $G(t, t')$, and boundary value problems with $G(\vec{r}, \vec{r}')$. But for a source which depends on both time and space, we need a Green's function $G = G(\vec{r}, \vec{r}'; t, t')$. We'll explore this with a series of examples.

Example 42.8 **A Pulse of Heat**

At time t', a solid bar at zero temperature is hit with a unit pulse of heat at position x'. The subsequent temperature distribution is the solution to the heat equation

$$\frac{\partial G}{\partial t} - \alpha \frac{\partial^2 G}{\partial x^2} = \delta(x - x')\,\delta(t - t'), \quad G(x, x'; t < t') \equiv 0. \tag{42.97}$$

Note that the cause (impulse at $t = t'$) precedes the effect (temperature response for $t > t'$), so this G is a causal Green's function. Indeed, since $G \equiv 0$ until t reaches t' and satisfies a homogeneous differential equation for $t > t'$, (42.97) has all the trappings of an initial value problem in $\tau \equiv t - t'$,

$$\frac{\partial G}{\partial \tau} - \alpha \frac{\partial^2 G}{\partial x^2} = 0, \quad \tau > 0, \quad G(x, x'; \tau = 0) = \delta(x - x'). \tag{42.98}$$

Expressed as a homogeneous initial value problem, G can be found by standard separation of variables. Equivalently, we can start with the expansion (42.68) and use the variable-separated ansatz

$$G(x, x'; \tau) = \sum_n c_n(\tau) \frac{\hat{\phi}_n(x')\hat{\phi}_n(x)}{-\lambda_n}, \tag{42.99}$$

with $\hat{\phi}_n$ the eigenfunctions of $\mathcal{L} = \partial_x^2$. Inserting this into (42.98) gives

$$\sum_n \frac{\hat{\phi}_n(x')}{-\lambda_n}\left(\dot{c}_n\hat{\phi}_n - \alpha c_n \hat{\phi}_n''\right) = \sum_n \frac{\hat{\phi}_n(x')\hat{\phi}_n(x)}{-\lambda_n}\left(\dot{c}_n + \alpha\lambda_n c_n\right) = 0. \tag{42.100}$$

Because the $\hat{\phi}_n$ are linearly independent, the expression in parentheses must vanish separately for each n; thus $c_n(\tau) = c_n(0)\exp(-\alpha\lambda_n\tau)$. Then because the $\hat{\phi}_n$ are complete, the initial condition in (42.98) requires $c_n(0) = -\lambda_n$. All together, the solution of the homogeneous initial value problem from an initial impulsive state $\delta(x - x')$ at $t = t'$ is

$$G(x, x'; t - t') = \sum_n e^{-\alpha\lambda_n(t-t')}\,\hat{\phi}_n(x')\hat{\phi}_n(x), \quad t > t'. \tag{42.101}$$

This, combined with $G(t < t') = 0$, gives the causal Green's function for (42.97). For instance, a bar of length L with ends held at zero temperature (homogeneous Dirichlet conditions) has the eigenfunctions and eigenvalues found in Example 41.3 — so for $t > t'$,

$$G_D(x, x'; t - t') = \frac{2}{L}\sum_{n=1}^{\infty} e^{-\alpha(n\pi/L)^2(t-t')}\,\sin(n\pi x'/L)\,\sin(n\pi x/L). \tag{42.102}$$

Homogeneous Neumann conditions can be accommodated using the appropriate eigenfunctions in (42.101).

Example 42.9 **This Old Heater**

Let's connect a bar to a heat source $q(x,t)$ — say, an old electric heater whose strength varies both along its length and in time. Then we need to solve the inhomogeneous heat equation

$$\frac{\partial T}{\partial t} - \alpha \frac{\partial^2 T}{\partial x^2} = q(x,t), \qquad T(x,0) = 0. \tag{42.103}$$

To apply a Green's function such as (42.102), we invoke Duhamel's principle introduced in Example 42.2: envision the source $q(x,t)$ as a continuous sequence of impulses $q(x,t')\delta(t-t')$,

$$q(x,t) = \int_0^t q(x,t')\delta(t-t')\,dt'. \tag{42.104}$$

Each one of these impulses is the source of its own initial value problem with $q(x,t')$ the initial condition at its unique value of t'. Each results in its own $t > t'$ temperature flow

$$J_q \equiv \left.\frac{\partial T}{\partial t}\right|_{t-t'} = \int_0^L q(x',t')\,G(x,x';t-t')\,dx', \qquad t-t' > 0. \tag{42.105}$$

So at each instant of time t' between 0 and t, a new homogeneous solution emerges and evolves as an initial value problem. Since the heat equation is linear, summing these freely propagating solutions gives the solution to the inhomogeneous problem (42.103) at time t,

$$T(x,t) = \int_0^t J_q(x,t-t')\,dt' = \int_0^t \int_0^L q(x',t')\,G(x,x';t,t')\,dx'\,dt'. \tag{42.106}$$

If we broaden (42.103) to have initial profile

$$T(x,0) = T_0(x), \tag{42.107}$$

the solution becomes (Problem 42.20)

$$T(x,t) = \int_0^L T_0(x')\,G(x,x';t,0)\,dx' + \int_0^t \int_0^L q(x',t')\,G(x,x';t,t')\,dx'\,dt'. \tag{42.108}$$

The first term is the homogeneous initial value solution, describing the diffusion of heat along the bar due only to the initial temperature profile $T_0(x)$. The impact of the heat source is governed by the second term, which is effectively (42.3) for a $(1+1)$-dimensional volume of space and time.

In addition to a physical source $q(x,t)$, an inhomogeneous term can be used to accommodate inhomogeneous boundary conditions. As we did in Examples 41.8–41.9, we can leverage linearity to construct a solution to the homogeneous equation $(\mathcal{L}-\lambda)\phi = 0$ with inhomogeneous conditions by recasting it as the sum of two pieces,

$$\phi(\vec{r}) = \tilde{\phi}(\vec{r}) + \xi(\vec{r}), \tag{42.109}$$

where $\tilde{\phi}$ is the solution with homogeneous conditions, and ξ is a function chosen to obey the inhomogeneous conditions. With this decomposition, the homogeneous differential equation becomes an inhomogeneous equation for $\tilde{\phi}$,

$$(\mathcal{L} - \lambda)\tilde{\phi} = \tilde{q}, \qquad \text{where } \tilde{q} = -\mathcal{L}\xi. \tag{42.110}$$

One then constructs the Green's function for $\mathcal{L} - \lambda$, and proceeds as before. In this way, a homogeneous equation with inhomogeneous boundary conditions can be reformulated as an inhomogeneous equation with homogeneous conditions. (You may wish to read that sentence more than once.) Though the solution ϕ is unique, it's clear from (42.110) that different ξ's produce different $\tilde{\phi}$'s — so there's some freedom in the choice of ξ. The simplest, physically inspired function which satisfies the required inhomogeneous boundary conditions is usually best.

Example 42.10 **Inhomogeneous Ends**

Consider again the homogeneous heat equation in 1+1 dimensions,

$$\frac{\partial T}{\partial t} - \alpha \frac{\partial^2 T}{\partial x^2} = 0, \tag{42.111a}$$

with initial condition

$$T(x, 0) = T_0(x), \tag{42.111b}$$

but this time with inhomogeneous Dirichlet conditions

$$T(0, t) = c_1(t) \qquad T(L, t) = c_2(t). \tag{42.111c}$$

Physically, this is the description of a rod of length L whose ends are maintained over time at temperatures $c_1(t)$ and $c_2(t)$. As in (42.109), we separate $T(x, t)$ into two pieces, $T = \xi + \tilde{T}$. The simplest choice for ξ which satisfies (42.111c) is the linear function

$$\xi(x, t) = c_1 + \frac{1}{L}(c_2 - c_1)x. \tag{42.112}$$

For time-independent $c_{1,2}$ as in Example 41.8, $\xi(x, t) \rightarrow \xi(x)$ is the temperature distribution after the initial profile $T_0(x)$ diffuses through the rod — that is, it's the steady-state solution to the heat equation once the transients die out.

Plugging $T(x, t) = \xi(x, t) + \tilde{T}(x, t)$ into (42.111a), we find that $\tilde{T}$ is a solution of the inhomogeneous heat equation

$$\frac{\partial \tilde{T}}{\partial t} - \alpha \frac{\partial^2 \tilde{T}}{\partial x^2} = \tilde{q}(x, t), \tag{42.113}$$

where

$$\tilde{q}(x, t) = -\frac{\partial \xi}{\partial t} + \alpha \frac{\partial^2 \xi}{\partial x^2} = -\left[\dot{c}_1 + \frac{1}{L}(\dot{c}_2 - \dot{c}_1)x \right]. \tag{42.114}$$

Thus our homogeneous heat equation with inhomogeneous boundary conditions has been recast as an inhomogeneous problem in $\tilde{T}$ with heat source $\tilde{q}$, homogeneous Dirichlet conditions

$$\tilde{T}(0, t) = \tilde{T}(L, t) = 0, \tag{42.115}$$

and initial temperature profile

$$\tilde{T}_0(x) = T_0(x) - \xi(x, 0). \tag{42.116}$$

Then with $T \to \tilde{T}$ and $q \to \tilde{q}$ in (42.108), the full solution to (42.111) becomes

$$T(x,t) = \xi(x,t) + \tilde{T}(x,t)$$

$$= c_1(t) + \frac{1}{L}\left[c_2(t) - c_1(t)\right]x + \int_0^L \tilde{T}_0(x')\, G_D(x,x';t,0)\, dx' \qquad (42.117)$$

$$+ \int_0^t \int_0^L \tilde{q}(x',t')\, G_D(x,x';t,t')\, dx'\, dt',$$

which is not so opaque once you identify each of the pieces as we did in (42.108). Moreover, this solution can be adapted to an inhomogeneous equation

$$\frac{\partial T}{\partial t} - \alpha\frac{\partial^2 T}{\partial x^2} = q(x,t) \qquad (42.118)$$

simply by adding q to $\tilde{q}$ in (42.114).

42.4 Green's Functions and Fourier Transforms

Our first example of a Green's function was that of the Coulomb potential, (42.5) — derived initially using our familiarity with the physics, and formally confirmed later in Example 42.3. But in fact, we derived the Coulomb Green's function earlier in Example 36.12, where we used Fourier transform techniques to solve Poisson's equation for charge density $\rho(\vec{r}) = q\,\delta(\vec{r} - \vec{r}\,')$. The applicability of Fourier transforms derives from the infinite range of the Coulomb potential — and hence the unbounded domain of its Poisson equation.

Nor is Poisson's equation the only time we've encountered a Fourier transform derivation of Green's functions. In Example 36.13, we mapped the differential equation for the forced damped harmonic oscillator,

$$\left[\frac{d^2}{dt^2} + 2\beta\frac{d}{dt} + \omega_0^2\right]u(t) = f(t), \qquad (42.119)$$

into an algebraic expression in frequency by decomposing $u(t)$ on the basis $\{e^{i\omega t}\}$,

$$\left[-\omega^2 + 2i\beta\omega + \omega_0^2\right]\tilde{u}(\omega) = \tilde{f}(\omega)\,. \qquad (42.120)$$

The solution $u(t)$ can then be expressed as a continuous expansion over the basis,

$$u(t) = \mathcal{F}^{-1}\left[\tilde{u}(\omega)\right] = \frac{1}{\sqrt{2\pi}}\int_{-\infty}^\infty \frac{\tilde{f}(\omega)\, e^{i\omega t}}{\omega_0^2 - \omega^2 + 2i\beta\omega}\, d\omega, \qquad (42.121)$$

with the conventional factor of $1/\sqrt{2\pi}$. The impulse $f(t) = f_0\,\delta(t - t')$ has Fourier transform $\tilde{f} = f_0/\sqrt{2\pi}$, so the Green's function is[7]

$$G(t,t') = \mathcal{F}^{-1}\left[\tilde{G}(\omega)\right] = \frac{f_0}{2\pi}\int_{-\infty}^\infty \frac{e^{i\omega(t-t')}}{\omega_0^2 - \omega^2 + 2i\beta\omega}\, d\omega\,. \qquad (42.122)$$

This is essentially what we did, albeit implicitly, to arrive at (22.8) — and, for that matter, all the other Green's functions in that Coda. We'll revisit some of them here.

---

[7] The identification of $\tilde{G}$ reveals this to be an application of the convolution theorem, (36.90a).

In this section, rather than the 2π convention used in (42.121), we'll instead place 2π's under each k and ω integral. Then decomposing on a basis of plane waves, our Fourier integrals become

$$\tilde{u}(\vec{k},\omega) = \mathcal{F}\left[u(\vec{r},t)\right] = \int d^3r \int dt\, u(\vec{r},t)\, e^{-i(\vec{k}\cdot\vec{r}-\omega t)}$$

$$u(\vec{r},t) = \mathcal{F}^{-1}\left[\tilde{u}(\vec{k},\omega)\right] = \int \frac{d^3k}{(2\pi)^3} \int \frac{d\omega}{2\pi}\, \tilde{u}(\vec{k},\omega)\, e^{+i(\vec{k}\cdot\vec{r}-\omega t)}. \tag{42.123}$$

Thus $\mathcal{F}\left[\delta(\vec{r})\delta(t)\right] = 1$. Note that the opposite sign convention for k and ω fixes the mappings between derivatives in t and x and multiplication by ω and k as

$$\mathcal{F}\left[\left(\frac{\partial}{\partial t}\right)^n\right] = (-i\omega)^n \qquad \mathcal{F}\left[\left(\frac{\partial}{\partial x}\right)^n\right] = (+ik)^n. \tag{42.124}$$

Example 42.11 The Heat Equation, Revisited

The heat equation Green's function in one space dimension is defined by

$$\left[\frac{\partial}{\partial t} - \alpha \frac{\partial^2}{\partial x^2}\right] G(x,x';t,t') = \delta(x-x')\delta(t-t'). \tag{42.125}$$

Unlike Example 42.8, we'll consider unbounded space and solve for the fundamental solution $G = G(x-x',t-t')$. Since it's translation invariant in both space and time, we can set $x' = t' = 0$ without loss of generality.

In Example 42.9, we found G for a finite-length bar as a sum over discrete eigenfunctions. For an infinite bar, that discrete spectrum becomes a continuum, and the Fourier sum in (42.102) becomes a Fourier integral. To find $G(x,t)$'s continuous spectrum $\tilde{G}(\omega,k)$, we use (42.124) to map (42.125) into

$$\left[(-i\omega) - \alpha\,(ik)^2\right]\tilde{G} = 1, \tag{42.126}$$

so that

$$\tilde{G}(\omega,k) = \frac{1}{\alpha k^2 - i\omega}. \tag{42.127}$$

This is a very simple, if complex, closed form expression for $\tilde{G}$. The inverse transform maps this back to G,

$$G(x,t) = \int \frac{dk\,d\omega}{2\pi\,2\pi} \frac{e^{+i(kx-\omega t)}}{\alpha k^2 - i\omega} = \frac{i}{(2\pi)^2} \int_{-\infty}^{\infty} dk\, e^{ikx} \int_{-\infty}^{\infty} d\omega\, \frac{e^{-i\omega t}}{\omega + i\alpha k^2}. \tag{42.128}$$

The ω integral is most-easily accomplished by extending it into the complex plane (Problem 22.6); the result is

$$G(x,t) = \frac{1}{2\pi} \int_{-\infty}^{\infty} dk\, e^{ikx} e^{-\alpha t k^2}, \qquad t > 0. \tag{42.129}$$

The k integral is a Gaussian which can be evaluated as in (8.70) to yield

$$G_R(x,t) = \Theta(t)\sqrt{\frac{1}{4\pi\alpha t}}\, e^{-x^2/4\alpha t}, \tag{42.130}$$

where $\Theta(t)$ is the step function of (9.27). The subscript R indicates that this is a retarded Green's function. (For an alternative derivation, see Problem 42.27.)

Example 42.12   **The Wave Equation, Revisited**

Since the three-dimensional wave equation over all space is translation invariant, the Green's function for the wave equation satisfies

$$\left(\frac{1}{v^2}\frac{\partial^2}{\partial t^2} - \nabla^2\right) G(\vec{r} - \vec{r}', t - t') = \delta(\vec{r} - \vec{r}')\,\delta(t - t'). \tag{42.131}$$

For simplicity and without loss of generality, we locate the impulse at the origin $\vec{r}' = 0$ at time $t' = 0$. The Fourier transform of (42.131) gives an algebraic expression for $\tilde{G}$ which is easily rearranged to find

$$\tilde{G}(\omega, \vec{k}) = \frac{1}{k^2 - \omega^2/v^2}, \tag{42.132}$$

where $k^2 = \vec{k}\cdot\vec{k}$. Then the Fourier representation of G is

$$G(\vec{r}, t) = -v^2 \int \frac{d^3k}{(2\pi)^3}\frac{d\omega}{2\pi}\frac{e^{i(\vec{k}\cdot\vec{r} - \omega t)}}{\omega^2 - v^2 k^2}. \tag{42.133}$$

Expressing the k-space integrals in spherical coordinates, this becomes

$$G(\vec{r}, t) = -\frac{v^2}{(2\pi)^3}\int_0^\infty dk\, k^2 \int_{-1}^1 d(\cos\theta)\, e^{ikr\cos\theta}\int_{-\infty}^\infty d\omega\,\frac{e^{-i\omega t}}{\omega^2 - v^2 k^2}$$

$$= -\frac{v^2}{4\pi^3}\frac{1}{r}\int_0^\infty dk\, k\sin(kr)\int_{-\infty}^\infty d\omega\,\frac{e^{-i\omega t}}{\omega^2 - v^2 k^2}. \tag{42.134}$$

The pole at $\omega = vk$ renders the ω integral ill-defined; obtaining a meaningful result requires specification of initial conditions. These were considered back in (22.23), where we found the retarded solution

$$\int_{-\infty}^\infty \frac{e^{-i\omega t}}{\omega^2 - v^2 k^2}\,d\omega = \begin{cases} -\frac{2\pi}{vk}\sin(kvt), & t > 0 \\ 0, & t < 0 \end{cases}. \tag{42.135}$$

To evaluate the remaining k integral, start by noting that $\sin(kr)\sin(kct)$ is even in k,

$$G(\vec{r}, t > 0) = +\frac{c}{4\pi^2 r}\int_{-\infty}^\infty dk\,\sin(kr)\sin(kvt)$$

$$= \frac{c}{8\pi r}\int_{-\infty}^\infty dk\,\cos[k(r - vt)] - \cos[k(r + vt)]$$

$$= \frac{c}{8\pi^2 r}\int_{-\infty}^\infty dk\left[e^{ik(r-vt)} - e^{-ik(r+vt)}\right]$$

$$= \frac{c}{4\pi r}\left[\delta(r - ct) - \delta(r + ct)\right]. \tag{42.136}$$

Of course for positive r and t, only the first delta contributes — and therefore

$$G_R(\vec{r}, t) = \frac{c}{4\pi r}\delta(r - vt)\,\Theta(t). \tag{42.137}$$

By translation invariance, this generalizes to

$$G_R(\vec{r} - \vec{r}', t - t') = \frac{v}{4\pi |\vec{r} - \vec{r}'|} \, \delta \left[|\vec{r} - \vec{r}'| - v(t - t') \right] \Theta(t - t')$$

$$= \frac{1}{4\pi |\vec{r} - \vec{r}'|} \, \delta(t' - t + |\vec{r} - \vec{r}'|/v) \, \Theta(t - t'). \qquad (42.138)$$

So not only is there no response before the impulse occurs at t' — so that cause precedes effect — no response manifests at $\vec{r}$ until sufficient time has elapsed for the signal to propagate a distance $|\vec{r} - \vec{r}'|$.

Example 42.13 **Retarded Potentials**

When you look at the stars at night, you are seeing them not as they are at that moment, but rather as they were at some time in the past. How far in the past is the time light takes to reach Earth; the further away the star is, the further in its past you are observing it. This sort of time delay must be true not only for starlight, but for any signal with finite speed — and finite speed, as relativity reveals, is a physical necessity.

This realization requires some adjustments to familiar static expressions. For instance, although Coulomb's law

$$\Phi(\vec{r}) = \frac{1}{4\pi} \int_V \frac{\rho(\vec{r}')}{|\vec{r} - \vec{r}'|} \, d^3 r' \qquad (42.139)$$

gives the electric potential at $\vec{r}$ of a charge distribution ρ, it presumes that the charges are at rest. But the effect on $\Phi(\vec{r})$ due to the motion of charge can propagate out from a point $\vec{r}'$ no faster than the speed of light, c. In other words, the motion of charge cannot affect Φ instantaneously; rather, the potential at $\vec{r}$ at time t must be due to the position of charges a distance $|\vec{r} - \vec{r}'| = c(t - t')$ at the *earlier* time $|\vec{r} - \vec{r}'|/c$. Thus for moving charges, (42.139) must be replaced by an integral over ρ at the *retarded time* $t_R \equiv t - |\vec{r} - \vec{r}'|/c$; the result is the *retarded potential*

$$\Phi(\vec{r}, t) = \frac{1}{4\pi} \int_V \frac{\rho(\vec{r}', t - |\vec{r} - \vec{r}'|/c)}{|\vec{r} - \vec{r}'|} \, d^3 r'. \qquad (42.140)$$

Physical reasoning has led us to insert t_R by hand. Recognizing that the wave equation (42.131) reduces to Poisson's equation for a unit static charge distribution (but beware the sign confusion — see footnote 2 in this chapter), a more formal approach starts with the retarded Green's function of (42.138) — which, notice, naturally accounts for the retarded time. And indeed, integrating G_R against a charge density $\rho(\vec{r}, t)$ gives

$$\Phi(\vec{r}, t) = \int_V \int \rho(\vec{r}', t') \, G_R(\vec{r} - \vec{r}', t - t') \, dt' \, d^3 r'$$

$$= \frac{1}{4\pi} \int_V \left[\int_0^\infty \frac{\rho(\vec{r}', t')}{|\vec{r} - \vec{r}'|} \delta(t' - t_R) \, dt' \right] d^3 r' = \frac{1}{4\pi} \int_V \frac{\rho(\vec{r}', t_R)}{|\vec{r} - \vec{r}'|} \, d^3 r', \qquad (42.141)$$

which is the retarded potential of (42.140).

A relatively straightforward example is that of a moving point charge with trajectory $\vec{s}(t)$, for which the density is simply

$$\rho(\vec{r}, t) = q \, \delta(\vec{r} - \vec{s}(t)). \qquad (42.142)$$

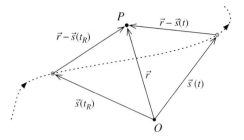

Figure 42.2 At time t, the potential at P depends not on a charge's position $\vec{s}(t)$, but on its earlier position $\vec{s}(t_R)$ along the trajectory (dotted) at the retarded time $t_R = t - |\vec{r} - \vec{s}(t_R)|/c$.

The retarded potential in this case is (Figure 42.2)

$$\Phi(\vec{r},t) = \int_V \left[\int \rho(\vec{r}',t') \, G_R(\vec{r} - \vec{r}',t - t') \, dt' \right] d^3 r'$$

$$= \frac{q}{4\pi} \int_0^\infty \int_V \frac{\delta\left[\vec{r}' - \vec{s}(t')\right]}{|\vec{r} - \vec{r}'|} \, \delta\left[t' - \left(t - \frac{1}{c}|\vec{r} - \vec{r}'| \right) \right] d^3 r' \, dt'$$

$$= \frac{q}{4\pi} \int_0^\infty \frac{1}{|\vec{r} - \vec{s}(t')|} \, \delta\left[t' - \left(t - \frac{1}{c}|\vec{r} - \vec{s}(t')| \right) \right] dt'. \tag{42.143}$$

We can recast the delta function using (9.17): with $g(t') = (t' - t) + |\vec{r} - \vec{s}(t')|/c$,

$$\frac{dg}{dt'} = \frac{d}{dt'} \left[(t' - t) + |\vec{r} - \vec{s}(t')|/c \right]$$

$$= 1 + \frac{1}{c} \vec{\nabla}_s |\vec{r} - \vec{s}(t')| \cdot \frac{d\vec{s}}{dt'}$$

$$= 1 - \frac{1}{c} \frac{\vec{r} - \vec{s}(t')}{|\vec{r} - \vec{s}(t')|} \cdot \frac{d\vec{s}}{dt'} \equiv 1 - \hat{R} \cdot \vec{u}/c, \tag{42.144}$$

where $\vec{\nabla}_s$ denotes the gradient with respect to $\vec{s}$, $\vec{R} \equiv \vec{r} - \vec{s}(t')$, and $\vec{u} \equiv \dot{\vec{s}}$ is the velocity of the charge. We need to evaluate this at the zero of g, which is that value of t' for which $t' = t - R(t')/c$. In other words, at the retarded time. All together then,

$$\Phi(\vec{r},t) = \frac{q}{4\pi} \int_0^\infty \frac{\delta(t' - t_R) \, dt'}{R(t') - \vec{R}(t') \cdot \vec{u}(t')/c} = \frac{q}{4\pi} \frac{1}{R - \vec{R} \cdot \vec{u}/c}\bigg|_{t_R}, \tag{42.145}$$

where both $\vec{R}$ and $\vec{u}$ are to be evaluated at t_R. This non-trivial, retarded potential of a moving charge is known as the *Liénard–Wiechert potential*.

As the previous example illustrates, a retarded Green's function takes an initial state at t_R and determines the future state of a system at $t > t_R$. One can think of G_R as "propagating" the solution forward in time. If this sounds familiar, it's because we've seen it before: in Chapter 22 we used contour integration in the complex plane to find impulse responses — Green's functions. For instance, (42.138) is the three-dimensional version of the propagator in (22.28) for a wave in one dimension. Although they differ functionally — one falls with distance, the other is constant — both clearly display a causal form in their δ and Θ functions. Flipping the sign of their arguments produces the advanced Green's function — just as in (22.29). So if G_R propagates forward in time from an initial state, G_A can be used to take a state backwards in time to determine whence it came.

Similar and yet distinct is the retarded Green's function for the heat equation (42.130). It too we found in Chapter 22 as the propagator (22.33) of the Schrödinger equation — which has the form of a diffusion equation with $\alpha = i\hbar/2m$. Although this G_R is causal in the sense that an impulse always produces its effect, it differs from the wave equation's in two essential ways. First, as noted below (22.33) the effect is instantaneous. Second, since the heat equation is first-order in time derivatives, it is not invariant under $t \rightarrow -t$. This is reflected in G_R's exponential decay, which diverges under time reversal. So while any forward-t evolution will smooth out a noisy initial temperature distribution, the process cannot be run in reverse. Indeed, many initial states correspond to the same final temperature distribution. This makes sense: a cup of hot coffee with cold milk will equilibrate over time to a uniform temperature distribution — but the process never happens in reverse. Diffusion thus identifies the direction into the future, the "thermodynamic arrow of time."

Problems

42.1 Find the Green's function for the Poisson equation in one dimension, $0 \le x \le L$, with the following boundary conditions:

(a) $\psi(0) = 0$, $\psi(L) = 0$
(b) $\psi(0) = 0$, $\psi'(L) = 0$
(c) $\psi'(0) = 0$, $\psi(L) = 0$.

42.2 For each of the Green's functions in Problem 42.1, find $\psi(x)$ for sources $\rho(x) = x^2$. Verify they are solutions of the differential equation with the required conditions.

42.3 Find the Green's function for the Poisson equation in one dimension, $0 \le x \le L$, with boundary conditions $\psi(0) = a$, $\psi'(L) = b$.

42.4 Rederive the Green's function of Example 42.6 using continuity and discontinuity conditions. Verify the singularity for $k \rightarrow n\pi/L$, and show that it reduces to (42.71) for $k \rightarrow 0$.

42.5 Find the Green's function for the Helmholtz equation in one dimension, $0 \le x \le L$, with (a) $\psi(0) = 0$, $\psi'(L) = 0$; (b) $\psi(0) = a$, $\psi(L) = b$.

42.6 Verify the expansion coefficient $a_{\ell m}$ in (42.30).

42.7 Use the delta function-imposed discontinuity to verify $c = -i/4$ in (42.42).

42.8 **An Electrostatic Classic**
A charge q is uniformly distributed on a ring of radius a. The ring lies in the xy-plane, inside and concentric with a grounded hollow sphere of radius R. Find the potential inside the sphere. [You may find Problem 9.10 helpful.]

42.9 Example 42.4 found the Green's function in polar coordinates for an outgoing wave in the plane. Repeat the derivation for the Green's function in spherical coordinates for an outgoing wave in three dimensions.

42.10 (a) Using (42.52), unpack Green's identity (42.51) to find (42.53).
(b) Use the divergence theorem to show that the fundamental solution in (42.52) is inconsistent with homogeneous Neumann conditions on a finite surface S. Show, however, that $\partial G/\partial n|_S = -1/A$, where A is the area of S, is consistent.
(c) Use this inhomogeneous Neumann condition to verify (42.55).

42.11 Consider the equation $\left(\frac{d^2}{dx^2} + 9\right)\psi = \rho(x)$ with Neumann conditions $\psi'(0) = \psi'(L) = 0$.

(a) Find the Green's function using continuity and discontinuity conditions.
(b) Find the Green's function using the eigenfunction expansion for G.
(c) Find the solution for $\rho(x) = x^2$.

42.12   Consider the equation $\frac{d^2}{dx^2}\psi - \psi = \rho(x)$ with ψ going to zero at $\pm\infty$.

  (a) Find the Green's function using continuity and discontinuity conditions.
  (b) Find the Green's function using Fourier transforms.
  (c) Find the solution for $\rho(x) = e^{-|x|}$.

42.13   Consider the one-dimensional equation $\left[\frac{d^2}{d\xi^2} - 2\frac{d}{d\xi} + 1\right]u(\xi) = 0$.

  (a) For $\xi \equiv x$, find the Green's function for the boundary value problem with $u(0) = u(1) = 0$. Use G to find the solution for source $f(x) = x$.
  (b) For $\xi \equiv t$, find the Green's function for the initial value problem with $u(0) = 0, u'(0) = 1$. Use G to find the solution for driving force $f(t) = t$.

42.14   Use the continuity and discontinuity conditions for G to find the retarded Green's function for the forced damped oscillator for underdamping, overdamping, and critical damping. Choosing reasonable values of the parameters, use Mathematica to make plots of each.

42.15   Duhamel's principle applied to the wave equation $\left(\partial_t^2 - c^2\partial_x^2\right)\psi(x,t) = f(x,t)$ should yield the general expression

$$\psi(x,t) = \int_0^t \int f(x',t')\,G(x,x';t,t')\,dx'\,dt'.$$

  Verify this for initial conditions $G(x,x',0) = 0$, $\partial_\tau G(x,x',0) = \delta(x - x')$. Then use the Green's function (22.28) to evaluate the integral.[8] Interpret the result in light of (41.140) from Problem 41.6. What's the full solution for $\psi(x,0) = \alpha(x), \partial_t\psi(x,0) = \beta(x)$?

42.16   (a) Derive the Green's function of the wave operator $\partial_t^2 - c^2\partial_x^2$ for a string held fixed at $x = 0$ and $x = L$.
  (b) Find an expression for the solution $\psi(x,t)$ with initial conditions $\psi(x,0) = \partial_t(x,0) = 0$ if the string has a driving force $f(x,t) = \sin x \cos 2t$.

42.17   Adapt the analysis on BTW 42.1 to show that a Green's function solution to the initial value problem is equivalent to the variation of constants approach.

42.18   Use the divergence theorem to show that for homogeneous Dirichlet and Neumann boundary conditions, the Green's function of the Sturm–Liouville operator $\mathcal{L} = \vec{\nabla} \cdot \left[p(\vec{r})\vec{\nabla}\right] - q(\vec{r})$ is symmetric, $G(\vec{r},\vec{r}') = G(\vec{r}',\vec{r})$.

42.19   Find the temperature distribution $T(x,t)$ of a rod of length L with homogeneous Dirichlet condition and initial profile $T_0(x) = \frac{1}{2}\cos x$, attached to a heat source $q(x,t) = \sin x \cos 2t$. Plot the result using Mathematica's `Plot3D` for $\alpha = 1$ and $L = \pi$.

42.20   Show that the Duhamel solutions (42.108) and (42.117) are solutions to the heat equation.

42.21   The eigenfunction expansion (42.64) shows that the Green's function emerges from orthogonal expansion. So in some sense, G is not really new. We'll demonstrate this for the concrete case of heat flow in a one-dimensional bar with homogeneous Dirichlet conditions $T(0,t) = T(L,t) = 0$ and an external source $q(x,t)$.

  (a) Expand both $T(x,t)$ and $q(x,t)$ in Fourier series with coefficients $a_n(t)$ and $b_n(t)$, respectively. [See Example 41.3.] If $T(x,0) = T_0(x)$, what's $a_n(0)$?
  (b) Insert the Fourier expansions into the heat equation to get a differential equation of $a_n(t)$. Solve this inhomogeneous equation for $a_n(t)$ in terms of an integral over $b_n(t)$.
  (c) Insert the solution for $a_n(t)$ into the Fourier series for $T(x,t)$, and show that it's equivalent to (42.108) with the Green's function (42.102).

---

[8] The difference between the inhomogeneous wave equation here and in (22.20) affects the form of G. Such variations in convention are not uncommon, but have no physical ramifications.

42.22 Example 41.8 found the temperature distribution of a rectangular plate with inhomogeneous boundary conditions. We want to show that this is equivalent to the Green's function method outlined in Example 42.10.

    (a) In what limit does the two-dimensional problem reduce to the one-dimensional configuration of (42.111)? Explain.

    (b) Show that in the one-dimensional limit, (41.112) simplifies to (42.112).

    (c) Construct the Green's function $G(x, x', y, y'; t, t')$ for the rectangular plate. Then recast the solution in the form of (42.117).

42.23 Consider an infinite bar, initially with zero temperature, attached to an external heat source $q(x, t) = \cos x \, e^{-t}$ at $t = 0$. Find the temperature distribution $T(x, t)$ of the bar. Use Mathematic's `Plot3D` command to visualize the solution, and describe it physically.

42.24 Consider a semi-infinite bar $0 \le x < \infty$, with $T(x, 0) = T_0(x)$, attached to an external heat source $q(x, t)$ at $t = 0$. Use the method of images to find the Green's function for (a) Dirichlet condition $T(0, t) = 0$; and (b) Neumann condition $\partial_x T(0, t) = 0$.

42.25 Use (22.28) to find G for a semi-infinite string $0 \le x < \infty$ fixed at $x = 0$.

42.26 Find G for a bar of length L, with $T(x, 0) = T_0(x)$ and Dirichlet conditions $T(0, t) = a$, $T(L, t) = b$. [Hint: Think images and periodicity.]

42.27 The heat equation Green's function (42.102) should yield the fundamental solution (42.130) as $L \to \infty$. The key feature emerging in this limit is translation invariance, $G(x, x'; t) \to G(x - x', \tau) \equiv G(\chi, \tau)$, leaving G with a dependence on two rather than three independent variables. But in fact, it depends on only one:

    (a) Use the initial value problem (42.98) to establish scale invariance of the Green's function: $\lambda G(\lambda \chi, \lambda^2 \tau) = G(\chi, \tau)$ for arbitrary dimensionless λ. Noting that the diffusivity α is the only dimensional scale in the problem, choose a value of λ to reduce the Green's function to the form $G = \lambda g(\xi)$ with a single dimensionless variable ξ and dimensionless g.

    (b) Find and solve the differential equation for g. Apply the initial condition in the $t \to 0$ limit to finally obtain G.

42.28 How would the analysis in Example 42.7 be affected with homogenous Dirichlet rather than Neumann conditions? Give both a mathematical and a physical explanation.

42.29 Use the (dis)continuity requirements to solve $\left(\frac{d^2}{dx^2} + \lambda \right) G(x, x') = \delta(x - x')$ with homogeneous Neumann conditions at $x = 0$ and L. Show that G is singular when λ is an eigenvalue of d^2/dx^2. For $\lambda = 0$, show that $\tilde{G} \equiv G - 1/\lambda$ is a modified Green's function which obeys (42.88), and reduces to the Green's function of Example 42.7 in the $\lambda \to 0$ limit.

42.30 Consider the equation $\left(\frac{d^2}{dx^2} + \frac{9\pi^2}{L^2} \right) u(x) = x + c$ for boundary conditions $u(0) = u(L) = 0$ and constant c. For what value of c does a solution exist? Find the modified Green's function for this value of c and solve for $u(x)$.

42.31 (a) Solve for the modified Green's function $\tilde{G}$ in Example 42.7.

    (b) Show that a steady-state solution $T(x)$ exists for a heat source $q(x) = \sin(2\pi x/L)$. Find the solution, and verify that it solves the differential equation with the required boundary conditions.

42.32 Solve the homogeneous heat equation of Example 42.10 with

$$T(0, t) = \cos t \qquad T(L, t) = \sin 2t.$$

Plot the result using Mathematica's `Plot3D` for $\alpha = 1$ and $L = 1$.

42.33 Deriving the Green's function of (42.24) in cylindrical coordinates is more elaborate than in spherical — but we can nonetheless follow the basic outline of Example 42.3:

(a) Use Table 41.1 to find the homogeneous solutions in cylindrical arguments. Show that the general solution on either side of $\vec{r}\,'$ is

$$G(\vec{r}, \vec{r}\,') = \sum_{m=-\infty}^{\infty} \int_{-\infty}^{\infty} dk\, e^{im\phi} e^{ikz} \begin{cases} a_m(k) I_m(kr), & 0 < r < r' \\ b_m(k) K_m(kr), & r' < r < \infty \end{cases}.$$

Why is the sum over k an integral?

(b) Use the continuity of G at $r = r'$ to relate the coefficients a_m and b_m.

(c) Using a 2ϵ-thick arc enclosing the source at $\vec{r}\,'$ with $d\vec{a} = \hat{r}\, r d\phi dz$, find a general expression for the discontinuity in G. Use this to show

$$a_m(k) = \frac{1}{4\pi^2} \frac{K_m(kr')}{kr'} \frac{e^{-i\phi'} e^{ikr'}}{W(kr')},$$

where $W(x)$ is the Wronskian of $I_m(x)$ and $K_m(x)$.

(d) Since a Wronskian can be evaluated at any point, use the asymptotic forms in (C.29) and (C.30) to find $W(kr')$.

(e) Bring all the pieces together to show that

$$G(\vec{r}, \vec{r}\,') = \frac{1}{2\pi^2} \sum_m \int dk\, I_m\,(kr_<) K_m\,(kr_>)\, e^{im(\phi - \phi')} \cos[k(z - z')].$$

43 Coda: Quantum Scattering

A plane wave of water hits a rock, producing an outgoing spherical wave. This picture of course is not special to water; sound and light waves exhibit analogous behavior, as do quantum mechanical wavefunctions. For quantum mechanical scattering of a particle moving through a region of non-zero potential V, the superposition of an incoming plane wave and an outgoing spherical wave has the form (Cohen-Tannoudji et al., 2019)

$$\phi(\vec{r}) = e^{i\vec{k}\cdot\vec{r}} + f(\theta,\phi)\frac{e^{ikr}}{r}, \qquad k^2 = 2mE/\hbar^2. \tag{43.1}$$

The first term represents a particle with momentum $\hbar\vec{k}$ before entering the region of V, the second the elastically scattered particle long after its interaction with the potential. Since V likely does not have a rock's distinct boundary, this superposition can only be valid asymptotically for large r — which, as a practical matter, is where an experiment's sources and detectors are located. At the very least, (43.1) requires the potential to fall off faster than the spherical wave's $1/r$.

The complex coefficient $f(\theta,\phi)$ is called the *scattering amplitude*. As a quantum mechanical expression, its magnitude-squared $|f|^2$ is directly related to the probability that the scattered particle arrives within $d\Omega$ of a detector oriented at $\hat{r} = \hat{r}(\Omega)$; integrating over solid angle gives the total *scattering cross section* σ,

$$\sigma \equiv \int \frac{d\sigma}{d\Omega}\, d\Omega = \int |f(\theta,\phi)|^2\, d\Omega. \tag{43.2}$$

The nomenclature derives from classical physics: in the deflection of water, we know intuitively that the amount of scattering depends upon the cross-sectional area of the rock. By analogy, the greater the cross section σ, the greater the probability of scattering.[1]

The form of (43.1) locates the details of the scattering in $f(\theta,\phi)$; within this function are encoded properties of the potential. Discovering and probing these properties are what experiments at particle accelerators seek to do. Thus solving for the scattering amplitude is crucial to the analysis of a scattering experiment; in this Coda, we'll examine two approaches.

43.1 The Born Approximation

Thus far, we've conceived of Green's functions as a vehicle for obtaining closed-form solutions to inhomogeneous differential equations. But under certain constraints, they can also be leveraged to find

---

[1] Scattering cross sections are measured in *barns*, where 1 barn = 10^{-28} m^2, the typical size in nuclear scattering. The coinage derives from the expression "you can't hit the broadside of a barn."

approximate series solutions. The time-independent Schrödinger equation for a particle of mass m in a potential V can be cast as an inhomogeneous Helmholtz equation (Problem 41.2)

$$\left(\nabla^2 + k^2\right)\phi(\vec{r}) = \rho(\vec{r}), \tag{43.3a}$$

where

$$k^2 = \frac{2mE}{\hbar^2} \quad \text{and} \quad \rho(\vec{r}) = \frac{2m}{\hbar^2}V(\vec{r})\phi(\vec{r}). \tag{43.3b}$$

Notice that ρ depends on the wavefunction ϕ — so despite superficial appearances, it is not actually a proper inhomogeneous term. As such, one might think a Green's function approach would be fruitless. Nonetheless, one can formally verify that a Helmholtz Green's function satisfying

$$\left(\nabla^2 + k^2\right)G(\vec{r} - \vec{r}') = \delta(\vec{r} - \vec{r}'), \tag{43.4}$$

can be used to construct the expression

$$\phi(\vec{r}) = \phi_0(\vec{r}) + \frac{2m}{\hbar^2}\int G(\vec{r} - \vec{r}')V(\vec{r}')\,\phi(\vec{r}')\,d^3r', \tag{43.5}$$

where $\phi_0 = e^{i\vec{k}\cdot\vec{r}}$ is a homogeneous Helmholtz (i.e., free-particle) solution.

With ϕ ensnared in the integration on the right, however, (43.5) is not actually a solution to (43.3); rather, it is an equivalent *integral* representation of Schrödinger's differential equation. Although it cannot be solved exactly, for a weak potential the solution ϕ should differ little from ϕ_0. The *Born approximation* leverages this by replacing ϕ in the integrand by ϕ_0 to get

$$\phi_1(\vec{r}) = \phi_0(\vec{r}) + \int G(\vec{r} - \vec{r}')V(\vec{r}')\phi_0(\vec{r}')\,d^3r'. \tag{43.6}$$

This approximate result can then be improved by returning to (43.5) and replacing ϕ by ϕ_1 to get

$$\begin{aligned}
\phi_2(\vec{r}) &= \phi_0(\vec{r}) + \int G(\vec{r} - \vec{r}')V(\vec{r}')\phi_1(\vec{r}')\,d^3r' \\
&= \phi_0(\vec{r}) + \int G(\vec{r} - \vec{r}')V(\vec{r}')\phi_0(\vec{r})\,d^3r' \\
&\quad + \int G(\vec{r} - \vec{r}')V(\vec{r}')G(\vec{r}' - \vec{r}'')V(\vec{r}'')\phi_0(\vec{r}'')\,d^3r'd^3r''.
\end{aligned} \tag{43.7}$$

Similarly,

$$\begin{aligned}
\phi_3(\vec{r}) &= \phi_0(\vec{r}) + \int G(\vec{r} - \vec{r}')V(\vec{r}')\phi_0(\vec{r})\,d^3r' \\
&\quad + \int G(\vec{r} - \vec{r}')V(\vec{r}')G(\vec{r}' - \vec{r}'')V(\vec{r}'')\phi_0(\vec{r}'')\,d^3r'd^3r'' \\
&\quad + \int G(\vec{r}-\vec{r}')V(\vec{r}')G(\vec{r}'-\vec{r}'')V(\vec{r}'')G(\vec{r}''-\vec{r}''')V(\vec{r}''')\phi_0(\vec{r}''')\,d^3r'd^3r''d^3r'''.
\end{aligned} \tag{43.8}$$

Continuing in this fashion generates successive approximations ϕ_n. For sufficiently weak V, the result is the convergent *Born series* solution to (43.5), $\phi \equiv \phi_{n\to\infty}$. The first-order approximation ϕ_1 is called the Born term; including higher orders allows one to obtain any level of accuracy desired. The weak potential is known as the "perturbation" to free particle propagation, and the technique of generating successive converging approximations to $\phi(\vec{r})$ starting from the "unperturbed" solution $\phi_0(\vec{r})$ is called *perturbation theory*.

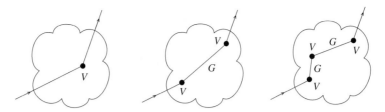

Figure 43.1 Single, double, and triple scattering. Only particles entering and exiting the interaction region are directly observed.

The iterative form of the Born series lends itself to a constructive and intuitive physical interpretation (Figure 43.1). The first integral represents a single scattering event by the potential. The second describes double scattering, with a factor of G responsible for "propagating" the particle between scattering events; having two factors of a weak V, this makes a smaller overall contribution than single scattering. Similarly, ϕ_3 includes the even-smaller effects of triple scattering. The full solution ϕ is the convergent sum over all possible multiple scatterings.

Though different potentials produce distinct Born series, what they have in common is the Helmholtz Green's function. Taking the Fourier transform of (43.4) gives

$$\left[-q^2 + k^2\right]\tilde{G}(q) = 1, \tag{43.9}$$

where we denote the Fourier frequency by q to distinguish it from k in the Helmholtz equation. The Green's function G is then the inverse transform

$$G(\vec{r} - \vec{r}') = \int \frac{d^3q}{(2\pi)^3} \frac{e^{i\vec{q}\cdot(\vec{r}-\vec{r}')}}{k^2 - q^2}. \tag{43.10}$$

As was the case for the ω integral in (42.134), we handle the pole at $q = k$ by applying boundary conditions. (Just as solving the differential equation requires specification of boundary conditions, so too for the integral equation.) Specifically, we want G to describe an outgoing wave due to an impulse; as we saw in Problem 22.7, with time dependence $e^{-i\omega t}$ this is

$$G_+(\vec{r} - \vec{r}') = -\frac{1}{4\pi} \frac{e^{ik|\vec{r}-\vec{r}'|}}{|\vec{r} - \vec{r}'|}, \tag{43.11}$$

describing a spherical wave of frequency k emerging from $\vec{r}'$. Inserting G_+ into (43.5) yields the *Lippmann–Schwinger equation*,

$$\phi(\vec{r}) = \phi_0(\vec{r}) - \frac{m}{2\pi\hbar^2} \int \frac{e^{ik|\vec{r}-\vec{r}'|}}{|\vec{r} - \vec{r}'|} V(\vec{r}')\,\phi(\vec{r}')\, d^3r'. \tag{43.12}$$

The similarity to (43.1) is clear. To extract the scattering amplitude, consider a potential $V(\vec{r}')$ of limited range centered at the origin from which a free-particle plane wave scatters (Figure 43.2).[2] In the potential-free region $r \gg r'$, we can use the first-order approximation (Problem 6.35)

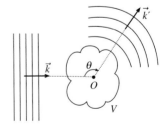

Figure 43.2 Plane wave scattering from a potential V.

$$|\vec{r} - \vec{r}'| \approx r - \hat{r}\cdot\vec{r}', \qquad r \gg r' \tag{43.13}$$

---

[2] A potential's limited range is distinct from the Born series requirement of a weak potential.

to expand the Lippman–Schwinger equation as

$$\phi(\vec{r}) = \phi_0(\vec{r}) - \frac{m}{2\pi\hbar^2}\frac{e^{ikr}}{r}\int e^{-i\vec{k}'\cdot\vec{r}'}V(\vec{r}')\phi(\vec{r}')\,d^3r' + \mathcal{O}(1/r^2), \tag{43.14}$$

where $\vec{k}' \equiv k\hat{r}$. Comparison with (43.1) identifies the scattering amplitude

$$f(\hat{r}) = -\frac{m}{2\pi\hbar^2}\int e^{-i\vec{k}'\cdot\vec{r}'}V(\vec{r}')\phi(\vec{r}')\,d^3r'. \tag{43.15}$$

Alas, finding f requires already knowing the solution $\phi(\vec{r})$. But for a weak potential, we can use the Born series to generate successive approximations. The Born approximation replaces ϕ in the integrand by $\phi_0 = e^{i\vec{k}\cdot\vec{r}}$ to give the first-order result

$$\begin{aligned} f_1(\hat{r}) &= -\frac{m}{2\pi\hbar^2}\int e^{-i\vec{k}'\cdot\vec{r}'}V(\vec{r}')\phi_0(\vec{r}')\,d^3r' \\ &= -\frac{m}{2\pi\hbar^2}\int e^{-i(\vec{k}'-\vec{k})\cdot\vec{r}'}V(\vec{r}')\,d^3r' = -\frac{m}{2\pi\hbar^2}\tilde{V}(\vec{q}), \end{aligned} \tag{43.16}$$

where $\hbar\vec{q} \equiv \hbar(\vec{k}' - \vec{k})$ is the *momentum transfer* between the incident and scattered waves. The momentum $\hbar\vec{k}$ is set by the experiment; then since $|\vec{k}| = |\vec{k}'|$, the only independent information is the scattering angle, which can be extracted from $\vec{q} \cdot \vec{q}$,

$$q^2 = 2k^2(1 - \cos\theta) = k^2\sin^2(\theta/2). \tag{43.17}$$

Thus measuring the angular distribution of the scattered particles determines $\tilde{V}(\vec{q})$. Moreover, the discovery that the scattering amplitude is given by the Fourier transform of the potential yields important physical insight: a potential $V(\vec{r})$ with a short-range distance scale $\sim r_0$ yields a $\tilde{V}(\vec{q})$ dominated by $q \sim 1/r_0$. So exploring short-distance behavior requires large momentum transfer. Physically, this is a consequence of the uncertainty principle.

Example 43.1　The Yukawa Potential

The Yukawa potential

$$V(r) = V_0\frac{e^{-\mu r}}{r} \tag{43.18}$$

is a model for the short-range nuclear attraction between nucleons (protons and neutrons). With an exponential drop-off, the Lippmann–Schwinger equation is clearly applicable. And as a central potential, the integrals are somewhat simplified: from (43.16), we find that in the Born approximation

$$f(\theta) \approx -\frac{m}{2\pi\hbar^2}\tilde{V}(\vec{q}) = -\frac{2V_0 m}{\hbar^2}\frac{1}{q^2 + \mu^2}. \tag{43.19}$$

Referring to (43.2) and using (43.17), the differential cross section is thus

$$\frac{d\sigma}{d\Omega} = |f(\theta)|^2 \approx \left(\frac{2V_0 m}{\hbar^2}\right)^2\frac{1}{[2k^2(1 - \cos\theta) + \mu^2]^2}. \tag{43.20}$$

So in a experiment in which protons are fired at a fixed nucleus, the Yukawa potential predicts that the protons predominantly scatter in the forward direction around $\theta = 0$. The total cross section includes scattering through all angles,

$$\sigma = \left(\frac{2V_0 m}{\hbar^2}\right)^2 \int_{-1}^{1} \frac{d(\cos\theta)\, d\phi}{[2k^2(1-\cos\theta)+\mu^2]^2} = \left(\frac{V_0 m}{\hbar^2}\right)^2 \frac{16\pi}{4k^2\mu^2 + \mu^4}. \qquad (43.21)$$

Both V_0 and μ can be experimentally determined from cross section data.

When Yukawa introduced this potential in 1935, he made a bold prediction. As seen in Problem 43.3, his potential is a Green's function for static solutions of the Klein–Gordon equation (22.35), implying that the nuclear force can be described as a source of spinless, relativistic particles with momentum q and mass $\hbar\mu/c$. Now the effective range of the nuclear force must be on the order of the size of a nucleus, $1\ \text{fm} = 10^{-15}$ m, so Yukawa knew that $\mu \sim 1\ \text{fm}^{-1}$. From this, he proposed the existence of a nuclear particle with mass of about

$$\hbar\mu/c = \hbar c\mu/c^2 = (200\ \text{MeV-fm})(1\ \text{fm}^{-1})/c^2 = 200\ \text{MeV}/c^2. \qquad (43.22)$$

In 1947 the pi meson (or pion) was discovered with a mass of about $140\ \text{MeV}/c^2$; Yukawa was awarded the physics Nobel prize two years later.

The Born expansion picture which emerges from Yukawa's work is of a pion with momentum $\hbar\vec{q} \equiv \hbar(\vec{k}' - \vec{k})$ being exchanged between nucleons, acting as the mediator of their nuclear interaction.[3] Such particle exchange is often depicted with ladder diagrams such as those in Figure 43.3. The portrayal of scattering as a converging sum of increasingly elaborate interactions as in Figures 43.1–43.3 are precursors to the Feynman diagrams of quantum field theory.

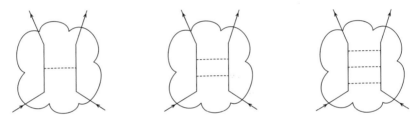

Figure 43.3 Particle scattering via single, double, and triple particle exchange (dotted lines).

43.2 The Method of Partial Waves

Consider a particle of mass m scattering off a short-range, spherically symmetric potential $V(\vec{r}) = V(r)$. Applying separation of variables $\Psi(\vec{r}) = R(r)Y(\theta,\phi)$ to the time-independent Schrödinger equation yields both the differential equation for spherical harmonics, as well as the radial equation

$$-\frac{\hbar^2}{2m}\frac{1}{r^2}\frac{d}{dr}\left(r^2\frac{dR}{dr}\right) + \left[V(r) + \frac{\hbar^2\ell(\ell+1)}{2mr^2}\right]R = ER. \qquad (43.23)$$

The term proportional to $\ell(\ell+1)$ is due to the particle's angular momentum — the quantum version of the centrifugal barrier $L^2/2I$ of classical physics. At very large distance r, neither it nor the short-range potential is relevant. But if V falls off faster than $1/r^2$, then at some intermediate distance the system effectively becomes that of a free particle with angular momentum ℓ, described by the radial equation of (41.69),

---

[3] In the formal $\mu \to 0$ limit, the Yukawa potential becomes the Coulomb potential, mediated by the exchange of a massless particle, the photon.

$$\frac{d}{dr}\left(r^2\frac{dR}{dr}\right) + \left[k^2 r^2 - \ell(\ell+1)\right]R(r) = 0, \tag{43.24}$$

with $k^2 = 2mE/\hbar^2$. As we know, the solutions are linear combinations of spherical Bessel and Neumann functions j_ℓ and n_ℓ,

$$R(r) = aj_\ell(kr) + bn_\ell(kr). \tag{43.25}$$

Note that since this is only valid for large r, the origin is not included — and so we cannot discard the Neumann function. In fact, more germane to scattering are the spherical Hankel functions $h_\ell^{(1,2)} \equiv j_\ell \pm in_\ell$. From their asymptotic form (C.39),

$$h_\ell^{(1,2)}(kr) \xrightarrow{\text{large } r} \frac{1}{\pm ikr}e^{\pm i(kr - \frac{\ell\pi}{2})} = (\mp i)^{\ell+1}\frac{e^{\pm ikr}}{kr}, \tag{43.26}$$

we see that these functions describe outgoing and incoming spherical waves. Thus it makes sense to recast the solution $R(r)$ as

$$R(r) = c_1 h_\ell^{(1)}(kr) + c_2 h_\ell^{(2)}(kr). \tag{43.27}$$

Since we're interested in outgoing waves scattered by the potential, we set $c_2 = 0$. (Compare this with Example 42.4 in cylindrical coordinates.) Thus the scattered wavefunction is a superposition,

$$\Psi_{\text{scat}} = \sum_\ell c_{\ell m}\psi_{\ell m} = \sum_{\ell m} c_{\ell m}h_\ell^{(1)}(kr)Y_{\ell m}(\theta, \phi), \tag{43.28}$$

each *partial wave* $\psi_{\ell m}$ having fixed angular momentum ℓ. Now with a spherically symmetric potential, there can be no ϕ dependence, so only $m = 0$ actually contributes; then (35.38) reduces the superposition of partial waves to

$$\Psi_{\text{scat}} = \sum_\ell \sqrt{\frac{2\ell+1}{4\pi}}\, c_{\ell 0}\, h_\ell^{(1)}(kr)P_\ell(\cos\theta). \tag{43.29}$$

We'll choose the incident wavefunction to propagate along z, so that the full scattering solution (including the incident wave) has the form

$$\Psi(\vec{r}) = e^{i\vec{k}\cdot\vec{r}} + \sum_\ell \sqrt{\frac{2\ell+1}{4\pi}}\, c_{\ell 0}\, h_\ell^{(1)}(kr)P_\ell(\cos\theta)$$

$$= e^{ikz} + k\sum_\ell i^{\ell+1}(2\ell+1)f_\ell(k)h_\ell^{(1)}(kr)P_\ell(\cos\theta), \tag{43.30}$$

where we've replaced $c_{\ell 0}$ by introducing the *partial wave coefficients* f_ℓ via

$$\sqrt{\frac{2\ell+1}{4\pi}}\, c_{\ell 0}(k) = i^{\ell+1}(2\ell+1)kf_\ell(k). \tag{43.31}$$

Including factors of i and k in the definition of f_ℓ cancels these same factors in the asymptotic form of $h^{(1)}$ in (43.26), so that

$$\Psi(\vec{r}) \xrightarrow{\text{large } r} e^{ikz} + \sum_\ell (2\ell+1)f_\ell(k)P_\ell(\cos\theta)\frac{e^{ikr}}{r}. \tag{43.32}$$

Comparing this with (43.1) identifies the scattering amplitude,

$$f(\theta) = \sum_\ell (2\ell + 1) f_\ell(k) P_\ell(\cos\theta). \tag{43.33}$$

Since Ψ_{scat} is expressed on a spherical basis, the same should be done for the incident wave; this is accomplished using the partial wave expansion (35.70) (see also BTW 41.2),

$$e^{ikz} = \sum_{\ell=0}^{\infty} \sum_{m=-\ell}^{\ell} (2\ell + 1) i^\ell j_\ell(kr) P_\ell(\cos\theta). \tag{43.34}$$

Using (43.26) gives

$$j_\ell(kr) = \frac{1}{2}\left(h^{(1)} + h^{(2)}\right) \xrightarrow{\text{large } r} \frac{1}{2ikr}\left[e^{i(kr - \frac{\ell\pi}{2})} - e^{-i(kr - \frac{\ell\pi}{2})}\right] = \frac{(-i)^\ell}{2ikr}\left[e^{ikr} - e^{-i(kr - \ell\pi)}\right], \tag{43.35}$$

which is the sum of outgoing and incoming spherical waves for each ℓ. All together then, (43.32) becomes

$$\Psi(\vec{r}) \xrightarrow{\text{large } r} \sum_\ell (2\ell + 1) P_\ell(\cos\theta)\left[\frac{1}{2ikr}\left(e^{ikr} - e^{-i(kr - \ell\pi)}\right) + f_\ell(k)\frac{e^{ikr}}{r}\right]$$
$$= \sum_\ell (2\ell + 1) P_\ell(\cos\theta)\frac{1}{2ik}\left[(1 + 2ikf_\ell)\frac{e^{ikr}}{r} + (-1)^{\ell+1}\frac{e^{-ikr}}{r}\right]. \tag{43.36}$$

Thus for each ℓ, the solution outside the range of V is the sum of an incoming spherical wave $\pm e^{-ikr}/r$ and an outgoing one e^{ikr}/r. The scattering itself manifests only in the coefficient $2ikf_\ell$ of the outgoing wave emerging from the potential region. So the effect of scattering is reflected only in non-vanishing partial wave coefficients; indeed, the total cross section depends only on the f_ℓ,

$$\sigma = \int |f(\theta)|^2 d\Omega = 4\pi \sum_\ell (2\ell + 1)|f_\ell(k)|^2, \tag{43.37}$$

as can be confirmed using (43.33). Of course, as an infinite sum, this expression may seem of limited utility — but a little physical reasoning gives insight into how quickly it may converge. Classically, the angular momentum of a particle of momentum p relative to the origin is $|\vec{L}| = |\vec{r} \times \vec{p}| = pb$, where b is the *impact parameter* (Figure 43.4). So for a potential of limited range $\sim r_0$, any particle with $b \gtrsim r_0$ should skate by without scattering. With $p = \hbar k$ and $L = pb \sim \hbar\ell$, this means that there is little to no scattering for $\ell \gtrsim kr_0$. So the shorter the range, the fewer the number of partial waves which contribute to σ. But the higher the incident momentum, the smaller the de Broglie wavelength $2\pi/k$, and thus the greater the resolution — at the cost of requiring contribution from higher-ℓ partial waves.

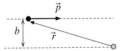

Figure 43.4 The impact parameter b.

Problems

43.1 Verify that (43.5) is a solution of the Helmholtz equation.

43.2 Show that the lowest order approximation (43.14) to the Lippmann–Schwinger equation emerges using the outgoing Green's function G_+ of (43.11).

43.3 Show that the Yukawa potential (43.18) is a Green's function of the static Klein–Gordon equation (22.35), $(\nabla^2 + \mu^2)G(\vec{r}) = -\delta(\vec{r} - \vec{r}')$.

43.4 Show that the total scattering cross section σ is given by the sum of partial wave contributions, (43.37).

43.5 Quantum mechanics conserves probability (see Problem 16.3), so the incoming probability flux in a scattering experiment must be equal to the outgoing flux.

(a) From (43.36), deduce that the quantity $1 + 2ikf_\ell$ must be a pure phase, $1 + 2ikf_\ell \equiv e^{2i\delta_\ell}$, so that the scattering amplitude can be expressed as

$$f(\theta) = \frac{1}{k} \sum_\ell (2\ell + 1)e^{i\delta_\ell} \sin \delta_\ell \, P_\ell(\cos \theta).$$

The quantity $\delta_\ell = \delta_\ell(k)$ is the phase shift acquired by the incoming wave as it scatters due to the potential, elegantly encoding the net effect of V on a particle with angular momentum ℓ.

(b) Show that in terms of phase shifts, the cross section is

$$\sigma = \frac{4\pi}{k^2} \sum_\ell (2\ell + 1) \sin^2 \delta_\ell.$$

43.6 **Hard-Sphere Scattering**

To calculate the scattering phase shifts δ_ℓ defined in Problem 43.5, we need a scattering potential. One of the simplest is the hard-sphere potential,

$$V(r) = \begin{cases} 0, & r \leq a \\ \infty, & r > a \end{cases}.$$

(a) For $r > a$, compare the general solution $R_\ell(r)$ of the radial equation (41.69) with the the large-r form in the brackets of (43.36) to find

$$R_\ell(r > a) = \cos \delta_\ell \, j_\ell(kr) - \sin \delta_\ell \, n_\ell(kr).$$

Apply a homogeneous Dirichlet condition at $r = a$ to solve for δ_0.

(b) For low-energy, large-wavelength $\lambda = 2\pi/k \gg a$, use the asymptotic forms of Bessel functions in Appendix C to conclude that for $ka \ll 1$, only $\ell = 0$ makes a significant contribution to the scattering — that is, only δ_0 is effectively non-zero. Then show that the total scattering cross section is

$$\sigma \approx 4\pi a^2, \qquad ka \ll 1,$$

four times the geometrical cross section πa^2 that we'd intuitively expect. Why do you think the quantum mechanical result is larger?

Appendix A

Curvilinear Coordinates

A curvilinear coordinate system (u_1, u_2, u_3) in $\mathbb{R}^3$ is specified by the line element

$$d\vec{r} = \frac{\partial \vec{r}}{\partial u_1} \, du_1 + \frac{\partial \vec{r}}{\partial u_2} \, du_2 + \frac{\partial \vec{r}}{\partial u_3} \, du_3$$

$$\equiv \hat{e}_1 \, h_1 du_1 + \hat{e}_2 \, h_2 du_2 + \hat{e}_3 \, h_3 du_3, \tag{A.1}$$

where the unit basis vectors are given by

$$\hat{e}_i \equiv \frac{1}{h_i} \frac{\partial \vec{r}}{\partial u_i}. \tag{A.2}$$

The scale factors $h_i \equiv |\partial \vec{r}/\partial u_i|$ are related to the cartesian (x_1, x_2, x_3) by

$$h_i^2 = \frac{\partial \vec{r}}{\partial u_i} \cdot \frac{\partial \vec{r}}{\partial u_i} = \left(\frac{\partial x_1}{\partial u_i}\right)^2 + \left(\frac{\partial x_2}{\partial u_i}\right)^2 + \left(\frac{\partial x_3}{\partial u_i}\right)^2. \tag{A.3}$$

The h_i allow straightforward expression of the infinitesimal distance

$$ds^2 \equiv d\vec{r} \cdot d\vec{r} = \sum_i h_i^2 \, du_i^2, \tag{A.4a}$$

as well as the area and volume elements

$$d\vec{a}_{ij} = \left(\sum_k \epsilon_{ijk} \hat{e}_k\right) h_i h_j \, du_i du_j \qquad d\tau = h_1 h_2 h_3 \, du_1 du_2 du_3. \tag{A.4b}$$

The gradient of a lone cartesian coordinate is just the unit vector along its axis — for instance, $\vec{\nabla} y \equiv \hat{\jmath}$. For curvilinear coordinates, however, (A.1) gives

$$\vec{\nabla} u_j = \left(\sum_i \frac{\hat{e}_i}{h_i} \frac{\partial}{\partial u_i}\right) u_j = \sum_i \frac{\hat{e}_i}{h_i} \delta_{ij} = \frac{\hat{e}_j}{h_j}. \tag{A.5}$$

Moreover, since the curl of a gradient vanishes, $\vec{\nabla} \times \hat{e}_j/h_j = 0$. We can use this to our advantage by writing $A_k \hat{e}_k = (h_k A_k)\left(\hat{e}_k/h_k\right)$ and using the $\vec{\nabla} \times (f\vec{A})$ identity in Table 12.1 to find the curl of the kth component of $\vec{A}$:

$$
\begin{aligned}
\vec{\nabla} \times \left(A_k \hat{e}_k\right) &= \vec{\nabla} \times \left[(h_k A_k)\left(\hat{e}_k/h_k\right)\right] \\
&= \vec{\nabla}(h_k A_k) \times \hat{e}_k/h_k + (h_k A_k)\,\underbrace{\vec{\nabla} \times (\hat{e}_k/h_k)}_{0} \\
&= \sum_j \frac{\hat{e}_j}{h_j} \frac{\partial}{\partial u_j}(h_k A_k) \times \hat{e}_k/h_k \\
&= \sum_j \frac{1}{h_j h_k} \frac{\partial(h_k A_k)}{\partial u_j}\left(\hat{e}_j \times \hat{e}_k\right) = \sum_{ij} \epsilon_{ijk} \frac{\hat{e}_i}{h_j h_k} \frac{\partial(h_k A_k)}{\partial u_j},
\end{aligned}
\tag{A.6}
$$

where we used (4.21a) in the last step. We can pretty this up using $\epsilon_{ijk} \frac{1}{h_j h_k} = \frac{1}{h_1 h_2 h_3} \epsilon_{ijk} h_i$. Summing (A.6) over k gives

$$
\vec{\nabla} \times \vec{A} = \frac{1}{h_1 h_2 h_3} \sum_{ijk} \epsilon_{ijk}\left(h_i \hat{e}_i\right) \frac{\partial(h_k A_k)}{\partial u_j},
\tag{A.7}
$$

which in determinant form is (12.49d),

$$
\vec{\nabla} \times \vec{A} = \frac{1}{h_1 h_2 h_3}
\begin{vmatrix}
h_1 \hat{e}_1 & h_2 \hat{e}_2 & h_3 \hat{e}_3 \\
\partial/\partial u_1 & \partial/\partial u_2 & \partial/\partial u_3 \\
h_1 A_1 & h_2 A_2 & h_3 A_3
\end{vmatrix}.
\tag{A.8}
$$

Note that the derivatives in the second row are to be applied to the elements in the third — so the determinant has terms $\partial_j(h_i A_i)$, *not* $(h_i A_i)\partial_j$.

What about the divergence? We start rather mysteriously by using (A.5) to write

$$
\begin{aligned}
\vec{\nabla} u_i \times \vec{\nabla} u_j &= \frac{\hat{e}_i}{h_i} \times \frac{\hat{e}_j}{h_j} = \epsilon_{ijk} \frac{\hat{e}_k}{h_i h_j} \\
&= \frac{1}{h_1 h_2 h_3} \epsilon_{ijk} h_k \hat{e}_k \equiv \epsilon_{ijk}\left(\hat{e}_k/w_k\right),
\end{aligned}
\tag{A.9}
$$

where $w_k \equiv h_1 h_2 h_3/h_k$. Our derivation of (A.7) relied on $\hat{e}_i/h_i$ having zero curl; this time, we want a corresponding expression for the divergence. As you're asked to verify in Problem 12.20, $\vec{\nabla} u_i \times \vec{\nabla} u_j$ is itself a curl, and so by (12.27b), has zero divergence. Therefore from (A.9) we get $\vec{\nabla} \cdot (\hat{e}_k/w_k) = 0$. With this, we're ready to find $\vec{\nabla} \cdot \vec{A}$: write $\vec{A} = \sum_k \hat{e}_k A_k = \sum_k \frac{\hat{e}_k}{w_k} w_k A_k$ and use the product rule (12.31),

$$
\begin{aligned}
\vec{\nabla} \cdot \vec{A} = \sum_k \vec{\nabla} \cdot \left(\frac{\hat{e}_k}{w_k} w_k A_k\right) &= \sum_k \left(\frac{\hat{e}_k}{w_k}\right) \cdot \vec{\nabla}(w_k A_k) + \underbrace{\vec{\nabla} \cdot \left(\frac{\hat{e}_k}{w_k}\right)}_{0}(w_k A_k) \\
&= \sum_{k,\ell} \frac{\hat{e}_k}{w_k} \cdot \frac{\hat{e}_\ell}{h_\ell} \frac{\partial}{\partial u_\ell}(w_k A_k) .
\end{aligned}
\tag{A.10}
$$

Since $\hat{e}_k \cdot \hat{e}_\ell = \delta_{k\ell}$, one of the sums collapses to yield (12.49c),

$$
\begin{aligned}
\vec{\nabla} \cdot \vec{A} &= \sum_k \frac{1}{w_k h_k} \frac{\partial}{\partial u_k}(w_k A_k) \\
&= \frac{1}{h_1 h_2 h_3}\left[\partial_1\left(h_2 h_3 A_1\right) + \partial_2\left(h_3 h_1 A_2\right) + \partial_3\left(h_1 h_2 A_3\right)\right],
\end{aligned}
\tag{A.11}
$$

where we've used the short-hand $\partial_k = \partial/\partial u_k$.

The expression (12.49b) for the Laplacian can be found from (A.1) and (A.11),

$$\nabla^2 f = \vec{\nabla} \cdot \vec{\nabla} f = \sum_k \frac{1}{h_k w_k} \frac{\partial}{\partial u_k} \left(\frac{w_k}{h_k} \frac{\partial}{\partial u_k} \right) f$$

$$= \frac{1}{h_1 h_2 h_3} \left[\partial_1 \left(\frac{h_2 h_3}{h_1} \partial_1 f \right) + \partial_2 \left(\frac{h_1 h_3}{h_2} \partial_2 f \right) + \partial_3 \left(\frac{h_1 h_2}{h_3} \partial_3 f \right) \right]. \tag{A.12}$$

Though these derivations are not simple, the results — (A.8), (A.11), and (A.12) — are not unappealing, requiring only knowledge of coordinate system scale factors.

A principal application of curvilinear coordinates is in solutions to the Helmholtz equation $(\vec{\nabla}^2 + k^2) f = 0$, which is separable not just in cartesian coordinates, but in the 10 orthogonal systems listed in Table A.1 as well.

Table A.1 Right-handed orthogonal curvilinear coordinates

System	Transformations	Domain	Scale factors	Coordinate surfaces
Spherical (r,θ,ϕ)	$x = r\sin\theta\cos\phi$ $y = r\sin\theta\sin\phi$ $z = r\cos\theta$	$0 \le r < \infty$ $0 \le \theta \le \pi$ $0 \le \phi < 2\pi$	$h_r = 1$ $h_\theta = r$ $h_\phi = r\sin\theta$	spheres, $r = $ const cones, $\theta = $ const planes, $\phi = $ const
Cylindrical (ρ,ϕ,z)	$x = \rho\cos\phi$ $y = \rho\sin\phi$ $z = z$	$0 \le r < \infty$ $0 \le \phi < 2\pi$ $-\infty < z < \infty$	$h_\rho = 1$ $h_\phi = \rho$ $h_z = 1$	cylinders, $\rho = $ const planes, $\phi = $ cnst, $z = $ const
Prolate spheroidal (r,θ,ϕ)	$x = a\sinh r\sin\theta\cos\phi$ $y = a\sinh r\sin\theta\sin\phi$ $z = a\cosh r\cos\theta$	$0 \le r < \infty$ $0 \le \theta \le \pi$ $0 \le \phi < 2\pi$	$h_r = a\sqrt{\sinh^2 r + \sin^2\theta}$ $h_\theta = a\sqrt{\sinh^2 r + \sin^2\theta}$ $h_\phi = a\sinh r\sin\theta$	prolate spheroids, $r = $ const hyperboloids, $\theta = $ const planes, $\phi = $ const
Oblate spheroidal (r,θ,ϕ)	$x = a\cosh r\sin\theta\cos\phi$ $y = a\cosh r\sin\theta\sin\phi$ $z = a\sinh r\cos\theta$	$0 \le r < \infty$ $0 \le \theta \le \pi$ $0 \le \phi < 2\pi$	$h_r = a\sqrt{\sinh^2 r + \cos^2\theta}$ $h_\theta = a\sqrt{\sinh^2 r + \cos^2\theta}$ $h_\phi = a\cosh r\sin\theta$	oblate spheroids, $r = $ const hyperboloids, $\theta = $ const planes, $\phi = $ const
Elliptic cylindrical (ρ,ϕ,z)	$x = a\cosh\rho\cos\phi$ $y = a\sinh\rho\sin\phi$ $z = z$	$0 \le \rho < \infty$ $0 \le \phi < 2\pi$ $-\infty < z < \infty$	$h_\rho = a\sqrt{\sinh^2\rho + \sin^2\phi}$ $h_\phi = a\sqrt{\sinh^2\rho + \sin^2\phi}$ $h_z = 1$	elliptic cylinders, $\rho = $ const hyperbolic cylinders, $\phi = $ const planes, $z = $ const
Parabolic cylindrical (η,ξ,z)	$x = (\eta^2 - \xi^2)/2$ $y = \eta\xi$ $z = z$	$0 \le \eta < \infty$ $-\infty < \xi < \infty$ $-\infty < z < \infty$	$h_\eta = \sqrt{\eta^2 + \xi^2}$ $h_v = \sqrt{\eta^2 + \xi^2}$ $h_z = 1$	parabolic cylinders, $\eta = $ const parabolic cylinders, $\xi = $ const planes, $z = $ const
Parabolic (η,ξ,ϕ)	$x = \eta\xi\cos\phi$ $y = \eta\xi\sin\phi$ $z = (\eta^2 - \xi^2)/2$	$0 \le \eta < \infty$ $0 \le \xi < \infty$ $0 \le \phi < 2\pi$	$h_\eta = \sqrt{\eta^2 + \xi^2}$ $h_\xi = \sqrt{\eta^2 + \xi^2}$ $h_\phi = \eta\xi$	paraboloids, $\eta = $ const paraboloids, $\xi = $ const planes, $\phi = $ const
Conical (η,ξ,χ)	$x^2 = \left(\dfrac{\eta\xi\chi}{bc}\right)^2$ $y^2 = \dfrac{\eta^2}{b^2}\dfrac{(\xi^2-b^2)(b^2-\chi^2)}{c^2-b^2}$ $z^2 = \dfrac{\eta^2}{c^2}\dfrac{(c^2-\xi^2)(c^2-\chi^2)}{c^2-b^2}$	$0 \le \eta < \infty$ $b^2 < \xi^2 < c^2$ $0 < \chi^2 < b^2$	$h_\eta^2 = 1$ $h_\xi^2 = \dfrac{\eta^2(\xi^2-\chi^2)}{(\xi^2-b^2)(c^2-\xi^2)}$ $h_\chi^2 = \dfrac{\eta^2(\xi^2-\chi^2)}{(b^2-\chi^2)(c^2-\chi^2)}$	spheres, $\eta = $ const elliptic cones, $\xi = $ const elliptic cones, $\chi = $ const
Confocal ellipsoidal (η,ξ,χ)	$x^2 = \dfrac{(a^2-\eta)(a^2-\xi)(a^2-\chi)}{(a^2-b^2)(a^2-c^2)}$ $y^2 = \dfrac{(b^2-\eta)(b^2-\xi)(b^2-\chi)}{(b^2-a^2)(b^2-c^2)}$ $z^2 = \dfrac{(c^2-\eta)(c^2-\xi)(c^2-\chi)}{(c^2-a^2)(c^2-b^2)}$	$-\infty < \eta < c^2$ $c^2 < \xi < b^2$ $b^2 < \chi < a^2$	$h_\eta^2 = \dfrac{1}{4}\dfrac{(\xi-\eta)(\chi-\eta)}{(a^2-\eta)(b^2-\eta)(c^2-\eta)}$ $h_\xi^2 = \dfrac{1}{4}\dfrac{(\chi-\xi)(\eta-\xi)}{(a^2-\xi)(b^2-\xi)(c^2-\xi)}$ $h_\chi^2 = \dfrac{1}{4}\dfrac{(\eta-\chi)(\xi-\chi)}{(a^2-\chi)(b^2-\chi)(c^2-\chi)}$	ellipsoids, $\eta = $ const hyperboloids, $\xi = $ const hyperboloids, $\chi = $ const
Confocal paraboloidal (η,ξ,χ)	$x^2 = \dfrac{(a^2-\eta)(a^2-\xi)(a^2-\chi)}{b^2-a^2}$ $y^2 = \dfrac{(b^2-\eta)(b^2-\xi)(b^2-\chi)}{a^2-b^2}$ $z^2 = \eta + \xi + \chi - a^2 - b^2$	$-\infty < \eta < b^2$ $b^2 < \xi < a^2$ $a^2 < \chi < \infty$	$h_\eta^2 = \dfrac{1}{4}\dfrac{(\xi-\eta)(\chi-\eta)}{(a^2-\eta)(b^2-\eta)}$ $h_\xi^2 = \dfrac{1}{4}\dfrac{(\chi-\xi)(\eta-\xi)}{(a^2-\xi)(b^2-\xi)}$ $\chi^2 = \dfrac{1}{4}\dfrac{(\eta-\chi)(\xi-\chi)}{(a^2-\chi)(b^2-\chi)}$	elliptic paraboloids, $\eta = $ const hyperbolic paraboloids, $\xi = $ const elliptic paraboloids, $\chi = $ const

Source: Moon and Spencer (1988); Morse and Feshbach (1953).

Appendix B
B

Rotations in $\mathbb{R}^3$

You might reasonably ask how we can be sure that every proper $\mathbb{R}^3$ rotation matrix has a direction $\hat{n}$ such that $R\hat{n} = \hat{n}$? The answer is that this is guaranteed by *Euler's rotation theorem* (BTW 27.1). But Euler's guarantee is not the end of the story. For one thing, $R\hat{n} = \hat{n}$ only gives the axis up to an overall sign; similarly, $\cos\varphi = \frac{1}{2}[\text{Tr}(R_n) - 1]$ cannot distinguish between $\pm\varphi$. So a sign convention is needed to ensure internal consistency. Moreover, though it's certainly reassuring to learn that every $\mathbb{R}^3$ rotation matrix has an axis, we've yet to develop techniques to actually *construct* a rotation matrix — that is, to find R for given $\hat{n}$ and φ. Of course, we know the canonical rotations (26.82) and (26.84) around coordinate axes; from them, any other rotation can be had by similarity. But unlike Problem 26.21, $R_n(\varphi) = SR_m(\varphi)S^{-1}$ generally requires construction of S from $\hat{n}$ and $\hat{m}$.

So thus far, we have no reliable method either to uniquely determine φ and $\hat{n}$ from a given 3×3 orthogonal matrix, or to construct a rotation matrix from specified φ and $\hat{n}$. Not surprisingly, there are several different approaches to constructing and deconstructing rotation matrices in $\mathbb{R}^3$; we'll look at three of the most common.

B.1 The Angle-Axis Representation

As you're asked to verify in Problem B.1, the rotation of an arbitrary vector $\vec{v}$ can be written directly in terms of the angle φ and axis $\hat{n}$ (Figure B.1),

$$R_n(\varphi)\,\vec{v} = \vec{v} + \sin\varphi\,(\hat{n}\times\vec{v}) + (1-\cos\varphi)\left[\hat{n}\times(\hat{n}\times\vec{v})\right]. \qquad (\text{B.1})$$

To express R_n as a matrix operator, $\vec{v}$ must be factored out to the far right. The first term, of course, is proportional to the identity; the other two must then somehow be related to the generators L_i of (26.136)–(26.137). For small $\varphi = \epsilon$, (26.138) suggests that the $\sin\varphi$ term reduces to $-i\epsilon\,\hat{n}\cdot\vec{L}$. To verify this, we need a matrix expression for the cross product:

$$(\hat{n}\times\vec{v})_i = \sum_{jk}\epsilon_{ijk}\hat{n}_j v_k = -\sum_{jk}\epsilon_{ikj}\hat{n}_j v_k$$

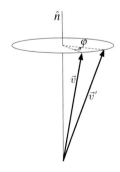

Figure B.1 Rotation around $\hat{n}$.

$$= \sum_k \left(-\sum_j \epsilon_{ikj}\hat{n}_j \right) v_k$$

$$\equiv \sum_k N_{ik}v_k , \tag{B.2}$$

where we have introduced the antisymmetric matrix $N = -N^T$,

$$N_{ij} \equiv -\sum_k \epsilon_{ijk}\hat{n}_k = \begin{pmatrix} 0 & -\hat{n}_z & \hat{n}_y \\ \hat{n}_z & 0 & -\hat{n}_x \\ -\hat{n}_y & \hat{n}_x & 0 \end{pmatrix}, \tag{B.3}$$

whose three independent elements are the three components of $\hat{n}$.[1]

The overall sign of N is chosen to be consistent with that of the generators L_i; this choice, coupled with the use of ϵ_{ijk}, fixes a right-handed convention. Since N is so closely related to $\hat{n}$, you may not be terribly surprised to learn that the double cross product in (B.1) is just N^2 (Problem B.2). Thus we have the matrix operator

$$R_n(\varphi) = \left[\mathbb{1} + \sin\varphi\, N + (1 - \cos\varphi)\, N^2 \right], \tag{B.4}$$

which is known as the *Rodrigues formula*. Written out as a matrix,

$$R_n(\varphi) = \begin{pmatrix} c + (1-c)\hat{n}_x^2 & (1-c)\hat{n}_x\hat{n}_y - s\,\hat{n}_z & (1-c)\hat{n}_x\hat{n}_z + s\,\hat{n}_y \\ (1-c)\hat{n}_x\hat{n}_y + s\,\hat{n}_z & c + (1-c)\hat{n}_y^2 & (1-c)\hat{n}_y\hat{n}_z - s\,\hat{n}_x \\ (1-c)\hat{n}_x\hat{n}_z - s\,\hat{n}_y & (1-c)\hat{n}_y\hat{n}_z + s\,\hat{n}_x & c + (1-c)\hat{n}_z^2 \end{pmatrix}, \tag{B.5}$$

where we write $c \equiv \cos\varphi$, $s \equiv \sin\varphi$ to reduce clutter. A quick consistency check shows that for $\hat{n} = \hat{\imath}, \hat{\jmath}$, or $\hat{k}$, the canonical rotations of (26.82) and (26.84) easily emerge. You can also verify that the rotation angle is given by (26.122). In practice, the components $(\hat{n}_x, \hat{n}_y, \hat{n}_z)$ are usually expressed in spherical coordinates (θ, ϕ) — which should not be confused with the rotation angle φ.

The angle-axis representation also shows how to determine the rotation axis from an arbitrary rotation matrix in $\mathbb{R}^3$. Using N's antisymmetry, it can be extracted from (B.4) as

$$N = \frac{1}{2\sin\varphi}(R - R^T). \tag{B.6}$$

With this, the axis $\hat{n}$ can be read off using (B.3) (note the indices are *anti*-cyclic),

$$\hat{n}_x = \frac{R_{32} - R_{23}}{2\sin\varphi} \qquad \hat{n}_y = \frac{R_{13} - R_{31}}{2\sin\varphi} \qquad \hat{n}_z = \frac{R_{21} - R_{12}}{2\sin\varphi}. \tag{B.7}$$

The rotation angle around $\hat{n}$ can then be determined using $\hat{n} \cdot \hat{n} = 1$. This of course presumes that $\varphi \neq 0$ — though in that case there's no rotation anyway. The only ambiguity occurs for $\varphi = \pi$, since $R_n(\pi) = R_n(-\pi) = R_{-n}(\pi)$. But for φ between 0 and π, (B.6) allows a full determination of $\hat{n}$. For example, from the matrix A in (26.124), we have

$$N = \frac{1}{2\sin(2\pi/3)}(A - A^T) = \frac{1}{\sqrt{3}} \begin{pmatrix} 0 & 1 & \sqrt{2} \\ -1 & 0 & 0 \\ -\sqrt{2} & 0 & 0 \end{pmatrix}. \tag{B.8}$$

Comparison with (B.3) shows that $\hat{n}$ is correctly given by the $+$ sign in (26.126).

---

[1] If this feels like déjà vu all over again, see Example 29.6.

B.2 Euler Angles

Though the angle-axis representation is very natural (axis + angle $\to$ rotation), it is by no means the only method, or even necessarily the most useful. Intuitive though it may be, the angle-axis representation requires that we know the rotation axis — which is a challenge for something like the precessing axis of a spinning top. Euler framed the problem differently by imagining axes $\hat{e}_1, \hat{e}_2$, and $\hat{e}_3$ fixed in a rotated rigid body. Before the rotation, this body-fixed frame coincides with a space-fixed frame $\hat{x}, \hat{y}$, and $\hat{z}$, sharing a common origin; after rotating, the orientation of the body is expressed by giving the body-fixed axes relative to the space-fixed coordinate axes.

If the rotation R is a function of time, the motion of a point $\vec{r}$ in the body is given by $r_i(t) = \sum_j R_{ij}(t) r_j(0)$. We can equivalently rotate the entire body-fixed frame using (26.19): $\hat{e}_i(t) = \sum_j R_{ij}^T(t) \hat{x}_j$. Either way, $R(t)$ describes the dynamics of the body as a path through the space of orthogonal matrices in $\mathbb{R}^3$ — that is, a path through the space $SO(3)$.[2]

Example B.1 The Instantaneous Angular Velocity

How is a rotating body's angular velocity $\vec{\omega}$ related to $R(t)$? The rate of change of the body's orientation relative to the space-fixed (lab) frame $\hat{x}_i$ is

$$\frac{d\hat{e}_i}{dt} = \frac{d}{dt}\left(\sum_j R_{ij}^T(t)\hat{x}_j\right) = \sum_j \dot{R}_{ij}^T \hat{x}_j, \tag{B.9}$$

since the space-fixed basis $\hat{x}_j$ is just that — fixed. Though the $\hat{e}_i$ are time dependent, they nevertheless form an orthonormal basis at every instant t. And since $d\hat{e}_i$ is itself a vector, we can expand it on the body-fixed basis,

$$\frac{d\hat{e}_i}{dt} = \sum_j \dot{R}_{ij}^T \hat{x}_j = \sum_{jk} \dot{R}_{ik}^T \left(R_{kj}\hat{e}_j\right) = \sum_j \left(\dot{R}^T R\right)_{ij} \hat{e}_j. \tag{B.10}$$

The matrix $\Omega \equiv \dot{R}^T R$ carries units of inverse time, and is obviously related to the angular velocity. In principle, Ω has nine independent elements — though since it's derived from an orthogonal R, it's likely to have fewer. Indeed, from the definition of orthogonal matrices we find

$$R^T R = \mathbb{1} \quad\Longrightarrow\quad \dot{R}^T R = -R^T \dot{R} = -\left(\dot{R}^T R\right)^T. \tag{B.11}$$

Thus Ω is an antisymmetric matrix, and so has only three independent components, which we'll suggestively parameterize as

$$\Omega = \begin{pmatrix} 0 & -\omega_3 & \omega_2 \\ \omega_3 & 0 & -\omega_1 \\ -\omega_2 & \omega_1 & 0 \end{pmatrix} = -i\sum_k \omega_k L_k = -i\vec{\omega}\cdot\vec{L}, \tag{B.12}$$

where $\vec{L}$ is the vector of generators in (26.137), and the angular velocity $\vec{\omega} = d\left(\varphi\,\hat{n}\right)/dt$. In fact, an antisymmetric matrix can be expressed explicitly as a vector $\vec{\omega}$ using the antisymmetry of ϵ_{ijk},

$$\Omega_{ij} = -\sum_k \epsilon_{ijk}\omega_k. \tag{B.13}$$

---

[2] A relevant feature of $SO(3)$ is that it's a continuous connected group — see BTW 26.5.

(The similarity with our angle-axis derivation is a clear reflection of the basic structure of $\mathbb{R}^3$ rotations.) Plugging this expression for Ω into (B.10), and using $\hat{e}_i \times \hat{e}_j = \sum_k \epsilon_{ijk}\hat{e}_k$ for right-handed systems, gives

$$\frac{d\hat{e}_j}{dt} = \sum_k \Omega_{jk}\hat{e}_k = -\sum_{ik} \epsilon_{jki}\omega_i\hat{e}_k$$

$$= \sum_i \omega_i \left(\sum_k \epsilon_{ijk}\hat{e}_k \right) = \sum_i \omega_i\hat{e}_i \times \hat{e}_j = \vec{\omega} \times \hat{e}_j, \qquad (B.14)$$

where $\vec{\omega} = \sum \omega_i\hat{e}_i$ identifies the ω_i as components on the body-fixed basis. Thus for a point fixed in the body with position $\vec{r}$ and components r_j on the $\hat{e}_j$ basis, we get the familiar result

$$\frac{d\vec{r}}{dt} = \vec{\omega} \times \vec{r}, \qquad (B.15)$$

verifying that $\vec{\omega}$ is the angular velocity.

As a vector expression, (B.15) is a basis-independent statement. But wait: although a space-fixed observer certainly sees the body rotating, a body-fixed observer should measure no angular velocity at all. So $\vec{\omega}$ cannot be the angular velocity directly *observed* by a rotating, non-inertial observer. Rather, the ω_i are the components of $\vec{\omega}$ at an instant of time in the body. That is, we imagine measuring ω through an infinity of successive inertial frames $\{\hat{e}_i\}$, each at rest with respect to the lab.

The challenge is to find $R(t)$. Euler's brilliant idea was to break down the transformation between these frames into successive canonical rotations around space-fixed axes. As a result, we do not need to define an overall axis of rotation; rather, we need only specify the rotation angles of Euler's canonical rotations. Because an orthogonal transformation in n dimensions needs $n(n - 1)/2$ independent parameters, Euler's construction requires three such rotations; the rotation angles are called the *Euler angles* — usually denoted α, β, and γ. Dynamic equations for these angles (also due to Euler) then allow one to track a body's changing orientation as a function of time.

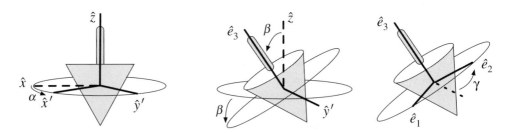

Figure B.2 Euler angles α, β, γ.

As shown in Figure B.2, the first Euler rotation of the body is by α around the space and body frames' initially common z-axis, $(0, 0, 1)$; this takes the body's $\hat{x}$- and $\hat{y}$-axes to $\hat{x}'$ and $\hat{y}'$. We then rotate by β around the $\hat{y}'$-axis, which has components $(-\sin\alpha, \cos\alpha, 0)$ on the space-fixed basis. A moment's thought shows that the resulting $\hat{e}_3$-axis has spherical coordinates $\theta \equiv \beta, \phi \equiv \alpha$ relative to the space-fixed frame, so $\hat{e}_3 = (\cos\alpha \sin\beta, \sin\alpha \sin\beta, \cos\beta)$. A final rotation by γ around $\hat{e}_3$ aligns the other two axes with $\hat{e}_1$ and $\hat{e}_2$. Writing this out in terms of coordinate rotations, reading from right to left, gives

$$R(\alpha, \beta, \gamma) = R_3(\gamma)R_{y'}(\beta)R_z(\alpha). \qquad (B.16)$$

The rotations in (B.16) are all around body axes — only the last of which, $R_3(\gamma)$, is one of the body-frame's final axes. And since rotations around different axes do not commute, the order matters. This makes it very cumbersome to work out the matrix elements of $R(\alpha, \beta, \gamma)$, since similarity transformations are needed to determine both $R_{y'}$ and R_3 from their canonical forms in (26.82) and (26.84). Fortunately, a similarity analysis demonstrates that $R(\alpha, \beta, \gamma)$ can be re-expressed as (Problem B.13)

$$R(\alpha, \beta, \gamma) = R_z(\alpha)R_y(\beta)R_z(\gamma). \tag{B.17}$$

In this form, all three Euler rotations are performed around *space-fixed* axes, allowing the full transformation to be easily written out in full glory:

$$R(\alpha, \beta, \gamma) = \begin{pmatrix} \cos\alpha & -\sin\alpha & 0 \\ \sin\alpha & \cos\alpha & 0 \\ 0 & 0 & 1 \end{pmatrix} \begin{pmatrix} \cos\beta & 0 & \sin\beta \\ 0 & 1 & 0 \\ -\sin\beta & 0 & \cos\beta \end{pmatrix} \begin{pmatrix} \cos\gamma & -\sin\gamma & 0 \\ \sin\gamma & \cos\gamma & 0 \\ 0 & 0 & 1 \end{pmatrix}$$

$$\tag{B.18}$$

$$= \begin{pmatrix} \cos\alpha\cos\beta\cos\gamma - \sin\alpha\sin\gamma & -\cos\alpha\cos\beta\sin\gamma - \sin\alpha\cos\gamma & \cos\alpha\sin\beta \\ \sin\alpha\cos\beta\cos\gamma + \cos\alpha\sin\gamma & \cos\alpha\cos\gamma - \sin\alpha\cos\beta\sin\gamma & \sin\alpha\sin\beta \\ -\sin\beta\cos\gamma & \sin\beta\sin\gamma & \cos\beta \end{pmatrix}.$$

These three successive Euler rotations applied to an object allow us to achieve any orientation relative to the space-fixed frame. Moreover, by restricting β to be between 0 and π, the orientation is unique — that is to say, any triplet of Euler angles in the range $-\pi \leq \alpha, \gamma < \pi$ and $0 \leq \beta \leq \pi$ corresponds to a single orientation of an object. Restriction to positive β, however, means that the inverse of $R^{-1}(\alpha, \beta, \gamma)$ is not $R(-\gamma, -\beta, -\alpha)$; rather,

$$R^{-1}(\alpha, \beta, \gamma) = R(-\gamma \pm \pi, +\beta, -\alpha \pm \pi), \tag{B.19}$$

where the sign is chosen to keep α and γ within $\pm\pi$. Exchanging $\alpha \longleftrightarrow -\gamma \pm \pi$ in (B.18) readily confirms (B.19) as R^{-1}.

BTW B.1 Gimbal Lock

There are two exceptions for which a specified Euler triplet of angles does *not* yield a unique orientation. One stipulation when deciding in what order to do the three Euler rotations is that no two consecutive rotations can be about the same axis; doing so would reduce the number of independent parameters from three to two. This is precisely what happens for $\beta = 0$ and π. In these cases, the freedom to choose a third Euler angle has been lost; only the sum $\alpha + \gamma$ is meaningful, rather than α and γ individually. And since $\alpha + \gamma$ alone does not provide a unique orientation, similar orientations can be described by vastly different triplets of Euler angles. For example, the position denoted by $(\pi/2, 0, -\pi/2)$ is mathematically speaking infinitesimally close to both $(\pi/2, \epsilon, -\pi/2)$ and $(0, \epsilon, 0)$ — but only the first one is physically close.

This effect, called *gimbal lock*, has practical consequences. Whether tracking a spacecraft (either in flight or as a computer graphic), or just directing a panning camera, the orientation can be thought of as a controlled path $(\alpha(t), \beta(t), \gamma(t))$ through "Euler space." The three angles give full control over the orientation. But whenever the β assumes its extreme values of 0 or π, there are suddenly only two rotation axes available, constraining the possible neighboring positions that can be achieved through the shortest and natural path. This difficulty is unavoidable in the Euler system, occurring for any choice of the three Euler rotation axes.

Incidentally, there's nothing fundamental about the *choice* of Euler rotation axes — other than not allowing two consecutive rotations around the same axis. Our ZYZ method is the one most commonly used in quantum mechanics, and has the added advantage that $\beta \equiv \theta$ and $\alpha \equiv \phi$ are the spherical coordinates of the final body-fixed z-axis. But there are $3 \times 2 \times 2 = 12$ possible conventions. In classical mechanics one often sees *ZXZ*, with an $\hat{x}'$ rotation replacing our $\hat{y}'$ rotation; in aeronautical engineering and computer graphics, a *ZYX* convention ("yaw-pitch-roll") is preferred.

The strength of the Euler formulation, however, is not in the construction of a rotation matrix, but rather in the analysis of a rotating body. That's because although rotations aren't vectors, angular velocity is (BTW 24.1). Thus in the progression of Figure B.2, an object's total angular velocity can be expressed as the sum of the angular velocities of each frame with respect to the previous one,

$$\vec{\omega} = \dot{\alpha}\hat{z} + \dot{\beta}\hat{y}' + \dot{\gamma}\hat{e}_3. \tag{B.20}$$

In spite of this being an expansion on a mixed basis, the physical interpretation is both intuitive and compelling (Figure B.3): $\dot{\gamma}$ gives the spin, $\dot{\beta}$ the nutation, and $\dot{\alpha}$ the precession of a rotating body. So the trio of functions $\alpha(t)$, $\beta(t)$, $\gamma(t)$ gives the full dynamic solution for a spinning object.

Figure B.3 Precession $\dot{\alpha}$, nutation $\dot{\beta}$, and spin $\dot{\gamma}$.

Although the expression for angular velocity in (B.20) may be physically appealing, the mixed basis can be difficult to work with. Of course, one need only go back to Example B.1 using $R(\alpha, \beta, \gamma)$ to find the components of $\vec{\omega}$ in either the space- or body-fixed frame. But Euler's construction offers an easier method. Consulting Figure B.2, we see that $\hat{e}_3$ has spherical coordinates β, α with respect to space-fixed frame (as noted earlier), and $\hat{y}'$ is given by rotation in the xy-plane,

$$\hat{e}_3 = \sin\beta\cos\alpha\,\hat{x} + \sin\beta\sin\alpha\,\hat{y} + \cos\beta\,\hat{z} \qquad \hat{y}' = -\sin\alpha\,\hat{x} + \cos\alpha\,\hat{y}. \tag{B.21}$$

Thus in the space-fixed frame, (B.20) becomes

$$\begin{aligned}
\vec{\omega} &= \dot{\alpha}\hat{z} + \dot{\beta}\left(-\sin\alpha\,\hat{x} + \cos\alpha\,\hat{y}\right) \\
&\quad + \dot{\gamma}\left(\sin\beta\cos\alpha\,\hat{x} + \sin\beta\sin\alpha\,\hat{y} + \cos\beta\,\hat{z}\right) \\
&= \left(-\dot{\beta}\sin\alpha + \dot{\gamma}\sin\beta\cos\alpha\right)\hat{x} \\
&\quad + \left(\dot{\beta}\cos\alpha + \dot{\gamma}\sin\beta\sin\alpha\right)\hat{y} + \left(\dot{\alpha} + \dot{\gamma}\cos\beta\right)\hat{z}.
\end{aligned} \tag{B.22}$$

Similarly, the components of $\vec{\omega}$ on the (instantaneous) body-fixed basis are

$$\begin{aligned}
\vec{\omega} &= \left(\dot{\beta}\sin\gamma - \dot{\alpha}\sin\alpha\cos\gamma\right)\hat{e}_1 \\
&\quad + \left(\dot{\beta}\cos\gamma + \dot{\alpha}\sin\alpha\sin\gamma\right)\hat{e}_2 + \left(\dot{\gamma} + \dot{\alpha}\cos\beta\right)\hat{e}_3.
\end{aligned} \tag{B.23}$$

These rather unappealing expressions stand in stark contrast with the intuitive mixed-basis description of (B.20).

B.3 Quaternions

A third representation for rotations in $\mathbb{R}^3$ begins with Example 26.8 and the curious observation that a $\mathbb{C}^2$ rotation also requires three parameters:[3]

$$U(\varphi) = n_0 \mathbb{1} - i\vec{n} \cdot \vec{\sigma}$$
$$= \mathbb{1} \cos(\varphi/2) - i\hat{n} \cdot \vec{\sigma} \sin(\varphi/2) = e^{-i\varphi \, \hat{n} \cdot \vec{\sigma}/2}, \tag{B.24}$$

where $n_0^2 + \vec{n} \cdot \vec{n} = 1$ — which the introduction of the angle φ in (26.50) accommodates naturally. If we can identify φ and $\hat{n}$ of this $\mathbb{C}^2$ representation with an angle and axis, then perhaps this parametrization can be used to rotate a vector $\vec{v}$ in $\mathbb{R}^3$.

Of course, the matrix $U(\varphi)$ cannot operate directly on a three-element column vector, so we'll need to repackage $\vec{v}$ into a two-dimensional form. We could try to write it as a unit-length, two-component complex vector — but that effort won't get very far. Instead, we try to find a 2×2 *matrix* representation V of $\vec{v}$. Rotation by U would then be implemented as a similarity transformation, $V' = UVU^\dagger$.

So how do we represent an $\mathbb{R}^3$ vector as a $\mathbb{C}^2$ matrix? Look back to (25.74), where we represented $\vec{v}$ by expanding its component on the basis of Pauli matrices,

$$V \equiv \vec{v} \cdot \vec{\sigma} = \begin{pmatrix} v_z & v_x - iv_y \\ v_x + iv_y & -v_z \end{pmatrix}. \tag{B.25}$$

This is an intriguing correspondence. As we discussed in Example 25.13, complex matrices form an inner product space, with

$$\langle V|W \rangle \equiv \frac{1}{2} \mathrm{Tr}(V^\dagger W). \tag{B.26}$$

For matrices $V = \vec{v} \cdot \vec{\sigma}$ and $W = \vec{w} \cdot \vec{\sigma}$, this is just the $\mathbb{R}^3$ inner product,

$$\langle V|W \rangle = \vec{v} \cdot \vec{w}. \tag{B.27}$$

So the correspondence $V \leftrightarrow \vec{v}$ is faithful in that the magnitudes and relative orientations of and between vectors is the same in both representations. V is called the *quaternion* representation of $\vec{v}$.

Note that since the Pauli matrices are traceless and Hermitian, so too is V. Moreover, both properties are unchanged under similarity transformations, so the matrix representation of a rotated vector will also be traceless and Hermitian. Thus even after a similarity transformation, we're assured that the components v_i' of the rotated vector are real. Also unchanged under similarity is the determinant; from (B.25) we find $\det V = -\|\vec{v}\|^2$ — which of course must be invariant under rotations.

So $UVU^\dagger$ is a proper rotation. To be a rotation of $\vec{v}$ it must yield a result of the form

$$V' = UVU^\dagger \equiv \begin{pmatrix} v_z' & v_x' - iv_y' \\ v_x' + iv_y' & -v_z' \end{pmatrix}, \tag{B.28}$$

where $\vec{v}' = (v_x', v_y', v_z')$. Let's work this out explicitly for a rotation around the z-axis. Using

$$U_z(\varphi) = \mathbb{1} \cos(\varphi/2) - i\sigma_3 \sin(\varphi/2) = \begin{pmatrix} e^{-i\varphi/2} & 0 \\ 0 & e^{i\varphi/2} \end{pmatrix} \tag{B.29}$$

---

[3] The sign choice here matches that of (26.142).

we find

$$V' = U_z(\varphi) \, V \, U_z^{\dagger}(\varphi) = \begin{pmatrix} v_z & e^{-i\varphi}(v_x - iv_y) \\ e^{i\varphi}(v_x + iv_y) & -v_z \end{pmatrix}. \tag{B.30}$$

Matching real and imaginary parts with (B.28) verifies that $\vec{v}\,'$ is the same vector as $R_z(\varphi)\vec{v}$. Though the algebra is a bit more tedious, similar calculations verify that the matrices

$$U_x(\varphi) = \begin{pmatrix} \cos\varphi/2 & -i\sin\varphi/2 \\ -i\sin\varphi/2 & \cos\varphi/2 \end{pmatrix} \qquad U_y(\varphi) = \begin{pmatrix} \cos\varphi/2 & -\sin\varphi/2 \\ \sin\varphi/2 & \cos\varphi/2 \end{pmatrix} \tag{B.31}$$

rotate $\vec{v}$ by φ around the x- and y-axes, respectively.

We can now begin to appreciate the algebraic forethought that went into the introduction of the half-angle in (26.51). Since a similarity transformation requires two factors of U, we end up with squares of trig functions multiplying the components of $\vec{v}$. Having started with half-angles, however, the familiar identities $\sin^2(\varphi/2) = \frac{1}{2}(1 - \cos\varphi)$ and $\cos^2(\varphi/2) = \frac{1}{2}(1 + \cos\varphi)$ can be invoked to express the transformations using the familar $\sin\varphi$ and $\cos\varphi$ of $\mathbb{R}^3$ rotation matrices R (Problem B.16). But the ramifications of needing two factors of U run deeper than mere algebraic convenience.

Example B.2   Double, Double Toil and Trouble

Although rotation by φ and $\varphi + 2\pi$ yields the same $\mathbb{R}^3$ rotation, they correspond to *distinct* $\mathbb{C}^2$ rotations $U(\varphi)$ and $-U(\varphi)$. In other words, every 3×3 orthogonal matrix R corresponds not to one, but rather *two* distinct 2×2 unitary matrices, $\pm U$. This two-to-one relation is, of course, mathematical; it doesn't really have any physical implications — at least for vectors in $\mathbb{R}^3$. On the other hand, if a physical system were discovered which forms, say, a two-dimensional complex vector . . . .

Consider a spin-1/2 system like an electron. The physical electron may live in $\mathbb{R}^3$, but as we saw in Example 26.6, its spin state is described by a $\mathbb{C}^2$ vector $\vec{\xi}$. Even though U and $-U$ correspond to the same R, they have *different* effects on $\vec{\xi}$. In particular,

$$U(2\pi) = e^{-i\pi\,\hat{n}\cdot\vec{\sigma}} = \mathbb{1}\cos(\pi) - i\hat{n}\cdot\vec{\sigma}\sin(\pi) = -\mathbb{1}. \tag{B.32}$$

In other words, a physical 2π rotation of an electron in the lab — for which, of course, $R = \mathbb{1}$ — only results in a π rotation of its spin state. A spin-1/2 system requires a 4π rotation to return to its original state! This is no mere mathematical curiosity: the distinction between a 2π and 4π rotation is observable by measuring the interference pattern produced by two electron beams, one rotated by 2π relative to the other.

BTW B.2   The Shape of Space

Though 3×3 proper orthogonal matrices and unit quaternions both describe rotations in $\mathbb{R}^3$, the different "shape" of their spaces — technically, the distinct topologies of $SO(3)$ and $SU(2)$ — can elucidate the unexpected subtlety of 2π rotation.

The shape of $SO(3)$ can be visualized as a solid ball of radius π. Each point within the ball corresponds to a unique rotation, its distance and direction from the origin denoting the angle and axis of rotation. Points on the surface correspond to π rotations, so antipodal points are identified as the same rotation. A solid ball, however, does not describe the quaternions of (B.24). Rather, the condition $\sum_{i=0}^{3} n_i^2 = 1$ reveals the shape of $SU(2)$ to be the three-dimensional spherical surface S^3 of a unit-radius four-dimensional ball. Every point on the three-sphere can be put in 1-to-1 correspondence with a quaternion rotation matrix (see (2.41), BTW 2.2). In both spaces, a continuous sequence of rotations can be visualized as a continuous path either through the ball of $SO(3)$ or along the surface of $SO(3)$ three-sphere. The connectedness of both spaces means that any continuous path can reach any point (BTW 26.5). Thus any finite rotation can thus be regarded as a path through the space of rotation operators.

Consider such a path for a 2π rotation. For the ball of $SO(3)$, begin at the origin ($\varphi = 0$), head along the z-axis to the north pole $\varphi = \pi$, and then continue upward from the south pole, sweeping out another π before closing the loop back at the origin. Although one can continuously deform this path, the two points on the surface must always be antipodal — otherwise the path will break. Thus there is no way to continuously shrink this loop to a point. Of course, whatever object you've rotated has indeed flipped by 2π, so this seems little more than a mathematical curiosity. Except that, remarkably, a 4π rotation *can* be shrunk to a point (Figure B.4). The rotated object in both cases looks to have undergone the same net change (that is, none), yet the processes are clearly distinct.

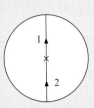

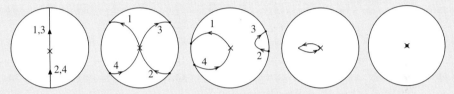

Figure B.4 The space of SO(3) rotations as points in a ball: continuously deforming a 4π rotation down to $\mathbb{1}$, maintaining the identity of antipodal points at every stage.

This behavior is reflected in a simpler way on the three-sphere of $SU(2)$: the path of a 2π rotation extends only halfway along a great circle of the three-sphere as $U \to -U$, so it's not a closed loop; for that, a full great circle is needed, corresponding to a 4π rotation (Problem B.18). And, as with all closed loops on a spherical surface, this great circle *can* be continuously shrunk to a point. Unlike $SO(3)$, the space $SU(2)$ is *simply connected* (Chapter 17).

Dirac's "belt trick" allows one to appreciate this topological distinction. Take a belt, and secure the end opposite the buckle. Hold the buckle end straight out, without any twist along the length of the belt. This is the $\varphi = 0$ state at the center of the ball. Now rotate the buckle by 2π. Though the buckle returns to its original orientation, the twist in the belt shows that 0 and 2π rotations are inequivalent in process — in particular, there is no way to undo the twist without additional rotation. They are topologically distinct. But now rotate the buckle by an additional 2π. The buckle again unremarkably returns to its original orientation. Remarkably, however, the 4π twist in the belt can be completely undone by guiding the buckle under the belt without ever changing its orientation — that is, *without rotation*. Thus 0 and 4π rotations are topologically equivalent, both in net rotation *and* process.

The correspondence between rotations of $\vec{v}$ and similarity transformations of V can be derived formally with a little index gymnastics. From (B.25),

$$V' = \sum_i v'_i \sigma_i = \sum_i \left(\sum_j R_{ij} v_j \right) \sigma_i$$

$$= \sum_{ij} v_j R_{ij} \sigma_i = \sum_j v_j \left(\sum_i R^T_{ji} \sigma_i \right), \tag{B.33}$$

or simply[4]

$$V' = \vec{v}\,' \cdot \vec{\sigma} = \left[R\vec{v} \right] \cdot \vec{\sigma} = \vec{v} \cdot \left[R^T \vec{\sigma} \right]. \tag{B.34}$$

We can think of $R^T \vec{\sigma}$ as a rotation of a three-element column vector, each component of which is a Pauli matrix. So in the last step, R went from mixing the components v_i of $\vec{v}$, to mixing the Pauli matrices σ_i of $\vec{\sigma}$.

So that's V' in terms of R. In terms of U,

$$V' = UVU^\dagger = U\,(\vec{v} \cdot \vec{\sigma})\,U^\dagger = \vec{v} \cdot \left(U\vec{\sigma}U^\dagger \right), \tag{B.35}$$

where we can pull out the v_i since U acts only in the matrix space of σ_i. Noting that $\vec{v}$ is arbitrary, comparison of (B.34) and (B.35) reveals $U\vec{\sigma}U^\dagger = R^T\vec{\sigma}$ — or equivalently,

$$U^\dagger \vec{\sigma} U = R\vec{\sigma} \qquad \Longleftrightarrow \qquad U^\dagger \sigma_j U = \sum_\ell R_{j\ell} \sigma_\ell. \tag{B.36}$$

The rotation matrix R_n corresponding to U_n is thus (Problem B.17)

$$(R_n)_{jk} = \frac{1}{2} \mathrm{Tr}\left(U_n^\dagger \sigma_j U_n \sigma_k \right). \tag{B.37}$$

Note that the two-to-one relationship between U and R is manifest.

BTW B.3 Just What Are Quaternions?

Vectors in the plane can be represented as complex numbers. And as we saw in Example 26.5, this allows any $\mathbb{R}^2$ vector to be regarded as an operator, its two parameters accounting for rotation and scaling of vectors in the plane. It all works so nicely that it's natural to seek an extension of the complex numbers which would similarly represent vectors and vector multiplication in $\mathbb{R}^3$. Inviting though this may sound, it cannot be done. A fairly straightforward way to see this is by counting parameters: one for magnification, and $n(n-1)/2$ for rotations in $\mathbb{R}^n$ (see BTW 26.4). So the numerical coincidence for $\mathbb{R}^2$ doesn't occur in $\mathbb{R}^3$: three parameters are required to specify a vector, whereas four are needed for rotation and scaling.

But let's see what we can do. One can "derive" complex numbers by starting with the one-dimensional set of reals, $\mathbb{R}$, and introduce a "basis vector" i such that $i^2 = -1$. Then the complex numbers $\mathbb{C}$ are defined as

$$\mathbb{C} = \mathbb{R} + \mathbb{R}i. \tag{B.38}$$

Now introduce another basis vector j such that $j^2 = -1$. The *quaternions* $\mathbb{Q}$ are defined as

$$\mathbb{Q} = \mathbb{C} + \mathbb{C}j = (\mathbb{R} + \mathbb{R}i) + (\mathbb{R} + \mathbb{R}i)\,j \tag{B.39}$$

$$= \mathbb{R} + \mathbb{R}i + \mathbb{R}j + \mathbb{R}k, \tag{B.40}$$

where $k = ij$ is another basis vector with $k^2 = -1$. Thus the most general form for a quaternion has one real and *three* imaginary parts,

$$Q = q_0 + iq_1 + jq_2 + kq_3, \tag{B.41}$$

with

$$i^2 = j^2 = k^2 = ijk = -1. \tag{B.42}$$

---

[4] In bra-ket notation, this is $\langle Rv | \sigma \rangle = \langle v | R^T \sigma \rangle$.

These conditions are sufficient to establish $ij = k$, $ji = -k$, as well as their cyclic permutations. So expanding from complex numbers to quaternions requires the loss of commutativity. Clearly then, ordinary numbers cannot represent quaternions. The simplest representation is given by the Pauli matrices, one each for i, j, and k. From Problem 25.24 and (25.137), we see that the quaternion basis vectors i, j, k can be represented by the σ's,

$$i \leftrightarrow -i\sigma_1 \qquad j \leftrightarrow -i\sigma_2 \qquad k \leftrightarrow -i\sigma_3. \tag{B.43}$$

So (26.47) is a 2×2 matrix representation of the quaternion U.

How can this extension beyond complex numbers be used to implement $\mathbb{R}^n$ rotations? Although the quaternion Q in (B.41) corresponds to a point (q_0, q_1, q_2, q_3) in $\mathbb{R}^4$, rotations in $\mathbb{R}^4$ require $n(n-1)/2 = 6$ parameters. On the other hand, *unit* quaternions, for which $q_0^2 + q_1^2 + q_2^2 + q_3^2 = 1$, have three free parameters — exactly what's needed to describe rotations in $\mathbb{R}^3$. From (B.24) and (B.43), a unit quaternion corresponds to $\det U = +1 = q_0^2 + q_1^2 + q_2^2 + q_3^2$, restricting the parameter space to the surface of the three-sphere, S^3.

As an example of their simplicity, consider successive $90°$ rotations around y and then x. The product is, of course, equivalent to a single rotation — but by how much and around what axis? The quickest way to determine $\hat{n}$ is using quaternions:

$$U_1(\pi/2) = \mathbb{1}\cos(\pi/4) - i\sigma_1 \sin(\pi/4) \rightarrow \frac{1}{\sqrt{2}}(1+i)$$

$$U_2(\pi/2) = \mathbb{1}\cos(\pi/4) - i\sigma_2 \sin(\pi/4) \rightarrow \frac{1}{\sqrt{2}}(1+j). \tag{B.44}$$

Then

$$U_1(\pi/2)U_2(\pi/2) \rightarrow \frac{1}{\sqrt{2}}(1+i)\frac{1}{\sqrt{2}}(1+j)$$

$$= \frac{1}{2}[1 + (i+j+k)]$$

$$= \frac{1}{2}\left[1 + \sqrt{3}\left(\frac{i+j+k}{\sqrt{3}}\right)\right]. \tag{B.45}$$

So the composite rotation is around $\hat{n} = +\frac{1}{\sqrt{3}}(\hat{i} + \hat{j} + \hat{k})$ by an angle given by $\cos(\varphi/2) = 1/2$ and $\sin(\varphi/2) = \sqrt{3}/2$ — that is, $\varphi = 120°$. Which is, of course, what we found in Example 26.13. (See Problem B.15.)

Quaternions and their applicability to rotations were first invented (or discovered, depending on your philosophical bent) by Hamilton. Historically, the standard use of $\hat{i}$, $\hat{j}$, $\hat{k}$ as unit vectors in $\mathbb{R}^3$ has its origins in his quaternion notation.

In principle, any of the three representations we've seen can and are used to implement rotations in $\mathbb{R}^3$; each has advantages and disadvantages relative one to another. The angle-axis is conceptually the simplest. Euler rotations allow for a straightforward connection between space-fixed and body-fixed frames. The quaternion approach is particularly popular in aeronautical engineering and computer graphics because the half-angle avoids the ambiguity that results when $\beta = 0$ or π (gimbal lock). Quaternions also have a numerical advantage. Numerical calculations inevitably introduce round-off errors, allowing the orthogonality of a matrix to numerically "leak away." And since the errors will compound, pretty soon all semblance of orthogonality will be lost — and thus we'll no longer working with rotations. Quaternions also suffer from round-off errors, eventually losing their unit magnitude. But as we saw in Example 26.8, any 2×2

complex matrix can be expressed in the form $U = n_0 \mathbb{1} + \vec{\sigma} \cdot \vec{n}$; a self-correcting algorithm need only divide U by $\sqrt{n_0^2 + \vec{n}^2}$ after every transformation to ensure orthogonality. Though rotation matrices can also be renormalized, calculating $R^T R$ is less computationally efficient.

Problems

B.1   Verify the decomposition $\vec{A} = \vec{A}_\parallel + \vec{A}_\perp = (\vec{A} \cdot \hat{e})\hat{e} + \hat{e} \times (\vec{A} \times \hat{e})$, and then use it to derive (B.1).

B.2   Show that $N^2 \vec{v} = \hat{n} \times (\hat{n} \times \vec{v})$.

B.3   By comparing N in (B.3) and the generators $\vec{L}$ of Section 26.6, express $R_n(\varphi)$ as an exponential of N; use this to verify that $N\hat{n} = 0$. Is R written this way still an orthogonal transformation?

B.4   (a) Derive the Rodrigues matrix (B.5) from (B.4).
(b) Verify that (B.5) is consistent with (26.122) and (26.123).

B.5   Here's an alternative derivation of the Rodrigues formula, (B.4):

(a) Show that around an arbitrary axis $\hat{n}$, (26.89) of Example 26.12 becomes

$$\dot{\vec{r}} = \omega N \vec{r},$$

with N given by (B.3). Assuming constant ω, solve this differential equation for $\vec{r}(t)$ and identify the rotation operator $R_n(t)$.
(b) Show that $R_n(t)$ is equivalent to (B.4).

B.6   Show that the Rodrigues formula yields a proper rotation. One could verify this directly from the matrix in (B.5). Tedious and inelegant. Instead, let's use index techniques. (Don't forget that the axis $\hat{n}$ is a unit vector.)

(a) Verify that N^2 is symmetric. Then show that $N^4 = -N^2$.
(b) Use these results to show that Rodrigues matrix (B.4) is orthogonal.
(c) Show that in odd-dimensional space, the determinant of an antisymmetric matrix vanishes — and hence that R_n is, in fact, a proper rotation.

B.7   For an infinitesimal rotation $\delta\vec{\varphi} \equiv \hat{n}\,\delta\varphi$, show (B.4) is consistent with (26.88).

B.8   Use the Rodrigues formula (B.4) to construct $R_z(\varphi)$.

B.9   Use the angle-axis representation to determine the angle φ and axis $\hat{n}$ of

$$R_n(\varphi) = \begin{pmatrix} -0.491958 & 0.695675 & -0.523464 \\ -0.804325 & -0.133042 & 0.579103 \\ 0.333224 & 0.70593 & 0.625 \end{pmatrix}.$$

How does φ compare with the angle emerging from the trace of R?

B.10   Find Euler angles $-\pi \le \alpha, \gamma < \pi$ and $0 \le \beta \le \pi$ for which $R(\alpha, \beta, \gamma)$ reduces to the canonical rotations R_x, R_y, and R_z.

B.11   Show the rotation $R_n(\varphi)$ corresponding to the Euler matrix (B.18) has angle

$$\cos\varphi = 2\cos^2\left(\frac{\alpha + \gamma}{2}\right)\cos^2\left(\frac{\beta}{2}\right) - 1$$

with axis $\hat{n} = (\sin\theta\cos\phi, \sin\theta\sin\phi, \cos\theta)$, where

$$\tan\theta = \frac{\tan\left(\frac{\beta}{2}\right)}{\sin\left(\frac{\alpha+\gamma}{2}\right)} \qquad \phi = \frac{1}{2}(\alpha - \gamma + \pi).$$

B.12 Show that the columns of the Euler matrix (B.18) give the representation of the body axes in the fixed frame. What do the rows give?

B.13 Derive the fixed-frame Euler matrix (B.17) from (B.16) using (26.121),

$$R_{S\hat{m}}(\theta) = SR_{\hat{m}}(\theta)S^{-1},$$

as follows. Write $R_{z'}(\gamma)$ as a similarity transformation around y'; substitute this into (B.16) and simplify. Then write $R_{y'}(\beta)$ as a similarity transformation around z; substitute this into the previous result and simplify. With a clever in the choice of angles for these similarity transformations, (B.17) will emerge.

B.14 Verify that $U_z(\varphi), U_x(\varphi)$ and $U_y(\varphi)$ in (B.29) and (B.31) are $\mathbb{C}^2$ representations of $\mathbb{R}^3$ rotations $R_z(\varphi), R_x(\varphi)$ and $R_y(\varphi)$, respectively.

B.15 (a) Use (B.24) and (25.138) from Problem 25.24 to find the effect of two consecutive rotations around the same axis $\hat{n}$, and verify that all is well.
(b) Show that a rotation by ϕ_1 around $\hat{n}_1$ followed by a rotation by ϕ_2 around $\hat{n}_2$ yields a rotation by θ around $\hat{n}$, where

$$\cos(\theta/2) = \cos(\phi_1/2)\cos(\phi_2/2) - \hat{n}_1 \cdot \hat{n}_2 \sin(\phi_1/2)\sin(\phi_2/2)$$

$$\hat{n}\sin(\theta/2) = \hat{n}_1\sin(\phi_1/2)\cos(\phi_2/2) + \hat{n}_2\cos(\phi_1/2)\sin(\phi_2/2)$$

$$+ \sin(\phi_1/2)\sin(\phi_2/2)\hat{n}_1 \times \hat{n}_2.$$

B.16 (a) Using only that the σ_i are Hermitian and satisfy (25.138), verify that the rotation matrix $U = \mathbb{1}\cos(\varphi/2) - i\hat{n} \cdot \vec{\sigma}\sin(\varphi/2)$ is unitary.
(b) Use (25.138) to show that the similarity transformation in (B.28) is equivalent to the angle-axis representation, (B.1).

B.17 (a) Use (B.36) to verify (B.37).
(b) Construct the matrix $R_z(\varphi)$ from (B.29) and (B.37).
(c) If you're up for some index calisthenics, derive the angle-axis matrix (B.5) from the quaternion representation, (B.24) and (B.37). You'll need (25.137) — and remember that individual Pauli matrices are traceless. [This is essentially Problem B.16b in index notation.]

B.18 In BTW B.2, we introduced the "shape" of the spaces $SU(2)$ and $SO(3)$.

(a) The unit quaternions of $SU(2)$ describe S^3, the surface of a unit sphere in three dimensions. Taking the north pole to represent the identity, use (2.41) in BTW 2.2 to find the correspondence between points on S^3 and rotations in $\mathbb{R}^3$. Show that a great circle corresponds to a 4π rotation.
(b) Unlike the ball with antipodal points identified, the surface of a sphere is simply connected. Discuss how, by analogy with quaternions on S^3, one might represent the space of $SO(3)$ rotation matrices as the surface of the two-sphere S^2, with the identity at the north pole. Show that even in this representation a 4π rather than a 2π rotation is equivalent to $\mathbb{1}$.

Appendix C

C The Bessel Family of Functions

Erf, Γ, H_n, L_m, P_ℓ^m, sinc, $Y_{\ell m}$ ... the list of so-called special functions extends from Ai to ζ (Abramowitz and Stegun, 1966). It would be a fool's errand to attempt to prioritize them by breadth of application, depth of insight, frequency of occurrence — or even inherent charm and allure. An exception might be sine and cosine, but their relative simplicity and overwhelming ubiquity generally excludes them from "special" status. Bessel functions, on the other hand, are certainly special functions — and as close cousins to "Fourier functions" sine and cosine, deserve special consideration.

Indeed, comparison with Fourier is illuminating. Just as with sine and cosine, Bessel functions have real and complex forms which can describe both standing waves and traveling waves. Both have oscillatory and decaying variants as well. With Sturm–Liouville boundary conditions, both constitute orthogonal bases with individual basis vectors distinguished by their zeros. Both have a continuum limit in which discrete expansion on the basis evolves into integral transforms. And like Fourier functions, the prominence of Bessel functions in physics and engineering derives from their appearance as part of the Helmholtz eigenfunction in cylindrical and spherical coordinates (see Table 41.1). So just as Fourier functions and spherical harmonics are fundamentally important in analyses on the circle and sphere, Bessel functions are indispensable when considering the interiors of disks, cylinders, and balls.

C.1 Cylindrical Bessel Functions

We start with Bessel's equation,

$$x \frac{d}{dx}\left(x \frac{du}{dx}\right) + \left(x^2 - v^2\right) u(x) = 0, \tag{C.1}$$

with real parameter v. As we derived in Example 39.7, this differential equation has the series solution

$$J_v(x) = \left(\frac{x}{2}\right)^v \sum_{m=0}^{\infty} \frac{(-1)^m}{m!\, \Gamma(m + v + 1)} \left(\frac{x}{2}\right)^{2m}, \tag{C.2}$$

which converges for all x. Recalling that $\Gamma(n+1) = n!$, we can immediately verify (35.18b) for small x and integer $v \equiv n$,

$$J_n(x) \xrightarrow{x\,\text{small}} \frac{x^n}{2^n n!}. \tag{C.3}$$

Beyond lowest order, the series for $n = 0$ and 1 gives

$$J_0(x) = 1 - \frac{x^2}{2^2(1!)^2} + \frac{x^4}{2^4(2!)^2} - \frac{x^6}{2^6(3!)^2} + \cdots \tag{C.4a}$$

$$J_1(x) = \frac{x}{2} - \frac{x^3}{2^3(2!)} + \frac{x^5}{2^5(2!)(3!)} - \frac{x^7}{2^7(3!)(4!)} + \cdots \tag{C.4b}$$

These are even and odd respectively, with $J_0(0) = 1$, $J_1(0) = 0$; similarly, $J_0'(x) = -J_1(x)$. In fact, (C.4) look like they could be series expansions of a damped cosine and sine.

Table C.1 Zeros of Bessel functions J_m and J_m'

J_0	J_1	J_2	J_3	J_0'	J_1'	J_2'	J_3'
2.4048	3.8317	5.1356	6.3802	0	1.8412	3.0542	4.2012
5.5201	7.0156	8.4172	9.7610	3.8317	5.3314	6.7061	8.0152
8.6537	10.174	11.620	13.015	7.0156	8.5363	9.9695	11.346

But they're not. At least, not precisely. For instance, as we saw in our discussion of the infinite circular well in Example 40.5, the zeros of J_0 do not occur at regular intervals; the same is true for all J_ν and their derivatives (see Table C.1). However, close scrutiny reveals that as x grows, the spacing of zeros converges to π (Problem C.2). We can confirm this by working out the large-x limit of J_n using one of the three integral representations in (35.17),

$$J_n(x) = \frac{1}{2\pi i^n} \int_0^{2\pi} e^{i(x \cos\varphi - n\varphi)} \, d\varphi. \tag{C.5}$$

The key is to note that since the integrand is oscillatory, the largest contributions to the integral come from regions in which the phase changes most slowly. So since $\cos\varphi$ is stationary at $\varphi = 0$ and π, we can use the stationary phase approximation explored in Problem 21.35 to expand around 0 and approximate the integral as

$$J_n(x) \approx \frac{1}{2\pi} 2\,\Re e \left[\frac{1}{i^n} \int_{-\infty}^{\infty} e^{ix\left(1 - \frac{1}{2}\varphi^2\right) - in\varphi} \, d\varphi \right], \tag{C.6}$$

where rapid oscillation of the integrand allows the extension to infinite limits. We've taken twice the real part to account for the stationary phase at both 0 and π since, as (C.5) shows, their contributions are complex conjugates of one another. The remaining integral can be treated as a Gaussian; from (8.70) we get

$$J_n(x) \approx \frac{1}{\pi} \Re e \left[e^{i(x - n\pi/2)} \left(\frac{2\pi}{ix} \right)^{1/2} e^{in^2/2x} \right]$$

$$= \sqrt{\frac{2}{\pi x}} \, \Re e \left[e^{i(x - n\pi/2 - \pi/4)} e^{in^2/2x} \right]. \tag{C.7}$$

Expanding the Gaussian to lowest order in $1/x$ reveals the asymptotic form

$$J_\nu(x) \xrightarrow{x\,\text{large}} \sqrt{\frac{2}{\pi x}} \cos\left(x - \frac{\nu\pi}{2} - \frac{\pi}{4} \right), \tag{C.8}$$

where we've reverted from n back to ν since this expression is valid for both integer and non-integer index. Plots for $n = 0$ to 2 are shown in Figure C.1a. Some common identities are presented in Table C.2.

Table C.2 Common Bessel J_n identities

$$J_n(-x) = (-1)^n J_n(x) \qquad\qquad \int x^{\pm n+1} J_n(x)\, dx = \pm x^{\pm n+1} J_{n+1}(x)$$

$$\int_0^\infty J_n(x)\, dx = 1 \qquad\qquad \sum_{n=-\infty}^{\infty} J_n(x) = 1$$

$$J_0^2 + 2\sum_{n=1}^{\infty} J_n^2 = 1 \qquad\qquad J_n(x+y) = \sum_{k=-\infty}^{\infty} J_k(x) J_{n-k}(x)$$

$$\cos(x\sin\varphi) = \sum_{n=-\infty}^{\infty} J_n(x)\cos n\varphi \qquad \sin(x\sin\varphi) = \sum_{n=-\infty}^{\infty} J_n(x)\sin n\varphi$$

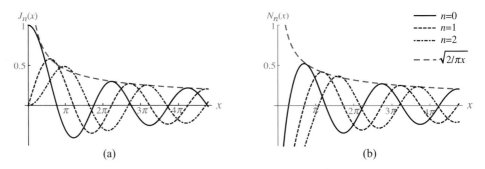

Figure C.1 (a) Bessel and (b) Neumann functions with $\sqrt{2/\pi x}$ large-x asymptote.

Bessel's equation depends on ν^2, implying that $J_{\pm\nu}(x)$ must both be solutions. Which sounds great since, as a second-order equation, Bessel's equation has two linearly independent solutions. But whereas *non*-integer ν yields two linearly independent series solutions, as we learned from (39.78)–(39.79), Bessel functions $J_{\pm n}$ for integer n are *not* independent. Indeed, the Wronskian (Problem C.1)

$$W[J_{\pm\nu}; x] = -\frac{2}{\pi x} \sin \nu\pi \tag{C.9}$$

vanishes for integer ν. This contrast between integer and non-integer index is irritating; what we'd like is a second solution which can be used for all ν. A glance at (C.9) suggests that a second solution can be constructed from $J_\nu / \sin \nu\pi$. The conventional choice is called a *Neumann* function, defined as[1]

$$N_\nu(x) \equiv \frac{\cos \nu\pi\, J_\nu(x) - J_{-\nu}(x)}{\sin \nu\pi}. \tag{C.10}$$

For non-integer index, N_ν is just a linearly combination of independent solutions $J_{\pm\nu}$ — and so is itself a solution independent of J_ν. But in the limit as ν goes to an integer, (C.10) still yields an independent solution,

$$N_n(x) \equiv \lim_{\nu \to n} N_\nu(x). \tag{C.11}$$

Thus the Neumann function $N_\nu(x)$ defined in (C.10) provides a second solution to Bessel's equation for both integer and non-integer ν, allowing the general solution of (C.1) to be written as

$$u_\nu(x) = A J_\nu(x) + B N_\nu(x). \tag{C.12}$$

---

[1] The notation $N_\nu(x)$ honors Carl Neumann, though one also sees $Y_\nu(x)$ — which should not be confused with spherical harmonics $Y_{\ell m}(\theta, \phi)$. In fact, Mathematica uses `BesselJ` and `BesselY`.

Plots of Neumann functions for $n = 0$ to 2 are shown in Figure C.1b. As expected from (39.23), Neumann functions diverge as $x \to 0$. But for large x, Neumann functions have the asymptotic form

$$N_\nu(x) \xrightarrow{x\,\text{large}} \sqrt{\frac{2}{\pi x}} \sin\left(x - \frac{\nu\pi}{2} - \frac{\pi}{4}\right) \tag{C.13}$$

which, together with (C.8), further underscores the parallel with cosine and sine.

The Sturm–Liouville orthogonality amongst the J_0's displayed in (40.43) applies to the J_ν's generally; the derivation is essentially an extension of Problem 40.23. What's not included in that analysis is a relationship between J_ν' and $J_{\nu+1}$. To find it, we start with (35.16) for integer index

$$e^{ix\cos\varphi} = \sum_{n=-\infty}^{\infty} i^n J_n(x)\, e^{in\varphi}. \tag{C.14}$$

With $t = ie^{i\varphi}$, this acquires the form of a generating function (Section 32.3)

$$G(x,t) \equiv e^{\frac{x}{2}\left(t - \frac{1}{t}\right)} = \sum_{n=-\infty}^{\infty} J_n(x)\, t^n. \tag{C.15}$$

Many of the identities in Table C.2 can be established from this generating function. But unlike previous generating functions we've seen, the negative powers of t preclude individual J_n's from being extracted from partials of G. But consider:

$$\sum_n n J_n(x) t^{n-1} = \frac{\partial G}{\partial t} = \frac{x}{2}\left(1 + \frac{1}{t^2}\right) G(x,t)$$

$$= \frac{x}{2}\left[\sum_{n=-\infty}^{\infty} J_n(x)\, t^n + \sum_{n=-\infty}^{\infty} J_n(x)\, t^{n-2}\right]$$

$$= \frac{x}{2} \sum_{n=-\infty}^{\infty} \left[J_{n-1}(x)\, t^{n-1} + J_{n+1}(x)\right] t^{n-1}. \tag{C.16}$$

The index shifts in the last step allows for straightforward matching of the power series coefficients to discover

$$2\frac{n}{x} J_n(x) = J_{n-1}(x) + J_{n+1}(x). \tag{C.17a}$$

With this, all J_n for $n \geq 2$ can be recursively generated from J_0 and J_1. A similar calculation gives

$$2J_n'(x) = J_{n-1}(x) - J_{n+1}(x). \tag{C.17b}$$

Remarkably, Bessel's equation can be derived from these two recursion relations for any index — not merely integer n (Problem C.7); thus general statements obtained from them should be valid for arbitrary ν. In particular, from (C.17) can be derived the derivative formulas

$$J_\nu(x) = \frac{\nu \pm 1}{x} J_{\nu\pm1}(x) \pm J_{\nu\pm1}'(x), \tag{C.18a}$$

and

$$\frac{d}{dx}\left[x^{\pm\nu} J_\nu(x)\right] = \pm x^{\pm\nu} J_{\nu\mp1}(x), \tag{C.18b}$$

which include the special case $J_0'(x) = -J_1(x)$. Using (C.18a), the Sturm–Liouville the statement of orthogonality for J_0 in (40.43) generalizes to

$$\int_0^a J_\nu(\alpha_{n\nu}r/a)J_\nu(\alpha_{m\nu}r/a)\,r\,dr = \frac{a^2}{2}J_\nu'^2(\alpha_{m\nu})\delta_{nm} = \frac{a^2}{2}J_{\nu\pm1}^2(\alpha_{m\nu})\delta_{nm}, \qquad \text{(C.19)}$$

where $\alpha_{m\nu}$ is the mth zero of J_ν.[2]

Incidentally, recall that it's the finite integration region in (C.19) which leads to discrete modes $J_\nu(\alpha_{m\nu}r/a)$. And just as the continuum limit which takes Fourier series to Fourier transforms, here too as $a \to \infty$ the discrete modes m in effect merge, discrete $k_{m\nu} = \alpha_{m\nu}/a$ becoming a continuous parameter k independent of both m and ν. This takes (C.19) to

$$\int_0^\infty J_\nu(k'r)J_\nu(kr)\,r\,dr = \frac{1}{k}\delta(k-k'), \qquad \nu > 1/2 \qquad \text{(C.20)}$$

which should be compared with Fourier orthogonality in (36.74). Clearly, this could just as easily be a completeness relation with $r \longleftrightarrow k$. Like the plane wave basis of Fourier transforms, this reflects the equivalence of k and r representations of a function decomposed on a basis of J_ν.

Table C.3 Bessel recursion relations ($\mathcal{J}_\nu \equiv J_\nu, N_\nu, H_\nu^{(1,2)}$)

$\frac{2\nu}{x}\mathcal{J}_\nu(x) = \mathcal{J}_{\nu-1}(x) + \mathcal{J}_{\nu+1}(x)$	$\frac{d}{dx}\left[x^{\pm\nu}\mathcal{J}_\nu(x)\right] = \pm x^{\pm\nu}\mathcal{J}_{\nu\mp1}(x)$
$2\mathcal{J}_\nu'(x) = \mathcal{J}_{\nu-1}(x) - \mathcal{J}_{\nu+1}(x)$	$x\mathcal{J}_\nu(x) = (\nu\pm1)\mathcal{J}_{\nu\pm1}(x) \pm x\mathcal{J}_{\nu\pm1}'(x)$

Since Bessel's equation can be derived from (C.17), any of its solutions — J_ν and N_ν for integer and non-integer ν, as well as any linear combinations — must obey these same recursion relations and those derived from them (Table C.3). Two of those linear combinations are clearly hinted at by the asymptotic forms of J_ν and N_ν: just as $\cos\nu x \pm i\sin\nu x = e^{\pm i\nu x}$, so too the *Hankel functions* of the first and second kind (occasionally referred to as Bessel functions of the third kind) are defined as

$$H_\nu^{(1,2)}(x) \equiv J_\nu(x) \pm iN_\nu(x), \qquad \text{(C.21)}$$

or equivalently,

$$J_\nu(x) = \frac{1}{2}\left[H_\nu^{(1)}(x) + H_\nu^{(2)}(x)\right] \qquad N_\nu(x) = \frac{1}{2i}\left[H_\nu^{(1)}(x) - H_\nu^{(2)}(x)\right]. \qquad \text{(C.22)}$$

These expressions are indeed reminiscent of the relationship between standing and traveling waves on a string. But the asymptotic forms

$$H_\nu^{(1,2)}(x) \xrightarrow{x\,\text{large}} \sqrt{\frac{2}{\pi x}}\,e^{\pm i(x - \frac{\nu\pi}{2} - \frac{\pi}{4})} \qquad \text{(C.23)}$$

reveal waves with intensity falling as $1/x$. This leads to the interpretation of Hankel functions $H_\nu^{(1,2)}(r)$ as outgoing and incoming waves in the plane. In fact, the Sturm–Liouville weight r in the orthogonality statements (C.19)–(C.20) can be construed as the Jacobian in cylindrical coordinates. For all these reasons, J_ν, N_ν and $H_\nu^{(1,2)}$ are often called "cylindrical" Bessel functions.

---

[2] In this more general notation, α_n in (40.40)–(40.45) and (41.32)–(41.35) becomes α_{0n}.

C.2 Modified Bessel Functions

So if $J_\nu(x)$ and $N_\nu(x)$ are analogs of $\cos x$ and $\sin x$, it's natural to ask about the Bessel parallels of $\cosh x$ and $\sinh x$. Using the integral representation in (C.5), we define the real-valued function

$$I_n(x) \equiv i^{-n} J_n(ix) = \frac{1}{2\pi} \int_0^{2\pi} e^{-x\cos\phi - in\varphi} \, d\varphi. \tag{C.24}$$

called the *modified Bessel function of the first kind*. We can generalize to non-integer index from the series in (C.2),

$$I_\nu(x) = \left(\frac{x}{2}\right)^\nu \sum_{m=0}^\infty \frac{1}{m! \, \Gamma(m+\nu+1)} \left(\frac{x}{2}\right)^{2m}, \tag{C.25}$$

which are solutions of the modified Bessel equation

$$x \frac{d}{dx} \left(x \frac{du}{dx}\right) - \left(x^2 + \nu^2\right) u(x) = 0. \tag{C.26}$$

As with J_ν, the modified Bessel functions $I_{\pm\nu}$ are only independent for non-integer ν. The *modified Bessel function of the second kind* $K_\nu(x)$ is defined as

$$K_\nu(x) = \frac{\pi}{2} \frac{I_{-\nu}(x) - I_\nu(x)}{\sin \nu x}, \tag{C.27}$$

with integral representation (Problem C.12)

$$K_\nu(x) = \frac{1}{2} \int_{-\infty}^\infty e^{-x\cosh t - \nu t} dt, \quad x > 0. \tag{C.28}$$

Though (C.27) is similar to (C.10) for Neumann functions, the modified function has large-x behavior

$$K_\nu(x) \xrightarrow{\;x\,\text{large}\;} \sqrt{\frac{\pi}{2x}} e^{-x}, \tag{C.29}$$

which complements the asymptotic form

$$I_\nu(x) \xrightarrow{\;x\,\text{large}\;} \sqrt{\frac{1}{2\pi x}} e^x. \tag{C.30}$$

Thus the modified Bessel functions I_ν and K_ν are independent for both integer and non-integer ν. So just as J_ν and N_ν are solutions to Bessel's equation, the modified Bessel functions span the general solution to (C.26),

$$u(x) = A I_\nu(x) + B K_\nu(x). \tag{C.31}$$

Moreover, the modified Bessel functions satisfy recursion relations similar to those for J_ν and N_ν, albeit with some sign flips — for instance,

$$X_{\nu+1} - X_{\nu-1} = \mp \frac{2\nu}{x} X_\nu(x) \qquad X_{\nu+1} + X_{\nu-1} = \mp 2 X_\nu'(x), \tag{C.32}$$

where the upper sign applies to $X_\nu \equiv I_\nu$, the lower for $X_\nu \equiv K_\nu$. Plots of I_n and K_n for $n = 0$ to 3 are shown in Figure C.2. Note that neither I_ν nor K_ν is oscillatory; in fact, both are divergent. But then, so are $\cosh$ and $\sinh$.

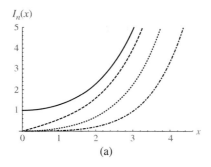
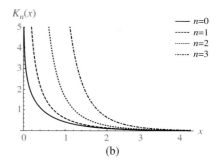

Figure C.2 Modified Bessel functions (a) $I_n(x)$ and (b) $K_n(x)$.

C.3 Spherical Bessel Functions

Though much of our attention thus far has been on Bessel functions with unspecified index, note that the asymptotic forms of J_ν and N_ν simplify for ν an odd half-integer. Defining integer $\ell \equiv \nu - 1/2$, (C.8) and (C.13) become

$$J_{\ell+\frac{1}{2}}(x) \xrightarrow{x\,\text{large}} \sqrt{\frac{2}{\pi x}} \sin\left(x - \frac{\ell\pi}{2}\right), \quad N_{\ell+\frac{1}{2}}(x) \xrightarrow{x\,\text{large}} -\sqrt{\frac{2}{\pi x}} \cos\left(x - \frac{\ell\pi}{2}\right). \tag{C.33}$$

We'll take this simplicity as motivation to introduce the *spherical Bessel functions*

$$j_\ell(x) \equiv \sqrt{\frac{\pi}{2x}} J_{\ell+\frac{1}{2}}(x) \qquad n_\ell(x) \equiv \sqrt{\frac{\pi}{2x}} N_{\ell+\frac{1}{2}}(x), \tag{C.34}$$

Table C.4 Spherical Bessel functions $x^2 \frac{d^2 v}{dx^2} + 2x \frac{dv}{dx} + \left[x^2 - \ell(\ell+1)\right] v = 0$

$j_0 = \frac{\sin x}{x}$	$n_0 = -\frac{\cos x}{x}$
$j_1 = \frac{\sin x}{x^2} - \frac{\cos x}{x}$	$n_1 = -\frac{\cos x}{x^2} - \frac{\sin x}{x}$
$j_2 = \left(\frac{3}{x^3} - \frac{1}{x}\right)\sin x - \frac{3}{x^2}\cos x$	$n_2 = -\left(\frac{3}{x^3} - \frac{1}{x}\right)\cos x - \frac{3}{x^2}\sin x$

for integer $\ell \geq 0$ (Table C.4). By itself this is hardly a compelling reason to define new functions — but let's explore their properties. First, either one multiplied by $\sqrt{x}$ must satisfy Bessel's equation; direct substitution of $u(x) = \sqrt{x}\,v(x)$ into (C.1) reveals the spherical Bessel equation

$$x^2 \frac{d^2 v}{dx^2} + 2x \frac{dv}{dx} + \left[x^2 - \ell(\ell+1)\right]v(x) = 0. \tag{C.35}$$

Despite the similarity to (C.1), this is a distinct differential equation. Distinct, but clearly not unrelated. Like their cylindrical upper-case cousins, spherical Bessel functions j_ℓ and n_ℓ are oscillatory though not periodic, damping out for large x. For small x, $j_\ell \sim x^\ell$ and $n_\ell \sim x^{-(\ell+1)}$. So like J_n, only $j_0(0)$ in non-zero at the origin, and like N_n, the n_ℓ diverge at the origin. In fact, as you can see in Figure C.3, the spherical and cylindrical versions have very similar profiles.

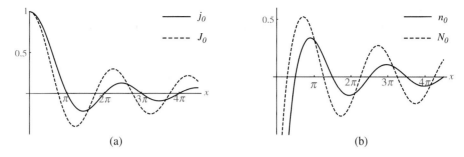

Figure C.3 Comparison of spherical and cylindrical Bessel functions.

Table C.5 Spherical Bessel function recursion relations ($\mathscr{J}_\ell \equiv j_\ell, n_\ell, h_\ell^{(1,2)}$)

$$\frac{2\ell+1}{x} \mathscr{J}_\ell = \mathscr{J}_{\ell-1} + \mathscr{J}_{\ell+1} \qquad\qquad \frac{d}{dx}\left[x^{\ell+1}\,\mathscr{J}_\ell(x)\right] = x^{\ell+1}\,\mathscr{J}_{\ell-1}$$

$$(2\ell+1)\,\mathscr{J}'_\ell = \ell\,\mathscr{J}_{\ell-1} - (\ell+1)\,\mathscr{J}_{\ell+1} \qquad\qquad \frac{d}{dx}\left[x^{-\ell}\,\mathscr{J}_\ell(x)\right] = -x^{-\ell}\,\mathscr{J}_{\ell+1}$$

Indeed, j_ℓ has essentially the same series expansion (C.2). Thus like J_n, the j_ℓ obey recursion relations — which can be found directly from (C.17)–(C.18) (see Table C.5),

$$\frac{2\ell+1}{x} j_\ell = j_{\ell-1} + j_{\ell+1} \tag{C.36a}$$

$$(2\ell+1)j'_\ell = \ell j_{\ell-1} - (\ell+1)j_{\ell+1}, \tag{C.36b}$$

and

$$\frac{d}{dx}\left[x^{\ell+1}j_\ell(x)\right] = x^{\ell+1}j_{\ell-1} \tag{C.37a}$$

$$\frac{d}{dx}\left[x^{-\ell}j_\ell(x)\right] = -x^{-\ell}j_{\ell+1}. \tag{C.37b}$$

Since the differential equation (C.35) can be derived from these relations, they hold also for n_ℓ — as well as any linear combinations of j_ℓ and n_ℓ.

So if j_ℓ and n_ℓ are so similar to J_n and N_n, why define them as distinct functions? The answer is in the adjective "spherical." Consider the two linear combinations

$$h_\ell^{(1,2)}(x) \equiv j_\ell(x) \pm in_\ell(x) = \sqrt{\frac{\pi}{2x}}\, H_{\ell+1/2}^{(1,2)}(x), \tag{C.38}$$

the *spherical Hankel functions* of the first and second kind (occasionally referred to as spherical Bessel functions of the third kind). As with the cylindrical versions, these expressions are clearly analogous of the relationship between standing and traveling waves. Their asymptotic forms

$$h_\ell^{(1,2)}(x) \xrightarrow{\ x\,\text{large}\ } \frac{(\mp i)^{\ell+1}}{x} e^{\pm ix} = \frac{1}{\pm ix} e^{\pm i(x-\ell\pi/2)} \tag{C.39}$$

reveal waves with intensity falling off as $1/x^2$. So the lower-case Hankel functions $h_\ell^{(1,2)}$ describe outgoing and incoming spherical waves.

Furthermore, comparing (40.12) with the spherical Bessel equation (C.35) shows that it can be viewed as a Sturm–Liouville system with weight $w = x^2$ — which is the radial part of the Jacobian in spherical coordinates. If we take $x \to kr$, (C.35) becomes

$$r^2 \frac{d^2 j_\ell}{dr^2} + 2r \frac{dj_\ell}{dr} + \left[k^2 r^2 - \ell(\ell + 1) \right] j_\ell(r) = 0. \tag{C.40}$$

Hence the finite solutions j_ℓ will be orthogonal if the surface term in (40.13) vanishes — and the usual Sturm–Liouville considerations lead to

$$\int_0^a j_\ell(\alpha_{\ell n} r/a) j_\ell(\alpha_{mv} r/a)\, r^2\, dr = \frac{a^3}{2} j_\ell'^2(\alpha_{\ell m}) \delta_{nm} = \frac{a^3}{2} j_{\ell \pm 1}^2(\alpha_{\ell m}) \delta_{nm}, \tag{C.41}$$

where $\alpha_{\ell n}$ is the nth zero of j_ℓ (Table C.6). There's also continuum orthogonality, similar to what we saw in (C.20),

$$\int_0^\infty j_\ell(k'r) j_\ell(kr)\, r^2\, dr = \frac{\pi}{2k^2}\, \delta(k - k'), \qquad v > -1. \tag{C.42}$$

Table C.6 Zeros of spherical Bessel j_ℓ and j_ℓ'

j_0	j_1	j_2	j_3	j_0'	j_1'	j_2'	j_3'
3.1416	4.4934	5.7635	6.9879	0	2.0816	3.3421	4.5141
6.2832	7.7253	9.0950	10.417	4.4934	5.9404	7.2899	8.5838
9.4248	10.904	12.323	13.698	7.7253	9.2058	10.614	11.973

Finally, note there are also modified spherical Bessel functions i_ℓ and k_ℓ, defined by analogy with those in Section C.2:

$$i_\ell(x) \equiv i^{-\ell} j_\ell(ix) = \sqrt{\frac{\pi}{2x}} I_{\ell+1/2}(x) \qquad k_\ell(x) \equiv \sqrt{\frac{2}{\pi x}} K_{\ell+1/2}(x). \tag{C.43}$$

These are independent solutions to the modified spherical Bessel equation

$$x^2 \frac{d^2 v}{dx^2} + 2x \frac{dv}{dx} - \left[x^2 + \ell(\ell + 1) \right] v(x) = 0. \tag{C.44}$$

Problems

C.1 Find the Wronskian $W(x)$ of $J_{\pm v}(x)$ to show that $J_{\pm v}$ are only independent for non-integer v. Then show that J_v and N_v are linearly independent for integer v as well. [You may find Problem 21.27 useful.]

C.2 Use Mathematica's `BesselJZero[n,k]` for the kth zero of $J_n(x)$ to show that the spacing between successive zeros converges to π. Do the same for Neumann functions $N_n(x)$ using `BesselYZero[n,k]` (see footnote 1, this appendix).

C.3 Derive the relations (C.18) from the recursion relations (C.17).

C.4 Expand the generating function $G(x, t)$ of (C.15) in powers of t and $1/t$. Then use the Cauchy product of Problem 6.20 to derive the series expansion for $J_n(x)$ in (C.2).

C.5 In a manner similar to that used to derive (C.17a), use the generating function (C.15) to derive (C.17b).

C.6   Use the generating function (C.15) and/or the recursion relations in (C.17) to verify the identities in Table C.2.

C.7   Use the recursion relations (C.17) to derive:

     (a) Bessel's equation (C.1)
     (b) the derivative formulas (C.18).

C.8   (a) Use (C.15) to find the generating function for the modified Bessel function $I_n(x)$.
     (b) Show that $I_{-n}(x) = I_n(x)$ and $I_n(-x) = (-1)^n I_n(x)$.

C.9   Show that:

     (a) $N_\nu(ix) = e^{i\pi(\nu+1)/2} I_\nu(x) - \frac{2}{\pi} e^{-i\pi\nu/2} K_\nu(x)$
     (b) $K_\nu(x) = \frac{\pi i^{\nu+1}}{2} H_\nu^{(1)}(ix) = \frac{\pi}{2i^{\nu+1}} H_\nu^{(2)}(-ix)$.

C.10   Starting from the integral representation of the Airy function,

     $\text{Ai}(x) = \frac{1}{2\pi} \int_{-\infty}^{\infty} e^{i\left(\frac{1}{3}t^3 + xt\right)} dt$ (Problem 36.14), show that:

     (a) $\text{Ai}(x) = \frac{1}{\sqrt{3\pi^2}} \sqrt{x} K_{\frac{1}{3}}\left(\frac{2}{3}x^{3/2}\right)$
     (b) $\text{Ai}'(x) = -\frac{1}{\sqrt{3\pi^2}} x K_{\frac{2}{3}}\left(\frac{2}{3}x^{3/2}\right)$.

C.11   Unlike Bessel functions, the zeros of sine and cosines are uniformly spaced, making them much easier to work with. But the underlying Sturm–Liouville orthogonality is the same. One way to appreciate this is to define, by analogy with $J_n(\alpha_{n\nu}r/a)$, the functions $\phi_n(x) = \sin(\alpha_m n x/a)$, where α_m for integer m is a zero of ϕ_n. Express the orthogonality of these functions in the form of (C.19). In what ways do the orthogonality relations differ?

C.12   Show that the integral (C.28) is a solution of the modified Bessel equation (C.26). Then formally identify this solution as $K_\nu(x)$ by examining the large x limit.

C.13   (a) Show that the Bessel equation (C.1) emerges as the radial equation of the Laplacian in cylindrical coordinates.
     (b) Show that the spherical Bessel equation (C.35) emerges as the radial equation of the Laplacian in spherical coordinates.

C.14   Find $j_0(x)$ and $n_0(x)$ from the series solution (C.2) and (C.10). With these in hand, use the recursion relations of Table C.5 to derive the rest of the spherical Bessel functions listed in Table C.4.

C.15   Show that $R(r) = a j_0(kr) + b n_0(kr)$ is the general solution to the spherical Bessel equation (C.35) with $\ell = 0$ by inserting the ansatz $u(r) \equiv R(r)/\sqrt{kr}$ into (C.1).

C.16   Show that spherical Bessel and Neumann functions have the asymptotic forms

$$j_\ell(x) \xrightarrow{x \text{ large}} \frac{1}{x} \sin\left(x - \frac{\ell\pi}{2}\right) \qquad n_\ell(x) \xrightarrow{x \text{ large}} -\frac{1}{x} \cos\left(x - \frac{\ell\pi}{2}\right)$$

and

$$j_\ell(x) \xrightarrow{x \text{ small}} \frac{2^\ell \ell!}{(2\ell+1)!} x^\ell \qquad n_\ell(x) \xrightarrow{x \text{ small}} -\frac{(2^\ell)!}{2^\ell \ell!} x^{-(\ell+1)}.$$

C.17   Find succinct expressions for $i_0(x)$ and $k_0(x)$ in terms of more-familiar functions.

C.18　In Section 35.4 the spherical Bessel functions were defined as the radial basis functions in the expansion of a plane wave in spherical coordinates (35.70a),

$$ e^{i\vec{k}\cdot\vec{r}} = 4\pi \sum_{\ell=0}^{\infty} \sum_{m=-\ell}^{\ell} i^{\ell} j_{\ell}(kr) Y^{*}_{\ell m}(\hat{k}) Y_{\ell m}(\hat{r}). $$

Use this expansion in the completeness relation $\int d^{3}r\, e^{i(\vec{k}-\vec{k}')\cdot\vec{r}} = (2\pi)^{3}\delta(\vec{k}-\vec{k}')$ to verify completeness (C.42) of the j_{ℓ}.

C.19　Use (C.2) to derive the Rodrigues-like formula for spherical Bessel functions (generally attributed to Rayleigh)

$$ j_{\ell}(x) \equiv \sqrt{\frac{\pi}{2x}} J_{\ell+1/2}(x) = (-x)^{\ell}\left(\frac{1}{x}\frac{d}{dx}\right)^{\ell}\frac{\sin x}{x}, \tag{C.45} $$

for integer $\ell \geq 0$. You'll want to use the identity $\Gamma\left(n+\frac{1}{2}\right) = \frac{\sqrt{\pi}}{2^{2n}}\frac{(2n)!}{n!}$ from (8.81a). You may also prefer to write

$$ \frac{1}{x}\frac{d}{dx} = 2\frac{d}{d(x^{2})} $$

and take derivatives with respect to x^{2}. Finally, keep an eye out for the opportunity to shift the summation's lower limit.

References

Abramowitz, M., and Stegun, I. A. (eds). 1966. *Handbook of Mathematical Functions, with Formulas, Graphs, and Mathematical Tables*. Dover.

Arfken, G. B., and Weber, H. J. 1995. *Mathematical Methods for Physicists*. 4th ed. Academic Press.

Backus, J. 1977. *The Acoustical Foundations of Music*. W.W. Norton.

Baierlein, R. 1999. *Thermal Physics*. Cambridge University Press.

Benade, A. H. 1990. *Fundamentals of Musical Acoustics*. 2nd revised ed. Dover.

Biedenharn, L. C., and Louck, J. D. 1981. *Angular Momentum in Quantum Phyiscs*. Encyclopedia of Mathematics and its Applications, vol. 8. Addison-Wesley.

Boas, M. L. 2005. *Mathematical Methods in the Physical Sciences*. 3rd ed. Wiley.

Churchill, R. V., and Brown, J. W. 1984. *Complex Variables and Applications*. McGraw-Hill.

Cohen-Tannoudji, C., Diu, B., and Laloë, F. 2019. *Quantum Mechanics*. Wiley.

de la Madrid, R. 2005. *Eur. J. Phys. (quant-ph/0502053)*, **26**, 287.

Dirac, P. A. M. 1931. *Proc. R. Soc. Lond. A*, **133**(60).

Dodelson, S., and Schmidt, F. 2020. *Modern Cosmology*. 2nd ed. Academic Press.

Feynman, R. P. 1972. *Statistical Mechanics*. Addison-Wesley.

Feynman, R. P. 2011. *The Feynman Lectures on Physics*. New millenium ed. Basic Books.

Georgi, H. 1993. *The Physics of Waves*. Prentice Hall.

Gieres, F. 2000. *Rep. Prog. Phys.*, **63**, 1893.

Gordan, C., and Webb, D. 1996. *Am. Scientist*, **84**, 46.

Gottlieb, H. P. W. 1979. *J. Acoust. Soc. Am.*, **66**(3), 647.

Griffiths, D. J. 2017. *Introduction to Electrodynamics*. 4th edn. Cambridge University Press.

Halmos, P. 2017. *Naïve Set Theory*. Dover.

Hand, L., and Finch, J. 1998. *Analytical Mechanics*. Cambridge University Press.

Jackson, J. D. 1998. *Classical Electrodynamics*. 3rd ed. Wiley.

Kac, M. 1966. *Am. Math. Monthly*, **73**(4), 1.

Kittel, C. 2012. *Introduction to Solid State Physics*. Wiley.

Kusse, B., and Westwig, E. 1998. *Mathematical Physics*. Wiley.

Lee, T. D. and Yang, C. N. 1956. *Phys. Rev.* **104**, 254.

Lighthill, M. J. 1958. *An Introduction to Fourier Analysis and Generalised Functions*. Cambridge University Press.

Lyons, R. 2011. *Understanding Digital Signal Processing*. 3rd ed. Prentice Hall.

Mathews, J. 1988. *Complex Variables for Mathematics and Engineering*. Wm.C. Brown.

Mathews, J., and Walker, R. L. 1970. *Mathematical Methods of Physics*. Addison-Wesley.

Moon, P., and Spencer, D. E. 1988. *Field Theory Handbook, Including Coordinate Systems, Differential Equations, and Their Solutions*. 3rd print ed. Springer-Verlag.

Morse, P. M., and Feshbach, H. 1953. *Methods of Theoretical Physics, Part I*. McGraw-Hill.

Namias, V. 1977. *Am. J. Phys.*, **45**, 773.

Needham, T. 1997. *Visual Complex Analysis*. Oxford University Press.

Peskin, M. E., and Schroeder, D. V. 1995. *An Introduction to Quantum Field Theory*. CRC Press.

Phelps, F. M., and Hunter, J. H. 1965. *Am. J. Phys.*, **33**, 285.

Proakis, J., and Manolakis, D. 2006. *Digital Signal Processing*. 4th ed. Pearson.

Rau, J. 2017. *Statistical Physics and Thermodynamics: An Introduction to Key Concepts*. Oxford.

Reed, M., and Simon, B. 1980. *Methods of Mathematical Physics, Vol I: Functional Analysis*. Academic Press.

Romer, R. H. 1982. *Am. J. Phys.*, **50**, 1089.

Russell, D. A. 2010. *Am. J. Phys.*, **78**, 549.

Saff, E. B., and Snider, A. D. 2003. *Fundamentals of Complex Analysis*. Prentice Hall.

Sakurai, J. J., and Napolitano, J. 2021. *Modern Quantum Mechanics*. Cambridge University Press.

Schutz, B. 2009. *A First Course in General Relativity*. 2nd ed. Cambridge University Press.

Sokal, A. 2010. *Am. J. Phys.*, **78**, 643.

Stakgold, I. 1979. *Green's Functions and Boundary Value Problems*. Wiley.

Taylor, J. R. 2005. *Classical Mechanics*. University Science Books.

Trott, M. 1996. *The Mathematica Journal*, **6**(4), 59.

Trott, M. 2004. *The Mathematica GuideBook for Graphics*. Springer.

Verbin, Y. 2015. *Eur. J. Phys.*, **36**, 015005.

Weinberg, S. 1972. *Gravitation and Cosmology*. Wiley.

Wong, C. W. 1991. *Introduction to Mathematical Physics*. Oxford University Press.

Wu, C. S. et al. 1957. *Phys. Rev.* 105, 1413.

Wunsch, A. 1983. *Complex Variables with Applications*. Addison-Wesley.

Index

Page numbers for BTWs are in **bold**, problems in *italics*

Representations of the Delta Function

- **Limit of sequence of functions**:

$$\delta(x) = \frac{1}{\sqrt{\pi}} \lim_{n\to\infty} n\, e^{-n^2 x^2}$$

$$= \frac{1}{2} \lim_{n\to\infty} n\, e^{-n|x|}$$

$$= \frac{1}{\pi} \lim_{n\to\infty} \frac{n}{1 + n^2 x^2}$$

$$= \lim_{n\to\infty} \frac{\sin nx}{\pi x} = \frac{1}{\pi} \lim_{n\to\infty} n\, \text{sinc}(nx)$$

$$= \lim_{n\to\infty} \frac{\sin^2 nx}{n\pi x^2}$$

- **Divergence**:

$$\vec{\nabla} \cdot \left(\hat{r}/r^2 \right) = 4\pi\, \delta(\vec{r})$$

- **Laplacian**:

$$\nabla^2 \left(1/r \right) = -4\pi\, \delta(\vec{r})$$

- **Fourier representation, ϕ periodic on 2π**:

$$\delta(\phi - \phi_0) = \frac{1}{2\pi} + \frac{1}{\pi} \sum_{m=1}^{\infty} [\cos(m\phi)\cos(m\phi_0) + \sin(m\phi)\sin(m\phi_0)]$$

$$= \frac{1}{2\pi} + \frac{1}{\pi} \sum_{m=1}^{\infty} \cos[m(\phi - \phi_0)]$$

$$= \frac{1}{2\pi} \sum_{m=-\infty}^{\infty} e^{im(\phi-\phi_0)}$$

- **Fourier representation, $|z| < \infty$**: $\displaystyle \delta(z - z_0) = \int_{-\infty}^{\infty} \frac{dk}{2\pi}\, e^{ik(z-z_0)}$

- **Cylindrical coordinates**:

$$\delta(\vec{r} - \vec{r}_0) = \frac{1}{r}\, \delta(r - r_0)\delta(\phi - \phi_0)\delta(z - z_0)$$

$$= \frac{1}{r}\, \delta(r - r_0) \frac{1}{2\pi} \sum_{n=-\infty}^{\infty} e^{im(\phi-\phi_0)} \int_{-\infty}^{\infty} \frac{dk}{2\pi}\, e^{ik(z-z_0)}$$

- **Spherical coordinates**:

$$\delta(\Omega - \Omega_0) = \delta(\cos\theta - \cos\theta_0)\, \delta(\phi - \phi_0) = \sum_{\ell m} Y_{\ell m}^*(\theta,\phi) Y_{\ell m}(\theta_0,\phi_0)$$